T0401932

Texts and Monographs in Physics

Series Editors: R. Balian W. Beiglböck H. Grosse E. H. Lieb
N. Reshetikhin H. Spohn W. Thirring

Springer
Berlin
Heidelberg
New York
Barcelona
Budapest
Hong Kong
London
Milan
Paris
Santa Clara
Singapore
Tokyo

Texts and Monographs in Physics

Series Editors: R. Balian W. Beiglböck H. Grosse E. H. Lieb
N. Reshetikhin H. Spohn W. Thirring

From Microphysics to Macrophysics I + II Methods and Applications of Statistical Physics By R. Balian

Variational Methods in Mathematical Physics A Unified Approach By P. Blanchard and E. Brüning

Quantum Mechanics: Foundations and Applications 3rd enlarged edition By A. Böhm

The Early Universe Facts and Fiction 3rd corrected and enlarged edition By G. Börner

Operator Algebras and Quantum Statistical Mechanics I + II 2nd edition By O. Bratteli and D. W. Robinson

Geometry of the Standard Model of Elementary Particles By A. Derdzinski

Scattering Theory of Classical and Quantum *N*-Particle Systems By J. Dereziński and C. Gérard

Effective Lagrangians for the Standard Model By A. Dobado, A. Gómez-Nicola, A. L. Maroto and J. R. Peláez

Quantum The Quantum Theory of Particles, Fields, and Cosmology By E. Elbaz

Quantum Relativity A Synthesis of the Ideas of Einstein and Heisenberg By D. R. Finkelstein

Quantum Mechanics I + II By A. Galindo and P. Pascual

The Elements of Mechanics By G. Gallavotti

Local Quantum Physics Fields, Particles, Algebras 2nd revised and enlarged edition By R. Haag

Supersymmetric Methods in Quantum and Statistical Physics By G. Junker

***CP* Violation Without Strangeness** Electric Dipole Moments of Particles, Atoms, and Molecules By I. B. Khriplovich and S. K. Lamoreaux

Quantum Groups and Their Representations By A. Klimyk and K. Schmüdgen

Inverse Schrödinger Scattering in Three Dimensions By R. G. Newton

Scattering Theory of Waves and Particles 2nd edition By R. G. Newton

Quantum Entropy and Its Use By M. Ohya and D. Petz

Generalized Coherent States and Their Applications By A. Perelomov

Essential Relativity Special, General, and Cosmological Revised 2nd edition By W. Rindler

Path Integral Approach to Quantum Physics An Introduction 2nd printing By G. Roepstorff

Finite Quantum Electrodynamics The Causal Approach 2nd edition By G. Scharf

From Electrostatics to Optics A Concise Electrodynamics Course By G. Scharf

The Mechanics and Thermodynamics of Continuous Media By M. Šilhavý

Large Scale Dynamics of Interacting Particles By H. Spohn

The Theory of Quark and Gluon Interactions 2nd completely revised and enlarged edition By F. J. Ynduráin

Relativistic Quantum Mechanics and Introduction to Field Theory By F. J. Ynduráin

Anatoli Klimyk Konrad Schmüdgen

Quantum Groups and Their Representations

Springer

Professor Dr. Anatoli Klimyk
Ukrainian Academy of Sciences
Institute for Theoretical Physics
Kiev 252143, Ukraine

Professor Dr.Konrad Schmüdgen
Universität Leipzig
Fakultät für Mathematik und Informatik
D-04109 Leipzig, Germany

Editors

Roger Balian
CEA
Service de Physique Théorique de Saclay
F-91191 Gif-sur-Yvette, France

Nicolai Reshetikhin
Department of Mathematics
University of California
Berkeley, CA 94720-3840, USA

Wolf Beiglböck
Institut für Angewandte Mathematik
Universität Heidelberg
Im Neuenheimer Feld 294
D-69120 Heidelberg, Germany

Herbert Spohn
Theoretische Physik
Ludwig-Maximilians-Universität München
Theresienstraße 37
D-80333 München, Germany

Harald Grosse
Institut für Theoretische Physik
Universität Wien
Boltzmanngasse 5
A-1090 Wien, Austria

Walter Thirring
Institut für Theoretische Physik
Universität Wien
Boltzmanngasse 5
A-1090 Wien, Austria

Elliott H. Lieb
Jadwin Hall
Princeton University, P. O. Box 708
Princeton, NJ 08544-0708, USA

Library of Congress Cataloging-in-Publication Data applied for.

Die Deutsche Bibliothek - CIP-Einheitsaufnahme

Klimyk, Anatolij U.:
Quantum groups and their representations / Anatoli Klimyk ; Konrad Schmüdgen. -
Berlin ; Heidelberg ; New York ; Barcelona ; Budapest ; Hong Kong ; London ; Milan ; Paris ;
Santa Clara ; Singapore ; Tokyo : Springer, 1997
(Texts and monographs in physics)

ISBN-13: 978-3-642-64601-0 e-ISBN-13: 978-3-642-60896-4

DOI: 10.1007/978-3-642-60896-4

This work is subject to copyright. All rights are reserved, whether the whole or part of the material is concerned, specifically the rights of translation, reprinting, reuse of illustrations, recitation, broadcasting, reproduction on microfilm or in any other way, and storage in data banks. Duplication of this publication or parts thereof is permitted only under the provisions of the German Copyright Law of September 9, 1965, in its current version, and permission for use must always be obtained from Springer-Verlag. Violations are liable for prosecution under the German Copyright Law.

© Springer-Verlag Berlin Heidelberg 1997
Softcover reprint of the hardcover 1st edition 1997

The use of general descriptive names, registered names, trademarks, etc. in this publication does not imply, even in the absence of a specific statement, that such names are exempt from the relevant protective laws and regulations and therefore free for general use.

Typesetting: Camera-ready copy from the authors using a Springer TEX macro package
Cover design: *design & production* GmbH, Heidelberg
SPIN: 10552562 55/3144-5 4 3 2 1 0 - Printed on acid-free paper

Preface

The invention of quantum groups is one of the outstanding achievements of mathematical physics and mathematics in the late twentieth century. The birth of the new theory and its rapid development are results of a strong interrelation between mathematics and physics.

Quantum groups arose in the work of L.D. Faddeev and the Leningrad school on the inverse scattering method in order to solve integrable models. The algebra $U_q(\mathrm{sl}_2)$ appeared first in 1981 in a paper by P.P. Kulish and N.Yu. Reshetikhin on the study of integrable XYZ models with highest spin. Its Hopf algebra structure was discovered later by E.K. Sklyanin. A major event was the discovery by V.G. Drinfeld and M. Jimbo around 1985 of a class of Hopf algebras which can be considered as one-parameter deformations of universal enveloping algebras of semisimple complex Lie algebras. These Hopf algebras will be called Drinfeld–Jimbo algebras in this book. Almost simultaneously, S.L. Woronowicz invented the quantum group $SU_q(2)$ and developed his theory of compact quantum matrix groups. An algebraic approach to quantized coordinate algebras was given about this time by Yu.I. Manin.

A striking feature of quantum group theory is the surprising connections with many, sometimes at first glance unrelated, branches of mathematics and physics. There are links with mathematical fields such as Lie groups, Lie algebras and their representations, special functions, knot theory, low-dimensional topology, operator algebras, noncommutative geometry, and combinatorics. On the physical side there are interrelations with the quantum inverse scattering method, the theory of integrable models, elementary particle physics, conformal and quantum field theories, and others. It is expected that quantum groups will lead to a deeper understanding of the concept of symmetry in physics.

Currently there is no satisfactory general definition of a quantum group. It is commonly accepted that quantum groups are certain "nice" Hopf algebras and that the standard deformations of the enveloping Hopf algebras of semisimple Lie algebras and of coordinate Hopf algebras of the corresponding Lie groups are guiding examples. Instead of searching for a rigorous definition of a quantum group it seems to be more fruitful to look for classes of Hopf algebras that give rise to a rich theory with important applications and contain enough interesting examples. In this book at least three such classes are extensively studied: quasitriangular Hopf algebras, coquasitriangular Hopf algebras, and compact quantum group algebras.

The aim of this book is to provide a treatment of the theory of quantum algebras (quantized universal enveloping algebras), quantum groups (quantized algebras of functions), their representations and corepresentations, and the noncommutative differential calculus on quantum groups. The exposition is organized such that different parts of the text can be read and used (almost) independently of others. Sections 1.2 and 1.3 contain the main general definitions and notions on Hopf algebras needed in the text. This book is divided into four parts.

Part I serves (among others) as an introduction to the theory of Hopf algebras, to the quantum algebra $U_q(\mathrm{sl}_2)$, the quantum group $SL_q(2)$, the q-oscillator algebra, and to their representations. The reader can use the corresponding chapters as first steps in order to learn the theory of quantum groups. A beginner might try to become aquainted with the language of Hopf algebras by reading Sect. 1.1 and portions of Sects. 1.2 and 1.3 and then passing immediately to Chaps. 3 or 4 (or start with Chaps. 3 or 4 and read parallel to them the relevant parts of Chap. 1). The main parts of the material of Chaps. 1, 3, and 4 can also be taken as a basis for an introductory course on quantum groups.

Parts II–IV cover some of the more advanced topics of the theory. In Part II (quantized universal enveloping algebras) and Part III (quantized algebras of functions) both fundamental approaches to quantum groups are developed in detail and as independently as possible, so readers interested in only one of these parts can restrict themselves to the corresponding chapters. Nevertheless the connections between both approaches appear to be very fruitful and instructive (see Sects. 4.4, 4.5.5, 9.4, 11.2.3, 11.5, and 11.6.6). A reader who is interested in only noncommutative differential calculus should pass directly to Part IV of the book and begin with Chaps. 12 or 14 (of course, some knowledge about the corresponding quantum groups from Chap. 9 and the L-functionals from Subsect. 10.1.3 is still required there). Together with Sects. 1.2 and 1.3, Parts II–IV form an advanced text on quantum groups. Selected material from these parts can also be used for graduate courses or seminars on quantum groups. Moreover, a large number of explicit formulas and new material (for instance, in Sects. 8.5, 10.1.3, 10.3.1, 13.2, and 14.3–5) are provided throughout the text, so we hope the book may be useful for experts as well.

Let us say a few words about the selected topics and the presentation in the book. Our objective in choosing the material was to cover important and useful tools and methods for (possible) applications in theoretical and mathematical physics (especially in representation theory and in noncommutative differential calculus). Of course, this depends on our personal view of the matter. We have tried to give a comprehensive treatment of the chosen topics at the price of not including some concepts (for instance, the quantum Weyl group). Although we develop a number of general concepts too, the emphasis in the book is always placed on the study of concrete quantum groups and quantum algebras and their representations. Most of the results are presented with complete (but sometimes concise) proofs. Often, missing proofs or gaps in the existing literature have been filled. For some rather technical proofs (in particular of advanced algebraic results) readers are referred to the original papers. In many cases we have omitted proofs that are similar to the classical case. Having the potential reader in mind, we have avoided abstract mathematical theories whenever it was possible. For instance, we do not use cohomology theory, category theory (apart from Subsect. 10.3.4), Poisson–Lie groups, deformation theory, and knot theory in the book. We assume, however, that the reader has some standard knowledge of Lie groups and Lie algebras and their representation theory.

The book is organized as follows. Formulas, results, definitions, examples, and remarks are numbered and quoted consecutively within the chapters. When a reference to an item in another chapter is made, the number of the chapter is added. For instance, (30) means formula (30) in the same chapter and Propostion 9.7 refers to Proposition 7 in Chap. 9. The end of a proof is marked by $\square$ and of an example or a remark by $\triangle$. The reader should also notice that often assumptions are fixed and kept in force throughout the whole chapter, section, or subsection. Bibliographical comments are usually gathered at the end of each chapter. There the sources of some results or notions are cited (as far as the authors are aware) and some related references are listed, but no attempt has been made to report the origins of all items.

We want to express our gratitude to A. Schüler and I. Heckenberger for their indispensible help and valuable suggestions in writing this book and to Mrs. K. Schmidt for typing parts of the manuscript. We also thank Yu. Bespalov, A. Gavrilik, and L. Vainerman for reading parts of the book.

Kiev and Leipzig, March 1997 *A.U. Klimyk, K. Schmüdgen*

Table of Contents

Part I. An Introduction to Quantum Groups

Part III. Quantized Algebras of Functions

Part IV. Noncommutative Differential Calculus

Part I

An Introduction to Quantum Groups

1. Hopf Algebras

The underlying mathematical notion for quantum groups is that of a *Hopf algebra*. These are associative algebras equipped with additional structures such as a comultiplication, a counit and an antipode. In some appropriate sense, these structures and their axioms reflect the multiplication, the unit element and the inverse elements of a group and their corresponding properties.

The purpose of this chapter is twofold. First, the reader unfamiliar with this topic may try to learn the basics of Hopf algebras here and then continue with Chaps. 3 or 4. Subsections marked by * should be omitted by beginners. Secondly, Sects. 1.2 and 1.3 collect general definitions and notation used throughout this book.

1.1 Prolog: Examples of Hopf Algebras of Functions on Groups

Probably the best way to take up a new concept and to understand the ideas behind its abstract definition might be to look at well-chosen examples. We shall do so before we enter the path through the wealth of necessary definitions. Our examples will only work with functions on "ordinary" groups, but they provide the motivation for later constructions on quantum groups.

To begin with, let G be an arbitrary group and let $\mathcal{F}(G)$ be the algebra of all complex-valued functions on G with pointwise algebraic operations. The group operations on G allow one to define the following mappings:

- comultiplication $\Delta : \mathcal{F}(G) \to \mathcal{F}(G \times G)$ by $(\Delta(f))(g_1, g_2) := f(g_1 g_2)$,
- counit $\varepsilon : \mathcal{F}(G) \to \mathbb{C}$ by $\varepsilon(f) := f(e)$,
- antipode $S : \mathcal{F}(G) \to \mathcal{F}(G)$ by $(S(f))(g) := f(g^{-1})$.

Here e denotes the unit element of G. From these definitions it is clear that

$$\Delta \text{ and } \varepsilon \text{ are algebra homomorphisms.} \tag{1}$$

The group axioms lead to the following identities (2)–(4) for the mappings Δ, ε and S. First, the associativity of the group multiplication implies that

$$(\Delta \otimes \mathrm{id}) \circ \Delta = (\mathrm{id} \otimes \Delta) \circ \Delta, \tag{2}$$

where id is the identity mapping on $\mathcal{F}(G)$. Indeed, the definition of Δ yields

$$(((\Delta \otimes \mathrm{id}) \circ \Delta)f)(g_1, g_2, g_3) = f((g_1g_2)g_3) \in \mathcal{F}(G \times G \times G),$$

$$(((\mathrm{id} \otimes \Delta) \circ \Delta)f)(g_1, g_2, g_3) = f(g_1(g_2g_3)) \in \mathcal{F}(G \times G \times G).$$

Since $(g_1g_2)g_3 = g_1(g_2g_3)$ by the associativity law, the two expressions on the right hand sides coincide and (2) follows. Secondly, we have the relation $f(eg) = f(ge) = f(g)$ which can be expressed as

$$(\varepsilon \otimes \mathrm{id}) \circ \Delta = (\mathrm{id} \otimes \varepsilon) \circ \Delta = \mathrm{id} \tag{3}$$

under the usual identifications of $\mathbb{C} \otimes \mathcal{F}(G)$ and $\mathcal{F}(G) \otimes \mathbb{C}$ with $\mathcal{F}(G)$. Define a mapping $m : \mathcal{F}(G \times G) \to \mathcal{F}(G)$ by $(mh)(g) := h(g, g)$, $h \in \mathcal{F}(G \times G)$, and a linear mapping $\eta : \mathbb{C} \to \mathcal{F}(G)$ such that $\eta(1)$ is the unit element of the algebra $\mathcal{F}(G)$. Then, thirdly, the relations $g^{-1}g = gg^{-1} = e$ for the inverse g^{-1} of a group element g yield the identities

$$m \circ (S \otimes \mathrm{id}) \circ \Delta = \eta \circ \varepsilon = m \circ (\mathrm{id} \otimes S) \circ \Delta. \tag{4}$$

Indeed, inserting the above definitions of m, Δ and S we obtain

$$((m \circ (\mathrm{id} \otimes S) \circ \Delta)f)(g) = (((\mathrm{id} \otimes S) \circ \Delta)f)(g, g) = f(gg^{-1}) = f(e) = (\eta \circ \varepsilon)(f)$$

which gives the second equality of (4). The first equality follows similarly.

Summarizing the preceding, we have seen that the group multiplication, the group unit and the group inverse of G induce mappings Δ, ε and S of functions on G such that the corresponding group axioms are expressed as identities (2)–(4). On the level of functions the properties of the group G are encoded in the three mappings Δ, ε and S defined above.

One disadvantage is that the algebra $\mathcal{F}(G \times G)$ of functions on $G \times G$ appeared in the definitions of the mappings Δ and m. In order to work with functions on G alone, we consider the tensor product $\mathcal{F}(G) \otimes \mathcal{F}(G)$ as a linear subspace of $\mathcal{F}(G \times G)$ by identifying $f_1 \otimes f_2 \in \mathcal{F}(G) \otimes \mathcal{F}(G)$ with the function $(f_1 \otimes f_2)(g_1, g_2) := f_1(g_1)f_2(g_2)$ on $G \times G$. Then we have $m(f_1 \otimes f_2)(g) = f_1(g)f_2(g)$, that is, $m(f_1 \otimes f_2)$ is just the product f_1f_2 of f_1 and f_2 in the algebra $\mathcal{F}(G)$. After the embedding $\mathcal{F}(G) \otimes \mathcal{F}(G) \subseteq \mathcal{F}(G \times G)$ one difficulty still remains: Δ does not map $\mathcal{F}(G)$ to $\mathcal{F}(G) \otimes \mathcal{F}(G)$ in general (for example, $\Delta(f) \notin \mathcal{F}(G) \otimes \mathcal{F}(G)$ if $f(t) = \sin t^2$ and $G = \mathbb{R}$). So it is natural to look for appropriate subalgebras $\mathcal{A}$ of $\mathcal{F}(G)$ for which $\Delta(\mathcal{A}) \subseteq \mathcal{A} \otimes \mathcal{A}$ and $S(\mathcal{A}) \subseteq \mathcal{A}$. Such a subalgebra $\mathcal{A}$ together with the linear mappings $\Delta : \mathcal{A} \to \mathcal{A} \otimes \mathcal{A}$, $\varepsilon : \mathcal{A} \to \mathbb{C}$ and $S : \mathcal{A} \to \mathcal{A}$ satisfying the above conditions (1)–(4) is called a *Hopf algebra* (by Definition 4 below). If the Hopf algebra $\mathcal{A}$ is sufficiently "large", one can expect that $\mathcal{A}$ and its structure maps Δ, ε and S store enough information about the group G. We consider three important examples of such Hopf algebras $\mathcal{A}$.

Example 1 (The Hopf algebra $\mathcal{F}(G)$ of a finite group). For a finite group G we have $\mathcal{F}(G) \otimes \mathcal{F}(G) = \mathcal{F}(G \times G)$, so by the preceding $\mathcal{F}(G)$ is a Hopf algebra. △

Example 2 (The Hopf algebra Rep(G) *of a compact group G).* Suppose that G is a compact topological group. Let T be a continuous representation of G on a finite-dimensional complex vector space V. Let $\mathcal{C}(T)$ denote the linear subspace of $\mathcal{F}(G)$ spanned by the matrix elements $t_{ij}(\cdot)$, $i,j = 1,2,\cdots,N$, of $T(\cdot)$ relative to some basis of V. Since $t_{ij}(g_1g_2) = \sum_k t_{ik}(g_1)t_{kj}(g_2)$, $g_1, g_2 \in G$, and $t_{ij}(e) = \delta_{ij}$, we have

$$\Delta(t_{ij}) = \sum\nolimits_k t_{ik} \otimes t_{kj}, \quad \varepsilon(t_{ij}) = \delta_{ij}. \tag{5}$$

The vector space $\mathcal{C}(T)$ equipped with the linear mappings $\Delta : \mathcal{C}(T) \to \mathcal{C}(T) \otimes \mathcal{C}(T)$ and $\varepsilon : \mathcal{C}(T) \to \mathbb{C}$ which satisfy the relations (2) and (3) is called a *coalgebra* (by Definition 2 below).

Let $\mathcal{A} = \mathrm{Rep}\,(G)$ be the linear span of spaces $\mathcal{C}(T)$ for all continuous finite-dimensional representations T of G. Since $\mathcal{C}(T_1 \oplus T_2) = \mathcal{C}(T_1) + \mathcal{C}(T_2)$ and $\mathcal{C}(T_1 \otimes T_2) = \mathcal{C}(T_1)\cdot\mathcal{C}(T_2)$ as is easily seen, $\mathcal{A}$ is a subalgebra of $\mathcal{F}(G)$. Recall that the contragredient representation T^c of T acts on the dual V' of the space V by $(T^c(g)v')(v) = v'(T(g^{-1})v)$, $g \in G$, $v \in V$, $v' \in V'$. If we choose the basis of V' dual to the basis of V, then T^c has the matrix elements $t^c_{ij} = S(t_{ji})$. Thus, we get $\mathcal{C}(T^c) = S(\mathcal{C}(T))$ and hence $S(\mathcal{A}) \subseteq \mathcal{A}$. Since $\Delta(\mathcal{C}(T)) \subseteq \mathcal{C}(T) \otimes \mathcal{C}(T)$ by (5), we also have $\Delta(\mathcal{A}) \subseteq \mathcal{A} \otimes \mathcal{A}$. Therefore $\mathcal{A} = \mathrm{Rep}\,(G)$ is a Hopf algebra. It is called the Hopf algebra of *representative functions* of G. Since Rep$\,(G)$ is dense in the algebra $C(G)$ of continuous functions on G by the Peter–Weyl theorem, Rep$\,(G)$ is sufficiently "large".

Let $(t_{ij})_{i,j=1}^N$ be a matrix of functions $t_{ij} \in \mathcal{F}(G)$. A simple consideration shows that the functions t_{ij} are matrix elements of some continuous finite-dimensional representation T of G if and only if they belong to the algebra Rep$\,(G)$ and satisfy the equations (5). This means that continuous finite-dimensional representations of G can be defined completely in terms of the Hopf algebra Rep$\,(G)$ without referring to the group G itself. Many other concepts of group representation theory can also be formulated in terms which are intrinsic to the Hopf algebra Rep$\,(G)$ (see Chap. 11). △

Example 3 (The coordinate Hopf algebras $\mathcal{O}(G)$ of simple matrix Lie groups). Let G denote one of the matrix groups $SL(N,\mathbb{C})$, $SO(N,\mathbb{C})$ or $Sp(N,\mathbb{C})$. Each element g of G is a complex $N \times N$ matrix $g = (g_{ij})$. Define the *coordinate functions* u^i_j on G by $u^i_j(g) := g_{ij}$, $g = (g_{ij}) \in G$. For $g, h \in G$ we have

$$\Delta(u^i_j)(g,h) = u^i_j(gh) = (gh)_{ij} = \sum\nolimits_k g_{ik}h_{kj} = \sum\nolimits_k u^i_k(g)u^k_j(h)$$

and $u^i_j(e) = \delta_{ij}$, so that

$$\Delta(u^i_j) = \sum\nolimits_k u^i_k \otimes u^k_j \quad \text{and} \quad \varepsilon(u^i_j) = \delta_{ij}. \tag{6}$$

Let $\mathcal{A} = \mathcal{O}(G)$ be the subalgebra of $\mathcal{F}(G)$ generated by the N^2 functions u^i_j, $i, j = 1, 2, \cdots, N$. Since $\Delta : \mathcal{F}(G) \to \mathcal{F}(G \times G)$ is an algebra homomorphism, we have $\Delta(\mathcal{A}) \subseteq \mathcal{A} \otimes \mathcal{A}$ by the first formula of (6). Any element $g \in G$ has determinant 1. Hence the function which is a constant equal to 1 on G belongs to $\mathcal{A}$, so $\mathcal{A}$ has a unit. Further, it follows that there are polynomials p_{ij} in N^2 indeterminates such that $(g^{-1})_{ij} = p_{ij}(g_{11}, g_{12}, \cdots, g_{NN})$, so that

$$\begin{aligned} S(u^i_j)(g) &= u^i_j(g^{-1}) = (g^{-1})_{ij} = p_{ij}(u^1_1(g), u^1_2(g), \cdots, u^N_N(g)) \\ &= p_{ij}(u^1_1, u^1_2, \cdots, u^N_N)(g). \end{aligned}$$

That is, we have $S(u^i_j) \in \mathcal{A}$ and hence $S(\mathcal{A}) \subseteq \mathcal{A}$. Therefore, by the discussion preceding Example 1, $\mathcal{A} = \mathcal{O}(G)$ is a Hopf algebra. It is called the *coordinate Hopf algebra* of G. $\triangle$

In all three examples of Hopf algebras the multiplication is *commutative* (being the ordinary multiplication of functions) and the antipode S is an algebra homomorphism such that $S^2 = \mathrm{id}$. These facts no longer remain true for general Hopf algebras and quantum groups.

Summarizing and slightly simplifying the above considerations, we have rephrased (certain) groups in terms of Hopf algebras of functions on the groups. This turns out to be the proper perspective that makes it possible to "deform" or to "quantize" the groups by passing to *noncommutative* algebras. Formally this is done by introducing Hopf algebras depending on say one parameter (usually denoted by q or e^h, h being thought of as Planck's constant) which specialize to the Hopf algebra associated with the group at a particular value (usually $q = 1$, resp. $h = 0$). The elements of such a deformed Hopf algebra will be viewed as functions on a "quantum group". The language of groups and representations will also be used for the deformed Hopf algebras as an inspiring source of motivation, though there are no groups around.

1.2 Coalgebras, Bialgebras and Hopf Algebras

Let us fix some general notation which will be kept throughout the book. The letter $\mathbb{K}$ always stands for a commutative ring with unit. If V and W are vector spaces over $\mathbb{K}$, $\mathcal{L}(V, W)$ denotes the $\mathbb{K}$-linear mappings of V to W and V' is the vector space of all $\mathbb{K}$-linear functionals on V. We set $\mathcal{L}(V) := \mathcal{L}(V, V)$. The symbol $\otimes$ means the tensor product $\otimes_{\mathbb{K}}$. We identify $\mathbb{K} \otimes V$ and $V \otimes \mathbb{K}$ with V by the isomorphisms $1 \otimes v \to v$ and $v \otimes 1 \to v$, respectively. If not specified otherwise, the letter τ denotes the flip operator given by $\tau(v \otimes w) = w \otimes v$.

1.2.1 Algebras

Recall that an *(associative) algebra* is a vector space $\mathcal{A}$ over $\mathbb{K}$ equipped with a mapping $(a, b) \to ab$ of $\mathcal{A} \times \mathcal{A}$ to $\mathcal{A}$ such that

$$a(bc) = (ab)c, \quad (a+b)c = ac+bc, \quad a(b+c) = ab+ac,$$

$$\alpha(ab) = (\alpha a)b = a(\alpha b) \tag{7}$$

for all $a, b \in \mathcal{A}$ and $\alpha \in \mathbb{K}$. An element 1 of an algebra $\mathcal{A}$ is called a *unit* of $\mathcal{A}$ if $1a = a1 = a$ for all $a \in \mathcal{A}$. If a unit element exists in $\mathcal{A}$, it is uniquely determined. Throughout this book, by an *algebra* we always mean an associative algebra with unit element.

The following slight reformulation of the algebra definition is the right perspective for its dualization in Subsect. 1.2.2.

Definition 1. *An* algebra *(that is, an associative algebra with unit) is a vector space $\mathcal{A}$ over $\mathbb{K}$ together with two linear maps $m : \mathcal{A} \otimes \mathcal{A} \to \mathcal{A}$, called the* multiplication *or the* product, *and $\eta : \mathbb{K} \to \mathcal{A}$, called the* unit, *such that*

$$m \circ (m \otimes \mathrm{id}) = m \circ (\mathrm{id} \otimes m), \tag{8}$$

$$m \circ (\eta \otimes \mathrm{id}) = \mathrm{id} = m \circ (\mathrm{id} \otimes \eta). \tag{9}$$

Given such an algebra $\mathcal{A}$, $ab := m(a \otimes b)$ is the product of a and b and the mapping η is determined by its value $\eta(1) \in \mathcal{A}$, which is the unit element of $\mathcal{A}$. Both definitions of an algebra are easily seen to be equivalent. Equation (8) is the associativity law, while (9) says that $\eta(1)$ is a unit element of $\mathcal{A}$ using the identification of $\mathbb{K} \otimes \mathcal{A}$ and $\mathcal{A} \otimes \mathbb{K}$ with $\mathcal{A}$. The conditions in (7) are built into the requirement that $m : \mathcal{A} \otimes \mathcal{A} \to \mathcal{A}$ is $\mathbb{K}$-linear.

The associativity (8) of the multiplication m means that the diagram

$$\begin{array}{ccc} \mathcal{A} \otimes \mathcal{A} \otimes \mathcal{A} & \xrightarrow{m \otimes \mathrm{id}} & \mathcal{A} \otimes \mathcal{A} \\ \Big\downarrow{\scriptstyle \mathrm{id} \otimes m} & & \Big\downarrow{\scriptstyle m} \\ \mathcal{A} \otimes \mathcal{A} & \xrightarrow{m} & \mathcal{A} \end{array} \tag{10}$$

is commutative. Likewise, the condition (9) of the unit can be expressed by the commutativity of the following diagram:

$$\begin{array}{ccccc} \mathbb{K} \otimes \mathcal{A} & \xrightarrow{\eta \otimes \mathrm{id}} & \mathcal{A} \otimes \mathcal{A} & \xleftarrow{\mathrm{id} \otimes \eta} & \mathcal{A} \otimes \mathbb{K} \\ \Big\downarrow{\scriptstyle \mathrm{id}} & & \Big\downarrow{\scriptstyle m} & & \Big\downarrow{\scriptstyle \mathrm{id}} \\ \mathbb{K} \otimes \mathcal{A} & \Longleftrightarrow & \mathcal{A} & \Longleftrightarrow & \mathcal{A} \otimes \mathbb{K}, \end{array} \tag{11}$$

where $\mathbb{K} \otimes \mathcal{A} \Longleftrightarrow \mathcal{A}$ and $\mathcal{A} \Longleftrightarrow \mathcal{A} \otimes \mathbb{K}$ denote the canonical identifications.

Let $\mathcal{A}$ and $\mathcal{B}$ be algebras. A $\mathbb{K}$-linear mapping $\varphi : \mathcal{A} \to \mathcal{B}$ is called an *algebra homomorphism* if $\varphi(aa') = \varphi(a)\varphi(a')$ for all $a, a' \in \mathcal{A}$ and $\varphi(1_{\mathcal{A}}) = 1_{\mathcal{B}}$. The two latter conditions can be rewritten as

$$\varphi \circ m_{\mathcal{A}} = m_{\mathcal{B}} \circ (\varphi \otimes \varphi) \quad \text{and} \quad \varphi \circ \eta_{\mathcal{A}} = \eta_{\mathcal{B}}.$$

There exists a *tensor product algebra* $\mathcal{A}\otimes\mathcal{B}$ whose vector space is the tensor product of vector spaces of $\mathcal{A}$ and $\mathcal{B}$ and whose multiplication is defined by $m_{\mathcal{A}\otimes\mathcal{B}} := (m_{\mathcal{A}}\otimes m_{\mathcal{B}})\circ(\mathrm{id}\otimes\tau\otimes\mathrm{id})$. That is, the algebra $\mathcal{A}\otimes\mathcal{B}$ has the product

$$(a\otimes b)(a'\otimes b') := aa'\otimes bb', \quad a,a'\in\mathcal{A},\ b,b'\in\mathcal{B}. \tag{12}$$

For each algebra $\mathcal{A}$ one can define the *opposite algebra* $\mathcal{A}^{\mathrm{op}}$. This is an algebra with the same underlying vector space as $\mathcal{A}$, but with the new multiplication $m_{\mathcal{A}^{\mathrm{op}}} := m_{\mathcal{A}}\circ\tau$. That is, we have $a\cdot_{\mathrm{op}} b = b\cdot a$, where $\cdot_{\mathrm{op}}$ and $\cdot$ denote the products of $\mathcal{A}^{\mathrm{op}}$ and $\mathcal{A}$, respectively.

Many algebras are constructed as quotients of free algebras or tensor algebras. Let us recall these notions in the case where $\mathbb{K}$ is the complex field $\mathbb{C}$. Let $\{x_i \mid i\in I\}$ be an indexed set of generators. We denote by I^∞ the union of all sets I^n, $n\in\mathbb{N}_0$, where I^0 consists of a single element o which is not in I. For $\mathrm{i}=(i_1,i_2,\cdots,i_k)\in I^k$ and $\mathrm{j}=(j_1,j_2,\cdots,j_n)\in I^n$, $k,n\in\mathbb{N}_0$, we set $(\mathrm{i},\mathrm{j})=(i_1,\cdots,i_k,j_1,\cdots,j_n)$ and $x_{\mathrm{i}} := x_{i_1}\cdots x_{i_k}$. For $\mathrm{o}\in I^0$ and $i\in I^k$, $k\in\mathbb{N}_0$, we put $(\mathrm{o},i)=(i,\mathrm{o}) := i$ and $x_{\mathrm{o}} := 1$. Then the *free algebra* $\mathbb{C}\langle x_i\rangle$ *with generators* x_i, $i\in I$, is defined as follows: it is a complex vector space with basis $\{x_{\mathrm{i}} \mid \mathrm{i}\in I^\infty\}$ and with the multiplication defined by $x_{\mathrm{i}}x_{\mathrm{j}} := x_{(\mathrm{i},\mathrm{j})}$, $\mathrm{i},\mathrm{j}\in I^\infty$. The algebra $\mathbb{C}\langle x_i\rangle$ possesses the following universal property: *for any indexed subset* $\{a_i \mid i\in I\}$ *of another algebra* $\mathcal{A}$ *there exists a unique algebra homomorphism* $\varphi:\mathbb{C}\langle x_i\rangle\to\mathcal{A}$ *such that* $\varphi(x_i)=a_i$ *for all* $i\in I$.

Next we define the tensor algebra $T(V)$ over a vector space V. Set $V^{\otimes 0} := \mathbb{K}$ and $V^{\otimes n} := V\otimes\cdots\otimes V$ (n times). The direct sum $T(V) := \bigoplus_{n=0}^{\infty} V^{\otimes n}$ of these vector spaces becomes an algebra, called the *tensor algebra* $T(V)$ over V, with multiplication determined by $x_n y_k := x_n\otimes y_k$ for $x_n\in V^{\otimes n}$ and $y_k\in V^{\otimes k}$. The algebra $T(V)$ also has a universal property: *for every linear map* $\varphi_1 : V\to\mathcal{A}$ *of* V *to an algebra* $\mathcal{A}$, *there exists a unique algebra homomorphism* $\varphi : T(V)\to\mathcal{A}$ *such that* $\varphi(v)=\varphi_1(v)$ *for* $v\in V$. The universal properties of the algebras $\mathbb{C}\langle x_i\rangle$ and $T(V)$ imply the following fact: for any basis $\{x_i \mid i\in I\}$ of the vector space V, there is a unique algebra isomorphism φ of the algebras $T(V)$ and $\mathbb{C}\langle x_i\rangle$ such that $\varphi(x_i)=x_i$, $i\in I$.

Often in this book an algebra $\mathcal{A}$ is introduced by stating that it is generated by elements $x_1,\cdots x_r$ with relations $f_k(x_1,\cdots,x_r)=0$, $k=1,2,\cdots,s$. By such a phrase we mean that $\mathcal{A}$ is the quotient algebra of the free algebra $\mathbb{C}\langle x_i\rangle$ with generators $x_1,\cdots,x_r$ by the two-sided ideal $\mathcal{I}$ of $\mathbb{C}\langle x_i\rangle$ generated by $f_k(x_1,\cdots,x_r)$, $k=1,2,\cdots,s$. For simplicity we shall denote the image of the elements $x_i\in\mathbb{C}\langle x_i\rangle$ in the quotient algebra $\mathcal{A}=\mathbb{C}\langle x_i\rangle/\mathcal{I}$ also by x_i.

1.2.2 Coalgebras

Recall that an algebra is a vector space $\mathcal{A}$ with two linear mappings $m : \mathcal{A}\otimes\mathcal{A}\to\mathcal{A}$ and $\eta:\mathbb{K}\to\mathcal{A}$ such that the diagrams (10) and (11) are commutative. We now dualize this definition by reversing all arrows and replacing all mappings by the corresponding dual ones. In doing so, the multiplication

$m : \mathcal{A} \otimes \mathcal{A} \to \mathcal{A}$ is replaced by the comultiplication $\Delta : \mathcal{A} \to \mathcal{A} \otimes \mathcal{A}$, the unit $\eta : \mathbb{K} \to \mathcal{A}$ by the counit $\varepsilon : \mathcal{A} \to \mathbb{K}$ and the diagrams (10) and (11) go into the following diagrams (13) and (14), which have to be commutative:

$$\begin{array}{ccc} \mathcal{A} \otimes \mathcal{A} \otimes \mathcal{A} & \xleftarrow{\Delta \otimes \mathrm{id}} & \mathcal{A} \otimes \mathcal{A} \\ \uparrow{\scriptstyle \mathrm{id} \otimes \Delta} & & \uparrow{\scriptstyle \Delta} \\ \mathcal{A} \otimes \mathcal{A} & \xleftarrow{\Delta} & \mathcal{A} \end{array} \tag{13}$$

$$\begin{array}{ccccc} \mathbb{K} \otimes \mathcal{A} & \xleftarrow{\varepsilon \otimes \mathrm{id}} & \mathcal{A} \otimes \mathcal{A} & \xrightarrow{\mathrm{id} \otimes \varepsilon} & \mathcal{A} \otimes \mathbb{K} \\ \uparrow{\scriptstyle \mathrm{id}} & & \uparrow{\scriptstyle \Delta} & & \uparrow{\scriptstyle \mathrm{id}} \\ \mathbb{K} \otimes \mathcal{A} & \Longleftrightarrow & \mathcal{A} & \Longleftrightarrow & \mathcal{A} \otimes \mathbb{K}. \end{array} \tag{14}$$

We thus obtain the following "dual" notion to an algebra.

Definition 2. *A* coalgebra *is a vector space $\mathcal{A}$ over $\mathbb{K}$ equipped with two linear mappings $\Delta : \mathcal{A} \to \mathcal{A} \otimes \mathcal{A}$, called the* comultiplication *or the* coproduct, *and $\varepsilon : \mathcal{A} \to \mathbb{K}$, called the* counit, *such that*

$$(\Delta \otimes \mathrm{id}) \circ \Delta = (\mathrm{id} \otimes \Delta) \circ \Delta, \tag{15}$$

$$(\varepsilon \otimes \mathrm{id}) \circ \Delta = \mathrm{id} = (\mathrm{id} \otimes \varepsilon) \circ \Delta. \tag{16}$$

Equation (15) is referred to as the *coassociativity* of the comultiplication Δ, because it dualizes the associativity (8) of the multiplication m.

Proceeding in a similar manner, most concepts on algebras can be dualized to coalgebra notions by giving diagrammatic formulations of the definitions and then reversing all arrows. We only list the outcomes of these dualizations in the following, but the reader unfamiliar with this procedure should carry out a few more examples in order to get used to working with "costructures".

Let $\mathcal{A}$ and $\mathcal{B}$ be coalgebras. A $\mathbb{K}$-linear mapping $\varphi : \mathcal{A} \to \mathcal{B}$ is said to be a *coalgebra homomorphism* if

$$\Delta_{\mathcal{B}} \circ \varphi = (\varphi \otimes \varphi) \circ \Delta_{\mathcal{A}} \quad \text{and} \quad \varepsilon_{\mathcal{A}} = \varepsilon_{\mathcal{B}} \circ \varphi.$$

The *tensor product coalgebra* $\mathcal{A} \otimes \mathcal{B}$ is the coalgebra built on the vector space $\mathcal{A} \otimes \mathcal{B}$ with comultiplication $\Delta_{\mathcal{A} \otimes \mathcal{B}} := (\mathrm{id} \otimes \tau \otimes \mathrm{id}) \circ (\Delta_{\mathcal{A}} \otimes \Delta_{\mathcal{B}})$ and counit $\varepsilon_{\mathcal{A} \otimes \mathcal{B}} := \varepsilon_{\mathcal{A}} \otimes \varepsilon_{\mathcal{B}}$. The *coopposite coalgebra* $\mathcal{A}^{\mathrm{cop}}$ is the coalgebra on the vector space $\mathcal{A}$ equipped with the new comultiplication $\Delta_{\mathcal{A}^{\mathrm{cop}}} := \tau \circ \Delta_{\mathcal{A}}$ and the counit $\varepsilon_{\mathcal{A}}$. A coalgebra $\mathcal{A}$ is said to be *cocommutative* if $\tau \circ \Delta = \Delta$. A linear subspace $\mathcal{B}$ of $\mathcal{A}$ is a *subcoalgebra* if $\Delta(\mathcal{B}) \subseteq \mathcal{B} \otimes \mathcal{B}$. A $\mathbb{K}$-linear subspace $\mathcal{I}$ of $\mathcal{A}$ is called a (two–sided) *coideal* if

$$\Delta(\mathcal{I}) \subseteq \mathcal{A} \otimes \mathcal{I} + \mathcal{I} \otimes \mathcal{A} \quad \text{and} \quad \varepsilon(\mathcal{I}) = \{0\}.$$

If $\mathcal{I}$ is a coideal of $\mathcal{A}$, the quotient vector space $\mathcal{A}/\mathcal{I}$ becomes a coalgebra with comultiplication and counit induced from $\mathcal{A}$.

Next we introduce the so-called *Sweedler notation* for the comultiplication. If a is an element of a coalgebra $\mathcal{A}$, the element $\Delta(a) \in \mathcal{A}\otimes\mathcal{A}$ is a finite sum

$$\Delta(a) = \sum\nolimits_i a_{1i} \otimes a_{2i}, \quad a_{1i}, a_{2i} \in \mathcal{A}. \tag{17}$$

Moreover, this representation of $\Delta(a)$ is not unique. For notational simplicity, we shall suppress the index i and write the sum (17) symbolically as

$$\Delta(a) = \sum a_{(1)} \otimes a_{(2)}. \tag{18}$$

Here the subscripts (1) and (2) refer to the corresponding tensor factors and the symbol $\sum$ reminds us that $\Delta(a)$ is actually a sum.

Let us define inductively mappings $\Delta^{(n)} : \mathcal{A} \to \mathcal{A}^{\otimes n+1}$ by

$$\Delta^{(n)} := (\mathrm{id} \otimes \Delta) \circ \Delta^{(n-1)}, \quad n > 1, \quad \text{and} \quad \Delta^{(1)} = \Delta.$$

(From the coassociativity (15) it follows that $\Delta^{(n)}$ is in fact equal to $n-1$ compositions of Δ independently of their order, that is, $\Delta^{(2)} = (\mathrm{id}\otimes\Delta)\circ\Delta = (\Delta\otimes\mathrm{id})\circ\Delta$, etc.) Then the element $\Delta^{(n)}(a) \in \mathcal{A}^{\otimes n+1}$ is denoted by

$$\Delta^{(n)}(a) := \sum a_{(1)} \otimes a_{(2)} \otimes \cdots \otimes a_{(n+1)}. \tag{19}$$

One can also replace the sums in (18) and (19) by the symbol $\sum_{(a)}$ in order to emphasize the element $a \in \mathcal{A}$. The reader may check the correct use of this notation by adding the corresponding elements to the sum.

The Sweedler notations (18) and (19) turn out to be very useful when working with general Hopf algebras, because they allow one to give elegant shorthand formulations of definitions, formulas and proofs. We strongly recommend the reader to practise its use (for instance, by carrying out proofs in Propositions 1, 5 and others below) in order to become acquainted with this powerful tool. The result will make the effort worthwhile.

The coassociativity (15) in terms of the notation in (18) and (19) yields

$$\begin{aligned}\sum (a_{(1)})_{(1)} \otimes (a_{(1)})_{(2)} \otimes a_{(2)} &= \sum a_{(1)} \otimes (a_{(2)})_{(1)} \otimes (a_{(2)})_{(2)} \\ &= \Delta^{(2)}(a) = \sum a_{(1)} \otimes a_{(2)} \otimes a_{(3)}.\end{aligned}$$

The counit property (16) reads in the Sweedler notation (18) as

$$\sum \varepsilon(a_{(1)})a_{(2)} = a = \sum a_{(1)}\varepsilon(a_{(2)}). \tag{20}$$

Proposition 1. *Let $\mathcal{A}$ be a coalgebra and $\mathcal{B}$ an algebra. Then the vector space $\mathcal{L}(\mathcal{A},\mathcal{B})$ of all $\mathbb{K}$-linear mappings of $\mathcal{A}$ to $\mathcal{B}$ equipped with the convolution product defined by*

$$(f * g)(a) := (m_{\mathcal{B}} \circ (f \otimes g)\Delta_{\mathcal{A}})(a) \equiv \sum f(a_{(1)})g(a_{(2)}), \quad a \in \mathcal{A}, \tag{21}$$

becomes an algebra with unit $\eta_{\mathcal{B}} \circ \varepsilon_{\mathcal{A}}$.

Proof. By the associativity of the multiplication in $\mathcal{B}$ and the coassociativity of the comultiplication in $\mathcal{A}$, we obtain

$$((f*g)*h)(a) = \sum f(a_{(1)})g(a_{(2)})h(a_{(3)}) = (f*(g*h))(a),$$

that is, the convolution product (21) is associative. The relation

$$((\eta_{\mathcal{B}} \circ \varepsilon_{\mathcal{A}}) * f)(a) = \sum \varepsilon_{\mathcal{A}}(a_{(1)})f(a_{(2)}) = f\Big(\sum \varepsilon_{\mathcal{A}}(a_{(1)})a_{(2)}\Big) = f(a)$$

by (20) shows that $\eta_{\mathcal{B}} \circ \varepsilon_{\mathcal{A}}$ is a left unit. It is similarly proved that $\eta_{\mathcal{B}} \circ \varepsilon_{\mathcal{A}}$ is a right unit. □

Corollary 2. *The dual vector space $\mathcal{A}'$ of a coalgebra $\mathcal{A}$ is an algebra with product*

$$(fg)(a) := (f*g)(a) = \sum f(a_{(1)})g(a_{(2)}), \quad a \in \mathcal{A}, \quad f,g \in \mathcal{A}'. \tag{22}$$

Proof. Apply Proposition 1 to the algebra $\mathcal{B} = \mathbb{K}$ (over $\mathbb{K}$). □

The dual space $\mathcal{B}'$ of an algebra $\mathcal{B}$ does not become a coalgebra in general by dualizing the multiplication (see the polynomial algebra $\mathbb{C}[x]$ for a counter-example), however it does if the algebra $\mathcal{B}$ is finite-dimensional. The reason is that we always have $\mathcal{A}' \otimes \mathcal{A}' \subseteq (\mathcal{A} \otimes \mathcal{A})'$ and hence $m_{\mathcal{A}'} : \mathcal{A}' \otimes \mathcal{A}' \to \mathcal{A}'$, while $\mathcal{B}' \otimes \mathcal{B}'$ is only a proper subspace of $(\mathcal{B} \otimes \mathcal{B})'$ if $\mathcal{B}$ is infinite-dimensional, so that $\Delta_{\mathcal{B}'}(f) := f \circ m_{\mathcal{B}}$ may not lie in $\mathcal{B}' \otimes \mathcal{B}'$. There are two possibilities to remedy this situation: to replace the dual space $\mathcal{B}'$ by a subspace (this will be done in Subsect. 1.2.8) or to enlarge the tensor product (for instance, by taking its completion in some topology).

Example 4 (The coalgebra $M_N(\mathbb{K})'$). Let u^i_j be the linear functional on the algebra $M_N(\mathbb{K})$ of $N \times N$ matrices over $\mathbb{K}$ defined by $u^i_j(g) := g_{ij}$ for $g = (g_{ij}) \in M_N(\mathbb{K})$. Then $M_N(\mathbb{K})' = \mathrm{Lin}\,\{u^i_j \mid i,j = 1,2,\cdots,N\}$ is a coalgebra with comultiplication $\Delta(u^i_j) = \sum_k u^i_k \otimes u^k_j$ and counit $\varepsilon(u^i_j) = \delta_{ij}$. △

1.2.3 Bialgebras

A bialgebra is an algebra and a coalgebra, where both structures are compatible in the sense of the following

Proposition 3. *If $\mathcal{A}$ is a vector space which is an algebra and a coalgebra, then the following two conditions are equivalent:*

(i) $\Delta : \mathcal{A} \to \mathcal{A} \otimes \mathcal{A}$ *and* $\varepsilon : \mathcal{A} \to \mathbb{K}$ *are algebra homomorphisms.*
(ii) $m : \mathcal{A} \otimes \mathcal{A} \to \mathcal{A}$ *and* $\eta : \mathbb{K} \to \mathcal{A}$ *are coalgebra homomorphisms.*

Proof. By the above definitions, the assertions of (i) mean that

$$\Delta \circ m = m_{\mathcal{A}\otimes\mathcal{A}} \circ (\Delta \otimes \Delta), \quad \Delta \circ \eta = \eta_{\mathcal{A}\otimes\mathcal{A}}, \quad \varepsilon \circ m = m_{\mathbb{K}} \circ (\varepsilon \otimes \varepsilon), \quad \varepsilon \circ \eta = \eta_{\mathbb{K}}, \tag{23}$$

while the statements of (ii) can be expressed as

$$\Delta \circ m = (m \otimes m) \circ \Delta_{\mathcal{A} \otimes \mathcal{A}}, \;\; \varepsilon_{\mathcal{A} \otimes \mathcal{A}} = \varepsilon \circ m, \;\; \Delta \circ \eta = (\eta \otimes \eta) \circ \Delta_{\mathbb{K}}, \;\; \varepsilon_{\mathbb{K}} = \varepsilon \circ \eta. \quad (24)$$

The right hand sides of the first relations of (23) and (24) are both equal to $(m \otimes m) \circ (\mathrm{id} \otimes \tau \otimes \mathrm{id}) \circ (\Delta \otimes \Delta)$. The second equality of (23) and the third of (24) are equivalent. The two last equations both mean that $\varepsilon(1) = 1$. Likewise, the third relation of (23) and the second of (24) are equivalent. (The reader is invited to express the relations of (23) and (24) as diagrams and to check their equivalence in this manner.) □

Definition 3. *A* bialgebra *is a vector space which is an algebra and a coalgebra such that the conditions in Proposition 3 hold. That is, if $\mathcal{A}$ is an algebra and a coalgebra, it is a bialgebra if and only if for any $a, b \in \mathcal{A}$ we have*

$$\Delta(ab) = \Delta(a)\Delta(b), \;\; \varepsilon(ab) = \varepsilon(a)\varepsilon(b), \;\; \Delta(1) = 1 \otimes 1, \;\; \varepsilon(1) = 1. \quad (25)$$

Let $\mathcal{A}$ and $\mathcal{B}$ be bialgebras. By a *bialgebra homomorphism* of $\mathcal{A}$ to $\mathcal{B}$ we mean a mapping that is both an algebra and a coalgebra homomorphism. The vector space $\mathcal{A} \otimes \mathcal{B}$ endowed with the tensor product algebra and coalgebra structures becomes a bialgebra. There are three bialgebras $\mathcal{A}^{\mathrm{op}}$, $\mathcal{A}^{\mathrm{cop}}$ and $\mathcal{A}^{\mathrm{op,cop}}$ which are obtained from $\mathcal{A}$ by taking the opposite of either the algebra or coalgebra structure or of both of them. That is, $\mathcal{A}^{\mathrm{op}}$ has the opposite multiplication $m_{\mathcal{A}^{\mathrm{op}}}$ and the comultiplication of $\mathcal{A}$, $\mathcal{A}^{\mathrm{cop}}$ has the multiplication of $\mathcal{A}$ and the opposite comultiplication $\Delta_{\mathcal{A}^{\mathrm{cop}}}$, and $\mathcal{A}^{\mathrm{op,cop}}$ carries the opposite multiplication $m_{\mathcal{A}^{\mathrm{op}}}$ and comultiplication $\Delta_{\mathcal{A}^{\mathrm{cop}}}$. (The reader may verify that the above structures are indeed bialgebras.)

A linear subspace $\mathcal{I}$ of a bialgebra $\mathcal{A}$ is called a *biideal* if it is both a two-sided ideal of the algebra $\mathcal{A}$ and a coideal of the coalgebra $\mathcal{A}$. The quotient $\mathcal{A}/\mathcal{I}$ of a bialgebra $\mathcal{A}$ by a biideal $\mathcal{I}$ is again a bialgebra with structures inherited from $\mathcal{A}$.

We now introduce two concepts of distinguished elements of a bialgebra $\mathcal{A}$. The motivation for this terminology stems from Examples 6 and 7 in Subsect. 1.2.6. A nonzero element $g \in \mathcal{A}$ is called *group-like* if $\Delta(g) = g \otimes g$. An element $x \in \mathcal{A}$ is called *primitive* if $\Delta(x) = x \otimes 1 + 1 \otimes x$.

Proposition 4. *Let $\mathcal{A}$ be a bialgebra. Then the product of group-like elements is group-like. If x and y are primitive elements of $\mathcal{A}$, then we have $\varepsilon(x) = \varepsilon(y) = 0$ and the element $[x, y] := xy - yx$ is also primitive.*

Proof. The first assertion is obvious. Let x and y be primitive elements. Since $(\mathrm{id} \otimes \varepsilon)\Delta(x) = x\varepsilon(1) + \varepsilon(x)1 = x$ by (16) and $\varepsilon(1) = 1$, we obtain $\varepsilon(x) = 0$. Using the fact that Δ is an algebra homomorphism, we get

$$\Delta(xy) = (x \otimes 1 + 1 \otimes x)(y \otimes 1 + 1 \otimes y) = xy \otimes 1 + x \otimes y + y \otimes x + 1 \otimes xy,$$

$$\Delta(yx) = yx \otimes 1 + y \otimes x + x \otimes y + 1 \otimes yx,$$

so that $\Delta([x, y]) = \Delta(xy) - \Delta(yx) = [x, y] \otimes 1 + 1 \otimes [x, y]$. □

By Proposition 4, the set of group-like elements of $\mathcal{A}$ is a semigroup with unit and the primitive elements of $\mathcal{A}$ form a Lie algebra with respect to the commutator bracket $[\cdot,\cdot]$.

Example 5 (The bialgebras $\mathcal{O}(M_N(\mathbb{K}))$ and $\mathbb{K}[x_1, x_2, \cdots, x_N]$). Let $\mathcal{A} = \mathcal{O}(M_N(\mathbb{K}))$ be the algebra generated by the functions u^i_j from Example 4. That is, $\mathcal{A}$ is just the polynomial algebra in N^2 indeterminates u^i_j, $i, j = 1, 2, \cdots, N$. Then $\mathcal{A}$ becomes a bialgebra with comultiplication and counit determined by $\Delta(u^i_j) = \sum_k u^i_k \otimes u^k_j$ and $\varepsilon(u^i_j) = \delta_{ij}$. The corresponding coordinate algebra of the diagonal matrices in $M_N(\mathbb{K})$ is the polynomial algebra $\mathbb{K}[x_1, x_2, \cdots, x_N]$ in the indeterminates $x_i = u^i_i$. It is a bialgebra with comultiplication $\Delta(x_i) = x_i \otimes x_i$ and counit $\varepsilon(x_i) = 1$. △

1.2.4 Hopf Algebras

The following notion is fundamental in quantum group theory.

Definition 4. *A bialgebra $\mathcal{A}$ is called a* Hopf algebra *if there exists a linear mapping $S : \mathcal{A} \to \mathcal{A}$, called the* antipode *or the* coinverse *of $\mathcal{A}$, such that*

$$m \circ (S \otimes \mathrm{id}) \circ \Delta = \eta \circ \varepsilon = m \circ (\mathrm{id} \otimes S) \circ \Delta. \tag{26}$$

Clearly, the equations (26) are just the requirement that the diagram

$$\begin{array}{ccccc}
\mathcal{A}\otimes\mathcal{A} & \xleftarrow{\ \Delta\ } & \mathcal{A} & \xrightarrow{\ \Delta\ } & \mathcal{A}\otimes\mathcal{A} \\
\Big\downarrow \mathrm{id}\otimes S & & \Big\downarrow \eta\circ\varepsilon & & \Big\downarrow S\otimes\mathrm{id} \\
\mathcal{A}\otimes\mathcal{A} & \xrightarrow{\ m\ } & \mathcal{A} & \xleftarrow{\ m\ } & \mathcal{A}\otimes\mathcal{A}
\end{array}$$

is commutative. In the Sweedler notation (18), the relations (26) say that

$$\sum S(a_{(1)})a_{(2)} = \varepsilon(a)1 = \sum a_{(1)}S(a_{(2)}), \tag{27}$$

where $\Delta(a) = \sum a_{(1)} \otimes a_{(2)}$. The equations (26) can be nicely interpreted by using the convolution product (21) of the algebra $\mathcal{L}(\mathcal{A}, \mathcal{A})$. Comparing the formulas (21) and (26) we see that (26) means nothing but that S is the convolution inverse of the identity mapping, that is, $S * \mathrm{id} = \mathrm{id} * S = \eta \circ \varepsilon$. Thus, *a bialgebra $\mathcal{A}$ is a Hopf algebra if and only if the identity map of $\mathcal{A}$ is invertible in the algebra $\mathcal{L}(\mathcal{A}, \mathcal{A})$*. Moreover, the preceding characterization implies that *an antipode of a Hopf algebra is uniquely determined.*

The next proposition gives some important properties of the antipode.

Proposition 5. *The antipode S of a Hopf algebra $\mathcal{A}$ is an algebra anti-homomorphism and a coalgebra anti-homomorphism of $\mathcal{A}$. This means that $S : \mathcal{A} \to \mathcal{A}^{\mathrm{op}}$ is an algebra homomorphism and $S : \mathcal{A} \to \mathcal{A}^{\mathrm{cop}}$ is a coalgebra homomorphism. That is, we have*

$$S(ab) = S(b)S(a), \quad a, b \in \mathcal{A}, \quad \textit{and} \quad S(1) = 1, \tag{28}$$

$$\Delta \circ S = \tau \circ (S \otimes S) \circ \Delta \quad \textit{and} \quad \varepsilon \circ S = \varepsilon. \tag{29}$$

Proof. The proof is a nice exercise in the use of the Sweedler notation. The reader who is unfamiliar with it might first try to carry out the proof itself and then compare it with the detailed presentation below. Using the Hopf algebra axioms and the facts that Δ and ε are algebra homomorphisms, we compute

$$\begin{aligned}
S(b)S(a) &= \sum S(b_{(1)}\varepsilon(b_{(2)}))S(a_{(1)}\varepsilon(a_{(2)})) \\
&= \sum S(b_{(1)})S(a_{(1)})\varepsilon(a_{(2)}b_{(2)}) \\
&= \sum S(b_{(1)})S(a_{(1)})(a_{(2)}b_{(2)})_{(1)}S((a_{(2)}b_{(2)})_{(2)}) \\
&= \sum S(b_{(1)})S(a_{(1)})a_{(2)}b_{(2)}S(a_{(3)}b_{(3)}) \\
&= \sum S(b_{(1)})(\varepsilon(a_{(1)})1)b_{(2)}S(a_{(2)}b_{(3)}) \\
&= \sum \varepsilon(a_{(1)})\varepsilon(b_{(1)})S(a_{(2)}b_{(2)}) = S(ab).
\end{aligned}$$

This proves the first equality of (28). The second one follows at once from the relation $m \circ (S \otimes \mathrm{id}) \circ \Delta(1) = \varepsilon(1)1$ combined with the facts that $\Delta(1) = 1 \otimes 1$ and $\varepsilon(1) = 1$. Similarly, we compute

$$\begin{aligned}
\sum S(a_{(2)}) \otimes S(a_{(1)}) &= \sum S(a_{(2)}\varepsilon(a_{(3)})) \otimes S(a_{(1)}) \\
&= \sum (S(a_{(2)}) \otimes S(a_{(1)}))(\varepsilon(a_{(3)})1 \otimes 1) \\
&= \sum (S(a_{(2)}) \otimes S(a_{(1)}))(\Delta(a_{(3)}S(a_{(4)}))) \\
&= \sum (S(a_{(2)}) \otimes S(a_{(1)}))(a_{(3)} \otimes a_{(4)})\Delta(S(a_{(5)})) \\
&= \sum (S(a_{(2)})a_{(3)} \otimes S(a_{(1)})a_{(4)})(\Delta(S(a_{(5)}))) \\
&= \sum (\varepsilon(a_{(2)})1 \otimes S(a_{(1)})a_{(3)})(\Delta(S(a_{(4)})) \\
&= \sum (1 \otimes S(a_{(1)})a_{(2)})(\Delta(S(a_{(3)})) \\
&= \sum (1 \otimes \varepsilon(a_{(1)})1)(\Delta(S(a_{(2)})) = \Delta(S(a)),
\end{aligned}$$

which is the first relation of (29). The second relation is derived by

$$\varepsilon(S(a)) = \varepsilon\Big(S\Big(\sum a_{(1)}\varepsilon(a_{(2)})\Big)\Big) = \varepsilon\Big(\sum S(a_{(1)})a_{(2)}\Big) = \varepsilon(\varepsilon(a)1) = \varepsilon(a).$$

□

In the Sweedler notation the first equality of (29) can be rewritten as

$$\sum S(a)_{(1)} \otimes S(a)_{(2)} = \sum S(a_{(2)}) \otimes S(a_{(1)}). \tag{30}$$

Let $\mathcal{A}$ be a Hopf algebra. Since the bialgebra $\mathcal{A}^{\text{op,cop}}$ has opposite multiplication and comultiplication, the antipode S of $\mathcal{A}$ satisfies the relations (26) for $\mathcal{A}^{\text{op,cop}}$ as well. Hence S is also an antipode of $\mathcal{A}^{\text{op,cop}}$ and the bialgebra $\mathcal{A}^{\text{op,cop}}$ is a Hopf algebra. However, for the bialgebras $\mathcal{A}^{\text{op}}$ and $\mathcal{A}^{\text{cop}}$ to be Hopf algebras S needs to be an invertible linear mapping of $\mathcal{A}$.

Proposition 6. *For any Hopf algebra $\mathcal{A}$ the following conditions are equivalent:*

(i) *The antipode S of $\mathcal{A}$ is invertible as a linear mapping of $\mathcal{A}$.*
(ii) *The bialgebra $\mathcal{A}^{\text{op}}$ is a Hopf algebra.*
(iii) *The bialgebra $\mathcal{A}^{\text{cop}}$ is a Hopf algebra.*

In this case, the inverse S^{-1} of S is the antipode of $\mathcal{A}^{\text{op}}$ and $\mathcal{A}^{\text{cop}}$.

Proof. (i)→(ii): The inverse S^{-1} of S is an algebra anti-homomorphism of $\mathcal{A}$ as S is by Proposition 5. Hence, by formula (28) we have

$$\begin{aligned}\sum S^{-1}(a_{(2)})a_{(1)} &= \sum S^{-1}(a_{(2)})(S^{-1}\circ S)(a_{(1)}) = S^{-1}\Big(\sum S(a_{(1)})a_{(2)}\Big)\\ &= S^{-1}(\varepsilon(a)1) = \varepsilon(a)1\end{aligned}$$

and similarly $\sum a_{(2)}S^{-1}(a_{(1)}) = \varepsilon(a)1$ for $a \in \mathcal{A}$. That is, S^{-1} satisfies condition (26) for the bialgebra $\mathcal{A}^{\text{op}}$, so that it is an antipode of $\mathcal{A}^{\text{op}}$.

(ii)→(i): Let $\hat{S}$ denote the antipode of $\mathcal{A}^{\text{op}}$. Since $\hat{S}$ is also an algebra anti-homomorphism of $\mathcal{A}$, we obtain

$$\begin{aligned}\hat{S}S(a) &= \hat{S}S\Big(\sum \varepsilon(a_{(2)})a_{(1)}\Big) = \sum \varepsilon(a_{(2)})\hat{S}S(a_{(1)})\\ &= \sum(a_{(3)}\hat{S}(a_{(2)}))\hat{S}S(a_{(1)}) = \sum a_{(3)}\hat{S}(S(a_{(1)})a_{(2)})\\ &= \sum(a_{(2)}\hat{S}(\varepsilon(a_{(1)})1) = a\hat{S}(1) = a\end{aligned}$$

and similarly $S\hat{S}(a) = a$ for $a \in \mathcal{A}$. Therefore, $\hat{S}$ is the inverse of S.

The equivalence of (i) and (iii) is proved in a similar manner. □

Corollary 7. *If a Hopf algebra $\mathcal{A}$ is commutative or cocommutative, then $S^2 = \text{id}$.*

Proof. In either case we have $\mathcal{A} = \mathcal{A}^{\text{op}}$ or $\mathcal{A} = \mathcal{A}^{\text{cop}}$, so $\mathcal{A}^{\text{op}}$ or $\mathcal{A}^{\text{cop}}$ is a Hopf algebra and the assertion follows from Proposition 6. □

The antipode S of a Hopf algebra $\mathcal{A}$ is said to be of *finite order* if $S^n = \text{id}$ for some $n \in \mathbb{N}$. The smallest such n is called the *order* of S. Otherwise, one says that S is of infinite order. By a result of D. E. Radford [Rad1], the antipode of any *finite-dimensional* Hopf algebra is of finite order (so, in particular, S is invertible and $\mathcal{A}^{\text{op}}$ and $\mathcal{A}^{\text{cop}}$ are Hopf algebras). Further, a result of R. G. Larson and D. E. Radford (see [LaR]) says that for any finite-dimensional complex cosemisimple Hopf algebra (see Sect. 11.2 for this notion) we even have $S^2 = \text{id}$. The latter solved an old conjecture of I. Kaplansky.

We shall see later that it is often easy to check the Hopf algebra axioms for a set of generators, so the following simple fact appears to be useful.

Proposition 8. *Let $\mathcal{A}_g$ be a subset of an algebra $\mathcal{A}$ which generates $\mathcal{A}$ as an algebra. Let $\Delta : \mathcal{A} \to \mathcal{A} \otimes \mathcal{A}$ and $\varepsilon : \mathcal{A} \to \mathbb{K}$ be homomorphisms and $S : \mathcal{A} \to \mathcal{A}$ an anti-homomorphism of the corresponding algebras. If the coassociativity condition (15) and the counit condition (16) (and the antipode condition (26)) are satisfied for elements in $\mathcal{A}_g$, then they are valid on the whole of $\mathcal{A}$ and hence $\mathcal{A}$ is a bialgebra (resp. a Hopf algebra).*

Proof. In the case of Δ and ε the assertions follow at once from the fact that $(\Delta \otimes \mathrm{id}) \circ \Delta$, $(\mathrm{id} \otimes \Delta) \circ \Delta$, $(\varepsilon \otimes \mathrm{id}) \circ \Delta$ and $(\mathrm{id} \otimes \varepsilon)\Delta$ are all algebra homomorphisms. In the case of S it suffices to show that (26) holds for a product ab provided that it is valid for a and for b. In order to do so, we compute

$$\begin{aligned}\sum S((ab)_{(1)})(ab)_{(2)} &= \sum S(a_{(1)}b_{(1)})a_{(2)}b_{(2)} = \sum S(b_{(1)})(S(a_{(1)})a_{(2)})b_{(2)} \\ &= \sum S(b_{(1)})(\varepsilon(a)1)b_{(2)} = \varepsilon(a)\varepsilon(b)1 = \varepsilon(ab)1.\end{aligned}$$

The other equality of (26) is verified similarly. □

We close this subsection with a few general definitions and facts. Let $\mathcal{A}$ and $\mathcal{B}$ be Hopf algebras. A linear mapping $\varphi : \mathcal{A} \to \mathcal{B}$ is called a *Hopf algebra homomorphism* if it respects the Hopf algebra structures, that is, if φ is a bialgebra homomorphism and $\varphi \circ S_{\mathcal{A}} = S_{\mathcal{B}} \circ \varphi$. Any bialgebra homomorphism between Hopf algebras is already a Hopf algebra homomorphism, that is, the condition $\varphi \circ S_{\mathcal{A}} = S_{\mathcal{B}} \circ \varphi$ is then automatically fulfilled.

A *Hopf ideal* of $\mathcal{A}$ is a biideal $\mathcal{I}$ such that $S(\mathcal{I}) \subseteq \mathcal{I}$. The quotient $\mathcal{A}/\mathcal{I}$ of a Hopf algebra $\mathcal{A}$ by a Hopf ideal $\mathcal{I}$ is again a Hopf algebra.

1.2.5* Dual Pairings of Hopf Algebras

Let us consider a finite-dimensional Hopf algebra $\mathcal{A}$. Recall that by Corollary 2 the dual vector space $\mathcal{A}'$ is an algebra with respect to the multiplication $fg(a) := (f \otimes g)\Delta(a) \equiv \sum f(a_{(1)})g(a_{(2)})$. For $f \in \mathcal{A}'$ we define a functional $\Delta(f) \in (\mathcal{A}\otimes\mathcal{A})'$ by $\Delta(f)(a\otimes b) := (f \circ m)(a\otimes b) \equiv f(ab)$, $a, b \in \mathcal{A}$. Since $\mathcal{A}$ is finite-dimensional, $(\mathcal{A}\otimes\mathcal{A})' = \mathcal{A}'\otimes\mathcal{A}'$ and so $\Delta(f) \in \mathcal{A}'\otimes\mathcal{A}'$. It is not difficult to show that the algebra $\mathcal{A}'$ equipped with the comultiplication Δ becomes a Hopf algebra. The antipode, the counit and the unit element of this Hopf algebra $\mathcal{A}'$ are given by $(Sf)(a) = f(S(a))$, $\varepsilon_{\mathcal{A}'}(f) = f(1)$ and $1_{\mathcal{A}'}(a) = \varepsilon(a)$, respectively. That is, the Hopf algebra $\mathcal{A}'$ is obtained by "dualizing" the structure maps of the Hopf algebra $\mathcal{A}$. This situation is generalized in

Definition 5. *A* dual pairing *of two bialgebras $\mathcal{U}$ and $\mathcal{A}$ is a bilinear mapping $\langle \cdot, \cdot \rangle : \mathcal{U} \times \mathcal{A} \to \mathbb{K}$ such that*

$$\langle \Delta_{\mathcal{U}}(f), a_1 \otimes a_2 \rangle = \langle f, a_1 a_2 \rangle, \quad \langle f_1 f_2, a \rangle = \langle f_1 \otimes f_2, \Delta_{\mathcal{A}}(a) \rangle, \tag{31}$$

$$\langle f, 1_{\mathcal{A}}\rangle = \varepsilon_{\mathcal{U}}(f), \quad \langle 1_{\mathcal{U}}, a\rangle = \varepsilon_{\mathcal{A}}(a) \tag{32}$$

for all $f, f_1, f_2 \in \mathcal{U}$ *and* $a, a_1, a_2 \in \mathcal{A}$. *A dual pairing* $\langle\cdot,\cdot\rangle$ *of* $\mathcal{U}$ *and* $\mathcal{A}$ *is called* nondegenerate *if* $\langle f, a\rangle = 0$ *for all* $f \in \mathcal{U}$ *implies that* $a = 0$ *and if* $\langle f, a\rangle = 0$ *for all* $a \in \mathcal{A}$ *implies* $f = 0$.

Proposition 9. *If* $\langle\cdot,\cdot\rangle$ *is a dual pairing of two Hopf algebras* $\mathcal{U}$ *and* $\mathcal{A}$, *then we have*

$$\langle S(f), a\rangle = \langle f, S(a)\rangle, \quad f \in \mathcal{U}, \quad a \in \mathcal{A}. \tag{33}$$

Proof. Consider the linear functionals F, G and H on the tensor product Hopf algebra $\mathcal{U} \otimes \mathcal{A}$ defined by $F(f \otimes a) = \langle f, a\rangle$, $G(f \otimes a) = \langle S(f), a\rangle$, $H(f \otimes a) = \langle f, S(a)\rangle$, $f \in \mathcal{U}$, $a \in \mathcal{A}$. We show that $FG = HF = \varepsilon$ in the algebra $(\mathcal{U} \otimes \mathcal{A})'$. Indeed, by (31) and (32), we get

$$FG(f \otimes a) = \sum \langle f_{(1)}, a_{(1)}\rangle\langle S(f_{(2)}), a_{(2)}\rangle = \sum \langle f_{(1)} S(f_{(2)}), a\rangle$$

$$= \langle \varepsilon(f)1, a\rangle = \varepsilon(f \otimes a),$$

that is, $FG = \varepsilon$, and similarly $HF = \varepsilon$. Thus, $G = (HF)G = H(FG) = H$ which yields (33). □

The discussion preceding Definition 5 has shown that for any finite-dimensional Hopf algebra $\mathcal{A}$ we have a dual pairing of the Hopf algebras $\mathcal{U} := \mathcal{A}'$ and $\mathcal{A}$ defined by $\langle f, a\rangle := f(a)$, $a \in \mathcal{A}$, $f \in \mathcal{A}'$. Obviously, this pairing is nondegenerate.

Let us discuss the above notion with some remarks. First note that in equation (31) the pairing of $\mathcal{U}$ and $\mathcal{A}$ must be extended to one of $\mathcal{U} \otimes \mathcal{U}$ and $\mathcal{A} \otimes \mathcal{A}$ by setting $\langle f_1 \otimes f_2, a_1 \otimes a_2\rangle := \langle f_1, a_1\rangle\langle f_2, a_2\rangle$.

Dual pairings of two Hopf algebras are by no means unique. For any couple of Hopf algebras $\mathcal{U}$ and $\mathcal{A}$ there is always the trivial pairing defined by $\langle f, a\rangle := \varepsilon_{\mathcal{U}}(f)\varepsilon_{\mathcal{A}}(a)$. Further, compositions with Hopf algebra homomorphisms in both variables again give a dual pairing.

Let $\langle\cdot,\cdot\rangle$ be a dual pairing of two Hopf algebras $\mathcal{U}$ and $\mathcal{A}$. Then $\mathcal{I}_{\mathcal{U}} := \{f \in \mathcal{U} \mid \langle f, a\rangle = 0 \text{ for all } a \in \mathcal{A}\}$ and $\mathcal{I}_{\mathcal{A}} := \{a \in \mathcal{A} \mid \langle f, a\rangle = 0 \text{ for all } f \in \mathcal{U}\}$ are Hopf ideals of $\mathcal{U}$ resp. $\mathcal{A}$ and the pairing $\langle\cdot,\cdot\rangle$ of $\mathcal{U}$ and $\mathcal{A}$ induces a *nondegenerate* dual pairing of the quotient Hopf algebras $\mathcal{U}/\mathcal{I}_{\mathcal{U}}$ and $\mathcal{A}/\mathcal{I}_{\mathcal{A}}$.

From the formulas (31) we see that any dual pairing of two bialgebras is completely determined by the values of the bilinear form $\langle\cdot,\cdot\rangle$ on sets of generators of the algebras $\mathcal{U}$ and $\mathcal{A}$. This allows one to use shorthand descriptions of pairings by means of matrices.

A nondegenerate dual pairing of two Hopf algebras $\mathcal{U}$ and $\mathcal{A}$ stores a lot of information about the links of both structures. (For instance, it follows then from (31) that $\mathcal{U}$ is commutative if and only if $\mathcal{A}$ is cocommutative.) In quantum group theory there is a general philosophy which says that then the Hopf algebras $\mathcal{U}$ and $\mathcal{A}$ provide *dual approaches* to the same "quantum object".

1.2.6 Examples of Hopf Algebras

Three important examples of Hopf algebras related to groups have already been developed in Sect. 1.1. Two others will be described below. The following example is the starting point for our first approach to quantum groups. The quantized universal enveloping algebras studied in Chaps. 3 and 6 are deformations of the Hopf algebras $U(\mathfrak{g})$ which are universal enveloping algebras of Lie algebras.

Example 6 (The universal enveloping algebra $U(\mathfrak{g})$). Let $\mathfrak{g}$ be a Lie algebra over $\mathbb{K}$ with Lie bracket $[\cdot,\cdot]$. The universal enveloping algebra $U(\mathfrak{g})$ is defined as the quotient algebra of the tensor algebra $T(\mathfrak{g})$ of the vector space $\mathfrak{g}$ by the two-sided ideal $\mathcal{I}$ generated by the elements $x\otimes y - y\otimes x - [x,y]$, where $x,y\in\mathfrak{g}$. That is, $U(\mathfrak{g})$ is the free algebra generated by a vector space basis of $\mathfrak{g}$ subject to the relations $[x,y]=xy-yx$ for $x,y\in\mathfrak{g}$. By its very definition, the algebra $U(\mathfrak{g})$ admits the following universal property: given a linear mapping $\varphi:\mathfrak{g}\to\mathcal{A}$ of $\mathfrak{g}$ to another algebra $\mathcal{A}$ satisfying

$$\varphi([x,y])=\varphi(x)\varphi(y)-\varphi(y)\varphi(x),\quad x,y\in\mathfrak{g}, \tag{34}$$

there is a unique algebra homomorphism $\tilde{\varphi}:U(\mathfrak{g})\to\mathcal{A}$ such that $\tilde{\varphi}(x)=\varphi(x)$ for $x\in\mathfrak{g}$. This fact is used in proving the following assertion.

There is a unique Hopf algebra structure on the algebra $U(\mathfrak{g})$ such that

$$\Delta(x)=x\otimes 1+1\otimes x,\quad \varepsilon(x)=0,\quad S(x)=-x,\quad x\in\mathfrak{g}. \tag{35}$$

In order to prove this, we define linear mappings $\varphi=\Delta$, ε and S of $\mathfrak{g}$ to the algebras $\mathcal{A}=U(\mathfrak{g})\otimes U(\mathfrak{g}),\mathbb{K},U(\mathfrak{g})^{\mathrm{op}}$ by (35). Condition (34) is obviously fulfilled for the mappings ε and S. One verifies (by arguing as in the proof of Proposition 4 above) that Δ also satisfies (34). Therefore, by the universal property of $U(\mathfrak{g})$, these mappings extend to algebra homomorphisms of $U(\mathfrak{g})$ to $\mathcal{A}$, denoted still by Δ, ε and S. By Proposition 8, it suffices to check the Hopf algebra axioms on the generators $x\in\mathfrak{g}$. This is easily done; as an example, we have $m\circ(S\otimes\mathrm{id})\circ\Delta(x)=m(-x\otimes 1+1\otimes x)=0=\varepsilon(x)$ for $x\in\mathfrak{g}$. By (35), any $x\in\mathfrak{g}$ is a primitive element of the Hopf algebra $U(\mathfrak{g})$. In the case $\mathbb{K}=\mathbb{C}$ it can be shown (see [Bou2]) that the elements of $\mathfrak{g}$ exhaust all primitive elements of $U(\mathfrak{g})$, so the Lie algebra of primitive elements of $U(\mathfrak{g})$ is just the Lie algebra $\mathfrak{g}$ itself. In particular, this gives an intrinsic characterization of the Lie algebra $\mathfrak{g}$ in terms of the Hopf algebra structure of $U(\mathfrak{g})$.

From (35) we see that the Hopf algebra $U(\mathfrak{g})$ is cocommutative.

Let $\mathcal{B}$ be either the Hopf algebra Rep(G) of Example 2 for a connected compact Lie group G or one of the Hopf algebras $\mathcal{O}(G)$ of Example 3. Suppose that $\mathbb{K}=\mathbb{C}$ and $\mathfrak{g}$ is the Lie algebra of the Lie group G. It is well-known that each element $a\in U(\mathfrak{g})$ acts as a left-invariant differential operator $\tilde{a}$ on G. If $a=x_1x_2\cdots x_n$ with $x_1,x_2,\cdots,x_n\in\mathfrak{g}$, the action of $\tilde{a}$ on a function $f\in C^\infty(G)$ is given by

$$(\tilde{a}f)(g) = \left.\frac{\partial^n}{\partial t_1 \cdots \partial t_n}\right|_{t_1=\cdots=t_n=0} f(g\exp(t_1x_1)\cdots\exp(t_nx_n)). \tag{36}$$

One can show that there is a nondegenerate dual pairing of the Hopf algebras $U(\mathfrak{g})$ and $\mathcal{B}$ determined by $\langle a, f\rangle := (\tilde{a}f)(e)$, where $a \in U(\mathfrak{g})$ and $f \in \mathcal{B}$. △

Example 7 (The group algebra $\mathbb{K}G$). Let $\mathbb{K}G$ be the vector space with basis given by the elements of G. It becomes an algebra by extending linearly the product of G. The unit of this algebra is the unit element of G. There is a unique Hopf algebra structure on the algebra $\mathbb{K}G$ such that $\Delta(g) = g \otimes g$, $\varepsilon(g) = 1$ and $S(g) = g^{-1}$ for $g \in G$.

The elements of G are obviously group-like. If $\mathbb{K} = \mathbb{C}$, it is easily checked that they exhaust all group-like elements of the Hopf algebra $\mathbb{K}G$. The Hopf algebra $\mathbb{K}G$ is cocommutative.

Let $\mathcal{A}$ denote one of the Hopf algebras $\mathcal{F}(G)$, $\mathrm{Rep}\,(G)$ or $\mathcal{O}(G)$ defined in Examples 1, 2 and 3, respectively. Then there is a nondegenerate dual pairing $\langle\cdot,\cdot\rangle$ of the Hopf algebras $\mathcal{A}$ and $\mathcal{B} := \mathbb{K}G$ defined by $\langle f, \sum_i \lambda_i g_i\rangle = \sum_i \lambda_i f(g_i)$, where $f \in \mathcal{A}$ and $\sum_i \lambda_i g_i \in \mathcal{B}$. △

Example 8 (The tensor algebra $T(V)$). Let V be a vector space and $T(V)$ the tensor algebra over V defined in Subsect. 1.2.1. Arguing similarly as in Example 6 (replacing thereby the universal property of $U(\mathfrak{g})$ by that of $T(V)$), it follows that there exists a unique Hopf algebra structure on $T(V)$ such that $\Delta(v) = v \otimes 1 + 1 \otimes v$, $\varepsilon(v) = 0$ and $S(v) = -v$ for $v \in V$. This Hopf algebra is cocommutative. For arbitrary elements $v_1, v_2, \cdots, v_n \in V$ we have

$$\Delta(v_1 \cdots v_n) = \sum_{k=0}^{n} \sum_{p\in\mathcal{P}_{nk}} v_{p(1)} \cdots v_{p(k)} \otimes v_{p(k+1)} \cdots v_{p(n)}, \tag{37}$$

$$S(v_1v_2 \cdots v_n) = (-1)^n v_n \cdots v_2v_1. \tag{38}$$

Here $\mathcal{P}_{nk}$ denotes the set of all permutations of $\{1, 2, \cdots, n\}$ such that $p(1) < \cdots < p(k)$ and $p(k+1) < \cdots < p(n)$. The elements of $\mathcal{P}_{nk}$ are called $(k, n-k)$-*shuffles*. In the cases $k = 0$ and $k = n$ the corresponding summands in (37) should be read as $1 \otimes v_1v_2 \cdots v_n$ and $v_1v_2 \cdots v_n \otimes 1$, respectively. Formula (37) can be proved by induction on n, and (38) follows immediately from (28).

Let V be a Lie algebra $\mathfrak{g}$. It is straightforward to check that the ideal $\mathcal{I} \equiv \langle x \otimes y - y \otimes x - [x, y] \mid x, y \in \mathfrak{g}\rangle$ is a Hopf ideal of the Hopf algebra $T(V)$. Hence, the Hopf algebra $U(\mathfrak{g})$ from Example 6 is just the quotient Hopf algebra $T(V)/\mathcal{I}$. In particular, the formulas (37) and (38) remain valid for elements $v_1, v_2, \cdots, v_n \in \mathfrak{g} = V$ in the Hopf algebra $U(\mathfrak{g})$.

Let us specialize to the case where V is an abelian Lie algebra with trivial Lie bracket $[x, y] = 0$, $x, y \in \mathfrak{g}$. Then the quotient algebra $T(V)/\mathcal{I}$ is the *symmetric algebra* $S(V)$ over V which is shown to be a Hopf algebra. △

Example 9 (M. E. Sweedler). Let $\mathcal{A}$ be the complex algebra with generators x and g and defining relations $g^2 = 1$, $x^2 = 0$ and $gxg = -x$. The elements

$1, g, x, gx$ form a vector space basis of $\mathcal{A}$. One verifies that the formulas $\Delta(g) = g \otimes g$, $\Delta(x) = x \otimes 1 + g \otimes x$, $\varepsilon(g) = 1$, $\varepsilon(x) = 0$, $S(g) = g = g^{-1}$ and $S(x) = -gx$ define a Hopf algebra structure on $\mathcal{A}$. It is the noncommutative noncocommutative complex Hopf algebra of smallest dimension. $\triangle$

1.2.7 $*$-Structures

When we consider $*$-structures we assume that $\mathbb{K}$ is the complex field $\mathbb{C}$.

By a *$*$-vector space* we mean a vector space V over the field $\mathbb{C}$ endowed with a mapping $a \to a^*$, called an *involution* and denoted by $*$, such that $(\alpha v + \beta w)^* = \overline{\alpha} v^* + \overline{\beta} w^*$ and $(v^*)^* = v$ for $v, w \in V$ and $\alpha, \beta \in \mathbb{C}$. Recall that a *$*$-algebra* is an (associative) algebra $\mathcal{A}$ (with unit) together with a mapping $a \to a^*$ such that $\mathcal{A}$ becomes a $*$-vector space and $(ab)^* = b^* a^*$ for all $a, b \in \mathcal{A}$. Such a mapping is called an *algebra involution* on $\mathcal{A}$. It follows then that $1^* = 1$.

A coalgebra $\mathcal{A}$ is called a *$*$-coalgebra* if $\mathcal{A}$ is equipped with an involution $*$ such that $\mathcal{A}$ is a $*$-vector space and $\Delta : \mathcal{A} \to \mathcal{A} \otimes \mathcal{A}$ is a $*$-homomorphism. The latter means that $\Delta(a^*) = \Delta(a)^*$ for $a \in \mathcal{A}$, where the involution of $\mathcal{A} \otimes \mathcal{A}$ is defined by $(a \otimes b)^* = a^* \otimes b^*$. In any $*$-coalgebra $\mathcal{A}$ we have $\varepsilon(a^*) = \overline{\varepsilon(a)}$, $a \in \mathcal{A}$. A *$*$-bialgebra* is a bialgebra $\mathcal{A}$ with an involution for which $\mathcal{A}$ is both a $*$-algebra and a $*$-coalgebra. A Hopf algebra which is a $*$-bialgebra is called a *Hopf $*$-algebra*. Though the latter definition does not contain a requirement concerning the behavior of the antipode with respect to the involution, the Hopf algebra structure implies the following compatibility relation.

Proposition 10. *In any Hopf $*$-algebra $\mathcal{A}$, we have $S(S(a^*)^*) = a$ for $a \in \mathcal{A}$, that is,*

$$S \circ * \circ S \circ * = \mathrm{id}. \tag{39}$$

In particular, S is invertible with inverse $S^{-1} = * \circ S \circ *$.

Proof. Put $T := * \circ S \circ *$. Setting $a = b^*$ in (27) and applying the involution yields

$$\sum b_{(2)} T(b_{(1)}) = \varepsilon(b)1 = \sum T(b_{(2)}) b_{(1)}, \quad b \in \mathcal{A}.$$

Hence T is an antipode for the Hopf algebra $\mathcal{A}^{\mathrm{op}}$. By Proposition 6, $T = S^{-1}$ and so $S \circ T = S \circ * \circ S \circ * = \mathrm{id}$. $\square$

Corollary 11. *The dual algebra $\mathcal{A}'$ of a Hopf $*$-algebra $\mathcal{A}$ is a $*$-algebra with involution defined by*

$$f^*(a) = \overline{f(S(a)^*)}, \quad f \in \mathcal{A}'. \tag{40}$$

Proof. Clearly, $f^* \in \mathcal{A}'$ and the involution is anti-linear. From (39) we obtain that $(f^*)^* = f$, while (29) implies that $(fg)^* = g^* f^*$. $\square$

Formula (40) provides the motivation for the following definition of dually paired Hopf $*$-algebras. A bilinear form $\langle \cdot, \cdot \rangle : \mathcal{U} \times \mathcal{A} \to \mathbb{C}$ is called a *dual* *pairing of two Hopf $*$-algebras $\mathcal{U}$ and $\mathcal{A}$* if it is a dual pairing of the Hopf algebras $\mathcal{U}$ and $\mathcal{A}$ (see Definition 5) and in addition we have

$$\langle f^*, a\rangle = \overline{\langle f, S(a)^*\rangle} \text{ and } \langle f, a^*\rangle = \overline{\langle S(f)^*, a\rangle}. \tag{41}$$

As for the other parts of Definition 5, condition (41) is symmetric in $\mathcal{U}$ and $\mathcal{A}$. But it suffices to require one of the relations in (41). Combined with other properties of dual pairings and Hopf $*$-algebras, the other one will follow. Indeed, by the first equality of (41) and the formulas (39) and (33), we get

$$\begin{aligned}\langle f, a^*\rangle &= \langle (f^*)^*, a^*\rangle = \overline{\langle f^*, S(a^*)^*\rangle} = \overline{\langle f^*, S^{-1}(a)\rangle}\\ &= \overline{\langle S^{-1}(f^*), a\rangle} = \overline{\langle S(f)^*, a\rangle}.\end{aligned}$$

Example 10 (Real forms of complex Lie algebras). Let $\mathfrak{g}_{\mathbb{R}}$ be a real form of a complex Lie algebra $\mathfrak{g}$. Since $\mathfrak{g}$ is a direct sum of $\mathfrak{g}_{\mathbb{R}}$ and $\mathrm{i}\mathfrak{g}_{\mathbb{R}}$, there is an involution of the complex vector space $\mathfrak{g}$ defined by $(x+\mathrm{i}y)^* = -x+\mathrm{i}y$, $x, y \in \mathfrak{g}_{\mathbb{R}}$. It extends uniquely to an algebra involution of $U(\mathfrak{g})$, and $U(\mathfrak{g})$ is then a Hopf $*$-algebra. Conversely, let $*$ be an involution such that $(U(\mathfrak{g}), *)$ becomes a Hopf $*$-algebra. Since $\mathfrak{g}$ is the set of primitive elements of $U(\mathfrak{g})$, this involution leaves $\mathfrak{g}$ invariant and hence $\mathfrak{g}_{\mathbb{R}} = \{x \in \mathfrak{g} \mid -x = x^*\}$ is a real form of $\mathfrak{g}$. That is, we have a one-to-one correspondence between real forms of the Lie algebra $\mathfrak{g}$ and Hopf $*$-structures on the Hopf algebra $U(\mathfrak{g})$. △

Example 11 (Real forms of connected complex Lie groups). Let $\mathcal{A}$ be the Hopf algebra $\mathcal{O}(G)$ from Example 3 and let $G_{\mathbb{R}}$ be a real connected Lie group which is a real form of the complex connected Lie group G. Then for any $f \in \mathcal{O}(G)$ there is a unique function $f^* \in \mathcal{O}(G)$ such that $f^*(g) = \overline{f(g)}$ for all $g \in G_{\mathbb{R}}$. Endowed with this mapping $*$, the Hopf algebra $\mathcal{O}(G)$ becomes a Hopf $*$-algebra. Let us illustrate this in the case $G = SL(N, \mathbb{C})$. For the real form $G_{\mathbb{R}} = SL(N, \mathbb{R})$ it is obvious that $\overline{u^i_j(g)} = u^i_j(g)$ for $g \in G_{\mathbb{R}}$, so $(u^i_j)^* = u^i_j$. For the compact real form $G_{\mathbb{R}} = SU(N)$ we have $(u^i_j)^*(g) = \overline{u^i_j(g)} = \overline{g_{ij}} = S(u^j_i)(g)$ for $g \in G_{\mathbb{R}}$ and hence $(u^i_j)^* = S(u^j_i)$.

Let $\mathfrak{g}_{\mathbb{R}}$ be the real Lie algebra of $G_{\mathbb{R}}$. It is a real form of the complex Lie algebra $\mathfrak{g}$ of the complex Lie group G, so $U(\mathfrak{g})$ is a Hopf $*$-algebra by the preceding example. Then the Hopf algebra pairing $\langle\cdot,\cdot\rangle$ from Example 6 (see (36)) is a dual pairing of the Hopf $*$-algebras $\mathcal{U} := U(\mathfrak{g})$ and $\mathcal{A} := \mathcal{O}(G)$. Let us sketch the proof of the latter assertion. As noted above, it is enough to prove the second relation of (41). Because of (31), it suffices to do so for elements $f = x \in \mathfrak{g}_{\mathbb{R}}$. Since $\exp(tx) \in G_{\mathbb{R}}$ for $t \in \mathbb{R}$, we then have

$$\langle x, a^*\rangle = \frac{d}{dt}\overline{a(\exp(tx))}\bigg|_{t=0} = \overline{\langle x, a\rangle} = \overline{\langle S(x)^*, a\rangle}. \qquad \triangle$$

The previous examples suggest the following terminology. A Hopf $*$-algebra structure on a Hopf algebra $\mathcal{A}$ is called a *real form* of $\mathcal{A}$. Two Hopf $*$-algebra structures $*$ and $*'$ on $\mathcal{A}$ are said to be *equivalent* if there exists a Hopf algebra automorphism φ of $\mathcal{A}$ such that $\varphi(a^*) = \varphi(a)^{*'}$ for all $a \in \mathcal{A}$.

1.2.8* The Dual Hopf Algebra $\mathcal{A}^\circ$

As discussed in Subsect. 1.2.2, the dual space $\mathcal{A}'$ of an infinite-dimensional algebra $\mathcal{A}$ does not become a coalgebra by dualizing the multiplication of $\mathcal{A}$. However, we shall see that $\mathcal{A}'$ contains a largest coalgebra $\mathcal{A}^\circ$ and that $\mathcal{A}^\circ$ is a Hopf algebra when $\mathcal{A}$ is so.

Suppose that $\mathcal{A}$ is an algebra. As usual, we consider $\mathcal{A}' \otimes \mathcal{A}'$ as a linear subspace of $(\mathcal{A}\otimes\mathcal{A})'$ by identifying $f\otimes g \in \mathcal{A}'\otimes\mathcal{A}'$ with the linear functional on $\mathcal{A}\otimes\mathcal{A}$ determined by $(f\otimes g)(a\otimes b) := f(a)g(b)$. For $f \in \mathcal{A}'$ let $\Delta(f)$ be the element of $(\mathcal{A}\otimes\mathcal{A})'$ defined by $\Delta(f)(a\otimes b) := f(ab)$, $a, b \in \mathcal{A}$. Let now $\mathcal{A}^\circ$ denote the set of all functionals $f \in \mathcal{A}'$ for which $\Delta(f)$ belongs to $\mathcal{A}'\otimes\mathcal{A}'$, that is, there are functionals $f_1, \cdots, f_r, g_1, \cdots, g_r \in \mathcal{A}'$, $r \in \mathbb{N}$, such that $f(ab) = \sum_{i=1}^r f_i(a)g_i(b)$ for all $a, b \in \mathcal{A}$. In this case, the element $\Delta(f) \in \mathcal{A}'\otimes\mathcal{A}'$ takes the form $\Delta(f) = \sum_i f_i \otimes g_i$.

Proposition 12. (i) *If $\mathcal{A}$ is an algebra, $\mathcal{A}^\circ$ is a coalgebra with comultiplication Δ defined above and counit ε given by $\varepsilon(f) = f(1)$.*

(ii) *If $\mathcal{A}$ is a bialgebra, then the coalgebra $\mathcal{A}^\circ$ equipped with the convolution multiplication (22) is also a bialgebra. The unit of the algebra $\mathcal{A}^\circ$ is then the counit ε of the coalgebra $\mathcal{A}$.*

(iii) *If $\mathcal{A}$ is a Hopf algebra, then so is $\mathcal{A}^\circ$ with antipode given by $S(f)(a) = f(S(a))$.*

(iv) *If $\mathcal{A}$ is a $*$-Hopf algebra, then the Hopf algebra $\mathcal{A}^\circ$ endowed with the involution (40) is a Hopf $*$-algebra too.*

Proof. (i): First we have to ensure that $\Delta(\mathcal{A}^\circ) \subseteq \mathcal{A}^\circ\otimes\mathcal{A}^\circ$. Let $f \in \mathcal{A}^\circ$ and $\Delta(f) = \sum_i f_i \otimes g_i$, where the functionals $\{f_i\}$ are chosen to be linearly independent. Then we can find elements $a_j \in \mathcal{A}$ such that $f_i(a_j) = \delta_{ij}1$. We have $g_j(ab) = \sum_i f_i(a_j)g_i(ab) = f(a_jab) = \sum_i f_i(a_ja)g_i(b)$ which implies that $g_j \in \mathcal{A}^\circ$. Similarly, $f_j \in \mathcal{A}^\circ$ and so $\Delta(f) \in \mathcal{A}^\circ\otimes\mathcal{A}^\circ$. Clearly, $\mathcal{A}^\circ$ is a linear subspace of $\mathcal{A}'$. The coassociativity of Δ is derived from the associativity of the multiplication of $\mathcal{A}$ by $((\Delta\otimes\mathrm{id})\circ\Delta(f))(a\otimes b\otimes c) = f((ab)c) = f(a(bc)) = ((\mathrm{id}\otimes\Delta)\circ\Delta(f))(a\otimes b\otimes c)$. The counit axiom is obvious.

(ii): We show that $\mathcal{A}^\circ$ is a subalgebra of $\mathcal{A}'$. Let $f, g \in \mathcal{A}^\circ$ with $\Delta(f) = \sum_i f_i \otimes g_i$ and $\Delta(g) = \sum_j h_j \otimes k_j$. Then we compute $fg(ab) = \sum_{i,j} f_ih_j(a)g_ik_j(b)$, so that $fg \in \mathcal{A}^\circ$. The compatibility condition of comultiplication and multiplication of $\mathcal{A}^\circ$ is easily derived from that of $\mathcal{A}$.

(iii): Let $\Delta(f) = \sum_i f_i \otimes g_i$ for $f \in \mathcal{A}^\circ$. Then we have $S(f)(ab) = \sum_i S(g_i)(a)S(f_i)(b)$ and hence $S(f) \in \mathcal{A}^\circ$. We check one relation of the antipode axiom (the other follows similarly) by

$$(m\circ(S\otimes\mathrm{id})\circ\Delta(f))(a) = \sum\nolimits_i (S(f_i)g_i)(a) = \sum\nolimits_{i,(a)} f_i(S(a_{(1)}))g_i(a_{(2)})$$

$$= \sum f(S(a_{(1)})a_{(2)}) = \varepsilon(a)f(1) = 1_{\mathcal{A}^\circ}(a)\varepsilon_{\mathcal{A}^\circ}(f) = ((\eta\circ\varepsilon)(f))(a).$$

(iv) is easily verified by using Corollary 11. □

The coalgebra (resp. bialgebra, Hopf algebra) $\mathcal{A}^\circ$ from Proposition 12 is called the *dual coalgebra* (resp. *bialgebra, Hopf algebra*) of the algebra (resp. bialgebra, Hopf algebra) $\mathcal{A}$.

Equivalent characterizations of $\mathcal{A}^\circ$ can be found in [Mon], Lemma 9.1.1, [Sw], Chap. 6, [A], Sects. 2.2 and 2.3, or [Jos], Sect. 1.4. In the Hopf algebra literature, $\mathcal{A}^\circ$ is usually introduced as the set of those elements in $\mathcal{A}'$ that vanish on a two-sided (or right or left) ideal of finite codimension.

Let $\mathcal{A}$ be a bialgebra. From the definition of the structure maps of $\mathcal{A}^\circ$ it is clear that we have a dual pairing of the bialgebras $\mathcal{A}^\circ$ and $\mathcal{A}$ given by $\langle f, a\rangle := f(a)$. If $\mathcal{A}^\circ$ separates the points of $\mathcal{A}$, then this pairing $\langle\cdot,\cdot\rangle$ is nondegenerate. There are groups $G \neq \{e\}$ such that the set $(\mathbb{C}G)^\circ$ for the group Hopf algebra $\mathbb{C}G$ (see Example 7) contains only multiples of ε (see, for instance, Example 2.5 in [A], p. 89). In this case the dual pairing of $(\mathbb{C}G)^\circ$ and $\mathbb{C}G$ is obviously denegenerate.

Let us consider a nondegenerate dual pairing $\langle\cdot,\cdot\rangle$ of two bialgebras (resp. Hopf algebras) $\mathcal{U}$ and $\mathcal{A}$. By the nondegeneracy of the pairing, we can consider $\mathcal{U}$ as a subspace of $\mathcal{A}'$ by identifying $f \in \mathcal{U}$ with the functional $\langle f,\cdot\rangle \in \mathcal{A}'$. Since $\langle f, aa'\rangle = \sum\langle f_{(1)}, a\rangle\langle f_{(2)}, a'\rangle$ by (31), $\mathcal{U}$ is then contained in $\mathcal{A}^\circ$. Comparing the structure maps of $\mathcal{A}^\circ$ with the properties of the dual pairing (see Definition 5) we see that $\mathcal{U}$ is even a subbialgebra (resp. Hopf subalgebra) of $\mathcal{A}^\circ$. Similarly, $\mathcal{A}$ becomes a subbialgebra (resp. Hopf subalgebra) of $\mathcal{U}^\circ$.

When dealing with the duality of Hopf algebras, one is faced with two problems. First, it may not be easy to determine all elements of $\mathcal{A}^\circ$, but one can often find Hopf subalgebras of $\mathcal{A}^\circ$. Secondly, in many cases one obtains a dual pairing quite naturally, but it may be degenerate or it is difficult to prove its nondegeneracy. For these reasons, the concept of dually paired Hopf algebras seems to be the proper setting of duality for quantum groups.

Example 12. Let $\mathcal{A} = \mathbb{C}[x]$ be the Hopf algebra of polynomials in x with comultiplication determined by $\Delta(x) = x\otimes 1+1\otimes x$. (This is actually the Hopf algebra $U(\mathfrak{g})$ of Example 6 with $\mathfrak{g} = \mathbb{R}$.) One verifies that the algebra $\mathcal{A}'$ is the algebra $\mathbb{C}[[h]]$ of formal power series, where a series $f = \sum_n a_n h^n \in \mathbb{C}[[h]]$ corresponds to the functional on $\mathcal{A}$ given by $f(\sum_k b_k x^k) := \sum_k k! b_k a_k$. Then it can be shown that $\mathcal{A}^\circ$ is just the linear span of functions $h^n e^{\lambda h}$, where $n \in \mathbb{N}_0$ and $\lambda \in \mathbb{C}$. See [PT], [LT] and [Mon], Example 9.1.7, for more details in this matter. △

1.2.9* Super Hopf Algebras

In the subsequent two subsections we give two variants of the above Hopf algebra definition. For a graded Hopf algebra $\mathcal{A}$ the structure maps have to preserve the graded structure and the algebra $\mathcal{A}\otimes\mathcal{A}$ carries a twisted product rather than the usual tensor product.

Definition 6. *An* $\mathbb{N}_0$-graded super Hopf algebra *is a direct sum* $\mathcal{A} = \bigoplus_{n\in\mathbb{N}_0} \mathcal{A}_n$ *of vector spaces* $\mathcal{A}_n$ *such that the following conditions are valid:*

(i) *$\mathcal{A}$ is an algebra such that $\mathcal{A}_m \cdot \mathcal{A}_n \subseteq \mathcal{A}_{m+n}$ for $m, n \in \mathbb{N}_0$ and $1 \in \mathcal{A}_0$.*

(ii) *$\mathcal{A}$ is a coalgebra such that $\Delta(\mathcal{A}_n) \subseteq \bigoplus_{i+j=n} \mathcal{A}_i \otimes \mathcal{A}_j$ for $n \in \mathbb{N}_0$ and $\varepsilon(\mathcal{A}_n) = \{0\}$ for $n \geq 1$.*

(iii) *$\Delta : \mathcal{A} \to \mathcal{A} \otimes \mathcal{A}$ and $\varepsilon : \mathcal{A} \to \mathbb{C}$ are algebra homomorphisms, where the product $\bullet$ of $\mathcal{A} \otimes \mathcal{A}$ is defined by*

$$(a \otimes a_k) \bullet (a'_n \otimes a') = (-1)^{kn} aa'_n \otimes a_k a', \quad a_k \in \mathcal{A}_k, \ a_n \in \mathcal{A}_n, \ a, a' \in \mathcal{A}. \quad (42)$$

(iv) *There is a linear mapping $S : \mathcal{A} \to \mathcal{A}$ such that condition (26) holds and $S(\mathcal{A}_n) \subseteq \mathcal{A}_n$, $n \in \mathbb{N}_0$.*

Let $\mathcal{A} = \bigoplus_{n \in \mathbb{N}_0} \mathcal{A}_n$ be such an $\mathbb{N}_0$-graded super Hopf algebra and set $\mathcal{A}_{\bar{0}} = \bigoplus_n \mathcal{A}_{2n}$ and $\mathcal{A}_{\bar{1}} = \bigoplus_n \mathcal{A}_{2n+1}$, where $\bar{0}$ and $\bar{1}$ are the two elements of $\mathbb{Z}_2 = \mathbb{Z}/2\mathbb{Z}$. Then we have:

(i)′ $\mathcal{A}$ is an algebra such that $\mathcal{A}_{\bar{k}} \cdot \mathcal{A}_{\bar{n}} \subseteq \mathcal{A}_{\bar{k}+\bar{n}}$ for $\bar{k}, \bar{n} \in \mathbb{Z}_2$ and $1 \in \mathcal{A}_{\bar{0}}$.

(ii)′ $\mathcal{A}$ is a coalgebra such that $\Delta(\mathcal{A}_{\bar{n}}) \subseteq \bigoplus_{\bar{i}+\bar{j}=\bar{n}} \mathcal{A}_{\bar{i}} \otimes \mathcal{A}_{\bar{j}}$ for $\bar{n} \in \mathbb{Z}_2$ and $\varepsilon(\mathcal{A}_{\bar{1}}) = \{0\}$.

(iii)′ $\Delta : \mathcal{A} \to \mathcal{A} \otimes \mathcal{A}$ and $\varepsilon : \mathcal{A} \to \mathbb{C}$ are algebra homomorphisms, where $\mathcal{A} \otimes \mathcal{A}$ obeys the product (42) with $k, n \in \{\bar{0}, \bar{1}\}$.

(iv)′ $\mathcal{A}$ admits an antipode S such that $S(\mathcal{A}_{\bar{0}}) \subseteq \mathcal{A}_{\bar{0}}$ and $S(\mathcal{A}_{\bar{1}}) \subseteq \mathcal{A}_{\bar{1}}$.

A direct sum $\mathcal{A} = \mathcal{A}_{\bar{0}} \oplus \mathcal{A}_{\bar{1}}$ of vector spaces $\mathcal{A}_{\bar{0}}$ and $\mathcal{A}_{\bar{1}}$ which satisfies the conditions (i)′–(iv)′ is called a *$\mathbb{Z}_2$-graded Hopf algebra* or a *super Hopf algebra*. Any $\mathbb{N}_0$-graded super Hopf algebra is in particular a super Hopf algebra. To be precise, Definition 6 is the definition of an $\mathbb{N}_0$-graded Hopf algebra, but we prefer to refer to it as an *$\mathbb{N}_0$-graded super Hopf algebra* in the following.

Remark 1. Let G be an abelian group and λ a fixed bicharacter of G, that is, a mapping $\lambda : G \times G \to \mathbb{C}$ such that $\lambda(n+m, k) = \lambda(n, k)\lambda(m, k)$ and $\lambda(k, n+m) = \lambda(k, n)\lambda(k, m)$ for $k, n, m \in G$. Then the concept of a G-graded Hopf algebra with respect to λ can be defined in a similar manner. The Hopf algebra $\mathcal{A}$ has to be graded by G and the factor $(-1)^{kn}$ in (42) must be replaced by $\lambda(n, k)$. In the case $\lambda(\bar{n}, \bar{k}) = (-1)^{kn}$ and $G = \mathbb{Z}_2$ we recover the $\mathbb{Z}_2$-graded Hopf algebra defined above. △

Example 13 (Tensor algebra $T(V)$ and exterior algebra $\Lambda(V)$). Let $T(V)$ be the tensor algebra of a vector space V. The quotient algebra of $T(V)$ by the two-sided ideal generated by the elements $v \otimes w + w \otimes v$, $v, w \in V$, is called the *exterior algebra* of V and denoted by $\Lambda(V)$. Both algebras $T(V)$ and $\Lambda(V)$ possess unique $\mathbb{N}_0$-graded super Hopf algebra structures such that $\Delta(v) = v \otimes 1 + 1 \otimes v$ and $\varepsilon(v) = 0$ for $v \in V$. For both algebras the comultiplication and the antipode on products of elements $v_1, v_2, \cdots, v_n \in V$ are given by

$$\Delta(v_1 v_2 \cdots v_n) = \sum_{k=0}^{n} \sum_{p \in \mathcal{P}_{nk}} (-1)^{l(p)} v_{p(1)} \cdots v_{p(k)} \otimes v_{p(k+1)} \cdots v_{p(n)}, \quad (43)$$

$$S(v_1 v_2 \cdots v_n) = (-1)^{n(n+1)/2} v_n \cdots v_2 v_1 \quad (44)$$

where $l(p)$ is the length of the permutation p. △

1.2.10* h-Adic Hopf Algebras

In this subsection we develop another variant of Hopf algebras. Roughly speaking, h-adic Hopf algebras are obtained by taking the ring $\mathbb{C}[[h]]$ of formal power series as $\mathbb{K}$ and requiring that the algebras and the corresponding tensor products are complete resp. completed in the so-called h-adic topology.

First let us describe the structure of the ring $\mathbb{C}[[h]]$. The elements of $\mathbb{C}[[h]]$ are formal power series $f = \sum_{n=0}^{\infty} a_n h^n$ in an indeterminate h with complex coefficients. We define an addition and a multiplication in $\mathbb{C}[[h]]$ by setting

$$f + g := \sum_{n=0}^{\infty} (a_n + b_n) h^n, \quad fg := \sum_{n=0}^{\infty} \left(\sum_{r+s=n} a_r b_s \right) h^n, \tag{45}$$

where $f = \sum_{n=0}^{\infty} a_n h^n$ and $g = \sum_{n=0}^{\infty} b_n h^n$. One easily verifies that these operations turn $\mathbb{C}[[h]]$ into a commutative ring with unit. The ring $\mathbb{C}[h]$ of all polynomials in h can be considered as a subring of $\mathbb{C}[[h]]$. Note that an element $f = \sum_{n=0}^{\infty} a_n h^n \in \mathbb{C}[[h]]$ is invertible in $\mathbb{C}[[h]]$ if and only if $a_0 \neq 0$. In particular, $\mathbb{C}[[h]]$ is not a field.

Let V be a complex vector space. Then the set $V[[h]]$ of all formal power series $f = \sum_{n=0}^{\infty} v_n h_n$ with coefficients v_n in V is a $\mathbb{C}[[h]]$-vector space with the addition of two elements $f, g \in V[[h]]$ defined by the first formula in (45) and the left multiplication of $f \in \mathbb{C}[[h]]$ and $g \in V[[h]]$ given by the second formula in (45).

If $\mathcal{A}$ is an algebra over $\mathbb{C}$, then $\mathcal{A}[[h]]$ becomes an algebra over $\mathbb{C}[[h]]$ with the product of two elements $f, g \in \mathcal{A}[[h]]$ defined by the second formula in (45).

Next we briefly develop the h-adic topology and its properties. Detailed treatments with proofs can be found in [Boul] and [Kas], Chap. XVI.

Let V and W be vector spaces over $\mathbb{C}[[h]]$. The topology on V for which the sets $\{h^n V + v \mid n \in \mathbb{N}_0\}$ are a neighborhood base of $v \in V$ is called the *h-adic topology*. We shall denote by $V \hat{\otimes} W$ the completion of the tensor product space $V \otimes_{\mathbb{C}[[h]]} W$ in the h-adic topology.

Let us make these notions more explicit. Let V and W be complex vector spaces. For any nonzero $f = \sum_{n=0}^{\infty} v_n h^n \in V[[h]]$ let $n(f)$ be the smallest integer n such that $v_n \neq 0$. For $f = 0$, we set $n(0) = +\infty$. Then $d(f, g) := 2^{-n(f-g)}$ defines a metric on $V[[h]]$ and the topology defined by that metric is just the h-adic topology of the vector space $V[[h]]$. A sequence $f^{(k)} = \sum_{n=0}^{\infty} v_n^{(k)} h^n$, $k \in \mathbb{N}$, converges to $f = \sum_{n=0}^{\infty} v_n h^n$ in the h-adic topology of $V[[h]]$ if and only if for any $r \in \mathbb{N}$ there exists an index k_r such that $v_n^{(k)} = v_n$ for $n = 1, 2, \cdots, r$ and all $k \geq k_r$. This characterization implies that in the h-adic topology the partial sums of any $f \in V[[h]]$ converge to f and that the polynomial space $V[h]$ is dense in $V[[h]]$. Further, it follows easily that the vector space $V[[h]]$ over $\mathbb{C}[[h]]$ is complete in the h-adic topology and that $V[[h]] \hat{\otimes} W[[h]] = (V \otimes W)[[h]]$.

If the complex vector space V is finite-dimensional, then $V[[h]]$ is canonically isomorphic to $V \otimes \mathbb{C}[[h]]$. However, the reader should be aware that for an infinite-dimensional space V the set $V[[h]]$ is strictly bigger than $V \otimes \mathbb{C}[[h]]$, because the latter contains only sums $\sum_{n=0}^{\infty} v_n h^n$ for which the linear span of all coefficients v_n has finite dimension.

Now we can define the h-adic variants of algebras, coalgebras, etc. by taking *h-adic complete* vector spaces over $\mathbb{C}[[h]]$ and replacing the algebraic tensor products $\otimes$ in the corresponding previous definitions by the *h-adic completions* $\hat{\otimes}$. That is, an *h-adic Hopf algebra* is a vector space $\mathcal{A}$ over $\mathbb{C}[[h]]$ which is complete in the h-adic topology and endowed with $\mathbb{C}[[h]]$-linear mappings $m : \mathcal{A} \hat{\otimes} \mathcal{A} \to \mathcal{A}$, $\eta : \mathbb{C}[[h]] \to \mathcal{A}$, $\Delta : \mathcal{A} \to \mathcal{A} \hat{\otimes} \mathcal{A}$, $\varepsilon : \mathcal{A} \to \mathbb{C}[[h]]$ and $S : \mathcal{A} \to \mathcal{A}$ which satisfy the Hopf algebra axioms with $\otimes$ replaced by $\hat{\otimes}$. h-Adic algebras, h-adic coalgebras and h-adic bialgebras are defined similarly. Simple examples of such h-adic structures are obtained as follows: if $\mathcal{A}$ is an algebra, coalgebra, etc., then $\mathcal{A}[[h]] = \mathcal{A} \hat{\otimes} \mathbb{C}[[h]]$ becomes an h-adic algebra, h-adic coalgebra, etc. with structure maps $\mathbb{C}[[h]]$-linearly extended from $\mathcal{A}$.

We illustrate the importance of the h-adic completeness by the following

Example 14 (Functional calculus in h-adic algebras). Let $\mathcal{A}$ be an h-adic algebra. Let a be an element of $\mathcal{A}$ and let $f = \sum_{n=0}^{\infty} c_n h^n \in \mathbb{C}[[h]]$. The elements $s_k := \sum_{n=0}^{k} c_n a^n h^n$, $k \in \mathbb{N}$, form a Cauchy sequence in the h-adic topology, because $s_k - s_{k'} \in h^n \mathcal{A}$ for $k, k' \geq n$. Since $\mathcal{A}$ is h-adic complete by the definition of an h-adic algebra, the sequence $\{s_n\}$ admits an h-adic limit $f(ha) := \sum_{n=0}^{\infty} c_n a^n h^n \in \mathcal{A}$. Setting (for instance) $f(h) = e^h$, we obtain an element $f(ha) = e^{ha}$ in $\mathcal{A}$. If $b \in \mathcal{A}$ commutes with a, we clearly have $e^{ha} e^{hb} = e^{h(a+b)}$. In particular, $(e^{ha})^{-1} = e^{h(-a)}$. △

In Sect. 6.1 h-adic algebras will be constructed in terms of generators and relations. In order to clarify what this means, let us consider a set of generators, say $x_1, x_2, \cdots, x_r$, and a set of relations, say $f_k(x_1, x_2, \cdots, x_r) = 0$. Then the h-adic algebra $\mathcal{A}$ with h-adic generators $x_1, x_2, \cdots, x_r$ and the set of relations $f_k(x_1, x_2, \cdots, x_r) = 0$ is defined as the quotient of the h-adic algebra $\mathbb{C}\langle x_i \rangle[[h]]$ by the closure in the h-adic topology of the two-sided ideal of $\mathbb{C}\langle x_i \rangle[[h]]$ generated by the elements $f_k(x_1, x_2, \cdots, x_r)$. Being the quotient of the complete metric space $\mathbb{C}\langle x_i \rangle[[h]]$ by a closed subspace, $\mathcal{A}$ is complete in the h-adic topology and hence is an h-adic algebra.

A technical argument for the need of h-adic Hopf algebras is that they are necessary for a mathematically rigorous treatment of universal R-matrices for quantized enveloping algebras. A conceptual reason is that they provide the proper setting of deformation theory. In this book the word "deformation" will be used only in the naive sense in order to describe structures depending on (complex) parameters. However, the precise definition is as follows.

Let $\mathcal{A}$ be a Hopf algebra over the complex field $\mathbb{C}$. An h-adic Hopf algebra $\mathcal{A}_d$ over $\mathbb{C}[[h]]$ is called a *formal deformation of the Hopf algebra $\mathcal{A}$* if $\mathcal{A}_d$ is isomorphic to $\mathcal{A}[[h]]$ as a vector space over $\mathbb{C}[[h]]$ and $m_{\mathcal{A}_d} \equiv m_{\mathcal{A}} \pmod h$ and $\Delta_{\mathcal{A}_d} \equiv \Delta_{\mathcal{A}} \pmod h$. This means that if a vector space isomorphism of

$\mathcal{A}_d$ and $\mathcal{A}[[h]]$ is fixed, then the multiplication and comultiplication of $\mathcal{A}_d$ can be regarded on $\mathcal{A}[[h]]$ and there exist linear mappings $m_n : \mathcal{A} \otimes \mathcal{A} \to \mathcal{A}$ and $\Delta_n : \mathcal{A} \to \mathcal{A} \otimes \mathcal{A}$ for any $n \in \mathbb{N}$ such that

$$m_{\mathcal{A}_d}(a \otimes b) \simeq m_{\mathcal{A}}(a \otimes b) + m_1(a \otimes b)h + m_2(a \otimes b)h^2 + \cdots,$$

$$\Delta_{\mathcal{A}_d}(a) \simeq \Delta_{\mathcal{A}}(a) + \Delta_1(a)h + \Delta_2(a)h^2 + \cdots.$$

The simplest example of a deformation of a Hopf algebra $\mathcal{A}$ is of course the h-adic Hopf algebra $\mathcal{A}[[h]]$ itself whose structure maps are obtained as $\mathbb{C}[[h]]$-linear extensions of the structure maps of $\mathcal{A}$. This deformation is called the null deformation of $\mathcal{A}$.

1.3 Modules and Comodules of Hopf Algebras

1.3.1 Modules and Representations

Let $\mathcal{A}$ be an arbitrary algebra and V a vector space. By a *representation* of $\mathcal{A}$ on V we mean an algebra homomorphism $\varphi \equiv \varphi_V$ of $\mathcal{A}$ to the algebra $\mathcal{L}(V)$ of linear operators on V. That is, we have $\varphi(\alpha a + \beta b) = \alpha\varphi(a) + \beta\varphi(b)$, $\varphi(ab) = \varphi(a)\varphi(b)$ and $\varphi(1) = I$ for $a, b \in \mathcal{A}$ and $\alpha, \beta \in \mathbb{K}$, where I is the identity operator on V. The vector space V is called a *left $\mathcal{A}$-module* if there exists a linear mapping $\varphi : \mathcal{A} \otimes V \to V$, written as $\varphi(a \otimes v) = a \triangleright v$, such that $(ab) \triangleright v = a \triangleright (b \triangleright v)$ and $1 \triangleright v = v$ for all $a, b \in \mathcal{A}$ and $v \in V$ or, equivalently, that

$$\varphi \circ (\mathrm{id} \otimes \varphi) = \varphi \circ (m \otimes \mathrm{id}) \quad \text{and} \quad \varphi \circ (\eta \otimes \mathrm{id}) = \mathrm{id}, \tag{46}$$

where the first equality acts on $\mathcal{A} \otimes \mathcal{A} \otimes V$ and the second on $\mathbb{K} \otimes V \equiv V$. The map φ is then called a *left action* of $\mathcal{A}$ on V. Sometimes we shall write $a.v$ or simply av for $a \triangleright v$. Clearly, there is a one-to-one correspondence, given by the formula $\varphi(a)v = \varphi(a \otimes v)$, between representations and left actions of $\mathcal{A}$ on V. Similarly, a *right $\mathcal{A}$-module* is a vector space V with a linear mapping $\varphi : V \otimes \mathcal{A} \to V$, called a *right action* of $\mathcal{A}$ on V and written as $\varphi(v \otimes a) = v \triangleleft a$, such that $v \triangleleft (ab) = (v \triangleleft a) \triangleleft b$ and $v = v \triangleleft 1$ for $a, b \in \mathcal{A}$ and $v \in V$. We also write $v.a$ or va instead of $v \triangleleft a$. The two conditions on a right module can be expressed as

$$\varphi \circ (\varphi \otimes \mathrm{id}) = \varphi \circ (\mathrm{id} \otimes m) \quad \text{and} \quad \varphi \circ (\mathrm{id} \otimes \eta) = \mathrm{id} \tag{47}$$

or as the commutativity of the diagrams

$$\begin{array}{ccc} V \otimes \mathcal{A} \otimes \mathcal{A} & \xrightarrow{\mathrm{id} \otimes m} & V \otimes \mathcal{A} \\ \downarrow{\scriptstyle \varphi \otimes \mathrm{id}} & & \downarrow{\scriptstyle \varphi} \\ V \otimes \mathcal{A} & \xrightarrow{\varphi} & V \end{array} \qquad \begin{array}{ccc} V \otimes \mathbb{K} & \xrightarrow{\mathrm{id}} & V \otimes \mathbb{K} \\ \downarrow{\scriptstyle \mathrm{id} \otimes \eta} & & \Updownarrow \\ V \otimes \mathcal{A} & \xrightarrow{\varphi} & V\,. \end{array} \tag{48}$$

Let V be a left $\mathcal{A}$-module with action φ denoted as $\varphi(a \otimes v) = av$. We say that a set $\{v_i \mid i \in I\}$ of elements $v_i \in V$ is a *left $\mathcal{A}$-module basis* of V if any $v \in V$ is a finite sum $v = \sum_i a_i v_i$ with uniquely determined elements $a_i \in \mathcal{A}$.

A vector space V is called an *$\mathcal{A}$-bimodule* if V is both a left $\mathcal{A}$-module with action av and a right $\mathcal{A}$-module with action va satisfying the compatibility condition $(av)b = a(vb)$ for $a, b \in \mathcal{A}$ and $v \in V$.

We shall freely use standard notions of representation theory of algebras. Let us recall a few of them. An *intertwining operator* of representations φ_V and φ_W on vector spaces V and W is a linear mapping $A : V \to W$ such that $A\varphi_V(a)v = \varphi_W(a)Av$ for $a \in \mathcal{A}$ and $v \in V$. If V and W are viewed as left $\mathcal{A}$-modules with actions $\varphi_V : \mathcal{A} \otimes V \to V$ and $\varphi_W : \mathcal{A} \otimes W \to W$, this is equivalent to

$$A \circ \varphi_V = \varphi_W \circ (\mathrm{id} \otimes A). \tag{49}$$

A is then called a *morphism* of left $\mathcal{A}$-modules. If A is invertible, it is an *isomorphism* of $\mathcal{A}$-modules. A representation φ_V on V is said to be *irreducible* (and the corresponding $\mathcal{A}$-module V is called *simple*) if V and $\{0\}$ are the only invariant subspaces of V. Representations which are direct sums of irreducible ones are said to be *completely reducible*. If $\mathcal{A}$ is a $*$-algebra and V carries a scalar product, then a representation φ of $\mathcal{A}$ on V is called a *$*$-representation* if $\langle \varphi(a)v, w\rangle = \langle v, \varphi(a^*)w\rangle$ for all $a \in \mathcal{A}$ and $v, w \in V$.

Suppose now that $\mathcal{A}$ is a bialgebra. Then the comultiplication of $\mathcal{A}$ enables us to define a *tensor product of representations*. Let φ_V and φ_W be representations of $\mathcal{A}$ on vector spaces V and W, respectively. There is a representation $\varphi_{V\otimes W}$ of $\mathcal{A}$ on the vector space $V \otimes W$, denoted also by $\varphi_V \otimes \varphi_W$ and called the tensor product of φ_V and φ_W, such that

$$\varphi_{V\otimes W}(a) := ((\varphi_V \otimes \varphi_W) \circ \Delta)(a) \equiv \sum \varphi_V(a_{(1)}) \otimes \varphi_W(a_{(2)}), \quad a \in \mathcal{A}.$$

The preceding formula defines a representation of the algebra $\mathcal{A}$, since we have $\varphi_{V\otimes W}(1) = (\varphi_V \otimes \varphi_W)(1 \otimes 1) = I_V \otimes I_W = I_{V\otimes W}$ and

$$\varphi_{V\otimes W}(ab) = \sum \varphi_V((ab)_{(1)}) \otimes \varphi_W((ab)_{(2)})$$

$$= \sum \varphi_V(a_{(1)})\varphi_V(b_{(1)}) \otimes \varphi_W(a_{(2)})\varphi_W(b_{(2)}) = \varphi_{V\otimes W}(a)\varphi_{V\otimes W}(b),$$

where we have used the relations (25).

If φ is a finite-dimensional representation of a complex algebra $\mathcal{A}$ on a complex vector space V, then there is a canonical extension of φ to a representation φ_h of the h-adic algebra $\mathcal{A}[[h]]$ on the space $V[[h]] = V \otimes \mathbb{C}[[h]]$ given by

$$\varphi_h\Big(\sum\nolimits_n a_n h^n\Big)(v \otimes f(h)) := \sum\nolimits_n (\varphi(a_n)v \otimes h^n f(h)). \tag{50}$$

The representation φ_h is reducible even if the representation φ is irreducible, since the subspace $hV[[h]]$ is invariant under the operators $\varphi_h(a)$, $a \in \mathcal{A}[[h]]$.

But if the representation φ of $\mathcal{A}$ is irreducible, then the representation φ_h of $\mathcal{A}[[h]]$ is indecomposable (that is, it cannot be represented as a direct sum of nontrivial irreducible representations).

1.3.2 Comodules and Corepresentations

Dualizing actions of algebras on modules leads to coactions of coalgebras on comodules. Let $\mathcal{A}$ be a coalgebra.

Definition 7. *A* corepresentation *or, equivalently, a* right coaction *of $\mathcal{A}$ on a vector space V, called then a* right $\mathcal{A}$-comodule, *is a linear mapping $\varphi : V \to V \otimes \mathcal{A}$ satisfying the relations*

$$(\varphi \otimes \mathrm{id}) \circ \varphi = (\mathrm{id} \otimes \Delta) \circ \varphi \quad \text{and} \quad (\mathrm{id} \otimes \varepsilon) \circ \varphi = \mathrm{id}. \tag{51}$$

A left coaction *of $\mathcal{A}$ on V is a linear mapping $\varphi : V \to \mathcal{A} \otimes V$ such that*

$$(\mathrm{id} \otimes \varphi) \circ \varphi = (\Delta \otimes \mathrm{id}) \circ \varphi \quad \text{and} \quad (\varepsilon \otimes \mathrm{id}) \circ \varphi = \mathrm{id}. \tag{52}$$

The vector space V equipped with the mapping φ is then called a left $\mathcal{A}$-comodule.

The two conditions of (51) are equivalent to the requirement that the two diagrams

$$\begin{array}{ccc} V \otimes \mathcal{A} \otimes \mathcal{A} & \xleftarrow{\mathrm{id} \otimes \Delta} & V \otimes \mathcal{A} \\ \uparrow{\scriptstyle \varphi \otimes \mathrm{id}} & & \uparrow{\scriptstyle \varphi} \\ V \otimes \mathcal{A} & \xleftarrow{\varphi} & V \end{array} \qquad \begin{array}{ccc} V \otimes \mathbb{K} & \xleftarrow{\mathrm{id}} & V \otimes \mathbb{K} \\ \uparrow{\scriptstyle \mathrm{id} \otimes \varepsilon} & & \Updownarrow \\ V \otimes \mathcal{A} & \xleftarrow{\varphi} & V \end{array} \tag{53}$$

are commutative. The diagrams (53) for a right coaction are obtained from the diagrams (48) for a right action by reversing arrows and replacing the multiplication m by the comultiplication Δ and the unit η by the counit ε. Note also that the equations of (51) and (52) which characterize right and left comodules are just the "duals" of the equations of (47) and (46) which define right and left modules, respectively.

Example 15 (The comultiplication). Clearly, Δ is a right coaction and also a left coaction of $\mathcal{A}$ on itself. Indeed, in the case $\varphi = \Delta$ the equations (51) and the equations (52) become the coalgebra axioms (15) and (16). The right coaction Δ of $\mathcal{A}$ on itself is called the *right regular corepresentation* of $\mathcal{A}$. △

Example 16 (Trivial corepresentation). The linear mapping $\varphi : V \to V \otimes \mathcal{A}$ defined by $\varphi(v) = v \otimes 1$ is the trivial corepresentation of $\mathcal{A}$ on V. △

Next we define some basic concepts on corepresentations. They may be viewed as dualizations of the corresponding notions on representations of algebras or as generalizations of concepts from group representation theory.

Let φ_V and φ_W be corepresentations of $\mathcal{A}$ on vector spaces V and W. A linear mapping $A : V \to W$ is called an *intertwining operator* of φ_V and φ_W (and a *morphism* of the corresponding right $\mathcal{A}$-comodules) if

$$\varphi_W \circ A = (A \otimes \mathrm{id}) \circ \varphi_V. \tag{54}$$

The set of such intertwiners A is denoted by $\mathrm{Mor}\,(\varphi_V, \varphi_W)$. We shall write $\mathrm{Mor}\,(\varphi_V)$ for $\mathrm{Mor}\,(\varphi_V, \varphi_V)$. If there exists an invertible intertwiner A of φ_V and φ_W, then the corepresentations φ_V and φ_W are said to be *equivalent* and A is called an *isomorphism* of the corresponding $\mathcal{A}$-comodules. A linear subspace U of V is called *invariant* under φ_V if $\varphi_V(U) \subseteq U \otimes \mathcal{A}$. The corepresentation φ_V on V is said to be *irreducible* if V and $\{0\}$ are the only invariant subspaces. Other concepts such as subcorepresentations, direct sums, etc. will have their obvious meanings.

Let φ be a left (resp. right) coaction of $\mathcal{A}$ on V. An element v of V is called *invariant under* φ if $\varphi(v) = 1 \otimes v$ (resp. $\varphi(v) = v \otimes 1$). We shall say that a linear functional f on V is *invariant under* φ if $(\mathrm{id} \otimes f)\varphi = f1$ (resp. $(f \otimes \mathrm{id})\varphi = f1$).

The next proposition characterizes finite-dimensional corepresentations in terms of matrices of elements of $\mathcal{A}$.

Proposition 13. *Let V be a finite-dimensional vector space with basis $e_1, e_2, \cdots, e_d$ and let $\varphi : V \to V \otimes \mathcal{A}$ be a linear mapping. Let $v = (v_{ij})$ be the $d \times d$ matrix of elements $v_{ij} \in \mathcal{A}$ defined by the equations*

$$\varphi(e_j) = \sum\nolimits_{k=1}^{d} e_k \otimes v_{kj}, \quad j = 1, 2, \cdots, d. \tag{55}$$

Then φ is a corepresentation of $\mathcal{A}$ on V if and only if

$$\Delta(v_{ij}) = \sum\nolimits_{k=1}^{d} v_{ik} \otimes v_{kj} \quad \text{and} \quad \varepsilon(v_{ij}) = \delta_{ij}, \quad i, j = 1, 2, \cdots, d. \tag{56}$$

Proof. Since $((\varphi \otimes \mathrm{id}) \circ \varphi)(e_j) = \sum_{i,k} e_i \otimes v_{ik} \otimes v_{kj}$, $((\mathrm{id} \otimes \Delta) \circ \varphi)(e_j) = \sum_i e_i \otimes \Delta(v_{ij})$ and $((\mathrm{id} \otimes \varepsilon) \circ \varphi)(e_j) = \sum_i e_i \varepsilon(v_{ij})$, it follows that φ satisfies the relations (51) if and only if the equations (56) hold. □

The elements $v_{ij} \in \mathcal{A}$ given by (55) are called the *matrix elements* of φ with respect to the basis $e_1, e_2, \cdots, e_d$. A matrix $v = (v_{ij})$ of elements of $\mathcal{A}$ satisfying (56) is called a *matrix corepresentation* of $\mathcal{A}$.

Conversely, let $v = (v_{ij})_{i,j=1,\dots,d}$ be a matrix corepresentation of $\mathcal{A}$ and let V be an arbitrary d-dimensional vector space with basis $e_1, e_2, \cdots, e_d$. Then, by Proposition 13, the linear mapping $\varphi : V \to V \otimes \mathcal{A}$ defined by (55) is a corepresentation of $\mathcal{A}$ on V.

It is often useful to express notions of corepresentations in terms of their coefficient matrices. As a sample, we shall do this in the case of an intertwiner. Let φ_V and φ_W be corepresentations on finite-dimensional vector spaces V and W. Let $v = (v_{ij})$ and $w = (w_{ij})$ be the corresponding matrix corepresentations with respect to some bases. Let $A : V \to W$ be a linear mapping and let (a_{rs}) be the matrix of A with respect to these bases. Then A is an intertwiner of φ_V and φ_W if and only if $\sum_i a_{ri}v_{ij} = \sum_s w_{rs}a_{sj}$ for all r and j. The set of such matrices (a_{rs}) is denoted by $\mathrm{Mor}\,(v, w)$.

A matrix corepresentation $v = (v_{ij})$ of a Hopf $*$-algebra with antipode S is called *unitary* if the matrix elements satisfy the relations $S(v_{ij}) = (v_{ji})^*$.

Example 17 (Corepresentation of $M_N(\mathbb{C})'$ on $\mathbb{C}^N$). Let $\mathcal{A}$ be the coalgebra $M_N(\mathbb{C})'$ from Example 4. The $N \times N$ matrix (u^i_j) of elements u^i_j introduced therein satisfies (56), hence it is a matrix corepresentation of $\mathcal{A}$. By (55), this matrix corepresentation defines a corepresentation of $\mathcal{A}$ on $\mathbb{C}^N$. $\triangle$

Example 18 (Group representations). Let $\mathcal{A}$ be one of the Hopf algebras of functions on a group G from Examples 1, 2 or 3. Then, by the definition of the comultiplication of $\mathcal{A}$, the equations (56) mean that $v_{ij}(g_1g_2) = \sum_k v_{ik}(g_1)v_{kj}(g_2)$ for all $g_1, g_2 \in G$ and $v_{ij}(e) = \delta_{ij}$. Thus, the matrix corepresentations of $\mathcal{A}$ are precisely the matrix representations of G with matrix coefficients in $\mathcal{A}$. Therefore, finite-dimensional corepresentations φ of the Hopf algebra $\mathcal{A}$ are in a one-to-one correspondence with finite-dimensional representations T of the group G whose matrix elements belong to $\mathcal{A}$. In order to make this correspondence more explicit, we consider $V \otimes \mathcal{A}$ as the vector space of V-valued functions on G by identifying an element $\sum v_i \otimes f_i$ of $V \otimes \mathcal{A}$ with the V-valued function $\sum_i f_i(\cdot)v_i$ on G. After this identification the correspondence $\varphi \to T$ is expressed by the formula

$$\varphi(v)(g) = T(g)v, \quad g \in G,\ v \in V. \tag{57}$$

Moreover, in the course of this the above notions on corepresentations reduce to the corresponding terms on group representations.

Finally, let us discuss the condition that the matrix coefficients of T are in $\mathcal{A}$. Clearly, it is redundant if $\mathcal{A} = \mathcal{F}(G)$ for a finite group G (Example 1). If $\mathcal{A} = \mathrm{Rep}\, G$ for a compact group G (Example 2), it ensures the continuity of T. In the case $\mathcal{A} = \mathcal{O}(G)$ of Example 3 this condition means precisely that the matrix elements of T are polynomial functions on G. Such representations of G are usually called *rational.* $\triangle$

There is also a *Sweedler notation for coactions* which has similar advantages as for the comultiplication (see Subsect. 1.2.2). If $\varphi : V \to V \otimes \mathcal{A}$ is a *right* coaction, one writes symbolically

$$\varphi(v) = \sum v_{(0)} \otimes v_{(1)}, \quad v \in V, \tag{58}$$

for the element $\varphi(v) \in V \otimes \mathcal{A}$. Then the first condition in (51) reads as

$$\sum (v_{(0)})_{(0)} \otimes (v_{(0)})_{(1)} \otimes v_{(1)} = \sum v_{(0)} \otimes (v_{(1)})_{(1)} \otimes (v_{(1)})_{(2)}.$$

This element will be written as $\sum v_{(0)} \otimes v_{(1)} \otimes v_{(2)}$. One has a similar notation for iterated applications of Δ and φ. The second condition in (51) says that

$$\sum v_{(0)}\varepsilon(v_{(1)}) = v, \quad v \in V. \tag{59}$$

The corresponding Sweedler notations for a *left* coaction $\varphi : V \to \mathcal{A} \otimes V$ are

$$\varphi(v) = \sum v_{(-1)} \otimes v_{(0)}, \tag{60}$$

$$(\mathrm{id} \otimes \varphi) \circ \varphi(v) \equiv (\Delta \otimes \mathrm{id}) \circ \varphi(v) = \sum v_{(-2)} \otimes v_{(-1)} \otimes v_{(0)}.$$

The reader should keep in mind that terms $v_{(k)}$ with $k \neq 0$ always belong to $\mathcal{A}$, while $v_{(0)}$ lies in V.

Let us assume now that $\mathcal{A}$ is a bialgebra. As in the case of representations, we then have a *tensor product of corepresentations.* If $\varphi_V : V \to V \otimes \mathcal{A}$ and $\varphi_W : W \to W \otimes \mathcal{A}$ are corepresentations, then it is easily verified that the linear mapping $\varphi_{V\otimes W} : V \otimes W \to (V \otimes W) \otimes \mathcal{A}$ defined by

$$\varphi_{V\otimes W}(v \otimes w) = \sum v_{(0)} \otimes w_{(0)} \otimes v_{(1)}w_{(1)}, \quad v \in V,\ w \in W, \tag{61}$$

is a corepresentation of $\mathcal{A}$ on the vector space $V \otimes W$. The corepresentation $\varphi_{V\otimes W}$ is also denoted by $\varphi_V \otimes \varphi_W$. Strictly speaking, we should have taken another symbol for the tensor product corepresentation rather than $\varphi_V \otimes \varphi_W$ in order to avoid confusion with the tensor product mapping $\varphi_V \otimes \varphi_W : V \otimes W \to V \otimes \mathcal{A} \otimes W \otimes \mathcal{A}$ which acts as $(\varphi_V \otimes \varphi_W)(v \otimes w) = \varphi_V(v) \otimes \varphi_W(w)$. But we prefer the notation $\varphi_V \otimes \varphi_W$ because it is commonly used in the physics literature. In any case, it will be clear from the context what is meant.

The tensor product of corepresentations is associative in the sense that the canonical isomorphism $(V \otimes W) \otimes U \simeq V \otimes (W \otimes U)$ of the tensor products of underlying vector spaces provides an equivalence of the corresponding corepresentations. However, the corepresentations $\varphi_V \otimes \varphi_W$ and $\varphi_W \otimes \varphi_V$ are not equivalent in general (but they are for our leading examples of quantum groups!). These statements also remain valid for representations.

Finally, let us note that the concepts and facts on right comodules developed above carry over to left comodules as well with a few necessary modifications. For instance, (55) must be replaced by $\varphi(e_j) = \sum_k v_{jk} \otimes e_k$. Moreover, if $\varphi : V \to V \otimes \mathcal{A}$ is a right coaction of a Hopf algebra $\mathcal{A}$ on V, then $\psi := \tau \circ (\mathrm{id} \otimes S) \circ \varphi : V \to \mathcal{A} \otimes V$ is a left coaction of $\mathcal{A}$ on V.

1.3.3 Comodule Algebras and Related Concepts

In many cases a Hopf algebra coacts on an algebra such that the coaction respects the algebra structure. This situation is described by the following notion. We suppose that $\mathcal{A}$ is a bialgebra.

Definition 8. *An algebra $\mathcal{X}$ is called a* right *(resp.* left*)* $\mathcal{A}$-comodule algebra *or, equivalently, a* right *(resp.* left*)* quantum space *for $\mathcal{A}$ if $\mathcal{X}$ is a right (resp. left) $\mathcal{A}$-comodule such that the multiplication $m : \mathcal{X} \otimes \mathcal{X} \to \mathcal{X}$ and the unit map $\eta : \mathbb{K} \to \mathcal{X}$ of $\mathcal{X}$ are $\mathcal{A}$-comodule homomorphisms.*

Let us be more explicit about these conditions. Suppose that $\mathcal{X}$ is an algebra and a right $\mathcal{A}$-comodule with coaction $\varphi_{\mathcal{X}} : \mathcal{X} \to \mathcal{X} \otimes \mathcal{A}$. Then $\mathcal{X} \otimes \mathcal{X}$ is a right $\mathcal{A}$-comodule with coaction $\varphi_{\mathcal{X}\otimes\mathcal{X}}$ given by (61) and $\mathbb{K}$ is a right $\mathcal{A}$-comodule with trivial coaction $\varphi_{\mathbb{K}}(\lambda) = \lambda \otimes 1$. That m and η are comodule homomorphisms means by (54) that $\varphi_{\mathcal{X}} \circ m = (m \otimes \mathrm{id}) \circ \varphi_{\mathcal{X}\otimes\mathcal{X}}$ and $\varphi_{\mathcal{X}} \circ \eta = (\eta \otimes \mathrm{id}) \circ \varphi_{\mathbb{K}}$. By (61), the two latter equations can be expressed as $\varphi_{\mathcal{X}}(xy) = \varphi_{\mathcal{X}}(x)\varphi_{\mathcal{X}}(y)$, $x, y \in \mathcal{X}$, and $\varphi_{\mathcal{X}}(1) = 1 \otimes 1$, where the product of $\mathcal{X} \otimes \mathcal{A}$ is defined by (12). That is, we have shown that *$\mathcal{X}$ is a right $\mathcal{A}$-comodule algebra if and only if the coaction $\varphi_{\mathcal{X}} : \mathcal{X} \to \mathcal{X} \otimes \mathcal{A}$ is an algebra homomorphism.* A similar statement is valid for left $\mathcal{A}$-comodule algebras.

Example 19 ($\mathbb{C}[x_1, x_2, \cdots, x_N]$ as comodule algebra of $\mathcal{O}(G)$ or $\mathcal{O}(M_N(\mathbb{C}))$). Let $\mathcal{A}$ be either the Hopf algebra $\mathcal{O}(G)$ from Example 3 or the bialgebra $\mathcal{O}(M_N(\mathbb{C}))$ from Example 5 and let $\mathcal{X}$ be the polynomial algebra $\mathbb{C}[x_1, x_2, \cdots, x_N]$. Define algebra homomorphisms $\varphi_L : \mathcal{X} \to \mathcal{A} \otimes \mathcal{X}$ and $\varphi_R : \mathcal{X} \to \mathcal{X} \otimes \mathcal{A}$ by $\varphi_L(x_i) = \sum_j u^i_j \otimes x_j$ and $\varphi_R(x_i) = \sum_j x_j \otimes u^j_i$. Then φ_L (resp. φ_R) is a left (resp. right) coaction of $\mathcal{A}$ on $\mathcal{X}$ and $\mathcal{X}$ is a left (resp. right) $\mathcal{A}$-comodule algebra, or equivalently, $\mathcal{X}$ is a left (resp. right) quantum space for $\mathcal{A}$. $\triangle$

In typical later applications, $\mathcal{A}$ will be the coordinate Hopf algebra of a quantum group and $\mathcal{X}$ will be the coordinate algebra of some quantum space. But a *quantum space* is a virtual object that exists only because of the algebraic structures of the $\mathcal{A}$-comodule algebra $\mathcal{X}$. For this reason, we shall abuse language and use the terms $\mathcal{A}$-comodule algebra and quantum space synonymously.

The corresponding notions of $*$-quantum spaces for a $*$-bialgebra $\mathcal{A}$ are easy to guess. A (*left* or *right*) *$*$-quantum space* for $\mathcal{A}$ or, equivalently, a (*left* or *right*) *$\mathcal{A}$-comodule $*$-algebra* is a $*$-algebra $\mathcal{X}$ which is a (left or right) $\mathcal{A}$-comodule algebra such that the coaction $\varphi_{\mathcal{X}}$ is $*$-preserving (that is, $\varphi_{\mathcal{X}}(x^*) = \varphi_{\mathcal{X}}(x)^*$ for $x \in \mathcal{X}$).

Although the previous Definition 8 is more abstract and less practical than its reformulation given afterwards, it provides the right setting to develop similar concepts such as $\mathcal{A}$-module algebras, $\mathcal{A}$-comodule coalgebras, $\mathcal{A}$-comodule bialgebras, etc. These notions are defined simply by requiring that all structure maps are $\mathcal{A}$-module or $\mathcal{A}$-comodule homomorphisms, respectively. Using the above formulas (49) resp. (54) and (58) resp. (60), it is not difficult to rewrite these abstract definitions in terms of the Sweedler notation. As a sample, we state these reformulations in the case of $\mathcal{A}$-module algebras.

A *left $\mathcal{A}$-module algebra* is an algebra $\mathcal{X}$ which is a left $\mathcal{A}$-module such that $m : \mathcal{X} \otimes \mathcal{X} \to \mathcal{X}$ and $\eta : \mathbb{K} \to \mathcal{X}$ are left $\mathcal{A}$-module homomorphisms. If $a \triangleright x$ denotes the action of $a \in \mathcal{A}$ on $x \in \mathcal{X}$, the two latter conditions mean that

$$a \triangleright (xy) = \sum (a_{(1)} \triangleright x)(a_{(2)} \triangleright y) \quad \text{and} \quad a \triangleright 1 = \varepsilon(a)1.$$

Likewise, an algebra $\mathcal{X}$ is called a *right $\mathcal{A}$-module algebra* if it is a right $\mathcal{A}$-module with action, denoted by $\triangleleft$, such that

$$(xy) \triangleleft a = \sum (x \triangleleft a_{(1)})(y \triangleleft a_{(2)}) \quad \text{and} \quad 1 \triangleleft a = \varepsilon(a)1.$$

Example 20. Let $\mathcal{X}$ be a left $\mathcal{A}$-module algebra for $\mathcal{A} = U(\mathfrak{g})$. Since $\Delta(a) = a \otimes 1 + 1 \otimes a$ for $a \in \mathfrak{g}$, we have $a \triangleright (xy) = (a \triangleright x)y + x(a \triangleright y)$ for $x, y \in \mathcal{X}$. That is, each $a \in \mathfrak{g}$ then acts as a derivation on $\mathcal{X}$. △

Example 21. Let $\mathcal{A} = \mathbb{C}G$. An algebra $\mathcal{X}$ is a left $\mathcal{A}$-module algebra if and only if the group G acts as a group of automorphisms of $\mathcal{X}$. △

1.3.4* Adjoint Actions and Coactions of Hopf Algebras

Let $\mathcal{A}$ be a Hopf algebra. For elements $a, b \in \mathcal{A}$ we define

$$\mathrm{ad}_L(a)b = \sum a_{(1)} b S(a_{(2)}), \quad \mathrm{ad}_R(a)b = \sum S(a_{(1)}) b a_{(2)},$$

$$\mathrm{Ad}_L(a) = \sum a_{(1)} S(a_{(3)}) \otimes a_{(2)}, \quad \mathrm{Ad}_R(a) = \sum a_{(2)} \otimes S(a_{(1)}) a_{(3)}.$$

The mappings ad_L and ad_R are called the *left* and *right adjoint actions*, respectively, and the mappings Ad_L and Ad_R are called the *left* and *right adjoint coactions*, respectively, of the Hopf algebra $\mathcal{A}$ on itself. That these are indeed actions and coactions follows from

Proposition 14. (i) ad_L *turns $\mathcal{A}$ into a left $\mathcal{A}$-module algebra and* ad_R *turns $\mathcal{A}$ into a right $\mathcal{A}$-module algebra.*

(ii) Ad_L *is a left coaction and* Ad_R *is a right coaction of $\mathcal{A}$ on itself.*

Proof. We carry out the proof in the case of ad_R. Set $a \triangleleft b = \mathrm{ad}_R(b)a$ for $a, b \in \mathcal{A}$. Clearly, we have $a \triangleleft 1 = a$ and $1 \triangleleft b = \varepsilon(b)1$. Further, we compute

$$a \triangleleft (bc) = \sum S((bc)_{(1)}) a (bc)_{(2)} = \sum S(c_{(1)})[S(b_{1)}) a b_{(2)}] c_{(2)} = (a \triangleleft b) \triangleleft c \quad (62)$$

and

$$\sum (a \triangleleft c_{(1)})(b \triangleleft c_{(2)}) = \sum S(c_{(1)}) a c_{(2)} S(c_{(3)}) b c_{(4)} = \sum S(c_{(1)}) a \varepsilon(c_{(2)}) b c_{(3)}$$

$$= \sum S(c_{(1)}) a b c_{(2)} = (ab) \triangleleft c.$$

These relations show that ad_R is a right action of $\mathcal{A}$ on itself which turns $\mathcal{A}$ into a right $\mathcal{A}$-module algebra. □

In general, the algebra $\mathcal{A}$ is not a left (resp. right) $\mathcal{A}$-comodule algebra with respect to the coaction Ad_L (resp. Ad_R). This matter will be taken up again in Sect. 10.3. However, $\mathcal{A}$ is a left (resp. right) $\mathcal{A}$-comodule coalgebra with respect to the coaction Ad_L (resp. Ad_R).

Remark 2. If V is a bimodule of $\mathcal{A}$, then the definitions of $\mathrm{ad}_L(a)b$ and $\mathrm{ad}_R(a)b$ make sense for $a \in \mathcal{A}$ and $b \in V$ and the reasoning of (62) remains valid.

Hence ad_L and ad_R are then left and right actions of $\mathcal{A}$ on the bimodule V, respectively. This general setting will be used in Chap. 13. △

Let us consider the above notions in two classical situations.

Example 22 (Conjugation representations of groups). Let $\mathcal{A}$ be the Hopf algebra $\mathcal{O}(G)$ from Example 3. Then we have

$$\mathrm{Ad}_L(f)(g,h) = f(ghg^{-1}) \quad \text{and} \quad \mathrm{Ad}_R(f)(g,h) = f(h^{-1}gh)$$

for $f \in \mathcal{O}(G)$ and $g, h \in G$. It suffices to verify these formulas on the generators. We shall do this in the case of Ad_R and compute

$$\begin{aligned} \mathrm{Ad}_R(u^i_j)(g,h) &= \sum\nolimits_{k,l} u^k_l(g)(S(u^i_k)u^l_j)(h) = \sum\nolimits_{k,l} g_{kl}(h^{-1})_{ik}h_{lj} \\ &= (h^{-1}gh)_{ij} = u^i_j(h^{-1}gh). \end{aligned}$$

The corepresentation Ad_R of $\mathcal{A}$ corresponds by (57) to the conjugation representation of G which is given by $(T(h)f)(g) := f(h^{-1}gh)$. △

Example 23 (Adjoint representations of Lie algebras). Let $\mathcal{A}$ be the Hopf algebra $U(\mathfrak{g})$ from Example 6. Since $\Delta(a) = a \otimes 1 + 1 \otimes a$ for $a \in \mathfrak{g}$, we get $\mathrm{ad}_L(a)b = [a,b]$ and $\mathrm{ad}_R(a)b = -[a,b]$ for $a, b \in \mathfrak{g}$. That is, on $\mathfrak{g}$ the mapping ad_L is just the adjoint representation of the Lie algebra $\mathfrak{g}$. △

1.3.5* Corepresentations and Representations of Dually Paired Coalgebras and Algebras

Let $\mathcal{A}$ be a coalgebra and $\mathcal{U}$ an algebra. We suppose in this subsection that $\langle\cdot,\cdot\rangle : \mathcal{U} \times \mathcal{A} \to \mathbb{C}$ is a (possibly degenerate) bilinear form such that

$$\langle f_1 \otimes f_2, \Delta(a)\rangle = \langle f_1 f_2, a\rangle \quad \text{and} \quad \langle 1, a\rangle = \varepsilon(a), \quad f_1, f_2 \in \mathcal{U}, \quad a \in \mathcal{A}. \tag{63}$$

Proposition 15. *If $\varphi : V \to V \otimes \mathcal{A}$ is a corepresentation of the coalgebra $\mathcal{A}$ on a vector space V, then*

$$\hat{\varphi}(f)v := ((\mathrm{id} \otimes f) \circ \varphi)(v) \equiv \sum v_{(0)}\langle f, v_{(1)}\rangle, \quad f \in \mathcal{U}, \ v \in V, \tag{64}$$

defines a representation of the algebra $\mathcal{U}$ on V.

Proof. By the second equality of (63) and (59), we have

$$\hat{\varphi}(1)v = \sum v_{(0)}\langle 1, v_{(1)}\rangle = \sum v_{(0)}\varepsilon(v_{(1)}) = v.$$

From the first equality of (63) we obtain

$$\hat{\varphi}(f_1 f_2)v = \sum v_{(0)}\langle f_1 f_2, v_{(1)}\rangle = \sum v_{(0)}\langle f_1 \otimes f_2, \Delta(v_{(1)})\rangle$$

$$= \sum v_{(0)}\langle f_1, v_{(1)}\rangle\langle f_2, v_{(2)}\rangle = \sum \hat{\varphi}(f_1)v_{(0)}\langle f_2, v_{(1)}\rangle = \hat{\varphi}(f_1)\hat{\varphi}(f_2)v. \qquad \square$$

There is an analogous result to Proposition 15 for left $\mathcal{A}$-comodules and right $\mathcal{U}$-modules. It says that any left $\mathcal{A}$-comodule with coaction ψ is also a right $\mathcal{U}$-module with action

$$\hat{\psi}(f)v = ((f \otimes \mathrm{id}) \circ \psi)v \equiv \sum v_{(0)} \langle f, v_{(-1)} \rangle.$$

Note that not every finite-dimensional representation of $\mathcal{U}$ is of the form $\hat{\varphi}$ for some corepresentation φ of $\mathcal{A}$ (see the end of Subsect. 4.4.2), not even when $\mathcal{U}$ and $\mathcal{A}$ are Hopf algebras and the pairing $\langle \cdot, \cdot \rangle$ is nondegenerate.

We now specialize the preceding to the case when the coactions φ and ψ are given by the comultiplication of $\mathcal{A}$. For $f \in \mathcal{U}$ and $a \in \mathcal{A}$, we define

$$f.a := \sum a_{(1)} \langle f, a_{(2)} \rangle \text{ and } a.f = \sum \langle f, a_{(1)} \rangle a_{(2)}. \tag{65}$$

Proposition 16. *$\mathcal{A}$ is a bimodule for the algebra $\mathcal{U}$ with left action $f.a$ and right action $a.f$ of $\mathcal{U}$ on $\mathcal{A}$. If $\langle \cdot, \cdot \rangle$ is a dual pairing of two bialgebras $\mathcal{U}$ and $\mathcal{A}$, then $\mathcal{A}$ becomes a left and right $\mathcal{U}$-module algebra in this manner.*

Proof. By Proposition 15 and the corresponding result for left comodules, the vector space $\mathcal{A}$ is a left and right $\mathcal{U}$-module. It is a $\mathcal{U}$-bimodule, since

$$(f_1.a).f_2 = \sum \langle f_2, a_{(1)} \rangle a_{(2)} \langle f_1, a_{(3)} \rangle = f_1.(a.f_2).$$

The second assertion follows easily from (31) and (32). □

Let us turn to another particular case. Suppose that $\mathcal{U}$ is the dual algebra $\mathcal{A}'$ and the pairing $\langle \cdot, \cdot \rangle$ is the evaluation of functionals $\langle f, a \rangle := f(a)$, $a \in \mathcal{A}$, $f \in \mathcal{A}'$. Then the conditions (63) hold by definition (22) of the product of $\mathcal{A}'$. Therefore, by Proposition 15, each corepresentation φ of the coalgebra $\mathcal{A}$ induces a representation $\hat{\varphi}$ of the dual algebra $\mathcal{A}'$. Finally, let us suppose in addition that the bilinear form $\langle \cdot, \cdot \rangle$ is *nondegenerate*. Then, by considering matrix coefficients with respect to some bases, we easily conclude that the corepresentation φ of $\mathcal{A}$ is uniquely determined by the associated representation $\hat{\varphi}$ of $\mathcal{U}$. That is, we obtain a one-to-one correspondence between corepresentations of $\mathcal{A}$ and *certain* (!) representations of $\mathcal{U}$. It is not difficult to see that this correspondence preserves standard notions of representation theory such as direct sums, tensor products, etc.

1.4 Notes

The notion of a Hopf algebra first occurred in the work of H. Hopf [Hop] on algebraic topology. The basic early monographs are [Sw] and [A]. A modern treatment is [Mon]. Hopf algebras appeared also in the work of G. I. Kac [K] in the 1960s. The concept of a dual pairing was introduced in [Ta1]. A nice discussion of dual pairings is in [VD1]. Only very few interesting examples of noncommutative noncocommutative Hopf algebras were known before quantum groups were invented (see [KP], [Sw], pp. 69–70, [Pa]).

2. q-Calculus

Quantum groups and their representations are closely related to the so-called q-calculus (q-numbers, q-factorials, q-differentiation, basic hypergeometric functions, and q-orthogonal polynomials). This chapter gives a brief introduction to these topics. The notions and facts developed below will be needed at various places in the book, but they are also of interest in themselves.

2.1 Main Notions of q-Calculus

2.1.1 q-Numbers and q-Factorials

For any nonzero complex number q, the *q-number* $[a]_q$, $a \in \mathbb{C}$, is defined by

$$[a]_q \equiv [a] := \frac{q^a - q^{-a}}{q - q^{-1}} = \frac{e^{ah} - e^{-ah}}{e^h - e^{-h}} = \frac{\sinh ah}{\sinh h}, \tag{1}$$

where $q = \exp h$. Clearly, $\lim_{q \to 1} [a]_q \to a$. We shall also use the expression

$$[[a]]_q \equiv [[a]] := \frac{1 - q^a}{1 - q} = q^{(a-1)/2} [a]_{q^{1/2}}. \tag{2}$$

A number of relations of these q-numbers can be derived by using properties of the exponential function. For $m, n \in \mathbb{N}$ and $a, b, c \in \mathbb{C}$ we have

$$\begin{aligned} [m] &= q^{m-1} + q^{m-3} + \cdots + q^{-(m-1)}, \\ [m+n] &= q^n[m] + q^{-m}[n] = q^{-n}[m] + q^m[n], \\ [m-n] &= q^n[m] - q^m[n] = q^{-n}[m] - q^{-m}[n], \\ 0 &= [a][b-c] + [b][c-a] + [c][a-b], \\ [n] &= [2][n-1] - [n-2]. \end{aligned}$$

The above equations also remain valid when q is considered as an indeterminate. Then any q-number $[m]_q$, $m \in \mathbb{Z}$, belongs to the space $\mathbb{Z}[q, q^{-1}]$ of Laurent polynomials in q with integral coefficients. Setting $q = \exp h$, we have that $[m]_q \equiv m \pmod{h}$.

If $m \in \mathbb{N}$, then we introduce the *q-factorial* $[m]_q! \equiv [m]!$ by setting

$$[m]! := [1][2]\cdots[m], \qquad [0]! := 1.$$

The expression

$$(a;q)_n := (1-a)(1-aq)(1-aq^2)\cdots(1-aq^{n-1}), \qquad (a;q)_0 = 1, \tag{3}$$

is extensively used in the theory of basic hypergeometric functions and in combinatorics. It is closely related to the q-factorials, since we have

$$[m]! = \frac{q^{-m(m-1)/2}}{(1-q^2)^m}(q^2;q^2)_m. \tag{4}$$

From (3) we see that $(q^{-m};q)_n = 0$ if $n = m+1, m+2, \cdots$. For $|q| < 1$, we define

$$(a;q)_\infty := \lim_{n\to\infty}(a;q)_n = \prod\nolimits_{j=1}^{\infty}(1-aq^{j-1}). \tag{5}$$

Since $|q| < 1$, the infinite product in (5) converges for all $a \in \mathbb{C}$ and defines an analytic function on $\mathbb{C}$. It is clear that

$$(a;q)_n = \frac{(a;q)_\infty}{(aq^n;q)_\infty}. \tag{6}$$

The latter relation allows us to define $(a;q)_{-n}$, $n \in \mathbb{N}$, by

$$(a;q)_{-n} := \frac{(a;q)_\infty}{(aq^{-n};q)_\infty} = \frac{1}{(aq^{-n};q)_n}. \tag{7}$$

By direct computations one verifies the following useful formulas:

$$(a;q)_{N+r} = (a;q)_N(aq^N;q)_r, \tag{8}$$

$$(a;q)_{N-r} = \frac{(a;q)_N}{(a^{-1}q^{1-N};q)_r}(-a^{-1})^r q^{r(r-1)/2} q^{-r(N-1)}, \tag{9}$$

$$(a;q)_{rN} = (a;q^r)_N(aq;q^r)_N\cdots(aq^{r-1};q^r)_N, \tag{10}$$

$$(aq^N;q)_r = \frac{(a;q)_r(aq^r;q)_N}{(a;q)_N}, \quad (aq^{-N};q)_r = \frac{(a;q)_r(qa^{-1};q)_N}{(a^{-1}q^{1-r};q)_N}q^{-Nr}, \tag{11}$$

where N and r are positive integers. Note that for any $a \in \mathbb{C}$ we get

$$(a;q^{-1})_n = (a^{-1};q)_n q^{-n(n-1)/2}(-a)^n. \tag{12}$$

This equation implies that

$$(q^{-2};q^{-2})_n = [n]!(-1)^n(1-q^2)^n q^{-n(n+3)/2}. \tag{13}$$

Next we turn to relations between expressions $(a;q)_n$ with different parameters q. We have $(a^2;q^2)_n = (a;q)_n(-a;q)_n$ and, more generally,

$$(a^r;q^r)_n = (a;q)_n(a\epsilon;q)_n(a\epsilon^2;q)_n\cdots(a\epsilon^{r-1};q)_n,$$

where $\epsilon := e^{2\pi i/r}$.

The equalities (8)–(11) for the expressions $(a;q)_n$ lead at once to the corresponding relations for q-factorials. Later the following equalities will be used:

$$[N+r]! = [N]!\frac{(q^{2N+2};q^2)_r}{(1-q^2)^r}q^{-r(r+2N-1)/2}, \tag{14}$$

$$[N-r]! = [N]!(-1)^r\frac{(1-q^2)^r}{(q^{-2N};q^2)_r}q^{r(r-2N-3)/2}. \tag{15}$$

By (3), (14) and (15), one derives asymptotics for $N \to +\infty$:

$$\frac{[N+r]!}{[N]!} \sim [N+1]^r q^{-r(r-1)/2}, \qquad \frac{[N-r]!}{[N]!} \sim [N]^{-r} q^{r(r-1)/2},$$

where $|q| < 1$ in the first relation and $|q| > 1$ in the second one.

There are particular properties of the q-numbers when q is a root of unity. Let d be the smallest positive integer such that $q^d = 1$ and let

$$e = d \quad \text{if } d \text{ is odd} \quad \text{and} \quad e = d/2 \quad \text{if } d \text{ is even.}$$

One easily verifies the assertion of the following

Proposition 1. *If d and e are as above and $n \in \mathbb{Z}$, then $[n]_q = 0$ if and only if $n \equiv 0 \pmod e$.*

2.1.2 q-Binomial Coefficients

The *q-binomial coefficients* are defined by the formula

$$\begin{bmatrix} n \\ m \end{bmatrix}_q \equiv \begin{bmatrix} n \\ m \end{bmatrix} := \frac{(q;q)_n}{(q;q)_m(q;q)_{n-m}} = \frac{[n]_{q^{1/2}}!q^{(n-m)m/2}}{[m]_{q^{1/2}}![n-m]_{q^{1/2}}!}, \tag{16}$$

where $n, m \in \mathbb{N}_0$ and $n \geq m$.

There is a close analogy between the usual binomial coefficients and their q-analogs. Many identities of the usual binomial coefficients have their counterparts for the q-binomial coefficients. For example,

$$\begin{bmatrix} n+1 \\ m \end{bmatrix}_q = \begin{bmatrix} n \\ m-1 \end{bmatrix}_q + q^m \begin{bmatrix} n \\ m \end{bmatrix}_q = q^{n-m+1}\begin{bmatrix} n \\ m-1 \end{bmatrix}_q + \begin{bmatrix} n \\ m \end{bmatrix}_q. \tag{17}$$

Proposition 2. *If v and w are noncommuting variables satisfying the relation $vw = qwv$, then we have*

$$(v+w)^n = \sum\nolimits_{m=0}^{n} \begin{bmatrix} n \\ m \end{bmatrix}_q w^m v^{n-m} = \sum\nolimits_{m=0}^{n} \begin{bmatrix} n \\ m \end{bmatrix}_{q^{-1}} v^m w^{n-m}. \tag{18}$$

If q is a primitive p-th root of unity and p is odd, then $(v+w)^p = v^p + w^p$.

Proof. Formula (18) can be proved by induction on n using the first identity in (17). The second assertion follows from (18). □

Proposition 3. *For $0 \le k \le n$, $k \in \mathbb{N}_0$, we have the identity*

$$\begin{bmatrix} n \\ m \end{bmatrix}_q = \sum_{s=0}^{\min(m,k)} q^{(k-s)(m-s)} \begin{bmatrix} k \\ s \end{bmatrix}_q \begin{bmatrix} n-k \\ m-s \end{bmatrix}_q . \tag{19}$$

Proof. Consider the free algebra generated by elements v and w subject to the relation $vw = qwv$. Using formula (18), we expand both sides of the identity $(v+w)^n = (v+w)^k(v+w)^{n-k}$ and compare coefficients of $v^m w^{n-m}$. □

Proposition 4. *The q-binomial coefficients satisfy the relation*

$$\sum_{k=1}^{n} (-1)^k \begin{bmatrix} n \\ k \end{bmatrix}_q q^{-nk} q^{(k+1)k/2} = 0. \tag{20}$$

Proof. By the second relation in (17), the left hand side of (20) is equal to

$$\sum_{k=1}^{n} (-1)^k q^{-nk} q^{(k+1)k/2} \left(\begin{bmatrix} n-1 \\ k \end{bmatrix}_q + q^{n-k} \begin{bmatrix} n-1 \\ k-1 \end{bmatrix}_q \right),$$

where the first summand in the brackets is set to zero if $k = n$ and the second summand vanishes if $k = 0$. This expression coincides with

$$\sum_{k=0}^{n-1} (-1)^k q^{-nk} q^{(k+1)k/2} \begin{bmatrix} n-1 \\ k \end{bmatrix}_q + \sum_{k=1}^{n} (-1)^k q^{-n(k-1)} q^{k(k-1)/2} \begin{bmatrix} n-1 \\ k-1 \end{bmatrix}_q ,$$

which clearly vanishes. □

2.1.3 Basic Hypergeometric Functions

In 1846 the German mathematician E. Heine introduced the series

$${}_2\varphi_1(a, b; c; q, z) = \sum_{n=0}^{\infty} \frac{(a;q)_n (b;q)_n}{(c;q)_n} \frac{z^n}{(q;q)_n}, \tag{21}$$

where $c \neq 0, -1, -2, \cdots$. If a or b is a negative integer, then this series is terminating and becomes a polynomial in z. The number q is called the *base* of the series (21). For $|q| < 1$, this series converges absolutely if $|z| < 1$. Since $\lim_{q \to 1}(1-q^a)(1-q)^{-1} = a$, the series (21) converges termwise to the classical hypergeometric series

$${}_2F_1(a, b; c; z) = \frac{\Gamma(c)}{\Gamma(a)\Gamma(b)} \sum_{n=0}^{\infty} \frac{\Gamma(a+n)\Gamma(b+n)}{\Gamma(c+n)} \frac{z^n}{n!}$$

when $q \to 1$ and $|q| < 1$. If $|q| > 1$, then using (12) we can transform (21) into the series ${}_2\varphi_1(a', b'; c'; q^{-1}, z')$ with base q^{-1}.

The function defined by the series (21) is called the *basic hypergeometric function* (or *q-hypergeometric function*) ${}_2\varphi_1$. It can be analytically continued outside of its domain of convergence.

More general basic hypergeometric functions are defined by the series

$$ {}_r\varphi_s(a_1, a_2, \cdots, a_r; b_1, b_2, \cdots, b_s; q, z) \equiv {}_r\varphi_s\left(\begin{matrix} a_1, a_2, \cdots, a_r \\ b_1, b_2, \cdots, b_s \end{matrix}; q, z\right) $$

$$ = \sum_{n=0}^{\infty} \frac{(a_1;q)_n(a_2;q)_n\cdots(a_r;q)_n}{(b_1;q)_n(b_2;q)_n\cdots(b_s;q)_n}\left((-1)^n q^{n(n-1)/2}\right)^{1+s-r}\frac{z^n}{(q;q)_n}. \tag{22} $$

In particular,

$$ {}_r\varphi_{r-1}(a_1, a_2, \cdots, a_r; b_1, b_2, \cdots, b_{r-1}; q, z) $$

$$ = \sum_{n=0}^{\infty} \frac{(a_1;q)_n(a_2;q)_n\cdots(a_r;q)_n}{(b_1;q)_n(b_2;q)_n\cdots(b_{r-1};q)_n}\frac{z^n}{(q;q)_n}. \tag{23} $$

Again, if at least one of the numbers a_1, a_2, $\cdots$, a_r is a negative integer, then the series (22) terminates. If $0 < |q| < 1$, then the series (22) converges absolutely for all z if $r \leq s$ and for all $|z| < 1$ if $r = s + 1$. It also converges absolutely if $|q| > 1$ and $|z| < |b_1 b_2 \cdots b_s|/|a_1 a_2 \cdots a_r|$. If the series (22) does not terminate, then it diverges for $z \neq 0$ if $0 < |q| < 1$ and $r > s + 1$ and if $|q| > 1$ and $|z| > |b_1 b_2 \cdots b_s|/|a_1 a_2 \cdots a_r|$.

2.1.4 The Function ${}_1\varphi_0(a; q, z)$

By (23), we have

$$ {}_1\varphi_0(a;q,z) = \sum_{n=0}^{\infty} \frac{(a;q)_n}{(q;q)_n} z^n. \tag{24} $$

From the expressions for ${}_1\varphi_0(a;q,z)$ and ${}_1\varphi_0(a;q,qz)$ we obtain

$$ {}_1\varphi_0(a;q,z) - {}_1\varphi_0(a;q,qz) = (1-a)z\ {}_1\varphi_0(aq;q,z), $$

$$ {}_1\varphi_0(a;q,z) - a\ {}_1\varphi_0(a;q,qz) = (1-a)\ {}_1\varphi_0(aq;q,z). $$

These relations yield the equation $(1-z)\ {}_1\varphi_0(a;q,z) = (1-az)\ {}_1\varphi_0(a;q,qz)$ and so

$$ {}_1\varphi_0(a;q,z) = \frac{(az;q)_n}{(z;q)_n}\ {}_1\varphi_0(a;q,q^n z). \tag{25} $$

If $|q| < 1$, then ${}_1\varphi_0(a;q,q^n z) \to 1$ when $n \to \infty$. Therefore, by (25),

$$ {}_1\varphi_0(a;q,z) = \sum_{n=0}^{\infty} \frac{(a;q)_n}{(q;q)_n} z^n = \prod_{r=0}^{\infty} \frac{1-azq^r}{1-zq^r} = \frac{(az;q)_\infty}{(z;q)_\infty}, \tag{26} $$

when $|q| < 1$. The latter formula has some interesting corollaries. One is

$$ {}_1\varphi_0(a;q,z)\,{}_1\varphi_0(b;q,az) = {}_1\varphi_0(ab;q,z). \tag{27} $$

Formula (26) implies that ${}_1\varphi_0(1;q,z) = 1$. Therefore, by (27),

$$ {}_1\varphi_0(a^{-1};q,az) = \{{}_1\varphi_0(a;q,z)\}^{-1}. \tag{28} $$

2.1.5 The Basic Hypergeometric Function ${}_2\varphi_1$

For many reasons it is useful to have formulas relating hypergeometric functions with different parameters. For the function ${}_2\varphi_1(a,b;c;q,z)$ the transformation formula

$$ {}_2\varphi_1(a,b;c;q,z) = \frac{(b;q)_\infty (az;q)_\infty}{(c;q)_\infty (z;q)_\infty} {}_2\varphi_1(c/b,z;az;q,b) \tag{29} $$

holds when $|q|<1$, $|z|<1$, $|b|<1$. Indeed, using the formulas (6) and then the equality (26), we get

$$ \begin{aligned} {}_2\varphi_1(a,b;c;q,z) &= \sum_{n=0}^{\infty} \frac{(a;q)_n (b;q)_n}{(c;q)_n (q;q)_n} z^n = \frac{(b;q)_\infty}{(c;q)_\infty} \sum_{n=0}^{\infty} \frac{(a;q)_n}{(q;q)_n} \frac{(cq^n;q)_\infty}{(bq^n;q)_\infty} z^n \\ &= \frac{(b;q)_\infty}{(c;q)_\infty} \sum_{n=0}^{\infty} \sum_{m=0}^{\infty} \frac{(a;q)_n}{(q;q)_n} z^n \frac{(c/b;q)_m b^m q^{mn}}{(q;q)_m} \\ &= \frac{(b;q)_\infty}{(c;q)_\infty} \sum_{m=0}^{\infty} \frac{(c/b;q)_m b^m}{(q;q)_m} \frac{(azq^m;q)_\infty}{(zq^m;q)_\infty} \\ &= \frac{(b;q)_\infty (az;q)_\infty}{(c;q)_\infty (z;q)_\infty} \sum_{m=0}^{\infty} \frac{(c/b;q)_m (z;q)_m}{(az;q)_m (q;q)_m} b^m. \end{aligned} $$

This proves (29). In a similar way, one obtains the relation

$$ {}_2\varphi_1(a,b;c;q,z) = \frac{(abz/c;q)_\infty}{(z;q)_\infty} {}_2\varphi_1(c/a,c/b;c;q,abz/c). \tag{30} $$

If $a = q^{-n}$, $n \in \mathbb{N}_0$, then we also have (see [GR])

$$ \begin{aligned} {}_2\varphi_1(q^{-n},b;c;q,z) &= (-1)^n q^{-n(n+1)/2} \frac{(b;q)_n}{(c;q)_n} z^n \\ &\quad \times {}_2\varphi_1(q^{-n}, q^{-n+1}c^{-1}; q^{-n+1}b^{-1}; q, cq^{n+1}/bz). \end{aligned} \tag{31} $$

There exist several summation formulas for the series ${}_2\varphi_1$. The most important one is

$$ {}_2\varphi_1(a,b;c;q,c/ab) = \frac{(c/a;q)_\infty (c/b;q)_\infty}{(c;q)_\infty (c/ab;q)_\infty}, \tag{32} $$

where $|c|<|ab|$, $|q|<1$. In order to prove this relation we first use (29) and then (26) and compute

$$ \begin{aligned} {}_2\varphi_1(a,b;c;q,c/ab) &= \sum_{n=0}^{\infty} \frac{(a;q)_n (b;q)_n}{(c;q)_n (q;q)_n} \left(\frac{c}{ab}\right)^n \\ &= \frac{(b;q)_\infty (c/b;q)_\infty}{(c;q)_\infty (c/ab;q)_\infty} \sum_{n=0}^{\infty} \frac{(c/ab;q)_n b^n}{(q;q)_n} \\ &= \frac{(b;q)_\infty (c/b;q)_\infty}{(c;q)_\infty (c/ab;q)_\infty} \frac{(c/a;q)_\infty}{(b;q)_\infty} = \frac{(c/a;q)_\infty (c/b;q)_\infty}{(c;q)_\infty (c/ab;q)_\infty}. \end{aligned} $$

In the case $b = q^{-n}$ the formula (32) can be written as

$$ {}_2\varphi_1(a, q^{-n}; c; q, q^n c/a) = \frac{(c/a; q)_n}{(c; q)_n}. \tag{33} $$

Further, it follows from (31) and (33) that

$$ {}_2\varphi_1(a, q^{-n}; c; q, q) = \frac{(c/a; q)_n}{(c; q)_n} a^n. \tag{34} $$

In what follows, we shall also use the notation

$$ {}_2\Phi_1(\alpha, \beta; \gamma; q, z) := {}_2\varphi_1(q^\alpha, q^\beta; q^\delta; q, z). \tag{35} $$

2.1.6 Transformation Formulas for ${}_3\varphi_2$ and ${}_4\varphi_3$

Transformation relations for the functions ${}_3\varphi_2$ and ${}_4\varphi_3$ are of a great importance for the theory of Clebsch–Gordan and Racah coefficients of the quantum algebra $U_q(\mathrm{su}_2)$. The fundamental transformation formula for the function ${}_3\varphi_2$ is

$$ {}_3\varphi_2(q^{-n}, a, b; c, d; q, q) = \frac{(c/a; q)_n}{(c; q)_n} a^n \; {}_3\varphi_2(q^{-n}, a, d/b; aq^{-n+1}/c, d; q, bq/c). \tag{36} $$

A repeated application of this formula leads to the relation

$$ {}_3\varphi_2(q^{-n}, q^{n-1}abcd, bz; bc, dc; q, q) = \left(\frac{b}{c}\right)^n \frac{(ac; q)_n (ad; q)_n}{(bc; q)_n (bd; q)_n} \; {}_3\varphi_2(q^{-n}, q^{n-1}abcd, az; ac, ad; q, q). \tag{37} $$

For the function ${}_4\varphi_3$, we have the transformation formula

$$ {}_4\varphi_3\left(\begin{matrix} q^{-n}, a, b, c \\ d,\ e,\ f \end{matrix}; q, q\right) = \left(\frac{bc}{d}\right)^n \frac{(aq^{1-n}/e; q)_n (aq^{1-n}/f; q)_n}{(e; q)_n (f; q)_n} \; {}_4\varphi_3\left(\begin{matrix} q^{-n},\ a,\ d/b,\ d/c \\ d, aq^{1-n}/e, aq^{1-n}/f \end{matrix}; q, q\right) \tag{38} $$

which holds when the condition $abcq^{1-n} = def$ is satisfied. The proofs of (36) and (38) are technically complicated and we refer the reader to the monograph [GR].

2.1.7 q-Analog of the Binomial Theorem

The classical binomial theorem says that $(1-x)^n = {}_2F_1(-n,c;c;x) = {}_1F_0(-n;x)$ for $n \in \mathbb{N}$. The expression

$$(1-x)_q^n := (1-q^{-1}x)(1-q^{-2}x)\cdots(1-q^{-n}x) = (q^{-n}x;q)_n. \tag{39}$$

can be regarded as a q-analog of $(1-x)^n$. In order to express $(1-x)_q^n$ by means of the q-hypergeometric function ${}_1\varphi_0$, we first state that

$$(z;q)_n = \sum\nolimits_{j=1}^{n} \begin{bmatrix} n \\ j \end{bmatrix}_q (-1)^j z^j q^{j(j-1)/2}. \tag{40}$$

Indeed, using (6) and (26), one derives

$$(z;q)_n = \frac{(z;q)_\infty}{(zq^n;q)_\infty} = \sum_{j=0}^{\infty} \frac{(q^{-n};q)_j z^j q^{jn}}{(q;q)_j} = \sum_{j=0}^{\infty} \frac{z^j q^{jn} \prod_{i=0}^{j-1}(1-q^{-n+i})}{(q;q)_j}$$

$$= \sum_{j=0}^{n} \frac{z^j q^{j(j-1)/2} \prod_{i=0}^{j-1}(1-q^{n-i})}{(-1)^j (q;q)_j} = \sum_{j=0}^{n} \frac{(q;q)_n}{(q;q)_j (q;q)_{n-j}} (-1)^j z^j q^{(j-1)j/2}.$$

This gives (40). From (39) and (40) we obtain

$$(1-x)_q^n = \sum\nolimits_{r=0}^{n} \begin{bmatrix} n \\ r \end{bmatrix}_q (-x)^r q^{r(r-2n-1)/2}. \tag{41}$$

Applying first the relation (9) for $a = q$ and then (24), we get

$$(1-x)_q^n = \sum\nolimits_{r=0}^{n} \frac{(q^{-n};q)_r}{(q;q)_r} x^r = {}_1\varphi_0(q^{-n};q,x). \tag{42}$$

Formula (42) is a q–analog of the binomial theorem.

Motivated by the equality (42), we define $(1-x)_q^a := {}_1\varphi_0(q^{-a};q,x)$ for arbitrary $a \in \mathbb{C}$ and $|q| < 1$. We also define $(b-x)_q^n$ by setting $(b-x)_q^n := b^n(1-x/b)_q^n$.

2.2 q-Differentiation and q-Integration

2.2.1 q-Differentiation

For a fixed complex number $q \neq 1$, the *q-differentiation operator* D_q is defined by

$$D_q f(x) = \frac{f(x) - f(qx)}{x - qx}. \tag{43}$$

Clearly, $D_q \to d/dx$ as $q \to 1$. Applying the Taylor formula to the right hand side of (43) we obtain

$$D_q f(x) = \sum_{n=0}^{\infty} \frac{(q-1)^n}{(n+1)!} x^n \frac{d^{n+1}}{dx^{n+1}} f(x), \tag{44}$$

provided that the expression on the right hand side of (44) exists.

One easily verifies that

$$D_q x^a = [[a]] x^{a-1}, \quad D_q (1-bx)_q^a = b[[a]](1-bqx)_q^{a-1}, \quad a, b \in \mathbb{C}, \tag{45}$$

where $[[a]] = (1-q^a)/(1-q) = q^{(a-1)/2}[a]_{q^{1/2}}$ as defined by (2).

For the product of two functions $f_1(x)$ and $f_2(x)$ we have

$$\begin{aligned} D_q\{f_1(x)f_2(x)\} &= \frac{f_1(x)f_2(x) - f_1(qx)f_2(qx)}{x(1-q)} \\ &= \frac{f_2(x)\{f_1(x) - f_1(qx)\} + f_1(qx)\{f_2(x) - f_2(qx)\}}{x(1-q)}. \end{aligned}$$

Thus we obtain the following q-analog of the Leibniz rule:

$$D_q\{f_1(x)f_2(x)\} = f_2(x) D_q f_1(x) + f_1(qx) D_q f_2(x). \tag{46}$$

From formula (46) it follows by induction on n that

$$D_q^n f(x) = \frac{x^{-n} q^{-n(n-1)/2}}{(q-1)^n} \sum_{m=0}^{n} \begin{bmatrix} n \\ m \end{bmatrix}_q (-1)^m q^{m(m-1)/2} f(q^{n-m} x).$$

In particular, we have

$$D_q^2 f(x) = (q-1)^{-2} x^{-2} q^{-1} \{f(q^2 x) - (1+q) f(qx) + q f(x)\}.$$

Using this formula one verifies that the function $y = {}_2\varphi_1(a, b; c;\ q, x)$ satisfies the second order q-differential equation

$$x(q^c - q^{a+b+1}x) D_q^2 y + \{[[c]] - (q^a[[b+1]] + q^b[[a]])x\} D_q y - [[a]][[b]] y = 0.$$

This is a q-analog of the second order differential equation for the ordinary hypergeometric function ${}_2F_1(a, b; c; x)$.

The equality

$$f(x) = f(a) + \frac{(x-a)_q^1}{[[1]]!} D_q f(a) + \frac{(x-a)_q^2}{[[2]]!} D_q^2 f(a) + \cdots, \tag{47}$$

is a q-analog of the classical Taylor series expansion. Here $(x-a)_q^n$ is defined in Subsect. 2.1.7, $[[m]]! := [[1]] \cdots [[m]]$ and $D_q^n f(a) = \{D_q^n f(x)\}_{x=a}$.

Sometimes, the q-differentiation operator is defined by

$$\tilde{D}_q f(x) = \frac{f(qx) - f(q^{-1}x)}{(q - q^{-1})x}. \tag{48}$$

Definition (48) of the q-differentiation operator corresponds to definition (1) of the q-numbers, because we have $\tilde{D}_q x^n = [n]_q x^{n-1}$. The q-derivative (48) can be expressed in terms of the q-derivative (43) by

$$\tilde{D}_q f(x) = D_{q^2} f(y), \qquad \text{where} \qquad y := q^{-1}x. \tag{49}$$

For the operator $\tilde{D}_q$ the q-analog of the Leibniz rule takes the form

$$\tilde{D}_q\{f_1(x)f_2(x)\} = f_2(q^{-1}x)\tilde{D}_q f_1(x) + f_1(qx)\tilde{D}_q f_2(x). \tag{50}$$

2.2.2 q-Integral

q-Integration will be defined as the inverse operation to q-differentiation. If $D_q F(x) = f(x)$, then we have $F(x) - F(qx) = (1-q)xf(x)$ and hence

$$F(q^j x) - F(q^{j+1}x) = (1-q)q^j x f(q^j x), \qquad j = 0, 1, 2, \cdots.$$

Summing these relations over $j = 0, 1, 2, \cdots, n-1$, one gets

$$F(x) - F(q^n x) = (1-q)x \sum_{j=0}^{n-1} q^j f(q^j x).$$

Suppose that $0 < q < 1$. Assuming that $F(q^n x) \to F(0)$ as $n \to \infty$ we obtain

$$F(x) - F(0) = (1-q)x \sum_{j=0}^{\infty} q^j f(q^j x).$$

Thus, for $0 < q < 1$ the *q-integral* of a function $f(x)$ on the interval $[0, c]$ is defined by

$$\int_0^c f(x) d_q x = c(1-q) \sum_{j=0}^{\infty} q^j f(q^j c) = \sum_{r=0}^{\infty} (x_r - x_{r+1}) f(x_r), \tag{51}$$

where $x_r := cq^r$. For the interval $[c, \infty)$ the definition of the q-integral is

$$\int_c^{\infty} f(x) d_q x = c(1-q) \sum_{j=1}^{\infty} q^{-j} f(q^{-j} c). \tag{52}$$

For arbitrary positive numbers a and b we set

$$\int_a^b f(x) d_q x = \int_0^b f(x) d_q x - \int_0^a f(x) d_q x.$$

Putting $c = 1$ into (51) and (52) and summing these equalities we obtain for the q-integral over an infinite interval the expression

$$\int_0^{\infty} f(x) d_q x = (1-q) \sum_{j=-\infty}^{\infty} q^j f(q^j). \tag{53}$$

The formula for q-integration by parts can be derived by q-integrating both sides of (46). In this manner we get

$$\int f_2(x)(D_q f_1(x)) d_q x = f_1(x) f_2(x) - \int f_1(qx)(D_q f_2(x)) d_q x. \tag{54}$$

2.2.3 q-Analog of the Exponential Function

There are the following three q-analogs of the exponential function:

$$e_q(z) := \sum_{n=0}^{\infty} \frac{z^n}{(q;q)_n} = {}_1\varphi_0(0;\; q,z) = (z;q)_\infty^{-1}, \quad |z| < 1, \tag{55}$$

$$E_q(z) := \sum_{n=0}^{\infty} \frac{q^{n(n-1)/2}}{(q;q)_n} z^n = {}_0\varphi_0(q,-z) = (-z;q)_\infty, \tag{56}$$

$$\tilde{E}_q(z) := \sum_{n=0}^{\infty} \frac{z^n}{[n]!}, \tag{57}$$

where $|q| < 1$. From (55) and (56) we derive that $e_q(z)E_q(-z) = 1$. If the variables x and y commute, then the relations $e_q(x)e_q(y) = e_q(x+y)$ and $E_q(x)E_q(y) = E_q(x+y)$ are not true in general, but one can show that

$$e_q(x+y)e_q^{-1}(x)e_q^{-1}(y) = \prod_{n=0}^{\infty} \left(1 + \frac{q^{2n}xy}{1 - q^n(x+y)}\right).$$

Multiplying the series of $e_q(x)$ and $E_q(y)$ and regrouping the corresponding terms it follows for commuting variables x and y that

$$e_q(x)E_q(y) = 1 + \frac{(x+y)_q^1}{(q;q)_1} + \frac{(x+y)_q^2}{(q;q)_2} + \cdots .$$

Proposition 5. *If x and y are variables such that $xy = qyx$, then we have*

$$e_q(x)e_q(y) = e_q(x+y) \quad \textit{and} \quad E_q(x)E_q(y) = E_q(x+y).$$

Proof. The first identity is easily obtained from (18) and (16). The second follows from the first, because the relation $e_q(z)E_q(-z) = 1$ implies that $E_q(x+y)$ is the inverse of $e_q(-x-y)$ and that $E_q(x)E_q(y)$ is the inverse of $e_q(-y)e_q(-x) = e_q(-x-y)$. □

Using the q-exponential functions we can introduce q-trigonometrical functions by

$$\sin_q x := \frac{1}{2\mathrm{i}}(e_q(\mathrm{i}x) - e_q(-\mathrm{i}x)) = \sum_{n=0}^{\infty} \frac{(-1)^n x^{2n+1}}{(q;q)_{2n+1}},$$

$$\cos_q x := \frac{1}{2}(e_q(\mathrm{i}x) + e_q(-\mathrm{i}x)) = \sum_{n=0}^{\infty} \frac{(-1)^n x^{2n}}{(q;q)_{2n}},$$

$$\mathrm{Sin}_q x := \frac{1}{2\mathrm{i}}(E_q(\mathrm{i}x) - E_q(-\mathrm{i}x)), \qquad \mathrm{Cos}_q x := \frac{1}{2}(E_q(\mathrm{i}x) + E_q(-\mathrm{i}x)).$$

Then we have the identities

$$\sin_q x \ \mathrm{Sin}_q x + \cos_q x \ \mathrm{Cos}_q x = 1, \quad \sin_q x \ \mathrm{Cos}_q x = \mathrm{Sin}_q x \cos_q x.$$

The functions $\cos_q(\omega x)$ and $\sin_q(\omega x)$ are linearly independent solutions of the second order q-differential equation

$$D_q^2 f + \omega^2 (1-q)^{-2} f = 0.$$

Similarly, the functions $\mathrm{Cos}_q(\omega x)$ and $\mathrm{Sin}_q(\omega x)$ are solutions of the q-differential equation $D_q^2 f(x) + \omega^2(1-q)^{-2} f(q^2 x) = 0$.

2.2.4 q-Analog of the Gamma Function

The *q-gamma function* $\Gamma_q(\nu)$ is defined by the q-integral

$$\int_0^\infty x^{\nu-1} E_q(-(1-q)x) d_q x = q^{-\nu(\nu-1)/2} \Gamma_q(\nu), \tag{58}$$

where $\mathrm{Re}\,\nu > 1$ and $|q| < 1$. Carrying out the q-integration by parts we get

$$\int_0^\infty x^{\nu-1} E_q(-(1-q)x) d_q x$$

$$= \frac{x^\nu}{[[\nu]]} E_q(-(1-q)x) \Big|_0^\infty + \frac{q^\nu}{[[\nu]]} \int_0^\infty x^\nu E_q(-(1-q)x) d_q x. \tag{59}$$

From the equality $e_q(-x)E_q(x) = 1$ we derive $E_q(-x) \to 0$ when $x \to +\infty$. Hence the first summand on the right hand side of (59) vanishes and we get

$$\Gamma_q(\nu+1) = [[\nu]] \Gamma_q(\nu).$$

In particular, if $\nu = n \in \mathbb{N}$, then

$$\Gamma_q(n) = [[n-1]]! = \frac{(q;q)_{n-1}}{(1-q)^{n-1}}. \tag{60}$$

If $\Gamma(\nu)$ is the "ordinary" gamma function, we have $\Gamma_q(\nu) \to \Gamma(\nu)$ when $q \nearrow 1$. The q-gamma function can also be represented as

$$\Gamma_q(\nu) = (1-q)^{1-\nu} \frac{(q;q)_\infty}{(q^\nu;q)_\infty}, \tag{61}$$

where $|q| < 1$. For $|q| > 1$ we define $\Gamma_q(\nu) := q^{(\nu-1)(\nu-2)/2} \Gamma_{1/q}(\nu)$.

For the study of Clebsch–Gordan coefficients of the quantum algebra $U_q(\mathrm{sl}_2)$ it is convenient to use the modified q-gamma function given by

$$\hat{\Gamma}_q(\nu) := q^{-(\nu-1)(\nu-2)/2} \Gamma_{q^2}(\nu). \tag{62}$$

This function has the properties $\hat{\Gamma}_q(\nu) = \hat{\Gamma}_{1/q}(\nu)$, $\hat{\Gamma}_q(n) = [n-1]!$, $n \in \mathbb{N}$, and

$$\hat{\Gamma}_q(\nu+1) = [\nu] \hat{\Gamma}_q(\nu). \tag{63}$$

By means of formula (63) one can show that if $\nu \to -m$, $m \in \mathbb{N}$, then

$$\hat{\Gamma}_q(\nu+1)/\hat{\Gamma}_q(\nu) \to -\hat{\Gamma}_q(m+1)/\hat{\Gamma}_q(m).$$

Consequently, for $\nu \to -m$, $m \in \mathbb{N}$, and $n \in \mathbb{N}$ we have

$$\frac{\hat{\Gamma}_q(\nu+1)}{\hat{\Gamma}_q(\nu-n+1)} \to (-1)^n \frac{\hat{\Gamma}_q(m+n)}{\hat{\Gamma}_q(m)}. \tag{64}$$

The Stirling formula describes the asymptotic behavior of the classical gamma function. Its analog for the q-gamma function is

$$\hat{\Gamma}_q(\nu+1) \sim q^{-\nu(\nu-1)/2}(1-q^2)^{-\nu} \exp(-C_q) \qquad \text{if} \qquad \nu \to +\infty, \tag{65}$$

where $0 < q < 1$ and $C_q = -\sum_{k=0}^{\infty} \ln\,(1-q^{2(k+1)})$. For the quotient of q-gamma functions one has the asymptotic formula

$$\frac{\hat{\Gamma}_q(\nu+a)}{\hat{\Gamma}_q(\nu)} \sim q^{-a(a+2\nu-3)/2}(1-q^2)^{-a} \quad \text{if} \quad \nu \to +\infty.$$

2.3 q-Orthogonal Polynomials

q-Analogs of classical orthogonal polynomials play an important role for the study of representations of quantum groups and q-oscillator algebras. In this section we consider some important q-orthogonal polynomials in the case when q is a real number.

2.3.1 Jacobi Matrices and Orthogonal Polynomials

Orthogonal polynomials satisfying three-term recurrence relations are connected with linear operators which can be represented by Jacobi matrices.

Let $\{a_n \mid n \in \mathbb{N}_0\}$, $\{b_n \mid n \in \mathbb{N}_0\}$, $\{c_n \mid n \in \mathbb{N}_0\}$ be complex sequences such that $a_n \neq 0$ for $n \in \mathbb{N}_0$. To these sequences we associate a linear operator L acting on a Hilbert space $\mathfrak{H}$ with orthonormal basis $|n\rangle$, $n = 0, 1, 2, \cdots$, by

$$L|n\rangle = a_n|n+1\rangle + b_n|n\rangle + c_n|n-1\rangle, \tag{66}$$

where $|-1\rangle := 0$. Suppose that $|z\rangle$ is an eigenvector of L for the eigenvalue z, that is, $L|z\rangle = z|z\rangle$. Then $|z\rangle$ can be written as $|z\rangle = \sum_{n=0}^{\infty} p_n(z)|n\rangle$ with certain coefficients $p_n(z)$. From (66) we get

$$\sum_{n=0}^{\infty} (p_n(z)a_n|n+1\rangle + p_n(z)b_n|n\rangle + p_n(z)c_n|n-1\rangle) = z\sum_{n=0}^{\infty} p_n(z)|n\rangle.$$

Equating coefficients of the vector $|n\rangle$ we obtain the following three-term recurrence relation for the coefficients $p_n(z)$:

$$c_{n+1}p_{n+1}(z) + b_n p_n(z) + a_{n-1}p_{n-1}(z) = zp_n(z). \tag{67}$$

Let us set $p_{-1}(z) = 0$ and $p_0(z) = 1$. Then (67) determines the coefficients $p_n(z)$ uniquely and each $p_n(z)$ is a polynomial in z of degree n.

Suppose now that the sequences (a_n) and (b_n) are real and $c_n = a_{n-1}$ for $n \in \mathbb{N}$. Then L is a symmetric operator. By Favard's theorem, there exists a nonnegative Borel measure μ on the real line such that the polynomials $p_n(z)$ satisfy the orthogonality relation

$$\int_a^b p_j(\lambda)p_k(\lambda)d\mu(\lambda) = \delta_{jk}, \quad j,k = 0,1,2,\cdots. \tag{68}$$

One can prove that the symmetric linear operator L always admits a self-adjoint extension, say $\tilde{L}$, in the Hilbert space $\mathfrak{H}$. If $E(\cdot)$ is the spectral measure of $\tilde{L}$ and $\mu(\lambda) = \langle 0 \mid E(\lambda) \mid 0\rangle$, then (68) holds. If the closure $\overline{L}$ of the operator L is self-adjoint, then any measure μ satisfying (68) is uniquely determined, that is, we have a unique orthogonality relation for the polynomials $p_n(z)$. This is in particular true if the symmetric operator L is bounded. In this case the measure μ has compact support. If the closure of L is not self-adjoint, then L has infinitely many self-adjoint extensions and to every such extension there corresponds a nonnegative Borel measure μ such that (68) is valid.

Special cases of the recurrence relation (67) lead to the so-called q-orthogonal polynomials.

2.3.2 q-Hermite Polynomials

The recurrence relation

$$2xp_n(x) = p_{n+1}(x) + (1-q^n)p_{n-1}(x) \tag{69}$$

determines the continuous q-Hermite polynomials $p_n(x) = H_n(x|q)$. They are given by

$$H_n(x|q) = \sum_{k=0}^{n} \frac{(q;q)_n}{(q;q)_k(q;q)_{n-k}} e^{i(n-2k)\theta},$$

where $x = \cos\theta$. The generating function for these polynomials is

$$|(ze^{i\theta};q)_\infty|^{-2} \equiv \prod_{n=0}^{\infty}(1-2xzq^n + z^2q^{2n})^{-1} = \sum_{n=0}^{\infty} \frac{H_n(x|q)}{(q;q)_n} z^n,$$

where again $x = \cos\theta$. If $-1 < q < 1$, then the polynomials $H_n(x|q)$ are orthogonal with orthogonality relation

$$\int_{-1}^{1} H_n(x|q)H_m(x|q)\left(\prod_{k=0}^{\infty}(1-2(2x^2-1)q^k + q^{2k})\right)(1-x^2)^{-1/2}dx$$

$$= \delta_{mn}2\pi(q;q)_n/(q;q)_\infty .$$

The corresponding self-adjoint operators are related to the position and momentum operators for the q-oscillator algebra (see [ChK] and [KlS]).

The polynomials

$$h_n(x|q) := \mathrm{i}^{-n} H_n(\mathrm{i}x|q), \qquad \mathrm{i} := \sqrt{-1}, \tag{70}$$

satisfy the recurrence relation

$$2x h_n(x|q) = h_{n+1}(x|q) + (q^n - 1) h_{n-1}(x|q). \tag{71}$$

For $0 < q < 1$, the orthogonality relation for these polynomials is

$$\int_{-\infty}^{\infty} \frac{h_n(\sinh u|q^{-1}) h_m(\sinh u|q^{-1})}{(-qe^{2u};q)_\infty (-qe^{-2u};q)_\infty} du = \frac{(-1)^n (q;q)_\infty (\log q^{-1})}{(q^{-1};q^{-1})_n^{-1}} \delta_{mn}.$$

The discrete q-Hermite polynomials $H_n(x;q)$, $n \in \mathbb{N}_0$, are defined by the recurrence relation

$$x p_n(x) = p_{n+1}(x) + q^n (1 - q^n) p_{n-1}(x). \tag{72}$$

In explicit form these polynomials are given by

$$H_n(x;q) = \sum_{k=0}^{[n/2]} \frac{(q;q)_n (-1)^k q^{k(k-1)}}{(q^2;q^2)_k (q;q)_{n-2k}} x^{n-2k},$$

where $[n/2]$ denotes the integral part of $n/2$. For $0 < q < 1$, the orthogonality relation for the polynomials $H_n(x;q)$ is of the form

$$\sum_{j=1}^{\infty} \{H_m(q^j;q) H_n(q^j;q) + H_m(-q^j;q) H_n(-q^j;q)\} \frac{q^j}{2} \frac{(q^{2j+2};q)_\infty (q;q^2)_\infty}{(q^2;q^2)_\infty}$$

$$+1/2\ (q;q^2)_\infty H_m(1;q) H_n(1;q) = q^{n(n-1)/2} (q;q)_n \delta_{mn},$$

that is, these polynomials are orthogonal on the set $x = \pm q^j$, $j \in \mathbb{N}_0$.

2.3.3 Little q-Jacobi Polynomials

The *little q-Jacobi polynomials* $p_n(x;a,b|q)$ are expressed in terms of the q-hypergeometric function ${}_2\varphi_1$ by

$$p_n(x;a,b|q) = {}_2\varphi_1(q^{-n}, abq^{n+1}; aq; q, xq). \tag{73}$$

One can easily verify that $\lim_{q\to 1} p_n((1-x)/2; q^\alpha, q^\beta|q) = c P_n^{(\alpha,\beta)}(x)$, where c is a constant and $P_n^{(\alpha,\beta)}(x)$ is the classical Jacobi polynomial.

The polynomials $p_n(x;a,b|q)$ satisfy the recurrence relation

$$x p_n(x) = A_n p_{n+1}(x) - (A_n + C_n) p_n(x) + C_n p_{n-1}(x), \tag{74}$$

where

$$A_n = (-q^n)(1-abq^{n+1})(1-aq^{n+1})\{(1-abq^{2n+1})(1-abq^{2n+2})\}^{-1},$$
$$C_n = (-aq^n)(1-q^n)(1-bq^n)\{(1-abq^{2n})(1-abq^{2n+1})\}^{-1}.$$

The orthogonality relation for the little q-Jacobi polynomials is

$$\sum_{x=0}^{\infty} p_m(q^x;a,b|q)p_n(q^x;a,b|q)\frac{(bq;q)_x}{(q;q)_x}(aq)^x = \frac{\delta_{mn}}{h_n(a,b;q)}, \tag{75}$$

for $0 < q < 1$, where

$$h_n(a,b;q) = (aq)^{-n}\frac{(abq;q)_n(1-abq^{2n+1})(aq;q)_n(aq;q)_\infty}{(q;q)_n(1-abq)(bq;q)_n(abq^2;q)_\infty}.$$

That is, these polynomials are orthogonal on a discrete set of points.

2.3.4 Big q-Jacobi Polynomials

Let α, β, c, d be fixed complex numbers. The functions

$$P_n^{(\alpha,\beta)}(x;\ c,d|q) = {}_3\varphi_2\left(\begin{matrix} q^{-n}, q^{\alpha+\beta+n+1}, q^{\alpha+1}x/c \\ q^{\alpha+1},\ -q^{\alpha+1}d/c \end{matrix} ;\ q,q\right),\quad n \in \mathbb{N}_0, \tag{76}$$

are polynomials in x, called the *big q-Jacobi polynomials.* If $\alpha, \beta \in \mathbb{R}$ and $c, d \in \mathbb{N}_0$, then they satisfy the orthogonality relation

$$\int_{-d}^{c} P_n^{(\alpha,\beta)}(x;c,d|q)P_m^{(\alpha,\beta)}(x;\ c,d|q)(qx/c;q)_\alpha(-qx/d;q)_\beta d_qx = h_n\delta_{nm}, \tag{77}$$

where

$$h_n = \frac{(q^{\beta+1};q)_{\alpha+1}(q;q)_n(1-q^{\alpha+\beta+1})(q^{\beta+1};q)_n(-q^{\beta+1}c/d;q)_n}{c(1-q)(q;q)_\alpha(-d/c;q)_{\alpha+1}(-qc/d;q)_\beta(q^{\alpha+\beta+1};q)_n}$$
$$\times\{(1-q^{\alpha+\beta+2n+1})(q^{\alpha+1};q)_n(-q^{\alpha+1}d/c;q)_n\}^{-1}.$$

Thus, the big q-Jacobi polynomials are also orthogonal on a discrete infinite set of points.

2.4 Notes

The expressions $(a;q)_n$ for $q \in \mathbb{N}$ have been known since the time of L. Euler. They have been used in combinatorics, Galois theory and finite geometry. The hypergeometric q-series ${}_2\varphi_1$ was invented in the nineteenth century by E. Heine. The q-integral was introduced and studied by F. H. Jackson [Jk] at the beginning of the twentieth century. A good exposition of the theory of basic hypergeometric functions and q-orthogonal polynomials can be found in the book [GR]. A nice survey on q-special functions is given in [Ko5].

3. The Quantum Algebra $U_q(\mathrm{sl}_2)$ and Its Representations

This chapter is devoted to a detailed study of the Hopf algebra $U_q(\mathrm{sl}_2)$. It can be considered as a one-parameter deformation of the universal enveloping algebra $U(\mathrm{sl}_2)$ (see Example 1.6) and is the simplest example of the quantized universal enveloping algebras $U_q(\mathfrak{g})$ which will be investigated in Chap. 6. Following common terminology in physics, we call $U_q(\mathrm{sl}_2)$ a *quantum algebra.*

3.1 The Quantum Algebras $U_q(\mathrm{sl}_2)$ and $U_h(\mathrm{sl}_2)$

3.1.1 The Algebra $U_q(\mathrm{sl}_2)$

Let q be a fixed complex number such that $q \neq 0$ and $q^2 \neq 1$. We denote by $U_q(\mathrm{sl}_2)$ the algebra (that is, the associative algebra with unit) over $\mathbb{C}$ with four generators E, F, K, K^{-1} satisfying the defining relations

$$KK^{-1} = K^{-1}K = 1, \quad KEK^{-1} = q^2E, \quad KFK^{-1} = q^{-2}F, \tag{1}$$

$$[E, F] = \frac{K - K^{-1}}{q - q^{-1}}. \tag{2}$$

It can be shown by induction that the relations (1) and (2) imply the formulas

$$[E, F^m] \equiv EF^m - F^mE = [m]F^{m-1}[K; 1-m], \tag{3}$$

$$[E^n, F] \equiv E^nF - FE^n = [n]E^{n-1}[K; n-1], \tag{4}$$

$$E^nF^m = \sum_{r=0}^{n} \frac{[n]![m]!}{[r]![n-r]![m-r]!} F^{m-r}E^{n-r} \times [K; n-m][K; n-m-1]\cdots[K; n-m-r+1], \quad n \le m, \tag{5}$$

where we set $[a] := [a]_q$ and $[K; n] := (Kq^n - K^{-1}q^{-n})/(q - q^{-1})$ for $n \in \mathbb{Z}$.

Proposition 1. *The set* $\{F^lK^mE^n \mid m \in \mathbb{Z},\ l, n \in \mathbb{N}_0\}$, *as well as the set* $\{E^nK^mF^l \mid m \in \mathbb{Z},\ l, n \in \mathbb{N}_0\}$, *is a vector space basis of* $U_q(\mathrm{sl}_2)$.

Proof. It follows from (1) and (2) that $U_q(\mathrm{sl}_2)$ is spanned by the monomials $F^lK^mE^n$, $m \in \mathbb{Z}$, $l, n \in \mathbb{N}_0$. To prove that they constitute a basis of $U_q(\mathrm{sl}_2)$ we consider the polynomial ring $\mathbb{C}[x, y, z, z^{-1}]$ in commuting indeterminates

x, y, z, z^{-1} satisfying $zz^{-1} = z^{-1}z = 1$. Define linear transformations $\tilde{E}$, $\tilde{F}$, $\tilde{K}$ and $\tilde{K}^{-1}$ of $\mathbb{C}[x, y, z, z^{-1}]$ by

$$\tilde{F}(x^r z^s y^t) = x^{r+1} z^s y^t, \quad \tilde{K}(x^r z^s y^t) = q^{-2r} x^r z^{s+1} y^t,$$

$$\tilde{E}(x^r z^s y^t) = q^{-2s} x^r z^s y^{t+1} + [r] x^{r-1} \frac{zq^{1-r} - z^{-1}q^{r-1}}{q - q^{-1}} z^s y^t.$$

Direct computations show that these transformations satisfy the relations (1) and (2). Therefore, the assignment $X \to \tilde{X}$, $X = E, F, K, K^{-1}$, determines a representation of $U_q(\mathrm{sl}_2)$ on $\mathbb{C}[x, y, z, z^{-1}]$. Since the monomials $x^r z^s y^t$, $s \in \mathbb{Z}$, $r, t \in \mathbb{N}_0$, form a basis of $\mathbb{C}[x, y, z, z^{-1}]$ and $\tilde{F}^l \tilde{K}^m \tilde{E}^n(1) = x^l z^m y^n$, the operators $\tilde{F}^l \tilde{K}^m \tilde{E}^n$, $m \in \mathbb{Z}$, $l, n \in \mathbb{N}_0$, are linearly independent. Thus, the set $\{F^l K^m E^n \mid m \in \mathbb{Z},\ l, n \in \mathbb{N}_0\}$ is a basis of $U_q(\mathrm{sl}_2)$. The second assertion is proved similarly. □

Using Proposition 1 one can show that the algebra $U_q(\mathrm{sl}_2)$ has no zero divisors, that is, we have $ab \neq 0$ in $U_q(\mathrm{sl}_2)$ whenever $a \neq 0$ and $b \neq 0$.

Proposition 2. *The quantum Casimir element*

$$C_q := EF + \frac{Kq^{-1} + K^{-1}q}{(q - q^{-1})^2} = FE + \frac{Kq + K^{-1}q^{-1}}{(q - q^{-1})^2} \tag{6}$$

lies in the center of the algebra $U_q(\mathrm{sl}_2)$. If q is not a root of unity, then the center of $U_q(\mathrm{sl}_2)$ is generated by C_q.

Proof. The second assertion is proved in Sect. 6.3. For the first assertion it is sufficient to verify that C_q commutes with E, F, K and K^{-1}. Using the relations (1) and (2) this is easily done. □

Proposition 3. *There exist automorphisms θ and $\vartheta_{\alpha,n,\nu}$, $\alpha \in \mathbb{C}\backslash\{0\}$, $n \in \mathbb{Z}$, $\nu = \pm 1$, of the algebra $U_q(\mathrm{sl}_2)$ such that*

$$\theta(E) = F, \quad \theta(F) = E, \quad \theta(K) = K^{-1},$$

$$\vartheta_{\alpha,n,\nu}(E) = \alpha K^n E, \quad \vartheta_{\alpha,n,\nu}(F) = \nu\alpha^{-1} q^{-2n} K^{-n} F, \quad \vartheta_{\alpha,n,\nu}(K) = \nu K.$$

Proof. First one shows by direct computation that the images of the generators under θ and $\vartheta_{\alpha,n,\nu}$ again satisfy the relations (1) and (2). As a sample, we check that $\vartheta \equiv \vartheta_{\alpha,n,\nu}$ preserves the relation (2) by computing

$$\vartheta(E)\vartheta(F) - \vartheta(F)\vartheta(E) = \nu(K^n E K^{-n} q^{-2n})F - \nu(K^{-n} F K^n q^{-2n})E$$
$$= \nu(EF - FE) = (q - q^{-1})^{-1}(\vartheta(K) - \vartheta(K^{-1})).$$

Hence there exist algebra homomorphisms θ and $\vartheta_{\alpha,n,\nu}$ given by the above formulas on the generators. Since $\theta \circ \theta = \mathrm{id}$ and $\vartheta_{\alpha,n,\nu} \circ \vartheta_{\nu^n\alpha^{-1},-n,\nu} = \mathrm{id}$, these homomorphisms are actually automorphisms. □

If q is not a root of unity, it can be shown (using the second assertion of Proposition 2) that the group of automorphisms of the algebra $U_q(\mathrm{sl}_2)$ is generated by the automorphisms θ and $\vartheta_{\alpha,n,\nu}$ (see [AlC]).

3.1.2 The Hopf Algebra $U_q(\mathrm{sl}_2)$

A key property of the algebra $U_q(\mathrm{sl}_2)$ is that it carries a Hopf algebra structure.

Proposition 4. *There exists a unique Hopf algebra structure on $U_q(\mathrm{sl}_2)$ with comultiplication Δ, counit ε and antipode S such that*

$$\Delta(E) = E \otimes K + 1 \otimes E, \quad \Delta(F) = F \otimes 1 + K^{-1} \otimes F, \quad \Delta(K) = K \otimes K,$$

$$S(K) = K^{-1},\ S(E) = -EK^{-1},\ S(F) = -KF,\ \varepsilon(K) = 1,\ \varepsilon(E) = \varepsilon(F) = 0.$$

Proof. First we extend Δ to an algebra homomorphism of the free algebra $\mathbb{C}\langle E, F, K, K^{-1}\rangle$ to $U_q(\mathrm{sl}_2) \otimes U_q(\mathrm{sl}_2)$. In order to prove that Δ passes to the quotient algebra $U_q(\mathrm{sl}_2)$ of $\mathbb{C}\langle E, F, K, K^{-1}\rangle$ it suffices to show that Δ preserves the relations (1) and (2). We shall do this in the case of (2). For the relations (1) the proof is similar and simpler. We have

$$\begin{aligned}[\Delta(E), \Delta(F)] \quad &= \quad K^{-1} \otimes EF + F \otimes E + EK^{-1} \otimes KF + EF \otimes K \\ &\quad\ -K^{-1} \otimes FE - K^{-1}E \otimes FK - F \otimes E - FE \otimes K\end{aligned}$$

$$= \frac{K^{-1} \otimes (K - K^{-1}) + (K - K^{-1}) \otimes K}{q - q^{-1}} = \{\Delta(K) - \Delta(K^{-1})\}/(q - q^{-1}).$$

It is proved similarly that S extends to an anti-homomorphism of $U_q(\mathrm{sl}_2)$ and ε to an algebra homomorphism of $U_q(\mathrm{sl}_2)$ to $\mathbb{C}$. By Proposition 1.8, it suffices to verify the Hopf algebra axioms on the generators. For instance, we have

$$(\Delta \otimes \mathrm{id}) \circ \Delta(E) = 1 \otimes 1 \otimes E + 1 \otimes E \otimes K + E \otimes K \otimes K = (\mathrm{id} \otimes \Delta) \circ \Delta(E).$$

The other conditions are checked in a similar manner. □

Definition 1. *The algebra $U_q(\mathrm{sl}_2)$ endowed with the Hopf algebra structure from Proposition 4 is called the* quantum algebra $U_q(\mathrm{sl}_2)$.

Some useful formulas about the Hopf algebra $U_q(\mathrm{sl}_2)$ are collected in

Proposition 5. *In the Hopf algebra $U_q(\mathrm{sl}_2)$ we have*

$$S^2(a) = KaK^{-1} \quad \text{for all} \quad a \in U_q(\mathrm{sl}_2), \tag{7}$$

$$\Delta(E^l K^m F^n) =$$

$$= \sum_{r=0}^{l} \sum_{s=0}^{n} \frac{[l]![n]!q^{r(l-r)-s(n-s)}}{[r]![l-r]![s]![n-s]!} E^{l-r} K^{m-n+s} F^s \otimes E^r K^{l-r+m} F^{n-s}, \tag{8}$$

$$S(F^n K^m E^l) = (-1)^{l+n} q^{n(n-1)-l(l-1)} E^l K^{n-l-m} F^n. \tag{9}$$

Proof. Since S^2 is an algebra homomorphism of $U_q(\mathrm{sl}_2)$ by Proposition 1.5, it suffices to check the equality (7) on the generators. We have $S^2(E) = -S(EK^{-1}) = -S(K^{-1})S(E) = KEK^{-1}$ and similarly $S^2(F) = KFK^{-1}$.

By induction one proves the formulas for $\Delta(E^l) = \Delta(E)^l$ and $\Delta(F^n) = \Delta(F)^n$ using that Δ is an algebra homomorphism. The formula (8) is obtained by multiplying the expressions of $\Delta(E^l)$, $\Delta(K^m) = K^m \otimes K^m$ and $\Delta(F^n)$. The relation (9) follows from the expressions for $S(F)$, $S(K)$ and $S(E)$ and the fact that S is an algebra anti-homomorphism. □

The next proposition deals with Hopf algebra isomorphisms of $U_q(\mathrm{sl}_2)$.

Proposition 6. (i) *For any $\alpha \in \mathbb{C}\backslash\{0\}$, there is a Hopf algebra automorphism ϑ_α of $U_q(\mathrm{sl}_2)$ such that*

$$\vartheta_\alpha(E) = \alpha E, \quad \vartheta_\alpha(F) = \alpha^{-1}F, \quad \vartheta_\alpha(K) = K.$$

Any Hopf algebra automorphism of $U_q(\mathrm{sl}_2)$ is of this form.

(ii) *Let α be a nonzero complex number. There exist Hopf algebra isomorphisms $\varphi_\alpha : U_q(\mathrm{sl}_2) \to U_{q^{-1}}(\mathrm{sl}_2)$ and $\psi_\alpha : U_q(\mathrm{sl}_2) \to U_{-q}(\mathrm{sl}_2)$ such that*

$$\varphi_\alpha(E) = \alpha FK, \quad \varphi_\alpha(F) = -\alpha^{-1}q^2EK^{-1}, \quad \varphi_\alpha(K) = K,$$

$$\psi_\alpha(E) = \alpha E, \quad \psi_\alpha(F) = -\alpha^{-1}F, \quad \psi_\alpha(K) = K.$$

Any Hopf algebra isomorphism $\varphi : U_q(\mathrm{sl}_2) \to U_{q^{-1}}(\mathrm{sl}_2)$, resp. $\psi : U_q(\mathrm{sl}_2) \to U_{-q}(\mathrm{sl}_2)$, is of the form $\varphi = \varphi_\alpha$, resp. $\psi = \psi_\alpha$.

(iii) *Two Hopf algebras $U_q(\mathrm{sl}_2)$ and $U_p(\mathrm{sl}_2)$ are isomorphic if and only if q is equal to $p, -p, p^{-1}$ or $-p^{-1}$.*

Proof. That the mappings defined above are indeed Hopf algebra isomorphisms is easily verified on the generators. We now prove the converse assertions and begin with two preliminary results. The first one is that the set of group-like elements of $U_q(\mathrm{sl}_2)$ is just the set of powers $K^n, n \in \mathbb{Z}$. Put $\mathcal{F}_q^\pm := \{x \in U_q(\mathrm{sl}_2) \,|\, \Delta(x) = 1 \otimes x + x \otimes K^{\pm 1}\}$. The second one is that $\mathcal{F}_q^+ = \mathrm{Lin}\{E, FK, 1-K\}$ and $\mathcal{F}_q^- = \mathrm{Lin}\{1-K^{-1}\}$. Both results are derived from formula (8) by writing an arbitrary element $x \in U_q(\mathrm{sl}_2)$ as a linear combination of basis elements $E^lK^mF^n$ and comparing coefficients in the expression of $\Delta(x)$. It is clear that the set of group-like elements and the sets $\mathcal{F}_q^\pm$ are preserved under Hopf algebra isomorphisms.

Now suppose that $\psi : U_q(\mathrm{sl}_2) \to U_p(\mathrm{sl}_2)$ is a Hopf algebra isomorphism. Since the image $\psi(K)$ of the group-like element K is again group-like, the first result above yields that $\psi(K) = K^n$ for some $n \in \mathbb{Z}$. Similarly, $\psi^{-1}(K) = K^m, m \in \mathbb{Z}$. Hence we have $\psi(K) = K$ or $\psi(K) = K^{-1}$. If $\psi(K) = K^{-1}$, then we have $\psi(\mathcal{F}_q^+) = \mathcal{F}_p^-$. But the latter is impossible, since dim $\mathcal{F}_q^+ = 3$ and dim $\mathcal{F}_p^- = 1$ by the second result. Thus, $\psi(K) = K$.

Next we note that the map $x \to KxK^{-1}$ leaves the space $\mathcal{F}_q^+$ invariant and has eigenvalues $1, q^2, q^{-2}$ with eigenvectors $1 - K, E, FK$, respectively. Therefore, we have either $q^2 = p^2$ and $\psi(E) = \alpha E, \psi(FK) = \beta FK$ or $q^2 = p^{-2}$ and $\psi(E) = \alpha FK, \psi(FK) = \beta E$ for some $\alpha, \beta \in \mathbb{C}$. Hence q is equal to $p, -p, p^{-1}$ or $-p^{-1}$. This completes the proof of (iii). Now suppose that $q = p$. Then, as shown by the preceding, $\psi(K) = K$, $\psi(E) = \alpha E$ and

$\psi(FK) = \beta FK$ and so $\psi(F) = \beta F$. By (2), $\beta = \alpha^{-1}$ and hence $\psi = \vartheta_\alpha$, which proves (i). The last assertion of (ii) follows similarly. $\square$

There is another important variant of the quantum algebra $U_q(\mathrm{sl}_2)$ which often occurs in the literature. It is defined as follows. Suppose that $q^4 \neq 0, 1$ and let $\breve{U}_q(\mathrm{sl}_2)$ denote the algebra over $\mathbb{C}$ with generators E, F, $K\,K^{-1}$ subject to the relations

$$KK^{-1} = K^{-1}K = 1, \quad KEK^{-1} = qE, \quad KFK^{-1} = q^{-1}F, \tag{10}$$

$$[E, F] = \frac{K^2 - K^{-2}}{q - q^{-1}}. \tag{11}$$

Arguing similarly as in the proof of Proposition 4, it follows that this new algebra $\breve{U}_q(\mathrm{sl}_2)$ becomes a Hopf algebra with structure maps given by

$$\Delta(K) = K \otimes K, \ \Delta(E) = E \otimes K + K^{-1} \otimes E, \ \Delta(F) = F \otimes K + K^{-1} \otimes F, \tag{12}$$

$$S(K) = K^{-1}, \ \ S(E) = -qE, \ \ S(F) = -q^{-1}F, \ \ \varepsilon(K) = 1, \ \ \varepsilon(E) = \varepsilon(F) = 0. \tag{13}$$

An advantage of the Hopf algebra $\breve{U}_q(\mathrm{sl}_2)$ over $U_q(\mathrm{sl}_2)$ is that the comultiplications of the generators E and F are given by the same formulas.

Using the above formulas and Proposition 1, one can show that there is an injective Hopf algebra homomorphism $\varphi : U_q(\mathrm{sl}_2) \to \breve{U}_q(\mathrm{sl}_2)$ determined by $\varphi(E) = EK$, $\varphi(F) = K^{-1}F$ and $\varphi(K) = K^2$. If we identify $U_q(\mathrm{sl}_2)$ with its image under φ, then $U_q(\mathrm{sl}_2)$ becomes a Hopf subalgebra of $\breve{U}_q(\mathrm{sl}_2)$.

Note that the algebra $\breve{U}_q(\mathrm{sl}_2)$ has four one-dimensional representations (sending K to 1, –1, i, –i and E, F to 0), while $U_q(\mathrm{sl}_2)$ has only two such representations. Hence the algebras $U_q(\mathrm{sl}_2)$ and $\breve{U}_q(\mathrm{sl}_2)$ are not isomorphic.

3.1.3 The Classical Limit of the Hopf Algebra $U_q(\mathrm{sl}_2)$

In this subsection we shall discuss how the enveloping algebra $U(\mathrm{sl}_2)$ is related to the classical limit $q \to 1$ of the Hopf algebra $U_q(\mathrm{sl}_2)$. Recall that $U(\mathrm{sl}_2)$ is the algebra with generators x_+, x_-, y and defining relations $[y, x_+] = 2x_+$, $[y, x_-] = -2x_-$, $[x_+, x_-] = y$ and that the Hopf algebra structure of $U(\mathrm{sl}_2)$ has been described by Example 1.6.

Let us formally write $q = e^h$, $K = e^{hH}$ and consider the limit $h \to 0$. Then the relations (1) and (2) imply (by differentiation at $h = 0$ resp. by taking the limit $h \to 0$) that $[H, E] = 2E$, $[H, F] = -2F$ and $[E, F] = H$. That is, we obtain the relations of $U(\mathrm{sl}_2)$. Similarly, the Hopf algebra structure of $U(\mathrm{sl}_2)$ is recovered from that of $U_q(\mathrm{sl}_2)$ as $h \to 0$. However, this viewpoint is not quite adequate if we want to compare the theory of $U_q(\mathrm{sl}_2)$ (at least for q not a root of unity) with the classical situation. In order to develop the proper setting we first give a slight reformulation of the definition of $U_q(\mathrm{sl}_2)$ that applies also in the case $q^2 = 1$.

Clearly, the elements E, F, K, K^{-1} and $G := (q-q^{-1})^{-1}(K-K^{-1})$ of the algebra $U_q(\mathrm{sl}_2)$ satisfy the relations

$$[G,E]=E(qK+q^{-1}K^{-1}),\quad [G,F]=-(qK+q^{-1}K^{-1})F, \tag{14}$$

$$[E,F]=G,\quad (q-q^{-1})G=K-K^{-1}. \tag{15}$$

Let $\tilde{U}_q(\mathrm{sl}_2)$ denote the algebra with generators E, F, K, K^{-1}, G and relations (1), (14) and (15). The algebras $\tilde{U}_q(\mathrm{sl}_2)$ and $U_q(\mathrm{sl}_2)$, $q^2 \neq 1$, are isomorphic (by mapping $E \to E$, $F \to F$, $K \to K$ and $G \to (q-q^{-1})(K-K^{-1})$). The Hopf algebra structure of $U_q(\mathrm{sl}_2)$ induces a Hopf algebra structure on $\tilde{U}_q(\mathrm{sl}_2)$ which is given by the formulas stated in Proposition 4 and the relations $\Delta(G) = G\otimes K + K^{-1}\otimes G$, $\varepsilon(G) = 0$ and $S(G) = -G$. It is easy to verify that these formulas endow the algebra $\tilde{U}_q(\mathrm{sl}_2)$ with a Hopf algebra structure also in the cases $q = \pm 1$ which were excluded for $U_q(\mathrm{sl}_2)$. We consider the Hopf algebra $\tilde{U}_1(\mathrm{sl}_2)$ as the *classical limit of the Hopf algebra* $\tilde{U}_q(\mathrm{sl}_2)$.

The two algebras $\tilde{U}_1(\mathrm{sl}_2)$ and $U(\mathrm{sl}_2)$ are closely related to each other. That is, there are canonical algebra isomorphisms

$$\tilde{U}_1(\mathrm{sl}_2) \simeq U(\mathrm{sl}_2)\otimes \mathbb{C}\mathbb{Z}_2 \quad\text{and}\quad U(\mathrm{sl}_2) \simeq \tilde{U}_1(\mathrm{sl}_2)/\langle K-1\rangle.$$

Indeed, putting $q = 1$ into (1), (14) and (15) we see that K belongs to the center of $\tilde{U}_1(\mathrm{sl}_2)$ and that $K^2 = 1$. Thus the defining relations of the algebra $\tilde{U}_1(\mathrm{sl}_2)$ read as $[E,F] = G$, $[G,E] = 2EK$ and $[G,F] = -2FK$. Hence there exists an algebra isomorphism of $\tilde{U}_1(\mathrm{sl}_2)$ and $U(\mathrm{sl}_2)\otimes\mathbb{C}\mathbb{Z}_2$ which maps $E \to x_+\chi$, $F \to x_-$ and $G \to y\chi$. Here χ is the generator of the two-dimensional algebra $\mathbb{C}\mathbb{Z}_2$ such that $\chi^2 = 1$ and we have simply written $ab := a\otimes b$ for $a \in U(\mathrm{sl}_2)$ and $b \in \mathbb{C}\mathbb{Z}_2$. That is, the algebra $\tilde{U}_1(\mathrm{sl}_2)$ is an extension of the algebra $U(\mathrm{sl}_2)$ by a single central element χ satisfying $\chi^2 = 1$. The appearance of this additional generator χ explains why we later have *two* irreducible representations in any dimension when q is not a root of unity.

Finally, we describe the Hopf algebra structure on $U(\mathrm{sl}_2)\otimes\mathbb{C}\mathbb{Z}_2$ that is induced from $\tilde{U}_1(\mathrm{sl}_2)$ by the above isomorphism. Tracing back the corresponding formulas, the structure maps on $U(\mathrm{sl}_2)\otimes\mathbb{C}\mathbb{Z}_2$ are given by

$$\Delta(x_+) = \chi x_+\otimes 1 + 1\otimes x_+,\quad \Delta(x_-) = \chi x_-\otimes 1 + 1\otimes x_-,\quad \Delta(y) = y\otimes 1 + 1\otimes y,$$

$$\varepsilon(x_+) = \varepsilon(x_-) = \varepsilon(y) = 0,\quad S(x_+) = -\chi x_+,\quad S(x_-) = -\chi x_-,\quad S(y) = -y.$$

3.1.4 Real Forms of the Quantum Algebra $U_q(\mathrm{sl}_2)$

Recall from Subsect. 1.2.7 that by a *real form* of $U_q(\mathrm{sl}_2)$ we mean this Hopf algebra together with an involution $* : U_q(\mathrm{sl}_2) \to U_q(\mathrm{sl}_2)$ which turns $U_q(\mathrm{sl}_2)$ into a Hopf $*$-algebra.

We do not give proofs in the subsequent discussion. The reader is invited to carry out one or two of them using the following general pattern. One first verifies that the corresponding formulas define an involution of the free algebra $\mathbb{C}\langle E, F, K, K^{-1}\rangle$. Next one shows that the defining relations (1) and (2) are invariant under this involution, hence it passes to the quotient algebra $U_q(\mathrm{sl}_2)$ to define an involution there. Finally, one checks on the generators that Δ and ε are $*$-preserving.

Case 1: $q \in \mathbb{R}$. For any $\gamma \in \mathbb{R}$, $\gamma \neq 0$, there exists a real form of $U_q(\mathrm{sl}_2)$ with involution determined by the equations

$$E^* = \gamma F K, \quad F^* = \gamma^{-1} K^{-1} E, \quad K^* = K.$$

Using the description of all Hopf algebra automorphisms of $U_q(\mathrm{sl}_2)$ from Proposition 6(i) we see that two such real forms with parameters γ and γ' are equivalent if and only if $\gamma\gamma' > 0$. Thus, there are precisely two inequivalent real forms of $U_q(\mathrm{sl}_2)$ in this case. The first one is given by

$$E^* = FK, \quad F^* = K^{-1}E, \quad K^* = K.$$

The corresponding Hopf $*$-algebra is denoted by $U_q(\mathrm{su}_2)$ and called the *compact real form* of $U_q(\mathrm{sl}_2)$. The second one is denoted $U_q(\mathrm{su}_{1,1})$ and defined by

$$E^* = -FK, \quad F^* = -K^{-1}E, \quad K^* = K.$$

Case 2: $|q| = 1$. For any $\gamma \in \mathbb{C}$, $|\gamma| = 1$, we then have a real form of $U_q(\mathrm{sl}_2)$ given by

$$E^* = -\gamma E, \quad F^* = -\gamma^{-1} F, \quad K^* = K.$$

But all such real forms are equivalent, so we can assume that $\gamma = 1$. This real form is denoted by $U_q(\mathrm{sl}_2(\mathbb{R}))$.

It is clear that the three real forms of $U_q(\mathrm{sl}_2)$ defined above correspond to the three real Lie algebras su_2, $\mathrm{su}_{1,1}$ and $\mathrm{sl}_2(\mathbb{R})$ of the complex Lie algebra $\mathrm{sl}(2, \mathbb{C})$. Note that the classical Lie algebras $\mathrm{su}_{1,1}$ and $\mathrm{sl}_2(\mathbb{R})$ are isomorphic. But their quantum analogs $U_q(\mathrm{su}_{1,1})$ and $U_q(\mathrm{sl}_2(\mathbb{R}))$ are not equivalent, simply because they are defined for different parameter domains.

Case 3: $q \in \sqrt{-1}\mathbb{R}$. The following sets of formulas define two inequivalent real forms of $U_q(\mathrm{sl}_2)$ which have no classical counterparts:

$$E^* = \sqrt{-1} F K, \quad F^* = \sqrt{-1} K^{-1} E, \quad K^* = K,$$
$$E^* = -\sqrt{-1} F K, \quad F^* = -\sqrt{-1} K^{-1} E, \quad K^* = K.$$

It can be shown that only for the above three parameter domains of q real forms of $U_q(\mathrm{sl}_2)$ exist and that the previous list is exhausting.

The Hopf algebra $\breve{U}_q(\mathrm{sl}_2)$ has, up to equivalence, precisely three real forms. They are denoted by $\breve{U}_q(\mathrm{su}_2)$, $\breve{U}_q(\mathrm{su}_{1,1})$, $\breve{U}_q(\mathrm{sl}_2(\mathbb{R}))$ and are determined respectively by the formulas

$$E^* = F, \quad F^* = E, \quad K^* = K \quad \text{for} \quad q \in \mathbb{R},$$
$$E^* = -F, \quad F^* = -E, \quad K^* = K \quad \text{for} \quad q \in \mathbb{R},$$
$$E^* = -E, \quad F^* = -F, \quad K^* = K \quad \text{for} \quad |q| = 1.$$

3.1.5 The h-Adic Hopf Algebra $U_h(\mathrm{sl}_2)$

Let h be an indeterminate. We denote by $U_h(\mathrm{sl}_2)$ the (complete) h-adic algebra (see Subsect. 1.2.10) with three generators E, F and H obeying the defining relations

$$[H,E]=2E, \quad [H,F]=-2F, \quad [E,F]=\frac{e^{hH}-e^{-hH}}{e^h-e^{-h}} \equiv \frac{\sinh hH}{\sinh h}, \tag{16}$$

where e^{hH} is the series $1+hH+(hH)^2/2!+\cdots$ belonging to $U_h(\mathrm{sl}_2)$.

The elements $E^lH^mF^n$, $l,m,n=0,1,2,\cdots$, constitute an h-adic vector space basis of $U_h(\mathrm{sl}_2)$, that is, these elements are linearly independent and the h-adic completion of their linear span coincides with $U_h(\mathrm{sl}_2)$. The element

$$C_h := EF+\frac{e^{h(H-1)}+e^{-h(H-1)}}{(e^h-e^{-h})^2} = FE+\frac{e^{h(H+1)}+e^{-h(H+1)}}{(e^h-e^{-h})^2} \tag{17}$$

commutes with all elements of $U_h(\mathrm{sl}_2)$ and generates the center of $U_h(\mathrm{sl}_2)$.

There exists a unique algebra automorphism θ of the algebra $U_h(\mathrm{sl}_2)$ such that

$$\theta(E)=F, \quad \theta(F)=E, \quad \theta(H)=-H, \quad \theta(h)=-h. \tag{18}$$

One can easily prove that there exists a unique h-adic Hopf algebra structure on $U_h(\mathrm{sl}_2)$ with comultiplication Δ, counit ε and antipode S defined on the generating elements by

$$\Delta(E)=E\otimes e^{hH}+1\otimes E, \qquad \Delta(F)=F\otimes 1+e^{-hH}\otimes F,$$

$$\Delta(H)=H\otimes 1+1\otimes H, \qquad \varepsilon(H)=\varepsilon(E)=\varepsilon(F)=0,$$

$$S(H)=-H, \quad S(E)=-Ee^{-hH}, \quad S(F)=-e^{hH}F.$$

The h-adic Hopf algebra obtained in this way is called the *h-adic quantum algebra* $U_h(\mathrm{sl}_2)$.

From the preceding formulas we see that the algebra $U_h(\mathrm{sl}_2)$ does not change and that the comultiplication Δ of $U_h(\mathrm{sl}_2)$ goes into the opposite comultiplication $\Delta^{\mathrm{cop}}=\tau\circ\Delta$ when h is replaced by $-h$.

We can also define the h-adic analog $\breve{U}_h(\mathrm{sl}_2)$ of the quantum algebra $\breve{U}_q(\mathrm{sl}_2)$ considered above. It is generated by elements E, F and H satisfying the defining relations

$$[H,E]=E, \quad [H,F]=F, \quad [E,F]=\frac{e^{2hH}-e^{-2hH}}{e^h-e^{-h}}$$

with the h-adic Hopf algebra structure

$$\Delta(E)=E\otimes e^{hH}+e^{-hH}\otimes E, \quad \Delta(F)=F\otimes e^{hH}+e^{-hH}\otimes F,$$

$$\Delta(H)=H\otimes 1+1\otimes H, \quad \varepsilon(H)=\varepsilon(E)=\varepsilon(F)=0,$$

$$S(H)=-H, \quad S(F)=-e^hE, \quad S(F)=-e^{-h}F.$$

An h-adic Hopf algebra $\mathcal{A}$ is called *quasitriangular* if there exists an invertible element $\mathcal{R}$ in the completion $\mathcal{A}\hat{\otimes}\mathcal{A}$ (in the meaning of Subsect. 1.2.10) such that $\Delta^{\mathrm{cop}}(a) = \mathcal{R}\Delta(a)\mathcal{R}^{-1}$, $a \in \mathcal{A}$, and

$$(\Delta \otimes \mathrm{id})\mathcal{R} = \mathcal{R}_{13}\mathcal{R}_{23} \quad \text{and} \quad (\mathrm{id} \otimes \Delta)\mathcal{R} = \mathcal{R}_{13}\mathcal{R}_{12} \quad \text{on} \quad \mathcal{A}\hat{\otimes}\mathcal{A}\hat{\otimes}\mathcal{A},$$

where $\mathcal{R}_{12} = \sum_i x_i \otimes y_i \otimes 1$, $\mathcal{R}_{13} = \sum_i x_i \otimes 1 \otimes y_i$, $\mathcal{R}_{23} = \sum_i 1 \otimes x_i \otimes y_i$ for $\mathcal{R} = \sum_i x_i \otimes y_i$. The element $\mathcal{R}$ is called a *universal R-matrix* of $\mathcal{A}$. It is useful for applications in physics, because it yields a solution of the *quantum Yang–Baxter equation* $R_{12}R_{13}R_{23} = R_{23}R_{13}R_{12}$ for any representation of the algebra $\mathcal{A}$. A detailed exposition of this theory will be given in Chap. 8. Here we state only the following proposition which is proved in Subsect. 8.3.1.

Proposition 7. *The h-adic quantum algebra $U_h(\mathrm{sl}_2)$ is quasitriangular and admits the universal R-matrix*

$$\mathcal{R} = e^{h(H \otimes H)/2} \sum_{n=0}^{\infty} \frac{q^{n(n+1)/2}(1-q^{-2})^n}{[n]_q!} E^n \otimes F^n, \quad q = e^h. \tag{19}$$

3.2 Finite-Dimensional Representations of $U_q(\mathrm{sl}_2)$ for q not a Root of Unity

Throughout this section we assume that q is not a root of unity.

3.2.1 The Representations $T_{\omega l}$

If T is a representation of the algebra $U_q(\mathrm{sl}_2)$, then the operator $T(K)$ is invertible and we have

$$T(K)T(E)T(K)^{-1} = q^2 T(E), \quad T(K)T(F)T(K)^{-1} = q^{-2}T(F), \tag{20}$$

$$T(E)T(F) - T(F)T(E) = \frac{T(K) - T(K)^{-1}}{q - q^{-1}}. \tag{21}$$

Conversely, since (1) and (2) are the defining relations of the algebra $U_q(\mathrm{sl}_2)$, any triple of operators $T(E)$, $T(F)$ and $T(K)$ on a vector space V such that $T(K)$ is invertible and (20) and (21) are fulfilled gives rise to a representation T of $U_q(\mathrm{sl}_2)$ on V.

Now let l be an integral or half-integral nonnegative number and let $\omega \in \{+1, -1\}$. Let V_l be a $(2l+1)$-dimensional complex vector space with basis $\mathbf{e}_m$, $m = -l, -l+1, \cdots, l$. For notational convenience, we set $\mathbf{e}_{l+1} = \mathbf{e}_{-l-1} = \mathbf{0}$. Define operators $T_{\omega l}(E)$, $T_{\omega l}(F)$, $T_{\omega l}(K)$ acting on V_l by

$$T_{\omega l}(K)\mathbf{e}_m = \omega q^{2m}\mathbf{e}_m, \quad T_{\omega l}(E)\mathbf{e}_m = ([l-m][l+m+1])^{1/2}\mathbf{e}_{m+1}, \tag{22}$$

$$T_{\omega l}(F)\mathbf{e}_m = \omega([l+m][l-m+1])^{1/2}\mathbf{e}_{m-1}, \tag{23}$$

where $[n] := [n]_q$ (see (2.1)). A simple computation shows that these operators satisfy the relations (20) and (21). Hence they define a representation $T_{\omega l}$ of the algebra $U_q(\mathrm{sl}_2)$ on V_l. Apart from the sign factor, the representation $T_{\omega l}$ is built in a similar manner as the $(2l+1)$-dimensional representation of the Lie algebra $\mathrm{sl}(2,\mathbb{C})$. The same method as in the classical case shows that $T_{\omega l}$ is irreducible. Therefore, the image $T_{\omega l}(C_q)$ of the Casimir element C_q (see Proposition 2) is a scalar multiple of the identity operator I on V_l. Since $T_{\omega l}(E)\mathbf{e}_l = 0$, we obtain from (6) that

$$T_{\omega l}(C_q) = \omega(q^{l+1} + q^{-l-1})(q - q^{-1})^{-2} I. \tag{24}$$

Let $T_{\omega,l}$ and $T_{\omega',l'}$ be two representations such that $(\omega, l) \neq (\omega', l')$. Since q is not a root of unity (!), the spectra of the operators $T_{\omega,l}(K)$ and $T_{\omega',l'}(K)$ do not coincide, so the representations $T_{\omega,l}$ and $T_{\omega',l'}$ are not equivalent. Moreover, by (24), the scalars of $T_{\omega,l}(C_q)$ and $T_{\omega',l'}(C_q)$ are different. Thus, we have proved

Proposition 8. *For any $l \in \frac{1}{2}\mathbb{N}_0$ and $\omega \in \{1, -1\}$, $T_{\omega l}$ is an irreducible representation of the algebra $U_q(\mathrm{sl}_2)$. If $(\omega, l) \neq (\omega', l')$, the representations $T_{\omega,l}$ and $T_{\omega',l'}$ are not equivalent and the values of the operators $T_{\omega,l}(C_q)$ and $T_{\omega',l'}(C_q)$ are different.*

There is a unique scalar product $\langle \cdot, \cdot \rangle$ on V_l such that $\langle \mathbf{e}_m, \mathbf{e}_n \rangle = \delta_{mn}$. If $q \in \mathbb{R}$, then the operators (22) and (23) satisfy the conditions $T_{\omega l}(E)^* = T_{\omega l}(F)$ and $T_{\omega l}(K)^* = T_{\omega l}(K)$. Therefore, by the definition of the involution of $U_q(\mathrm{su}_2)$, $T_{\omega l}$ is a $*$-representation of the $*$-algebra $U_q(\mathrm{su}_2)$ on V_l.

The family $T_{\omega l}$, $\omega \in \{+1, -1\}$, $l \in \frac{1}{2}\mathbb{N}_0$, exhausts all finite-dimensional irreducible representations of $U_q(\mathrm{sl}_2)$ as shown by

Proposition 9. *Any irreducible finite-dimensional representation T of $U_q(\mathrm{sl}_2)$ is equivalent to one of the representations $T_{\omega l}$, $\omega \in \{+1, -1\}$, $l = 0, \frac{1}{2}, 1, \cdots$.*

Proof. Because the underlying vector space V is finite-dimensional, the operator $T(K)$ has at least one eigenvector $\mathbf{e}$, say $T(K)\mathbf{e} = \mu\mathbf{e}$. Since $KE^r = q^{2r}E^rK$ by (1) and hence $T(K)T(E^r)\mathbf{e} = \mu q^{2r}T(E^r)\mathbf{e}$, $T(E^r)\mathbf{e}$ is either zero or an eigenvector of $T(K)$ with the eigenvalue μq^{2r}. Since these numbers are pairwise distinct (because q is not a root of unity) and V is finite-dimensional, there exists $r \in \mathbb{N}_0$ such that $\mathbf{e}_0 := T(E^r)\mathbf{e} \neq 0$ and $T(E)\mathbf{e}_0 = 0$. Let $T(K)\mathbf{e}_0 = \lambda\mathbf{e}_0$. Set $F^{(s)} := F^s/[s]!$ and $\mathbf{e}_s := T(F^{(s)})\mathbf{e}_0$, $s = 0, 1, 2, \cdots$. Since $KF = q^{-2}FK$, we have $T(K)\mathbf{e}_s = \lambda q^{-2s}\mathbf{e}_s$. There exists a smallest $p \in \mathbb{N}$ such that $T(F^{(p)})\mathbf{e}_0 \neq 0$ and $T(F^{(p+1)})\mathbf{e}_0 = 0$. By (3), for any $r \in \{0, 1, 2, \cdots\}$ we have

$$\begin{aligned} T(E)\mathbf{e}_r &= T(E)T(F^{(r)})\mathbf{e}_0 = T(EF^{(r)})\mathbf{e}_0 \\ &= \frac{q^{-(r-1)}\lambda - q^{r-1}\lambda^{-1}}{q - q^{-1}} T(F^{(r-1)})\mathbf{e}_0 = \frac{q^{-(r-1)}\lambda - q^{r-1}\lambda^{-1}}{q - q^{-1}} \mathbf{e}_{r-1}. \end{aligned}$$

Thus, the subspace V_1 spanned by the vectors $\mathbf{e}_0, \mathbf{e}_1, \cdots, \mathbf{e}_p$ is invariant under the operators $T(E)$, $T(F)$, $T(K)$. Since the representation T is irreducible, $V_1 = V$ and the vectors $\mathbf{e}_0, \mathbf{e}_1, \cdots, \mathbf{e}_p$ form a basis of V.

Let us find the possible eigenvalues λ. Since $T(F^{(p+1)})\mathbf{e}_0 = 0$, we have $\{T(E)T(F^{(p+1)}) - T(F^{(p+1)})T(E)\}\mathbf{e}_0 = 0$, so $(q^{-p}\lambda - q^p\lambda^{-1})/(q - q^{-1}) = 0$ by (3). Therefore, $\lambda = \pm q^p$, and the representation T is given by

$$T(E)\mathbf{e}_r = \omega[p - r + 1]\mathbf{e}_{r-1}, \tag{25}$$

$$T(F)\mathbf{e}_r = [r + 1]\mathbf{e}_{r+1}, \quad T(K)\mathbf{e}_r = \omega q^{p-2r}\mathbf{e}_r, \tag{26}$$

where $\omega \in \{\pm 1\}$. We set $l := \frac{p}{2}$ and $\mathbf{e}'_r := \omega^{-r}([l-r]!/[l+r+1]!)^{1/2}\mathbf{e}_{l-r}$ for $r = -l, -l+1, \cdots, l$. Then the formulas (25) and (26) turn into

$$T(E)\mathbf{e}'_s = ([l + s + 1][l - s])^{1/2}\mathbf{e}'_{s+1},$$

$$T(F)\mathbf{e}'_s = \omega([l - s + 1][l + s])^{1/2}\mathbf{e}'_{s-1}, \quad T(K)\mathbf{e}'_s = \omega q^s\mathbf{e}'_s.$$

This shows that the representations T and $T_{\omega l}$ are equivalent. □

The representations $T_{\omega l}$ with $\omega = 1$ are denoted by T_l. They are called *representations of type 1*. Obviously, the representation $T_{\omega l}$ is the tensor product of T_l and the one-dimensional representation $T_{\omega 0}$.

3.2.2 Weight Representations and Complete Reducibility

We begin with a preliminary result.

Lemma 10. *Let T be a finite-dimensional representation of $U_q(\mathrm{sl}_2)$. Then the operators $T(E)$ and $T(F)$ are nilpotent, that is, there is an $r \in \mathbb{N}_0$ such that $T(E)^r = T(F)^r = 0$.*

Proof. The finite-dimensional representation T takes in some appropriate basis the following matrix form:

$$\begin{pmatrix} T^{(1)} & * & * & \cdots & * \\ 0 & T^{(2)} & * & \cdots & * \\ 0 & 0 & T^{(3)} & \cdots & * \\ \cdots & \cdots & \cdots & \cdots & \cdots \\ 0 & 0 & 0 & \cdots & T^{(n)} \end{pmatrix}, \tag{27}$$

where $T^{(1)}, \cdots, T^{(n)}$ are irreducible representations of $U_q(\mathrm{sl}_2)$ and asterisks mean certain submatrices. By Proposition 9, each representation $T^{(k)}$ in (27) is equivalent to some $T_{\omega l}$. Therefore, since the operators $T_{\omega l}(E)$ and $T_{\omega l}(F)$ are nilpotent by (22) and (23), there exists $m \in \mathbb{N}_0$ such that $T^{(k)}(E)^m = T^{(k)}(F)^m = 0$ for $k = 1, 2, \cdots, n$. Hence, by (27), $T^{(k)}(E)^r = T^{(k)}(F)^r = 0$ for some r. □

Next we give the analogs for $U_q(\mathrm{sl}_2)$ of some notions from classical representation theory of Lie algebras. Let T be a representation of $U_q(\mathrm{sl}_2)$ on a finite-dimensional vector space V. For any complex number λ, we set $V_\lambda := \{\mathbf{x} \in V \mid T(K)\mathbf{x} = \lambda\mathbf{x}\}$. If $V_\lambda \neq \{0\}$, then V_λ is called a *weight space* and λ is a *weight* of the representation T. The nonzero vectors of V_λ are the *weight vectors.* A vector $\mathbf{x} \in V$ such that $T(E)\mathbf{x} = 0$ and $T(K)\mathbf{x} = \mu\mathbf{x}$, is called a *highest weight vector* and μ is called a *highest weight* of T.

If V is the linear span of weight spaces of T, then T is called a *weight representation.* Then V is the direct sum of weight spaces. In other words, T is a weight representation if and only if the operator $T(K)$ is diagonal with respect to some basis of V. Each representation $T_{\omega l}$ is a weight representation.

Proposition 11. *Every finite-dimensional representation of* $U_q(\mathrm{sl}_2)$ *is a weight representation.*

Proof. Let T be a representation of $U_q(\mathrm{sl}_2)$ on a finite-dimensional vector space V. It is known from linear algebra that the operator $T(K)$ is diagonalizable if there exist pairwise distinct numbers $a_1, a_2, \cdots, a_k$ such that

$$(T(K) - a_1)(T(K) - a_2)\cdots(T(K) - a_k)V = \{0\}.$$

By Lemma 10, there exists $m \in \mathbb{N}$ such that $T(F)^m V = \{0\}$. Recall that $[K; n] = (Kq^n - K^{-1}q^{-n})/(q - q^{-1})$. Put $K_0 := I$ and

$$K_r = \prod\nolimits_{j=-(r-1)}^{r-1} [K; r - m + j], \quad r = 1, 2, \cdots, m.$$

We prove by induction on r that

$$T(F^{m-r})T(K_r)V = \{0\}, \quad r = 0, 1, 2, \cdots, m. \tag{28}$$

For $r = 0$ this is true because $T(F)^m V = \{0\}$. Suppose that (28) holds for $i = 0, 1, 2, \cdots, n-1$. Setting $B = E^n F^m \prod_{j=1}^{n-1} [K; n - m + j]$, we have $T(B)V = \{0\}$ and

$$B = \sum_{i=0}^{n} b_i F^{m-i} \Big(\prod_{j=1}^{i} [K; i - n - m + j] \Big) \Big(\prod_{j=1}^{n-1} [K; 2i - n - m + j] \Big) E^{n-i}$$

by (5), where $b_i = [n]![m]!/[i]![n-i]![m-i]!$. Further, we get

$$\begin{aligned} B &= \sum\nolimits_{i=0}^{n} b_i F^{m-i} \Big(\prod\nolimits_{j=1}^{i+n-1} [K; i - n - m + j] \Big) E^{n-i} \\ &= \sum\nolimits_{i=0}^{n} b_i F^{m-i} K_i \Big(\prod\nolimits_{j=0}^{n-i-1} [K; -m - j] \Big) E^{n-i}. \end{aligned}$$

Since $T(B)V = \{0\}$ as noted above and $T(F^{m-r})T(K_r)V = \{0\}$ for $r = 0, 1, 2, \cdots, n-1$ by the induction hypothesis, the last equation implies the relation $T(F^{m-n})T(K_n)V = \{0\}$. This completes the proof of (28).

In the case $r = m$ formula (28) yields

$$T(K_m)V = \Big(\prod\nolimits_{j=-(m-1)}^{m-1} \frac{q^j T(K^{-1}) T(K^2 - q^{-2j})}{q - q^{-1}} \Big) V = \{0\}.$$

Applying some appropriate power of $T(K)$ we get

$$\Big(\prod_{j=-(m-1)}^{m-1}\{T(K)-q^{-j}\}\{T(K)+q^{-j}\}\Big)V=\{0\}.$$

The numbers in the set $q^{-j}, -q^{-j}$, $j\in\mathbb{Z}$, are pairwise distinct since q is not a root of unity. □

Proposition 12. *Any finite-dimensional representation T of $U_q(\mathrm{sl}_2)$ is completely reducible, that is, T is a direct sum of irreducible representations.*

Proof. Let T be a finite-dimensional representation of $U_q(\mathrm{sl}_2)$ on a space V. We first suppose that T reduces to the matrix form (27) with all $T^{(i)}$ equivalent to a fixed irreducible representation $T_{\omega l}$. Let $V_1=\{\mathbf{v}\in V \mid T(E)\mathbf{v}=0\}$. From the special form of T it is clear that all $\mathbf{v}\in V_1$ are highest weight vectors belonging to the same weight. In particular, V_1 is a linear subspace of V. Let $\mathbf{v}_0^1,\cdots,\mathbf{v}_0^k$ be a basis of V_1. For each $i=1,2,\cdots,k$, let V^i be the span of vectors $\mathbf{v}_s^i:=F^s\mathbf{v}_0^i$, $s=0,1,\cdots,2l$. Repeating the reasoning of the proof of Proposition 9, it follows that on each space V^i the irreducible representation $T_{\omega l}$ is realized. We show that the set of vectors $\mathbf{v}_s^i$ is linearly independent. Indeed, only vectors of the same weight may be linearly dependent. If $\sum_i\alpha_i\mathbf{v}_s^i=0$, then the action by the operator $T(E)^s$ yields $\sum_i\alpha_i\mathbf{v}_0^i=0$ and so $\alpha_i=0$ for all $i=1,2,\cdots,k$. Hence the vectors $\mathbf{v}_s^i$ are linearly independent and we have a direct sum $\tilde V:=\bigoplus_i V^i$. Since T is a weight representation by Proposition 11, we have $V=\tilde V$. Thus, T is completely reducible.

Now let T be an arbitrary representation of the form (27). On the block diagonals the Casimir operator $T(C_q)$ acts as $T^{(i)}(C_q)=c_iI$. The constants c_i are the generalized eigenvalues of the operator $T(C_q)$. It is known from linear algebra that the space V decomposes into the direct sum $V=\bigoplus_j V_j$ of generalized eigenspaces $V_j:=\{\mathbf{v}\in V\mid (T(C_q)-c_j)^r\mathbf{v}=0 \text{ for some } r\}$ of $T(C_q)$ corresponding to distinct generalized eigenvalues c_j. Since C_q belongs to the center of $U_q(\mathrm{sl}_2)$ we have

$$(T(C_q)-c_j)^rT(a)\mathbf{v}=T(a)(T(C_q-c_j))^r\mathbf{v}=0 \ \text{ for }\ a\in U_q(\mathrm{sl}_2),\quad \mathbf{v}\in V_j.$$

That is, each V_j is an invariant subspace for the representation T. By Proposition 8, the restrictions $T\lceil V_j$ of T are of the form treated in the preceding paragraph. Hence $T\lceil V_j$ and so T is completely reducible. □

3.2.3 Finite-Dimensional Representations of $\breve U_q(\mathrm{sl}_2)$ and $U_h(\mathrm{sl}_2)$

The previous considerations carry over with a few modifications to the algebras $\breve U_q(\mathrm{sl}_2)$ and $U_h(\mathrm{sl}_2)$ as well. In this subsection we state without proof the corresponding classification results for these algebras.

Theorem 13. (i) *For each $\omega\in\{+1,-1,\mathrm{i},-\mathrm{i}\}$ and $l=\frac12\mathbb{N}_0$ there exists up to equivalence a unique $(2l+1)$-dimensional irreducible representation $T_{\omega l}$ of*

$\breve{U}_q(\mathrm{sl}_2)$. *This representation acts on basis vectors* $\mathbf{e}_i$, $i = -l, -l+1, \cdots, l$, *of the representation space by*

$$T_{\omega l}(E)\mathbf{e}_m = \omega([l-m][l+m+1])^{1/2}\mathbf{e}_{m+1}, \tag{29}$$

$$T_{\omega l}(F)\mathbf{e}_m = \omega([l+m][l-m+1])^{1/2}\mathbf{e}_{m-1}, \quad T_{\omega l}(K)\mathbf{e}_m = \omega q^m \mathbf{e}_m, \tag{30}$$

where $\mathbf{e}_{l+1} = \mathbf{e}_{-l-1} := 0$. *Any irreducible finite-dimensional representation of* $\breve{U}_q(\mathrm{sl}_2)$ *is equivalent to one of these representations* $T_{\omega l}$.

(ii) *Every finite-dimensional representation of* $\breve{U}_q(\mathrm{sl}_2)$ *is completely reducible.*

Since $U_h(\mathrm{sl}_2)$ is an algebra over $\mathbb{C}[[h]]$, we consider representations of $U_h(\mathrm{sl}_2)$ on linear spaces over the ring $\mathbb{C}[[h]]$. Let l be an integral or half-integral nonnegative number and let V_l be a complex vector space with a basis $\mathbf{e}_i$, $i = -l, -l+1, \cdots, l$. Then the formulas

$$T_l(E)\mathbf{e}_m = ([l-m][l+m+1])^{1/2}\mathbf{e}_{m+1}, \tag{31}$$

$$T_l(F)\mathbf{e}_m = ([l+m][l-m+1])^{1/2}\mathbf{e}_{m-1}, \quad T_l(H)\mathbf{e}_m = 2m\mathbf{e}_m \tag{32}$$

define an indecomposable representation T_l of $U_h(\mathrm{sl}_2)$ on the vector space $V_l[[h]]$ (see Subsect. 1.2.10) over the ring $\mathbb{C}[[h]]$. (Recall that a representation T is called indecomposable if the underlying space cannot be represented as a direct sum of nontrivial invariant subspaces.) The representation T_l is not irreducible, since any subspace $h^k V_l[[h]]$ is invariant.

Proposition 14. *Every indecomposable representation of* $U_h(\mathrm{sl}_2)$ *on a* $\mathbb{C}[[h]]$-*vector space of the form* $V[[h]]$, *where* V *is a finite-dimensional complex vector space, is equivalent to one of the representations* T_l.

3.3 Representations of $U_q(\mathrm{sl}_2)$ for q a Root of Unity

Throughout this section we assume that the parameter q is a primitive p-th root of unity, where $p \geq 3$. The representation theory of $U_q(\mathrm{sl}_2)$ is then very different from the case when q is not a root of unity.

We set $p' = p$ if p is odd and $p' = p/2$ if p is even.

3.3.1 The Center of $U_q(\mathrm{sl}_2)$

Let $\mathfrak{z}$ denote the center of the algebra $U_q(\mathrm{sl}_2)$. If q is not a root of unity, we know from Proposition 2 that $\mathfrak{z}$ is generated by the Casimir element C_q defined by (6). However, in the root of unity case the center $\mathfrak{z}$ is much larger.

Proposition 15. (i) *The elements* $E^{p'}$, $F^{p'}$, $K^{p'}$, $K^{-p'}$ *belong to* $\mathfrak{z}$.

(ii) $\mathfrak{z}$ *is generated by* $E^{p'}$, $F^{p'}$, $K^{p'}$, $K^{-p'}$ *and the Casimir element* C_q.

Proof. (i): We give the proof for $F^{p'}$. By (1), $KF^{p'} = q^{-2p'}F^{p'}K = F^{p'}K$. Since $[p']_q = 0$ by Proposition 2.1, we have $EF^{p'} = F^{p'}E$ by (3). Thus, $F^{p'} \in \mathfrak{z}$. The proof for $E^{p'}$, $K^{p'}$ and $K^{-p'}$ is similar.

(ii): A proof can be found, for example, in [DCK]. □

Corollary 16. *Any irreducible representation of* $U_q(\mathrm{sl}_2)$ *is finite-dimensional.*

Proof. Since an irreducible representation T maps central elements into scalar operators, it follows from Proposition 15(ii) that $T(U_q(\mathrm{sl}_2))$ coincides with the linear span of operators $T(E^rK^sF^t)$, $r,s,t \le p'-1$. Therefore, if $\mathbf{v}$ is a nonzero vector of the representation space V, then $T(U_q(\mathrm{sl}_2))\mathbf{v}$ is a finite-dimensional invariant subspace, so that $T(U_q(\mathrm{sl}_2))\mathbf{v} = V$. □

Let $\mathfrak{z}_0$ be the subalgebra of $\mathfrak{z}$ generated by the elements $E^{p'}, F^{p'}, K^{p'}, K^{-p'}$ and let J be the two-sided ideal of $U_q(\mathrm{sl}_2)$ generated by $E^{p'}, F^{p'}, K^{p'}-1$. Using the assumption $q^p = 1$ it is easy to calculate that

$$\Delta(K^{p'}) = K^{p'} \otimes K^{p'}, \quad \Delta(E^{p'}) = E^{p'} \otimes K^{p'} + 1 \otimes E^{p'},$$

$$\Delta(F^{p'}) = F^{p'} \otimes 1 + K^{-p'} \otimes F^{p'}.$$

Since $S(E^{p'})$ and $S(F^{p'})$ are multiples of $E^{p'}K^{-p'}$ and $K^{p'}F^{p'}$, respectively, it is clear from these formulas that $\mathfrak{z}_0$ is a Hopf subalgebra and J is a Hopf ideal of $U_q(\mathrm{sl}_2)$. Therefore, the quotient $U_q^{\mathrm{res}}(\mathrm{sl}_2) = U_q(\mathrm{sl}_2)/J$ is a finite-dimensional Hopf algebra.

Using Proposition 15 it can be shown that the algebra $U_q(\mathrm{sl}_2)$ is a $\mathfrak{z}_0$-bimodule with free left (and right) $\mathfrak{z}_0$-module basis

$$E^rK^sF^t, \quad r,s,t = 0,1,2,\cdots,p'-1. \tag{33}$$

Hence the elements (33) also constitute a vector space basis of $U_q^{\mathrm{res}}(\mathrm{sl}_2)$.

3.3.2 Representations of $U_q(\mathrm{sl}_2)$

In this subsection T denotes an irreducible representation of $U_q(\mathrm{sl}_2)$ on a vector space V. Then the images $T(E^{p'})$ and $T(F^{p'})$ of the central elements $E^{p'}$ and $F^{p'}$ are scalar multiples of the identity operator. Concerning the values of these scalars we distinguish three possible cases.

Case 1: $T(E^{p'}) = T(F^{p'}) = 0$.

In order to find such representations T let us first look at the representations $T_{\omega l}$ from Subsect. 3.2.1. For any $\omega \in \{+1,-1\}$ and $l \in \frac{1}{2}\mathbb{N}_0$ the operators $T_{\omega l}(E)$, $T_{\omega l}(F)$ and $T_{\omega l}(K)$ given by (22) and (23) also define a representation $T_{\omega l}$ of the algebra $U_q(\mathrm{sl}_2)$ in the root of unity case.

Next we construct a three-parameter family of representations $T_{ab\lambda}$ which plays a crucial role in the following. Let a, b, and λ be complex numbers such that $\lambda \ne 0$ and let V denote a p'-dimensional vector space with basis $\mathbf{e}_i$, $i = 0,1,2,\cdots,p'-1$. One immediately checks that the operators

$$T_{ab\lambda}(K)\mathbf{e}_i = q^{-2i}\lambda\mathbf{e}_i,\ T_{ab\lambda}(F)\mathbf{e}_{p'-1} = b\mathbf{e}_0,\ T_{ab\lambda}(F)\mathbf{e}_i = \mathbf{e}_{i+1},\ i < p'-1, \tag{34}$$

$$T_{ab\lambda}(E)\mathbf{e}_0 = a\mathbf{e}_{p'-1},\ \ T_{ab\lambda}(E)\mathbf{e}_i = \left(ab + [i]_q\frac{\lambda q^{1-i} - \lambda^{-1}q^{i-1}}{q-q^{-1}}\right)\mathbf{e}_{i-1},\ i > 0, \tag{35}$$

satisfy the defining relations (1) and (2) of the algebra $U_q(\mathrm{sl}_2)$. Hence these operators determine a representation $T_{ab\lambda}$ of $U_q(\mathrm{sl}_2)$ on V. Clearly, we have $T_{00\lambda}(E^{p'}) = T_{00\lambda}(F^{p'}) = 0$ for any $\lambda \neq 0$.

Proposition 17. (i) *The representation $T_{\omega l}$ is irreducible if and only if $2l < p'$. The representations $T_{\omega l}$, $\omega \in \{+1, -1\}$, $l = 0, \frac{1}{2}, 1, \cdots, \frac{1}{2}(p'-1)$, are pairwise nonequivalent and satisfy the condition $T_{\omega l}(E^{p'}) = T_{\omega l}(F^{p'}) = 0$.*

(ii) *The representation $T_{00\lambda}$ is irreducible if and only if $\lambda \neq \pm q^n$ for $n = 0, 1, \cdots, p'-2$. The irreducible representations $T_{00\lambda}$ are pairwise inequivalent.*

(iii) *The p'-dimensional irreducible representations $T_{00\lambda}$, $\lambda = \omega q^{p'-1}$, and $T_{\omega,(p'-1)/2}$ for $\omega = \pm 1$ are equivalent. Any two other irreducible representations of the form $T_{\omega l}$ and $T_{00\lambda}$ are not equivalent.*

(iv) *Each irreducible representation T of the algebra $U_q(\mathrm{sl}_2)$ such that $T(E^{p'}) = T(F^{p'}) = 0$ is equivalent to one of the irreducible representations from* (i) *or* (ii).

Proof. (i): By the formulas (25) and (26), the action of the representation $T_{\omega l}$ on the generators E, F, K can be written as

$$T_{\omega l}(K)\mathbf{e}'_r = \omega q^{2l-2r}\mathbf{e}'_r,\ T_{\omega l}(E)\mathbf{e}'_r = \omega[2l-r+1]\mathbf{e}'_{r-1},\ T_{\omega l}(F)\mathbf{e}'_r = \omega[r+1]\mathbf{e}'_{r+1},$$

where $\mathbf{e}'_{-1} = \mathbf{e}'_{2l+1} := 0$ and $\mathbf{e}'_r, r = 0, 1, \cdots, 2l$, is a basis of the representation space V_l. First suppose that $2l \geq p'$. Since $[p'] = 0$, $T_{\omega l}(F)\mathbf{e}'_{p'-1} = [p']\mathbf{e}'_{p'} = 0$. Hence the subspace W spanned by the vectors $\mathbf{e}'_i, i = 0, 1, 2, \cdots, p'-1$, is invariant for the representation $T_{\omega l}$. Since $W \neq V_l$, $T_{\omega l}$ is not irreducible.

Now assume that $2l < p'$. It is obvious that $T_{\omega l}(E^{p'}) = T_{\omega l}(F^{p'}) = 0$. Since q is a primitive p-th root of unity, all q-numbers $[1], [2], \cdots, [p'-1]$ are nonzero. Using this fact the remaining assertions of (i) follow completely similar to the classical case.

(ii): The reducibility of the representations $T_{00\lambda}$, $\lambda = \pm q^n$, $0 \leq n \leq p'-2$, and the irreducibility of the others are proved as in (i). Suppose that two irreducible representations $T_{00\lambda}$ and $T_{00\mu}$ are equivalent. Then there exists an invertible operator B such that $BT_{00\lambda}(a) = T_{00\mu}(a)B$ for all $a \in U_q(\mathrm{sl}_2)$. Let $\{\mathbf{e}_i^\lambda\}$ and $\{\mathbf{e}_i^\mu\}$ be the corresponding bases in the representation spaces of $T_{00\lambda}$ and $T_{00\mu}$, respectively. Since $\ker T_{00\lambda}(F) = \mathbb{C}\mathbf{e}_0^\lambda$ and $\ker T_{00\mu}(F) = \mathbb{C}\mathbf{e}_0^\mu$, it follows that $c\mathbf{e}_0^\mu = B\mathbf{e}_0^\lambda$ for some $c \in \mathbb{C}$. But then we have $\lambda B\mathbf{e}_0^\lambda = BT_{00\lambda}(K)\mathbf{e}_0^\lambda = T_{00\mu}(K)B\mathbf{e}_0^\lambda = cT_{00\mu}(K)\mathbf{e}_0^\mu = c\mu\mathbf{e}_0^\mu = \mu B\mathbf{e}_0^\lambda$, so that $\lambda = \mu$.

(iii): The equivalence of $T_{00\lambda}, \lambda = \omega q^{p'-1}$, and $T_{\omega,(p'-1)/2}$ can be seen at once from the above formulas. Other irreducible representations $T_{00\lambda}$ and $T_{\omega l}$ cannot be equivalent, because they act on spaces of different dimensions.

(iv) is proved in the same way as Proposition 9. □

Note that any irreducible representation from Proposition 17 is a representation with highest weight and with lowest weight.

Case 2: $T(E^{p'}) \neq 0$ and $T(F^{p'}) \neq 0$.

Such irreducible representations are called *cyclic*. The reason for this name will be seen below.

Proposition 18. (i) *The representation $T_{ab\lambda}$ is irreducible and cyclic if and only if all coefficients $ab+[i]_q(\lambda q^{1-i}-\lambda^{-1}q^{i-1})(q-q^{-1})^{-1}$, $i=0,1,\cdots,p'-1$, (in particular, the parameters a and b) are nonzero.*

(ii) *Two irreducible cyclic representations $T_{ab\lambda}$ and $T_{a'b'\lambda'}$ are equivalent if and only if $a'=a+b^{-1}[i](\lambda q^{1-i}-\lambda^{-1}q^{i-1})(q-q^{-1})^{-1}$, $b'=b$ and $\lambda'=q^{-2i}\lambda$ for some $i\in\{0,1,2,\cdots,p'-1\}$.*

(iii) *Any irreducible cyclic representation of $U_q(\mathrm{sl}_2)$ is equivalent to a representation $T_{ab\lambda}$.*

Proof. The assertion of (i) is obvious.

(ii): Let $T_{ab\lambda}$ and $T_{a'b'\lambda'}$ be equivalent irreducible cyclic representations. Then the spectra of the operators $T_{ab\lambda}(K)$ and $T_{a'b'\lambda'}(K)$ coincide, so $\lambda'=q^{-2i}\lambda$ for some $i\in\{0,1,2,\cdots,p'-1\}$. For the central elements $F^{p'}$ and C_q we have

$$T_{ab\lambda}(F^{p'})=bI,\quad T_{ab\lambda}(C_q)=((\lambda q+\lambda^{-1}q^{-1})(q-q^{-1})^{-2}+ab)I.$$

Since these values must be the same for the equivalent representation $T_{a'b'\lambda'}$, we get $b=b'$ and

$$\frac{\lambda q+\lambda^{-1}q^{-1}}{(q-q^{-1})^2}+ab=\frac{\lambda q^{-2i+1}+\lambda^{-1}q^{2i-1}}{(q-q^{-1})^2}+a'b,$$

that is, $a'=a+b^{-1}[i](\lambda q^{1-i}-\lambda^{-1}q^{i-1})(q-q^{-1})^{-1}$. Conversely, a direct calculation shows that for such values of a',b',λ' the representations $T_{ab\lambda}$ and $T_{a'b'\lambda'}$ are indeed equivalent.

(iii): Let T be an irreducible cyclic representation of $U_q(\mathrm{sl}_2)$ on a vector space V. By Corollary 16, V is finite-dimensional. Hence the operator $T(K)$ has an eigenvector $\mathbf{e}_0\in V$, say $T(K)\mathbf{e}_0=\lambda\mathbf{e}_0$. We set $\mathbf{e}_i:=T(F^i)\mathbf{e}_0$, $i=0,1,,\cdots,p'-1$. Since T is cyclic, $T(F^{p'})=bI$ for some nonzero $b\in\mathbb{C}$. Then we have $T(F^{p'-i})\mathbf{e}_i=T(F^{p'})\mathbf{e}_0=b\mathbf{e}_0\neq 0$ and so $\mathbf{e}_i\neq 0$. The relations (1) imply that $T(K)\mathbf{e}_i=q^{-2i}\lambda\mathbf{e}_i$, $i=0,1,2,\cdots,p'-1$. Because the numbers $q^{-2i}\lambda$, $i=0,1,2,\cdots,p'-1$, are pairwise distinct, the vectors $\mathbf{e}_i$, $i=0,1,2,\cdots,p-1$, are linearly independent. Since the Casimir operator (6) acts on V as a scalar operator, there is an $a'\in\mathbb{C}$ such that

$$T(FE)\mathbf{e}_0=T(C_q)\mathbf{e}_0-\frac{T(K)q+T(K^{-1})q^{-1}}{q-q^{-1}}\mathbf{e}_0=a'\mathbf{e}_0.$$

Therefore, $bT(E)\mathbf{e}_0=T(F^{p'})T(E)\mathbf{e}_0=T(F^{p'-1})a'\mathbf{e}_0=a'\mathbf{e}_{p'-1}$, that is, $T(E)\mathbf{e}_0=a\mathbf{e}_{p'-1}$, where $a:=a'b^{-1}$. Applying relation (3) we find that

$$T(E)\mathbf{e}_i = T(E)T(F^i)\mathbf{e}_0 = \left(ab + [i]_q \frac{\lambda q^{1-i} - \lambda^{-1}q^{i-1}}{q-q^{-1}}\right)\mathbf{e}_{i-1},\ i>0.$$

Thus we see that $W := \mathrm{Lin}\{\mathbf{e}_i \,|\, i = 0, 1, \cdots, p'-1\}$ is an invariant subspace. Since T is irreducible, $W = V$. The preceding formulas show that T is equivalent to the representation $T_{ab\lambda}$. □

Let us see what is special about the cyclic representations $T_{ab\lambda}$. The above formulas show that one can move from an arbitrary weight vector to a nonzero complex multiple of any other by repeated application of the operator $T_{ab\lambda}(E)$ (or of $T_{ab\lambda}(F)$). This explains why these representations are called cyclic. Recall that in contrast to this property, for the representations $T_{\omega l}, 2l < p'$, the action of the operator $T_{\omega l}(E)$ (or of $T_{\omega l}(F)$) defines a *linear* order on the weight vectors.

Case 3: $T(E^{p'}) = 0$ and $T(F^{p'}) \neq 0$ or $T(E^{p'}) \neq 0$ and $T(F^{p'}) = 0$.

In this case the irreducible representation T will be called *semicyclic*.

We first treat the case where $T(E^{p'}) = 0$ and $T(F^{p'}) \neq 0$. Examples of such semicyclic representations are obtained by setting $a = 0$ in the three parameter family $T_{ab\lambda}$. More precisely, we have

Proposition 19. (i) *$T_{0b\lambda}$ is an irreducible representation such that $T_{0b\lambda}(E^{p'}) = 0$ and $T_{0b\lambda}(F^{p'}) \neq 0$ if and only if $b \neq 0$ and $\lambda^{p'} \neq \pm 1, 0$. Two such representations $T_{0b\lambda}$ and $T_{0b'\lambda'}$ are equivalent if and only if $b = b'$ and $\lambda = \lambda'$.*

(ii) *Every irreducible representation T of $U_q(\mathrm{sl}_2)$ such that $T(E^{p'}) = 0$ and $T(F^{p'}) \neq 0$ is equivalent to one of the representations $T_{0b\lambda}$, $b \neq 0, \lambda \neq 0$.*

Proof. The proof of (i) is straightforward. In order to prove (ii), we repeat the reasoning of the proof of Proposition 18(ii). Since $T(E^{p'}) = 0$ by assumption, the constant a must be zero, that is, T is equivalent to $T_{0b\lambda}$. □

From the formulas (34) and (35) we see that $\mathbf{e}_0$ is a highest weight vector for the representation $T_{0b\lambda}$ and that any representation $T_{0b\lambda}$ with $b \neq 0$ and $\lambda^{p'} \neq \pm 1, 0$ has no lowest weight vector. Thus a semicyclic representation T such that $T(E^{p'}) = 0$ and $T(F^{p'}) \neq 0$ always admits a highest weight vector, but no lowest weight vector. Similarly, every semicyclic representation T for which $T(E^{p'}) \neq 0$ and $T(F^{p'}) = 0$ has a lowest weight vector, but no highest weight vector. It is clear from their definition that cyclic representations have neither highest nor lowest weights.

Now we turn to the semicyclic representations T such that $T(E^{p'}) \neq 0$ and $T(F^{p'}) = 0$. Let θ be the algebra automorphism of $U_q(\mathrm{sl}_2)$ determined by $\theta(E) = F, \theta(F) = E$ and $\theta(K) = K^{-1}$. Then $T' := T \circ \theta$ is also an irreducible representation. The representation $T'_{0b\lambda} := T_{0b\lambda} \circ \theta$ on V acts as

$$T'_{0b\lambda}(K)\mathbf{e}_i = q^{2i}\lambda^{-1}\mathbf{e}_i,\ T'_{0b\lambda}(E)\mathbf{e}_{p'-1} = b\mathbf{e}_0,\ T'_{0b\lambda}(E)\mathbf{e}_i = \mathbf{e}_{i+1},\ i < p'-1,$$

$$T'_{0b\lambda}(F)\mathbf{e}_0 = 0,\ \ T'_{0b\lambda}(F)\mathbf{e}_i = [i]_q \frac{\lambda q^{1-i} - \lambda^{-1}q^{i-1}}{q-q^{-1}}\mathbf{e}_{i-1},\ i>0.$$

Clearly, we have $T(E^{p'}) = 0$ and $T(F^{p'}) \neq 0$ if and only if $T'(E^{p'}) \neq 0$ and $T'(F^{p'}) = 0$. Therefore, the assertions of Proposition 19 remain valid if $T_{0b\lambda}$ is replaced by $T'_{0b\lambda}$ and if the elements $E^{p'}$ and $F^{p'}$ are interchanged. In particular, this shows that *any semicyclic irreducible representation T such that $T(E^{p'}) \neq 0$ and $T(F^{p'}) = 0$ is equivalent to one of the representations $T'_{0b\lambda}$, $b \neq 0$, $\lambda^{p'} \neq \pm 1, 0$.*

Since the Cases 1–3 exhaust all possible cases for the irreducible representation T, the preceding results can be summarized in the following

Theorem 20. *Every irreducible representation of $U_q(\mathrm{sl}_2)$ has dimension $\leq p'$ and is a weight representation. The irreducible representations T of dimension $< p'$ are up to equivalence the representations $T_{\omega l}$, $2l < p' - 1$, from Proposition 17. Any irreducible representation of dimension p' is equivalent to one of the representations $T_{ab\lambda}$ or $T'_{0b\lambda}$ from Propositions 17–19.*

Finally, let us return once more to the representation $T_{00\lambda}$ for $\lambda := \omega q^n$, $\omega = \pm 1$, where n is one of the numbers $0, 1, \cdots, p' - 2$. Recall that these representations are not irreducible and the linear span of basis elements $\mathbf{e}_j$, $n < j \leq p' - 1$, is a nontrivial invariant subspace. Comparing the corresponding formulas we see that the restriction of $T_{00\lambda}$ to this subspace is equivalent to the irreducible representation $T_{\omega l}$, $2l = p' - n - 1$, from Proposition 17. A closer look at the operators $T_{00\lambda}(E)$, $T_{00\lambda}(F)$ and $T_{00\lambda}(K)$ shows that the representation $T_{00\lambda}$ is *not completely reducible.* That is, the assertion of Proposition 12 is no longer true if q is a root of unity. The existence of finite-dimensional representations that are not completely reducible is another new and interesting phenomenon of the root of unity case.

3.3.3 Representations of $U_q^{\mathrm{res}}(\mathrm{sl}_2)$

In this subsection we determine the irrreducible representations of the finite-dimensional quotient algebra $U_q^{\mathrm{res}}(\mathrm{sl}_2) = U_q(\mathrm{sl}_2)/J$ defined in Subsect. 3.3.1. Clearly, the irreducible representations of $U_q^{\mathrm{res}}(\mathrm{sl}_2)$ are in one-to-one correspondence to those irreducible representations of $U_q(\mathrm{sl}_2)$ which annihilate the elements $E^{p'}, F^{p'}, K^{p'} - 1$. Thus, only the representations from Case 1 can give rise to irreducible representations of $U_q^{\mathrm{res}}(\mathrm{sl}_2)$. For notational simplicity let us denote the corresponding representations of $U_q^{\mathrm{res}}(\mathrm{sl}_2)$ and $U_q(\mathrm{sl}_2)$ by the same symbol. From the results developed above we obtain

Proposition 21. *The algebra $U_q^{\mathrm{res}}(\mathrm{sl}_2)$ has the following irreducible representations:*

(i) *$T_{1,l}$ with $0 \leq 2l + 1 \leq p$ if p is odd;*

(ii) *$T_{\pm 1,l}$, $2l + 1 < p'$, with $2l$ even if p and p' are even;*

(iii) *$T_{1,l}$, $2l + 1 < p'$, with $2l$ even and $T_{-1,l}$, $2l + 1 \leq p'$, with $2l$ odd if p is even and p' is odd.*

Any irreducible representation of $U_q^{\mathrm{res}}(\mathrm{sl}_2)$ is equivalent to one of these representations.

3.4 Tensor Products of Representations. Clebsch–Gordan Coefficients

During the study of representations in the two previous sections only the algebra structure of $U_q(\mathrm{sl}_2)$ was needed. In the subsequent sections we essentially use the comultiplication $U_q(\mathrm{sl}_2)$ by considering tensor products of representations.

Throughout Sects. 3.4–3.6 we assume that q is not a root of unity.

3.4.1 Tensor Products of Representations T_l

Let T and T' be representations of the quantum algebra $U_q(\mathrm{sl}_2)$ resp. $\breve{U}_q(\mathrm{sl}_2)$ on vector spaces V and W, respectively. Recall from Subsect. 1.3.1 that the tensor product $T \otimes T'$ is the representation on the vector space $V \otimes W$ defined by

$$(T \otimes T')(a) := \sum T(a_{(1)}) \otimes T'(a_{(2)}). \tag{36}$$

In what follows, let T_{l_1} and T_{l_2}, $l_1, l_2 \in \mathbb{N}_0$, be two irreducible representations of $U_q(\mathrm{sl}_2)$ (resp. $\breve{U}(\mathrm{sl}_2)$). The next result is the Clebsch–Gordan formula on the decomposition of the tensor product $T_{l_1} \otimes T_{l_2}$ into a direct sum of irreducible representations.

Proposition 22. *If q is not a root of unity, then we have*

$$T_{l_1} \otimes T_{l_2} \simeq T_{|l_1-l_2|} \oplus T_{|l_1-l_2|+1} \oplus \cdots \oplus T_{|l_1+l_2|}. \tag{37}$$

Proof. We carry the proof for $U_q(\mathrm{sl}_2)$. By Propositions 9 and 12, $T_{l_1} \otimes T_{l_2}$ is equivalent to a direct sum $T = \bigoplus m_{\omega l} T_{\omega l}$ of multiples of irreducible representations $T_{\omega l}$. The operator $(T_{l_1} \otimes T_{l_2})(K) = T_{l_1}(K) \otimes T_{l_2}(K)$ has precisely the eigenvectors $\mathbf{e}_m \otimes \mathbf{e}_n$ with eigenvalues $q^{2(m+n)}$, where $m = -l_1, \cdots, l_1$ and $n = -l_2, \cdots, l_2$. Comparing with the eigenvalues of $T_{\omega l}(K)$ and their corresponding multiplicities, it follows that $m_{1l} = 1$ if $l = |l_1 - l_2|, |l_1 - l_2| + 1, \cdots, |l_1 + l_2|$ and $m_{\omega l} = 0$ otherwise. □

Note that formula (37) is also valid for the h-adic algebra $U_h(\mathrm{sl}_2)$.

In the classical case, the flip operator $\tau : V_{l_1} \otimes V_{l_2}$ given by $\tau(v_1 \otimes v_2) = v_2 \otimes v_1$ provides an equivalence of the representations $T_{l_1} \otimes T_{l_2}$ and $T_{l_2} \otimes T_{l_1}$ of $U(\mathrm{sl}_2)$. This is no longer true for $U_q(\mathrm{sl}_2)$, but there is another "natural" operator $\hat{R}_{l_1 l_2}$ that realizes this equivalence at least if q is not a root of unity.

We first consider the quasitriangular h-adic Hopf algebra $U_h(\mathrm{sl}_2)$ which is equipped with the universal R-matrix $\mathcal{R}$ from Proposition 7. Let

$$R_{l_1 l_2} := (T_{l_1} \otimes T_{l_2})(\mathcal{R}) \quad \text{and} \quad \hat{R}_{l_1 l_2} := \tau \circ (T_{l_1} \otimes T_{l_2})(\mathcal{R}).$$

Then $\hat{R}_{l_1 l_2}$ maps $V_{l_1} \otimes V_{l_2}$ to $V_{l_2} \otimes V_{l_1}$ and, by the relation $\Delta^{\mathrm{cop}}(a)\mathcal{R} = \mathcal{R}\Delta(a)$ for $a \in U_h(\mathrm{sl}_2)$, we have

$$\begin{aligned}\hat{R}_{l_1l_2}((T_{l_1}\otimes T_{l_2})(a)) &= \tau(T_{l_1}\otimes T_{l_2})(\mathcal{R})(T_{l_1}\otimes T_{l_2})(\Delta(a))\\ &= \tau(T_{l_1}\otimes T_{l_2})(\mathcal{R}(\Delta(a)) = \tau(T_{l_1}\otimes T_{l_2})(\Delta^{\mathrm{cop}}(a))(T_{l_1}\otimes T_{l_2})(\mathcal{R})\\ &= (T_{l_2}\otimes T_{l_1})(\Delta(a))\tau(T_{l_1}\otimes T_{l_2})(\mathcal{R}) = (T_{l_2}\otimes T_{l_1})(a)\hat{R}_{l_1l_2}. \end{aligned} \tag{38}$$

Since $\hat{R}_{l_1l_2}$ is invertible with inverse $(T_{l_1}\otimes T_{l_2})(\mathcal{R}^{-1})\circ\tau$, the operator $\hat{R}_{l_1l_2}$ realizes the equivalence of the representations $T_{l_1}\otimes T_{l_2}$ and $T_{l_2}\otimes T_{l_1}$ of $U_h(\mathrm{sl}_2)$.

Using (19) we compute the matrix elements $(R_{l_1l_2})^{ij}_{mn}$ of the operator $R_{l_1l_2}$ with respect to the basis $\{\mathbf{e}_m\otimes\mathbf{e}_n\}$ of $V_{l_1}\otimes V_{l_2}$. For this reason we abbreviate $q=e^h$. By (31) and (32), we obtain

$$\begin{aligned}(R_{l_1l_2})^{n_1,n_2}_{n_1+n,n_2-n} &:= \langle \mathbf{e}_{n_1+n}\otimes\mathbf{e}_{n_2-n}\mid R_{l_1l_2}\mid \mathbf{e}_{n_1}\otimes\mathbf{e}_{n_2}\rangle\\ &= c\langle \mathbf{e}_{n_1+n}\otimes\mathbf{e}'_{n_2-n}\mid e^{h(H\otimes H)/2}E^n\otimes F^n\mid \mathbf{e}_{n_1}\otimes\mathbf{e}'_{n_2}\rangle,\end{aligned}$$

where $c=q^{n(n+1)/2}(1-q^{-2})^n([n]_q!)^{-1}$. The other matrix elements of $R_{l_1l_2}$ vanish. A direct calculation yields

$$\begin{aligned}(R_{l_1l_2})^{n_1,n_2}_{n_1+n,n_2-n} &= \frac{(1-q^{-2})^n}{[n]!}\left(\frac{[l_1-n_1]![l_1+n_1+n]!}{[l_1+n_1]![l_1-n_1-n]!}\right)^{1/2}\\ &\times\left(\frac{[l_2+n_2]![l_2-n_2+n]!}{[l_2-n_2]![l_2+n_2-n]!}\right)^{1/2} q^{2(n_1+n)(n_2-n)}q^{n(n+1)/2}.\end{aligned} \tag{39}$$

The expression in (39) is a formal power series in the indeterminate h. But it is also well-defined for fixed $q\in\mathbb{C}$, when q is not a root of unity. The corresponding matrix $\hat{R}_{l_1l_2}$ realizes an equivalence of the representations $T_{l_1}\otimes T_{l_2}$ and $T_{l_2}\otimes T_{l_1}$ of the algebra $U_q(\mathrm{sl}_2)$ as seen by the following reasoning. Equation (38) holds in particular for the elements $e^{\pm hH}$, E and F of the algebra $U_h(\mathrm{sl}_2)$ and for vectors in the complex linear span of the basis elements $\mathbf{e}_m\otimes\mathbf{e}_n$. Now we put $q=e^h$ in the matrix entries of $\hat{R}_{l_1l_2}$ and $K^{\pm1}:=e^{\pm hH}$ in (38). Then, by the definitions of the representations T_l for $U_h(\mathrm{sl}_2)$ and $U_q(\mathrm{sl}_2)$ and by the formulas of the comultiplications, (38) is just the intertwining property for the generators of $U_q(\mathrm{sl}_2)$. Hence $\hat{R}_{l_1l_2}$ intertwines the representations $T_{l_1}\otimes T_{l_2}$ and $T_{l_2}\otimes T_{l_1}$ of $U_q(\mathrm{sl}_2)$ and gives their equivalence. A similar result holds for the algebra $\breve{U}_q(\mathrm{sl}_2)$.

Let us turn to the special case $l_1=l_2=\frac{1}{2}$ and denote the operators $R_{1/2,1/2}$ and $\hat{R}_{1/2,1/2}$ by R and $\hat{R}$, respectively. With respect to the basis $\mathbf{e}_{-1/2}\otimes\mathbf{e}_{-1/2}$, $\mathbf{e}_{-1/2}\otimes\mathbf{e}_{1/2}$, $\mathbf{e}_{1/2}\otimes\mathbf{e}_{-1/2}$, $\mathbf{e}_{1/2}\otimes\mathbf{e}_{1/2}$ these matrices take the form

$$R=\begin{pmatrix} q & 0 & 0 & 0\\ 0 & 1 & 0 & 0\\ 0 & q-q^{-1} & 1 & 0\\ 0 & 0 & 0 & q\end{pmatrix},\quad \hat{R}=\begin{pmatrix} q & 0 & 0 & 0\\ 0 & q-q^{-1} & 1 & 0\\ 0 & 1 & 0 & 0\\ 0 & 0 & 0 & q\end{pmatrix}. \tag{40}$$

3.4.2 Clebsch–Gordan Coefficients

Let $\tilde{V}_l$ denote the subspace of $V_{l_1} \otimes V_{l_2}$ on which the representation T_l in the decomposition (37) is realized. Let $\{\mathbf{e}_j\}$, $\{\mathbf{e}_k\}$, $\{\mathbf{e}^l_m\}$ be bases of the vector spaces V_{l_1}, V_{l_2}, $\tilde{V}_l$ such that the corresponding representations of $U_q(\mathrm{sl}_2)$ (resp. $\breve{U}_q(\mathrm{sl}_2)$) are given by the formulas (22) and (23) (resp. (29) and (30)). Since both sets

$$\{\mathbf{e}_j \otimes \mathbf{e}_k \mid j = -l_1, -l_1+1, \cdots, l_1,\ k = -l_2, -l_2+1, \cdots, l_2\},$$

$$\{\mathbf{e}^l_m \mid l = |l_1 - l_2|, |l_1 - l_2| + 1, \cdots, l_1 + l_2,\ m = -l, -l+1, \cdots, l\}$$

are bases of $V_{l_1} \otimes V_{l_2}$, there exists an invertible matrix C with entries $C_q(l_1, l_2, l; j, k, m)$ such that

$$\mathbf{e}^l_m = \sum\nolimits_{j,k} C_q(l_1, l_2, l; j, k, m)\mathbf{e}_j \otimes \mathbf{e}_k. \tag{41}$$

The numbers $C_q(l_1, l_2, l; j, k, m)$ are called the *Clebsch–Gordan coefficients* (abbreviated, CGC's) of the tensor product $T_{l_1} \otimes T_{l_2}$. We often write

$$C_q(\ell, \mathbf{j}) \equiv C_q(l_1, l_2, l; j, k, m).$$

In the subsequent treatment we shall see that the main properties of the CGC's known for the classical group $SU(2)$ (as described, for instance, in [VK1], Chap. 8) remain valid for the CGC's of $\breve{U}_q(\mathrm{su}_2)$ as well. In particular,

$$C_q(\ell, \mathbf{j}) = 0 \quad \text{if} \quad j + k \neq m.$$

Let $t^{(l)}_{mn}$, $m, n = -l, -l+1, \cdots, l$, denote the matrix coefficients of the representation T_l of $U_q(\mathrm{sl}_2)$ (resp. $\breve{U}_q(\mathrm{sl}_2)$) with respect to the basis $\{\mathbf{e}_m\}$ of V_l. That is, $t^{(l)}_{mn}$ are linear functionals on $U_q(\mathrm{sl}_2)$ (resp. $\breve{U}_q(\mathrm{sl}_2)$) such that

$$T_l(\cdot)\mathbf{e}_n = \sum\nolimits_m t^{(l)}_{mn}(\cdot)\mathbf{e}_m. \tag{42}$$

We apply the representation $T_{l_1} \otimes T_{l_2}$ to both sides of (41). On the left hand side, we take into account that $T_{l_1} \otimes T_{l_2}$ acts on $\mathbf{e}^l_m$ as T_l does and use the formulas (41) and (42). On the right hand side, we use (42) and the definition of the product in the dual algebra $U_q(\mathrm{sl}_2)'$ (resp. $\breve{U}_q(\mathrm{sl}_2)'$). Equating the coefficients of basis elements, we obtain the identity

$$C_q(l_1, l_2, l; i, j, m)t^{(l)}_{mn} = \sum\nolimits_{r,s} C_q(l_1, l_2, l; r, s, n)t^{(l_1)}_{ir}t^{(l_2)}_{js}$$

for $i = -l_1, -l_1+1, \cdots, l_1$, $j = -l_2, -l_2+1, \cdots, l_2$. Since the matrix C_q of CGC's is invertible, we derive from this identity that

$$t^{(l_1)}_{ir}t^{(l_2)}_{js} = \sum\nolimits_l C_q(l_1, l_2, l; i, j, m)C_q^{-1}(l_1, l_2, l; r, s, n)t^{(l)}_{mn}. \tag{43}$$

Before we continue the general treatment of CGC's, let us illustrate the preceding in the case $l_1 = l_2 = \frac{1}{2}$. We abbreviate $t_{11} := t^{(1/2)}_{-1/2,-1/2}, t_{12} := t^{(1/2)}_{-1/2,1/2}$, $t_{21} := t^{(1/2)}_{1/2,-1/2}$, $t_{(22)} := t^{1/2}_{1/2,1/2}$. In the above notation, we have

$$\mathbf{e}^1_1 = \mathbf{e}_{1/2} \otimes \mathbf{e}_{1/2}, \quad \mathbf{e}^1_{-1} = \mathbf{e}_{-1/2} \otimes \mathbf{e}_{-1/2},$$

$$\mathbf{e}^1_0 = (q^{-1/2}/[2]^{1/2})\mathbf{e}_{1/2} \otimes \mathbf{e}_{-1/2} + (q^{1/2}/[2]^{1/2})\mathbf{e}_{-1/2} \otimes \mathbf{e}_{1/2},$$

$$\mathbf{e}^0_0 = (q^{1/2}/[2]^{1/2})\mathbf{e}_{1/2} \otimes \mathbf{e}_{-1/2} - (q^{-1/2}/[2]^{1/2})\mathbf{e}_{-1/2} \otimes \mathbf{e}_{1/2}.$$

In order to prove these formulas, it suffices to check that the actions of the representations T_l, $l = 0$, 1, and $T_{1/2} \otimes T_{1/2}$ on both sides are the same. Comparing with (41), we see that

$$C_q\left(\tfrac{1}{2},\tfrac{1}{2},1;\tfrac{1}{2},-\tfrac{1}{2},0\right) = -C_q\left(\tfrac{1}{2},\tfrac{1}{2},0;-\tfrac{1}{2},\tfrac{1}{2},0\right) = q^{-1/2}/[2]^{1/2},$$

$$C_q\left(\tfrac{1}{2},\tfrac{1}{2},1;-\tfrac{1}{2},\tfrac{1}{2},0\right) = C_q\left(\tfrac{1}{2},\tfrac{1}{2},0;\tfrac{1}{2},-\tfrac{1}{2},0\right) = q^{1/2}/[2]^{1/2},$$

$$C_q\left(\tfrac{1}{2},\tfrac{1}{2},1;\tfrac{1}{2},\tfrac{1}{2},1\right) = C_q\left(\tfrac{1}{2},\tfrac{1}{2},0;-\tfrac{1}{2},-\tfrac{1}{2},-1\right) = 1.$$

Inserting this into (43), we obtain

$$t_{11}t_{22} = q(t^1_{00} + q^{-2}t^0_{00})/[2], \quad t_{12}t_{21} = (t^1_{00} - t^0_{00})/[2].$$

If we denote the unit element $t^0_{00} = \varepsilon$ of the dual algebra by 1, the two previous equations yield $t_{11}t_{22} - qt_{12}t_{21} = 1$. Proceeding in a similar manner we derive from (43) at $l_1 = l_2 = 1/2$ that

$$t_{11}t_{12} = qt_{12}t_{11}, \quad t_{11}t_{21} = qt_{21}t_{11}, \quad t_{12}t_{22} = qt_{22}t_{12}, \quad t_{21}t_{22} = qt_{22}t_{21}, \tag{44}$$

$$t_{12}t_{21} = t_{21}t_{12}, \quad t_{11}t_{22} - qt_{12}t_{21} = 1, \quad t_{22}t_{11} - q^{-1}t_{12}t_{21} = 1. \tag{45}$$

These relations define the coordinate algebra of the quantum group $SL_q(2)$ which will be extensively studied in the next chapter.

In the remainder of this chapter we assume that q is a real number such that $q \neq 0, \pm 1$ and we consider the compact real form $\breve{U}_q(\mathrm{su}_2)$ of the quantum algebra $\breve{U}_q(\mathrm{sl}_2)$. Then, as noted in Subsect. 3.2.1, any T_l is a $*$-representation of the $*$-algebra $\breve{U}_q(\mathrm{su}_2)$, where the scalar product of the representation space V_l is defined by requiring that the basis $\{\mathbf{e}_m\}$ is orthonormal. Since $\breve{U}_q(\mathrm{su}_2)$ is a Hopf $*$-algebra, $T_{l_1} \otimes T_{l_2}$ is also a $*$-representation of $\breve{U}_q(\mathrm{su}_2)$. Therefore, (37) gives an orthogonal decomposition of the Hilbert space $V_{l_1} \otimes V_{l_2}$ and we can choose the basis $\{\mathbf{e}^l_m\}$ of $\tilde{V}_l$ to be orthonormal. Hence we have

$$C_q(l_1, l_2, l; j, k, m) = \langle \mathbf{e}_j \otimes \mathbf{e}_k, \mathbf{e}^l_m \rangle.$$

The unitarity of the matrix C_q leads to the orthogonality relations for the CGC's:

$$\sum\nolimits_j C_q(l_1, l_2, l; j, m-j, m)\overline{C_q(l_1, l_2, l'; j, m-j, m)} = \delta_{ll'}, \tag{46}$$

$$\sum_l C_q(l_1,l_2,l;j,m-j,m)\overline{C_q(l_1,l_2,l;j',m-j',m)} = \delta_{jj'}. \tag{47}$$

As in the case of $SU(2)$, the CGC's for the quantum algebra $\breve{U}_q(\mathrm{su}_2)$ can be calculated by the method of highest weights. Applying the operator $T_l(E)$ to both sides of (41) at $m=l$ and taking into account that $T_l(E)\mathbf{e}_l^l = 0$ and

$$T_l(E) \sum_{j+k=m} C_q(\ell;\mathbf{j})(\mathbf{e}_j\otimes\mathbf{e}_k) = \sum_{j+k=m} C_q(\ell;\mathbf{j})T_l(E)(\mathbf{e}_j\otimes\mathbf{e}_k)$$

$$= \sum_{j+k=m} C_q(\ell;\mathbf{j})\{q^k(T_{l_1}(E)\mathbf{e}_j)\otimes\mathbf{e}_k + q^{-j}\mathbf{e}_j\otimes T_{l_2}(E)\mathbf{e}_k\},$$

we derive a recurrence relation for $C_q(\ell;\mathbf{j})$, $j+k=l$. After some simple calculations this leads to the solution

$$C_q(\ell;\mathbf{j}) = (-1)^{l_1-j} q^{-(j+1)(l_1-j)+j(l+1)/2}\left(\frac{[l_1+j]![l_2+k]!}{[l_1-j]![l_2-k]!}\right)^{1/2}\alpha,$$

where $j+k=l$ and α is a constant independent of j and k. By the orthogonality relation $\sum_{j+k=l}|C_q(\ell;\mathbf{j})|^2 = 1$, we find

$$\alpha^{-2} = \sum_{j+k=l} q^{-2(j+1)(l_1-j)+j(l+1)/2}\frac{[l_1+j]![l_2+k]!}{[l_1-j]![l_2-k]!}.$$

The right hand side can be summed by means of the basic hypergeometric functions ${}_2\varphi_1$. Indeed, by the relations (2.14) and (2.15), we have

$$\begin{aligned}\alpha^{-2} &= [2l_1]![l+l_2-l_1]![l_1+l_2-l]!^{-1}q^{2l_1(l+1)}\\ &\quad\times{}_2\Phi_1(l+l_2-l_1+1, l-l_1-l_2; -2l_1;\ q^2, q^{-2(2l+1)}),\end{aligned}$$

where the notation (2.35) is used. We apply the formula (2.33) to this hypergeometric series. As in the classical case (see [VK1], Chap. 8), the CGC's $C_q(\ell;\mathbf{j})$ are uniquely determined only up to a number $c(l)$ of modulus 1, which does not depend on j and k. Hence we can assume that $\alpha>0$. Thus,

$$\alpha^{-1} = q^{-(l_1+l_2-l)(l+l_2-l_1+1)}\left(\frac{[2l+1]![l_1+l_2-l]!q^{2l_1(l+1)}}{[l_1+l_2+l+1]![l+l_1-l_2]![l-l_1+l_2]!}\right)^{1/2}$$

and for $j+k=l$ we obtain

$$\begin{aligned}C_q(l_1,l_2,l;j,k,l) &= (-1)^{l_1-j}q^{\frac{1}{2}\{l_2(l_2+1)-l_1(l_1+1)-l(l+1)+2j(l+1)\}}\\ &\quad\times\left(\frac{[2l+1]![l_1+l_2-l]![l_1+j]![l_2+k]!}{[l_1+l_2+l+1]![l+l_1-l_2]![l-l_1+l_2]![l_1-j]![l_2-k]!}\right)^{1/2}.\end{aligned} \tag{48}$$

From the relation $T_l(E)^* = T_l(F)$ and the equality

$$T_l(F)^{l-m}\mathbf{e}_l^l = ([2l]![l-m]![l+m]!^{-1})^{1/2}\mathbf{e}_m^l$$

we derive

$$C_q(\ell;\mathbf{j}) = N_{l,m}\langle \mathbf{e}_j \otimes \mathbf{e}_k, T_l(F)^{l-m}\mathbf{e}_l^l\rangle = N_{l,m}\langle T_l(E)^{l-m}(\mathbf{e}_j\otimes\mathbf{e}_k), \mathbf{e}_l^l\rangle, \quad (49)$$

where $j+k=m$, $T_l(E) = T_{l_1}(E)\otimes T_{l_2}(K) + T_{l_1}(K^{-1})\otimes T_{l_2}(E)$ and $N_{l,m} = ([l+m]!)^{1/2}([2l]![l-m]!)^{-1/2}$. Setting $E_{l_i} := T_{l_i}(E)$ and $K_{l_i} := T_{l_i}(K)$, we have $(E_{l_1}\otimes K_{l_2})((K_{l_1})^{-1}\otimes E_{l_2}) = q^2((K_{l_1})^{-1}\otimes E_{l_2})(E_{l_1}\otimes K_{l_2})$. Therefore, the ooperator $E_l^{l-m} \equiv T_l(E)^{l-m}$ can be represented in the form

$$E_l^{l-m} = \sum_{k=0}^{l-m} \begin{bmatrix} l-m \\ k \end{bmatrix}_{q^{-2}} (E_{l_1}\otimes K_{l_2})^k \left((K_{l_1})^{-1}\otimes E_{l_2}\right)^{l-m-k} \quad (50)$$

by Proposition 2.2. Combining (49) and (50) we get

$$C_q(\ell,\mathbf{j}) = \sum_{r=0}^{l-m} q^{\beta}\frac{[l-m]!}{[r]![l-m-r]!}\left(\frac{[l_1-j]![l_2-k]!}{[l_1+j]![l_2+k]!}\frac{[l+m]!}{[2l]![l-m]!}\right)^{1/2}$$

$$\times\left(\frac{[l_2+l-j-r]![l_1+j+r]!}{[l_2-l+j+r]![l_1-j-r]!}\right)^{1/2} C_q(l_1,l_2,l;j+r,k+l-m-r,l),$$

where $j+k=m$ and $\beta = rk - j(l-m-r)$. From the latter equation and (48) we obtain the following expression for the CGC's of $\breve{U}_q(\mathrm{su}_2)$:

$$C_q(\ell,\mathbf{j}) = \frac{(-1)^{l_1-j}q^{\beta_1}([2l+1][\ell;\mathbf{j}])^{1/2}[l_1+l_2-l]!}{\Delta(\ell)[l_1+l_2+l+1]![l_1+j]![l_2+k]!}$$

$$\times\sum_r \frac{(-1)^r q^{r(l+m+1)}[l_1+j+r]![l_2+l-j-r]!}{[r]![l-m-r]![l_1-j-r]![l_2-l+j+r]!}, \quad (51)$$

where $m=j+k$, $\beta_1 = \frac{1}{2}\{l_2(l_2+1) - l_1(l_1+1) - l(l+1) + 2j(m+1)\}$,

$$\Delta(\ell) \equiv \Delta(l_1,l_2,l) = \left(\frac{[l_1+l_2-l]![l_1-l_2+l]![l-l_1+l_2]!}{[l_1+l_2+l+1]!}\right)^{1/2}, \quad (52)$$

$$[\ell;\mathbf{j}] = [l_1+j]![l_1-j]![l_2+k]![l_2-k]![l+m]![l-m]!, \quad (53)$$

and the summation is over all integral values of r for which the numbers in the q-factorials are nonnegative. Equation (51) is a q-analog of the Racah formula for the CGC's of the group $SU(2)$.

By (2.14) and (2.15), the sum in (51) reduces to the hypergeometric series ${}_3\varphi_2$. Using the notation ${}_3\varphi_2(q^a,q^b,q^c;q^d,q^e;\ q,x) := {}_3\Phi_2(a,b,c;d,e;\ q,x)$ we have

$$C_q(\ell;\mathbf{j}) = \frac{(-1)^{l_1-j}q^{\beta_1}\Delta(\ell)[l+l_2-j]!([\ell;\mathbf{j}][2l+1])^{1/2}}{[l_1-l_2+l]![l+l_2-l_1]!]l_2-l+j]![l_1-j]![l_2+k]![l-m]!}$$

$$\times{}_3\Phi_2\left(\begin{matrix} j-l_1,\ l_1+j+1,\ -l+m \\ l_2-l+j+1,\ -l-l_2+j\end{matrix};\ q^2,q^2\right), \quad (54)$$

where β_1 is as in (51).

3.4.3 Other Expressions for Clebsch–Gordan Coefficients

A number of other expressions for the CGC's of $\check{U}_q(\mathrm{su}_2)$ can be derived from (54) by means of the relations (2.36) and (2.37). Setting

$$n = l_1 - j,\ a = q^{2(j+l_1+1)},\ b = q^{2(j-l+k)},\ c = q^{2(j-l+l_2+1)},\ d = q^{2(j-l-l_2)}$$

in (2.36) it follows from (54) that

$$C_q(\ell;\mathbf{j}) = \frac{q^{\beta_2}\Delta(\ell)[l+l_2-j]!([\ell;\mathbf{j}][2l+1])^{1/2}}{[l-l_2+j]![l_1+l_2-l]!]l-l_1+l_2]![l_1-j]![l_2+k]![l-m]!} \times {}_3\Phi_2\left(\begin{matrix} j+l_1+1,\ j-l_1,\ -l_2-k \\ j-l-l_2,\ l-l_2+j+1 \end{matrix};\ q^2, q^{2(k-l_2)}\right), \tag{55}$$

where $\beta_2 = \frac{1}{2}\{l_1(l_1+1)+l_2(l_2+1)-l(l+1)+2jk\}$. Inserting here

$$n = l_1 - j,\ a = q^{-2(k+l_2)},\ b = q^{2(l-l_1-l_2)},\ c = q^{2(l-l_1-k+1)},\ d = q^{2(l-l_2+j+1)}$$

and using (2.36) we get a q-analog of the van der Waerden formula for the CGC's of $SU(2)$:

$$C_q(\ell;\mathbf{j}) = \frac{q^{\beta_3}\Delta(\ell)([\ell;\mathbf{j}][2l+1])^{1/2}}{[l-l_1-k]![l_1+l_2-l]!][l-l_2+j]![l_1-j]![l_2+k]!} \times {}_3\Phi_2\left(\begin{matrix} j-l_1,\ l-l_1-l_2,\ -l_2-k \\ l-l_2+j+1,\ l-l_1-k+1 \end{matrix};\ q^2, q^2\right), \tag{56}$$

where $\beta_3 = \frac{1}{2}\{(l_1+l_2-l)(l_1+l_2+l+1)+2(kl_1-jl_2)\}$. Putting here

$$n = l_2 + k,\ a = q^{2(l-l_1-l_2)},\ b = q^{2(j-l_1)},\ c = q^{2(l-l_1-k+1)},\ d = q^{2(l-l_2+j+1)},$$

and applying again (2.36) we obtain

$$C_q(\ell;\mathbf{j}) = \frac{q^{\beta_4}\Delta(\ell)[2l_2]!([\ell;\mathbf{j}][2l+1])^{1/2}}{[l-l_2+j]![l_1+l_2-l]!][l+l_2-l_1]![l_1-j]![l_2-k]![l_2+k]!} \times {}_3\Phi_2\left(\begin{matrix} -k-l_2,\ l-l_1-l_2,\ l+l_1-l_2+1 \\ l-l_2+j+1,\ \ -2l_2 \end{matrix};\ q^2, q^{2(m-l)}\right), \tag{57}$$

where $\beta_4 = \frac{1}{2}\{(l_1-l+l_2)(l_1+l-l_2+1)+2(kl-kl_2-jl_2)\}$. Setting

$$n = l_1 - j,\ a = q^{-2(l-m)},\ b = q^{2(l_1+j+1)},\ c = q^{2(l_2-l+j+1)},\ d = q^{2(j-l-l_2)}$$

into formula (54) and using relation (2.36) we derive a q-analog of the Jusys–Bandzaitis formula for the CGC's of $SU(2)$:

$$C_q(\ell;\mathbf{j}) = \frac{(-1)^{l_1-j}q^{\beta_5}[l+l_2-j]![l_1+l_2-m]![l_1+j]![l+m]!([2l+1])^{1/2}}{\Delta(\ell)([\ell;\mathbf{j}])^{1/2}[l_1+l_2+l+1]!]}$$

$$\times {}_3\Phi_2\left(\begin{matrix} m-l, & -l-l_1-l_2-1, & j-l_1 \\ -l-l_2+j, & -l_1-l_2+m & \end{matrix}; q^2, q^{2(l+l_1-l_2+1)}\right), \tag{58}$$

where $\beta_5 = \frac{1}{2}\{(l_2-l_1-l)(l_2+l+1)+2j(l+l_1+1)+2kl_1\}$. Letting

$$n = l+l_1+l_2+1,\ a = q^{2(m-l)},\ b = q^{2(l_1-l_2-l)},\ c = q^{-4l},\ d = q^{2(j-l-l_2)}$$

and applying (2.36) we derive

$$C_q(\ell;\mathbf{j}) = \frac{(-1)^{l_1-l+k}q^{\beta_6}[l+l_2-j]![l_1+l_2-l]![2l]![l_1+j]!([2l+1])^{1/2}}{\Delta(\ell)([\ell;\mathbf{j}])^{1/2}[l_1+l_2+l+1]!]}$$

$$\times {}_3\Phi_2\left(\begin{matrix} m-l, & -l-l_1-l_2-1, & l_1-l_2-l \\ & -l-l_2+j, & -2l \end{matrix}; q^2, q^2\right), \tag{59}$$

where $\beta_6 = \frac{1}{2}\{(l+l_2-l_1)(l+l_1+l_2+1)-2l_2m-2kl-2k\}$. If we set here

$$n = l+l_2-l_1,\ a = q^{-2(l_1+l_2+l+1)},\ b = q^{2(m-l)},\ c = q^{2(j-l-l_2)},\ d = q^{-4l}$$

and use (2.36), we get

$$C_q(\ell;\mathbf{j}) = \frac{(-1)^{l_2+k}q^{\beta_7}\Delta(\ell)[l+l_2+j]![2l]![l_1-j]!([2l+1])^{1/2}}{[l_1-l_2+l]![l_2-l_1+l]!([\ell;\mathbf{j}])^{1/2}}$$

$$\times {}_3\Phi_2\left(\begin{matrix} -m-l, & -l-l_1-l_2-1, & l_1-l_2-l \\ & -l-l_2-j, & -2l \end{matrix}; q^2, q^{2(l_2+k+1)}\right), \tag{60}$$

where $\beta_7 = \frac{1}{2}\{(l_1-l_2-l)(l_1+l_2+l+1)-2k(l+l_2+1)+2jl_2\}$. If we set

$$n = l+l_2-l_1,\ a = q^{-2(l+m)},\ b = q^{2(l_1-j+1)},\ c = q^{2(l_1-l_2-m+1)},\ d = q^{-2(l_2+l+j)}$$

and use again (2.36), we derive a q-analog of the Wigner formula for the CGC's of $SU(2)$:

$$C_q(\ell;\mathbf{j}) = \frac{(-1)^{l_2+k}q^{\beta_8}\Delta(\ell)[l+l_2+j]![l-m]![l_1-j]!([2l+1])^{1/2}}{[l_1-l_2-m]![l_2-l_1+l]!([\ell;\mathbf{j}])^{1/2}}$$

$$\times {}_3\Phi_2\left(\begin{matrix} -m-l, & l_1-j+1, & l_1-l_2-l \\ -l-l_2-j, & l_1-l_2-m+1 & \end{matrix}; q^2, q^2\right), \tag{61}$$

where $\beta_8 = \frac{1}{2}\{(l_1-l_2-l)(l_1+l_2-l+1)+2j(l-l_1)-2k(l_1+1)\}$. Setting

$$n = l+m,\ a = q^{2(l_1-l-l_2)},\ b = q^{2(l_1+l_2-l+1)},\ c = q^{2(l_1-l-k+1)},\ d = q^{-4l}$$

into (60), we obtain a q-analog of the Majumdar formula:

$$C_q(\ell;\mathbf{j}) = \frac{(-1)^{l_2+k}q^{\beta_9}\Delta(\ell)[2l]!([\ell;\mathbf{j}][2l+1])^{1/2}}{[l_1-l_2+l]![l-l_1+l_2]![l_1-l-k]![l-m]![l+m]![l_2+k]!}$$

$$\times {}_3\Phi_2\left(\begin{matrix} -m-l, & l_1+l_2-l+1, & l_1-l_2-l \\ & l_1-l-k+1, & -2l \end{matrix}; q^2, q^2\right), \tag{62}$$

where $\beta_9 = \frac{1}{2}\{(l_1 - l_2 - l)(l_1 + l_2 - l + 1) + 2j(l - l_1) - 2k(l_1 + 1)\}$.

Using the relations (2.14) and (2.15), formula (56) can be rewritten as

$$C_q(\ell; \mathbf{j}) = q^{\beta_3} \Delta(\ell)([\ell; \mathbf{j}][2l + 1])^{1/2}$$
$$\times \sum_r \frac{(-1)^r q^{-r(l_1+l_2+l+1)} [l - l_1 - k + r]!^{-1}}{[r]![l_1 + l_2 - l - r]![l_1 - j - r]![l_2 + k - r]![l - l_2 + j + r]!}. \tag{63}$$

In a similar manner, formula (58) can be expressed as

$$C_q(\ell; \mathbf{j}) = \frac{(-1)^{l_1 - j} q^{\beta_5}}{\Delta(\ell)} \frac{([\ell; \mathbf{j}][2l + 1])^{1/2}}{[l_2 - k]![l_2 + k]!}$$
$$\times \sum_r \frac{(-1)^r q^{r(l_1 - l_2 + l)} [l_1 + l_2 - m - r]![l_2 + l - j - r]!}{[r]![l_1 + l_2 + l - r + 1]![l_1 - j - r]![l - m - r]!}. \tag{64}$$

For formula (60) we obtain the expression

$$C_q(\ell; \mathbf{j}) = (-1)^{l_2 + k} q^{\beta_7} \Delta(\ell) \frac{[l_1 - j]![l + m]![l_1 + l_2 + l + 1]!}{[l_1 - l_2 + 1]!} \left(\frac{[2l + 1]}{([\ell; \mathbf{j}]} \right)^{1/2}$$
$$\times \sum_r \frac{(-1)^r q^{r(l_2 + k)} [2l - r]![l_2 + l + j - r]!}{[r]![l + l_2 - l_1 - r]![l_1 + l_2 + l - r + 1]![l + m - r]!}. \tag{65}$$

The q-analog of the Wigner formula (61) can be written as

$$C_q(\ell; \mathbf{j}) = (-1)^{l_2 + k} q^{\beta_8} \Delta(\ell)[l - m]![l + m]! \left(\frac{[2l + 1]}{([\ell; \mathbf{j}]} \right)^{1/2}$$
$$\times \sum_r \frac{(-1)^r q^{r(l_1 + l_2 - l + 1)} [l_1 - j + r]![l_2 + l + j - r]!}{[r]![l + l_2 - l_1 - r]![l_1 - l_2 - m + r]![l + m - r]!}. \tag{66}$$

Formula (62) can be represented as

$$C_q(\ell; \mathbf{j}) = (-1)^{l_2 + k} q^{\beta_9} \frac{\Delta(\ell)([2l + 1]([\ell; \mathbf{j}])^{1/2}}{[l_1 - l_2 + l]![l_1 + l_2 - l]![l_2 + k]![l - m]!}$$
$$\times \sum_r \frac{(-1)^r q^{r(l_1 - j + 2)} [2l - r]![l_1 + l_2 - l + r]!}{[r]![l + l_2 - l_1 - r]![l_1 - l - k + r]![l + m - r]!}. \tag{67}$$

The expressions for β_s in formulas (63)–(67) are the same as in (56), (58), (60)–(62), respectively.

Formula (54) for $l = 0$ gives

$$C_q(l_1, l_2, 0; j, -j, 0) = (-1)^{l_1 - j} q^j [2l_1 + 1]^{-1/2} \delta_{l_1 l_2}.$$

For $l = l_1 + l_2$, (56) yields

$$C_q(l_1, l_2, l_1 + l_2; j, k, m) = q^{(kl_1 - jl_2)}$$

$$\times \left(\frac{[2l_1]![2l_2]![l_1+l_2-m]![l_1+l_2+m]!}{[2l_1+2l_2]![l_1-j]![l_1+j]![l_2-k]![l_2+k]!} \right)^{1/2}$$

where $m = j + k$. Setting $l = l_1 - l_2$ into (61) we get

$$C_q(l_1, l_2, l_1 - l_2; j, k, m) = (-1)^{l_2+k} q^{-(kl_1+jl_2+k)}$$

$$\times \left(\frac{[2l_2]![2l_1-2l_2+1]![l_1-j]![l_1+j]!}{[2l_1+1]![l_2-k]![l_2+k]![l_1-l_2+m]![l_1-l_2-m]!} \right)^{1/2},$$

where $m = j + k$. From the last two formulas we easily derive the expressions for the CGC's of the tensor product $T_l \otimes T_{1/2}$:

$$C_q(l, \tfrac{1}{2}, l + \tfrac{1}{2}; j, \pm\tfrac{1}{2}, j \pm \tfrac{1}{2}) = q^{\pm(l \mp j)/2}([l \pm j + \tfrac{1}{2}]/[2l+1])^{1/2}, \qquad (68)$$

$$C_q(l, \tfrac{1}{2}, l - \tfrac{1}{2}; j, \pm\tfrac{1}{2}, j \pm \tfrac{1}{2}) = \mp q^{\mp(l \pm j + 1)/2}([l \mp j]/[2l+1])^{1/2}. \qquad (69)$$

3.4.4 Symmetries of Clebsch–Gordan Coefficients

From formula (56) it follows that

$$C_q(\ell, \mathbf{j}) \equiv C_q(l_1, l_2, l; j, k, m) = C_q(l_2, l_1, l; -k, -j, -m). \qquad (70)$$

Equation (58) leads to the formula for permutation symmetry of the CGC's:

$$C_q(\ell, \mathbf{j}) = (-1)^{l-l_1-k} q^{-k} \left(\frac{[2l+1]}{[2l_1+1]} \right)^{1/2} C_q(l, l_2, l_1; m, -k, j). \qquad (71)$$

The expression (56) is symmetric with respect to the transformation

$$C_q(\ell, \mathbf{j}) \to C_q \left(\frac{l^+ + m}{2}, \frac{l^+ - m}{2}, l; \frac{l^- + j - k}{2}, \frac{l^- - j + k}{2}, l_1 - l_2 \right), \qquad (72)$$

where $l^+ = l_1 + l_2$ and $l^- = l_1 - l_2$. This is the q-analog of Regge's symmetry for the CGC's of the group $SU(2)$. Using (2.13), we derive from (51) that

$$C_q(\ell, \mathbf{j}) = (-1)^{l_1+l_2-l} C_{q^{-1}}(l_1, l_2, l; -j, -k, -m). \qquad (73)$$

This formula relates the CGC's of the algebras $\breve{U}_q(\mathrm{su}_2)$ and $\breve{U}_{q^{-1}}(\mathrm{su}_2)$.

The classical analogs of the relations (70)–(73) generate a symmetry group for the CGC's of $SU(2)$ which contains 72 elements (see Subsect. 8.2.2 in [VK1]). The relations (70)–(73) for $\breve{U}_q(\mathrm{su}_2)$ generate a symmetry group for the quantum CGC's also containing 72 elements, but now relation (73) changes the parameter q to q^{-1}.

Along with the above mentioned symmetries, there exist symmetries of the expressions for the CGC's related to the transition $l \to \bar{l} \equiv -l - 1$. This substitution leads to expressions of the type $[-k]!/[-m]!$, where $k, m \in \mathbb{N}$. By (2.64), such expressions have to be replaced by $(-1)^{k-m}[m+1]!/[k+1]!$. We collect here only a few of these symmetries:

$$\begin{aligned} C_q(l_1,l_2,l;j,k,m) &= (-1)^{l_2+k}C_q(\bar{l}_1,l_2,\bar{l};j,k,m) \\ &= (-1)^{l_1-j}C_q(l_1,\bar{l}_2,\bar{l};j,k,m) \\ &= (-1)^{l_2-l-j}C_q(\bar{l}_1,l_2,l;j,k,m) \\ &= (-1)^{l_1+l_2-l}C_q(\bar{l}_1,\bar{l}_2,\bar{l};j,k,m). \end{aligned}$$

3.5 Racah Coefficients and $6j$ Symbols of $U_q(\mathrm{su}_2)$

3.5.1 Definition of the Racah Coefficients

Consider three irreducible representations T_{l_1}, T_{l_2}, T_{l_3} of $\breve{U}_q(\mathrm{su}_2)$. Let $\{\mathbf{e}_j\}$, $\{\mathbf{f}_k\}$, $\{\mathbf{h}_m\}$ denote the orthonormal bases of their respective representation spaces such that the formulas (22) and (23) hold. The tensor product of these three representations is associative, that is, under the canonical identification of the vector spaces $(V_{l_1}\otimes V_{l_2})\otimes V_{l_3}$ and $V_{l_1}\otimes(V_{l_2}\otimes V_{l_3})$ we have

$$(T_{l_1}\otimes T_{l_2})\otimes T_{l_3} = T_{l_1}\otimes(T_{l_2}\otimes T_{l_3}). \tag{74}$$

Proposition 22 yields the decompositions

$$T_{l_1}\otimes T_{l_2} = \bigoplus_{l_{12}=|l_1-l_2|}^{l_1+l_2} T_{l_{12}}, \qquad T_{l_2}\otimes T_{l_3} = \bigoplus_{l_{23}=|l_2-l_3|}^{l_2+l_3} T_{l_{23}}. \tag{75}$$

Inserting the latter into (74), we get

$$\bigoplus_{l_{12}=|l_1-l_2|}^{l_1+l_2} (T_{l_{12}}\otimes T_{l_3}) = \bigoplus_{l_{23}=|l_2-l_3|}^{l_2+l_3} (T_{l_1}\otimes T_{l_{23}}).$$

The carrier spaces of the representations $T_{l_{12}}$ and $T_{l_{23}}$ in (75) have orthonormal bases consisting of the elements

$$\mathbf{a}_m^{l_{12}} = \sum_{i=-l_1}^{l_1}\sum_{j=-l_2}^{l_2} C^{l_1 l_2 l_{12}}_{i\,j\,m}(\mathbf{e}_i\otimes\mathbf{f}_j), \quad \mathbf{b}_n^{l_{23}} = \sum_{j=-l_2}^{l_2}\sum_{n=-l_3}^{l_3} C^{l_2 l_3 l_{23}}_{j\,k\,n}(\mathbf{f}_j\otimes\mathbf{h}_k),$$

respectively, where $C^{abc}_{\alpha\beta\gamma} := C_q(a,b,c;\alpha,\beta,\gamma)$ denote the CGC's of $\breve{U}_q(\mathrm{su}_2)$. By Proposition 22, we have

$$T_{l_{12}}\otimes T_{l_3} = \bigoplus\nolimits_{l=|l_{12}-l_3|}^{l_{12}+l_3} T_l. \tag{76}$$

Let $\{\mathbf{e}_p^{l_1 l_2(l_{12}),l_3,l}\}$ be an orthonormal basis of the carrier space of the subrepresentation T_l in (76) such that the formulas (22) and (23) hold. Note that the upper indices indicate the order of the tensor products: first the tensor product $T_{l_1}\otimes T_{l_2}$ is formed and then the components $T_{l_{12}}$ of $T_{l_1}\otimes T_{l_2}$ are tensored by T_{l_3}. We have

$$\mathbf{e}_p^{l_1l_2(l_{12}),l_3,l} = \sum_m C^{l_{12}l_3l}_{m\ k\ p}(\mathbf{a}^{l_{12}}_m \otimes \mathbf{h}_k) = \sum_{i,j,k} C^{l_1l_2l_{12}}_{i\ j\ m} C^{l_{12}l_3l}_{m\ k\ p}(\mathbf{e}_i \otimes \mathbf{f}_j) \otimes \mathbf{h}_k, \quad (77)$$

where $i+j=m$, $m+k=p$.

Analogously, if $\{\mathbf{e}_p^{l_1,l_2l_3(l_{23}),l}\}$ is the corresponding orthonormal basis of the carrier space of T_l in the decomposition $T_{l_1} \otimes T_{l_{23}} = \bigoplus_{l=|l_1-l_{23}|}^{l_1+l_{23}} T_l$, then we have

$$\mathbf{e}_p^{l_1,l_2l_3(l_{23}),l} = \sum\nolimits_{i,j,k} C^{l_2l_3l_{23}}_{j\ k\ n} C^{l_1l_{23}l}_{i\ n\ p} \mathbf{e}_i \otimes (\mathbf{f}_j \otimes \mathbf{h}_k), \quad (78)$$

where $j+k=n$ and $i+n=p$. Since the vectors $\mathbf{e}_p^{l_1l_2(l_{12}),l_3,l}$ with

$$|l_1-l_2| \le l_{12} \le l_1+l_2,\ |l_{12}-l_3| \le l \le l_{12}+l_3,\ p=-l,-l+1,\cdots,l,$$

as well as the vectors $\mathbf{e}_p^{l_1,l_2l_3(l_{23}),l}$ with

$$|l_1-l_{23}| \le l \le l_1+l_{23},\ |l_2-l_3| \le l_{23} \le l_2+l_3,\ p=-l,-l+1,\cdots,l,$$

are orthonormal bases of the space $V_{l_1} \otimes V_{l_2} \otimes V_{l_3}$, by Schur's lemma there is a unitary matrix R with entries $R_q(l_1l_2l_3, l_{12}l_{23}, l)$ such that

$$\mathbf{e}_p^{l_1l_2(l_{12}),l_3,l} = \sum\nolimits_{l_{23}} R_q(l_1l_2l_3, l_{12}l_{23}, l)\mathbf{e}_p^{l_1,l_2l_3(l_{23}),l}. \quad (79)$$

The complex numbers $R_q(l_1l_2l_3, l_{12}l_{23}, l)$ are called the *Racah coefficients* of the quantum algebra $\breve{U}_q(\mathrm{su}_2)$ or the *q-Racah coefficients*. Note that the matrix R is block-diagonal and that the numbers $R_q(l_1l_2l_3, l_{12}l_{23}, l)$ do not depend on the indices i, j, k, m, n, p of basis elements in (77) and (78). The unitarity of the matrix R leads to the orthogonality relations

$$\sum\nolimits_{l_{23}} R_q(l_1l_2l_3, l_{12}l_{23}, l)\overline{R_q(l_1l_2l_3, l'_{12}l_{23}, l)} = \delta_{l_{12}l'_{12}}, \quad (80)$$

$$\sum\nolimits_{l_{12}} R_q(l_1l_2l_3, l_{12}l_{23}, l)\overline{R_q(l_1l_2l_3, l_{12}l'_{23}, l)} = \delta_{l_{23}l'_{23}}. \quad (81)$$

By the orthogonality relations for the CGC's, (78) implies that

$$\mathbf{e}_i \otimes (\mathbf{f}_j \otimes \mathbf{h}_k) = \sum_{l_{23}} \sum_l \overline{C^{l_2l_3l_{23}}_{j\ k\ n} C^{l_1l_{23}l}_{i\ n\ p}} \mathbf{e}_p^{l_1,l_2l_3(l_{23}),l}. \quad (82)$$

Since $(\mathbf{e}_i \otimes \mathbf{f}_j) \otimes \mathbf{h}_k = \mathbf{e}_i \otimes (\mathbf{f}_j \otimes \mathbf{h}_k)$, (82) can be inserted into (77). Comparing with (79) we obtain

$$R_q(l_1l_2l_3, l_{12}l_{23}, l) = \sum_{i=-l_1}^{l_1} \sum_{j=-l_2}^{l_2} \sum_{k=-l_3}^{l_3} C^{l_1l_2l_{12}}_{i\ j\ k} C^{l_{12}l_3l}_{m\ k\ p} \overline{C^{l_2l_3l_{23}}_{j\ k\ n} C^{l_1l_{23}l}_{i\ n\ p}}, \quad (83)$$

where $i+j=m$, $m+k=p$, $j+k=n$, $i+n=p$. Note that (83) holds for each value of the index p. We see from (83) that the Racah coefficients are real if the CGC's are real.

Recall that the CGC $C^{l_1l_2l}_{i,j,i+j}$ vanishes if the *triangle condition* $|l_1-l_2| \le l \le l_1+l_2$ does not hold. Therefore, by (83), $R_q(l_1l_2l_2, l_{12}l_{23}, l) = 0$ if at least one of the triples (l_1, l_2, l_{12}), (l_{12}, l_3, l), (l_2, l_3, l_{23}) and (l_1, l_{23}, l) does not satisfy this condition.

3.5.2 Relations Between Racah and Clebsch–Gordan Coefficients

In what follows we assume that the CGC's and the Racah coefficients are real. The converse relation to (77) has the form

$$(\mathbf{e}_i \otimes \mathbf{f}_j) \otimes \mathbf{h}_k = \sum_{l_{12}=|l_1-l_2|}^{l_1+l_2} \sum_{l=|l_{12}-l_3|}^{l_{12}+l_3} C_{i\ j\ m}^{l_1 l_2 l_{12}} C_{m\ k\ p}^{l_{12} l_3 l} \mathbf{e}_p^{l_1 l_2 (l_{12}), l_3, l}.$$

We now insert (79) into this equation. Comparing with the right hand side of (82) and equating the coefficients of $\mathbf{e}_p^{l_1, l_2 l_3 (l_{23}), l}$, we obtain

$$C_{j,k,j+k}^{l_2 l_3 l_{23}} C_{i,j+k,i+j+k}^{l_1 l_{23} l} = \sum_{l_{12}=|l_1-l_2|}^{l_1+l_2} R_q(l_1 l_2 l_3, l_{12} l_{23}, l) C_{i,j,i+j}^{l_1 l_2 l_{12}} C_{i+j,k,i+j+k}^{l_{12} l_3 l}.$$

Using the orthogonality relation (46) for the CGC's one derives

$$\sum_{i,j} C_{i,j,i+j}^{l_1 l_2 l_{12}} C_{j,k,j+k}^{l_2 l_3 l_{23}} C_{i,j+k,i+j+k}^{l_1 l_{23} l} = R_q(l_1 l_2 l_3, l_{12} l_{23}, l) C_{i+j,k,i+j+k}^{l_{12} l_3 l}, \tag{84}$$

where the summations are over the values of i and j such that $i + j = \text{const}$.

3.5.3 Symmetry Relations

It is more convenient to formulate symmetry properties in terms of the Wigner $6j$ symbols which are defined by

$$\begin{Bmatrix} l_1 & l_2 & l_{12} \\ l_3 & l & l_{23} \end{Bmatrix}_q$$

$$= (-1)^{l_1+l_2+l_3+l}([2l_{12}+1][2l_{23}+1])^{-1/2}\, R_q(l_1 l_2 l_3, l_{12} l_{23}, l). \tag{85}$$

As in the classical case, symmetry relations for the Racah coefficients follow from those for the CGC's by means of (83). For this reason, we give these symmetries without proof.

The $6j$ symbol $\begin{Bmatrix} l_1 & l_2 & l_{12} \\ l_3 & l & l_{23} \end{Bmatrix}_q$ is invariant under all permutations of its columns as well as under simultaneous permutations of l_1 and l_3, l_2 and l. Further, the following q-analog of the Regge symmetry holds:

$$\begin{Bmatrix} l_1 & l_2 & l_{12} \\ l_3 & l & l_{23} \end{Bmatrix}_q = \begin{Bmatrix} l_1 & s-l_2 & s-l_{12} \\ l_3 & s-l & s-l_{23} \end{Bmatrix}_q,$$

where $s = \frac{1}{2}(l + l_{12} + l_2 + l_{23})$. The described symmetries generate the symmetry group of the Wigner $6j$ symbols. This group contains 144 elements.

Besides, as follows from (73), we have a new symmetry relation replacing q by q^{-1}:

$$\begin{Bmatrix} l_1 & l_2 & l_{12} \\ l_3 & l & l_{23} \end{Bmatrix}_q = \begin{Bmatrix} l_1 & l_2 & l_{12} \\ l_3 & l & l_{23} \end{Bmatrix}_{q^{-1}} .$$

As noted in Subsect. 3.4.4, the expressions for the CGC's possess symmetry properties related to the transitions $l \to \bar{l} := -l-1$. These relations and formula (83) lead to symmetries of expressions for Racah coefficients. Some of them are given by the formulas

$$\begin{aligned} \begin{Bmatrix} l_1 & l_2 & l_{12} \\ l_3 & l & l_{23} \end{Bmatrix}_q &= (-1)^m \begin{Bmatrix} \bar{l}_1 & l_2 & l_{12} \\ l_3 & l & l_{23} \end{Bmatrix}_q = (-1)^n \begin{Bmatrix} \bar{l}_1 & l_2 & l_{12} \\ \bar{l}_3 & l & l_{23} \end{Bmatrix}_q \\ &= (-1)^{2(l_{12}+l_{23})} \begin{Bmatrix} l_1 & l_2 & \bar{l}_{12} \\ l_3 & l & \bar{l}_{23} \end{Bmatrix}_q , \end{aligned}$$

where $m = l_2 - l_{12} + l_{23} - l$ and $n = 2(l_1 + l_3)$.

3.5.4 Calculation of Racah Coefficients

The Racah coefficients can be computed by formula (84). Setting $i + j = l_{12}$ and $k = l - l_{12}$ therein we obtain

$$R_q(l_1 l_2 l_3, l_{12} l_{23}, l) = \left(C^{l_{12} l_3 l}_{l_{12}, l-l_{12}, l} \right)^{-1} \times \sum_{i,j} C^{l_1 l_2 l_{12}}_{i,j,l_{12}} C^{l_2 l_3 l_{23}}_{j,l-l_{12},j+l-l_{12}} C^{l_1 l_{23} l}_{i,j+l-l_{12},l} . \quad (86)$$

Utilizing here the expression

$$C^{l_1 l_2 l}_{j\ k\ l} = (-1)^{l_1 - j} q^{\alpha} \frac{\Delta(l_1, l_2, l)}{[l_1 - l_2 + l]!} \left(\frac{[l_1 + j]![l_2 + k]![2l+1]}{[l_2 - l_1 + l]!^2 [l_1 - j]![l_2 - k]!} \right)^{1/2}$$

from Subsect. 3.4.2, where $\alpha = \frac{1}{2}\{l_2(l_2+1) - l_1(l_1+1) - l(l+1) + 2j(l+1)\}$ and $j + k = l$, we get

$$\begin{aligned} \begin{Bmatrix} l_1 & l_2 & l_{12} \\ l_3 & l & l_{23} \end{Bmatrix}_q &= \gamma q^m \sum_i q^{i(l+l_{12}+2)} \frac{[l_1 + i]!}{[l_1 - i]!} \\ &\times \left(\frac{[l + l_{23} - i]![l_2 + l_{12} - i]!}{[l_{23} - l + i]![l_2 - l_{12} + i]!} \right)^{1/2} C^{l_2 l_3 l_{23}}_{l_{12}-i, l-l_{12}, l-i} , \end{aligned} \quad (87)$$

where $m = \frac{1}{2}\{c(l_2) + c(l_{23}) - c(l_3)\} - c(l_1) - l_{12}(l+1)$, $c(l) := l(l+1)$, and

$$\begin{aligned} \gamma = (-1)^{l - l_1 + l_2 + l_3} \Bigg(& \frac{[l + l_3 + l_{12} + 1]![l_1 + l_2 - l_{12}]![l_1 + l_{23} - l]!}{[l_1 + l_2 + l_{12} + 1]![l_1 + l + l_{23} + 1]![l_1 - l_2 + l_{12}]!} \\ & \times \frac{[l - l_3 + l_{12}]![2l_{23} + 1]^{-1}}{[l_2 - l_1 + l_{12}]![l_1 + l - l_{23}]![l - l_1 + l_{23}]!} \Bigg)^{1/2} . \end{aligned}$$

Substituting the expression (56) for the CGC's into (87) we derive

$$\begin{Bmatrix} l_1 & l_2 & l_{12} \\ l_3 & l & l_{23} \end{Bmatrix}_q = \frac{(-1)^{l-l_1+2l_2+l_3-l_{12}}\Delta(l_1,l_2,l_{12})\Delta(l_1,l_{23},l)}{[l+l_3-l_{12}]![l_1-l_2+l_{12}]![l_2-l_1+l_{12}]![l_1-l_{23}+l]!}$$

$$\times\frac{\Delta(l_2,l_3,l_{23})\Delta(l_{12},l_3,l)[l_{12}+l_3+l+1]!}{[l_{23}-l_1+l]![l_2-l_3+l_{23}]![l_3-l_2+l_{23}]!}\sum_{i,r}\frac{(-1)^{i+r}q^{\nu}[l_1+i]!}{[r]![l_1-i]!}$$

$$\times\frac{[l_2+l_{12}-i+r]![l_{23}+l-i]![l_3+l_{23}-l_{12}+i-r]}{[l_2-l_{12}+i-r]![l_3+l_{12}-l_{23}-i+r]![l_{23}-l+i-r]!},$$

where $\nu = c(i)+r(l-i+l_{23}+1)-c(l_1)$. Now we change the summation from r and i to $k=i-r$ and i. The sum over i reduces to ${}_2\Phi_1(-l_1+k,\ l_1+k+1;\ -l_{23}-l+k;\ q^2,q^2)$ and can be summed by using formula (2.34). Putting $s=l_2-l_{12}+k$ we get

$$\begin{Bmatrix} l_1 & l_2 & l_{12} \\ l_3 & l & l_{23} \end{Bmatrix}_q = \frac{(-1)^{a}\Delta(l_1,l_2,l_{12})\Delta(l_1,l_{23},l)\Delta(l_2;l_3,l_{23})\Delta(l_{12},l_3,l)}{[l+l_3-l_{12}]![l_1-l_2+l_{12}]![l_2-l_1+l_{12}]![l_1-l_{23}+l]!}$$

$$\times\frac{[l_{12}+l_3+l+1]![l_1+l+l_{23}+1]!}{[l_2-l_3+l_{23}]![l_3-l_2+l_{23}]!}\sum_{s}\frac{(-1)^{s}[2l_2-s]![l_1-l_2+l_{12}+s]!}{[s]![l_1+l_2-l_{12}-s]!}$$

$$\times\frac{[l_3+l_{23}-l_2+s]![l_2+l_3-l_{23}-s]!^{-1}}{[l_{23}+l_{12}-l_2-l+s]![l_{23}+l_{12}-l_2+l+s+1]!}, \tag{88}$$

where $a=l-l_1+l_2+l_3$ and the summation is over all integral values of s for which the numbers in the q-factorials are nonnegative. The sum over s can be expressed in terms of the hypergeometric series ${}_4\varphi_3(\cdots;\ q^2,q^2)$. Using first the symmetry relation

$$\begin{Bmatrix} l_1 & l_2 & l_{12} \\ l_3 & l & l_{23} \end{Bmatrix}_q = (-1)^{2(l_{12}+l_{23})}\begin{Bmatrix} l_1 & l_2 & \bar{l}_{12} \\ l_3 & l & \bar{l}_{23} \end{Bmatrix}_q,$$

where $\bar{l}=-l-1$, and applying formula (2.23) for ${}_4\varphi_3(\cdots;\ q,q)$ we obtain

$$\begin{Bmatrix} l_1 & l_2 & l_{12} \\ l_3 & l & l_{23} \end{Bmatrix}_q = \frac{(-1)^{l+l_2+l_{12}+l_{23}}\Delta(l_1,l_2,l_{12})\Delta(l_1,l_{23},l)\Delta(l_2;l_3,l_{23})}{[l-l_3+l_{12}]![l_2-l_1+l_{12}]![l_1+l_2-l_{12}]![l_{12}-l+l_3]!}$$

$$\times\frac{\Delta(l_{12},l_3,l)[2l_2]![l_2+l_{12}-l+l_{23}]![l_2+l+l_{12}+l_{23}+1]!}{[l_1-l+l_{23}]![l-l_1+l_{23}]![l_2+l_3-l_{23}]![l_2-l_3+l_{23}]!}\times$$

$${}_4\Phi_3\left(\begin{matrix} l_1-l_2-l_{12}, l_3-l_2-l_{23}, -l_1-l_2-l_{12}-1, -l_2-l_3-l_{23}-1 \\ -2l_2,\ -l_2-l_{12}+l-l_{23},\ -l_2-l_{12}-l-l_{23}-1 \end{matrix};q^2,q^2\right) \tag{89}$$

where ${}_4\Phi_3$ is related to the function ${}_4\varphi_3$ as ${}_3\Phi_2$ is to ${}_3\varphi_2$ (see Subsect. 3.4.3).

Other expressions for the Racah coefficients of $U_q(\mathrm{su}_2)$ can be derived from (88) and (89) by symmetry relations. Symmetries of expressions for

Racah coefficients are equivalent to relation (2.38) for the basic hypergeometric function ${}_4\varphi_3$. Applying (2.38) to the right hand side of (89), we obtain distinct expressions for the Racah coefficients of $\breve{U}_q(\mathrm{su}_2)$ in terms of ${}_4\Phi_3$. In particular, in this way we get q-analogs of all known expressions for the Racah coefficients of the group $SU(2)$. *All these q-analogs are obtained from the corresponding classical expressions by replacing the factorials m! by the q-factorials [m]! and, if the Racah coefficients are expressed in terms of hypergeometric series, the function*

$$ {}_4F_3(a,b,c,d;\ e,f,g;\ 1) \quad by \quad {}_4\Phi_3(a,b,c,d;\ e,f,g;\ q^2,q^2). $$

Next we give two other relations, which are analogous to (87). Putting $k = l_3$, $i + j + k = l$ in (86) we find

$$ \begin{Bmatrix} l_1 & l_2 & l_{12} \\ l_3 & l & l_{23} \end{Bmatrix}_q = (-1)^{l_{12}+l_3+l_{23}} q^s $$

$$ \times \frac{([l+l_3-l_{12}![l+l_3+l_{12}+1]!)^{1/2}}{\Delta(l_1,l,l_{23})\Delta(l_2,l_3,l_{23})[2l_{12}+1]!^{1/2}} \sum_m (-1)^m q^{m(l+l_3+1)/2} $$

$$ \times \frac{[l+l_{23}+m]!}{l_{23}-l+m]!} \left(\frac{l_1+m]![l_2-l+l_3+m]!}{l_1-m]![l_2+l-l_3-m]!} \right)^{1/2} C^{l_2\ l_2\ l_{12}}_{m,l-l_3-m,l-l_3}, \tag{90} $$

where $s = \frac{1}{2}\{l_{12}(l_{12}+1) - l_1(l_1+1) - (l_2(l_2+1) - 2l(l+1) + 2l_{23}(l_{23}+1)\}$. Another such relation is

$$ \begin{Bmatrix} l_1 & l_2 & l_{12} \\ l_3 & l & l_{23} \end{Bmatrix}_q = q^s \frac{(-1)^{l+l_1+l_2+l_3} \Delta(l_1,l,l_{23})\Delta(l_2,l_3,l_{23})}{([l+l_3-l_{12}]![l+l_3+l_{12}+1]![2l_{12}+1])^{1/2}} $$

$$ \times \sum_m \frac{q^{-m(l+l_3+1)/2}([l_1+m]![l_2+l-l_3+m]!)^{1/2}[l-l_{23}+m]!^{-1}}{[l+l_{23}+m+1]!([l_1-m]![l_2-l+l_3-m]!)^{1/2}} C^{l_1\ l_2\ l_{12}}_{m,j,l_3-l}, \tag{91} $$

where $j = l_3 - l - m$ and s is as in (90).

We close this subsection by giving the Racah coefficients for some special values of indices. If $l_{12} = 0$ in (89), then the triangular conditions yield $l_1 = l_2$ and $l_1 - l_2 - l_{12} = 0$, so we get

$$ \begin{Bmatrix} l_1 & l_2 & 0 \\ l_3 & l & l_{23} \end{Bmatrix}_q = \frac{(-1)^{l+l_1+l_{23}}}{([2l_1+1][2l_3+1])^{1/2}} \delta_{l_1 l_2} \delta_{l l_3}. $$

From (89) and the symmetry relations, for $l_1 + l_2 = l_{12}$ we derive

$$ \begin{Bmatrix} a & b & a+b \\ d & e & f \end{Bmatrix}_q = (-1)^s \left(\frac{[2a]![2b]![a+b+d+e+1]![a+b-d+e]!}{[2a+2b+1]![d+e-a-b]![a+e-f]!} \right. $$

$$ \left. \times \frac{[a+b+d-e]![e+f-a]![d+f-b]![a-e+f]!^{-1}}{[a+e+f+1]![b+d-f]![b-d+f]![b+d+f+1]!} \right)^{1/2}, $$

where $s = a+b+d+e$.

For the Racah coefficients with $l_3 = \frac{1}{2}$ we have the expressions

$$\begin{Bmatrix} l_1 & l_2 & l \\ 1/2 & l+1/2 & l_2 \pm 1/2 \end{Bmatrix}_q$$

$$= (-1)^{l_1+l_2+l+\frac{1}{2}\pm\frac{1}{2}} \left(\frac{[l_1+l_2 \pm l+1 \pm 1][\mp l_1 \pm l_2 + l + 1]}{[2l_2+1][2l+1][2l_2+1\pm 1][2l+2]} \right)^{1/2}, \tag{92}$$

$$\begin{Bmatrix} l_1 & l_2 & l \\ 1/2 & l-1/2 & l_2 \pm 1/2 \end{Bmatrix}_q$$

$$= (-1)^{l_1+l_2+l} \left(\frac{[l_1+l_2 \mp l+1][\pm l_1 \mp l_2 + l]}{[2l_2+1][2l+1][2l_2+1\pm 1][2l]} \right)^{1/2}. \tag{93}$$

3.5.5 The Biedenharn–Elliott Identity

The Racah coefficients of $\breve{U}_q(\mathrm{su}_2)$ satisfy the Biedenharn–Elliott identity

$$\sum_{l_{23}} R_q(l_1l_2l_3, l_{12}l_{23}, l_{123}) R_q(l_1l_{23}l_4, l_{123}l_{234}, l) R_q(l_2l_3l_4, l_{23}l_{34}, l_{234})$$
$$= R_q(l_{12}l_3l_4, l_{123}l_{34}, l) R_q(l_1l_2l_{34}, l_{12}l_{234}, l). \tag{94}$$

Its proof is similar to the case of the group $SU(2)$. The starting point is the observation that there are five possibilities to decompose the tensor product $T_{l_1} \otimes T_{l_2} \otimes T_{l_3} \otimes T_{l_4}$ into irreducible representations by Proposition 22:

$$\{(T_{l_1} \otimes T_{l_2}) \otimes T_{l_3}\} \otimes T_{l_4}, \tag{I}$$
$$\{T_{l_1} \otimes (T_{l_2} \otimes T_{l_3})\} \otimes T_{l_4}, \tag{II}$$
$$T_{l_1} \otimes \{(T_{l_2} \otimes T_{l_3}) \otimes T_{l_4}\}, \tag{III}$$
$$T_{l_1} \otimes \{T_{l_2} \otimes (T_{l_3} \otimes T_{l_4})\}, \tag{IV}$$
$$(T_{l_1} \otimes T_{l_2}) \otimes (T_{l_3} \otimes T_{l_4}). \tag{V}$$

We can go from decomposition (I) to decomposition (IV) by the chain (I)→(II)→(III)→(IV) and also by the chain (I)→(V)→(IV), using Racah coefficients at each step. Since the final formulas connect the same decompositions (I) and (IV), the corresponding matrices are equal in both cases. Writing down the equality of matrix elements, we get the identity (94).

In terms of Wigner $6j$ symbols the Biedenharn–Elliott identity has the form

$$\begin{Bmatrix} c' & a & a' \\ e & b' & b \end{Bmatrix}_q \begin{Bmatrix} a' & e & b' \\ d & d' & c \end{Bmatrix}_q = \sum_f (-1)^p [2f+1]$$
$$\times \begin{Bmatrix} a & b & e \\ d & c & f \end{Bmatrix}_q \begin{Bmatrix} c' & b & b' \\ d & d' & f \end{Bmatrix}_q \begin{Bmatrix} c' & a & a' \\ c & d' & f \end{Bmatrix}_q,$$

where $p = a' + b' - c' - d' - a - b - d + e - f$.

By the orthogonality relation for the Racah coefficients, it follows from (94) that

$$R_q(aa'f, c'c, d') = \sum_{e,e'} R_q(abd, ef, c) R_q(c'bd, e'f, d') R_q(a'ab, c'c, e') R_q(a'ed, e'c, d'). \quad (94')$$

Putting here $d = \frac{1}{2}$, $b = f - \frac{1}{2}$ and using the expressions (92) and (93), we obtain a recurrence relation for the Racah coefficients. Using the notation

$$R^{abc}_{jkm}(r) = R_q(r-m, a, b;\ r-k, c;\ r), \qquad j + k = m,$$

this recurrence relation can be written as

$$\begin{aligned}
R^{abc}_{jkm}(r) = {} & \left(\frac{[a+b-c][a-b+c+1][b-k][2r-b-k+1][c+m+1]}{[2b]^2[2c+1][2r+1]^2[2c+2][2r+c-m+2]^{-1}} \right)^{1/2} \\
& \times R^{a,b-1/2,c+1/2}_{j,k+1/2,m+1/2}(r+1/2) \\
& + \left(\frac{[b-a+c][a+b+c+1][b-k][2r-b-k+1][c-m]}{[2b]^2[2c][2c+1][2r+1]^2[2r-c-m+1]^{-1}} \right)^{1/2} \\
& \times R^{a,b-1/2,c-1/2}_{j,k+1/2,m+1/2}(r+1/2) \\
& + \left(\frac{[a-b+c][a-b+c+1][b+k][2r+b-k+1][c-m+1]}{[2b]^2[2c+1][2c+2][2r+1]^2[2r-c-m]^{-1}} \right)^{1/2} \\
& \times R^{a,b-1/2,c+1/2}_{j,k-1/2,m-1/2}(r-1/2) \\
& + \left(\frac{[b-a+c][a+b+c+1][b+k][2r+b-k+1][c+m]}{[2b]^2[2c][2c+1][2r+1]^2[2r+c-m+1]^{-1}} \right)^{1/2} \\
& \times R^{a,b-1/2,c-1/2}_{j,k-1/2,m-1/2}(r-1/2). \qquad (95)
\end{aligned}$$

Starting with the Racah coefficients $R^{a,1/2,a\pm1/2}_{j,\pm1/2,j\pm1/2}$, given by (92) and (93), equation (95) allows us to compute inductively all Racah coefficients.

Our next result is the relation

$$\sum_r (-1)^{c+f+r} q^{\alpha}[2r+1] \begin{Bmatrix} a & d & r \\ e & b & c \end{Bmatrix}_q \begin{Bmatrix} a & d & r \\ b & e & f \end{Bmatrix}_q = q^{\beta} \begin{Bmatrix} a & b & c \\ d & e & f \end{Bmatrix}_q, \quad (96)$$

where we abbreviated $\alpha := -\{f(f+1) + c(c+1) + r(r+1)\}$ and $\beta := -\{a(a+1) + b(b+1) + d(d+1) + e(e+1)\}$. Note that (96) can be interpreted as an addition theorem for Racah coefficients.

In order to prove formula (96) we replace the first and the second Racah coefficients on the left hand side by the expressions (90) and (91), respectively, and use the orthogonality relation for the CGC's. Then the left hand side of (96) coincides with the expression (88) multiplied by q^{β}.

3.5.6 The Hexagon Relation

The expression

$$\begin{Bmatrix} a & b & c \\ e & d & f \\ h & k & g \end{Bmatrix}_q = \sum_n c \begin{Bmatrix} a & e & h \\ k & g & n \end{Bmatrix}_q \begin{Bmatrix} b & d & k \\ e & n & f \end{Bmatrix}_q \begin{Bmatrix} c & f & g \\ n & a & b \end{Bmatrix}_q, \tag{97}$$

where $c = (-1)^{2n}[2n+1]q^{-\{n(n+1)+h(h+1)+d(d+1)+c(c+1)\}}$, will be called the $9j$ *symbol* of $\breve{U}_q(\mathrm{su}_2)$. These symbols possess the following symmetry relation permuting the second and the third columns:

$$\begin{Bmatrix} a & b & c \\ e & d & f \\ h & k & g \end{Bmatrix}_q = (-1)^{\beta} q^{\tau(\beta)} \begin{Bmatrix} a & c & b \\ e & f & d \\ h & g & k \end{Bmatrix}_q, \tag{98}$$

where

$$\beta = \sum_{i=1}^{9} a_i \equiv a+b+c+d+e+f+h+k+g, \quad \tau(\beta) = \sum_{i=1}^{9} a_i(a_i+1).$$

We sketch the proof of (98). First we express both sides in terms of Racah coefficients. Next we insert the expression for the Racah coefficient from (96), with q replaced by q^{-1} by (86), instead of the first Racah coefficient of the left hand side of (98). Summing over n by using the identity (94′) we finally arrive at the right hand side of (98).

Writing (98) in terms of Racah coefficients, we obtain the *hexagon relation*

$$\sum_n c \begin{Bmatrix} a & g & n \\ k & e & h \end{Bmatrix}_q \begin{Bmatrix} a & c & b \\ f & n & g \end{Bmatrix}_q \begin{Bmatrix} k & b & d \\ f & e & n \end{Bmatrix}_q$$

$$= \sum_m (-1)^{2m} q^{\tau} \begin{Bmatrix} k & c & m \\ f & h & g \end{Bmatrix}_q \begin{Bmatrix} a & m & d \\ f & e & h \end{Bmatrix}_q \begin{Bmatrix} a & c & b \\ k & d & m \end{Bmatrix}_q, \tag{99}$$

where c is as in (97) and $\tau = -\{m(m+1)+g(g+1)+e(e+1)+b(b+1)\}$.

3.5.7 Clebsch–Gordan Coefficients as Limits of Racah Coefficients

The main aim of this subsection is to show that the CGC's of $\breve{U}_q(\mathrm{su}_2)$ are the limit cases of the Racah coefficients:

$$\lim_{r\to\infty} R^{abc}_{jkm}(r) = C^{abc}_{jkm}, \qquad j+k=m, \quad |q|>1. \tag{100}$$

In order to prove this formula we first note that the relation

$$\lim_{r\to\infty} \frac{[r+\alpha]}{[r+\beta]} = q^{(\alpha-\beta)/2}, \qquad |q|>1, \tag{101}$$

together with (68), (69), (92) and (93) imply that

$$\lim_{r\to\infty} R^{a,1/2,a+m}_{j,k,j+k}(r) = C^{a,1/2,a+m}_{j,k,j+k}, \qquad |q|>1, \tag{102}$$

where $m=\pm\frac{1}{2}$, $k=\pm\frac{1}{2}$. This is the assertion (100) for $b=\frac{1}{2}$. In order to derive (100) in the general case we take the limit $r\to\infty$ in the recurrence relation (95). Setting $\hat{C}^{abc}_{jkm} = \lim_{r\to\infty} R^{abc}_{jkm}(r)$, $s=j+2k$, $t=c-b$, $p=b+c$ and using (101), we obtain

$$\begin{aligned}
&\hat{C}^{abc}_{jkm}\\
&= q^{(t-s+1)/4}\left(\frac{[a+b-c][a-b+c+1][b-k][c+m+1]}{[2b]^2[2c+1][2c+2]}\right)^{\frac{1}{2}}\hat{C}^{a,b-1/2,c+1/2}_{j,k+1/2,m+1/2}\\
&= q^{-(p+s)/4}\left(\frac{[a-b+c][a+b+c+1][b-k][c-m]}{[2b]^2[2c+1][2c]}\right)^{\frac{1}{2}}\hat{C}^{a,b-1/2,c-1/2}_{j,k+1/2,m+1/2}\\
&= q^{(t-s-1)/4}\left(\frac{[a+b-c][a-b+c+1][b+k][c-m+1]}{[2b]^2[2c+1][2c+2]}\right)^{\frac{1}{2}}\hat{C}^{a,b-1/2,c+1/2}_{j,k-1/2,m-1/2}\\
&= q^{(p-s)/4}\left(\frac{[b-a+c][a+b+c+1][b+k][c+m]}{[2b]^2[2c+1][2c]}\right)^{\frac{1}{2}}\hat{C}^{a,b-1/2,c-1/2}_{j,k-1/2,m-1/2}.
\end{aligned} \tag{103}$$

On the other hand, formula (86) leads to the identity

$$C^{a\ f\ e}_{j,k,j+k} = \sum\nolimits_{i,p,c} R_q(abd,cf,e)C^{a\ b\ c}_{j,i,j+i}C^{c\ d\ e}_{j+i,p,j+k}C^{b\ d\ f}_{i,p,k}.$$

Substituting $d=\frac{1}{2}$, $b=f-\frac{1}{2}$ and using the expressions (68), (69), (92) and (93) for special values of CGC's and Racah coefficients, we conclude that (103) remains valid if all $\hat{C}^{abc}_{jkm}$ are replaced by the CGC's C^{abc}_{jkm}. That is, both families of coefficients satisfy the same recurrence relation (103). Since $\hat{C}^{a,1/2,c}_{j,k,m} = C^{a,1/2,c}_{j,k,m}$ as shown by (102), it follows that $\hat{C}^{abc}_{jkm} = C^{abc}_{jkm}$ for all values of indices. This completes the proof of formula (100).

Relation (100) yields the following asymptotic formula for large r:

$$\begin{Bmatrix} a & b & c\\ d+r & e+r & f+r\end{Bmatrix}_q \sim \frac{(-1)^{a+b+d+e}q^{-f/2}}{([2r+1][2c+1])^{1/2}}C^{a\ b\ c}_{f-e,d-f,d-e}, \qquad |q|>1,$$

If $|q|<1$, then the same reasoning shows that

$$\lim_{r\to\infty} R^{a\ b\ c}_{j\ k\ m}(r) = (-1)^{a+b-c}C^{a\ b\ c}_{-j,-k,-m}, \qquad |q|<1. \tag{104}$$

By (73), the latter can also be written as

$$\lim_{r\to\infty} R^{a\ b\ c}_{j\ k\ m}(r) = C^{a\ b\ c}_{j\ k\ m}(q^{-1}), \qquad |q|<1.$$

For the Wigner $6j$ symbols, (104) gives the asymptotic formula

$$\left\{\begin{matrix} a & b & c \\ d+r & e+r & f+r \end{matrix}\right\}_q \sim \frac{(-1)^{a+b-c}q^{f/2}}{([2r+1][2c+1])^{1/2}} C^{a\ b\ c}_{f-e,d-f,d-e},$$

where $|q| < 1$ and $r \to \infty$.

The preceding consideration shows that, as in the classical case, *the five-term recurrence relation (95) is of fundamental importance for the theory of CGC's and Racah coefficients of the quantum algebra $\breve{U}_q(\mathrm{su}_2)$. This relation together with the expressions (92) and (93) for the simplest Racah coefficients determine uniquely all Racah coefficients of $\breve{U}_q(\mathrm{su}_2)$. By (100) and (104), the Racah coefficients in turn determine all CGC's of $\breve{U}_q(\mathrm{su}_2)$.*

We also note the equation

$$\lim_{r\to\infty} (-1)^{2r} \frac{[2r+c+f+1]!}{[2r+b+e]!} \left\{\begin{matrix} a & b+r & c+r \\ d & e+r & f+r \end{matrix}\right\}_q$$
$$= (-1)^{a+b+d+e} T(a,d,f-e,b-f,b-c,c-e) \quad (105)$$

for $b+e \le c+f$, where

$$T(a,d,a',d',a'',d'') = \frac{1}{[a'-a'']!} \left(\frac{[a+a']![a-a'']![d-d']![d+d'']!}{[a-a']![a+a'']![d+d']![d-d'']!} \right)^{1/2}.$$

If $b+e > c+f$, the limit in (105) vanishes.

3.6 Tensor Operators and the Wigner–Eckart Theorem

In this section we define tensor operators transforming under irreducible representations of $\breve{U}_q(\mathrm{su}_2)$ and prove the Wigner–Eckart theorem for these operators. First we recall the corresponding result for compact Lie groups.

3.6.1 Tensor Operators for Compact Lie Groups

Let T_σ be an irreducible representation of a compact Lie group G acting on a vector space V_σ and let t^σ_{mn} be the matrix elements of T_σ with respect to some basis of V_σ. We say that a set $\mathbf{R}^\sigma \equiv \{R^\sigma_m \mid m = 1,2,\cdots,\dim\ T_\sigma\}$ of operators acting on a Hilbert space $\mathfrak{H}$ is a *tensor operator*, transforming under the representation T_σ, if there exists a continuous representation T of G on $\mathfrak{H}$ such that

$$T(g)R^\sigma_m T(g^{-1}) = \sum\nolimits_{n=1}^{\dim T_\sigma} t^\sigma_{nm}(g) R^\sigma_n, \quad g \in G, \quad m = 1,2,\cdots,\dim T_\sigma. \quad (106)$$

Let us see what the latter means in terms of the associated representation dT of the Lie algebra $\mathfrak{g}$ of G. Let $g(t)$ be a one-parameter subgroup in G with the tangent vector $X \in \mathfrak{g}$. Setting $g = g(t)$ into (106) and differentiating at $t = 0$, we obtain

$$[dT(X), R_m^\sigma] = \sum_{n=1}^{\dim T_\sigma} t_{nm}^\sigma(X) R_n^\sigma.$$

This equality defines tensor operators in infinitesimal form.

Let $T = \bigoplus_i T_{\lambda_i}$ be a decomposition of T into irreducible components. For simplicity we assume that this decomposition has no multiple irreducible representations. Let $\mathbf{e}_s^{\lambda_i}$, $s = 1, 2, \cdots, \dim T_{\lambda_i}$, be an orthonormal basis in the carrier space of the representation T_{λ_i}. Then the Wigner–Eckart theorem asserts that the matrix elements $\langle \mathbf{e}_s^{\lambda_i} \mid R_m^\sigma \mid \mathbf{e}_r^{\lambda_j} \rangle$ of the operators R_m^σ with respect to the basis $\{\mathbf{e}_s^{\lambda_i}\}$ are of the form

$$\langle \mathbf{e}_s^{\lambda_i} \mid R_m^\sigma \mid \mathbf{e}_r^{\lambda_j} \rangle = \langle \lambda_i, s \mid \sigma, m;\ \lambda_j, r \rangle \langle \lambda_i \| \mathbf{R}^\sigma \| \lambda_j \rangle.$$

Here $\langle \lambda_i, s \mid \sigma, m;\ \lambda_j, r \rangle$ are the CGC's of the tensor product of the representations T_σ and T_{λ_j} of G (recall that we assumed that this tensor product does not contain multiple irreducible representations in its decomposition) and $\langle \lambda_i \| \mathbf{R}^\sigma \| \lambda_j \rangle$ are the so-called *reduced matrix elements* of the tensor operator $\mathbf{R}^\sigma$. The reduced matrix elements $\langle \lambda_i \| \mathbf{R}^\sigma \| \lambda_j \rangle$ are independent of the basis elements.

Thus, the Wigner–Eckart theorem represents the matrix elements of tensor operators as products of two quantities, where the first is determined by the group structure and the second is independent of the group.

3.6.2 Tensor Operators and the Wigner–Eckart Theorem for $\breve{U}_q(\mathrm{su}_2)$

Let T_k be the $(2k+1)$-dimensional irreducible representation of $\breve{U}_q(\mathrm{su}_2)$. We say that a set of operators $\mathbf{R}^k = \{R_n^k \mid n = -k, -k+1, \cdots, k\}$ acting on a Hilbert space $\mathfrak{H}$ is a *tensor operator*, transforming under the representation T_k, if there exists a representation T of $\breve{U}_q(\mathrm{su}_2)$ on $\mathfrak{H}$ such that

$$T(K) R_n^k T(K)^{-1} = q^n R_n^k, \tag{107}$$

$$T(E) R_n^k - q^n R_n^k T(E) = ([k-n][k+n+1])^{1/2} R_{n+1}^k T(K)^{-1}, \tag{108}$$

$$T(F) R_n^k - q^n R_n^k T(F) = ([k+n][k-n+1])^{1/2} R_{n-1}^k T(K)^{-1}. \tag{109}$$

Note that the coefficients on the right hand sides are just the matrix elements of the operators $T_k(K)$, $T_k(E)$ and $T_k(F)$, respectively.

Suppose that there is a subset $I \subseteq \frac{1}{2}\mathbb{N}_0$ such that the representation T decomposes into an orthogonal direct sum

$$T = \bigoplus_{l \in I} T_l \tag{110}$$

of irreducible representations T_l, $l \in I$. In other words, the multiplicities of all irreducible representations T_n, $n \in \frac{1}{2}\mathbb{N}_0$, in the decomposition (110) do not exceed 1. Let $\{\mathbf{e}_m^l\}$ be the corresponding orthonormal basis of the carrier space of the representation T_l in (110).

Proposition 23. *If $\mathbf{R}^k$ is a tensor operator as above, then there exist numbers $\langle l'\|\mathbf{R}^k\|l\rangle$ such that the matrix elements of the operators R^k_n with respect to the bases $\{\mathbf{e}^l_m\}$ are*

$$\langle \mathbf{e}^{l'}_{m'} \mid R^k_n \mid \mathbf{e}^l_m\rangle = C_q(l,k,l';m,n,m')\langle l'\|\mathbf{R}^k\|l\rangle, \tag{111}$$

where $C_q(\cdot)$ are the CGC's of $\breve{U}_q(\mathrm{su}_2)$. The reduced matrix elements $\langle l'\|\mathbf{R}^k\|l\rangle$ are independent of n, m, m'.

Proof. We consider the vectors $\mathbf{e}^{kl}_{nm} := R^k_n|\mathbf{e}^l_m\rangle$. By (107), $T(K)\mathbf{e}^{kl}_{nm} = T(K)R^k_n|\mathbf{e}^l_m\rangle = q^{n+m}R^k_n\mathbf{e}^{kl}_{nm}$. Similarly, by (108) and (109), we get

$$T(E)\mathbf{e}^{kl}_{nm} = q^n([l-m][l+m+1])^{1/2}\mathbf{e}^{kl}_{n,m+1} + q^{-m}([k-n][k+n+1])^{1/2}\mathbf{e}^{kl}_{n+1,m},$$

$$T(F)\mathbf{e}^{kl}_{nm} = q^n([l+m][l-m+1])^{1/2}\mathbf{e}^{kl}_{n,m-1} + q^{-m}([k+n][k-n+1])^{1/2}\mathbf{e}^{kl}_{n-1,m}.$$

These formulas show that the representation T acts on the vectors $\mathbf{e}^{kl}_{nm'}$ as the tensor product $T_l \otimes T_k$ does. Therefore, we can apply formula (41) and express them as sums of vectors transforming under irreducible representations:

$$\mathbf{e}^{kl}_{nm} = \sum\nolimits_{l'} C_q(l,k,l';m,n,m')\tilde{\mathbf{e}}^{l'}_{n+m}$$

(we assume that the CGC's are real). Now we take the scalar product of both sides with the basis vector $\mathbf{e}^{l'}_{m'}$. Using the definition of $\mathbf{e}^{kl}_{nm}$ and the fact that $\langle \mathbf{e}^{l'}_{m'} \mid \tilde{\mathbf{e}}^{l'}_r\rangle = \delta_{m'r}\langle l'\|\mathbf{R}^k\|l\rangle$, we obtain the relation (111), where the numbers $\langle l'\|\mathbf{R}^k\|l\rangle$ are independent of n, m, m'. □

3.7 Applications

The aim of this section is to give a short sketch of some applications of representations of $U_q(\mathrm{sl}_2)$ in nuclear physics.

3.7.1 The $U_q(\mathrm{sl}_2)$ Rotator Model of Deformed Nuclei

In nuclear physics it was suggested that rotational spectra of deformed nuclei can be described by a q-deformed rotator which corresponds to the Casimir element of $U_q(\mathrm{sl}_2)$. This rotator is defined by the Hamiltonian $H = (2I)^{-1}C'_q + E_0$, where I is the moment of inertia, E_0 is the bandhead energy and C'_q is the quadratic Casimir element chosen such that its eigenvalues on the representations T_l are $[l][l+1]$. For real q, $q = e^\mu$, the energy levels of the q-rotator are

$$E(j) = \frac{1}{2I}[j][j+1] + E_0 = \frac{1}{2I}\frac{\sinh(\mu j)\sinh(\mu(j+1))}{\sinh^2\mu} + E_0. \tag{112}$$

Following the physics literature the number $l \in \frac{1}{2}\mathbb{N}_0$ parametrizing the irreducible finite-dimensional type 1 representations of $U_q(\mathrm{sl}_2)$ is here denoted by j. If $q = e^{\mathrm{i}\mu}$, $\mu \in \mathbb{R}$, is of modulus one, then one has

$$E(j) = \frac{1}{2I}\frac{\sin(\mu j)\sin(\mu(j+1))}{\sin^2\mu} + E_0. \tag{113}$$

The expressions (112) and (113) can be represented as

$$E(j) = E_0 + Aj(j+1) + B(j(j+1))^2 + C(j(j+1))^3 + D(j(j+1))^4 + \cdots, \tag{114}$$

where the coefficients A, B, C, D depend on μ and are expressed in terms of the spherical Bessel functions of the first kind in the case (113) and of the modified Bessel functions of the first kind in the case (112). Since rotational spectra in the first order approximation are described by the usual Lie algebra sl_2, it is natural to expect μ to be near zero. (It was found that μ is around the value 0.03.) Empirically it is known that for nuclear rotational spectra the coefficients A, B, C, D have alternating signs starting with positive A. This implies that rotational spectra can be described only by the case (113).

Numerical expressions for ground state bands of rare earth and actinide nuclei given by the formula (113) have led to very good agreement with experimental data. There is also satisfactory agreement for the β- and γ-bands of these nuclei.

The variable moment of inertia (VMI) model of rotational nuclear spectra is described by the formula

$$E(j) = \frac{1}{2\Theta_0}\left(j(j+1) - \frac{\sigma}{2}(j(j+1))^2 + \sigma^2(j(j+1))^3 - 3\sigma^3(j(j+1))^4 + \cdots\right), \tag{115}$$

where Θ_0 is a free parameter of the model corresponding to the ground state moment of inertia and σ is related to the softness of the nucleus. Both expressions (114) and (115) have similar properties (alternating signs, successive coefficients falling by about 3 orders of magnitude, etc.). However, the expansion (114) gives better agreement with the empirical values of the ratios $AC/(4B^2) = 1/10$ and $A^2D/(24B^7) = 1/280$.

3.7.2 Electromagnetic Transitions in the $U_q(\mathrm{sl}_2)$ Model

In the case of the VMI model no method has been found for making predictions for the $B(E2)$ transition probabilities connecting the levels of a collective band. The $U_q(\mathrm{sl}_2)$ model provides such a link.

In rotational bands, the collective model of Bohr and Mottelson gives the transition probabilities described by the formula

$$B(E2; j+1 \to j) = \frac{5}{16\pi}Q_0^2\left|C^{j+2,2,j}_{K,0,-K}\right|^2, \tag{116}$$

where $C^{j+2,2,j}_{K,0,-K}$ is the CGC of the group $SU(2)$, Q_0^2 is the intrinsic electric quadrupole moment and K is the projection of the angular momentum j on the symmetry axis of the nucleus in the body-fixed frame. For the $K = 0$ bands we have

$$B(E2;\, j+1 \to j) = \frac{5}{16\pi} Q_0^2 \frac{3}{2} \frac{(j+1)(j+2)}{(2j+3)(2j+5)}.$$

From the preceding it is clear that the $B(E2)$ values should saturate with increasing value of j.

In the case of the $U_q(\mathrm{sl}_2)$ model we have to take the corresponding CGC of the quantum algebra $U_q(\mathrm{sl}_2)$ instead of that of $SU(2)$ in formula (116). Using the explicit form of these coefficients, we find for $K = 0$ bands that

$$B(E2;\, j+1 \to j) = \frac{5}{16\pi} Q_0^2 \frac{[3][4][j+1]^2[j+2]^2}{[2][2j+2][2j+3][2j+4][2j+5]}.$$

Expanding these functions and keeping corrections of the leading order in μ (it is assumed that the range of values of μ is the same as in the description of ground state spectra), one gets

$$B(E2;\, j+1 \to j) = \frac{5}{16\pi} Q_0^2 \frac{3}{2} \frac{(j+1)(j+2)}{(2j+3)(2j+5)} \left(1 + \frac{\mu^2}{3}(6j^2 + 22j + 12)\right).$$

Thus we see that the additional factor depending on μ^2 leads to an increase with j, while the Bohr–Mottelson model reaches saturation at high j. A discussion of this effect is contained in [BDKL].

3.8 Notes

The associative algebra $U_q(\mathrm{sl}_2)$ first appeared in the paper by P. P. Kulish and N. Yu. Reshetikhin [KuR] devoted to the construction of integrable models of type XXZ with higher spin. Its Hopf algebra structure was found later by E. K. Sklyanin [Skl2]. The universal R-matrix of $U_h(\mathrm{sl}_2)$ was discovered by V. G. Drinfeld [Dr2]. The irreducible representations of $U_q(\mathrm{sl}_2)$ when q is not a root of unity were constructed in [Skl1], see also [Wor2]. The irreducible representations of $U_q(\mathrm{sl}_2)$ in the root of unity case are given in [RA]. The expressions for the Clebsch–Gordan coefficients of $U_q(\mathrm{su}_2)$ were derived in [KR1], [KK], [GKK], [Vak] (see also [BL], [VK1], Chap. 14, and the references therein). The formulas for the Racah coefficients of $U_q(\mathrm{su}_2)$ have been obtained in [KR1] and [KlK] (see [BL] and [VK1], Chap. 14). Tensor operators for $U_q(\mathrm{sl}_2)$ were introduced by Biedenharn (see [BL]).

There is a large variety of applications of the quantum algebra $U_q(\mathrm{sl}_2)$ in physics. The relations of quantum groups to the theory of integrable models and ideas of applications in quantum field theory are discussed in [F1], [F2], [AGS], [DV], GoS], [MR]. A good review and extensive references on applications of representations of $U_q(\mathrm{sl}_2)$ in nuclear physics can be found in [BDKL].

4. The Quantum Group $SL_q(2)$ and Its Representations

In this chapter we investigate the coordinate Hopf algebra of the quantum group $SL_q(2)$ and develop its corepresentation theory. It is the simplest example from the series of quantum groups associated with simple complex Lie groups which will be studied extensively in the second part of the book.

In this chapter q denotes a fixed nonzero complex number.

4.1 The Hopf Algebra $\mathcal{O}(SL_q(2))$

In Sect. 1.1 we have seen that the algebraic properties of the group $SL(2,\mathbb{C})$ are stored in its coordinate Hopf algebra $\mathcal{O}(SL(2))$. Let us briefly recall the structure of this Hopf algebra from Example 1.3. As an algebra, $\mathcal{O}(SL(2))$ is the quotient of the commutative polynomial algebra $\mathbb{C}[u_1^1, u_2^1, u_1^2, u_2^2]$ in four indeterminates $u_1^1, u_2^1, u_1^2, u_2^2$ (the coordinate functions on $SL(2,\mathbb{C})$) by the two-sided ideal generated by the element $u_1^1u_2^2 - u_2^1u_1^2 - 1$. On the generators the comultiplication Δ, the counit ε and the antipode S are given by $\Delta(u_j^i) = \sum_k u_k^i \otimes u_j^k$, $\varepsilon(u_j^i) = \delta_{ij}$, $S(u_1^1) = u_2^2$, $S(u_2^1) = -u_2^1$, $S(u_1^2) = -u_1^2$, $S(u_2^2) = u_1^1$. The Hopf algebras $\mathcal{O}(SL_q(2))$ introduced below are a one-parameter deformation of this Hopf algebra $\mathcal{O}(SL(2))$.

4.1.1 The Bialgebra $\mathcal{O}(M_q(2))$

Let $\mathcal{O}(M_q(2))$ be the complex (associative) algebra with generators a, b, c, d satisfying the following relations:

$$ab = qba, \quad ac = qca, \quad bd = qdb, \quad cd = qdc, \quad bc = cb \tag{1}$$

$$ad - da = (q - q^{-1})bc. \tag{2}$$

In order to shorten formulas we also write $u_1^1 = a$, $u_2^1 = b$, $u_1^2 = c$, $u_2^2 = d$.

Proposition 1. *There exists a unique bialgebra structure on the algebra $\mathcal{O}(M_q(2))$ with comultiplication Δ and counit ε such that*

$$\Delta(a) = a \otimes a + b \otimes c, \quad \Delta(b) = a \otimes b + b \otimes d, \tag{3}$$

$$\Delta(c) = c \otimes a + d \otimes c, \quad \Delta(d) = c \otimes b + d \otimes d, \tag{4}$$

$$\varepsilon(a) = \varepsilon(d) = 1, \quad \varepsilon(b) = \varepsilon(c) = 0. \tag{5}$$

Note that the relations (3)–(5) can be written as $\Delta(u^i_j) = \sum_k u^i_k \otimes u^k_j$ and $\varepsilon(u^i_j) = \delta_{ij}$, $i, j = 1, 2$.

Proof. The algebra $\mathcal{O}(M_q(2))$ is the quotient of the free algebra $\mathbb{C}\langle u^i_j\rangle$ with generators u^i_j by the two-sided ideal generated by the elements obtained from the relations (1) and (2). By the universal property of the free algebra $\mathbb{C}\langle u^i_j\rangle$ (see Subsect. 1.2.1), there exist algebra homomorphisms $\Delta : \mathbb{C}\langle u^i_j\rangle \to \mathbb{C}\langle u^i_j\rangle \otimes \mathbb{C}\langle u^i_j\rangle$ and $\varepsilon : \mathbb{C}\langle u^i_j\rangle \to \mathbb{C}$ such that $\Delta(u^i_j) = \sum_k u^i_k \otimes u^k_j$ and $\varepsilon(u^i_j) = \delta_{ij}$. In order to show that Δ and ε pass to algebra homomorphisms of the algebra $\mathcal{O}(M_q(2))$ it is sufficient to prove that Δ and ε preserve the relations (1) and (2), that is, $\Delta(a)\Delta(b) = q\Delta(b)\Delta(a)$, etc. As a sample, we check this for the first relation. By (3) and (4),

$$\Delta(a)\Delta(b) = aa \otimes ab + ab \otimes ad + ba \otimes cb + bb \otimes cd,$$

$$q\Delta(b)\Delta(a) = aa \otimes qba + qba \otimes da + ab \otimes qbc + bb \otimes qdc.$$

Indeed, by (1) and (2), both terms coincide. Since Δ and ε are algebra homomorphisms of $\mathcal{O}(M_q(2))$ as just shown, by Proposition 1.8 it suffices to verify the coassociativity and the counit axiom on the generators u^i_j. We have $((\Delta \otimes \mathrm{id}) \circ \Delta)(u^i_j) = \sum_{k,l} u^i_k \otimes u^k_l \otimes u^l_j = ((\mathrm{id} \otimes \Delta) \circ \Delta)(u^i_j)$ and $((\varepsilon \otimes \mathrm{id}) \circ \Delta)(u^i_j) = \sum_k \varepsilon(u^i_k)u^k_j = u^i_j = ((\mathrm{id} \otimes \varepsilon) \circ \Delta)(u^i_j)$. □

Definition 1. *The bialgebra $\mathcal{O}(M_q(2))$ is called the* coordinate algebra of the quantum matrix space $M_q(2)$.

The elements $\{a^ib^jc^kd^l \mid i, j, k, l \in \mathbb{N}_0\}$ form a vector space basis of $\mathcal{O}(M_q(2))$. The formulas (3) and (4) give the comultiplication Δ only on the generators. The action of Δ on products of generators can be calculated by taking into account that Δ is an algebra homomorphism. We find that

$$\Delta(a^{2l}) = (a \otimes a + b \otimes c)^{2l} = \sum_{i=-l}^{l} \begin{bmatrix} 2l \\ l+i \end{bmatrix}_{q^{-2}} a^{l-i}b^{l+i} \otimes a^{l-i}c^{l+i}, \tag{6}$$

$$\Delta(c^{2l}) = \sum_{i=-l}^{l} \begin{bmatrix} 2l \\ l+i \end{bmatrix}_{q^{-2}} c^{l-i}d^{l+i} \otimes a^{l-i}c^{l+i}, \tag{7}$$

where $l \in \frac{1}{2}\mathbb{N}_0$ and the summation is over integral (resp. half-integral) values of i if l is integral (resp. half-integral). We also have

$$\Delta(a^{l-i}c^{l+i}) = \sum_{j=-l}^{l} \sum_{\mu \geq 0} \begin{bmatrix} l-i \\ \mu \end{bmatrix}_{q^{-2}} \begin{bmatrix} l+i \\ l+j-\mu \end{bmatrix}_{q^{-2}}$$
$$\times q^{-\mu(\mu+i-j)} a^{l-i-\mu}b^{\mu}c^{i-j+\mu}d^{l+j-\mu} \otimes a^{l-j}c^{l+j}. \tag{8}$$

We conclude this subsection by giving a reformulation of the six quadratic relations (1) and (2). For this purpose we take the matrix $R = (R^{ij}_{kl})_{i,j,k,l=1,2}$ with $R^{11}_{11} = R^{22}_{22} = q$, $R^{12}_{12} = R^{21}_{21} = 1$, $R^{21}_{12} = q - q^{-1}$ and $R^{ij}_{kl} = 0$ otherwise. That is, R is just the matrix (3.40) derived in Subsect. 3.4.1 from the universal R-matrix of the h-adic Hopf algebra $U_h(\mathrm{sl}_2)$. Let us consider the following equations

$$\sum\nolimits_{k,l} R^{ij}_{kl} u^k_n u^l_m = \sum\nolimits_{k,l} u^j_l u^i_k R^{kl}_{nm}, \quad i,j,n,m = 1,2. \tag{9}$$

Inserting the values of R^{ij}_{kl}, these 16 equations reduce precisely to the 6 relations (1) and (2). The equations (9) are the starting point for the FRT approach to quantum groups which will be elaborated in Chap. 9.

4.1.2 The Hopf Algebra $\mathcal{O}(SL_q(2))$

By relation (2), we have

$$ad - qbc = da - q^{-1}bc. \tag{10}$$

This element of $\mathcal{O}(M_q(2))$ is denoted $\mathcal{D}_q$ and called the *quantum determinant.*

Proposition 2. *The quantum determinant $\mathcal{D}_q$ is a group-like element (that is, $\Delta(\mathcal{D}_q) = \mathcal{D}_q \otimes \mathcal{D}_q$ and $\varepsilon(\mathcal{D}_q) = 1$) belonging to the center of the algebra $\mathcal{O}(M_q(2))$.*

Proof. The proof is given by direct computation using the relations (1) and (2). □

If q is not a root of unity, then it can be shown that the center of $\mathcal{O}(M_q(2))$ is generated by $\mathcal{D}_q$.

Since $\mathcal{D}_q$ is group-like, the two-sided ideal $\langle \mathcal{D}_q - 1\rangle$ generated by the element $\mathcal{D}_q - 1$ is a biideal of $\mathcal{O}(M_q(2))$. Hence the quotient $\mathcal{O}(SL_q(2)) := \mathcal{O}(M_q(2))/\langle \mathcal{D}_q - 1\rangle$ is again a bialgebra.

Proposition 3. *The bialgebra $\mathcal{O}(SL_q(2))$ is a Hopf algebra. The antipode S of $\mathcal{O}(SL_q(2))$ is determined by*

$$S(a) = d,\ S(b) = -q^{-1}b,\ S(c) = -qc,\ S(d) = a. \tag{11}$$

Proof. Put $a' := d$, $b' := -q^{-1}b$, $c' := -qc$, $d' := a$. One verifies that the elements a', b', c', d' satisfy the defining relations of the algebra $\mathcal{O}(SL_{q^{-1}}(2))$. But $\mathcal{O}(SL_{q^{-1}}(2))$ is the opposite algebra of $\mathcal{O}(SL_q(2))$. Hence there exists an algebra anti-homomorphism $S : \mathcal{O}(SL_q(2)) \to \mathcal{O}(SL_q(2))$ such that $S(a) = a'$, $S(b) = b'$, $S(c) = c'$ and $S(d) = d'$. To prove that S is an antipode, by Proposition 1.8 it is enough to check the antipode axiom (1.26) on the four generators. We omit this straightforward verification. □

A direct computation shows that for the algebra $\mathcal{O}(SL_q(2))$ the matrices

$$\mathbf{u} = \begin{pmatrix} a & b \\ c & d \end{pmatrix} \quad \text{and} \quad S(\mathbf{u}) = \begin{pmatrix} d & -q^{-1}b \\ -qc & a \end{pmatrix}$$

are inverse to each other. This fact is actually equivalent to the validity of the antipode condition (1.26) for the generators a, b, c, d.

Definition 2. *The Hopf algebra $\mathcal{O}(SL_q(2))$ is called the* coordinate algebra of the quantum group $SL_q(2)$.

As motivated at the end of Sect. 1.1, we shall think of and treat elements of the Hopf algebra $\mathcal{O}(SL_q(2))$ as functions on the "quantum group" $SL_q(2)$. However, this Hopf algebra is not a group Hopf algebra (it is neither commutative nor cocommutative if $q \neq 1$) and the quantum group $SL_q(2)$ exists only in terms of the Hopf algebra $\mathcal{O}(SL_q(2))$ and its structure.

Recall that the defining relations of the algebra $\mathcal{O}(SL_q(2))$ are the five equations (1) and the two equations

$$ad - qbc = da - q^{-1}bc = 1. \tag{12}$$

These are precisely the relations (3.44) and (3.45) which were derived in Subsect. 3.4.1 for the matrix elements of the representation $T_{1/2}$ of the quantum algebra $U_q(\mathrm{sl}_2)$. A geometric approach to these relations will be given in Subsect. 4.1.3.

We now begin to study algebraic properties of the Hopf algebra $\mathcal{O}(SL_q(2))$. Using the formulas (2.17) and (12) one proves by induction on m that

$$a^m d^m = \sum_{k=0}^{m} \begin{bmatrix} m \\ k \end{bmatrix}_{q^{-2}} q^{2km-k^2} (bc)^k, \quad d^m a^m = \sum_{k=0}^{m} \begin{bmatrix} m \\ k \end{bmatrix}_{q^{-2}} q^{-k^2} (bc)^k. \tag{13}$$

The following result is often used in the subsequent sections.

Proposition 4. *The set $\{a^n b^m c^r,\ b^m c^r d^s \mid m, r, s \in \mathbb{N}_0,\ n \in \mathbb{N}\}$ is a vector space basis of $\mathcal{O}(SL_q(2))$.*

Proof. By the relations (1) and (2), this set obviously spans $\mathcal{A}(SL_q(2))$. Using one of the representations π_u^∞ from Corollary 20 below, it is not difficult to show that they are linearly independent. In Subsect. 4.1.5 we use the diamond lemma in order to give another proof of the linear independence. □

Proposition 5. (i) *There are algebra automorphisms θ and $\vartheta_{\alpha,\beta}$, $\alpha, \beta \in \mathbb{C}\backslash\{0\}$, of the algebra $\mathcal{O}(SL_q(2))$ such that*

$$\theta(a) = a, \quad \theta(b) = c, \quad \theta(c) = b, \quad \theta(d) = d,$$

$$\vartheta_{\alpha,\beta}(a) = \alpha a, \quad \vartheta_{\alpha,\beta}(b) = \beta b, \quad \vartheta_{\alpha,\beta}(c) = \beta^{-1} c, \quad \vartheta_{\alpha,\beta}(d) = \alpha^{-1} d.$$

(ii) *For $\beta \in \mathbb{C}\backslash\{0\}$, $\vartheta_{1,\beta}$ is a Hopf algebra automorphism of $\mathcal{O}(SL_q(2))$.*

(iii) *There is a Hopf algebra isomorphism $\rho : \mathcal{O}(SL_q(2)) \to \mathcal{O}(SL_{q^{-1}}(2)) = \mathcal{O}(SL_q(2))^{\mathrm{op}}$ such that $\rho(a) = d$, $\rho(b) = c$, $\rho(c) = b$, and $\rho(d) = a$.*

Proof. These results follow by direct verification of the defining relations. □

4.1.3 A Geometric Approach to $SL_q(2)$

Matrices with complex entries act as linear transformations on complex spaces. We shall see now that the defining relations (1) and (2) of $\mathcal{O}(M_q(2))$ occur quite naturally if we think of quantum matrices of $M_q(2)$ as transformations of the quantum plane $\mathbb{C}_q^2$.

Let $\mathcal{O}(\mathbb{C}_q^2)$ be the algebra with generators x and y satisfying the relation

$$xy = qyx. \tag{14}$$

We call $\mathcal{O}(\mathbb{C}_q^2)$ the *coordinate algebra of the quantum plane* $\mathbb{C}_q^2$.

Let a, b, c, d be elements of an algebra $\mathcal{A}$. We "transform" the vectors $\begin{pmatrix} x \\ y \end{pmatrix}$ and (x, y) by the matrix $\begin{pmatrix} a & b \\ c & d \end{pmatrix}$ from the left resp. from the right, that is, we set

$$\begin{pmatrix} a & b \\ c & d \end{pmatrix} \otimes \begin{pmatrix} x \\ y \end{pmatrix} = \begin{pmatrix} a \otimes x + b \otimes y \\ c \otimes x + d \otimes y \end{pmatrix} =: \begin{pmatrix} x' \\ y' \end{pmatrix}, \tag{15}$$

$$(x, y) \otimes \begin{pmatrix} a & b \\ c & d \end{pmatrix} = (x \otimes a + y \otimes c,\ x \otimes b + y \otimes d) =: (x'', y''). \tag{16}$$

Proposition 6. *Suppose $q^2 + 1 \neq 0$. The couples (x', y') and (x'', y'') satisfy (14) if and only if the elements a, b, c, d fulfill the relations (1) and (2).*

Proof. First suppose that $x'y' = qy'x'$ and $x''y'' = qy''x''$. The first equation says $(a \otimes x + b \otimes y)(c \otimes x + d \otimes y) = q(c \otimes x + d \otimes y)(a \otimes x + b \otimes y)$, that is,

$$(ac - qca) \otimes x^2 + (ad - da + q^{-1}bc - qcb) \otimes xy + (bd - qdb)) \otimes y^2 = 0.$$

Since the elements x^2, xy and y^2 are linearly independent in $\mathcal{O}(\mathbb{C}_q^2)$, we get

$$ac = qca, \quad bd = qdb, \quad ad - da - qcb + q^{-1}bc = 0. \tag{17}$$

Similarly, the relation $x''y'' = qy''x''$ leads to

$$ab = qba, \quad cd = qdc, \quad ad - da - qbc + q^{-1}cb = 0. \tag{18}$$

The last relations in (17) and (18) imply that $bc = cb$. Inserting the latter into (17) and (18) we obtain (1) and (2). The same computations read in reversed order prove the opposite implication. □

In Hopf algebra terminology, the second implication of Proposition 6 yields the following (see also Example 1.19).

Proposition 7. *$\mathcal{O}(\mathbb{C}_q^2)$ is a left and right comodule algebra of the bialgebra $\mathcal{O}(M_q(2))$ with left coaction φ_L and right coaction φ_R determined by*

$$\varphi_L(x) = a \otimes x + b \otimes y, \quad \varphi_L(y) = c \otimes x + d \otimes y, \tag{19}$$

$$\varphi_R(x) = x \otimes a + y \otimes c, \quad \varphi_R(y) = x \otimes b + y \otimes d. \tag{20}$$

Proof. It follows at once from the converse implication of Proposition 6 that there are algebra homomorphisms $\varphi_L : \mathcal{O}(\mathbb{C}_q^2) \to \mathcal{O}(M_q(2)) \otimes \mathcal{O}(\mathbb{C}_q^2)$ and $\varphi_R : \mathcal{O}(\mathbb{C}_q^2) \to \mathcal{O}(\mathbb{C}_q^2) \otimes \mathcal{O}(M_q(2))$ such that (19) and (20) hold. To show that φ_L and φ_R are coactions, it suffices to verify the conditions (1.51) and (1.52) for the generators x and y which is easily done. □

Since $\mathcal{O}(SL_q(2))$ is a quotient bialgebra of $\mathcal{O}(M_q(2))$, the assertions of Proposition 7 remain valid for $\mathcal{O}(SL_q(2))$.

In order to express the quantum determinant also in geometric terms, we introduce the *exterier algebra* $\Lambda(\mathbb{C}_q^2)$ *of the quantum plane.* It is the algebra generated by elements ξ and η and relations

$$\xi^2 = 0, \quad \eta^2 = 0, \quad \xi\eta = -q^{-1}\eta\xi.$$

Setting $\xi' := a\otimes\xi + b\otimes\eta$ and $\eta' := c\otimes\xi + d\otimes\eta$, we have $\xi'\eta' = (ad - qbc)\otimes\xi\eta$. In the classical case $q = 1$, the determinant appears in the first factor, so it is natural to take $\mathcal{D}_q := ad - qbc$ as the quantum determinant.

4.1.4 Real Forms of $\mathcal{O}(SL_q(2))$

In the following we list three real forms of $\mathcal{O}(SL_q(2))$. That these formulas give well-defined Hopf $*$-algebra structures on $\mathcal{O}(SL_q(2))$ can be proved by the method sketched in Subsect. 3.1.3.

Case 1: $q \in \mathbb{R}$. There is an involution of the algebra $\mathcal{O}(SL_q(2))$ determined by

$$a^* = d, \quad b^* = -qc, \quad c^* = -q^{-1}b, \quad d^* = a. \tag{21}$$

The corresponding Hopf $*$-algebra is called the *coordinate algebra of the real quantum group* $SU_q(2)$ and is denoted by $\mathcal{O}(SU_q(2))$. It is called the *compact real form* of $SL_q(2)$. Inserting the formulas (21) for the involution into the relations (1) and (12) for $\mathcal{O}(SL_q(2))$ we see that $\mathcal{O}(SU_q(2))$ is the $*$-algebra generated by two elements a and c subject to the relations

$$ac = qca, \quad ac^* = qc^*a, \quad cc^* = c^*c, \quad a^*a + c^*c = 1, \quad aa^* + q^2c^*c = 1. \tag{22}$$

The Hopf $*$-algebra $\mathcal{O}(SU_q(1,1))$ defined by the involution

$$a^* = d, \quad b^* = qc, \quad c^* = q^{-1}b, \quad d^* = a$$

on $\mathcal{O}(SL_q(2))$ is the *coordinate algebra of the real quantum group* $SU_q(1,1)$.

Case 2: $|q| = 1$. There exists a real form $\mathcal{O}(SL_q(2,\mathbb{R}))$ of $\mathcal{O}(SL_q(2))$ with $*$-structure given by $a^* = a$, $b^* = b$, $c^* = c$ and $d^* = d$.

4.1.5 The Diamond Lemma

The diamond lemma [Ber] provides a general method in order to prove that certain sets are bases of algebras which are defined in terms of generators and relations. For instance, the Poincaré–Birkhoff–Witt theorem for enveloping algebras $U(\mathfrak{g})$ can be derived from it. In this subsection we develop a simplified variant of this lemma.

Let $\mathbb{C}\langle x_i\rangle$ be the free algebra with generators $x_1, x_2, \cdots, x_r$. Setting $I := \{1, 2, \cdots, r\}$, the set $X := \{x_\mathfrak{i} \mid \mathfrak{i} \in I^\infty\}$ of all monomials (see Subsect. 1.2.1 for the corresponding notation) is a basis of $\mathbb{C}\langle x_i\rangle$. We define an ordering "$\leq$" on X as follows. Two monomials of different length are ordered according to their length and two monomials of equal length are ordered lexicographically with respect to their indices. Let Π be a set of pairs $(x_{\mathfrak{i}_k}, f_k)$, $k = 1, 2, \cdots, s$, where $x_{\mathfrak{i}_k} \in X$ and $f_k \in \mathbb{C}\langle x_i\rangle$. We assume that Π is *compatible* with the ordering "$\leq$", that is, each f_k is a linear combination of basis elements $x_\mathfrak{i} \leq x_{\mathfrak{i}_k}$. For $x, y \in X$, let r_{xky} be the linear map of $\mathbb{C}\langle x_i\rangle$ which sends the basis element $x x_{\mathfrak{i}_k} y$ to $x f_k y$ and leaves the others fixed. The maps r_{xky} are called *reductions.* A monomial $x \in X$ is called *reduced* if $r(x) = x$ for all reductions r. Let R be the set of all compositions of reductions.

We say that two elements $x_{\mathfrak{i}_k}$ and $x_{\mathfrak{i}_l}$ and three elements $x, y, z \in X$ form an *overlap ambiguity* if $x_{\mathfrak{i}_k} = xy$ and $x_{\mathfrak{i}_l} = yz$. Such an ambiguity is called *resolvable* if there are $r, r' \in R$ such that $r(x_{\mathfrak{i}_k} z) = r'(x x_{\mathfrak{i}_l})$. Similarly, we say that elements $x_{\mathfrak{i}_k}$ and $x_{\mathfrak{i}_l}$, $k \neq l$, and $x, y \in X$ form an *inclusion ambiguity* if $x_{\mathfrak{i}_l} = x x_{\mathfrak{i}_k} y$. It is said to be *resolvable* if there exist $r, r' \in R$ such that $r(x_{\mathfrak{i}_l}) = r'(x x_{\mathfrak{i}_k})$.

Lemma 8. *Let $\mathcal{X}$ denote the algebra with generators $x_1, x_2, \cdots, x_r$ and defining relations $x_{\mathfrak{i}_k} - f_k = 0$, $k = 1, 2, \cdots, s$. If the set Π of pairs $(x_{\mathfrak{i}_k}, f_k)$ is compatible with the ordering "$\leq$" and if all ambiguities are resolvable, then the set of reduced monomials is a vector space basis of $\mathcal{X}$.*

Proof. This lemma is a special case of Theorem 1.2 in [Ber]. □

Despite the many definitions the basic idea behind this lemma becomes clear if one thinks of the passage from the free algebra $\mathbb{C}\langle x_i\rangle$ to its quotient $\mathcal{X}$ as follows. Terms of $\mathbb{C}\langle x_i\rangle$ are reduced to smaller ones according to the order "$\leq$" by inserting repeatedly the defining relations $x_{\mathfrak{i}_k} = f_k$. The monomials that remain unchanged during this procedure form a basis of $\mathcal{X}$.

The following proof of Proposition 4 shows the usefulness of Lemma 8. A number of similar later results (see Chap. 9) can be proved analogously.

Second Proof of Proposition 4. Put $x_1 = b$, $x_2 = a$, $x_3 = d$ and $x_4 = c$. Let Π be the set of seven pairs (ab, qba), $(ad, qbc + 1)$, $(ca, q^{-1}ac)$, (cb, bc), (cd, qdc), $(da, q^{-1}bc+1)$, $(db, q^{-1}bd)$. Clearly, $\mathcal{X}$ is just the algebra $\mathcal{O}(SL_q(2))$. A moment's consideration shows that Π is compatible with the ordering "$\leq$" and that the monomials $b^i a^j d^k c^l$ with $j = 0$ or $k = 0$ are precisely the reduced monomials. Thus, by Lemma 8 and by the relations (1), the proof is complete as soon as we have shown that all ambiguities are resolvable.

From the form of the set Π we see that there are no inclusion ambiguities and we have eight overlap ambiguities. Written as pairs, the latter are (ad, da), (ad, db), (ca, ab), (ca, ad), (cd, da), (cd, db), (da, ab), (da, ad). We check the resolvability of (ad, da) and denote reduction maps by arrows. We have

$$(ad)a \to (qbc + 1)a = qb(ca) + a \to qb(q^{-1}ac) + a = bac + a,$$

$$a(da) \to a(q^{-1}bc + 1) = q^{-1}(ab)c + a \to q^{-1}(qba)c + a = bac + a.$$

That is, there are $r, r' \in R$ such that $r((qbc + 1)a) = r'(a(q^{-1}bc + 1))$, so (ad, da) is resolvable. The other ambiguities are treated similarly. □

4.2 Representations of the Quantum Group $SL_q(2)$

Throughout this section we assume that the parameter q is not a root of unity and that $\mathcal{A}$ denotes the Hopf algebra $\mathcal{O}(SL_q(2))$.

4.2.1 Finite-Dimensional Corepresentations of $\mathcal{O}(SL_q(2))$: Main Results

First we recall that the finite-dimensional irreducible representations of the group $SL(2, \mathbb{C})$ with polynomial coefficients are labeled by nonnegative half-integers. For $l \in \frac{1}{2}\mathbb{N}_0$, the corresponding representation T_l acts on the vector space of homogeneous polynomials $f \in \mathbb{C}[x, y]$ of degree $2l$ by

$$(T_l(g)f)(x, y) = f(ax + cy, bx + dy), \quad g = \begin{pmatrix} a & b \\ c & d \end{pmatrix} \in SL(2, \mathbb{C}).$$

Expressed in the language of Hopf algebras this picture carries over directly to the quantum case.

Let $\mathcal{O}(\mathbb{C}_q^2)_n$ be the subspace of $\mathcal{O}(\mathbb{C}_q^2)$ formed by the homogeneous elements of degree n. Recall that the right coaction φ_R of the Hopf algebra $\mathcal{O}(SL_q(2))$ on its comodule algebra $\mathcal{O}(\mathbb{C}_q^2)$ is, by Proposition 7, an algebra homomorphism. Therefore, by (20), $\mathcal{O}(\mathbb{C}_q^2)_n$ is invariant under φ_R. Hence the restriction of φ_R to $\mathcal{O}(\mathbb{C}_q^2)_n$ is a corepresentation, denoted $T_{n/2}$, of the Hopf algebra $\mathcal{O}(SL_q(2))$ on the $(n+1)$-dimensional complex vector space $\mathcal{O}(\mathbb{C}_q^2)_n$. We call $T_{n/2}$ the *spin* $\frac{n}{2}$ representation of the quantum group $SL_q(2)$.

Theorem 9. *The corepresentations T_l, $l \in \frac{1}{2}\mathbb{N}_0$, are irreducible and pairwise inequivalent. Moreover, any finite-dimensional irreducible corepresentation of $\mathcal{O}(SL_q(2))$ is equivalent to one of these corepresentations.*

Proof. The proof of this theorem will be given in Subsect. 4.2.5 below. □

In the subsequent subsections we will develop some harmonic analysis on the quantum group $SL_q(2)$. The corepresentations T_l will be studied further in this context. The reader is invited to compare these considerations and formulas with the classical situation (see, for instance, [VK1], Chap. 6).

4.2.2 A Decomposition of $\mathcal{O}(SL_q(2))$

In this subsection we give a direct sum decomposition of $\mathcal{O}(SL_q(2))$ with respect to the coaction of the quantum subgroup K of diagonal matrices.

Let K be the subgroup of diagonal matrices of the group $SL(2,\mathbb{C})$. Obviously, the coordinate algebra $\mathcal{O}(K)$ is the Hopf algebra $\mathbb{C}[z,z^{-1}]$ of Laurent polynomials in z with comultiplication, counit and antipode determined by $\Delta(z) = z\otimes z$, $\varepsilon(z) = 1$ and $S(z) = z^{-1}$.

A *quantum subgroup* of a Hopf algebra $\mathcal{A}$ is defined by a Hopf algebra $\mathcal{B}$ together with a Hopf algebra homomorphism ϕ of $\mathcal{A}$ *onto* $\mathcal{B}$. (The corresponding classical picture is as follows: if H is a subgroup of a group G and $\mathcal{A}$ and $\mathcal{B}$ are Hopf algebras of functions on G and H, respectively, then $\phi(f)$ is just the restriction of the function $f \in \mathcal{A}$ to H.)

Clearly, there is a Hopf algebra homomorphism ϕ_K of $\mathcal{O}(SL_q(2))$ onto $\mathcal{O}(K)$ such that

$$\phi_K(a) = z, \quad \phi_K(b) = \phi_K(c) = 0, \quad \phi_K(d) = z^{-1}, \tag{23}$$

so the Hopf algebra $\mathcal{O}(K)$ is a quantum subgroup of the Hopf algebra $\mathcal{O}(SL_q(2))$. We express this by saying that K is a quantum subgroup of $SL_q(2)$. Now we define mappings $L_K : \mathcal{A} \to \mathcal{O}(K)\otimes\mathcal{A}$ and $R_K : \mathcal{A} \to \mathcal{A}\otimes\mathcal{O}(K)$ by

$$L_K = (\phi_K \otimes \mathrm{id})\circ\Delta \quad \text{and} \quad R_K = (\mathrm{id}\otimes\phi_K)\circ\Delta. \tag{24}$$

Then L_K and R_K are algebra homomorphisms and also left and right coactions of the Hopf algebra $\mathcal{O}(K)$ on $\mathcal{A} \equiv \mathcal{O}(SL_q(2))$, respectively. In other words, $\mathcal{A}$ becomes a left (resp. right) $\mathcal{O}(K)$-comodule algebra.

For $m,n \in \mathbb{Z}$ we introduce a subspace $\mathcal{A}[m,n]$ of $\mathcal{A}$ by setting

$$\mathcal{A}[m,n] = \{x\in\mathcal{A} \mid L_K(x) = z^m\otimes x,\ R_K(x) = x\otimes z^n\}. \tag{25}$$

We have $L_K(a) = z\otimes a$, $R_K(a) = a\otimes z$, that is, $a \in \mathcal{A}[1,1]$. Similarly, $b \in \mathcal{A}[1,-1]$, $c\in\mathcal{A}[-1,1]$, $d\in\mathcal{A}[-1,-1]$. Since L_K and R_K are homomorphisms,

$$\mathcal{A}[m,n]\cdot\mathcal{A}[p,r] \subset \mathcal{A}[m+p,n+r]. \tag{26}$$

Consequently, for the basis elements of $\mathcal{A}$ from Proposition 4 we have

$$b^pc^ld^r \in \mathcal{A}[p-l-r,l-p-r], \quad a^rb^pc^l \in \mathcal{A}[r+p-l,r+l-p]. \tag{27}$$

The elements of $\mathcal{A}[0,0]$ are called *K-biinvariant*. By (26), $\mathcal{A}[0,0]$ is a subalgebra of $\mathcal{A}$ and each vector space $\mathcal{A}[m,n]$ is an $\mathcal{A}[0,0]$-bimodule. In the case when $m-n$ is even we introduce an element $e_{mn}\in\mathcal{A}[m,n]$ by setting

$$e_{mn} = a^{(m+n)/2} c^{(n-m)/2} \quad \text{if} \quad m+n \geq 0,\ m \leq n, \tag{28}$$

$$e_{mn} = a^{(m+n)/2} b^{(m-n)/2} \quad \text{if} \quad m+n \geq 0,\ m \geq n, \tag{29}$$

$$e_{mn} = b^{(m-n)/2} d^{(-n-m)/2} \quad \text{if} \quad m+n \leq 0,\ m \geq n, \tag{30}$$

$$e_{mn} = c^{(n-m)/2} d^{(-n-m)/2} \quad \text{if} \quad m+n \leq 0,\ m \leq n. \tag{31}$$

The next proposition collects the main properties of the spaces $\mathcal{A}[m,n]$. All assertions follow immediately from (27) combined with the fact that the set $\{a^n b^m c^r,\ b^m c^r d^s\}$ is a vector space basis of $\mathcal{A}$.

Proposition 10. (i) $\mathcal{A}[0,0]$ *is the algebra* $\mathbb{C}[\zeta]$ *of all polynomials in* $\zeta = -qbc$.
(ii) *We have the decomposition*

$$\mathcal{A} \equiv \mathcal{O}(SL_q(2)) = \bigoplus\nolimits_{m,n\in\mathbb{Z}} \mathcal{A}[m,n]. \tag{32}$$

(iii) *If* $m,n \in \mathbb{Z}$ *and* $m-n$ *is odd, then* $\mathcal{A}[m,n] = \{0\}$.
(iv) *If* $m,n \in \mathbb{Z}$ *and* $m-n$ *is even, then* $\mathcal{A}[m,n] = \mathbb{C}[\zeta]e_{mn} = e_{mn}\mathbb{C}[\zeta]$. *Moreover, if* $p(\zeta)e_{mn} = 0$ *in* $\mathcal{A}[m,n]$ *for a polynomial* $p \in \mathbb{C}[\zeta]$, *then* $p = 0$.

4.2.3 Finite-Dimensional Subcomodules of $\mathcal{O}(SL_q(2))$

First we identify the left and right $\mathcal{A}$-comodule algebras $\mathcal{O}(\mathbb{C}_q^2)$ from Proposition 7 with certain subalgebras of $\mathcal{A}$.

Let $\mathcal{A}(a,b)$ denote the subalgebra of $\mathcal{A}$ generated by the elements a and b. Since $ab = qba$ by (1), there is an algebra homomorphism $\vartheta_R : \mathcal{O}(\mathbb{C}_q^2) \to \mathcal{A}(a,b)$ such that $\vartheta_R(x) = a$ and $\vartheta_R(y) = b$. Proposition 4 implies that ϑ_R is an isomorphism. Comparing (3) and (20) we obtain $(\vartheta_R \otimes \mathrm{id}) \circ \varphi_R = \Delta \circ \vartheta_R$, that is, under the isomorphism ϑ_R the right coaction φ_R of $\mathcal{A}$ on $\mathcal{O}(\mathbb{C}_q^2)$ goes into the restriction of the comultiplication to $\mathcal{A}(a,b)$. Thus, ϑ_R provides an isomorphism of the right $\mathcal{A}$-comodule algebra $\mathcal{O}(\mathbb{C}_q^2)$ with right coaction φ_R (by Proposition 7) to the right $\mathcal{A}$-comodule algebra $\mathcal{A}(a,b)$ with the comultiplication as right coaction. Similarly, the left $\mathcal{A}$-comodule algebra $\mathcal{O}(\mathbb{C}_q^2)$ with coaction φ_L defined by (19) is isomorphic to the left $\mathcal{A}$-comodule algebra $\mathcal{A}(a,c)$ with an isomorphism ϑ_L such that $\vartheta_L(x) = a$ and $\vartheta_L(y) = c$.

Let $l \in \mathbb{N}_0$. The image of $\mathcal{O}(\mathbb{C}_q^2)_{2l}$ under the isomorphisms ϑ_L and ϑ_R are

$$V_l^L = \bigoplus\nolimits_{i=-l}^{l} \mathbb{C}e_i^{(l)}, \qquad V_l^R = \bigoplus\nolimits_{i=-l}^{l} \mathbb{C}f_i^{(l)},$$

where the elements $e_i^{(l)}$ and $f_i^{(l)}$ are given by

$$e_i^{(l)} = \begin{bmatrix} 2l \\ l+i \end{bmatrix}_{q^{-2}}^{1/2} e_{-2i,2l} = \begin{bmatrix} 2l \\ l+i \end{bmatrix}_{q^{-2}}^{1/2} a^{l-i} c^{l+i}, \tag{33}$$

$$f_i^{(l)} = \begin{bmatrix} 2l \\ l+i \end{bmatrix}_{q^{-2}}^{1/2} e_{2l,-2i} = \begin{bmatrix} 2l \\ l+i \end{bmatrix}_{q^{-2}}^{1/2} a^{l-i} b^{l+i}. \tag{34}$$

Since $\mathcal{O}(\mathbb{C}_q^2)$ is invariant under φ_L and φ_R, we have

$$\Delta : V_l^L \to \mathcal{A} \otimes V_l^L, \quad \Delta : V_l^R \to V_l^R \otimes \mathcal{A}. \tag{35}$$

Hence V_l^L is a left $\mathcal{A}$-comodule and V_l^R is a right $\mathcal{A}$-comodule with coactions denoted by T_l^L and T_l^R, respectively. Let $t_{ij}^{(l)}$, $i, j = -l, -l+1, \cdots, l$, be the matrix coefficients with respect to the basis $\{f_i^{(l)}\}$ of V_l^R, that is,

$$T_l^R(f_i^{(l)}) \equiv \Delta(f_i^{(l)}) = \sum_j f_j^{(l)} \otimes t_{ji}^{(l)}. \tag{36}$$

Since the corepresentations T_l^R and T_l are equivalent by construction, $t_{ij}^{(l)}$ are also the matrix elements of T_l.

Lemma 11. (i) $e_i^{(l)} = t_{i,-l}^{(l)}$ *and* $f_i^{(l)} = t_{-l,i}^{(l)}$.
(ii) $t_{ij}^{(l)} \in \mathcal{A}[-2i, -2j]$.

Proof. (i): By (27), $f_i^{(l)} \in \mathcal{A}[2l, -2i]$ and hence $L_K(f_i^{(l)}) = z^{2l} \otimes f_i^{(l)}$. On the other hand, (23) implies that $\phi_K(f_i^{(l)}) = \delta_{i,-l} z^{2l}$, so $L_K(f_i^{(l)}) = z^{2l} \otimes t_{-l,i}^{(l)}$ by (36). Comparing both expressions, we get $f_i^{(l)} = t_{-l,i}^{(l)}$. Further, since $f_{-l}^{(l)} = a^{2l}$ by (34), formula (6) yields

$$\Delta(f_{-l}^{(l)}) = \sum_{j=-l}^{l} \begin{bmatrix} 2l \\ l+j \end{bmatrix}_{q^{-2}}^{1/2} f_j^{(l)} \otimes a^{l-j} c^{l+j}. \tag{37}$$

Comparing (36) with (37) and using (33) we obtain $e_j^{(l)} = t_{j,-l}^{(l)}$.

(ii): First note that $(\mathrm{id} \otimes R_K) \circ \Delta = (\Delta \otimes \mathrm{id}) \circ R_K$ by the definition (24) of R_K. Since $f_j^{(l)} \in \mathcal{A}[2l, -2j]$, we obtain from (36) and (25) that

$$((\Delta \circ \mathrm{id}) \circ R_K)(f_j^{(l)}) = \Delta(f_j^{(l)}) \otimes z^{-2j} = \sum_i f_i^{(l)} \otimes t_{ij}^{(l)} \otimes z^{-2j},$$

$$((\mathrm{id} \otimes R_K) \circ \Delta)(f_j^{(l)}) = \sum_i f_i^{(l)} \otimes R_K(t_{ij}^{(l)}).$$

Hence $R_K(t_{ij}^{(l)}) = t_{ij}^{(l)} \otimes z^{-2j}$. Similar reasoning shows that $L_K(t_{ij}^{(l)}) = z^{-2i} \otimes t_{ij}^{(l)}$. Thus, $t_{ij}^{(l)} \in \mathcal{A}[-2i, -2j]$. □

By Proposition 1.13, $t_{ij}^{(l)}$ satisfies the condition $\Delta(t_{ij}^{(l)}) = \sum_{k=-l}^{l} t_{ik}^{(l)} \otimes t_{kj}^{(l)}$. Since $e_i^{(l)} = t_{i,-l}^{(l)}$ by Lemma 11(i), we have

$$T_l^L(e_i^{(l)}) \equiv \Delta(e_i^{(l)}) = \sum_k t_{ik}^{(l)} \otimes e_k^{(l)}, \tag{38}$$

that is, $t_{ij}^{(l)}$ are the matrix coefficients for the left coaction T_l^L as well. The explicit form of $t_{ij}^{(l)}$ can be derived if one inserts the definition (33) of $e_i^{(l)}$ into (38), then uses (8) and takes on the right hand side the coefficient of $e_j^{(l)}$.

We close this subsection with two remarkable symmetry properties of the matrix elements $t_{ij}^{(l)}$. The first one will be used in Subsect. 4.3.1. Recall from Proposition 5 that θ is the algebra automorphism and ρ is the algebra anti-automorphism of $\mathcal{A} = \mathcal{O}(SL_q(2))$ given by $\theta(a) = a$, $\theta(b) = c$, $\theta(c) = b$, $\theta(d) = d$ and $\rho(a) = d$, $\rho(b) = c$, $\rho(c) = b$, $\rho(d) = a$, respectively.

Proposition 12. *$\theta(t_{ij}^{(l)}) = t_{ji}^{(l)}$ and $\rho(t_{ij}^{(l)}) = t_{-i,-j}^{(l)}$.*

Proof. Clearly, θ is a coalgebra anti-homomorphism of $\mathcal{A}$. From this fact and the formulas (36) and (38) we get

$$\begin{aligned}\sum\nolimits_j f_j^{(l)} \otimes t_{ji}^{(l)} &= \Delta(f_i^{(l)}) = (\Delta \circ \theta)(e_i^{(l)}) = (\tau \circ (\theta \otimes \theta) \circ \Delta)(e_i^{(l)}) \\ &= \sum\nolimits_j f_j^{(l)} \otimes \theta(t_{ij}^{(l)}),\end{aligned}$$

which yields the first assertion.

Let σ be the algebra anti-automorphism of $\mathcal{O}(\mathbb{C}_q^2)$ such that $\sigma(x) = y$ and $\sigma(y) = x$. We put $\alpha_k := \begin{bmatrix} 2l \\ l+k \end{bmatrix}_{q^{-2}}^{1/2}$. Recalling the definition of the corepresentation T_l on $\mathcal{O}(\mathbb{C}_q^2)_{2l}$ we have

$$\begin{aligned} T_l(\alpha_i x^{l-i} y^{l+i}) &= \alpha_i \varphi_R(x)^{l-i} \varphi_R(y)^{l+i} \\ = \alpha_i (x \otimes a + y \otimes c)^{l-i} (x \otimes b + y \otimes d)^{l+i} &= \sum \alpha_j x^{l-j} y^{l+j} \otimes t_{ji}^{(l)}. \end{aligned} \tag{39}$$

We now apply the algebra anti-automorphism $\sigma \otimes \rho$ of $\mathcal{O}(\mathbb{C}_q^2)_{2l} \otimes \mathcal{A}$ to the last two parts of (39) and replace (i, j) by $(-i, -j)$. Since $\alpha_k = \alpha_{-k}$, the third part of (39) and the first tensor factor of the fourth part remain unchanged. Equating the coefficients of $\alpha_j x^{l-j} y^{l+j}$ we get $t_{-i,-j}^{(l)} = \rho(t_{ij}^{(l)})$. □

4.2.4 Calculation of the Matrix Coefficients

In this subsection we express the matrix coefficients $t_{ij}^{(l)}$ explicitly in terms of the little q-Jacobi polynomials.

Since the basis element $e_i^{(l)}$ of V_l^L is a complex multiple of $a^{l-i} c^{l+i}$, we compute the matrix elements $t_{ij}^{(l)}$ by using formula (8). We carry out this calculation in the case where $i + j \leq 0$ and $i \geq j$. Applying formula (13) to the expression $a^{l+j-\mu} d^{l+j-\mu}$ in (8), it follows that the first tensor factor of the term $a^{l-j} c^{l+j}$ in (8) is

$$\begin{aligned}\tilde{t}_{ij}^{(l)} = \sum_{\mu \geq 0} \sum_{k \geq 0} a^{-i-j} c^{i-j} \begin{bmatrix} l-i \\ \mu \end{bmatrix}_{q^{-2}} \begin{bmatrix} l+i \\ l+j-\mu \end{bmatrix}_{q^{-2}} \begin{bmatrix} l+j-\mu \\ k \end{bmatrix}_{q^{-2}} \\ \times q^{\mu(\mu+i-j)-(l+j-\mu)(2\mu+i-j)+k^2-2k(l+j-\mu)} (bc)^{\mu+k}.\end{aligned}$$

(Note that $\tilde{t}_{ij}^{(l)}$ is a complex multiple of $t_{ij}^{(l)}$.) Changing the summation from k to $r = \mu + k$ we reduce this expression to

$$\tilde{t}_{ij}^{(l)} = \sum_{\mu,r} a^{-i-j} c^{i-j} q^{(l+j)(j-i)} \begin{bmatrix} l-i \\ \mu \end{bmatrix}_{q^{-2}} \begin{bmatrix} l+i \\ l+j-\mu \end{bmatrix}_{q^{-2}} \begin{bmatrix} l+j-\mu \\ r-\mu \end{bmatrix}_{q^{-2}}$$
$$\times q^{2\mu(\mu+i-j)-2(l+j)r+r^2} (bc)^r.$$

Let I_r denote the sum over μ on the right hand side of the last formula. Regrouping terms in q-binomial coefficients, we get

$$\begin{aligned} I_r &= \sum_{\mu} q^{2\mu(\mu+i-j)-2(l+j)r+r^2} \begin{bmatrix} l-i \\ \mu \end{bmatrix}_{q^{-2}} \begin{bmatrix} l+i \\ l+j-\mu \end{bmatrix}_{q^{-2}} \begin{bmatrix} l+j-\mu \\ r-\mu \end{bmatrix}_{q^{-2}} \\ &= \sum_{\mu} q^{2\mu(\mu+i-j)-2(l+j)r+r^2} \begin{bmatrix} l-i \\ \mu \end{bmatrix}_{q^{-2}} \begin{bmatrix} l+i \\ l+j-\mu \end{bmatrix}_{q^{-2}} \begin{bmatrix} i-j+r \\ r-\mu \end{bmatrix}_{q^{-2}}. \end{aligned}$$

This sum reduces to a hypergeometric series of type (2.33) and we have

$$I_r = (-q)^r \begin{bmatrix} l+i \\ i-j \end{bmatrix}_{q^{-2}} \frac{(q^{2(l+j)}; q^{-2})_r (q^{-2(l-j+1)}; q^{-2})_r}{(q^{-2}; q^{-2})_r (q^{-2(i-j+1)}; q^{-2})_r}.$$

Therefore,

$$\tilde{t}_{ij}^{(l)} = a^{-i-j} c^{i-j} q^{-(l+j)(j-i)} \begin{bmatrix} l+i \\ i-j \end{bmatrix}_{q^{-2}}$$
$$\times {}_2\varphi_1(q^{2(l+j)}, q^{-2(l-j+1)};\ q^{2(j-i-1)};\ q^{-2}, -q^{-1}bc).$$

Passing to $t_{ij}^{(l)}$ and using the notation (2.35) for ${}_2\varphi_1$ we obtain

$$t_{ij}^{(l)} = q^{-(l+j)(j-i)} a^{-i-j} c^{i-j} \begin{bmatrix} l+i \\ i-j \end{bmatrix}_{q^{-2}} \begin{bmatrix} l-j \\ i-j \end{bmatrix}_{q^{-2}}$$
$$\times {}_2\Phi_1(-l-j, l-j+1;\ i-j+1;\ q^{-2}, q^{-2}\zeta),$$

where $\zeta = -qbc$. By (2.73), $t_{ij}^{(l)}$ can be expressed in terms of the little q-Jacobi polynomials by

$$t_{ij}^{(l)} = N_{ij}^l a^{-i-j} c^{i-j} p_{l+j}(\zeta; q^{-2(i-j)}, q^{2(i+j} \mid q^{-2}), \qquad (40)$$

where

$$N_{ij}^l = q^{-(l+j)(j-i)} \begin{bmatrix} l+i \\ i-j \end{bmatrix}_{q^{-2}}^{1/2} \begin{bmatrix} l-j \\ i-j \end{bmatrix}_{q^{-2}}^{1/2}. \qquad (41)$$

Recall that we assumed that $i+j \le 0$ and $i \ge j$. In the same way the other cases can be treated. Retaining the notation (41), we then obtain

$$t_{ij}^{(l)} = N_{ji}^l a^{-i-j} b^{j-i} p_{l+i}(\zeta; q^{-2(j-i)}, q^{2(i+j)} \mid q^{-2}), \quad i+j \le 0,\ j \ge i, \quad (42)$$

$$t_{ij}^{(l)} = N_{-i,-j}^l p_{l-j}(\zeta; q^{-2(j-i)}, q^{-2(i+j)} \mid q^{-2}) b^{j-i} d^{i+j}, \quad i+j \ge 0,\ j \ge i, \quad (43)$$

$$t_{ij}^{(l)} = N_{-j,-i}^l p_{l-i}(\zeta; q^{-2(i-j)}, q^{-2(i+j)} \mid q^{-2}) c^{i-j} d^{i+j}, \quad i+j \ge 0,\ i \ge j. \quad (44)$$

4.2.5 The Peter–Weyl Decomposition of $\mathcal{O}(SL_q(2))$

Let $\mathcal{C}(T_l^R)$ denote the linear span of matrix elements $t_{ij}^{(l)}$, $i,j = -l, -l+1, \cdots, l$. It is clear from Proposition 1.13 that $\mathcal{C}(T_l^R)$ is a subcoalgebra of $\mathcal{O}(SL_q(2))$.

Theorem 13. *The set $\{t_{ij}^{(l)} \mid l \in \mathbb{N}_0,\ i,j = -l, -l+1, \cdots, l\}$ is a vector space basis of $\mathcal{O}(SL_q(2))$. The Hopf algebra $\mathcal{O}(SL_q(2))$ is a direct sum of the coalgebras $\mathcal{C}(T_l^R)$, that is,*

$$\mathcal{O}(SL_q(2)) = \bigoplus\nolimits_{l \in \frac{1}{2}\mathbb{N}} \mathcal{C}(T_l^R). \tag{45}$$

Proof. To prove the first assertion, we use essentially the results on the matrix elements $t_{ij}^{(l)}$ established in the previous subsection. From the expressions for $t_{ij}^{(l)}$ derived there we get

$$t_{ij}^{(l)} = e_{-2i,-2j} F_{ij}^{(l)}(\zeta) = \tilde{F}_{ij}^{(l)}(\zeta) e_{-2i,-2j}, \tag{46}$$

where $F_{ij}^{(l)}$ and $\tilde{F}_{ij}^{(l)}$ are elements of $\mathcal{A}[0,0] = \mathbb{C}[\zeta]$. It follows from (32), Proposition 10(iv) and (46) that for fixed l the matrix elements $t_{ij}^{(l)}$, $-l \le i,j \le l$, are linearly independent. From (40)–(44) we conclude that the function $F_{ij}^{(l)}(\zeta)$ is a polynomial of degree $l - \max\ (|i|,|j|)$. Therefore, the functions $F_{ij}^{(l)}$, $l = \max\ (|i|,|j|) + n$, $n = 0,1,2,\cdots$, form a basis of $\mathcal{A}[0,0] = \mathbb{C}[\zeta]$. It follows from Proposition 10(iv) and (46) that

$$\mathcal{A}[-2i,-2j] = \bigoplus_{l=\max(|i|,|j|)}^{\infty} \mathbb{C} t_{ij}^{(l)}. \tag{47}$$

Using once more formula (32) the first assertion follows.

The second assertion is an obvious consequence of the first one. □

The direct sum decomposition (45) is called the *Peter–Weyl decomposition* of the Hopf algebra $\mathcal{O}(SL_q(2))$.

The coalgebra $\mathcal{C}(T_l^R)$ carries the full information about the corepresentation $T_l^R \simeq T_l$. All basis elements $e_i^{(l)} = t_{i,-l}^{(l)}$ and $f_i^{(l)} = t_{-l,i}^{(l)}$ (by Lemma 11(i)) belong to $\mathcal{C}(T_l^R)$. Hence the left coaction T_l^L and the right coaction T_l^R are realized on the coalgebra $\mathcal{C}(T_l^R)$ and are given there by the comultiplication.

Theorem 13 is a basic result in the representation theory of the quantum group $SL_q(2)$. A first application of it is the

Proof of Theorem 9. We show that T_l^R is irreducible. Let $V \neq \{0\}$ be an invariant subspace for T_l^R and let $f = \sum_i \alpha_i f_i^{(l)}$ be a nonzero element of V. There is an index r with $\alpha_r \neq 0$. Since the matrix elements $t_{ij}^{(l)}$ are linearly independent, there are linear functionals F_n on $\mathcal{A}$ such that $F_n(t_{ij}^{(l)}) = \delta_{in}\delta_{jr}$.

We have $(\mathrm{id} \otimes F_n)T_l^R(f) = \alpha_r f_n^{(l)} \in V$. Thus, $f_n^{(l)} \in V$ for each n and hence $V = V_l^R$. This proves the irreducibility of T_l^R.

It is obvious that two corepresentations T_l^R and $T_{l'}^R$, $l \neq l'$, are not equivalent, because they act on spaces of different dimensions.

Now let $T : V \to V \otimes \mathcal{A}$ be an arbitrary irreducible corepresentation on a vector space V. Let $F \in V'$, $F \not\equiv 0$. Then the linear mapping $A := (F \otimes \mathrm{id})T : V \to \mathcal{A}$ intertwines the right coactions T and Δ. Let P_l^i be the projection of $\mathcal{A}$ onto the irreducible invariant subspace $\mathcal{C}_i(T_l^R) = \mathrm{Span}\,\{t_{ij}^{(l)}, j = -l, -l+1, \cdots, l\}$ of $\mathcal{A}$, given by the decomposition (45). Since $\varepsilon \circ A = F \neq 0$, A is not the null operator. Thus, $P_l^i \circ A \not\equiv 0$ for some l and i. Since $P_l^i \circ A \in \mathrm{Mor}\,(T, T_l^R)$, then $\ker P_l^i \circ A$ and $\mathrm{im}\, P_l^i \circ A$ are invariant subspaces for T and T_l^R, respectively. But both corepresentations T and T_l^R are irreducible, so that $\ker P_l^i \circ A = \{0\}$ and $\mathrm{im}\, P_l^i \circ A = V_l^R$. Hence $P_l^i \circ A : V \to V_l^R$ is bijective and T_l and T_l^R are equivalent. □

Since for fixed l the matrix elements $t_{ij}^{(l)}$ are linearly independent and $\Delta(t_{ij}^{(l)}) = \sum_k t_{ik}^{(l)} \otimes t_{kj}^{(l)}$, we have $\dim \mathcal{C}(T_l^R) = (2l+1)^2$ and the right (and also the left) coaction of Δ on $\mathcal{C}(T_l^R)$ is the direct sum of $2l+1$ copies of the irreducible corepresentation T_l^R (resp. T_l^L).

4.2.6 The Haar Functional of $\mathcal{O}(SL_q(2))$

In order to motivate the subsequent definition, we begin with an example.

Example 1. Let $\mathcal{A}$ be a Hopf algebra of functions on a group G. Let μ be a measure on G. We set $h(f) := \int f(g)d\mu(g)$, $f \in \mathcal{A}$. (For simplicity we suppress questions on the existence of integrals, but everything works if $\mathcal{A} = \mathrm{Rep}\, G$ for some compact group G and μ is a Borel measure.) The linear functional h on $\mathcal{A}$ is called left- and right-invariant if

$$h(f) \equiv \int f(g)d\mu(g) = \int f(g_0 g)d\mu(g) = \int f(g g_0)d\mu(g), \quad f \in \mathcal{A}, \tag{48}$$

for all $g_0 \in G$. In terms of the comultiplication on $\mathcal{A}$ these equalities read as $h(f)1 = ((\mathrm{id} \otimes h) \circ \Delta)(f) = ((h \otimes \mathrm{id}) \circ \Delta)(f)$, where 1 is the unit of $\mathcal{A}$. Note that (48) holds if μ is the Haar measure of a compact group G. △

Definition 3. *A linear functional h on a Hopf algebra $\mathcal{A}$ is called* invariant *if it satisfies the condition*

$$((\mathrm{id} \otimes h) \circ \Delta)(x) = h(x)I = ((h \otimes \mathrm{id}) \circ \Delta)(x), \quad x \in \mathcal{A}. \tag{49}$$

As a second application of Theorem 13 we get a simple proof for the existence and uniqueness of an invariant functional h on the Hopf algebra $\mathcal{A} = \mathcal{O}(SL_q(2))$ such that $h(1) = 1$. Since the matrix elements $t_{ij}^{(l)}$ form a vector space basis of $\mathcal{A}$, we can define a linear functional h on $\mathcal{A}$ by setting

$$h(t^{(0)}_{00}) \equiv h(I) = 1 \quad \text{and} \quad h(t^{(l)}_{ij}) = 0, \quad l > 0. \tag{50}$$

Since $\Delta(t^{(l)}_{ij}) = \sum_k t^{(l)}_{ik} \otimes t^{(l)}_{kj}$ by Proposition 1.13, one immediately sees that h is invariant. Conversely, if h is an invariant linear functional on $\mathcal{A}$, then condition (49), applied to $x = t^{(l)}_{nm}$, implies that $h(t^{(l)}_{nm}) = 0$ for $l > 0$. This proves that there is a unique invariant linear functional h on $\mathcal{A}$ satisfying $h(1) = 1$. This functional will be called the *Haar functional* of $\mathcal{A}$.

Next we want to find the explicit form of the Haar functional h.

Let $x \in \mathcal{A}[m,n]$. From the definition (24) of L_K and R_K and from the invariance of h we obtain $z^m h(x) = h(x)1 = h(x)z^n$. Therefore, $h(x) = 0$ if $(m,n) \neq (0,0)$. It remains to compute h on $\mathcal{A}[0,0] = \mathbb{C}[\zeta]$. We will show that

$$h(\zeta^n) = \frac{1-q^{-2}}{1-q^{-2(n+1)}}, \qquad n = 0,1,2,\cdots. \tag{51}$$

Let $P : \mathcal{A} \to \mathcal{A}[0,0]$ be the projection determined by the direct sum decomposition $\mathcal{A} = \bigoplus_{m,n} \mathcal{A}[m,n]$. By (3), (4), (13) and (2.17), we have

$$(\mathrm{id} \otimes P) \circ \Delta(\zeta^n) = \sum_{i+j=n} \begin{bmatrix} n \\ i \end{bmatrix}^2_{q^{-2}} q^{2ij} \zeta^j (\zeta; q^2)_i \otimes \zeta^i (q^{-2}\zeta; q^{-2})_j,$$

where the expressions $(a;q)_m$ are given by (2.3). Hence, the first equality in (49) yields

$$h(\zeta^n) \cdot I = \sum_{i+j=n} \begin{bmatrix} n \\ i \end{bmatrix}^2_{q^{-2}} q^{2ij} \zeta^j (\zeta; q^2)_i \, h(\zeta^i (q^{-2}\zeta; q^{-2})_j).$$

This is an equality of polynomials in $\zeta = -qbc$. Equating the coefficients of ζ we obtain the recurrence formula

$$h(\zeta^n) = \frac{1-q^{-2n}}{1-q^{-2(n+1)}} h(\zeta^{n-1}), \qquad n \geq 1.$$

Since $h(I) = 1$ by assumption, the latter implies (51).

Formula (51) can be used for the calculation of $h(f)$, where f is a polynomial in ζ. For example, we have

$$h(\zeta^r (\zeta; q^2)_s) = q^{-2(r+1)} \frac{(q^{-2}; q^{-2})_r (q^{-2}; q^{-2})_s (q^{-2}; q^{-2})_1}{(q^{-2}; q^{-2})_{r+s+1}}, \tag{52}$$

$$h(\zeta^r (q^{-2}\zeta; q^{-2})_s) = \frac{(q^{-2}; q^{-2})_r (q^{-2}; q^{-2})_s (q^{-2}; q^{-2})_1}{(q^{-2}; q^{-2})_{r+s+1}}. \tag{52'}$$

It follows from (11) that $S : \mathcal{A}[m,n] \to \mathcal{A}[-n,-m]$ and $S(\zeta) = \zeta$. Hence we get $h(S(x)) = h(x)$, $x \in \mathcal{O}(SL_q(2))$.

We summarize some of the preceding results in

Theorem 14. *There exists a unique invariant linear functional h on the Hopf algebra $\mathcal{O}(SL_q(2))$ such that $h(1) = 1$. It is called the Haar functional of $\mathcal{O}(SL_q(2))$. The Haar functional vanishes on the space $\mathcal{A}[m,n]$ if $(m,n) \neq (0,0)$ and on the basis elements $a^r b^k c^l$ and $b^k c^l d^r$ if $r \neq 0$ or if $r = 0$ and $k \neq l$. On the space $\mathcal{A}[0,0]$ and on the elements $b^k c^k$ it is given by (51) and by*

$$h(b^k c^k) = (-1)^k \frac{q - q^{-1}}{q^{k+1} - q^{-(k+1)}}, \quad k \in \mathbb{N}.$$

Remark 1. It can be shown (see Theorem 11.13 in Subsect. 11.2.1) that if a Hopf algebra $\mathcal{A}$ possesses a Haar functional (that is, an invariant linear functional h satisfying $h(1) = 1$), then any corepresentation of $\mathcal{A}$ is completely reducible. Taking this result for granted, it follows from Theorem 14 that *corepresentations of $\mathcal{O}(SL_q(2))$ are always completely reducible.* △

The Haar functional h on $\mathcal{O}(SL_q(2))$ is not *central*, that is, we do not have $h(xy) = h(yx)$ in general. (For instance, since q is not a root of unity, $h(ad - da) = (q - q^{-1})h(bc) \neq 0$.) The next proposition is a substitute for the missing centrality. Let ϑ be the algebra automorphism $\vartheta_{q^2,1}$ from Proposition 5, that is, ϑ is determined by the equations $\vartheta(a) = q^2 a$, $\vartheta(b) = b$, $\vartheta(c) = c$, $\vartheta(d) = q^{-2} d$. In particular, by (27), we have $\vartheta(x) = q^{m+n} x$ for $x \in \mathcal{A}[m,n]$.

Proposition 15. *For all $x, y \in \mathcal{O}(SL_q(2))$, we have $h(xy) = h(\vartheta(y)x)$.*

Proof. Since ϑ is an algebra homomorphism, it is clear that the assertion holds for $y_1 y_2$ (and all $x \in \mathcal{O}(SL_q(2))$) if it is true for y_1 and for y_2. Thus, it suffices to treat the cases $y = a, b, c, d$. Taking the basis elements for x and using the explicit description of h given in Theorem 14, this reduces to a number of straighforward verifications. We omit the details. □

4.3 The Compact Quantum Group $SU_q(2)$ and Its Representations

Recall that the quantum group $SU_q(2)$ is described by the Hopf $*$-algebra $\mathcal{O}(SU_q(2))$ which is just the Hopf algebra $\mathcal{O}(SL_q(2))$ for real q equipped with the $*$-structure defined by (21). Therefore, the theory developed in Sect. 4.2 applies in particular to $SU_q(2)$. But it still remains to deal with all questions and properties related to the $*$-structure.

Throughout this section q is a real number such that $q \neq 0, \pm 1$ and $\mathcal{A}$ denotes the Hopf $*$-algebra $\mathcal{O}(SU_q(2))$.

4.3.1 Unitary Representations of the Quantum Group $SU_q(2)$

Let us say that a matrix corepresentation $\mathbf{v} = (v_{ij})_{i,j=1}^d$ of a Hopf $*$-algebra $\mathcal{A}$ is *unitary* if

$$\sum_k v_{ki}^* v_{kj} = \sum_k v_{ik} v_{jk}^* = \delta_{ij} \cdot 1, \quad i,j = 1,2,\cdots,d. \tag{53}$$

Setting $\mathbf{v}^* = ((v^*)_{ij}) := (v_{ji}^*)$, (53) is equivalent to the equations $\mathbf{v}^*\mathbf{v} = \mathbf{v}\mathbf{v}^* = I$.

In what follows let $T^{(l)}$ denote the matrix corepresentation $(t_{ij}^{(l)})$ of $\mathcal{A}$.

Proposition 16. *For any $l \in \frac{1}{2}\mathbb{N}_0$, $T^{(l)}$ is a unitary matrix corepresentation of $\mathcal{O}(SU_q(2))$.*

Proof. We shall use the algebra automorphism θ of $\mathcal{A}$ which keeps a and d fixed and interchanges b and c. By the formulas (11) for the antipode S and (21) for the involution of $\mathcal{O}(SU_q(2))$, for any $n, m, r, s \in \mathbb{N}_0$ we have

$$(a^n b^m c^r d^s)^* = a^s(-q^{-1}b)^r(-qc)^m d^n = (S \circ \theta)(a^n b^m c^r d^s). \tag{54}$$

Since q is real, it is clear from the construction (compare (8), (33) and (38)) that each matrix element $t_{ij}^{(l)}$ is a *real* linear combination of terms $a^n b^m c^r d^s$. Therefore, by (54) and the first relation in Proposition 12, we have $(t_{ij}^{(l)})^* = (S \circ \theta)(t_{ij}^{(l)}) = S(t_{ji}^{(l)})$. This implies the equations (53), since, by (1.27), $\sum_k S(t_{ik}^{(l)})t_{kj}^{(l)} = \sum_k t_{ik}^{(l)} S(t_{kj}^{(l)}) = \delta_{ij}1$. □

A corepresentation $T : V \to V \otimes \mathcal{A}$ of a Hopf $*$-algebra $\mathcal{A}$ on a finite-dimensional Hilbert space V is said to be *unitary* if there is an orthonormal basis of V such that the corresponding matrix corepresentation of T (see Proposition 1.13) is unitary. A number of equivalent conditions will be given later in Proposition 11.11.

Let us define a scalar product on the vector space V_l^R by requiring that the basis $\{f_i^{(l)}\}$ of V_l^R from (34) is orthonormal. Then, by the previous definition and by Proposition 16, each corepresentation T_l^R of $\mathcal{O}(SU_q(2))$ on the Hilbert space V_l^R in unitary.

4.3.2 The Haar State and the Peter–Weyl Theorem for $\mathcal{O}(SU_q(2))$

Let h be the Haar functional of the Hopf algebra $\mathcal{O}(SL_q(2))$. We introduce two Hermitian forms on $\mathcal{A} \equiv \mathcal{O}(SU_q(2))$ by setting

$$\langle x, y \rangle_R = h(xy^*), \quad \langle x, y \rangle_L = h(x^*y), \qquad x, y \in \mathcal{O}(SU_q(2)). \tag{55}$$

Note that in our terminology scalar products are conjugate linear in the first variable and linear in the second one. We shall see below (see Theorem 17(i)) that $\langle \cdot, \cdot \rangle_L$ and $\overline{\langle \cdot, \cdot \rangle_R}$ are both scalar products on the vector space $\mathcal{A}$. In particular, we have $h(x^*x) = \langle x, x \rangle_L \geq 0$ for $x \in \mathcal{A}$ and $h(1) = 1$. Since a linear functional on a $*$-algebra with the two latter properties is usually called a state, h is said to be the *Haar state* of the quantum group $SU_q(2)$.

From the definition (55) it is obvious that $\langle xz, y \rangle_R = \langle x, yz^* \rangle_R$ and $\langle zx, y \rangle_L = \langle x, z^*y \rangle_L$. Moreover, by Proposition 15, $\langle x, y \rangle_L = \langle \vartheta(y), x \rangle_R$.

The next theorem can be considered as an analog of the classical Peter–Weyl theorem for the quantum group $SU_q(2)$.

Theorem 17. (i) $\langle\cdot,\cdot\rangle_L$ *and* $\overline{\langle\cdot,\cdot\rangle_R}$ *are scalar products on the vector space* $\mathcal{O}(SU_q(2))$. *The Peter–Weyl decomposition (45) of* $\mathcal{O}(SU_q(2))$ *is an orthogonal decomposition with respect to both scalar products.*

(ii) *The matrix elements* $t^{(l)}_{mn}$ *of the irreducible corepresentations* $T^{(l)}$ *satisfy the orthogonality relations*

$$\langle t^{(l)}_{mn}, t^{(l')}_{m'n'}\rangle_R = [2l+1]_q^{-1} q^{2n}\delta_{ll'}\delta_{mm'}\delta_{nn'}, \tag{56}$$

$$\langle t^{(l)}_{mn}, t^{(l')}_{m'n'}\rangle_L = [2l+1]_q^{-1} q^{-2m}\delta_{ll'}\delta_{mm'}\delta_{nn'}. \tag{57}$$

In the proof of this theorem we require the following lemma.

Lemma 18. *Let* $k,l \in \frac{1}{2}\mathbb{N}_0$ *and let* M *be a* $(2l+1)\times(2k+1)$ *matrix with complex entries. Let* $\tilde{M} := h(T^{(l)}MT^{(k)*})$ *and* $\tilde{M}' := h(T^{(l)*}MT^{(k)})$. *Then we have* $\tilde{M}=0$, $\tilde{M}'=0$ *if* $l \neq k$ *and* $\tilde{M}=\alpha I$, $\tilde{M}'=\alpha' I$ *for some* $\alpha,\alpha' \in \mathbb{C}$ *if* $l=k$.

Proof. We prove the assertion for the matrix $\tilde{M}$. The proof for $\tilde{M}'$ is analogous. We denote the matrix $T^{(l)}\otimes I$ by $T^{(l)}_1$ and the matrix $I\otimes T^{(l)}$ by $T^{(l)}_2$. Using the invariance condition (49) of the Haar functional h, we obtain

$$\begin{aligned} T^{(l)}\tilde{M}T^{(k)*} &= (\mathrm{id}\otimes h)(T^{(l)}_1 T^{(l)}_2 M T^{(k)*}_2 T^{(k)*}_1) \\ &= (\mathrm{id}\otimes h)(T^{(l)}_1 T^{(l)}_2 M (T^{(k)}_1 T^{(k)}_2)^*) \\ &= ((\mathrm{id}\otimes h)\circ\Delta)(T^{(l)}MT^{(k)*}) = h(T^{(l)}MT^{(k)*}) = \tilde{M}. \end{aligned}$$

Since $T^{(k)}$ is unitary by Proposition 16 and hence $T^{(k)*}T^{(k)} = I$, we have $T^{(l)}\tilde{M} = \tilde{M}T^{(k)}$. Therefore, $\tilde{M}$ defines a linear mapping, still denoted by $\tilde{M}$, of V^R_l to V^R_k which intertwines the irreducible corepresentations $T^{(l)}$ and $T^{(k)}$. Hence the invariant subspaces $\ker\tilde{M}$ and $\mathrm{im}\,\tilde{M}$ are either $\{0\}$ or the whole space. Thus, we get $\ker\tilde{M} = 0$ if $l\neq k$ and $\tilde{M}=\alpha I$ if $l=k$. (The latter reasoning is just Schur's lemma for Hopf algebras.) □

Proof of Theorem 17. (ii): Let E_{ij} be the $(2l+1)\times(2k+1)$ matrix having 1 in the (i,j)-position and 0 elsewhere. Let $\tilde{E}_{ij}$ be the corresponding matrix from Lemma 18. Then $(\tilde{E}_{ij})_{rs} = h(t^{(l)}_{ri}t^{(k)*}_{sj}) = \langle t^{(l)}_{ri}, t^{(k)}_{sj}\rangle_R$. By Lemma 18, we have $\langle t^{(l)}_{ri}, t^{(k)}_{sj}\rangle_R = 0$ if $l\neq k$. Let $l=k$. Since $t^{(l)}_{ri}t^{(l)*}_{sj} \in \mathcal{A}[2s-2r, 2j-2i]$ and $h(a)=0$ for all $a\in\mathcal{A}[m,n]$, $(m,n)\neq(0,0)$, then $\langle t^{(l)}_{ri}, t^{(k)}_{sj}\rangle_R = 0$ for $(r,i)\neq(s,j)$. Since $t^{(k)}_{rs}\in\mathcal{A}[-2r,-2s]$, we have $\vartheta(t^{(k)}_{rs}) = q^{-2(r+s)}t^{(k)}_{rs}$ and so

$$\langle t^{(l)}_{ij}, t^{(m)}_{rs}\rangle_L = \langle\vartheta(t^{(m)}_{rs}), t^{(l)}_{ij}\rangle_R = q^{-2(r+s)}\langle t^{(m)}_{rs}, t^{(l)}_{ij}\rangle_R. \tag{58}$$

Hence $\langle t^{(l)}_{ij}, t^{(k)}_{rs}\rangle_L = 0$ if $l\neq k$ and if $l=k$, $(i,j)\neq(r,s)$.

Let now $\tilde{E}_{jj} = h(T^{(l)}E_{jj}T^{(l)*})$. By Lemma 18, there exists a constant $\alpha_j \in \mathbb{C}$ such that $(\tilde{E}_{jj})_{ii} = h(t_{ij}^{(l)}t_{ij}^{(l)*}) = \langle t_{ij}^{(l)}, t_{ij}^{(l)} \rangle_R = \alpha_j$ for all i and j $(-l \le i, j \le l)$. In just the same way, considering $\tilde{E}'_{ij} = h(T^{(l)*}E_{ij}T^{(l)})$, we find that there exists a constant $\alpha'_i \in \mathbb{C}$ such that $\langle t_{ij}^{(l)}, t_{ij}^{(l)} \rangle_L = \alpha'_i$ for all i and j. By (58), we have $\alpha'_i = q^{-2(i+j)}\alpha_j$. Consequently, there exists a constant $\alpha \in \mathbb{C}$ such that $\alpha = q^{2i}\alpha'_i = q^{-2j}\alpha_j$ for all i and j. Since $t_{i,-l}^{(l)} = e_i^{(l)}$, then by the formulas (33), (13), (2.17) and (52) we have $\alpha_{-l} = q^{-4l}(1-q^{-2})/(1-q^{-4l-2})$. Therefore, $\alpha = q^{-2l}(1-q^{-2})/(1-q^{-4l-2})$ and so

$$\alpha_j = [2l+1]_q^{-1}q^{2j}, \qquad \alpha'_i = [2l+1]_q^{-1}q^{-2i}. \tag{59}$$

(i): By (56) and (57), the numbers $\langle t_{mn}^{(l)}, t_{mn}^{(l)} \rangle_L$ and $\langle t_{mn}^{(l)}, t_{mn}^{(l)} \rangle_R$ are positive for all real $q \ne 0, \pm 1$. Since the matrix elements $t_{mn}^{(l)}$ form a vector space basis of $\mathcal{A}$, the equations (56) and (57) imply that $\langle x, x \rangle_L > 0$ and $\langle x, x \rangle_R > 0$ for any nonzero $x \in \mathcal{A}$. Hence $\langle \cdot, \cdot \rangle_L$ and $\overline{\langle \cdot, \cdot \rangle_R}$ are scalar products on $\mathcal{A}$. The orthogonality of the Peter–Weyl decomposition follows at once from (ii). □

Let us turn once more to the Haar state. Formula (51) enables us to write the Haar state on $\mathbb{C}[\zeta]$ as a q-integral (see Subsect. 2.2.2). For $f(\zeta) \in \mathbb{C}[\zeta]$, we have

$$h(f) = \int_0^1 f(\zeta) d_{q^{-2}}\zeta = (1-q^{-2}) \sum_{j=0}^{\infty} q^{-2j} f(q^{-2j}) \ \text{ if } \ q > 1, \tag{60}$$

$$h(f) = \int_0^1 f(q^2\zeta) d_{q^2}\zeta = (1-q^2) \sum_{j=0}^{\infty} q^{2j} f(q^{2j+2}) \ \text{ if } \ 0 < q < 1. \tag{61}$$

Indeed, since these formulas are obviously true for $f(\zeta) = \zeta^n$ by (51), they are valid for any $f \in \mathbb{C}[\zeta]$.

Next we give the scalar product $\overline{\langle \cdot, \cdot \rangle_R}$ on a fixed space $\mathcal{A}[m,n]$. By Proposition 10(iv), $\mathcal{A}[m,n] = \mathbb{C}[\zeta]e_{mn}$, where the elements e_{mn} are defined by (28)–(31). From the formulas (21), (2.17) and (28)–(31) it is easy to compute that $e_{mn}e_{mn}^* = \Phi_{mn}(\zeta)$, where $\Phi_{mn}(\zeta)$ is the weight function given by

$$\Phi_{mn}(\zeta) = q^{(m-n)(2-m-n)/2}\zeta^{(n-m)/2}(\zeta; q^2)_{(m+n)/2}, \quad m+n \ge 0, \ m \le n,$$

$$\Phi_{mn}(\zeta) = q^{(m-n)(m+n)/2}\zeta^{(m-n)/2}(\zeta; q^2)_{(m+n)/2}, \quad m+n \ge 0, \ m \ge n,$$

$$\Phi_{mn}(\zeta) = \zeta^{(m-n)/2}(q^{-2}\zeta; q^{-2})_{-(m+n)/2}, \quad m+n \le 0, \ m \ge n,$$

$$\Phi_{mn}(\zeta) = q^{m-n}\zeta^{(n-m)/2}(q^{-2}\zeta; q^{-2})_{-(m+n)/2}, \quad m+n \le 0, \ m \le n.$$

In terms of these functions Φ_{mn} the scalar product $\overline{\langle \cdot, \cdot \rangle_R}$ for elements of $\mathcal{A}[m,n]$ can be derived from the formulas

$$\langle x, y \rangle_R = h(f_1(\zeta)\Phi_{mn}(\zeta)f_2(\zeta)^*), \qquad x = f_1(\zeta)e_{mn}, \ y = f_2(\zeta)e_{mn}. \tag{62}$$

The expression on the right hand side of (62) can be computed explicitly by means of the formulas (52) and (52′).

4.3.3 The Fourier Transform on $SU_q(2)$

By Theorem 17, the matrix elements $t_{ij}^{(l)}$ form a complete orthogonal system in $\mathcal{O}(SU_q(2))$ with respect to the scalar product $\overline{\langle\cdot,\cdot\rangle_R}$. To each element $f \in \mathcal{O}(SU_q(2))$ we associate a matrix $\hat{f}^{(l)} = (\hat{f}_{mn}^{(l)})$ by $\hat{f}_{mn}^{(l)} := h(ft_{mn}^{(l)*})$, $l \in \frac{1}{2}\mathbb{N}_0$, $m,n = -l,-l+1,\cdots,l$. The mapping $f \to \hat{f}^{(l)}$ is called the *Fourier transform of the Hopf $*$-algebra $\mathcal{O}(SU_q(2))$* or the *Fourier transform on the quantum group $SU_q(2)$*. Using the formula (56), it is easy to show that the inverse transform is

$$f = \sum\nolimits_{l\in\frac{1}{2}\mathbb{N}_0} [2l+1]_q \sum\nolimits_{i,j=-l}^{l} q^{-2i} \hat{f}_{ij}^{(l)} t_{ji}^{(l)},$$

and that the following Plancherel formula holds:

$$\langle f,g\rangle_R = \sum\nolimits_{l\in\frac{1}{2}\mathbb{N}_0} [2l+1]_q \sum\nolimits_{i,j=-l}^{l} q^{-2i} \hat{f}_{ij}^{(l)} \overline{\hat{g}_{ij}^{(l)}},$$

where $[2l+1]_q$ is given by (2.1).

4.3.4 $*$-Representations and the C^*-Algebra of $\mathcal{O}(SU_q(2))$

The Hopf $*$-algebra $\mathcal{A} \equiv \mathcal{O}(SU_q(2))$ is in particular a $*$-algebra, so it is natural to look for its $*$-representations. Let $\tilde{\pi}$ be a $*$-representation of $\mathcal{A}$ on a vector space V equipped with a scalar product $\langle\cdot,\cdot\rangle$. By (22), we have

$$\begin{aligned}\|\tilde{\pi}(a)v\|^2 + \|\tilde{\pi}(c)v\|^2 &= \langle\tilde{\pi}(a)^*\tilde{\pi}(a)v,v\rangle + \langle\tilde{\pi}(c)^*\tilde{\pi}(c)v,v\rangle \\ &= \langle\tilde{\pi}(a^*a+c^*c)v,v\rangle = \langle\tilde{\pi}(1)v,v\rangle = \|v\|^2\end{aligned}$$

for $v \in V$, so the operators $\tilde{\pi}(a)$ and $\tilde{\pi}(c)$ are bounded on V with norms not exceeding 1. This implies that the operators $\tilde{\pi}(a^*)$ and $\tilde{\pi}(c^*)$ are also bounded and have norms less than or equal to 1. Since any $x \in \mathcal{A}$ is a linear combination of elements $a^i(c^*)^j c^k (a^*)^l$, it follows that $\tilde{\pi}(x)$ is bounded on V and that there is a constant C_x not depending on the $*$-representation π (!) such that $\|\tilde{\pi}(x)\| \le C_x$. Thus any operator $\tilde{\pi}(x)$ extends by continuity to an operator $\pi(x)$ on the Hilbert space $\mathfrak{H}$ obtained by completing $(V,\langle\cdot,\cdot\rangle)$. Then π is a $*$-representation of $\mathcal{A}$ by bounded operators on the Hilbert space $\mathfrak{H}$.

For $x \in \mathcal{A}$, we set

$$\|x\| := \sup_{\pi} \|\pi(x)\|,$$

where π runs through the $*$-representations of $\mathcal{A}$ on Hilbert spaces. Since $\|x\| = \sup_\pi \|\pi(x)\| \le C_x$, $\|\cdot\|$ is finite on $\mathcal{A}$. We shall show that $\|x\| = 0$ implies $x = 0$. Let us consider the vector space $\mathcal{A}$ endowed with the scalar product $\langle\cdot,\cdot\rangle_L$ defined by (55). It is easily verified that $\tilde{\pi}(x)y := xy$, $x,y \in \mathcal{A}$, is a $*$-representation of $\mathcal{A}$ on $(\mathcal{A},\langle\cdot,\cdot\rangle_L)$. Let π be the corresponding $*$-representation on the Hilbert space completion of $(\mathcal{A},\langle\cdot,\cdot\rangle_L)$. If $\|x\| = 0$, then

$\pi(x) = 0$ and so $\|\pi(x)1\|_L^2 = \langle x, x\rangle_L = 0$. Because $\langle \cdot, \cdot \rangle_L$ is a scalar product on $\mathcal{A}$, we get $x = 0$. Hence $\|\cdot\|$ is a norm on $\mathcal{A}$ which obviously has the C^*-property $\|x^*x\| = \|x\|^2$, $x \in \mathcal{A}$. Therefore, the completion of $\mathcal{A}$ in the norm $\|\cdot\|$ is a C^*-algebra. It is called the *C^*-algebra of the compact quantum group $SU_q(2)$*.

Next we describe the structure of $*$-representations of $\mathcal{A}$ by means of the following operator-theoretic model. We suppose that $q \in \mathbb{R}$, $|q| < 1$.

Let $\mathfrak{G}$ and $\mathfrak{H}_0$ be Hilbert spaces and let v and w be unitary operators on $\mathfrak{G}$ and $\mathfrak{H}_0$, respectively. Let $\mathfrak{H} = \bigoplus_{n=0}^{\infty} \mathfrak{H}_n$, where $\mathfrak{H}_n := \mathfrak{H}_0$. On the Hilbert space $\mathfrak{G} \oplus \mathfrak{H}$ we define operators a' and c' by

$$a' = v, \quad c' = 0 \quad \text{on} \quad \mathfrak{G},$$

$$a'(\eta_n) = ((1-q^2)^{1/2}\eta_1, (1-q^4)^{1/2}\eta_2, \cdots), \quad c'(\eta_n) = (q^n w \eta_n) \quad \text{for} \quad (\eta_n) \in \mathfrak{H}.$$

Then the adjoint operators a'^* and c'^* act on $\mathfrak{H}$ by

$$a'^*(\eta_n) = (0, (1-q^2)^{1/2}\eta_0, (1-q^4)^{1/2}\eta_1, \cdots), \quad c'^*(\eta_n) = (q^n w^* \eta_n).$$

Using these formulas one immediately verifies that a' and c' satisfy the relations (22), so there exists a unique $*$-representation π of the $*$-algebra $\mathcal{O}(SU_q(2))$ on the Hilbert space $\mathfrak{G} \oplus \mathfrak{H}$ such that $\pi(a) = a'$ and $\pi(c) = c'$.

Proposition 19. *Any $*$-representation of the $*$-algebra $\mathcal{O}(SU_q(2))$ on a Hilbert space is unitarily equivalent to one of the above form.*

Proof. Consider an arbitrary $*$-representation of $\mathcal{O}(SU_q(2))$ on a Hilbert space $\mathcal{E}$. For notational simplicity we denote the images of a, b, c, d by the same symbols. Throughout this proof we freely use the relations (22) and the polar decomposition $a = u|a|$ of the operator a (see, for instance, [RS]). First we show that a^* is injective. Let $a^*\eta = 0$. Then $q^2c^*c\eta = \eta$ and so $|q|^2\|c\eta\| = \|\eta\|$. Since $\|c\eta\|^2 \le \|c\eta\|^2 + \|a\eta\|^2 = \langle (a^*a + c^*c)\eta, \eta\rangle = \|\eta\|^2$ and $|q| < 1$, we conclude that $\eta = 0$. Thus, a^* is injective and hence $uu^* = I$. Since $aa^* = u|a|^2u^* = ua^*au^*$, the two last equations of (22) imply that

$$I = aa^* + q^2c^*c = ua^*au^* + q^2c^*c = u(I - c^*c)u^* + q^2c^*c.$$

Therefore, since $uu^* = I$ as just shown, we obtain

$$uc^*cu^* = q^2c^*c. \tag{63}$$

Now let $\mathfrak{H}$ be the closure of the linear span of subspaces $\mathfrak{D}_n := d^n(\ker a)$, $n \in \mathbb{N}_0$. By (22), $\mathfrak{H}$ is invariant under the operators a, b, c, d, u, u^*. Hence its orthogonal complement $\mathfrak{G}$ in $\mathcal{E}$ has the same property. Let us consider the restriction of the operators to $\mathfrak{G}$. Since $\ker a \subseteq \mathfrak{H}$, a is injective on $\mathfrak{G}$ and hence u is unitary on $\mathfrak{G}$. Therefore, by (63), the spectrum of the operator c^*c is invariant under multiplication by q^{-2}. Since c^*c is bounded and $|q| < 1$, this implies that $c^*c = 0$ and so $c = 0$ on $\mathfrak{G}$. Hence $aa^* = a^*a = I$ on $\mathfrak{G}$

by (22). That is, on $\mathfrak{G}$ the operators a and c are of the same form as in the above model.

Next we determine the operators on the subspace $\mathfrak{H}$. Let $\zeta \in \mathfrak{D}_n$. Then $\zeta = d^n\eta$ with $a\eta = 0$ and $c^*c\zeta = c^*cd^n\eta = q^{2n}d^nc^*c\eta = q^{2n}d^n(I - a^*a)\eta = q^{2n}\zeta$. Hence the closure $\mathfrak{H}_n$ of $\mathfrak{D}_n$ is the eigenspace of c^*c with eigenvalue q^{2n} and we have $\mathfrak{H} = \bigoplus_{n=0}^{\infty} \mathfrak{H}_n$. Here we used the assumption $q(q^2 - 1) \neq 0$, because then $\mathfrak{H}_n$ and $\mathfrak{H}_m$, $n \neq m$, belong to different eigenvalues of c^*c and so are mutually orthogonal. For $\zeta \in \mathfrak{H}_n$, $|a|^2\zeta = a^*a\zeta = (I - c^*c)\zeta = (1 - q^{2n})\zeta$ by (22) and hence $|a|\zeta = (1 - q^{2n})^{1/2}\zeta$. Since $d = a^* = |a|u^*$ maps $\mathfrak{D}_n$ onto $\mathfrak{D}_{n+1}$ and $uu^* = I$, it follows that u^* is an isometric mapping of $\mathfrak{H}_n$ onto $\mathfrak{H}_{n+1}$. We have

$$au^{*n}\eta = u|a|u^{*n}\eta = (1 - q^{2n})^{1/2}u^{*n-1}\eta, \quad \eta \in \mathfrak{H}_0. \tag{64}$$

Further, we have $|a|^2c = dac = cda = c|a|^2$ and so $|a|c = c|a|$. Thus, $cd = c|a|u^* = |a|cu^* = qdc = q|a|u^*c$. Since $|a|$ is injective on $cu^*\mathfrak{H}$ and $u^*c\mathfrak{H}$, we obtain $cu^* = qu^*c$ and hence

$$cu^{*n}\eta = q^nu^{*n}c\eta, \quad \eta \in \mathfrak{H}_0. \tag{65}$$

The operator c maps $\mathfrak{H}_0 = \ker |a|$ to itself (because of $|a|c = c|a|$) and satisfies $c^*c = cc^* = I$ on $\mathfrak{H}_0$. That is, the restriction of c to $\mathfrak{H}_0$ is a unitary operator on $\mathfrak{H}_0$. Thus, if we identify $\mathfrak{H}_n$ and $\mathfrak{H}_0$ via the isometry u^{*n}, we see from (64) and (65) that the operators a and c on $\mathcal{E}$ have the required structure. □

From Proposition 19 we easily derive the following

Corollary 20. *For any number $u \in \mathbb{C}$, $|u| = 1$, there are two irreducible $*$-representations π_u^1 and π_u^∞ of the $*$-algebra $\mathcal{O}(SU_q(2))$. The $*$-representation π_u^1 is one-dimensional and is determined by the formulas $\pi_u^1(a) = u$ and $\pi_u^1(c) = 0$. The $*$-representation π_u^∞ acts on a Hilbert space with orthonormal basis $\{\mathbf{e}_n \mid n = 0, 1, 2 \cdots\}$ by*

$$\pi_u^\infty(a)\mathbf{e}_n = (1 - q^{2n})^{1/2}\mathbf{e}_{n-1}, \quad \pi_u^\infty(c)\mathbf{e}_n = q^nu\mathbf{e}_n, \quad n \in \mathbb{N}_0,$$

where $\mathbf{e}_{-1} := \mathbf{0}$. These $*$-representations are mutually inequivalent. Any irreducible $*$-representation of $\mathcal{O}(SU_q(2))$ is unitarily equivalent to one of these representations.

4.4 Duality of the Hopf Algebras $U_q(\mathrm{sl}_2)$ and $\mathcal{O}(SL_q(2))$

4.4.1 Dual Pairing of the Hopf Algebras $U_q(\mathrm{sl}_2)$ and $\mathcal{O}(SL_q(2))$

Recall from Example 1.6 that there is a nondegenerate dual pairing of the Hopf algebras $U(\mathfrak{g})$ and $\mathcal{O}(G)$ given by formula (1.36) with $g = e$. Let us specialize this to the case $\mathfrak{g} = \mathrm{sl}(2, \mathbb{C})$ and $G = SL(2, \mathbb{C})$. In terms of the standard generators H, E, F of $\mathrm{sl}(2, \mathbb{C})$ satisfying the relations $[H, E] = 2E$,

$[H,F]=-2F$, $[E,F]=H$ and the coordinate functions $a,\ b,\ c,\ d$ of $SL(2,\mathbb{C})$, the dual pairing $\langle\cdot,\cdot\rangle$ of $U(\mathrm{sl}_2)$ and $\mathcal{O}(SL(2,\mathbb{C}))$ is expressed by the equations

$$-\langle H,a\rangle=\langle H,d\rangle=\langle E,c\rangle=\langle F,b\rangle=1 \quad \text{and zero otherwise.} \tag{66}$$

The next theorem says that there exists a similar dual pairing in the quantum case.

Theorem 21. *There exists a unique dual pairing $\langle\cdot,\cdot\rangle^{\smile}$ of the Hopf algebras $\breve{U}_q(\mathrm{sl}_2)$ and $\mathcal{O}(SL_q(2))$ such that*

$$\langle K,a\rangle^{\smile}=q^{-1/2},\quad \langle K,d\rangle^{\smile}=q^{1/2},\quad \langle E,c\rangle^{\smile}=\langle F,b\rangle^{\smile}=1, \tag{67}$$

$$\langle K,b\rangle^{\smile}=\langle K,c\rangle^{\smile}=\langle E,a\rangle^{\smile}=\langle E,b\rangle^{\smile}=\langle E,d\rangle^{\smile}=\langle F,a\rangle^{\smile}=\langle F,c\rangle^{\smile}=\langle F,d\rangle^{\smile}=0. \tag{68}$$

Likewise, there exists a unique dual pairing $\langle\cdot,\cdot\rangle$ of the Hopf algebras $U_q(\mathrm{sl}_2)$ and $\mathcal{O}(SL_q(2))$ such that the nonvanishing pairings between the generators are

$$\langle K,a\rangle=q^{-1},\quad \langle K,d\rangle=q,\quad \langle E,c\rangle=\langle F,b\rangle=1. \tag{69}$$

If q is not a root of unity, both pairings $\langle\cdot,\cdot\rangle$ and $\langle\cdot,\cdot\rangle^{\smile}$ are nondegenerate.

The proof of Theorem 21 is given below. The dual pairings are completely determined by their values on the generators, but for computations it is often useful to know them on the whole algebras.

Proposition 22. *If q is not a root of unity, the dual pairing $\langle\cdot,\cdot\rangle^{\smile}$ of the Hopf algebras $\breve{U}_q(\mathrm{sl}_2)$ and $\mathcal{O}(SL_q(2))$ is given by the formulas*

$$\langle K^mE^nF^l,d^sc^rb^t\rangle^{\smile}=q^{(n-r)^2}\begin{bmatrix} s \\ n-r\end{bmatrix}_{q^2}\gamma^{srt}_{mnl},$$

if $0\le n-r=l-t\le s$, $\langle K^mE^nF^l,d^sc^rb^t\rangle^{\smile}=0$ otherwise, and

$$\langle K^mE^nF^l,a^sc^rb^t\rangle^{\smile}=\delta_{rn}\delta_{tl}\gamma^{-s,r,t}_{mnl},$$

where

$$\gamma^{srt}_{mnl}=\frac{q^{m(s+r-t)/2}q^{-s(n+l)/2}}{q^{n(n-1)/2}q^{l(l-1)/2}}\,\frac{(q^2;q^2)_l(q^2;q^2)_n}{(1-q^2)^{l+n}}.$$

Proof. A proof can be given by induction using the relations (1.31) and (1.32). We do not carry out the technical details (see [Koe]). □

Let us discuss the previous results. First we note that both pairings $\langle\cdot,\cdot\rangle$ and $\langle\cdot,\cdot\rangle^{\smile}$ yield the equations (66) in the limit $q\to 1$ if we put formally $K=e^{hH}$ and $q=e^h$. Further, both pairings become degenerate when q is a root of unity. For instance, if $q^k=1$, one verifies that the nonzero element $a^{2k}-1$ is annihilated by the whole algebra $U_q(\mathrm{sl}_2)$ resp. $\breve{U}_q(\mathrm{sl}_2)$.

Let φ be the embedding of the Hopf algebra $U_q(\mathrm{sl}_2)$ into $\breve{U}_q(\mathrm{sl}_2)$ defined in Subsect. 3.1.2. Then the two pairings $\langle\cdot,\cdot\rangle$ and $\langle\cdot,\cdot\rangle\breve{}$ are related by the formulas

$$\langle x,y\rangle = \langle\varphi(x),\vartheta_{1,q^{-1/2}}(y)\rangle\breve{} = \langle(\varphi\circ\vartheta_{q^{1/2}})(x),y\rangle\breve{}, \tag{70}$$

where $x\in U_q(\mathrm{sl}_2)$, $y\in\mathcal{O}(SL_q(2))$ and $\vartheta_{q^{1/2}}$ and $\vartheta_{1,q^{-1/2}}$ are Hopf algebra automorphisms of $U_q(\mathrm{sl}_2)$ and $\mathcal{O}(SL_q(2))$, respectively, from Propositions 3.6 and 5. Recall that $\vartheta_{1,q^{-1/2}}$ (resp. $\vartheta_{q^{1/2}}$) leaves a, d (resp. K) unchanged and maps b to $q^{-1/2}b$, c to $q^{1/2}c$ (resp. E to $q^{1/2}E$ and F to $q^{-1/2}F$). Combined with (70), Proposition 22 gives also the full dual pairing $\langle\cdot,\cdot\rangle$ on the whole algebras $U_q(\mathrm{sl}_2)$ and $\mathcal{O}(SL_q(2))$.

Next we turn to the relationship between pairings and $*$-structures. For real q the bilinear form $\langle\cdot,\cdot\rangle\breve{}$ is a dual pairing of the Hopf $*$-algebras $\breve{U}_q(\mathrm{su}_2)$ and $\mathcal{O}(SU_q(2))$ and also of the Hopf $*$-algebras $\breve{U}_q(\mathrm{su}_{1,1})$ and $\mathcal{O}(SU_q(1,1))$. Indeed, it suffices to check condition (1.41) on the generators which is easily done. But $\langle\cdot,\cdot\rangle\breve{}$ is not a dual pairing of the Hopf $*$-algebras $\breve{U}_q(\mathrm{sl}(2,\mathbb{R}))$ and $\mathcal{O}(SL_q(2,\mathbb{R}))$ for $|q|=1$. However, one can define another involution $*'$ on $\mathcal{O}(SL_q(2))$ by transposing the involution of $\breve{U}_q(\mathrm{sl}_2)$ according to formula (1.41) such that $(\mathcal{O}(SL_q(2)),*')$ becomes a Hopf $*$-algebra and $\langle\cdot,\cdot\rangle\breve{}$ is a pairing of the Hopf algebras $\breve{U}_q(\mathrm{sl}(2,\mathbb{R}))$ and $(\mathcal{O}(SL_q(2)),*')$. The involution $*'$ is determined by the relations $a^{*'}=a$, $b^{*'}=q^{-1}b$, $c^{*'}=qc$, $d^{*'}=d$.

Similarly, the bilinear form $\langle\cdot,\cdot\rangle$ on $U_q(\mathrm{sl}_2)\times\mathcal{O}(SL_q(2))$ is not a dual pairing of Hopf $*$-algebras for all three real forms. But, as just explained, it becomes a pairing of the corresponding Hopf $*$-algebras when passing to other involutions either of $U_q(\mathrm{sl}_2)$ or of $\mathcal{O}(SL_q(2))$.

Proof of Theorem 21. Let $\mathcal{B}$ be the subalgebra of $U_q(\mathrm{sl}_2)'$ generated by the matrix elements t_{ij}, $i,j=1,2$, of the representation $T_{1/2}$ of $U_q(\mathrm{sl}_2)$. Since $t_{ij}(xy)=\sum_k t_{ik}(x)t_{kj}(y)$ for $x,y\in U_q(\mathrm{sl}_2)$ and $t_{ij}(1)=\delta_{ij}$, it is not difficult to show that $\mathcal{B}$ becomes a Hopf algebra with comultiplication, counit and antipode induced by duality from $U_q(\mathrm{sl}_2)$. (This means that $\mathcal{B}$ is a Hopf subalgebra of the dual Hopf algebra $U_q(\mathrm{sl}_2)^\circ$, see Subsect. 1.2.8.) That is, by construction, we have a dual pairing $\langle x,f\rangle := f(x)$, $x\in U_q(\mathrm{sl}_2)$, $f\in\mathcal{B}$, of $U_q(\mathrm{sl}_2)$ and $\mathcal{B}$. As shown in Subsect. 3.4.2, the elements t_{ij} satisfy the relations (3.44) and (3.45). Since the latter are precisely the defining relations for the generators u^i_j of $\mathcal{A}:=\mathcal{O}(SL_q(2))$, there exists an algebra homomorphism $\psi:\mathcal{A}\to\mathcal{B}$ such that $\psi(u^i_j)=t_{ij}$, $i,j=1,2$. The formulas for the comultiplication and the counit of t_{ij} and u^i_j are the same. Thus, ψ is a bialgebra (and hence a Hopf algebra) homomorphism. Therefore, $\langle x,y\rangle := \langle x,\psi(y)\rangle=\psi(y)(x)$ is a dual pairing of the Hopf algebras $U_q(\mathrm{sl}_2)$ and $\mathcal{A}=\mathcal{O}(SL_q(2))$. By definition, $\langle x,u^i_j\rangle$ is just the matrix coefficient $t_{ij}(x)$. Inserting these values from (3.22) and (3.23) at $x=K,E,F$, we obtain the above formulas. The same proof works for the pairing $\langle\cdot,\cdot\rangle\breve{}$ as well.

Now we prove that the pairings are nondegenerate when q is not a root of unity. This can be derived from (70) and Proposition 22. As an example, suppose that $\langle x,y\rangle\check{} = 0$ for all $x \in \check{U}_q(\mathrm{sl}_2)$. We write the element y of $\mathcal{O}(SL_q(2))$ as a finite sum $y = \sum(\alpha_{lmn}a^l c^m b^n + \alpha_{-lmn}d^l c^m b^n)$. Assume that $y \neq 0$. We choose the smallest $m \in \mathbb{N}_0$ such that $\alpha_{lmn} \neq 0$ for some $l \in \mathbb{Z}$, $n \in \mathbb{N}_0$ and then the smallest $n \in \mathbb{N}_0$ such that $\alpha_{lmn} \neq 0$ for minimal m. By Proposition 22, $\langle K^l E^m F^n, y\rangle$ is a nonzero multiple of α_{lmn} which is the desired contradiction.

There is another proof of the nondegeneracy based on representation theory. We sketch it in the case of $U_q(\mathrm{sl}_2)$. For $n \in \mathbb{N}$, let $\mathcal{A}_n$ be the linear span of basis elements from $\mathcal{A} = \mathcal{O}(SL_q(2))$ of degree less than or equal to n. Obviously, $\dim \mathcal{A}_n = \sum_{k=0}^n (k+1)^2$. From the decomposition of the tensor products of representations in Subsect. 3.4.2 it is clear that $\mathcal{B}_n := \psi(\mathcal{A}_n)$ is spanned by the matrix elements of all irreducible representations T_l, $l = 0, \frac{1}{2}, \cdots, \frac{n}{2}$. From the representation theory of associative algebras (see, for example, [CR]) it is known that the set of these matrix elements is linearly independent. Thus, we have $\dim \mathcal{B}_n = \sum_{k=0}^n (k+1)^2$. Hence the restriction of ψ to $\mathcal{A}_n$ coincides with $\mathcal{B}_n$ for $n \in \mathbb{N}$ and ψ is bijective. Therefore, if $\langle x,y\rangle = \psi(y)(x) = 0$ for all $x \in U_q(\mathrm{sl}_2)$, then $\psi(y) = 0$ as a linear functional on $U_q(\mathrm{sl}_2)$ and so $y = 0$. This proves the nondegeneracy in the second variable.

If $\langle x,y\rangle = \psi(y)(x) = 0$ for all $y \in \mathcal{O}(SL_q(2))$, then we have in particular that $t^l_{mn}(x) = 0$ for all matrix elements t^l_{mn} (see Subsect. 3.4.2) of irreducible representations T_l of $U_q(\mathrm{sl}_2)$. Since the irreducible representations T_l, $l \in \frac{1}{2}\mathbb{N}_0$, separate the elements of $U_q(\mathrm{sl}_2)$ (this assertion will be proved in a more general context in Subsect. 7.1.5), we conclude that $x = 0$. □

Corollary 23. *If q is not a root of unity, then there is a Hopf algebra isomorphism ψ of $\mathcal{O}(SL_q(2))$ onto the (Hopf) subalgebra of the dual Hopf algebra $U_q(\mathrm{sl}_2)^\circ$ generated by the matrix coefficients t_{ij} of the representation $T_{1/2}$ such that $\psi(u^i_j) = t_{ij}$, $i,j = 1, 2$.*

Proof. As shown in the above proof, ψ is a Hopf algebra isomorphism. □

Recall that in the above approach in Sect. 4.1 the relations (3.44) and (3.45) for the matrix elements of the representation $T_{1/2}$ in the dual algebra $U_q(\mathrm{sl}_2)'$ have been taken as the defining relations of $\mathcal{O}(SL_q(2))$. Likewise, one can start with the Hopf algebra $\mathcal{O}(SL_q(2))$ and develop the theory of the Hopf algebra $U_q(\mathrm{sl}_2)$ in the dual of $\mathcal{O}(SL_q(2))$. We do not carry out the details and sketch only the idea. Let K and K^{-1} be the characters on the algebra $\mathcal{O}(SL_q(2))$ determined by $K(b) = K(c) = K^{-1}(b) = K^{-1}(c) = 0$, $K(a) = K^{-1}(d) = q^{-1}$, $K(d) = K^{-1}(a) = q$. Next one checks that there are linear functionals E, F on $\mathcal{O}(SL_q(2))$ such that $E(xy) = E(x)K(y) + \varepsilon(x)E(y)$ and $F(xy) = F(x)\varepsilon(y) + K^{-1}(x)F(y)$ for $x, y \in \mathcal{O}(SL_q(2))$ and $E(a) = E(b) = E(d) = E(1) = F(a) = F(c) = F(d) = F(1) = 0$, $E(c) = 1$, $F(b) = 1$. Then it is not difficult to verify that the functionals $E, F, K, K^{-1} \in \mathcal{O}(SL_q(2))'$ satisfy the defining relations of the algebra $U_q(\mathrm{sl}_2)$. The preceding formulas lead also to the coalgebra structure of $U_q(\mathrm{sl}_2)$, so that the Hopf algebra $U_q(\mathrm{sl}_2)$ can be recovered in this manner.

4.4.2 Corepresentations of $\mathcal{O}(SL_q(2))$ and Representations of $U_q(\mathrm{sl}_2)$

Since we have a dual pairing of the Hopf algebras $U_q(\mathrm{sl}_2)$ (resp. $\breve{U}_q(\mathrm{sl}_2)$) and $\mathcal{O}(SL_q(2))$, by Proposition 1.15 each corepresentation φ of $\mathcal{O}(SL_q(2))$ gives rise to a representation $\hat{\varphi}$ of $U_q(\mathrm{sl}_2)$ (resp. $\breve{U}_q(\mathrm{sl}_2)$) by formula (1.64). That is, we have $\hat{\varphi}(x)v = \sum v_{(0)}\langle x, v_{(1)}\rangle$ for $x \in U_q(\mathrm{sl}_2)$, when $\varphi(v) = \sum v_{(0)} \otimes v_{(1)}$. We will consider the passage from φ to $\hat{\varphi}$ in three interesting special cases. For the first two let us assume that q is not a root of unity.

First let φ be the corepresentation T_l^R from Sect. 4.2. Since $T_l^R(f_i^{(l)}) = \sum_j f_j^{(l)} \otimes t_{ji}^{(l)}$ by (36), we obtain $\hat{T}_l^R(x)f_i^{(l)} = \sum_j f_j^{(l)}\langle x, t_{ji}^{(l)}\rangle$. Using the expressions (40)–(44) for $t_{ij}^{(l)}$ and Theorem 21, one derives that

$$\hat{T}_l^R(K)f_i^{(l)} = q^{-2i}f_i^{(l)}, \quad \hat{T}_l^R(F)f_i^{(l)} = q^{i+1}([l+i+1][l-i])^{1/2}f_{i+1}^{(l)},$$

$$\hat{T}_l^R(E)f_i^{(l)} = q^{-i}([l-i+1][l+i])^{1/2}f_{i-1}^{(l)}.$$

Comparing these formulas with (3.22) and (3.23) we get the assertion of

Proposition 24. *For any $l \in \frac{1}{2}\mathbb{N}_0$, the representation $\hat{T}_l^R$ of $U_q(\mathrm{sl}_2)$ is equivalent to the representation $T_{1,l}$ defined in Subsect. 3.2.1.*

Because q is not a root of unity, the dual pairing $\langle\cdot,\cdot\rangle$ of $U_q(\mathrm{sl}_2)$ and $\mathcal{O}(SL_q(2))$ is nondegenerate. Therefore, any element $y \in \mathcal{O}(SL_q(2))$ can be identified with the functional $\langle\cdot, y\rangle$ on $U_q(\mathrm{sl}_2)$. As seen from the above formulas, under this identification the matrix elements $t_{ij}^{(l)} \in \mathcal{O}(SL_q(2))$ go into the functionals $t_{ij}^{(l)} \in U_q(\mathrm{sl}_2)'$ defined by (3.42). In particular this justifies denoting both quantities by $t_{ij}^{(l)}$. By Corollary 23, $\mathcal{O}(SL_q(2))$ becomes a Hopf subalgebra of $U_q(\mathrm{sl})^\circ$ in this manner. This subalgebra is just the linear span of coefficient functionals $t_{ij}^{(l)}$ for the representations $T_{1,l}$ of $U_q(\mathrm{sl}_2)$. Proposition 24 and the Hopf algebra embedding $\mathcal{O}(SL_q(2)) \subseteq U_q(\mathrm{sl}_2)^\circ$ have the following important consequence: *the results on decompositions of tensor product representations and on Clebsch–Gordan and Racah coefficients of $U_q(\mathrm{sl}_2)$ obtained in Sects. 3.4 and 3.5 remain valid for corepresentations of $\mathcal{O}(SL_q(2))$ when q is not a root of unity.*

Next let us consider the corepresentation $\varphi = \Delta$ of $\mathcal{O}(SL_q(2))$. By Theorem 13, $\mathcal{O}(SL_q(2))$ decomposes as a direct sum of subcoalgebras $\mathcal{C}(T_l^R)$, $l \in \frac{1}{2}\mathbb{N}_0$. As we have seen in Subsect. 4.2.5, the restriction of $\varphi = \Delta$ to $\mathcal{C}(T_l^R)$ is a direct sum of $2l+1$ corepresentations T_l^R. Therefore, by Proposition 24 and formula (3.24), the image $\hat{\varphi}(C_q)$ of the Casimir element C_q of $U_q(\mathrm{sl}_2)$ acts on $\mathcal{C}(T_l^R)$ as the multiple of the identity by the complex number $(q^{l+1}+q^{-l-1})/(q-q^{-1})^{-2}$. That is, *the Peter–Weyl decomposition (45) of $\mathcal{O}(SL_q(2))$ is just the eigenspace decomposition of the operator $\hat{\varphi}(C_q)$.*

Further, if φ is a coaction of $\mathcal{O}(SL_q(2))$ on a right comodule algebra $\mathcal{X}$, it is easy to verify that $\mathcal{X}$ becomes a left module algebra of $U_q(\mathrm{sl}_2)$ with action

$\hat{\varphi}$. Let us specialize this to the right comodule algebra $\mathcal{O}(\mathbb{C}_q^2)$ with coaction $\varphi := \varphi_R$ given by (20). Then we conclude that *$\mathcal{O}(\mathbb{C}_q^2)$ is a left module algebra of $U_q(\mathrm{sl}_2)$ with action $\hat{\varphi}$ determined by the equations*

$$\hat{\varphi}(K)x = qx, \quad \hat{\varphi}(K)y = q^{-1}y, \quad \hat{\varphi}(E)x = \hat{\varphi}(F)y = 0,$$

$$\hat{\varphi}(E)y = x, \quad \hat{\varphi}(F)x = y.$$

Finally, let us note that the one-dimensional representation of $U_q(\mathrm{sl}_2)$ with $K \to -1$, $E \to 0$, $F \to 0$ obviously is not of the form $\hat{\varphi}$ for some corepresentation φ of $\mathcal{O}(SL_q(2))$.

4.5 Quantum 2-Spheres

In this section we are concerned with a one-parameter family of quantum homogeneous spaces for the quantum group $SL_q(2)$. For special parameter values they are quantum analogs of two-dimensional spheres. Throughout this section we assume that q is not a root of unity.

4.5.1 A Family of Quantum Spaces for $SL_q(2)$

Let α and β be complex numbers such that $(\alpha, \beta) \neq (0,0)$. We denote by $\mathcal{X}_{q\alpha\beta}$ the algebra with three generators x_{-1}, x_0, x_1 and defining relations

$$x_0^2 - qx_1x_{-1} - q^{-1}x_{-1}x_1 = \beta 1, \tag{71}$$

$$(1-q^2)x_0^2 + qx_{-1}x_1 - qx_1x_{-1} = (1-q^2)\alpha x_0, \tag{72}$$

$$x_{-1}x_0 - q^2x_0x_{-1} = (1-q^2)\alpha x_{-1}, \tag{73}$$

$$x_0x_1 - q^2x_1x_0 = (1-q^2)\alpha x_1. \tag{74}$$

Further, let us recall from Sect. 4.2 that the three-dimensional irreducible corepresentation T_1 of $\mathcal{O}(SL_q(2))$ can be given by the matrix corepresentation

$$(t_{ij}^{(1)})_{i,j=-1,0,1} = \begin{pmatrix} a^2 & q'ab & b^2 \\ q'ac & 1+(q+q^{-1})bc & q'bd \\ c^2 & q'cd & d^2 \end{pmatrix}, \tag{75}$$

where $q' := (1+q^{-2})^{1/2}$.

Proposition 25. *Each algebra $\mathcal{X}_{q\alpha\beta}$ is a left quantum space and a right quantum space for $\mathcal{A} = \mathcal{O}(SL_q(2))$ with left coaction φ_L and right coaction φ_R determined by*

$$\varphi_L(x_i) = \sum_{j=-1}^{1} t_{ij}^{(1)} \otimes x_j \quad \text{and} \quad \varphi_R(x_i) = \sum_{j=-1}^{1} x_j \otimes t_{ji}^{(1)}, \quad i = -1, 0, 1. \tag{76}$$

Proof. We sketch the reasoning for φ_L. We first define an algebra homomorphism φ_L of the free algebra $\mathbb{C}\langle x_i\rangle$ with generators x_{-1}, x_0, x_1 to $\mathcal{A}\otimes\mathbb{C}\langle x_i\rangle$ by (76) and show then that φ_L passes to the quotient algebra $\mathcal{X}_{q\alpha\beta}$ of $\mathbb{C}\langle x_i\rangle$ by checking that φ_L preserves the relations (71)–(74). We omit these straightforward verifications and mention only the reason why it works: if we rewrite the equations (71)–(74) as $\sum_{i,j} a_{0,ij}x_ix_j = \beta 1$ and $\sum_{i,j} b_{k,ij}x_ix_j = (1-q^2)\alpha x_k$, $k = -1, 0, 1$, then the 1×3^2-matrix $(a_{0,ij})$ intertwines the tensor product corepresentation $T_1 \otimes T_1$ and the one-dimensional corepresentation T_0 and the 3×3^2-matrix $(b_{k,ij})$ intertwines the corepresentations $T_1 \otimes T_1$ and T_1. □

Each of the sets $\{x_{-1}^i x_0^j,\ x_0^i x_1^k \mid i,j \in \mathbb{N}_0,\ k \in \mathbb{N}\}$ and $\{x_0^i x_1^k,\ x_0^i x_{-1}^j \mid i,j \in \mathbb{N}_0,\ k \in \mathbb{N}\}$ forms a vector space basis of $\mathcal{X}_{q\alpha\beta}$. It can be seen from the relations (71)–(74) that these elements span $\mathcal{X}_{q\alpha\beta}$. Their linear independence can be derived from Lemma 8. One may also use the algebra homomorphism ψ into $\mathcal{O}(SL_q(2))$ defined below combined with the fact that the elements $a^n b^m c^r$, $b^m c^r d^s$, $m,n,r \in \mathbb{N}_0$, $s \in \mathbb{N}$, are linearly independent.

Let us say that two quantum spaces $\mathcal{X}$ and $\mathcal{Y}$ of a Hopf algebra $\mathcal{A}$ are *equivalent* (or *isomorphic*) if there exists an algebra isomorphism of $\mathcal{X}$ and $\mathcal{Y}$ which intertwines the coaction of $\mathcal{A}$ on $\mathcal{X}$ and $\mathcal{Y}$. Clearly, the quantum spaces $\mathcal{X}_{q\alpha\beta}$ and $\mathcal{X}_{q,c\alpha,c^2\beta}$ are isomorphic for any $c \in \mathbb{C}$, $c \neq 0$. (Indeed, under the scaling $x_i \to c^{-1}x_i$ of the generators the defining relations of $\mathcal{X}_{q\alpha\beta}$ go into that of $\mathcal{X}_{q,c\alpha,c^2\beta}$. Hence, there is an isomorphism θ of both quantum spaces such that $\theta(x_i) = c^{-1}x_i$.)

Next we remove one superfluous parameter by choosing a "normal form", denoted by $\mathcal{O}(S^2_{q\rho})$ for $\rho \in \mathbb{C}\bigcup\{\infty\}$, of the quantum space $\mathcal{X}_{q\alpha\beta}$. In the case when $\beta - \alpha^2 \neq 0$ we take numbers ρ and c such that $\rho^2 = \alpha^2(\beta-\alpha^2)^{-1}$, $c^2 = \beta - \alpha^2$ and $\alpha = c\rho$. Since then $(\alpha,\beta) = (c\rho, c^2(1+\rho^2))$, the quantum spaces $\mathcal{X}_{q\alpha\beta}$ and $\mathcal{X}_{q,\rho,(1+\rho^2)}$ are isomorphic. We denote the quantum space $\mathcal{X}_{q,\rho,(1+\rho^2)}$, $\rho \in \mathbb{C}$, by $\mathcal{O}(S^2_{q\rho})$. If $\beta - \alpha^2 = 0$, then $\mathcal{X}_{q\alpha\beta}$ is isomorphic to $\mathcal{X}_{q11}$. Let us denote the quantum space $\mathcal{X}_{q11}$ by $\mathcal{O}(S^2_{q\infty})$.

Now we realize the quantum spaces $\mathcal{O}(S^2_{q\rho})$ as subalgebras of $\mathcal{O}(SL_q(2))$. For $\rho \in \mathbb{C}$, we put

$$\tilde{x}_{-1} := (1+q^2)^{-1/2}a^2 + \rho(1+q^{-2})^{1/2}ac - q(1+q^2)^{-1/2}c^2, \tag{77}$$

$$\tilde{x}_0 := ba + \rho(1+(q+q^{-1})bc) - cd, \tag{78}$$

$$\tilde{x}_1 := (1+q^2)^{-1/2}b^2 + \rho(1+q^2)^{1/2}db - q(1+q^2)^{-1/2}d^2. \tag{79}$$

In the case $\rho = \infty$ we set

$$\tilde{x}_{-1} := (1+q^2)^{1/2}ac, \quad \tilde{x}_0 := 1 + (q+q^{-1})bc, \quad \tilde{x}_1 := (1+q^2)^{1/2}db. \tag{80}$$

Note that the elements $\tilde{x}_i$ are linear combinations of matrix coefficients $t^{(1)}_{ij}$. One verifies that the elements $\tilde{x}_{-1}, \tilde{x}_0, \tilde{x}_1$ of $\mathcal{O}(SL_q(2))$ satisfy the defining relations (71)–(74) of the algebra $\mathcal{O}(S^2_{q\rho})$. Hence there exists an algebra homomorphism $\psi : \mathcal{O}(S^2_{q\rho}) \to \mathcal{O}(SL_q(2))$ such that $\psi(x_i) = \tilde{x}_i$. By Proposition

31 below, ψ is injective, so that we can consider $\mathcal{O}(S^2_{q\rho})$ as a subalgebra of $\mathcal{O}(SL_q(2))$ by identifying x_i with $\tilde{x}_i$. Under this identification the coaction φ_R of $\mathcal{O}(SL_q(2))$ on $\mathcal{O}(S^2_{q\rho})$ corresponds to the comultiplication of $\mathcal{O}(SL_q(2))$.

If the parameters q and ρ are real, then the algebra $\mathcal{O}(S^2_{q\rho})$ becomes a $*$-algebra with involution determined by

$$x^*_{-1} = -q^{-1}x_1, \quad x^*_0 = x_0, \quad x^*_1 = -qx_{-1}. \tag{81}$$

The coactions φ_L and φ_R are then $*$-preserving for the real form $\mathcal{O}(SU_q(2))$, so the $*$-algebra $\mathcal{O}(S^2_{q\rho})$ is a left and right $*$-quantum space for the Hopf $*$-algebra $\mathcal{O}(SU_q(2))$. Moreover, the embedding ψ of $\mathcal{O}(S^2_{q\rho})$ into $\mathcal{O}(SL_q(2))$ is then a $*$-homomorphism. We call this $*$-algebra $\mathcal{O}(S^2_{q\rho})$ the *coordinate algebra of the quantum 2-sphere* $S^2_{q\rho}$.

Finally, let us look at the classical counterpart of the $*$-algebra $\mathcal{O}(S^2_{q\rho})$. In the limit $q \to 1$, the relations (72)–(74) mean nothing but that the generators x_i mutually commute and the equations (71) and (81) read as $x_0^2 + 2x_1x_1^* = (1+\rho^2)1$. Taking hermitian variables by setting $y_1 := x_0$, $\sqrt{2}y_2 := x_1 + x_1^*$ and $\sqrt{2}y_3 := -\sqrt{-1}(x_1 - x_1^*)$, the latter becomes the equation $y_1^2 + y_2^2 + y_3^2 = 1 + \rho^2$. That is, in the limit $q \to 1$ the $*$-algebra $\mathcal{O}(S^2_{q\rho})$ is just the coordinate $*$-algebra of a 2-sphere in $\mathbb{R}^3$.

4.5.2 Decomposition of the Algebra $\mathcal{O}(S^2_{q\rho})$

First we proceed in a similar manner as in Subsect. 4.2.2 and decompose the right quantum space $\mathcal{O}(S^2_{q\rho})$ into a direct sum with respect to the coaction of the quantum subgroup K. For this reason, we define an algebra homomorphism

$$R_K := (\mathrm{id} \otimes \phi_K)\varphi_R : \mathcal{O}(S^2_{q\rho}) \to \mathcal{O}(S^2_{q\rho}) \otimes \mathcal{O}(K),$$

where φ_R is the right coaction of $\mathcal{O}(SL_q(2))$ on $\mathcal{O}(S^2_{q\rho})$ and ϕ_K is the Hopf algebra homomorphism of $\mathcal{O}(SL_q(2))$ to $\mathcal{O}(K)$ given by (23). For $n \in 2\mathbb{Z}$ we set

$$\mathcal{O}(S^2_{q\rho})[n] := \{x \in \mathcal{O}(S^2_{q\rho}) \mid R_K(x) = x \otimes z^n\}.$$

By (76), (75) and (23), we get $x_0 \in \mathcal{O}(S^2_{q\rho})[0]$, $x_{-1} \in \mathcal{O}(S^2_{q\rho})[2]$ and $x_1 \in \mathcal{O}(S^2_{q\rho})[-2]$. Because R_K is an algebra homomorphism, the latter yields that

$$x^i_{-1}x^j_0x^k_1 \in \mathcal{O}(S^2_{q\rho})[2i-2k] \quad \text{for} \quad i,j,k \in \mathbb{N}_0. \tag{82}$$

As noted in Subsect. 4.5.1, the set $\{x^i_{-1}x^j_0, x^i_0x^k_1\}$ is a basis of $\mathcal{O}(S^2_{q\rho})$. Therefore (82) implies that

$$\mathcal{O}(S^2_{q\rho})[2n] = \begin{cases} x^n_{-1}\mathbb{C}[x_0] & \text{if} \quad n \geq 0, \\ \mathbb{C}[x_0]x_1^{-n} & \text{if} \quad n \leq 0. \end{cases}$$

In particular, the space $\mathcal{O}(S^2_{q\rho})[0]$ of K-invariant elements of $\mathcal{O}(S^2_{q\rho})$ coincides with the algebra $\mathbb{C}[x_0]$ of polynomials in the generator x_0. Further, by (82) we have the decomposition

$$\mathcal{O}(S^2_{q\rho}) = \bigoplus\nolimits_{m\in\mathbb{Z}} \mathcal{O}(S^2_{q\rho})[2m].$$

Our next result determines the decomposition of the coaction φ_R into irreducible components.

Theorem 26. *The corepresentation φ_R of $\mathcal{O}(SL_q(2))$ on $\mathcal{O}(S^2_{q\rho})$ is equivalent to the direct sum of irreducible corepresentations T_l, $l \in \mathbb{N}_0$, with multiplicity 1. That is, we have*

$$\varphi_R \simeq \bigoplus\nolimits_{l\in\mathbb{N}_0} T_l \quad on \quad \mathcal{O}(S^2_{q\rho}) = \bigoplus\nolimits_{l=0}^{\infty} V_l. \tag{83}$$

Proof. Let W^n be the linear span of elements $x^i_{-1}x^j_0x^k_1$, $i,j,k \in \mathbb{N}_0$, $i+j+k \le n$. We prove by induction on n that the restriction $\varphi_{R,n}$ of φ_R to W^n is equivalent to a direct sum of corepresentations T_l, $l = 0,1,\cdots,n$. Obviously, this in turn implies Theorem 26.

For $n = 0$ the assertion is trivially true. Suppose that it holds for some $n \in \mathbb{N}_0$. Since corepresentations of $\mathcal{O}(SL_q(2))$ are completely reducible (see Remark 1 in Subsect. 4.2.6) and any irreducible corepresentation of $\mathcal{O}(SL_q(2))$ is equivalent to some T_l, there exists a φ_R-invariant subspace W_{n+1} of W^{n+1} such that $W^{n+1} = W^n \oplus W_{n+1}$ and $\varphi_R\lceil W_{n+1} \simeq \bigoplus_{l'} T_{l'}$. On the other hand, since the basis elements $f^{(l)}_i$ (see (34)) of the representation space of $T_l \simeq T^R_l$ are in $\mathcal{A}[2l,-2i]$ by Lemma 11, we have $(\mathrm{id}\otimes\phi_K)T_l(f^{(l)}_i) = f^{(l)}_i \otimes z^{-2i}$ and so $\phi_K(t^{(l)}_{ji}) = \delta_{ij}z^{-2i}$. Since $\varphi_{R,n} \simeq \bigoplus_{l=0}^n T_l$ by the induction hypothesis, it follows that $R_K(W^n) = (\mathrm{id}\otimes\phi_K)\varphi_{R,n}(W^n) \subseteq \bigoplus_{i=-n}^n \mathcal{O}(S^2_{q\rho})[2i]$.

By (82), $x^{n+1}_{-1} \in W^{n+1}$ belongs to $\mathcal{O}(S^2_{q\rho})[2n+2]$ and up to multiples it is the only element of W^{n+1} in $\mathcal{O}(S^2_{q\rho})[2n+2]$. Hence there occurs at least one corepresentation $T_{l'}$ in the decomposition $\varphi_R\lceil W_{n+1} \simeq \bigoplus_{l'} T_{l'}$ such that $l' > n$. But the latter is possible only when $\varphi_R\lceil W_{n+1} \simeq T_{n+1}$, because we have $\dim W_{n+1} = \dim W^{n+1} - \dim W^n \le \dim \mathrm{Lin}\,\{x^i_{-1}x^j_0x^k_1 \mid i+j = m+k = n+1\} = 2n+1$. This completes the induction proof.

In the preceding reasoning only corepresentation theory of $\mathcal{O}(SL_q(2))$ was used. Another proof can be given by decomposing the representation $\hat{\varphi}_R$ of the algebra $U_q(\mathrm{sl}_2)$ which is induced by the dual pairing from Theorem 21. The key observation in this proof is the fact that the highest weight vectors of $\hat{\varphi}_R$ are precisely the complex multiples of x^n_{-1}, $n \in \mathbb{N}_0$. □

Remark 2. The following fact from the preceding proof will be used later. Up to multiples, x^l_{-1} and $f^{(l)}_{-l}$ are the only elements in the corresponding carrier spaces of the corepresentation T_l such that $(\mathrm{id}\otimes\phi_K)\varphi_R(x^l_{-1}) = x^l_{-1}\otimes z^{2l}$ and $(\mathrm{id}\otimes\phi_K)\Delta(f^{(l)}_{-l}) = f^{(l)}_{-l}\otimes z^{2l}$, respectively. Moreover, both elements are highest weight vectors for the irreducible representation $\hat{T}_l$ of $U_q(\mathrm{sl}_2)$. △

The quantum spaces $\mathcal{O}(S^2_{q\rho}) \equiv \mathcal{X}_{q,\rho,1+\rho^2}$ and $\mathcal{O}(S^2_{q,-\rho}) \equiv \mathcal{X}_{q,-\rho,1+\rho^2}$ are obviously isomorphic for $\rho \in \mathbb{C}$. As a first application of Theorem 26, we show that these are the only isomorphisms of the quantum spheres $\mathcal{O}(S^2_{q,\rho})$.

Proposition 27. *Let $\rho, \rho' \in \mathbb{C}\bigcup\{\infty\}$. The quantum spaces $\mathcal{O}(S^2_{q,\rho})$ and $\mathcal{O}(S^2_{q,\rho'})$ are isomorphic if and only if $\rho = \rho'$ or $\rho = -\rho'$.*

Proof. Suppose that θ is an isomorphism of the quantum spaces $\mathcal{O}(S^2_{q,\rho})$ and $\mathcal{O}(S^2_{q,\rho'})$. Since θ intertwines the coactions of $\mathcal{O}(SL_q(2))$, Theorem 26 implies that θ maps $V_1 \simeq \mathrm{Lin}\,\{x_i\}$ to $V_1' \simeq \mathrm{Lin}\,\{x_i'\}$, where $\{x_i\}$ and $\{x_i'\}$ denote the generators of $\mathcal{O}(S^2_{q,\rho})$ and $\mathcal{O}(S^2_{q,\rho'})$, respectively. Since the corepresentation T_1 is irreducible, it follows that $\theta(x_i) = \lambda x_i'$ for some $\lambda \in \mathbb{C}$.

From the relations (71)–(74) we see that the one-dimensional representations π of the algebra $\mathcal{O}(S^2_{q,\rho})$ are given by numbers $\pi(x_i)$, $i = -1, 0, 1$, such that $\pi(x_1)\pi(x_{-1}) = 0$, $\pi(x_0) = 1$ if $\rho = \infty$ and $(q + q^{-1})\pi(x_1)\pi(x_{-1}) = -1$, $\pi(x_0) = \rho$ if $\rho \in \mathbb{C}$. Being an algebra isomorphism, θ preserves the one-dimensional representations. Therefore, the previous description implies that $\mathcal{O}(S^2_{q,\infty})$ is not isomorphic to $\mathcal{O}(S^2_{q,\rho'})$ for $\rho' \in \mathbb{C}$ and that $\lambda^2 = 1$ and hence $\rho' = \rho$ or $\rho' = -\rho$ for $\rho, \rho' \in \mathbb{C}$. □

As a second application of Theorem 26 we obtain the existence of an invariant functional on $\mathcal{O}(S^2_{q,\rho})$. A linear functional h on the algebra $\mathcal{O}(S^2_{q\rho})$ is called *right-invariant* if

$$((h \otimes \mathrm{id}) \circ \varphi_R)(x) = h(x)1, \quad x \in \mathcal{O}(S^2_{q,\rho}).$$

Proposition 28. *There exists a unique right-invariant functional h on $\mathcal{O}(S^2_{q\rho})$ such that $h(1) = 1$. The functional h vanishes on all subspaces $\mathcal{O}(S^2_q)[2m]$, $m \neq 0$.*

Proof. The proof proceeds similarly to the case of the Haar functional in Subsect. 4.2.6. By Theorem 26, we can define a linear functional h on $\mathcal{O}(S^2_{q\rho})$ by $h(1) = 1$ and $h(x) = 0$ if $x \in V_l$, $l \neq 0$. As in Subsect. 4.2.6, h is right-invariant and any right-invariant linear functional on $\mathcal{O}(S^2_{q\rho})$ with $h(1) = 1$ is of this form.

For $x \in \mathcal{O}(S^2_{q\rho})[2n]$, we have $((h \otimes \phi_K)\varphi_R)(x) = h(x)z^{2n} = h(x)\phi_K(1)$ and so $h(x) = 0$ if $n \neq 0$. □

By Proposition 31 below the algebra $\mathcal{O}(S^2_{q,\rho})$ can be identified with a subalgebra of $\mathcal{O}(SL_q(2))$ such that the right coaction φ_R corresponds to the comultiplication. Then, by the uniqueness assertion of Proposition 28, the right-invariant functional h must be the restriction of the Haar functional of $\mathcal{O}(SL_q(2))$ to $\mathcal{O}(S^2_{q,\rho})$.

Finally, let us suppose that q is real. We define a Hermitian form $\langle \cdot, \cdot \rangle$ on $\mathcal{O}(S^2_{q\rho})$ by setting

$$\langle x, y \rangle = \overline{h(xy^*)}, \quad x, y \in \mathcal{O}(S^2_{q\rho}).$$

Because h is the restriction of the Haar state of $\mathcal{O}(SU_q(2))$ to its $*$-subalgebra $\mathcal{O}(S^2_{q,\rho})$, it follows from Theorem 17 that $\langle \cdot, \cdot \rangle$ is a scalar product on $\mathcal{O}(S^2_{q,\rho})$ and that the subspaces V_l are mutually orthogonal with respect to $\langle \cdot, \cdot \rangle$.

4.5.3 Spherical Functions on $S^2_{q\rho}$

There exists a basis $e_i^{(l)}$, $i = -l, -l+1, \cdots, l$, in the carrier space V_l of the corepresentation T_l in the decomposition (83) such that

$$T_l(e_i^{(l)}) = \sum\nolimits_{j=-l}^{l} e_j^{(l)} \otimes t_{ji}^{(l)} \quad i = -l, -l+1, \cdots, l,$$

where $t_{ij}^{(l)}$ are the matrix elements of the corepresentation T_l of $\mathcal{O}(SU_q(2))$ from Subsect. 4.2.4. By Remark 2 after Theorem 26, we can assume without loss of generality that $e_{-l}^{(l)} = x_{-1}^l$. If q is real, then the elements $e_i^{(l)} \in \mathcal{O}(S^2_{q\rho})$ are called *spherical functions on the quantum 2-sphere* $S^2_{q\rho}$.

Proposition 29. (i) *If* $0 \le i \le l$, *then*

$$e_i^{(l)} = N_i(-q^{-2(1+i)}\beta/\alpha; q^{-2})_{l-i} P_{l-i}^{(i,i)}(y;\ \alpha, \beta | q^{-2}) x_{-1}^i,$$

and if $-l \le i \le 0$, *then*

$$e_i^{(l)} = N_{-i}(-q^{2(i-1)}\beta/\alpha; q^{-2})_{l+i} x_1^{-i} P_{l+i}^{(-i,-i)}(y;\ \alpha, \beta | q^{-2}),$$

where $P_n^{(a,b)}(x;\ \alpha, \beta | q)$ *are big* q*-Jacobi polynomials (see Subsect. 2.3.4) and*

$$N_i = (-\alpha)^{l-i} q^{(l-i)(l+3i+3)/2} \begin{bmatrix} 2l \\ l-i \end{bmatrix}_{q^{-2}}^{-1/2} \begin{bmatrix} l \\ l-i \end{bmatrix}_{q^{-2}},$$

$$\alpha = ((q+q^{-1})^{-2} + \rho^2/4)^{1/2} + \rho/2, \quad \beta = \alpha - \rho.$$

(ii) *If* q *is real, the spherical functions* $e_i^{(l)}$ *satisfy the orthogonality relation*

$$\langle e_i^{(m)}, e_j^{(n)} \rangle = \delta_{mn}\delta_{ij} \frac{1-q^{-2}}{1-q^{-2(2m+1)}} \begin{bmatrix} 2m \\ m \end{bmatrix}_{q^{-2}}^{-1} \prod_{r=1}^{m} (\alpha + q^{-2r}\beta)(\beta + q^{-2r}\alpha).$$

Proof. The proof can be found in [NM]. □

4.5.4 An Infinitesimal Characterization of $\mathcal{O}(S^2_{q\rho})$

In this subsection we give an elegant description of $\mathcal{O}(S^2_{q\rho})$ as the subalgebra of elements of $\mathcal{O}(SL_q(2))$ which are infinitesimally invariant with respect to a certain twisted primitive element of $\breve{U}_q(\mathrm{sl}_2)$.

An element x of a bialgebra $\mathcal{A}$ is called *twisted primitive* with respect to a group-like element g of $\mathcal{A}$ if $\Delta(x) = g^{-1} \otimes x + x \otimes g$. If x is twisted primitive with respect to g, then we have $S(x) = -g^{-1}xg$ and $\varepsilon(x) = 0$.

From the formula (12) for the comultiplication of the Hopf algebra $\breve{U}_q(\mathrm{sl}_2)$ we see that any linear combination of $K - K^{-1}$, E and F is twisted primitive with respect to the group-like element K of $\breve{U}_q(\mathrm{sl}_2)$. We put

$$X_\rho := \rho q^{1/2}\frac{q+q^{-1}}{q^{-1}-q}(K-K^{-1})+qE+F, \quad \rho\in\mathbb{C}, \tag{84}$$

$$X_\infty := K-K^{-1}. \tag{85}$$

In the notation of Subsect. 1.3.5, we write $x.f = \sum x_{(2)}\langle f, x_{(1)}\rangle^{\vee}$ for $x \in \mathcal{O}(SL_q(2))$ and $f \in \breve{U}_q(\mathrm{sl}_2)$, where $\langle\cdot,\cdot\rangle^{\vee}$ is the dual pairing from Theorem 21. As noted in Subsect. 1.3.5, this defines a right action of the algebra $\breve{U}_q(\mathrm{sl}_2)$ on $\mathcal{O}(SL_q(2))$. Using the formulas (40)–(44) for the matrix coefficients and (67)–(68) for the pairing $\langle\cdot,\cdot\rangle^{\vee}$ we calculate

$$t^{(l)}_{ij}.K = q^{-i}t^{(l)}_{ij}, \quad t^{(l)}_{ij}.E = \sqrt{[l+i+1][l-i]}\,t^{(l)}_{i+1,j}, \tag{86}$$

$$t^{(l)}_{ij}.F = \sqrt{[l+i][l-i+1]}\,t^{(l)}_{i-1,j}. \tag{87}$$

Further, let

$$\tilde{\mathcal{O}}(S^2_{q\rho}) := \{x\in\mathcal{O}(SL_q(2)) \mid x.X_\rho = 0\} \tag{88}$$

denote the set of right X_ρ-invariant elements of $\mathcal{O}(SL_q(2))$. Using the fact that X_ρ is twisted primitive one immediately checks that $\tilde{\mathcal{O}}(S^2_{q\rho})$ is a subalgebra of $\mathcal{O}(SL_q(2))$ and $\Delta(\tilde{\mathcal{O}}(S^2_{q\rho}))\subseteq\tilde{\mathcal{O}}(S^2_{q\rho})\otimes\mathcal{O}(SL_q(2))$.

Lemma 30. $\dim\tilde{\mathcal{O}}(S^2_{q\rho})\bigcap\mathcal{C}(T^R_l) = 2l+1$ *for* $l\in\mathbb{N}_0$ *and* $\tilde{\mathcal{O}}(S^2_{q\rho})\bigcap\mathcal{C}(T^R_l) = \{0\}$ *for* $l\in\frac{1}{2}+\mathbb{N}_0$.

Proof. Let $\mathcal{C}_j(T^R_l)$ denote the linear span of matrix elements $t^{(l)}_{ij}$, $i=-l,-l+1,\cdots,l$. Clearly, $\mathcal{C}(T^R_l) = \bigoplus_{j=-l}^{l}\mathcal{C}_j(T^R_l)$. Since $\Delta(t^{(l)}_{ij}) = \sum_k t^{(l)}_{ik}\otimes t^{(l)}_{kj}$, the map $x\rightarrow x.X_\rho$ leaves each subspace $\mathcal{C}_j(T^R_l)$ invariant. Moreover, from (86) and (87) we see that X_ρ acts on the subspaces $\mathcal{C}_j(T^R_l)$, $j=-l,-l+1,\cdots,l$, by the same formulas.

In the case $\rho=\infty$ the assertion follows immediately from (86) and (87). Suppose now that $\rho<\infty$. We consider the equation $x.X_\rho = 0$ for $x = \sum_i \alpha_i t^{(l)}_{ij}\in\mathcal{C}_j(T^R_l)$, where $\alpha_i\in\mathbb{C}$. Inserting the formulas (84), (86) and (87) and comparing the coefficients of $t^{(l)}_{ij}$, it follows that this equation is equivalent to the recurrence relations

$$\alpha_i(q^{-i}-q^i)\gamma+\alpha_{i-1}q\sqrt{[l-i][l+i+1]}+\alpha_{i+1}\sqrt{[l-i+1][l+i]} = 0 \tag{89}$$

with $\alpha_{-l-1}=\alpha_{l+1}:=0$ for the coefficients α_i, where we set $\gamma=\rho q^{1/2}(q+q^{-1})/(q^{-1}-q)$. If α_{-l} is given, then (89) determines uniquely the coefficients $\alpha_{-l+1},\alpha_{-l+2},\cdots,\alpha_l,\alpha_{l+1}$. This procedure yields a solution of the equation $x.X_\rho=0$ if and only if $\alpha_{l+1}=0$. Therefore, since the recurrence relation (89) does not depend on j, it follows that either $\dim\tilde{\mathcal{O}}(S^2_{q\rho})\bigcap\mathcal{C}(T^R_l)=2l+1$ or $\tilde{\mathcal{O}}(S^2_{q\rho})\bigcap\mathcal{C}(T^R_l)=\{0\}$.

We find by a direct computation that $\dim\tilde{\mathcal{O}}(S^2_{q\rho})\bigcap\mathcal{C}(T^R_1) = 3$ and $\tilde{\mathcal{O}}(S^2_{q\rho})\bigcap\mathcal{C}(T^R_{1/2}) = \{0\}$. In fact, the solutions for the corepresentation T^R_1 are precisely the elements $\tilde{x}_{-1},\tilde{x}_0,\tilde{x}_1$ from (77)–(79), that is,

$$\tilde{x}_j = \beta_{-1} t^{(1)}_{-1,j} + \beta_0 t^{(1)}_{0,j} + \beta_1 t^{(1)}_{1,j},$$

where $\beta_{-1} = (1+q^2)^{-1/2}$, $\beta_0 = \rho$ and $\beta_1 = -q(1+q^2)^{-1/2}$.

Now we prove by induction on l that dim $\tilde{\mathcal{O}}(S^2_{q\rho}) \bigcap \mathcal{C}(T^R_l) = 2l+1$ for integral l. For $l=0$ this is trivially true. Suppose that it holds for a fixed $l \in \mathbb{N}$. Take a nonzero element $x = \sum_i \alpha_i t^{(l)}_{il} \in \tilde{\mathcal{O}}(S^2_{q\rho}) \bigcap \mathcal{C}(T^R_l)$. By the discussion after Proposition 24 in Subsect. 4.4.2, the product $t^{l_1}_{ij} t^{l_2}_{mn} \in \mathcal{O}(SL_q(2))$ can be computed by means of formula (3.43). In this way we obtain

$$x\tilde{x}_1 = \sum_{i,m} \alpha_i \beta_m t^{(l)}_{il} t^{(1)}_{m1} = \sum_{i,m} \alpha_i \beta_m C^{l,1,l+1}_{i,m,i+m} C^{l,1,l+1}_{l,1,l+1} t^{l+1}_{i+m,l+1}.$$

Thus, $x\tilde{x}_1 \in \mathcal{C}_{l+1}(T^R_{l+1})$. Since $\tilde{\mathcal{O}}(S^2_{q\rho})$ is an algebra and $x \in \tilde{\mathcal{O}}(S^2_{q\rho})$ by assumption and $\tilde{x}_1 \in \tilde{\mathcal{O}}(S^2_{q\rho})$, we have $x\tilde{x}_1 \in \tilde{\mathcal{O}}(S^2_{q\rho}) \bigcap \mathcal{C}(T^R_{l+1})$. Because the algebra $\mathcal{O}(SL_q(2))$ has no zero divisors, $x\tilde{x}_1 \neq 0$ and so $\tilde{\mathcal{O}}(S^2_{q\rho}) \bigcap \mathcal{C}(T^R_{l+1}) \neq \{0\}$. Hence we have dim $\tilde{\mathcal{O}}(S^2_{q\rho}) \bigcap \mathcal{C}(T^R_{l+1}) = 2(l+1)+1$, which completes the induction proof.

Next we show that $\tilde{\mathcal{O}}(S^2_{q\rho}) \bigcap \mathcal{C}(T^R_l) = \{0\}$ for half-integral l. Assume the contrary and let l be the smallest half-integer such that $\tilde{\mathcal{O}}(S^2_{q\rho}) \bigcap \mathcal{C}(T^R_l) \neq \{0\}$. As noted above, $l \neq \frac{1}{2}$, so $l \geq \frac{3}{2}$. Let $x = \sum_i \alpha_i t^{(l)}_{ik} \in \tilde{\mathcal{O}}(S^2_{q\rho}) \bigcap \mathcal{C}(T^R_l)$. From formula (3.43) it follows that for $k+j \leq l-1$ the coefficient of $x\tilde{x}_j$ at $t^{(l-1)}_{s,k+j}$ is

$$(\alpha_{s+1}\beta_{-1} C^{l,1,l-1}_{s+1,-1,s} + \alpha_s \beta_0 C^{l,1,l-1}_{s,0,s} + \alpha_{s-1}\beta_1 C^{l,1,l-1}_{s-1,1,s}) C^{l,1,l-1}_{k,j,k+j}.$$

Since l is minimal, the expression in brackets must vanish and we obtain the system of equations

$$\alpha_{s+1}\beta_{-1} C^{l,1,l-1}_{s+1,-1,s} + \alpha_s \beta_0 C^{l,1,l-1}_{s,0,s} + \alpha_{s-1}\beta_1 C^{l,1,l-1}_{s-1,1,s} = 0$$

for $s = -l+1, -l+2, \cdots, l-1$. Substituting the explicit form of the coefficients β_i and $C^{l,1,l-1}_{m,n,m+n}$ from Subsect. 3.4.3 and comparing with the recurrence relations (89) for α_i, we conclude that $\alpha_i = 0$ for all i. Thus we have $x = 0$ and $\tilde{\mathcal{O}}(S^2_{q\rho}) \bigcap \mathcal{C}(T^R_l) = \{0\}$. This contradiction completes the proof. □

Proposition 31. *There is an isomorphism ψ of the right quantum spaces $\mathcal{O}(S^2_{q\rho})$ (with respect to the coaction φ_R) and $\tilde{\mathcal{O}}(S^2_{q\rho})$ (with respect to the comultiplication) for the quantum group $\mathcal{O}(SL_q(2))$ such that $\psi(x_i) = \tilde{x}_i$, $i = -1, 0, 1$. Here $\tilde{x}_i$ are the elements of $\mathcal{O}(SL_q(2))$ defined by (77)–(80).*

Proof. As already noted above, the elements $\tilde{x}_j$ are in $\tilde{\mathcal{O}}(S^2_{q\rho})$ and satisfy the relations (71)–(74). Hence there is an algebra homomorphism $\psi : \mathcal{O}(S^2_{q\rho}) \to \tilde{\mathcal{O}}(S^2_{q\rho})$ such that $\psi(x_i) = \tilde{x}_i$. One also verifies that $\Delta(\tilde{x}_i) = \sum_j \tilde{x}_j \otimes t^{(l)}_{ji}$. Hence ψ intertwines the corresponding right coactions φ_R and Δ of $\mathcal{O}(SL_q(2))$ on $\mathcal{O}(S^2_{q\rho})$ and $\tilde{\mathcal{O}}(S^2_{q\rho})$, respectively.

The direct sum $\mathcal{C}(T_l^R) = \bigoplus_{i=-l}^{l} \mathrm{Lin}\, \{t_{ij}^{(l)} \mid j = -l, -l+1, \cdots, l\}$ gives a decomposition of the corepresentation $\Delta\lceil\mathcal{C}(T_l^R)$ into a direct sum of $2l+1$ irreducible corepresentations T_l. Therefore, by Lemma 30, the comultplication Δ considered as a corepresentation on the vector space $\tilde{\mathcal{O}}(S^2_{q\rho}) \bigcap \mathcal{C}(T_l^R)$ is equivalent to T_l for $l \in \mathbb{N}_0$. Hence there is a direct sum decomposition $\tilde{\mathcal{O}}(S^2_{q\rho}) = \bigoplus_{l\in\mathbb{N}_0} \tilde{V}_l$ such that $\Delta\lceil\tilde{V}_l \simeq T_l$. On the other hand, we have $\mathcal{O}(S^2_{q\rho}) = \bigoplus_{l\in\mathbb{N}_0} V_l$ and $\varphi_R\lceil V_l \simeq T_l$ by Theorem 26. Since ψ intertwines both coactions, ψ maps V_l to $\tilde{V}_l$ for any $l \in \mathbb{N}_0$. Because T_l is irreducible, $\psi\lceil V_l$ is either zero or bijective. Assume that $\psi(V_l) = \{0\}$. From Remark 2 in Subsect. 4.5.2 we know that $x_{-1}^l \in V_l$. Hence $\psi(x_{-1}^l) = \psi(x_{-1})^l = \tilde{x}_{-1}^l = 0$. By (77) and (80), this is impossible, because the elements $a^i c^j$ are linearly independent. Thus, the map $\psi : \mathcal{O}(S^2_{q\rho}) \to \tilde{\mathcal{O}}(S^2_{q\rho})$ is bijective. □

4.6 Notes

The quantum group $SL_q(2)$ and its corepresentation theory were introduced and developed independently by S. L. Woronowicz [Wor2] and L. L. Vaksman and Y. S. Soibelman [VS1]. The approach to $SL_q(2)$ via symmetries of the quantum plane is from [Man]. The matrix coefficients of irreducible corepresentations of $\mathcal{O}(SU_q(2))$ were determined in [Ko1] and [MMN]. Our treatment of harmonic analysis on $SL_q(2)$ followed mainly the paper [MMN], see also [KV]. The explicit pairing in Proposition 22 is taken from [Koe]. The quantum 2-spheres were invented by P. Podleś [Pod1]. The harmonic analysis on these spheres was developed in [NM] and [DK2]. The infinitesimal characterization of the quantum 2-spheres was discovered by T. Koornwinder (see [DK2] and [Ko4]). It has been a key result for the construction of quantum homogeneous spaces (see Sect. 11.6).

5. The q-Oscillator Algebras and Their Representations

The q-oscillator algebras are deformations of the harmonic oscillator algebra from quantum mechanics and of its central extension. This chapter gives an introduction into these algebras and their representations. Throughout we suppose that $q^2 \neq 1$.

5.1 The q-Oscillator Algebras $\mathcal{A}_q^c$ and $\mathcal{A}_q$

5.1.1 Definitions and Algebraic Properties

Recall that the one-dimensional harmonic oscillator algebra $\mathcal{A}$ is generated by two elements a and a^+ satisfying the commutation relation

$$[a, a^+] \equiv aa^+ - a^+a = 1. \tag{1}$$

In the Fock representation the generators a^+ and a act as the creation and annihilation operators, respectively, and the element $N := a^+a$ corresponds to the number operator. From relation (1) we immediately obtain that

$$[N, a^+] = a^+ \quad \text{and} \quad [N, a] = -a \tag{2}$$

in the algebra $\mathcal{A}$. Let $\mathcal{A}^c$ be the algebra with three generators a, a^+, N and defining relations (1) and (2). It is clear that the element $c := N - a^+a$ belongs to the center of the algebra $\mathcal{A}^c$ and that the algebra $\mathcal{A}^c$ is just the extension of the algebra $\mathcal{A}$ by an additional generator c which commutes with a and a^+. The q-oscillator algebras $\mathcal{A}_q$ and $\mathcal{A}_q^c$ defined below can be viewed as q-deformations of these two algebras $\mathcal{A}$ and $\mathcal{A}^c$, respectively.

Definition 1. *The* centrally extended q-oscillator algebra $\mathcal{A}_q^c$ *is the complex (associative unital) algebra with four generators a^+, a, q^N, q^{-N} and relations*

$$[a, a^+]_q \equiv aa^+ - qa^+a = q^{-N}, \quad q^{-N}q^N = q^Nq^{-N} = 1, \tag{3}$$

$$q^Na^+ = qa^+q^N, \quad q^Na = q^{-1}aq^N. \tag{4}$$

It might be necessary to emphasize that there is no element N in the algebra $\mathcal{A}_q^c$ and that q^N and q^{-N} are only symbols for two of the four abstract generators of the algebra $\mathcal{A}_q^c$.

The defining relations (3) and (4) of the algebra $\mathcal{A}_q^c$ are *formally* equivalent to the following relations:

$$[a, a^+]_q = q^{-N}, \quad Na^+ = a^+(N+1), \quad aN = (N+1)a. \tag{5}$$

Indeed, the second equation $Na^+ = a^+(N+1)$ of (5) implies that $f(N)a^+ = a^+f(N+1)$ for any polynomial f, so that formally $q^N a^+ = a^+ q^{N+1} = qa^+q^N$. Conversely, if we write $q = e^h$ and $q^N = e^{hN}$, then the relation $q^N a^+ = qa^+q^N$ means that $e^{hN}a^+ = e^h a^+ e^{hN}$. Differentiating the latter at $h = 0$ yields the equality $Na^+ = a^+(N+1)$. The equations $aN = (N+1)a$ and $q^N a = q^{-1}aq^N$ are related in a similar manner. (Note that for N-finite representations (see Sect. 5.2) such formal manipulations have a rigorous meaning.) In the limit $q \to 1$ the three relations of (5) go into the defining relations (1) and (2) of the algebra $\mathcal{A}^c$.

For $k \in \mathbb{Z}$ and $\alpha \in \mathbb{C}$, we introduce the notation

$$q^{kN+\alpha} := q^\alpha (q^N)^k \quad \text{and} \quad [N+\alpha] \equiv [N+\alpha]_q := \frac{q^{N+\alpha} - q^{-N-\alpha}}{q - q^{-1}}.$$

Using the relations (3) and (4) one easily verifies that the element

$$c_q := q^{1-N}([N]_q - a^+a) = q^{1-N}[N]_q + q^{-2N} - q^{-N}aa^+ \tag{6}$$

belongs to the center of the algebra $\mathcal{A}_q^c$. In the limit $q \to 1$ this element gives just the central element $c = N - a^+a$ of the algebra $\mathcal{A}^c$. If q is not a root of unity, then the center of $\mathcal{A}_q^c$ is generated by the element c_q. (This can be shown by modifying the proof of Theorem 6.45′ below.)

The defining relations (3) and (4) of $\mathcal{A}_q^c$ are not symmetric with respect to q and q^{-1}. The following algebra $\mathcal{A}_q$ can be regarded as a symmetrized version of the algebra $\mathcal{A}_q^c$.

Definition 2. *The* symmetric q-oscillator algebra $\mathcal{A}_q$ *is the complex (associative unital) algebra generated by four elements a^+, a, q^N, q^{-N} subject to the relations*

$$[a, a^+]_q \equiv aa^+ - qa^+a = q^{-N}, \quad [a, a^+]_{q^{-1}} \equiv aa^+ - q^{-1}a^+a = q^N, \tag{7}$$

$$q^{-N}q^N = q^N q^{-N} = 1, \quad q^N a^+ = qa^+q^N, \quad q^N a = q^{-1}aq^N. \tag{8}$$

We shall refer to both algebras $\mathcal{A}_q^c$ and $\mathcal{A}_q$ simply as *q-oscillator algebras*. From the relations (7) we derive at once that

$$a^+a = [N]_q \quad \text{and} \quad aa^+ = [N+1]_q. \tag{9}$$

This in turn leads to

$$[N]_q a^+ = a^+aa^+ = a^+[N+1]_q \quad \text{and} \quad a[N]_q = aa^+a = [N+1]_q a.$$

All such formulas are q-deformations of the corresponding formulas for the harmonic oscillator algebra $\mathcal{A}$. The two relations (7) in Definition 2 can be replaced by any two of the four relations (7) and (9).

In the limit $q \to 1$ the relations of the algebra $\mathcal{A}_q$ reduce to (1) and the relation $N = a^+a$. Hence the q-oscillator algebra $\mathcal{A}_q$ can be considered as a q-deformation of the oscillator algebra $\mathcal{A}$.

By definition, the algebra $\mathcal{A}_q$ is the quotient of the algebra $\mathcal{A}_q^c$ by the two-sided ideal $\mathcal{J}$ generated by the element $aa^+ - q^{-1}a^+a - q^N$. Since the second condition of (7) can be replaced by the first of (9), the ideal $\mathcal{J}$ is also generated by the central element c_q. Thus the element c_q given by (6) is zero in the algebra $\mathcal{A}_q$ and the algebra $\mathcal{A}_q$ has only the trivial center $\mathbb{C}\cdot 1$ if q is not a root of unity.

Proposition 1. *The following sets are bases of the complex vector spaces:*

$$\mathcal{A}_q^c: \quad \{(a^+)^k a^n q^{mN} \mid k,n \in \mathbb{N},\ m \in \mathbb{Z}\},\ \{a^n (a^+)^k q^{mN} \mid k,n \in \mathbb{N},\ m \in \mathbb{Z}\},$$

$$\mathcal{A}_q: \quad \{(a^+)^k q^{mN}, q^{mN} a^n \mid k,n \in \mathbb{N},\ m \in \mathbb{Z},\ (m,n) \neq (0,0)\}.$$

Proof. All assertions can be derived from the diamond Lemma 4.8. □

Proposition 2. (i) *There exist automorphisms ϑ_β, $\beta \in \mathbb{C}$, $\beta \neq 0$, and θ_n, $n \in \mathbb{Z}$, of the algebras $\mathcal{A}_q^c$ and $\mathcal{A}_q$ such that*

$$\vartheta_\beta(a) = \beta a, \quad \vartheta_\beta(a^+) = \beta^{-1}a^+, \quad \vartheta_\beta(q^{-N}) = q^{-N},$$

$$\theta_n(a) = q^{nN}a, \quad \theta_n(a^+) = a^+q^{-nN}, \quad \theta_n(q^{-N}) = q^{-N}.$$

(ii) *There are algebra automorphisms $\varphi_{\alpha,r}$, $\alpha, r \in \mathbb{C}$, $\alpha \neq 0$, of $\mathcal{A}_q^c$ determined by*

$$\varphi_{\alpha,r}(a) = q^r a, \quad \varphi_{\alpha,r}(a^+) = \alpha q^r a^+, \quad \varphi_{\alpha,r}(q^{-N}) = \alpha q^{2r} q^{-N}.$$

(iii) *There exists an automorphism ψ of the algebra $\mathcal{A}_q$ such that*

$$\psi(a) = a^+, \quad \psi(a^+) = a, \quad \psi(q^{-N}) = -qq^N.$$

Proof. One checks that the images of the generators under ϑ_β, θ_n, $\varphi_{\alpha,r}$ and ψ satisfy the defining relations of the algebras $\mathcal{A}_q^c$ and $\mathcal{A}_q$, respectively. Hence the above formulas indeed define algebra homomorphisms. Since the mappings are invertible, these homomorphisms are actually automorphisms. □

If q is not a root of unity, it can be shown that the automorphisms from Proposition 2 generate the groups of algebra automorphisms of $\mathcal{A}_q$ and $\mathcal{A}_q^c$, respectively.

We close this subsection by looking at $*$-structures on the q-oscillator algebras. If the parameter q is real, then both algebras $\mathcal{A}_q^c$ and $\mathcal{A}_q$ become $*$-algebras with involutions determined by $a^* := a^+$ and $(q^N)^* := q^N$. In the case $|q| = 1$ the algebra $\mathcal{A}_q^c$ is a $*$-algebra with involution such that $a^* := a^+$ and $(q^N)^* := q^{-N}$.

5.1.2 Other Forms of the q-Oscillator Algebra

In this subsection we list some other variants of the q–oscillator algebra that occur in the literature. They are derived by simple *formal* algebraic manipulations from the algebra $\mathcal{A}_q^c$. As abstract algebras they are of course different. But for N-finite representations of the algebra $\mathcal{A}_q^c$ these formal replacements lead to representations of the corresponding algebras.

First, inserting the formal substitution

$$b := q^{-N/2}a, \qquad b^+ := a^+q^{-N/2}$$

into (3) and using (4) we obtain the relation $bb^+ - b^+b = q^{-2N}$. Replacing formally q^{-2} by q in this equation and in (4), we obtain the relations

$$[b, b^+] \equiv bb^+ - b^+b = q^N, \quad q^N b^+ = qb^+q^N, \quad q^N b = q^{-1}bq^N. \tag{10}$$

Let $\mathcal{A}_b$ denote the algebra with generators $b,\ b^+,\ q^N,\ q^{-N}$ subject to the relations (10) and $q^Nq^{-N} = q^{-N}q^N = 1$.

Similarly, putting

$$A := q^{N/2}a, \qquad A^+ := a^+q^{N/2}$$

and then replacing q^2 by q, the relations (3) and (4) lead formally to the equations

$$[A, A^+]_q \equiv AA^+ - qA^+A = 1, \quad q^N A^+ = qA^+q^N, \quad q^N A = q^{-1}Aq^N. \tag{11}$$

The algebra with generators $A,\ A^+,\ q^N,\ q^{-N}$ and defining relations (11) and $q^Nq^{-N} = q^{-N}q^N = 1$ is denoted by $\mathcal{A}_A$. Some authors use the name q-oscillator algebra for the algebra with generators $A,\ A^+$ subject to the relation $AA^+ - qA^+A = 1$.

A two-parameter deformation of the harmonic oscillator algebra is derived from the q-oscillator algebra $\mathcal{A}_q^c$ in the following way. First we replace the parameter q by r. Setting

$$\hat{a} := r^{\alpha N}a, \qquad \hat{a}^+ := a^+r^{\alpha N} \tag{12}$$

for $\alpha \in \mathbb{C}$ and $q = r^{2\alpha+1}$, $p = r^{-2\alpha+1}$ in the defining equations (3) and (4) of the algebra $\mathcal{A}_r^c$ we formally obtain the relations

$$\hat{a}\hat{a}^+ - q\hat{a}^+\hat{a} = p^{-N}, \quad p^N\hat{a}^+ = p\hat{a}^+p^N, \quad p^N\hat{a} = p^{-1}\hat{a}p^N. \tag{13}$$

The algebra $\mathcal{A}_{p,q}^c$ with generators $\hat{a}$, $\hat{a}^+$, p^N, p^{-N} and relations (13) is called the *(p, q)-oscillator algebra*. Of course, the inverse of the formal substitution (12), that is, $a = r^{-\alpha N}\hat{a}$ and $a^+ = \hat{a}^+r^{-\alpha N}$ with $r^2 = pq$, $r^{4\alpha} = qp^{-1}$ transforms the two-parameter deformation $\mathcal{A}_{p,q}^c$ back to the algebra $\mathcal{A}_r^c$.

5.1.3 The q-Oscillator Algebra and the Quantum Algebra $\breve{U}_q(\mathrm{sl}_2)$

The quantum algebra $\breve{U}_q(\mathrm{sl}_2)$ admits various homomorphisms into the q-oscillator algebra $\mathcal{A}_q$. More precisely, we have the following

Proposition 3. (i) *For any $\alpha \in \mathbb{C}$, there exists the algebra homomorphism $\mathcal{T}_\alpha : \breve{U}_q(\mathrm{sl}_2) \to \mathcal{A}_q$ such that*

$$\mathcal{T}_\alpha(E) = a, \quad \mathcal{T}_\alpha(F) = a^+[N-2\alpha]_q, \quad \mathcal{T}_\alpha(K) = q^{N-\alpha}.$$

(ii) *There is an algebra homomorphism $\mathcal{T} : \breve{U}_{q^2}(\mathrm{sl}_2) \to \mathcal{A}_q$ determined by*

$$\mathcal{T}(E) = -\frac{(a^+)^2}{q+q^{-1}}, \quad \mathcal{T}(F) = \frac{a^2}{q+q^{-1}}, \quad \mathcal{T}(K) = q^{N+1/2}.$$

Proof. (i): As usual, it suffices to check the defining relations (3.10) and (3.11) of the algebra $\breve{U}_q(\mathrm{sl}_2)$. As a sample, we carry out this for the relation (3.11). By (8) and (9), we have

$$\begin{aligned}
[\mathcal{T}_\alpha(E), \mathcal{T}_\alpha(F)] &= aa^+[N-2\alpha]_q - a^+[N-2\alpha]_q a \\
&= aa^+[N-2\alpha]_q - a^+a[N-2\alpha-1]_q \\
&= [N+1]_q[N-2\alpha]_q - [N]_q[N-2\alpha-1]_q \\
&= (q-q^{-1})^{-2}\{(q^{N+1}-q^{-N-1})(q^{N-2\alpha}-q^{-N+2\alpha}) \\
&\quad -(q^N-q^{-N})(q^{N-2\alpha-1}-q^{-N+2\alpha+1})\} \\
&= \frac{q^{2(N-\alpha)}-q^{-2(N-\alpha)}}{q-q^{-1}} = \frac{\mathcal{T}_\alpha(K)^2-\mathcal{T}_\alpha(K)^{-2}}{q-q^{-1}}.
\end{aligned}$$

(ii): The proof follows by similar calculations using the relations

$$(a^+)^2a^2 = [N]_q[N-1]_q, \quad a^2(a^+)^2 = [N+1]_q[N+2]_q. \qquad \square$$

Our next aim is to realize the algebra $\breve{U}_q(\mathrm{sl}_2)$ in terms of two commuting q-oscillator algebras. In order to do this we need an algebra $\mathcal{A}_q^{\mathrm{ext}}$ which is obtained by adjoining formally elements $q^{N/2}$ and $q^{-N/2}$ to $\mathcal{A}_q$. More precisely, $\mathcal{A}_q^{\mathrm{ext}}$ is the associative unital algebra with generators a, a^+, $q^{N/2}$ and $q^{-N/2}$ subject to the relations

$$[a,a^+]_q = q^{-N} := (q^{-N/2})^2, \quad [a,a^+]_{q^{-1}} = q^N := (q^{N/2})^2,$$

$$q^{-N/2}q^{N/2} = q^{N/2}q^{-N/2} = 1, \quad q^{N/2}a^+ = q^{1/2}a^+q^{N/2}, \quad q^{N/2}a = q^{-1/2}aq^{N/2}.$$

It is clear that the algebra $\mathcal{A}_q$ from Definition 2 is isomorphic to the subalgebra of $\mathcal{A}_q^{\mathrm{ext}}$ generated by the elements a, a^+, q^N and q^{-N}.

Let $\mathcal{A}_q^{\mathrm{ext},2}$ be the tensor product of two q-oscillator algebras $\mathcal{A}_q^{\mathrm{ext}}$. The corresponding generators of the algebra $\mathcal{A}_q^{\mathrm{ext},2}$ are denoted by $a_1, a_1^+, q^{N_1/2}$, $q^{-N_1/2}, a_2, a_2^+, q^{N_2/2}q^{-N_2/2}$. Note that every element of the set $a_1, a_1^+, q^{\pm N_1/2}$ commutes with any element from $a_2, a_2^+, q^{\pm N_2/2}$.

Proposition 4. *There exists a unique algebra homomorphism $\varphi : \breve{U}_q(\mathrm{sl}_2) \to \mathcal{A}_q^{\mathrm{ext},2}$ such that*

$$\varphi(E) = a_1^+ a_2, \quad \varphi(F) = a_2^+ a_1, \quad \varphi(K) = q^{(N_1 - N_2)/2}. \tag{14}$$

Proof. It is enough to verify that the elements $\varphi(E)$, $\varphi(F)$, $\varphi(K)$ and $\varphi(K^{-1})$ satisfy the defining relations (3.10) and (3.11) of $\breve{U}_q(\mathrm{sl}_2)$. As an example, we check the relation (3.11). By (9), we obtain

$$\varphi(E)\varphi(F) = a_1^+ a_2 a_2^+ a_1 = [N_1]_q [N_2 + 1]_q.$$

Similarly, $\varphi(F)\varphi(E) = [N_1 + 1]_q [N_2]_q$. A straightforward computation shows that

$$[N_1]_q [N_2 + 1]_q - [N_1 + 1]_q [N_2]_q = [N_1 - N_2]_q.$$

Hence $\varphi(E)\varphi(F) - \varphi(F)\varphi(E) = (\varphi(K)^2 - \varphi(K^{-1})^2)/(q - q^{-1})$. □

The algebra homomorphism φ from Proposition 4 is called the *Jordan–Schwinger realization* of the algebra $\breve{U}_q(\mathrm{sl}_2)$.

An algebra homomorphism ψ of a Hopf algebra $\mathcal{H}$ into another algebra $\mathcal{X}$ might be a useful tool for the study of the Hopf algebra. First the composition of ψ with representations of $\mathcal{X}$ provides us with representations of $\mathcal{H}$. For the Jordan–Schwinger homomorphism $\varphi : \breve{U}_q(\mathrm{sl}_2) \to \mathcal{A}_q^{\mathrm{ext},2}$ this will be carried out in Subsect. 5.3.4 where the irreducible representations of $\breve{U}_q(\mathrm{sl}_2)$ are realized on the q-analog of the Bargmann–Fock space. Another application is given by the following lemma. It is a slight generalization of Proposition 1.14 and shows that such a homomorphism turns the algebra $\mathcal{X}$ into an $\mathcal{H}$-module algebra.

Lemma 5. *Let $\psi : \mathcal{H} \to \mathcal{X}$ be an algebra homomorphism of a Hopf algebra $\mathcal{H}$ into an algebra $\mathcal{X}$. Then $\mathcal{X}$ is a left $\mathcal{H}$-module algebra with respect to the left action of $\mathcal{H}$ on $\mathcal{X}$ defined by*

$$a \triangleright x \equiv \mathrm{ad}_{\psi,L}(a)x := \sum \psi(a_{(1)}) \cdot x \cdot \psi(S(a_{(2)})), \quad a \in \mathcal{H},\ x \in \mathcal{X}. \tag{15}$$

Proof. Since ψ is an algebra homomorphism, it is clear that there are left and right actions of $\mathcal{H}$ on $\mathcal{X}$ given by $ax := \psi(a) \cdot x$ and $xa := x \cdot \psi(a)$, respectively, where the dot means the multiplication of $\mathcal{X}$. By the associativity of this multiplication, $\mathcal{X}$ becomes then an $\mathcal{H}$-bimodule. Therefore, as noted in Remark 2 in Subsect. 1.3.4, $\mathrm{ad}_{\psi,L}$ is a left action of $\mathcal{H}$ on $\mathcal{X}$. Since $a \triangleright 1 = \varepsilon(a)1$ and

$$\sum (a_{(1)} \triangleright x)(a_{(2)} \triangleright y) = \sum \psi(a_{(1)}) x \psi(S(a_{(2)})) \psi(a_{(3)}) y \psi(S(a_{(4)}))$$

$$= \sum \psi(a_{(1)}) x \psi(S(a_{(2)}) a_{(3)}) y \psi(S(a_{(4)})) = a \triangleright (xy),$$

$\mathcal{X}$ is indeed an $\mathcal{H}$-module algebra. □

Lemma 5 applies in particular to the algebra homomorphisms from Propositions 3 and 4. Let us write down the corresponding action in the case of the

Jordan–Schwinger homomorphism φ. Setting $\psi = \varphi$ in (15) and using (14), (3.12) and (3.13), the left action (15) for the generators of $\breve{U}_q(\mathrm{sl}_2)$ yields

$$\mathrm{ad}_{\psi,L}(E)x = a_1^+ a_2 x q^{(N_2-N_1)/2} - q q^{(N_2-N_1)/2} x a_1^+ a_2,$$

$$\mathrm{ad}_{\psi,L}(F)x = a_2^+ a_1 x q^{(N_2-N_1)/2} - q^{-1} q^{(N_2-N_1)/2} x a_2^+ a_1,$$

$$\mathrm{ad}_{\psi,L}(K)x = q^{(N_1-N_2)/2} x q^{(N_2-N_1)/2}, \quad x \in \mathcal{A}_q^{\mathrm{ext},2}.$$

Our next example describes a few irreducible submodules of the $\breve{U}_q(\mathrm{sl}_2)$-module $\mathcal{A}_q^{\mathrm{ext},2}$.

Example 1. A direct computation shows that the elements

$$r_{1/2}^{1/2} := a_1^+ q^{-N_2/2}, \quad r_{-1/2}^{1/2} := a_2^+ q^{N_1/2}$$

of $\mathcal{A}_q^{\mathrm{ext},2}$ satisfy the relations

$$\mathrm{ad}_{\psi,L}(E) r_{-1/2}^{1/2} = r_{1/2}^{1/2}, \ \mathrm{ad}_{\psi,L}(E) r_{1/2}^{1/2} = 0,$$

$$\mathrm{ad}_{\psi,L}(F) r_{1/2}^{1/2} = r_{-1/2}^{1/2}, \ \mathrm{ad}_{\psi,L}(F) r_{-1/2}^{1/2} = 0, \ \mathrm{ad}_{\psi,L}(K) r_{\pm 1/2}^{1/2} = q^{\pm 1} r_{\pm 1/2}^{1/2}.$$

This means that $r_{1/2}^{1/2}$ and $r_{-1/2}^{1/2}$ form a basis of an irreducible submodule $T_{1/2}$ of the $\breve{U}_q(\mathrm{sl}_2)$-module $\mathcal{A}_q^{\mathrm{ext},2}$. Similarly,

$$\hat{r}_{1/2}^{1/2} := -a_2 q^{(N_1+1)/2}, \quad \hat{r}_{-1/2}^{1/2} := a_1 q^{-(N_2+1)/2}$$

are basis elements of another irreducible $\breve{U}_q(\mathrm{sl}_2)$-submodule $T_{1/2}$ of $\mathcal{A}_q^{\mathrm{ext},2}$.

Using the Clebsch–Gordan coefficients of the algebra $\breve{U}_q(\mathrm{sl}_2)$ from Sect. 3.4 one can find basis elements of polynomials of $r_{\pm 1/2}^{1/2}$ and $\hat{r}_{\pm 1/2}^{1/2}$ transforming under any representation T_l, $l \in \frac{1}{2}\mathbb{N}_0$, of $\breve{U}_q(\mathrm{sl}_2)$. For instance, the element

$$r_0^0 := q^{1/2} r_{1/2}^{1/2} \hat{r}_{-1/2}^{1/2} - q^{-1/2} r_{-1/2}^{1/2} \hat{r}_{1/2}^{1/2} = [N_1 + N_2]$$

is invariant under the representation $\mathrm{ad}_{\psi,L}$, that is, $\mathrm{ad}_{\psi,L}(a) r_0^0 = \varepsilon(a) r_0^0$ for all $a \in \breve{U}_q(\mathrm{sl}_2)$. Further, the three elements

$$r_1^1 := r_{1/2}^{1/2} \hat{r}_{1/2}^{1/2} = \frac{1}{\sqrt{[2]}} q^{(N_1-N_2)/2} a_1^+ a_2,$$

$$r_{-1}^1 := r_{1/2}^{1/2} \hat{r}_{-1/2}^{1/2} = -\frac{1}{\sqrt{[2]}} q^{(N_1-N_2)/2} a_1 a_2^+,$$

$$r_0^1 = \frac{1}{\sqrt{[2]}} (q^{-1/2} r_{1/2}^{1/2} \hat{r}_{-1/2}^{1/2} + q^{1/2} r_{-1/2}^{1/2} \hat{r}_{1/2}^{1/2}) = \frac{1}{[2]} (q^{-N_2-\frac{1}{2}}[N_1] - q^{N_1+\frac{1}{2}}[N_2]),$$

form a basis of an irreducible $\breve{U}_q(\mathrm{sl}_2)$-submodule T_1 of $\mathcal{A}_q^{\mathrm{ext},2}$. △

5.1.4 The q-Oscillator Algebras and the Quantum Space $M_{q^2}(2)$

Let $\mathcal{A}_q^{\mathrm{ext},4}$ denote the 4-fold tensor product of the q-oscillator algebra $\mathcal{A}_q^{\mathrm{ext}}$ defined in the preceding subsection. The generators of this algebra $\mathcal{A}_q^{\mathrm{ext},4}$ are denoted by $a_{ij}, a_{ij}^+, q^{N_{ij}/2}, q^{-N_{ij}/2}$, where $i,j=1,2$. Recall that the algebra $\mathcal{O}(M_{q^2}(2))$ is generated by elements $u_1^1, u_2^1, u_1^2, u_2^2$ satisfying the relations (4.1) and (4.2) with q replaced by q^2.

Proposition 6. *There exists an algebra automorphism $\gamma : \mathcal{O}(M_{q^2}(2)) \to \mathcal{A}_q^{\mathrm{ext},4}$ such that*

$$\gamma(u_1^1) = a_{11}^+ q^{(N_{12}-N_{11}+N_{21}+N_{22})/2}, \quad \gamma(u_2^1) = a_{12}^+ q^{(N_{21}-3N_{11}-N_{12}+N_{22})/2},$$

$$\gamma(u_1^2) = a_{21}^+ q^{(N_{12}-3N_{11}-N_{21}+N_{22})/2},$$

$$\gamma(u_2^2) = a_{22}^+ q^{(N_{11}-3N_{12}-3N_{21}-N_{22})/2} + (1-q^2) a_{21}^+ a_{12}^+ a_{11} q^{(N_{22}-N_{11}-N_{12}-N_{21})/2},$$

$$\gamma(\mathcal{D}_q) = q^{3/2} q^{-(N_{12}+N_{21}+1)} a_{11}^+ a_{22}^+ - q^{3/2} q^{(N_{11}+N_{22}+1)} a_{21}^+ a_{12}^+.$$

Proof. The proof is given by direct computation. We omit the details. □

There is also an algebra homomorphism $\nu : \mathcal{O}(SL_q(2)) \to \mathcal{A}_q$ such that

$$\nu(u_1^1) = a^+, \quad \nu(u_2^1) = q^{-N}, \quad \nu(u_1^2) = -q^{-N-1}, \quad \nu(u_2^2) = (q-q^{-1}) q^{-N-1} a.$$

5.2 Representations of q-Oscillator Algebras

5.2.1 N-Finite Representations

We shall use the bra and ket notation as is common in physics. If $T(N)$ is an operator on a vector space V and $w \in \mathbb{C}$, then the symbol $|w\rangle$ denotes a vector from V such that $T(N)|w\rangle = w|w\rangle$.

Suppose that V is a vector space and $T(a^+)$, $T(a)$ and $T(N)$ are three operators on V such that the eigenvectors of $T(N)$ span V and the relations

$$[T(a), T(a^+)]_q = q^{-T(N)}, \quad T(N)T(a^+) = T(a^+)(T(N)+I), \tag{16}$$

$$T(a)T(N) = (T(N)+I)T(a) \tag{17}$$

are fulfilled on V. Here the operators $q^{\pm T(N)}$ are defined by $q^{\pm T(N)}|w\rangle = q^{\pm w}|w\rangle$ when $T(N)|w\rangle = w|w\rangle$. Then it is clear that the operators $T(a^+)$, $T(a)$, and $T(q^{\pm N}) := q^{\pm T(N)}$ satisfy the relations (3) and (4) of the algebra $\mathcal{A}_q^c$. Hence these operators define a unique algebra homomorphism T of $\mathcal{A}_q^c$ to the algebra $\mathcal{L}(V)$ of linear operators on V, that is, T is a representation of $\mathcal{A}_q^c$ on the vector space V.

Definition 3. *A representation T of the algebra $\mathcal{A}_q^c$ on a vector space V of the above form is called* N-finite.

The following observations are crucial in the following.

Proposition 7. *Let T be an N-finite representation of $\mathcal{A}_q$ and let $|w\rangle$ be an eigenvector of $T(N)$ with eigenvalue w.*

(i) *Either $T(a^+)|w\rangle = 0$ or $T(a^+)|w\rangle$ is an eigenvector of $T(N)$ with eigenvalue $w+1$.*

(ii) *Either $T(a)|w\rangle = 0$ or $T(a)|w\rangle$ is an eigenvector of $T(N)$ with eigenvalue $w-1$.*

(iii) *If T is irreducible, then $|w\rangle$ is an eigenvector of $T(a^+)T(a)$ and $T(a)T(a^+)$.*

Proof. It follows from (16) that

$$T(N)T(a^+)|w\rangle = T(a^+)T(N)|w\rangle + T(a^+)|w\rangle = (w+1)T(a^+)|w\rangle,$$

which proves (i). The proof of (ii) is similar. If T is irreducible, then $T(c_q) = \alpha{\cdot}I$ for some $\alpha \in \mathbb{C}$ and hence

$$T(c_q)|w\rangle = q^{1-w}[w]|w\rangle - T(a^+)T(a)|w\rangle$$
$$= (q^{1-w}[w] + q^{-2w})|w\rangle - T(q^{-N})T(a)T(a^+)|w\rangle = \alpha|w\rangle.$$

This implies that $|w\rangle$ is an eigenvector of $T(a^+)T(a)$ and $T(a)T(a^+)$. □

Now let T be an irreducible N-finite representation of $\mathcal{A}_q^c$ and let $|w\rangle$ be an eigenvector of $T(N)$ with eigenvalue w. Concerning the actions of the operators $T(a)^n$ and $T(a^+)^n$ on $|w\rangle$, there are three possible cases:

Case 1: There exists a natural number n such that $T(a)^n|w\rangle = 0$. Then the representation T is called a *representation with lowest weight.*

Case 2: There is a natural number n such that $T(a^+)^n|w\rangle = 0$. Then T is called a *representation with highest weight.*

Case 3: $T(a)^n|w\rangle \neq 0$ and $T(a^+)^n|w\rangle \neq 0$ for any natural number n.

5.2.2 Irreducible Representations with Highest (Lowest) Weights

Throughout this subsection we suppose that q is not a root of unity.

The next proposition says that the irreducible N-finite representations of $\mathcal{A}_q^c$ with highest weights are parametrized by complex numbers.

Proposition 8. *To every complex number w there corresponds an irreducible N-finite representation T_w^+ of $\mathcal{A}_q^c$ with lowest weight. It acts on a vector space V with basis elements $|w+m\rangle$, $m \in \mathbb{N}_0$, and the operators $T_w^+(N)$, $T_w^+(a^+)$ and $T_w^+(a)$ are given by*

$$T_w^+(N)|w+m\rangle = (w+m)|w+m\rangle, \quad T_w^+(a^+)|w+m\rangle = |w+m+1\rangle, \tag{18}$$

$$T_w^+(a)|w+m\rangle = q^{-w}[m]|w+m-1\rangle, \tag{19}$$

where we have set $|w-1\rangle := 0$ *and* $[m]$ *denotes the q-number (2.1).*

Every irreducible N*-finite representation of* $\mathcal{A}_q^c$ *with lowest weight is equivalent to one of the representations* T_w^+. *Two representations* T_w^+ *and* $T_{w'}^+$ *are equivalent if and only if* $q^w = q^{w'}$.

Proof. One immediately checks that the operators $T_w^+(N)$, $T_w^+(a^+)$ and $T_w^+(a)$ defined by (18) and (19) satisfy (16) and (17), hence they indeed determine an N-finite representation T_w^+ of $\mathcal{A}_q^c$. We have $T_w^+(a)|w\rangle = 0$, so T_w^+ has a lowest weight. Since q is not a root of unity, one easily verifies that the representation T_w^+ has no nontrivial invariant subspace. Thus, T_w^+ is irreducible. Since the formulas for the operators $T_w^+(q^N)$, $T_w^+(a^+)$ and $T_w^+(a)$ depend only on q^w, T_w^+ and $T_{w'}^+$ are equivalent if $q^w = q^{w'}$. Conversely, if $q^w \neq q^{w'}$, then the operators $T_w^+(q^N)$ and $T_{w'}^+(q^N)$ have different spectra, because q is not a root of unity. Hence the representations T_w^+ and $T_{w'}^+$ are not equivalent.

Let T be an arbitrary irreducible N-finite representation of $\mathcal{A}_q^c$ with lowest weight on a vector space V. Then there exist a number $w \in \mathbb{C}$ and a nonzero vector $|w\rangle \in V$ such that $T(N)|w\rangle = w|w\rangle$ and $T(a)|w\rangle = 0$. We denote the vector $T(a^+)^m|w\rangle$ by $|w+m\rangle$, $m \in \mathbb{N}$. Since $aa^+ = qa^+a + q^{-N}$, we have

$$T(a)|w+m\rangle = T(aa^+)|w+m-1\rangle$$

$$= qT(a^+a)|w+m-1\rangle + q^{-(w+m-1)}|w+m-1\rangle. \tag{20}$$

In particular, $T(a)|w+1\rangle = q^{-w}|w\rangle$. We show by induction on m that

$$T(a)|w+m\rangle = q^{-w}[m]|w+m-1\rangle. \tag{21}$$

Suppose that (21) holds for $m = 2, 3, \cdots, n-1$. Then, by (20), we obtain

$$T(a)|w+n\rangle = \frac{q^{n-1} - q^{-n+1}}{q^{w-1}(q-q^{-1})} T(a^+)|w+n-2\rangle + q^{-(w+n-1)}|w+n-1\rangle$$

$$= q^{-w}[n]|w+n-1\rangle,$$

so (21) holds for any $m \in \mathbb{N}$. Thus, the operators $T(N)$, $T(a^+)$ and $T(a)$ act on the vectors $|w+m\rangle$ as described by (18) and (19). Each vector $|w+m\rangle$ is nonzero. Indeed, otherwise $V_m := \mathrm{Lin}\,\{|w+k\rangle \mid k>m\}$ is an invariant subspace for the irreducible representation T by (21) and Proposition 7(i), so that $V_m = V$. This is impossible, since then $|w\rangle$ is not in V. Similarly, we get $V = \mathrm{Lin}\,\{|w+k\rangle \mid k \in \mathbb{N}_0\}$. Since the (nonzero!) vectors $|w+m\rangle$ are eigenvectors of $T(N)$ for pairwise distinct eigenvalues (by Proposition 7(i)), they are linearly independent. Thus we have shown that T is equivalent to T_w. □

Proposition 9. *To any complex number* w *there corresponds an irreducible* N*-finite representation* T_w^- *of* $\mathcal{A}_q^c$ *with highest weight. It acts on a vector space* V *with basis elements* $|w-m\rangle$, $m \in \mathbb{N}_0$, *by the formulas*

$$T_w^-(N)|w-m\rangle = (w-m)|w-m\rangle, \quad T_w^-(a)|w-m\rangle = |w-m-1\rangle,$$

$$T_w^-(a^+)|w-m\rangle = -q^{-w-1}[m]|w-m+1\rangle,$$

where $|w+1\rangle := 0$. *Every irreducible N-finite representation of $\mathcal{A}_q^c$ with highest weight is equivalent to one of these representations T_w^-. Two such representations T_w^- and $T_{w'}^-$ are equivalent if and only if $q^w = q^{w'}$.*

Note that the action of the central element c_q of $\mathcal{A}_q^c$ is given by

$$T_w^+(c_q) = q^{1-w}[w]I \quad \text{and} \quad T_w^-(c_q) = q^{-w}[w+1]I. \tag{22}$$

5.2.3 Representations Without Highest and Lowest Weights

In this subsection we assume that q is not a root of unity.

Let α and w be complex numbers such that $0 \le \operatorname{Re} w < 1$. Let V be a vector space with basis $|w+m\rangle$, $m \in \mathbb{Z}$. A straightforward verification shows that the operators $T_{\alpha w}(N)$, $T_{\alpha w}(a^+)$ and $T_{\alpha w}(a)$ on V defined by

$$T_{\alpha w}(N)|w+m\rangle = (w+m)|w+m\rangle, \quad T_{\alpha w}(a^+)|w+m\rangle = |w+m+1\rangle, \tag{23}$$

$$T_{\alpha w}(a)|w+m\rangle = (\alpha q^m + q^{-w}[m])|w+m-1\rangle \tag{24}$$

satisfy the relations (16) and (17). Hence these operators define an N-finite representation $T_{\alpha w}$ of the algebra $\mathcal{A}_q^c$ on V. For the central element c_q we get

$$T_{\alpha w}(c_q) = q^{1-w}([w]-\alpha)I. \tag{25}$$

Proposition 10. *If $\alpha q^m + q^{-w}[m] \ne 0$ for all $m \in \mathbb{Z}$, then the representation $T_{\alpha w}$ is irreducible and has neither a highest weight nor a lowest weight. Two such representations $T_{\alpha w}$ and $T_{\alpha' w'}$ are equivalent if and only if $\alpha = \alpha'$ and $q^w = q^{w'}$. Every irreducible N-finite representation of the algebra $\mathcal{A}_q^c$ without highest and lowest weights is equivalent to one of these representations $T_{\alpha w}$.*

Proof. The properties of the representations $T_{\alpha w}$ asserted above follow from the formulas (23) and (24). In order to prove the last assertion, let T be an N-finite irreducible representation of $\mathcal{A}_q^c$ without lowest weight and without highest weight. If $|x\rangle$ is an eigenvector of $T(N)$ with eigenvalue x, then it follows from the assertions (i) and (ii) of Proposition 7 that the vectors $T(a)^r|x\rangle$, $T(a^+)^s|x\rangle$, $r \in \mathbb{N}_0, s \in \mathbb{N}$, are linearly independent. Since T is irreducible, their linear span coincides with V by Proposition 7(iii). Thus, the spectrum of the operator $T(N)$ is simple and formed by the eigenvalues $x+m$, where $m \in \mathbb{Z}$.

There is an eigenvalue w of $T(N)$ such that $0 \le \operatorname{Re} w < 1$. Let $|w\rangle$ be a corresponding eigenvector. Let $|w-1\rangle, |w\rangle, \cdots, |w+m\rangle, \cdots$ be eigenvectors for $T(N)$ such that $T(N)|w+i\rangle = (w+i)|w+i\rangle$ and $T(a^+)|w+i\rangle = |w+i+1\rangle$, $i = -1, 0, 1, 2, \cdots$. By Proposition 7(i), there is a complex number α such that $T(a)|w\rangle = \alpha|w-1\rangle$. Since T has no lowest weight, $\alpha \ne 0$. Then we get

$$T(a)|w+1\rangle = T(a)T(a^+)|w\rangle = T(qa^+a + q^{-N})|w\rangle = (q\alpha + q^{-w})|w\rangle.$$

Using similar equalities, we prove by induction on m that

$$\begin{aligned} T(a)|w+m\rangle &= (\alpha q^m + q^{-w}(q^{m-1}+q^{m-3}+\cdots+q^{-(m-1)}))|w+m-1\rangle \\ &= (\alpha q^m + q^{-w}[m])|w+m-1\rangle, \end{aligned}$$

for $m \in \mathbb{N}_0$. That is, the operators $T(N)$, $T(a^+)$ and $T(a)$ act on the vectors $|w+m\rangle$, $m=-1,0,1,\cdots$, by the formulas (23) and (24).

If $|w\rangle' := \alpha^{-w}|w\rangle$ and $|w-1\rangle' := \alpha^{-w+1}|w-1\rangle$, then $T(a^+)|w-1\rangle' := \alpha|w\rangle'$ and $T(a)|w\rangle' := |w-1\rangle'$. Setting $|w-m-1\rangle' := T(a)|w-m\rangle'$, $m \in \mathbb{N}_0$, and using the relation $a^+a = q^{-1}aa^+ - q^{-N-1}$, we find by a simple induction argument that

$$T(a^+)|w-m\rangle' = (\alpha q^{-m+1} - q^{-w}(q^{-m+2}+q^{-m+4}+\cdots+q^{m-2}))|w-m+1\rangle'$$

$$= (\alpha q^{-m+1} - q^{-w}[m-1])|w-m+1\rangle'.$$

By assumption, the representation T has no lowest weight. Therefore, the preceding implies that $|w-m\rangle' \neq 0$ for $m \in \mathbb{N}_0$ and that $\alpha q^m - q^{-w}[m] \neq 0$ for any $m \in \mathbb{Z}$. Setting

$$|w-m\rangle := \prod\nolimits_{i=0}^{m}(\alpha q^{-i+1} - q^{-w}[i-1])^{-1}|w-m\rangle',$$

it is straightforward to check that the operators $T(N)$, $T(a^+)$ and $T(a)$ act on the vectors $|w-m\rangle$, $m=-1,-2,\cdots$, by the formulas (23) and (24). □

Summarizing the preceding, if q is not a root of unity, then the above Propositions 8–10 classify all irreducible N-finite representations of the algebra $\mathcal{A}_q^c$ up to equivalence.

Let us briefly turn to the symmetric q-oscillator algebra $\mathcal{A}_q$. As noted in Subsect. 5.1.1, the algebra $\mathcal{A}_q$ is the quotient of the algebra $\mathcal{A}_q^c$ by the two-sided ideal generated by the central element c_q. Thus, the (irreducible) representations of $\mathcal{A}_q$ are in one-to-one correspondence to those (irreducible) representations of $\mathcal{A}_q^c$ that annihilate c_q. A representation of $\mathcal{A}_q$ is called *N-finite* if the corresponding representation of $\mathcal{A}_q^c$ is so.

Proposition 11. (i) *Let w_+ and w_- denote fixed complex numbers such that $q^{w_+} = -1$ and $q^{w_-+1} = -1$. Then the representations T_0^+, $T_{w_+}^+$, T_{-1}^- and $T_{w_-}^-$ are pairwise inequivalent irreducible N-finite representations of the algebra $\mathcal{A}_q$ with highest or lowest weight.*

(ii) *Let $w \in \mathbb{C}$ be such that $0 \leq \operatorname{Re} w < 1$ and $[w]q^m + q^{-w}[m] \neq 0$ for all $m \in \mathbb{Z}$. Then $T_{[w],w}$ is an irreducible N-finite representation of $\mathcal{A}_q$. Two such representations $T_{[w],w}$ and $T_{[w'],w'}$ are equivalent if and only if $q^w = q^{w'}$.*

(iii) *Any irreducible N-finite representation of $\mathcal{A}_q$ is equivalent to one of the representations from* (i) *or* (ii).

Proof. (i): From (22) we see that the representation T_w^+ (resp. T_w^-) of $\mathcal{A}_q^c$ maps c_q into the zero operator and so passes to a representation of $\mathcal{A}_q$ if and only if $[w]=0$ (resp. $[w+1]=0$) or equivalently $q^{2w}=1$ (resp. $q^{2(w+1)}=1$). Since two representations T_w^+ and $T_{w'}^+$ (resp. T_w^- and $T_{w'}^-$) are equivalent if and only if $q^w = q^{w'}$, there are precisely two equivalence classes with representatives T_0^+ and $T_{w_+}^+$ (resp. T_{-1}^- and $T_{w_-}^-$). Two representations T_w^+ and $T_{w'}^-$ cannot be equivalent, because T_w^+ has a lowest weight and $T_{w'}^-$ does not.

(ii): If $T_{\alpha w}(c_q) = 0$, then $\alpha = [w]$ by (25). The other assertions of (ii) follow from Proposition 10.

(iii): Each irreducible N-finite representation of $\mathcal{A}_q$ yields an irreducible N-finite representation of $\mathcal{A}_q^c$ that annihilates the element c_q. Hence the assertion follows from the fact that the representations of (i) and (ii) exhaust all representations T from Propositions 8–10 such that $T(c_q) = 0$. □

5.2.4 Irreducible Representations of $\mathcal{A}_q^c$ for q a Root of Unity

In this subsection q is a root of unity. More precisely, we assume that $q^p = 1$, where $p \in \mathbb{N}$ is odd and minimal.

Lemma 12. *The elements $(a^+)^p$, $(a)^p$, $(q^N)^p$ and $(q^{-N})^p$ belong to the center of $\mathcal{A}_q^c$.*

Proof. It follows from the relations (3) and (4) that

$$a(a^+)^n = [n](a^+)^{n-1}q^{-N} + q^{-n}(a^+)^n a, \quad (a^+)^n q^{-N} = q^n q^{-N}(a^+)^n.$$

Putting $n = p$ and using the relations $q^p = 1$ and $[p] = 0$, we see that $(a^+)^p$ commutes with the generators and so with all elements of $\mathcal{A}_q^c$. The proof for a^p, $(q^N)^p$ and $(q^{-N})^p$ is similar. □

Definition 4. *Let T be an irreducible representation of $\mathcal{A}_q^c$. T is said to be* cyclic *if $T(a^p) \neq 0$ and $T((a^+)^p) \neq 0$. If $T(a^p) = 0$ and $T((a^+)^p) \neq 0$ or if $T((a^+)^p) = 0$ and $T(a^p) \neq 0$, then T is called* semicyclic.

Let μ and ξ be complex numbers and $\xi \neq 0$. Let V be a p-dimensional vector space with basis $|m\rangle$, $m = 0, 1, \cdots, p-1$. We define operators on V by the formulas

$$T_{\mu\xi}(a^+)|m\rangle = |m+1\rangle, \quad 0 \le m \le p-2, \qquad T_{\mu\xi}(a^+)|p-1\rangle = \xi|0\rangle,$$

$$T_{\mu\xi}(a)|m\rangle = [m+\mu]|m-1\rangle, \quad 1 \le m \le p-1, \qquad T_{\mu\xi}(a)|0\rangle = [\mu]\xi^{-1}|p-1\rangle,$$

$$T_{\mu\xi}(q^{\pm N})|m\rangle = q^{\pm(m+\mu)}|m\rangle$$

and

$$T'_{\mu\xi}(a)|m\rangle = |m-1\rangle, \quad 1 \le m \le p-1, \qquad T'_{\mu\xi}(a)|0\rangle = \xi|p-1\rangle,$$

$$T'_{\mu\xi}(a^+)|m\rangle = [m+\mu+1]|m+1\rangle, \quad 0 \le m \le p-2, \qquad T'_{\mu\xi}(a^+)|p-1\rangle = [\mu]\xi^{-1}|0\rangle,$$

$$T'_{\mu\xi}(q^{\pm N})|m\rangle = q^{\pm(m+\mu)}|m\rangle.$$

Both sets of operators fulfill the relations (3) and (4), so they define representations $T_{\mu\xi}$ and $T'_{\mu\xi}$ of the algebra $\mathcal{A}_q^c$. One easily checks that the representations $T_{\mu\xi}$ and $T'_{\mu\xi}$ are irreducible. Moreover, if $\mu \neq 0$ and $\xi \neq 0$, then the representation $T'_{\mu\xi}$ is equivalent to $T_{\mu,[\mu]\xi^{-1}}$.

The formulas (18) and (19) also define a representation of $\mathcal{A}_q^c$ on V when q is a root of unity. But this representation is not irreducible. The subspace V_p spanned by the basis vectors $|w+m\rangle$, $m = p, p+1, \cdots$, is invariant. The quotient space V/V_p is irreducible. The corresponding representation operators $T_w(q^{\pm N})$, $T_w(a^+)$ and $T_w(a)$ are then given by

$$T_w(q^{\pm N})|w+m\rangle = q^{\pm(w+m)}|w+m\rangle, \quad T_w(a^+)|w+m\rangle = |w+m+1\rangle,$$

$$T_w(a)|w+m\rangle = q^{-w}[m]|w+m-1\rangle,$$

where we set $|w-1\rangle = |w+p\rangle := 0$.

Proposition 13. (i) *Suppose that $q^p = 1$ for an odd nonnegative integer p. Then every irreducible representation of the algebra $\mathcal{A}_q^c$ is p-dimensional and equivalent to one of the representations $T_{\mu\xi}$ for $\mu, \xi \in \mathbb{C}$, $\xi \neq 0$, $T'_{0,\xi}$ for $\xi \in \mathbb{C}$, $\xi \neq 0$, or T_w, $w \in \mathbb{C}$. The representations $T_{\mu\xi}$, $\mu \neq 0$, $\xi \neq 0$, are cyclic. The representations $T_{0,\xi}$, $\xi \neq 0$, are semicyclic with lowest weight and the representations $T'_{0,\xi}$, $\xi \neq 0$, are semicyclic with highest weight. The representations T_w have highest and lowest weights.*

(ii) *The representations $T_{\mu\xi}$ and $T_{\mu'\xi'}$, $\mu, \mu' \neq 0$, are equivalent if and only if $\xi = \xi'$ and $q^\mu = q^{\mu'+k}$ for some $k = 0, 1, 2, \cdots, p-1$. The representations $T_{0,\xi}$ (and the representations $T'_{0,\xi}$) are pairwise nonequivalent. The representations T_w and $T_{w'}$ are equivalent if and only if $q^w = q^{w'}$.*

Proof. (i): The stated properties of the representations $T_{\mu\xi}$, $T'_{0\xi}$ and T_w are immediately verified.

Let T be an arbitrary irreducible representation of the algebra $\mathcal{A}_q^c$ on a vector space V. Since T maps elements of the center to scalar operators, it follows from Lemma 12 that $T(\mathcal{A}_q^c)$ is spanned by the operators $T(a^i(a^+)^j(q^N)^k)$, $i, j, |k| = 0, 1, \cdots, p-1$. In particular, we conclude that $\dim T(\mathcal{A}_q^c) < \infty$ and hence $\dim V < \infty$. Let q^μ be an eigenvalue of $T(q^N)$ and $|0\rangle$ an eigenvector for this eigenvalue. Since $(a^+)^p$ belongs to the center of $\mathcal{A}_q^c$, there is a complex number ξ such that $T((a^+)^p) = \xi \cdot I$.

First let us suppose that $\xi \neq 0$. Then each vector $|m\rangle := T(a^+)^m|0\rangle$, $m = 0, 1, \cdots, p-1$, is nonzero. From the first relation of (4) we obtain that $T(q^N)|m\rangle = q^{\mu+m}|m\rangle$. Hence the vectors $|m\rangle$, $m = 0, 1, \cdots, p-1$, are linearly independent. Since their linear span is invariant under T by Proposition 7(iii), it coincides with the carrier space V. From (3), (4), and (6) it follows that the operator $T(a)$ is given by the formulas $T(a)|m\rangle = [m+\mu]|m-1\rangle$, $1 \leq m \leq p-1$ and $T(a)|0\rangle = [\mu]\xi^{-1}|p-1\rangle$. Thus we have shown that T is equivalent to the representation $T_{\mu\xi}$. In the case $\xi = 0$ one proves analogously that T is equivalent to one of the representations $T'_{0,\zeta}$ or T_w.

(ii): Since the values of two equivalent irreducible representations $T_{\mu\xi}$ and $T_{\mu'\xi'}$ at the central element $(a^+)^p$ are equal, we get $\xi = \xi'$. Further, the spectra of the operators $T_{\mu\xi}(q^N)$ and $T_{\mu'\xi'}(q^N)$ must coincide, so that $q^\mu = q^{\mu'+k}$ for some $k = 0, 1, 2, \cdots, p-1$. The proofs of the converse and of the remaining assertions of (ii) are straightforward and will be omitted. □

It is easy to check that all representations $T_{\mu\xi}$ and $T'_{\mu\xi}$ annihilate the element c_q and therefore give representations of the algebra $\mathcal{A}_q$. For the representation T_w we have $T_w(c_q) = 0$ if and only if $q^{2w} = 1$. Thus, only T_w with $w = 0, p/2$ gives a representation of $\mathcal{A}_q$. That is, we have

Corollary 14. *Any irreducible representation of $\mathcal{A}_q$ is p-dimensional and equivalent to one of the representations $T_{\mu\xi}$, $T'_{0,\xi}$, T_0 and $T_{p/2}$.*

5.2.5 Irreducible $*$-Representations of $\mathcal{A}_q^c$ and $\mathcal{A}_q$

In this subsection we suppose that $q > 0$ and $q \neq 1$. Then, as noted in Subsect. 5.1.1, the algebras $\mathcal{A}_q^c$ and $\mathcal{A}_q$ are $*$-algebras with involutions such that $a^* = a^+$ and $(q^N)^* = q^N$. The next proposition decides which of the irreducible N-finite representations of $\mathcal{A}_q^c$ introduced in Subsects. 5.2.2 and 5.2.3 are $*$-representations. Recall that a $*$-representation of a $*$-algebra $\mathcal{A}$ is a representation T on a vector space V, equipped with a scalar product $\langle\cdot,\cdot\rangle$, such that

$$\langle T(x)v, v'\rangle = \langle v, T(x^*)v'\rangle, \qquad x \in \mathcal{A}, \quad v, v' \in V. \tag{26}$$

Proposition 15. (i) *T_w^+ (resp. T_w^-) is a $*$-representation of $\mathcal{A}_q^c$ if and only if $q^w > 0$ (resp. $q^w < 0$).*

(ii) *If $0 < q < 1$, then $T_{\alpha w}$ is a $*$-representation of $\mathcal{A}_q^c$ if and only if $q^w > 0$ and $\alpha(q - q^{-1}) + q^{-w} \leq 0$. If $q > 1$, then $T_{\alpha w}$ is a $*$-representation if and only if $q^w < 0$ and $\alpha q^w + (q - q^{-1})^{-1} \leq 0$.*

Proof. We carry out the proof of (ii). The proof of (i) is similar. Suppose that $T_{\alpha w}$ is a $*$-representation. Applied to $x = q^N$, (26) means that the operator $T(q^N)$ is symmetric, so that its eigenvalues q^{w+m}, $m \in \mathbb{Z}$, are real. If $q^w > 0$, then applying (26) to $x = a$, $v = |w+m\rangle$, $v' = |w+m-1\rangle$, we find that $\alpha q^m + q^{-w}[m] > 0$ for all $m \in \mathbb{Z}$. This is equivalent to the inequality

$$\alpha + (q - q^{-1})^{-1}q^{-w} > q^{-w}q^{-2m}(q - q^{-1})^{-1} \quad \text{for all} \quad m \in \mathbb{Z}.$$

Since $q^{-w} > 0$, this is impossible if $q > 1$. If $0 < q < 1$, it leads to the condition $\alpha(q - q^{-1}) + q^{-w} \leq 0$. If $q^w < 0$, then it follows in a similar manner that $q > 1$ and $\alpha q^w + (q - q^{-1})^{-1} \leq 0$.

Conversely, suppose that $0 < q < 1$, $q^w > 0$ and $\alpha(q - q^{-1}) + q^{-w} \leq 0$. Then one easily checks that there is a scalar product $\langle\cdot,\cdot\rangle$ on the space V defined by $\langle w+m+1|w+m+1\rangle = (\alpha q^m + q^{-w}[m])\langle w+m|w+m\rangle$ and $\langle w+n|w+m\rangle = 0$ for $m \neq n$ such that (26) holds for the generators of $\mathcal{A}_q^c$.

Hence (26) is true for all $x \in \mathcal{A}_q^c$ and $T_{\alpha w}$ is a $*$-representation. The case $q > 1$ is treated similarly. $\square$

Comparing Propositions 11 and 15 we obtain

Corollary 16. *T_0^+ and $T_{w_-}^-$ are the only N-finite irreducible $*$-representations of the $*$-algebra $\mathcal{A}_q$.*

The operator $T(N)$ is usually considered as a q-analog of the number operator. Hence it is natural to require that it has a real spectrum. Let us call an N-finite $*$-representation T of $\mathcal{A}_q^c$ or $\mathcal{A}_q$ a *physical $*$-representation* if there exists an operator $T(N)$ as in Subsect. 5.2.1 with real eigenvalues. The $*$-representations T_w^+ and $T_{\alpha w}$, $0<q<1$, of $\mathcal{A}_q^c$ from Proposition 15 are physical $*$-representations. The $*$-representations T_w^- and $T_{\alpha w}$, $q>1$, of $\mathcal{A}_q^c$ are not, since there is no real w such that $q^w < 0$. The $*$-representation T_0^+ of $\mathcal{A}_q$ is physical, while $T_{w_-}^-$ is not. Thus, by Corollary 16, *the q-oscillator algebra $\mathcal{A}_q$ admits a unique irreducible N-finite physical $*$-representation T_0^+.* It is called the Fock representation and will be studied in Sect. 5.3 below.

5.2.6 Irreducible $*$-Representations of Another q-Oscillator Algebra

In this short subsection we list the irreducible $*$-representations of another q-oscillator algebra without proof. We assume that $q > 0$ and $q \neq 1$.

Let $\mathcal{B}_q$ denote the algebra with generators b and b^+ satisfying the relation

$$bb^+ - qb^+b = 1. \tag{27}$$

The algebra $\mathcal{B}_q$ is a $*$-algebra with involution given by $b^* := b^+$. Clearly, $\mathcal{B}_q$ is a q-deformation of the q-oscillator algebra $\mathcal{A}$ defined by relation (1).

Next we introduce three families of $*$-representations of the $*$-algebra $\mathcal{B}_q$. Let $|n\rangle$ denote elements of an orthonormal basis of a Hilbert space.

For arbitrary positive $q \neq 1$ we define linear operators

$$T_0(b)|n\rangle = [[n]]^{1/2}|n-1\rangle,\ T_0(b^+)|n\rangle = [[n+1]]^{1/2}|n+1\rangle,\quad n \in \mathbb{N}_0,$$

acting on the vector space $V := \text{Lin}\,\{|n\rangle \mid n \in \mathbb{N}_0\}$, where $|-1\rangle := 0$ and $[[n]] := (q^n - 1)(q - q^{-1})^{-1}$ (see (2.2)).

For $0 < q < 1$ and $q \le \gamma < 1$, we set

$$T_\gamma(b)|n\rangle = \gamma_n^{1/2}|n-1\rangle,\ T_\gamma(b^+)|n\rangle = \gamma_{n+1}^{1/2}|n+1\rangle,\ n \in \mathbb{Z},$$

on $V := \text{Lin}\,\{|n\rangle \mid n \in \mathbb{Z}\}$, where $\gamma_n := (1-q)^{-1}(q^n\gamma + 1)$.

For $0 < q < 1$ and $\beta \in \mathbb{C}$, $|\beta| = (1-q)^{-1/2}$, we define operators

$$T_\beta^1(b) = \beta,\ \ T_\beta^1(b^+) = \overline{\beta}$$

on the one-dimensional vector space $V := \mathbb{C}$.

One immediately verifies that each of these families of linear operators satisfy the relation (27) and condition (26) for $x = b, b^+$. Hence they define $*$-representations T_0, T_γ, T^1_β of the $*$-algebra $\mathcal{B}_q$ on V. It is easily seen that these representations are irreducible and pairwise inequivalent.

The representation T_0 admits a unit vector $|0\rangle$ such that $T_0(b)|0\rangle = 0$. Therefore, T_0 is called the Fock representation of $\mathcal{B}_q$. Clearly, T_0 goes into the Fock representation of the oscillator algebra $\mathcal{A}$ as $q \to 1$, while the representations T_γ and T^1_β have no counterparts for $q = 1$.

In the case $q < 1$ the operators $T_0(b)$ and $T_0(b^+)$ are bounded and their norms are given by

$$\| T_0(b)\| = \| T_0(b^+)\| = (1-q)^{-1/2}.$$

Both operators $T_0(b)$ and $T_0(b^+)$ are unbounded if $q > 1$. The operators $T_\gamma(b)$ and $T_\gamma(b^+)$ are unbounded for any $\gamma \in [q, 1)$.

5.3 The Fock Representation of the q-Oscillator Algebra

The most important representation of the q-oscillator algebra is the Fock representation. The aim of this section is to study a q-analog of the Bargmann–Fock realization of this representation and coherent states. Throughout this section we suppose that $q > 0$ and $q \neq 1$.

5.3.1 The Fock Representation

First let us recall the structure of the $*$-representation $T \equiv T^+_0$ of the $*$-algebra $\mathcal{A}_q$. The corresponding representation operators for the generators are determined by the formulas (18) and (19) with $w = 0$. If we write $|m\rangle'$ instead of $|m\rangle$ therein and then $|m\rangle := [m]!^{-1/2}|m\rangle'$, these formulas go into the equations

$$T(a)|n\rangle = \sqrt{[n]}|n-1\rangle, \quad T(a^+)|n\rangle = \sqrt{[n+1]}|n+1\rangle, \quad T(N)|n\rangle = n|n\rangle. \quad (28)$$

Let $\mathfrak{H}$ be a Hilbert space with orthonormal basis $|m\rangle$, $m \in \mathbb{N}_0$, and let $\mathcal{D}$ be the dense linear subspace of $\mathfrak{H}$ spanned by the vectors $|m\rangle$. Then the representation T given by (28) is a $*$-representation of the $*$-algebra $\mathcal{A}_q$ on $\mathcal{D}$. Indeed, it is straightforward to check that equation (26) holds for the four generators of $\mathcal{A}_q$, hence (26) is valid for all elements of the algebra $\mathcal{A}_q$. This $*$-representation T is called the *Fock representation* of the q-oscillator algebra $\mathcal{A}_q$.

Each of the other variants of the q-oscillator algebras mentioned in Subsect. 5.1.2 also possesses a Fock representation. For the algebra $\mathcal{A}_b$ it is given by

$$T(b)|n\rangle = [[n]]^{1/2}|n-1\rangle, \; T(b^+)|n\rangle = [[n+1]]^{1/2}|n+1\rangle, \; T(N)|n\rangle = n|n\rangle,$$

where $[[n]] = (q^n - 1)/(q - 1)$ (see (2.2)). The Fock representation for the algebra $\mathcal{A}_A$ is given by the same formulas if b is replaced by A and b^+ by A^+. The Fock representation of the two-parameter algebra $\mathcal{A}^c_{p,q}$ takes the form

$$T(\hat{a})|n\rangle = [n]_{qp}^{1/2}|n-1\rangle, \quad T(\hat{a}^+)|n\rangle = [n+1]_{qp}^{1/2}|n+1\rangle, \quad T(N)|n\rangle = n|n\rangle,$$

where $[n]_{qp} = (q^n - p^{-n})/(q - p^{-1})$.

Let us return to the Fock representation of the algebra $\mathcal{A}_q$. It describes a q-deformed harmonic oscillator with Hamiltonian $H = (h\omega/2)(aa^+ + a^+a)$, where h and ω are as in case of the "usual" quantum harmonic oscillator. The vectors $|n\rangle$ are eigenvectors of this Hamiltonian with eigenvalues

$$E(n) = \frac{h\omega}{2}([n] + [n+1]) = \frac{h\omega}{2}\frac{\sinh(\nu(n+1/2))}{\sinh(\nu/2)},$$

where $q = e^\nu$. In the limit $q \to 1$ these numbers give the eigenvalues of the Hamiltonian of the usual quantum harmonic oscillator.

5.3.2 The Bargmann–Fock Realization

In this subsection we study a realization of the Fock representation of $\mathcal{A}_q$ on a Hilbert space of entire holomorphic functions.

Recall that $\tilde{D}_q$ is the q-derivative given by formula (2.48). For polynomials $f, g \in \mathbb{C}[z]$ we define

$$(f, g) = \overline{f}(\tilde{D}_q)g(z)|_{z=0}, \tag{29}$$

where $\overline{f} = \sum_n \overline{\alpha_n} z^n$ for $f = \sum_n \alpha_n z^n$. Since $\tilde{D}_q z^n = [n]z^{n-1}$, it follows that the form $(\cdot,\cdot)$ is a scalar product on the vector space $\mathbb{C}[z]$ such that the polynomials

$$u_n(z) = \frac{z^n}{\sqrt{[n]!}}, \qquad n = 0, 1, 2, \cdots, \tag{30}$$

constitute an orthonormal basis. Let $\mathfrak{F}$ denote the Hilbert space completion of $(\mathbb{C}[z], (\cdot,\cdot))$. Since $\{u_n\}$ is an orthonormal basis, the elements of $\mathfrak{F}$ are precisely those entire holomorphic functions $f(z) = \sum_{n=0}^\infty \alpha_n z^n$ on the complex plane for which $\sum_{n=0}^\infty |\alpha_n|^2 [n]! < \infty$. In particular, the q-exponential function $\tilde{E}_q(z) = \sum_{n=0}^\infty z^n/[n]!$ (see (2.57)) belongs to $\mathfrak{F}$ and satisfies the equation

$$\sum_{n=0}^\infty u_n(\xi)u_n(\overline{z}) = \tilde{E}_q(\xi\overline{z}). \tag{31}$$

The function $\tilde{E}_q(z)$ fulfills the inequality

$$|\tilde{E}_q(z)| \le \tilde{E}_q(|z|) \le \exp|z|, \quad z \in \mathbb{C}. \tag{32}$$

For positive values of x, the function $\tilde{E}_q(x)$ is positive, while for negative x it is an alternating series which oscillates within the bounds (32).

It is clear that there is an isomorphism of the Hilbert spaces $\mathfrak{H}$ and $\mathfrak{F}$ such that $|n\rangle$ is mapped to u_n. Under this isomorphism the operators $T(a)$, $T(a^+)$, $T(N)$ from (28) are transformed into the operators

$$T(a) = \tilde{D}_q, \quad T(a^+) = z, \quad T(N) = z\frac{d}{dz} \tag{33}$$

on the space $\mathfrak{F}$. This form of the Fock representation of $\mathcal{A}_q$ is called the *Bargmann–Fock realization.*

In the case of the q-oscillator algebra $\mathcal{A}_b$ we define a scalar product $(\cdot,\cdot)$ on $\mathbb{C}[z]$ by the formula

$$(f,g) = \overline{f}(D_q)g(z)|_{z=0},$$

where D_q is given by (2.43). Then the polynomials

$$v_n(z) = \frac{z^n}{\sqrt{[[n]]!}}, \qquad n = 0, 1, 2, \cdots,$$

form an orthonormal basis, where $[[n]]! := [[1]] \cdot [[2]] \cdots [[n]]$. Let $\mathfrak{F}_b$ denote the Hilbert space which is obtained by completing the space $(\mathbb{C}[z], (\cdot,\cdot))$. The Hilbert space $\mathfrak{F}_b$ consists of all functions $f(z) = \sum_{n=0}^{\infty} \alpha_n z^n$ on the complex plane $\mathbb{C}$ satisfying the condition $\sum_{n=0}^{\infty} |\alpha_n|^2 [[n]]! < \infty$. In the case $q > 1$ the exponential function

$$e_q((1-q)z) = \sum_{n=0}^{\infty} \frac{z^n}{[[n]]!},$$

from (2.55) is contained in $\mathfrak{F}_b$ and we have

$$\sum\nolimits_{n=0}^{\infty} v_n(\xi) v_n(\bar{z}) = e_q((1-q)\xi\bar{z}).$$

Note that for $q > 1$ the series defining $e_q((1-q)z)$ converges uniformly and absolutely for all finite z. But $e_q((1-q)z)$ has an essential singularity at infinity. For positive values of z, the function $e_q((1-q)z)$ is positive.

It is clear that there exists an isomorphism ψ of the Hilbert spaces $\mathfrak{H}$ and $\mathfrak{F}_b$ such that $\psi(|n\rangle) = v_n$, $n = 0, 1, 2, \cdots$. Under this isomorphism the operators $T(b)$, $T(b^+)$ and $T(N)$ of the Fock representation are transformed into the operators

$$T(b) = D_q, \quad T(b^+) = z, \quad T(N) = z\frac{d}{dz}$$

on the space $\mathfrak{F}_b$.

5.3.3 Coherent States

Let z be a complex number. A unit vector $|z\rangle$ of the carrier space of the Fock representation of $\mathcal{A}_q$ is called a *q-coherent state* if

$$T(a)|z\rangle = z|z\rangle.$$

Any complex number is an eigenvalue of $T(a)$ with unit eigenvector

$$|z\rangle = \tilde{E}_q(z\bar{z})^{-1/2} \sum_{n=0}^{\infty} \frac{z^n}{[n]!^{1/2}} |n\rangle = \tilde{E}_q(z\bar{z})^{-1/2} \tilde{E}_q(zT(a^+))|0\rangle, \tag{34}$$

where $\tilde{E}_q(zT(a^+)) := \sum_{n=0}^{\infty} (zT(a^+))^n/[n]!$. The factor $\tilde{E}_q(z\bar{z})^{-1/2}$ in (34) is only taken in order to ensure that $|z\rangle$ is a unit vector of $\mathcal{F}$. Note that the operator $T(a)$ is not symmetric and that its eigenvectors $|z\rangle$, $z \in \mathbb{C}$, are not mutually orthogonal. More precisely, we have

$$\langle z'|z\rangle = \tilde{E}_q(z'\overline{z'})^{-1/2} \tilde{E}_q(z\bar{z})^{-1/2} \tilde{E}_q(z\overline{z'}).$$

In the case of the Fock representation of the algebra $\mathcal{A}_b$ a q-coherent state is defined by $T(b)|z\rangle = z|z\rangle$. For arbitrary $z, z' \in \mathbb{C}$ we have

$$|z\rangle = e_q((1-q)z\bar{z})^{-1/2} \sum_{n=0}^{\infty} \frac{z^n}{[[n]]!^{1/2}} |n\rangle$$

and

$$\langle z'|z\rangle = e_q((1-q)z'\overline{z"})^{-1/2} e_q((1-q)z\bar{z})^{-1/2} e_q((1-q)z\overline{z'}).$$

It can be shown (see, for example, [DKu]) that in the case $0 < q < 1$ the q-coherent states of the algebra $\mathcal{A}_b$ have the property that

$$\int |z\rangle\langle z| d\mu(z) = 1, \tag{35}$$

where

$$d\mu(z) = \frac{1}{2\pi} e_q((1-q)|z|^2) e_{1/q}(-(1-q)q|z|^2) d_q|z|^2 d\theta. \tag{36}$$

In equation (36), $\theta = \arg z$ and $d_q|z|^2$ means the q-integration from Subsect. 2.2.2. The integration in (35) is over $|z|^2 \leq \eta_0$ and $0 \leq \theta < 2\pi$, where η_0 is the smallest pole of $e_q((1-q)z)$ on the positive real half-line.

There exists a similar decomposition for the q-coherent states of the algebra $\mathcal{A}_q$. But in this case the measure $d\mu(z)$ is more complicated.

5.3.4 Bargmann–Fock Space Realization of Irreducible Representations of $\breve{U}_q(\mathrm{sl}_2)$

In this subsection we compose the Jordan–Schwinger realization (see Proposition 4) with the Fock representation of the algebra $\mathcal{A}_q^{\mathrm{ext},2}$ and develop an approach to the irreducible type 1 representations of $\breve{U}_q(\mathrm{sl}_2)$ in this manner.

The Fock representation T of the algebra $\mathcal{A}_q^{\mathrm{ext},2}$ acts on the Hilbert space $\mathfrak{H}^{\otimes 2} := \mathfrak{H} \otimes \mathfrak{H}$ with orthonormal basis $|m,n\rangle$, $m,n = 0,1,2,\cdots$, and is determined by the formulas (28). Let $\mathcal{D}^{\otimes 2}$ denote the span of basis elements $|m,n\rangle$. The composition $\pi := T \circ \varphi$ defines an infinite-dimensional representation of the algebra $\breve{U}_q(\mathrm{sl}_2)$ by linear operators on the space $\mathcal{D}^{\otimes 2}$:

$$\breve{U}_q(\mathrm{sl}_2) \xrightarrow{\varphi} \mathcal{A}_q^{\mathrm{ext},2} \xrightarrow{T} \mathcal{L}(\mathcal{D}^{\otimes 2}). \tag{37}$$

The decomposition of this representation π into irreducible components will be given by formula (38) below. In order to achieve this we first note that the basis elements $|m,n\rangle$ of $\mathfrak{H}^{\otimes 2}$ can be represented in the form

$$|m,n\rangle = \frac{T(a_1^+)^m}{\sqrt{[m]!}}\frac{T(a_2^+)^n}{\sqrt{[n]!}}|0,0\rangle.$$

By (9), we have

$$\pi(E)|m,n\rangle = T(a_1^+)T(a_2)\frac{T(a_1^+)^m}{\sqrt{[m]!}}\frac{T(a_2^+)^n}{\sqrt{[n]!}}|0,0\rangle$$

$$= ([m+1][n])^{1/2}\frac{T(a_1^+)^{m+1}}{\sqrt{[m+1]!}}\frac{T(a_2^+)^{n-1}}{\sqrt{[n-1]!}}|0,0\rangle = ([m+1][n])^{1/2}|m+1,n-1\rangle.$$

Similarly, we obtain

$$\begin{aligned}\pi(F)|m,n\rangle &= ([m][n+1])^{1/2}\frac{T(a_1^+)^{m-1}}{\sqrt{[m-1]!}}\frac{T(a_2^+)^{n+1}}{\sqrt{[n+1]!}}|0,0\rangle \\ &= ([m][n+1])^{1/2}|m-1,n+1\rangle.\end{aligned}$$

Thus, for each $l \in \frac{1}{2}\mathbb{N}$ the linear subspace V_l spanned by the basis elements $|m,n\rangle$ with $m+n = 2l$ is invariant under the representation π. Let us redenote $m+n$, m and n by $2l$, $l+k$ and $l-k$, respectively. Then the invariant subspace V_l of $\mathfrak{H}^{\otimes 2}$ is spanned by the vectors

$$\mathbf{e}_k^l := |l+k,l-k\rangle = \frac{T(a_1^+)^{l+k}}{\sqrt{[l+k]!}}\frac{T(a_2^+)^{l-k}}{\sqrt{[l-k]!}}|0,0\rangle, \quad k = -l,-l+1,\cdots,l.$$

The operators $\pi(E)$, $\pi(F)$ and $\pi(K)$ act on these vectors by the formulas

$$\pi(E)\mathbf{e}_k^l = ([l+k+1][l-k])^{1/2}\mathbf{e}_{k+1}^l,$$

$$\pi(F)\mathbf{e}_k^l = ([l+k][l-k+1])^{1/2}\mathbf{e}_{k-1}^l, \quad \pi(K)\mathbf{e}_k^l = q^k \mathbf{e}_k^l.$$

That is, the restriction of T to the invariant subspace V_l is equivalent to the irreducible representation $T_{1,l}$ of $\breve{U}_q(\mathrm{sl}_2)$ from Subsect. 3.2.3. Thus we have

$$\pi = \bigoplus\nolimits_{l\in\frac{1}{2}\mathbb{N}_0} T_{1,l}. \tag{38}$$

If we consider the Bargmann–Fock realization on the space $\mathfrak{H}^{\otimes 2}$, then V_l becomes the space $\mathfrak{F}_l$ of homogeneous polynomials in two variables z_1 and z_2 of degree $2l$. To the basis element $\mathbf{e}_k^l \in V_l$ there corresponds the monomial

$$z_1^{l+k} z_2^{l-k} / ([l+k]![l-k]!)^{1/2}$$

in $\mathfrak{F}_l$. These monomials are orthonormal in $\mathcal{F}^{\otimes 2}$, since the elements $\mathbf{e}_k^l$ are orthonormal with respect to the scalar product (29). By (14) and (33), the operators $\pi(E)$, $\pi(F)$ and $\pi(K)$ are realized on $\mathfrak{F}_l$ as

$$\pi(E) = z_1 \tilde{D}_q^{(2)}, \quad \pi(F) = z_2 \tilde{D}_q^{(1)}, \quad \pi(K) = q^{(z_1\partial_1 - z_2\partial_2)/2},$$

where the symbol $\tilde{D}^{(i)}$ means the q-derivative $\tilde{D}_q$ (see (2.48)) with respect to the variable z_i. Note that the operator $q^{(z_1\partial_1 - z_2\partial_2)/2}$ on $\mathfrak{F}_l$ is well-defined, because $N_1 - N_2 = z_1\partial_1 - z_2\partial_2$ is a diagonal operator on $\mathfrak{F}_l$.

Finally, the results of this subsection will be used to prove the following

Proposition 17. *The Jordan–Schwinger homomorphism $\varphi : \breve{U}_q(\mathrm{sl}_2) \to \mathcal{A}_q^{\mathrm{ext},2}$ from Proposition 4 is an algebra isomorphism of $\breve{U}_q(\mathrm{sl}_2)$ into $\mathcal{A}_q^{\mathrm{ext},2}$.*

Proof. Since q is not a root of unity, the dual pairing of the Hopf algebras $\breve{U}_q(\mathrm{sl}_2)$ and $\mathcal{O}(SL_q(2))$ is nondegenerate. From this it follows easily that for any nonzero element $x \in \breve{U}_q(\mathrm{sl}_2)$ there exists an irreducible finite-dimensional representation $T_{1,l}$ of $\breve{U}_q(\mathrm{sl}_2)$ such that $T_{1,l}(x) \neq 0$. (In fact, this assertion is a special case of Proposition 7.21 below.)

Now let us suppose that $\varphi(x) = 0$ for $x \in \breve{U}_q(\mathrm{sl}_2)$. For the representation $\pi = T \circ \varphi$ of the algebra $\breve{U}_q(\mathrm{sl}_2)$ defined by (37) we then have $\pi(x) = 0$. From (38) we obtain $T_{1,l}(x) = 0$ for all $l \in \frac{1}{2}\mathbb{N}_0$, so $x = 0$ by the above result. □

5.4 Notes

The q-oscillator algebra and its Fock representation were introduced by L. C. Biedenharn [Bid] and A. Macfarlane [Macf]. The irreducible representations of q-oscillator algebras for q not a root of unity have been determined by P. P. Kulish [Kul] and G. Rideau [Rid]. The description of q-coherent states and the Bargmann–Fock realization of the Fock representation of $\mathcal{A}_q^{\mathrm{c}}$ can be found in the paper [DKu]. The Jordan–Schwinger realization of the quantum algebra $\breve{U}_q(\mathrm{sl}_2)$ and its irreducible representations were given in [Bid]. Other references, as well as further results, on q-oscillator algebras are contained in [BL], [DKu], [ChK] and [Ge]. Physical applications of q-oscillators are described in [DKu] and [BDKL].

Part II

Quantized Universal Enveloping Algebras

6. Drinfeld–Jimbo Algebras

The aim of this chapter is to define the quantized universal enveloping algebras, called Drinfeld–Jimbo algebras, and to develop basic algebraic structures and results on these algebras such as the Poincaré–Birkhoff–Witt theorem, braid group actions, Verma modules, quantum Killing forms, quantum Casimir elements, centers and Harish-Chandra homomorphisms.

6.1 Definitions of Drinfeld–Jimbo Algebras

6.1.1 Semisimple Lie Algebras

In this subsection we collect general notions, facts and notation on semisimple Lie algebras that will be freely used in the rest of the book.

Let $\mathfrak{g}$ be a finite-dimensional complex semisimple Lie algebra. The operators $\operatorname{ad} X$, $X \in \mathfrak{g}$, acting on $\mathfrak{g}$ as $\operatorname{ad} X(Y) = [X, Y]$, $Y \in \mathfrak{g}$, define the adjoint representation of $\mathfrak{g}$. A maximal commutative Lie subalgebra $\mathfrak{h}$ in $\mathfrak{g}$ consisting of semisimple elements is called a *Cartan subalgebra* of $\mathfrak{g}$. Recall that an element $X \in \mathfrak{g}$ is *semisimple* if the operator $\operatorname{ad} X$ can be diagonalized.

Let $\mathfrak{h}$ be a Cartan subalgebra of $\mathfrak{g}$. Since the operators $\operatorname{ad} H$, $H \in \mathfrak{h}$, are semisimple and commute with each other, the Lie algebra $\mathfrak{g}$ is a direct sum of joint eigenspaces of these operators:

$$\mathfrak{g} = \mathfrak{h} \oplus \bigoplus\nolimits_{\alpha \neq 0} \mathfrak{g}_\alpha, \tag{1}$$

where α are nonvanishing linear forms on $\mathfrak{h}$, called *roots*, and $\mathfrak{g}_\alpha$ are the corresponding eigenspaces, called *root subspaces*. The Cartan subalgebra $\mathfrak{h}$ coincides with the eigenspace of the eigenvalue 0. The set of all roots α is denoted by Δ. The dimension of $\mathfrak{h}$ is called the *rank* of $\mathfrak{g}$. Let us collect the main properties of roots and root subspaces:

(i) If α is a root of $\mathfrak{g}$, then so is $-\alpha$. There exist no other roots of the form $c\alpha$, $c \in \mathbb{C}$.

(ii) If $\mathfrak{g}_\alpha$ and $\mathfrak{g}_\beta$, $\alpha \neq -\beta$, are root subspaces of $\mathfrak{g}$ and $E_\alpha \in \mathfrak{g}_\alpha$, $E_\beta \in \mathfrak{g}_\beta$, then $[E_\alpha, E_\beta] \in \mathfrak{g}_{\alpha+\beta}$ if $\alpha + \beta$ is a root of $\mathfrak{g}$ and $[E_\alpha, E_\beta] = 0$ otherwise. For $\alpha = -\beta$ we have $[E_\alpha, E_\beta] \in \mathfrak{h}$.

(iii) All root subspaces $\mathfrak{g}_\alpha$, $\alpha \in \Delta$, are one-dimensional.

(iv) If $\alpha + \beta \neq 0$, then the corresponding root subspaces $\mathfrak{g}_\alpha$ and $\mathfrak{g}_\beta$ are orthogonal with respect to the Killing form $B(X, Y) := \mathrm{Tr}\,(\mathrm{ad}\, X \circ \mathrm{ad}\, Y)$.

(v) The restriction of $B(X, Y)$ to $\mathfrak{h} \times \mathfrak{h}$ is nondegenerate. For every root α there exists a unique element $H_\alpha \in \mathfrak{h}$ such that $B(H_\alpha, H) = \alpha(H)$ for all $H \in \mathfrak{h}$. Let $\mathfrak{h}_R = \sum_{\alpha \in \Delta} \mathbb{R} H_\alpha$. The complex dimension of $\mathfrak{h}$ coincides with the real dimension of $\mathfrak{h}_R$. The Killing form is real and positive definite on $\mathfrak{h}_R$.

(vi) The Killing form is nondegenerate on $\mathfrak{g} \times \mathfrak{g}$. One can choose elements $E_{\pm\alpha} \in \mathfrak{g}_{\pm\alpha}$ such that $B(E_\alpha, E_{-\alpha}) = 1$. Then, $[E_\alpha, E_{-\alpha}] = H_\alpha$.

We choose a basis $H_1, H_2, \cdots, H_l$ of the Cartan subalgebra $\mathfrak{h}$ such that $\alpha(H_k) \in \mathbb{R}$ for all $\alpha \in \Delta$ and take a basis vector E_α in each root subspace $\mathfrak{g}_\alpha$ such that $[E_\alpha, E_{-\alpha}] = H_\alpha$. Then the elements

$$H_1, H_2, \cdots, H_l, \quad E_\alpha,\ \alpha \in \Delta, \tag{2}$$

form a vector space basis of $\mathfrak{g}$. We have the commutation relations

$$[H_i, H_j] = 0, \quad [H_i, E_\alpha] = \alpha(H_i) E_\alpha, \quad \alpha \in \Delta, \tag{3}$$

$$[E_\alpha, E_{-\alpha}] = H_\alpha, \quad \alpha \in \Delta, \qquad [E_\alpha, E_\beta] = N_{\alpha\beta} E_{\alpha+\beta}, \quad \alpha \neq -\beta, \tag{4}$$

where $N_{\alpha\beta} = 0$ if $\alpha + \beta$ is not a root. A set of elements (2) satisfying the relations (3) and (4) is called a *Cartan–Weyl basis* of $\mathfrak{g}$.

We fix an *ordered* basis $H_1, H_2, \cdots, H_l$ of the Cartan subalgebra $\mathfrak{h}$. A root $\alpha \in \Delta$ is called *positive* (resp. *negative*) if the first nonzero number in the sequence $\alpha(H_1), \alpha(H_2), \cdots, \alpha(H_l)$ is positive (resp. negative). The set of positive (resp. negative) roots will be denoted by Δ_+ (resp. Δ_-).

Let $\mathfrak{h}'_R = \sum_{\alpha \in \Delta} \mathbb{R}\alpha$ be the subspace of the dual space $\mathfrak{h}'$ for $\mathfrak{h}$. If $\gamma \in \mathfrak{h}'_R$, then the formula $\gamma(H) = B(H, H_\gamma)$, $H \in \mathfrak{h}$, defines uniquely an element $H_\gamma \in \mathfrak{h}_R$. There is a one-to-one correspondence $\gamma \leftrightarrow H_\gamma$ between elements from $\mathfrak{h}'_R$ and $\mathfrak{h}_R$. We define a symmetric bilinear form $(\gamma, \gamma') = cB(H_\gamma, H_{\gamma'})$, $\gamma, \gamma' \in \mathfrak{h}'_R$, on $\mathfrak{h}'_R \times \mathfrak{h}'_R$, where c is a fixed constant. If $\alpha \in \Delta$, then the formula

$$w_\alpha : \gamma \to \gamma - \frac{2(\gamma, \alpha)}{(\alpha, \alpha)} \alpha, \qquad \gamma \in \mathfrak{h}'_R, \tag{5}$$

defines a reflection of the space $\mathfrak{h}'_R$. Clearly, $w_\alpha = w_{-\alpha}$. The group W generated by all reflections w_α, $\alpha \in \Delta$, is called the *Weyl group* of the Lie algebra $\mathfrak{g}$ with respect to $\mathfrak{h}$. W is a finite group which acts transitively on Δ.

A root $\alpha \in \Delta_+$ is said to be *simple* if it is not the sum of two other positive roots. We state some properties of simple roots:

(i) Any positive (negative) root is a linear combination of simple roots with nonnegative (nonpositive) integral coefficients.

(ii) The number of simple roots is equal to the rank of the Lie algebra $\mathfrak{g}$, that is, it is the dimension l of the Cartan subalgebra $\mathfrak{h}$. Simple roots are linearly independent.

(iii) If $\alpha_1, \alpha_2, \cdots, \alpha_l$ are simple roots of $\mathfrak{g}$, then the reflections w_{α_i}, $i = 1, 2, \cdots, l$, generate the Weyl group W of $\mathfrak{g}$.

Note that the set of simple roots of $\mathfrak{h}$ is not uniquely determined. If $\alpha_1, \alpha_2, \cdots, \alpha_l$ are simple roots, then $w\alpha_1, w\alpha_2, \cdots, w\alpha_l$ for any fixed $w \in W$ are also simple roots. In particular, the set $-\alpha_1, -\alpha_2, \cdots, -\alpha_l$ (which corresponds to the basis $-H_1, -H_2, \cdots, -H_l$ of $\mathfrak{h}$) is a set of simple roots.

Formula (1) gives a direct sum decomposition $\mathfrak{g} = \mathfrak{n}_+ \oplus \mathfrak{h} \oplus \mathfrak{n}_-$, where $\mathfrak{n}_+ := \sum_{\alpha>0} \mathfrak{g}_\alpha$ and $\mathfrak{n}_- := \sum_{\alpha>0} \mathfrak{g}_{-\alpha}$.

Proposition 1. *The subalgebras $\mathfrak{n}_+$ and $\mathfrak{n}_-$ are maximal nilpotent subalgebras of $\mathfrak{g}$. Every maximal nilpotent subalgebra of $\mathfrak{g}$ can be mapped by an inner automorphism onto $\mathfrak{n}_+$. The subalgebras $\mathfrak{h} + \mathfrak{n}_+$ and $\mathfrak{h} + \mathfrak{n}_-$ are maximal solvable subalgebras in $\mathfrak{g}$.*

Let us fix an ordered sequence $\alpha_1, \cdots, \alpha_l$ of simple roots. Let $E_{\alpha_1}, \cdots, E_{\alpha_l}$ be the corresponding root elements of $\mathfrak{g}$. By (4), successive application of the commutator $[\cdot, \cdot]$ gives all root elements E_α, $\alpha \in \Delta_+$. That is, the elements $E_{\alpha_1}, \cdots, E_{\alpha_l}$ generate the Lie subalgebra $\mathfrak{n}_+$. Analogously, the root elements $E_{-\alpha_1}, \cdots, E_{-\alpha_l}$ generate the Lie subalgebra $\mathfrak{n}_-$. The elements $[E_{\alpha_i}, E_{-\alpha_i}] = H_{\alpha_i}$, $i = 1, 2, \cdots, l$, form a basis of the Cartan subalgebra $\mathfrak{h}$. Thus, *the complex semisimple Lie algebra $\mathfrak{g}$ is generated by the root elements $E_{\alpha_1}, \cdots, E_{\alpha_l}$, $E_{-\alpha_1}, \cdots, E_{-\alpha_l}$ corresponding to the simple roots $\alpha_1, \cdots, \alpha_l$.*

The $l \times l$ matrix $A = (a_{ij})$ with entries

$$a_{ij} = 2(\alpha_i, \alpha_j)/(\alpha_i, \alpha_i)$$

is called the *Cartan matrix* of $\mathfrak{g}$. It determines the semisimple Lie algebra $\mathfrak{g}$ uniquely up to isomorphisms. It can be proved that

(i) $\det A \neq 0$;
(ii) a_{ij} are integers, $a_{ij} \leq 0$ for $i \neq j$ and $a_{ii} = 2$, $i = 1, 2, \cdots, l$;
(iii) there exists a diagonal matrix $D = \text{diag}\ (d_1, \cdots, d_l)$ with entries in the set $\{1, 2, 3\}$ such that the matrix $DA = (d_i a_{ij})$ is symmetric and positive definite. The matrix DA is called the *symmetrized Cartan matrix* of $\mathfrak{g}$.

We assume throughout that the length (α, α) of the shortest root α is 2. Then $d_i = (\alpha_i, \alpha_i)/2$ and the matrix DA has the entries $(DA)_{ij} = (\alpha_i, \alpha_j)$. We also use the symmetric bilinear form $(\cdot, \cdot)$ on $\mathfrak{h} \times \mathfrak{h}$ defined by $(H_{\alpha_i}, H_{\alpha_j}) = d_j^{-1} a_{ij}$. It is a multiple of the restriction of the Killing form to $\mathfrak{h} \times \mathfrak{h}$.

Let us collect the Cartan matrices for the classical simple complex Lie algebras $\mathrm{sl}(l+1, \mathbb{C})$, $\mathrm{so}(2l+1, \mathbb{C})$, $\mathrm{sp}(2l, \mathbb{C})$, and $\mathrm{so}(2l, \mathbb{C})$:

$\mathrm{sl}(l+1, \mathbb{C})$: $a_{ii} = 2$, $i = 1, 2, \cdots, l$, $a_{i,i+1} = a_{i+1,i} = -1$, $i = 1, 2, \cdots, l-1$, $a_{ij} = 0$ otherwise;

$\mathrm{so}(2l+1, \mathbb{C})$: as for $\mathrm{sl}(l+1, \mathbb{C})$ with $a_{l,l-1} = -1$ replaced by $a_{l,l-1} = -2$;

$\mathrm{sp}(2l, \mathbb{C})$: as for $\mathrm{sl}(l+1, \mathbb{C})$ with $a_{l-1,l} = -1$ replaced by $a_{l-1,l} = -2$;

so$(2l,\mathbb{C})$: as for sl$(l+1,\mathbb{C})$ with $a_{l-1,l}=a_{l,l-1}=-1$ and $a_{l-2,l}=a_{l,l-2}=0$ replaced by $a_{l-1,l}=a_{l,l-1}=0$ and $a_{l-2,l}=a_{l,l-2}=-1$, respectively.

The corresponding matrices $D=\mathrm{diag}\,(d_1,d_2,\cdots,d_l)$ are

$$
\begin{aligned}
D&=\mathrm{diag}\,(1,1,\cdots,1) \quad \text{for} \quad \mathrm{sl}(l+1,\mathbb{C}) \quad \text{and} \quad \mathrm{so}(2l,\mathbb{C}),\\
D&=\mathrm{diag}\,(2,\cdots,2,1) \quad \text{for} \quad \mathrm{so}(2l+1,\mathbb{C}),\\
D&=\mathrm{diag}\,(1,\cdots,1,2) \quad \text{for} \quad \mathrm{sp}(2l,\mathbb{C}).
\end{aligned}
$$

The *universal enveloping algebra* $U(\mathfrak{g})$ of the Lie algebra $\mathfrak{g}$ is the quotient of the tensor algebra $T(\mathfrak{g})$ over $\mathfrak{g}$ by the two-sided ideal generated by the elements $x\otimes y-y\otimes x-[x,y]$, $x,y\in\mathfrak{g}$. The next result characterizes the algebra $U(\mathfrak{g})$ in terms of generators and relations (see [Ser] for a proof).

Theorem 2 (*Serre's theorem*). *Let $\mathfrak{g}$ be a complex semisimple Lie algebra with Cartan matrix $A=(a_{ij})$ and simple roots $\alpha_1,\alpha_2,\cdots,\alpha_l$, $l=\mathrm{rank}\,\mathfrak{g}$. Then $E_i\equiv E_{\alpha_i}$, $F_i\equiv E_{-\alpha_i}$, $H_i:=[E_{\alpha_i},F_{\alpha_i}]$, $i=1,2,\cdots,l$, can be chosen in such a way that the universal enveloping algebra $U(\mathfrak{g})$ is generated by the elements E_i,F_i,H_i, $i=1,2,\cdots,l$, subject to the relations*

$$[H_i,H_j]=0,\quad [E_i,F_i]=H_i,\quad [E_i,F_j]=0,\quad i\neq j, \tag{6}$$

$$[H_i,E_j]=a_{ij}E_j,\quad [H_i,F_j]=-a_{ij}F_j, \tag{7}$$

$$(\mathrm{ad}\,E_i)^{1-a_{ij}}E_j\equiv\sum_{k=0}^{1-a_{ij}}(-1)^k\binom{1-a_{ij}}{k}E_i^{1-a_{ij}-k}E_jE_i^k=0,\quad i\neq j, \tag{8}$$

$$(\mathrm{ad}\,F_i)^{1-a_{ij}}F_j\equiv\sum_{k=0}^{1-a_{ij}}(-1)^k\binom{1-a_{ij}}{k}F_i^{1-a_{ij}-k}F_jF_i^k=0,\quad i\neq j. \tag{9}$$

The relations (8) and (9) are usually called the *Serre relations.*

Let $U(\mathfrak{n}_\pm)$ and $U(\mathfrak{h})$ be the universal enveloping algebras of the Lie subalgebras $\mathfrak{n}_\pm$ and $\mathfrak{h}$ of $\mathfrak{g}$, respectively. Then the map $U(\mathfrak{n}_+)\otimes U(\mathfrak{h})\otimes U(\mathfrak{n}_-)\ni n_+\otimes H\otimes n_-\to n_+Hn_-\in U(\mathfrak{g})$ is bijective. In particular, we have

$$U(\mathfrak{g})=U(\mathfrak{n}_+)U(\mathfrak{h})U(\mathfrak{n}_-)=U(\mathfrak{n}_-)U(\mathfrak{h})U(\mathfrak{n}_+). \tag{10}$$

The elements E_α, $\alpha\in\Delta_+$, form a basis of $\mathfrak{n}_+$ and $E_{-\alpha}$, $\alpha\in\Delta_+$, constitute a basis of $\mathfrak{n}_-$. We choose an order $\beta_1>\beta_2>\cdots>\beta_n$ in the set of positive roots, where n is the number of elements in Δ_+. Then the elements $E_{\beta_1}^{r_1}\cdots E_{\beta_n}^{r_n}$, $r_j\in\mathbb{N}_0$, form a basis of $U(\mathfrak{n}_+)$. The same procedure yields a basis of $U(\mathfrak{n}_-)$. The elements $H_1^{s_1}\cdots H_l^{s_l}$, $s_j\in\mathbb{N}_0$, are a basis of $U(\mathfrak{h})$.

Theorem 3 (*Poincaré–Birkhoff–Witt theorem*). *The elements*

$$E_{\beta_1}^{r_1}E_{\beta_2}^{r_2}\cdots E_{\beta_n}^{r_n}H_1^{s_1}H_2^{s_2}\cdots H_l^{s_l}E_{-\beta_n}^{p_n}\cdots E_{-\beta_2}^{p_2}E_{-\beta_1}^{p_1},\quad r_j,s_j,p_j\in\mathbb{N}_0, \tag{11}$$

form a basis of the universal enveloping algebra $U(\mathfrak{g})$.

Since the Weyl group W of $\mathfrak{g}$ is generated by the reflections $w_{\alpha_i} \equiv w_i$, $i = 1, 2, \cdots, l$, any element $w \in W$ is a certain product of reflections $w_1, \cdots, w_l$. Such a representation of w is not unique. A decomposition $w = w_{i_1} w_{i_2} \cdots w_{i_k}$ is called *reduced* if the number k is minimal. This number is called the *length* of w and is denoted by $l(w)$. Clearly, $l(ww') \leq l(w) + l(w')$.

Proposition 4. (i) *There exists a unique element w_0 of maximal length in the Weyl group W and $l(w_0)$ is the number of positive roots of $\mathfrak{g}$.*

(ii) *Let $w_0 = w_{i_1} \cdots w_{i_n}$ be a reduced decomposition of w_0. Define roots $\beta_1, \beta_2, \cdots, \beta_n$, $n = l(w_0)$, by setting*

$$\beta_r = w_{i_1} \cdots w_{i_{r-1}}(\alpha_{i_r}), \qquad r = 1, 2, \cdots, n,$$

where α_{i_r} are the corresponding simple roots. Then the roots $\beta_1, \beta_2, \cdots, \beta_n$ are positive and pairwise distinct. They exhaust all positive roots of $\mathfrak{g}$.

6.1.2 The Drinfeld–Jimbo Algebras $U_q(\mathfrak{g})$

Let $\mathfrak{g}$ be a finite-dimensional complex semisimple Lie algebra of rank l with Cartan matrix (a_{ij}) and $d_i := (\alpha_i, \alpha_i)/2$. Let q be a fixed nonzero complex number and let $q_i = q^{d_i}$. Suppose that $q_i^2 \neq 1$ for $i = 1, 2, \cdots, l$.

Let $U_q(\mathfrak{g})$ be the (associative unital complex) algebra with $4l$ generators E_i, F_i, K_i, K_i^{-1}, $1 \leq i \leq l$, and defining relations

$$K_i K_j = K_j K_i, \quad K_i K_i^{-1} = K_i^{-1} K_i = 1, \tag{12}$$

$$K_i E_j K_i^{-1} = q_i^{a_{ij}} E_j, \quad K_i F_j K_i^{-1} = q_i^{-a_{ij}} F_j, \tag{13}$$

$$E_i F_j - F_j E_i = \delta_{ij} \frac{K_i - K_i^{-1}}{q_i - q_i^{-1}}, \tag{14}$$

$$\sum_{r=0}^{1-a_{ij}} (-1)^r \left[\!\left[\begin{matrix} 1 - a_{ij} \\ r \end{matrix} \right]\!\right]_{q_i} E_i^{1-a_{ij}-r} E_j E_i^r = 0, \quad i \neq j, \tag{15}$$

$$\sum_{r=0}^{1-a_{ij}} (-1)^r \left[\!\left[\begin{matrix} 1 - a_{ij} \\ r \end{matrix} \right]\!\right]_{q_i} F_i^{1-a_{ij}-r} F_j F_i^r = 0, \quad i \neq j, \tag{16}$$

where

$$\left[\!\left[\begin{matrix} n \\ r \end{matrix} \right]\!\right]_q = \frac{[n]_q!}{[r]_q![n-r]_q!} = q^{(r-n)r} \left[\begin{matrix} n \\ r \end{matrix} \right]_{q^2}, \quad [n]_q = \frac{q^n - q^{-n}}{q - q^{-1}}.$$

Note that the relations (15) and (16) are obtained from (8) and (9) by replacing the usual binomial coefficients by the q-binomial coefficients. Setting $q = e^h$ and $K_i = q_i^{hH_i}$, the relations (12) and (13) are *formally* equivalent to (7) and the first relation in (6).

Proposition 5. *There is a unique Hopf algebra structure on the algebra $U_q(\mathfrak{g})$ with comultiplication Δ, counit ε and antipode S such that*

$$\Delta(K_i) = K_i \otimes K_i, \quad \Delta(K_i^{-1}) = K_i^{-1} \otimes K_i^{-1}, \tag{17}$$

$$\Delta(E_i) = E_i \otimes K_i + 1 \otimes E_i, \quad \Delta(F_i) = F_i \otimes 1 + K_i^{-1} \otimes F_i, \tag{18}$$

$$\varepsilon(K_i) = 1, \quad \varepsilon(E_i) = \varepsilon(F_i) = 0, \tag{19}$$

$$S(K_i) = K_i^{-1}, \quad S(E_i) = -E_i K_i^{-1} \quad S(F_i) = -K_i F_i. \tag{20}$$

Proof. Let $\mathcal{A} = \mathbb{C}\langle E_i, F_i, K_i, K_i^{-1}\rangle$ be the free algebra generated by E_i, F_i, K_i, K_i^{-1}, $i = 1, 2, \cdots, l$. Then $U_q(\mathfrak{g})$ is the quotient algebra $\mathcal{A}/\mathcal{J}$, where $\mathcal{J}$ is the two-sided ideal generated by the elements corresponding to the relations (12)–(16). We proceed as in the proof of Proposition 3.4 and define first algebra homomorphisms $\Delta : \mathcal{A} \to U_q(\mathfrak{g}) \otimes U_q(\mathfrak{g})$ and $\varepsilon : \mathcal{A} \to \mathbb{C}$ and an algebra anti-homomorphism $S : \mathcal{A} \to U_q(\mathfrak{g})$ such that (17)–(20) hold. Then we verify that Δ, ε and S preserve the defining relations (12)–(16). We carry out this proof for Δ and the first Serre relation (15). That is, we shall show that

$$\sum\nolimits_{r=0}^{1-a_{ij}} (-1)^r \begin{bmatrix}\begin{bmatrix} 1-a_{ij} \\ r \end{bmatrix}\end{bmatrix}_{q_i} \Delta(E_i)^{1-a_{ij}-r} \Delta(E_j) \Delta(E_i)^r = 0. \tag{21}$$

Using Proposition 2.2 and the relation (2.17), we easily verify that

$$\Delta(E_i)^s = \sum\nolimits_{k=0}^{s} q_i^{-k(s-k)} \begin{bmatrix}\begin{bmatrix} s \\ k \end{bmatrix}\end{bmatrix}_{q_i} E_i^k \otimes K_i^k E_i^{s-k}. \tag{22}$$

Therefore, the left hand side in (21) can be represented as

$$\sum_{r=0}^{1-a_{ij}} \sum_{s=0}^{r} \sum_{t=0}^{1-a_{ij}-r} (-1)^r q_i^{b_{rs}} \begin{bmatrix}\begin{bmatrix} 1-a_{ij} \\ r \end{bmatrix}\end{bmatrix}_{q_i} \begin{bmatrix}\begin{bmatrix} r \\ s \end{bmatrix}\end{bmatrix}_{q_i} \begin{bmatrix}\begin{bmatrix} 1-a_{ij}-r \\ t \end{bmatrix}\end{bmatrix}_{q_i}$$

$$\times (E_i^s \otimes K_i^s E_i^{r-s})(E_j \otimes K_j + 1 \otimes E_j)(E_i^t \otimes K_i^t E_i^{1-a_{ij}-r-t}), \tag{23}$$

where $b_{rs} := -s(r-s) - t(1-a_{ij}-r-t)$. We split this sum into two parts. The first part contains summands with E_j in the first factor and the second one contains summands with E_j in the second factor. Both parts vanish. Let us check this for the second part. It can be written as

$$\sum_{r=0}^{1-a_{ij}} \sum_{s=0}^{r} \sum_{t=0}^{1-a_{ij}-r} (-1)^r q_i^{-s(r-s)+t(2s+t-r-1)} \begin{bmatrix}\begin{bmatrix} 1-a_{ij} \\ r \end{bmatrix}\end{bmatrix}_{q_i} \begin{bmatrix}\begin{bmatrix} r \\ s \end{bmatrix}\end{bmatrix}_{q_i}$$

$$\times \begin{bmatrix}\begin{bmatrix} 1-a_{ij}-r \\ t \end{bmatrix}\end{bmatrix}_{q_i} (E_i^{t+s} \otimes K_i^{t+s} E_i^{r-s} E_j E_i^{1-a_{ij}-r-t}). \tag{24}$$

Changing summation to $n = s + t$ and $p = r - s$ and using the identity

$$\begin{bmatrix}\begin{bmatrix} b \\ p+s \end{bmatrix}\end{bmatrix}_{q_i} \begin{bmatrix}\begin{bmatrix} p+s \\ s \end{bmatrix}\end{bmatrix}_{q_i} \begin{bmatrix}\begin{bmatrix} b-p-s \\ n-s \end{bmatrix}\end{bmatrix}_{q_i} = \begin{bmatrix}\begin{bmatrix} b-n \\ p \end{bmatrix}\end{bmatrix}_{q_i} \begin{bmatrix}\begin{bmatrix} b \\ n \end{bmatrix}\end{bmatrix}_{q_i} \begin{bmatrix}\begin{bmatrix} n \\ s \end{bmatrix}\end{bmatrix}_{q_i}$$

where $b = 1 - a_{ij}$, we can write the expression (24) in the form

$$\sum_{n=0}^{b}\sum_{p=0}^{b-n}\left(\sum_{s=0}^{n}(-1)^s\left[\!\left[\begin{matrix}n\\ s\end{matrix}\right]\!\right]_{q_i} q_i^{-s(n-1)}\right)(-1)^p q^{n(n-p-1)}\left[\!\left[\begin{matrix}b-n\\ p\end{matrix}\right]\!\right]_{q_i}\left[\!\left[\begin{matrix}b\\ n\end{matrix}\right]\!\right]_{q_i}$$

$$\times E_i^n \otimes K_i^n E_i^p E_j E_i^{b-p-n}.$$

By Proposition 2.4, the interior sum over s vanishes if $n \neq 0$. The summand for $n = 0$ reduces to (15), hence it also vanishes. This shows that the second part of the sum (23) is zero. The proof for the first part of the sum is similar. This completes the proof of (21).

Since Δ, ε, and S preserve the relations (12)–(16), they pass to algebra homomorphisms resp. an anti-homomorphism of $U_q(\mathfrak{g}) = \mathcal{A}/\mathcal{J}$. By Proposition 1.8, it is enough to check the Hopf algebra axioms on the generators which is easily done. (It is in fact the same verification as in the case of $U_q(\mathrm{sl}_2)$.) □

Definition 1. *The Hopf algebra $U_q(\mathfrak{g})$ of Proposition 5 is called the* Drinfeld–Jimbo quantized universal enveloping algebra *(or, briefly, the* Drinfeld–Jimbo algebra*) corresponding to the Lie algebra $\mathfrak{g}$ and the complex number q.*

Remark 1. One may also define the Drinfeld–Jimbo algebra $U_q(\mathfrak{g})$ by omitting first the Serre relations (15) and (16) that are unpleasant to deal with. That is, one first verifies that the algebra $\hat{U}_q(\mathfrak{g})$ with generators E_i, F_i, K_i, K_i^{-1} and defining relations (12)–(14) is a Hopf algebra with structure maps given by the formulas (17)–(20). Then one shows that the two-sided ideal of $\hat{U}_q(\mathfrak{g})$ generated by the elements on the left hand sides of (15) and (16) is a Hopf ideal of $\hat{U}_q(\mathfrak{g})$. The corresponding quotient Hopf algebra of $\hat{U}_q(\mathfrak{g})$ is then the Drinfeld–Jimbo algebra $U_q(\mathfrak{g})$. △

Let $\mathfrak{g}$ be a simple Lie algebra of type A_l, D_l or E_l. Then the Cartan matrix (a_{ij}) is symmetric and we have $a_{ii} = 2$ and $a_{ij} = 0$ or -1 if $i \neq j$. Hence, $(\alpha_i, \alpha_i)/2 = 1$ for all simple roots α_i and all q_i coincide with q. In these cases the relations (15) and (16) for $a_{ij} = -1$ take the form

$$E_i^2 E_j - (q+q^{-1})E_iE_jE_i + E_jE_i^2 = 0, \quad F_i^2 F_j - (q+q^{-1})F_iF_jF_i + F_jF_i^2 = 0.$$

For many reasons (see, for instance, Subsect. 11.5.1) it is convenient to have also the Hopf algebra $U_q(\mathfrak{g})$ for $\mathfrak{g} = \mathrm{gl}(N, \mathbb{C})$. Most of the theory developed later for $U_q(\mathfrak{g})$ with a semisimple Lie algebra $\mathfrak{g}$ remains valid with a few minor modifications for $U_q(\mathrm{gl}_N)$. It is defined as follows. The algebra $U_q(\mathrm{gl}_N)$ is generated by elements E_i, F_i, K_j, K_j^{-1}, $i = 1, 2, \cdots, N-1$, $j = 1, 2, \cdots, N$, subject to the relations

$$K_iK_j = K_jK_i, \qquad K_iK_i^{-1} = K_i^{-1}K_i = 1,$$

$$K_iE_jK_i^{-1} = q^{\delta_{ij}-\delta_{i,j+1}}E_j, \quad K_iF_jK_i^{-1} = q^{-\delta_{ij}+\delta_{i,j+1}}F_j,$$

$$E_iF_j - F_jE_i = \delta_{ij}\frac{K_iK_{i+1}^{-1} - K_i^{-1}K_{i+1}}{q-q^{-1}},$$

$$E_iE_j = E_jE_i, \quad F_iF_j = F_jF_i, \quad |i-j| \le 2,$$
$$E_i^2E_{i\pm1} - (q+q^{-1})E_iE_{i\pm1}E_i + E_{i\pm1}E_i^2 = 0,$$
$$F_i^2F_{i\pm1} - (q+q^{-1})F_iF_{i\pm1}F_i + F_{i\pm1}F_i^2 = 0.$$

The algebra $U_q(\mathrm{gl}_N)$ becomes a Hopf algebra with structure maps given on the generators by

$$\Delta(K_i^{\pm1}) = K_i^{\pm1} \otimes K_i^{\pm1}, \quad \Delta(E_i) = E_i \otimes K_iK_{i+1}^{-1} + 1 \otimes E_i,$$
$$\Delta(F_i) = F_i \otimes 1 + K_i^{-1}K_{i+1} \otimes F_i, \quad \varepsilon(K_i) = 1, \quad \varepsilon(E_i) = \varepsilon(F_i) = 0,$$
$$S(K_i) = K_i^{-1}, \quad S(E_i) = -E_iK_i^{-1}K_{i+1} \quad S(F_i) = -K_iK_{i+1}^{-1}F_i.$$

The element $K_1K_2\cdots K_N$ is obviously group-like and central in $U_q(\mathrm{gl}_N)$. Hence the two-sided ideal $\langle K_1K_2\cdots K_N - 1\rangle$ is a Hopf ideal of $U_q(\mathrm{gl}_N)$. The quotient Hopf algebra $U_q(\mathrm{gl}_N)/\langle K_1K_2\cdots K_N - 1\rangle$ is the extended Drinfeld–Jimbo algebra $U_q^{\mathrm{ext}}(\mathrm{sl}_N)$ used in Sect. 8.5. Moreover, the subalgebra of $U_q(\mathrm{gl}_N)$ generated by the elements $K_iK_{i+1}^{-1}$, E_i and F_i is a Hopf subalgebra that is isomorphic to $U_q(\mathrm{sl}_N)$ with Hopf algebra isomorphism π determined by $\pi(K_iK_{i+1}^{-1}) = K_i$, $\pi(E_i) = E_i$ and $\pi(F_i) = F_i$, $i = 1, 2, \cdots, N-1$.

Let us return to the Drinfeld–Jimbo algebra $U_q(\mathfrak{g})$ for a semisimple Lie algebra $\mathfrak{g}$. Let $\alpha_1, \cdots, \alpha_l$ be the simple roots of $\mathfrak{g}$ and let

$$Q = \{\textstyle\sum_i n_i\alpha_i \mid n_i \in \mathbb{Z}\} \quad \text{and} \quad Q_+ = \{\textstyle\sum_i n_i\alpha_i \in Q \mid n_i \in \mathbb{N}_0\}. \tag{25}$$

For every form $\lambda = \sum_i n_i\alpha_i \in Q$ we define an element

$$K_\lambda = K_1^{n_1}K_2^{n_2}\cdots K_l^{n_l} \tag{26}$$

of the algebra $U_q(\mathfrak{g})$. From (13) one derives that

$$K_\lambda E_i K_\lambda^{-1} = q^{(\lambda,\alpha_i)}E_i, \quad K_\lambda F_i K_\lambda^{-1} = q^{-(\lambda,\alpha_i)}F_i. \tag{27}$$

Proposition 6. *Let ρ be the half-sum of positive roots of $\mathfrak{g}$. Then we have*

$$S^2(a) = K_{2\rho}aK_{2\rho}^{-1}, \quad a \in U_q(\mathfrak{g}). \tag{28}$$

The antipode S is a bijective mapping of $U_q(\mathfrak{g})$.

Proof. Since the maps $a \to K_{2\rho}aK_{2\rho}^{-1}$ and S^2 are both algebra homomorphisms of $U_q(\mathfrak{g})$, it is sufficient to prove the equality (28) for the generators of $U_q(\mathfrak{g})$. This is done by direct verification. Obviously, (28) implies that S is injective. Since $a = S(K_{2\rho}S(a)K_{2\rho}^{-1})$ by (20) and (28), S is surjective. □

In the case $\mathfrak{g} = \mathrm{sl}(2,\mathbb{C})$ the Drinfeld–Jimbo algebra $U_q(\mathfrak{g})$ is just the Hopf algebra $U_q(\mathrm{sl}_2)$ studied in Chap. 3. We now introduce the counterpart to the second quantum algebra $\breve{U}_q(\mathrm{sl}_2)$. Let $\breve{U}_q(\mathfrak{g})$ be the algebra with generators E_i, F_i, K_i, K_i^{-1}, $i = 1, 2, \cdots, l$, and the following defining relations:

$$K_iK_j = K_jK_i, \quad K_iK_i^{-1} = K_i^{-1}K_i = 1, \tag{29}$$

$$K_iE_jK_i^{-1} = q_i^{a_{ij}/2}E_j, \quad K_iF_jK_i^{-1} = q_i^{-a_{ij}/2}F_j, \tag{30}$$

$$E_iF_j - F_jE_i = \delta_{ij}\frac{K_i^2 - K_i^{-2}}{q_i - q_i^{-1}}, \tag{31}$$

$$\sum_{r=0}^{1-a_{ij}}(-1)^r \begin{bmatrix}\begin{bmatrix} 1-a_{ij} \\ r \end{bmatrix}\end{bmatrix}_{q_i} E_i^{1-a_{ij}-r}E_jE_i^r = 0, \quad i \neq j, \tag{32}$$

$$\sum_{r=0}^{1-a_{ij}}(-1)^r \begin{bmatrix}\begin{bmatrix} 1-a_{ij} \\ r \end{bmatrix}\end{bmatrix}_{q_i} F_i^{1-a_{ij}-r}F_jF_i^r = 0, \quad i \neq j. \tag{33}$$

The algebra $\breve{U}_q(\mathfrak{g})$ becomes a Hopf algebra with structure maps determined on the generators as follows:

$$\Delta(K_i) = K_i \otimes K_i, \quad \Delta(K_i^{-1}) = K_i^{-1} \otimes K_i^{-1}, \tag{34}$$

$$\Delta(E_i) = E_i \otimes K_i + K_i^{-1} \otimes E_i, \quad \Delta(F_i) = F_i \otimes K_i + K_i^{-1} \otimes F_i, \tag{35}$$

$$\varepsilon(K_i) = 1, \quad \varepsilon(E_i) = \varepsilon(F_i) = 0, \tag{36}$$

$$S(K_i) = K_i^{-1}, \quad S(E_i) = -q_iE_i \quad S(F_i) = -q_i^{-1}F_i. \tag{37}$$

Similarly as in the case of sl_2, there is an injective Hopf algebra homomorphism $\varphi : U_q(\mathfrak{g}) \to \breve{U}_q(\mathfrak{g})$ such that

$$\varphi(E_i) = E_iK_i, \quad \varphi(F_i) = K_i^{-1}F_i, \quad \varphi(K_i) = K_i^2.$$

That is, $U_q(\mathfrak{g})$ can be considered as a Hopf subalgebra of $\breve{U}_q(\mathfrak{g})$ if we identify $X \in U_q(\mathfrak{g})$ with $\varphi(X) \in \breve{U}_q(\mathfrak{g})$. Note that the algebras $U_q(\mathfrak{g})$ and $\breve{U}_q(\mathfrak{g})$ are not isomorphic, because they have different numbers of one-dimensional representations. The quantum algebra $\breve{U}_q(\mathfrak{g})$ for $\mathfrak{g} = \mathrm{gl}(N, \mathbb{C})$ will be defined and studied in Sect. 7.3.

For some investigations it is convenient to treat q as an indeterminate and to work with the field $\mathbb{Q}(q)$ of rational functions in q over the field $\mathbb{Q}$ of rational numbers. Replacing in the above definition of $U_q(\mathfrak{g})$ the base field $\mathbb{C}$ by $\mathbb{Q}(q)$, we obtain a Hopf algebra over the field $\mathbb{K} = \mathbb{Q}(q)$. It will be denoted by $U'_q(\mathfrak{g})$ and called the *rational form* of the Drinfeld–Jimbo algebra.

6.1.3 The h-Adic Drinfeld–Jimbo Algebras $U_h(\mathfrak{g})$

The Drinfeld–Jimbo algebras $U_h(\mathfrak{g})$ are h-adic Hopf algebras over the ring $\mathbb{C}[[h]]$. Recall that h-adic algebras and Hopf algebras were defined in Subsect. 1.2.10.

Let $\mathfrak{g}$, (a_{ij}) and d_i be as in Subsect. 6.1.2, and let h be an indeterminate. Let $U_h(\mathfrak{g})$ be the h-adic algebra over the ring $\mathbb{C}[[h]]$ generated by $3l$ elements E_i, F_i, H_i, $i = 1, 2, \cdots, l$, subject to the relations

$$[H_i, H_j] = 0, \quad [H_i, E_j] = a_{ij}E_j, \quad [H_i, F_j] = -a_{ij}F_j, \tag{38}$$

$$E_iF_j - F_jE_i = \delta_{ij}\frac{e^{d_ihH_i} - e^{-d_ihH_i}}{e^{d_ih} - e^{-d_ih}}, \tag{39}$$

$$\sum_{r=0}^{1-a_{ij}} (-1)^r \left[\left[\begin{matrix} 1-a_{ij} \\ r \end{matrix}\right]\right]_{q_i} E_i^{1-a_{ij}-r} E_j E_i^r = 0, \quad i \neq j, \tag{40}$$

$$\sum_{r=0}^{1-a_{ij}} (-1)^r \left[\left[\begin{matrix} 1-a_{ij} \\ r \end{matrix}\right]\right]_{q_i} F_i^{1-a_{ij}-r} F_j F_i^r = 0, \quad i \neq j, \tag{41}$$

where $q_i := e^{d_ih}$.

Proposition 7. *There exists a unique h-adic Hopf algebra structure on $U_h(\mathfrak{g})$ with comultiplication $\Delta : U_h(\mathfrak{g}) \to U_h(\mathfrak{g})\hat{\otimes}U_h(\mathfrak{g})$, counit ε and antipode S such that*

$$\Delta(H_i) = H_i \otimes 1 + 1 \otimes H_i, \tag{42}$$

$$\Delta(E_i) = E_i \otimes e^{d_ihH_i} + 1 \otimes E_i, \quad \Delta(F_i) = F_i \otimes 1 + e^{-d_ihH_i} \otimes F_i, \tag{43}$$

$$\varepsilon(H_i) = \varepsilon(E_i) = \varepsilon(F_i) = 0, \tag{44}$$

$$S(H_i) = -H_i, \quad S(E_i) = -E_ie^{-d_ihH_i} \quad S(F_i) = -e^{d_ihH_i}F_i. \tag{45}$$

Proof. The proof is similar to that for the Hopf algebras $U_q(\mathfrak{g})$ in Subsect. 6.1.2. □

Definition 2. *The h-adic Hopf algebra $U_h(\mathfrak{g})$ is called the* h-adic Drinfeld–Jimbo algebra.

One can also define the h-adic Hopf algebra $U_h(\mathfrak{g})$ for $\mathfrak{g} = \mathrm{gl}(N, \mathbb{C})$. It is generated by elements E_i, F_i, H_j, $i = 1, 2, \cdots, N-1$, $j = 1, 2, \cdots, N$, subject to the relations

$$[H_i, H_j] = 0, \quad [H_i, E_j] = (\delta_{ij} - \delta_{i,j+1})E_j,$$

$$[H_i, F_j] = (-\delta_{ij} + \delta_{i,j+1})F_j, \quad [E_i, F_j] = \delta_{ij}\frac{e^{h(H_i - H_{i+1})} - e^{-h(H_i - H_{i+1})}}{e^h - e^{-h}}$$

and the same Serre relations for E_i and F_i as in the case $U_q(\mathrm{gl}_N)$ with $q := e^h$. The Hopf algebra structure of $U_h(\mathrm{gl}_N)$ is determined by

$$\Delta(H_i) = H_i \otimes 1 + 1 \otimes H_i, \quad \Delta(E_i) = E_i \otimes e^{h(H_i - H_{i+1})} + 1 \otimes E_i,$$

$$\Delta(F_i) = F_i \otimes 1 + e^{-h(H_i - H_{i+1})} \otimes F_i, \quad \varepsilon(H_j) = \varepsilon(E_i) = \varepsilon(F_i) = 0,$$

$$S(H_j) = -H_j, \quad S(E_i) = -E_ie^{-h(H_i - H_{i+1})}, \quad S(F_i) = -e^{h(H_i - H_{i+1})}F_i.$$

Remarks: 2. The element $e^{d_ih} - e^{-d_ih} \in \mathbb{C}[[h]]$ is not invertible in $\mathbb{C}[[h]]$, because its constant term is zero. But the expression on the right hand side of (39) is a formal power series $\sum_n p_n(H_i)h^n$ with certain polynomials $p_n(H_i)$, so it is a well-defined element of the h-adic algebra generated by E_j, F_j, H_j.

3. The algebra $U_q(\mathfrak{g})$ is *formally* derived from the h-adic Drinfeld–Jimbo algebra $U_h(\mathfrak{g})$ if e^h and $e^{d_jhH_j}$ are replaced by q and K_j, respectively.

4. The constant terms in the defining relations (38)–(41) give just the defining relations (6)–(9) for the universal enveloping algebra $U(\mathfrak{g})$ according to Serre's theorem. Hence there is an isomorphism

$$U_h(\mathfrak{g})/hU_h(\mathfrak{g}) \simeq U(\mathfrak{g})$$

of complex algebras. This fact is sometimes expressed by saying that the algebra $U(\mathfrak{g})$ is the classical limit $h \to 0$ of the h-adic algebra $U_h(\mathfrak{g})$.

5. Let us adopt the following notational convention in the h-adic case. For $\alpha \in \mathbb{C}$, we write q^α for the formal power series $e^{\alpha h} \in \mathbb{C}[[h]]$ and $q_i^{\alpha H_i}$ for the element $e^{\alpha d_i h H_i} \in U_h(\mathfrak{g})$. △

Proposition 6 extends at once to the Hopf algebra $U_h(\mathfrak{g})$. That is, we have

$$S^2(a) = e^{H_{2\rho} h} a e^{-H_{2\rho} h}, \quad a \in U_h(\mathfrak{g}) \tag{46}$$

where $H_{2\rho} = \sum_{i=1}^{l} d_i n_i H_i$ if $2\rho = \sum_i n_i \alpha_i$.

As an abstract h-adic algebra, $U_h(\mathfrak{g})$ is isomorphic to $U(\mathfrak{g})[[h]]$, see Subsect. 1.2.10 for the latter notion. More precisely, we have

Proposition 8. *There is an isomorphism φ of the h-adic algebra $U_h(\mathfrak{g})$ onto $U(\mathfrak{g})[[h]]$ which coincides with the identity map modulo h. Further, if φ and φ' are two such isomorphisms of $U_h(\mathfrak{g})$ onto $U(\mathfrak{g})[[h]]$, then there exists an invertible element $F \in U(\mathfrak{g})[[h]]$ such that $F \equiv 1 \pmod h$ and $\varphi'(a) = F\varphi(a)F^{-1}$ for $a \in U_h(\mathfrak{g})$.*

Proof. See [Dr2] or [ShS], Chap. 11. □

We give the explicit form of such an isomorphism, denoted by φ, for the algebra $U_h(\mathrm{sl}_2)$ (see [Jim2]). Recall from Subsect. 3.1.4 that $U_h(\mathrm{sl}_2)$ is the h-adic algebra generated by elements H, E, F such that $[H, E] = 2E$, $[H, F] = -2F$ and $[E, F] = (e^{hH} - e^{-hH})/(e^h - e^{-h})$. The isomorphism φ of the algebras $U_h(\mathrm{sl}_2)$ and $U(\mathrm{sl}_2)[[h]]$ over $\mathbb{C}[[h]]$ is uniquely determined by its action on the generating elements and is given by

$$\varphi(H) = H', \quad \varphi(F) = F', \quad \varphi(E) = 2\left(\frac{\cosh h(H'-1) - \cosh 2h\sqrt{C'}}{[H'-1]^2 - 4C' \sinh^2 h}\right) E'.$$

Here H', E', F' are the generators of $U(\mathrm{sl}_2)$ satisfying the relations $[H', E'] = 2E'$, $[H', F'] = -2F'$, $[E', F'] = H'$, and $C' = \frac{1}{4}(H'-1)^2 + E'F'$ is the Casimir element of $U(\mathrm{sl}_2)$.

6.1.4 Some Algebra Automorphisms of Drinfeld–Jimbo Algebras

The algebra automorphisms of $U_q(\mathrm{sl}_2)$ from Proposition 3.3 have the following counterparts for general $U_q(\mathfrak{g})$.

Proposition 9. *Let $\epsilon = (\epsilon_1, \cdots, \epsilon_l)$, $\epsilon_i = \pm 1$, and $\alpha = (\alpha_1, \cdots, \alpha_l)$, $\alpha_i \in \mathbb{C}$, $\alpha_i \neq 0$. There exist algebra automorphisms θ and $\vartheta_{\epsilon\alpha}$ of $U_q(\mathfrak{g})$ such that*

$$\theta(E_i) = F_i, \quad \theta(F_i) = E_i, \quad \theta(K_i) = K_i^{-1},$$

$$\vartheta_{\epsilon\alpha}(E_i) = \alpha_i E_i, \quad \vartheta_{\epsilon\alpha}(F_i) = \epsilon_i \alpha_i^{-1} F_i, \quad \vartheta_{\epsilon\alpha}(K_i) = \epsilon_i K_i.$$

Proof. The proof is given by direct verification. □

Other algebra automorphisms of $U_q(\mathfrak{g})$ will be given in Subsect. 6.2.2.

Proposition 10. *Let α be as in Proposition 9. There exist automorphisms θ and ϑ_α of the h-adic algebra $U_h(\mathfrak{g})$ such that*

$$\theta(E_i) = F_i, \quad \theta(F_i) = E_i, \quad \theta(H_i) = -H_i, \tag{47}$$

$$\vartheta_\alpha(E_i) = \alpha_i E_i, \quad \vartheta_\alpha(F_i) = \alpha_i^{-1} F_i, \quad \vartheta_\alpha(H_i) = H_i.$$

There exists a $\mathbb{C}$-algebra automorphism ϕ of $U_h(\mathfrak{g})$ such that

$$\phi(E_i) = F_i, \qquad \phi(F_i) = E_i, \qquad \phi(H_i) = -H_i, \qquad \phi(h) = -h. \tag{48}$$

There is also a $\mathbb{C}$-algebra anti-automorphism ω of $U_h(\mathfrak{g})$ such that

$$\omega(E_i) = F_i, \qquad \omega(F_i) = E_i, \qquad \omega(H_i) = H_i, \qquad \omega(h) = -h. \tag{49}$$

Below we shall use the $\mathbb{Q}$-algebra automorphism ψ of $U_q'(\mathfrak{g})$ defined by

$$\psi(E_i) = E_i, \quad \psi(F_i) = F_i, \qquad \psi(K_i) = K_i^{-1}, \quad \psi(q) = q^{-1}. \tag{50}$$

6.1.5 Triangular Decomposition of $U_q(\mathfrak{g})$

In this subsection we develop an analog of the decomposition (10) for $U_q(\mathfrak{g})$. Throughout we assume that q is not a root of unity.

Let $U_q(\mathfrak{n}_+)$, $U_q(\mathfrak{n}_-)$ and $U_q(\mathfrak{h})$ be the subalgebras of $U_q(\mathfrak{g})$ generated by the elements $E_1, \cdots, E_l$, the elements $F_1, \cdots, F_l$ and the elements $H_1, \cdots, H_l$, respectively. The subalgebras generated by $E_1, \cdots, E_l, H_1, \cdots, H_l$ resp. $F_1, \cdots, F_l, H_1, \cdots, H_l$ will be denoted by $U_q(\mathfrak{b}_+)$ resp. $U_q(\mathfrak{b}_-)$. From (17)–(20) we see that $U_q(\mathfrak{h})$, $U_q(\mathfrak{b}_+)$ and $U_q(\mathfrak{b}_-)$ are Hopf subalgebras of $U_q(\mathfrak{g})$.

The action of the elements K_λ, $\lambda \in Q$, on $U_q(\mathfrak{g})$ by conjugation determines a Q-gradation of the algebra $U_q(\mathfrak{g})$ and its subalgebras. A nonzero element $a \in U_q(\mathfrak{g})$ is said to be of *degree* $\alpha = \sum_{i=1}^l n_i \alpha_i$, $n_i \in \mathbb{Z}$, if $K_\lambda a K_\lambda^{-1} = q^{(\lambda,\alpha)} a$ for all $\lambda \in Q$. If the element has degrees α and β, then $q^{(\lambda,\alpha)} = q^{(\lambda,\beta)}$ for all $\lambda \in Q$ and hence $\alpha = \beta$, because we assumed that q is not a root of unity and the form $(\cdot,\cdot)$ on Q is nondegenerate. Thus, the degree is well-defined. Further, if a is a monomial in the generators with n_i factors E_i and m_i factors F_i, then it follows from (27) that a is of degree $\sum_i (n_i - m_i)\alpha_i$. Since such monomials span $U_q(\mathfrak{g})$, we have a direct sum decomposition

$$U_q(\mathfrak{g}) = \bigoplus_{\beta \in Q} U_q^\beta(\mathfrak{g}), \quad U_q^\beta(\mathfrak{g}) = \{a \in U_q(\mathfrak{g}) \mid K_\lambda a K_\lambda^{-1} = q^{(\lambda,\beta)} a\}. \tag{51}$$

Since $U_q^\alpha(\mathfrak{g}) \cdot U_q^\beta(\mathfrak{g}) \subseteq U_q^{\alpha+\beta}(\mathfrak{g})$, the preceding shows that $U_q(\mathfrak{g})$ is indeed a *Q-graded algebra*. This gradation induces obviously Q-gradations on the subalgebras $U_q(\mathfrak{n}_+)$ and $U_q(\mathfrak{n}_-)$ as well. In particular, we have

$$U_q(\mathfrak{n}_+) = \bigoplus\nolimits_{\beta\in Q_+} U_q^\beta(\mathfrak{n}_+), \quad U_q(\mathfrak{n}_-) = \bigoplus\nolimits_{\beta\in Q_+} U_q^{-\beta}(\mathfrak{n}_-), \tag{52}$$

where

$$U_q^{\pm\beta}(\mathfrak{n}_\pm) = \{a \in U_q(\mathfrak{n}_\pm) \mid K_\lambda a K_\lambda^{-1} = q^{\pm(\lambda,\beta)}a\}. \tag{53}$$

Lemma 11. *$E_1^{n_1}\cdots E_l^{n_l} \neq 0$ for every l-tuple $(n_1,\cdots,n_l)\in\mathbb{N}_0^l$.*

Proof. Let φ be the algebra homomorphism of $U_q(\mathfrak{g})$ to $U_{q_i}(\mathrm{sl}_2)$ defined by $\varphi(E_i) = E$, $\varphi(F_i) = F$, $\varphi(K_i) = K$ and $\varphi(E_j) = \varphi(F_j) = 0$, $\varphi(K_j) = 1$, $j\neq i$. Since $\varphi(E_i^n) = E^n \neq 0$ by Proposition 3.1, $E_i^n \neq 0$ for any $n\in\mathbb{N}$.

The algebra $U_q(\mathfrak{g})^{\otimes n}$ is Q^n-graded. Let $\Delta^{(r)}$ be as in Subsect. 1.2.2. From (18) we easily get $\Delta^{(n-1)}(E_i) = \sum_{j=0}^{n-1} 1^{\otimes j}\otimes E_i \otimes K_i^{\otimes(n-1-j)}$. From this formula it follows that the component of degree $(n_1\alpha_1,\cdots,n_l\alpha_l)$ of the element $\Delta^{(l-1)}(E_1^{n_1}\cdots E_l^{n_l}) \in U_q(\mathfrak{g})^{\otimes l}$ is a nonzero multiple of

$$E_1^{n_1}\otimes E_2^{n_2}K_1^{n_1}\otimes E_2^{n_3}K_1^{n_1}K_2^{n_2}\otimes\cdots\otimes E_l^{n_l}K_1^{n_1}\cdots K_{l-1}^{n_{l-1}}.$$

This element is nonzero, since $E_i^n\neq 0$. Hence $E_1^{n_1}\cdots E_l^{n_l}\neq 0$. □

Lemma 12. *The elements K_γ, $\gamma\in Q$, form a vector space basis of $U_q(\mathfrak{h})$.*

Proof. Suppose that $\sum_\gamma a_\gamma K_\gamma = 0$, where the sum is finite and $a_\gamma\in\mathbb{C}$. Using the left adjoint representation ad_L of $U_q(\mathfrak{g})$ (see Subsect. 1.3.4) we find

$$\mathrm{ad}_L\Big(\sum\nolimits_\gamma a_\gamma K_\gamma\Big)(E_1^{n_1}\cdots E_l^{n_l}) = \Big(\sum\nolimits_\gamma a_\gamma q^{(\gamma,\sum n_i\alpha_i)}\Big)E_1^{n_1}\cdots E_l^{n_l} = 0.$$

Since $E_1^{n_1}\cdots E_l^{n_l}\neq 0$ by Lemma 11, we have $\sum_\gamma a_\gamma q^{(\gamma,\sum n_i\alpha_i)} = 0$ for all $n_1,\cdots,n_l\in\mathbb{N}$. Since the roots $\alpha_1,\cdots,\alpha_l$ form a basis in the dual space $\mathfrak{h}'$ and q is not a root of unity, it is not difficult to show that the latter is only possible if all a_γ vanish. Thus, the set $\{K_\gamma\}$ is linearly independent. □

From the definition of $U_q(\mathfrak{n}_+)$ it is clear that it has a vector space basis consisting of certain monomials $E_{i_1}^{n_1}\cdots E_{i_l}^{n_l}$, where $i_j\in\{1,2,\cdots,l\}$ and $n_j\in\mathbb{N}_0$. Let us fix such a basis and denote it by $\{\mathbf{E}_r \mid r\in I\}$, where I is an appropriate set of indices.

Lemma 13. *The elements $\mathbf{E}_rK_\gamma$, $r\in I$, $\gamma\in Q$, form a basis of $U_q(\mathfrak{b}_+)$.*

Proof. These elements generate $U_q(\mathfrak{b}_+)$. We have to show that they are linearly independent. Let $a \equiv \sum a_{r\gamma}\mathbf{E}_rK_\gamma = 0$. Without loss of generality we assume that all summands of a are of the same Q-degree β. The summand of degree $(\beta,0)$ in $\Delta(\sum a_{r\gamma}\mathbf{E}_rK_\gamma)$ must be 0. This implies that $\sum a_{r\gamma}\mathbf{E}_rK_\gamma\otimes K_\beta K_\gamma = 0$. The latter equality can be written as

$$\sum\nolimits_\gamma\Big(\sum\nolimits_{r\in I} a_{r\gamma}\mathbf{E}_rK_\gamma\Big)\otimes K_\beta K_\gamma = 0.$$

By Lemma 12, the expression in the parentheses vanishes. Hence we get $\sum a_{r\gamma}\mathbf{E}_r = 0$ for every $\gamma \in Q$. Since $\{\mathbf{E}_r \mid r \in I\}$ is a basis of $U_q(\mathfrak{n}_+)$, all coefficients $a_{r\gamma}$ vanish. □

Let θ be the algebra automorphism from Proposition 9. We set $\mathbf{F}_r := \theta(\mathbf{E}_r)$. Since $\mathbf{F}_r K_\gamma := \theta(\mathbf{E}_r K_{-\gamma})$, it follows from Lemma 13 that the set $\{\mathbf{F}_r K_\gamma \mid r \in I, \gamma \in Q\}$ is a basis of $U_q(\mathfrak{b}_-)$.

Theorem 14. *The elements $\mathbf{E}_r K_\gamma \mathbf{F}_{r'}$, $r, r' \in I$, $\gamma \in Q$, form a basis of the vector space $U_q(\mathfrak{g})$. The mapping $\mathbf{E}_r \otimes K_\gamma \otimes \mathbf{F}_{r'} \to \mathbf{E}_r K_\gamma \mathbf{F}_{r'}$ gives an isomorphism of the vector spaces $U_q(\mathfrak{n}_+) \otimes U_q(\mathfrak{h}) \otimes U_q(\mathfrak{n}_-)$ and $U_q(\mathfrak{g})$.*

Proof. Since the algebra $U_q(\mathfrak{g})$ is spanned by the elements $\mathbf{E}_r K_\gamma \mathbf{F}_{r'}$, $r, r' \in I$, $\gamma \in Q$, it suffices to prove that the elements $\mathbf{E}_r \mathbf{F}_{r'} K_\gamma$ are linearly independent. Suppose that a finite sum $a \equiv \sum a_{rr'\gamma}\mathbf{E}_r\mathbf{F}_{r'}K_\gamma$ is zero, where $a_{rr'\gamma} \in \mathbb{C}$. If $\mathbf{E}_r$ and $\mathbf{F}_{r'}$ are of Q-degrees λ_r and $-\lambda_{r'}$, respectively, then $\mathbf{E}_r\mathbf{F}_{r'}K_\gamma$ is of Q-degree $\lambda_r - \lambda_{r'}$. Without loss of generality we may assume that all summands of a are of fixed Q-degree $\lambda_r - \lambda_{r'}$.

We introduce an order in Q. If $\alpha = \sum n_i\alpha_i$ and $\alpha' = \sum n_i'\alpha_i$, then we write $\alpha < \alpha'$ if $\sum n_i < \sum n_i'$ or if $\sum n_i = \sum n_i'$ and $n_i < n_i'$ for the smallest index i such that $n_i \neq n_i'$. Let J be the set of $r \in I$ for which the degree γ_r of the term $\mathbf{E}_r$ in the sum a is maximal with respect to this order. Clearly, the sum of all terms in $\Delta(a)$ which have maximal Q-degree in the first tensor factor and minimal Q-degree in the second one must vanish, that is,

$$\sum_{\gamma, r\in J, r'\in J'} a_{rr'\gamma}(\mathbf{E}_r K_{-\gamma_{r'}} \otimes K_{\gamma_r}\mathbf{F}_{r'})(K_\gamma \otimes K_\gamma) = 0,$$

where γ_r and $\gamma_{r'}$ are fixed. Hence $\sum_{\gamma,r\in J,r'\in J'} a_{rr'\gamma}(\mathbf{E}_r K_\gamma \otimes \mathbf{F}_{r'}K_\gamma) = 0$. Since the sets $\{\mathbf{F}_{r'}K_\gamma\}$ and $\{\mathbf{E}_r K_\gamma\}$ are linearly independent, we get $a_{rr'\gamma} = 0$ for $r \in J$ and $r' \in J'$. Replacing a by $a - \sum_{\gamma,r\in J,r'\in J'} a_{rr'\gamma}(\mathbf{E}_r\mathbf{F}_{r'}K_\gamma)$ and repeating this procedure it follows that all coefficients $a_{rr'\gamma}$ are zero. □

Applying the automorphism θ to the basis elements in Theorem 14 we obtain the following assertion: *The elements $\mathbf{F}_r K_\gamma \mathbf{E}_{r'}$, $r, r' \in I$, $\gamma \in Q$, are also a basis of the vector space $U_q(\mathfrak{g})$. The mapping $\mathbf{F}_r \otimes K_\gamma \otimes \mathbf{E}_{r'} \to \mathbf{F}_r K_\gamma \mathbf{E}_{r'}$ defines a vector space isomorphism of $U_q(\mathfrak{n}_-) \otimes U_q(\mathfrak{h}) \otimes U_q(\mathfrak{n}_+)$ and $U_q(\mathfrak{g})$.*

Remark 6. Recall that $U_q(\mathfrak{n}_+)$, $U_q(\mathfrak{n}_-)$ and $U_q(\mathfrak{h})$ have been defined as subalgebras of $U_q(\mathfrak{g})$. Using the approach sketched in Remark 1 it can be shown that $U_q(\mathfrak{n}_+)$ (resp. $U_q(\mathfrak{n}_-)$) is isomorphic to the algebra with generators $E_1, \dots, E_l$ (resp. $F_1, \dots, F_l$) and defining relations (15) (resp. (16)) and that the elements K_γ, $\gamma \in Q$, form a vector space basis of $U_q(\mathfrak{h})$. These facts hold for any $q \in \mathbb{C}$ such that $q_i^2 \neq 1$. For such $q \in \mathbb{C}$ the multiplication map gives vector space isomorphisms of $U_q(\mathfrak{n}_+) \otimes U_q(\mathfrak{h}) \otimes U_q(\mathfrak{n}_-)$ and $U_q(\mathfrak{n}_-) \otimes U_q(\mathfrak{h}) \otimes U_q(\mathfrak{n}_+)$ to $U_q(\mathfrak{g})$. However, if q is a root of unity, then the decompositions (51) and (52) are no longer direct sums. △

6.1.6 Hopf Algebra Automorphisms of $U_q(\mathfrak{g})$

In this subsection we suppose that q is not a root of unity. The classification of Hopf algebra automorphisms of $U_q(\mathfrak{g})$ is derived from

Lemma 15. *Let $\psi : U_q(\mathfrak{g}) \to U_q(\mathfrak{g})$ be a coalgebra homomorphism. Then there are elements $\gamma_i \in Q_+$ and finite subsets $I_{i1}, I_{i2}, J_{i1}, J_{i2}$ of the set I from Theorem 14 such that*

$$\psi(K_i) = K_{\gamma_i}, \qquad \psi(E_i) = \sum\nolimits_{r \in I_{i1}} a_r \mathbf{E}_r + \sum\nolimits_{s \in I_{i2}} b_s K_{\gamma_i} \mathbf{F}_s, \tag{54}$$

$$\psi(F_i) = \sum\nolimits_{r \in J_{i1}} c_r \mathbf{E}_r K_{\gamma_i}^{-1} + \sum\nolimits_{s \in J_{i2}} d_s \mathbf{F}_s, \tag{55}$$

where $a_r, b_s, c_r, d_s \in \mathbb{C}$. For $r \in I_{i1} \cup J_{i1}$ and $s \in I_{i2} \cup J_{i2}$ we have

$$\mathbf{E}_r \in U_q^{\gamma_i}(\mathfrak{n}_+) \quad \text{and} \quad \mathbf{F}_s \in U_q^{-\gamma_i}(\mathfrak{n}_-). \tag{56}$$

Proof. By Theorem 14, the element $\psi(E_i)$ can be written as a finite sum $\sum a_{r\gamma r'} \mathbf{E}_r K_\gamma \mathbf{F}_{r'}$ with $a_{r\gamma r'} \in \mathbb{C}$. Since ψ is a coalgebra homomorphism, we have $\Delta(\psi(E_i)) = (\psi \otimes \psi) \circ \Delta(E_i)$. Using (17) and (18), we expand $\Delta(\psi(E_i))$ and $(\psi \otimes \psi) \circ \Delta(E_i)$ in terms of the basis elements $\mathbf{E}_r K_\gamma \mathbf{F}_{r'}$. Comparing coefficients we conclude that the term $\psi(E_i) \otimes \psi(K_i)$ appearing in the expression $(\psi \otimes \psi) \circ \Delta(E_i)$ must be of the form given by (54) with $\mathbf{E}_r$ and $\mathbf{F}_s$ as in (56). Formula (55) is proved analogously. □

Let $A = (a_{ij})$ be the Cartan matrix of the Lie algebra $\mathfrak{g}$. A permutation μ of the set $\{1, 2, \cdots, l\}$ of vertices of the Dynkin diagram of $\mathfrak{g}$ such that $a_{\mu(i),\mu(j)} = a_{ij}$, $i, j = 1, 2, \cdots, l$, is called a *diagram automorphism of* $\mathfrak{g}$. From the explicit form of Dynkin diagrams it is easily seen that the Lie algebras B_l, $l > 2$, and C_l, $l > 2$, have no nontrivial diagram automorphisms. The Lie algebras A_l and D_l, $l > 4$, have precisely one nontrivial diagram automorphism μ and μ has order 2. The Lie algebra D_4 has nontrivial diagram automorphisms of orders 2 and 3.

Theorem 16. *Let $c_1, c_2, \cdots, c_l$ be nonzero complex numbers and let μ be a diagram automorphism of $\mathfrak{g}$. Then there exists a unique Hopf algebra automorphism ψ of $U_q(\mathfrak{g})$ such that*

$$\psi(K_i) = K_{\mu(i)}, \quad \psi(E_i) = c_i E_{\mu(i)}, \quad \psi(F_i) = c_i^{-1} F_{\mu(i)}.$$

Every Hopf algebra automorphism of $U_q(\mathfrak{g})$ is of this form.

Proof. It is easy to verify that each such ψ is a Hopf automorphism. In order to prove the last assertion we apply Lemma 15. Let ψ be a Hopf algebra automorphism of $U_q(\mathfrak{g})$. Since ψ is invertible, the element γ_i satisfying $K_{\gamma_i} = \psi(K_i)$ must be a simple root $\alpha_{\mu(i)}$. Moreover, μ defines a permutation of the roots in the Dynkin diagram of $\mathfrak{g}$. By (54), $\psi(E_i)$ is of the form $a_i E_{\mu(i)} + b_i K_{\gamma_i} F_{\mu(i)}$. Since ψ is also an algebra automorphism, we have $\psi(K_i)\psi(E_i)\psi(K_i)^{-1} = q_i^2 \psi(E_i)$. This equation implies that $b_i = 0$. The assertion concerning $\psi(F_i)$ is derived from (55) in a similar manner. □

6.1.7 Real Forms of Drinfeld–Jimbo Algebras

As discussed in Example 1.10, real forms of the complex Lie algebra $\mathfrak{g}$ correspond to Hopf $*$-structures on the Hopf algebra $U(\mathfrak{g})$. By a *real form* of a Drinfeld–Jimbo algebra $U_q(\mathfrak{g})$ we mean an involution on $U_q(\mathfrak{g})$ such that the Hopf algebra $U_q(\mathfrak{g})$ becomes a Hopf $*$-algebra.

As for $U_q(\mathrm{sl}_2)$, we distinguish three domains of parameter values.

Case 1: $q \in \mathbb{R}$. In this case the real forms of $U_q(\mathfrak{g})$ are described by

Proposition 17. *Let μ be a diagram automorphism of $\mathfrak{g}$ and let c_i, $i = 1, 2, \cdots, l$, be nonzero complex numbers such that $\mu^2 = \mathrm{id}$ and $c_{\mu(i)} = \overline{c_i}$. There exists a real form of $U_q(\mathfrak{g})$ with involution determined by*

$$K_i^* = K_{\mu(i)}, \quad E_i^* = c_i K_{\mu(i)} F_{\mu(i)}, \quad F_i^* = c_i^{-1} E_{\mu(i)} K_{\mu(i)}^{-1}.$$

Two real forms defined by the data (μ, c_i) and (μ', c_i') are equivalent if and only if there exists a diagram automorphism ν such that $\nu\mu = \mu'\nu$ and $c_i' c_{\nu(i)}^{-1} > 0$ for all i satisfying $\mu(i) = i$.

Proof. The proof is given by a direct verification. □

A closer look at the equivalence conditions shows that up to equivalence there are only the following standard real forms:

$$K_i^* = K_{\mu(i)}, \quad E_i^* = \sigma_i K_{\mu(i)} F_{\mu(i)}, \quad F_i^* = \sigma_i E_{\mu(i)} K_{\mu(i)}^{-1}, \tag{57}$$

where μ is a diagram automorphism of $\mathfrak{g}$ such that $\mu^2 = \mathrm{id}$, $\sigma_i = 1$ for $\mu(i) \neq i$ and $\sigma_i = \pm 1$ for $\mu(i) = i$. The real form of $U_q(\mathfrak{g})$ defined by (57) with $\mu = \mathrm{id}$ and $\sigma_i = 1$, $i = 1, 2, \cdots, l$, is called the *compact real form*. The compact real form of $U_q(\mathrm{sl}_n)$ is denoted by $U_q(\mathrm{su}_n)$. Putting $\mu = \mathrm{id}$ in (57) we obtain a real form of $U_q(\mathrm{sl}_n)$ denoted by $U_q(\mathrm{su}(\sigma_1, \cdots, \sigma_l))$.

Case 2: $|q| = 1$.

Proposition 18. *Let μ be a diagram automorphism of $\mathfrak{g}$ and let c_i, $i = 1, 2, \cdots, l$, be nonzero complex numbers such that $\mu^2 = \mathrm{id}$ and $c_{\mu(i)}\overline{c_i} = 1$. Then there is a real form of $U_q(\mathfrak{g})$ with involution determined by*

$$K_i^* = K_{\mu(i)}, \quad E_i^* = c_i E_{\mu(i)}, \quad F_i^* = c_i^{-1} F_{\mu(i)}.$$

If q is not a root of unity, then two real forms defined by the data (μ, c_i) and (μ', c_i') are equivalent if and only if there exists a diagram automorphism ν such that $\nu\mu = \mu'\nu$.

The equivalence conditions allow us to restrict ourselves to the following standard real forms:

$$K_i^* = K_{\mu(i)}, \quad E_i^* = -E_{\mu(i)}, \quad F_i^* = -F_{\mu(i)}, \tag{58}$$

where μ is as in Proposition 18. For $U_q(\mathrm{sl}_n)$ and $\mu = \mathrm{id}$ this real form is denoted by $U_q(\mathrm{sl}_n(\mathbb{R}))$. It is an analog of the universal enveloping algebra $U(\mathrm{sl}(n, \mathbb{R}))$.

Case 3: $q \in \sqrt{-1}\mathbb{R}$, $q \neq \pm\sqrt{-1}$.

Proposition 19. *Then the Hopf algebra $U_q(\mathfrak{g})$ has real forms if and only if $\mathfrak{g} = \mathrm{sp}(2n, \mathbb{C})$. In this case, the corresponding involutions are given by*

$$K_i^* = K_i, \qquad E_i^* = (-1)^{d_i} c_i^{-1} K_i F_i, \qquad F_i^* = c_i E_i K_i^{-1},$$

where c_i are nonzero complex numbers such that $c_i = (-1)^{d_i}\overline{c_i}$. Two real forms with data c_i and c_i' are equivalent if and only if $c_i' c_i^{-1} > 0$ for all i.

In this case we may restrict ourselves up to equivalence to the real forms

$$K_i^* = K_i, \qquad E_i^* = (-1)^{d_i} s_i K_i F_i, \qquad F_i^* = s_i E_i K_i^{-1}$$

of $U_q(\mathrm{sp}_{2n})$, where $s_i = 1$ for even d_i and $s_i = \pm\sqrt{-1}$ for odd d_i.

Using the form of Hopf algebra automorphisms of $U_q(\mathfrak{g})$ obtained in Subsect. 6.1.6 it can be shown that the above lists are exhausting.

Theorem 20. *If q is not a root of unity, then the Drinfeld–Jimbo algebra $U_q(\mathfrak{g})$ has real forms if and only if $q \in \mathbb{R}$ or $|q| = 1$ or $q \in \sqrt{-1}\mathbb{R}$, $\mathfrak{g} = \mathrm{sp}(2n, \mathbb{C})$. All real forms of $U_q(\mathfrak{g})$ are then described in Propositions 17–19.*

6.2 Poincaré–Birkhoff–Witt Theorem and Verma Modules

In this section we construct explicitly a vector space basis for the Drinfeld–Jimbo algebra $U_q(\mathfrak{g})$ in terms of general root vectors. The main technical tool for doing that is an action of the braid group of the Lie algebra $\mathfrak{g}$ as an automorphism group on the algebra $U_q(\mathfrak{g})$.

6.2.1 Braid Groups

If $A = (a_{ij})$ is the Cartan matrix of $\mathfrak{g}$, then the numbers $a_{ij}a_{ji}$ may be equal to 0, 1, 2 or 3. Let m_{ij} be equal to 2, 3, 4, 6 when $a_{ij}a_{ji}$ is equal to $0, 1, 2, 3$, respectively. Recall (see [Bou2]) that the Weyl group W of $\mathfrak{g}$ can be defined as the group generated by the reflections $w_1, w_2, \cdots, w_l$ (corresponding to the simple roots of $\mathfrak{g}$) satisfying the defining relations

$$w_i^2 = 1, \quad i = 1, 2, \cdots, l, \quad w_i w_j w_i w_j \cdots = w_j w_i w_j w_i \cdots, \quad i \neq j, \tag{59}$$

where in the latter equations there are m_{ij} w's on each side.

Example 1. The Weyl group W of $\mathrm{sl}(l+1, \mathbb{C})$ is the permutation group $\mathcal{P}_{l+1}$. It is generated by elements $w_1, w_2, \cdots, w_l$ with defining relations

$$w_i^2 = 1, \quad w_i w_{i+1} w_i = w_{i+1} w_i w_{i+1}, \quad \text{and} \quad w_i w_j = w_j w_i \ \text{ if } \ |i-j| > 1. \quad \triangle$$

Definition 3. *The* braid group *associated with* $\mathfrak{g}$ *is the group* $\mathfrak{B}_\mathfrak{g}$ *generated by elements* $s_1, \cdots, s_l$ *subject to the relations*

$$s_i s_j s_i s_j \cdots = s_j s_i s_j s_i \cdots, \qquad i \neq j, \tag{60}$$

where there are m_{ij} *s's on each side.*

Example 2. The braid group $\mathfrak{B}_\mathfrak{g}$ for $\mathfrak{g} = \mathrm{sl}(l+1, \mathbb{C})$ has l generators and the defining relations

$$s_i s_{i+1} s_i = s_{i+1} s_i s_{i+1} \quad \text{and} \quad s_i s_j = s_j s_i \ \text{ if } \ |i-j| > 1.$$

This group is usually called Artin's braid group on l strands. △

Clearly, a braid group contains an infinite number of elements, because the elements s_i^n, $n \in \mathbb{Z}$, are pairwise distinct. The correspondence $s_i^n \to n$, $n \in \mathbb{Z}$, $i = 1, 2, \cdots, l$, defines a homomorphism from $\mathfrak{B}_\mathfrak{g}$ to $\mathbb{Z}$.

If J denotes the normal subgroup of $\mathfrak{B}_\mathfrak{g}$ generated by the elements s_i^2, $i = 1, 2, \cdots, l$, then the quotient group $\mathfrak{B}_\mathfrak{g}/J$ is isomorphic to the Weyl group W of $\mathfrak{g}$. Thus, we have a natural homomorphism φ from $\mathfrak{B}_\mathfrak{g}$ onto W such that $\varphi(s_i) = w_i$.

The braid groups $\mathfrak{B}_\mathfrak{g}$ have the following important property.

Proposition 21. *Let* $w = w_{i_1} w_{i_2} \cdots w_{i_k}$, $k = l(w)$, *be a reduced decomposition of an element* $w \in W$. *Then the element* $s_w = s_{i_1} s_{i_2} \cdots s_{i_k}$ *of* $\mathfrak{B}_\mathfrak{g}$ *depends only on* w *and not on the choice of reduced decomposition for* w.

Proof. The proof is given in [Bou2], Chap. 5, § 1.5, Proposition 5. □

6.2.2 Action of Braid Groups on Drinfeld–Jimbo Algebras

The importance of the braid group $\mathfrak{B}_\mathfrak{g}$ in the present context stems from the fact that it acts on the algebra $U_q(\mathfrak{g})$.

Theorem 22. *To every* i, $i = 1, 2, \cdots, l$, *there corresponds an algebra automorphism* $\mathcal{T}_i$ *of* $U_q(\mathfrak{g})$ *which acts on the generators* H_j, E_j, F_j *as*

$$\mathcal{T}_i(K_j) = K_j K_i^{-a_{ij}}, \quad \mathcal{T}_i(E_i) = -F_i K_i, \quad \mathcal{T}_i(F_i) = -K_i^{-1} E_i, \tag{61}$$

$$\mathcal{T}_i(E_j) = \sum\nolimits_{r=0}^{-a_{ij}} (-1)^{r-a_{ij}} q_i^{-r} (E_i)^{(-a_{ij}-r)} E_j (E_i)^{(r)}, \quad i \neq j, \tag{62}$$

$$\mathcal{T}_i(F_j) = \sum\nolimits_{r=0}^{-a_{ij}} (-1)^{r-a_{ij}} q_i^{r} (F_i)^{(r)} F_j (F_i)^{(-a_{ij}-r)}, \quad i \neq j, \tag{63}$$

where

$$(E_i)^{(n)} = E_i^n/[n]_{q_i}!, \qquad (F_i)^{(n)} = F_i^n/[n]_{q_i}!.$$

The mapping $s_i \to \mathcal{T}_i$ *determines a homomorphism of the braid group* $\mathfrak{B}_\mathfrak{g}$ *into the group of algebra automorphisms of* $U_q(\mathfrak{g})$.

Proof. The proof is given by direct and lengthy calculations. First one shows that the elements $\mathcal{T}_i(K_j)$, $\mathcal{T}_i(E_j)$, $\mathcal{T}_i(F_j)$ given by (61)–(63) satisfy the defining relations (12)–(16) of $U_q(\mathfrak{g})$. Thus, there exists an algebra automorphism $\mathcal{T}_i$ as stated above. Then one verifies that the automorphisms $\mathcal{T}_i$ satisfy the defining relations (60) of the braid group $\mathfrak{B}_\mathfrak{g}$. Hence the map $s_i \to \mathcal{T}_i$ defines an action of $\mathfrak{B}_\mathfrak{g}$ on $U_q(\mathfrak{g})$. □

The braid group $\mathfrak{B}_\mathfrak{g}$ acts also as a group of algebra automorphisms on the h-adic Drinfeld–Jimbo algebra $U_h(\mathfrak{g})$. The corresponding algebra automorphisms $\mathcal{T}_i$ are given on the generating elements H_i, E_i and F_i, $i = 1, 2, \cdots, l$, by the formulas (62) and (63), with q_i replaced by $e^{d_i h}$, and by

$$\mathcal{T}_i(H_j) = H_j - a_{ji}H_i, \quad \mathcal{T}_i(E_i) = -F_i e^{d_i h H_i}, \quad \mathcal{T}_i(F_i) = -e^{-d_i h H_i} E_i. \quad (64)$$

Clearly, this braid group action induces the classical action of the Weyl group W on the subalgebra $U_h(\mathfrak{h})$ of $U_h(\mathfrak{g})$ which is generated by $H_1, \cdots, H_l$, .

6.2.3 Root Vectors and Poincaré–Birkhoff–Witt Theorem

In the universal enveloping algebra $U(\mathfrak{g})$, there are root elements for every root of $\mathfrak{g}$. In $U_q(\mathfrak{g})$ we have so far only root elements E_i and F_i corresponding to the simple roots of $\mathfrak{g}$. Using the braid group action on $U_q(\mathfrak{g})$, we shall define root elements of $U_q(\mathfrak{g})$ for arbitrary positive and negative roots of $\mathfrak{g}$.

Let w_0 be the longest element of the Weyl group W of $\mathfrak{g}$ and let $w_0 = w_{i_1} w_{i_2} \cdots w_{i_n}$ be a fixed reduced decomposition of w_0. Let $\alpha_1, \cdots, \alpha_l$ be the simple roots of $\mathfrak{g}$. Recall that by Proposition 4 the sequence

$$\beta_1 = \alpha_{i_1}, \quad \beta_2 = w_{i_1}(\alpha_{i_2}), \cdots, \beta_n = w_{i_1} \cdots w_{i_{n-1}}(\alpha_{i_n})$$

exhausts all positive roots of $\mathfrak{g}$. The corresponding root elements of $U_q(\mathfrak{g})$ are obtained by

Definition 4. *The elements*

$$E_{\beta_r} = \mathcal{T}_{i_1} \mathcal{T}_{i_2} \cdots \mathcal{T}_{i_{r-1}}(E_{i_r}) \quad \text{and} \quad F_{\beta_r} = \mathcal{T}_{i_1} \mathcal{T}_{i_2} \cdots \mathcal{T}_{i_{r-1}}(F_{i_r}) \quad (65)$$

from $U_q(\mathfrak{g})$ *are called* root vectors *of* $U_q(\mathfrak{g})$ *corresponding to the roots* β_r *and* $-\beta_r$*, respectively.*

Root vectors for the h-adic algebra $U_h(\mathfrak{g})$ are defined similarly.

The root vectors of the universal enveloping algebra $U(\mathfrak{g})$ can be defined in the same way using the Weyl group W instead of the braid group. Then different reduced decompositions of w_0 give, up to signs, the same root vectors of $U(\mathfrak{g})$. This is no longer true for the root vectors (65) of $U_q(\mathfrak{g})$.

Example 3. For the Lie algebra $\mathfrak{g} = \mathrm{sl}(3, \mathbb{C})$ there are precisely two reduced decompositions $w_0 = w_1 w_2 w_1$ and $w_0 = w_2 w_1 w_2$ of the longest element w_0 in W. The corresponding sequences of positive root vectors of $U_q(\mathrm{sl}_3)$ are

$$E_1, \quad \mathcal{T}_1(E_2) = -E_1E_2 + q^{-1}E_2E_1, \quad \mathcal{T}_1\mathcal{T}_2(E_1) = E_2,$$

$$E_2, \quad \mathcal{T}_2(E_1) = -E_2E_1 + q^{-1}E_1E_2, \quad \mathcal{T}_2\mathcal{T}_1(E_2) = E_1.$$

Note that $\mathcal{T}_1(E_2)$ is not proportional to $\mathcal{T}_2(E_1)$. △

Proposition 23. (i) *If $w \in W$ and α_i is a simple root such that $w(\alpha_i) \in \Delta_+$, then $\mathcal{T}_w(E_i) \in U_q(\mathfrak{n}_+)$ and $\mathcal{T}_w(F_i) \in U_q(\mathfrak{n}_-)$. All root elements E_{β_r} and F_{β_r}, $r = 1, 2, \cdots, n$, from (65) belong to $U_q(\mathfrak{n}_+)$ and $U_q(\mathfrak{n}_-)$, respectively.*

(ii) *If $w(\alpha_i) = \alpha_j$ is a simple root, then $\mathcal{T}_w(E_i) = E_j$ and $\mathcal{T}_w(F_i) = F_j$.*

(iii) *For the root elements E_{β_r} and F_{β_r} we have*

$$K_\lambda E_{\beta_r} K_\lambda^{-1} = q^{(\lambda,\beta_r)} E_{\beta_r}, \qquad K_\lambda F_{\beta_r} K_\lambda^{-1} = q^{-(\lambda,\beta_r)} F_{\beta_r}.$$

Proof. The proof is given by direct calculation using the definition of $\mathfrak{B}_\mathfrak{g}$, Proposition 4 and the formulas (62), (63) and (65). □

Having root elements in $U_q(\mathfrak{g})$ corresponding to all roots of $\mathfrak{g}$ we can state an analog of the *Poincaré–Birkhoff–Witt theorem* for $U_q(\mathfrak{g})$:

Theorem 24. *Let E_{β_r}, F_{β_r}, $r = 1, 2, \cdots, n$, be the root elements from Definition 4. Then the following set of elements is a vector space basis of $U_q(\mathfrak{g})$:*

$$F_{\beta_1}^{r_1} \cdots F_{\beta_n}^{r_n} K_1^{t_1} \cdots K_l^{t_l} E_{\beta_n}^{s_n} \cdots E_{\beta_1}^{s_1}, \quad r_j, s_j \in \mathbb{N}_0, \; t_j \in \mathbb{Z}. \tag{66}$$

Proof. The proof of this theorem can be found in [Ros2] and [Yam] for $U_q(\mathrm{sl}_n)$ and in [Lus] for the general case. □

Corollary 25. *The elements $E_{\beta_1}^{r_1} \cdots E_{\beta_n}^{r_n}$, $r_1, \cdots, r_n \in \mathbb{N}_0$, form a vector space basis of the subalgebra $U_q(\mathfrak{n}_+)$ of $U_q(\mathfrak{g})$. Likewise, the set of elements $F_{\beta_n}^{s_n} \cdots F_{\beta_1}^{s_1}$, $s_n, \cdots, s_1 \in \mathbb{N}_0$, is a vector space basis of $U_q(\mathfrak{n}_-)$.*

In particular we see that there is a natural one-to-one correspondence between basis elements of $U_q(\mathfrak{n}_\pm)$ and $U(\mathfrak{n}_\pm)$.

It can be shown by using Theorem 24 that the algebra $U_q(\mathfrak{g})$ has no zero divisors, that is, we have $ab \neq 0$ in $U_q(\mathfrak{g})$ if $a \neq 0$ and $b \neq 0$ (see [DCK]).

An analog of the Poincaré–Birkhoff–Witt theorem holds also for the h-adic algebra $U_h(\mathfrak{g})$.

Theorem 24′. *Let E_{β_r}, F_{β_r}, $r = 1, 2, \cdots, n$, be the root elements of the algebra $U_h(\mathfrak{g})$. Then the set of elements*

$$F_{\beta_1}^{r_1} \cdots F_{\beta_n}^{r_n} H_1^{t_1} \cdots H_l^{t_l} E_{\beta_n}^{s_n} \cdots E_{\beta_1}^{s_1}, \quad r_j, t_j, s_j \in \mathbb{N}_0,$$

is a basis of the $\mathbb{C}[[h]]$-vector space $U_h(\mathfrak{g})$.

From Corollary 25 we obtain additional information about the decomposition (52) of the subalgebra $U_q(\mathfrak{n}_\pm)$. The dimension of the subspace $U_q^{\pm\beta}(\mathfrak{n}_\pm)$ from (53) is equal to the number of basis elements in $U_q(\mathfrak{n}_\pm)$ of degree $\pm\beta$. As in the classical case, the latter number coincides with the *Kostant partition function* $K(\beta)$. If $\beta \neq 0$, then $K(\beta)$ is defined to be the number of partitions of the linear form β into a sum of positive roots of $\mathfrak{g}$, where roots may enter into partitions with multiplicities. If $\beta = 0$, then $K(\beta) := 0$. Thus, we have

$$\dim U_q^{-\beta}(\mathfrak{n}_-) = \dim U_q^{\beta}(\mathfrak{n}_+) = K(\beta). \tag{67}$$

6.2.4 Representations with Highest Weights

Let T be a representation of the Drinfeld–Jimbo algebra $U_q(\mathfrak{g})$ on a vector space V. For any function λ on the root lattice Q we set

$$V_\lambda = \{\mathbf{x} \in V \mid T(K_\alpha)\mathbf{x} = \lambda(\alpha)\mathbf{x} \text{ for all } \alpha \in Q\},$$

where $K_\alpha = K_1^{r_1} \cdots K_l^{r_l}$ for $\alpha = r_1\alpha_1 + \cdots + r_l\alpha_l$. That is, each V_λ is a joint eigenspace of the commuting operators $T(K_i)$, $i = 1, 2, \cdots, l$.

Definition 5. *If $V_\lambda \neq \{0\}$, then we say that the function λ is a* weight, *the number $m_\lambda = \dim V_\lambda$ is the* multiplicity *of the weight λ and V_λ is a* weight subspace *of the representation T. The nonzero vectors in V_λ are called* weight vectors. *A representation T of $U_q(\mathfrak{g})$ is called a* weight representation *if its underlying space V decomposes into a direct sum of weight subspaces.*

Weight representations are the most important representations of $U_q(\mathfrak{g})$. If q is not a root of unity, then every finite-dimensional representation of $U_q(\mathfrak{g})$ is a weight representation (see Subsect. 7.1.1 below).

Definition 6. *A weight representation T of $U_q(\mathfrak{g})$ on a vector space V is called a* representation with highest weight *if there exists a weight vector $\mathbf{e}_{\Lambda'} \in V$ such that $T(K_\alpha)\mathbf{e}_{\Lambda'} = \Lambda'(\alpha)\mathbf{e}_{\Lambda'}$, $\alpha \in Q$,*

$$T(E_i)\mathbf{e}_{\Lambda'} = 0 \ \ for \ \ i = 1, 2, \cdots, l \ \ and \ \ T(U_q(\mathfrak{g}))\mathbf{e}_{\Lambda'} = V. \tag{68}$$

We then call the function Λ' on Q a highest weight *and the vector $\mathbf{e}_{\Lambda'}$ a* highest weight vector *of the representation T.*

Since $U_q(\mathfrak{n}_+)$ contains the unit element 1, the first condition in (68) implies that $T(U_q(\mathfrak{n}_+))\mathbf{e}_{\Lambda'} = \mathbb{C}\mathbf{e}_{\Lambda'}$. The last formula in (68) means that the vector $\mathbf{e}_{\Lambda'}$ is *cyclic* for the representation T. From (68) we easily derive the relations

$$V = T(U_q(\mathfrak{n}_-))T(U_q(\mathfrak{h}))T(U_q(\mathfrak{n}_+))\mathbf{e}_{\Lambda'} = T(U_q(\mathfrak{n}_-))\mathbf{e}_{\Lambda'}, \tag{69}$$

$$V = \sum\nolimits_{\beta \in Q_+} T(U_q^{-\beta}(\mathfrak{n}_-))\mathbf{e}_{\Lambda'}. \tag{70}$$

Remark 7. The definition of highest weight representations depends on the positive roots and so on the corresponding ordered sequence $\alpha_1, \alpha_2, \cdots, \alpha_l$ of simple roots of $\mathfrak{g}$. As noted in Subsect. 6.1.1, we may also take the ordered sequence of simple roots $-\alpha_1, -\alpha_2, \cdots, -\alpha_l$. In the latter case the elements K_i^{-1}, F_i play the role of the elements K_i, E_i in Definition 6. That is, T is a highest weight representation with respect to $\alpha_1, \alpha_2, \cdots, \alpha_l$ if and only if $T \circ \theta$ is a highest weight representation with respect to $-\alpha_1, -\alpha_2, \cdots, -\alpha_l$, where θ is the algebra automorphism of $U_q(\mathfrak{g})$ from Proposition 9. Moreover, $\Lambda'(\alpha)$ is a highest weight of T if and only if $\Lambda'(-\alpha)$ is a highest weight of $T \circ \theta$. All results of Subsects. 6.2.4–6 and Chap. 7 remain valid (under appropriate reformulations) for highest weight representations with respect to $-\alpha_1, -\alpha_2, \cdots, -\alpha_l$. We shall use this setting in Subsects. 8.4.1, 11.2.3, 11.5.3 and 11.6.4 below. △

In the rest of this section we suppose that q is not a root of unity.

Proposition 26. *Let T be a representation of $U_q(\mathfrak{g})$ with highest weight Λ' on a vector space V. Then the sum (70) is a direct sum and gives the weight subspace decomposition of the space V. We have*

$$V_\mu = T(U_q^{-\beta}(\mathfrak{n}_-))\mathbf{e}_{\Lambda'}, \qquad \text{where} \qquad \mu(\alpha) = q^{-(\alpha,\beta)}\Lambda'(\alpha). \tag{71}$$

All weights of the representation T are of the form $\mu(\alpha) = q^{-\left(\alpha, \sum n_i \alpha_i\right)}\Lambda'(\alpha)$, where $n_i \in \mathbb{N}_0$ and α_i are the simple roots of $\mathfrak{g}$.

Proof. Let $x \in U_q^{-\beta}(\mathfrak{n}_-)$. By (53), we have

$$T(K_\alpha)T(x)\mathbf{e}_{\Lambda'} = T(K_\alpha)T(x)T(K_\alpha^{-1})T(K_\alpha)\mathbf{e}_{\Lambda'} = q^{-(\alpha,\beta)}\Lambda'(\alpha)T(x)\mathbf{e}_{\Lambda'},$$

that is, $T(x)\mathbf{e}_{\Lambda'}$ is a weight vector with respect to the weight $q^{-(\alpha,\beta)}\Lambda'(\alpha)$. Because q is not a root of unity, the sum (70) is a direct sum. Since all forms $\beta \in Q_+$ in (70) are sums of simple roots with nonnegative integral coefficients, all weights of T are of the form stated in the proposition. □

A representation of $U_q(\mathfrak{g})$ is called a *representation of type 1* if it has a highest weight of the form $\Lambda'(\alpha) = q^{(\alpha,\Lambda)}$, where Λ is a linear form on the Cartan subalgebra $\mathfrak{h}$ of $\mathfrak{g}$. We shall prove in Subsect. 7.1.2 that any irreducible representation is a tensor product of a representation of type 1 and a one-dimensional representation. Therefore, we shall restrict ourselves in this chapter to representations of type 1. If T is a representation of type 1 with highest weight $\Lambda'(\alpha) = q^{(\alpha,\Lambda)}$, then (with a slight abuse of language) we call Λ also a highest weight of T and denote T by T_Λ. The corresponding highest weight vector $\mathbf{e}_{\Lambda'}$ will be denoted by $\mathbf{e}_\Lambda$. By Proposition 26, for any type 1 representation T_Λ we have a direct sum decomposition $V = \bigoplus_{\beta\in Q_+} V_{\Lambda-\beta}$, where

$$V_{\Lambda-\beta} = T(U_q^{-\beta}(\mathfrak{n}_-))\mathbf{e}_\Lambda = \{\mathbf{x} \in V \mid T(K_\alpha)\mathbf{x} = q^{(\Lambda-\beta,\alpha)}\mathbf{x},\ K_\alpha \in U_q(\mathfrak{h})\}. \tag{72}$$

Note that the functions $q^{(\Lambda-\beta,\alpha)}$ and $q^{(\Lambda-\beta',\alpha)}$ on Q do not coincide if $\beta \neq \beta'$, because q is not a root of unity.

6.2.5 Verma Modules

The formulas (67) and (72) imply that for a type 1 representation T_Λ with highest weight Λ on a space V one has the inequality

$$\dim V_{\Lambda-\beta} \leq K(\beta), \tag{73}$$

where K is the Kostant partition function (see (67)).

Definition 7. *Let T_Λ be a type 1 representation of $U_q(\mathfrak{g})$ with highest weight Λ on a vector space V_Λ. The corresponding $U_q(\mathfrak{g})$-module V_Λ is called a* Verma module *if*

$$\dim V_{\Lambda-\beta} = K(\beta), \tag{74}$$

for all $\beta \in \Delta_+$, where $V_{\Lambda-\beta}$ is the weight subspace (72). A Verma module with highest weight Λ is denoted by M_Λ and the corresponding representation of $U_q(\mathfrak{g})$ by T_Λ^V.

It follows from (67), (72) and (74) that the map $U_q(\mathfrak{n}_-) \ni x \to T_\Lambda^V(x)\mathbf{e}_\Lambda \in M_\Lambda$ is bijective. That is, a Verma module is freely generated by the action of the algebra $U_q(\mathfrak{n}_-)$ on the vector $\mathbf{e}_\Lambda$.

Any Verma module M_Λ can be realized directly on the vector space $U_q(\mathfrak{n}_-)$. In this case, the unit 1 of this algebra is a highest weight vector and the representation operators are given by the formulas

$$T_\Lambda^V(K_\alpha)1 = q^{(\Lambda,\alpha)}1, \qquad K_\alpha \in U_q(\mathfrak{h}), \tag{75}$$

$$T_\Lambda^V(F_i)1 = F_i, \quad T_\Lambda^V(E_i)1 = 0, \qquad i = 1, 2, \cdots, l, \tag{76}$$

$$T_\Lambda^V(K_\alpha)F_\beta = q^{(\Lambda-\beta,\alpha)}F_\beta, \qquad F_\beta \in U_q^{-\beta}(\mathfrak{n}_-), \tag{77}$$

$$T_\Lambda^V(F_i)F_\beta = F_iF_\beta, \quad T_\Lambda^V(E_i)F_\beta = E_iF_\beta, \qquad F_\beta \in U_q^{-\beta}(\mathfrak{n}_-). \tag{78}$$

Let us explain why the right hand side of the last equality belongs to $U_q(\mathfrak{n}_-)$ and is well-defined. Since the element E_iF_β is contained in $U_q(\mathfrak{n}_-)U_q(\mathfrak{h})U_q(\mathfrak{n}_+)$, it is a sum of products x_-hx_+, $x_- \in U_q(\mathfrak{n}_-)$, $h \in U_q(\mathfrak{h})$, $x_+ \in U_q(\mathfrak{n}_+)$. To every such summand there corresponds a vector of the underlying space $U_q(\mathfrak{n}_-)$ of the representation T_Λ^V. Namely, if the degree of x_+ is nonzero, then this vector is $T_\Lambda^V(x_-hx_+)1 = 0$. If the degree of x_+ is 0 and $x_-hx_+ \in U_q(\mathfrak{n}_-)U_q(\mathfrak{h})$, the corresponding vector is $T_\Lambda^V(x_-h)1 = T_\Lambda^V(x_-)T_\Lambda^V(h)1$. This vector is uniquely determined by the formulas (75)–(78).

Definition 8. *A linear form λ on $\mathfrak{h}$ is called* integral *if the numbers $(\lambda, \alpha_i^\vee)$, $\alpha_i^\vee = 2\alpha_i/(\alpha_i, \alpha_i)$, $i = 1, 2, \cdots, l$, are integers. The set of integral linear forms is denoted by P. A linear form λ on $\mathfrak{h}$ is called* dominant *if $(\lambda, \alpha_i) \geq 0$ for $i = 1, 2, \cdots, l$. The set of integral dominant linear forms is denoted by P_+.*

Verma modules of $U_q(\mathfrak{g})$ have similar properties as in the classical case. Let us describe some of them. Let $\Lambda \in P$ be such that $(\Lambda, \alpha_i^\vee) \geq 0$ for fixed $i \in \{1, 2, \cdots, l\}$ and put $n_i := (\Lambda, \alpha_i^\vee)$. Let M_Λ be a Verma module with highest weight vector $\mathbf{e}_\Lambda$. Using the relations (14) and repeating the reasoning of Subsect. 3.2.1 we find that

$$E_i F_i^{n_i+1} \mathbf{e}_\Lambda = 0, \qquad E_i F_i^{n_i} \mathbf{e}_\Lambda \neq 0, \qquad E_j F_i^{n_i+1} \mathbf{e}_\Lambda = 0, \quad j \neq i.$$

The vector $\mathbf{e}_{\Lambda'} = F_i^{n_i+1}\mathbf{e}_\Lambda$ has the weight $\Lambda' = \Lambda - (n_i+1)\alpha_i$. We have

$$T_\Lambda^V(U_q(\mathfrak{g}))\mathbf{e}_{\Lambda'} = T_\Lambda^V(U_q(\mathfrak{n}_-))T_\Lambda^V(U_q(\mathfrak{h}))T_\Lambda^V(U_q(\mathfrak{n}_+))\mathbf{e}_{\Lambda'} = T_\Lambda^V(U_q(\mathfrak{n}_-))\mathbf{e}_{\Lambda'},$$

that is, $T_\Lambda^V(U_q(\mathfrak{n}_-))\mathbf{e}_{\Lambda'}$ is an invariant subspace and the restriction of T_Λ^V to $T_\Lambda^V(U_q(\mathfrak{n}_-))\mathbf{e}_{\Lambda'}$ is a representation with highest weight $\Lambda' = \Lambda - (n_i+1)\alpha_i$. One can see that this submodule is isomorphic to the Verma module $M_{\Lambda'}$.

Thus, we have shown that *if a Verma module* M_Λ *has a highest weight* $\Lambda \in P$ *such that* $(\Lambda, \alpha_i^\vee) \geq 0$, *then it has a* $U_q(\mathfrak{g})$*-submodule isomorphic to the Verma module* $M_{\Lambda'}$. *Moreover, we have* $(\Lambda', \alpha_i^\vee) < 0$. If $\Lambda \in P_+$, then such a submodule $M_{\Lambda'}$ exists for every $i = 1, 2, \cdots, l$. In the latter case we have the following stronger result.

Proposition 27. *Let* M_Λ *be a Verma module with* $\Lambda \in P_+$. *Then for every element* w *of the Weyl group* W *of* $\mathfrak{g}$ *there exists the Verma submodule in* M_Λ *with highest weight*

$$\Lambda_w = w(\Lambda + \rho) - \rho, \tag{79}$$

where ρ *is the half-sum of all positive roots of* $\mathfrak{g}$. *Every irreducible* $U_q(\mathfrak{g})$*-module contained in* M_Λ *determines a type 1 representation with highest weight. These highest weights are of the form (79).*

Proof. The proof of this proposition is analogous to that of the corresponding assertion in the classical theory [Dix]. □

6.2.6 Irreducible Representations with Highest Weights

A Verma module M_Λ has nontrivial $U_q(\mathfrak{g})$-submodules in general. It can be proved that if M_Λ is reducible, then there is always a unique *maximal* proper $U_q(\mathfrak{g})$-submodule M of M_Λ. That is, any other proper $U_q(\mathfrak{g})$-submodule of M_Λ is contained in M. We denote by L_Λ the representation of $U_q(\mathfrak{g})$ determined by the quotient module M_Λ/M.

Proposition 28. (i) *L_Λ is an irreducible representation.*

(ii) *If T_Λ is an irreducible type 1 representation of $U_q(\mathfrak{g})$ with highest weight Λ, then T_Λ is equivalent to L_Λ. In particular, for any linear form Λ on the Cartan subalgebra $\mathfrak{h}$ of $\mathfrak{g}$ there exists, up to equivalence, a unique irreducible type 1 representation of $U_q(\mathfrak{g})$ with highest weight Λ.*

(iii) *If $\Lambda \in P_+$, then the representation L_Λ is finite-dimensional.*

Proof. (i) is clear by the maximality of M.

(ii): Let V_Λ be the underlying space of the representation T_Λ and let $\mathbf{e}_\Lambda$ be its highest weight vector. By Proposition 26, V_Λ is spanned by the vectors $T_\Lambda(F_{i_1} \cdots F_{i_n})\mathbf{e}_\Lambda$, $n \in \mathbb{N}_0$. It is easy to see that the operators $T_\Lambda(X)$,

$X \in U_q(\mathfrak{g})$, act on these vectors by the formulas (75)–(78). Hence the linear mapping $\mathcal{T} : U_q(\mathfrak{n}_-) \to V_\Lambda$ given by $\mathcal{T}(F_{i_1} \cdots F_{i_n}) = T_\Lambda(F_{i_1} \cdots F_{i_n})\mathbf{e}_\Lambda$ intertwines the representations T_Λ^V and T_Λ. Since $\mathcal{T}(M)$ is an invariant subspace of V_Λ and T_Λ is the irreducible representation of $U_q(\mathfrak{g})$ with highest weight Λ, we have $\mathcal{T}(M) = \{0\}$, so that $T_\Lambda \simeq L_\Lambda$.

(iii): Let i be one of the numbers $1, 2, \cdots, l$. As noted before Proposition 27, M_Λ has the submodule $M_{\Lambda'}$, $\Lambda' = \Lambda-(n_i+1)\alpha_i$, since $\Lambda \in P_+$ by assumption. The set of weights (counted with multiplicities) of the quotient module $M_\Lambda/M_{\Lambda'}$ is invariant with respect to the element $w_i \in W$ corresponding to the simple root α_i. Indeed, the multiplicity m_λ of a weight λ in $M_\Lambda/M_{\Lambda'}$ coincides with $m_\lambda^\Lambda - m_\lambda^{\Lambda'}$, where m_λ^Λ (resp. $m_\lambda^{\Lambda'}$) is the multiplicity of λ in the module M_Λ (resp. in $M_{\Lambda'}$). The multiplicities of weights for Verma modules of $U_q(\mathfrak{g})$ coincide with the corresponding multiplicities for Verma modules of the Lie algebra $\mathfrak{g}$. Therefore, the multiplicities of weights in the $U_q(\mathfrak{g})$-module $M_\Lambda/M_{\Lambda'}$ coincide with those in the quotient $M_\Lambda/M_{\Lambda'}$ of the corresponding Verma modules for $\mathfrak{g}$. In the classical case, the set of weights of the module $M_\Lambda/M_{\Lambda'}$ is invariant with respect to the element w_i of the Weyl group W. Therefore, this invariance holds in the quantum case as well.

Since M_Λ contains the submodule $M_{\Lambda'}$, $\Lambda' = \Lambda - (n_i + 1)\alpha_i$, for any $i = 1, \cdots, l$, the set of weights of the representation L_Λ is invariant under the whole Weyl group W. Since M_Λ (and hence L_Λ) has only a finite number of dominant weights, it follows that the set of all weights of L_Λ is finite, that is, L_Λ is finite-dimensional. □

Proposition 29. *Let L_Λ be the representation from Proposition 28 with highest weight $\Lambda \in P_+$ and highest weight vector* $\mathbf{e}$. *Let $\beta = \sum_{i=1}^l m_i\alpha_i$ be an element from Q_+ such that $(\Lambda, \alpha_i^\vee) \geq m_i$, $i = 1, 2, \cdots, l$, and let $V_\Lambda \simeq M_\Lambda/M$ be the carrier space of the representation L_Λ. Then the linear mapping $U_q^{-\beta}(\mathfrak{n}_-) \ni x \to V_\Lambda(x)\mathbf{e}$ is injective.*

Proof. By the definition of a Verma module, it is enough to show that β is not a weight of the maximal $U_q(\mathfrak{g})$-submodule M of M_Λ. This follows from Proposition 27, because no set of weights $\{\Lambda_w - \sum_{i=1}^l n_i\alpha_i \mid n_i \in \mathbb{N}_0\}$, $w \in W$, contains β. □

6.2.7 The Left Adjoint Action of $U_q(\mathfrak{g})$

Recall that by Proposition 1.14(i) the left adjoint action

$$\mathrm{ad}_L(a)b = \sum a_{(1)}bS(a_{(2)})$$

defines a representation of the algebra $\check{U}_q(\mathfrak{g})$ on itself. In the classical case, this representation is the adjoint action $\mathrm{ad}(X)b = Xb - bX$, $X \in \mathfrak{g}$, $b \in U(\mathfrak{g})$, and it decomposes into a direct sum of irreducible finite-dimensional representations over the center of $U(\mathfrak{g})$ (see [Dix], Subsects. 2.3.3 and 8.2.4). The last statement is not true for Drinfeld–Jimbo algebras.

An element $b \in \check{U}_q(\mathfrak{g})$ is called *locally finite* if the space $\mathrm{ad}_L(\check{U}_q(\mathfrak{g}))b$ is finite-dimensional. Let $\mathcal{F}$ be the set of all locally finite elements of $\check{U}_q(\mathfrak{g})$. It is obvious that $\mathcal{F}$ is a vector space which is invariant under the representation ad_L and that $\mathcal{F}$ decomposes into a sum of finite-dimensional ad_L-invariant subspaces. Since $\mathrm{ad}_L(b)cd = \sum \mathrm{ad}_L(b_{(1)})c \cdot \mathrm{ad}_L(b_{(2)})d$, the set $\mathcal{F}$ is a subalgebra of $\check{U}(\mathfrak{g})$.

Throughout this subsection we assume that q is not a root of unity. Our aim is to describe the algebra $\mathcal{F}$. We give a brief exposition for the case $\check{U}_q(\mathrm{sl}_2)$ and state the main results for the general case without proofs. Let $T_q(\mathfrak{h})$ denote the multiplicative group of the algebra $\check{U}_q(\mathfrak{h})$, that is, $T_q(\mathfrak{h})$ is the group of elements $K_\lambda, \lambda \in Q$.

Let us begin with the case $\mathfrak{g} = \mathrm{sl}_2$. From the formulas (3.12) and (3.13) for the comultiplication and the antipode of $\check{U}_q(\mathrm{sl}_2)$ we obtain that

$$\mathrm{ad}_L(E)a = EaK^{-1} - qK^{-1}aE, \quad \mathrm{ad}_L(F)a = FaK^{-1} - q^{-1}K^{-1}aF, \tag{80}$$

$$\mathrm{ad}_L(K)a = KaK^{-1}, \qquad \mathrm{ad}_L(K^{-1})a = K^{-1}aK \tag{81}$$

for any $a \in \check{U}_q(\mathrm{sl}_2)$.

Proposition 30. *The set $\mathcal{F} \subset \check{U}_q(\mathrm{sl}_2)$ is not trivial and $\mathcal{F} \neq \check{U}_q(\mathrm{sl}_2)$.*

Proof. Let $b = EK$. Since $\mathrm{ad}_L(E)b = 0$ and $\mathrm{ad}_L(K)b = qb$ by (80) and (81), b is a highest weight vector of a weight subrepresentation of ad_L. A direct calculation shows that $\mathrm{ad}_L(F^3)b = 0$, so that $b \in \mathcal{F}$. Similarly, $c = FK \in \mathcal{F}$. Thus, $\mathcal{F}$ is not trivial. Using (80) we easily find that

$$\mathrm{ad}_L(E^n)E = \prod\nolimits_{i=1}^{n} (1 - q^{-2i+1}) E^{n+1} K^{-n}.$$

Therefore, since q is not a root of unity, $\dim \mathrm{ad}_L(\check{U}_q(\mathrm{sl}_2))E = \infty$ and so $E \notin \mathcal{F}$. $\square$

Proposition 31. $\mathcal{S} := T_q(\mathfrak{h}) \bigcap \mathcal{F} = \{K^{2r} \mid r \in \mathbb{N}_0\}$.

Proof. From the first formula of (80) we derive that

$$\mathrm{ad}_L(E^n)K^s = \prod\nolimits_{i=1}^{n} (1 - q^{s-2i+2}) E^n K^{s-n}.$$

Therefore, if $s \notin 2\mathbb{N}_0$, then we have $1 - q^{s-2i+2} \neq 0$ for all $i \in \mathbb{N}$ and hence $\dim \mathrm{ad}_L(\check{U}_q(\mathrm{sl}_2))K^s = \infty$, so that $K^s \notin \mathcal{F}$.

Suppose now that $s = 2r \in 2\mathbb{N}_0$. Then the above formula implies that $\mathrm{ad}_L(E^{r+1})K^{2r} = 0$. Similarly, $\mathrm{ad}_L(F^{r+1})K^{2r} = 0$. Further, by (81), $\mathrm{ad}_L(K^n)K^{2r} = K^{2r}$ for any $n \in \mathbb{Z}$. From the preceding facts and formula (3.5) (more precisely, the corresponding formula for $\check{U}_q(\mathrm{sl}_2)$) it follows that the vector space $\mathrm{ad}_L(\check{U}_q(\mathrm{sl}_2))K^{2r}$ coincides with the span of elements $\mathrm{ad}_L(E^iF^j)K^{2r}$, $0 \leq i, j \leq r+1$. Thus, $K^{2r} \in \mathcal{F}$ for $r \in \mathbb{N}_0$. $\square$

Let $\mathcal{S}^{-1} := \{s^{-1} \mid s \in \mathcal{S}\}$. Clearly, $\mathcal{S}\mathcal{S}^{-1}$ is a subgroup of the group $T_q(\mathfrak{h})$. By Proposition 31, the coset space $T_q(\mathfrak{h})/\mathcal{S}\mathcal{S}^{-1}$ consists of two elements and we have

$$T_q(\mathfrak{h}) = \mathcal{S}\mathcal{S}^{-1} + \mathcal{S}\mathcal{S}^{-1}K^{-1}.$$

Since EK and FK belong to $\mathcal{F}$ (see the proof of Proposition 30) and $K^{-2} \in \mathcal{S}^{-1}$, FK^{-1} and EK^{-1} lie in $\mathcal{F}\mathcal{S}^{-1}$. Since $\mathcal{F}$ is an algebra and $\mathcal{S}^{-1} = \{K^{-2r} \mid r \in \mathbb{N}_0\}$, the set $\mathcal{F}\mathcal{S}^{-1}$ is also an algebra. Therefore, the polynomial algebras $\mathbb{C}[FK^{-1}]$ and $\mathbb{C}[EK^{-1}]$ are contained in $\mathcal{F}\mathcal{S}^{-1}$. Hence $\mathbb{C}[FK^{-1}] \cdot \mathcal{S}\mathcal{S}^{-1} \cdot \mathbb{C}[EK^{-1}] \subset \mathcal{F}\mathcal{S}^{-1}$. Since $\mathbb{C}[FK^{-1}] \cdot T_q(\mathfrak{h}) \cdot \mathbb{C}[EK^{-1}] = \breve{U}_q(\mathrm{sl}_2)$, we conclude that

$$\breve{U}_q(\mathrm{sl}_2) = \mathcal{F}\mathcal{S}^{-1} + \mathcal{F}\mathcal{S}^{-1}K^{-1}.$$

This equality shows that although the algebra $\mathcal{F}$ does not coincide with $\breve{U}_q(\mathrm{sl}_2)$ it is nevertheless rather large. Note that $\mathcal{S}^{-1} \bigcap \mathcal{F}$ contains only the unit element. An explicit description of the set $\mathcal{F}$ is given by

Proposition 32. (i) *For any $r \in \mathbb{N}_0$, the set $\mathcal{F}(r) := \mathrm{ad}_L(\breve{U}_q(\mathrm{sl}_2))K^{2r}$ is an ad_L-invariant vector space of dimension $(r+1)^2$ with a basis formed by the elements $(FK^{-1})^m K^{2r} (EK^{-1})^n$, $0 \le m, n \le r$.*

(ii) $\mathcal{F} = \bigoplus_{r=0}^{\infty} \mathcal{F}(r) = \mathrm{ad}_L(\breve{U}_q(\mathrm{sl}_2))\mathcal{S}$.

Proof. (i): Using (80), (81) and the formula (3.5) for the algebra $\breve{U}_q(\mathrm{sl}_2)$, a direct computation shows that the repeated action of the operators $\mathrm{ad}_L(E)$, $\mathrm{ad}_L(F)$ and $\mathrm{ad}_L(K)$ on K^{2r} gives all elements $(FK^{-1})^m K^{2r} (EK^{-1})^n$, $0 \le m, n \le r$, and that the span of these elements is ad_L-invariant.

(ii): See [JL]. □

Now we turn to the corresponding results for an arbitrary Drinfeld–Jimbo algebra $\breve{U}_q(\mathfrak{g})$. By (34), (35) and (37), the formulas (80) and (81) then remain valid for any $a \in \breve{U}_q(\mathfrak{g})$ if one adds the index i to the elements E, F, K, K^{-1} and to the parameter q therein.

Let φ be the isomorphism from the additive group $Q = \sum_{i=1}^{l} \mathbb{Z}\alpha_i$ to the multiplicative group $T_q(\mathfrak{h})$ defined by $\varphi(\alpha_i) = K_i$.

Theorem 33. *Suppose that q is not a root of unity. If $\mathcal{F}$ is the set of locally finite elements of $\breve{U}_q(\mathfrak{g})$, then we have:*

(i) $\mathcal{S} := T_q(\mathfrak{h}) \bigcap \mathcal{F} = \{\varphi(\alpha) \mid \alpha \in 4P_+ \bigcap Q\}$.

(ii) *The set $\mathcal{S}\mathcal{S}^{-1}$ is a subgroup of $T_q(\mathfrak{h})$ and the coset space $T_q(\mathfrak{h})/\mathcal{S}\mathcal{S}^{-1}$ is finite. If $K^{(i)}$, $i = 1, 2, \cdots, m$, is a complete set of coset representatives for $\mathcal{S}\mathcal{S}^{-1}$ in $T_q(\mathfrak{h})$, then $\breve{U}_q(\mathfrak{g}) = \sum_{i=1}^{m} \mathcal{F}\mathcal{S}^{-1}K^{(i)}$.*

(iii) $\mathcal{F} = \mathrm{ad}_L(\breve{U}_q(\mathfrak{g}))\mathcal{S}$.

Proof. The proof of these assertions can be found in [JL] or [Jos]. □

6.3 The Quantum Killing Form and the Center of $U_q(\mathfrak{g})$

Except for Proposition 34 and Lemma 35 we assume in Subsects. 6.3.1–4 that q is not a root of unity.

6.3.1 A Dual Pairing of the Hopf Algebras $U_q(\mathfrak{b}_+)$ and $U_q(\mathfrak{b}_-)^{\mathrm{op}}$

The following bilinear form $\langle\cdot,\cdot\rangle$ is the key ingredient for the construction of the quantum Killing form on $U_q(\mathfrak{g})$ in the next subsection.

Proposition 34. *There exists a unique dual pairing* $\langle\cdot,\cdot\rangle : U_q(\mathfrak{b}_+)\times U_q(\mathfrak{b}_-)\to \mathbb{C}$ *of the Hopf algebras* $U_q(\mathfrak{b}_+)$ *and* $U_q(\mathfrak{b}_-)^{\mathrm{op}}$ *such that*

$$\langle K_i, K_j\rangle = q^{-(\alpha_i,\alpha_j)},\quad \langle K_j, F_i\rangle = \langle E_j, K_i\rangle = 0,\quad \langle E_j, F_i\rangle = \frac{\delta_{ij}}{q_i^{-1}-q_i} \tag{82}$$

for $i,j=1,2,\cdots,l$. *Moreover, we have* $\langle S(a), S(b)\rangle = \langle a,b\rangle$ *for* $a,b\in U_q(\mathfrak{g})$ *and*

$$\langle aK_\lambda, bK_\mu\rangle = q^{-(\lambda,\mu)}\langle ab\rangle,\quad \textit{for}\quad \lambda,\mu\in Q,\ a\in U_q(\mathfrak{b}_+),\ b\in U_q(\mathfrak{b}_-), \tag{83}$$

$$\langle a,b\rangle = 0\quad \textit{for}\quad a\in U_q^{\alpha}(\mathfrak{b}_+),\ b\in U_q^{-\beta}(\mathfrak{b}_-),\ \alpha,\beta\in Q_+,\ \alpha\neq\beta. \tag{84}$$

Proof. The uniqueness assertion is clear, since any dual pairing of bialgebras is determined by the values on the generators. We prove the existence of the pairing $\langle\cdot,\cdot\rangle$. We define linear functionals φ_i, $\psi_i\in U_q(\mathfrak{b}_-)'$, $i=1,2,\cdots,l$, by

$$\varphi_i(aK_\mu) = q^{-(\alpha_i,\mu)}\varepsilon(a),\quad a\in U_q(\mathfrak{n}_-),\qquad \psi_i(F_iK_\mu) = (q_i^{-1}-q_i)^{-1},$$

$$\psi_i(aK_\mu) = 0\quad\text{for}\quad \mu\in Q,\quad a\in U_q^{-\beta}(\mathfrak{n}_-),\ \beta\neq\alpha_i.$$

Then one verifies that the elements φ_i and ψ_i of the algebra $U_q(\mathfrak{b}_-)'$ satisfy the defining relations of the generating elements of $U_q(\mathfrak{b}_+)$ (see Remark 6). Hence there exists an algebra homomorphism $\Phi : U_q(\mathfrak{b}_+)\to U_q(\mathfrak{b}_-)'$ such that $\Phi(K_i)=\varphi_i$ and $\Phi(E_i)=\psi_i$. We now define $\langle a,b\rangle := \Phi(a)(b)$ for $a\in U_q(\mathfrak{b}_+)$, $b\in U_q(\mathfrak{b}_-)$. By construction, (82) is fulfilled and we have

$$\langle aa', b\rangle = (\Phi(a)\Phi(a'))(b) = \sum \Phi(a)(b_{(1)})\Phi(a')(b_{(2)}) = \sum \langle a, b_{(1)}\rangle\langle a', b_{(2)}\rangle. \tag{85}$$

Next we prove that

$$\langle a, bb'\rangle = \sum\langle a_{(1)}, b'\rangle\langle a_{(2)}, b\rangle,\qquad a\in U_q(\mathfrak{b}_+),\ \ b,b'\in U_q(\mathfrak{b}_-). \tag{86}$$

It is straightforward to check that (86) holds for $a=a_1a_2$ provided that it holds separately for a_1 and for a_2 (and arbitrary b and b'). Similarly, (86) is valid for $b=b_1b_2$ when it is valid for b_1 and for b_2 (and arbitrary a and b'). Hence it suffices to check (86) on the generators which is easily done. By (85) and (86), the bilinear form $\langle\cdot,\cdot\rangle$ satisfies the condition (1.31) for $\mathcal{U}=U_q(\mathfrak{b}_+)$

and $\mathcal{A} = U_q(\mathfrak{b}_-)^{\mathrm{op}}$. Since $\Phi(1) = \varepsilon$ by definition, (1.32) also holds. Hence $\langle\cdot,\cdot\rangle$ is a dual pairing of $U_q(\mathfrak{b}_+)$ and $U_q(\mathfrak{b}_-)^{\mathrm{op}}$.

The formulas (83) and (84) can be verified by induction. □

Lemma 35. *For $a \in U_q(\mathfrak{b}_+)$ and $b \in U_q(\mathfrak{b}_-)$ we have*

$$ab = \sum \langle S(a_{(1)}), b_{(1)}\rangle \langle a_{(3)}, b_{(3)}\rangle b_{(2)} a_{(2)}, \tag{87}$$

$$ba = \sum \langle a_{(1)}, b_{(1)}\rangle \langle S(a_{(3)}), b_{(3)}\rangle a_{(2)} b_{(2)}. \tag{88}$$

Proof. Again one checks that the formulas are valid for $a = a_1 a_2$ (resp. $b = b_1 b_2$) when they hold for a_1 and for a_2 (resp. for b_1 and for b_2). Thus it suffices to prove these formulas for the generators. We omit these verifications (a sample will be carried out later in the proof of Proposition 8.13). □

Each of the relations (87) and (88) means in fact that the Drinfeld–Jimbo algebra $U_q(\mathfrak{g})$ is a quotient of the quantum double of $U_q(\mathfrak{b}_+)$ and $U_q(\mathfrak{b}_-)$ (see Subsect. 8.2.1). We shall state this fact later as Corollary 8.15.

In what follows we assume that q is not a root of unity.

Lemma 36. *If an element $a \in U_q^{-\beta}(\mathfrak{b}_-)$ satisfies the relations $E_i a = a E_i$ for $i = 1, 2, \cdots, l$, then we have $a = 0$. If $F_i b = b F_i$, $i = 1, 2, \cdots, l$, for some $b \in U_q^{\beta}(\mathfrak{b}_+)$, then $b = 0$.*

Proof. Let L_Λ be the irreducible representation from Subsect. 6.2.6 with highest weight $\Lambda \in P_+$ such that the conditions of Proposition 29 are satisfied for β. Since $L_\Lambda(E_i a)\mathbf{e} = L_\Lambda(a E_i)\mathbf{e} = 0$, $i = 1, 2, \cdots, l$, for a highest weight vector $\mathbf{e}$ of the representation L_Λ, the vector $L_\Lambda(a)\mathbf{e}$ generates a proper subrepresentation of L_Λ. Since L_Λ is irreducible, $L_\Lambda(a)\mathbf{e} = 0$. Hence $a = 0$ by Proposition 29. In order to prove the second assertion, we apply the automorphism θ from Proposition 9 to the equation $E_i a = a E_i$. □

Proposition 37. *For any $\beta \in Q_+$, the restriction of the bilinear form $\langle\cdot,\cdot\rangle$ from Proposition 34 to $U_q^{\beta}(\mathfrak{b}_+) \times U_q^{-\beta}(\mathfrak{b}_-)$ is nondegenerate.*

Proof. We have to show that $\langle b, a\rangle = 0$, $a \in U_q^{-\beta}(\mathfrak{b}_-)$, for some $b \in U_q^{\beta}(\mathfrak{b}_+)$ implies that $b = 0$. This will be proved by induction with respect to the usual ordering of Q_+. If β is a simple root, then it is true by (82). Let $\beta > 0$ and suppose that it holds for all $\gamma \in Q_+$ such that $\beta - \gamma \in Q_+$.

Fix $b \in U_q^{\beta}(\mathfrak{b}_+)$ such that $\langle b, a\rangle = 0$ for all $a \in U_q^{-\beta}(\mathfrak{b}_-)$. By (83), it is sufficient to assume that $b \in U_q^{\beta}(\mathfrak{n}_+)$. By the formulas (17) and (18) for the comultiplication we can write

$$\Delta(b) = \sum\nolimits_{0 \le \gamma \le \beta} b_\gamma (1 \otimes K_\gamma), \qquad b_\gamma \in U_q^{\gamma}(\mathfrak{n}_+) \otimes U_q^{\beta-\gamma}(\mathfrak{n}_+), \tag{89}$$

where $b_0 = 1 \otimes b$ and $b_\beta = b \otimes 1$. Let $\gamma \in Q_+$, $0 < \gamma < \beta$, $x \in U_q^{-(\beta-\gamma)}(\mathfrak{b}_-)$ and $y \in U_q^{-\gamma}(\mathfrak{b}_-)$. By (1.31) and (83), we have

$$0 = \langle b, xy\rangle = \langle \Delta(b), y \otimes x\rangle = \langle b_\gamma (1 \otimes K_\gamma), y \otimes x\rangle = \langle b_\gamma, y \otimes x\rangle. \tag{90}$$

Since for any $\gamma' < \beta$ the restriction of $\langle\cdot,\cdot\rangle$ to $U_q^{\gamma}(\mathfrak{b}_+) \times U_q^{-\gamma}(\mathfrak{b}_-)$ is nondegenerate, so is its extension to a bilinear form on

$$[U_q^{\gamma}(\mathfrak{b}_+) \otimes U_q^{\beta-\gamma}(\mathfrak{b}_+)] \times [U_q^{-\gamma}(\mathfrak{b}_-) \otimes U_q^{-(\beta-\gamma)}(\mathfrak{b}_-)].$$

Hence it follows from (90) that $b_\gamma = 0$. Because of (89) this means that $\Delta(b) = b \otimes K_\beta + 1 \otimes b$. Applying the formulas (87) and (88) to $F_i b$ and bF_i and using the expressions $\Delta^{(2)}(b) = b \otimes K_\beta \otimes K_\beta + 1 \otimes b \otimes K_\beta + 1 \otimes 1 \otimes b$, $\Delta^{(2)}(F_i) = F_i \otimes 1 \otimes 1 + K_i^{-1} \otimes F_i \otimes 1 + K_i^{-1} \otimes K_i^{-1} \otimes F_i$, we derive that $F_i b = bF_i$ for $i = 1, 2, \cdots, l$. Thus, by Lemma 36, $b = 0$.

Similar reasoning shows that $\langle a, b\rangle = 0$, $a \in U_q^{\beta}(\mathfrak{b}_+)$, for $b \in U_q^{-\beta}(\mathfrak{b}_-)$ implies that $b = 0$. □

In Subsect. 8.3.2 the h-adic counterpart to the above bilinear form $\langle\cdot,\cdot\rangle$ will be needed. Replacing formally K_r by $e^{hd_rH_r}$ and q by e^h in the relation $\langle K_i, K_j\rangle = q^{-(\alpha_i,\alpha_j)}$ of Proposition 34, we obtain $\langle e^{hd_iH_i}, e^{hd_jH_j}\rangle = e^{-h(\alpha_i,\alpha_j)}$. In order to get this equality we should define $hd_id_j\langle H_i, H_j\rangle = -(\alpha_i,\alpha_j) \equiv -d_ia_{ij}$ and so $h\langle H_i, H_j\rangle = -d_j^{-1}a_{ij}$. But, since h is not invertible in the ring $\mathbb{C}[[h]]$, there is no value $\langle H_i, H_j\rangle \in \mathbb{C}[[h]]$ such that the latter holds. In order to circumvent this difficulty, we shall work with the h-adic Hopf algebra $\tilde{U}_h(\mathfrak{b}_-) := \mathbb{C}[[h]]\cdot 1 + hU_h(\mathfrak{b}_-)$ instead of $U_h(\mathfrak{b}_-)$. The elements

$$\tilde{H}_i := hH_i \quad\text{and}\quad \tilde{F}_i := hF_i, \quad i = 1, 2, \cdots, l,$$

belong to $\tilde{U}_h(\mathfrak{b}_-)$ and satisfy

$$\Delta(\tilde{H}_i) = \tilde{H}_i \otimes 1 + 1 \otimes \tilde{H}_i, \quad \Delta(\tilde{F}_i) = \tilde{F}_i \otimes 1 + e^{-d_i\tilde{H}_i} \otimes \tilde{F}_i.$$

The element $e^{-d_i\tilde{H}_i} = 1 + \sum_{k\geq 1} \frac{1}{k!}(-d_ih)^k H_i^k$ is also in $\tilde{U}_h(\mathfrak{b}_-)$. Note that $e^{-d_i\tilde{H}_i}$ is not in the h-adic subalgebra of $U_h(\mathfrak{g})$ generated by $\tilde{H}_i$.

Proposition 38. *There exists a dual pairing* $\langle\cdot,\cdot\rangle : U_h(\mathfrak{b}_+) \times \tilde{U}_h(\mathfrak{b}_-) \to \mathbb{C}[[h]]$ *of the h-adic Hopf algebras* $U_h(\mathfrak{b}_+)$ *and* $\tilde{U}_h(\mathfrak{b}_-)^{\rm op}$ *such that*

$$\langle H_i, \tilde{H}_j\rangle = -d_i^{-1}a_{ji}, \;\; \langle H_j, \tilde{F}_i\rangle = \langle E_j, \tilde{H}_i\rangle = 0, \;\; \langle E_i, \tilde{F}_j\rangle = \frac{\delta_{ij}h}{e^{-d_ih} - e^{d_ih}}$$

for $i, j = 1, 2, \cdots, l$.

For the study of the Drinfeld–Jimbo algebras $U_q(\mathfrak{g})$ and $U_h(\mathfrak{g})$ as quotients of quantum doubles in Sect. 8.3, we shall also use the bilinear form $\langle\cdot,\cdot\rangle'$ on $U_q(\mathfrak{b}_+) \times U_q(\mathfrak{b}_-)$ resp. $U_h(\mathfrak{b}_+) \times \tilde{U}_h(\mathfrak{b}_-)$ defined by

$$\langle a, b\rangle' := \langle S(a), b\rangle.$$

Clearly, the form $\langle\cdot,\cdot\rangle'$ is then a dual pairing of the Hopf algebras $U_q(\mathfrak{b}_+)^{\rm op}$ and $U_q(\mathfrak{b}_-)$ resp. of $U_h(\mathfrak{b}_+)^{\rm op}$ and $\tilde{U}_h(\mathfrak{b}_-)$. From the formulas for $\langle\cdot,\cdot\rangle$ we get

$$\langle K_i, K_j\rangle' = q^{(\alpha_i,\alpha_j)}, \quad \langle K_j, F_i\rangle' = \langle E_j, K_i\rangle' = 0, \quad \langle E_j, F_i\rangle' = \frac{\delta_{ij}}{q_i - q_i^{-1}},$$

$$\langle H_i, \tilde{H}_j\rangle' = d_i^{-1}a_{ji}, \;\; \langle H_j, \tilde{F}_i\rangle' = \langle E_j, \tilde{H}_i\rangle' = 0, \;\; \langle E_i, \tilde{F}_j\rangle' = \frac{\delta_{ij}h}{e^{d_ih} - e^{-d_ih}}.$$

6.3.2 The Quantum Killing Form on $U_q(\mathfrak{g})$

Let $\mathcal{A}$ be a Hopf algebra and let ρ be a representation of $\mathcal{A}$ on a vector space V. We say that a linear functional f on V is *invariant* with respect to ρ if $f(\rho(a)v) = \varepsilon(a)f(v)$ for all $a \in \mathcal{A}$ and $v \in V$. A bilinear form $\langle\cdot,\cdot\rangle$ on $\mathcal{A} \times \mathcal{A}$ is called ad_L-*invariant* if the form $\langle\cdot,\cdot\rangle$ considered as a linear functional on $\mathcal{A} \otimes \mathcal{A}$ is invariant under the representation $\mathrm{ad}_L \otimes \mathrm{ad}_L$ of $\mathcal{A}$ on $\mathcal{A} \otimes \mathcal{A}$, that is,

$$\sum \langle \mathrm{ad}_L(c_{(1)})a, \mathrm{ad}_L(c_{(2)})b\rangle = \varepsilon(c)\langle a, b\rangle, \quad a, b, c \in \mathcal{A}.$$

It is easily seen that the latter is equivalent to the condition

$$\langle \mathrm{ad}_L(c)a, b\rangle = \langle a, \mathrm{ad}_L(S(c))b\rangle, \quad a, b, c \in \mathcal{A}. \tag{91}$$

Let us turn now to the Drinfeld–Jimbo algebra $U_q(\mathfrak{g})$. From Subsect. 6.1.5 we recall that the multiplication map $a \otimes k \otimes b \to akb$ of $U_q(\mathfrak{n}_-) \otimes U_q(\mathfrak{h}) \otimes U_q(\mathfrak{n}_+)$ to $U_q(\mathfrak{g})$ is an isomorphism of vector spaces. Hence there exists a well-defined bilinear form $\langle\cdot,\cdot\rangle_U$ on $U_q(\mathfrak{g}) \times U_q(\mathfrak{g})$ given by

$$\langle a_1 K_\alpha b_1, a_2 K_\beta b_2\rangle_U := q^{-(\alpha,\beta)/2} \langle S(b_1), a_2\rangle \langle b_2, S(a_1)\rangle \tag{92}$$

for $a_1, a_2 \in U_q(\mathfrak{n}_-)$, $b_1, b_2 \in U_q(\mathfrak{n}_+)$ and $\alpha, \beta \in Q$. Here $\langle\cdot,\cdot\rangle$ is the dual pairing of $U_q(\mathfrak{b}_+)$ and $U_q(\mathfrak{n}_-)^{\mathrm{op}}$ from the preceding subsection.

Theorem 39. *The bilinear form $\langle\cdot,\cdot\rangle_U$ on $U_q(\mathfrak{g}) \times U_q(\mathfrak{g})$ is ad_L-invariant, that is,*

$$\langle \mathrm{ad}_L(a)b_1, b_2\rangle_U = \langle b_1, \mathrm{ad}_L(S(a))b_2\rangle_U, \quad a, b_1, b_2 \in U_q(\mathfrak{g}). \tag{93}$$

Proof. By Proposition 1.14, ad_L is a representation of $U_q(\mathfrak{g})$. Hence it suffices to prove (93) for elements K_α, $a_\beta \in U_q^\beta(\mathfrak{n}_+)$, $b_\beta \in U_q^{-\beta}(\mathfrak{n}_-)$. First let $a = K_\alpha$. We may assume that $b_1 = vK_\lambda S(w)$, $v \in U_q^{-\theta}(\mathfrak{n}_-)$, $w \in U_q^\gamma(\mathfrak{n}_+)$, and $b_2 = S(v')K_\mu w'$, $v' \in U_q^{-\omega}(\mathfrak{n}_-)$, $w' \in U_q^\delta(\mathfrak{n}_+)$. A direct computation shows that

$$\langle \mathrm{ad}_L(a)b_1, b_2\rangle_U = \delta_{\theta\delta}\delta_{\omega\gamma} q^{-(\alpha,\theta-\omega)} \langle b_1, b_2\rangle_U,$$

$$\langle b_1, \mathrm{ad}_L(S(a))b_2\rangle_U = \delta_{\theta\delta}\delta_{\omega\gamma} q^{-(\alpha,\theta-\omega)} \langle b_1, b_2\rangle_U.$$

That is, in this case condition (93) is fulfilled.

Now let $a \in U_q^{-\beta}(\mathfrak{n}_-)$ and let b_1 and b_2 be as before. Using (88) we get

$$\mathrm{ad}_L(a)b_1 = \sum a_{(1)} vK_\lambda S(w) S(a_{(2)}) = \sum a_{(1)} vK_\lambda S(a_{(2)} w)$$

$$= \sum a_{(1)} vK_\lambda S\Big(\sum \langle w_{(1)}, a_{(2)}\rangle \langle S(w_{(3)}), a_{(4)}\rangle w_{(2)} a_{(3)}\Big)$$

$$= \sum \langle w_{(1)}, a_{(2)}\rangle \langle S(w_{(3)}), a_{(4)}\rangle a_{(1)} vK_\lambda S(a_{(3)}) S(w_{(2)}). \tag{94}$$

Since a is a product of the generators F_i and

$$\Delta^{(3)}(F_i) = F_i \otimes 1 \otimes 1 \otimes 1 + K_i^{-1} \otimes F_i \otimes 1 \otimes 1 + K_i^{-1} \otimes K_i^{-1} \otimes F_i \otimes 1$$
$$+ K_i^{-1} \otimes K_i^{-1} \otimes K_i^{-1} \otimes F_i,$$

it follows that

$$\Delta^{(3)}(a) = \sum\nolimits_{(\beta_n)\in I} a^{(\beta_n)}(K^{-1}_{\beta_1+\beta_2+\beta_3} \otimes K^{-1}_{\beta_2+\beta_3} \otimes K^{-1}_{\beta_3} \otimes 1),$$

where $(\beta_n) \equiv (\beta_0, \beta_1, \beta_2, \beta_3) \in Q_+^4$, $I = \{(\beta_0, \beta_1, \beta_2, \beta_3) \mid \beta_0 + \beta_1 + \beta_2 + \beta_3 = \beta\}$ and

$$a^{(\beta_n)} = \sum\nolimits_r a^{(\beta_n)}_{0,r} \otimes a^{(\beta_n)}_{1,r} \otimes a^{(\beta_n)}_{2,r} \otimes a^{(\beta_n)}_{3,r} \in U_q^{-\beta_0}(\mathfrak{n}_-) \otimes \cdots \otimes U_q^{-\beta_3}(\mathfrak{n}_-).$$

Similarly, for $w \in U_q^{\gamma}(\mathfrak{n}_+)$ we obtain

$$\Delta^{(2)}(w) = \sum\nolimits_{(\gamma_m)\in J} w^{(\gamma_m)}(1 \otimes K_{\gamma_0} \otimes K_{\gamma_0+\gamma_1}),$$

where $(\gamma_m) \equiv (\gamma_0, \gamma_1, \gamma_2) \in Q_+^3$, $J = \{(\gamma_0, \gamma_1, \gamma_2) \mid \gamma_0 + \gamma_1 + \gamma_2 = \gamma\}$ and

$$w^{(\gamma_m)} = \sum\nolimits_s w^{(\gamma_m)}_{0,s} \otimes w^{(\gamma_m)}_{1,s} \otimes w^{(\gamma_m)}_{2,s} \in U_q^{\gamma_0}(\mathfrak{n}_+) \otimes U_q^{\gamma_1}(\mathfrak{n}_+) \otimes U_q^{\gamma_2}(\mathfrak{n}_+).$$

From (94) we conclude that

$$\mathrm{ad}_L(a)b_1 = \sum_{(\beta_n),(\gamma_m),r,s} \langle w^{(\gamma_m)}_{0,s}, a^{(\beta_n)}_{1,r} K^{-1}_{\beta_2+\beta_3}\rangle \langle S(w^{(\gamma_m)}_{2,s} K_{\gamma_0+\gamma_1}), a^{(\beta_n)}_{3,r}\rangle D, \tag{95}$$

where

$$D := a^{(\beta_n)}_{0,r} K^{-1}_{\beta_1+\beta_2+\beta_3} v K_\lambda K_{\beta_3} S(a^{(\beta_n)}_{2,r}) S(w^{(\gamma_m)}_{1,s} K_{\gamma_0})$$
$$= q^{-(\theta,\beta_1+\beta_2+\beta_3)-(\beta_2,-\lambda+\beta_1+\beta_2)} a^{(\beta_n)}_{0,r} v S(a^{(\beta_n)}_{2,r}) K^{-1}_{\beta_2} K_{\lambda-\gamma_0-\beta_1} S(w^{(\gamma_m)}_{1,s})$$

with $a^{(\beta_n)}_{0,r} v S(a^{(\beta_n)}_{2,r}) K^{-1}_{\beta_2} \in U_q(\mathfrak{n}_-)$. The expressions in parentheses on the right hand side of (95) vanish unless $\gamma_0 = \beta_1$ and $\gamma_2 = \beta_3$. Therefore,

$$\langle \mathrm{ad}_L(a)b_1, b_2\rangle_U = \sum_{(\beta_n),r} q^{N(\beta_n)} \langle \Delta^{(2)}(w), a^{(\beta_n)}_{1,r} K^{-1}_{\beta_2+\beta_3} \otimes S^{-1}(v') \otimes S^{-1}(a^{(\beta_n)}_{3,r})\rangle$$
$$\times \langle S^{-1}(w'), a^{(\beta_n)}_{0,r} v S(a^{(\beta_n)}_{2,r}) K^{-1}_{\beta_2}\rangle$$
$$= \sum_{(\beta_n),r} q^{N(\beta_n)} \langle w, S^{-1}(v' a^{(\beta_n)}_{3,r}) a^{(\beta_n)}_{1,r} K^{-1}_{\beta_2+\beta_3}\rangle \langle S^{-1}(w'), a^{(\beta_n)}_{0,r} v S(a^{(\beta_n)}_{2,r}) K^{-1}_{\beta_2}\rangle$$
$$= \sum_{(\beta_n),r} q^{N(\beta_n)} \langle w, S^{-1}(v' a^{(\beta_n)}_{3,r}) a^{(\beta_n)}_{1,r}\rangle \langle S^{-1}(w'), a^{(\beta_n)}_{0,r} v S(a^{(\beta_n)}_{2,r})\rangle,$$

where $N(\beta_n) = -(\theta, \beta_1 + \beta_2 + \beta_3) + (\beta_2, -\lambda + \beta_1 + \beta_2) - (\beta_1 - \lambda, \mu)/2$. By a similar calculation we show that the right hand side of (93) takes the same value. Thus, the assertion is proved for $a \in U_q^{-\beta}(\mathfrak{n}_-)$.

The proof of (93) for $a \in U_q^{\beta}(\mathfrak{n}_+)$ is completely analogous. □

Definition 9. *The bilinear form* $\langle\cdot,\cdot\rangle_U$ *defined by (92) is called the* quantum Killing form *or the* Rosso form *of the Drinfeld–Jimbo algebra* $U_q(\mathfrak{g})$.

It can be shown that the quantum Killing form $\langle\cdot,\cdot\rangle_U$ is nondegenerate on $U_q(\mathfrak{g}) \times U_q(\mathfrak{g})$ if q is not a root of unity.

One may also define a bilinear form $\langle\cdot,\cdot\rangle'_U$ on $U_q(\mathfrak{g}) \times U_q(\mathfrak{g})$ by

$$\langle a_1 K_\alpha S(b_1), b_2 K_\beta S(a_2)\rangle'_U := q^{-(\alpha,\beta)/2}\langle b_2, a_1\rangle\langle b_1, a_2\rangle, \tag{96}$$

where $a_1, a_2 \in U_q(\mathfrak{n}_-)$, $b_1, b_2 \in U_q(\mathfrak{n}_+)$ and $\alpha, \beta \in Q$. The form (96) is related to (92) by

$$\langle a, b\rangle_U = \langle a, S(b)\rangle'_U.$$

Since the adjoint actions ad_L and ad_R are connected by the formula

$$\mathrm{ad}_L(S(a))S(b) = S(\mathrm{ad}_R(a)b),$$

the ad_L-invariance of the form (92) means that the form (96) satisfies the condition

$$\langle \mathrm{ad}_L(c)a, b\rangle'_U = \langle a, \mathrm{ad}_R(c)b\rangle'_U.$$

6.3.3 A Quantum Casimir Element

The Casimir element Ω constructed in this subsection does not belong to the algebra $U_q(\mathfrak{g})$. It is an infinite sum of elements from $U_q^\beta(\mathfrak{b}_+)U_q^{-\beta}(\mathfrak{n}_-)$, $\beta \in Q_+$. In order to deal with this infinite sum we first define certain completions $\bar{U}_q^\pm(\mathfrak{g})$ of the algebra $U_q(\mathfrak{g})$. Consider the vector spaces

$$\bar{U}_q^\pm(\mathfrak{g}) := \prod\nolimits_{\beta\in Q_+} U_q(\mathfrak{b}_\pm)U_q^{\mp\beta}(\mathfrak{n}_\mp).$$

The elements of $\bar{U}_q^\pm(\mathfrak{g})$ are sequences $x = (x_\beta)_{\beta\in Q_+}$ of elements x_β from $U_q(\mathfrak{b}_\pm)U_q^{\mp\beta}(\mathfrak{n}_\mp)$. Let us write such a sequence formally as an infinite sum $x = \sum_\beta x_\beta$. Then, by this definition, the components x_β are uniquely determined by the element x. Since

$$U_q(\mathfrak{g}) = \bigoplus\nolimits_{\beta\in Q_+} U_q(\mathfrak{b}_+)U_q^{-\beta}(\mathfrak{n}_-)$$

by Theorem 14, $U_q(\mathfrak{g})$ can be considered as the subspace of $\bar{U}_q^+(\mathfrak{g})$ formed by the sums $x = \sum_\beta x_\beta$ for which all but finitely many terms x_β vanish. From the commutation relations of the generators of $U_q(\mathfrak{g})$ it follows that for any $\beta, \gamma \in Q_+$, $x_\beta \in U_q^{-\beta}(\mathfrak{n}_-)$ and $y_\gamma \in U_q^\gamma(\mathfrak{n}_+)$ the element $x_\beta y_\gamma$ belongs to the sum of spaces $U_q(\mathfrak{b}_+)U_q^{-\delta}(\mathfrak{n}_-)$, $|\beta| - |\gamma| \le |\delta| \le |\beta|$, where $|\alpha| := \sum_i n_i$ for $\alpha = \sum_i n_i\alpha_i \in Q_+$. From this fact one easily concludes that the multiplication of $U_q(\mathfrak{g})$ extends canonically to the vector space $\bar{U}_q^+(\mathfrak{g})$ such that $\bar{U}_q^+(\mathfrak{g})$ is also an algebra. In a similar way, $\bar{U}_q^-(\mathfrak{g})$ and

$$\bar{U}_q^\pm(\mathfrak{g})\bar{\otimes}\cdots\bar{\otimes}\bar{U}_q^\pm(\mathfrak{g}) := \bar{U}_q^\pm(\mathfrak{g}\oplus\cdots\oplus\mathfrak{g}) \quad (n \text{ times})$$

become algebras which contain $U_q(\mathfrak{g})$ and $U_q(\mathfrak{g}) \otimes \cdots \otimes U_q(\mathfrak{g})$, respectively, as subalgebras.

By Proposition 37, the pairing $\langle \cdot,\cdot \rangle$ of $U_q^\beta(\mathfrak{n}_+)$ and $U_q^{-\beta}(\mathfrak{n}_-)$ is nondegenerate. Hence there are vector space bases $\{a_r^\beta\}$ and $\{b_s^\beta\}$ of $U_q^{-\beta}(\mathfrak{n}_-)$ and $U_q^\beta(\mathfrak{n}_+)$, respectively, such that $\langle b_s^\beta, a_r^\beta \rangle = \delta_{rs}$. Let θ be the algebra automorphism of $U_q(\mathfrak{g})$ from Proposition 9. We set

$$C_\beta = \sum\nolimits_r a_r^\beta \otimes b_r^\beta, \quad \Omega_\beta := \sum\nolimits_r S(b_r^\beta) a_r^\beta, \quad \Omega'_\beta := \theta(\Omega_\beta), \tag{97}$$

$$\Omega := \sum\nolimits_{\beta \in Q_+} \Omega_\beta, \quad \text{and} \quad \Omega' := \sum\nolimits_{\beta \in Q_+} \Omega'_\beta. \tag{98}$$

Because $C_\beta \in U_q^{-\beta}(\mathfrak{n}_-) \otimes U_q^\beta(\mathfrak{n}_+)$ is the canonical element with respect to the pairing $\langle \cdot,\cdot \rangle$, C_β and so Ω_β, Ω and Ω' are independent of the choice of bases $\{a_r^\beta\}$ and $\{b_s^\beta\}$. By construction, we have $\Omega_\beta \in U_q^\beta(\mathfrak{b}_+) U_q^{-\beta}(\mathfrak{n}_-)$ and $\Omega'_\beta \in U_q^{-\beta}(\mathfrak{b}_-) U_q^\beta(\mathfrak{n}_+)$, so that $\Omega \in \bar{U}_q^+(\mathfrak{g})$ and $\Omega' \in \bar{U}_q^-(\mathfrak{g})$.

Definition 10. *The element $\Omega \in \bar{U}_q^+(\mathfrak{g})$ is called a* quantum Casimir element *for the Drinfeld–Jimbo algebra $U_q(\mathfrak{g})$.*

The reason for this terminology lies in Proposition 41 below. It says that up to an algebra automorphism ψ, which corresponds to the identity in the case $q = 1$, Ω commutes with all elements of $U_q(\mathfrak{g})$.

Lemma 40. *The elements C_β satisfy the relations*

$$[1 \otimes F_i, C_{\beta+\alpha_i}] = C_\beta (F_i \otimes K_i) - (F_i \otimes K_i^{-1}) C_\beta, \tag{99}$$

$$[E_i \otimes 1, C_{\beta+\alpha_i}] = C_\beta (K_i^{-1} \otimes E_i) - (K_i \otimes E_i) C_\beta. \tag{100}$$

Proof. These two relations are proved in the same manner. For this reason, we prove only (99). Since both sides of (99) lie in $U_q^{-\beta-\alpha_i}(\mathfrak{n}_-) \otimes U_q(\mathfrak{g})$, by Proposition 37 it is enough to show that they coincide when the first tensor factor is paired against an arbitrary element $a \in U_q^{\beta+\alpha_i}(\mathfrak{n}_+)$. For the left hand side of (99) we then obtain

$$\sum\nolimits_r \langle a, a_r^{\beta+\alpha_i} \rangle (F_i b_r^{\beta+\alpha_i} - b_r^{\beta+\alpha_i} F_i)$$

$$= F_i \Big(\sum\nolimits_r \langle a, a_r^{\beta+\alpha_i} \rangle b_r^{\beta+\alpha_i} \Big) - \Big(\sum\nolimits_r \langle a, a_r^{\beta+\alpha_i} \rangle b_r^{\beta+\alpha_i} \Big) F_i = F_i a - a F_i.$$

Here we used the relation $\sum_r \langle a, a_r^\beta \rangle b_r^\beta = a$ which follows from $\langle b_s^\beta, a_r^\beta \rangle = \delta_{rs}$. Since

$$\Delta(a) = \sum_{\substack{\gamma,\delta \in Q_+ \\ \gamma+\delta=\beta+\alpha_i}} a_{\gamma\delta}(1 \otimes K_\gamma), \qquad a_{\gamma\delta} \in U_q^\gamma(\mathfrak{n}_+) \otimes U_q^\delta(\mathfrak{n}_+),$$

with $a_{0,\beta+\alpha_i} = 1 \otimes a$, $a_{\beta+\alpha_i,0} = a \otimes 1$, $a_{\alpha_i,\beta} = E_i \otimes u$, $a_{\beta,\alpha_i} = v \otimes E_i$ for some $u, v \in U_q^\beta(\mathfrak{n}_+)$, the pairing of the right hand side of (99) against a gives

$$\sum_r \{\langle a, a_r^\beta F_i\rangle b_r^\beta K_i - \langle a, F_i a_r^\beta\rangle K_i^{-1} b_r^\beta\}$$
$$= \sum_r \{\langle \Delta(a), F_i \otimes a_r^\beta\rangle b_r^\beta K_i - \langle \Delta(a), a_r^\beta \otimes F_i\rangle K_i^{-1} b_r^\beta\}$$
$$= \sum_r \{\langle E_i, F_i\rangle \langle uK_i, a_r^\beta\rangle b_r^\beta K_i - \langle v, a_r^\beta\rangle \langle E_i K_\beta, F_i\rangle K_i^{-1} b_r^\beta\}$$
$$= \frac{1}{q_i - q_i^{-1}} \left\{ K_i^{-1} \sum_r \langle v, a_r^\beta\rangle b_r^\beta - \left(\sum_r \langle u, a_r^\beta\rangle b_r^\beta\right) K_i \right\} = \frac{K_i^{-1} v - uK_i}{q_i - q_i^{-1}}.$$

Therefore, the proof is complete as soon as we have shown that

$$F_i a - aF_i = (K_i^{-1} v - uK_i)/(q_i - q_i^{-1}). \tag{101}$$

We have $\Delta^{(2)}(F_i) = F_i \otimes 1 \otimes 1 + K_i^{-1} \otimes F_i \otimes 1 + K_i^{-1} \otimes K_i^{-1} \otimes F_i$ and

$$\Delta^{(2)}(a) = \sum_{\substack{\gamma_1, \gamma_2, \gamma_3 \in Q_+ \\ \gamma_1 + \gamma_2 + \gamma_3 = \beta + \alpha_i}} a_{\gamma_1 \gamma_2 \gamma_3} (1 \otimes K_{\gamma_1} \otimes K_{\gamma_1 + \gamma_2}),$$

where $a_{0, \beta + \alpha_i, 0} = 1 \otimes a \otimes 1$, $a_{\alpha_i, \beta, 0} = E_i \otimes u \otimes 1$, $a_{0, \beta, \alpha_i} = 1 \otimes v \otimes E_i$ and $a_{\gamma_1 \gamma_2 \gamma_3} \in U_q^{\gamma_1}(\mathfrak{n}_+) \otimes U_q^{\gamma_2}(\mathfrak{n}_+) \otimes U_q^{\gamma_3}(\mathfrak{n}_+)$. Now relation (101) follows immediately from (87) and (88). □

Proposition 41. *Let ψ and φ be the algebra automorphisms of $U_q(\mathfrak{g})$ such that $\psi(K_\alpha) = K_\alpha$, $\psi(E_i) = E_i K_i^{-2}$, $\psi(F_i) = K_i^2 F_i$ and $\varphi(K_\alpha) = K_\alpha$, $\varphi(E_i) = K_i^{-2} E_i$, $\varphi(F_i) = F_i K_i^2$. Then*

$$\psi(a)\Omega = \Omega a \quad \text{and} \quad \varphi(a)\Omega' = \Omega' a \quad \text{for} \quad a \in U_q(\mathfrak{g}).$$

Proof. Since ψ is an algebra automorphism, it is enough to prove the first assertion for the generators $a = K_i, E_i, F_i$. For $a = K_i$ it is obviously true. Applying the mapping $m \circ (S \otimes 1) \circ \tau$ to both sides of (99) and (100) and summing over $\beta \in Q_+$ we obtain $E_i K_i^{-2} \Omega = \Omega E_i$ and $K_i^2 F_i \Omega = \Omega F_i$. This means that $\psi(E_i)\Omega = \Omega E_i$ and $\psi(F_i)\Omega = \Omega F_i$. Applying the automorphism θ we get the assertion for Ω'. □

Next we apply the element Ω' to the study of Verma modules. Let V be the carrier space of a weight representation. We define a linear operator ω on V by setting $\omega \mathbf{x} = q^{(\lambda + \rho, \lambda + \rho)} \mathbf{x}$, $\mathbf{x} \in V_\lambda$, on a weight subspace V_λ of V. Here, as earlier, ρ is the half-sum of all positive roots of $\mathfrak{g}$.

Proposition 42. *If M_Λ is a Verma module with highest weight Λ, then the operator $T_\Lambda^V(\Omega')\omega$ is a multiple of the identity operator, that is,*

$$T_\Lambda^V(\Omega')\omega = q^{(\Lambda + \rho, \Lambda + \rho)} I. \tag{102}$$

Proof. Let $\mathbf{e}_\Lambda$ be a highest weight vector of the Verma module M_Λ. Then

$$T_\Lambda^V(U_q(\mathfrak{g}))\mathbf{e}_\Lambda = T_\Lambda^V(U_q(\mathfrak{n}_-))\mathbf{e}_\Lambda = \sum_{\beta \in Q_+} T_\Lambda^V(U_q^{-\beta}(\mathfrak{n}_-))\mathbf{e}_\Lambda.$$

If $F_\beta \in U_q^{-\beta}(\mathfrak{n}_-)$, then $\mathbf{e}_{\Lambda-\beta} := T_\Lambda^V(F_\beta)\mathbf{e}_\Lambda$ is a weight vector for the weight $\Lambda - \beta$. We assert that

$$(T_\Lambda^V(\Omega')\omega)T_\Lambda^V(F_i)\mathbf{e}_{\Lambda-\beta} = T_\Lambda^V(F_i)(T_\Lambda^V(\Omega')\omega)\mathbf{e}_{\Lambda-\beta} \tag{103}$$

for any $\beta \in Q_+$ and any $i = 1, 2, \cdots, l$. Indeed, we have

$$(T_\Lambda^V(\Omega')\omega)T_\Lambda^V(F_i)\mathbf{e}_{\Lambda-\beta} = T_\Lambda^V(\Omega')T_\Lambda^V(F_i)q^\mu\omega\mathbf{e}_{\Lambda-\beta}$$

$$= T_\Lambda^V(F_iK_i^2)T_\Lambda^V(\Omega')q^\mu\omega\mathbf{e}_{\Lambda-\beta} = T_\Lambda^V(F_i)q^{2(\Lambda-\beta,\alpha_i)}q^\mu T_\Lambda^V(\Omega')\omega\mathbf{e}_{\Lambda-\beta}, \tag{104}$$

where

$$\mu = (\Lambda - \beta - \alpha_i + \rho, \Lambda - \beta - \alpha_i + \rho) - (\Lambda - \beta + \rho, \Lambda - \beta + \rho).$$

Denoting $\Lambda - \beta$ by σ, we have

$$(\sigma - \alpha_i + \rho, \sigma - \alpha_i + \rho) - (\sigma + \rho, \sigma + \rho) + 2(\sigma, \alpha_i) = (\alpha_i, \alpha_i) - 2(\alpha_i, \rho) = 0,$$

since $2(\rho, \alpha_i)/(\alpha_i, \alpha_i) = 1$. Therefore, the right hand side of (104) coincides with $T_\Lambda^V(F_i)T_\Lambda^V(\Omega')\omega\mathbf{e}_{\Lambda-\beta}$ and (103) is proved. The relation (103) yields

$$\begin{aligned}(T_\Lambda^V(\Omega')\omega)T_\Lambda^V(F_\beta)\mathbf{e}_\Lambda &= T_\Lambda^V(F_\beta)(T_\Lambda^V(\Omega')\omega)\mathbf{e}_\Lambda \\ &= q^{(\Lambda+\rho,\Lambda+\rho)}T_\Lambda^V(F_\beta)T_\Lambda^V(\Omega')\mathbf{e}_\Lambda.\end{aligned}$$

By (97) and (98),

$$T_\Lambda^V(\Omega')\mathbf{e}_\Lambda = T_\Lambda^V(\theta(\Omega))\mathbf{e}_\Lambda = T_\Lambda^V(\theta(\Omega_0))\mathbf{e}_\Lambda = T_\Lambda^V(1)\mathbf{e}_\Lambda = \mathbf{e}_\Lambda$$

and the relation (102) follows. □

Corollary 43. *For the irreducible representation L_Λ, realized on the quotient module M_Λ/M of the Verma module M_Λ by its maximal submodule M, we have $L_\Lambda(\Omega')\omega = q^{(\Lambda+\rho,\Lambda+\rho)}I$.*

6.3.4 The Center of $U_q(\mathfrak{g})$ and the Harish-Chandra Homomorphism

The aim of this subsection is to give a quantum analog of Harish-Chandra's description of the center of a universal enveloping algebra $U(\mathfrak{g})$.

Let $\mathfrak{z}_q$ be the center of the Drinfeld–Jimbo algebra $U_q(\mathfrak{g})$. Since the elements of $\mathfrak{z}_q$ commute in particular with all K_α, $\alpha \in Q_+$, it follows from (51) that $\mathfrak{z}_q \subseteq U_q^0(\mathfrak{g})$. By Theorem 24 and Corollary 25, any $Z \in \mathfrak{z}_q$ is a finite sum

$$Z = \sum\nolimits_{\mathbf{r},\mathbf{s}} F_{\beta_1}^{r_1} \cdots F_{\beta_n}^{r_n} k_{\mathbf{r},\mathbf{s}} E_{\beta_n}^{s_n} \cdots E_{\beta_1}^{s_1}, \qquad k_{\mathbf{r},\mathbf{s}} \in U_q(\mathfrak{h}), \tag{105}$$

where $\beta_1, \cdots, \beta_n$ are the positive roots of $\mathfrak{g}$, $\mathbf{r} = (r_1, \cdots, r_n)$, $\mathbf{s} = (s_1, \cdots, s_n)$ and the summation is over all $\mathbf{r}, \mathbf{s} \in \mathbb{N}_0^n$ such that $r_1\beta_1 + \cdots + r_n\beta_n = s_1\beta_1 + \cdots + s_n\beta_n$.

Let $\bar{\mathfrak{z}}_q$ denote the center of the algebra $\bar{U}_q^+(\mathfrak{g})$. Any $Z \in \bar{\mathfrak{z}}_q$ is also of the form (105) with a possibly infinite sum.

To define the Harish-Chandra homomorphism for $U_q(\mathfrak{g})$ we introduce an algebra homomorphism $\gamma : U_q(\mathfrak{h}) \to U_q(\mathfrak{h})$ such that

$$\gamma(K_\alpha) = q^{-(\alpha,\rho)} K_\alpha, \qquad \alpha \in Q,$$

and mappings $\Phi : \mathfrak{z}_q \to U_q(\mathfrak{h})$, $\bar{\Phi} : \bar{\mathfrak{z}}_q \to U_q(\mathfrak{h})$ defined by

$$\Phi(Z) = k_{\mathbf{0}\mathbf{0}}, \quad \bar{\Phi}(Z) = k_{\mathbf{0}\mathbf{0}}, \tag{106}$$

where Z is of the form (105). Proceeding as in the classical case (see [Dix]), one proves

Proposition 44. *The mappings $\gamma \circ \Phi : \mathfrak{z}_q \to U_q(\mathfrak{h})$ and $\gamma \circ \bar{\Phi} : \bar{\mathfrak{z}}_q \to U_q(\mathfrak{h})$ are algebra homomorphisms.*

Definition 11. *The mappings $\gamma \circ \Phi$ and $\gamma \circ \bar{\Phi}$ from Proposition 44 are called the* Harish-Chandra homomorphisms.

Put $T = \sum_{\lambda \in 2P \cap Q} \mathbb{C} K_\lambda \subset U_q(\mathfrak{h})$, where P is the set of all integral weights and Q is given by (25). The Weyl group W of $\mathfrak{g}$ acts on T by $wK_\lambda = K_{w\lambda}$. We denote by T^W the set of all W-invariant elements of T.

Theorem 45. *The Harish-Chandra homomorphisms $\gamma \circ \Phi$ and $\gamma \circ \bar{\Phi}$ are injective and their images coincide with the set T^W.*

This theorem says that an element $K \in U_q(\mathfrak{h})$ is the "$k_{\mathbf{0}\mathbf{0}}$ part" of a central element $Z \in \mathfrak{z}_q$ if and only if $\gamma(K) \in T^W$. Such an element $K \in U_q(\mathfrak{h})$ determines uniquely an element Z of $\mathfrak{z}_q$ such that $(\gamma \circ \Phi)(Z) = K$. There exists an inductive procedure which allows one to calculate the other summands in (105).

We give the proof of Theorem 45 only for the case of $U_q(\mathrm{sl}_2)$. A proof in the general case can be found in [Tan3] (see also [Lus] and [JL]). For $U_q(\mathrm{sl}_2)$, Theorem 45 can be reformulated in the following form.

Theorem 45$'$. *For $U_q(\mathrm{sl}_2)$ we have $\bar{\mathfrak{z}}_q = \mathfrak{z}_q$ and the center $\mathfrak{z}_q$ is generated by the Casimir element C_q from (3.6).*

Proof. We first show that $\bar{\mathfrak{z}}_q = \mathfrak{z}_q$. Clearly, $\mathfrak{z}_q = U_q(\mathrm{sl}_2) \bigcap \bar{\mathfrak{z}}_q$. If $Z \in \bar{\mathfrak{z}}_q$, then

$$Z = \sum\nolimits_{m=0}^{\infty} F^m k_m E^m, \qquad k_m \in U_q(\mathfrak{h}). \tag{107}$$

Let ϑ be the algebra automorphism of $U_q(\mathfrak{h})$ determined by the formula $\vartheta(K^n) = q^{-n} K^n$, $n \in \mathbb{Z}$. Substituting the expression (107) into the relation $aZ = Za$ for $a = F, E, K$, we find by direct calculation that

$$k_m = \vartheta(k_m) + \frac{q^{m+1} - q^{-m-1}}{(q - q^{-1})^2} (q^{-m} K - q^m K^{-1}) k_{m+1}, \quad m \in \mathbb{N}. \tag{108}$$

Assume that $Z \in \bar{\mathfrak{z}}$ and $Z \notin U_q(\mathrm{sl}_2)$. By (108), $k_m \neq 0$ for $m \in \mathbb{N}$. For $n \in \mathbb{Z}$ we set $U_q(\mathfrak{h})_n = \{k \in U_q(\mathfrak{h}) \mid \vartheta(k) = q^n k\}$. Then $U_q(\mathfrak{h}) = \bigoplus_{n\in\mathbb{Z}} U_q(\mathfrak{h})_n$. For any nonzero element $k = \sum_{n\in\mathbb{Z}} k^{(n)}$, $k^{(n)} \in U_q(\mathfrak{h})_n$, we define a number

$$a(k) = \max\{n \in \mathbb{Z} \mid k^{(n)} \neq 0\} - \min\{n \in \mathbb{Z} \mid k^{(n)} \neq 0\}.$$

By the definition of ϑ, one has $a(\vartheta(k_m) - k_m) \leq a(k_m)$. It follows from (108) that $a(\vartheta(k_m) - k_m) = a(k_{m+1}) + 2$. Thus, $a(k_m) \geq a(k_{m+1}) + 2$ for all $m \in \mathbb{N}$. Hence we obtain that $a(k_0) \geq a(k_m) + 2m \geq 2m$, $m \in \mathbb{N}$. This is impossible, since all $a(k_j)$ are finite. Therefore, $\bar{\mathfrak{z}}_q = \mathfrak{z}_q$ and the sum (107) is finite.

We now show that $\mathfrak{z}_q = \mathbb{C}[C_q]$, where C_q is given by (3.6). Let

$$Z = \sum\nolimits_{m=0}^{r} F^m k_m E^m \in \mathfrak{z}_q, \qquad k_m \in U_q(\mathfrak{h}). \tag{109}$$

Suppose that all $Z \in \mathfrak{z}_q$ of this form with $r < s$ are in $\mathbb{C}[C_q]$. Let $Z \in \mathfrak{z}_q$ be of the form (109) with $r = s$. As shown above, the elements k_m have to satisfy the relation (108). Since $k_{r+1} = 0$, it follows from (108) that $k_r = \alpha \cdot 1$ for some $\alpha \in \mathbb{C}$. Hence $Z - \alpha C_q^s$ is a central element of the form (109) with $r = s - 1$, so $Z - \alpha C_q^s \in \mathbb{C}[C_q]$ by the induction hypothesis. Thus, $Z \in \mathbb{C}[C_q]$ and the induction proof is complete. □

For every integral weight $\lambda \in P$ we define a character λ on $U_q(\mathfrak{h})$, that is, an algebra homomorphism $\lambda : U_q(\mathfrak{h}) \to \mathbb{C}$, by setting

$$\lambda(K_\alpha) \equiv \lambda(K_{\alpha_1}^{n_1} \cdots K_{\alpha_l}^{n_l}) := K_\alpha(\lambda) \equiv q^{n_1(\alpha_1,\lambda)} \cdots q^{n_l(\alpha_l,\lambda)} = q^{(\alpha,\lambda)}$$

for $\alpha = n_1\alpha_1 + \cdots + n_l\alpha_l \in Q$. Clearly, we have $\gamma(K_\alpha)(\lambda + \rho) = K_\alpha(\lambda)$. The composition $\xi_\lambda = \lambda \circ \gamma \circ \Phi$ of the two algebra homomorphisms $\gamma \circ \Phi : \mathfrak{z}_q \to U_q(\mathfrak{h})$ and $\lambda : U_q(\mathfrak{h}) \to \mathbb{C}$ is a character of the center $\mathfrak{z}_q$, called a *central character* of $U_q(\mathfrak{g})$. The following theorem is the quantum analog of Harish-Chandra's classical result (see [Dix]) on central characters.

Theorem 46. *For $\lambda, \mu \in P$ we have $\xi_\lambda = \xi_\mu$ if and only if $\lambda = w(\mu)$ for some $w \in W$.*

Proof. The proof can be found in [Ros3]. □

6.3.5 The Center of $U_q(\mathfrak{g})$ for q a Root of Unity

In this subsection we describe the center $\mathfrak{z}_q$ of $U_q(\mathfrak{g})$ when q is a root of unity. The proofs of the results stated below are complicated and can be found in the papers [L3] and [DCK], see also [Lus]. We assume that q is a primitive p-th root of unity, where p is an odd integer such that $p > d_i$, $i = 1, 2, \cdots, l$.

As it is expected from the case of $U_q(\mathrm{sl}_2)$ (see Proposition 3.15), there are many additional elements in the center $\mathfrak{z}_q$.

Proposition 47. *Let E_α and F_α, $\alpha \in \Delta$, be the root elements from Definition 4. Then the elements E_α^p, F_α^p, $\alpha \in \Delta$, and K_i^p, $i = 1, 2, \cdots, l$, belong to $\mathfrak{z}_q$.*

Outline of proof. For K_i^p the assertion follows immediately from the relations $K_iE_jK_i^{-1} = q^{d_ia_{ij}}E_j$ and $K_iF_jK_i^{-1} = q^{-d_ia_{ij}}F_j$. Similarly, the assertion for E_i^p and F_i^p is obtained at once from the defining relations. The proof for the elements $(E_\alpha)^p$ and $(F_\alpha)^p$, $\alpha \neq \alpha_i$, is more involved and is omitted. □

We denote the elements E_α^p, F_α^p, and K_i^p by e_α, f_α and k_i, respectively. Clearly, $k_i^{-1} \in \mathfrak{z}_q$. For the simple roots α_i we write e_{α_i} and f_{α_i} as e_i and f_i, respectively. Let $\mathfrak{z}_0$ be the subalgebra of $\mathfrak{z}_q$ generated by e_α, f_α, $\alpha \in \Delta_+$, and k_i, k_i^{-1}, $i = 1, 2, \cdots, l$. We also consider the subalgebras $\mathfrak{z}_0^0$, $\mathfrak{z}_0^+$, $\mathfrak{z}_0^-$ of $\mathfrak{z}_0$ generated by the elements k_i, k_i^{-1}, $1 \leq i \leq l$, the elements e_α, $\alpha \in \Delta_+$, and the elements f_α, $\alpha \in \Delta_+$, respectively. The sets $\mathfrak{z}_0^{0+} := \mathfrak{z}_0^0\mathfrak{z}_0^+$ and $\mathfrak{z}_0^{0-} = \mathfrak{z}_0^-\mathfrak{z}_0^0$ are also subalgebras of $\mathfrak{z}_0$.

Proposition 48. (i) *The algebras $\mathfrak{z}_0$, $\mathfrak{z}_0^0$, $\mathfrak{z}_0^{0+}$, $\mathfrak{z}_0^{0-}$ are Hopf subalgebras of $U_q(\mathfrak{g})$. The comultiplication of $U_q(\mathfrak{g})$ acts on the elements e_i, f_i, k_i as*

$$\Delta(e_i) = e_i \otimes k_i + 1 \otimes e_i, \quad \Delta(f_i) = f_i \otimes 1 + k_i^{-1} \otimes f_i, \quad \Delta(k_i) = k_i \otimes k_i.$$

(ii) *We have $\mathfrak{z}_0^\pm = U_q(\mathfrak{n}_\pm) \bigcap \mathfrak{z}_q$. The multiplication map defines an algebra isomorphism of $\mathfrak{z}_0^- \otimes \mathfrak{z}_0^0 \otimes \mathfrak{z}_0^+$* onto $\mathfrak{z}_0$.

(iii) *The action of the braid group $\mathfrak{B}_\mathfrak{g}$ on $U_q(\mathfrak{g})$ leaves $\mathfrak{z}_0$ invariant.*

It is quite remarkable that the formulas for the comultiplications of the elements $e_i = E_i^p$ and $f_i = F_i^p$ are the same as the corresponding formulas (18) for the generators E_i and F_i, respectively.

The following theorem determines the structure of $U_q(\mathfrak{g})$ in terms of the subalgebra $\mathfrak{z}_0$ of the center $\mathfrak{z}_q$.

Theorem 49. *The algebra $U_q(\mathfrak{g})$ is a finite-dimensional linear space over $\mathfrak{z}_0$ with basis*

$$E_{\beta_1}^{r_1} \cdots E_{\beta_n}^{r_n} K_1^{t_1} \cdots K_l^{t_l} F_{\beta_n}^{s_n} \cdots F_{\beta_1}^{s_1}, \qquad 1 \leq r_j, t_j, s_j \leq p.$$

Let Spec $(\mathfrak{z}_q)$ and Spec $(\mathfrak{z}_0)$ denote the sets of all characters on the commutative algebras $\mathfrak{z}_q$ and $\mathfrak{z}_0$, respectively. That is, Spec $(\mathfrak{z}_q)$ and Spec $(\mathfrak{z}_0)$ consist of all algebra homomorphisms of $\mathfrak{z}_q$ and $\mathfrak{z}_0$, respectively, to $\mathbb{C}$. These sets are of importance for the study of irreducible representations of $U_q(\mathfrak{g})$. For any irreducible representation of $U_q(\mathfrak{g})$, the elements of $\mathfrak{z}_q$ act as scalar operators. Hence they determine a unique algebra homomorphism of $\mathfrak{z}_q$ to $\mathbb{C}$, called the *central character* of the representation. The following two important results allow us to describe the structure of Spec $(\mathfrak{z}_q)$ and Spec $(\mathfrak{z}_0)$.

Theorem 50. *The elements e_α, f_α, $\alpha \in \Delta_+$, k_i, $i = 1, 2, \cdots, l$, are algebraically independent (that is, they do not satisfy a nontrivial algebraic equation with complex coefficients). The center $\mathfrak{z}_q$ of $U_q(\mathfrak{g})$ is algebraic over $\mathfrak{z}_0$ (that is, every element of $\mathfrak{z}_q$ fulfills a nontrivial algebraic equation with coefficients in $\mathfrak{z}_0$).*

Example 4 ($U_q(\mathrm{sl}_2)$). The center of the algebra $U_q(\mathrm{sl}_2)$ is generated by the elements E^p, F^p, K^p, K^{-p} and C_q, where C_q is the Casimir element (3.6).

It follows from (3.6) that $C_q - (q-q^{-1})^{-2}(Kq+K^{-1}q^{-1}) = FE$. Using this relation, one easily computes that

$$\prod_{j=0}^{p-1}(C_q - (q-q^{-1})^{-2}(Kq^{j+1}+K^{-1}q^{-j-1})) = E^pF^p,$$

that is,

$$C_q^p + \gamma_1 C_q^{p-1} + \cdots + \gamma_{p-1}C_q + X = E^pF^p, \tag{110}$$

where $\gamma_i \in \mathbb{C}$ and $X = (-1)^p(q-q^{-1})^{-2p}(K^p - K^{-p}) \in \mathfrak{z}_0$. △

Since the elements e_α, f_α and k_i are algebraically independent, the set $\mathrm{Spec}\,(\mathfrak{z}_0)$ is isomorphic to $\mathbb{C}^{2n} \times (\mathbb{C}^\times)^l$, where $\mathbb{C}^\times := \mathbb{C}\backslash\{0\}$, $l = \mathrm{rank}\,\mathfrak{g}$ and n is the number of roots in Δ_+. Thus, the set $\mathrm{Spec}\,(\mathfrak{z}_0)$ is characterized by $2n + l = \dim\mathfrak{g}$ complex parameters.

The restriction of characters on $\mathfrak{z}_q$ to its subalgebra $\mathfrak{z}_0$ defines a map $\nu : \mathrm{Spec}\,(\mathfrak{z}_q) \to \mathrm{Spec}\,(\mathfrak{z}_0)$. The sets

$$\nu^{-1}(\xi) = \{\eta \in \mathrm{Spec}\,(\mathfrak{z}_q) \mid \nu(\eta) = \xi\}$$

for $\xi \in \mathrm{Spec}\,(\mathfrak{z}_0)$ are called the *fibers* of the map ν. The second part of Theorem 50 implies that for every point $\xi \in \mathrm{Spec}\,(\mathfrak{z}_0)$ the fiber $\nu^{-1}(\xi)$ contains only a finite number of points.

Proposition 51. *The fibers of ν have at most p^l points and the generic fiber has precisely p^l points, where l is the rank of $\mathfrak{g}$.*

6.4 Notes

The quantized universal enveloping algebras were discovered independently by V. G. Drinfeld [Dr1] and M. Jimbo [Jim1]. The triangular decomposition of $U_q(\mathfrak{g})$ is proved in [Ros1]. The Hopf algebra automorphisms and the real forms of Drinfeld–Jimbo algebras were described in [Tw].

The Poincaré–Birkhoff–Witt theorem is from [Ros2] and [Yam] for $\mathfrak{g} = \mathrm{sl}_N$ and from [Lus] in the general case. Braid group actions and general root elements were invented by G. Lusztig [L1] (see also [L2] and [Lus]) and S. Z. Levendorskii and Y. S. Soibelman [LS1]. Verma modules for $U_q(\mathfrak{g})$ and the corresponding descriptions of finite-dimensional representations of $U_q(\mathfrak{g})$ appeared in [L1]. The adjoint action of Drinfeld–Jimbo algebras was extensively studied in [JL], see also the book [Jos] and the references therein.

The existence of a dual pairing of $U_q(\mathfrak{b}_+)$ and $U_q(\mathfrak{b}_-)^{\mathrm{op}}$ was observed by Drinfeld [Dr2]. In our exposition we followed the paper of T. Tanisaki [Tan3]. The description of the quantum analog of the Casimir element is also taken from [Tan3]. The results on the Harish-Chandra homomorphism in Subsect. 6.3.4 appeared in [Ros3]. The center of $U_q(\mathfrak{g})$ in the root of unity case was investigated in [L3], [Lus], and [DCK].

7. Finite-Dimensional Representations of Drinfeld–Jimbo Algebras

Weight representations and Verma modules of Drinfeld–Jimbo algebras $U_q(\mathfrak{g})$ appeared in Subsects. 6.2.5–7. The present chapter is devoted to a detailed study of finite-dimensional representations of these algebras.

In Sects. 7.1–4 we assume that q is not a root of unity. As we have seen in Sect. 3.3 in the case of $U_q(\mathrm{sl}_2)$, the corresponding representation theory of $U_q(\mathfrak{g})$ is similar in many aspects to the classical theory. In Sect. 7.5 we investigate representations of Drinfeld–Jimbo algebras $U_q(\mathfrak{g})$ in the root of unity case. Then the representation theory strongly differs from the case when q is not a root of unity.

7.1 General Properties of Finite-Dimensional Representations of $U_q(\mathfrak{g})$

7.1.1 Weight Structure and Classification

The aim of this subsection is to prove the following theorem:

Theorem 1. *Any irreducible finite-dimensional representation of a Drinfeld–Jimbo algebra $U_q(\mathfrak{g})$ is a weight representation and a representation with highest weight. Such a representation is uniquely determined, up to equivalence, by its highest weight.*

Proof. The proof will be given by several steps stated as propositions.

Proposition 2. *Every irreducible finite-dimensional representation of a Drinfeld–Jimbo algebra $U_q(\mathfrak{g})$ is a weight representation.*

Proof. Let T be a nontrivial irreducible representation of $U_q(\mathfrak{g})$ on a finite-dimensional vector space V. Since the operators $T(K_i)$, $i = 1, 2, \cdots, l = \operatorname{rank} \mathfrak{g}$, commute with each other, they possess a nonzero common eigenvector. Let V' be a maximal subspace of V on which all operators $T(K_i)$ are diagonalizable. Then we have $\dim V' \geq 1$ and $V' = \bigoplus_\mu V'_\mu$, where $V'_\mu = \{\mathbf{v} \in V' \mid T(K_i)\mathbf{v} = \mu_i \mathbf{v}\}$, $\mu = (\mu_1, \cdots, \mu_l)$. Assume on the contrary that $\dim V' < \dim V$. Since T is irreducible, V' cannot be invariant under all operators $T(E_i)$, $T(F_i)$, $i = 1, 2, \cdots, l$. (Indeed, if $T(E_i)\mathbf{v}' \in V'$ and $T(F_i)\mathbf{v}' \in V'$ for all weight vectors $\mathbf{v}' \in V'$, then V' is an invariant subspace of V.) Let $\mathbf{v}$ be an element of V' such that $T(E_j)\mathbf{v}$ is not in V' for some $j \in \{1, 2, \cdots, l\}$. Without loss of generality we can assume that $\mathbf{v}$ is a weight vector. Then there is a μ such that $\mathbf{v} \in V'_\mu$. Then we have $T(E_j)\mathbf{v} \neq 0$ and $T(K_i)T(E_j)\mathbf{v} = q_i^{a_{ij}} T(E_j)T(K_i)\mathbf{v} = \mu_i q_i^{a_{ij}} T(E_j)\mathbf{v}$, that is, $\mathbf{v}'' \equiv T(E_j)\mathbf{v}$ is a common eigenvector for all $T(K_i)$ and all operators $T(K_i)$ are diagonalizable on $V'' = V' \oplus \mathbb{C}\mathbf{v}''$. Since $\dim V'' > \dim V'$, this is a contradiction. □

Lemma 3. *If T is a finite-dimensional weight representation of a Drinfeld–Jimbo algebra $U_q(\mathfrak{g})$, then there exists $N \in \mathbb{N}$ such that $T(E_{i_1}) \cdots T(E_{i_p}) = 0$ for all $i_1, \cdots i_p \in \{1, 2, \cdots, l\}$, $p \geq N$.*

Proof. Let V be the carrier space of T and let $V = \bigoplus_\mu V_\mu$, be its weight decomposition, where $V_\mu = \{\mathbf{v} \in V \mid T(K_i)\mathbf{v} = \mu_i \mathbf{v}\}$. It is enough to show that for every $\mathbf{v} \in V_\mu$, $\mu = (\mu_1, \cdots, \mu_l)$, we have $T(E_{i_1}) \cdots T(E_{i_p})\mathbf{v} = 0$ for sufficiently large p. Set $\mathbf{v}' = T(E_{i_1}) \cdots T(E_{i_p})\mathbf{v}$. Then $\mathbf{v}' \in V_{\mu'}$, where $\mu' = (\mu'_1, \cdots, \mu'_l)$ and $\mu'_i = \mu_i q_i^{\sum n_k a_{ik}}$. Here n_k is the number of $T(E_k)$ appearing in the product $T(E_{i_1}) \cdots T(E_{i_p})$. The representation T has a finite number of weights. We denote them by $\mu, \mu^{(1)}, \cdots, \mu^{(r)}$. We shall prove that there exists $N \in \mathbb{N}$ such that for $p \geq N$, μ' does not occur in this list. Let $\lambda_i^{(s)} = \mu_i^{(s)}/\mu_i$, $\quad i = 1, 2, \cdots, l$. We show that there exists a number $i_0 \in \{1, 2, \cdots, l\}$ such that

$$q_{i_0}^{\sum n_k a_{i_0 k}} \notin \{1, \lambda_{i_0}^{(1)}, \cdots, \lambda_{i_0}^{(r)}\}. \tag{1}$$

Let us write the complex numbers q and $\lambda_i^{(s)}$ as $q = \exp 2\mathrm{i}\pi\nu$ and $\lambda_i^{(s)} = \exp\,(2\mathrm{i}\pi\nu y_i^{(s)})$. Since $q_i = q^{(\alpha_i, \alpha_i)/2}$, the equality $q_i^{\sum n_k a_{ik}} = \lambda_i^{(s)}$ implies that there is an integer m such that

$$\frac{(\alpha_i, \alpha_i)}{2} \sum_{k=1}^{l} n_k a_{ik} = y_i^{(s)} + \frac{m}{\nu}.$$

By the definition of a_{ik}, the latter writes as $\sum_k n_k(\alpha_i, \alpha_k) = y_i^{(s)} + m/\nu$. Since $\sum_k n_k(\alpha_i, \alpha_k) \in \mathbb{Z}$, we conclude that $y_i^{(s)} + m/\nu \in \mathbb{Z}$. Since $\nu \notin \mathbb{Q}$, there exists at most one integer m such that $y_i^{(s)} + m/\nu \in \mathbb{Z}$. Set $z_i^{(s)} := y_i^{(s)} + m/\nu$. Assume that for every $i \in \{1, 2, \cdots, l\}$ there exists $s \in \{0, 1, \cdots, r\}$ such that $\sum_k n_k(\alpha_i, \alpha_k) = z_i^{(s)}$ with integral $z_i^{(s)}$. This is a system of linear equations with unknowns $n_1, n_2, \cdots, n_l$. Since the $l \times l$ matrix consisting of the entries (α_i, α_k) is invertible, for a given $z_1^{(s_1)}, \cdots, z_l^{(s_l)}$ there exists at most one integral solution of this system. However, the number of possible sets $\{z_1^{(s_1)}, \cdots, z_l^{(s_l)}\}$ is finite. Therefore, if the integers $n_1, n_2, \cdots, n_l$ do not belong to a certain finite set $\mathfrak{M}$, there exists an index i_0 such that (1) is satisfied. Let $N := \sup(|n_1| + \cdots + |n_l|) + 1$, where the supremum is taken over $\mathfrak{M}$. Then the assertion of Lemma 3 holds. □

Let T be a finite-dimensional representation of $U_q(\mathfrak{g})$ on a vector space V. Recall that $\mathbf{v} \in V$ is a *highest weight vector* of T if $\mathbf{v}$ is a weight vector such that $T(E_i)\mathbf{v} = 0$ for $i = 1, 2, \cdots, l$ and $T(U_q(\mathfrak{g}))\mathbf{v} = V$.

Proposition 4. *Every irreducible finite-dimensional representation of* $U_q(\mathfrak{g})$ *is a representation with highest weight.*

Proof. Let T be an irreducible representation of $U_q(\mathfrak{g})$ on a finite-dimensional vector space V and let $V_0 = \bigcap_i \ker T(E_i)$. Let N be the smallest number in $\mathbb{N}$ for which the assertion of Lemma 3 holds. Then there exist a vector $\mathbf{v} \in V$ and indices $i_1, \cdots, i_{N-1} \in \{1, 2, \cdots, l\}$ such that $\mathbf{v}_0 = T(E_{i_1} \cdots E_{i_{N-1}})\mathbf{v} \neq 0$ and $T(E_i)\mathbf{v}_0 = 0$ for $i = 1, 2, \cdots, l$. Thus, $\mathbf{v}_0 \in V_0$ and so $V_0 \neq \{0\}$. From the formula (6.13) it is clear that V_0 is invariant under all operators $T(K_i)$. Since $V_0 \neq \{0\}$, there exists a common eigenvector $\mathbf{e}_0 \in V_0$ for the operators $T(K_i)$, $i = 1, 2, \cdots, l$. Because T is irreducible, the invariant subspace $T(U_q(\mathfrak{g}))\mathbf{e}_0$ is equal to V. Since $\mathbf{e}_0 \in V_0$, we have $T(E_i)\mathbf{e}_0 = 0$ for all i. The preceding shows that $\mathbf{e}_0$ is a highest weight vector for T. □

Let $\omega = (\omega_1, \cdots, \omega_l)$, $\omega_i = \pm 1$. Then there is a one-dimensional representation T_ω of $U_q(\mathfrak{g})$ such that

$$T_\omega(E_i) = T_\omega(F_i) = 0, \quad T_\omega(K_i) = \omega_i, \qquad i = 1, 2, \cdots, l.$$

Clearly, every one-dimensional representation of $U_q(\mathfrak{g})$ is of this form.

If T is an irreducible representation of $U_q(\mathfrak{g})$ with highest weight $\lambda = (\lambda_1, \cdots, \lambda_l)$, then the tensor product $T \otimes T_\omega$ is an irreducible representation of $U_q(\mathfrak{g})$ with highest weight $\omega \cdot \lambda \equiv (\omega_1\lambda_1, \cdots, \omega_l\lambda_l)$. In particular, the irreducible finite-dimensional representation T_λ with highest weight $\lambda = \omega \cdot q^{\mathbf{n}} = (\omega_1 q^{n_1}, \cdots, \omega_l q^{n_l})$ is the tensor product of representations T_ω and $T_{\lambda'}$ with $\lambda' = q^{\mathbf{n}}$.

Proposition 5. $\lambda = (\lambda_1, \cdots, \lambda_l)$ *is a highest weight of an irreducible finite-dimensional representation* T *of* $U_q(\mathfrak{g})$ *if and only if it is of the form* $\lambda = \omega \cdot q^{\mathbf{n}} = (\omega_1 q^{n_1}, \cdots, \omega_l q^{n_l})$ *with* $\omega_i = \pm 1$ *and* $n_i \in \mathbb{N}_0$.

Proof. Let $U_q^i(\mathrm{sl}_2)$ be the subalgebra of $U_q(\mathfrak{g})$ generated by the elements E_i, F_i, K_i, K_i^{-1}. It is isomorphic to $U_{q_i}(\mathrm{sl}_2)$. We restrict the representation T to $U_q^i(\mathrm{sl}_2)$. Since finite-dimensional representations of $U_{q_i}(\mathrm{sl}_2)$ are completely reducible (see Proposition 3.12), this restriction decomposes into a direct sum of irreducible representations of $U_q^i(\mathrm{sl}_2)$. The highest weight vector $\mathbf{e}$ of T belongs to the carrier space of one of these irreducible representations and is its highest weight vector with weight λ_i. By the results of Subsect. 3.2.1, λ_i is of the form stated in the proposition. Conversely, let $\mathbf{n} = (n_1, \cdots, n_l)$, $n_i \in \mathbb{N}_0$, and $\omega = (\omega_1, \cdots, \omega_l)$, $\omega_i = \pm 1$. By Proposition 6.28, there is an irreducible representation $L_{\mathbf{n}}$ with highest weight $q^{\mathbf{n}}$. Then $T_\omega \otimes L_{\mathbf{n}}$ is an irreducible representation with highest weight $\lambda = \omega \cdot q^{\mathbf{n}}$. □

Proposition 6. *Every irreducible finite-dimensional representation* T *of* $U_q(\mathfrak{g})$ *is uniquely determined, up to equivalence, by its highest weight.*

Proof. Let $\lambda = \omega \cdot q^{\mathbf{n}}$ and $\lambda' = \omega' \cdot q^{\mathbf{n}'}$ be highest weights of the representation T. From the form of eigenvalues of the operators $T(K_i)$ we see that $\omega = \omega'$. Since $\omega = \omega'$, the irreducible representation $T_\omega \otimes T$ has the highest weights $q^{\mathbf{n}}$ and $q^{\mathbf{n}'}$, so $q^{\mathbf{n}} = q^{\mathbf{n}'}$ by Proposition 6.28. □

Remark 1. By Remark 6.7, we may take the roots $-\alpha_1, -\alpha_2, \cdots, -\alpha_l$ as a set of simple roots of $\mathfrak{g}$. In this case, the elements F_i, $i = 1, 2, \cdots, l$, correspond to simple roots. Then highest weight vectors are taken with respect to the operators $T(F_i)$ instead of $T(E_i)$, and weights $\lambda = (\lambda_1, \cdots, \lambda_l)$ and the corresponding weight vectors $\mathbf{v}$ are defined by $T(K_i^{-1})\mathbf{v} = \lambda_i \mathbf{v}$. All results of this chapter are true (under appropriate reformulations if necessary) for this setting. This approach will be used in Sect. 8.4.

7.1.2 Properties of Representations

Let T be an irreducible representation of $U_q(\mathfrak{g})$ on a finite-dimensional vector space V. Then T is a representation with highest weight $\lambda = (\lambda_1, \cdots, \lambda_l)$ and corresponding highest weight vector $\mathbf{e}$. We repeat the main properties of such representations from Subsects. 6.2.5–7:

(i) *V is spanned by the vectors* $\mathbf{e}$ *and* $T(F_{i_1}) \cdots T(F_{i_p})\mathbf{e}$, $i_1, \cdots, i_p \in \{1, 2, \cdots, l\}$. *Moreover,* $\dim V_\lambda = 1$.

(ii) *The vector* $T(F_{i_1}) \cdots T(F_{i_p})\mathbf{e}$ *in* V *is of weight* $\mu = (\mu_1, \cdots, \mu_l)$ *with* $\mu_k = \lambda_k q_k^{-\sum_j a_{k i_j}}$, *where* a_{ki} *are the entries of the Cartan matrix of* $\mathfrak{g}$. *Every weight of the representation* T *is of this form.*

The counterpart to Proposition 5 for the algebras $\check{U}_q(\mathfrak{g})$ is

Proposition 7. *If* $\omega = (\omega_1, \cdots, \omega_l)$, $\omega_k \in \{1, -1, \sqrt{-1}, -\sqrt{-1}\}$, *and* $\mathbf{n} = (n_1, \cdots, n_l)$ *is a dominant integral weight for the Lie algebra* $\mathfrak{g}$, *then* $\lambda = \omega \cdot q^{\mathbf{n}}$ *is the highest weight of some irreducible finite-dimensional representation of* $\check{U}_q(\mathfrak{g})$. *Moreover, every highest weight of an irreducible finite-dimensional representation of* $\check{U}_q(\mathfrak{g})$ *is of this form.*

The classical H. Weyl theorem on complete reducibility has the following quantum analog.

Theorem 8. *Each finite-dimensional representation of* $U_q(\mathfrak{g})$ *or of* $\check{U}_q(\mathfrak{g})$ *is completely reducible.*

Proof. A proof of this theorem can be found in each of the papers [Ros1], [APW] and [JL]. □

Any dominant integral weight $\mathbf{n} = (n_1, n_2, \cdots, n_l)$ corresponds uniquely to a dominant integral form Λ on the Cartan subalgebra of $\mathfrak{g}$ by $n_i = 2(\Lambda, \alpha_i)/(\alpha_i, \alpha_i)$. We shall denote the irreducible finite-dimensional representation T of $U_q(\mathfrak{g})$ with highest weight $q^{\mathbf{n}}$ by $T_{\mathbf{n}}$ and also by T_Λ with a slight abuse of notation. Such representations $T_{\mathbf{n}} \equiv T_\Lambda$ are called of *type 1.*

Since any irreducible representation is a tensor product of a representation $T_{\mathbf{n}}$ and a one-dimensional representation T_ω, we study in the following mainly type 1 representations. The following properties of representations $T_{\mathbf{n}} \equiv T_\Lambda$ are similar to those of irreducible finite-dimensional representations of $\mathfrak{g}$.

Proposition 9. (i) *The Weyl group W of the Lie algebra $\mathfrak{g}$ acts naturally on the set Π_Λ of weights of the type 1 irreducible finite-dimensional representation T_Λ of $U_q(\mathfrak{g})$. This action of W leaves Π_Λ invariant and preserves the dimensions of weight spaces.*

(ii) *If Ω' is the element and ω is the operator from Subsect. 6.3.3, then we have $T_\Lambda(\Omega')\omega = q^{(\Lambda+\rho,\Lambda+\rho)}I$.*

(iii) *For any type 1 irreducible representation T_Λ of $U_q(\mathfrak{g})$ we have $T_\Lambda(Z) = \xi_\Lambda(Z)I$, $Z \in \mathfrak{z}_q$, where $\mathfrak{z}_q$ is the center of $U_q(\mathfrak{g})$ and ξ_Λ, $\Lambda \in P_+$, is the central character of $U_q(\mathfrak{g})$ from Subsect. 6.3.4. The central characters separate the finite-dimensional irreducible representations of $U_q(\mathfrak{g})$, that is, for any two representations $T_{\Lambda'}$ and $T_{\Lambda''}$, $\Lambda' \neq \Lambda''$, there exists an element $Z \in \mathfrak{z}_q$ such that $\xi_{\Lambda'}(Z) \neq \xi_{\Lambda''}(Z)$.*

Proof. Since T_Λ is equivalent to the representation L_Λ from Subsect. 6.2.6, the proof of (i) is in fact given by the proof of Proposition 6.28(iii). The assertion of (ii) follows from Corollary 6.43. The first part of (iii) follows from the formula $\xi_\Lambda = \Lambda \circ \gamma \circ \Phi$ (see Subsect. 6.3.4) and from the expression (6.105) for the elements of $\mathfrak{z}_q$. The second part of (iii) is a consequence of Theorem 6.46. □

Let us describe the highest weights Λ of type 1 irreducible representations of Drinfeld–Jimbo algebras $U_q(\mathfrak{g})$ corresponding to the simple Lie algebras $\mathrm{sl}(l+1,\mathbb{C})$, $\mathrm{so}(2l+1,\mathbb{C})$, $\mathrm{sp}(2l,\mathbb{C})$, and $\mathrm{so}(2l,\mathbb{C})$. As in the classical case, it is convenient to characterize the highest weights Λ by the following numbers m_i. Suppose that $\Lambda \equiv \mathbf{n} = (n_1, n_2, \cdots, n_l)$, where $n_i = 2(\Lambda,\alpha_i)/(\alpha_i,\alpha_i)$. Then the relations between the numbers n_i and m_i are

$$n_i = m_i - m_{i+1},\ \ i = 1,2,\cdots,l,\ \text{ for }\ U_q(\mathrm{sl}_{l+1}),$$
$$n_i = m_i - m_{i+1},\ \ i = 1,2,\cdots,l-1,\ \ n_l = 2m_l\ \text{ for }\ U_q(\mathrm{so}_{2l+1}),$$
$$n_i = m_i - m_{i+1},\ \ i = 1,2,\cdots,l-1,\ \ n_l = m_l\ \text{ for }\ U_q(\mathrm{sp}_{2l}),$$
$$n_i = m_i - m_{i+1},\ \ i = 1,2,\cdots,l-1,\ \ n_l = m_{l-1} + m_l\ \text{ for }\ U_q(\mathrm{so}_{2l}).$$

Note that $n_{l-1} = n_l$ for $U_q(\mathrm{so}_{2l})$ if and only if $m_l = 0$.

For $U_q(\mathrm{sl}_{l+1})$ and $U_q(\mathrm{sp}_{2l})$ all numbers m_i are integers, while for $U_q(\mathrm{so}_N)$ they are all integers or all half-integers. They satisfy the dominantness conditions

$$m_1 \geq m_2 \geq \cdots \geq m_{l+1} \quad \text{for} \quad U_q(\mathrm{sl}_{l+1}), \tag{2}$$

$$m_1 \geq m_2 \geq \cdots \geq m_l \geq 0 \quad \text{for} \quad U_q(\mathrm{so}_{2l+1}), U_q(\mathrm{sp}_{2l}) \tag{3}$$

$$m_1 \geq m_2 \geq \cdots \geq m_{l-1} \geq |m_l| \quad \text{for} \quad U_q(\mathrm{so}_{2l}). \tag{4}$$

For $\mathfrak{g} = \mathrm{so}_{2l}$ and $\mathfrak{g} = \mathrm{sp}_{2l}$ there is a one-to-one correspondence between irreducible finite-dimensional representations of $U_q(\mathfrak{g})$ and collections of numbers $(m_1, m_2, \cdots, m_l)$ as described above. This is not true for the algebra $U_q(\mathrm{sl}_{l+1})$. Two such sets $(m_1, m_2, \cdots, m_{l+1})$ and $(\hat{m}_1, \hat{m}_2, \cdots, \hat{m}_{l+1})$ describe the same irreducible finite-dimensional representation of $U_q(\mathrm{sl}_{l+1})$ if and only if there exists an integer m such that $\hat{m}_i = m_i + m$, $i = 1, 2, \cdots, l+1$.

7.1.3 Representations of h-Adic Drinfeld–Jimbo Algebras

By Proposition 6.8, the h-adic algebra $U_h(\mathfrak{g})$ is isomorphic to $U(\mathfrak{g})[[h]]$, where $U(\mathfrak{g})$ is the universal enveloping algebra of $\mathfrak{g}$. Therefore, as noted in Subsect. 1.3.1, if T is a finite-dimensional representation of the (complex) algebra $U(\mathfrak{g})$ on a (complex) vector space V, then formula (1.50) defines a representation T_h of the h-adic algebra $U_h(\mathfrak{g}) \simeq U(\mathfrak{g})[[h]]$ on the $\mathbb{C}[[h]]$-vector space $V[[h]]$. Conversely, let V be a finite-dimensional complex vector space and let T' be a representation of $U_h(\mathfrak{g})$ on $V_h := V[[h]]$. Then it is clear that the equation

$$T(x + hU_h(\mathfrak{g}))(v + hV_h) := T'(x)v + hV_h, \quad x \in U(\mathfrak{g}), v \in V,$$

defines a representation T of $U_h(\mathfrak{g})/hU_h(\mathfrak{g}) \simeq U(\mathfrak{g})$ on $V_h/hV_h \simeq V$ such that $T_h = T'$. It is easily seen that T is irreducible if and only if T_h is indecomposable. Thus, we have proved the following

Proposition 10. *The map $T \to T_h$ determines a one-to-one correspondence between finite-dimensional representations of the complex Lie algebra $\mathfrak{g}$ and representations of the h-adic algebra $U_h(\mathfrak{g})$ on $\mathbb{C}[[h]]$-vector spaces of the form $V[[h]]$, where V is a finite-dimensional complex vector space.*

The carrier space $V[[h]]$ of an indecomposable representation $(T_\Lambda)_h$ of $U_h(\mathfrak{g})$ with highest weight Λ decomposes into weight subspaces

$$V[[h]] = \bigoplus_\mu V[[h]]_\mu, \quad V[[h]]_\mu = V_\mu \otimes \mathbb{C}[[h]],$$

where V_μ is the corresponding weight subspace of the representation T_Λ of $\mathfrak{g}$ on V. In particular, we get $\dim_{\mathbb{C}[[h]]} V[[h]]_\mu = \dim_{\mathbb{C}} V_\mu$. A similar assertion is true for the representations T_Λ of the Drinfeld–Jimbo algebra $U_q(\mathfrak{g})$.

Proposition 11. *A type 1 irreducible finite-dimensional representation T_Λ of $U_q(\mathfrak{g})$ acts on a space of the same dimension as the corresponding irreducible representation T_Λ of the Lie algebra $\mathfrak{g}$. Moreover, the dimensions of weight subspaces, corresponding to the same weight in these representations of $U_q(\mathfrak{g})$ and $\mathfrak{g}$, coincide.*

Sketch of proof. For $\mathfrak{g} = \mathrm{sl}(n, \mathbb{C})$ this will be shown in Subsect. 7.3.3. For a general Drinfeld–Jimbo algebra $U_q(\mathfrak{g})$ it can be proved by means of Verma modules. If M'_Λ and M_Λ are the Verma modules of $\mathfrak{g}$ and $U_q(\mathfrak{g})$, respectively, with highest weight Λ, then by (6.74) the dimensions of their weight subspaces, corresponding to the same weight, coincide. We can construct the irreducible finite-dimensional representations T_Λ of $U_q(\mathfrak{g})$ and of $\mathfrak{g}$ by means of their Verma modules M_Λ and M'_Λ. By Proposition 6.28, the representations T_Λ are realized on M_Λ/M and M'_Λ/M', respectively, where M and M' are the maximal proper submodules. By Proposition 6.27, the submodule M (resp. M') coincides with the sum of all Verma submodules M_{Λ_w} (resp. M'_{Λ_w}) with $\Lambda_w = w(\Lambda+\rho)-\rho$, $w \in W$, $w \neq 1$. This implies that the weight subspaces of M and M', belonging to the same weight, have the same dimensions. Therefore, the dimensions of the corresponding weight subspaces in M_Λ/M and M'_Λ/M' coincide. □

Most of the above considerations and facts remain valid almost verbatim for representations of the quantum algebra $U'_q(\mathfrak{g})$ over the field $\mathbb{Q}(q)$. In this case the representation spaces are vector spaces over $\mathbb{Q}(q)$.

7.1.4 Characters of Representations and Multiplicities of Weights

Proposition 11 allows us to define characters of finite-dimensional representations of $U_q(\mathfrak{g})$ which characterize these representations up to equivalence.

If T is a type 1 finite-dimensional representation of $U_q(\mathfrak{g})$ on a vector space V with weight subspace decomposition $V = \bigoplus_{\mu \in P} V_\mu$, then the function

$$\chi(T) = \sum\nolimits_{\mu \in P} (\dim V_\mu) e^\mu$$

on the Cartan subalgebra $\mathfrak{h}$ of $\mathfrak{g}$ is called the *character* of T. Here e^μ is the function on $\mathfrak{h}$ defined by $e^\mu(h) = e^{\mu(h)}$, $h \in \mathfrak{h}$. Recall that the character of a finite-dimensional representation of the Lie algebra $\mathfrak{g}$ is defined in the same way. Therefore, by Proposition 11, the characters of type 1 finite-dimensional representations of $U_q(\mathfrak{g})$ are in a one-to-one correspondence with characters of finite-dimensional representations of $\mathfrak{g}$. This leads to

Proposition 12. (i) *Type 1 irreducible finite-dimensional representations of $U_q(\mathfrak{g})$ are determined uniquely, up to equivalence, by their characters.*

(ii) *The character $\chi(T_\Lambda)$ of the irreducible finite-dimensional representation T_Λ with highest weight Λ is given by the classical Weyl formula*

$$\chi(T_\Lambda) = \frac{\sum_{w \in W} (-1)^{l(w)} e^{w(\Lambda+\rho)}}{\sum_{w \in W} (-1)^{l(w)} e^{w(\rho)}},$$

where W is the Weyl group of $\mathfrak{g}$, ρ is the half-sum of positive roots of $\mathfrak{g}$ and $l(w)$ is the length of the element $w \in W$.

Recall that the dimensions of weight subspaces V_μ of the underlying space V of an irreducible finite-dimensional representation T_Λ of $U_q(\mathfrak{g})$ are called *multiplicities* of weights μ in T_Λ and are denoted by m^Λ_μ. By Proposition 11, they coincide with the corresponding weight multiplicities in the irreducible representation of $\mathfrak{g}$ with highest weight Λ. There exist several formulas for the calculation of weight multiplicities (see [Hum] or [Zhe]).

7.1.5 Separation of Elements of $U_q(\mathfrak{g})$

The aim of this subsection is to prove the following theorem.

Theorem 13. *Let $a \in U_q(\mathfrak{g})$. If $T(a) = 0$ for all irreducible finite-dimensional type 1 representations of $U_q(\mathfrak{g})$, then $a = 0$.*

Proof. Assume on the contrary that a nonzero element $a \in U_q(\mathfrak{g})$ is annihilated by all such representations T. As shown in Subsect. 6.1.5, a can be represented as

$$a = \sum\nolimits_{i,\mu,j} c_{i\mu j} f_i K_\mu e_j, \quad f_i \in U_q^{-\beta'(i)}(\mathfrak{n}_-), \ e_j \in U_q^{\beta(j)}(\mathfrak{n}_+), \ c_{i\mu j} \in \mathbb{C},$$

where only finitely many coefficients are nonvanishing. Let β_0 be the maximal element in the set of all $\beta(j)$ for which there exists a coefficient $c_{i\mu j} \neq 0$.

For $\Lambda \in P_+$, let T_Λ be the irreducible type 1 representation of $U_q(\mathfrak{g})$ on a space V_Λ with highest weight Λ and highest weight vector $\mathbf{e}_\Lambda$. If θ is the algebra automorphism of $U_q(\mathfrak{g})$ from Proposition 6.9, then $T_\Lambda^\theta := T_\Lambda \circ \theta$ is also an irreducible type 1 representation of $U_q(\mathfrak{g})$ on V_Λ. Since $\theta(K_i) = K_i^{-1}$ and $\theta(E_i) = F_i$, we have

$$T_\Lambda^\theta(K_\mu)\mathbf{e}_\Lambda = q^{-(\mu,\Lambda)}\mathbf{e}_\Lambda \quad \text{and} \quad T_\Lambda^\theta(F_i)\mathbf{e}_\Lambda = 0.$$

Let us denote the vector $\mathbf{e}_\Lambda$ and the vector space V_Λ considered for the representation T_Λ^θ by $\mathbf{e}_\Lambda^\theta$ and V_Λ^θ, respectively. Since $T_\Lambda^\theta \otimes T_{\Lambda'}$ is a direct sum of irreducible type 1 representations by Theorem 8, we have $(T_\Lambda^\theta \otimes T_{\Lambda'})(a) = 0$ for all $\Lambda, \Lambda' \in P_+$.

By the definition of the comultiplication of $U_q(\mathfrak{g})$, $\Delta(e_j) = e_j \otimes K_{\beta(j)} + b$, where b is a sum of terms $x \otimes yE_i$. Therefore, since $T_\Lambda(E_i)\mathbf{e}_\Lambda = 0$, we have $(T_\Lambda^\theta \otimes T_{\Lambda'})(b)(\mathbf{e}_\Lambda^\theta \otimes \mathbf{e}_{\Lambda'}) = 0$ and so

$$(T_\Lambda^\theta \otimes T_{\Lambda'})(e_j)(\mathbf{e}_\Lambda^\theta \otimes \mathbf{e}_{\Lambda'}) = q^{(\beta(j),\Lambda')} T_\Lambda^\theta(e_j)\mathbf{e}_\Lambda^\theta \otimes \mathbf{e}_{\Lambda'}.$$

Since $\Delta(K_\mu) = K_\mu \otimes K_\mu$, we obtain

$$(T_\Lambda^\theta \otimes T_{\Lambda'})(K_\mu e_j)(\mathbf{e}_\Lambda^\theta \otimes \mathbf{e}_{\Lambda'}) = q^{(\beta(j),\Lambda')+(\mu,\Lambda'-\Lambda+\beta(j))} T_\Lambda^\theta(e_j)\mathbf{e}_\Lambda^\theta \otimes \mathbf{e}_{\Lambda'}.$$

Further, we have $\Delta(f_i) = K_{\beta'(i)}^{-1} \otimes f_i + d$, where d is a sum of terms $xF_j \otimes y$. Thus we get

$$(T_\Lambda^\theta \otimes T_{\Lambda'})(f_i K_\mu e_j)(\mathbf{e}_\Lambda^\theta \otimes \mathbf{e}_{\Lambda'}) =$$
$$q^{(\beta(j),\Lambda')+(\mu,\Lambda'-\Lambda+\beta(j))-(\beta'(i),-\Lambda+\beta(j))} T_\Lambda^\theta(e_j)\mathbf{e}_\Lambda^\theta \otimes T_{\Lambda'}(f_i)\mathbf{e}_{\Lambda'} + g, \quad (5)$$

where g is a sum of terms from $(V_\Lambda^\theta)_{-\Lambda+\beta} \otimes V_{\Lambda'}$ with $\beta < \beta(j)$ (recall that $(V_\Lambda^\theta)_{-\Lambda+\beta}$ is the weight subspace of V_Λ^θ for the weight $-\Lambda + \beta$). Since β_0 is the maximal weight in the set $\{\beta(j)\}$, the term $(T_\Lambda^\theta \otimes T_{\Lambda'})(f_i K_\mu e_j)(\mathbf{e}_\Lambda^\theta \otimes \mathbf{e}_{\Lambda'})$ has a component in $(V_\Lambda^\theta)_{-\Lambda+\beta_0} \otimes V_{\Lambda'}$ only for $\beta(j) = \beta_0$ and in this case this component coincides with the right hand side of (5) when g is omitted. Thus, the canonical projection of $(T_\Lambda^\theta \otimes T_{\Lambda'})(a)(\mathbf{e}_\Lambda^\theta \otimes \mathbf{e}_{\Lambda'})$ onto $(V_\Lambda^\theta)_{-\Lambda+\beta_0} \otimes V_{\Lambda'}$ is

$$\sum_{\substack{i,\mu,j\\ \beta(j)=\beta_0}} c_{i\mu j} q^{(\beta_0,\Lambda')+(\mu,\Lambda'-\Lambda+\beta_0)-(\beta'(i),-\Lambda+\beta_0)} T_\Lambda^\theta(e_j)\mathbf{e}_\Lambda^\theta \otimes T_{\Lambda'}(f_i)\mathbf{e}_{\Lambda'}. \qquad (6)$$

Since $(T_\Lambda^\theta \otimes T_{\Lambda'})(a)(\mathbf{e}_\Lambda^\theta \otimes \mathbf{e}'_{\Lambda'}) = 0$, this projection is zero.

Let N be a positive integer such that

$$\beta_0 < N(\alpha_1 + \cdots + \alpha_l), \quad \beta'(i) < N(\alpha_1 + \cdots + \alpha_l)$$

for all i for which there exists a coefficient $c_{i\mu j} \neq 0$ in (6). Let P_N be the set of highest weights λ for $U_q(\mathfrak{g})$ such that $2(\lambda, \alpha_i)/(\alpha_i, \alpha_i) > N$, $i = 1, 2, \cdots, l$.

By Proposition 6.29, the mapping $e \to T_\Lambda^\theta(e)\mathbf{e}_\Lambda^\theta$ of $U_q^{\beta_0}(\mathfrak{n}_+)$ to $(V_\Lambda^\theta)_{-\Lambda+\beta_0}$ is bijective. Thus, the vectors $T_\Lambda^\theta(e_j)\mathbf{e}_\Lambda^\theta$ with $\beta(j) = \beta_0$ are linearly independent. Therefore, the vanishing of the sum (6) implies that

$$\sum\nolimits_{i,\mu} c_{i\mu j} q^{(\beta_0,\Lambda')+(\mu,\Lambda'-\Lambda+\beta_0)-(\beta'(i),-\Lambda+\beta_0)} T_{\Lambda'}(f_i)\mathbf{e}_{\Lambda'} = 0 \qquad (7)$$

for all $\Lambda \in P_N$ and all j with $\beta(j) = \beta_0$.

It is shown similarly that for all $\Lambda' \in P_N$ the vectors $T_{\Lambda'}(f_i)\mathbf{e}_{\Lambda'}$ that occur in (7) with $c_{i\mu j} \neq 0$ are linearly independent. The factors $q^{(\beta_0,\Lambda')-(\beta'(i),-\Lambda-\beta_0)}$ in (7) are independent of μ. If we cancel these factors, we get

$$\sum\nolimits_{\mu} c_{i\mu j} q^{(\mu,\beta_0-\Lambda)} q^{(\mu,\Lambda')} = 0 \qquad (8)$$

for all i, j with $\beta(j) = \beta_0$ and all $\Lambda, \Lambda' \in P_N$. For fixed Λ, we consider the left hand side of (8) as a linear combination of the distinct characters $\Lambda' \to q^{(\mu,\Lambda')}$ on the semigroup P_N. Now we apply Artin's theorem on the linear independence of characters which is also true for semigroups. Thus, all coefficients $c_{i\mu j} q^{(\mu,\beta_0-\Lambda)}$ in (8) vanish. Hence $c_{i\mu j} = 0$ for all i, j and μ with $\beta(j) = \beta_0$. This contradicts the choice of β_0 and completes the proof. □

The counterpart to Theorem 13 in the h-adic case is

Theorem 13′. *Let $a \in U_h(\mathfrak{g})$. If $T(a) = 0$ for all indecomposable finite-dimensional representations T of $U_h(\mathfrak{g})$, then $a = 0$.*

Proof. The result follows from the discussion preceding Proposition 10 and from the corresponding result for representations of the universal enveloping algebra of the Lie algebra $\mathfrak{g}$ (see [Dix], Theorem 2.5.7). □

7.1.6 The Quantum Trace of Finite-Dimensional Representations

If T is a finite-dimensional representation of a *cocommutative* Hopf algebra $\mathcal{A}$, then we have

$$\operatorname{Tr} T(\operatorname{ad_L}(a)b) = \operatorname{Tr} \sum T(a_{(1)})T(bS(a_{(2)}))$$

$$= \operatorname{Tr} \sum T(a_{(2)})T(bS(a_{(1)})) = \operatorname{Tr} \sum T(bS(a_{(1)}))T(a_{(2)}) = \varepsilon(a) \operatorname{Tr} T(b)$$

for $a, b \in \mathcal{A}$, where Tr denotes the "usual" trace of a linear mapping on a finite-dimensional vector space. In this subsection we give a generalization of this relation to the Drinfeld–Jimbo algebra $\breve{U}_q(\mathfrak{g})$.

Let ρ denote the half-sum of positive roots of the Lie algebra $\mathfrak{g}$. Suppose that T is a finite-dimensional representation of the algebra $\breve{U}_q(\mathfrak{g})$. We define the *left* and *right quantum traces* of T by

$$\mathrm{Tr}_{q,L}\, T(a) = \mathrm{Tr}\, T(aK_{2\rho}) \quad \text{and} \quad \mathrm{Tr}_{q,R}\, T(a) = \mathrm{Tr}\, T(aK_{2\rho}^{-1}), \quad a \in \breve{U}_q(\mathfrak{g}).$$

The main properties of these quantum traces are given by

Proposition 14. (i) *For arbitrary elements $a, b \in \breve{U}_q(\mathfrak{g})$ we have*

$$\mathrm{Tr}_{q,L}\, T(\mathrm{ad}_L(a)b) = \varepsilon(a)\, \mathrm{Tr}_{q,L}\, T(b), \quad \mathrm{Tr}_{q,R}\, T(\mathrm{ad}_R(a)b) = \varepsilon(a)\, \mathrm{Tr}_{q,R}\, T(b),$$

$$\mathrm{Tr}_{q,L}\, T(ab) = \mathrm{Tr}_{q,L}\, T(ba), \quad \mathrm{Tr}_{q,R}\, T(ab) = \mathrm{Tr}_{q,R}\, T(ba).$$

(ii) *If T is a direct sum of type 1 representations of $\breve{U}_q(\mathfrak{g})$, then*

$$\mathrm{Tr}_{q,L}\, T(1) = \mathrm{Tr}_{q,R}\, T(1).$$

Proof. (i): We prove the first formula. Since ad_L is an algebra homomorphism by Proposition 1.14, it suffices to verify this equality for the generators $a = K_i, E_i, F_i$ of $\breve{U}_q(\mathfrak{g})$. First we let $a = E_i$. Since $(\rho, \alpha_i)/(\alpha_i, \alpha_i) = 1$ (see, for example, [Hum]), we have $(\rho, \alpha_i) = (\alpha_i, \alpha_i)$. Thus, from (6.30) we obtain

$$K_i^{-1} K_{2\rho} E_i = q^{-(\alpha_i, \alpha_i)/2 + (\rho, \alpha_i)} E_i K_i^{-1} K_{2\rho} = q_i E_i K_{2\rho} K_i^{-1}.$$

Using the formulas (6.34)–(6.37) and the preceding relation we compute

$$\begin{aligned}
\mathrm{Tr}_{q,L}\, T(\mathrm{ad}_L(E_i)b) &= \mathrm{Tr}\, T((\mathrm{ad}_L(E_i)b)K_{2\rho}) \\
&= \mathrm{Tr}\, T(E_i)T(bK_i^{-1}K_{2\rho}) - q_i \mathrm{Tr}\, T(K_i^{-1})T(bE_iK_{2\rho}) \\
&= \mathrm{Tr}\, T(bK_i^{-1}K_{2\rho})T(E_i) - q_i \mathrm{Tr}\, T(bE_iK_{2\rho})T(K_i^{-1}) \\
&= \mathrm{Tr}\, T(b)T(K_i^{-1}K_{2\rho}E_i - q_iE_iK_{2\rho}K_i^{-1}) = 0 = \varepsilon(E_i)\, \mathrm{Tr}_{q,L}\, T(b).
\end{aligned}$$

The proof for the generator $a = F_i$ is similar. In the case $a = K_i$ we get

$$\begin{aligned}
\mathrm{Tr}_{q,L}\, T(\mathrm{ad}_L(K_i)b) &= \mathrm{Tr}\, T(K_i)T(bK_i^{-1}K_{2\rho}) \\
&= \mathrm{Tr}\, T(bK_i^{-1}K_{2\rho})T(K_i) = \varepsilon(K_i)\, \mathrm{Tr}_{q,L}\, T(b).
\end{aligned}$$

The proof of the second formula is similar. The third and the fourth formulas are obvious.

(ii): It suffices to assume that T is an irreducible type 1 representation. Then T is a highest weight representation. Let $T_{\mathfrak{g}}$ be the corresponding representation of the Lie algebra $\mathfrak{g}$ with the same highest weight. If G is the simply connected connected complex linear group with Lie algebra $\mathfrak{g}$, then there is a representation T_G of G which is the exponential of $T_{\mathfrak{g}}$. From the structure of the representations T and $T_{\mathfrak{g}}$ it follows that there is an element

$h \in \exp \mathfrak{h} \subset G$ such that $T(K_{2\rho}) = T_G(h)$. The Weyl group W of $\mathfrak{g}$ can be considered as a subgroup of G. If w_0 is the longest element of W, then we have $K_{2\rho}^{-1} = K_{-2\rho} = K_{w_0(2\rho)}$ and hence $h = w_0 h^{-1} w_0^{-1}$ (see, for instance, [Zhe]). Thus we obtain

$$\begin{aligned}\operatorname{Tr} T(K_{2\rho}) &= \operatorname{Tr} T_G(h) = \operatorname{Tr} T_G(w_0) T_G(h^{-1}) T_G(w_0)^{-1} \\ &= \operatorname{Tr} T_G(h)^{-1} = \operatorname{Tr} T(K_{2\rho})^{-1} = \operatorname{Tr} T(K_{-2\rho}),\end{aligned}$$

which gives the assertion of (ii). □

Clearly, we have

$$\operatorname{Tr}_{q,L} \big(\bigoplus\nolimits_i T_i\big)(a) = \sum\nolimits_i \operatorname{Tr}_{q,L} T_i(a), \quad \operatorname{Tr}_{q,R} \big(\bigoplus\nolimits_i T_i\big)(a) = \sum\nolimits_i \operatorname{Tr}_{q,R} T_i(a).$$

Since $\Delta(K_{2\rho}) = K_{2\rho} \otimes K_{2\rho}$, it follows at once from the definition of the tensor product $T_1 \otimes T_2$ of two representations T_1 and T_2 (see Subsect. 1.3.1) that

$$\operatorname{Tr}_{q,L} (T_1 \otimes T_2)(a) = \sum \operatorname{Tr}_{q,L} T_1(a_{(1)}) \operatorname{Tr}_{q,L} T_2(a_{(2)}),$$

$$\operatorname{Tr}_{q,R} (T_1 \otimes T_2)(a) = \sum \operatorname{Tr}_{q,R} T_1(a_{(1)}) \operatorname{Tr}_{q,R} T_2(a_{(2)}).$$

If T is a direct sum of type 1 representations of $\breve{U}_q(\mathfrak{g})$, then the number

$$\dim_q T := \operatorname{Tr}_{q,L} T(1) = \operatorname{Tr}_{q,R} T(1)$$

is called the *quantum dimension* of T. If T' and T'' are two such representations, then the preceding formulas yield

$$\dim_q (T_1 \otimes T_2) = (\dim_q T_1)(\dim_q T_2).$$

Example 1. Let $T_{1,l}$ be the type 1 irreducible $(2l+1)$-dimensional representation of $\breve{U}_q(\mathrm{sl}_2)$ from Theorem 3.13. Since $K_{2\rho} = K$ for $\breve{U}_q(\mathrm{sl}_2)$, we get

$$\dim_q T_{1,l} = \sum_{i=0}^{2l} q^{l-i} = \frac{q^{(2l+1)/2} - q^{-(2l+1)/2}}{q^{-1/2} - q^{-1/2}} = [2l+1]_{q^{1/2}}. \qquad \triangle$$

7.2 Tensor Products of Representations

The investigation of tensor products and Clebsch–Gordan coefficients for representations of Drinfeld–Jimbo algebras $U_q(\mathfrak{g})$ of higher ranks is much more complicated than in the case of $U_q(\mathrm{sl}_2)$. The reason is that multiple irreducible representations appear in the decompositions of tensor products. By Proposition 12, the multiplicities of irreducible components in tensor products are determined by the same formulas as in the classical case. We develop these results on multiplicities in Subsect. 7.2.1. Let us assume that all representations in this section are of type 1.

7.2.1 Multiplicities in Tensor Products of Representations

By Theorem 8, finite-dimensional representations of $U_q(\mathfrak{g})$ are completely reducible. Therefore, the tensor product $T_\Lambda \otimes T_{\Lambda'}$ of two irreducible finite-dimensional representations of $U_q(\mathfrak{g})$ decomposes into a direct sum of irreducible components:

$$T_\Lambda \otimes T_{\Lambda'} = \bigoplus\nolimits_\lambda m_\lambda^{\Lambda,\Lambda'} T_\lambda, \tag{9}$$

where $m_\lambda^{\Lambda,\Lambda'}$ denotes the multiplicity of the irreducible representation T_λ in $T_\Lambda \otimes T_{\Lambda'}$. This multiplicity $m_\lambda^{\Lambda,\Lambda'}$ can be expressed in terms of weight multiplicities of one of the representations T_Λ and $T_{\Lambda'}$. In order to give the corresponding formulas we introduce some notation. If μ is a weight of some representation of $\mathfrak{g}$, let $\{\mu\}$ be the dominant weight lying on the orbit $W\mu$. The element w of W for which $w\mu = \{\mu\}$ is denoted by $w_{\{\mu\}}$. If m_μ^Λ is the multiplicity of the weight μ in the representation T_Λ, then we have (see [Kl1])

$$m_\lambda^{\Lambda,\Lambda'} = \sum\nolimits_\mu m_\mu^\Lambda (\det w_{\{\mu+\Lambda'+\rho\}}), \tag{10}$$

where $\det w = (-1)^{l(w)}$ and the summation is over all weights μ of T_Λ such that $\{\mu+\Lambda'+\rho\} = \lambda+\rho$. Since any such weight μ can be represented in the form $w(\lambda+\rho) - \Lambda' - \rho$, $w \in W$, formula (10) can be written as

$$m_\lambda^{\Lambda,\Lambda'} = \sum\nolimits_{w\in W} (-1)^{l(w)} m^\Lambda_{w(\lambda+\rho)-\Lambda'-\rho} = \sum\nolimits_{w\in W} (-1)^{l(w)} m^\Lambda_{\lambda+\rho-w(\Lambda'+\rho)}.$$

From (10) one derives the following

Proposition 15. *The decomposition of the tensor product $T_\Lambda \otimes T_{\Lambda'}$ of two irreducible finite-dimensional representations T_Λ and $T_{\Lambda'}$ of $U_q(\mathfrak{g})$ is given by*

$$T_\Lambda \otimes T_{\Lambda'} = \sum\nolimits_{\mu\in\Pi_\Lambda} m_\mu^\Lambda \beta_{\mu+\Lambda'+\rho} T_{\{\mu+\Lambda'+\rho\}-\rho}, \tag{11}$$

where Π_Λ is the set of all weights of the representation T_Λ and β_ν is the number defined as follows: $\beta_\nu = 0$ if there exists $w \in W$, $w \neq 1$, such that $w\nu = \nu$ and $\beta_\nu = \det w_{\{\nu\}}$ if no such element exists.

Note that the relation $w(\mu+\Lambda'+\rho) = \mu+\Lambda'+\rho$, $w \neq 1$, means that the linear form $\{\mu+\Lambda'+\rho\}-\rho$ is not dominant, hence this linear form cannot be a highest weight and the symbol $T_{\{\mu+\Lambda'+\rho\}-\rho}$ has no meaning. But in this case the corresponding coefficient β in (11) is equal to 0, so the summand in (11) does not occur. In (11), there are negative and nonnegative coefficients. Adding the coefficients of the same irreducible representation we obtain a nonnegative number which is the multiplicity of this representation.

Formula (11) shows that the set of weights of one of the irreducible representations T_Λ and $T_{\Lambda'}$ already determines the decomposition (9). Some special cases of the formula (11) are stated separately in

Corollary 16. *If the form $w\Lambda + \Lambda' + \rho$ is dominant for any $w \in W$, then*

$$T_\Lambda \otimes T_{\Lambda'} = \bigoplus_\mu m_\mu^\Lambda T_{\mu+\Lambda'}, \tag{12}$$

where the summation is over all weights $\mu \in \Pi_\Lambda$ for which $\mu+\Lambda'$ is dominant. If the linear form $w\Lambda + \Lambda'$ is dominant for any $w \in W$, then the summation in (12) is over all weights $\mu \in \Pi_\Lambda$. In particular, if all forms $w\Lambda+\Lambda'$, $w \in W$, are dominant, then to every weight $\mu \in \Pi_\Lambda$ there corresponds a representation $T_{\mu+\Lambda'}$ on the right hand side of (9) and the multiplicity of $T_{\mu+\Lambda'}$ in $T_\Lambda \otimes T_{\Lambda'}$ is equal to the multiplicity of the weight μ in T_Λ.

If the weights Λ are given by the numbers $(\lambda_1, \cdots, \lambda_l)$ with $\lambda_i = 2(\Lambda, \alpha_i)/(\alpha_i, \alpha_i)$, then a direct calculation shows that for the classical complex Lie algebras A_l, B_l, C_l, D_l the forms $w\Lambda+\Lambda'$ are dominant for all $w \in W$ if and only if

$$\begin{aligned}
&\text{for } A_l: && \lambda_i' \geq \lambda_1 + \lambda_2 + \cdots + \lambda_l, && 1 \leq i \leq l,\\
&\text{for } B_l: && \lambda_i' \geq \lambda_1 + 2\lambda_2 + \cdots + 2\lambda_{l-1} + \lambda_l, && 1 \leq i \leq l-1,\\
& && \lambda_l' \geq \lambda_1 + 2\lambda_2 + \cdots + 2\lambda_{l-1} + \lambda_l, && \\
&\text{for } C_l: && \lambda_i' \geq \lambda_1 + 2\lambda_2 + \cdots + 2\lambda_{l-1} + 2\lambda_l, && 1 \leq i \leq l-1,\\
& && \lambda_l' \geq \lambda_1 + \lambda_2 + \cdots + \lambda_{l-1} + \lambda_l, && \\
&\text{for } D_l: && \lambda_i' \geq \lambda_1 + 2\lambda_2 + \cdots + 2\lambda_{l-2} + \lambda_{l-1} + \lambda_l, && 1 \leq i \leq l.
\end{aligned}$$

If the decomposition (9) is of the form (12), then the set of highest weights λ of irreducible representations which occur in (12) is contained in the set $\Pi_\Lambda + \Lambda'$. This statement is also true in the general case.

Proposition 17. *The highest weights of irreducible representations in the decomposition (9) belong to the set $\Pi_\Lambda + \Lambda'$. Moreover, the multiplicity of an irreducible representation $T_{\mu+\Lambda'}$, $\mu \in \Pi_\Lambda$, in $T_\Lambda \otimes T_{\Lambda'}$ does not exceed the multiplicity of the weight μ in T_Λ.*

Using the embeddings of irreducible finite-dimensional representations of the Lie algebra $\mathfrak{g}$ into its infinite-dimensional representations of the principal nonunitary series one proves the following result (see [VK2], Sect. 4.2).

Proposition 18. *The multiplicity $m_\lambda^{\Lambda\Lambda'}$ of an irreducible representation T_λ in the decomposition (9) does not exceed the multiplicity of the weight $\Lambda - \overline{\Lambda'}$ in T_λ, where $\overline{\Lambda'}$ is the highest weight of the contragredient representation $T_{\overline{\Lambda'}}$ of $T_{\Lambda'}$. Moreover, $m_\lambda^{\Lambda,\Lambda'} = m_{\overline{\Lambda'}}^{\Lambda,\overline{\lambda}} = m_{\overline{\Lambda}}^{\Lambda',\overline{\lambda}}$.*

Propositions 17 and 18 admit the following useful corollary.

Corollary 19. (i) *If all weight multiplicities of an irreducible representation T_λ are at most one, then T_λ appears in the decomposition (9) of any tensor product $T_\Lambda \otimes T_{\Lambda'}$ with multiplicity less than or equal to one.*

(ii) *If the weight multiplicities of one of the representations T_Λ and $T_{\Lambda'}$ do not exceed 1, then all multiplicities of irreducible representations in the decomposition (9) are at most one.*

Next we develop formulas for the decomposition of the tensor product of an irreducible representation and the first fundamental representation (also called the *vector representation*) of a Drinfeld–Jimbo algebra. We characterize the highest weights by the numbers m_i given in (2)–(4). Let $T_{\mathbf{m}}$ be an irreducible representation of $U_q(\mathfrak{g})$, $\mathfrak{g} = A_{l-1}, B_l, C_l, D_l$, with highest weight $\mathbf{m} \equiv (m_1, m_2, \cdots, m_l)$ and let T_1 be the first fundamental representation of $U_q(\mathfrak{g})$ (that is, with highest weight $(1, 0, \cdots, 0)$). Then the tensor product $T_{\mathbf{m}} \otimes T_1$ decomposes into irreducible components as

$$T_{\mathbf{m}} \otimes T_1 = \bigoplus\nolimits_{i=1}^{l} T_{\mathbf{m}+\mathbf{e}_i} \tag{13}$$

for $U_q(\mathrm{sl}_l)$, where $\mathbf{e}_i = (0, \cdots, 0, 1, 0, \cdots, 0)$ (1 is in the i-th place), and as

$$T_{\mathbf{m}} \otimes T_1 = \bigoplus\nolimits_{i=1}^{l} T_{\mathbf{m}+\mathbf{e}_i} \oplus \bigoplus\nolimits_{i=1}^{l} T_{\mathbf{m}-\mathbf{e}_i} \tag{14}$$

for $U_q(\mathrm{so}_{2l+1})$ if $m_l = 0$ and also for $U_q(\mathrm{sp}_{2l})$ and $U_q(\mathrm{so}_{2l})$. If $m_l \neq 0$, then for $U_q(\mathrm{so}_{2l+1})$ we have

$$T_{\mathbf{m}} \otimes T_1 = T_{\mathbf{m}} \oplus \bigoplus\nolimits_{i=1}^{l} T_{\mathbf{m}+\mathbf{e}_i} \oplus \bigoplus\nolimits_{i=1}^{l} T_{\mathbf{m}-\mathbf{e}_i}. \tag{15}$$

If for some $\mathbf{m} \pm \mathbf{e}_i$ in (13)–(15) the dominantness condition in (2)–(4) is not fulfilled, then the corresponding representation $T_{\mathbf{m}\pm\mathbf{e}_i}$ must be omitted.

Let $\mathfrak{P}(\mathfrak{g})$ be the set of highest weights of irreducible representations of $U_q(\mathfrak{g})$ which are contained in some r-fold tensor product $T_1^{\otimes r}$, $r \in \mathbb{N}_0$, of the vector representation. Using the decompositions (13)–(15), we derive the following

Proposition 20. *If the highest weights are given by the numbers* $\mathbf{m} := (m_1, m_2, \cdots, m_l)$ *satisfying the corresponding conditions (2)–(4), then*

$$\mathfrak{P}(\mathrm{sl}_l) = P_+, \quad \mathfrak{P}(\mathrm{sp}_{2l}) = P_+,$$

$$\mathfrak{P}(\mathrm{so}_{2l+1}) = \{\mathbf{m} \in P_+ \mid m_i \in \mathbb{Z}\}, \quad \mathfrak{P}(\mathrm{so}_{2l}) = \{\mathbf{m} \in P_+ \mid m_i \in \mathbb{Z}\}.$$

Note that in terms of $\Lambda = (\lambda_1, \lambda_2, \cdots, \lambda_l)$, $\lambda_i = 2(\lambda, \alpha_i)/(\alpha_i, \alpha_i)$, the sets $\mathfrak{P}(\mathrm{so}_N)$ are described as

$$\mathfrak{P}(\mathrm{so}_{2l+1}) = \{\Lambda \in P_+ \mid \lambda_l \in 2\mathbb{Z}\}, \quad \mathfrak{P}(\mathrm{so}_{2l}) = \{\Lambda \in P_+ \mid \lambda_{l-1} + \lambda_l \in 2\mathbb{Z}\}.$$

If $\mathfrak{g}$ is a classical simple Lie algebra, a closer look at the proof of Theorem 13 shows that it remains valid if we replace the set of all type 1 irreducible finite-dimensional representations of $U_q(\mathfrak{g})$ by the set of representations T_Λ, $\Lambda \in \mathfrak{P}(\mathfrak{g})$. Therefore, we get

Proposition 21. *Let* $\mathfrak{g}$ *be one of the complex simple Lie algebras* A_l, B_l, C_l, D_l *and let* $a \in U_q(\mathfrak{g})$. *If* $T_\Lambda(a) = 0$ *for all irreducible representations* T_Λ *with* $\Lambda \in \mathfrak{P}(\mathfrak{g})$, *then* $a = 0$.

7.2.2 Clebsch–Gordan Coefficients

Let T_Λ and $T_{\Lambda'}$ be type 1 irreducible representations of $U_q(\mathfrak{g})$ acting on finite-dimensional vector spaces V and V', respectively. As already noted above, their tensor product decomposes into a direct sum of irreducible representations:

$$T_\Lambda \otimes T_{\Lambda'} = \bigoplus\nolimits_s m_s^{\Lambda,\Lambda'} T_{\Lambda_s}. \tag{16}$$

The corresponding decomposition of the carrier space $V \otimes V'$ is $V \otimes V' = \bigoplus_s m_s^{\Lambda,\Lambda'} V_s$, where $m_s V_s := V_s \oplus \cdots \oplus V_s$ (m_s times). In order to distinguish different subspaces V_s with the same index s we equip them with an additional index r and write V_{sr}, $r = 1, 2, \cdots, m_s^{\Lambda,\Lambda'}$.

The Clebsch–Gordan coefficients of the tensor product (16) will be defined as for the algebra $U_q(\mathrm{sl}_2)$ (see Subsect. 3.4.2), but now multiple irreducible representations may appear in the decomposition (16). As in Subsect. 3.4.2, we consider two bases of the space $V \otimes V'$. The first one consists of the vectors $\mathbf{e}_i \otimes \mathbf{e}'_j$, $i = 1, 2, \cdots, \dim V$, $j = 1, 2, \cdots, \dim V'$, where $\{\mathbf{e}_i\}$ and $\{\mathbf{e}'_j\}$ are bases of V and V', respectively. The second is formed by bases $\{\mathbf{e}_k^{sr}\}$ of the subspaces V_{sr}. We suppose that the bases $\{\mathbf{e}_k^{sr}\}$, $r = 1, 2, \cdots, m_s^{\Lambda,\Lambda'}$, with fixed s are such that the representations T_{Λ_s} are given by the same matrices with respect to these bases. Both bases $\{\mathbf{e}_i \otimes \mathbf{e}'_j\}$ and $\{\mathbf{e}_k^{sr}\}$ are connected by an invertible matrix U with complex entries $\langle \mathbf{e}_i, \mathbf{e}'_j \mid \mathbf{e}_k^{sr} \rangle$ such that

$$\mathbf{e}_k^{sr} = \sum\nolimits_{i,j} \langle \mathbf{e}_i, \mathbf{e}'_j \mid \mathbf{e}_k^{sr} \rangle \mathbf{e}_i \otimes \mathbf{e}'_j. \tag{17}$$

The numbers $\langle \mathbf{e}_i, \mathbf{e}'_j \mid \mathbf{e}_k^{sr} \rangle$ are called the *Clebsch–Gordan coefficients* (briefly, the CGC's) of the tensor product $T_\Lambda \otimes T_{\Lambda'}$.

Let q be real. Then we can assume that V and V' are Hilbert spaces and $\{\mathbf{e}_i\}$, $\{\mathbf{e}'_j\}$ and $\{\mathbf{e}_k^{sr}\}$ are orthonormal bases. Then the matrix U is unitary and $U^{-1} = U^*$ transforms the basis $\{\mathbf{e}_k^{sr}\}$ into $\{\mathbf{e}_i \otimes \mathbf{e}'_j\}$. The entries $\overline{\langle \mathbf{e}_i, \mathbf{e}'_j \mid \mathbf{e}_k^{sr} \rangle}$ of the matrix U^* will be denoted by $\langle \mathbf{e}_k^{sr} \mid \mathbf{e}_i, \mathbf{e}'_j \rangle$. The CGC's $\langle \mathbf{e}_i, \mathbf{e}'_j \mid \mathbf{e}_k^{sr} \rangle$ are then equal to the scalar products of basis vectors, that is, we have

$$\langle \mathbf{e}_i, \mathbf{e}'_j \mid \mathbf{e}_k^{sr} \rangle = (\mathbf{e}_i \otimes \mathbf{e}'_j, \mathbf{e}_k^{sr}), \qquad \langle \mathbf{e}_k^{sr} \mid \mathbf{e}_i, \mathbf{e}'_j \rangle = (\mathbf{e}_k^{sr}, \mathbf{e}_i \otimes \mathbf{e}'_j).$$

Since the matrix U of CGC's is unitary, one has the orthogonality relations

$$\sum\nolimits_{i,j} \langle \mathbf{e}_k^{sr} \mid \mathbf{e}_i, \mathbf{e}'_j \rangle \langle \mathbf{e}_i, \mathbf{e}'_j \mid \mathbf{e}_{k'}^{s'r'} \rangle = \delta_{ss'} \delta_{rr'} \delta_{kk'}, \tag{18}$$

$$\sum\nolimits_{s,k,r} \langle \mathbf{e}_i, \mathbf{e}'_j \mid \mathbf{e}_k^{sr} \rangle \langle \mathbf{e}_k^{sr} \mid \mathbf{e}_{i'}, \mathbf{e}'_{j'} \rangle = \delta_{ii'} \delta_{jj'}. \tag{19}$$

As in the case of the quantum algebra $U_q(\mathrm{sl}_2)$ we then have the following relations between matrix coefficients of the representations and CGC's:

$$t_{ij}^{\Lambda} t_{i'j'}^{\Lambda'} = \sum\nolimits_{\Lambda_s,r,k,k'} \langle \mathbf{e}_i, \mathbf{e}'_{i'} \mid \mathbf{e}_k^{sr} \rangle t_{kk'}^{\Lambda_s} \langle \mathbf{e}_{k'}^{sr} \mid \mathbf{e}_j, \mathbf{e}'_{j'} \rangle, \tag{20}$$

$$t_{kk'}^{\Lambda_s} = \sum\nolimits_{i,j,i',j'} \langle \mathbf{e}_k^{sr} \mid \mathbf{e}_i, \mathbf{e}'_{i'} \rangle t_{ij}^{\Lambda} t_{i'j'}^{\Lambda'} \langle \mathbf{e}_j, \mathbf{e}'_{j'} \mid \mathbf{e}_{k'}^{sr} \rangle. \tag{21}$$

For general complex q, the above formulas (20) and (21) remain valid if the numbers $\langle \mathbf{e}_k^{sr} \mid \mathbf{e}_i, \mathbf{e}'_{i'} \rangle$ therein are replaced by the corresponding entries of the matrix U^{-1}.

Repeated application of formula (21) leads to the following

Proposition 22. *The matrix coefficients t_{ij}^{Λ} of any irreducible representation T_Λ with highest weight Λ from the set $\mathfrak{P}(\mathfrak{g})$, described in Proposition 20, are polynomials of the matrix coefficients of the vector representation.*

7.3 Representations of $\breve{U}_q(\mathrm{gl}_n)$ for q not a Root of Unity

7.3.1 The Hopf Algebra $\breve{U}_q(\mathrm{gl}_n)$

The aim of this subsection is to introduce the Hopf algebra $\breve{U}_q(\mathrm{gl}_n)$.

The algebra $\breve{U}_q(\mathrm{gl}_n)$ is generated by elements E_i, F_i, K_j, K_j^{-1}, $i = 1, 2, \cdots n-1$, $j = 1, 2, \cdots, n$, subject to the relations

$$K_iK_j = K_jK_i, \quad K_iK_i^{-1} = K_i^{-1}K_i = 1,$$

$$K_iE_jK_i^{-1} = q^{\delta_{ij}/2}q^{-\delta_{i,j+1}/2}E_j, \quad K_iF_jK_i^{-1} = q^{-\delta_{ij}/2}q^{\delta_{i,j+1}/2}F_j,$$

$$[E_i, F_r] = \delta_{ir}\frac{K_i^2K_{i+1}^{-2} - K_i^{-2}K_{i+1}^2}{q - q^{-1}}, \quad [E_i, E_j] = [F_i, F_j] = 0, \quad |i-j| \geq 2,$$

$$E_i^2E_{i\pm1} - (q+q^{-1})E_iE_{i\pm1}E_i + E_{i\pm1}E_i^2 = 0,$$

$$F_i^2F_{i\pm1} - (q+q^{-1})F_iF_{i\pm1}F_i + F_{i\pm1}F_i^2 = 0.$$

The algebra $\breve{U}_q(\mathrm{gl}_n)$ is a Hopf algebra with structure maps given by

$$\Delta(E_i) = E_i \otimes K_iK_{i+1}^{-1} + K_i^{-1}K_{i+1} \otimes E_i,$$

$$\Delta(F_i) = F_i \otimes K_iK_{i+1}^{-1} + K_i^{-1}K_{i+1} \otimes F_i,$$

$$\Delta(K_i) = K_i \otimes K_i, \quad \varepsilon(E_i) = \varepsilon(F_i) = 0, \quad \varepsilon(K_i) = 1.$$

$$S(E_i) = -qE_i, \quad S(F_i) = -q^{-1}F_i, \quad S(K_i) = K_i^{-1}.$$

For $\breve{U}_q(\mathrm{gl}_n)$ we also have triangular decompositions

$$\breve{U}_q(\mathrm{gl}_n) = U_q(\mathfrak{n}_+) \otimes \breve{U}_q(\mathfrak{h}) \otimes U_q(\mathfrak{n}_-) = U_q(\mathfrak{n}_-) \otimes \breve{U}_q(\mathfrak{h}) \otimes U_q(\mathfrak{n}_+), \tag{22}$$

where $U_q(\mathfrak{n}_+)$ and $U_q(\mathfrak{n}_-)$ are the subalgebras of $\breve{U}_q(\mathrm{gl}_n)$ generated by E_i, $i = 1, 2, \cdots, n-1$, and by F_i, $i = 1, 2, \cdots, n-1$, respectively, and $\breve{U}_q(\mathfrak{h})$ is generated by K_i, $i = 1, 2, \cdots, n$.

Root vectors of $\breve{U}_q(\mathrm{gl}_n)$ corresponding to positive and negative roots of the Lie algebra $\mathrm{gl}(n, \mathbb{C})$ can be introduced by means of the braid group action (see Subsect. 6.2.1). In the present case we prefer to define them explicitly. We set $E_{i,i+1} := E_i$ and $E_{i+1,i} := F_i$. Then the formulas

$$E_{i,j+1} = [E_{ij}, E_{j,j+1}]_q \equiv E_{ij}E_{j,j+1} - qE_{j,j+1}E_{ij}, \qquad i<j,$$
$$E_{j+1,i} = [E_{j+1,j}, E_{ji}]_{q^{-1}} \equiv E_{j+1,j}E_{ji} - q^{-1}E_{ji}E_{j+1,j}, \qquad i<j,$$

determine recursively elements E_{ij} and E_{ji}, $1 \le i < j \le n$, of $\breve{U}_q(\mathrm{gl}_n)$. A direct computation shows that

$$K_jE_{jk}K_j^{-1} = q^{1/2}E_{jk}, \quad K_kE_{jk}K_k^{-1} = q^{-1/2}E_{jk},$$

that is, the elements E_{jk}, $j \neq k$, indeed have properties of root vectors. If $i<k<l$ or $i>k>l$, then we have

$$[E_{ik}, E_{ki}] = \frac{K_i^2K_k^{-2} - K_i^{-2}K_k^2}{q-q^{-1}},$$
$$E_{ik}E_{kl}^2 - (q+q^{-1})E_{kl}E_{ik}E_{kl} + E_{kl}^2E_{ik} \equiv [[E_{ik}, E_{kl}]_q, E_{kl}]_{q^{-1}} = 0.$$

7.3.2 Finite-Dimensional Representations of $\breve{U}_q(\mathrm{gl}_n)$

Since $K_1K_2\cdots K_n$ belongs to the center of $\breve{U}_q(\mathrm{gl}_n)$, the results of Subsects. 7.1.1–4 can be extended to the algebra $\breve{U}_q(\mathrm{gl}_n)$ as follows.

Theorem 23. *Let T be a representation of $\breve{U}_q(\mathrm{gl}_n)$ on a finite-dimensional vector space V such that its restriction to the subalgebra $\breve{U}_q(\mathfrak{h})$ is completely reducible. Then we have:*

(i) *V is a direct sum of weight subspaces with respect to $\breve{U}_q(\mathfrak{h})$.*

(ii) *The representation T is completely reducible.*

(iii) *If T is irreducible, then it is a highest weight representation and the highest weight subspace is one-dimensional.*

(iv) *If T is irreducible, then its highest weight is of the form $\omega\cdot q^{\mathbf{m}} = (\omega_1q^{m_1},\cdots,\omega_nq^{m_n})$, where $\omega_i \in \{1,-1,\mathrm{i},-\mathrm{i}\}$ and $\mathbf{m} = (m_1,\cdots,m_n)$ is a highest weight of an irreducible finite-dimensional representation of the Lie algebra $\mathrm{gl}(n,\mathbb{C})$. Moreover, every weight of this form is the highest weight of an irreducible finite-dimensional representation of $\breve{U}_q(\mathrm{gl}_n)$.*

(v) *If two irreducible finite-dimensional representations of $\breve{U}_q(\mathrm{gl}_n)$ have the same highest weights, then they are equivalent.*

The highest weights of irreducible finite-dimensional representations of $\mathrm{gl}(n,\mathbb{C})$ are given by n integers $\mathbf{m} = (m_1,\cdots,m_n)$ such that

$$m_1 \ge m_2 \ge \cdots \ge m_n.$$

As above, the irreducible representations $T_{\mathbf{m}}$ of $\breve{U}_q(\mathrm{gl}_n)$ with highest weights $q^{\mathbf{m}} = (q^{m_1},\cdots,q^{m_n})$ are called *representations of type 1*.

The characters of finite-dimensional representations of $\breve{U}_q(\mathrm{gl}_n)$ can be defined as in Subsect. 7.1.4. They coincide with the characters of the corresponding representations of the Lie algebra $\mathrm{gl}(n,\mathbb{C})$ and the analog of Proposition 12 is also true for $\breve{U}_q(\mathrm{gl}_n)$.

7.3.3 Gel'fand–Tsetlin Bases and Explicit Formulas for Representations

The Gel'fand–Tsetlin bases of carrier spaces of irreducible representations of $\breve{U}_q(\mathrm{gl}_n)$ are formed by successive restrictions of the representations to the subalgebras $\breve{U}_q(\mathrm{gl}_{n-1})$, $\breve{U}_q(\mathrm{gl}_{n-2})$, $\cdots, \breve{U}_q(\mathrm{gl}_1) \equiv U(\mathrm{gl}_1)$. From the character theory we know that the decomposition of the representation $T_{\mathbf{m}}$ of $U_q(\mathrm{gl}_n)$ into irreducible representations of $\breve{U}_q(\mathrm{gl}_{n-1})$ is the same as for the corresponding representation $T_{\mathbf{m}}$ of $\mathrm{gl}(n,\mathbb{C})$. Hence the restriction of $T_{\mathbf{m}}$, $\mathbf{m} = (m_1, \cdots, m_n)$, to $\breve{U}_q(\mathrm{gl}_{n-1})$ decomposes into the irreducible representations $T_{\mathbf{m}_{n-1}}$, $\mathbf{m}_{n-1} = (m_{1,n-1}, \cdots, m_{n-1,n-1})$, such that

$$m_1 \geq m_{1,n-1} \geq m_2 \geq m_{2,n-1} \geq \cdots \geq m_{n-1,n-1} \geq m_n \tag{23}$$

and each of these representations enters into the decomposition exactly once. Since the irreducible representations of $U(\mathrm{gl}_1)$ are one-dimensional, we obtain a basis of the carrier space $V_{\mathbf{m}}$ of the representation $T_{\mathbf{m}}$ of $\breve{U}_q(\mathrm{gl}_n)$ labeled by the Gel'fand–Tsetlin tableaux

$$M = \begin{pmatrix} m_{1,n} & & m_{2,n} & & \cdots & \cdots & & m_{n,n} \\ & m_{1,n-1} & & m_{2,n-1} & \cdots & & m_{n-1,n-1} & \\ & & \cdots & \cdots & \cdots & \cdots & & \\ & & & & m_{11} & & & \end{pmatrix}, \tag{24}$$

where $m_{i,n} \equiv m_i$. The entries in (24) are integers satisfying the betweenness conditions

$$m_{i,j+1} \geq m_{ij} \geq m_{i+1,j+1}, \qquad i = 1, 2, \cdots, j, \quad j = 1, 2, \cdots, n. \tag{25}$$

The set of all tableaux (24), satisfying these conditions, labels the basis elements of the carrier space of $T_{\mathbf{m}}$. The corresponding basis element will be denoted by $|M\rangle$, where M is the tableau (24).

Theorem 24. *Let q be a positive number and let $T_{\mathbf{m}}$ be the irreducible representation of $\breve{U}_q(\mathrm{gl}_n)$ with highest weight $\mathbf{m}$. Then the generators of $\breve{U}_q(\mathrm{gl}_n)$ act on the Gel'fand–Tsetlin basis of this representation by*

$$T_{\mathbf{m}}(K_k)|M\rangle = q^{a_k/2}|M\rangle, \quad a_k = \sum_{i=1}^{k} m_{i,k} - \sum_{i=1}^{k-1} m_{i,k-1}, \quad 1 \leq k \leq n, \tag{26}$$

$$T_{\mathbf{m}}(E_k)|M\rangle = \sum_{j=1}^{k} A_k^j(M)|M_k^j\rangle, \quad T_{\mathbf{m}}(F_k)|M\rangle = \sum_{j=1}^{k} A_k^j(M_k^{-j})|M_k^{-j}\rangle, \quad 1 \leq k \leq n-1. \tag{27}$$

Here $M_k^{\pm j}$ is the Gel'fand–Tsetlin tableau obtained from the tableau (24) if m_{jk} is replaced by $m_{j,k} \pm 1$, and $A_k^j(M)$ is the expression

$$A_k^j(M) = \left(- \frac{\prod_{i=1}^{k+1} [l_{i,k+1} - l_{j,k}] \prod_{i=1}^{k-1} [l_{i,k-1} - l_{j,k} - 1]}{\prod_{i \neq j} [l_{i,k} - l_{j,k}][l_{i,k} - l_{j,k} - 1]} \right)^{1/2},$$

where $l_{ir} = m_{ir} - i$, the positive value of the square root is taken and $[m]$ is the q-number defined by (2.1).

Multiplying the basis elements by appropriate factors, we obtain from (27) that

$$T_{\mathbf{m}}(E_k)|M\rangle' = \sum_{j=1}^{k} a_k^j(M)|M_k^j\rangle', \quad T_{\mathbf{m}}(F_k)|M\rangle' = \sum_{j=1}^{k} b_k^j(M)|M_k^{-j}\rangle', \quad (28)$$

where

$$a_k^j(M) = - \frac{\prod_{i=1}^{k+1} [l_{i,k+1} - l_{j,k}]}{\prod_{i \neq j} [l_{i,k} - l_{j,k}]}, \quad b_k^j(M) = \frac{\prod_{i=1}^{k-1} [l_{i,k-1} - l_{j,k}]}{\prod_{i \neq j} [l_{i,k} - l_{j,k}]}.$$

These formulas are valid for any complex q which is not a root of unity.

Since the proof of Theorem 24 is long and technically complicated, we give only a sketch of the proof. A complete proof can be found in the paper [UTS].

Sketch of proof of Theorem 24. Let $T_{\mathbf{m}}$ be a type 1 irreducible representation of $\breve{U}_q(\mathrm{gl}_n)$ with highest weight $\mathbf{m} = (m_1, \cdots, m_n)$ acting on the vector space $V_{\mathbf{m}}$. For notational simplicity, we consider $V_{\mathbf{m}}$ as a left $\breve{U}_q(\mathrm{gl}_n)$-module and write av instead of $T_{\mathbf{m}}(a)v$, $a \in \breve{U}_q(\mathrm{gl}_n)$, $v \in V_{\mathbf{m}}$. We denote by $|0\rangle$ the highest weight vector of $V_{\mathbf{m}}$ (that is, $E_j|0\rangle = 0$, $j = 1, 2, \cdots, n-1$).

The left action of $\breve{U}_q(\mathrm{gl}_n)$ on $V_{\mathbf{m}}$ induces a right action of $\breve{U}_q(\mathrm{gl}_n)$ on the dual space $V'_{\mathbf{m}}$ by $v'(Xv) = \overline{(v'X^*)(v)}$, $v \in V_{\mathbf{m}}$, $v' \in V'_{\mathbf{m}}$, where $*$ is the involution uniquely defined by $(E_i)^* = F_i$ and $(K_i)^* = K_i$. Let $\langle 0|$ be a vector of $V'_{\mathbf{m}}$ such that $\langle 0|F_j = 0$, $j = 1, 2, \cdots, n-1$. For $a, b \in \breve{U}_q(\mathrm{gl}_n)$, let $(\langle 0|a, b|0\rangle)$ denote the value of the functional $\langle 0|a \in V'_{\mathbf{m}}$ on the vector $b|0\rangle \in V_{\mathbf{m}}$.

We define inductively lowering and raising operators d_{ki} and c_{ik} by

$$d_{kk} = 1, \quad d_{k,k-1} = F_k, \qquad c_{kk} = 1, \quad c_{k-1,k} = E_k,$$

$$d_{ki} = \langle K_{i+1}^2 K_k^{-2} q^{k-i} \rangle F_k d_{k-1,i} - \langle K_{i+1}^2 K_k^{-2} q^{k-i-1} \rangle d_{k-1,i} F_k,$$

$$c_{ik} = c_{i,k-1} E_k \langle K_{i+1}^2 K_k^{-2} q^{k-i} \rangle - E_k c_{i,k-1} \langle K_{i+1}^2 K_k^{-2} q^{k-i-1} \rangle,$$

where $\langle K_j^2 K_s^{-2} q^m \rangle$ denotes the expression defined by

$$\langle K_j^2 K_s^{-2} q^m \rangle := \frac{K_j^2 K_s^{-2} q^m - K_j^{-2} K_s^2 q^{-m}}{q - q^{-1}}.$$

One can verify that the elements d_{ki}, $i = 1, 2, \cdots, k$, as well as the c_{ik}, $i = 1, 2, \cdots, k$, commute with each other and satisfy the relations

$$F_k^2 d_{k-1,i} - (q+q^{-1})F_k d_{k-1,i} F_k + d_{k-1,i} F_k^2 = 0,$$

$$d_{k-1,i}^2 F_k - (q+q^{-1}) d_{k-1,i} F_k d_{k-1,i} + F_k d_{k-1,i}^2 = 0,$$

$$F_k d_{k-2,i} = d_{k-2,i} F_k, \qquad F_k d_{k,i} = d_{k,i} F_k.$$

For each multi-index $\mathbf{r} = (r_0, r_1, \cdots r_{k-1})$, $r_i \in \mathbb{N}_0$, we set

$$d_k^{\mathbf{r}} = d_{k,0}^{r_0} \cdots d_{k,k-1}^{r_{k-1}}, \qquad c_k^{\mathbf{r}} = c_{k-1,k}^{r_{k-1}} \cdots c_{0,k}^{r_0}.$$

The main relation, used for the construction of the Gel'fand–Tsetlin basis elements $|M\rangle$ from the highest weight vector, is

$$c_k^{\mathbf{r}} d_k^{\mathbf{s}} = \delta_{\mathbf{r},\mathbf{s}} [\mathbf{r}]! \prod_{i=0}^{k-1} \Big\{ \prod_{p=1}^{r_i} \prod_{t=1}^{k-i} \langle K_{i+1}^2 K_{i+t+1}^{-2} q^{t-p} \rangle$$

$$\times \prod_{p=1}^{r_i} \prod_{t=1}^{k-i-1} \langle K_{i+1}^2 K_{i+t+1}^{-2} q^{t-p+1+r_{i+t}} \rangle \Big\} \mod J_k,$$

where $[\mathbf{r}]! = [r_0]![r_1]! \cdots [r_{k-1}]!$ and J_k is the left ideal of $\breve{U}_q(\mathrm{gl}_n)$ generated by the elements $E_1, \cdots, E_{k-1}$.

Now with every Gel'fand–Tsetlin tableau (24) we associate the elements

$$d^M = d_1^{\mathbf{m}_2 - \mathbf{m}_1} d_2^{\mathbf{m}_3 - \mathbf{m}_2} \cdots d_{n-1}^{\mathbf{m}_n - \mathbf{m}_{n-1}}, \qquad c^M = c_{n-1}^{\mathbf{m}_n - \mathbf{m}_{n-1}} \cdots c_2^{\mathbf{m}_3 - \mathbf{m}_2} c_1^{\mathbf{m}_2 - \mathbf{m}_1}, \tag{29}$$

where $\mathbf{m}_k - \mathbf{m}_{k-1} \equiv (m_{1,k} - m_{1,k-1}, \cdots, m_{k-1,k} - m_{k-1,k-1})$. One can prove the following facts:

(i) The weight of the vector $d^M|0\rangle$ coincides with the weight of the basis vector corresponding to the Gel'fand–Tsetlin tableau (24).

(ii) For any Gel'fand–Tsetlin tableaux M and M' we have

$$(\langle 0|c^{M'}, d^M|0\rangle) = \delta_{M,M'} N_M^2,$$

where $N_M^2 = \prod_{k=1}^{n-1} \tau_k(\mathbf{m}_k, \mathbf{m}_{k+1})$ and $\tau_k(\mathbf{m}_k, \mathbf{m}_{k+1})$ denotes the expression

$$\prod_{i \le j} \frac{[m_{i,k+1} - m_{j,k} + j - i]!}{[m_{i,k} - m_{j,k} + j - i]!} \prod_{i \le j} \frac{[m_{i,k+1} - m_{j,k+1} + j - i - 1]!}{[m_{i,k} - m_{j,k+1} + j - i - 1]!}.$$

(iii) The vectors $|M\rangle = d^M|0\rangle$ form a basis of the space $V_{\mathbf{m}}$ if M runs over the set of Gel'fand–Tsetlin tableaux for the representation $T_{\mathbf{m}}$.

Finally, using (29) and the actions of E_k and F_k on $d^M|0\rangle$, one proves that the actions of the operators $T_{\mathbf{m}}(E_k)$ and $T_{\mathbf{m}}(F_k)$ on the vectors $|M\rangle$ are given by the formulas (28). □

7.3.4 Representations of Class 1

In most applications only the simplest irreducible representations of $\breve{U}_q(\mathrm{gl}_n)$ are needed. An irreducible representation $T_{\mathbf{m}}$ of $\breve{U}_q(\mathrm{gl}_n)$ is said to be of *class 1* with respect to $\breve{U}_q(\mathrm{gl}_{n-1})$ if the restriction of $T_{\mathbf{m}}$ to $\breve{U}_q(\mathrm{gl}_{n-1})$ contains the trivial irreducible representation of this subalgebra, that is, the representation with highest weight $(0, \cdots, 0)$.

Proposition 25. *An irreducible finite-dimensional representation $T_{\mathbf{m}}$ of $\breve{U}_q(\mathrm{gl}_n)$ with highest weight $\mathbf{m} = (m_1, \cdots, m_n)$ is of class 1 with respect to $\breve{U}_q(\mathrm{gl}_{n-1})$ if and only if $m_2 = m_3 = \cdots = m_{n-1} = 0$.*

Proof. The assertion follows from (23). □

If an irreducible representation $T_{ll'} := T_{(l,0,\cdots,0,l')}$ of $\breve{U}_q(\mathrm{gl}_n)$ has a highest weight $(l, 0, \cdots, 0, l')$, $l \geq 0 \geq l'$, then its Gel'fand–Tsetlin tableaux are of the form

$$M_{jj'}^{mm'} \equiv \begin{pmatrix} l & & 0 & & 0 & & \cdots & & 0 & & 0 & & l' \\ & m & & 0 & & 0 & \cdots & 0 & & 0 & & m' & \\ & & j & & 0 & & \cdots & & 0 & & j' & & \\ & & & & \cdots & & \cdots & & \cdots & & & & \end{pmatrix}, \tag{30}$$

where $l \geq m \geq j \geq \cdots \geq 0 \geq \cdots \geq j' \geq m' \geq l'$. The operators $T_{ll'}(K_n)$, $T_{ll'}(E_{n-1})$ and $T_{ll'}(F_{n-1})$ act on the corresponding basis vectors $|M_{jj'}^{mm'}\rangle$ by

$$T_{ll'}(K_n)|M_{jj'}^{mm'}\rangle = q^{(l+l'-m-m')/2}|M_{jj'}^{mm'}\rangle, \tag{31}$$

$$\begin{aligned} &T_{ll'}(E_{n-1})|M_{jj'}^{mm'}\rangle \\ &= \left(\frac{[l-m][m-l'+n-1][m-j+1][m-j'+n-2]}{[m-m'+n-1][m-m'+n-2]}\right)^{\frac{1}{2}} |M_{jj'}^{m+1,m'}\rangle \\ &+ \left(\frac{[l-m'+n-2][m'-l'+1][j-m'+n-3][j'-m']}{[m-m'+n-2][m-m'+n-3]}\right)^{\frac{1}{2}} |M_{jj'}^{m,m'+1}\rangle, \end{aligned} \tag{32}$$

$$\begin{aligned} &T_{ll'}(F_{n-1})|M_{jj'}^{mm'}\rangle \\ &= \left(\frac{[l-m+1][m-l'+n-2][m-j][m-j'+n-3]}{[m-m'+n-3][m-m'+n-2]}\right)^{\frac{1}{2}} |M_{jj'}^{m-1,m'}\rangle \\ &+ \left(\frac{[l-m'+n-1][m'-l'][j-m'+n-2][j'-m'+1]}{[m-m'+n-2][m-m'+n-1]}\right)^{\frac{1}{2}} |M_{jj'}^{m,m'-1}\rangle. \end{aligned} \tag{33}$$

For representations with highest weights $(l, 0, \cdots, 0)$, these formulas turn into

$$T_{l,0}(K_n)|M_{j,0}^{m,0}\rangle = q^{(l-m)/2}|M_{j,0}^{m,0}\rangle, \tag{34}$$

$$T_{l,0}(E_{n-1})|M_{j,0}^{m,0}\rangle = ([l-m][m-j+1])^{1/2}|M_{j,0}^{m+1,0}\rangle, \tag{35}$$

$$T_{l,0}(F_{n-1})|M_{j,0}^{m,0}\rangle = ([l-m+1][m-j])^{1/2}|M_{j,0}^{m-1,0}\rangle. \tag{36}$$

The formulas (31)–(33) show that $|M_{jj'}^{mm'}\rangle$ is an eigenvector of the operator $T_{ll'}(K_n)$ and that the operators $T_{ll'}(E_{n-1})$ and $T_{ll'}(F_{n-1})$ change only the indices j and j' in the tableaux (30). Similarly, the matrix elements of the operators $T_{ll'}(K_{k+1})$, $T_{ll'}(E_k)$ and $T_{ll'}(F_k)$ depend only on the $(k+1)$-th, k-th and $(k-1)$-th rows of the Gel'fand–Tsetlin basis elements (30).

We see from the formulas (34)–(36) that the matrix elements of the operators $T_{l,0}(K_n)$, $T_{l,0}(E_{n-1})$ and $T_{l,0}(F_{n-1})$ do not depend explicitly on n. The next proposition shows that this is the case for all irreducible representations of $\breve{U}_q(\mathrm{gl}_n)$ with highest weights $(m_1, m_2, \cdots, m_i, 0, \cdots, 0)$.

Proposition 26. *If an irreducible representation $T_{\mathbf{m}}$ of $\breve{U}_q(\mathrm{gl}_n)$ has a highest weight $\mathbf{m} = (m_1, m_2, \cdots, m_i, 0, \cdots, 0)$ with more than two zeros, then the matrix elements of the operators $T_{\mathbf{m}}(K_n)$, $T_{\mathbf{m}}(E_{n-1})$ and $T_{\mathbf{m}}(F_{n-1})$ with respect to the Gel'fand–Tsetlin basis are independent of n. The corresponding formulas (27) for these operators contain only i summands.*

Proof. The proof follows from the expressions for the coefficients $A_{n-1}^{j}(M)$ in (27) at $m_{i+1} = \cdots = m_n = 0$. □

7.3.5 Tensor Products of Representations

In this subsection we consider tensor products of irreducible representations of $\breve{U}_q(\mathrm{gl}_n)$ for which the multiplicities of irreducible components do not exceed 1. The most important such examples are the tensor products $T_{\mathbf{m}} \otimes T_1$, where $T_{\mathbf{m}}$ is an irreducible representation and T_1 is the *vector representation* of $\breve{U}_q(\mathrm{gl}_n)$. Recall that the decomposition of this tensor product was given in Subsect. 7.2.1.

Other tensor products with multiplicities not exceeding 1 are $T_{\mathbf{m}} \otimes T_p$, where $\mathbf{m}$ is any highest weight and T_p is the *symmetric irreducible representation* with highest weight $(p, 0, \cdots, 0)$. From (11) one can derive that

$$T_{\mathbf{m}} \otimes T_p = \bigoplus\nolimits_{\mathbf{r}} T_{\mathbf{m}+\mathbf{r}}, \tag{37}$$

where the summation is over all $\mathbf{r} = (r_1, \cdots, r_n)$ such that $r_j \in \mathbb{N}_0$ and

$$r_1 + \cdots + r_n = p, \quad m_1 + r_1 \ge m_1 \ge m_2 + r_2 \ge m_2 \ge \cdots \ge m_n + r_n \ge m_n. \tag{38}$$

If the multiplicities of irreducible components in the decomposition of a tensor product $T_{\mathbf{m}_n} \otimes T_{\mathbf{m}'_n}$ are at most one, then the CGC's with respect to the Gel'fand–Tsetlin bases can be written in the form

$$\langle M\ M' \,|\, M'' \rangle \equiv \left\langle \begin{matrix} \mathbf{m}_n & \mathbf{m}'_n \\ \mathbf{m}_{n-1} & \mathbf{m}'_{n-1} \\ \cdots & \cdots \end{matrix} \,\middle|\, \begin{matrix} \mathbf{m}''_n \\ \mathbf{m}''_{n-1} \\ \cdots \end{matrix} \right\rangle, \tag{39}$$

where $\mathbf{m}_j$ are the rows of the corresponding Gel'fand–Tsetlin tableaux. As in the classical case (see [VK2], Chap. 4), it can be proved that the CGC (39) is a product of the so-called $\breve{U}_q(\mathrm{gl}_{n-i})$-*scalar factors* (or *reduced CGC's*)

$$\langle M\ M' \mid M'' \rangle = \left(\begin{matrix} \mathbf{m}_n & \mathbf{m}'_n \\ \mathbf{m}_{n-1} & \mathbf{m}'_{n-1} \end{matrix} \middle| \begin{matrix} \mathbf{m}''_n \\ \mathbf{m}''_{n-1} \end{matrix} \right) \left(\begin{matrix} \mathbf{m}_{n-1} & \mathbf{m}'_{n-1} \\ \mathbf{m}_{n-2} & \mathbf{m}'_{n-2} \end{matrix} \middle| \begin{matrix} \mathbf{m}''_{n-1} \\ \mathbf{m}''_{n-2} \end{matrix} \right)$$

$$\times \cdots \times \left(\begin{matrix} \mathbf{m}_2 & \mathbf{m}'_2 \\ \mathbf{m}_1 & \mathbf{m}'_1 \end{matrix} \middle| \begin{matrix} \mathbf{m}''_2 \\ \mathbf{m}''_1 \end{matrix} \right), \tag{40}$$

which depend only on two rows of the CGC.

7.3.6 Tensor Operators and the Wigner–Eckart Theorem

In Subsects. 7.3.6–9 we suppose that q is a positive number.

Let $T_{\mathbf{m}}$ be an irreducible finite-dimensional representation of $\breve{U}_q(\mathrm{gl}_n)$ with highest weight $\mathbf{m}$ on a space $V_{\mathbf{m}}$ with Gel'fand–Tsetlin basis $\{|M\rangle\}$. Suppose that $\{R^{\mathbf{m}}_M\}$ is a set of operators acting on a Hilbert space $\mathfrak{H}$ and indexed by the Gel'fand–Tsetlin tableaux $|M\rangle$ of the representation $T_{\mathbf{m}}$. Let T be a representation of $\breve{U}_q(\mathrm{gl}_n)$ acting on $\mathfrak{H}$, which is a direct sum of irreducible finite-dimensional representations. We say that the set of operators $\{R^{\mathbf{m}}_M\}$ is a *tensor operator* transforming under the representation $T_{\mathbf{m}}$ of $\breve{U}_q(\mathrm{gl}_n)$ if for all generators E_i, F_i, K_i we have

$$T(E_i)R^{\mathbf{m}}_M - q^{(a_i - a_{i+1})/2} R^{\mathbf{m}}_M T(E_i) = \sum_{k=1}^{i} A^k_i(M) R^{\mathbf{m}}_{M^k_i} T(K_i^{-1}K_{i+1}), \tag{41}$$

$$T(F_i)R^{\mathbf{m}}_M - q^{(a_i - a_{i+1})/2} R^{\mathbf{m}}_M T(F_i) = \sum_{k=1}^{i} A^k_i(M^{-k}_i) R^{\mathbf{m}}_{M^{-k}_i} T(K_i^{-1}K_{i+1}), \tag{42}$$

$$T(K_i)R^{\mathbf{m}}_M T(K_i^{-1}) = q^{a_i/2} R^{\mathbf{m}}_M, \tag{43}$$

where $A^k_i(M)$ and $a_i \equiv a_i(M)$ are as in (26) and (27).

Repeating the arguments of Subsect. 3.6.2 one proves the following *Wigner–Eckart theorem* for the matrix elements of the operators $R^{\mathbf{m}}_M$.

Theorem 27. *If* $\mathfrak{H} = \bigoplus_{\mathbf{m}'} V_{\mathbf{m}'}$ *is a decomposition of* $\mathfrak{H}$ *into irreducible subspaces for the representation* T *and* $\{|\mathbf{m}', M'\rangle\}$ *are Gel'fand–Tsetlin bases of the subspaces* $V_{\mathbf{m}'}$, *then the matrix elements of the operators* $R^{\mathbf{m}}_M$ *with respect to these bases are expressed in terms of the CGC's of* $\breve{U}_q(\mathrm{gl}_n)$ *by*

$$\langle \mathbf{m}', M' \mid R^{\mathbf{m}}_M \mid \mathbf{m}'', M'' \rangle = \sum_r \langle \mathbf{m}' \,\|R^{\mathbf{m}}\|\, \mathbf{m}'' \rangle_r \left\langle \begin{matrix} \mathbf{m}' \\ M' \end{matrix} \middle| \begin{matrix} \mathbf{m} & \mathbf{m}'' \\ M & M'' \end{matrix} \right\rangle_r . \tag{44}$$

Here $\langle \mathbf{m}' \| R^{\mathbf{m}} \| \mathbf{m}'' \rangle_r$ *are the so-called reduced matrix elements of the tensor operator which do not depend on* M', M, M''. *The summation index* r *in*

(44) distinguishes multiple irreducible representations in the tensor product $T_{\mathbf{m}} \otimes T_{\mathbf{m}''}$.

Note that if the tensor product $T_{\mathbf{m}} \otimes T_{\mathbf{m}''}$ contains only irreducible components with multiplicities not exceeding 1, then the right hand side of (44) contains only one summand and the index r can be omitted.

7.3.7 Clebsch–Gordan Coefficients for the Tensor Product $T_{\mathbf{m}} \otimes T_1$

These CGC's are factorized into products of reduced CGC's of the form

$$\left(\begin{matrix} \mathbf{m}_n & (1,\mathbf{0}) \\ \mathbf{m}_{n-1} & (0,\mathbf{0}) \end{matrix}\middle| \begin{matrix} \mathbf{m}_n + \mathbf{e}_i \\ \mathbf{m}_{n-1} \end{matrix}\right), \qquad \left(\begin{matrix} \mathbf{m}_n & (1,\mathbf{0}) \\ \mathbf{m}_{n-1} & (1,\mathbf{0}) \end{matrix}\middle| \begin{matrix} \mathbf{m}_n + \mathbf{e}_i \\ \mathbf{m}_{n-1} + \mathbf{e}_j \end{matrix}\right),$$

where $\mathbf{0} = (0,\cdots,0)$. The corresponding reduced CGC's are given by the formulas

$$\left(\begin{matrix} \mathbf{m}_n & (1,\mathbf{0}) \\ \mathbf{m}_{n-1} & (0,\mathbf{0}) \end{matrix}\middle| \begin{matrix} \mathbf{m}_n + \mathbf{e}_i \\ \mathbf{m}_{n-1} \end{matrix}\right) = q^{-1/2\left(i+1+\sum_j m_{j,n-1} - \sum_{j\neq i} m_{jn}\right)}$$

$$\times \left(\frac{\prod_{j=1}^{n-1} [m_{j,n-1} - m_{i,n} - j + i - 1]}{\prod_{j\neq i} [m_{j,n} - m_{i,n} - j + i]} \right)^{1/2}, \tag{45}$$

$$\left(\begin{matrix} \mathbf{m}_n & (1,\mathbf{0}) \\ \mathbf{m}_{n-1} & (1,\mathbf{0}) \end{matrix}\middle| \begin{matrix} \mathbf{m}_n + \mathbf{e}_i \\ \mathbf{m}_{n-1} + \mathbf{e}_j \end{matrix}\right) = \vartheta(j-i) q^{-(m_{j,n-1} - m_{i,n} - j + i)/2}$$

$$\times \left(\prod_{k\neq i} \frac{[m_{k,n} - m_{j,n-1} - k + j]}{[m_{k,n} - m_{i,n} - k + i]} \prod_{k\neq j} \frac{[m_{k,n-1} - m_{i,n} - k + i - 1]}{[m_{k,n-1} - m_{j,n-1} - k + j - 1]} \right)^{1/2}, \tag{46}$$

where $[m] \equiv [m]_q$ is the q-number (2.1), $m_{s,n}$ are the components of the highest weight $\mathbf{m}_n$ and we have abbreviated $\vartheta(j-i) := 1$ if $j - i \geq 0$ and $\vartheta(j-i) := -1$ if $j - i < 0$.

In the case of the tensor product $T_1 \otimes T_{\mathbf{m}}$, the corresponding expressions for the reduced CGC's

$$\left(\begin{matrix} (1,\mathbf{0}) & \mathbf{m}_n \\ (0,\mathbf{0}) & \mathbf{m}_{n-1} \end{matrix}\middle| \begin{matrix} \mathbf{m}_n + \mathbf{e}_i \\ \mathbf{m}_{n-1} \end{matrix}\right), \qquad \left(\begin{matrix} (1,\mathbf{0}) & \mathbf{m}_n \\ (1,\mathbf{0}) & \mathbf{m}_{n-1} \end{matrix}\middle| \begin{matrix} \mathbf{m}_n + \mathbf{e}_i \\ \mathbf{m}_{n-1} + \mathbf{e}_j \end{matrix}\right)$$

are obtained from (45) and (46), respectively, if we replace q by q^{-1}.

Both expressions (45) and (46) are special cases of CGC's considered in the next subsection.

7.3.8 Clebsch–Gordan Coefficients for the Tensor Product $T_{\mathbf{m}} \otimes T_p$

The corresponding CGC's factorize into products of reduced CGC's of the form

$$\left(\begin{matrix} \mathbf{m}_n & (p,\mathbf{0}) \\ \mathbf{m}_{n-1} & (0,\mathbf{0}) \end{matrix}\middle| \begin{matrix} \mathbf{m}'_n \\ \mathbf{m}_{n-1} \end{matrix}\right), \qquad \left(\begin{matrix} \mathbf{m}_n & (p,\mathbf{0}) \\ \mathbf{m}_{n-1} & (r,\mathbf{0}) \end{matrix}\middle| \begin{matrix} \mathbf{m}'_n \\ \mathbf{m}'_{n-1} \end{matrix}\right). \tag{47}$$

The expression for the first reduced CGC is

$$\left(\begin{matrix} \mathbf{m}_n & (p,\mathbf{0}) \\ \mathbf{m}_{n-1} & (0,\mathbf{0}) \end{matrix}\middle| \begin{matrix} \mathbf{m}'_n \\ \mathbf{m}_{n-1} \end{matrix}\right) = q^a \sqrt{[p]} \prod_{i<j} [l'_{i,n} - l'_{j,n}]$$

$$\times \left(\prod_{i \le j} \frac{[l'_{i,n} - l_{j,n-1}]!}{[l_{i,n} - l_{j,n-1}]!} \prod_{i<j} \frac{[l_{i,n-1} - l_{j,n} - 1]!}{[l_{i,n-1} - l'_{j,n} - 1]!} \frac{\prod_{i<j} [l_{i,n} - l'_{j,n} - 1]!}{\prod_{i \le j} [l'_{i,n} - l_{j,n}]!} \right)^{1/2}, \tag{48}$$

where $l_{i,n} = m_{i,n} - i$, $l'_{i,n} = m'_{i,n} - i$, $l_{j,n-1} = m_{j,n-1} - j$ and

$$a = \frac{p^2}{4} - \frac{1}{4} \sum_{i=1}^{n} (m'_{in} - m_{in})(m_{in} - m'_{in} - 2i + 2) + \frac{p}{2} \left(\sum_{i=1}^{n} m_{in} - \sum_{j=1}^{n-1} m_{j,n-1} \right).$$

In order to describe the second reduced CGC in (47) we introduce the notation

$$S_{pr}(\mathbf{m}_p, \mathbf{m}_r) = \left(\frac{\prod_{i \le j} [m_{ip} - m_{jr} - i + j]!}{\prod_{i<j} [m_{ir} - m_{jp} - i + j - 1]!} \right)^{1/2},$$

where $\mathbf{m}_n = (m_{1n}, \cdots, m_{nn})$. It is not difficult to verify that if $\mathbf{m}_n$ is the highest weight of an irreducible representation $T_{\mathbf{m}_n}$, then

$$S_{nn}(\mathbf{m}_n, \mathbf{m}_n)^2 = \prod\nolimits_{i<j} [l_{in} - l_{jn}]. \tag{49}$$

For the second reduced CGC in (47) we obtain the expression

$$\left(\begin{matrix} \mathbf{m}_n & (p,\mathbf{0}) \\ \mathbf{m}_{n-1} & (r,\mathbf{0}) \end{matrix}\middle| \begin{matrix} \mathbf{m}'_n \\ \mathbf{m}'_{n-1} \end{matrix}\right) = q^{b/2} \sqrt{[p-r]!}\, \frac{S_{nn}(\mathbf{m}'_n, \mathbf{m}'_n) S_{n,n-1}(\mathbf{m}_n, \mathbf{m}_{n-1})}{S_{nn}(\mathbf{m}'_n, \mathbf{m}_n) S_{n,n-1}(\mathbf{m}'_n, \mathbf{m}'_{n-1})}$$

$$\times S_{n-1,n-1}(\mathbf{m}'_{n-1}, \mathbf{m}_{n-1}) S_{n-1,n-1}(\mathbf{m}_{n-1}, \mathbf{m}_{n-1}) \sum_{\mathbf{m}} \Bigg((-1)^{\sum_i (m_i - m_{i,n-1})}$$

$$\times \frac{q^{(r-p-1)\sum_i (m_i - m_{i,n-1})} S^2_{n,n-1}(\mathbf{m}'_n, \mathbf{m}) S^2_{n-1,n-1}(\mathbf{m}, \mathbf{m})}{S^2_{n,n-1}(\mathbf{m}_n, \mathbf{m}) S^2_{n-1,n-1}(\mathbf{m}'_{n-1}, \mathbf{m}) S^2_{n-1,n-1}(\mathbf{m}, \mathbf{m}_{n-1})} \Bigg), \tag{50}$$

where the summation is over all integers $\mathbf{m} = (m_1, \cdots, m_{n-1})$ such that

$$\max(m_{i+1,n+1}, m'_{i+1,n}) \le m_i \le \min(m'_{i,n-1}, m_{in})$$

and

$$b=\sum_{i<j}(m'_{in}-m_{in})(m'_{jn}-m_{jn})-\sum_i(m'_{in}-m_{in})(m_{in}-i+1)$$
$$+\sum_i(m'_{i,n-1}-m_{i,n-1})(m_{i,n-1}-i+1)+(p-r)\sum_i m_{in}$$
$$-(p-r)\sum_j m_{j,n-1}-\sum_{i<j}(m'_{i,n-1}-m_{i,n-1})(m'_{j,n-1}-m_{j,n-1}).$$

We sketch the proof of the formulas (48) and (50). If T_{p_n} is the symmetric irreducible representation of $\breve{U}_q(\mathrm{gl}_n)$, then its Gel'fand–Tsetlin tableaux are of the form

$$P=\begin{pmatrix} p_n & & 0 & & 0 & & \cdots & & 0 & & 0 & & 0 \\ & p_{n-1} & & 0 & & 0 & \cdots & 0 & & 0 & & 0 & \\ & & p_{n-2} & & 0 & & \cdots & & 0 & & 0 & & \\ & & & & \cdots & & \cdots & & \cdots & & & & \\ & & & & & & p_1 & & & & & & \end{pmatrix}, \qquad (51)$$

where $p_n\ge p_{n-1}\ge\cdots\ge p_1\ge 0$. With such a tableau we associate the element

$$R^{p_n}_{p_{n-1},\cdots,p_1}=\left(\frac{[p_n]!}{\prod_{i=1}^n[p_i-p_{i-1}]!}\right)^{1/2}\prod_{i=1}^n E^{p_i-p_{i-1}}_{i,n+1}\prod_{j=1}^n K_j^{p_j-p_{j-1}} \qquad (52)$$

of the algebra $\breve{U}_q(\mathrm{gl}_{n+1})$. Using the relations $E^n_{i,i+1}E^m_{i,k}=q^{-nm}E^m_{i,k}E^n_{i,i+1}$ and

$$E^n_{i,k-1}E^m_{k-1,k}=\sum_s\frac{[n]![m]!q^{(n-s)(m-s)}}{[s]![n-s]![m-s]!}E^{m-s}_{k-1,k}E^s_{ik}E^{n-s}_{i,k-1},$$

$$E^n_{i,i+1}E^m_{k,i}=\sum_s\frac{(-1)^s[n]![m]!q^{-s(n-m+1)}}{[s]![n-s]![m-s]!}E^{m-s}_{k,i}E^s_{k,i-1}E^{n-s}_{i,i+1}K_i^{-2s}K_{i+1}^{2s},$$

where $i+1<k$, one proves that the elements (52) satisfy the relations

$$E_{i,i+1}R^{p_n}_{p_{n-1},\cdots,p_1}K_iK_{i+1}^{-1}-q^{-1}K_iK_{i+1}^{-1}R^{p_n}_{p_{n-1},\cdots,p_1}E_{i,i+1}$$
$$=([p_{i+1}-p_i][p_i-p_{i-1}+1])^{1/2}R^{p_n}_{p_{n-1},\cdots,p_i+1,\cdots,p_1}, \qquad (53)$$
$$E_{i+1,i}R^{p_n}_{p_{n-1},\cdots,p_1}K_iK_{i+1}^{-1}-qK_iK_{i+1}^{-1}R^{p_n}_{p_{n-1},\cdots,p_1}E_{i+1,i}$$
$$=([p_{i+1}-p_i+1][p_i-p_{i-1}])^{1/2}R^{p_n}_{p_{n-1},\cdots,p_i-1,\cdots,p_1}. \qquad (54)$$

This means that for any finite-dimensional representation T of $\breve{U}_q(\mathrm{gl}_{n+1})$, the operators $T(R^{p_n}_{p_{n-1},\cdots,p_1})$ form a tensor operator transforming under the symmetric representation T_{p_n} of $\breve{U}_q(\mathrm{gl}_n)$. Therefore, by Theorem 27 and formula (44) therein, for any irreducible representation $T_{\mathbf{m}_{n+1}}$ of $\breve{U}_q(\mathrm{gl}_{n+1})$ we have

$$\langle M\mid R^{p_n}_{p_{n-1},\cdots,p_1}\mid M'\rangle=B_{p_n,p_{n-1}}(\mathbf{m}_{n+1})\langle\overline{M}\mid P\ \overline{M'}\rangle.$$

Here M and M' are the Gel'fand–Tsetlin tableaux given by $M = \begin{pmatrix} \mathbf{m}_{n+1} \\ \overline{M} \end{pmatrix}$ and $M' = \begin{pmatrix} \mathbf{m}_{n+1} \\ \overline{M'} \end{pmatrix}$, respectively, P is the Gel'fand–Tsetlin tableau (51), $\langle \overline{M} \mid P \ \ \overline{M'}\rangle$ is the Clebsch–Gordan coefficient and $B_{p_n,p_{n-1}}(\mathbf{m}_{n+1})$ is the reduced matrix element depending on $p_n, p_{n-1}, \mathbf{m}_n, \mathbf{m}'_n$, where $\mathbf{m}_n$ and $\mathbf{m}'_n$ are the first rows in the tableaux $\overline{M}$ and $\overline{M'}$, respectively.

As noted in Subsect. 7.2.2, the CGC $\langle \overline{M} \ \ P \mid \overline{M'}\rangle$ is equal to the scalar product $\langle \overline{M} \otimes P \ , \ \overline{M'}\rangle$. Since $q > 0$, $T_{\mathbf{m}_n} \otimes T_{p_n}$ is a $*$-representation with respect to the involution of $\breve{U}_q(\mathrm{gl}_n)$ given by $(E_i)^* := F_i$ and $(K_i)^* := K_i$, so that

$$\langle \overline{M} \otimes P, (T_{\mathbf{m}_n} \otimes T_{p_n})(E_{n-1})\overline{M'}\rangle = \langle (T_{\mathbf{m}_n} \otimes T_{p_n})(F_{n-1})(\overline{M} \otimes P), \overline{M'}\rangle.$$

Inserting the formulas (26) and (27) we derive the recurrence relation

$$\sum_{j=1}^{n-1} A^j_{n-1}(\overline{M'})\langle \overline{M} \ \ P \mid (\overline{M'})^j_{n-1}\rangle = A^j_{n-1}(P^{-1}_{n-1})q^c\langle \overline{M} \ \ P^{-1}_{n-1} \mid \overline{M'}\rangle$$
$$+\sum_{j=1}^{n-1} A^j_{n-1}(\overline{M}^{-j}_{n-1})q^{p_n-p_{n-1}}\langle \overline{M}^{-j}_{n-1} \ \ P \mid \overline{M'}\rangle,$$

where A^j_{n-1} and $M^{\pm j}_{n-1}$ are as in (27) and $c = m_{1n} + \cdots + m_{nn} - m_{1,n-1} - \cdots - m_{n-1,n-1}$. Setting $p_{n-1} = p_{n-2} = \cdots = p_1 = 0$, we obtain a recurrence relation connecting the reduced CGC's

$$\left(\begin{array}{cc|c} \mathbf{m}_n & (p,\mathbf{0}) & \mathbf{m}'_n \\ \mathbf{m}_{n-1} & (0,\mathbf{0}) & \mathbf{m}_{n-1} \end{array}\right) \quad \text{and} \quad \left(\begin{array}{cc|c} \mathbf{m}_n & (p,\mathbf{0}) & \mathbf{m}'_n \\ \mathbf{m}^{+s}_{n-1} & (0,\mathbf{0}) & \mathbf{m}^{+s}_{n-1} \end{array}\right),$$

where s takes one of the values $1, 2, \cdots, n-1$ and $\mathbf{m}^{+s}_{n-1}$ is obtained from $\mathbf{m}_{n-1}$ if we replace $m_{s,n-1}$ by $m_{s,n-1}+1$. These recurrence relations determine the reduced CGC's (48) up to a normalization constant. This constant can be computed by using the orthogonality relations for the reduced CGC's.

Using (53) one can easily verify that $R^p_{r,0,\cdots,0}$ coincides with

$$\left(\frac{[p-r]![r]!}{[p]!}\right)^{\frac{1}{2}} \sum_{s=0}^{r} \frac{(-1)^s q^{s(r-1-p/2)}}{[s]![r-s]!} E^{r-s}_{n-1,n} R^p_{0,\cdots,0} E^s_{n-1,n} K^r_{n-1} K^{-r}_n,$$

where $p = p_n$ and $r \equiv p_{n-1}$. From the matrix form of this relation the second reduced CGC in (47) can be expressed in terms of the first reduced CGC's and the matrix elements of the operators $T_{\mathbf{m}}(E^k_{n-1,n})$. The latter matrix elements are given by the formula

$$\left\langle \begin{array}{c} \mathbf{m}'_{n-1} \\ \mathbf{m}_{n-2} \\ \cdots \end{array} \middle| T_{\mathbf{m}_n}(E^p_{n-1,n}) \middle| \begin{array}{c} \mathbf{m}_{n-1} \\ \mathbf{m}_{n-2} \\ \cdots \end{array} \right\rangle = \left\langle \begin{array}{c} \mathbf{m}_{n-1} \\ \mathbf{m}_{n-2} \\ \cdots \end{array} \middle| T_{\mathbf{m}_n}(E^p_{n,n-1}) \middle| \begin{array}{c} \mathbf{m}'_{n-1} \\ \mathbf{m}_{n-2} \\ \cdots \end{array} \right\rangle$$

$$= \frac{[p]! d_{n-1}(\mathbf{m}_{n-1}) d_{n-1}(\mathbf{m}'_{n-1}) S_{n,n-1}(\mathbf{m}_n, \mathbf{m}_{n-1}) S_{n-1,n-2}(\mathbf{m}'_{n-1}, \mathbf{m}_{n-2})}{(S_{n-1,n-1}(\mathbf{m}'_{n-1}, \mathbf{m}_{n-1}))^2 S_{n,n-1}(\mathbf{m}_n, \mathbf{m}'_{n-1}) S_{n-1,n-2}(\mathbf{m}_{n-1}, \mathbf{m}_{n-2})}, \tag{55}$$

where the first rows $\mathbf{m}_n$ in the Gel'fand–Tsetlin tableaux are omitted and

$$d_{n-1}(\mathbf{m}_{n-1}) = S_{n-1,n-1}(\mathbf{m}_{n-1}, \mathbf{m}_{n-1}).$$

This leads to the expression (50) for the reduced CGC's. Formula (55) is proved by induction on p. For this, we use the recurrence relation

$$\langle M' \mid T_{\mathbf{m}_n}(E^p_{n-1,n}) \mid M\rangle = \sum\nolimits_{i=1}^{n-1} A^i_{n-1}(M) \langle M' \mid T_{\mathbf{m}_n}(E^{p-1}_{n-1,n}) \mid M^{+i}_{n-1}\rangle$$
$$\times \langle M^{+i}_{n-1} \mid T_{\mathbf{m}_n}(E_{n-1,n}) \mid M\rangle,$$

where $A^i_{n-1}(M)$ is as in (27). In this formula we substitute the expression (55) for the operator $T_{\mathbf{m}_n}(E^{p-1}_{n-1,n})$ and use the equality

$$1 - x_1 x_2 \cdots x_n = \sum\nolimits_{s=1}^{n} (1 - x_s) \prod\nolimits_{i \neq s} \frac{y_s - x_i y_i}{y_s - y_i}, \tag{56}$$

which is well-known in combinatorics (see [Mac]). After some simplifications we obtain then formula (55) for $T_{\mathbf{m}_n}(E^p_{n-1,n})$.

7.3.9 The Tensor Product $T_{\mathbf{m}} \otimes T_1$ for $q^{\pm 1} \to 0$

Let us consider $U_q(\mathrm{gl}_n)$ as a one-parameter Hopf algebra deformation of the universal enveloping algebra $U(\mathrm{gl}_n)$ and assume that $q > 0$. The CGC's of $U_q(\mathrm{gl}_n)$ are then functions of the parameter q and we can look at their limits for $q \to 0$ and $q \to \infty$. If $q \to 0$, then a direct calculation shows that the reduced CGC's (45) and (46) vanish except for the following cases:

$$\lim_{q\to 0} \left(\begin{matrix} \mathbf{m}_n & (1,0) \\ \mathbf{m}_{n-1} & (0,0) \end{matrix} \middle| \begin{matrix} \mathbf{m}^{+i}_n \\ \mathbf{m}_{n-1} \end{matrix} \right) = 1 \quad \text{if} \quad m_{k,n-1} = m_{k+1,n}, \quad i \le k < n,$$

$$\lim_{q\to 0} \left(\begin{matrix} \mathbf{m}_n & (1,0) \\ \mathbf{m}_{n-1} & (1,0) \end{matrix} \middle| \begin{matrix} \mathbf{m}^{+i}_n \\ \mathbf{m}^{+j}_{n-1} \end{matrix} \right) = 1 \quad \text{if} \quad i < j, \quad m_{k,n-1} = m_{k+1,n}, \quad i \le k < n,$$

$$\lim_{q\to 0} \left(\begin{matrix} \mathbf{m}_n & (1,0) \\ \mathbf{m}_{n-1} & (1,0) \end{matrix} \middle| \begin{matrix} \mathbf{m}^{+i}_n \\ \mathbf{m}^{+j}_{n-1} \end{matrix} \right) = 1 \quad \text{if} \quad i = j,$$

where $\mathbf{m}^{+i}_n := \mathbf{m}_n + \mathbf{e}_i$. Similarly, when $q \to \infty$, then the reduced CGC's (45) and (46) vanish except for

$$\lim_{q\to \infty} \left(\begin{matrix} \mathbf{m}_n & (1,0) \\ \mathbf{m}_{n-1} & (0,0) \end{matrix} \middle| \begin{matrix} \mathbf{m}^{+i}_n \\ \mathbf{m}_{n-1} \end{matrix} \right) = 1 \quad \text{if} \quad i = 1,$$

$$\lim_{q\to \infty} \left(\begin{matrix} \mathbf{m}_n & (1,0) \\ \mathbf{m}_{n-1} & (1,0) \end{matrix} \middle| \begin{matrix} \mathbf{m}^{+i}_n \\ \mathbf{m}^{+j}_{n-1} \end{matrix} \right) = 1 \text{ if } i = j+1 \text{ or } i = j \text{ and } m_{i,n-1} = m_{in}.$$

Each CGC of the tensor product $T_{\mathbf{m}_n} \otimes T_1$ is a product of several reduced CGC's of the type (46) and one reduced CGC of the type (45). A closer analysis of these reduced CGC's shows that for fixed i and M' all CGC's $\begin{pmatrix} \mathbf{m}_n & (1,\mathbf{0}) & \mathbf{m}_n^{+i} \\ M & M'' & M' \end{pmatrix}$ vanish in the limits $q \to 0$ and $q \to \infty$ except for one. Thus, by (17), for every basis vector $|\mathbf{m}_n^{+i}, M'\rangle$ of the representation space of $T_{\mathbf{m}_n^{+i}}$ from (13), there exists a unique basis vector $|\mathbf{m}_n, M''\rangle \otimes |(1,\mathbf{0}), M\rangle$ of the carrier space of $T_{\mathbf{m}_n} \otimes T_1$ such that in both limits $q \to 0$ and $q \to \infty$ we have

$$|\mathbf{m}_n^{+i}, M'\rangle = |\mathbf{m}_n, M''\rangle \otimes |(1,\mathbf{0}), M\rangle.$$

Now we apply this procedure inductively starting from the tensor product $T_1 \otimes T_1$ of two vector representations T_1 of $U_q(\mathrm{gl}_n)$. Let $\mathbf{v}_1, \mathbf{v}_2, \cdots, \mathbf{v}_n$ denote the elements of the Gel'fand–Tsetlin basis of T_1. Since the irreducible representation $T_{\mathbf{m}_n}$ with highest weight $\mathbf{m}_n = (m_{1n}, m_{2n}, \cdots, m_{nn})$, $m_{nn} \geq 0$, is contained in the tensor product of $m = m_{1n} + \cdots + m_{nn}$ copies of the vector representation T_1, then *at $q = 0$ and at $q = \infty$ every Gel'fand–Tsetlin basis vector of the representation space of $T_{\mathbf{m}_n}$ takes the following simple form:*

$$\mathbf{v}_{i_1} \otimes \mathbf{v}_{i_2} \otimes \cdots \otimes \mathbf{v}_{i_m}, \quad m = m_{1n} + \cdots + m_{nn}, \quad i_j \in \{1, 2, \cdots, n\}.$$

This observation is a starting point for the construction of crystal bases for spaces of finite-dimensional representations of Drinfeld–Jimbo algebras $U'_q(\mathfrak{g})$.

7.4 Crystal Bases

The construction of convenient bases in representation spaces is one of the main problems of applied representation theory. For example, Gel'fand–Tsetlin bases for irreducible representations of the Lie algebras $\mathrm{gl}(N, \mathbb{C})$ and $\mathrm{so}(N, \mathbb{C})$ have been proved to be a useful tool for a number of applications of representations in theoretical physics. In the preceding section an analog of such bases has been constructed for representations of the algebras $\breve{U}_q(\mathrm{gl}_N)$. However, it is not known how to get Gel'fand–Tsetlin bases for $U_q(\mathrm{so}_N)$, because there is no Hopf algebra embedding $U_q(\mathrm{so}_{N-1}) \subseteq U_q(\mathrm{so}_N)$.

This section deals with the construction of Kashiwaras's crystal bases for finite-dimensional representations of the Drinfeld–Jimbo algebra $U'_q(\mathfrak{g})$ over the field $\mathbb{Q}(q)$ defined in Subsect. 6.1.2. Roughly speaking, these are bases of vector spaces at "$q = 0$" that behave well simultaneously with respect to the action of all generators E_i and F_i of $U'_q(\mathfrak{g})$. (These statements are made precise in Definitions 1 and 2 below.) In applications to exactly solvable models in statistical physics the parameter q plays the role of the absolute temperature. Hence crystal bases desribe the behavior of the model at absolute temperature zero which explains the origin of their name.

In this section we give a brief exposition, without proofs, of the main results on crystal bases. The complete proofs of the results are quite involved and can be found in Kashiwara's original paper [Kas2], see also [Kas1] and [Kas3]. For notational simplicity, we shall use the language of left modules rather than of representations, thus following the terminology used in [Kas2].

7.4.1 Crystal Bases of Finite-Dimensional Modules

Recall that $\mathbb{Q}(q)$ is the field of rational functions in one variable q with coefficients in the rational field $\mathbb{Q}$. Let $\mathbb{A}$ be the subring of $\mathbb{Q}(q)$ consisting of all rational functions without poles at $q = 0$. For $a \in \mathbb{A}$ we identify the coset $a + q\mathbb{A}$ with the rational number $a(0)$. Then we have $\mathbb{A}/q\mathbb{A} = \mathbb{Q}$.

Definition 1. *A basis of a vector space V over $\mathbb{Q}(q)$ at $q = 0$ is a pair (L, B) such that L is a free left $\mathbb{A}$-module satisfying $V \simeq \mathbb{Q}(q) \otimes_{\mathbb{A}} L$ and B is a basis of the $\mathbb{Q}$-vector space L/qL.*

Let V_1 and V_2 be vector spaces over $\mathbb{Q}(q)$ with bases (L_1, B_1) and (L_2, B_2) at $q = 0$, respectively. It is easily verified that the pair $(L_1, B_1) \oplus (L_2, B_2) := (L_1 \oplus L_2, B_1 \cup B_2)$ is a basis of the vector space $V_1 \oplus V_2$ at $q = 0$ and that $(L_1, B_1) \otimes (L_2, B_2) = (L_1 \otimes_{\mathbb{A}} L_2, B_1 \otimes B_2)$ is a basis of $V_1 \otimes V_2$ at $q = 0$, where $B_1 \otimes B_2 := \{b_1 \otimes b_2 \,|\, b_1 \in B_1, b_2 \in B_2\}$.

The following example will play a crucial role in what follows.

Example 2. For $l \in \frac{1}{2}\mathbb{N}_0$, let V_l denote the $(2l+1)$-dimensional simple left $U_q'(\mathrm{sl}_2)$-module defined by the formulas (3.22) and (3.23) with $\omega = 1$, and let $f_0 := \mathbf{e}_l \in V_l$ be its highest weight vector. Set $F^{(i)} := F^i/[i]!$ and $f_i^{(l)} := F^{(i)} f_0$ for $i \in \mathbb{N}_0$. The vectors $f_i^{(l)}$, $i = 0, 1, \cdots, 2l$, form a basis of V_l such that

$$K f_i^{(l)} = q^{2(l-i)} f_i^{(l)}, \quad E f_i^{(l)} = [2l - i + 1] f_{i-1}^{(l)}, \quad F f_i^{(l)} = [i] f_{i+1}^{(l)}. \tag{57}$$

If

$$L_l := \bigoplus\nolimits_i \mathbb{A} f_i^{(l)} \quad \text{and} \quad B_l := \{f_i^{(l)} \bmod qL_l \,|\, i = 0, 1, \cdots, 2l\},$$

then the pair (L_l, B_l) is a basis of the vector space V_l at $q = 0$. △

Next we define linear operators $\tilde{E}_i$, $\tilde{F}_i$, $i = 1, 2, \cdots, \mathrm{rank}\, \mathfrak{g}$, on each finite-dimensional type 1 left $U_q'(\mathfrak{g})$-module V. If $\mathfrak{g} = \mathrm{sl}_2$ and $V = V_l$ is as in Example 2, we set $\tilde{E} f_i = f_{i-1}$ and $\tilde{F} f_i = f_{i+1}$ for $i = 0, 1, \cdots, 2l$, where $f_{-1} = f_{2l+1} := 0$. If V is an arbitrary finite-dimensional type 1 left $U_q'(\mathrm{sl}_2)$-module, then we decompose V as a direct sum of modules V_l and define $\tilde{E}$ and $\tilde{F}$ on V as direct sums of the corresponding operators for the modules V_l. Now let V be a finite-dimensional type 1 left $U_q'(\mathfrak{g})$-module. The subalgebra $U_q'(\mathfrak{g})_i$ generated by the elements E_i, F_i, K_i, K_i^{-1} is obviously isomorphic to $U_{q_i}'(\mathrm{sl}_2)$. Hence V is in particular a left $U_{q_i}'(\mathrm{sl}_2)$-module and the operators $\tilde{E}_i$ and $\tilde{F}_i$ are defined by the preceding. Summarizing, if $V = \bigoplus_\lambda V_\lambda$ is the weight space decomposition, then every element $f \in V_\lambda$ is of the form $f = \sum_k F_i^{(k)} f_k$ with $f_k \in V_{\lambda + k\alpha_i}$, $E_i f_k = 0$ and we have defined

$$\tilde{E}_i f = \sum_k F_i^{(k-1)} f_k, \quad \tilde{F}_i f = \sum_k F_i^{(k+1)} f_k.$$

Finally, we can give the definition of crystal bases.

Definition 2. *A* crystal basis *of a finite-dimensional type 1 left* $U_q'(\mathfrak{g})$*-module* V *is a basis* (L, B) *of* V *at* $q = 0$ *such that the following conditions hold:*

(i) $L = \bigoplus_{\lambda \in P} L_\lambda$ *and* $B = \bigcup_{\lambda \in P} B_\lambda$*, where* $L_\lambda = L \cap V_\lambda$ *and* $B_\lambda = B \cap (L_\lambda / qL_\lambda)$.

(ii) $\tilde{E}_i(L) \subseteq L$, $\tilde{F}_i(L) \subseteq L$, $\tilde{E}_i(B) \subseteq B \cup \{0\}$, $\tilde{F}_i(B) \subseteq B \cup \{0\}$ *for* $i = 1, 2, \cdots, n = \operatorname{rank} \mathfrak{g}$.

(iii) *For any* $b_1, b_2 \in B$ *and* $i \in \{1, 2, \cdots, n\}$ *we have* $\tilde{E}_i b_2 = b_1$ *if and only if* $\tilde{F}_i b_1 = b_2$.

Note that because of $\tilde{E}_i(L) \subseteq L$ and $\tilde{F}_i(L) \subseteq L$ the mappings $\tilde{E}_i$ and $\tilde{F}_i$ pass to the quotient L/qL and define mappings, again denoted by $\tilde{E}_i$ and $\tilde{F}_i$, on L/qL, so that $\tilde{E}_i(b)$ and $\tilde{F}_i(b)$ are indeed well-defined for $b \in B$.

Conditions (i)–(iii) express the "nice" behavior of the basis (L, B) with respect to the actions of the generators K_i, E_i, F_i on V. For instance, if $b \in B$, then either $\tilde{E}_i(b) = 0$ or $\tilde{E}_i(b) \in B$ by (ii). Further, if $\tilde{E}_i(b) \in B$, then $b = \tilde{F}_i \tilde{E}_i(b)$ by (iii).

The following assertion follows easily from Definition 2.

Lemma 28. *If* (L_j, B_j) *is a crystal basis of the* $U_q'(\mathfrak{g})$*-modules* V_j, $j = 1, 2, \cdots, r$*, so is the pair* $(L, B) := (\bigoplus_j L_j, \bigcup_j B_j)$ *for the direct sum* $\bigoplus_j V_j$.

7.4.2 Existence and Uniqueness of Crystal Bases

Let us begin with the simplest case $\mathfrak{g} = \mathrm{sl}_2$. The pair (L_l, B_l) from Example 2 is obviously a crystal basis of the $U_q'(\mathrm{sl}_2)$-module V_l. Since any finite-dimensional type 1 $U_q'(\mathrm{sl}_2)$-module V is a direct sum of such modules V_l, it follows at once from Lemma 28 that V has a crystal basis.

The existence of crystal bases in the general case is much more subtle. For $\lambda \in P_+$, let $V(\lambda)$ be the irreducible left $U_q'(\mathfrak{g})$-module with highest weight λ and highest weight vector e_λ. We set

$$L(\lambda) = \sum_{r=0}^{\infty} \sum_{i_1, \cdots, i_r = 1}^{n} \mathbb{A} \tilde{F}_{i_1} \cdots \tilde{F}_{i_r}(e_\lambda),$$

$$B(\lambda) = \{\tilde{F}_{i_1} \cdots \tilde{F}_{i_r}(e_\lambda) \mod qL(\lambda) \mid i_1, \cdots, i_r \in \{1, 2, \cdots, n\}\}.$$

Theorem 29. $(L(\lambda), B(\lambda))$ *is a crystal basis of* $V(\lambda)$. *Any finite-dimensional type 1 left* $U_q'(\mathfrak{g})$*-module has a crystal basis.*

Proof. By Lemma 28, the first assertion implies the second one. It is obvious that $B(\lambda)$ spans the $\mathbb{Q}$-vector space $L(\lambda)/qL(\lambda)$ and that $\tilde{F}_i(B) \subseteq B \cup \{0\}$. The complete proof that $(L(\lambda), B(\lambda))$ is a crystal basis is rather long. It is given in [Kas2] by an induction procedure called the grand loop. □

The following proposition says that crystal bases are unique up to isomorphism.

Proposition 30. *Let (L, B) be a crystal basis of a finite-dimensional type 1 left $U_q'(\mathfrak{g})$-module V. Then there is an isomorphism γ of V to a direct sum $\bigoplus_\lambda V(\lambda)$ of highest weight modules $V(\lambda)$ (possibly with multiplicities) which maps (L, B) to the crystal basis $(\bigoplus_\lambda L(\lambda), \bigcup_\lambda B(\lambda))$ of $\bigoplus_\lambda V(\lambda)$.*

Proof. A proof is given in [Kas2], p. 478, using Theorem 29. □

7.4.3 Crystal Bases of Tensor Product Modules

One advantage of crystal bases is their behavior under tensor products. Before we state the corresponding result, we consider an example.

Example 3. From Subsect. 3.4.1 we know that the tensor product left $U_q'(\mathrm{sl}_2)$-module $V_l \otimes V_{1/2}$ (see Example 2) decomposes as a direct sum $V_{l+1/2} \oplus V_{l-1/2}$. The vectors

$$w_i^{(l+1/2)} = f_i^{(l)} \otimes f_0^{(1/2)} + q^{2l-i+1} f_{i-1}^{(l)} \otimes f_1^{(1/2)}, \qquad i = 0, 1, \cdots, 2l+1,$$

$$w_i^{(l-1/2)} = \frac{q^{-i}[2l-i]}{[2l]} f_i^{(l)} \otimes f_1^{(1/2)} - \frac{q[i+1]}{[2l]} f_{i+1}^{(l)} \otimes f_0^{(1/2)}, \quad i = 0, 1, \cdots, 2l-1,$$

are bases of $V_{l+1/2}$ and $V_{l-1/2}$, respectively, on which K, E and F act by (57). It follows from these formulas that

$$w_i^{(l+1/2)} = f_i^{(l)} \otimes f_0^{(1/2)} \ (\mathrm{mod}\, q(L^{(l)} \otimes L^{(1/2)})), \ i \le 2l, \ w_{2l+1}^{(l+1/2)} = f_{2l}^{(l)} \otimes f_1^{(1/2)} \tag{58}$$

and

$$w_i^{(l-1/2)} = f_i^{(l)} \otimes f_1^{(1/2)} \ (\mathrm{mod}\ q(L^{(l)} \otimes L^{(1/2)})). \tag{59}$$

The pair (L, B), where $L = L^{(l+1/2)} \oplus L^{(l-1/2)}$ and B consists of the vectors (58) and (59), is a crystal basis of $V_l \otimes V_{1/2}$. △

Let (L, B) be a crystal basis of a $U_q'(\mathfrak{g})$-module V. For $b \in B$ we set

$$\epsilon_i(b) = \max\{k \in \mathbb{N} \mid \tilde{E}_i^k(b) \in B\}, \qquad \phi_i(b) = \max\{k \in \mathbb{N} \mid \tilde{F}_i^k(b) \in B\}.$$

Theorem 31. (i) *Let (L_1, B_1) and (L_2, B_2) be crystal bases of two $U_q'(\mathfrak{g})$-modules V_1 and V_2, respectively. Then $(L_1, B_1) \otimes (L_2, B_2)$ is a crystal basis of $V_1 \otimes V_2$.*

(ii) *If $b_1 \in B_1$ and $b_2 \in B_2$, then for any $i \in \{1, 2, \cdots, n\}$ we have*

$$\tilde{E}_i(b_1 \otimes b_2) = \begin{cases} \tilde{E}_i b_1 \otimes b_2 & \text{for} \quad \phi_i(b_1) \ge \epsilon_i(b_2), \\ b_1 \otimes \tilde{E}_i b_2 & \text{for} \quad \phi_i(b_1) < \epsilon_i(b_2), \end{cases}$$

$$\tilde{F}_i(b_1 \otimes b_2) = \begin{cases} \tilde{F}_i b_1 \otimes b_2 & \text{for} \quad \phi_i(b_1) > \epsilon_i(b_2), \\ b_1 \otimes \tilde{F}_i b_2 & \text{for} \quad \phi_i(b_1) \le \epsilon_i(b_2). \end{cases}$$

Sketch of Proof. (i) follows easily from (ii). To prove (ii), it suffices to assume that $\mathfrak{g} = \mathrm{sl}_2$. In this case the assertion can be derived from the formulas for the Clebsch–Gordan coefficients. See [Kas2] for details. □

To a crystal basis one can associate its crystal graph. A *directed graph* consists of a set of *vertices* and a set of ordered pairs of vertices, called *arrows.* A directed graph is called *colored* if it is equipped with a map from the set of arrows to another set, the so-called set of *colors.*

Let (L, B) be a crystal basis of a $U'_q(\mathfrak{g})$-module V. We define the *crystal graph* of (L, B) as the colored graph with set of vertices B and set of colors $\{1, 2, \cdots, n\}$, $n = \operatorname{rank} \mathfrak{g}$, such that elements $b, b' \in B$ are connected by an arrow, directed from b to b', with color i if and only if $b' = \tilde{F}_i(b)$ (and hence $b = \tilde{E}_i(b')$ by Definition 2(iii)).

The crystal graph of a crystal basis $(L(\lambda), B(\lambda))$, $\lambda \in P_+$, is a connected graph, because every vertex $b \in B(\lambda)$ is obviously connected with the highest weight vector.

Suppose that (L, B) is a crystal basis of a type 1 finite-dimensional $U'_q(\mathfrak{g})$-module V which is isomorphic to a direct sum $\bigoplus_i V(\lambda_i)$ of irreducible modules $V(\lambda_i)$, $\lambda_i \in P_+$. It can be shown that then the crystal graph of (L, B) is (isomorphic to) the disjoint union of the crystal graphs of all $(L(\lambda_i), B(\lambda_i))$. Thus, the decomposition of V into irreducible components is equivalent to the decomposition of the crystal graph of (L, B) into connected components.

From Theorem 31 we can read off the crystal graph of the tensor product $V_1 \otimes V_2$ of two $U'_q(\mathfrak{g})$-modules V_1 and V_2. Without going further into technical details we only state that it is obtained as the union of tensor products of each component of the crystal graph of (L_1, B_1) with each component of the crystal graph of (L_2, B_2). Combined with the result of the preceding paragraph, one obtains in this manner an interesting graphic algorithm for the irreducible decomposition of the tensor product $V_1 \otimes V_2$ (see [KN] and [Na]). We illustrate this by a simple

Example 4. The crystal graph of the crystal basis of the $U'_q(\mathrm{sl}_2)$-module V_1 is $f_0^{(1)} \longrightarrow f_1^{(1)} \longrightarrow f_2^{(1)}$. Hence the crystal graph of $V_1 \otimes V_1$ is

$$\begin{array}{ccccc} f_0^{(1)} \otimes f_0^{(1)} & \rightarrow & f_1^{(1)} \otimes f_0^{(1)} & \rightarrow & f_2^{(1)} \otimes f_0^{(1)} \\ & & & & \downarrow \\ f_0^{(1)} \otimes f_1^{(1)} & \rightarrow & f_1^{(1)} \otimes f_1^{(1)} & & f_2^{(1)} \otimes f_1^{(1)} \\ & & \downarrow & & \downarrow \\ f_0^{(1)} \otimes f_2^{(1)} & & f_1^{(1)} \otimes f_2^{(1)} & & f_2^{(1)} \otimes f_2^{(1)} \end{array}$$

Therefore, as discussed above, $V_1 \otimes V_1$ decomposes as $V_2 \oplus V_1 \oplus V_0$. △

7.4.4 Globalization of Crystal Bases

Let $(L(\lambda), B(\lambda))$ be the crystal basis of the $U'_q(\mathfrak{g})$-module $V(\lambda)$, $\lambda \in P_+$. Then $B(\lambda)$ is a basis of the $\mathbb{Q}$-vector space $L(\lambda)/qL(\lambda)$. In this subsection we shall show how to obtain a basis of the $\mathbb{Q}(q)$-vector space $V(\lambda)$ from $B(\lambda)$. Let us fix an irreducible finite-dimensional $U_q'(\mathfrak{g})$-module $V(\lambda)$, $\lambda \in P_+$, with highest weight vector $\mathbf{e}_\lambda$.

There exists a $\mathbb{Q}$-algebra automorphism η of $U_q'(\mathfrak{g})$ defined by

$$\eta(E_i) = E_i, \quad \eta(F_i) = F_i, \quad \eta(K_i) = K_i^{-1}, \quad \eta(q) = q^{-1}.$$

From Proposition 6.28 it follows easily that there is a $\mathbb{Q}$-linear mapping $\eta : V(\lambda) \to V(\lambda)$ such that $\eta(a\mathbf{e}_\lambda) = \eta(a)\mathbf{e}_\lambda$, $a \in U_q'(\mathfrak{g})$. Clearly, $\bar{\mathbb{A}} := \eta(\mathbb{A})$ is the subring of functions in $\mathbb{Q}(q)$ without poles at $q = \infty$ and $\overline{L(\lambda)} = \eta(L(\lambda))$ is a free $\bar{\mathbb{A}}$-module such that $V(\lambda) = \mathbb{Q}(q) \otimes_{\bar{\mathbb{A}}} \overline{L(\lambda)}$. Further, we set

$$V_{\mathbb{Q}}(\lambda) = \sum \mathbb{Q}[q, q^{-1}] F_{i_1}^{(n_1)} \cdots F_{i_r}^{(n_r)} \mathbf{e}_\lambda.$$

The crucial step for a globalization of crystal bases is

Proposition 32. *The natural map $\zeta : L(\lambda) \cap \overline{L(\lambda)} \cap V_{\mathbb{Q}}(\lambda) \to L(\lambda)/qL(\lambda)$ is an isomorphism of $\mathbb{Q}$-vector spaces.*

Let $\mathcal{B}(\lambda)$ be the set $\zeta^{-1}(B(\lambda))$. Then we have

Theorem 33. *$\mathcal{B}(\lambda)$ is a basis of the $\mathbb{Q}(q)$-vector space $V(\lambda)$.*

The basis $\mathcal{B}(\lambda)$ of $V(\lambda)$ is called a *global crystal basis* of $V(\lambda)$. One can also prove that

$$F_i^m V(\lambda) = \bigoplus_{b \in \mathcal{B}(\lambda), \epsilon_i(b) \geq m} \mathbb{Q}(q) b, \qquad E_i^m V(\lambda) = \bigoplus_{b \in \mathcal{B}(\lambda), \phi_i(b) \geq m} \mathbb{Q}(q) b,$$

where ϵ_i and ϕ_i are as in Subsect. 7.4.3.

7.4.5 Crystal Bases of $U_q'(\mathfrak{n}_-)$

In Subsects. 7.4.1–4 crystal bases of type 1 finite-dimensional $U_q'(\mathfrak{g})$-modules have been investigated. A similar theory can be developed for the subalgebra $U_q'(\mathfrak{n}_-)$ of $U_q'(\mathfrak{g})$ which is generated by the elements F_i, $i = 1, 2, \cdots, n = \operatorname{rank} \mathfrak{g}$. This will be sketched in this subsection.

In order to imitate Definition 2 we first need to define the counterpart to the operators $\tilde{E}_i$ and $\tilde{F}_i$ for $U_q'(\mathfrak{n}_-)$. Putting

$$E_i'(x) := -(q_i - q_i^{-1}) K_i (E_i x - x E_i), \quad x \in U_q'(\mathfrak{n}_-),$$

one easily verifies that

$$E_i'(F_j x) = q^{-(\alpha_i, \alpha_j)} F_j E_i'(x) + \delta_{ij} x, \tag{60}$$

$$E_i'(xy) = E_i'(x) y + \sigma_i(x) E_i'(y), \tag{61}$$

where $\sigma_i(x) := K_i x K_i^{-1}$ and $x, y \in U_q'(\mathfrak{n}_-)$. Equation (61) means that E_i' is a *skew-derivation* of the algebra $U_q'(\mathfrak{n}_-)$ with respect to the automorphism σ_i.

By (60) and (61), we have $E_i'(F_i) = 1$ and $E_i'(F_i^m) = q_i^{1-m}[m]F_i^{m-1}$. Further, the linear mapping E_i' leaves $U_q'(\mathfrak{n}_-)$ invariant and for any $x \in U_q'(\mathfrak{n}_-)$ there exists $m \in \mathbb{N}$ such that $(E_i')^m(x) = 0$. Using the preceding facts it can be shown (see [Kas2], Subsect. 3.2.1) that any $x \in U_q'(\mathfrak{n}_-)$ is of the form

$$x = \sum\nolimits_{m=0}^{k} F_i^{(m)}(y_m), \quad E_i'(y_m) = 0, \quad k \in \mathbb{N},$$

with uniquely determined elements $y_m \in U_q'(\mathfrak{n}_-)$. Then we define

$$\tilde{E}_i(x) = \sum\nolimits_{m=1}^{k} F_i^{(m-1)}(y_m), \quad \tilde{F}_i(x) = \sum\nolimits_{m=0}^{k} F_i^{(m+1)}(y_m).$$

Now Definition 2 can be adapted to the present situation as follows. A *crystal basis of* $U_q'(\mathfrak{n}_-)$ is a basis of the $\mathbb{Q}(q)$-vector space $U_q'(\mathfrak{n}_-)$ at $q = 0$ (see Definition 1) such that the conditions (ii) and (iii) of Definition 2 are fulfilled. Note that condition (i) of Definition 2 is not needed here. Since obviously $\tilde{E}_i\tilde{F}_i = I$, it follows from (ii) that $\tilde{F}_i(b) \in B$ for any $b \in B$. One can prove that $\tilde{F}_i\tilde{E}_i$ is the projection of $U_q'(\mathfrak{n}_-)$ onto $F_i \cdot U_q'(\mathfrak{n}_-)$ with respect to the direct sum decomposition $U_q'(\mathfrak{n}_-) = F_i \cdot U_q'(\mathfrak{n}_-) \oplus \ker E_i'$.

Let $L(\infty)$ be the $\mathbb{A}$-linear span of elements $\tilde{F}_{i_1} \cdots \tilde{F}_{i_r} \cdot 1$ for $i_1, \cdots, i_r \in \{1, 2, \cdots, n\}$, and let $B(\infty)$ be the set of images of these elements in $L(\infty)/qL(\infty)$. Further, by Proposition 6.28, every irreducible $U_q'(\mathfrak{g})$-module $V(\lambda)$, $\lambda \in P_+$, is isomorphic to a quotient of the Verma module M_λ, which can be realized as a $U_q'(\mathfrak{g})$-module $U_q'(\mathfrak{n}_-)$ by Subsect. 6.2.5. Therefore, there exists a homomorphism π_λ of the $U_q'(\mathfrak{g})$-module $U_q'(\mathfrak{n}_-)$ to $V(\lambda)$.

Proposition 34. (i) *$(L(\infty), B(\infty))$ is a crystal basis of $U_q'(\mathfrak{n}_-)$.*

(ii) *The crystal basis $(L(\lambda), B(\lambda))$ of the $U_q'(\mathfrak{g})$-module $V(\lambda)$ (see Subsect. 7.4.2) is obtained by $L(\lambda) = \pi_\lambda(L(\infty))$ and $B(\lambda) = \{x \in B(\infty) \mid \overline{\pi}_\lambda(b) \neq 0\}$, where $\overline{\pi}_\lambda : L(\infty)/qL(\infty) \to L(\lambda)/qL(\lambda)$ is the induced map of $\pi_\lambda : L(\infty) \to L(\lambda)$.*

Also the globalization of the crystal basis $(L(\infty), B(\infty))$ of $U_q'(\mathfrak{n}_-)$ proceeds in a similar manner as in the case of $U_q'(\mathfrak{g})$-modules. If $U_{\mathbb{Q}}'(\mathfrak{n}_-)$ denotes the $\mathbb{Q}[q, q^{-1}]$-subalgebra of $U_q'(\mathfrak{n}_-)$ generated by the elements $F_i^{(m)}$, then it can be proved that the canonical map

$$\zeta : U_{\mathbb{Q}}'(\mathfrak{n}_-) \bigcap L(\infty) \bigcap \overline{L(\infty)} \to L(\infty)/qL(\infty)$$

is an isomorphism of $\mathbb{Q}$-vector spaces.

Theorem 35. (i) *The set $\mathcal{B}(\infty) := \zeta^{-1}(B(\infty))$ is a basis of the $\mathbb{Q}(q)$-vector space $U_q'(\mathfrak{n}_-)$.*

(ii) *For each $\lambda \in P_+$, the homomorphism π_λ maps the set of all $b \in \mathcal{B}(\infty)$ such that $\pi_\lambda(b) \neq 0$ to a basis of the $\mathbb{Q}(q)$-vector space $V(\lambda)$.*

The set $\mathcal{B}(\infty)$ is called Kashiwara's *global crystal basis of* $U_q'(\mathfrak{n}_-)$. It can be shown (see [L5] and [GL]) that it coincides with Lusztig's *canonical basis* which was constructed in [L4] and [L5] using a different approach. A detailed exposition of the latter can be found in the book [Lus].

7.5 Representations of $U_q(\mathfrak{g})$ for q a Root of Unity

In this section we study representations of Drinfeld–Jimbo algebras $U_q(\mathfrak{g})$ when q is a primitive p-th root of unity, where p is odd and $p > d_i$, $i = 1, 2, \cdots, l$ (l is a rank of $\mathfrak{g}$). We denote the parameter q in this case by ϵ.

7.5.1 General Results

In what follows, we shall use the notation and some facts from Subsect. 6.3.5. Recall that $\mathfrak{z}_\epsilon$ is the center of the algebra $U_\epsilon(\mathfrak{g})$ and $\mathrm{Spec}\,(\mathfrak{z}_\epsilon)$ is the set of all algebra homomorphisms $\chi : \mathfrak{z}_\epsilon \to \mathbb{C}$. Let $\mathrm{Rep}\,(U_\epsilon(\mathfrak{g}))$ denote the set of all equivalence classes of irreducible representations of $U_\epsilon(\mathfrak{g})$.

If T is an irreducible representation of $U_\epsilon(\mathfrak{g})$ on V, then any $T(Z)$, $Z \in \mathfrak{z}_\epsilon$, is a scalar operator on V, that is, there is a complex number $\chi_T(Z)$ such that $T(Z) = \chi_T(Z)I$. Obviously, $\chi_T : \mathfrak{z}_\epsilon \to \mathbb{C}$ is an algebra homomorphism, called the *central character* of T. Clearly, equivalent representations have the same central character. Therefore, the assignment $T \to \chi_T$ defines a map

$$\Psi : \mathrm{Rep}\,(U_\epsilon(\mathfrak{g})) \to \mathrm{Spec}\,(\mathfrak{z}_\epsilon). \tag{62}$$

Proposition 36. *Every irreducible representation of the algebra $U_\epsilon(\mathfrak{g})$ is finite-dimensional.*

Proof. Let $\mathcal{I}^{\chi_T}$ be the two-sided ideal of $U_\epsilon(\mathfrak{g})$ generated by the set $\ker \chi_T \equiv \{Z - \chi_T(Z)1 \mid Z \in \mathfrak{z}_\epsilon\}$. Since $\ker \chi_T \subseteq \ker T$, the irreducible representation T of $U_\epsilon(\mathfrak{g})$ induces an irreducible representation $\tilde{T}$ of the quotient algebra $U_\epsilon^{\chi_T} := U_\epsilon(\mathfrak{g})/\mathcal{I}^{\chi_T}$. By Theorem 6.49, the algebra $U_\epsilon^{\chi_T}$ is finite-dimensional. Hence its irreducible representation $\tilde{T}$ is finite-dimensional (see [CR]). □

Proposition 37. *The mapping Ψ is surjective, that is, for any $\chi \in \mathrm{Spec}\,(\mathfrak{z}_\epsilon)$ there exists an irreducible representation T of $U_\epsilon(\mathfrak{g})$ such that $\chi_T = \chi$.*

Idea of proof. As in the preceding proof, $U_\epsilon^\chi := U_\epsilon(\mathfrak{g})/\mathcal{I}^\chi$ is a finite-dimensional algebra, where $\mathcal{I}^\chi$ is the two-sided ideal generated by $\ker \chi$. It can be shown that $U_\epsilon^\chi \neq \{0\}$. Any irreducible subrepresentation of the regular representation of U_ϵ^χ gives an irreducible representation, say T, of $U_\epsilon(\mathfrak{g})$. Since T annihilates $\ker \chi$, we get $\chi_T = \chi$. □

The mapping Ψ is not injective in general. The main properties of this mapping are contained in the following

Theorem 38. *There exists a nonempty closed proper subset D of* $\mathrm{Spec}\,(\mathfrak{z}_\epsilon)$ *such that:*

(i) *If $\chi \in \mathrm{Spec}\,(\mathfrak{z}_\epsilon)$ and $\chi \notin D$, then $\Psi^{-1}(\chi)$ consists of one irreducible representation. This representation is of dimension p^n, where n is the number of positive roots of $\mathfrak{g}$.*

(ii) *If $\chi \in D$, then $\Psi^{-1}(\chi)$ consists of a finite number of irreducible representations and their dimensions are strictly less than p^n.*

Proof. A proof is given in the paper [DCK]. During this proof it is shown that for any $\chi \in \mathrm{Spec}\,(\mathfrak{z}_\epsilon)$, $\chi \notin D$, the quotient algebra $U_\epsilon^\chi = U_\epsilon(\mathfrak{g})/\mathcal{I}^\chi$ is isomorphic to the matrix algebra $M_{p^n}(\mathbb{C})$. This algebra has, up to equivalence, a unique irreducible representation which is of dimension p^n. □

Recall from Subsect. 6.3.5 that the center $\mathfrak{z}_\epsilon$ is algebraic over $\mathfrak{z}_0$ (that is, each element of $\mathfrak{z}_\epsilon$ satisfies a nontrivial algebraic equation with coefficients in $\mathfrak{z}_0$) and that the set Spec $(\mathfrak{z}_0)$ is isomorphic to $\mathbb{C}^{2n} \times (\mathbb{C}^\times)^l$, where $\mathbb{C}^\times = \mathbb{C}\backslash\{0\}$, n is the number of positive roots of the Lie algebra $\mathfrak{g}$ and l is the rank of $\mathfrak{g}$. From these facts we conclude that Theorem 38 gives a parametrization of the irreducible representations T of $U_\epsilon(\mathfrak{g})$ such that $\chi_T \notin D$ by $2n+l = \dim \mathfrak{g}$ complex parameters. If $\chi_T \in D$, then additional parameters are needed.

Note that the set Spec $(\mathfrak{z}_\epsilon)$ is a complex algebraic variety of dimension p^{2n+l}. It can be shown (see [DCK]) that D is a proper subvariety. This means that almost all irreducible representations T of $U_\epsilon(\mathfrak{g})$ are of dimension p^n.

Let us illustrate Theorem 38 by the simplest example.

Example 5 ($U_\epsilon(\mathrm{sl}_2)$). By Proposition 3.15, the center $\mathfrak{z}_\epsilon$ of the algebra $U_\epsilon(\mathrm{sl}_2)$ is generated by the elements E^p, F^p, K^p, K^{-p} and C_ϵ. (Note that q has been replaced by ϵ.) All irreducible representations of $U_\epsilon(\mathrm{sl}_2)$ have been classified in Sect. 3.3. Recall that these are the cyclic representations $T_{ab\lambda}$, the semicyclic representations $T_{0b\lambda}$ and $T'_{0b\lambda}$, and the irreducible representations $T_{\omega l}$, $2l < p$ (see Propositions 3.17–19). All representations $T_{ab\lambda}$, $T'_{0b\lambda}$ and $T_{\omega,(p-1)/2}$ are p-dimensional. The only irreducible representations of dimensions less than p are $T_{\omega l}$, $l < (p-1)/2$. For these representations we have $T_{\omega l}(E^p) = T_{\omega l}(F^p) = 0$, $T_{\omega l}(K^p) = \omega I$ and $T_{\omega l}(C_\epsilon) = \omega c_l := \omega(\epsilon^{2l+1} + \epsilon^{-2l-1})(\epsilon - \epsilon^{-1})^{-2} I$. The set Spec $(\mathfrak{z}_\epsilon)$ is determined by the points $(x, y, z, c) \in \mathbb{C}^4$, where $T(E^p) = xI$, $T(F^p) = yI$, $T(K^p) = zI$, $T(C_\epsilon) = cI$ and T is an arbitrary irreducible representation of $U_\epsilon(\mathrm{sl}_2)$. The 3-tuple (x, y, z) runs over the set $\mathbb{C}^2 \times \mathbb{C}^\times$ and for each such 3-tuple (x, y, z) the corresponding values c are related to x, y, z by formula (6.110). Thus the subset $D \subset$ Spec $(\mathfrak{z}_\epsilon)$ consists of the points

$$(0, 0, \omega, \omega c_l), \qquad \omega = \pm 1, \quad l = 0, \tfrac{1}{2}, 1, \cdots \tfrac{p}{2} - 1. \qquad \triangle$$

Proposition 39. *For any nonzero $a \in U_\epsilon(\mathfrak{g})$ there exists an irreducible representation T of $U_\epsilon(\mathfrak{g})$ such that $T(a) \neq 0$.*

Proof. By Theorem 6.49, the algebra $U_\epsilon(\mathfrak{g})$ is a finite-dimensional vector space over $\mathfrak{z}_0$ with a certain basis, say $x_1, \cdots, x_d$. Hence we can write a as $a = \sum_i x_i z_i$ with $z_i \in \mathfrak{z}_0$. Since $a \neq 0$, there is an index $k \in \{1, \cdots, d\}$ such that $z_k \neq 0$. By the remarks after Theorem 6.50, Spec $(\mathfrak{z}_0)$ separates the elements of $\mathfrak{z}_0$. Hence there exists $\chi' \in$ Spec $(\mathfrak{z}_0)$ such that $\chi'(z_k) \neq 0$. We choose a character $\chi \in$ Spec $(\mathfrak{z}_\epsilon)$ such that $\chi' = \chi$ on $\mathfrak{z}_0$. Since D is a proper subvariety of Spec $(\mathfrak{z}_\epsilon)$, we can choose χ' such that $\chi \notin D$.

Now we show that a has a nonzero image in $U_\epsilon(\mathfrak{g})/\mathcal{I}^\chi$. Assume on the contrary that $a \in \mathcal{I}^\chi$. Since $\mathcal{I}^\chi$ is the two-sided ideal of $U_\epsilon(\mathfrak{g})$ generated by ker χ, it follows then that a can be expressed as $a = \sum_j b_j y_j$, where

$b_j \in U_\epsilon(\mathfrak{g})$, $y_j \in \mathfrak{z}_\epsilon$ and $\chi(y_j) = 0$. If we represent the elements b_j also as $b_j = \sum_i x_i z_{ij}$ with $z_{ij} \in \mathfrak{z}_0$, we obtain $a = \sum_i x_i z_i = \sum_{i,j} x_i z_{ij} y_j$. Comparing the coefficients of the basis element x_k we get $z_k = \sum_j z_{kj} y_j$. Since $\chi(z_k) = \chi'(z_k) \neq 0$ and $\chi(y_j) = 0$ for all j by construction, this is a contradiction. Hence the canonical image of a in $U_\epsilon(\mathfrak{g})/\mathcal{I}^\chi$ is indeed nonzero.

As noted in the proof of Theorem 38, $U_\epsilon(\mathfrak{g})/\mathcal{I}^\chi$ is isomorphic to the matrix algebra $M_{p^n}(\mathbb{C})$. Hence the composition of the quotient map $U_\epsilon(\mathfrak{g}) \to U_\epsilon(\mathfrak{g})/\mathcal{I}^\chi$ with the isomorphism $U_\epsilon(\mathfrak{g})/\mathcal{I}^\chi \simeq M_{p^n}(\mathbb{C})$ defines a p^n-dimensional irreducible representation T of $U_\epsilon(\mathfrak{g})$ such that $T(a) \neq 0$. □

Let T be an irreducible representation of $U_\epsilon(\mathfrak{g})$ on V. As in the case when q is not a root of unity, one can prove that T is a weight representation, that is, $V = \bigoplus_\lambda V_\lambda$, where $V_\lambda = \{\mathbf{v} \in V \mid T(K_i)\mathbf{v} = \epsilon^{(\lambda,\alpha_i)}\mathbf{v},\ i = 1, 2, \cdots, l\}$. However, T is not necessarily a highest or lowest weight representation.

The problems of classification and explicit descriptions of all irreducible representations of $U_\epsilon(\mathfrak{g})$ are only partially solved. For instance, the dimensions of the representations T with $\chi_T \in D$ are not even known (see [DCKP]).

7.5.2 Cyclic Representations

An important class of irreducible representations of the Drinfeld–Jimbo algebras $U_\epsilon(\mathfrak{g})$ are the so-called cyclic representations.

Definition 3. *An irreducible representation T of $U_\epsilon(\mathfrak{g})$ is called* cyclic *if $T(E_i)^p$ and $T(F_i)^p$, $i = 1, 2, \cdots, l$, are nonzero scalar operators.*

For Proposition 40 and Corollary 41 we assume that p and the determinant of the symmetrized Cartan matrix $(d_i a_{ij})$ have no common prime factor.

Proposition 40. *Let T be a cyclic representation of $U_\epsilon(\mathfrak{g})$ on a vector space V. Then all weight subspaces of V are of the same dimension and the dimension of V is divisible by p^l, where l is the rank of $\mathfrak{g}$. Moreover, T is neither a highest nor a lowest weight representation.*

Sketch of proof. If V_λ and V_μ are two weight subspaces of V, then $\lambda - \mu = \sum_i k_i \alpha_i$, where $k_i \in \mathbb{Z}$ and $\alpha_1, \cdots, \alpha_l$ are the simple roots of $\mathfrak{g}$. Set $X_i^{k_i} = E_i^{k_i}$ if $k_i > 0$ and $X_i^{k_i} = F_i^{k_i}$ for $k_i < 0$. By the definition of a cyclic representation, $T(X_1^{k_1} X_2^{k_2} \cdots X_l^{k_l})$ is an invertible operator from V_μ to V_λ. Therefore, $\dim V_\lambda = \dim V_\mu$. Since $\ker T(E_i) = \ker T(F_i) = \{0\}$ by Definition 3, T is not a highest and not a lowest weight representation.

Finally, we prove that $\dim V$ is divisible by p^l. Because $V = \bigoplus_\lambda V_\lambda$ as noted above and $\dim V_\lambda = \dim V_\mu$ as just shown, it suffices to prove that $V_\lambda = V_\mu$ if and only if $\lambda - \mu \in pQ$, where Q is the set of all $\sum_i n_i \alpha_i$, $n_i \in \mathbb{Z}$. By (6.13), $V_\lambda = V_\mu$ means that $\sum_j d_i a_{ij} k_j \equiv 0 \pmod{p}$, $i = 1, 2, \cdots, l$. Since p is coprime to $\det(d_i a_{ij})$ by assumption, this is true if and only if all k_j are divisible by p. Hence $\dim V$ is divisible by p^l. □

Corollary 41. *If T is a cyclic representation of $U_\epsilon(\mathfrak{g})$ on a vector space V, then $p^l \le \dim V \le p^n$, where l is the rank of $\mathfrak{g}$ and n is the number of positive roots of $\mathfrak{g}$.*

In order to construct explicitly cyclic representations of $U_\epsilon(\mathfrak{g})$, one uses compositions of certain algebra homomorphisms of $U_\epsilon(\mathfrak{g})$ to an auxiliary algebra $\mathfrak{W}_{q,m}$ defined below with irreducible representations of $\mathfrak{W}_{q,m}$.

Let $\mathfrak{W}_\epsilon$ denote the algebra with generators x, x^{-1}, z, z^{-1} and defining relations

$$zx = \epsilon xz, \quad xx^{-1} = x^{-1}x = zz^{-1} = z^{-1}z = 1.$$

In the literature the algebra $\mathfrak{W}_\epsilon$ is occasionally refered to as the *q-Weyl algebra.*

Let X and Z be the operators acting on a p-dimensional vector space V with basis $\mathbf{e}_0, \mathbf{e}_1, \cdots, \mathbf{e}_{p-1}$ by the matrices

$$X = \begin{pmatrix} 0 & 1 & 0 & \cdots & 0 \\ 0 & 0 & 1 & \cdots & 0 \\ \cdots & \cdots & \cdots & \cdots & \cdots \\ 0 & 0 & 0 & \cdots & 1 \\ 1 & 0 & 0 & \cdots & 0 \end{pmatrix}, \quad Z = \begin{pmatrix} \epsilon^0 & 0 & 0 & \cdots & 0 \\ 0 & \epsilon^1 & 0 & \cdots & 0 \\ 0 & 0 & \epsilon^2 & \cdots & 0 \\ \cdots & \cdots & \cdots & \cdots & \cdots \\ 0 & 0 & 0 & \cdots & \epsilon^{p-1} \end{pmatrix},$$

that is,

$$Z\mathbf{e}_i = \epsilon^i \mathbf{e}_i, \qquad X\mathbf{e}_i = \mathbf{e}_{i+1}, \qquad \text{where} \qquad \mathbf{e}_p \equiv \mathbf{e}_0. \tag{63}$$

Since $\epsilon^p = 1$ by assumption, we have $ZX = \epsilon XZ$ and $X^p = Z^p = I$. Thus there is an irreducible representation π of $\mathfrak{W}_\epsilon$ on the vector space V such that $\pi(x) = X$ and $\pi(z) = Z$.

For $m \in \mathbb{N}$, let $\mathfrak{W}_{\epsilon,m}$ be the m-fold tensor product of m (commuting) copies of the algebra $\mathfrak{W}_\epsilon$. The corresponding generators of $\mathfrak{W}_{\epsilon,m}$ are denoted by x_i and z_i, $i = 1, 2, \cdots, m$. Let $\mathbf{g} = (g_1, g_2, \cdots, g_m)$ and $\mathbf{h} = (h_1, h_2, \cdots, h_m)$ be m-tuples of nonzero complex numbers g_i, h_i. We define an irreducible representation $\pi_{\mathbf{gh}}$ of the algebra $\mathfrak{W}_{\epsilon,m}$ on the vector space $V^{\otimes m}$ by

$$\pi_{\mathbf{gh}}(x_i) = 1 \otimes \cdots \otimes 1 \otimes g_i X \otimes 1 \otimes \cdots \otimes 1,$$

$$\pi_{\mathbf{gh}}(z_i) = 1 \otimes \cdots \otimes 1 \otimes h_i Z \otimes 1 \otimes \cdots \otimes 1,$$

where X and Z both act on the i-th tensor factor.

7.5.3 Cyclic Representations of the Algebra $U_\epsilon(\mathrm{sl}_{l+1})$

Set $m := l(l+1)/2$. We shall label the generators of $\mathfrak{W}_{\epsilon,m}$ and the entries of the m-tuples $\mathbf{g}$ and $\mathbf{h}$ from Subsect. 7.5.2 as x_{ij}, z_{ij} and g_{ij}, h_{ij}, respectively, with $i \le j$, $i, j = 1, 2, \cdots, l$. The main step for the construction of cyclic representations of $U_\epsilon(\mathrm{sl}_{l+1})$ is the algebra homomorphism $\rho_{\mathbf{rs}}$ from

Proposition 42. *Let* $\mathbf{r} = (r_1, r_2, \cdots, r_l)$ *and* $\mathbf{s} = (s_1, s_2, \cdots, s_l)$ *be two l-tuples of nonzero complex numbers. Then there exists a unique algebra homomorphism* $\rho_{\mathbf{rs}} : U_\epsilon(\mathrm{sl}_{l+1}) \to \mathfrak{W}_{\epsilon,m}$ *such that* $\rho_{\mathbf{rs}}(K_i) = r_i s_i^{-1} z_{il}^2 z_{i-1,l}^{-1} z_{i+1,l}^{-1}$ *and*

$$\rho_{\mathbf{rs}}(E_i) = \sum\nolimits_{k=i}^{l} \{r_i z_{ik} z_{i,k-1} z_{i-1,k-1}^{-1} z_{i+1,k}^{-1}\} x_{ik} z_{i,k+1} \cdots x_{il},$$

$$\rho_{\mathbf{rs}}(F_i) = \sum\nolimits_{k=1}^{i} \{s_i z_{i+1-k,l-k} z_{i+1-k,l+1-k}^{-1} z_{i-k,l+1-k} z_{i-k,l-k}^{-1}\}$$

$$\times x_{i+1-k,l+1-k}^{-1} x_{i+2-k,l+2-k}^{-1} \cdots x_{il}^{-1},$$

where $\{z\}$ *denotes the expression* $\{z\} = (z - z^{-1})(\epsilon - \epsilon^{-1})^{-1}$.

Proof. A proof is given in [DJMM2]. □

Now we define a representation $\sigma \equiv \sigma_{\mathbf{rsgh}}$ of $U_\epsilon(\mathrm{sl}_{l+1})$ on $V^{\otimes m}$ by composing the homomorphism $\rho_{\mathbf{rs}} : U_\epsilon(\mathrm{sl}_{l+1}) \to \mathfrak{W}_{\epsilon,m}$ and the representation $\pi_{\mathbf{gh}} : \mathfrak{W}_{\epsilon,m} \to \mathcal{L}(V^{\otimes m})$. The representation $\sigma_{\mathbf{rsgh}}$ depends then on $2l + 2m$ complex parameters r_i, s_i, g_{ij}, h_{ij}. However, not all of these parameters are independent. It can be shown that the same set of representations $\sigma_{\mathbf{rsgh}}$ is obtained if we put $s_i = 1$, $i = 1, 2, \cdots, l$. Then $\sigma \equiv \sigma_{\mathbf{rgh}}$ depends on $\dim \mathrm{sl}_{l+1} = l(l+2)$ complex parameters.

Theorem 43. *For a generic choice of the parameters* $\mathbf{r}, \mathbf{g}, \mathbf{h}$ *(that is, except for a set of Lebesgue measure zero in* $\mathbb{C}^{l(l+2)}$*) the representations* $\sigma = \sigma_{\mathbf{rgh}}$ *of* $U_\epsilon(\mathrm{sl}_{l+1})$ *defined above are irreducible and cyclic.*

Proof. The proof can be found in [DJMM2]. □

Recall that the dimension of the carrier space $V^{\otimes m}$ of any representation $\sigma_{\mathbf{rgh}}$ is p^m, where $m = l(l+1)/2$ is the number of positive roots of the Lie algebra $\mathrm{sl}(l+1, \mathbb{C})$. That is, all irreducible representations $\sigma_{\mathbf{rgh}}$ obtained from Theorem 43 have the *maximal* dimension (see Corollary 41).

For special values of the parameters g_{ij}, h_{ij}, r_i, the representations σ are reducible (but not completely reducible in general) and we get invariant subspaces of the carrier space $V^{\otimes l(l+1)/2}$. Let us treat such an example.

Let i and j be integers such that $0 \le i \le p-1$ and $1 \le j \le l-1$. In (63) we used the basis $\mathbf{e}_k$, $k = 0, 1, 2, \cdots, p-1$, of the space V. Let $\mathbf{e}_{i,j}$ be the vector from the tensor product of $j(j+1)/2$ copies of the space V of the form $\mathbf{e}_{i,j} = \mathbf{e}_i \otimes \mathbf{e}_i \otimes \cdots \otimes \mathbf{e}_i$, where the vectors $\mathbf{e}_i$ belong to the tensor factors V labeled by all pairs (k, n), $1 \le k \le n \le j$. We consider the linear subspace $V_{i,j} \equiv \mathbb{C}\mathbf{e}_{i,j} \otimes V \otimes \cdots \otimes V$ of $V^{\otimes l(l+1)/2}$, where the tensor factors V correspond to labels (k, n) such that $1 \le k \le n$, $j < n \le l$. We choose values of the parameters r_k, $k = 1, 2, \cdots, j$, and h_{kn}, $1 \le k \le n \le j$, such that $r_k = \epsilon^{-2}$, $k = 1, 2, \cdots, j$, and

$$h_{kn} h_{k,n-1} h_{k-1,n-1}^{-1} h_{k+1,n}^{-1} = \epsilon^{1 - i\delta_{k,1}}, \qquad 1 \le k \le n \le j, \tag{64}$$

$$h_{n+1-k,n}^{-1} h_{n-k,n-1}^{-1} h_{n+1-k,n-1} h_{n-k,k} = \epsilon^{-1 + i\delta_{k,1}}, \qquad 1 \le k \le n \le j. \tag{65}$$

It can be shown that the system of equations (64) and (65) has a unique solution h_{kn}, $1 \le k \le n \le j$. The parameters r_k, $k = j+1, \cdots, l$, g_{kn}, $1 \le k \le n \le l$, and h_{kn}, $1 \le k \le n$, $l < n \le l$, are independent.

Proposition 44. *The representations σ for the values of g_{kn}, h_{kn}, r_k just described are reducible and the subspace $V_{i,j}$ is invariant.*

Proof. The proof follows easily from the explicit expressions for the elements $\rho_{\mathbf{r},\mathbf{s}}(X)$, $X = E_k, F_k, K_k$, $s_n = 1$, and the operators $\sigma(X)$. □

Since $V_{0,l-1} = V^{\otimes l}$, it follows easily that the corresponding representation of $U_\epsilon(\mathrm{sl}_{n+1})$ realized on the invariant subspace $V_{0,l-1}$ is irreducible and cyclic. It has the minimal dimension p^l.

7.5.4 Representations of Minimal Dimensions

Recall from Corollary 41 that the dimensions of cyclic representations of $U_\epsilon(\mathfrak{g})$ are greater than or equal to p^l, where l is the rank of $\mathfrak{g}$. In this subsection we construct for $\mathfrak{g}$ of type A_l, B_l or C_l cyclic representations of $U_\epsilon(\mathfrak{g})$ of dimensions p^l. The main tool for doing this is the algebra homomorphism ρ obtained in the next proposition.

Proposition 45. *Let us write $\langle y\rangle$ for the expression $y + y^{-1}$. There exists an algebra homomorphism*

(i) $\rho : U_\epsilon(\mathrm{sl}_{l+1}) \to \mathfrak{W}_{\epsilon,l+1}$ *such that* $\rho(K_i) = z_i^{-1} z_{i+1}$ *and*

$$\rho(E_i) = \frac{x_i^{-1} x_{i+1}}{(\epsilon^{1/2}+\epsilon^{-1/2})^2}\langle \epsilon^{-1/2} z_i\rangle, \quad \rho(F_i) = \frac{x_i x_{i+1}^{-1}}{(\epsilon^{1/2}+\epsilon^{-1/2})^2}\langle \epsilon^{-1/2} z_{i+1}\rangle;$$

(ii) $\rho : U_\epsilon(\mathrm{so}_{2l+1}) \to \mathfrak{W}_{\epsilon,l}$ *such that* $\rho(F_i), \rho(E_i), \rho(K_i)$, $i < l$, *are as in* (i) *and*

$$\rho(E_l) = \frac{x_l^{-1}}{(\epsilon^{1/2}+\epsilon^{-1/2})^2}\langle \epsilon^{-1/2} z_l\rangle, \qquad \rho(F_l) = x_l, \qquad \rho(K_l) = z_l^{-1};$$

(iii) $\rho : U_\epsilon(\mathrm{sp}_{2l}) \to \mathfrak{W}_{\epsilon^2,l}$ *such that* $\rho(F_i), \rho(E_i), \rho(K_i)$, $i < l$, *are as in* (i) *and*

$$\rho(E_l) = \frac{x_l^{-2}}{(\epsilon^{8}+\epsilon^{-8})^2}\langle \epsilon^{-8} z_l\rangle\langle \epsilon^{-24} z_l\rangle, \qquad \rho(F_l) = x_l^2, \qquad \rho(K_l) = z_l^{-2}.$$

Proof. The proof follows by direct verifications of the defining relations. □

As in the preceding subsection, we define a representation $\sigma_{\mathbf{gh}} := \pi_{\mathbf{gh}} \circ \rho$ of $U_\epsilon(\mathfrak{g})$ for $\mathfrak{g} = \mathrm{sl}_{l+1}, \mathrm{so}_{2l+1}, \mathrm{sp}_{2l}$.

Proposition 46. (i) *The representation $\sigma_{\mathbf{gh}}$ of $U_\epsilon(\mathrm{sl}_{l+1})$ is a direct sum of p irreducible representations of minimal dimension p^l. The carrier subspaces of these p subrepresentations are just the eigenspaces of the operator $\sigma_{\mathbf{hg}}(z_1 z_2 \cdots z_{l+1})$.*

(ii) *The representations* $\sigma_{\mathbf{gh}}$ *of* $U_\epsilon(\mathrm{so}_{2l+1})$ *and* $U_\epsilon(\mathrm{sp}_{2l})$ *are irreducible and of dimension* p^l.

(iii) *Any cyclic representation of* $U_\epsilon(\mathfrak{g})$, $\mathfrak{g} = \mathrm{sl}_{l+1}, \mathrm{so}_{2l+1}, \mathrm{sp}_{2l}$, *of dimension* p^l *is equivalent to a representation from* (i) *or* (ii), *respectively.*

(iv) *The dimension of any cyclic representation of* $U_\epsilon(\mathrm{so}_{2l})$ *is strictly greater than* p^l.

Proof. The proof is given in [CP2]. □

All cyclic irreducible representations obtained in parts (i) and (ii) of Proposition 46 have dimension p^l. Therefore, by Corollary 41, it follows that for $U_\epsilon(\mathrm{sl}_{l+1})$, $U_\epsilon(\mathrm{so}_{2l+1})$ and $U_\epsilon(\mathrm{sp}_{2l})$ the number p^l is indeed the minimal dimension of cyclic representations. It can be shown (see [CP3]) that the minimal dimension of cyclic representations for the algebra $U_\epsilon(\mathrm{so}_8)$ is p^5.

7.5.5 Representations of $U_\epsilon(\mathrm{sl}_{l+1})$ in Gel'fand–Tsetlin Bases

In this subsection it will be shown that the method of Gel'fand–Tsetlin bases works for the Drinfeld–Jimbo algebra $U_\epsilon(\mathrm{sl}_{l+1})$ with ϵ a root of unity as well if the parameters are chosen suitably.

We fix complex numbers $m_{i,l+1}$, $i = 1, 2, \cdots, l+1$, and c_{jk}, h_{jk}, $1 \le j \le k \le l$, and suppose that the differences $h_{ik} - h_{jk}$ are not integers. Let V be a complex vector space with a basis labeled by the tableaux (M) of the form

$$\begin{pmatrix} m_{1,l+1} & & m_{2,l+1} & & \cdots & & \cdots & & m_{l+1,l+1} \\ & m_{1l} & & m_{2l} & & \cdots & & m_{ll} & \\ & & m_{1,l-1} & & \cdots & & m_{l-1,l-1} & & \\ & & & \cdots & & \cdots & & & \\ & & & & m_{11} & & & & \end{pmatrix},$$

where the first row is fixed by the above numbers and the m_{ij}, $1 \le i \le j \le l$, run independently over the values $h_{ij}, h_{ij}+1, \cdots, h_{ij}+p-1$. Thus, the dimension of the space V is $p^{l(l+1)/2}$. Recall that $l(l+1)/2$ is the number of positive roots of $\mathrm{sl}(l+1, \mathbb{C})$.

Let $T(K_k)$, $T(E_k)$ and $T(F_k)$ be the operators on V determined by

$$T(K_k)|\,M\rangle = q^{a_k(M)-a_{k+1}(M)}|\,M\rangle, \quad a_k(M) = \sum\nolimits_i m_{ik} - \sum\nolimits_i m_{i,k-1}, \tag{66}$$

$$T(E_k)|\,M\rangle = \sum_{j=1}^{k} \frac{1}{c_{jk}} A_k^j(M)|\,M_k^{+j}\rangle, \; T(F_k)|\,M\rangle = \sum_{j=1}^{k} c_{jk} A_k^{-j}(M)|\,M_k^{-j}\rangle, \tag{67}$$

where $(M_k^{\pm j})$ is the tableau (M) with m_{jk} replaced by $m_{jk} \pm 1$,

$$A_k^j(M) = \left(\frac{\prod_{i=1}^{k+1} [m_{i,k+1} - m_{jk} - i + j][m_{i,k-1} - m_{jk} - i + j - 1]}{\prod_{\substack{i=1 \\ i \le j}}^{k} [m_{ik} - m_{jk} - i + j][m_{ik} - m_{jk} - i + j - 1]} \right)^{1/2} \tag{68}$$

and $[m] := [m]_\epsilon$ is defined by (2.1).

Proposition 47. *The operators $T(E_k)$, $T(F_k)$ and $T(K_k)$ given by the formulas (66)–(68) determine a representation T of $U_\epsilon(\mathrm{sl}_{l+1})$. For generic values of the parameters $m_{i,l+1}$, c_{jk}, h_{jk} the representation T is irreducible.*

Outline of proof. In order to prove that the operators (66)–(68) give a representation it suffices to show that they satisfy the defining relations of the algebra $U_\epsilon(\mathrm{sl}_{l+1})$. This verification is done either by a long direct calculation or by the method used in Subsect. 7.3.3. As usual, the irreducibility is proved by showing that every nonzero vector of V is cyclic. The corresponding assertion is then derived from the fact that for generic values of $m_{i,l+1}$, c_{jk}, h_{jk} the coefficients in (66) and (67) do not vanish. □

Any representation T contains $l(l+1)$ parameters c_{jk}, h_{ik} and the numbers $m_{i,l+1}$, $i = 1, 2, \cdots, l+1$. However, they are not all independent, since if we add to all numbers m_{jk} of a tableaux (M) some fixed number, then this does not change the formulas (66)–(68). Thus, the representations T have in fact $l(l+2) = \dim \mathrm{sl}(l+1, \mathbb{C})$ parameters. It can be shown (see [ACh2]) that these parameters are independent.

Proposition 48. *For generic values of the parameters $m_{i,l+1}$, c_{jk}, h_{jk} the representation T determined by formulas (66)–(68) is cyclic.*

Proof. In order to prove this assertion we calculate the operators $T(E_k)^p$, $T(F_k)^p$, $k = 1, 2, \cdots, l$. First of all, the matrix element $\langle M' \,|\, T(F_k)^r \mid M\rangle$ can be nonzero only if $m_{is} = m'_{is}$ for all $s \le k$. For $s = k$, the k-th row in M' is of the form $(m'_{1k}, m'_{2k}, \cdots, m'_{kk}) = (m_{1k} - p_1, m_{2k} - p_2, \cdots, m_{kk} - p_k)$, where $p_1 + \cdots + p_k = r$. Using (56), one proves by induction on r that

$$\langle M' \mid T(F_k)^r \mid M\rangle = \left(\prod_{i\le j} \frac{[l'_{ik} - l'_{jk}]}{[l_{ik} - l_{jk}]} \frac{[r]!}{[p_1]![p_2]! \cdots [p_k]!}\right)^{1/2}$$

$$\times \prod_{j=1}^{k} c_{jk}^{p_j} \prod_{s=1}^{p_j} \left(- \frac{\prod_{i=1}^{k+1} [l_{i,k+1} - l_{jk} + s] \prod_{i=1}^{k-1} [l_{i,k-1} - l_{jk} + s - 1]}{\prod_{i\ne j} [l_{ik} - l_{jk} + s]^2} \right)^{\frac{1}{2}}, \quad (69)$$

where $l_{ir} = m_{ir} - i$ and $l'_{ir} = m'_{ir} - i$.

We put $r = p$ in (69). Then all coefficients $[p]!/[p_1]! \cdots [p_k]!$ vanish except for the cases when one of the p_i is equal to p. Hence we get

$$T(F_k)^p|M\rangle$$

$$= \prod_{j=1}^{k} c_{jk}^{p} \prod_{s=1}^{p} \left(- \frac{\prod_{i=1}^{k+1} [l_{i,k+1} - l_{jk} + s - 1] \prod_{i=1}^{k-1} [l_{i,k-1} - l_{jk} + s - 2]}{\prod_{i\ne j} [l_{ik} - l_{jk} + s - 1]^2} \right)^{\frac{1}{2}} |M\rangle,$$

that is, $T(F_k)^p$ is a scalar operator. Since we have $\langle M' \mid T(F_k)^p \mid M\rangle = \langle M \mid T(E_k)^p \mid M'\rangle$, $T(E_k)^p$ is also a scalar operator. The preceding formulas imply that for generic values of $m_{i,l+1}$, c_{jk}, h_{jk} the operators $T(F_k)^p$ and $T(E_k)^p$ are nonzero scalar operators, so the representation T is cyclic. □

One can also construct the representation T when some of the numbers h_{jk} are equal to each other. Then T becomes reducible and we can find subrepresentations of T with dimensions smaller than $p^{l(l+1)/2}$. For example, let all m_{jk}, $j \geq 2$, in (M) be equal. Then we obtain an irreducible representations T' of $U_\epsilon(\mathrm{sl}_{l+1})$ of dimension p^l given by $T'(K_k)|M\rangle = q^{a_k - a_{k+1}}|M\rangle$ and

$$T'(F_k)|M\rangle = c_k(-[m_{k+1} - m_k + 1][m_{k-1} - m_k])^{1/2}|M_k^{-1}\rangle,$$
$$T'(E_k)|M\rangle = c_k^{-1}(-[m_{k+1} - m_k][m_{k-1} - m_k - 1])^{1/2}|M_k^{+1}\rangle,$$

where a_k is as in (66) and $m_{1,k+1}$, $m_{1,k}$, $m_{1,k-1}$ are denoted by m_{k+1}, m_k, m_{k-1}, respectively. Each m_i takes the values $m_i = h_i, h_i + 1, \cdots, h_i + p - 1$. The representation T' depends on $2l + 1$ parameters h_i, $i = 1, 2, \cdots, l + 1$, and c_i, $i = 1, 2, \cdots, l$.

7.6 Applications

The aim of this section is to sketch in a short review how Drinfeld–Jimbo algebras and their representations are used in physics.

The Drinfeld–Jimbo algebras $U_q(\mathrm{sl}_n)$ have been used in order to replace the unitary groups $SU(n)$ and their irreducible representations in describing flavor symmetries of hadrons, in particular, of vector mesons (see [Gav]). In this case, the algebra $U_q(\mathrm{sl}_{n+1})$ is considered as a dynamical symmetry. The latter realizes the necessary breaking of flavor symmetries up to exact isospin symmetry $U_q(\mathrm{su}_2)_I$ and produces some q-analogs of mass relations. This application of $U_q(\mathrm{sl}_n)$ uses the generators corresponding to nonsimple root elements.

The well-known octet mass sum rules of Gell-Mann and Okubo are

$$3m_\Lambda + m_\Sigma = 2(m_N + m_\Xi) \tag{70}$$

for baryons $(1/2)^+$ and

$$3m_{\omega_0} + m_\rho = 4m_{K^*} \tag{71}$$

for vector mesons 1^-. They have been derived on the basis of an $SU(3)$ symmetry breaking. If instead the quantum algebra $U_q(\mathrm{sl}_3)$ with $q = e^{i\pi/6}$ is used (note that then $[2]_q = \pm\sqrt{3}$), the q-analog of formula (70) becomes

$$m_N + \frac{1+\sqrt{3}}{2} m_\Xi = \frac{2}{\sqrt{3}} m_\Lambda + \frac{9-\sqrt{3}}{6} m_\Sigma. \tag{72}$$

The equality (72) gives a surprisingly good mass formula and holds with an accuracy of 0.22%. For a comparison, the Gell-Mann–Okubo relation (70) is satisfied only with an accuracy of 0.57%.

The q-analog of the mass relation (71) in the case of three flavors takes the form

$$m_{\omega_8} + \left(2\frac{[2]}{[3]} - 1\right) m_\rho = 2\frac{[2]}{[3]} m_{K^*}. \tag{73}$$

In the case $[3]_q = [2]_q$ this formula simplifies and reads as $m_{\omega_8} + m_\rho = 2m_{K^*}$. Setting $m_{\omega_8} = m_\phi$, one recognizes here the nonet formula of Okubo. This relation agrees with the corresponding experimental data (up to errors of the experiment and of an averaging over isoplets).

Applying the same approach in the cases of more flavors, new mass sum rules (higher analogs of Okubo's nonet mass relation (73)) are obtained that are also consistent with experimental data. These mass formulas are closely related to the Alexander polynomials of certain torus knots.

One of the main physical motivations for the representation theory of Drinfeld–Jimbo algebras in the root of unity case originates from the chiral Potts model. This is a solvable lattice model built on solutions of the quantum Yang–Baxter equation whose spectral parameters live on certain algebraic curves of genus greater than 1. These solutions can be derived as intertwiners of tensor product representations of the Drinfeld–Jimbo algebra for the affine Lie algebra of sl_2 with q a root of unity (see [BSt]). Attempts for extending these results to the case of sl_3 have been made in [BK] and [DJMM1].

It turned out that some properties of the chiral Potts model are closely related to the underlying structure of nonhighest weight representations of Drinfeld–Jimbo algebras. This understanding has proved to be very fruitful and allowed the construction of infinitely many new exactly solvable models as generalizations of the chiral Potts model (see [BK], [DJMM1]). The free parameters which characterize these nonhighest weight representations appear in the form of spectral parameters in these generalizations.

Among others, Drinfeld–Jimbo algebras have been invented for the construction of solvable models of statistical mechanics. In this context, the parameter q appears as the temperature, and the case $q = 0$ corresponds to the absolute temperature zero. For this reason, the passage to $q = 0$ is called "crystallization" (with a naive faith that any material is crystallized at temperature zero).

The theory of crystal bases was used for the computation of 1-point functions for the 6 vertex model and its generalizations [JMMO], [KKMMNN].

The 1-point functions are basic macroscopic quantities that describe the multiphase structure of a given lattice model of statistical mechanics. For the two-dimensional solvable models there is a method for computing the 1-point functions, called the corner transfer matrix method. It reduces the two-dimensional statistical sums of the 1-point functions to the one-dimensional statistical sums over certain paths. In the vertex models given by the R-matrices of the Drinfeld–Jimbo algebras, the parameter q behaves like the temperature, and the sum over the paths leads to series expansions at low temperature for the 1-point functions.

Under certain conditions the set $\mathcal{P}$ of paths can be identified with the crystal basis B of a $U_q'(\mathfrak{g})$-module. The sets of paths corresponding to the multiphases are in a one-to-one correspondence with the crystals of certain irreducible highest weight $U_q'(\mathfrak{g})$-modules. As a result, closed expressions of the one-dimensional statistical sums (and of the 1-point functions) in terms of the string functions of the corresponding affine Lie algebras have been derived in [KKMMNN].

Each line of a two-dimensional lattice bears a finite-dimensional module. Multiple lines are equipped with the tensor products of the associated modules. In this manner, the termodynamic limit refers to an infinite tensor product. An exact treatment of this infinite product is quite difficult. It turns out to be tractable when only the situation at $q = 0$ is considered. The result is then summarized as an isomorphism between the set of paths $\mathcal{P}$ and the crystal B.

7.7 Notes

The classification of finite-dimensional irreducible representations of Drinfeld–Jimbo algebras for q not a root of unity is due to G. Lusztig [L1] and M. Rosso [Ros1]. Among others, Theorem 8 on complete reducibility of finite-dimensional representations was proved in these papers. Theorem 13 on separation of elements by irreducible representations is from [Lus]. We followed the proof given in [Jan].

Theorem 24 on irreducible finite-dimensional representations of the algebra $U_q(\mathrm{gl}_n)$ is stated without proof in [Jim3]. It is proved in [UTS]. The general formulation of the Wigner–Eckart theorem for Drinfeld–Jimbo algebras is given in [BT] (see also [Kl2] and [RiS]). Expressions for the Clebsch–Gordan coefficients of tensor products $T_{\mathbf{m}} \otimes T_p$ of $U_q(\mathrm{gl}_n)$ have been derived in [AlS]. Clebsch–Gordan coefficients of $U_q(\mathrm{gl}_n)$ were also studied in [GLB], [Gou] and [Kl3]. The results of Subsect. 7.3.9 are taken from [DJM].

Crystal bases of representations of Drinfeld–Jimbo algebras were invented and developed by M. Kashivara [Kas1], [Kas2]. More material on crystal bases can be found in [Jan], [JMM], [Kas3], [KN] and [Var]. For an application of crystal bases to representations of classical Lie algebras see [Na].

Basic results on irreducible representations of Drinfeld–Jimbo algebras $U_q(\mathfrak{g})$ for q a root of unity have been obtained by C. De Concini, V. G. Kac and C. Procesi [DCK], [DCKP]. The results of Subsect. 7.5.2 on cyclic representations have been taken from the paper [DJMM2]. A recursive method for the construction of such representations was developed in [Schn1], [Schn2]. Proposition 46 was obtained in [CP2]. Gel'fand–Tsetlin bases for irreducible representations of $U_q(\mathrm{gl}_n)$ for q a root of unity were constructed in [ACh2]. Further results on representations in the root of unity case can be found in [ACh1], [Bec], [DCL] and [L3].

8. Quasitriangularity and Universal R-Matrices

A quasitriangular Hopf algebra is a Hopf algebra equipped with a so-called universal R-matrix which induces solutions of the quantum Yang–Baxter equation on their representations. The Drinfeld–Jimbo algebras $U_h(\mathfrak{g})$ are h-adic quasitriangular Hopf algebras. This emerges from Drinfeld's quantum double construction. The universal R-matrix of $U_h(\mathfrak{g})$ allows one also to introduce the so-called L-functionals. Suppressing mathematical subtleties they provide an alternative approach to the Drinfeld–Jimbo algebras which turns out to be very useful for applications (see Sects. 11.5–6 and 14.5–6). This chapter is devoted to a detailed study of these topics. The universal R-matrices, the R-matrices and the main L-functionals for the vector representations of Drinfeld–Jimbo algebras are explicitly determined.

8.1 Quasitriangular Hopf Algebras

8.1.1 Definition and Basic Properties

Recall that a bialgebra $\mathcal{A}$ is cocommutative if $\Delta^{\mathrm{cop}} = \Delta$, where $\Delta^{\mathrm{cop}} = \tau \circ \Delta$ and τ is the flip. Now we consider bialgebras $\mathcal{A}$ which are cocommutative up to conjugation by an element $\mathcal{R} \in \mathcal{A} \otimes \mathcal{A}$.

Definition 1. *A bialgebra (resp. Hopf algebra) $\mathcal{A}$ is called* quasitriangular *if there exists an invertible element $\mathcal{R}$ of $\mathcal{A} \otimes \mathcal{A}$ such that*

$$\Delta^{\mathrm{cop}}(a) = \mathcal{R}\Delta(a)\mathcal{R}^{-1}, \qquad a \in \mathcal{A}, \tag{1}$$

$$(\Delta \otimes \mathrm{id})(\mathcal{R}) = \mathcal{R}_{13}\mathcal{R}_{23}, \qquad (\mathrm{id} \otimes \Delta)(\mathcal{R}) = \mathcal{R}_{13}\mathcal{R}_{12}, \tag{2}$$

where $\mathcal{R}_{12} = \sum_i x_i \otimes y_i \otimes 1$, $\mathcal{R}_{13} = \sum_i x_i \otimes 1 \otimes y_i$, $\mathcal{R}_{23} = \sum_i 1 \otimes x_i \otimes y_i$ for $\mathcal{R} = \sum_i x_i \otimes y_i$. An invertible element $\mathcal{R} \in \mathcal{A} \otimes \mathcal{A}$ satisfying conditions (1) and (2) is called a universal R-matrix *of $\mathcal{A}$. A quasitriangular bialgebra (resp. Hopf algebra) with universal R-matrix $\mathcal{R}$ is said to be* triangular *if $\mathcal{R}_{21} = \mathcal{R}^{-1}$, where $\mathcal{R}_{21} = \tau(\mathcal{R}) \equiv \sum_i y_i \otimes x_i$.*

The preceding definitions carry over almost verbatim to h-adic bialgebras and Hopf algebras. The only modification is that the element $\mathcal{R}$ then belongs to the h-adic completion $\mathcal{A}\hat{\otimes}\mathcal{A}$ of $\mathcal{A} \otimes \mathcal{A}$.

We rewrite (1) by means of the Sweedler notation. If $\mathcal{R}$ is of the form $\mathcal{R} = \sum_i x_i \otimes y_i$ and $\Delta(a) = \sum a_{(1)} \otimes a_{(2)}$, then (1) is equivalent to

$$\sum\nolimits_{i,(a)} a_{(2)}x_i \otimes a_{(1)}y_i = \sum\nolimits_{i,(a)} x_i a_{(1)} \otimes y_i a_{(2)}, \quad a \in \mathcal{A}. \tag{3}$$

The relations (2) mean that

$$\sum\nolimits_i \sum (x_i)_{(1)} \otimes (x_i)_{(2)} \otimes y_i = \sum\nolimits_{i,j} x_i \otimes x_j \otimes y_i y_j, \tag{4}$$

$$\sum\nolimits_i \sum x_i \otimes (y_i)_{(1)} \otimes (y_i)_{(2)} = \sum\nolimits_{i,j} x_i x_j \otimes y_j \otimes y_i. \tag{5}$$

Example 1. The element $\mathcal{R} = 1 \otimes 1$ is a universal R-matrix for any cocommutative bialgebra. △

Example 2 (Sweedler's Hopf algebra). Let $\mathcal{A}$ be the Hopf algebra from Example 1.9. For any complex number α, the element

$$\mathcal{R}(\alpha) := \frac{1}{2}(1\otimes 1 + 1\otimes g + g\otimes 1 - g\otimes g) + \frac{\alpha}{2}(x\otimes x - x\otimes gx + gx\otimes x + gx\otimes gx)$$

is a universal R-matrix of $\mathcal{A}$ (see [Rad2] for details). △

Important examples of quasitriangular Hopf algebras will be obtained later in Subsects. 8.2.2 and 8.3.2.

Leg numbering notations such as $\mathcal{R}_{13}\mathcal{R}_{23}$, $\mathcal{R}_{12}$, etc., will often be used in the following. The lower indices always refer to the numbers of factors of a tensor product on which the corresponding element, matrix or transformation acts.

Let $\mathcal{A}$ be a bialgebra. Recall that $\mathcal{A}^{\mathrm{op}}$ (resp. $\mathcal{A}^{\mathrm{cop}}$) denotes the bialgebra with coalgebra (resp. algebra) structure of $\mathcal{A}$ but with opposite product (resp. coproduct).

Proposition 1. *Let $\mathcal{R}$ be a universal R-matrix of a bialgebra $\mathcal{A}$.*

(i) *$\mathcal{R}_{21}\mathcal{R}\Delta(a) = \Delta(a)\mathcal{R}_{21}\mathcal{R}$ for all $a \in \mathcal{A}$.*

(ii) *$\mathcal{R}_{21}^{-1}$ is also a universal R-matrix of $\mathcal{A}$.*

(iii) *$\mathcal{R}_{21}$ and $\mathcal{R}_{12}^{-1}$ are both universal R-matrices for the bialgebras $\mathcal{A}^{\mathrm{op}}$ and $\mathcal{A}^{\mathrm{cop}}$. In particular, if a bialgebra $\mathcal{A}$ is quasitriangular, then so are $\mathcal{A}^{\mathrm{op}}$ and $\mathcal{A}^{\mathrm{cop}}$.*

Proof. (i): The flip transforms (1) into $\Delta(a) = \mathcal{R}_{21}\Delta^{\mathrm{cop}}(a)\mathcal{R}_{21}^{-1}$, so that $\Delta(a)\mathcal{R}_{21}\mathcal{R} = \mathcal{R}_{21}\Delta^{\mathrm{cop}}(a)\mathcal{R} = \mathcal{R}_{21}\mathcal{R}\Delta(a)$.

(ii): From $\Delta(a) = \mathcal{R}_{21}\Delta^{\mathrm{cop}}(a)\mathcal{R}_{21}^{-1}$ we get $\Delta^{\mathrm{cop}}(a) = \mathcal{R}_{21}^{-1}\Delta(a)\mathcal{R}_{21}$. Since Δ is an algebra homomorphism, we have $((\Delta\otimes \mathrm{id})(\mathcal{R}))^{-1} = (\Delta\otimes\mathrm{id})(\mathcal{R}^{-1})$. Therefore, by (2), $(\Delta\otimes\mathrm{id})(\mathcal{R}^{-1}) = \mathcal{R}_{23}^{-1}\mathcal{R}_{13}^{-1}$. Permuting the order of tensor factors, we derive that $(\mathrm{id}\otimes\Delta)(\mathcal{R}_{21}^{-1}) = \mathcal{R}_{31}^{-1}\mathcal{R}_{21}^{-1} = (\mathcal{R}_{21}^{-1})_{13}(\mathcal{R}_{21}^{-1})_{12}$. Likewise, $(\mathrm{id}\otimes\Delta)(\mathcal{R}^{-1}) = \mathcal{R}_{12}^{-1}\mathcal{R}_{13}^{-1}$ leads to the relation $(\Delta\otimes\mathrm{id})(\mathcal{R}_{21}^{-1}) = (\mathcal{R}_{21}^{-1})_{13}(\mathcal{R}_{21}^{-1})_{23}$, that is, $\mathcal{R}_{21}^{-1}$ is indeed a universal R-matrix for $\mathcal{A}$.

(iii) is proved similarly. □

Some basic properties of the universal R-matrix are described in

Proposition 2. *Let $\mathcal{A}$ be a quasitriangular bialgebra with universal R-matrix $\mathcal{R}$. Then we have*

$$\mathcal{R}_{12}\mathcal{R}_{13}\mathcal{R}_{23} = \mathcal{R}_{23}\mathcal{R}_{13}\mathcal{R}_{12}, \tag{6}$$

$$(\varepsilon \otimes \mathrm{id})(\mathcal{R}) = (\mathrm{id} \otimes \varepsilon)(\mathcal{R}) = 1. \tag{7}$$

If $\mathcal{A}$ is a Hopf algebra, then we also have

$$(S \otimes \mathrm{id})(\mathcal{R}) = \mathcal{R}^{-1}, \quad (\mathrm{id} \otimes S)(\mathcal{R}^{-1}) = \mathcal{R}, \quad (S \otimes S)(\mathcal{R}) = \mathcal{R}. \tag{8}$$

Proof. Using the first equality in (2) twice and relation (1) we obtain

$$\mathcal{R}_{12}\mathcal{R}_{13}\mathcal{R}_{23} = \mathcal{R}_{12}\cdot(\Delta \otimes \mathrm{id})(\mathcal{R}) = (\Delta^{\mathrm{cop}} \otimes \mathrm{id})(\mathcal{R})\cdot\mathcal{R}_{12}$$

$$= (\tau \otimes \mathrm{id})(\Delta \otimes \mathrm{id})(\mathcal{R})\cdot\mathcal{R}_{12} = (\tau \otimes \mathrm{id})(\mathcal{R}_{13}\mathcal{R}_{23})\cdot\mathcal{R}_{12} = \mathcal{R}_{23}\mathcal{R}_{13}\mathcal{R}_{12}.$$

From $(\varepsilon \otimes \mathrm{id}) \circ \Delta = \mathrm{id}$ and (2) one derives that

$$\mathcal{R} = (\varepsilon \otimes \mathrm{id} \otimes \mathrm{id})(\Delta \otimes \mathrm{id})(\mathcal{R}) = (\varepsilon \otimes \mathrm{id} \otimes \mathrm{id})(\mathcal{R}_{13}\mathcal{R}_{23}) = (\varepsilon \otimes \mathrm{id})(\mathcal{R})\cdot\mathcal{R},$$

that is, $(\varepsilon \otimes \mathrm{id})(\mathcal{R}) = 1$. The second part in (7) is proved analogously. Since $m(S \otimes \mathrm{id})\Delta(a) = \varepsilon(a)1$ by the antipode axiom, we have

$$(m \otimes \mathrm{id})(S \otimes \mathrm{id} \otimes \mathrm{id})(\Delta \otimes \mathrm{id})(\mathcal{R}) = (\varepsilon \otimes \mathrm{id})(\mathcal{R}) = 1. \tag{9}$$

Since S is an algebra anti-homomorphism, (9) implies that

$$1 = (m \otimes \mathrm{id})(S \otimes \mathrm{id} \otimes \mathrm{id})(\mathcal{R}_{13}\mathcal{R}_{23}) = (S \otimes \mathrm{id})(\mathcal{R})\cdot\mathcal{R},$$

so that $(S \otimes \mathrm{id})(\mathcal{R}) = \mathcal{R}^{-1}$. Using the fact that $\mathcal{R}^{-1}$ is a universal R-matrix for $\mathcal{A}^{\mathrm{op}}$ (by Proposition 1), similar reasoning yields the second equality of (8). Finally, the third equality follows by

$$(S \otimes S)(\mathcal{R}) = (\mathrm{id} \otimes S)(S \otimes \mathrm{id})(\mathcal{R}) = (\mathrm{id} \otimes S)(\mathcal{R}^{-1}) = \mathcal{R}. \tag{10}$$

Definition 2. *The relation (6) is called the* quantum Yang–Baxter equation *(abbreviated, the* QYBE*).*

Let $\mathcal{R} = \sum_i x_i \otimes y_i$. Define mappings $\psi_{\mathcal{R}}$ and $\psi^{\mathcal{R}}$ from the dual bialgebra $\mathcal{A}^\circ$ to $\mathcal{A}$ by setting

$$\psi_{\mathcal{R}}(f) = \sum\nolimits_i f(x_i)y_i, \quad \psi^{\mathcal{R}}(f) = \sum\nolimits_i x_i f(y_i), \qquad f \in \mathcal{A}^\circ.$$

From relations (2) one easily proves the following assertions.

Proposition 3. *Let $\mathcal{R}$ be a universal R-matrix of a quasitriangular bialgebra $\mathcal{A}$. Then $\psi_{\mathcal{R}} : \mathcal{A}^\circ \to \mathcal{A}^{\mathrm{cop}}$ and $\psi^{\mathcal{R}} : \mathcal{A}^\circ \to \mathcal{A}^{\mathrm{op}}$ are bialgebra homomorphisms. That is, $\psi_{\mathcal{R}} : \mathcal{A}^\circ \to \mathcal{A}$ is an algebra homomorphism and a coalgebra anti-homomorphism and $\psi^{\mathcal{R}} : \mathcal{A}^\circ \to \mathcal{A}$ is an algebra anti-homomorphism and a coalgebra homomorphism.*

8.1.2 R-Matrices for Representations

If T_V and T_W are representations of a bialgebra $\mathcal{A}$, one may ask whether or not the tensor product representations $T_V \otimes T_W$ and $T_W \otimes T_V$ are equivalent. We shall see that for quasitriangular bialgebras this is indeed true and the equivalence is realized by an operator derived from the universal R-matrix.

Suppose that $\mathcal{A}$ is a quasitriangular bialgebra with universal R-matrix $\mathcal{R} = \sum_i x_i \otimes y_i$. Let $T_V : \mathcal{A} \to \mathcal{L}(V)$ and $T_W : \mathcal{A} \to \mathcal{L}(W)$ be representations of $\mathcal{A}$ on vector spaces V and W, respectively, where $\mathcal{L}(V)$ denotes the algebra of linear operators on V. We define linear mappings R_{VW} and $\hat{R}_{VW}$ by

$$R_{VW} := (T_V \otimes T_W)(\mathcal{R}) \quad \text{and} \quad \hat{R}_{VW} := \tau \circ (T_V \otimes T_W)(\mathcal{R}). \tag{11}$$

That is, by definition we have

$$\hat{R}_{VW}(v \otimes w) = \tau \circ (T_V \otimes T_W)(\mathcal{R})(v \otimes w) = \sum\nolimits_i T_W(y_i)w \otimes T_V(x_i)v.$$

The inverse of $\hat{R}_{VW}$ acts as $\hat{R}^{-1}_{VW}(w \otimes v) = (T_V \otimes T_W)(\mathcal{R}^{-1})(v \otimes w)$.

Proposition 4. (i) *The mapping $\hat{R}_{VW}$ defined by (11) provides an equivalence of the representations $T_V \otimes T_W$ and $T_W \otimes T_V$ of $\mathcal{A}$.*

(ii) *For any triple T_V, T_W, T_U of representations of $\mathcal{A}$ on vector spaces V, W and U, respectively, we have*

$$\hat{R}_{V\otimes W,U} = (\hat{R}_{VU} \otimes \mathrm{id}_W)(\mathrm{id}_V \otimes \hat{R}_{WU}), \tag{12}$$

$$\hat{R}_{V,W\otimes U} = (\mathrm{id}_W \otimes \hat{R}_{VU})(\hat{R}_{VW} \otimes \mathrm{id}_U), \tag{13}$$

$$\begin{aligned}(\hat{R}_{WU} \otimes \mathrm{id}_V)&(\mathrm{id}_W \otimes \hat{R}_{VU})(\hat{R}_{VW} \otimes \mathrm{id}_U)\\ &= (\mathrm{id}_U \otimes \hat{R}_{VW})(\hat{R}_{VU} \otimes \mathrm{id}_W)(\mathrm{id}_V \otimes \hat{R}_{WU}).\end{aligned} \tag{14}$$

Proof. Since $\hat{R}_{VW}$ is invertible as noted above and

$$\begin{aligned}\hat{R}_{VW}(T_V \otimes T_W)(a) &= (\tau \circ (T_V \otimes T_W)(\mathcal{R}))(T_V \otimes T_W)(\Delta(a))\\ &= \tau \circ ((T_V \otimes T_W)(\mathcal{R}\Delta(a)))\\ &= \tau \circ ((T_V \otimes T_W)(\Delta^{\mathrm{cop}}(a))(T_V \otimes T_W)(\mathcal{R}))\\ &= (T_W \otimes T_V)(\Delta(a))(\tau \circ (T_V \otimes T_W))(\mathcal{R})\\ &= (T_W \otimes T_V)(a)\hat{R}_{VW},\end{aligned}$$

$\hat{R}_{VW}$ realizes an equivalence of the representations $T_V \otimes T_W$ and $T_W \otimes T_V$.

Equation (12) is a consequence of the first formula in (2). Indeed, we have

$$\begin{aligned}\hat{R}_{V\otimes W,U} &= (\tau \circ (T_{V\otimes W} \otimes T_U))(\mathcal{R}) = (\tau_{12,3} \circ (T_{V\otimes W} \otimes T_U)((\Delta \otimes \mathrm{id})(\mathcal{R}))\\ &= \tau_{12,3} \circ (T_V \otimes T_W \otimes T_U)(\mathcal{R}_{13}\mathcal{R}_{23}))\\ &= \sum\nolimits_{i,j} T_U(y_j y_i) \otimes T_V(x_j) \otimes T_W(x_i)\\ &= \Big(\sum\nolimits_j T_U(y_j) \otimes T_V(x_j) \otimes \mathrm{id}_W\Big)\Big(\sum\nolimits_i T_U(y_i) \otimes \mathrm{id}_V \otimes T_W(x_i)\Big)\\ &= (\hat{R}_{VU} \otimes \mathrm{id}_W)(\mathrm{id}_V \otimes \hat{R}_{WU}),\end{aligned}$$

where $\tau_{12,3}(v \otimes w \otimes u) = u \otimes v \otimes w$. Similarly, (13) follows from the second formula in (2), and (14) can be derived from the QYBE (6) or by combining (12) and (13). □

Let us consider (14) in the special case when $V = W = U$ and $T_V = T_W = T_U$. Setting $\hat{R} := \hat{R}_{VV}$, (14) reads as

$$\hat{R}_{12}\hat{R}_{23}\hat{R}_{12} = \hat{R}_{23}\hat{R}_{12}\hat{R}_{23}. \tag{15}$$

Equation (15) is called the *braid relation* for the linear transformation $\hat{R}$. It is not difficult to check that an arbitrary linear transformation R satisfies the quantum Yang–Baxter equation (6) if and only if the transformation $\hat{R} := \tau \circ R$ fulfills the braid relation (15). In this sense the QYBE and the braid relation are equivalent. Therefore, by Proposition 4(ii), any transformation $R_{VV} = \tau \circ \hat{R}_{VV} = (T_V \otimes T_V)(\mathcal{R})$ satisfies the quantum Yang–Baxter equation (6). That is, the universal R-matrix $\mathcal{R}$ gives rise to a solution R_{VV} of the QYBE for every representation T_V of the quasitriangular bialgebra $\mathcal{A}$. Since solutions of the QYBE are usually referred to as *R-matrices*, this fact is the reason why $\mathcal{R}$ is called the *universal* R-matrix of $\mathcal{A}$. Finding all solutions of the QYBE turns out to be a difficult task. To a large extent, quantum groups were invented as structures which allow one to produce such R-matrices.

8.1.3 Square and Inverse of the Antipode

Recall that by Corollary 1.7 the square of the antipode of a cocommutative Hopf algebra is the identity map. For quasitriangular Hopf algebras this is not longer true, but S^2 is still an inner automorphism as stated in

Proposition 5. *Suppose that $\mathcal{A}$ is a quasitriangular Hopf algebra with universal R-matrix $\mathcal{R}$. Then $u := m(S \otimes \mathrm{id})(\mathcal{R}_{21})$ is an invertible element of $\mathcal{A}$ with inverse $u^{-1} = m(\mathrm{id} \otimes S^2)(\mathcal{R}_{21})$. The element $uS(u) = S(u)u$ is central. The antipode S of $\mathcal{A}$ is invertible and we have*

$$S^2(a) = uau^{-1} \quad \textit{and} \quad S^{-1}(a) = u^{-1}S(a)u \quad \textit{for all} \;\; a \in \mathcal{A}.$$

Proof. Let $\mathcal{R} = \sum_i x_i \otimes y_i$. We first prove that $S^2(a)u = ua$ for $a \in \mathcal{A}$. By (1) we have the relation $(\mathcal{R} \otimes 1)\Delta^{(2)}(a) = (\Delta^{\mathrm{cop}} \otimes \mathrm{id})\Delta(a)(\mathcal{R} \otimes 1)$. It gives

$$\sum\nolimits_{i,(a)} x_i a_{(1)} \otimes y_i a_{(2)} \otimes a_{(3)} = \sum\nolimits_{i,(a)} a_{(2)} x_i \otimes a_{(1)} y_i \otimes a_{(3)}. \tag{16}$$

We now apply the mapping $\mathrm{id} \otimes S \otimes S^2$ to both sides and multiply the terms in reversed order. By (1.20), (1.27) and (1.28), the two sides of (16) yield

$$\begin{aligned}
\sum\nolimits_{i,(a)} S^2(a_{(3)})S(y_i a_{(2)})x_i a_{(1)} &= \sum\nolimits_{i,(a)} S(a_{(2)}S(a_{(3)}))S(y_i)x_i a_{(1)} \\
&= \sum\nolimits_i S(y_i)x_i a = ua, \\
\sum\nolimits_{i,(a)} S^2(a_{(3)})S(a_{(1)}y_i)a_{(2)}x_i &= \sum\nolimits_{i,(a)} S^2(a_{(3)})S(y_i)S(a_{(1)})a_{(2)}x_i \\
&= \sum\nolimits_i S^2(a)S(y_i)x_i = S^2(a)u,
\end{aligned}$$

respectively, so that $S^2(a)u = ua$.

Let $v = m(\mathrm{id} \otimes S^2)(\mathcal{R}_{21}) = \sum_i y_i S^2(x_i)$. Setting $\mathcal{R}^{-1} = \sum_i x_i^{(-1)} \otimes y_i^{(-1)}$ and using the relations $(S \otimes S)(\mathcal{R}) = \mathcal{R}$ and $(S \otimes \mathrm{id})(\mathcal{R}) = \mathcal{R}^{-1}$, we derive

$$\begin{aligned} vu &= \sum\nolimits_i y_i S^2(x_i)u = \sum\nolimits_i y_i u x_i = \sum\nolimits_{i,j} y_i S(y_j) x_j x_i \\ &= \sum\nolimits_{i,j} S(y_i) S(y_j) x_j S(x_i) = \sum\nolimits_{i,j} S(y_i^{(-1)}) S(y_j) x_j x_i^{(-1)}. \end{aligned}$$

Applying $\mathrm{id} \otimes S$ to the identity $\mathcal{R}\mathcal{R}^{-1} = \sum_{i,j} x_j x_i^{(-1)} \otimes y_j y_i^{(-1)} = 1 \otimes 1$ and multiplying again in reversed order we obtain

$$vu = \sum\nolimits_{i,j} S(y_i^{(-1)}) S(y_j) x_j x_i^{(-1)} = S(1)1 = 1.$$

Similarly,

$$\begin{aligned} uv &= \sum\nolimits_i u y_i S^2(x_i) = \sum\nolimits_i S^2(y_i) u S^2(x_i) \\ &= \sum\nolimits_i S(y_i) u S(x_i) = \sum\nolimits_i y_i u x_i = 1. \end{aligned}$$

By the preceding we have shown that v is the inverse of u and that $S^2(a) = uau^{-1}$, $a \in \mathcal{A}$.

Since $S^2(u^{\pm 1}) = uu^{\pm 1}u^{-1} = u^{\pm 1}$, we have

$$S^2(u^{-1}au) = S^2(u^{-1})S^2(a)S^2(u) = u^{-1}uau^{-1}u = a, \quad a \in \mathcal{A},$$

so that $\mathcal{A} = S(\mathcal{A})$. Applying S to the relation $ua = S^2(a)u$ we get $S(a)S(u) = S(u)S^3(a) = S(u)uS(a)u^{-1}$ and so $S(a)S(u)u = S(u)uS(a)$ for $a \in \mathcal{A}$. Since $S(\mathcal{A}) = \mathcal{A}$, $S(u)u$ belongs to the center of $\mathcal{A}$. Setting $u = S(a)$, the latter identity yields $uS(u)u = S(u)uu$ and hence $uS(u) = S(u)u$.

The mapping T defined by $T(a) := u^{-1}S(a)u$ is the inverse of S, since

$$ST(a) = S(u^{-1}S(a)u) = S(u)S^2(a)S(u^{-1}) = S(u)uau^{-1}S(u^{-1}) = a,$$

$$TS(a) = u^{-1}S^2(a)u = u^{-1}uau^{-1}u = a. \qquad \square$$

Note that the square of the antipode for the Drinfeld–Jimbo algebras has been described as an inner automorphism in Subsect. 6.1.2.

The central element $uS(u)$ is occasionally called the *quantum Casimir element* of the quasitriangular Hopf algebra $\mathcal{A}$.

Remark 1. A Hopf algebra $\mathcal{A}$ is said to be *quasi-cocommutative* if its antipode is bijective and if there exists an invertible element $\mathcal{R} \in \mathcal{A} \otimes \mathcal{A}$ satisfying (1). By some slight modifications of the first part of the above proof (use $m \circ (S^{-1} \otimes \mathrm{id})(\mathcal{R}_{21}^{-1})$ as v) it follows that the formula $S^2(a) = uau^{-1}$, $a \in \mathcal{A}$, is valid for quasi-cocommutative Hopf algebras. △

Corollary 6. $(\mathrm{id} \otimes S^{-1})(\mathcal{R}) = \mathcal{R}^{-1}$.

Proof. The third identity of (8) implies that $(S^{-1} \otimes S^{-1})(\mathcal{R}^{-1}) = \mathcal{R}^{-1}$. Applying $S^{-1} \otimes S^{-1}$ to the first formula of (8), we get $(\mathrm{id} \otimes S^{-1})(\mathcal{R}) = (S^{-1} \otimes S^{-1})(\mathcal{R}^{-1}) = \mathcal{R}^{-1}$. □

Proposition 7. *If $\mathcal{A}$, $\mathcal{R}$ and u are as in Proposition 5, then the element $w := uS(u^{-1})$ is group-like and*

$$S^4(a) = waw^{-1}, \quad a \in \mathcal{A}.$$

Moreover, $\Delta(u) = (\mathcal{R}_{21}\mathcal{R})^{-1}(u \otimes u)$ and $\Delta(S(u)) = (\mathcal{R}_{21}\mathcal{R})^{-1}(S(u) \otimes S(u))$.

Proof. We first prove the relation $\Delta(u)\mathcal{R}_{21}\mathcal{R} = u \otimes u$. By (1), we have

$$\begin{aligned}\Delta(u)\mathcal{R}_{21}\mathcal{R} &= \sum_i \Delta(S(y_i))\Delta(x_i)\mathcal{R}_{21}\mathcal{R} = \sum_i (S \otimes S)(\Delta^{\mathrm{cop}}(y_i))\Delta(x_i)\mathcal{R}_{21}\mathcal{R} \\ &= \sum_i (S \otimes S)(\Delta^{\mathrm{cop}}(y_i))\mathcal{R}_{21}\mathcal{R}\Delta(x_i).\end{aligned}$$

Defining a right action $\triangleleft$ of the Hopf algebra $\mathcal{A}^{\otimes 4}$ on $\mathcal{A} \otimes \mathcal{A}$ by

$$(a \otimes b) \triangleleft (X \otimes Y) := (S \otimes S)(Y)(a \otimes b)X, \quad a, b \in \mathcal{A}, \quad X, Y \in \mathcal{A} \otimes \mathcal{A},$$

and using the formulas (2) we can rewrite the previous relation as

$$\Delta(u)\mathcal{R}_{21}\mathcal{R} = \mathcal{R}_{21} \triangleleft (\mathcal{R}_{12}\mathcal{R}_{13}\mathcal{R}_{23}\mathcal{R}_{14}\mathcal{R}_{24}).$$

By the QYBE, the right hand side is equal to $\mathcal{R}_{21} \triangleleft (\mathcal{R}_{23}\mathcal{R}_{13}\mathcal{R}_{12}\mathcal{R}_{14}\mathcal{R}_{24})$. Since $(\mathrm{id} \otimes S^{-1})(\mathcal{R}) = \mathcal{R}^{-1}$ by Corollary 6, we derive

$$\begin{aligned}\mathcal{R}_{21} \triangleleft \mathcal{R}_{23} &= \sum\nolimits_{i,j} S(y_j)y_i \otimes x_i x_j = (S \otimes \mathrm{id}) \sum\nolimits_{i,j} S^{-1}(y_i)y_j \otimes x_i x_j \\ &= (S \otimes \mathrm{id})(\mathcal{R}_{21}^{-1}\mathcal{R}_{21}) = 1 \otimes 1.\end{aligned}$$

Thus, $\mathcal{R}_{21} \triangleleft (\mathcal{R}_{23}\mathcal{R}_{13}) = (1 \otimes 1) \triangleleft \mathcal{R}_{13} = \sum_i S(y_i)x_i \otimes 1 = u \otimes 1$ and hence

$$\begin{aligned}\mathcal{R}_{21} \triangleleft (\mathcal{R}_{23}\mathcal{R}_{13}\mathcal{R}_{12}\mathcal{R}_{14}) &= (u \otimes 1)\Big(\sum\nolimits_{i,j} x_i x_j \otimes S(y_j)y_i\Big) \\ &= (u \otimes 1)(\mathrm{id} \otimes S)\sum\nolimits_{i,j} x_i x_j \otimes S^{-1}(y_i)y_j \\ &= (u \otimes 1)(\mathrm{id} \otimes S)(\mathcal{R}^{-1}\mathcal{R}) = u \otimes 1,\end{aligned}$$

so that

$$\Delta(u)\mathcal{R}_{21}\mathcal{R} = (u \otimes 1) \triangleleft \mathcal{R}_{24} = (u \otimes 1)(1 \otimes u) = u \otimes u. \tag{17}$$

Because $\Delta(a)\mathcal{R}_{21}\mathcal{R} = \mathcal{R}_{21}\mathcal{R}\Delta(a)$ by Proposition 1(i), this yields $\Delta(u) = (\mathcal{R}_{21}\mathcal{R})^{-1}(u \otimes u)$.

From the latter relation and the third formula of (8) we obtain

$$\begin{aligned}\Delta(S(u)) &= (S\otimes S)(\tau\circ\Delta)(u) = (S\otimes S)(\tau\circ(\mathcal{R}_{21}\mathcal{R})^{-1}(u\otimes u))\\ &= (S\otimes S)(u\otimes u)(S\otimes S)(\mathcal{R}\mathcal{R}_{21})^{-1}\\ &= ((S\otimes S)(u\otimes u))(\mathcal{R}_{21}\mathcal{R})^{-1}.\end{aligned}$$

Therefore, by (17), we have

$$\Delta(w) = \Delta(uS(u^{-1})) = (u\otimes u)(\mathcal{R}_{21}\mathcal{R})^{-1}\mathcal{R}_{21}\mathcal{R}(S(u^{-1})\otimes S(u^{-1})) = w\otimes w,$$

that is, w is group-like. Since $S^2(S(a)) = S(uau^{-1}) = S(u^{-1})S(a)S(u)$ and hence $S^2(b) = S(u^{-1})bS(u)$, $b\in\mathcal{A}$, we get $S^4(b) = S^3(u^{-1})S^2(b)S^3(u) = S(u^{-1})ubu^{-1}S(u) = wbw^{-1}$. □

8.2 The Quantum Double and Universal R-Matrices

The quantum double is one of the most celebrated Hopf algebra constructions. We develop this method here for general skew-paired bialgebras. A number of other approaches to the double will be treated in Sect. 10.2. In the present context the quantum double is used in order to derive explicit formulas for the universal R-matrices of the h-adic Drinfeld–Jimbo algebras.

8.2.1 The Quantum Double of Skew-Paired Bialgebras

If not stated otherwise, $\mathcal{A}$ and $\mathcal{B}$ are bialgebras in this subsection.

Definition 3. *A bilinear mapping $\sigma:\mathcal{A}\times\mathcal{B}\to\mathbb{C}$ is called* a skew-pairing *of the bialgebras $\mathcal{A}$ and $\mathcal{B}$ if $\sigma(\cdot,\cdot)$ is a dual pairing of the bialgebras $\mathcal{A}$ and $\mathcal{B}^{\mathrm{op}}$, that is, for all $a,a'\in\mathcal{A}$ and $b,b'\in\mathcal{B}$ we have*

$$\sigma(a,1)=\varepsilon(a),\quad \sigma(1,b)=\varepsilon(b), \tag{18}$$

$$\sigma(aa',b)=\sum\sigma(a,b_{(1)})\sigma(a',b_{(2)}),\quad \sigma(a,bb')=\sum\sigma(a_{(2)},b)\sigma(a_{(1)},b'). \tag{19}$$

The mapping σ is said to be convolution invertible *(or, briefly,* invertible*) if there exists another bilinear mapping $\overline{\sigma}:\mathcal{A}\times\mathcal{B}\to\mathbb{C}$ such that $\sigma\overline{\sigma}=\overline{\sigma}\sigma=\varepsilon_{\mathcal{A}}\otimes\varepsilon_{\mathcal{B}}$, that is, for all $a\in\mathcal{A}$ and $b\in\mathcal{B}$ we have*

$$\sum\sigma(a_{(1)},b_{(1)})\overline{\sigma}(a_{(2)},b_{(2)}) = \sum\overline{\sigma}(a_{(1)},b_{(1)})\sigma(a_{(2)},b_{(2)}) = \varepsilon_{\mathcal{A}}(a)\varepsilon_{\mathcal{B}}(b).$$

The inverse of σ is always denoted by $\overline{\sigma}$. It is easily seen that σ is a skew-pairing of $\mathcal{A}$ and $\mathcal{B}$ if and only if $\overline{\sigma}_{21}$ is a skew-pairing of $\mathcal{B}$ and $\mathcal{A}$, where $\overline{\sigma}_{21}(b,a):=\overline{\sigma}(a,b)$. A skew-pairing σ of $\mathcal{A}$ and $\mathcal{B}$ is invertible if $\mathcal{A}$ is a Hopf algebra or if $\mathcal{B}$ is a Hopf algebra with invertible antipode. In these cases the inverse $\overline{\sigma}$ is given by

$$\overline{\sigma}(a,b)=\sigma(S(a),b),\quad\text{resp.}\quad \overline{\sigma}(a,b)=\sigma(a,S^{-1}(b)),\quad a\in\mathcal{A},\quad b\in\mathcal{B}. \tag{20}$$

We verify, for instance, the second formula. By (18) and (19), we have

$$\sum \sigma(a_{(1)}, S^{-1}(b_{(1)}))\sigma(a_{(2)}, b_{(2)}) = \sum \sigma(a, b_{(2)}S^{-1}(b_{(1)}))$$

$$= \sum \sigma(a, S^{-1}(b_{(1)}S(b_{(2)}))) = \sigma(a, \varepsilon(b)1) = \varepsilon(a)\varepsilon(b).$$

Similarly we get $\sum \sigma(a_{(1)}, b_{(1)})\sigma(a_{(2)}, S^{-1}(b_{(2)})) = \varepsilon(a)\varepsilon(b)$.

Proposition 8. (i) *Let $\mathcal{A}$ and $\mathcal{B}$ be bialgebras equipped with an invertible skew-pairing $\sigma : \mathcal{A} \times \mathcal{B} \to \mathbb{C}$. Then the vector space $\mathcal{B} \otimes \mathcal{A}$ becomes an algebra with product defined by*

$$(b \otimes a)(b' \otimes a') = \sum bb'_{(2)} \otimes a_{(2)}a'\overline{\sigma}(a_{(1)}, b'_{(1)})\sigma(a_{(3)}, b'_{(3)}) \tag{21}$$

for $a, a' \in \mathcal{A}$ and $b, b' \in \mathcal{B}$. With the tensor product coalgebra structure of $\mathcal{B} \otimes \mathcal{A}$ (that is, with the coproduct $\Delta_{\mathcal{B}\otimes\mathcal{A}}(b \otimes a) = \sum b_{(1)} \otimes a_{(1)} \otimes b_{(2)} \otimes a_{(2)}$ and counit $\varepsilon_{\mathcal{B}\otimes\mathcal{A}}(b \otimes a) = \varepsilon_{\mathcal{B}}(b)\varepsilon_{\mathcal{A}}(a)$), this algebra is a bialgebra.

(ii) *If $\mathcal{A}$ and $\mathcal{B}$ are Hopf algebras, this bialgebra is a Hopf algebra with antipode S given by $S(b \otimes a) = (1 \otimes S_{\mathcal{A}}(a))(S_{\mathcal{B}}(b) \otimes 1)$.*

Proof. First we prove the associativity of the product. Since σ is a dual pairing of $\mathcal{A}$ and $\mathcal{B}^{\mathrm{op}}$, $\overline{\sigma}$ is a dual pairing of $\mathcal{A}^{\mathrm{op}}$ and $\mathcal{B}$. Using appropriate properties of σ and $\overline{\sigma}$, we compute that both expressions $((b \otimes a)(b' \otimes a'))(b'' \otimes a'')$ and $(b \otimes a)((b' \otimes a')(b'' \otimes a''))$ are equal to

$$\sum bb'_{(2)}b''_{(3)} \otimes a_{(3)}a'_{(2)}a''\overline{\sigma}(a_{(1)}, b'_{(1)})\overline{\sigma}(a_{(2)}, b''_{(2)})\sigma(a_{(4)}, b''_{(4)})\sigma(a_{(5)}, b'_{(3)})$$
$$\times\overline{\sigma}(a'_{(1)}, b''_{(1)})\sigma(a'_{(3)}, b''_{(5)}).$$

This proves the associativity. It is obvious that $(b\otimes a)(1\otimes 1) = (1\otimes 1)(b\otimes a) = b \otimes a$ and that $\varepsilon_{\mathcal{B}\otimes\mathcal{A}}$ is an algebra homomorphism. We verify that $\Delta_{\mathcal{B}\otimes\mathcal{A}}$ is an algebra homomorphism by computing

$$\Delta_{\mathcal{B}\otimes\mathcal{A}}(b\otimes a)\Delta_{\mathcal{B}\otimes\mathcal{A}}(b'\otimes a') = \sum (b_{(1)}\otimes a_{(1)})(b'_{(1)}\otimes a'_{(1)})\otimes(b_{(2)}\otimes a_{(2)})(b'_{(2)}\otimes a'_{(2)})$$

$$= \sum b_{(1)}b'_{(2)} \otimes a_{(2)}a'_{(1)} \otimes b_{(2)}b'_{(5)} \otimes a_{(5)}a'_{(2)}\overline{\sigma}(a_{(1)}, b'_{(1)})\sigma(a_{(3)}, b'_{(3)})$$
$$\times\overline{\sigma}(a_{(4)}, b'_{(4)})\sigma(a_{(6)}, b'_{(6)})$$

$$= \sum b_{(1)}b'_{(2)} \otimes a_{(2)}a'_{(1)} \otimes b_{(2)}b'_{(3)} \otimes a_{(3)}a'_{(2)}\overline{\sigma}(a_{(1)}, b'_{(1)})\sigma(a_{(4)}, b'_{(4)})$$

$$= \Delta_{\mathcal{B}\otimes\mathcal{A}}((b \otimes a)(b' \otimes a')).$$

Finally, we show that $S(b \otimes a) := (1 \otimes S_{\mathcal{A}}(a))(S_{\mathcal{B}}(b) \otimes 1)$ is an antipode if $\mathcal{A}$ and $\mathcal{B}$ are Hopf algebras. Using (21) we get

$$\begin{aligned}\sum S(b_{(1)} \otimes a_{(1)})(b_{(2)} \otimes a_{(2)}) \\ &= \sum (1 \otimes S_{\mathcal{A}}(a_{(1)}))(S_{\mathcal{B}}(b_{(1)}) \otimes 1)(b_{(2)} \otimes 1)(1 \otimes a_{(2)}) \\ &= \sum (1 \otimes S_{\mathcal{A}}(a_{(1)}))(S_{\mathcal{B}}(b_{(1)})b_{(2)} \otimes 1)(1 \otimes a_{(2)}) \\ &= \varepsilon(b) \sum 1 \otimes S_{\mathcal{A}}(a_{(1)})a_{(2)} = \varepsilon(a)\varepsilon(b)(1 \otimes 1)\end{aligned}$$

and similarly $\sum (b_{(1)} \otimes a_{(1)})S(b_{(2)} \otimes a_{(2)}) = \varepsilon(a)\varepsilon(b)(1 \otimes 1)$. □

Definition 4. *The bialgebra from Proposition 8 is called the* generalized quantum double *of the bialgebras $\mathcal{A}$ and $\mathcal{B}$ with respect to the skew-pairing σ or, briefly, the* quantum double *of $\mathcal{A}$ and $\mathcal{B}$. It is denoted by $\mathcal{D}(\mathcal{A}, \mathcal{B}; \sigma)$ or simply by $\mathcal{D}(\mathcal{A}, \mathcal{B})$ if no confusion can arise.*

Let us discuss some basic properties of the quantum double. From the corresponding definitions in Proposition 8 it is clear that both mappings $\mathcal{B} \ni b \to b \otimes 1 \in \mathcal{D}(\mathcal{A}, \mathcal{B}; \sigma)$ and $\mathcal{A} \ni a \to 1 \otimes a \in \mathcal{D}(\mathcal{A}, \mathcal{B}; \sigma)$ are injective bialgebra homomorphisms. Let us identify $b \otimes 1$ with b and $1 \otimes a$ with a. Then $\mathcal{B}$ and $\mathcal{A}$ are subbialgebras of $\mathcal{D}(\mathcal{A}, \mathcal{B}; \sigma)$. Note that the expression $ba \equiv (b \otimes 1)(1 \otimes a) = b \otimes a$ is different from $ab \equiv (1 \otimes a)(b \otimes 1)$ in general. By (21), we have the following cross commutation relations between elements $a \equiv 1 \otimes a \in \mathcal{A}$ and $b \equiv b \otimes 1 \in \mathcal{B}$ in the algebra $\mathcal{D}(\mathcal{A}, \mathcal{B}; \sigma)$:

$$ab = \sum \overline{\sigma}(a_{(1)}, b_{(1)})b_{(2)}a_{(2)}\sigma(a_{(3)}, b_{(3)}), \tag{22}$$

$$\sum \sigma(a_{(1)}, b_{(1)})a_{(2)}b_{(2)} = \sum b_{(1)}a_{(1)}\sigma(a_{(2)}, b_{(2)}). \tag{23}$$

In fact, as an *algebra* the double $\mathcal{D}(\mathcal{A}, \mathcal{B}; \sigma)$ is the free algebra generated by the algebras $\mathcal{B}$ and $\mathcal{A}$ with cross relations (22) or, equivalently, (23). As a *coalgebra*, $\mathcal{D}(\mathcal{A}, \mathcal{B}; \sigma)$ is just the tensor product coalgebra of $\mathcal{B}$ and $\mathcal{A}$.

We give another isomorphic variant of the quantum double. Since $\overline{\sigma}_{21}$ is an invertible skew-pairing of $\mathcal{B}$ and $\mathcal{A}$, as noted above, the quantum double $\mathcal{D}(\mathcal{B}, \mathcal{A}; \overline{\sigma}_{21})$ is also well-defined. The bialgebra $\mathcal{D}(\mathcal{B}, \mathcal{A}; \overline{\sigma}_{21})$ is realized on the vector space $\mathcal{A} \otimes \mathcal{B}$. Its coalgebra structure is inherited from the tensor product of the coalgebras $\mathcal{A}$ and $\mathcal{B}$ and its algebra structure is given by the product

$$(a \otimes b)(a' \otimes b') = \sum aa'_{(2)} \otimes b_{(2)}b'\sigma(a'_{(1)}, b_{(1)})\overline{\sigma}(a'_{(3)}, b_{(3)}) \tag{24}$$

for $a, a' \in \mathcal{A}$ and $b, b' \in \mathcal{B}$. The cross commutation relations of $b \equiv 1 \otimes b \in \mathcal{B}$ and $a \equiv a \otimes 1 \in \mathcal{A}$ in the algebra $\mathcal{D}(\mathcal{B}, \mathcal{A}; \overline{\sigma}_{21})$ are

$$ba = \sum \sigma(a_{(1)}, b_{(1)})a_{(2)}b_{(2)}\overline{\sigma}(a_{(3)}, b_{(3)}). \tag{25}$$

In the literature both variants $\mathcal{D}(\mathcal{A}, \mathcal{B}; \sigma)$ and $\mathcal{D}(\mathcal{B}, \mathcal{A}; \overline{\sigma}_{21})$ occur as generalized quantum doubles of the bialgebras $\mathcal{A}$ and $\mathcal{B}$. The bialgebras $\mathcal{D}(\mathcal{A}, \mathcal{B}; \sigma)$ and $\mathcal{D}(\mathcal{B}, \mathcal{A}; \overline{\sigma}_{21})$ are indeed isomorphic with isomorphism θ defined by

$$\theta(b\otimes a)=\sum a_{(2)}\otimes b_{(2)}\sigma(a_{(1)},b_{(1)})\overline{\sigma}(a_{(3)},b_{(3)}),\quad a\in\mathcal{A},\ b\in\mathcal{B}. \tag{26}$$

Remark 2. All quantum doubles occurring in this book are of the form $\mathcal{D}(\mathcal{A},\mathcal{B};\sigma)$ with a skew-pairing σ of $\mathcal{A}$ and $\mathcal{B}$ and built over the vector space $\mathcal{B}\otimes\mathcal{A}$. The reader should notice that there are also variants of the double in the literature (see [Jos]) where the tensor factors are interchanged. △

Let us specialize the preceding to the following situation. We suppose that $\mathcal{A}$ is a Hopf algebra with invertible antipode and let $\mu(f,a)$ denote the evaluation $\langle f,a\rangle:=f(a)$ of the functional $f\in\mathcal{A}^\circ$ at $a\in\mathcal{A}$. Since the antipode of $\mathcal{A}$ is invertible, $\mathcal{A}^{\rm op}$ and $\mathcal{A}^{\rm cop}$ are Hopf algebras by Proposition 1.6 and so are $(\mathcal{A}^{\rm op})^\circ\equiv(\mathcal{A}^\circ)^{\rm cop}$ and $(\mathcal{A}^{\rm cop})^\circ\equiv(\mathcal{A}^\circ)^{\rm op}$ as well by Proposition 1.12. As noted in Subsect. 1.2.8, μ is a dual pairing of $\mathcal{A}^\circ$ and $\mathcal{A}$. Hence μ is a skew-pairing of $(\mathcal{A}^\circ)^{\rm cop}$ and $\mathcal{A}$, and by (20) its convolution inverse $\overline{\mu}$ is given by $\overline{\mu}(f,a)=\langle f,S^{-1}(a)\rangle$. Therefore, by Proposition 8, the quantum doubles $\mathcal{D}((\mathcal{A}^\circ)^{\rm cop},\mathcal{A};\mu)$ and $\mathcal{D}(\mathcal{A},(\mathcal{A}^\circ)^{\rm cop};\overline{\mu}_{21})$ are well-defined Hopf algebras. Further, since μ is a dual pairing of $\mathcal{A}^\circ$ and $\mathcal{A}$, $\nu:=\mu_{21}$ is a skew-pairing of $\mathcal{A}$ and $(\mathcal{A}^\circ)^{\rm op}$ with convolution inverse $\overline{\nu}$ such that $\overline{\nu}(a,f)=\langle f,S(a)\rangle$ (again by (20)). Thus, the corresponding quantum doubles $\mathcal{D}(\mathcal{A},(\mathcal{A}^\circ)^{\rm op};\nu)$ and $\mathcal{D}((\mathcal{A}^\circ)^{\rm op},\mathcal{A};\overline{\nu}_{21})$ are also well-defined. Each of the four Hopf algebras

$$\mathcal{D}((\mathcal{A}^\circ)^{\rm cop},\mathcal{A};\mu),\ \mathcal{D}(\mathcal{A},(\mathcal{A}^\circ)^{\rm cop};\overline{\mu}_{21}),\ \mathcal{D}(\mathcal{A},(\mathcal{A}^\circ)^{\rm op};\nu),\ \mathcal{D}((\mathcal{A}^\circ)^{\rm op},\mathcal{A};\overline{\nu}_{21})$$

is called a *quantum double of the Hopf algebra* $\mathcal{A}$. Although their products look quite different at first glance, all four Hopf algebras are in fact isomorphic. It was already noted above that the first and second ones and the third and fourth ones are isomorphic with isomorphism θ given by (26). One immediately verifies that $S_{\mathcal{A}^\circ}\otimes{\rm id}$ is an isomorphism of the Hopf algebras $\mathcal{D}(\mathcal{A},(\mathcal{A}^\circ)^{\rm op};\nu)$ and $\mathcal{D}(\mathcal{A},(\mathcal{A}^\circ)^{\rm cop};\overline{\mu}_{21})$. The reader should notice that $\overline{\mu}_{21}$ is different from $\overline{\nu}\equiv\overline{\mu_{21}}$ in general, because the convolution inverses of μ and $\nu=\mu_{21}$ are formed with respect to different pairs of Hopf algebras!

In Subsect. 8.2.2 we shall use the double $\mathcal{D}(\mathcal{A},(\mathcal{A}^\circ)^{\rm cop};\overline{\mu}_{21})$, so we restate here its structure explicitly. By definition, $\mathcal{D}(\mathcal{A},(\mathcal{A}^\circ)^{\rm cop};\overline{\mu}_{21})$ is the quantum double $\mathcal{D}(\mathcal{A},\mathcal{B};\sigma)$ with product (24) and $\mathcal{B}=(\mathcal{A}^\circ)^{\rm cop}$, $\sigma=\overline{\mu}_{21}$. By (20), the convolution inverse of $\overline{\mu}_{21}$ is $\mu_{21}(a,f)=\langle f,a\rangle$. Let us adopt the convention that the Sweedler notation $\Delta_{\mathcal{A}^\circ}(f)=\sum f_{(1)}\otimes f_{(2)}$ refers to the comultiplication of $\mathcal{A}^\circ$, so that $\Delta_{(\mathcal{A}^\circ)^{\rm cop}}(f)=\sum f_{(2)}\otimes f_{(1)}$. Then the Hopf algebra $\mathcal{D}(\mathcal{A},(\mathcal{A}^\circ)^{\rm cop};\overline{\mu}_{21})$ is realized on the vector space $\mathcal{A}^\circ\otimes\mathcal{A}$ with product

$$(f\otimes a)(g\otimes b)=\sum fg_{(2)}\otimes a_{(2)}b\,\langle g_{(3)},a_{(1)}\rangle\langle g_{(1)},S_{\mathcal{A}}^{-1}(a_{(3)})\rangle, \tag{27}$$

unit $\varepsilon\otimes 1$, counit $1\otimes\varepsilon$, comultiplication

$$\Delta(f\otimes a)=\sum f_{(2)}\otimes a_{(1)}\otimes f_{(1)}\otimes a_{(2)} \tag{28}$$

and antipode

$$S(f\otimes a)=(1\otimes S_{\mathcal{A}^\circ}^{-1}(f))(S_{\mathcal{A}}(a)\otimes 1).$$

Finally, let us note that the previous definitions and results carry over almost verbatim to h-adic Hopf algebras. The only modifications are that $\sigma : \mathcal{A}\hat{\otimes}\mathcal{B} \to \mathbb{C}[[h]]$ is $\mathbb{C}[[h]]$-linear and that the quantum double $\mathcal{D}(\mathcal{A},\mathcal{B})$ is built over the completion $\mathcal{B}\hat{\otimes}\mathcal{A}$ of $\mathcal{B}\otimes\mathcal{A}$ in the h-adic topology.

8.2.2 Quasitriangularity of Quantum Doubles of Finite-Dimensional Hopf Algebras

Suppose that $\mathcal{A}$ is a finite-dimensional Hopf algebra. Let $\{e_i \mid i = 1,2,\cdots,n\}$ and $\{f_i \mid i = 1,2,\cdots,n\}$ be bases of the vector spaces $\mathcal{A}$ and $\mathcal{A}' \equiv \mathcal{A}^\circ$, respectively, such that $\langle f_j, e_i\rangle = \delta_{ij}$. We retain the notation from Subsect. 8.2.1.

Theorem 9. *The quantum double $\mathcal{D}(\mathcal{A}) := \mathcal{D}(\mathcal{A},(\mathcal{A}^\circ)^{\rm cop};\overline{\mu}_{21})$ is a quasitriangular Hopf algebra with the universal R-matrix*

$$\mathcal{R} = \sum\nolimits_i (\varepsilon\otimes e_i)\otimes(f_i\otimes 1) \in \mathcal{D}(\mathcal{A})\otimes\mathcal{D}(\mathcal{A}). \tag{29}$$

Proof. Since the antipode of the finite-dimensional Hopf algebra $\mathcal{A}^\circ$ is always bijective ([Sw], Corollary 5.1.6), $(\mathcal{A}^\circ)^{\rm cop}$ is a Hopf algebra. Therefore, by Proposition 8, $\mathcal{D}(\mathcal{A}) \equiv \mathcal{D}(\mathcal{A},(\mathcal{A}^\circ)^{\rm cop};\overline{\mu}_{21})$ is a Hopf algebra. The main part of this proof consists of showing that $\mathcal{R}$ is a universal R-matrix of $\mathcal{D}(\mathcal{A})$.

First we prove that $(\Delta\otimes{\rm id})(\mathcal{R}) = \mathcal{R}_{13}\mathcal{R}_{23}$. Since $\{e_i\}$ and $\{f_i\}$ are dual bases, we have $a = \sum\langle f_i, a\rangle e_i$ and $g = \sum\langle g, e_i\rangle f_i$ for $a\in\mathcal{A}$ and $g\in\mathcal{A}^\circ$. Using these identities we compute the expression $\langle gh, a\rangle$ in two possible ways and obtain

$$\sum\nolimits_i \sum \langle g, e_{i,(1)}\rangle\langle h, e_{i,(2)}\rangle\langle f_i, a\rangle = \sum\nolimits_{i,j} \langle g, e_i\rangle\langle h, e_j\rangle\langle f_i f_j, a\rangle$$

for $a\in\mathcal{A}$ and $g,h\in\mathcal{A}^\circ$. Since the pairing $\langle\cdot,\cdot\rangle$ of $\mathcal{A}^\circ$ and $\mathcal{A}$ is nondegenerate (because $\mathcal{A}$ is finite-dimensional), the latter equation implies that

$$\sum\nolimits_i \sum \varepsilon\otimes e_{i,(1)}\otimes\varepsilon\otimes e_{i,(2)}\otimes f_i\otimes 1 = \sum\nolimits_{i,j} \varepsilon\otimes e_i\otimes\varepsilon\otimes e_j\otimes f_i f_j\otimes 1.$$

This means that $(\Delta\otimes{\rm id})(\mathcal{R}) = \mathcal{R}_{13}\mathcal{R}_{23}$. The relation $({\rm id}\otimes\Delta)(\mathcal{R}) = \mathcal{R}_{13}\mathcal{R}_{12}$ is derived in a similar manner.

Next we show that $\mathcal{R}$ is invertible. More precisely, we prove that

$$\mathcal{R}^{-1} := \sum\nolimits_i (\varepsilon\otimes S(e_i))\otimes(f_i\otimes 1) \tag{30}$$

is the inverse of $\mathcal{R}$ in the algebra $\mathcal{D}(\mathcal{A})\otimes\mathcal{D}(\mathcal{A})$. The relation $\mathcal{R}^{-1}\mathcal{R} = 1\otimes 1$ is obviously equivalent to the equation

$$\sum\nolimits_{i,j} S(e_i)e_j\otimes f_i f_j = 1\otimes\varepsilon. \tag{31}$$

Pairing the second tensor factor of the left hand side against $a\in\mathcal{A}$ and using the identity $b = \sum_i\langle f_i, b\rangle e_i$, we get

$$\sum_{i,j}\sum S(e_i)e_j\langle f_i,a_{(1)}\rangle\langle f_j,a_{(2)}\rangle$$

$$=\sum S\Big(\sum_i\langle f_i,a_{(1)}\rangle e_i\Big)\sum_j\langle f_j,a_{(2)}\rangle e_j=\sum S(a_{(1)})a_{(2)}=\varepsilon(a)1.$$

This implies (31) and so the relation $\mathcal{R}^{-1}\mathcal{R}=1\otimes 1$. It is proved similarly that $\mathcal{R}\mathcal{R}^{-1}=1\otimes 1$.

Our next aim is to prove that $\mathcal{R}\Delta(g\otimes a)=\Delta^{\rm cop}(g\otimes a)\mathcal{R}$ for $g\in\mathcal{A}^\circ$ and $a\in\mathcal{A}$. Inserting the definitions of the R-matrix (29), the product (27) and the coproduct (28) of the quantum double $\mathcal{D}(\mathcal{A})$ we get

$$\mathcal{R}\Delta(g\otimes a)=\sum_i\sum(\varepsilon\otimes e_i)(g_{(2)}\otimes a_{(1)})\otimes(f_i\otimes 1)(g_{(1)}\otimes a_{(2)})$$

$$=\sum_i\sum g_{(3)}\otimes e_{i,(2)}a_{(1)}\otimes f_ig_{(1)}\otimes a_{(2)}\langle g_{(4)},e_{i,(1)}\rangle\langle g_{(2)},S^{-1}(e_{i,(3)})\rangle, \quad (32)$$

$$\Delta^{\rm cop}(g\otimes a)\mathcal{R}=\sum_i\sum(g_{(1)}\otimes a_{(2)})(\varepsilon\otimes e_i)\otimes(g_{(2)}\otimes a_{(1)})(f_i\otimes 1)$$

$$=\sum_i\sum g_{(1)}\otimes a_{(4)}e_i\otimes g_{(2)}f_{i,(2)}\otimes a_{(2)}\langle f_{i,(3)},a_{(1)}\rangle\langle f_{i,(1)},S^{-1}(a_{(3)})\rangle. \quad (33)$$

It suffices to prove that the expressions (32) and (33) are equal when paired in the third tensor factor against $b\in\mathcal{A}$. Applying the mapping $(\Delta\otimes{\rm id})\circ\Delta$ to the identity $c=\sum_i\langle f_i,c\rangle e_i$, $c\in\mathcal{A}$, yields that

$$\sum c_{(1)}\otimes c_{(2)}\otimes c_{(3)}=\sum_i\sum\langle f_i,c\rangle e_{i,(1)}\otimes e_{i,(2)}\otimes e_{i,(3)}. \quad (34)$$

We first pair the third tensor factor of (32) against b. Using the relation (34) and properties of the dual pairing $\langle\cdot,\cdot\rangle$ of the Hopf algebras $\mathcal{A}^\circ$ and $\mathcal{A}$ we obtain

$$\sum_i\sum g_{(3)}\otimes e_{i,(2)}a_{(1)}\otimes a_{(2)}\langle f_i,b_{(1)}\rangle\langle g_{(1)},b_{(2)}\rangle\langle g_{(4)},e_{i,(1)}\rangle\langle g_{(2)},S^{-1}(e_{i,(3)})\rangle$$

$$=\sum g_{(3)}\otimes b_{(2)}a_{(1)}\otimes a_{(2)}\langle g_{(1)},b_{(4)}\rangle\langle g_{(4)},b_{(1)}\rangle\langle g_{(2)},S^{-1}(b_{(3)})\rangle$$

$$=\sum g_{(2)}\otimes b_{(2)}a_{(1)}\otimes a_{(2)}\langle g_{(1)},b_{(4)}S^{-1}(b_{(3)})\rangle\langle g_{(3)},b_{(1)}\rangle$$

$$=\sum g_{(1)}\otimes b_{(2)}a_{(1)}\otimes a_{(2)}\langle g_{(2)},b_{(1)}\rangle.$$

Pairing the third tensor factor of (33) against b it follows that

$$\sum_i\sum g_{(1)}\otimes a_{(4)}e_i\otimes a_{(2)}\langle g_{(2)},b_{(1)}\rangle\langle f_{i,(2)},b_{(2)}\rangle\langle f_{i,(3)},a_{(1)}\rangle\langle f_{i,(1)},S^{-1}(a_{(3)})\rangle$$

$$=\sum g_{(1)}\otimes a_{(4)}\Big(\sum_i e_i\langle f_i,S^{-1}(a_{(3)})b_{(2)}a_{(1)}\rangle\Big)\otimes a_{(2)}\langle g_{(2)},b_{(1)}\rangle$$

$$=\sum g_{(1)}\otimes a_{(4)}S^{-1}(a_{(3)})b_{(2)}a_{(1)}\otimes a_{(2)}\langle g_{(2)},b_{(1)}\rangle$$

$$=\sum g_{(1)}\otimes b_{(2)}a_{(1)}\otimes a_{(2)}\langle g_{(2)},b_{(1)}\rangle.$$

Thus, we have shown that $\mathcal{R}\Delta(g\otimes a)=\Delta^{\rm cop}(g\otimes a)\mathcal{R}$. Since $\mathcal{R}$ is invertible, we conclude that $\Delta^{\rm cop}=\mathcal{R}\Delta\mathcal{R}^{-1}$ on $\mathcal{D}(\mathcal{A})\otimes\mathcal{D}(\mathcal{A})$. □

Corollary 10. *Any finite-dimensional Hopf algebra $\mathcal{A}$ is isomorphic to a Hopf subalgebra of a quasitriangular Hopf algebra.*

Proof. $\mathcal{A}$ is isomorphic to the Hopf subalgebra $1\otimes\mathcal{A}$ of $\mathcal{D}(\mathcal{A})$. □

Remarks: 3. Since the map θ defined by (26) is an isomorphism of the quantum doubles $\mathcal{D}(\mathcal{A},(\mathcal{A}^\circ)^{\rm cop};\overline{\mu}_{21})$ and $\mathcal{D}((\mathcal{A}^\circ)^{\rm cop},\mathcal{A};\mu)$, the element

$$(\theta\otimes\theta)(\mathcal{R})=\sum\nolimits_i(e_i\otimes\varepsilon)\otimes(1\otimes f_i)$$

is a universal R-matrix of the Hopf algebra $\mathcal{D}((\mathcal{A}^\circ)^{\rm cop},\mathcal{A};\mu)$.

4. The structure of the universal R-matrix (29) becomes more transparent by the following interpretation. To each element $x=\sum_r a_r\otimes b_r\in\mathcal{A}\otimes\mathcal{A}^\circ$ we associate a linear mapping $\tilde{x}$ of $\mathcal{A}^\circ$ by setting $\tilde{x}g:=\sum_r\langle g,a_r\rangle b_r$, $g\in\mathcal{A}^\circ$. Then, $x\to\tilde{x}$ is an injective map of $\mathcal{A}\otimes\mathcal{A}^\circ$ into $\mathcal{L}(\mathcal{A}^\circ)$. Since $\sum_r\langle g,e_i\rangle f_i=g$ for any $g\in\mathcal{A}^\circ$, the so-called *canonical element* $\sum_i e_i\otimes f_i$ of $\mathcal{A}\otimes\mathcal{A}^\circ$ corresponds to the identity map of $\mathcal{A}^\circ$. Under the embeddings $a\to\varepsilon\otimes a$ of $\mathcal{A}$ and $f\to f\otimes 1$ of $\mathcal{A}^\circ$ into the double $\mathcal{D}(\mathcal{A})$, *the canonical element goes into the universal R-matrix $\mathcal{R}$.* Therefore, the canonical element of $\mathcal{A}\otimes\mathcal{A}^\circ$ and so the R-matrix (29) are independent of the choice of bases. △

Finally, we return to the general quantum double $\mathcal{D}(\mathcal{A},\mathcal{B};\sigma)$. Let $\mathcal{A}$ and $\mathcal{B}$ be (not necessarily finite-dimensional) Hopf algebras equipped with an invertible skew-pairing σ such that its convolution inverse $\overline{\sigma}$ is *nondegenerate*. As discussed at the end of Subsect. 8.2.1, the Hopf algebra $\mathcal{D}(\mathcal{A},(\mathcal{A}^\circ)^{\rm cop};\overline{\mu}_{21})$ is the double $\mathcal{D}(\mathcal{A},\mathcal{B};\sigma)$ with $\mathcal{B}=(\mathcal{A}^\circ)^{\rm cop}$, $\sigma=\overline{\mu}_{21}$ and $\overline{\sigma}(a,f)=\langle f,a\rangle$. That is, for the general quantum double $\mathcal{D}(\mathcal{A},\mathcal{B};\sigma)$ the convolution inverse $\overline{\sigma}$ corresponds to the evaluation bilinear form $\langle\cdot,\cdot\rangle$. Therefore, if we want to translate the assertion of Theorem 9 to the double $\mathcal{D}(\mathcal{A},\mathcal{B};\sigma)$, we have to take dual bases $\{e_i\}$ and $\{f_i\}$ of the vector spaces $\mathcal{A}$ and $\mathcal{B}$, respectively, with respect to the inverse form $\overline{\sigma}$. The element $\mathcal{R}=\sum_i(1\otimes e_i)\otimes(f_i\otimes 1)$ is then an infinite sum in general, so it is not a universal R-matrix of $\mathcal{D}(\mathcal{A},\mathcal{B};\sigma)$ in the strict sense. However, when the tensor factors are paired against elements of $\mathcal{B}$ resp. $\mathcal{A}$, all sums occurring in the proof of Theorem 9 become finite and the whole reasoning remains valid. This means that the sum $\mathcal{R}$ is considered as bilinear form on some appropriate dual. An elaboration of this idea leads to the concept of a universal r-form (see Sect. 10.1). Another possibility to remedy this situation is to work with topological completions of the tensor products.

The reasoning of the above proof of Theorem 9 also gives the following h-adic version which will be used in Subsect. 8.3.2.

Proposition 11. *Suppose that $\mathcal{A}$ and $\mathcal{B}$ are h-adic Hopf algebras equipped with an invertible skew-pairing σ of $\mathcal{A}$ and $\mathcal{B}$ such that its inverse $\overline{\sigma}$ is nondegenerate. Let $\{e_i\}$ and $\{f_i\}$ be dual bases of the vector spaces $\mathcal{A}$ and $\mathcal{B}$,* *respectively, with respect to the form $\overline{\sigma}$. If $\mathcal{R} = \sum_i (1 \otimes e_i) \otimes (f_i \otimes 1)$ belongs to the h-adic completion $\mathcal{D}(\mathcal{A}, \mathcal{B}; \sigma) \hat{\otimes} \mathcal{D}(\mathcal{A}, \mathcal{B}; \sigma)$, then $\mathcal{R}$ is a universal R-matrix of the h-adic Hopf algebra $\mathcal{D}(\mathcal{A}, \mathcal{B}; \sigma)$.*

8.2.3 The Rosso Form of the Quantum Double

Another interesting property of the quantum double is that it possesses a natural ad_L-invariant bilinear form (see Subsect. 6.3.2 for the latter notion).

Let $\mathcal{A}$ and $\mathcal{B}$ be Hopf algebras and let σ be a skew-pairing of $\mathcal{A}$ and $\mathcal{B}$. The bilinear form $\langle \cdot, \cdot \rangle_R$ on $\mathcal{D}(\mathcal{A}, \mathcal{B}; \sigma) \times \mathcal{D}(\mathcal{A}, \mathcal{B}; \sigma)$ defined by

$$\langle b \otimes a, b' \otimes a' \rangle_R = \sigma(S(a), b')\sigma(a', S(b)), \quad a, a' \in \mathcal{A}, \ b, b' \in \mathcal{B}, \tag{35}$$

is called the *Rosso form* of the quantum double $\mathcal{D}(\mathcal{A}, \mathcal{B}; \sigma)$.

Proposition 12. *The Rosso form $\langle \cdot, \cdot \rangle_R$ on $\mathcal{D}(\mathcal{A}, \mathcal{B}; \sigma)$ is ad_L-invariant.*

Proof. As noted in Subsect. 6.3.2, it suffices to show that $\langle \cdot, \cdot \rangle_R$ satisfies condition (6.91). Since ad_L is a homomorphism and the algebra $\mathcal{D}(\mathcal{A}, \mathcal{B}; \sigma)$ is generated by the subalgebras $\mathcal{B}$ and $\mathcal{A}$, it is enough to verify (6.91) for elements $1 \otimes c$ and $d \otimes 1$, where $c \in \mathcal{A}$ and $d \in \mathcal{B}$. By (21), we have

$$\begin{aligned} \mathrm{ad}_L(d \otimes 1)(b \otimes a) &= \sum (d_{(1)} \otimes 1)(b \otimes a)(S(d_{(2)}) \otimes 1) \\ &= \sum d_{(1)} b S(d_{(3)}) \otimes a_{(2)} \overline{\sigma}(a_{(1)}, S(d_{(4)})) \sigma(a_{(3)}, S(d_{(2)})), \\ \mathrm{ad}_L(S(d) \otimes 1)(b' \otimes a') &= \sum (S(d_{(2)}) \otimes 1)(b' \otimes a')(S^2(d_{(1)}) \otimes 1) \\ &= \sum S(d_{(4)}) b' S^2(d_{(2)}) \otimes a'_{(2)} \overline{\sigma}(a'_{(1)}, S^2(d_{(1)})) \sigma(a'_{(3)}, S^2(d_{(3)})). \end{aligned}$$

Recall that $\overline{\sigma}(\cdot, \cdot) = \sigma(S(\cdot), \cdot)$ and $\sigma(\cdot, \cdot) = \overline{\sigma}(\cdot, S(\cdot))$. Using these relations, formula (35) and the fact that $\sigma(\cdot, \cdot)$ is a skew-pairing, we compute

$$\begin{aligned} &\langle \mathrm{ad}_L(d \otimes 1)(b \otimes a), b' \otimes a' \rangle_R \\ &= \sum \sigma(S(a_{(2)}), b') \sigma(a', S((d_{(1)}) b S(d_{(3)}))) \sigma(S(a_{(1)}), S(d_{(4)})) \sigma(a_{(3)}, S(d_{(2)})) \\ &= \sum \sigma(S(a), S(d_{(4)}) b' S(d_{(2)})) \sigma(a', S^2(d_{(3)}) S(b) S(d_{(1)})) \end{aligned}$$

and

$$\begin{aligned} &\langle b \otimes a, \mathrm{ad}_L(S(d) \otimes 1)(b' \otimes a') \rangle_R \\ &= \sum \sigma(S(a), S(d_{(4)}) b' S^2(d_{(2)})) \sigma(a'_{(2)}, S(b)) \sigma(a'_{(1)}, S(d_{(1)})) \sigma(a'_{(3)}, S(d_{(3)})) \\ &= \sum \sigma(S(a), S(d_{(4)}) b' S^2(d_{(2)})) \sigma(a', S^2(d_{(3)}) S(b) S(d_{(1)})). \end{aligned}$$

Thus,

$$\langle \mathrm{ad}_L(d \otimes 1)(b \otimes a), b' \otimes a' \rangle_R = \langle b \otimes a, \mathrm{ad}_L(S(d) \otimes 1)(b' \otimes a') \rangle_R.$$

In a similar way one verifies that both expressions

$$\langle \mathrm{ad}_L(1 \otimes c)(b \otimes a), b' \otimes a' \rangle_R \ \text{ and } \ \langle b \otimes a, \mathrm{ad}_L(1 \otimes S(c))(b' \otimes a') \rangle_R$$

are equal to

$$\sum \sigma(S^2(c_{(4)}) S(a) S(c_{(2)}), b') \sigma(S(c_{(3)}) a' S(c_{(1)}), S(b)).$$

This completes the proof of the ad_L-invariance of $\langle \cdot, \cdot \rangle_R$. □

8.2.4 Drinfeld–Jimbo Algebras as Quotients of Quantum Doubles

The bilinear forms $\sigma(\cdot,\cdot) := \langle\cdot,\cdot\rangle$ from Propositions 6.34 and 6.38 are dual pairings of the Hopf algebras $U_q(\mathfrak{b}_+)$ and $U_q(\mathfrak{b}_-)^{\mathrm{op}}$ resp. $U_h(\mathfrak{b}_+)$ and $\tilde{U}_h(\mathfrak{b}_-)^{\mathrm{op}}$, hence they are skew-pairings of $\mathcal{A} := U_q(\mathfrak{b}_+)$ and $\mathcal{B} := U_q(\mathfrak{b}_-)$ resp. $\mathcal{A} := U_h(\mathfrak{b}_+)$ and $\mathcal{B} := \tilde{U}_h(\mathfrak{b}_-)$. Therefore, the corresponding quantum doubles $\mathcal{D}(\mathcal{A},\mathcal{B};\sigma)$ are well-defined Hopf algebras by Proposition 8. We denote these doubles by $\mathcal{D}(U_q(\mathfrak{b}_+),U_q(\mathfrak{b}_-))$ and $\mathcal{D}(U_h(\mathfrak{b}_+),\tilde{U}_h(\mathfrak{b}_-))$, respectively. Their products are defined by formula (24). By (20), the convolution inverse $\overline{\sigma}$ of σ is just the bilinear form $\langle\cdot,\cdot\rangle' \equiv \langle S(\cdot),\cdot\rangle$ defined at the end of Subsect. 6.3.1. We shall show that these two quantum doubles are closely related to the Drinfeld–Jimbo algebras $U_q(\mathfrak{g})$ and $U_h(\mathfrak{g})$.

Proposition 13. *As an algebra, $\mathcal{D}(U_q(\mathfrak{b}_+),U_q(\mathfrak{b}_-))$ is generated by $6l$ elements E_i, K_i, K_i^{-1}, F_i, K_i', $K_i'^{-1}$. The defining relations are the relations for the generators E_i, K_i, K_i^{-1}, resp. F_i, K_i', $K_i'^{-1}$, of the algebra $U_q(\mathfrak{b}_+)$, resp. $U_q(\mathfrak{b}_-)$, and the following cross relations:*

$$K_i'E_jK'^{-1}_i = q_i^{a_{ij}}E_j, \quad K_iF_jK_i^{-1} = q_i^{-a_{ij}}F_j,$$

$$K_iK_j' = K_j'K_i, \quad E_iF_j - F_jE_i = \delta_{ij}\frac{K_i - K'^{-1}_i}{q_i - q_i^{-1}}.$$

Proof. As noted above, $\mathcal{D}(\mathcal{A},\mathcal{B};\sigma)$ is the free algebra generated by the algebras $\mathcal{A}$ and $\mathcal{B}$ with cross commutation relations (23). Further, it suffices to require the cross relations (23) for the generators $a = E_i, K_i$ of $\mathcal{A} \equiv U_q(\mathfrak{b}_+)$ and $b = F_i, K_i'$ of $\mathcal{B} \equiv U_q(\mathfrak{b}_-)$. For example, let $a = E_i$ and $b = F_j$. Recall that $\Delta(E_i) = E_i \otimes K_i + 1 \otimes E_i$ and $\Delta(F_j) = F_j \otimes 1 + K'^{-1}_j \otimes F_j$. Since $\langle E_i, 1\rangle = \langle K_i, F_j\rangle = \langle E_i, K'^{-1}_i\rangle = \langle 1, F_j\rangle = 0$ by (6.82), relation (23) gives

$$K_i{\cdot}1\langle E_i,F_j\rangle + E_iF_j\langle 1, K'^{-1}_j\rangle = F_jE_i\langle K_i,1\rangle + K'^{-1}_j{\cdot}1\langle E_i,F_j\rangle.$$

Inserting the values $\langle K_i,1\rangle = \langle 1, K'^{-1}_j\rangle = 1$ and $\langle E_i,F_j\rangle = \delta_{ij}(q_i^{-1} - q_i)^{-1}$ from (6.82), we get $[E_i,F_j] = \delta_{ij}(q_i - q_i^{-1})^{-1}(K_i - K'^{-1}_i)$. The other three cross relations are derived in a similar manner. (Note that the equivalent cross relations (22) have already been stated without proof as formula (6.87).) □

Proposition 14. *$\mathcal{D}(U_h(\mathfrak{b}_+),\tilde{U}_h(\mathfrak{b}_-))$ is the free h-adic algebra generated by the h-adic algebras $U_h(\mathfrak{b}_+)$ and $\tilde{U}_h(\mathfrak{b}_-)$ subject to the commutation relations (23). For the generators E_i, H_i of $U_h(\mathfrak{b}_+)$ and the elements $\tilde{F}_i$, $\tilde{H}_i$ (see Subsect. 6.3.1) of $\tilde{U}_h(\mathfrak{b}_-)$ these cross relations (23) are*

$$[H_i,\tilde{H}_j'] = 0, \quad [H_i,\tilde{F}_j] = -a_{ij}\tilde{F}_j, \qquad [E_j,\tilde{H}_i'] = -a_{ij}hE_j,$$

$$[E_i,\tilde{F}_j] = \delta_{ij}h\frac{e^{d_ihH_i} - e^{-d_i\tilde{H}_i'}}{e^{d_ih} - e^{-d_ih}}.$$

Proof. The proof is analogous to the preceding proof. We verify (for instance) the relation $[E_j, \tilde{H}_i'] = -a_{ij}hE_j$. Setting $a = E_j$ and $b = \tilde{H}_i$ in (23) and using (6.42) and (6.43), we get

$$\langle 1,1\rangle E_j\tilde{H}_i = \langle e^{d_jhH_j}, 1\rangle \tilde{H}_iE_j + \langle e^{d_jhH_j}, \tilde{H}_i\rangle E_j.$$

By the formulas in Proposition 6.38, $\langle e^{d_jhH_j}, \tilde{H}_i\rangle = \langle d_jhH_j, \tilde{H}_i\rangle = -a_{ij}h$ and $\langle 1,1\rangle = \langle e^{d_jhH_j}, 1\rangle = 1$, so the relation follows. □

Comparing the doubles $\mathcal{D}(U_q(\mathfrak{b}_+),\ U_q(\mathfrak{b}_-))$ and $\mathcal{D}(U_h(\mathfrak{b}_+),\ \tilde{U}_h(\mathfrak{b}_-))$ with the algebras $U_q(\mathfrak{g})$ and $U_h(\mathfrak{g})$, we see that the generators of the Drinfeld–Jimbo algebras satisfy similar relations, but the number of generators corresponding to the Cartan parts is doubled. Quotienting out a half of these generators we indeed get the Drinfeld–Jimbo algebras.

Corollary 15. *The two-sided ideal $\mathcal{I}_q(\mathfrak{g})$, resp. $\mathcal{I}_h(\mathfrak{g})$, of $\mathcal{D}(U_q(\mathfrak{b}_+), U_q(\mathfrak{b}_-))$, resp. $\mathcal{D}(U_h(\mathfrak{b}_+), \tilde{U}_h(\mathfrak{b}_-))$, generated by the elements $K_i - K'_i$, resp. $hH_i - \tilde{H}_i'$, $i = 1, 2, \cdots, l$, is a Hopf ideal and we have canonical isomorphisms*

$$\mathcal{D}(U_q(\mathfrak{b}_+), U_q(\mathfrak{b}_-))/\mathcal{I}_q(\mathfrak{g}) \simeq U_q(\mathfrak{g}), \quad \mathcal{D}(U_h(\mathfrak{b}_+), \tilde{U}_h(\mathfrak{b}_-))/\mathcal{I}_h(\mathfrak{g}) \simeq U_h(\mathfrak{g})$$

of Hopf algebras resp. h-adic Hopf algebras.

Proof. The proof follows from Proposition 13, resp. 14, combined with the defining relations of the algebras $U_q(\mathfrak{g})$ and $U_h(\mathfrak{g})$. □

8.3 Explicit Form of Universal R-Matrices

8.3.1 The Universal R-Matrix for $U_h(\mathrm{sl}_2)$

Let us recall that $U_h(\mathrm{sl}_2)$ is the h-adic Hopf algebra with generators E, F and H satisfying the relations

$$[H, E] = 2E, \quad [H, F] = -2F, \quad [E, F] = \frac{e^{hH} - e^{-hH}}{e^h - e^{-h}}.$$

The comultiplication Δ of $U_h(\mathrm{sl}_2)$ is uniquely determined by the formulas

$$\Delta(E) = E \otimes e^{hH} + 1 \otimes E, \;\; \Delta(F) = F \otimes 1 + e^{-hH} \otimes F, \;\; \Delta(H) = H \otimes 1 + 1 \otimes H.$$

Theorem 16. *The h-adic Drinfeld–Jimbo algebra $U_h(\mathrm{sl}_2)$ is quasitriangular with universal R-matrix given by*

$$\mathcal{R} = e^{h(H\otimes H)/2} \sum\nolimits_{n=0}^{\infty} R_n(h)(E^n \otimes F^n), \tag{36}$$

where $R_n(h) := q^{n(n+1)/2}(1 - q^{-2})^n([n]_q!)^{-1}$ and $q = e^h$.

Proof. We begin with the relation $(\Delta \otimes \mathrm{id})(\mathcal{R}) = \mathcal{R}_{13}\mathcal{R}_{23}$. First we remark that

$$(\Delta \otimes \mathrm{id})(\mathcal{R}) = \sum\nolimits_{m,n=0}^{\infty} a_{mn}(E^m \otimes E^n \otimes F^{m+n}), \tag{37}$$

where a_{mn} is the element of $U_h(\mathfrak{h})^{\otimes 3}$ given by

$$q^{-mn}\frac{[m+n]_q!}{[m]_q![n]_q!}R_{m+n}(h)\{(\Delta \otimes \mathrm{id})(e^{h(H\otimes H)/2})\}(1 \otimes e^{mhH} \otimes 1). \tag{38}$$

The summand in $\mathcal{R}_{13}\mathcal{R}_{23}$ containing the term $E^m \otimes E^n \otimes F^{m+n}$ is equal to

$$R_m(h)R_n(h)e^{h(H\otimes 1\otimes H)/2}(E^m \otimes 1 \otimes F^m)e^{h(1\otimes H\otimes H)/2}(1 \otimes E^n \otimes F^n). \tag{39}$$

By induction on m one shows that

$$[1 \otimes H \otimes H, E^m \otimes 1 \otimes F^m] = -2m(E^m \otimes H \otimes F^m).$$

Since $e^{bhH}Ee^{-bhH} = e^{bh}E$ and $e^{bhH}Fe^{-bhH} = e^{-bh}F$, the latter relation leads to the equality

$$e^{-h(1\otimes H\otimes H)/2}(E^m \otimes 1 \otimes F^m)e^{h(1\otimes H\otimes H)/2} = (1 \otimes e^{mhH} \otimes 1)(E^m \otimes 1 \otimes F^m). \tag{40}$$

On the other hand,

$$(\Delta \otimes \mathrm{id})(e^{h(H\otimes H)/2}) = e^{h(\Delta(H)\otimes H)/2} = e^{h(H\otimes 1\otimes H + 1\otimes H\otimes H)/2}. \tag{41}$$

From formulas (37)–(41) we see that the assertion $(\Delta \otimes \mathrm{id})(\mathcal{R}) = \mathcal{R}_{13}\mathcal{R}_{23}$ is valid provided we know that

$$q^{-mn}\frac{[m+n]_q!}{[m]_q![n]_q!}R_{m+n}(h) = R_m(h)R_n(h).$$

This identity is proved by an easy computation.

The equation $(\mathrm{id} \otimes \Delta)(\mathcal{R}) = \mathcal{R}_{13}\mathcal{R}_{12}$ is proved in a similar manner.

Next we turn to the relation $\Delta^{\mathrm{cop}}(a) = \mathcal{R}\Delta(a)\mathcal{R}^{-1}$. Since both sides of this equality are algebra homomorphisms from $U_h(\mathrm{sl}_2)$ to $U_h(\mathrm{sl}_2)\hat{\otimes}U_h(\mathrm{sl}_2)$, it is enough to prove that

$$\Delta^{\mathrm{cop}}(a)\mathcal{R} = \mathcal{R}\Delta(a) \quad \text{for} \quad a = E, F, H. \tag{42}$$

We carry out the proof for $a = E$. The cases of F and H are treated similarly. For $a = E$, (42) can be rewritten as

$$(E \otimes 1 + e^{hH} \otimes E)\sum_{n=0}^{\infty} R_n(h)e^{h(H\otimes H)/2}(E^n \otimes F^n)$$

$$= \left(\sum_{n=0}^{\infty} R_n(h)e^{h(H\otimes H)/2}(E^n \otimes F^n)\right)(E \otimes e^{hH} + 1 \otimes E).$$

The difference between the left and the right hand sides is

$$\sum_{n=0}^{\infty} e^{h(H\otimes H)/2}\Big\{R_n(h)(1\otimes(e^{-hH}-q^{2n}e^{hH}))$$

$$+R_{n+1}(h)[n+1]_q\left(1\otimes\frac{q^n e^{hH}-q^{-n}e^{-hH}}{q-q^{-1}}\right)\Big\}(E^{n+1}\otimes F^n). \tag{43}$$

Since $[n+1]_q R_{n+1}(h) = q^n(q-q^{-1})R_n(h)$, the coefficient at $E^{n+1}\otimes F^n$ in (43) is zero and hence (43) vanishes.

Finally, a direct computation shows that the element

$$e^{-h(H\otimes H)/2}\sum_{n=0}^{\infty} e^{-hn}R_n(h)((e^{hH}E)^n\otimes(e^{hH}F)^n)$$

of $U_h(\mathrm{sl}_2)\hat{\otimes}U_h(\mathrm{sl}_2)$ is the inverse of $\mathcal{R}$. □

8.3.2 The Universal R-Matrix for $U_h(\mathfrak{g})$

Let $\mathfrak{g}$ be a finite-dimensional simple complex Lie algebra. The following theorem provides an explicit expression for the universal R-matrix of the Drinfeld–Jimbo algebra $U_h(\mathfrak{g})$.

Theorem 17. *$U_h(\mathfrak{g})$ is a quasitriangular h-adic Hopf algebra. Let E_{β_r} and F_{β_r} be the root elements of $U_h(\mathfrak{g})$ from (6.65) corresponding to the sequence $\beta_1, \beta_2, \cdots, \beta_n$ of positive roots of the Lie algebra $\mathfrak{g}$. Let (B_{ij}) be the inverse of the matrix $(C_{ij}) := (d_j^{-1}a_{ij})$. Then the element*

$$\mathcal{R}=\exp\Big(h\sum_{i,j}B_{ij}(H_i\otimes H_j)\Big)\sum_{r_1,\cdots,r_n=0}^{\infty}\prod_{j=1}^{n} q_{\beta_j}^{\frac{1}{2}r_j(r_j+1)}\frac{(1-q_{\beta_j}^{-2})^{r_j}}{[r_j]_{q_{\beta_j}}!}E_{\beta_j}^{r_j}\otimes F_{\beta_j}^{r_j}$$

$$=\exp\Big(h\sum_{i,j}B_{ij}(H_i\otimes H_j)\Big)\prod_{\beta\in\Delta_+}\exp_{q_\beta}\Big((1-q_\beta^{-2})(E_\beta\otimes F_\beta)\Big) \tag{44}$$

of $U_h(\mathfrak{g})\hat{\otimes}U_h(\mathfrak{g})$ is a universal R-matrix of $U_h(\mathfrak{g})$, where $q_\beta := q_i$ if β and the simple root α_i lie in the same orbit with respect to the Weyl group W of $\mathfrak{g}$, the factors in the product (44) appear in the order $\beta_n, \beta_{n-1}, \cdots, \beta_1$ and the q-exponential function $\exp_q x$ is defined by $\exp_q x = \sum_{r=0}^{\infty} q^{r(r+1)/2}x^r/[r]_q!$.

Sketch of proof. The key step of the proof is to determine dual bases of the h-adic vector spaces $U_h(\mathfrak{b}_+)$ and $\tilde{U}_h(\mathfrak{b}_-)$ with respect to the bilinear form $\overline{\sigma} = \langle\cdot,\cdot\rangle'$ from Subsect. 6.3.1. By Theorem 6.24′, the set of elements

$$H_1^{m_1}\cdots H_l^{m_l}E_{\beta_n}^{r_n}\cdots E_{\beta_1}^{r_1},\quad m_1,\cdots,m_l,r_1,\cdots,r_n\in\mathbb{N}_0, \tag{45}$$

constitute a basis of $U_h(\mathfrak{b}_+)$. Similarly, the set of elements

$$\breve{H}_1^{m_1}\cdots\breve{H}_l^{m_l}\tilde{F}_{\beta_n}^{s_n}\cdots\tilde{F}_{\beta_1}^{s_1},\quad m_1,\cdots,m_l,s_1,\cdots,s_n\in\mathbb{N}_0, \tag{46}$$

is a basis of $\tilde{U}_h(\mathfrak{b}_-)$, where $\breve{H}_i := \sum_k B_{ik}\tilde{H}_k$.

We compute the bilinear form $\langle \cdot,\cdot \rangle'$ at the basis elements (45) and (46). For this we essentially use the formulas at the end of Subsect. 6.3.1 and the fact that $\langle \cdot,\cdot \rangle'$ is a dual pairing of the Hopf algebras $\tilde{U}_h(\mathfrak{b}_+)^{\mathrm{op}}$ and $\tilde{U}_h(\mathfrak{b}_-)$. Since

$$\langle H_j, \breve{H}_i\rangle' = \sum_k B_{ik}\langle H_j, \tilde{H}_k\rangle' = \sum_k B_{ik} d_j^{-1} a_{kj} = \delta_{ij}$$

and $\Delta^{(n-1)}(H_j) = \sum_{i=0}^{n-1} 1^{n-i-1} \otimes H_j \otimes 1^i$, we obtain

$$\langle H_j^{m_j}, \breve{H}_i^{n_i}\rangle = \langle (\Delta^{(n_i-1)}(H_j))^{m_j}, (\breve{H}_i \otimes \cdots \otimes \breve{H}_i)^{n_i}\rangle' = \delta_{ij}\delta_{m_j n_i} m_i!. \tag{47}$$

A similar slightly longer reasoning shows that

$$\langle E_j^{m_j}, \tilde{F}_i^{n_i}\rangle' = \delta_{ij}\delta_{m_j n_i} \frac{[n_i]_{q_i}! h^{n_i}}{(q_i - q_i^{-1})^{n_i}} q_i^{-n_i(n_i-1)/2}. \tag{48}$$

In order to determine the values of $\langle \cdot,\cdot \rangle'$ at arbitrary elements (45) and (46), we need an expression for $\Delta(E_\beta)$, $\beta \in \Delta_+$. Clearly, the Hopf subalgebra $U_h(\mathfrak{g}_i)$ of $U_h(\mathfrak{g})$ generated by E_i, F_i, H_i is isomorphic to $U_h(\mathrm{sl}_2)$. By Theorem 16, the universal R-matrix for $U_h(\mathfrak{g}_i)$ is, up to a certain factor, of the form

$$\tilde{\mathcal{R}}_i = \sum_{m=0}^{\infty} \frac{(1-q_i^{-2})^m}{[m]_{q_i}!} q_i^{m(m+1)/2} (E_i^m \otimes F_i^m),$$

where $q_i = e^{d_i h}$. Recall that T_i is the algebra automorphism of $U_h(\mathfrak{g})$ defined by the braid group action (6.62)–(6.64) on $U_h(\mathfrak{g})$. One can prove that

$$\Delta(T_i(a)) = \tilde{\mathcal{R}}_i^{-1}((T_i \otimes T_i)\Delta(a))\tilde{\mathcal{R}}_i, \qquad a \in U_h(\mathfrak{g}). \tag{49}$$

By Definition 6.4, the root element E_{β_r} for a positive root $\beta_r \in \Delta_+$ is given by $E_{\beta_r} = T_{i_1}T_{i_2}\cdots T_{i_{r-1}}(E_{i_r})$. We set

$$\tilde{\mathcal{R}}_{\beta_r} := (T_{i_1}T_{i_2}\cdots T_{i_{r-1}} \otimes T_{i_1}T_{i_2}\cdots T_{i_{r-1}})\tilde{\mathcal{R}}_{i_r}.$$

One can derive from (49) that

$$\Delta(E_{\beta_r}) = \tilde{\mathcal{R}}^{-1}(\beta_r)(E_{\beta_r} \otimes e^{hH_{\beta_r}} + 1 \otimes E_{\beta_r})\tilde{\mathcal{R}}(\beta_r),$$

where $\tilde{\mathcal{R}}(\beta_r) := \tilde{\mathcal{R}}_{\beta_{r-1}}\tilde{\mathcal{R}}_{\beta_{r-2}}\cdots\tilde{\mathcal{R}}_{\beta_1}$ and $H_{\beta_r} = \sum_{i=1}^{l} d_i r_i H_i$ if $\beta_r = \sum_i r_i\alpha_i$, $r_i \geq 0$. Then we have

$$\Delta(E_{\beta_r}^n) = \tilde{\mathcal{R}}^{-1}(\beta_r)(E_{\beta_r} \otimes e^{hH_{\beta_r}} + 1 \otimes E_{\beta_r})^n\tilde{\mathcal{R}}(\beta_r).$$

From the latter formula we derive that

$$\Delta(E_{\beta_r}^n) - (E_{\beta_r} \otimes e^{hH_{\beta_r}} + 1 \otimes E_{\beta_r})^n \in U_h(\mathfrak{b}_+) \otimes U_h^{\beta_r}(\mathfrak{n}_+). \tag{50}$$

One verifies that any braid group automorphism T_i satisfies the condition $\langle T_i(a), T_i(b)\rangle' = \langle a, b\rangle'$ for $a \in U_h(\mathfrak{n}_+)$ and $b \in \tilde{U}_h(\mathfrak{n}_-)$. Therefore, if a positive root β and a simple root α_i of $\mathfrak{g}$ belong to the same W-orbit, then

$$\langle E_\beta, \tilde{F}_\beta\rangle' = \langle E_i, \tilde{F}_i\rangle' = h(q_i^{-1} - q_i)^{-1}.$$

Further, from (50) one derives that $\langle E_\beta^m, \tilde{F}_\beta^n\rangle' = \langle E_i^m, \tilde{F}_i^n\rangle'$ for $n, m \in \mathbb{N}_0$.

Using the above relations, especially (47) and (48), one proves by induction that

$$\begin{aligned}
&\langle H_1^{m_1} \cdots H_l^{m_l} E_{\beta_n}^{r_n} \cdots E_{\beta_1}^{r_1}, \breve{H}_1^{n_1} \cdots \breve{H}_l^{n_l} \tilde{F}_{\beta_n}^{s_n} \cdots \tilde{F}_{\beta_1}^{s_1}\rangle' \\
&= \prod\nolimits_{i=1}^{l} \langle H_i^{m_i}, \breve{H}_i^{n_i}\rangle' \cdot \prod\nolimits_{j=1}^{n} \langle E_{\beta_j}^{r_j}, F_{\beta_j}^{s_j}\rangle' \\
&= \prod\nolimits_{i=1}^{l} \delta_{n_i m_i} m_i! \cdot \prod\nolimits_{j=1}^{n} \delta_{s_j r_j} \frac{[r_j]_{q_{\beta_j}}! h^{r_j}}{(1 - q_{\beta_j}^{-2})^{r_j}} q_{\beta_j}^{-r_j(r_j+1)/2}.
\end{aligned} \tag{51}$$

That is, the elements (46) divided by the scaling factor appearing in (51) and the elements (45) form dual bases of $U_h(\mathfrak{b}_+)$ and $\tilde{U}_h(\mathfrak{b}_-)$, respectively, with respect to $\langle \cdot, \cdot \rangle'$. For brevity, let us denote these sets by $\{e_i\}$ resp. $\{f_i\}$. From the particular form of these elements we see that $\mathcal{R}_\mathcal{D} := \sum_i (1 \otimes e_i) \otimes (f_i \otimes 1)$ is an element of the h-adic completion $\mathcal{D}(U_h(\mathfrak{b}_+), \tilde{U}_h(\mathfrak{b}_-)) \hat{\otimes} \mathcal{D}(U_h(\mathfrak{b}_+), \tilde{U}_h(\mathfrak{b}_-))$. Hence, by Proposition 11, $\mathcal{R}_\mathcal{D}$ is a universal R-matrix of the quantum double $\mathcal{D}(U_h(\mathfrak{b}_+), \tilde{U}_h(\mathfrak{b}_-))$. Under the usual identification of e_i with $1 \otimes e_i$ and of f_i with $f_i \otimes 1$, we get $\mathcal{R}_\mathcal{D} = \sum_i e_i \otimes f_i$. Let π be the canonical homomorphism of the double $\mathcal{D}(U_h(\mathfrak{b}_+), \tilde{U}_h(\mathfrak{b}_-))$ to $U_h(\mathfrak{g})$ given by Corollary 15. Inserting the elements (45) and (46) into $\mathcal{R}_\mathcal{D}$, one computes that $\mathcal{R} := (\pi \otimes \pi)\mathcal{R}_\mathcal{D}$ takes the form stated in the theorem. Since $\mathcal{R}_\mathcal{D}$ is a universal R-matrix of the double $\mathcal{D}(U_h(\mathfrak{b}_+), \tilde{U}_h(\mathfrak{b}_-))$ and π is a Hopf algebra homomorphism, $\mathcal{R}$ is a universal R-matrix of $U_h(\mathfrak{g})$. □

Remarks: 5. Recall from Subsect. 6.2.3 that the sequences $\{E_{\beta_i}\}$ and $\{F_{\beta_i}\}$ of root vectors of $U_h(\mathfrak{g})$ depend on the reduced representation $w_0 = w_{i_1} \cdots w_{i_n}$ of the longest element w_0 of the Weyl group of $\mathfrak{g}$. Nevertheless, it can be shown (see [KR2]) that the universal R-matrix (44) is independent of the choice of the reduced decomposition of w_0.

6. The sum $\sum_{i,j} B_{ij} H_i \otimes H_j$ occurring in formula (44) is just the canonical element with respect to the symmetric bilinear form $\langle .,. \rangle$ on $\mathfrak{h} \otimes \mathfrak{h}$ given by $\langle H_i, H_j\rangle = d_j^{-1} a_{ij}$ (see Subsect. 6.1.1). Indeed, setting $G_j := \sum_i B_{ij} H_i$, we have $\sum_{i,j} B_{ij} H_i \otimes H_j = \sum_k G_k \otimes H_k$ and

$$\langle G_j, H_k\rangle = \sum\nolimits_i B_{ij} \langle H_i, H_k\rangle = \sum\nolimits_i B_{ij} \langle H_k, H_i\rangle = \sum\nolimits_i C_{ki} B_{ij} = \delta_{jk}.$$

Sometimes it is also convenient to choose other dual bases in $\mathfrak{h}$ or to take a self-dual basis of $\mathfrak{h}$. Then only the first factor $\exp\left(h \sum_{i,j} B_{ij}(H_i \otimes H_j)\right)$ in (44) has to be correspondingly changed.

7. The h-adic Hopf algebra $U_h(\mathrm{gl}_{l+1})$ is also quasitriangular. A universal R-matrix of it is given by the universal R-matrix (44) for $U_h(\mathrm{sl}_{l+1})$ with $\exp\left(h \sum_{i,j} B_{ij}(H_i \otimes H_j)\right)$ replaced by $\exp\left(h \sum_i (H_i' \otimes H_i')\right)$, where H_i', $i = 1, 2, \cdots, l+1$, is a self-dual basis of the Cartan subalgebra with respect to the bilinear form $\langle .,. \rangle$. △

8.3.3 R-Matrices for Representations of $U_q(\mathfrak{g})$

Recall that the h-adic algebra $U_h(\mathfrak{g})$ is formally transformed into the complex algebra $U_q(\mathfrak{g})$ if we replace $e^{d_i h}$ by q_i and $e^{d_i h H_i}$ by K_i. But the universal R-matrix (44) of $U_h(\mathfrak{g})$ does not yield a universal R-matrix of $U_q(\mathfrak{g})$ in this manner. The first difficulty is that the second factor in (44) is not in $U_q(\mathfrak{g}) \otimes U_q(\mathfrak{g})$, because it is an infinite sum. However, it belongs to the completion $\bar{U}_q^+(\mathfrak{g})\bar{\otimes}\bar{U}_q^+(\mathfrak{g})$ defined in Subsect. 6.3.3, and it yields a finite sum when a highest (lowest) weight representation acts on the first (second) tensor factor. The second problem is that the first factor in (44) has no counterpart in $\bar{U}_q^+(\mathfrak{g})\bar{\otimes}\bar{U}_q^+(\mathfrak{g})$. But the inner algebra automorphism of $U_h(\mathfrak{g})\hat{\otimes}U_h(\mathfrak{g})$ generated by this factor does and we shall work with this algebra automorphism of $\bar{U}_q^+(\mathfrak{g})\bar{\otimes}\bar{U}_q^+(\mathfrak{g})$ instead.

We assume throughout this subsection that q is not a root of unity.

One verifies that there is an automorphism Φ of the algebra $\bar{U}_q^+(\mathfrak{g})\bar{\otimes}\bar{U}_q^+(\mathfrak{g})$ such that

$$\Phi(K_\lambda \otimes 1) = K_\lambda \otimes 1, \qquad \Phi(1 \otimes K_\lambda) = 1 \otimes K_\lambda, \tag{52}$$

$$\Phi(E_i \otimes 1) = E_i \otimes K_i, \qquad \Phi(1 \otimes E_i) = K_i \otimes E_i, \tag{53}$$

$$\Phi(F_i \otimes 1) = F_i \otimes K_i^{-1}, \qquad \Phi(1 \otimes F_i) = K_i^{-1} \otimes F_i. \tag{54}$$

Recall that $\bar{U}_q^+(\mathfrak{g})$ and $\bar{U}_q^+(\mathfrak{g})\bar{\otimes}\bar{U}_q^+(\mathfrak{g})$ are the completions defined in Subsect. 6.3.3 and $C_\beta \in U_q^{-\beta}(\mathfrak{n}_-)\otimes U_q^\beta(\mathfrak{n}_+)$ denotes the canonical element with respect to the pairing $\langle\cdot,\cdot\rangle$ (see (6.82), (6.97) and Proposition 6.37). We define

$$\mathfrak{R} := \sum_{\beta\in Q_+} q^{(\beta,\beta)}(K_\beta \otimes K_\beta^{-1})C_\beta \in \bar{U}_q^+(\mathfrak{g})\bar{\otimes}\bar{U}_q^+(\mathfrak{g}).$$

A direct computation shows that $\mathfrak{R}$ coincides with the sum occurring in formula (44) for the universal R-matrix of $U_h(\mathfrak{g})$, that is,

$$\mathfrak{R} = \sum_{r_1,\cdots,r_n=0}^{\infty} \prod_{j=1}^{n} \frac{(1-q_{\beta_j}^{-2})^{r_j}}{[r_j]_{q_{\beta_j}}!} q_{\beta_j}^{r_j(r_j+1)/2} E_{\beta_j}^{r_j} \otimes F_{\beta_j}^{r_j}.$$

Theorem 18. *The element $\mathfrak{R}$ and the above algebra automorphism Φ of $\bar{U}_q^+(\mathfrak{g})\bar{\otimes}\bar{U}_q^+(\mathfrak{g})$ satisfy the following conditions:*

$$\mathfrak{R}\Delta(a) = \Phi(\Delta^{\mathrm{cop}}(a))\mathfrak{R}, \qquad a \in \bar{U}_q^+(\mathfrak{g}), \tag{55}$$

$$(\Phi_{23} \circ \Phi_{13})(\mathfrak{R}_{12}) = \mathfrak{R}_{12}, \qquad (\Phi_{12} \circ \Phi_{13})(\mathfrak{R}_{23}) = \mathfrak{R}_{23}, \tag{56}$$

$$\Phi_{23}(\mathfrak{R}_{13})\mathfrak{R}_{23} = (\Delta \otimes 1)(\mathfrak{R}), \quad \Phi_{12}(\mathfrak{R}_{13})\mathfrak{R}_{12} = (1 \otimes \Delta)(\mathfrak{R}). \tag{57}$$

$\mathfrak{R}$ is invertible in the algebra $\bar{U}_q^+(\mathfrak{g})\bar{\otimes}\bar{U}_q^+(\mathfrak{g})$.

Proof. Our first aim is to prove that

$$\mathfrak{R}^{-1} := \sum_{\beta\in Q_+} q^{(\beta,\beta)}(1 \otimes K_\beta^{-1})(S \otimes 1)(C_\beta)$$

is the inverse of $\mathfrak{R}$. Clearly, the product $\mathfrak{R}\mathfrak{R}^{-1}$ is a sum over $\beta \in Q_+$ of terms

$$\sum_{\substack{\gamma,\delta \in Q_+ \\ \gamma+\delta=\beta}} C_\gamma \cdot (K_\delta^{-1} \otimes 1) \cdot (S \otimes 1)(C_\delta). \tag{58}$$

For $\beta = 0$ this expression is obviously equal to $1 \otimes 1$. Thus, in order to prove that $\mathfrak{R}\mathfrak{R}^{-1} = 1 \otimes 1$ it suffices to show that all terms in (58) for $\beta \neq 0$ are zero. The sum (58) belongs to $U_q^{-\beta}(\mathfrak{n}_-) \otimes U_q(\mathfrak{g})$. Since the bilinear form $\langle \cdot, \cdot \rangle$ is nondegenerate on $U_q^\beta(\mathfrak{b}_+) \times U_q^{-\beta}(\mathfrak{b}_-)$ by Proposition 6.37, it is enough to prove that any term (58) for $\beta \neq 0$ vanishes when the first tensor factor is paired against an arbitrary element $a \in U_q^\beta(\mathfrak{n}_+)$. We can write

$$\Delta(a) = \sum_{\substack{\gamma,\delta \in Q_+ \\ \gamma+\delta=\beta}} a_{\delta,\gamma}(1 \otimes K_\delta), \qquad a_{\delta,\gamma} \in U_q^\delta(\mathfrak{n}_+) \otimes U_q^\gamma(\mathfrak{n}_+),$$

$$a_{\delta,\gamma} = \sum_m a_{\delta,m}^{\delta,\gamma} \otimes a_{\gamma,m}^{\delta,\gamma}, \qquad a_{\delta,m}^{\delta,\gamma} \in U_q^\delta(\mathfrak{n}_+), \quad a_{\gamma,m}^{\delta,\gamma} \in U_q^\gamma(\mathfrak{n}_+).$$

If a_r^β and b_r^β are as in the proof of Lemma 6.40, then pairing (58) in the first tensor factor against a yields

$$\begin{aligned}
(\langle a, \cdot \rangle \otimes 1)\Big(\sum_{\gamma,\delta,r,s} a_r^\gamma K_\delta^{-1} S(a_s^\delta) \otimes b_r^\gamma b_s^\delta\Big) &= \sum_{\gamma,\delta,r,s} \langle a, a_r^\gamma K_\delta^{-1} S(a_s^\delta)\rangle b_r^\gamma b_s^\delta \\
&= \sum_{\gamma,\delta,r,s} \langle \Delta(a), K_\delta^{-1} S(a_s^\delta) \otimes a_r^\gamma \rangle b_r^\gamma b_s^\delta \\
&= \sum_{\gamma,\delta,r,s,m} \langle a_{\delta,m}^{\delta,\gamma}, K_\delta^{-1} S(a_s^\delta)\rangle \langle a_{\gamma,m}^{\delta,\gamma} K_\delta, a_r^\gamma \rangle b_r^\gamma b_s^\delta \\
&= \sum_{\gamma,\delta,m} \Big(\sum_r \langle a_{\gamma,m}^{\delta,\gamma} K_\delta, a_r^\gamma\rangle b_r^\gamma\Big)\Big(\sum_s \langle a_{\delta,m}^{\delta,\gamma}, K_\delta^{-1} S(a_s^\delta)\rangle b_s^\delta\Big) \\
&= \sum_{\gamma,\delta,m} \Big(\sum_r (a_{\gamma,m}^{\delta,\gamma}, a_r^\gamma) b_r^\gamma\Big)\Big(\sum_s (K_\delta S^{-1}(a_{\delta,m}^{\delta,\gamma}), a_s^\delta) b_s^\delta\Big) \\
&= \sum_{\gamma,\delta,m} a_{\gamma,m}^{\delta,\gamma} K_\delta S^{-1}(a_{\delta,m}^{\delta,\gamma}) = S^{-1}\Big(\sum_{\gamma,\delta,m} a_{\delta,m}^{\delta,\gamma} S(a_{\gamma,m}^{\delta,\gamma} K_\delta)\Big) \\
&= (S^{-1} \circ m \circ (1 \otimes S) \otimes \Delta)(a) = \varepsilon(a)1 = 0,
\end{aligned}$$

where we used various properties of the dual pairing $\langle \cdot, \cdot \rangle$ of the Hopf algebras $U_q(\mathfrak{b}_+)$ and $U_q(\mathfrak{b}_-)^{\mathrm{op}}$. Thus, we have proved that $\mathfrak{R}\mathfrak{R}^{-1} = 1 \otimes 1$. Similar reasoning shows that $\mathfrak{R}^{-1}\mathfrak{R} = 1 \otimes 1$.

Next we turn to the relation (55). Since Δ and $\mathfrak{R}^{-1} \cdot \Phi \circ \Delta^{\mathrm{cop}} \cdot \mathfrak{R}$ are both algebra automorphisms, it is sufficient to prove (55) for the generators $a = K_\lambda, E_i, F_i$. The case $a = K_\lambda$ is easily done. The cases $a = E_i$ and $a = F_i$ follow from (6.99) and (6.100). The relations (56) are proved by straightforward verification. The relations (57) are equivalent to

$$(\Delta\otimes 1)(C_\beta) = \sum_{\substack{\gamma,\delta\in Q_+\\ \gamma+\delta=\beta}} q^{-(\gamma,\delta)}(K_\delta^{-1}\otimes 1\otimes 1)\cdot(C_\gamma)_{13}\cdot(C_\delta)_{23}$$

and

$$(1\otimes\Delta)(C_\beta) = \sum_{\substack{\gamma,\delta\in Q_+\\ \gamma+\delta=\beta}} q^{-(\gamma,\delta)}(1\otimes 1\otimes K_\delta)\cdot(C_\gamma)_{13}\cdot(C_\delta)_{12}.$$

These identities are proved in a similar way as the fact that the terms (58) vanish for $\beta\neq 0$. We omit the details. □

Suppose that T_V and T_W are two weight representations of $U_q(\mathfrak{g})$ on vector spaces V and W, respectively, with weight space decompositions $V=\bigoplus_\mu V_\mu$ and $W=\bigoplus_{\mu'} W_{\mu'}$. Let B_{VW} denote the linear operator on $V\otimes W$ given by

$$B_{VW}(v\otimes w) := q^{(\mu,\mu')} v\otimes w \quad \text{for} \quad v\in V_\mu,\ w\in W_{\mu'}.$$

Now we define the R-matrix $R_{VW}: V\otimes W\to V\otimes W$ and the linear mapping $\hat{R}_{VW}: V\otimes W\to W\otimes V$ for the two representations T_V and T_W by

$$R_{VW} := B_{VW}\circ(T_V\otimes T_W)(\mathfrak{R}) \quad \text{and} \quad \hat{R}_{VW} := \tau\circ R_{VW}. \tag{59}$$

Clearly, $\hat{R}_{VW}$ is invertible with inverse $\hat{R}_{VW}^{-1} = (T_V\otimes T_W)(\mathfrak{R}^{-1})\circ B_{VW}^{-1}\circ\tau$.

Proposition 19. *The mapping $\hat{R}_{VW}$ intertwines the representations $T_V\otimes T_W$ and $T_W\otimes T_V$. The family of mappings $\hat{R}_{VW}$ has the properties stated in Proposition 4.*

Proof. A direct computation shows that on weight vectors we have

$$((T_V\otimes T_W)\circ\Phi)(a_1\otimes a_2) = B_{VW}^{-1}(T_V\otimes T_W)(a_1\otimes a_2)B_{VW},\quad a_1,a_2\in U_q(\mathfrak{g}).$$

Therefore, this relation holds on the whole space $V\otimes W$. Applying the tensor product representation $T_V\otimes T_W$ to both sides of (55) we derive

$$B_{VW}^{-1}R_{VW}(T_V\otimes T_W)(\Delta(a)) = ((T_V\otimes T_W)\circ\Phi)(\Delta^{\rm cop}(a))B_{VW}^{-1}R_{VW}.$$

Using the previous equation and applying the flip to both sides, we obtain $\hat{R}_{VW}(T_V\otimes T_W)(a) = (T_W\otimes T_V)(a)\hat{R}_{VW}$.

Next we prove (12). It follows from the first relation in (57) that

$$B_{WU}^{-1}((T_V\otimes T_U)(\mathfrak{R}_{13}))B_{WU}(T_W\otimes T_U)(\mathfrak{R}_{23}) = (T_V\otimes T_W\otimes T_U)((\Delta\otimes{\rm id})(\mathfrak{R})).$$

This implies that $B_{WU}^{-1}B_{VU}^{-1}R_{VU}R_{WU} = B_{V\otimes W,U}^{-1}R_{V\otimes W,U}$. Clearly, we have $B_{WU}^{-1}B_{VU}^{-1} = B_{V\otimes W,U}^{-1}$. Therefore, applying the flip permuting elements of $V\otimes W$ and U, we obtain (12). The relation (13) is derived in a similar manner from the second relation in (57). □

8.4 Vector Representations and *R*-Matrices

This section is concerned with the R-matrices $R_{1,1} := R_{VV}$ for the vector representations $T_1 = T_V$ of the Drinfeld–Jimbo algebras $U_h(\mathfrak{g})$ and $U_q(\mathfrak{g})$. These matrices will be the starting point for the definitions of quantized simple Lie groups in Chap. 9. Using the notation $q^\alpha = e^{\alpha h}$, $\alpha \in \mathbb{C}$, in the h-adic case, we shall obtain expressions for the R-matrices that apply to both algebras $U_h(\mathfrak{g})$ and $U_q(\mathfrak{g})$ simultaneously. In the case $\mathfrak{g} = \mathrm{so}_{2n+1}$ we shall use the algebras $U_{q^{1/2}}(\mathrm{so}_{2n+1})$ and $U_{h/2}(\mathrm{so}_{2n+1})$.

8.4.1 Vector Representations of Drinfeld–Jimbo Algebras

The "usual" realizations of the Lie algebras $\mathrm{gl}(N, \mathbb{C})$, $\mathrm{sl}(N, \mathbb{C})$, $\mathrm{so}(N, \mathbb{C})$ and $\mathrm{sp}(N, \mathbb{C})$ by complex $N \times N$ matrices are irreducible representations with highest weight $(1, 0, \cdots, 0)$. They are commonly referred to as the first fundamental representations or as the vector representations. We now generalize these representations to Drinfeld–Jimbo algebras.

Definition 5. *The* vector representation T_1 *of the Drinfeld–Jimbo algebra* $U_q(\mathfrak{g})$ *or* $U_h(\mathfrak{g})$ *is the irreducible type 1 representation with highest weight* $\lambda = (1, 0, \cdots, 0)$ *with respect to the ordered sequence* $-\alpha_1, \cdots, -\alpha_l$ *of simple roots.*

The simple roots $-\alpha_1, \cdots, -\alpha_l$ are chosen only for technical convenience. Note that a representation T of $U_q(\mathfrak{g})$ or $U_h(\mathfrak{g})$ has the highest weight λ with respect to the simple roots $-\alpha_1, \cdots, -\alpha_l$ if and only if the representation $T \circ \theta$ has the highest weight λ with respect to the simple roots $\alpha_1, \cdots, \alpha_l$ (recall that in the first case weights are considered with respect to the elements K_i^{-1} and in the second case with respect to K_i). Here θ denotes the algebra automorphism of $U_q(\mathfrak{g})$ or $U_h(\mathfrak{g})$ from Proposition 6.9 or from formula (6.47), respectively.

Let E_{ij} be the $N \times N$ matrix with 1 in the (i, j)-position and 0 elsewhere. The vector representation T_1 of $U_q(\mathfrak{g})$ is described by the following list:

$U_q(\mathrm{gl}_N)$:

$$T_1(K_i) = q^{-1}E_{ii} + \sum\nolimits_{j \neq i} E_{jj}, \qquad i = 1, 2, \cdots, N,$$

$$T_1(E_i) = E_{i+1,i}, \quad T_1(F_i) = E_{i,i+1}, \qquad i = 1, 2, \cdots, N-1.$$

$U_q(\mathrm{sl}_N)$:

$$T_1(K_i) = q^{-1}E_{ii} + qE_{i+1,i+1} + \sum\nolimits_{j \neq i, i+1} E_{jj}, \qquad i = 1, 2, \cdots, N-1,$$

$$T_1(E_i) = E_{i+1,i}, \quad T_1(F_i) = E_{i,i+1}, \qquad i = 1, 2, \cdots, N-1.$$

$U_{q^{1/2}}(\mathrm{so}_N)$, $N = 2n+1$:

$$T_1(K_i) = D_i^{-1} D_{i+1} D_{2n-i+1}^{-1} D_{2n-i+2}, \quad i = 1, 2, \cdots, n-1,$$

$$T_1(F_i) = E_{i,i+1} - E_{2n-i+1,2n-i+2}, \qquad i = 1,2,\cdots n-1,$$
$$T_1(E_i) = E_{i+1,i} - E_{2n-i+2,2n-i+1}, \qquad i = 1,2,\cdots n-1,$$
$$T_1(F_n) = c(E_{n,n+1} - q^{-1/2}E_{n+1,n+2}), \;\; T_1(E_n) = c(E_{n+1,n} - q^{1/2}E_{n+2,n+1}),$$
$$T_1(K_n) = D_n^{-1}D_{n+2},$$

where $c = [2]_{q^{1/2}}^{1/2} = (q^{1/2}+q^{-1/2})^{1/2}$ and D_j is the diagonal matrix $\sum_i q^{\delta_{ij}} E_{ii}$.

$U_q(\mathrm{sp}_N)$, $N = 2n$:

$$T_1(K_i) = D_i^{-1}D_{i+1}D_{2n-i}^{-1}D_{2n-i+1}, \quad i = 1,2,\cdots,n-1,$$
$$T_1(F_i) = E_{i,i+1} - E_{2n-i,2n-i+1}, \qquad i = 1,2,\cdots,n-1,$$
$$T_1(E_i) = E_{i+1,i} - E_{2n-i+1,2n-i}, \qquad i = 1,2,\cdots,n-1,$$
$$T_1(F_n) = E_{n,n+1}, \;\; T_1(E_n) = E_{n+1,n} \;\; T_1(K_n) = D_n^{-2}D_{n+1}^2.$$

$U_q(\mathrm{so}_N)$, $N = 2n$:

$$T_1(K_i) = D_i^{-1}D_{i+1}D_{2n-i}^{-1}D_{2n-i+1}, \quad i = 1,2,\cdots,n-1,$$
$$T_1(F_i) = E_{i,i+1} - E_{2n-i,2n-i+1}, \qquad i = 1,2,\cdots,n-1,$$
$$T_1(E_i) = E_{i+1,i} - E_{2n-i+1,2n-i}, \qquad i = 1,2,\cdots,n-1,$$
$$T_1(F_n) = -E_{n,n+2} + E_{n-1,n+1}, \;\; T_1(E_n) = -E_{n+2,n} + E_{n+1,n-1},$$
$$T_1(K_n) = D_{n-1}^{-1}D_n^{-1}D_{n+1}D_{n+2}.$$

By straightforward calculations one checks that the preceding formulas define an irreducible weight representation T_1 of the algebra $U_q(\mathfrak{g})$ on the vector space $\mathbb{C}^N$. For the basis vector $\mathbf{e}_1 = (1,0,\cdots,0)$ we easily verify that $T_1(U_q(\mathfrak{g}))\mathbf{e}_1 = \mathbb{C}^N$, $T_1(K_i^{-1})\mathbf{e}_1 = q^{\delta_{1i}}\mathbf{e}_1$ and $T_1(F_i)\mathbf{e}_1 = 0$ for all $i = 1,\cdots,l$. Hence T_1 is the type 1 representation with highest weight $\lambda = (1,0,\cdots,0)$ with respect to the simple roots $-\alpha_1,\cdots,-\alpha_l$ (see also Definition 6.6 and Remark 6.7 in Subsect. 6.2.4). Thus, T_1 is indeed the vector representation of $U_q(\mathfrak{g})$.

If a reduced decomposition of the longest element w_0 of the Weyl group W of $\mathfrak{g}$ is fixed, then the action of the corresponding nonsimple root elements can be derived from the above formulas. For example, let $\mathfrak{g} = \mathrm{sl}(N,\mathbb{C})$. We take the reduced decomposition

$$w_0 = (w_1w_2\cdots w_{N-1})(w_1w_2\cdots w_{N-2})\cdots(w_1w_2)w_1$$

of the longest element w_0 of the Weyl group. Here w_i is the reflection corresponding to the i-th simple root. Then the roots from Subsect. 6.2.3 are

$$\beta_{12},\beta_{13},\cdots,\beta_{1N},\beta_{23},\cdots,\beta_{2N},\cdots,\beta_{N-2,N-1},\beta_{N-2,N},\beta_{N-1,N},$$

where $\beta_{ij} = \alpha_i + \alpha_{i+1} + \cdots + \alpha_{j-1}$. One easily calculates that

$$T_1(E_{\beta_{ij}}) = (-1)^{i-j+1}E_{ji}, \qquad T_1(F_{\beta_{ij}}) = (-1)^{i-j+1}E_{ij}.$$

This formula holds also for the root elements of $U_q(\mathrm{gl}_N)$.

The vector representation T_1 of the h-adic algebra $U_h(\mathfrak{g})$ is determined in a similar manner. It acts on the vector space $\mathbb{C}^N[[h]]$ over the ring $\mathbb{C}[[h]]$. The formulas for the representation operators $T_1(E_i)$ and $T_1(F_i)$ are the same as above. The actions of the Cartan generators H_i are given as follows:

$U_h(\mathrm{gl}_N)$:

$$T_1(H_i) = -E_{ii}, \qquad i = 1, 2, \cdots, N.$$

$U_h(\mathrm{sl}_N)$:

$$T_1(H_i) = -E_{ii} + E_{i+1,i+1}, \qquad i = 1, 2, \cdots, N-1.$$

$U_{h/2}(\mathrm{so}_{2n+1})$:

$$T_1(H_i) = -E_{ii} + E_{i+1,i+1} - E_{2n-i+1,2n-i+1} + E_{2n-i+2,2n-i+2}, \; i = 1, \cdots, n-1$$

$$T_1(H_n) = 2(-E_{nn} + E_{n+2,n+2}).$$

$U_h(\mathrm{sp}_{2n})$:

$$T_1(H_i) = -E_{ii} + E_{i+1,i+1} - E_{2n-i,2n-i} + E_{2n-i+1,2n-i+1}, \; i = 1, 2, \cdots, n-1,$$

$$T_1(H_n) = -E_{nn} + E_{n+1,n+1}.$$

$U_h(\mathrm{so}_{2n})$:

$$T_1(H_i) = -E_{ii} + E_{i+1,i+1} - E_{2n-i,2n-i} + E_{2n-i+1,2n-i+1}, \; i = 1, 2, \cdots, n-1,$$

$$T_1(H_n) = -E_{n-1,n-1} - E_{nn} + E_{n+1,n+1} + E_{n+2,n+2}.$$

8.4.2 R-Matrices for Vector Representations

In this subsection we compute the R-matrices $R_{1,1} := R_{V_1V_1}$ for the vector representations $T_1 = T_{V_1}$ of the Drinfeld–Jimbo algebras $U_h(\mathfrak{g})$ and $U_q(\mathfrak{g})$. As noted above, for $\mathfrak{g} = \mathrm{so}_{2n+1}$ we shall take the algebras $U_{q^{1/2}}(\mathrm{so}_{2n+1})$ and $U_{h/2}(\mathrm{so}_{2n+1})$ instead of $U_q(\mathrm{so}_{2n+1})$ and $U_h(\mathrm{so}_{2n+1})$, respectively. Recall that $R_{1,1} = R_{V_1V_1}$ was defined in Subsects. 8.1.2 and 8.3.3. Let us set

$$\mathsf{R} := q^{1/N}R_{1,1} \text{ for } \mathfrak{g} = \mathrm{gl}_N, \mathrm{sl}_N \;\text{ and }\; \mathsf{R} := R_{1,1} \text{ for } \mathfrak{g} = \mathrm{so}_N, \mathrm{sp}_N.$$

As earlier, we put $\hat{\mathsf{R}} := \tau \circ \mathsf{R}$. The notation R and $\hat{\mathsf{R}}$ (with sans serif letter R) will be kept throughout the remainder of the book in order to distinguish between these particular R-matrices and general R-matrices. The latter are usually denoted by R and $\hat{R}$.

We begin with the h-adic algebras $U_h(\mathfrak{g})$. Then, by (11), we have $R_{1,1} = (T_1 \otimes T_1)(\mathcal{R})$, where $\mathcal{R}$ is the universal R-matrix (44) of $U_h(\mathfrak{g})$.

First let $\mathfrak{g} = \mathrm{gl}_N, \mathrm{sl}_N$. Using the description of the vector representation T_1 from Subsect. 8.4.1 we find that

$$(T_1 \otimes T_1)\Big\{\exp\Big(h \sum\nolimits_{i,j}^{N-1} B_{ij}(H_i \otimes H_j)\Big)\Big\}$$
$$= e^{-h/N}\Big(e^h \sum\nolimits_{i=1}^{N} E_{ii} \otimes E_{ii} + \sum\nolimits_{i \neq j} E_{ii} \otimes E_{jj}\Big).$$

For all positive roots β we have $T_1(E_\beta)^2 = 0$. From the expressions for the operators $T_1(E_{\beta_{ij}})$ and $T_1(F_{\beta_{ij}})$ obtained in Subsect. 8.4.1 we derive that

$$(T_1 \otimes T_1)\{\exp_q((1-q^{-2})(E_{\beta_{ij}} \otimes F_{\beta_{ij}}))\} = 1 + (e^h - e^{-h})(E_{ji} \otimes E_{ij}).$$

Multiplying both expressions and inserting the convention $q^\alpha = e^{\alpha h}$ we get

$$\mathsf{R} = q \sum_i (E_{ii} \otimes E_{ii}) + \sum_{i \neq j} (E_{ii} \otimes E_{jj}) + (q - q^{-1}) \sum_{i > j} (E_{ij} \otimes E_{ji}). \tag{60}$$

The operator $(T_1 \otimes T_1)(\mathcal{R})$ for $U_{h/2}(\mathrm{so}_{2n+1})$, $U_h(\mathrm{so}_{2n})$ and $U_h(\mathrm{sp}_{2n})$ can be calculated similarly. We state only the outcome of these computations. In order to do so, some more notation is needed. Let $(\rho_1, \rho_2, \cdots, \rho_N)$ denote the N-tuple

$$(n - \tfrac{1}{2}, n - \tfrac{3}{2}, \cdots \tfrac{1}{2}, 0, -\tfrac{1}{2}, \cdots, -n + \tfrac{1}{2}) \quad \text{for} \quad \mathfrak{g} = \mathrm{so}_{2n+1},$$
$$(n-1, n-2, \cdots, 1, 0, 0, -1, \cdots, -n+1) \quad \text{for} \quad \mathfrak{g} = \mathrm{so}_{2n},$$
$$(n, n-1, \cdots, 1, -1, \cdots, -n) \quad \text{for} \quad \mathfrak{g} = \mathrm{sp}_{2n}.$$

We set $\epsilon_i := 1$, $i = 1, 2, \cdots, N$, for $\mathfrak{g} = \mathrm{so}_N$ and $\epsilon_i := 1$ if $i \leq n$, $\epsilon_i := -1$ if $i > n$ for $\mathfrak{g} = \mathrm{sp}_N$, $N = 2n$. Put $i' := N + 1 - i$. For $U_{h/2}(\mathrm{so}_{2n+1})$, $U_h(\mathrm{sp}_{2n})$ and $U_h(\mathrm{so}_{2n})$ we then derive that

$$\mathsf{R} = q \sum_{\substack{i \\ i \neq i'}} (E_{ii} \otimes E_{ii}) + \sum_{\substack{i,j \\ i \neq j, j'}} (E_{ii} \otimes E_{jj}) + q^{-1} \sum_{\substack{i \\ i \neq i'}} (E_{i'i'} \otimes E_{ii})$$
$$+ (q - q^{-1})\Big\{\sum_{\substack{i,j \\ i > j}} (E_{ij} \otimes E_{ji}) - \sum_{\substack{i,j \\ i > j}} q^{\rho_i - \rho_j} \epsilon_i \epsilon_j (E_{ij} \otimes E_{i'j'})\Big\}$$
$$+ E_{(N+1)/2,(N+1)/2} \otimes E_{(N+1)/2,(N+1)/2}, \tag{61}$$

where the last summand must be omitted for $\mathfrak{g} = \mathrm{sp}_{2n}, \mathrm{so}_{2n}$ and the summations are over $1, 2, \cdots, N$.

Now let us turn to the Drinfeld–Jimbo algebras $U_q(\mathfrak{g})$. In this case, we have $R_{1,1} = B_{VV} \circ (T_1 \otimes T_1)(\mathfrak{R})$ by (59), where $\mathfrak{R}$ and B_{VV} are defined in Subsect. 8.3.3. Carrying out similar computations as above we find that the matrix R is given also by the formulas (60) for $U_q(\mathrm{gl}_N)$ and $U_q(\mathrm{sl}_N)$ and by (61) for $U_{q^{1/2}}(\mathrm{so}_{2n+1})$, $U_q(\mathrm{sp}_{2n})$ and $U_q(\mathrm{so}_{2n})$ with q being now a complex number rather than an abbreviation for e^h. Strictly speaking, this derivation of R is justified only if q is not a root of unity, because $\mathfrak{R}$ was defined under this assumption. But the expressions (60) and (61) make sense for any q, so we take them as definitions in the root of unity case. By continuity, the intertwining property $\hat{\mathsf{R}}(T_1 \otimes T_1) = (T_1 \otimes T_1)\hat{\mathsf{R}}$, the quantum Yang–Baxter equation for R and the braid relation for $\hat{\mathsf{R}}$ hold for all q.

We conclude this subsection with an important symmetry property of the matrix $\mathsf{R} = R_{1,1}$ for $\mathfrak{g} = \mathrm{so}_{2n+1}, \mathrm{so}_{2n}, \mathrm{sp}_{2n}$ which emerges from the self-duality of the corresponding vector representations T_1.

Let T be a representation of a Hopf algebra $\mathcal{A}$ on a finite-dimensional vector space V. Then the formula

$$(T'(a)v')(v) := v'(T(S(a))v), \quad a \in \mathcal{A},\ v \in V,\ v' \in V',$$

defines a representation T' of $\mathcal{A}$ on the dual vector space V' called the *contragredient* or the *dual representation* to T. If T is equivalent to T', then T is called a *self-dual representation.* Let us identify the vector spaces V and V' by fixing an isomorphism. For linear mapping $A \in \mathcal{L}(V)$, we denote by A^t the transposed map to A defined by $(A^t v)(w) = v(Aw)$, $v, w \in V$. One easily checks that *a representation T is self-dual if and only if there is an invertible linear mapping $A \in \mathcal{L}(V)$ such that*

$$T(S(a)) = AT(a)^t A^{-1}, \qquad a \in \mathcal{A}.$$

Now we turn to the vector representation T_1 of $U_q(\mathfrak{g})$ resp. $U_h(\mathfrak{g})$ on V_1. We identify V_1 with its dual V_1' by taking the standard bases of $\mathbb{C}^N$ resp. $\mathbb{C}[[h]]^N$ as dual bases of V_1 and V_1'. For $\mathfrak{g} = \mathrm{so}_{2n+1}, \mathrm{so}_{2n}, \mathrm{sp}_{2n}$, we introduce the $N \times N$ matrix

$$\mathsf{C} = (\mathsf{C}^i_j) \quad \text{with} \quad \mathsf{C}^i_j := \epsilon_i \delta_{ij'} q^{-\rho_i}. \tag{62}$$

Note that $\mathsf{C}^2 = I$ for $\mathfrak{g} = \mathrm{so}_{2n+1}, \mathrm{so}_{2n}$ and $\mathsf{C}^2 = -I$ for $\mathfrak{g} = \mathrm{sp}_{2n}$.

Proposition 20. *Let $\mathfrak{g}$ be so_{2n+1}, so_{2n} or sp_{2n}. Then the vector representations T_1 of $U_q(\mathfrak{g})$ and $U_h(\mathfrak{g})$ are self-dual and we have*

$$T_1(S(a)) = \mathsf{C}T_1(a)^t\mathsf{C}^{-1}, \quad a \in U_q(\mathfrak{g}) \textit{ resp. } a \in U_h(\mathfrak{g}), \tag{63}$$

$$\mathsf{R} = \mathsf{C}_1(\mathsf{R}^{t_1})^{-1}\mathsf{C}_1^{-1} = \mathsf{C}_2(\mathsf{R}^{-1})^{t_2}\mathsf{C}_2^{-1}, \tag{64}$$

where t_1 and t_2 are the transpositions in the first resp. second tensor factors.

Proof. Using the formulas from Subsect. 8.4.1 one verifies the relation (63) on the generators. Hence it is valid on the whole of $U_q(\mathfrak{g})$ resp. $U_h(\mathfrak{g})$.

We prove (64) for $U_h(\mathfrak{g})$. Using the first formula of (8) we compute

$$R^{-1}_{V_1V_1} = (T_1 \otimes T_1)(\mathcal{R}^{-1}) = (T_1 \otimes T_1)(S \otimes \mathrm{id})(\mathcal{R})$$

$$= (T_1 \circ S \otimes T_1)(\mathcal{R}) = (\mathsf{C}T_1^t\mathsf{C}^{-1} \otimes T_1)(\mathcal{R}) = \mathsf{C}_1 R^{t_1}_{V_1V_1}\mathsf{C}_1^{-1}.$$

This in turn implies that $\mathsf{R} = \mathsf{C}_1(\mathsf{R}^{t_1})^{-1}\mathsf{C}_1^{-1}$. The second equality of (64) follows similarly using the second formula of (8).

By construction, the matrices R and C for $U_h(\mathfrak{g})$ go into the corresponding matrices for $U_q(\mathfrak{g})$ if $q = e^h \in \mathbb{C}[[h]]$ is replaced by the complex number q. Therefore, formula (64) holds also for $U_q(\mathfrak{g})$. □

8.4.3 Spectral Decompositions of R-Matrices for Vector Representations

We begin with two general results for the h-adic algebra $U_h(\mathfrak{g})$.

Proposition 21. *Let u be the element from Proposition 5 for the h-adic quasitriangular Hopf algebra $U_h(\mathfrak{g})$ and let ρ be the half-sum of all positive roots of $\mathfrak{g}$. Then the element $e^{-hH_\rho}u = ue^{-hH_\rho}$ belongs to the center of $U_h(\mathfrak{g})$. If T_λ is an irreducible type 1 representation of $U_h(\mathfrak{g})$ with highest weight λ, then $T_\lambda(e^{-hH_\rho}u) = \exp(-h(\lambda, \lambda + 2\rho))I$. The comultiplication of $e^{-hH_\rho}u$ is given by*

$$\Delta(e^{-hH_\rho}u) = (e^{-hH_\rho}u \otimes e^{-hH_\rho}u)\cdot(\mathcal{R}_{21}\mathcal{R})^{-1}. \tag{65}$$

Proof. The proof is given in [Dr4]. □

Proposition 22. *Suppose that T_λ and $T_{\lambda'}$ are two irreducible type 1 representations of $U_h(\mathfrak{g})$ with highest weights λ and λ', respectively. Let T_μ be an irreducible representation with highest weight μ which occurs in the decomposition of the tensor product $T_\lambda \otimes T_{\lambda'}$ into irreducible components. Then the operator $(T_\lambda \otimes T_{\lambda'})(\mathcal{R}_{21}\mathcal{R})$ acts on the corresponding subspace V_μ of $V_\lambda \otimes V_{\lambda'}$ as*

$$q^{-(\lambda,\lambda+2\rho)-(\lambda',\lambda'+2\rho)+(\mu,\mu+2\rho)}I_\mu, \tag{66}$$

where I_μ is the identity map of V_μ.

Proof. Since $\Delta(a)\mathcal{R}_{21}\mathcal{R} = \mathcal{R}_{21}\mathcal{R}\Delta(a)$ by Proposition 1, $(T_\lambda \otimes T_{\lambda'})(\mathcal{R}_{21}\mathcal{R})$ commutes with all operators $(T_\lambda \otimes T_{\lambda'})(a) \equiv (T_\lambda \otimes T_{\lambda'})(\Delta(a))$, $a \in U_h(\mathfrak{g})$. Therefore, the restriction of $(T_\lambda \otimes T_{\lambda'})(\mathcal{R}_{21}\mathcal{R})$ to the carrier space V_μ of any irreducible component T_μ of $T_\lambda \otimes T_{\lambda'}$ is a multiple of the unit operator I_μ. The explicit expression for the operator $(T_\lambda \otimes T_{\lambda'})(\mathcal{R}_{21}\mathcal{R})$ on V_μ can be found by means of Proposition 21. Indeed, by (65), we have

$$(T_\lambda \otimes T_{\lambda'})(\mathcal{R}_{21}\mathcal{R})(T_\lambda \otimes T_{\lambda'})(\Delta(e^{-hH_\rho}u)) = T_\lambda(e^{-hH_\rho}u) \otimes T_{\lambda'}(e^{-hH_\rho}u).$$

On the irreducible subspace V_μ the operator $(T_\lambda \otimes T_{\lambda'})(\Delta(e^{-hH_\rho}u))$ acts as $e^{-h(\mu,\mu+2\rho)}I$. Hence $(T_\lambda \otimes T_{\lambda'})(\mathcal{R}_{21}\mathcal{R})$ takes the value (66) on V_μ. □

Corollary 23. *Let T_λ be as in Proposition 22. Then all eigenvalues of the operator $\hat{R}_{\lambda\lambda} := \tau \circ (T_\lambda \otimes T_\lambda)(\mathcal{R})$ are of the form*

$$\pm q^{-(2(\lambda,\lambda+2\rho)-(\mu,\mu+2\rho))/2}, \tag{67}$$

where μ is the highest weight of an irreducible component in the decomposition of $T_\lambda \otimes T_\lambda$.

Proof. For the operator $\hat{R}_{\lambda\lambda} := \tau \circ (T_\lambda \otimes T_\lambda)(\mathcal{R})$, we have

$$\hat{R}^2_{\lambda\lambda} = \tau \circ (T_\lambda \otimes T_\lambda)(\mathcal{R}) \cdot \tau \circ (T_\lambda \otimes T_\lambda)(\mathcal{R}) = (T_\lambda \otimes T_\lambda)(\mathcal{R}_{21}\mathcal{R}).$$

Thus, the squares of eigenvalues of $\hat{R}_{\lambda\lambda}$ are eigenvalues of $(T_\lambda \otimes T_\lambda)(\mathcal{R}_{21}\mathcal{R})$. Hence the form (67) follows from (66). □

Let $\mathcal{S}(V_\lambda \otimes V_\lambda)$ be the span of eigenspaces for the eigenvalues (67) with sign $+$ for any possible μ. We call $\mathcal{S}(V_\lambda \otimes V_\lambda)$ the *quantum symmetric subspace* of $V_\lambda \otimes V_\lambda$. The subspace $\Lambda(V_\lambda \otimes V_\lambda)$ generated by the eigenspaces for all possible eigenvalues (67) with sign $-$ is called the *quantum antisymmetric subspace* of $V_\lambda \otimes V_\lambda$. Since $T_\lambda \otimes T_\lambda$ is completely reducible, we have

$$V_\lambda \otimes V_\lambda = \mathcal{S}(V_\lambda \otimes V_\lambda) \oplus \Lambda(V_\lambda \otimes V_\lambda). \tag{68}$$

Now we apply these results to the vector representation T_1 of $U_h(\mathfrak{g})$, $\mathfrak{g} = \mathrm{gl}_N, \mathrm{sl}_N, \mathrm{so}_N, \mathrm{sp}_N$. The corresponding R-matrices $R_{1,1} := (T_1 \otimes T_1)(\mathcal{R})$ have been computed in Subsect. 8.4.2. As noted there, the R-matrix R for the algebra $U_q(\mathfrak{g})$ is obtained from the R-matrix R for $U_h(\mathfrak{g})$ by replacing the power series $q = e^h \in \mathbb{C}[[h]]$ by the complex number q. Under this replacement, the subspaces $\mathcal{S}(V_\lambda \otimes V_\lambda)$ and $\Lambda(V_\lambda \otimes V_\lambda)$ are also well-defined for $U_q(\mathfrak{g})$ and the subsequent results are valid for both algebras $U_h(\mathfrak{g})$ and $U_q(\mathfrak{g})$.

First let $\mathfrak{g} = \mathrm{gl}_N, \mathrm{sl}_N$. By (7.13), the tensor product $T_1 \otimes T_1$ decomposes into two irreducible representations with highest weights $(2, 0, \cdots, 0)$ and $(1, 1, 0, \cdots, 0)$. Inserting these highest weights into (67), we find that the matrix $\hat{R}_{1,1}$ has the eigenvalues $q^{1-1/N}$ on the first component $\mathcal{S}(V_1 \otimes V_1)$ and $-q^{-1-1/N}$ on the second subspace $\Lambda(V_1 \otimes V_1)$. That is, we have

Proposition 24. *The matrix* $\hat{\mathsf{R}} = \tau \circ \mathsf{R} = q^{1/N}\hat{R}_{1,1}$ *for the algebras* $U_h(\mathrm{gl}_N)$, $U_h(\mathrm{sl}_N)$, $U_q(\mathrm{gl}_N)$ *and* $U_q(\mathrm{sl}_N)$ *satisfies the quadratic equation*

$$(\hat{\mathsf{R}} - qI)(\hat{\mathsf{R}} + q^{-1}I) = 0 \quad \text{or} \quad \hat{\mathsf{R}}^2 = (q - q^{-1})\hat{\mathsf{R}} + I. \tag{69}$$

The matrix $\hat{\mathsf{R}}$ has the spectral decomposition

$$\hat{\mathsf{R}} = q\mathsf{P}_+ - q^{-1}\mathsf{P}_-, \tag{70}$$

where the projections P_+ and P_- are of the form

$$\mathsf{P}_+ = \frac{\hat{\mathsf{R}} + q^{-1}I}{q + q^{-1}}, \qquad \mathsf{P}_- = \frac{-\hat{\mathsf{R}} + qI}{q + q^{-1}}. \tag{71}$$

The operators (71) are quantum analogs of the symmetrization and antisymmetrization operators on $\mathbb{C}^N \otimes \mathbb{C}^N$, respectively.

Now we consider the cases $\mathfrak{g} = \mathrm{so}_{2n+1}$ and $\mathfrak{g} = \mathrm{so}_{2n}$. By (7.14),

$$T_1 \otimes T_1 = T_2 \oplus T_0 \oplus T_{(1,1)}. \tag{72}$$

The irreducible representation $T_{(1,1)}$ with highest weight $(1,1,0,\cdots,0)$ acts on the subspace $\Lambda(V_1 \otimes V_1)$. The subspace $\mathcal{S}(V_1 \otimes V_1)$ is the direct sum of two invariant spaces. On the first one, denoted by $\mathcal{S}'(V_1 \otimes V_1)$, the symmetric irreducible representation T_2 with highest weight $(2,0,\cdots,0)$ is realized. The second subspace is one-dimensional and the corresponding subrepresentation T_0 is trivial. From Corollary 23 we obtain the eigenvalues of R stated in the following

Proposition 25. *The operator* $\hat{\mathsf{R}} = \tau \circ \mathsf{R} = \hat{R}_{1,1}$ *for the algebras* $U_{h/2}(\mathrm{so}_{2n+1})$, $U_h(\mathrm{so}_{2n})$ *and* $U_{q^{1/2}}(\mathrm{so}_{2n+1})$, $U_q(\mathrm{so}_{2n})$ *acts on* $\Lambda(V_1 \otimes V_1)$ *as* $-q^{-1}I$, *on* $\mathcal{S}'(V_1 \otimes V_1)$ *as* qI *and on the one-dimensional invariant subspace as* $q^{1-N}I$, *where* $N = 2n$ *resp.* $N = 2n+1$. $\hat{\mathsf{R}}$ *satisfies the cubic equation*

$$(\hat{\mathsf{R}} - qI)(\hat{\mathsf{R}} + q^{-1}I)(\hat{\mathsf{R}} - q^{1-N}I) = 0. \tag{73}$$

If $N > 2$ and $(1+q^2)(1+q^{2-N})(1-q^{-N}) \neq 0$, it follows from (73) that the operator $\hat{\mathsf{R}}$ for $U_{q^{1/2}}(\mathrm{so}_{2n+1})$ and $U_q(\mathrm{so}_{2n})$ has the spectral decomposition

$$\hat{\mathsf{R}} = q\mathsf{P}_+ - q^{-1}\mathsf{P}_- + q^{1-N}\mathsf{P}_0,$$

where

$$\mathsf{P}_+ = \frac{\hat{\mathsf{R}}^2 - (q^{1-N} - q^{-1})\hat{\mathsf{R}} - q^{-N}I}{(q+q^{-1})(q-q^{1-N})}, \tag{74}$$

$$\mathsf{P}_- = \frac{\hat{\mathsf{R}}^2 - (q + q^{1-N})\hat{\mathsf{R}} + q^{-N+2}I}{(q+q^{-1})(q^{-1}+q^{1-N})}, \tag{75}$$

$$\mathsf{P}_0 = \frac{\hat{\mathsf{R}}^2 - (q - q^{-1})\hat{\mathsf{R}} - I}{(q^{1-N} - q)(q^{-1} + q^{1-N})}. \tag{76}$$

Here P_+, P_- and P_0 are the projections onto $\Lambda(V_1 \otimes V_1)$, $\mathcal{S}'(V_1 \otimes V_1)$ and the one-dimensional invariant subspace, respectively.

One can easily verify that the matrix $\hat{\mathsf{R}}$ for $U_{q^{1/2}}(\mathrm{so}_{2n+1})$ and $U_q(\mathrm{so}_{2n})$ also satisfies the relations

$$\hat{\mathsf{R}}^{-1} = \tau \circ \hat{\mathsf{R}}(q^{-1}) \circ \tau, \qquad \hat{\mathsf{R}} - \hat{\mathsf{R}}^{-1} = (q - q^{-1})(I - \mathsf{K}) \tag{77}$$

with

$$\mathsf{K} := \sum\nolimits_{i,j=1}^{N} q^{\rho_i - \rho_j} E_{N+1-i,j} \otimes E_{i,N+1-j}, \tag{78}$$

where the numbers ρ_i have been defined in Subsect. 8.4.2 and $\hat{\mathsf{R}}(q^{-1})$ denotes the matrix $\hat{\mathsf{R}}$ with q replaced by q^{-1}. The matrix K is a complex multiple of P_0 and

$$\mathsf{K}^2 = (q - q^{-1} - q^{1-N} + q^{N-1})(q - q^{-1})^{-1}\mathsf{K}.$$

Finally, we turn to the case $\mathfrak{g} = \mathrm{sp}_N$. Then we also have the tensor product decomposition (72), but now the restriction of $T_1 \otimes T_1$ to $\mathcal{S}(V_1 \otimes V_1)$ is irreducible and the space $\Lambda(V_1 \otimes V_1)$ splits into the direct sum of a one-dimensional invariant subspace and its irreducible complement $\Lambda'(V_1 \otimes V_1)$.

Proposition 26. *The operator* $\hat{\mathsf{R}} = \tau \circ \mathsf{R} = \hat{\mathsf{R}}_{1,1}$ *for the algebras* $U_h(\mathrm{sp}_N)$ *and* $U_q(\mathrm{sp}_N)$ *acts as* $-q^{-1}I$ *on* $\Lambda'(V_1 \otimes V_1)$*, as* qI *on* $\mathcal{S}(V_1 \otimes V_1)$ *and as* $-q^{-(1+N)}I$ *on the one-dimensional invariant subspace. We have*

$$(\hat{\mathsf{R}} - qI)(\hat{\mathsf{R}} + q^{-1}I)(\hat{\mathsf{R}} + q^{-N-1}I) = 0.$$

If $N > 4$ and $(1+q^2)(1-q^{-N})(1+q^{-N-2}) \neq 0$, then the matrix $\hat{\mathsf{R}}$ for $U_q(\mathrm{sp}_N)$ admits the spectral decomposition

$$\hat{\mathsf{R}} = q\mathsf{P}_+ - q^{-1}\mathsf{P}_- - q^{-N-1}\mathsf{P}_0,$$

where

$$\mathsf{P}_+ = \frac{\hat{\mathsf{R}}^2 + (q^{-1} + q^{-N-1})\hat{\mathsf{R}} + q^{-N-2}I}{(q+q^{-1})(q+q^{-N-1})}, \tag{79}$$

$$\mathsf{P}_- = \frac{\hat{\mathsf{R}}^2 - (q - q^{-N-1})\hat{\mathsf{R}} - q^{-N}I}{(q+q^{-1})(q^{-1} - q^{-N-1})}, \tag{80}$$

$$\mathsf{P}_0 = \frac{\hat{\mathsf{R}}^2 - (q - q^{-1})\hat{\mathsf{R}} - I}{(-q - q^{-N-1})(q^{-1} - q^{-N-1})}$$

are the projections onto $\mathcal{S}(V_1 \otimes V_1)$, $\Lambda'(V_1 \otimes V_1)$ and the one-dimensional invariant subspace, respectively.

The matrix $\hat{\mathsf{R}}$ for the algebra $U_q(\mathrm{sp}_N)$ also satisfies the relations (77) with

$$\mathsf{K} := \sum\nolimits_{i,j=1}^{N} q^{\rho_i - \rho_j} \epsilon_i \epsilon_j (E_{N+1-i,j} \otimes E_{i,N+1-j}). \tag{81}$$

As above, K is a multiple of P_0, but now we have

$$\mathsf{K}^2 = (q - q^{-1} + q^{-N-1} - q^{N+1})(q - q^{-1})^{-1}\mathsf{K}.$$

8.5 L-Operators and L-Functionals

8.5.1 L-Operators and L-Functionals

Throughout this subsection, let T_V be a representation of the h-adic Drinfeld–Jimbo algebra $U_h(\mathfrak{g})$ on a finite-dimensional vector space V. The elements

$$L_V^+ = (\mathrm{id} \otimes T_V)(\mathcal{R}), \quad L_V^- = (T_V \otimes \mathrm{id})(\mathcal{R}^{-1}) \tag{82}$$

of $U_h(\mathfrak{g}) \otimes \mathcal{L}(V)$ and $\mathcal{L}(V) \otimes U_h(\mathfrak{g})$, respectively, are called the *L-operators* of the algebra $U_h(\mathfrak{g})$ with respect to the representation T_V.

Proposition 27. *The L-operators* L_V^+ *and* L_V^- *satisfy the relations*

$$(L_V^\pm)_1 (L_V^\pm)_2 R_{VV} = R_{VV} (L_V^\pm)_2 (L_V^\pm)_1, \tag{83}$$

$$(L_V^-)_1 (L_V^+)_2 R_{VV} = R_{VV} (L_V^+)_2 (L_V^-)_1, \tag{84}$$

$$(\Delta\otimes \mathrm{id})L_V^+ = (L_V^+)_{13}(L_V^+)_{23}, \quad (\mathrm{id}\otimes\Delta)L_V^- = (L_V^-)_{12}(L_V^-)_{13}, \tag{85}$$

$$(\varepsilon\otimes \mathrm{id})L_V^+ = I, \quad (\mathrm{id}\otimes\varepsilon)L_V^- = I. \tag{86}$$

Proof. Applying the operator $\mathrm{id}\otimes T_V\otimes T_V$ to both sides of the QYBE $\mathcal{R}_{12}\mathcal{R}_{13}\mathcal{R}_{23} = \mathcal{R}_{23}\mathcal{R}_{13}\mathcal{R}_{12}$ we obtain the relation (83) with sign $+$. The QYBE is equivalent to the identity $\mathcal{R}_{13}^{-1}\mathcal{R}_{23}^{-1}\mathcal{R}_{12} = \mathcal{R}_{12}\mathcal{R}_{23}^{-1}\mathcal{R}_{13}^{-1}$. Applying $T_V\otimes T_V\otimes \mathrm{id}$ to the latter relation we get formula (83) with sign $-$. Equation (84) is obtained by applying the operator $T_V\otimes\mathrm{id}\otimes T_V$ to the identity $\mathcal{R}_{12}^{-1}\mathcal{R}_{23}\mathcal{R}_{13} = \mathcal{R}_{13}\mathcal{R}_{23}\mathcal{R}_{12}^{-1}$. The formulas (85) and (86) are immediate consequences of (2) and (7), respectively. For example, using the second relation of (2) and the fact that T_V is a representation, we derive that

$$(\mathrm{id}\otimes\Delta)L_V^- = (T_V\otimes\mathrm{id})((\mathrm{id}\otimes\Delta)(\mathcal{R}^{-1}))$$

$$= (T_V\otimes\mathrm{id})(\mathcal{R}_{12}^{-1}\mathcal{R}_{13}^{-1}) = (L_V^-)_{12}(L_V^-)_{13}. \qquad \square$$

Let us choose a basis $\{e_i\}$ of V. Then there exist linear functionals t_{kl} on $U_h(\mathfrak{g})$, called the *matrix coefficients* of the representation T_V, such that

$$T_V(\cdot)e_l = \sum\nolimits_k t_{kl}(\cdot)e_k. \tag{87}$$

Further, by (82), there are uniquely determined elements $l_{ij}^\pm$ of $U_h(\mathfrak{g})$ such that

$$L_V^+(1\otimes e_j) = \sum\nolimits_i l_{ij}^+\otimes e_i \quad\text{and}\quad L_V^-(e_j\otimes 1) = \sum\nolimits_i e_i\otimes l_{ij}^-.$$

These elements $l_{ij}^\pm\in U_h(\mathfrak{g})$ are called *L-functionals* associated with the representation T_V. This terminology stems from the fact that the elements $l_{ij}^\pm$ are commonly treated as functionals on the corresponding quantum matrix groups (see Subsect. 10.1.3). Let (R_{jl}^{ik}) and $((R^{-1})_{jl}^{ik})$ be the matrices of the transformations R_{VV} and R_{VV}^{-1}, respectively, with respect to the basis $\{e_i\otimes e_k\}$ of $V\otimes V$. Since $(T_V\otimes\mathrm{id})L_V^+ = R_{VV}$ and $(\mathrm{id}\otimes T_V)L_V^- = R_{VV}^{-1}$ by (82), we have

$$t_{kl}(l_{ij}^+) = R_{lj}^{ki} \quad\text{and}\quad t_{kl}(l_{ij}^-) = (R^{-1})_{jl}^{ik}. \tag{88}$$

In terms of L-functionals the formulas (85) and (86) read as

$$\Delta(l_{ij}^\pm) = \sum\nolimits_k l_{ik}^\pm\otimes l_{kj}^\pm \quad\text{and}\quad \varepsilon(l_{ij}^\pm) = \delta_{ij}. \tag{89}$$

In the remainder of this subsection, T_{V_1} will be vector representation T_1 of $U_h(\mathfrak{g})$ on V_1. We fix the basis of V_1 used in Subsect. 8.4.1. Let R be the corresponding matrix from Subsect. 8.4.2 and let $\mathsf{L}^+ = (l_{ij}^+)$ and $\mathsf{L}^- = (l_{ij}^-)$ denote the $N\times N$ matrices of L-functionals in the vector representation T_1. Then the relations (83) and (84) yield the following matrix equations

$$\mathsf{L}_1^+\mathsf{L}_2^+\mathsf{R} = \mathsf{R}\mathsf{L}_2^+\mathsf{L}_1^+, \quad \mathsf{L}_1^-\mathsf{L}_2^-\mathsf{R} = \mathsf{R}\mathsf{L}_2^-\mathsf{L}_1^-, \quad \mathsf{L}_1^-\mathsf{L}_2^+\mathsf{R} = \mathsf{R}\mathsf{L}_2^+\mathsf{L}_1^-. \tag{90}$$

Properties of L-functionals in the vector representation are contained in

Proposition 28. (i) $l^+_{ij} = l^-_{ji} = 0$ *for* $i > j$ *and* $l^+_{ii} l^-_{ii} = l^-_{ii} l^+_{ii} = 1$ *for* $i = 1, 2, \cdots, N$.

(ii) $l^+_{11} l^+_{22} \cdots l^+_{NN} = 1$ *for* $\mathfrak{g} = \mathrm{sl}_N$.

(iii) $\mathsf{L}^\pm \mathsf{C}^t (\mathsf{L}^\pm)^t (\mathsf{C}^{-1})^t = \mathsf{C}^t (\mathsf{L}^\pm)^t (\mathsf{C}^{-1})^t \mathsf{L}^\pm = I$ *for* $\mathfrak{g} = \mathrm{so}_N, \mathrm{sp}_N$, *where* C *is the matrix from (62).*

Proof. (i): The relations $l^+_{ij} = l^-_{ji} = 0$, $i > j$, follow from the particular form (44) of the matrix $\mathcal{R}$ and its inverse $\mathcal{R}^{-1}$ combined with the fact that $T_1(E_i)$ and $T_1(F_i)$ are lowering and raising operators, respectively (see Subsect. 8.4.1). For the same reason, only the Cartan parts of $\mathcal{R}$ and $\mathcal{R}^{-1}$ contribute to the diagonal L-functionals. From this we derive that $l^+_{ii} l^-_{ii} = l^-_{ii} l^+_{ii} = 1$.

(ii) follows from the explicit formulas in Subsect. 8.4.1 and (iii) is obtained from Proposition 20. □

8.5.2 L-Functionals for Vector Representations

The first aim of this subsection is to derive explicit expressions of the L-functionals l^+_{ii}, $l^+_{i,i+1}$, $l^-_{i+1,i}$ in terms of the generators of $U_h(\mathfrak{g})$. For this we use the formula (44) for the universal R-matrix $\mathcal{R}$ of $U_h(\mathfrak{g})$. In order to treat the first factor in (44), we change the standard basis $\{H_1, H_2, \cdots, H_l\}$ of the Cartan subalgebra $\mathfrak{h}$ which appears in the defining relations of $U_h(\mathfrak{g})$ to an orthonormal basis $\{H'_1, H'_2, \cdots, H'_l\}$ with respect to the symmetric bilinear form $\langle .,. \rangle$ on $\mathfrak{h} \times \mathfrak{h}$ (see Subsect. 6.1.1 and Remark 6 in Subsect. 8.3.2). Such a basis will be called *self-dual.* If $\{H'_1, H'_2, \cdots, H'_l\}$ is a self-dual basis of $\mathfrak{h}$, then we have $\sum_{i,j} B_{ij} H_i \otimes H_j = \sum_i H'_i \otimes H'_i$. The L-functionals can be calculated by inserting this expression into (44) and using the formulas for the vector representations from Subsect. 8.4.1. We collect the results of these computations and the corresponding self-dual bases in the following list:

$U_h(\mathrm{gl}_N)$:

$$H'_i = H_i, \quad i = 1, 2, \cdots, N,$$

$$l^+_{ii} = q^{-H'_i}, \quad l^+_{i,i+1} = (q - q^{-1}) q^{-H'_i} E_i, \quad l^-_{i+1,i} = -(q - q^{-1}) F_i q^{H'_i},$$

$$l^+_{i,j} = (q - q^{-1})(-1)^{i-j+1} q^{-H'_i} E_{\beta_{ij}}, \quad l^-_{j,i} = -(q - q^{-1}) F_{\beta_{ij}} q^{H'_i},$$

where $j > i + 1$ and the positive roots are as in Subsect. 8.4.1.

$U_h(\mathrm{sl}_N)$:

$$H'_i = -\sum\nolimits_{k=1}^{N-1} \frac{k}{N} H_k + \sum\nolimits_{s=i}^{N-1} H_s, \quad i = 1, 2, \cdots, N-1,$$

$$H_j = H'_j - H'_{j+1}, \quad j = 1, \cdots, N-2,$$

$$l^+_{ii} = q^{-H'_i}, \quad i = 1, 2, \cdots, N-1, \quad l^+_{NN} = q^{\sum_{k=1}^{N-1} k H_k / N},$$

$$l^+_{i,i+1} = (q - q^{-1}) l^+_{ii} E_i, \quad l^-_{i+1,i} = -(q - q^{-1}) F_i l^-_{ii},$$

$$l^+_{i,j} = (q-q^{-1})(-1)^{i-j+1} l^+_{ii} E_{\beta_{ij}}, \quad l^-_{j,i} = -(q-q^{-1}) F_{\beta_{ij}} l^-_{ii},$$

where $j > i+1$ and the positive roots are as in Subsect. 8.4.1.

$U_{h/2}(\mathrm{so}_{2n+1})$:

$$H'_i = H_i + H_{i+1} + \cdots + H_{n-1} + \tfrac{1}{2} H_n, \quad i = 1, 2, \cdots, n,$$

$$l^+_{ii} = q^{-H'_i}, \quad l^+_{i'i'} = q^{H'_i}, \quad i = 1, 2, \cdots, n, \quad l^+_{n+1,n+1} = l^-_{n+1,n+1} = 1,$$

$$l^+_{k,k+1} = (q-q^{-1}) q^{-H'_k} E_k, \quad l^+_{2n-k+1,2n-k+2} = -(q-q^{-1}) q^{H'_{k+1}} E_k,$$

$$l^-_{k+1,k} = -(q-q^{-1}) F_k q^{H'_k}, \quad l^-_{2n-k+2,2n-k+1} = (q-q^{-1}) F_k q^{-H'_{k+1}},$$

$$1 \le k \le n-1,$$

$$l^+_{n,n+1} = c(q^{1/2}-q^{-1/2}) q^{-H'_n} E_n, \quad l^+_{n+1,n+2} = -c q^{-1/2}(q^{1/2}-q^{-1/2}) E_n,$$

$$l^-_{n+1,n} = -c(q^{1/2}-q^{-1/2}) F_n q^{H'_n}, \quad l^-_{n+2,n+1} = c q^{1/2}(q^{1/2}-q^{-1/2}) F_n,$$

where $c = [2]^{1/2}_{q^{1/2}}$, $q = e^h$, and $i' = 2n+2-i$.

$U_h(\mathrm{sp}_{2n})$:

$$H'_i = H_i + H_{i+1} + \cdots + H_n, \quad i = 1, 2, \cdots, n,$$

$$l^+_{ii} = q^{-H'_i}, \quad l^+_{i'i'} = q^{H'_i}, \quad i = 1, 2, \cdots, n,$$

$$l^+_{k,k+1} = (q-q^{-1}) q^{-H'_k} E_k, \quad l^+_{2n-k,2n-k+1} = -(q-q^{-1}) q^{H'_{k+1}} E_k,$$

$$l^-_{k+1,k} = -(q-q^{-1}) F_k q^{H'_k}, \quad l^-_{2n-k+1,2n-k} = (q-q^{-1}) F_k q^{-H'_{k+1}},$$

$$1 \le k \le n-1,$$

$$l^+_{n,n+1} = (q^2-q^{-2}) q^{-H'_n} E_n, \quad l^-_{n+1,n} = -(q^2-q^{-2}) F_n q^{H'_n},$$

where $i' = 2n+1-i$.

$U_h(\mathrm{so}_{2n})$:

$$H'_i = H_i + H_{i+1} + \cdots + H_{n-2} + (H_{n-1} + H_n)/2, \quad i = 1, 2, \cdots, n-1,$$

$$H'_n = (-H_{n-1} + H_n)/2$$

$$l^+_{ii} = q^{-H'_i}, \quad l^+_{i'i'} = q^{H'_i}, \quad i = 1, 2, \cdots, n,$$

$$l^+_{k,k+1} = (q-q^{-1}) q^{-H'_k} E_k, \quad l^+_{2n-k,2n-k+1} = -(q-q^{-1}) q^{H'_{k+1}} E_k,$$

$$l^-_{k+1,k} = -(q-q^{-1}) F_k q^{H'_k}, \quad l^-_{2n-k+1,2n-k} = (q-q^{-1}) F_k q^{-H'_{k+1}},$$

$$1 \le k \le n-1,$$

$$l^+_{n-1,n+1} = (q-q^{-1}) q^{-H'_{n-1}} E_n, \quad l^+_{n,n+2} = -(q-q^{-1}) q^{-H'_n} E_n,$$

$$l^-_{n+1,n-1} = -(q-q^{-1}) F_n q^{H'_{n-1}}, \quad l^-_{n+2,n} = (q-q^{-1}) F_n q^{H'_n}.$$

Expressions for the other L-functionals l^+_{ij}, l^-_{ji}, $i \le j$, can be derived from the previous formulas combined with the relations (90).

Proposition 29. *The L-functionals satisfy the recurrence relations*

$$(q-q^{-1})l_{ij}^+ = (l_{rj}^+ l_{ir}^+ - l_{ir}^+ l_{rj}^+) l_{rr}^-, \quad i<r<j, \tag{91}$$

$$(q-q^{-1})l_{ji}^- = l_{rr}^+ (l_{jr}^- l_{ri}^- - l_{ri}^- l_{jr}^-), \quad i<r<j, \tag{92}$$

for $\mathfrak{g} = \mathrm{gl}_N, \mathrm{sl}_N$ *and, when in addition* $r \neq i'$, $r \neq j'$, *for* $\mathfrak{g} = \mathrm{so}_N, \mathrm{sp}_N$.

Proof. By (90) we have the commutation relations

$$\sum\nolimits_{k,l} \mathsf{R}_{lk}^{pm} l_{ks}^\pm l_{lt}^\pm = \sum\nolimits_{k,l} l_{pl}^\pm l_{mk}^\pm \mathsf{R}_{ts}^{lk}. \tag{93}$$

We apply this relation in case $i=p<r=m=t<j=s$ with sign $+$ and in case $i=t<r=p=s<j=m$ with sign $-$. Inserting the corresponding values of the matrix entries R_{lk}^{pm} from (60) and (61) and using the relations $l_{kk}^+ l_{kk}^- = l_{kk}^- l_{kk}^+ = 1$ we derive (91) and (92). □

Note that the condition $r=i'$, $r=j'$ excludes only the functionals $l_{n-1,n+1}^+$, $l_{n,n+2}^+$ resp. $l_{n+1,n-1}^-$, $l_{n+2,n}^-$ for $\mathfrak{g} = \mathrm{sp}_{2n}$ and $\mathfrak{g} = \mathrm{so}_{2n}$ on the left hand side of (91) and (92), because in these cases there is no such r. But these functionals for $\mathfrak{g} = \mathrm{so}_{2n}$ are already known by the above list. In order to find them for $\mathfrak{g} = \mathrm{sp}_{2n}$ we use the relations (93). Setting $p=n-1$, $m=r=n$, $s=n+1$ in (93) with sign $+$ and using the values of the matrix entries R_{lk}^{pm}, we obtain

$$(q^2-q^{-2})l_{n-1,n+1}^+ = (q l_{n,n+1}^+ l_{n-1,n}^+ - l_{n-1,n}^+ l_{n,n+1}^+) l_{nn}^-.$$

Inserting the expressions for $l_{n-1,n}^+$, $l_{n,n+1}^+$, $l_{n,n}^-$ in terms of E_{n-1}, E_n, $q^{-H'_{n-1}}$, $q^{\pm H'_n}$, we derive

$$l_{n-1,n+1}^+ = (q-q^{-1})(E_n q^{-H'_{n-1}} E_{n-1} q^{-H'_n} - q^{-H'_{n-1}} E_{n-1} q^{-H'_n} E_n) q^{H'_n}.$$

Similarly, we get

$$l_{n+1,n-1}^- = -(q-q^{-1}) q^{-H'_n} (q^{H'_n} F_{n-1} q^{-H'_{n-1}} F_n - F_n q^{H'_n} F_{n-1} q^{H'_{n-1}}),$$

$$l_{n,n+2}^+ = (q^2-q^{-2})(q^{-H'_n} E_n q^{H'_n} E_{n-1} q^{-H'_n} - E_{n-1} q^{-H'_n} E_n),$$

$$l_{n+2,n}^- = -(q^2-q^{-2})(q^{H'_n} F_{n-1} q^{-H'_n} F_n q^{H'_n} - F_n q^{H'_n} F_{n-1}).$$

Corollary 30. *Let* $U_{\mathbb{Z}}(\mathfrak{g})$ *denote the subalgebra over the ring* $\mathbb{Z}$ *of integers of* $U_h(\mathfrak{g})$ *resp.* $U_{h/2}(\mathrm{so}_{2n+1})$ *generated by the elements* E_i, F_i $q^{H'_j}$, $q^{-H'_j}$. *Except for the functionals* l_{ij}^+, l_{ji}^- *with* $i=n<j$ *for* $\mathfrak{g} = \mathrm{sp}_{2n}$ *and* l_{ij}^+, l_{ji}^- *with* $i=n<j$ *or* $i=n+1<j$ *for* $\mathfrak{g} = \mathrm{so}_{2n+1}$, *we have*

$$l_{ij}^\pm \in (q-q^{-1}) U_{\mathbb{Z}}(\mathfrak{g}) \quad for \quad i \neq j.$$

In the excluded cases we have the relations

$$l_{ij}^+, l_{ji}^- \in (q^2-q^{-2}) U_{\mathbb{Z}}(\mathfrak{g}) \quad for \quad \mathfrak{g} = \mathrm{sp}_{2n}, \quad i=n<j,$$

$$l^+_{nj}, l^-_{jn} \in (q^{1/2}-q^{-1/2})(q^{1/2}+q^{-1/2})^{1/2}U_{\mathbb{Z}}(\mathfrak{g}),$$
$$l^+_{n+1,j}, l^-_{j,n+1} \in q^{\mp 1/2}(q^{1/2}-q^{-1/2})(q^{1/2}+q^{-1/2})^{1/2}U_{\mathbb{Z}}(\mathfrak{g})$$

for $\mathfrak{g} = \mathrm{so}_{2n+1}$.

Proof. By the formulas of the above list, the assertion holds for all functionals $l^+_{i,i+1}, l^-_{i+1,i}$ and in addition for $l^+_{n-1,n+1}, l^+_{n,n+2}, l^-_{n+1,n-1}, l^-_{n+2,n}$ in the cases $\mathfrak{g} = \mathrm{sp}_{2n}, \mathrm{so}_{2n+1}$. Therefore, the assertion follows by induction using the recurrence formulas (91) and (92). □

Finally, let us specialize to the case $\mathfrak{g} = \mathrm{gl}_N, \mathrm{sl}_N$. Then, by (60), the relations (90) (or (93)) imply the commutation rules

$$l^+_{jj}l^+_{ij} = ql^+_{ij}l^+_{jj},\quad l^+_{ii}l^+_{ij} = q^{-1}l^+_{ij}l^+_{ii},\quad l^-_{ji}l^-_{ii} = ql^-_{ii}l^-_{ji},\quad l^-_{jj}l^-_{ji} = ql^-_{ji}l^-_{jj},\quad i<j.$$
$$l^+_{is}l^+_{jj} = l^+_{jj}l^+_{is},\quad l^-_{si}l^-_{jj} = l^-_{jj}l^-_{si}\quad \text{if}\quad j\neq i,\ j\neq s,\ i\neq s.$$

Combining the preceding formulas with the recurrence relations (91) and (92) one derives the following explicit expressions for the L-functionals for $m<k$:

$$(q-q^{-1})^{k-m-1}l^+_{mk}$$
$$= \Big(\sum_{\Omega\subseteq I}(-1)^{|\Omega|}l^+_{i_1-1,i_1}l^+_{i_2-1,i_2}\cdots l^+_{i_s-1,i_s}l^+_{k-1,k}l^+_{j_r-1,j_r}\cdots l^+_{j_2-1,j_2}l^+_{j_1-1,j_1}\Big)$$
$$\times l^-_{m+1,m+1}l^-_{m+2,m+2}\cdots l^-_{k-1,k-1},$$
$$(q-q^{-1})^{k-m-1}l^-_{km} = l^+_{m+1,m+1}l^+_{m+2,m+2}\cdots l^+_{k-1,k-1}$$
$$\times\Big(\sum_{\Omega\subseteq I}(-1)^{|\Omega|}l^-_{i_1,i_1-1}l^-_{i_2,i_2-1}\cdots l^-_{i_s,i_s-1}l^-_{k,k-1}l^-_{j_r,j_r-1}\cdots l^-_{j_2,j_2-1}l^-_{j_1,j_1-1}\Big),$$

where the summations are over all subsets Ω (including the empty set) of $I=\{m+1,m+2,\cdots,k-1\}$. The numbers in Ω are denoted by $i_1,i_2,\cdots,i_s$, $i_1<i_2<\cdots<i_s$, and the numbers in $I\backslash\Omega$ are denoted by $j_1,j_2,\cdots j_r$, $j_1<j_2<\cdots<j_r$. As usual, $|\Omega|$ denotes the number of elements in Ω.

Example 3 ($U_h(\mathrm{gl}_2)$ *and* $U_h(\mathrm{sl}_2)$). Then the L-operators are of the form

$$L^+ = \begin{pmatrix} q^{-H_1'} & (q-q^{-1})q^{-H_1'}E \\ 0 & q^{-H_2'}\end{pmatrix},\quad L^- = \begin{pmatrix} q^{H_1'} & 0 \\ (q^{-1}-q)Fq^{H_1'} & q^{H_2'}\end{pmatrix}$$

for $U_h(\mathrm{gl}_2)$ and

$$L^+ = \begin{pmatrix} q^{-H/2} & (q-q^{-1})q^{-H/2}E \\ 0 & q^{H/2}\end{pmatrix},\quad L^- = \begin{pmatrix} q^{H/2} & 0 \\ (q^{-1}-q)Fq^{H/2} & q^{-H/2}\end{pmatrix}$$

for $U_h(\mathrm{sl}_2)$. △

Example 4. For $U_h(\mathrm{gl}_3)$, the L-functionals l^+_{13} and l^-_{31} are given by

$$l^+_{13} = -(q-q^{-1})q^{-H_1'}(q^{-1}E_1E_2-E_2E_1),\quad l^-_{31} = (q-q^{-1})(F_1F_2-qF_2F_1)q^{H_1'}.$$

△

8.5.3 The Extended Hopf Algebras $U_q^{\text{ext}}(\mathfrak{g})$

In the preceding subsections L-functionals for the vector representation of the h-adic Hopf algebra $U_h(\mathfrak{g})$ have been investigated. The rest of this section will be concerned with L-functionals for the vector representation in the case of fixed q. The idea to define such L-functionals is to replace in the expressions of $l_{ij}^{\pm} \in U_h(\mathfrak{g})$ functions of generators of $U_h(\mathfrak{g})$ by their counterparts in $U_q(\mathfrak{g})$. Because not all main diagonal functionals $l_{ii}^{\pm}$ have counterparts in $U_q(\mathfrak{g})$, we need extensions $U_q^{\text{ext}}(\mathfrak{g})$ of the Drinfeld–Jimbo algebras $U_q(\mathfrak{g})$ for $\mathfrak{g} = \text{sl}_N, \text{sp}_{2n}, \text{so}_{2n}$. They are constructed by adjoining formally certain products of the elements $K_i^{\pm 1/N}$, $i = 1, 2, \cdots, N-1$, to $U_q(\text{sl}_N)$, the elements $K_n^{\pm 1/2}$ to $U_q(\text{sp}_{2n})$ and $K_{n-1}^{\pm 1/2} K_n^{\pm 1/2}$, $K_{n-1}^{\pm 1/2} K_n^{\mp 1/2}$ to $U_q(\text{so}_{2n})$. Let us give the precise definitions of these extended Hopf algebras.

The algebra $U_q^{\text{ext}}(\text{sl}_N)$ is generated by elements E_i, F_i, $i = 1, 2, \cdots, N-1$, $\hat{K}_j$, $\hat{K}_j^{-1}$, $j = 1, 2, \cdots, N$, subject to the defining relations

$$\hat{K}_i \hat{K}_j = \hat{K}_j \hat{K}_i, \quad \hat{K}_i \hat{K}_i^{-1} = \hat{K}_i^{-1} \hat{K}_i = 1, \quad \hat{K}_1 \hat{K}_2 \cdots \hat{K}_N = 1,$$

$$\hat{K}_i E_{i-1} \hat{K}_i^{-1} = q^{-1} E_{i-1}, \quad \hat{K}_i E_i \hat{K}_i^{-1} = q E_i, \quad \hat{K}_i E_j \hat{K}_i^{-1} = E_j, \ j \neq i, i-1,$$
$$\hat{K}_i F_{i-1} \hat{K}_i^{-1} = q F_{i-1}, \quad \hat{K}_i F_i \hat{K}_i^{-1} = q^{-1} F_i, \quad \hat{K}_i F_j \hat{K}_i^{-1} = F_j, \ j \neq i, i-1, \tag{94}$$

$$E_i F_j - F_j E_i = \delta_{ij} \frac{\hat{K}_i \hat{K}_{i+1}^{-1} - \hat{K}_i^{-1} \hat{K}_{i+1}}{q - q^{-1}}, \tag{95}$$

and the Serre relations (6.15) and (6.16) for the elements E_i and F_i of $U_q(\text{sl}_N)$. As in the case of the Drinfeld–Jimbo algebra $U_q(\text{sl}_N)$ one verifies that the algebra $U_q^{\text{ext}}(\text{sl}_N)$ becomes a Hopf algebra with comultiplication Δ, counit ε and antipode S such that

$$\Delta(\hat{K}_i) = \hat{K}_i \otimes \hat{K}_i, \quad \Delta(\hat{K}_i^{-1}) = \hat{K}_i^{-1} \otimes \hat{K}_i^{-1},$$
$$\Delta(E_i) = E_i \otimes \hat{K}_i \hat{K}_{i+1}^{-1} + 1 \otimes E_i, \quad \Delta(F_i) = F_i \otimes 1 + \hat{K}_i^{-1} \hat{K}_{i+1} \otimes F_i,$$
$$\varepsilon(\hat{K}_i) = 1, \quad \varepsilon(E_i) = \varepsilon(F_i) = 0,$$
$$S(\hat{K}_i) = \hat{K}_i^{-1}, \quad S(E_i) = -E_i \hat{K}_i^{-1} \hat{K}_{i+1}, \quad S(F_i) = -\hat{K}_i \hat{K}_{i+1}^{-1} F_i.$$

Note that the Hopf algebra $U_q^{\text{ext}}(\text{sl}_N)$ is just the quotient of the Hopf algebra $U_q(\text{gl}_N)$ (see Subsect. 6.1.2) by the Hopf ideal $\langle \hat{K}_1 \hat{K}_2 \cdots \hat{K}_N - 1 \rangle$.

The algebra $U_q^{\text{ext}}(\text{sp}_{2n})$ is generated by elements E_i, F_i, $i = 1, 2, \cdots, n$, K_i, K_i^{-1}, $i = 1, 2, \cdots, n-1$, and $\hat{K}_n, \hat{K}_n^{-1}$ subject to all defining relations of the algebra $U_q(\text{sp}_{2n})$ involving E_j, F_j, K_i, K_i^{-1}, $i = 1, 2, \cdots, n-1$, and

$$\hat{K}_n K_i = K_i \hat{K}_n, \ \hat{K}_n \hat{K}_n^{-1} = \hat{K}_n^{-1} \hat{K}_n = 1, \ E_n F_n - F_n E_n = \frac{\hat{K}_n^2 - \hat{K}_n^{-2}}{q^2 - q^2}, \tag{96}$$

$$\hat{K}_n E_j \hat{K}_n^{-1} = q^{a_{nj}/2} E_j, \quad \hat{K}_n F_j \hat{K}_n^{-1} = q^{-a_{nj}/2} F_j. \tag{97}$$

The Hopf algebra structure for $U_q^{\rm ext}(\mathrm{sp}_{2n})$ is given by the formulas of $U_q(\mathrm{sp}_{2n})$ for E_i, F_i, K_i, $i = 1, 2, \cdots, n-1$, and

$$\Delta(\hat{K}_n) = \hat{K}_n \otimes \hat{K}_n, \quad \Delta(\hat{K}_n^{-1}) = \hat{K}_n^{-1} \otimes \hat{K}_n^{-1}, \tag{98}$$

$$\Delta(E_n) = E_n \otimes \hat{K}_n^2 + 1 \otimes E_n, \quad \Delta(F_n) = F_n \otimes 1 + \hat{K}_n^{-2} \otimes F_n, \tag{99}$$

$$\varepsilon(\hat{K}_n) = 1, \quad \varepsilon(E_n) = \varepsilon(F_n) = 0, \tag{100}$$

$$S(\hat{K}_n) = \hat{K}_n^{-1}, \quad S(E_n) = -E_n\hat{K}_n^{-2}, \quad S(F_n) = -\hat{K}_n^2 F_n.$$

The algebra $U_q^{\rm ext}(\mathrm{so}_{2n})$ is generated by elements E_i, F_i, $i = 1, 2, \cdots, n$, K_i, K_i^{-1}, $i = 1, 2, \cdots, n-2$, and $\hat{K}_j, \hat{K}_j^{-1}$, $j = n-1, n$. The defining relations are the relations of the algebra $U_q(\mathrm{so}_{2n})$ containing E_i, F_i, K_i, K_i^{-1}, as well as the relations

$$\hat{K}_i\hat{K}_i^{-1} = \hat{K}_i^{-1}\hat{K}_i = 1, \quad \hat{K}_{n-1}\hat{K}_n = \hat{K}_n\hat{K}_{n-1}, \quad \hat{K}_iK_j = K_j\hat{K}_i,$$

$$\hat{K}_{n-1}E_j\hat{K}_{n-1}^{-1} = q^{(a_{nj}+a_{n-1,j})/2}E_j, \quad \hat{K}_nE_j\hat{K}_n^{-1} = q^{(a_{nj}-a_{n-1,j})/2}E_j,$$

$$\hat{K}_{n-1}F_j\hat{K}_{n-1}^{-1} = q^{-(a_{nj}+a_{n-1,j})/2}F_j, \quad \hat{K}_nF_j\hat{K}_n^{-1} = q^{-(a_{nj}-a_{n-1,j})/2}F_j.$$

The Hopf algebra structure for $U_q^{\rm ext}(\mathrm{so}_{2n})$ is determined by the formulas of $U_q(\mathrm{so}_{2n})$ for the generators E_i, F_i, K_i, $i = 1, 2, \cdots, n-2$, by the formulas (98) and (100) taken for n and $n-1$ instead of n and by

$$\Delta(E_{n-1}) = E_{n-1} \otimes \hat{K}_{n-1}\hat{K}_n^{-1} + 1 \otimes E_{n-1},$$

$$\Delta(F_{n-1}) = F_{n-1} \otimes 1 + \hat{K}_{n-1}^{-1}\hat{K}_n \otimes F_{n-1},$$

$$\Delta(E_n) = E_n \otimes \hat{K}_{n-1}\hat{K}_n + 1 \otimes E_n, \quad \Delta(F_n) = F_n \otimes 1 + \hat{K}_{n-1}^{-1}\hat{K}_n^{-1} \otimes F_n,$$

$$S(E_{n-1}) = -E_{n-1}\hat{K}_{n-1}^{-1}\hat{K}_n, \quad S(E_n) = -E_n\hat{K}_{n-1}^{-1}\hat{K}_n^{-1},$$

$$S(F_{n-1}) = -\hat{K}_{n-1}\hat{K}_n^{-1}F_{n-1}, \quad S(F_n) = -\hat{K}_{n-1}\hat{K}_nF_n.$$

The "usual" Drinfeld–Jimbo algebras $U_q(\mathrm{sl}_N), U_q(\mathrm{sp}_{2n})$ and $U_q(\mathrm{so}_{2n})$ can be considered as Hopf subalgebras of $U_q^{\rm ext}(\mathrm{sl}_N), U_q^{\rm ext}(\mathrm{sp}_{2n})$ and $U_q^{\rm ext}(\mathrm{so}_{2n})$, respectively. Indeed, we shall identify $U_q(\mathrm{sl}_N)$ with its image under the injective Hopf algebra homomorphism $\varphi : U_q(\mathrm{sl}_N) \to U_q^{\rm ext}(\mathrm{sl}_N)$ such that $\varphi(E_i) = E_i$, $\varphi(F_i) = F_i$, $\varphi(K_i) = \hat{K}_i\hat{K}_{i+1}^{-1}$. Similarly, the Hopf algebras $U_q(\mathrm{sp}_{2n})$ and $U_q(\mathrm{so}_{2n})$ become Hopf subalgebras in $U_q^{\rm ext}(\mathrm{sp}_{2n})$ and $U_q^{\rm ext}(\mathrm{so}_{2n})$, respectively. The embedding $\varphi : U_q(\mathfrak{g}) \to U_q^{\rm ext}(\mathfrak{g})$, $\mathfrak{g} = \mathrm{sp}_{2n}, \mathrm{so}_{2n}$, is uniquely determined by $\varphi(E_i) = E_i$, $\varphi(F_i) = F_i$, $\varphi(K_j) = K_j$, $j = 1, 2, \cdots, n-1$ for $U_q^{\rm ext}(\mathrm{sp}_{2n})$ and $j = 1, 2, \cdots, n-2$ for $U_q^{\rm ext}(\mathrm{so}_{2n})$, $\varphi(K_n) = \hat{K}_n^2$ for $U_q^{\rm ext}(\mathrm{sp}_{2n})$ and $\varphi(K_{n-1}) = \hat{K}_{n-1}\hat{K}_n^{-1}$, $\varphi(K_n) = \hat{K}_{n-1}\hat{K}_n$ for $U_q^{\rm ext}(\mathrm{so}_{2n})$.

The algebra $U_q^{\rm ext}(\mathrm{sp}_{2n})$ was defined above by adjoining the square root $K_n^{1/2}(= \hat{K}_n)$ and its inverse to $U_q(\mathrm{sp}_{2n})$. Similarly, the algebra $U_q^{\rm ext}(\mathrm{sl}_N)$ can be obtained by adjoining only the N-th root $(K_1K_2^2 \cdots K_{N-1}^{N-1})^{1/N}$ (which corresponds to $\hat{K}_N^{-1}$ in the above notation) and its inverse to the algebra

$U_q(\mathrm{sl}_N)$. That is, $U_q^{\mathrm{ext}}(\mathrm{sl}_N)$ is generated by the algebra $U_q(\mathrm{sl}_N)$ with generators $E_i, F_i, K_i, K_i^{-1}, i = 1, \dots, N-1$, and $\hat{K}_N, \hat{K}_N^{-1}$ satisfying the relations

$$\hat{K}_N E_i \hat{K}_N^{-1} = q^{-\delta_{i,N-1}} E_i, \quad \hat{K}_N F_i \hat{K}_N^{-1} = q^{\delta_{i,N-1}} F_i, \quad \hat{K}_N^{-N} = K_1 K_2^2 \cdots K_{N-1}^{N-1}.$$

In a similar way, $U_q^{\mathrm{ext}}(\mathrm{so}_{2n})$ can be constructed by adjoining only the square root $(K_{n-1}K_n^{-1})^{1/2}$ $(= \hat{K}_n^{-1})$ and its inverse to $U_q(\mathrm{so}_{2n})$.

Most constructions and results for the Drinfeld–Jimbo algebras $U_q(\mathfrak{g})$ carry over in a straightforward manner to the extended algebras $U_q^{\mathrm{ext}}(\mathfrak{g})$. For instance, the vector representation T_1 for $U_q^{\mathrm{ext}}(\mathrm{sl}_N)$ is obtained from that of $U_q(\mathrm{sl}_N)$ if we leave the operators $T_1(E_i)$ and $T_1(F_i)$ unchanged and set

$$T_1(\hat{K}_i) = T_1(K_1)^{-1/N} T_1(K_2)^{-2/N} \cdots T_1(K_{i-1})^{-(i-1)/N} T_1(K_i)^{(N-i)/N}$$
$$\times T_1(K_{i+1})^{(N-i-1)/N} \cdots T_1(K_{N-1})^{1/N}, \quad i = 1, 2, \cdots, N-1,$$

where $T_1(K_i)$ are the representation operators for $U_q(\mathrm{sl}_N)$. Similarly, we set $T_1(\hat{K}_n) = T_1(K_n)^{1/2}$ for $U_q^{\mathrm{ext}}(\mathrm{sp}_{2n})$ and $T_1(\hat{K}_{n-1}) = (T_1(K_{n-1})T_1(K_n))^{1/2}$, $T_1(\hat{K}_n) = (T_1(K_{n-1}^{-1})T_1(K_n))^{1/2}$ for $U_q^{\mathrm{ext}}(\mathrm{so}_{2n})$. The matrices R, C and K for both algebras $U_q^{\mathrm{ext}}(\mathfrak{g})$ and $U_q(\mathfrak{g})$ are the same.

8.5.4 L-Functionals for Vector Representations of $U_q(\mathfrak{g})$

A direct translation of the definition of L-functionals for the h-adic Hopf algebra $U_h(\mathfrak{g})$ to $U_q(\mathfrak{g})$ does not lead to well-defined elements $l_{ij}^{\pm}$ of $U_q(\mathfrak{g})$ or $U_q^{\mathrm{ext}}(\mathfrak{g})$, because the universal R-matrix does not make sense for fixed complex q. The key for getting L-functionals $l_{ij}^{\pm}$ for $q \in \mathbb{C}$ is the algebra homomorphism ψ_q from the next proposition. For notational simplicity, we write $U_q^{\mathrm{ext}}(\mathrm{gl}_N) := U_q(\mathrm{gl}_N)$ and $U_q^{\mathrm{ext}}(\mathrm{so}_{2n+1}) := U_{q^{1/2}}(\mathrm{so}_{2n+1})$.

Let $\mathbb{C}[e^h, e^{-h}]$ be the subring of $\mathbb{C}[[h]]$ formed by the Laurent polynomials in e^h and let $\mathbb{K}$ denote the ring $\mathbb{C}[e^h, e^{-h}]$ for $\mathfrak{g} = \mathrm{gl}_N, \mathrm{sl}_N, \mathrm{so}_{2n}, \mathrm{sp}_{2n}$ and $\mathbb{C}[e^{h/2}, e^{-h/2}]$ for $\mathfrak{g} = \mathrm{so}_{2n+1}$. Further, we denote by $U_{\mathbb{K}}(\mathfrak{g})$ the $\mathbb{K}$-subalgebra of $U_h(\mathfrak{g})$ resp. $U_{h/2}(\mathrm{so}_{2n+1})$ generated by the elements $E_i' := (q_i - q_i^{-1})E_i = (e^{d_i h} - e^{-d_i h})E_i$, F_i, $q^{H_i'} = \exp hH_i'$ and $q^{-H_i'}$ for $i = 1, 2, \cdots, l = \mathrm{rank}\ \mathfrak{g}$. Note that the generators $q_i^{H_i} = \exp d_i h H_i$ of the algebra $U_{\mathbb{K}}(\mathfrak{g})$ are just the counterparts to the generators K_i of $U_q^{\mathrm{ext}}(\mathfrak{g})$ for $\mathfrak{g} = \mathrm{sl}_N, \mathrm{so}_{2n}, \mathrm{sp}_{2n}$.

Proposition 31. *Let q be a nonzero complex number such that $q_i \equiv q^{d_i} \neq 1$, resp. $q^{d_i/2} \neq 1$ if $\mathfrak{g} = \mathrm{so}_{2n+1}$, $i = 1, 2, \cdots, l$. There exists a unique $\mathbb{C}$-algebra homomorphism $\psi_q : U_{\mathbb{K}}(\mathfrak{g}) \to U_q^{\mathrm{ext}}(\mathfrak{g})$ such that $\psi_q(e^h \cdot 1) = q \cdot 1$, $\psi_q(E_i') = (q_i - q_i^{-1})E_i$, $\psi_q(F_i) = F_i$ and $\psi_q\left(q_j^{H_j}\right) = K_j$ resp. $\psi_q\left(q^{H_j'}\right) = \hat{K}_j$ for all i, j in the case $\mathfrak{g} = \mathrm{gl}_N, \mathrm{sl}_N, \mathrm{so}_{2n}, \mathrm{sp}_{2n}$. For $\mathfrak{g} = \mathrm{so}_{2n+1}$ the homomorphism ψ is given by the same formulas with h replaced by $h/2$ and q_i by $q_i^{1/2}$. The map ψ_q intertwines the comultiplications, counits and antipodes of the Hopf algebras $U_{\mathbb{K}}(\mathfrak{g})$ and $U_q^{\mathrm{ext}}(\mathfrak{g})$, respectively.*

Proof. We carry out the proof in the case $\mathfrak{g} = \mathrm{gl}_N, \mathrm{sl}_N, \mathrm{so}_{2n}, \mathrm{sp}_{2n}$. The case $\mathfrak{g} = \mathrm{so}_{2n+1}$ is treated similarly. Let E'_{β_i} denote the elements of $U_h(\mathfrak{n}_+)$ which are obtained when in definition (6.65) of the root vectors the generators E_j are replaced by E'_j. From Theorem 6.24′ it follows that the elements

$$F_{\beta_1}^{r_1} \cdots F_{\beta_n}^{r_n} (q^{H'_1})^{k_i} \cdots (q^{H'_l})^{k_l} E'^{s_n}_{\beta_n} \cdots E'^{s_1}_{\beta_1}, \quad r_i, s_i \in \mathbb{N}_0, \quad k_i \in \mathbb{Z}, \qquad (101)$$

are linearly independent in $U_h(\mathfrak{g})$ over the ring $\mathbb{C}[[h]]$. All elements (101) are contained in $U_\mathbb{K}(\mathfrak{g})$. This set is also linearly independent in $U_\mathbb{K}(\mathfrak{g})$ over $\mathbb{K} = \mathbb{C}[e^h, e^{-h}]$. Since the relation $E'_i F_j - F_j E'_i = \delta_{ij}(q_i^{H_i} - q_i^{-H_i})$ holds in the algebra $U_h(\mathfrak{g})$ and the $q_i^{H_i}$, $i = 1, 2, \cdots, l$, can be expressed in terms of $q^{H'_j}$, the $\mathbb{K}$-linear span of elements (101) coincides with the algebra $U_\mathbb{K}(\mathfrak{g})$. Thus, these elements form a $\mathbb{K}$-vector space basis of $U_\mathbb{K}(\mathfrak{g})$.

Let $a \in U_\mathbb{K}(\mathfrak{g})$ be a multiple of an element (101) by some coefficient from $\mathbb{K} = \mathbb{C}[e^h, e^{-h}]$. We then define $\psi_q(a)$ by replacing in the corresponding expressions the power series $e^h \in \mathbb{K}$ by the number $q \in \mathbb{C}$ and the elements E_i, F_i, $q_j^{H_j}$, $q^{H'_i}$ of $U_\mathbb{K}(\mathfrak{g})$ by the corresponding generators E_i, F_i, K_i, $\hat{K}_i$ of $U_q^{\mathrm{ext}}(\mathfrak{g})$. From the commutation rules in both algebras $U_h(\mathfrak{g})$ and $U_q^{\mathrm{ext}}(\mathfrak{g})$ it is clear that ψ_q is a $\mathbb{C}$-algebra homomorphism. Comparing the corresponding formulas for the comultiplications, counits and antipodes in $U_\mathbb{K}(\mathfrak{g})$ and $U_q^{\mathrm{ext}}(\mathfrak{g})$ we see that ψ_q intertwines these operations. □

We now apply Proposition 31 to the definition of L-functionals of $U_q^{\mathrm{ext}}(\mathfrak{g})$. First we set $\gamma(q) = q - q^{-1}$ for $\mathfrak{g} = \mathrm{gl}_N, \mathrm{sl}_N, \mathrm{so}_{2n}$, $\gamma(q) = q^2 - q^{-2}$ for $\mathfrak{g} = \mathrm{sp}_{2n}$ and $\gamma(q) = ((q - q^{-1})(q^{1/2} - q^{-1/2}))^{1/2}$ for $\mathfrak{g} = \mathrm{so}_{2n+1}$. From Corollary 30 it follows that for some $k \in \mathbb{N}$ all functionals $\gamma(e^h)^k l_{ij}^{\pm}$ belong to the algebra $U_\mathbb{K}(\mathfrak{g})$. We define the *$L$-functionals $l_{ij}^{\pm}$ for the vector representation* of $U_q^{\mathrm{ext}}(\mathfrak{g})$ by

$$l_{ij}^{\pm} := \gamma(q)^{-k} \psi_q(\gamma(e^h)^k l_{ij}^{\pm}).$$

Note that $\gamma(q)^{-k}$ is only a complex constant. Although this definition may look artificial at first glance, its meaning is clear: any L-functional $l_{ij}^{\pm} \in U_q^{\mathrm{ext}}(\mathfrak{g})$ is obtained if in the expression for the L-functional $l_{ij}^{\pm} \in U_h(\mathfrak{g})$ the elements $e^{\alpha h} \in \mathbb{C}[[h]]$ and $E_i, F_i, q_i^{H_i}, q^{H'_i} \in U_h(\mathfrak{g})$ are replaced by $q^\alpha \in \mathbb{C}$ and the corresponding generators $E_i, F_i, K_i, \hat{K}_i \in U_q^{\mathrm{ext}}(\mathfrak{g})$, respectively. Under these replacements the lists in Subsect. 8.5.2 become the lists for $U_q(\mathfrak{g})$ resp. $U_q^{\mathrm{ext}}(\mathfrak{g})$. Some main diagonal L-functionals l_{ii}^- then take the following form:

$U_q(\mathrm{gl}_N): \ l_{ii}^- = K_i; \quad U_q^{\mathrm{ext}}(\mathrm{sl}_N): \ l_{ii}^- = \hat{K}_i; \ U_q(\mathrm{sl}_N): \ K_i = l_{ii}^- l_{i+1,i+1}^+;$

$U_{q^{1/2}}(\mathrm{so}_{2n+1}): \ l_{ii}^- = K_i K_{i+1} \cdots K_{n-1} K_n, \ i \le n;$

$U_q^{\mathrm{ext}}(\mathrm{sp}_{2n}): \ l_{ii}^- = K_i \cdots K_{n-1} \hat{K}_n, \ i \le n;$

$U_q(\mathrm{sp}_{2n}): \ K_i = l_{ii}^- l_{i+1,i+1}^+, \ i \le n; \ K_n = (l_{nn}^-)^2;$

$U_q^{\mathrm{ext}}(\mathrm{so}_{2n}): \ l_{ii}^- = K_i \cdots K_{n-2} \hat{K}_{n-1} \hat{K}_n, \ i \le n-2; \ l_{ii}^- = \hat{K}_i, \ i = n-1, n;$

$U_q(\mathrm{so}_{2n}):\ K_i = l_{ii}^- l_{i+1,i+1}^+,\ i < n-2,\ K_{n-2} = l_{n-2,n-2}^- l_{n-1,n-1}^+ l_{nn}^+,$
$K_{n-1} = l_{n-1,n-1}^- l_{nn}^+,\ K_n = l_{n-1,n-1}^- l_{nn}^-.$

Some of these formulas express the generators K_i of $U_q(\mathfrak{g})$ in terms of L-functionals $l_{jj}^{\pm}$. Since the elements $l_{jj}^{\pm}$ are in $U_q^{\mathrm{ext}}(\mathfrak{g})$ and not in $U_q(\mathfrak{g})$, they cannot be expressed by the generators of $U_q(\mathfrak{g})$ for $\mathfrak{g} = \mathrm{sl}_N, \mathrm{sp}_{2n}, \mathrm{so}_{2n}$.

Recall that the matrices R and C for $U_h(\mathfrak{g})$, resp. $U_{h/2}(\mathrm{so}_{2n+1})$, go into the corresponding matrices R and C for $U_q(\mathfrak{g})$, resp. $U_{q^{1/2}}(\mathrm{so}_{2n+1})$, and $U_q^{\mathrm{ext}}(\mathfrak{g})$ if the power series e^h is replaced by q. Therefore, because ψ_q is a $\mathbb{C}$-algebra homomorphism, all algebraic properties of the L-functionals $l_{ij}^{\pm}$ for the vector representation developed above in the h-adic case remain valid for fixed q as well. That is, *the formulas (88)–(90) and the assertions of Proposition 28 hold also for the L-functionals $l_{ij}^{\pm}$ of $U_q^{\mathrm{ext}}(\mathfrak{g})$*. We shall essentially use this fact in the proof of Theorem 33 below.

Example 5. The L-operators for the algebra $U_q^{\mathrm{ext}}(\mathrm{sl}_2)$ are of the form

$$L^+ = \begin{pmatrix} \hat{K}^{-1} & (q-q^{-1})\hat{K}^{-1}E_1 \\ 0 & \hat{K}^{-1} \end{pmatrix},\quad L^- = \begin{pmatrix} \hat{K} & 0 \\ (q^{-1}-q)F_1\hat{K} & \hat{K} \end{pmatrix}. \quad \triangle$$

8.5.5 The Hopf Algebras $\mathcal{U}(\mathsf{R})$ and $U_q^L(\mathfrak{g})$

In Subsect. 8.5.1 we have seen that the L-functionals $l_{ij}^{\pm}$ for the vector representation of $U_h(\mathfrak{g})$ satisfy the equations (90) and the relations stated in Proposition 28. We now reverse the viewpoint and take these properties as the defining relations of an abstract algebra. We carry out this only for fixed complex q and show that this algebra is isomorphic to the extended Drinfeld–Jimbo algebra $U_q^{\mathrm{ext}}(\mathfrak{g})$. With some technical modifications a similar result holds also in the h-adic case.

We begin by being a bit more general. Let $R = (R_{jl}^{ik})_{i,j,k,l=1,\cdots,N}$ be an arbitrary matrix with complex entries. Let $\mathcal{U}(R)$ denote the algebra with $N(N+1)$ generators $\ell_{ij}^+, \ell_{ji}^-, i \le j, i,j = 1,2,\cdots,N$, and defining relations

$$\mathcal{L}_1^{\pm}\mathcal{L}_2^{\pm}R = R\mathcal{L}_2^{\pm}\mathcal{L}_1^{\pm},\quad \mathcal{L}_1^{-}\mathcal{L}_2^{+}R = R\mathcal{L}_2^{+}\mathcal{L}_1^{-}, \tag{102}$$

$$\ell_{ii}^+\ell_{ii}^- = \ell_{ii}^-\ell_{ii}^+ = 1,\quad i = 1,2,\cdots,N, \tag{103}$$

where we set $\mathcal{L}^{\pm} := (\ell_{ij}^{\pm})$ and $\ell_{ij}^+ = \ell_{ji}^- = 0$, $i > j$, in (102).

Proposition 32. *$\mathcal{U}(R)$ is a Hopf algebra with comultiplication Δ, counit ε and antipode S determined by*

$$\Delta(\ell_{ij}^{\pm}) = \sum\nolimits_k \ell_{ik}^{\pm} \otimes \ell_{kj}^{\pm},\quad \varepsilon(\ell_{ij}^{\pm}) = \delta_{ij},\quad S(\mathcal{L}^{\pm}) = (\mathcal{L}^{\pm})^{-1}. \tag{104}$$

Proof. That $\mathcal{U}(R)$ becomes a bialgebra is easily verified. (The proof of Proposition 9.1 below contains the details of this reasoning.) The matrices $\mathcal{L}^+$ and

$\mathcal{L}^-$ are upper resp. lower triangular and the entries of their main diagonals are invertible in $\mathcal{U}(R)$. Hence both matrices are invertible. The entries $\tilde{\ell}^{\pm}_{ij}$ of the inverses $(\mathcal{L}^{\pm})^{-1}$ obviously satisfy the opposite commutation relations to (102) and (103). Hence there is an algebra anti-homomorphism $S : \mathcal{U}(R) \to \mathcal{U}(R)$ such that $S(\ell^{\pm}_{ij}) = \tilde{\ell}^{\pm}_{ij}$. We then have the relations $S(\mathcal{L}^{\pm})\mathcal{L}^{\pm} = I$ which mean that the antipode axiom is fulfilled on the generators. Therefore, by Proposition 1.8, $\mathcal{U}(R)$ is a Hopf algebra. □

Now we set $R = \mathsf{R}$, where R is one of the matrices from Subsect. 8.4.2. Let $\mathcal{I}(\mathfrak{g})$ be the two-sided ideal in $\mathcal{U}(\mathsf{R})$ generated by the relations

$$\ell^{\pm}_{11}\ell^{\pm}_{22}\cdots\ell^{\pm}_{NN} = 1 \quad \text{for} \quad \mathfrak{g} = \mathrm{sl}_N,$$

$$\mathcal{L}^{\pm}\mathsf{C}^t(\mathcal{L}^{\pm})^t(\mathsf{C}^{-1})^t = \mathsf{C}^t(\mathcal{L}^{\pm})^t(\mathsf{C}^{-1})^t\mathcal{L}^{\pm} = I \;\text{ for }\; \mathfrak{g} = \mathrm{so}_N, \mathrm{sp}_N.$$

Since $\mathcal{I}(\mathfrak{g})$ is a Hopf ideal of $\mathcal{U}(\mathsf{R})$ as easily checked, the quotient $\mathcal{U}(\mathsf{R})/\mathcal{I}(\mathfrak{g})$ is a Hopf algebra which will be denoted by $U_q^L(\mathfrak{g})$. The following important result says that this Hopf algebra $U_q^L(\mathfrak{g})$ is isomorphic to the extended Drinfeld–Jimbo algebra $U_q^{\mathrm{ext}}(\mathfrak{g})$ defined in Subsect. 8.5.3.

Theorem 33. *Let $\mathfrak{g}$ be one of the Lie algebras gl_N, sl_N, so_N or sp_N and set $U_q^{\mathrm{ext}}(\mathrm{gl}_N) := U_q(\mathrm{gl}_N)$ and $U_q^{\mathrm{ext}}(\mathrm{so}_{2n+1}) := U_{q^{1/2}}(\mathrm{so}_{2n+1})$. Then the Hopf algebras $U_q^L(\mathfrak{g})$ and $U_q^{\mathrm{ext}}(\mathfrak{g})$ are isomorphic with a Hopf algebra isomorphism $\theta : U_q^L(\mathfrak{g}) \to U_q^{\mathrm{ext}}(\mathfrak{g})$ such that $\theta(\ell^+_{ij}) = l^+_{ij}$ and $\theta(\ell^-_{ji}) = l^-_{ji}$, $i \le j$, where l^+_{ij} and l^-_{ji} are the elements of $U_q^{\mathrm{ext}}(\mathfrak{g})$ defined in Subsect. 8.5.4.*

Outline of proof. As noted at the end of Subsect. 8.5.4, the elements $l^+_{ij}, l^-_{ji} \in U_q^{\mathrm{ext}}(\mathfrak{g})$ satisfy the defining relations of the algebra $U_q^L(\mathfrak{g})$. Hence there is an algebra homomorphism $\theta : U_q^L(\mathfrak{g}) \to U_q^{\mathrm{ext}}(\mathfrak{g})$ such that $\theta(\ell^+_{ij}) = l^+_{ij}$ and $\theta(\ell^-_{ji}) = l^-_{ji}$, $i \le j$. Comparing the formulas (89) and (104) for the comultiplications and the counits, it is clear that θ is a bialgebra and hence a Hopf algebra homomorphism. From the formulas listed in Subsects. 8.5.2 and 8.5.4 we check easily that the image of θ contains all generators of $U_q^{\mathrm{ext}}(\mathfrak{g})$. Therefore, θ is surjective.

It remains to show that θ is injective. We carry out this proof in the case $\mathfrak{g} = \mathrm{sl}_N$. The proof in the other cases is similar. First we solve the formulas of l^+_{jj}, $l^+_{i,i+1}$ $l^-_{i+1,i}$ in Subsects. 8.5.2 and 8.5.4 for the generators of $U_q^{\mathrm{ext}}(\mathrm{sl}_N)$ and set

$$E'_i := (q-q^{-1})^{-1}\ell^-_{ii}\ell^+_{i,i+1}, \quad F'_i := -(q-q^{-1})^{-1}\ell^-_{i+1,i}\ell^+_{ii}, \tag{105}$$

$$K'_j := \ell^-_{jj}, \quad {K'}_j^{-1} := \ell^+_{jj}, \quad j = 1, 2, \cdots, N. \tag{106}$$

Using the relations of $U_q^L(\mathrm{sl}_N)$, we next show that the elements E'_i, F'_i, K'_i of $U_q^L(\mathrm{sl}_N)$ satisfy the defining relations of the extended Drinfeld–Jimbo algebra $U_q^{\mathrm{ext}}(\mathrm{sl}_N)$. In order to do this, we shall write down the relations (102) for $U_q^L(\mathrm{sl}_N)$ explicitly. The equation $\mathcal{L}_1^+\mathcal{L}_2^+\mathsf{R} = \mathsf{R}\mathcal{L}_2^+\mathcal{L}_1^+$ for $U_q^L(\mathrm{sl}_N)$ is equivalent to the following relations:

$$\ell_{ij}^{+} l_{rs}^{+} = \ell_{rs}^{+} \ell_{ij}^{+} - (q-q^{-1}) \ell_{is}^{+} \ell_{rj}^{+}, \quad i<r,\ j<s, \tag{107}$$

$$\ell_{ij}^{+} \ell_{rj}^{+} = q^{-1} \ell_{rj}^{+} \ell_{ij}^{+}, \quad i<r, \tag{108}$$

$$\ell_{ij}^{+} \ell_{rs}^{+} = \ell_{rs}^{+} \ell_{ij}^{+}, \quad i<r,\ j>s, \tag{109}$$

$$\ell_{ij}^{+} \ell_{is}^{+} = q^{-1} \ell_{is}^{+} \ell_{ij}^{+}, \quad j<s, \tag{110}$$

$$\ell_{ii}^{+} \ell_{jj}^{+} = \ell_{jj}^{+} \ell_{ii}^{+},$$

where $i \le j$ and $r \le s$. The relation $\mathcal{L}_1^- \mathcal{L}_2^- \mathsf{R} = \mathsf{R} \mathcal{L}_2^- \mathcal{L}_1^-$ can be written as

$$\ell_{ij}^{-} \ell_{rs}^{-} = \ell_{rs}^{-} \ell_{ij}^{-} + (q-q^{-1}) \ell_{is}^{-} \ell_{rj}^{-}, \quad i>r,\ j>s, \tag{111}$$

$$\ell_{ij}^{-} \ell_{rj}^{-} = q \ell_{rj}^{-} \ell_{ij}^{-}, \quad i>r, \tag{112}$$

$$\ell_{ij}^{-} \ell_{rs}^{-} = \ell_{rs}^{-} \ell_{ij}^{-}, \quad i>r,\ j<s, \tag{113}$$

$$\ell_{ij}^{-} \ell_{is}^{-} = q \ell_{is}^{-} \ell_{ij}^{-}, \quad j>s, \tag{114}$$

$$\ell_{ii}^{-} \ell_{jj}^{-} = \ell_{jj}^{-} \ell_{ii}^{-},$$

where $i \ge j$ and $r \ge s$. The matrix equation $\mathcal{L}_1^- \mathcal{L}_2^+ \mathsf{R} = \mathsf{R} \mathcal{L}_2^+ \mathcal{L}_1^-$ is equivalent to the relations

$$\begin{gathered} \ell_{ij}^{-} \ell_{rs}^{+} = \ell_{rs}^{+} \ell_{ij}^{-}, \quad i<r,\ j<s, \quad \ell_{ij}^{-} \ell_{is}^{+} = q \ell_{is}^{+} \ell_{ij}^{-}, \quad j<s, \\ \ell_{ij}^{-} \ell_{rs}^{+} = \ell_{rs}^{+} \ell_{ij}^{-} + (q-q^{-1})(\ell_{is}^{+} \ell_{rj}^{-} - \ell_{rj}^{+} \ell_{is}^{-}), \quad i>r,\ j<s, \\ \ell_{ij}^{-} \ell_{rj}^{+} = q^{-1} \ell_{rj}^{+} \ell_{ij}^{-}, \quad i>r, \quad \ell_{ij}^{-} \ell_{rs}^{+} = \ell_{rs}^{+} \ell_{ij}^{-}, \quad i>r,\ j>s, \\ \ell_{ii}^{-} \ell_{jj}^{+} = \ell_{jj}^{+} \ell_{ii}^{-}, \quad \ell_{jj}^{-} \ell_{ii}^{+} = \ell_{ii}^{+} \ell_{jj}^{-}, \end{gathered} \tag{115}$$

where $i \ge j$ and $r \le s$.

Relation (107) for $(i,j) = (i,i+1)$ and $(r,s) = (i+1,i+2)$ gives

$$\ell_{i,i+1}^{+} \ell_{i+1,i+2}^{+} - \ell_{i+1,i+2}^{+} \ell_{i,i+1}^{+} = -(q-q^{-1}) \ell_{i,i+2}^{+} \ell_{i+1,i+1}^{+}.$$

The latter equation expresses $\ell_{i,i+2}^{+}$ in terms of $\ell_{i,i+1}^{+}$, $\ell_{i+1,i+2}^{+}$ and $\ell_{i+1,i+1}^{-}$. Inserting this expression into the relation $\ell_{i,i+2}^{+} \ell_{i,i+1}^{+} = q \ell_{i,i+1}^{+} \ell_{i,i+2}^{+}$ (by (110)) we obtain

$$\begin{aligned} &(\ell_{i,i+1}^{+} \ell_{i+1,i+2}^{+} - \ell_{i+1,i+2}^{+} \ell_{i,i+1}^{+}) \ell_{i+1,i+1}^{-} \ell_{i,i+1}^{+} \\ &\qquad - q \ell_{i,i+1}^{+} (\ell_{i,i+1}^{+} \ell_{i+1,i+2}^{+} - \ell_{i+1,i+2}^{+} \ell_{i,i+1}^{+}) \ell_{i+1,i+1}^{-} = 0. \end{aligned}$$

If we multiply this equation by $(q-q^{-1})^{-3} q (l_{ii}^{-})^2$ and substitute the elements E_i' and E_{i+1}' from (105), then a slight simplification yields the Serre relation

$$E'^{2}_{i+1} E_i' - (q+q^{-1}) E_{i+1}' E_i' E_{i+1}' + E_i' E'^{2}_{i+1} = 0.$$

The second Serre relation for E_j' follows from (107) and (109). The Serre relations for F_j' can be derived similarly from the formulas (111)–(114).

Equation (108) for $i = j-1$ and $r = j$ and equation (110) for $j = i+1$ and $s = i$ lead to the first two defining relations (94). The third relation in (94) follows from (109). The corresponding defining relations for the elements F_i' can be obtained from (112)–(114). Relation (95) follows easily from (115).

Thus, there exists an algebra homomorphism $\vartheta : U_q^{\mathrm{ext}}(\mathrm{sl}_N) \to U_q^L(\mathrm{sl}_N)$ such that $\vartheta(E_i) = E_i'$, $\vartheta(F_i) = F_i'$, $\vartheta(\hat{K}_j) = K_j'$ for $i = 1,2,\cdots,N-1$, $j = 1,2,\cdots,N$. The formulas (105) and (106) imply that $\theta(E_i') = E_i$, $\theta(F_i') = F_i$, $\theta(K_j') = \hat{K}_j$. Since the sets of elements $E_i, F_i, \hat{K}_j$ and E_i', F_i', K_j' generate the algebras $U_q^{\mathrm{ext}}(\mathrm{sl}_N)$ and $U_q^L(\mathrm{sl}_N)$, respectively, ϑ is the inverse of θ. $\square$

8.6 An Analog of the Brauer–Schur–Weyl Duality

In this section we develop an analog of the Brauer–Schur–Weyl duality for the Drinfeld–Jimbo algebras $U_q(\mathfrak{g})$, $\mathfrak{g} = \mathrm{sl}_N, \mathrm{so}_N, \mathrm{sp}_N$. For this we need to extend the Hopf algebras $U_q(\mathrm{so}_N)$ by the group algebra $\mathbb{C}\mathbb{Z}_2$. Throughout this section we assume that q is not a root of unity.

8.6.1 The Algebras $\tilde{U}_q(\mathrm{so}_N)$

Let $\mathbb{C}\mathbb{Z}_2$ be the group algebra of the group $\mathbb{Z}_2$ consisting of the two elements 1 and χ. We denote by $\tilde{U}_q(\mathrm{so}_{2n+1})$ the tensor product $\mathbb{C}\mathbb{Z}_2 \otimes U_{q^{1/2}}(\mathrm{so}_{2n+1})$ of the Hopf algebras $\mathbb{C}\mathbb{Z}_2$ and $U_{q^{1/2}}(\mathrm{so}_{2n+1})$.

The group $\mathbb{Z}_2$ acts as a group of Hopf algebra automorphisms of $U_q(\mathrm{so}_{2n})$ such that $1(\cdot) = \mathrm{id}$ and $\chi(\cdot)$ is the Hopf algebra automorphism defined by

$$\chi(E_i) = E_{\chi(i)}, \quad \chi(F_i) = F_{\chi(i)}, \quad \chi(K_i) = K_{\chi(i)},$$

$$\chi(i) = i \ \text{ for } \ i = 1,2,\cdots,n-2, \quad \chi(n-1) = n, \ \ \chi(n) = n-1.$$

Since obviously $\chi \circ \chi = \mathrm{id}$, this defines indeed an action of the group $\mathbb{Z}_2$. Let $\tilde{U}_q(\mathrm{so}_{2n}) = \mathbb{C}\mathbb{Z}_2 \ltimes U_q(\mathrm{so}_{2n})$ be the corresponding right crossed product algebra (see Subsect. 10.2.1). That is, $\tilde{U}_q(\mathrm{so}_{2n})$ is the algebra generated by the algebra $U_q(\mathrm{so}_{2n})$ and an additional element χ such that $\chi^2 = 1$ with the commutation rule $\chi(a) = \chi a \chi$, $a \in U_q(\mathrm{so}_{2n})$. One easily checks that $\tilde{U}_q(\mathrm{so}_{2n})$ is a Hopf algebra with comultiplication defined by the comultiplication of $U_q(\mathrm{so}_{2n})$ and the relation $\Delta(\chi) = \chi \otimes \chi$.

The finite-dimensional irreducible type 1 representations of the algebra $\tilde{U}_q(\mathrm{so}_{2n})$ are described by the following proposition.

Proposition 34. (i) *Let* $\lambda = (\lambda_1, \lambda_2, \cdots, \lambda_n)$, $\lambda_i = 2(\lambda, \alpha_i)/(\alpha_i, \alpha_i)$, *be the highest weight of a type 1 irreducible finite-dimensional representation of the algebra* $U_q(\mathrm{so}_{2n})$.

Case 1: $\lambda_{n-1} = \lambda_n$. *There exist exactly two nonequivalent irreducible representations* $\tilde{T}_\lambda$ *and* $\tilde{T}_\lambda^\circ$ *of* $\tilde{U}_q(\mathrm{so}_{2n})$ *such that their restrictions to* $U_q(\mathrm{so}_{2n})$ *are equivalent to the irreducible representation* T_λ. *If* $\mathbf{v}$ *and* $\mathbf{v}^\circ$ *are highest weight vectors for* $\tilde{T}_\lambda$ *and* $\tilde{T}^\circ_\lambda$, *respectively, then* $\tilde{T}_\lambda(\chi)\mathbf{v} = \mathbf{v}$ *and* $\tilde{T}^\circ_\lambda(\chi)\mathbf{v}^\circ = -\mathbf{v}^\circ$.

Case 2: $\lambda_{n-1} \neq \lambda_n$. *Then there exists a unique irreducible representation* $\tilde{T}_\lambda$ *of* $\tilde{U}_q(\mathrm{so}_{2n})$ *such that its restriction to* $U_q(\mathrm{so}_{2n})$ *is equivalent to the direct sum representation* $T_\lambda \oplus T_{\chi(\lambda)}$, *where* $\chi(\lambda) = (\lambda_1, \cdots, \lambda_{n-2}, \lambda_n, \lambda_{n-1})$ *if* $\lambda = (\lambda_1, \cdots, \lambda_{n-2}, \lambda_{n-1}, \lambda_n)$.

(ii) *Every type 1 irreducible finite-dimensional representation of* $\tilde{U}_q(\mathrm{so}_{2n})$ *is equivalent to one of the representations from Case 1 or Case 2. Every finite-dimensional representation of* $\tilde{U}_q(\mathrm{so}_{2n})$ *is completely reducible.*

Proof. The proof of this proposition is similar to the classical case and we omit it. □

The vector representation $\tilde{T}_1$ of the algebra $\tilde{U}_q(\mathrm{so}_N)$ is given by the same formulas of $\tilde{T}_1(E_i)$, $\tilde{T}_1(F_i)$ and $\tilde{T}_1(K_i)$ as for the vector representation of $U_q(\mathrm{so}_N)$ and by

$$\tilde{T}_1(\chi) = -I \quad \text{if} \quad N = 2n+1,$$

$$\tilde{T}_1(\chi) = E_{n,n+1} + E_{n+1,n} + I - E_{nn} - E_{n+1,n+1} \ \text{ if } \ N = 2n.$$

8.6.2 Tensor Products of Vector Representations

In this subsection we decompose the r-fold tensor product of the vector representations of $\tilde{U}_q(\mathrm{so}_N)$ and $U_q(\mathfrak{g})$, $\mathfrak{g} = \mathrm{sl}_N, \mathrm{sp}_N$, into irreducible components. This result will be used in the proofs of Theorems 38 and 11.22 below. In order to treat all needed cases at once, we write $\tilde{U}_q(\mathfrak{g})$ instead of $U_q(\mathfrak{g})$ for $\mathfrak{g} = \mathrm{sl}_N, \mathrm{sp}_N$. In what follows, $\mathfrak{g}$ denotes one of the Lie algebras $\mathrm{sl}_N, \mathrm{so}_N, \mathrm{sp}_N$.

A set of integers $\mathbf{n} = (n_1, n_2, \cdots, n_k)$ such that $n_1 \geq n_2 \geq \cdots \geq n_k \geq 0$ is called a *partition.* Every partition is representable by a *Young diagram* consisting of boxes placed in k rows (if $n_k \neq 0$) with n_i boxes in the i-th row. If $\mathbf{n}$ is a partition, then its *transpose* is the partition $\mathbf{n}' = (n'_1, n'_2, \cdots, n'_l)$, where n'_i is the number of boxes in the i-th column of the Young diagram of the partition $\mathbf{n}$. In particular, n'_1 is equal to the number of rows in the Young diagram corresponding to $\mathbf{n}$. The number $|\mathbf{n}| := \sum_i n_i$ is called the *length* of the partition $\mathbf{n}$.

Let $\mathcal{P}$ be the set of all partitions (including the partition $\mathbf{0} = (0)$). For $A_n \equiv \mathrm{sl}_{n+1}$, $B_n \equiv \mathrm{so}_{2n+1}$, $C_n \equiv \mathrm{sp}_{2n}$ and $D_n \equiv \mathrm{so}_{2n}$ we define the following sets of partitions (see, for instance, [Wey]):

$$\mathcal{P}(A_n) = \{\mathbf{n} \in \mathcal{P} \mid n'_1 \leq n+1\}, \quad \mathcal{P}(C_n) = \{\mathbf{n} \in \mathcal{P} \mid n'_1 \leq n\},$$

$$\mathcal{P}(B_n) = \{\mathbf{n} \in \mathcal{P} \mid n'_1 + n'_2 \leq 2n+1\}, \quad \mathcal{P}(D_n) = \{\mathbf{n} \in \mathcal{P} \mid n'_1 + n'_2 \leq 2n\}$$

and

$$\mathcal{P}_r(A_n) = \{\mathbf{n} \in \mathcal{P}(A_n) \bigm| |\mathbf{n}| = r\},$$

$$\mathcal{P}_r(X_n) = \{\mathbf{n} \in \mathcal{P}(X_n) \bigm| |\mathbf{n}| \leq r, |\mathbf{n}| \equiv r \ (\mathrm{mod}\ 2)\}, \quad X_n = B_n, C_n, D_n.$$

For A_n and C_n, the sets $\mathcal{P}(A_n)$ and $\mathcal{P}(C_n)$ coincide with the sets of highest weights $(m_1, m_2, \cdots)$ described in Subsect. 7.1.2 (see (7.2)–(7.4)). Note that different partitions of $\mathcal{P}(A_n)$ may correspond to equivalent representations of $U_q(\mathrm{sl}_{n+1})$. If $\mathbf{n} \in \mathcal{P}(B_n)$ or $\mathbf{n} \in \mathcal{P}(D_n)$, we set $\mathbf{n}^\circ := (N - n'_1, n'_2, n'_3 \cdots)'$, where $N = 2n+1$ for B_n and $N = 2n$ for D_n. In fact, $\mathbf{n}^\circ$ coincides with $\mathbf{n}$ when the first column is appropriately changed. If $n'_1 > n$ in $\mathbf{n}$, then $n^{\circ\prime}_1 \le n$. Clearly, $\mathbf{n}^{\circ\circ} = \mathbf{n}$. Further, we have $\mathbf{n}^\circ = \mathbf{n}$ if and only if $\mathbf{n} \in \mathcal{P}(D_n)$ and $n'_1 = n$.

With every partition $\mathbf{n} \in \mathcal{P}(\mathfrak{g})$ we associate a type 1 irreducible finite-dimensional representation $\tilde{T}(\mathbf{n})$ of $\tilde{U}_q(\mathfrak{g})$. It is defined as follows. For $\mathfrak{g} = \mathrm{sl}_{n+1}, \mathrm{sp}_{2n}$, it is just the irreducible representation of $U_q(\mathfrak{g})$ with highest weight $\mathbf{n}$. For $\mathfrak{g} = \mathrm{so}_{2n+1}$, the restriction of $\tilde{T}(\mathbf{n})$ to $U_{q^{1/2}}(\mathrm{so}_{2n+1})$ coincides with the irreducible representation with highest weight $\mathbf{n}$ if $n'_1 \le n$ and with highest weight $\mathbf{n}^\circ$ if $n'_1 > n$, and the operator $\tilde{T}(\mathbf{n})(\chi)$ is given by $\tilde{T}(\mathbf{n})(\chi) = (-1)^{|\mathbf{n}|}$. In the case $\mathfrak{g} = \mathrm{so}_{2n}$ we set $\tilde{T}(\mathbf{n}) := \tilde{T}_{\lambda(\mathbf{n})}$ if $n'_1 \le n$ and $\tilde{T}(\mathbf{n}) := \tilde{T}^\circ_{\lambda(\mathbf{n})}$ if $n'_1 > n$, where $\tilde{T}_\lambda$ and $\tilde{T}^\circ_\lambda$ are as in Proposition 34 and

$$\lambda(\mathbf{n}) = (n_1 - n_2, \cdots, n_{n-1} - n_n, n_{n-1} + n_n) \quad \text{if} \quad n'_1 \le n,$$

$$\lambda(\mathbf{n}) := \lambda(\mathbf{n}^\circ) \quad \text{if} \quad n'_1 > n.$$

Except for the case where $\mathfrak{g} = \mathrm{so}_{2n}$ and $n'_1 = n$, the restriction of the representation $\tilde{T}(\mathbf{n})$ to $U_q(\mathfrak{g})$, $\mathfrak{g} = \mathrm{sl}_N, \mathrm{sp}_N, \mathrm{so}_{2n}$, resp. $U_{q^{1/2}}(\mathrm{so}_{2n+1})$ is irreducible. By Proposition 34, if $\mathfrak{g} = \mathrm{so}_{2n}$ and $n'_1 = n$, then the restriction of $\tilde{T}(\mathbf{n})$ is the direct sum of two nonequivalent irreducible representations of $U_q(\mathrm{so}_{2n})$.

Proposition 35. *If $\mathfrak{g} = \mathrm{so}_{2n+1}, \mathrm{sp}_{2n}, \mathrm{so}_{2n}$ and if $\mathbf{n}, \mathbf{m} \in \mathcal{P}(\mathfrak{g})$, $\mathbf{n} \ne \mathbf{m}$, then the representations $\tilde{T}(\mathbf{n})$ and $\tilde{T}(\mathbf{m})$ of $\tilde{U}(\mathfrak{g})$ are not equivalent.*

Proof. Since $\tilde{U}_q(\mathrm{sp}_{2n}) = U_q(\mathrm{sp}_{2n})$ and $\mathcal{P}(\mathrm{sp}_{2n})$ is the set of all highest weights of type 1 irreducible representations of $U_q(\mathrm{sp}_{2n})$, the assertion for $\mathfrak{g} = \mathrm{sp}_{2n}$ follows. If $\mathfrak{g} = \mathrm{so}_{2n+1}$, then $\tilde{T}(\mathbf{n})(\chi) \ne \tilde{T}(\mathbf{n}^\circ)(\chi)$ and hence $\tilde{T}(\mathbf{n}) \not\simeq \tilde{T}(\mathbf{n}^\circ)$. It follows from the definition of $\tilde{T}(\mathbf{n})$ for $\tilde{U}_q(\mathrm{so}_{2n+1})$ that $\tilde{T}(\mathbf{n})$ and $\tilde{T}(\mathbf{m})$ are not equivalent if $n'_1 \le n$ and $m'_1 \le n$ or if $n'_1 > n$ and $m'_1 > n$. This implies the assertion for $\mathfrak{g} = \mathrm{so}_{2n+1}$. For $\mathfrak{g} = \mathrm{so}_{2n}$, the assertion follows from the results of Subsect. 7.1.2 and Proposition 34. □

For partitions $\mathbf{m}, \mathbf{n} \in \mathcal{P}(\mathfrak{g})$ let us write $\mathbf{m} \sim \mathbf{n}$ if $\mathbf{m} = \mathbf{n} + \mathbf{e}_i$ for some i in case $\mathfrak{g} = \mathrm{sl}_N$ and $\mathbf{m} = \mathbf{n} \pm \mathbf{e}_i$ for some i in cases $\mathfrak{g} = \mathrm{so}_N, \mathrm{sp}_N$. Here $\mathbf{e}_i$ is the vector with 1 in the i-th component and 0 otherwise.

Proposition 36. *The tensor product of an irreducible representation $\tilde{T}(\mathbf{n})$, $\mathbf{n} \in \mathcal{P}(\mathfrak{g})$, and the vector representation $\tilde{T}_1$ of $\tilde{U}_q(\mathfrak{g})$ decomposes into irreducible representations as*

$$\tilde{T}(\mathbf{n}) \otimes \tilde{T}_1 \simeq \bigoplus_{\mathbf{m} \sim \mathbf{n}, \mathbf{m} \in \mathcal{P}(\mathfrak{g})} \tilde{T}(\mathbf{m}). \tag{116}$$

Proof. For $\mathfrak{g} = \mathrm{so}_N$, one easily verifies that both sides of (116) coincide for the element χ. Thus it suffices to prove (116) for elements of the algebras $U_q(\mathfrak{g}), \mathfrak{g} = \mathrm{sl}_N, \mathrm{sp}_N, \mathrm{so}_{2n}$, resp. $U_{q^{1/2}}(\mathrm{so}_{2n+1})$. If $\mathfrak{g} = \mathrm{sl}_N, \mathrm{so}_{2n}, \mathrm{sp}_N$ or if $\mathfrak{g} = \mathrm{so}_{2n+1}$ and $n'_1 \neq n, n+1$, then the decomposition (116) follows from the formulas (7.13)–(7.15) and Propositions 34 and 35. If $\mathfrak{g} = \mathrm{so}_{2n+1}$ and $n'_1 = n$ (resp. $n'_1 = n+1$), then the representation corresponding to the first summand on the right hand side of (7.15) appears when a box in the $(n+1)$-th row is added to (resp. subtracted from) the Young diagram of $\mathbf{n}$. □

Corollary 37. (i) *The r-fold tensor product $\tilde{T}_1^{\otimes r}$ of the vector representation $\tilde{T}_1$ of $\tilde{U}_q(\mathfrak{g})$ decomposes into a direct sum of irreducible representations as*

$$\tilde{T}_1^{\otimes r} \simeq \bigoplus\nolimits_{\mathbf{n}\in\mathcal{P}_r(\mathfrak{g})} m_{\mathbf{n}} \tilde{T}(\mathbf{n}), \tag{117}$$

where $m_{\mathbf{n}}$, $m_{\mathbf{n}} > 0$, is the multiplicity of $\tilde{T}(\mathbf{n})$ in the decomposition.

(ii) *If $\mathfrak{g} = \mathrm{sl}_N$, then $\tilde{T}_1^{\otimes r}$ contains the trivial irreducible representation (that is, the representation with highest weight $(0,0,\cdots,0)$) in the decomposition (117) if and only if $r = kN$, $k \in \mathbb{N}_0$. If $\mathfrak{g} = \mathrm{so}_N$ or $\mathfrak{g} = \mathrm{sp}_N$, then $\tilde{T}_1^{\otimes r}$ contains the trivial representation if and only if $r \in 2\mathbb{N}_0$.*

8.6.3 The Brauer–Schur–Weyl Duality for Drinfeld–Jimbo Algebras

First we briefly describe the corresponding classical results. Let G be one of the groups $SL(N,\mathbb{C})$, $O(N,\mathbb{C})$ or $Sp(N,\mathbb{C})$ and let T_1 be the vector (first fundamental) representation of G on $V_1 = \mathbb{C}^N$. It is well-known that the problem of decomposing the tensor product representation $T_1^{\otimes r}$ of G into irreducible components is closely related to the structure of its centralizer algebra $T_1^{\otimes r}(G)' = \{A \in \mathcal{L}(V_1^{\otimes r}) \mid AT_1^{\otimes r}(g) = T_1^{\otimes r}(g)A,\ g \in G\}$. For $G = SL(N,\mathbb{C})$, a classical result of I. Schur says that the algebra $T_1^{\otimes r}(G)'$ is generated by the flip operators $\tau_{i,i+1}$, $i = 1,2,\cdots,r-1$, of the i-th and $(i+1)$-th tensor factors in $V_1^{\otimes r}$. In the cases $G = O(N,\mathbb{C}), Sp(N,\mathbb{C})$ the Brauer–Weyl duality theorem asserts that $T_1^{\otimes r}(G)'$ is generated by $\tau_{i,i+1}$ and $\mathsf{K}_{i,i+1}$, $i = 1,2,\cdots,r-1$. Here K is the projection of $V_1 \otimes V_1$ onto its one-dimensional $T_1^{\otimes 2}$-invariant subspace and $\mathsf{K}_{i,i+1}$ denotes the operator K acting in the i-th and $(i+1)$-th factors of $V_1^{\otimes r}$.

We now turn to the quantum algebras. Let $\tilde{T}_1$ be the vector representation of $\tilde{U}_q(\mathfrak{g})$ on the vector space V_1, where $\mathfrak{g} = \mathrm{sl}_N, \mathrm{so}_N, \mathrm{sp}_N$. The image $\tilde{T}_1^{\otimes r}(\tilde{U}_q(\mathfrak{g}))$ of $\tilde{U}_q(\mathfrak{g})$ under the r-fold tensor product representation $\tilde{T}_1^{\otimes r}$ is a subalgebra of the algebra $\mathcal{L}(V_1^{\otimes r})$ of linear operators on $V_1^{\otimes r}$. Let R and K be the matrices for $U_q(\mathfrak{g})$, resp. $U_{q^{1/2}}(\mathrm{so}_{2n+1})$, from the formulas (60), (61), (78) and (81). As usual, $\hat{\mathsf{R}}_{i,i+1}$ and $\mathsf{K}_{i,i+1}$ are the operators on $V_1^{\otimes r}$ acting as $\hat{\mathsf{R}} = \tau \circ \mathsf{R}$ and K, respectively, in the i-th and $(i+1)$-th tensor factors and as the identity elsewhere. We denote by $B(r)$ the subalgebra of $\mathcal{L}(V_1^{\otimes r})$ generated by the operators $\hat{\mathsf{R}}_{i,i+1}$, $i = 1,2,\cdots,r-1$, for $\mathfrak{g} = \mathrm{sl}_N$ and by $\hat{\mathsf{R}}_{i,i+1}$ and

$\mathsf{K}_{i,i+1}$, $i = 1, 2, \cdots, r-1$, for $\mathfrak{g} = \mathrm{so}_N, \mathrm{sp}_N$. Let $\tilde{T}_1^{\otimes r}(\tilde{U}_q(\mathfrak{g}))'$ and $B(r)'$ be the sets of operators in $\mathcal{L}(V_1^{\otimes r})$ commuting with all operators from $\tilde{T}_1^{\otimes r}(\tilde{U}_q(\mathfrak{g}))$ and $B(r)$, respectively. Recall that a complex number is called *transcendental* if it is not a root of a nontrivial polynomial with integral coefficients.

The q-analog of the classical Brauer–Schur–Weyl duality is stated as

Theorem 38. *Let $r \in \mathbb{N}$, $r \geq 2$, and let q be transcendental. Then we have*

$$\tilde{T}_1^{\otimes r}(\tilde{U}_q(\mathfrak{g}))' = B(r) \quad \text{and} \quad B(r)' = \tilde{T}_1^{\otimes r}(\tilde{U}_q(\mathfrak{g})).$$

Moreover, the algebra $\tilde{T}_1^{\otimes r}(\tilde{U}_q(\mathfrak{g}))$ decomposes as a direct sum of algebras

$$\tilde{T}_1^{\otimes r}(\tilde{U}_q(\mathfrak{g})) = \bigoplus\nolimits_{\mathbf{n}\in\mathcal{P}_r(\mathfrak{g})} \mathcal{L}(V(\mathbf{n})), \tag{118}$$

where $V(\mathbf{n})$ is the space of the irreducible representation $\tilde{T}(\mathbf{n})$ of $\tilde{U}_q(\mathfrak{g})$.

Proof. We carry out the proof for $\mathfrak{g} = \mathrm{so}_{2n+1}$. The other cases are treated in a similar manner. Since the irreducible representations $\tilde{T}(\mathbf{n})$, $\mathbf{n} \in \mathcal{P}_r(\mathfrak{g})$, are mutually inequivalent by Proposition 35, the decomposition (117) of the representation $\tilde{T}_1^{\otimes r}$ implies (118). Further, since the multiplicities $m_\mathbf{n}$ in (117) are independent of q, so is $m := \dim \tilde{T}_1^{\otimes r}(\tilde{U}_q(\mathfrak{g}))' = \sum_{\mathbf{n}\in\mathcal{P}_r(\mathfrak{g})} m_\mathbf{n}^2$.

On the other hand, let $\hat{T}_1$ be the vector representation of the classical group $G = O(2n+1, \mathbb{C})$. The r-fold tensor product $\hat{T}_1^{\otimes r}$ of this representation decomposes also into a direct sum of irreducible components according to the formula (117). Therefore,

$$\dim \hat{T}_1^{\otimes r}(G)' = \sum\nolimits_{\mathbf{n}\in\mathcal{P}_r(\mathfrak{g})} m_\mathbf{n}^2 = m.$$

Let us express the dependence of R and K on q by writing $\mathsf{R}(q)$ and $\mathsf{K}(q)$, respectively. By the Brauer–Weyl duality theorem for the group $G = O(2n+1, \mathbb{C})$ (see [Bra], [Wey]), there exists a basis v_k, $k = 1, 2, \cdots, m$, of the vector space $\hat{T}_1^{\otimes r}(G)'$ consisting of monomials of the operators $\tau_{i,i+1}$ and $\mathsf{K}_{i,i+1}$, $i = 1, 2, \cdots, r-1$. Replacing $\tau_{i,i+1}$ by $\hat{\mathsf{R}}(q)_{i,i+1}$ and $\mathsf{K}_{i,i+1}$ by $\mathsf{K}(q)_{i,i+1}$ in these monomials v_k we obtain vectors denoted by $v_k(q)$. Consider the matrix of coefficients of the set $\{v_k(q) \mid k = 1, 2, \cdots, m\}$ with respect to the standard basis $E_{i_1j_1} \otimes \cdots \otimes E_{i_rj_r}$ of $\mathcal{L}(V_1^{\otimes r})$. By (61) and (78), the entries of this matrix are Laurent polynomials in $q^{1/2}$ with integral coefficients. For $q = 1$, the vectors $v_k = v_k(1)$, $k = 1, 2, \cdots, m$, are linearly independent. Hence there is a regular $m \times m$ submatrix of the coefficient matrix. For general q, the determinant of this submatrix is a Laurent polynomial, say f, in $q^{1/2}$ with integral coefficients. Since $f(1) \neq 0$, f is nontrivial. Because q is transcendental, so is $q^{1/2}$ and hence $f(q^{1/2}) \neq 0$. Therefore, the vectors $\{v_k(q) \mid k = 1, 2, \cdots, m\}$ are linearly independent.

By Proposition 19, the matrix $\hat{\mathsf{R}}(q)$ and so the polynomial $\mathsf{K}(q)$ of $\hat{\mathsf{R}}(q)$ intertwine the representation $\tilde{T}_1 \otimes \tilde{T}_1$. This implies that $B(r) \subseteq \tilde{T}_1^{\otimes r}(\tilde{U}_q(\mathfrak{g}))'$.

Since the vectors $\{v_k(q) \,|\, k = 1, 2, \cdots, m\}$ of $B(r)$ are linearly independent and $m = \dim \tilde{T}_1^{\otimes r}(\tilde{U}_q(\mathfrak{g}))'$, as noted above, we conclude that $B(r) = \tilde{T}_1^{\otimes r}(\tilde{U}_q(\mathfrak{g}))'$. Hence $B(r)' = \tilde{T}_1^{\otimes r}(\tilde{U}_q(\mathfrak{g}))'' = \tilde{T}_1^{\otimes r}(\tilde{U}_q(\mathfrak{g}))$, where the second equality follows immediately from (118). □

8.6.4 Hecke and Birman–Wenzl–Murakami Algebras

In this subsection we introduce Hecke and Birman–Wenzl–Murakami algebras and show that the algebras $B(r)$ appearing in Theorem 38 are images of representations of these algebras.

Definition 6. *Let $q \in \mathbb{C}$, $q \neq 0$, and $r \in \mathbb{N}$, $r \geq 2$. The* Hecke algebra $H_r(q)$ *is the complex unital algebra with generators $g_1, g_2, \cdots g_{r-1}$ and defining relations*

$$g_i g_{i+1} g_i = g_{i+1} g_i g_{i+1}, \tag{119}$$

$$g_i g_j = g_j g_i \quad for \quad |i - j| \geq 2, \tag{120}$$

$$g_i^2 = (q - q^{-1}) g_i + 1. \tag{121}$$

In the case $q = 1$ the equations (119)–(121) are the defining relations for the permutation group $\mathcal{P}_r$. Hence $H_r(1)$ is just the group algebra $\mathbb{C}\mathcal{P}_r$.

Definition 7. *Let $p, q \in \mathbb{C}\backslash\{0\}$, $p, q \neq 0$, $q^2 \neq 1$, and $r \in \mathbb{N}$, $r \geq 2$. The* Birman–Wenzl–Murakami algebra $\mathrm{BWM}_r(p, q)$ *is the complex unital algebra with invertible generators $g_1, g_2, \cdots, g_{r-1}$ subject to the relations (119), (120) and*

$$e_i g_i = p^{-1} e_i, \tag{122}$$

$$e_i g_{i-1} e_i = p e_i, \quad e_i g_{i-1}^{-1} e_i = p^{-1} e_i, \tag{123}$$

where

$$e_i := 1 - (q - q^{-1})^{-1}(g_i - g_i^{-1}). \tag{124}$$

From the preceding relations it follows in particular that e_i is a complex multiple of a projection and that g_i satisfies a cubic equation. More precisely, we have

$$e_i^2 = (1 + (p - p^{-1})(q - q^{-1})^{-1}) e_i, \quad (g_i - q)(g_i + q^{-1})(g_i - p^{-1}) = 0.$$

With another set of relations the algebra $\mathrm{BWM}_r(p, q)$ can also be defined in the cases $q = \pm 1$ which have been excluded above.

In order to give vector space bases of both algebras, we consider the following sets of monomials:

$$M_n = \{1, g_n, g_n g_{n-1}, \cdots, g_n g_{n-1} \cdots g_1\},$$

$$M_{-n} = \{e_{1,n+1}, e_{2,n+1}, \cdots, e_{n,n+1}\}$$

for $n = 1, 2, \cdots, r - 1$, where

$$e_{ij} := g_{j-1}g_{j-2}\cdots g_{i+1}e_i g_{i+1}\cdots g_{j-1},\ i+1<j, \quad \text{and} \quad e_{i,i+1} := e_i.$$

Further, we set $M_0 = \{1\}$.

Proposition 39. (i) *The set* $B_r := \{x_1x_2\cdots x_{r-1} \mid x_i \in M_i\}$ *is a basis of the vector space* $H_r(q)$.

(ii) *The set* $B_r := \{x_{n_1}\cdots x_{n_r} | -r+1 \le n_1 < n_2 < \cdots < n_r \le r-1,\ n_i + n_j \neq 0$ if $i \neq j$, $x_{n_i} \in M_{n_i}\}$ *is a vector space basis of* $\mathrm{BWM}_r(p,q)$.

Proof. See [Bou2], pp. 54–56, [Wen1] or [HKW] for $H_r(q)$ and [BW] or [Wen2] for $\mathrm{BWM}_r(p,q)$. □

The above description of bases is recursive, that is, we have $B_{r+1} = B_r{\cdot}M_r$ for the Hecke algebras and $B_{r+1} = M_{-r}{\cdot}B_r \cup B_r{\cdot}M_r$ for the Birman–Wenzl–Murakami algebras. Since the sets M_r and M_{-r} consist of $r+1$ and r elements, respectively, it follows that

$$\dim H_r(q) = r! \quad \text{and} \quad \dim \mathrm{BWM}_r(p,q) = (2r-1)(2r-3)\cdots 3{\cdot}1.$$

Both algebras play an important role in knot theory (see, for instance, [HKW], [BW], [Wen2]). The interest in these algebras in the present context stems from the following

Proposition 40. *Let* $\hat{\mathsf{R}}$ *and* $\hat{\mathsf{R}}_{i,i+1}$ *be as in Subsects. 8.4.3 and 8.6.3. Set* $\epsilon := 1$ *for* so_N *and* $\epsilon := -1$ *for* sp_N. *There exists a unique representation* π_r *of the Hecke algebra* $H_r(q)$ *for* $\mathfrak{g} = \mathrm{sl}_N$ *and of the Birman–Wenzl–Murakami algebra* $\mathrm{BWM}_r(q, \epsilon q^{N-\epsilon})$ *for* $\mathfrak{g} = \mathrm{so}_N, \mathrm{sp}_N$ *on the vector space* $(\mathbb{C}^N)^{\otimes r}$ *such that* $\pi_r(g_i) = \hat{\mathsf{R}}_{i,i+1}$, $i = 1,2,\cdots,r-1$. *That is, the algebra* $B(r)$ *defined in Subsect. 8.6.3 is the image of* $H_r(q)$ *resp.* $\mathrm{BWM}_r(q,\epsilon q^{N-\epsilon})$ *under this representation* π_r.

Proof. We have to show that the matrices $\hat{\mathsf{R}}_{i,i+1}$, $i = 1,2,\cdots,r-1$, satisfy the defining relations of the algebras $H_r(q)$ and $\mathrm{BWM}_r(q,\epsilon q^{N-\epsilon})$, respectively. Since R satisfies the QYBE by Proposition 4, $\hat{\mathsf{R}}$ fulfills the braid relation (15). This in turn implies (119). Condition (120) is obviously fulfilled. Equation (121) for $\mathfrak{g} = \mathrm{sl}_N$ follows from (69). The relations (122) and (124) of the algebra $\mathrm{BWM}_r(q,\epsilon q^{N-\epsilon})$ for $\mathfrak{g} = \mathrm{so}_N, \mathrm{sp}_N$ follow from Propositions 25 and 26 and the second equality of (77). Relation (123) can be easily derived (for instance) from (64). □

8.7 Applications

The universal R-matrices are the main bridge connecting quantum groups with integrable systems. However, in the theory of integrable systems one needs R-matrices that depend on a spectral parameter. There is a procedure, called Baxterization, for obtaining solutions $R(\lambda)$ of the quantum Yang–Baxter equation from the R-matrices considered in previous sections. In this section we give a brief review of this method and the relations between R-matrices and integrable systems.

8.7.1 Baxterization

The parametrized quantum Yang–Baxter equation gives an approach to solvable (integrable) systems of statistical mechanics. A state of such a system is defined by a configuration at the vertices of a square lattice in the plane (for example, by spins). Each edge of the lattice is to be thought of as an interaction and contributes an energy $E(\sigma, \sigma')$ to the total energy, where σ and σ' are the spins at the ends of the edge. For instance, in the Ising model the total energy of a state σ is $\sum_{i,j}(k_1\sigma_{ij}\sigma_{i+1,j} + k_2\sigma_{ij}\sigma_{i,j+1})$. One of the most important characteristics of the system is given by the *partition function*

$$Z = \sum_{\sigma} \exp(-E(\sigma)/kT)$$

depending on parameters k_1, k_2, k, T. Usually one collects this dependence of the function Z by a single parameter called the *spectral parameter* λ.

The partition function $Z \equiv Z(\lambda)$ does not make sense for an infinite system. For that reason, one considers rectangular approximations of the system with N vertices. If Z_N is the corresponding partition function, then the free energy of a site is $F(\lambda) = \lim_{N\to\infty}(\log Z_N)/N$, where the approximating rectangles have to tend to the whole lattice. One needs to calculate $F(\lambda)$ as a function of λ which means the model is solved.

The usual technique of finding $F(\lambda)$ is the method of transfer matrices. A *transfer matrix* is a matrix $T(\lambda)$, depending on the horizontal dimension m of the approximating rectangle, such that $\mathrm{Tr}\,(T(\lambda)^n)$ is the partition function for the $m \times n$ rectangle. The limit of $F(\lambda)$ when $n \to \infty$ is given by the largest eigenvalue of $T(\lambda)$. But the size of $T(\lambda)$ grows exponentially with m and its diagonalization is not straightforward. R. Baxter proposed considering systems for which the transfer matrices commute among themselves for different values of λ. Then the matrices $T(\lambda)$ have a common eigenvector, say $T(\lambda)e = f(\lambda)e$. It turns out that one can then determine all possible functions $f(\lambda)$ including the largest one, that is, one can solve the model. In order to find systems for which the transfer matrices $T(\lambda)$ commute it is natural to factorize the matrix $T(\lambda)$ as $R_m(\lambda)R_{m-1}(\lambda)\cdots R_1(\lambda)$, where R_i corresponds to the contribution obtained by adding energy at the i-th position. Thus, it is necessary to look for local conditions on the matrices R_i, which guarantee that the matrices $T(\lambda)$ mutually commute. It can be shown (see [Bax]) that if for each λ and λ' there exists λ'' such that

$$R_i(\lambda)R_{i+1}(\lambda')R_i(\lambda'') = R_{i+1}(\lambda'')R_i(\lambda')R_{i+1}(\lambda), \tag{125}$$

$$R_i(\lambda)R_j(\lambda') = R_j(\lambda')R_i(\lambda), \quad |i-j| \geq 2, \tag{126}$$

then under certain boundary conditions we have $T(\lambda)T(\lambda') = T(\lambda')T(\lambda)$.

If the matrices R_i are independent of λ, then (125) is just the braid relation (15) which is known to be equivalent to the QYBE (6). The method of obtaining solutions $R_i(\lambda)$ of (125) from solutions of the constant(!) QYBE is called *Baxterization.* Recall from Subsects. 8.1.2 and 8.3.3 that representations of Drinfeld–Jimbo algebras give rise to solutions of the latter equation. In this subsection we carry out the Baxterization procedure for the R-matrices of the vector representations of the Drinfeld–Jimbo algebras $U_q(\mathfrak{g})$, $\mathfrak{g} = \mathrm{sl}_N, \mathrm{so}_N, \mathrm{sp}_N$.

First let R be the R-matrix for the vector representation of $U_q(\mathrm{sl}_N)$ from Proposition 24. By Proposition 19, $\hat{\mathsf{R}} := \tau \circ \mathsf{R}$ (recall that, as usual, τ is the flip operator) satisfies the braid relation

$$\hat{\mathsf{R}}_{12}\hat{\mathsf{R}}_{23}\hat{\mathsf{R}}_{12} = \hat{\mathsf{R}}_{23}\hat{\mathsf{R}}_{12}\hat{\mathsf{R}}_{23}, \tag{127}$$

which is equivalent to the QYBE (6). We shall consider the Baxterization of this R-matrix when the parameters λ, λ' and λ'' are related by the condition $\lambda'' = \lambda^{-1}\lambda'$. Then the Baxterized R-matrix is a solution of the equation

$$\hat{\mathsf{R}}_{12}(x)\hat{\mathsf{R}}_{23}(xy)\hat{\mathsf{R}}_{12}(y) = \hat{\mathsf{R}}_{23}(y)\hat{\mathsf{R}}_{12}(xy)\hat{\mathsf{R}}_{23}(x). \tag{128}$$

We look for solutions of this equation of the form

$$\hat{\mathsf{R}}(x) = g(x)(I + f(x)\hat{\mathsf{R}}),$$

where $f(x)$ and $g(x)$ are functions of the parameter x. Inserting this expression into (128), and using the braid relation (127) and the Hecke condition $\hat{\mathsf{R}}^2 = (q - q^{-1})\hat{\mathsf{R}} + I$ for the matrix $\hat{\mathsf{R}}$ (see (69)), (128) turns out to be equivalent to the equation

$$f(x) + f(y) + (q - q^{-1})f(x)f(y) = f(xy) \tag{129}$$

for the function f. The function $g(x)$ is arbitrary. Substituting $f(x) = (q - q^{-1})^{-1}(\tilde{f}(x) - 1)$, we find that the general solution of equation (129) is given by $f(x) = (q - q^{-1})^{-1}(x^\gamma - 1)$, where γ is an arbitrary parameter. An important particular solution is obtained for $\gamma = -2$ and $g(x) = (q - q^{-1})x$. In this case we get

$$\hat{\mathsf{R}}(x) = x^{-1}\hat{\mathsf{R}} - x\hat{\mathsf{R}}^{-1}.$$

Now we turn to the Drinfeld–Jimbo algebras $U_q(\mathrm{so}_N)$ and $U_q(\mathrm{sp}_N)$. In this case we represent the Baxterized matrix $\hat{\mathsf{R}}(x)$ in the form

$$\hat{\mathsf{R}}(x) = h(x)\big(I + f(x)\hat{\mathsf{R}} + g(x)\mathsf{K}\big),$$

where $\hat{\mathsf{R}}$ and K are given by (61), (78) and (81), respectively. As in the previous case, inserting this expression into (128) we find that

$$\hat{\mathsf{R}}(x) = h(x)\Big(I + \frac{x^\gamma - 1}{q - q^{-1}}\hat{\mathsf{R}} + \frac{x^\gamma + 1}{\alpha_\pm x^\gamma + 1}\mathsf{K}\Big),$$

where $h(x)$ is an arbitrary function, γ is a parameter and $\alpha_\pm = \pm\epsilon q^{N-\epsilon\pm 1}$. In the special case when $h(x) = (q - q^{-1})x$ and $\gamma = -2$ we have

$$\begin{aligned}\hat{\mathsf{R}}(x) &= x^{-1}\hat{\mathsf{R}} - x\hat{\mathsf{R}}^{-1} + \frac{(q-q^{-1})(\alpha_\pm + 1)}{\alpha_\pm x^{-1}+1}\mathsf{K} \\ &= \frac{x-x^{-1}}{x+\alpha_\pm x^{-1}}\Big(-x\hat{\mathsf{R}}^{-1} - \alpha_\pm x^{-1}\hat{\mathsf{R}} + \frac{(q-q^{-1})(\alpha_\pm+1)}{x-x^{-1}}\Big) \\ &= (x^{-1}q - xq^{-1})\mathsf{P}_+ + (xq-(xq)^{-1})\mathsf{P}_- + a(x)\mathsf{P}_0,\end{aligned}$$

where $a(x) = (x+\alpha_\pm x^{-1})^{-1}(\alpha_\pm + x^{-1})(\epsilon q^{\epsilon - N}\alpha_\pm + (q-q^{-1} - \epsilon q^{\epsilon N}\alpha_\pm)x^2)$ and ϵ, P_+, P_-, P_0 are as in Subsect. 8.4.3. The last equation determines the spectral decomposition of $\hat{\mathsf{R}}(x)$.

Another relation between the values λ, λ' and λ'' is obtained under the following change of parameters:

$$x = e^{(q^{-1}-q)(\theta-\theta')/2}, \qquad y = e^{(q^{-1}-q)\theta'/2}. \tag{130}$$

Then instead of (128) we have the equation

$$\hat{\mathsf{R}}_{12}(\theta-\theta')\hat{\mathsf{R}}_{23}(\theta)\hat{\mathsf{R}}_{12}(\theta') = \hat{\mathsf{R}}_{23}(\theta')\hat{\mathsf{R}}_{12}(\theta)\hat{\mathsf{R}}_{23}(\theta-\theta'). \tag{131}$$

The substitution (130) transforms a matrix $\hat{\mathsf{R}}(x)$ satisfying (128) to a matrix $\hat{\mathsf{R}}(\theta)$ that fulfills (131).

8.7.2 Elliptic Solutions of the Quantum Yang–Baxter Equation

The previous solutions of the QYBE are called *rational* (dependence on x) and *trigonometric* (dependence on θ). Elliptic solutions are expressed in terms of elliptic functions of the spectral parameter. In this subsection we shall give elliptic solutions which are symmetric with respect to the commutative group $\mathbb{Z}_p \otimes \mathbb{Z}_p$, $p \in \mathbb{N}$, where $\mathbb{Z}_p := \mathbb{Z}/p\mathbb{Z}$.

Let X and Z be the matrices from Subsect. 7.5.2, where $\epsilon = \exp(2\pi \mathrm{i}/p)$. We set

$$T(\alpha) \equiv T(\alpha_1, \alpha_2) := Z^{\alpha_1}X^{\alpha_2}, \quad \alpha_1, \alpha_2 = 0, 1, 2, \cdots, p-1.$$

Clearly, this equation defines a projective representation of the group $\mathbb{Z}_p \otimes \mathbb{Z}_p$, that is, we have $T(\alpha)T(\beta) = \epsilon^{\alpha_2\beta_1}T(\alpha+\beta)$.

Since the algebra generated by X and Z coincides with the matrix algebra $M_p(\mathbb{C})$, any solution of the QYBE on $\mathbb{C}^p \otimes \mathbb{C}^p$ can be written as $R(\theta) = w_{\alpha\beta}(\theta)T(\alpha)\otimes T(\beta)$ with $w_{\alpha\beta} \in \mathbb{C}$. We are looking for solutions of the following form

$$R(\theta) = w_\alpha(\theta)T(\alpha) \otimes T(\alpha)^{-1},\ w_\alpha(\theta) \in \mathbb{C}. \tag{132}$$

One immediately checks that such solutions are invariant under the projective representation T of the group $\mathbb{Z}_p \otimes \mathbb{Z}_p$ in the sense that they satisfy the condition

$$R(\theta)(T\otimes T)(\alpha) = (T\otimes T)(\alpha)R(\theta) \quad \text{for} \quad \alpha \in \mathbb{Z}_p \otimes \mathbb{Z}_p.$$

In order to obtain elliptic solutions we impose the additional conditions

$$R(\theta+1)=Z_1^{-1}R(\theta)Z_1=Z_2R(\theta)Z_2^{-1},\quad R(0)=T_\alpha\otimes T_\alpha^{-1}, \tag{133}$$

$$R(\theta+\tau)=e^{-\mathrm{i}\pi\tau}e^{-2\pi\mathrm{i}\theta}X_1^{-1}R(\theta)X_1=e^{-\mathrm{i}\pi\tau}e^{-2\pi\mathrm{i}\theta}X_2R(\theta)X_2^{-1}, \tag{134}$$

where τ is a complex parameter. These equations are in fact consistent with the quantum Yang–Baxter equation. If we substitute (132) into (133) and (134), we obtain the equations

$$w_\alpha(\theta+1)=\epsilon^{\alpha_2}w_\alpha(\theta),\quad w_\alpha(\theta+\tau)=e^{-\mathrm{i}\pi\tau}e^{-2\pi\mathrm{i}\theta}\epsilon^{-\alpha_1}w_\alpha(\theta),\quad w_\alpha(0)=1.$$

These equations are satisfied by the elliptic function

$$w_\alpha(\theta)=\frac{\Theta_\alpha(\theta+\eta)}{\Theta_\alpha(\eta)},$$

where η is an arbitrary parameter and $\Theta_\alpha(\theta)$ is the following Θ-function:

$$\Theta_\alpha(\theta)=\sum_{m=-\infty}^{\infty}\exp\Big(\mathrm{i}\pi\tau\Big(m+\frac{\alpha_2}{N}\Big)^2+2\pi\mathrm{i}\Big(m+\frac{\alpha_2}{N}\Big)\Big(\theta+\frac{\alpha_1}{N}\Big)\Big).$$

8.7.3 R-Matrices and Integrable Systems

In order to construct integrable systems with periodic boundary conditions corresponding to a given Baxterized R-matrix one uses relation (131) written in the equivalent form

$$R_{12}(\theta-\theta')R_{13}(\theta)R_{23}(\theta')=R_{23}(\theta')R_{13}(\theta)R_{12}(\theta-\theta'), \tag{135}$$

where $R_{ij}:=\tau\circ\hat{R}_{ij}$. We construct a transfer matrix $(T(\lambda)^i_j)$ by setting

$$T(\theta)^{i\,k_1\,k_2\,\cdots\,k_m}_{j\,l_1\,l_2\,\cdots\,l_m}=\sum\nolimits_{s_1,\cdots,s_{m-1}}R(\theta)^{i\,k_1}_{s_1\,l_1}R(\theta)^{s_1\,k_2}_{s_2\,l_2}\cdots R(\theta)^{s_{m-1}k_m}_{j\,l_m}.$$

Here m denotes the number of rows in the rectangle. The matrix entries $T(\theta)^i_j$ are linear transformations of $\mathcal{L}(V_1^{\otimes m})$. The partition function $Z_{R(\theta)}$ is defined by

$$Z_{R(\theta)}=\mathrm{Tr}_{V_1^{\otimes m}}(\mathrm{Tr}\,T(\theta))^n,$$

where n is the number of columns of the rectangle.

If we represent $T(\theta)$ as a product of R-matrices $R_i(\theta)$ and use the quantum Yang–Baxter equation, we easily derive the relation

$$\sum\nolimits_{s,r}R(\theta-\theta')^{i\,k}_{s\,r}T(\theta)^s_jT(\theta')^r_l=\sum\nolimits_{s,r}T(\theta')^k_rT(\theta)^i_sR(\theta-\theta')^{s\,r}_{j\,l}.$$

The latter can be expressed by the matrix equation

$$R(\theta-\theta')_{12}T_1(\theta)T_2(\theta')=T_2(\theta')T_1(\theta)R(\theta-\theta')_{12} \tag{136}$$

in the vector space $V_1 \otimes V_2 \otimes V_1^{\otimes m}$, where $T_1 := R_{13}R_{14}\cdots R_{1,m+2}$ and $T_2 := R_{23}R_{24}\cdots R_{2,m+2}$. Equation (136) is called the *fundamental relation* of the quantum inverse scattering method.

Now we multiply both sides of (136) on the right by $R(\theta-\theta')_{12}^{-1}$ and take the trace Tr in $\mathcal{L}(V_1 \otimes V_1)$. Then we obtain

$$\begin{aligned}\mathrm{Tr}_{V_1\otimes V_1} T_2(\theta')T_1(\theta) &= \mathrm{Tr}_{V_1\otimes V_1}(R(\theta-\theta')_{12}T_1(\theta)T_2(\theta')R(\theta-\theta')_{12}^{-1}) \\ &= \mathrm{Tr}_{V_1\otimes V_1} T_1(\theta)T_2(\theta'),\end{aligned}$$

that is, we have

$$\mathrm{Tr}\, T_1(\theta)\mathrm{Tr}\, T_2(\theta') = \mathrm{Tr}\, T_2(\theta')\mathrm{Tr}\, T_1(\theta).$$

This means that the system is solvable.

To every two-dimensional lattice system of statistical mechanics there corresponds an associated one-dimensional quantum chain. This is a quantum system for a particle in the space consisting of a chain of n sites. In the continuum limit $n \to \infty$ we obtain a $(1+1)$-dimensional quantum integrable system. This quantum system is the quantization of an integrable system of the equations of motion in 1+1 dimensions. In such a way we also obtain the conformal field theory.

As an example, let us consider the elliptic R-matrix

$$R(\theta) = \begin{pmatrix} \dfrac{\mathrm{sn}\,(\theta\tau+\tau)}{\mathrm{sn}\,\tau} & 0 & 0 & k\mathrm{sn}\,(\theta\tau)\mathrm{sn}\,(\theta\tau+\tau) \\ 0 & \dfrac{\mathrm{sn}\,(\theta\tau)}{\mathrm{sn}\,\tau} & 1 & 0 \\ 0 & 1 & \dfrac{\mathrm{sn}\,(\theta\tau)}{\mathrm{sn}\,\tau} & 0 \\ k\mathrm{sn}\,(\theta\tau)\mathrm{sn}\,(\theta\tau+\tau) & 0 & 0 & \dfrac{\mathrm{sn}\,(\theta\tau+\tau)}{\mathrm{sn}\,\tau} \end{pmatrix},$$

where τ is a complex parameter and $\mathrm{sn}\,\tau$ is the Jacobi elliptic function depending on the parameter k. This R-matrix describes the so-called eight vertex or XYZ model. The term XYZ originates from the existence of anisotropies in the associated quantum model in all three space dimensions. In the limit $k \to 0$ we have $\mathrm{sn}\,\tau \to \sin\tau$ and then $R(\theta)$ describes the trigonometric six vertex or XXZ model. In the limit $\tau \to 0$ we obtain the R-matrix for the rational six vertex or XXX model.

Equation (135) can be rewritten as

$$S_{23}(\theta-\theta')S_{13}(\theta)S_{12}(\theta') = S_{12}(\theta')S_{13}(\theta)S_{23}(\theta-\theta').$$

Under the conditions of unitarity and crossing symmetry

$$S_{12}(\theta)S_{21}(-\theta) = I, \qquad S_{12} = (S_{21}(\mathrm{i}\pi-\theta))^{t_1},$$

this equation determines factorized S-matrices describing the scattering of particle-like excitations in $(1+1)$-dimensional integrable relativistic models. The term $S_{j_1j_2}^{i_1i_2}(\theta)$ is interpreted as the S-matrix for the scattering of two particles with isotopic spins i_1 and i_2 to two particles with isotopic spins j_1 and j_2, and the spectral parameter θ is the difference of the rapidities of these particles. The many-particle S-matrices in such models factorize into products of two-particle S-matrices.

8.8 Notes

The fundamental concepts of a quasitriangular Hopf algebra and of the quantum double and their basic properties were established by V. G. Drinfeld [Dr1], [Dr2]. Various general approaches to the double have been given by S. Majid (see [Maj1] and [Maj], Sect. 10.2). The importance of the quantum double for the construction of universal R-matrices was also observed by Drinfeld who obtained the universal R-matrix for $U_h(\mathrm{sl}_2)$ and also Theorem 9. The universal R-matrix for $U_h(\mathrm{sl}_N)$ was computed in [Ros2]. The formula (44) for the universal R-matrices of arbitrary h-adic Drinfeld–Jimbo algebras was determined in [KR2] and [LS1].

The R-matrices and L-functionals for the vector representations of the Drinfeld–Jimbo algebras and their properties have been derived and studied in [RTF]. The construction of R-matrices for fixed q given in Subsect. 8.3.3 is due to T. Tanisaki [Tan3]. The corresponding approach to L-functionals for fixed q developed in Sect. 8.5 appears here for the first time. The results on the Brauer–Schur–Weyl duality in Subsect. 8.6.3 are taken from [H3]. The Birman–Wenzl–Murakami algebra was invented independently in [Mur] and [BW].

The Yang–Baxter equation arose in the work of C. N. Yang [Yan] and R. J. Baxter [Bax] on exactly solvable models in statistical mechanics. Finding solutions of this equation was one of the main reasons for the development of quantum groups. There is now an extensive literature on solutions of the QYBE and on its algebraic aspects (see, for instance, [DV], [Gur], [Hi], [Is1], [Jim5], [LR], [Maj1], [Rad3], [Res1], [S1]). The braid relation first came up in E. Artin's theory of braids [Ar].

Part III

Quantized Algebras of Functions

9. Coordinate Algebras of Quantum Groups and Quantum Vector Spaces

With this chapter we take up the detailed study of quantized algebras of functions. The rest of this book will be mainly devoted to this topic, although many results are obtained for general Hopf algebras. The aim of this chapter is to develop basic definitions and facts on the standard quantized simple Lie groups and their quantum vector spaces.

Following common terminology in the physics literature, let us adopt the convention from now on to sum over repeated indices belonging to different terms. However, if confusion is possible or if a formula looks too unfamiliar, we shall write the sum sign.

9.1 The Approach of Faddeev–Reshetikhin–Takhtajan

9.1.1 The FRT Bialgebra $\mathcal{A}(R)$

Let V be a finite-dimensional complex vector space and let R be a linear mapping of $V \otimes V$ to itself. Let $\hat{R} := \tau \circ R$, where τ is the flip of $V \otimes V$.

We fix a basis $\{x_i\}_{i=1,2,\cdots,N}$ of V. Then there are complex numbers $R^{ij}_{nm} = \hat{R}^{ji}_{nm}$ such that

$$R(x_i \otimes x_j) = R^{nm}_{ij} x_n \otimes x_m \quad \text{and} \quad \hat{R}(x_i \otimes x_j) = \hat{R}^{nm}_{ij} x_n \otimes x_m.$$

Let $\mathbb{C}\langle u^i_j \rangle$ denote the free algebra with N^2 generators $u^i_j, i, j = 1, 2, \cdots, N$, and let $\mathcal{J}(R)$ be the two-sided ideal of $\mathbb{C}\langle u^i_j \rangle$ generated by the N^4 elements

$$I^{ij}_{mn} := R^{ji}_{kl} u^k_m u^l_n - u^i_k u^j_l R^{lk}_{mn}, \qquad i, j, m, n = 1, 2, \cdots, N.$$

Let $\mathcal{A}(R)$ denote the quotient algebra $\mathbb{C}\langle u^i_j \rangle / \mathcal{J}(R)$, that is, $\mathcal{A}(R)$ is the algebra generated by elements u^i_j, $i, j = 1, 2, \cdots, N$, subject to the relations

$$R^{ji}_{kl} u^k_m u^l_n = u^i_k u^j_l R^{lk}_{mn}, \qquad i, j, m, n = 1, 2, \cdots, N. \tag{1}$$

The $N \times N$ matrix (u^i_j) of the generators u^i_j is called the *fundamental matrix* of $\mathcal{A}(R)$ and is denoted by $\mathbf{u}$. In matrix notation the defining relations (1) of $\mathcal{A}(R)$ can be reformulated as

$$R\mathbf{u}_1\mathbf{u}_2 = \mathbf{u}_2\mathbf{u}_1 R \tag{2}$$

or equivalently as

$$\hat{R}\mathbf{u}_1\mathbf{u}_2 = \mathbf{u}_1\mathbf{u}_2\hat{R}, \tag{3}$$

where $\mathbf{u}_1$ and $\mathbf{u}_2$ are the $N^2 \times N^2$ matrices defined by $\mathbf{u}_1 = \mathbf{u} \otimes I$ and $\mathbf{u}_2 = I \otimes \mathbf{u}$ (Kronecker product of matrices). Then all products in (2) and (3) are just matrix products. The reader should notice that $\mathbf{u}_1\mathbf{u}_2 \neq \mathbf{u}_2\mathbf{u}_1$ unless $\mathcal{A}(R)$ is commutative. In this matrix notation we shall write the algebra $\mathcal{A}(R)$ as

$$\mathcal{A}(R) = \mathbb{C}\langle u^i_j\rangle / \langle \hat{R}\mathbf{u}_1\mathbf{u}_2 - \mathbf{u}_1\mathbf{u}_2\hat{R}\rangle. \tag{4}$$

Similar notation as in (2), (3) and (4) will often be used in the following.

From the defining relations it is clear that $\mathcal{A}(\tau \circ R \circ \tau)$ is the opposite algebra $\mathcal{A}(R)^{\mathrm{op}}$ and that $\mathcal{A}(\tau \circ R^{-1} \circ \tau) = \mathcal{A}(R)$ if R is invertible.

Example 1 ($R =$ id *and* $R = \tau$). If R is the identity map, then (1) reads as $u^j_m u^i_n = u^i_n u^j_m$ and $\mathcal{A}(R)$ is the *commutative* polynomial algebra in N^2 indeterminates, that is, $\mathcal{A}(R)$ is the algebra of coordinate functions of the matrix algebra $M_N(\mathbb{C})$. At the other extreme, if R is the flip τ, then $\mathcal{A}(R)$ is the free algebra $\mathbb{C}\langle u^i_j\rangle$ with generators u^i_j. △

Proposition 1. *There is a unique bialgebra structure on the algebra $\mathcal{A}(R)$ such that*

$$\Delta(u^i_j) = u^i_k \otimes u^k_j \quad \textit{and} \quad \varepsilon(u^i_j) = \delta_{ij}, \quad i,j = 1,2,\cdots,N. \tag{5}$$

In matrix form the equations (5) are commonly written as $\Delta(\mathbf{u}) = \mathbf{u}\otimes\mathbf{u}$ and $\varepsilon(\mathbf{u}) = I$.

Proof. Since $\mathbb{C}\langle u^i_j\rangle$ is the free algebra with generators u^i_j, there are unique algebra homomorphisms $\Delta : \mathbb{C}\langle u^i_j\rangle \to \mathbb{C}\langle u^i_j\rangle \otimes \mathbb{C}\langle u^i_j\rangle$ and $\varepsilon : \mathbb{C}\langle u^i_j\rangle \to \mathbb{C}$ such that (5) holds. Then $\mathbb{C}\langle u^i_j\rangle$ becomes a bialgebra with comultiplication Δ and counit ε. Indeed, by Proposition 1.8, it suffices to check the axioms (1.15) and (1.16) for the generators u^i_j which is easily done.

Since $\Delta(I^{ij}_{mn}) = I^{ij}_{kl} \otimes u^k_m u^l_n + u^i_k u^j_l \otimes I^{kl}_{mn}$ and $\varepsilon(I^{ij}_{mn}) = 0$, we get $\Delta(\mathcal{J}(R)) \subseteq \mathcal{J}(R) \otimes \mathbb{C}\langle u^i_j\rangle + \mathbb{C}\langle u^i_j\rangle \otimes \mathcal{J}(R)$ and $\varepsilon(\mathcal{J}(R)) = \{0\}$. That is, $\mathcal{J}(\mathcal{R})$ is a biideal, so the quotient algebra $\mathcal{A}(R) = \mathbb{C}\langle u^i_j\rangle / \mathcal{J}(R)$ is also a bialgebra. □

Definition 1. *The bialgebra $\mathcal{A}(R)$ from Proposition 1 is called* the coordinate algebra of the quantum matrix space *associated with the linear transformation R or, briefly, the* FRT bialgebra.

Of course, the preceding definition of the bialgebra $\mathcal{A}(R)$ depends on the matrix coefficients R^{ij}_{nm} of R and so on the basis $\{x_i\}$ of V. The next proposition gives a basis-free characterization of the bialgebra $\mathcal{A}(R)$. In particular, this implies that the bialgebra $\mathcal{A}(R)$ is uniquely determined up to isomorphism by the linear transformation R which justifies the notation $\mathcal{A}(R)$.

Proposition 2. *There is a linear map* $\varphi_V : V \to V \otimes \mathcal{A}(R)$ *such that* V *is a right comodule of* $\mathcal{A}(R)$ *with coaction* φ_V *and* $\hat{R} \in \mathrm{Mor}\,(\varphi_V \otimes \varphi_V)$. *If* $\tilde{\mathcal{A}}$ *is another bialgebra and* $\psi_V : V \to V \otimes \tilde{\mathcal{A}}$ *is a right coaction of* $\tilde{\mathcal{A}}$ *on* V *such that* $\hat{R} \in \mathrm{Mor}\,(\psi_V \otimes \psi_V)$, *then there exists a unique bialgebra homomorphism* $\theta : \mathcal{A}(R) \to \tilde{\mathcal{A}}$ *such that* $(\mathrm{id} \otimes \theta) \circ \varphi_V = \psi_V$.

Proof. By Proposition 1.13 and (5), the linear map $\varphi_V : V \to V \otimes \mathcal{A}(R)$ defined by

$$\varphi_V(x_i) = x_j \otimes u^j_i, \quad i = 1, 2, \cdots, N, \tag{6}$$

is a right coaction of $\mathcal{A}(R)$ on V. Using (1), we obtain

$$\begin{aligned}
(\varphi_V \otimes \varphi_V)\hat{R}(x_m \otimes x_n) &= (\varphi_V \otimes \varphi_V)(\hat{R}^{kl}_{mn} x_k \otimes x_l) \\
&= x_i \otimes x_j \otimes u^i_k u^j_l R^{lk}_{mn} = x_i \otimes x_j \otimes R^{ji}_{kl} u^k_m u^l_n \\
&= (\hat{R} \otimes \mathrm{id})(x_k \otimes x_l \otimes u^k_m u^l_n) \\
&= (\hat{R} \otimes \mathrm{id})(\varphi_V \otimes \varphi_V)(x_m \otimes x_n),
\end{aligned}$$

so that $\hat{R} \in \mathrm{Mor}\,(\varphi_V \otimes \varphi_V)$. This proves the first part of the proposition.

To prove the second assertion, let $\tilde{\mathcal{A}}$ and ψ_V be as above. There are elements $t^i_j \in \tilde{\mathcal{A}}$, $i, j = 1, 2, \cdots, N$, such that $\psi_V(x_i) = x_j \otimes t^j_i$. Since ψ_V is a right coaction, we have $\Delta(t^i_j) = t^i_k \otimes t^k_j$ and $\varepsilon(t^i_j) = \delta_{ij}$ by Proposition 1.13. From the computation carried out in the first part of this proof it follows that $\hat{R} \in \mathrm{Mor}\,(\psi_V \otimes \psi_V)$ implies that $\hat{R}^{ij}_{kl} t^k_m t^l_n = t^i_k t^j_l \hat{R}^{kl}_{mn}$ for $i, j = 1, 2, \cdots, N$. Therefore, the map $\theta : u^i_j \to t^i_j$ extends to an algebra homomorphism θ of $\mathcal{A}(R)$ to $\tilde{\mathcal{A}}$ which obviously satisfies the condition $(\mathrm{id} \otimes \theta) \circ \varphi_V = \psi_V$. On the other hand, the latter relation implies that $\theta(u^i_j) = t^i_j$, hence the uniqueness assertion follows. □

By (6), the elements u^i_j are the matrix coefficients of the corepresentation φ_V with respect to the basis $\{x_i\}$. We call the coaction φ_V and likewise the matrix $\mathbf{u} = (u^i_j)$ the *fundamental corepresentation* of the FRT bialgebra $\mathcal{A}(R)$. A similar terminology will be used for the quantum groups $GL_q(N), SL_q(N), O_q(N), SO_q(N)$, and $Sp_q(N)$ which are derived below from FRT bialgebras $\mathcal{A}(R)$.

Remark 1. The FRT bialgebra $\mathcal{A}(R)$ is a special case of the following more general construction. Let $\mathcal{C}$ be an arbitrary coalgebra and let $\mathbf{r}$ be a linear functional on $\mathcal{C} \otimes \mathcal{C}$. Let $\mathcal{M}(\mathcal{C}, \mathbf{r})$ be the quotient algebra of the tensor algebra $T(\mathcal{C})$ by the two-sided ideal $\mathcal{J}$ generated by the elements

$$I(x, y) := \sum \left(\mathbf{r}(x_{(1)} \otimes y_{(1)}) x_{(2)} \otimes y_{(2)} - y_{(1)} \otimes x_{(1)} \mathbf{r}(x_{(2)} \otimes y_{(2)}) \right), \quad x, y \in \mathcal{C}.$$

By the universal property of $T(\mathcal{C})$, the comultiplication and the counit of $\mathcal{C}$ extend uniquely to algebra homomorphisms $\Delta : T(\mathcal{C}) \to T(\mathcal{C}) \otimes T(\mathcal{C})$ and $\varepsilon : T(\mathcal{C}) \to \mathbb{C}$. Then, $T(\mathcal{C})$ becomes a bialgebra. Since $\varepsilon(I(x, y)) = 0$ and

$$\Delta(I(x, y)) = \sum \left(I(x_{(1)}, y_{(1)}) \otimes x_{(2)} \otimes y_{(2)} + y_{(1)} \otimes x_{(1)} \otimes I(x_{(2)}, y_{(2)}) \right),$$

$\mathcal{J}$ is a biideal of $T(\mathcal{C})$. Hence $\mathcal{M}(\mathcal{C}, \mathbf{r})$ *is a bialgebra* with structures induced from $T(\mathcal{C})$.

Let us consider the special case $\mathcal{C} = M_N(\mathbb{C})'$, see Example 1.4. Let $u_l^k \in \mathcal{C}$ be given by $u_l^k(g) := g_{kl}$ for $g = (g_{kl}) \in M_N(\mathbb{C})$. As above, let R be a linear transformation of $V \otimes V$ with matrix coefficients R^{ij}_{mn} and define $\mathbf{r} \in (\mathcal{C} \otimes \mathcal{C})'$ by $\mathbf{r}(u_m^i \otimes u_n^j) := R^{ij}_{mn}$. Then the bialgebra $\mathcal{M}(\mathcal{C}, \mathbf{r})$ is nothing but the FRT bialgebra $\mathcal{A}(R)$. △

We close this subsection by studying $*$-structures on FRT bialgebras $\mathcal{A}(R)$ and on Hopf algebra quotients of it. Recall that a $*$-bialgebra (resp. a Hopf $*$-algebra) is a bialgebra (resp. a Hopf algebra) $\mathcal{A}$ which is a $*$-algebra such that $\Delta(a^*) = \Delta(a)^*$ and $\varepsilon(a^*) = \overline{\varepsilon(a)}$ for $a \in \mathcal{A}$.

Suppose that the vector space V is equipped with an involution, that is, with an anti-linear mapping $v \to v^*$ such that $(v^*)^* = v$ for $v \in V$. Define an involution on $V \otimes V$ by $(v \otimes w)^* := v^* \otimes w^*$ and assume that the basis elements x_i of V satisfy $x_i^* = x_i$. Then the transformation R of $V \otimes V$ is called *real* if $R((v \otimes w)^*) = (R(v \otimes w))^*, v, w \in V$, and *inverse real* if $R((v \otimes w)^*) = (R^{-1}(v \otimes w))^*$, $v, w \in V$. Clearly, R is real if and only if $\overline{R^{mn}_{ij}} = R^{mn}_{ij}$ for all i,j,m,n (that is, $\bar{R} = R$) and R is *inverse real* if and only if $\overline{R^{mn}_{ij}} = (R^{-1})^{mn}_{ij}$ for all i,j,m,n (that is, $\bar{R} = R^{-1}$).

Since $\mathbb{C}\langle u_j^i \rangle$ is the free algebra with generators u_j^i, there is a unique anti-linear multiplicative mapping θ of $\mathbb{C}\langle u_j^i \rangle$ (that is, $\theta(\lambda a + \mu b) = \bar{\lambda}\theta(a) + \bar{\mu}\theta(b)$, $\theta(ab) = \theta(a)\theta(b)$ for $\lambda, \mu \in \mathbb{C}$, $a, b \in \mathbb{C}\langle u_j^i \rangle$) such that $\theta(u_j^i) = u_i^j$. The next proposition will be used for the treatment of real forms in Subsects. 9.2.4 and 9.3.5 below.

Proposition 3. (i) *Suppose that R is real. Let $\mathcal{I}$ be a subset of $\mathbb{C}\langle u_j^i \rangle$ such that the two-sided ideal $\mathcal{J} := \langle \mathcal{J}(R), \mathcal{I} \rangle$ of $\mathbb{C}\langle u_j^i \rangle$ is a biideal and the bialgebra $\mathcal{A} = \mathbb{C}\langle u_j^i \rangle)/\mathcal{J}$ is a Hopf algebra. Assume that $\theta(\mathcal{I}) \subseteq \mathcal{J}$. Let ψ be an automorphism of the Hopf algebra $\mathcal{A}$ such that $\psi^2 = \mathrm{id}$. Then $\mathcal{A}$ is a Hopf $*$-algebra with involution determined by $(u_j^i)^* = S(\psi(u_i^j))$, $i, j = 1, 2, \cdots, N$.*

(ii) *If R is inverse real, then $\mathcal{A}(R)$ is a $*$-bialgebra with involution defined by $(u_j^i)^* = u_j^i$, $i, j = 1, 2, \cdots, N$.*

Proof. (i): We first prove the assertion in the case $\psi = \mathrm{id}$. Define an anti-linear anti-multiplicative mapping $\vartheta : \mathbb{C}\langle u_j^i \rangle \to \mathcal{A}$ by $\vartheta(u_j^i) = S(u_i^j)$. Let $I : \mathbb{C}\langle u_j^i \rangle \to \mathcal{A} = \mathbb{C}\langle u_j^i \rangle / \mathcal{J}$ be the canonical map. Using the assumption that R is real and the properties of S, θ and ϑ we get $\vartheta(I^{ij}_{nm}) = S(I(\theta(I^{ij}_{nm}))) = 0$, so $\vartheta(\mathcal{J}(R)) = \{0\}$, and $\vartheta(x) = S(I(\theta(x)))$ for $x \in \mathbb{C}\langle u_j^i \rangle$. Since $\theta(\mathcal{I}) \subseteq \mathcal{J}$ by assumption, it follows that $\vartheta(\mathcal{J}) = \{0\}$. Hence ϑ passes to an anti-linear anti-multiplicative map $a \to a^* := \vartheta(a)$ of $\mathcal{A} = \mathbb{C}\langle u_j^i \rangle / \mathcal{J}$ to $\mathcal{A}$. By definition we then have $(u_j^i)^* = S(u_i^j)$. For $\mathcal{A}$ being a Hopf $*$-algebra we still have to show that $a^{**} = a$ for $a \in \mathcal{A}$ and that Δ and ε are $*$-preserving. It suffices to do this on the generators u_j^i. The relations $\Delta((u_j^i)^*) = \Delta(u_j^i)^*$ and $\varepsilon((u_j^i)^*) = \delta_{ij}$ are easily verified. Applying the $*$-operation to the identity $\sum_n (u_i^n)^* u_j^n =$

$\sum_n S(u^i_n)u^n_j = \delta_{ij}$ we get $\sum_n (u^n_j)^*(u^n_i)^{**} = \sum_n S(u^j_n)(u^n_i)^{**} = \delta_{ij}$. This implies that $(u^n_i)^{**} = u^n_i$ and completes the proof in the case $\psi = \mathrm{id}$.

Let us turn to the general case. Put $a^{*\prime} := \psi(a)^*$, $a \in \mathcal{A}$. Since ψ is a Hopf algebra automorphism, $\psi \circ S = S \circ \psi$ and hence $\psi(a^*) = \psi(a)^*$, $a \in \mathcal{A}$. Using the latter and the assumption $\psi^2 = \mathrm{id}$ one verifies that $(\mathcal{A}, *')$ is also a Hopf $*$-algebra.

(ii): The bialgebra $\mathbb{C}\langle u^i_j\rangle$ is a $*$-bialgebra with involution defined by $(u^i_j)^* = u^i_j$. By $\bar{R} = R^{-1}$, we obtain $(I^{ij}_{nm})^* = -\sum_{k,l,r,s}(R^{-1})^{ji}_{kl} I^{kl}_{rs}(R^{-1})^{rs}_{nm}$, so the involution of $\mathbb{C}\langle u^i_j\rangle$ passes to the quotient $\mathcal{A}(R) = \mathbb{C}\langle u^i_j\rangle/\mathcal{J}(R)$. □

9.1.2 The Quantum Vector Spaces $\mathcal{X}_L(f;R)$ and $\mathcal{X}_R(f;R)$

We retain the notation from the preceding subsection. Let $f = \{f_1, \cdots, f_n\}$ be a set of polynomials f_i in one variable with complex coefficients. We first give a basis-free definition of the algebra $\mathcal{X}_L(f;R)$. Namely, $\mathcal{X}_L(f;R)$ is the quotient of the tensor algebra $T(V')$ over the dual V' of the vector space V by the two-sided ideal $\mathcal{J}_L(f;R)$ of $T(V')$ which is generated by the ranges of the mappings $f_m(\hat{R})^t$, $m = 1, 2, \cdots, n$, of $(V \otimes V)'$. Here $f_m(\hat{R})^t$ denotes the transpose of the linear transformation $f_m(\hat{R})$ of $V \otimes V$.

Now we describe the algebra $\mathcal{X}_L(f;R)$ in terms of coordinates. Let us identify V and V' and so $V \otimes V$ and $(V \otimes V)'$ by identifying $v \in V$ with the linear functional $\tilde{v}$ on V given by $\tilde{v}(w) = \sum_i \alpha^i\beta^i$, where $v = \sum_i \alpha^i x_i$ and $w = \sum_i \beta^i x_i \in V$. Then the tensor algebra $T(V')$ is (isomorphic to) the free associative algebra $\mathbb{C}\langle x_i\rangle$ with generators $x_1, \cdots, x_N$ and $\mathcal{J}_L(f;R)$ is the two-sided ideal of $\mathbb{C}\langle x_i\rangle$ generated by the nN^2 elements $f_m(\hat{R})^{ij}_{kl}x_k x_l$, where $m = 1, 2, \cdots, n$ and $i, j = 1, 2, \cdots, N$. That is, $\mathcal{X}_L(f;R)$ is the algebra with generators $x_1, \cdots, x_N$ and defining relations

$$f_m(\hat{R})^{ij}_{kl}x_k x_l = 0, \qquad m = 1, 2, \cdots, n; \quad i, j = 1, 2, \cdots, N. \tag{7}$$

Let $\mathbf{x}$ denote the column vector $(x_1, \cdots, x_N)^t$ of generators. In matrix form the relations (7) can be written as

$$f_m(\hat{R})\mathbf{x}_1\mathbf{x}_2 = 0, \qquad m = 1, 2, \cdots, n. \tag{8}$$

Indeed, if $\mathbf{x}_1$ denotes the matrix $\mathbf{x} \otimes I$ and $\mathbf{x}_2$ is the matrix $I \otimes \mathbf{x}$, then the expression $f_m(\hat{R})\mathbf{x}_1\mathbf{x}_2$ in (8) is just the product of the three matrices $f_m(\hat{R})$, $\mathbf{x}_1$ and $\mathbf{x}_2$.

Similarly, the algebra $\mathcal{X}_R(f;R)$ is defined as the quotient of the tensor algebra $T(V)$ over V by the two-sided ideal generated by the images of the transformations $f_m(\hat{R})$, $m = 1, 2, \cdots, n$, of $V \otimes V$. Clearly, $\mathcal{X}_R(f;R)$ is the algebra with generators $x_1, \cdots, x_N$ subject to the relations

$$f_m(\hat{R})^{kl}_{ij}x_k x_l = 0, \qquad m = 1, 2, \cdots, n; \quad i, j = 1, 2, \cdots, N. \tag{9}$$

Proposition 4. *Let $\mathcal{A}$ be a bialgebra which is a quotient of the bialgebra $\mathcal{A}(R)$. Then there are algebra homomorphisms $\varphi_L : \mathcal{X}_L(f;R) \to \mathcal{A}\otimes\mathcal{X}_L(f;R)$ and $\varphi_R : \mathcal{X}_R(f;R) \to \mathcal{X}_R(f;R)\otimes\mathcal{A}$ such that*

$$\varphi_L(x_i) = u^i_j \otimes x_j \quad and \quad \varphi_R(x_i) = x_j \otimes u^j_i. \tag{10}$$

These mappings φ_L and φ_R turn the algebras $\mathcal{X}_L(f;R)$ and $\mathcal{X}_R(f;R)$ into left and right $\mathcal{A}$-quantum spaces, respectively.

Proof. We carry out the proof for $\mathcal{X}_L(f;R)$. By the universal property of the free algebra $\mathbb{C}\langle x_i\rangle$, there is a unique algebra homomorphism $\varphi_L : \mathbb{C}\langle x_i\rangle \to \mathcal{A}\otimes\mathbb{C}\langle x_i\rangle$ such that $\varphi_L(x_i) = u^i_j \otimes x_j$. Since

$$\varphi_L(f_m(\hat{R})^{ij}_{kl}x_kx_l) = f_m(\hat{R})^{ij}_{kl}u^k_r u^l_s \otimes x_r x_s = u^i_k u^j_l \otimes f_m(\hat{R})^{kl}_{rs}x_r x_s$$

by (3), $\varphi_L(\mathcal{J}_L(f;R)) \subseteq \mathcal{A}\otimes\mathcal{J}_L(f;R)$. Hence φ_L passes to an algebra homomorphism, still denoted by φ_L, of $\mathcal{X}_L(f;R)$ to $\mathcal{A}\otimes\mathcal{X}_L(f;R)$. To prove that φ_L is a left coaction, it suffices to check the conditions (1.52) on the generators x_i. We omit this simple verification. □

Definition 2. *The algebras $\mathcal{X}_L(f;R)$ and $\mathcal{X}_R(f;R)$ are called the* coordinate algebras of the left and right quantum vector spaces, *respectively, associated with the set of polynomials f and the transformation R.*

Suppose that the matrix $\hat{R}$ is symmetric, that is, $\hat{R}^{ij}_{kl} = \hat{R}^{kl}_{ij}$. The matrices $\hat{R}$ for the quantum groups $GL_q(N)$, $SL_q(N)$, $O_q(N)$ and $Sp_q(N)$ considered in Sects. 9.2 and 9.3 will have this property (see formulas (13) and (30) below). Then the defining relations (7) and (9) are the same, so the algebras $\mathcal{X}_L(f;R)$ and $\mathcal{X}_R(f;R)$ coincide and are denoted by $\mathcal{X}(f;R)$. By Proposition 4, $\mathcal{X}(f;R)$ is a left and right quantum space for the bialgebra $\mathcal{A}$.

Example 2 (R = id). Let R be the identity map and $f(t) = 1 - t$. Then $\mathcal{X}(f;R)$ is the commutative polynomial algebra in N indeterminates and the left coaction φ_L is given by the matrix multiplication of $\mathbf{u}$ and $\mathbf{x}$. △

Next we consider some quantum spaces which are obtained by modifying the preceding construction. For simplicity we restrict ourselves to the right handed coordinate versions and to the case of single polynomial relations. For any triple of polynomials f, g, h in one variable and any complex number γ, let $\mathcal{X}(f,g,h,\gamma;R)$ denote the unital algebra with $2N$ generators $x_1,\cdots,x_N,y_1,\cdots,y_N$ and defining relations

$$f(\hat{R})^{kl}_{ij}x_kx_l = g(\hat{R})^{ji}_{lk}y_ky_l = h(\hat{R})^{ik}_{jl}x_ky_l - y_ix_j + \gamma\delta_{ij} = 0, \quad i,j = 1,\cdots,N. \tag{11}$$

Proposition 5. *Let $\mathcal{A}$ be a Hopf algebra which is a quotient of the FRT bialgebra $\mathcal{A}(R)$. Then the algebra $\mathcal{X} \equiv \mathcal{X}(f,g,h,\gamma;R)$ is a right quantum space for $\mathcal{A}$ with respect to the coaction φ_R determined by the equations*

$$\varphi_R(x_i) = x_j \otimes u^j_i \quad and \quad \varphi_R(y_i) = y_j \otimes S(u^i_j). \tag{12}$$

Proof. The proof is similar to the proof of Proposition 4. One first defines an algebra homomorphism $\varphi_R : \mathbb{C}\langle x_i, y_j\rangle \to \mathbb{C}\langle x_i, y_j\rangle \otimes \mathcal{A}$ by (12). As usual, $\mathbb{C}\langle x_i, y_j\rangle$ denotes the free algebra with generators $x_i, y_j, i, j = 1, \cdots, N$. Using (3) one then derives that φ_R passes to the quotient algebra $\mathcal{X}$ of $\mathbb{C}\langle x_i, y_j\rangle$ to give a right coaction there. □

The defining equations (11) may look strange at first glance. However, they become very natural in terms of corepresentation theory (see Sect. 11.1). Set $A^{kl}_{ij} = f(\hat{R})^{kl}_{ij}, B^{kl}_{ij} = g(\hat{R})^{ji}_{lk}$ and $C^{kl}_{ij} = h(\hat{R})^{ik}_{jl}$. Then the matrices A, B and C belong to the intertwining spaces $\mathrm{Mor}\,(\mathbf{u} \otimes \mathbf{u}), \mathrm{Mor}\,(\mathbf{u}^c \otimes \mathbf{u}^c)$ and $\mathrm{Mor}\,(\mathbf{u}^c \otimes \mathbf{u}, \mathbf{u} \otimes \mathbf{u}^c)$, respectively, and it is only these properties which are used during the proof of Proposition 5. The term $y_i x_j$ in (11) may also be replaced by $T^{kl}_{ij} y_k x_l$ with $T \in \mathrm{Mor}\,(\mathbf{u}^c \otimes \mathbf{u})$. The second equation of (12) means that φ_R acts as the contragredient corepresentation $\mathbf{u}^c$ on the generators y_i.

In the following Sects. 9.2 and 9.3 we realize the constructions $\mathcal{A}(R)$ and $\mathcal{X}(f; R)$ by taking as R the matrices R which have been derived in Subsect. 8.4.2 for the vector representations of the Drinfeld–Jimbo algebras $U_q(\mathfrak{g})$, $\mathfrak{g} = \mathrm{gl}_N, \mathrm{sl}_N, \mathrm{so}_{2n}, \mathrm{sp}_N$, and $U_{q^{1/2}}(\mathrm{so}_{2n+1})$. Our aim is to develop the coordinate Hopf algebras of the quantum groups $GL_q(N)$, $SL_q(N)$, $O_q(N), SO_q(N)$ and $Sp_q(N)$. Since $\mathcal{A}(R)$ is only a bialgebra, this requires additional constructions.

9.2 The Quantum Groups $GL_q(N)$ and $SL_q(N)$

In this section R is the matrix $q^{1/N} R_{1,1}$ obtained in Subsect. 8.4.2 for the Drinfeld–Jimbo algebra $U_q(\mathrm{sl}_N)$. By (8.60), $\hat{\mathsf{R}} = \tau \circ \mathsf{R}$ has the matrix entries

$$\hat{\mathsf{R}}^{ij}_{mn} \equiv \mathsf{R}^{ji}_{mn} = q^{\delta_{ij}} \delta_{in} \delta_{jm} + (q - q^{-1}) \delta_{im} \delta_{jn} \theta(j - i), \tag{13}$$

where θ is the Heaviside symbol, that is, $\theta(k) = 1$ if $k > 0$ and $\theta(k) = 0$ if $k \leq 0$.

We suppose that $\lambda_+ := q + q^{-1} \neq 0$. By (13), the matrix $\hat{\mathsf{R}}$ is symmetric. From Proposition 8.24 we know that $\hat{\mathsf{R}}$ satisfies the quadratic equation

$$(\hat{\mathsf{R}} - qI)(\hat{\mathsf{R}} + q^{-1}I) = 0 \tag{14}$$

and that

$$\mathsf{P}_+ := \lambda_+^{-1}(\hat{\mathsf{R}} + q^{-1}I) \quad \text{and} \quad \mathsf{P}_- := \lambda_+^{-1}(-\hat{\mathsf{R}} + qI) \tag{15}$$

are projections (that is, $\mathsf{P}_\pm^2 = \mathsf{P}_\pm$) such that $\mathsf{P}_+\mathsf{P}_- = \mathsf{P}_-\mathsf{P}_+ = 0$ and

$$\hat{\mathsf{R}} = q\mathsf{P}_+ - q^{-1}\mathsf{P}_-. \tag{16}$$

9.2.1 The Quantum Matrix Space $M_q(N)$ and the Quantum Vector Space $\mathbb{C}_q^N$

We begin by describing the FRT bialgebra $\mathcal{A}(\mathsf{R})$ explicitly.

Definition 3. *The bialgebra $\mathcal{A}(\mathsf{R})$, with $\mathsf{R} = (\mathsf{R}^{ij}_{mn})$ given by (13), is called the* coordinate algebra of the quantum matrix space $M_q(N)$ *and is denoted by $\mathcal{O}(M_q(N))$.*

Inserting the matrix entries R^{ji}_{nm} from (13) into the relations (1), we see that the algebra $\mathcal{O}(M_q(N))$ is generated by the elements u^i_j, $i, j = 1, 2, \cdots, N$, with defining relations:

$$u^i_k u^j_k = q u^j_k u^i_k, \quad u^k_i u^k_j = q u^k_j u^k_i, \quad i < j, \tag{17a}$$

$$u^i_l u^j_k = u^j_k u^i_l, \quad i < j, \ k < l, \tag{17b}$$

$$u^i_k u^j_l - u^j_l u^i_k = (q - q^{-1}) u^j_k u^i_l, \quad i < j, \ k < l. \tag{17c}$$

These relations are equivalent to the requirement that for any k, l, m, n with $k < l$, $m < n$ the entries $a = u^k_m$, $b = u^k_n$, $c = u^l_m$, $d = u^l_n$ of the corresponding 2×2 submatrix of $\mathbf{u} = (u^i_j)$ satisfy the defining relations of $M_q(2)$ (see the formulas (4.1) and (4.2)).

From the above defining relations for $M_q(N)$ it easily follows that there are an automorphism θ and an anti-automorphism ϑ of the algebra $\mathcal{O}(M_q(N))$ such that $\theta(u^i_j) = u^j_i$ and $\vartheta(u^i_j) = u^{N+1-i}_{N+1-j}$. We have $\Delta \circ \theta = \tau \circ (\theta \otimes \theta) \circ \Delta$, $\Delta \circ \vartheta = (\vartheta \otimes \vartheta) \circ \Delta$ and $\theta \circ \vartheta = \vartheta \circ \theta$. Moreover, for any nonzero $\alpha \in \mathbb{C}$ the mapping $u^i_j \to \alpha^{j-i} u^i_j$ extends to an automorphism of the bialgebra $\mathcal{O}(M_q(N))$.

Definition 4. *Let $f_s(t) = t - q$ and $f_a(t) = t + q^{-1}$. The algebras $\mathcal{O}(\mathbb{C}_q^N) := \mathcal{X}(f_s; \mathsf{R})$ and $\Lambda(\mathbb{C}_q^N) := \mathcal{X}(f_a; \mathsf{R})$ are called the* coordinate algebra *and the* exterior algebra of the quantum vector space $\mathbb{C}_q^N$, *respectively.*

Note that $f_s(\hat{\mathsf{R}})$ and $f_a(\hat{\mathsf{R}})$ are just the two factors in equation (14).

In order to avoid confusion, we denote the generators of $\Lambda(\mathbb{C}_q^N)$ by $y_1, \cdots, y_N$ and the column vector $(y_1, \cdots, y_N)^t$ by $\mathbf{y}$. Since $f_s(\hat{\mathsf{R}}) = -\lambda_+ \mathsf{P}_-$ and $f_a(\hat{\mathsf{R}}) = \lambda_+ \mathsf{P}_+$ by (15), the algebras $\mathcal{O}(\mathbb{C}_q^N)$ and $\Lambda(\mathbb{C}_q^N)$ are defined by the matrix equations $\mathsf{P}_- \mathbf{x}_1 \mathbf{x}_2 = 0$ and $\mathsf{P}_+ \mathbf{y}_1 \mathbf{y}_2 = 0$, respectively. We insert the matrix entries (13) into these equations and obtain the following explicit forms. The algebra $\mathcal{O}(\mathbb{C}_q^N)$ has generators $x_1, \cdots, x_N$ with defining relations

$$x_i x_j = q x_j x_i, \qquad i < j.$$

The quantum exterior algebra $\Lambda(\mathbb{C}_q^N)$ is generated by elements $y_1, \cdots, y_N$ subject to the relations

$$y_i^2 = 0 \quad \text{and} \quad y_i y_j = -q^{-1} y_j y_i, \quad i < j.$$

Obviously, in the "classical" case $q = 1$ the algebra $\mathcal{O}(\mathbb{C}_q^N)$ is the polynomial algebra in N commuting indeterminates, that is, the symmetric algebra of the vector space $\mathbb{C}^N$, and $\Lambda(\mathbb{C}_q^N)$ is the exterior algebra of $\mathbb{C}^N$.

Proposition 6. *The following sets of monomials are bases of the corresponding complex vector spaces:*

$$\mathcal{O}(M_q(N)): \quad \{(u_1^1)^{k_{11}}(u_2^1)^{k_{12}}\cdots(u_N^N)^{k_{NN}} \mid k_{11}, k_{12}, \cdots, k_{NN} \in \mathbb{N}_0\},$$

$$\mathcal{O}(\mathbb{C}_q^N): \quad \{x_1^{k_1}x_2^{k_2}\cdots x_N^{k_N} \mid k_1, k_2, \cdots, k_N \in \mathbb{N}_0\},$$

$$\Lambda(\mathbb{C}_q^N): \quad \{y_1^{i_1}y_2^{i_2}\cdots y_N^{i_N} \mid i_1, i_2, \cdots, i_N \in \{0,1\}\}.$$

Proof. The proof can be found in [PW], Sects. 3.4 and 3.5. The assertions can also be derived from Lemma 4.8. For the first assertion this is carried out in [NYM], Sect. 1.5. □

Proposition 6 implies that for all three algebras the subspaces spanned by the monomials of a fixed degree have the same dimensions as in the classical case $q = 1$.

Let $\mathcal{O}(\mathbb{C}_q^N)_n$ and $\Lambda(\mathbb{C}_q^N)_n$ denote the subspaces of $\mathcal{O}(\mathbb{C}_q^N)$ and $\Lambda(\mathbb{C}_q^N)$, respectively, consisting of homogeneous elements of degree n. Since the matrix $\hat{\mathsf{R}}$ is symmetric (see (13)), we know from Proposition 4 that $\mathcal{O}(\mathbb{C}_q^N)$ and $\Lambda(\mathbb{C}_q^N)$ are both left and right quantum spaces for the bialgebra $\mathcal{O}(M_q(N))$ with coactions φ_L and φ_R given by (10). Since the coactions are algebra homomorphisms, the subspaces $\mathcal{O}(\mathbb{C}_q^N)_n$ and $\Lambda(\mathbb{C}_q^N)_n$ are invariant under φ_L and φ_R and the restrictions $\varphi_{L,n}$ of φ_L and $\varphi_{R,n}$ of φ_R to these subspaces are corepresentations of the bialgebra $\mathcal{O}(M_q(N))$. We call the mapping

$$\varphi_{R,n} : \mathcal{O}(\mathbb{C}_q^N)_n \to \mathcal{O}(\mathbb{C}_q^N)_n \otimes \mathcal{O}(M_q(N))$$

the *symmetric tensor corepresentation* of $\mathcal{O}(M_q(N))$ and the mapping

$$\varphi_{R,n} : \Lambda(\mathbb{C}_q^N)_n \to \Lambda(\mathbb{C}_q^N)_n \otimes \mathcal{O}(M_q(N))$$

the *alternating tensor corepresentation* of $\mathcal{O}(M_q(N))$.

9.2.2 Quantum Determinants

We first introduce some multi-index notation. Let Ω_n denote the set of all subsets of $\{1, 2, \cdots, N\}$ consisting of n elements. For subsets $I, J \in \Omega_n$ we write I and J as $I = \{i_1, \cdots, i_n\}$ and $J = \{j_1, \cdots, j_n\}$ with $i_1 < i_2 < \cdots < i_n$ and $j_1 < j_2 < \cdots < j_n$. We set $y_I := y_{i_1} \cdots y_{i_n}$, $u_J^I := u_{j_1}^{i_1} \cdots u_{j_n}^{i_n}$ and

$$\mathcal{D}_J^I := \sum\nolimits_{p \in \mathcal{P}_n} (-q)^{l(p)} u_{j_1}^{i_{p(1)}} \cdots u_{j_n}^{i_{p(n)}}, \tag{18}$$

where $\mathcal{P}_n$ is the permutation group of the set $\{1, 2, \cdots, n\}$ and $l(p)$ is the number of inversions in p. If $p \in \mathcal{P}_n$ and $p' = p^{-1}$, then

$$u^{i_{p(1)}}_{j_1} \cdots u^{i_{p(n)}}_{j_n} = u^{i_1}_{j_{p'(1)}} \cdots u^{i_n}_{j_{p'(n)}}. \tag{19}$$

Indeed, both expressions contain the same factors, but in a different order. Using (17b), one side of (19) can be rearranged to get the other side. Since $l(p') = l(p)$, (19) yields the alternative expression

$$\mathcal{D}^I_J = \sum\nolimits_{p \in \mathcal{P}_n} (-q)^{l(p)} u^{i_1}_{j_{p(1)}} \cdots u^{i_n}_{j_{p(n)}}. \tag{20}$$

Definition 5. *The element* $\mathcal{D}^I_J$ *of* $\mathcal{O}(M_q(N))$ *is called the* quantum n-minor determinant *of the matrix* $\mathbf{u} = (u^i_j)_{i,j=1,\cdots,N}$ *with respect to the row* I *and column* J. *In the case* $I = J = \{1, 2, \cdots, N\}$ *the element* $\mathcal{D}^I_J$ *is called the* quantum determinant *of* $\mathbf{u} = (u^i_j)_{i,j=1,\cdots,N}$ *and is denoted by* $\mathcal{D}_q$.

Proposition 7. (i) $\varphi_{R,n}(y_I) = y_J \otimes \mathcal{D}^J_I$ *and* $\varphi_{L,n}(y_I) = \mathcal{D}^I_J \otimes y_J$ *for* $I \in \Omega_n$.
(ii) $\Delta(\mathcal{D}^I_J) = \mathcal{D}^I_K \otimes \mathcal{D}^K_J$ *and* $\varepsilon(\mathcal{D}^I_J) = \delta_{IJ}$ *for* $I, J \in \Omega_n$.

(Note that there are summations in (i) and (ii) over $J \in \Omega_n$ resp. $K \in \Omega_n$.)

Proof. (i): We carry out the proof for $\varphi_{R,n}$. From (10) we obtain $\varphi_{R,n}(y_I) = y_{j_1} \cdots y_{j_n} \otimes u^{j_1}_{i_1} \cdots u^{j_n}_{i_n}$. The defining relations for $\Lambda(\mathcal{O}^N_q)$ imply that $y_{j_1} \cdots y_{j_n} = 0$ if $j_k = j_l$ for some $k \neq l$ and that $y_{j_{p(1)}} \cdots y_{j_{p(n)}} = (-q)^{l(p)} y_{j_1} \cdots y_{j_n}$ for $p \in \mathcal{P}_n$ and $J = \{j_1, \cdots, j_n\} \in \Omega_n$. Using these facts we get

$$\varphi_{R,n}(y_I) = \sum_{J \in \Omega_n} \sum_{p \in \mathcal{P}_n} y_{j_{p(1)}} \cdots y_{j_{p(n)}} \otimes u^{j_{p(1)}}_{i_1} \cdots u^{j_{p(n)}}_{i_n} = \sum_{J \in \Omega_n} y_J \otimes \mathcal{D}^J_I.$$

(ii): By Proposition 6, $\{y_J \mid J \in \Omega_n\}$ is a basis of the vector space $\Lambda(\mathbb{C}^N_q)_n$. Hence, by (i), $\mathcal{D}^I_J$ are the matrix elements of the corepresentation $\varphi_{R,n}$ of $\mathcal{O}(M_q(N))$ and the assertion follows from Proposition 1.13. □

We reformulate the above results for the quantum determinant $\mathcal{D}_q$. Let ζ be a nonzero element of the one-dimensional space $\Lambda(\mathbb{C}^N_q)_N$. Then, by Proposition 7(i), $\mathcal{D}_q$ is the unique element of $\mathcal{O}(M_q(N))$ which satisfies the equations

$$\varphi_{R,N}(\zeta) = \zeta \otimes \mathcal{D}_q \quad \text{and} \quad \varphi_{L,N}(\zeta) = \mathcal{D}_q \otimes \zeta.$$

Applying this characterization to $\zeta = y_1 \cdots y_N$ and $\zeta' = y_{r(1)} \cdots y_{r(N)} = (-q)^{l(r)} \zeta$ for $r \in \mathcal{P}_N$, we conclude that

$$\mathcal{D}_q = \sum_{p \in \mathcal{P}_N} (-q)^{l(p)-l(r)} u^{p(1)}_{r(1)} \cdots u^{p(N)}_{r(N)} = \sum_{p \in \mathcal{P}_N} (-q)^{l(p)-l(r)} u^{r(1)}_{p(1)} \cdots u^{r(N)}_{p(N)} \tag{21}$$

for any permutation $r \in \mathcal{P}_N$. By Proposition 7(ii) applied with $I = J = \{1, 2, \cdots, N\}$, we have

$$\Delta(\mathcal{D}_q) = \mathcal{D}_q \otimes \mathcal{D}_q \quad \text{and} \quad \varepsilon(\mathcal{D}_q) = 1, \tag{22}$$

so the quantum determinant is a nonzero group-like element of $\mathcal{O}(M_q(N))$.

For $i,j \in \{1,2,\cdots,N\}$, let $A_i^j := \mathcal{D}_I^J$, where $J=\{1,\cdots,j-1,j+1,\cdots,N\}$ and $I=\{1,\cdots,i-1,i+1,\cdots,N\}$.

Proposition 8. *For $i,k \in \{1,2,\cdots,N\}$ we have*

$$\sum\nolimits_j (-q)^{i-j} A_i^j u_k^j = \sum\nolimits_j (-q)^{j-k} u_i^j A_k^j = \delta_{ik}\mathcal{D}_q, \tag{23}$$

$$\sum\nolimits_j (-q)^{j-k} u_j^i A_j^k = \sum\nolimits_j (-q)^{i-j} A_j^i u_j^k = \delta_{ik}\mathcal{D}_q. \tag{24}$$

Proof. For subsets I and J from $\{1,2,\cdots,N\}$, let $\mathrm{sgn}_q(I,J)=0$ if $I\bigcap J\neq\emptyset$ and $\mathrm{sgn}_q(I,J)=(-q)^{l(I,J)}$ if $I\bigcap J=\emptyset$, where $l(I,J)$ is the number of pairs (i,j), $i\in I$, $j\in J$, such that $i>j$. We then have $y_I y_J = \mathrm{sgn}_q(I,J) y_K$, $K=I\bigcup J$. Applying φ_R to this equation with appropriate sets I and J, using Proposition 7(i) and comparing the coefficients of $y_1\cdots y_N$, (23) follows. Using φ_L the identity (24) is derived in a similar way. □

Set $\tilde{u}_j^i := (-q)^{i-j} A_i^j$, $i,j=1,2,\cdots,N$. The matrix $\tilde{\mathbf{u}} := (\tilde{u}_j^i)_{i,j=1,\cdots,N}$ is called the *cofactor matrix* of $\mathbf{u}=(u_j^i)_{i,j=1,\cdots,N}$. The first formulas of (23) and (24) can be rewritten in matrix form as

$$\tilde{\mathbf{u}}\mathbf{u} = \mathbf{u}\tilde{\mathbf{u}} = \mathcal{D}_q \cdot I_N, \tag{25}$$

where I_N is the unit matrix. Multiplying (25) by the matrix $\mathbf{u}$ on the left and on the right and comparing both equations we obtain $\mathcal{D}_q\cdot\mathbf{u} = \mathbf{u}\tilde{\mathbf{u}}\mathbf{u} = \mathbf{u}\cdot\mathcal{D}_q$ which means that $\mathcal{D}_q u_j^i = u_j^i \mathcal{D}_q$ for all $i,j=1,2,\cdots,N$. This yields the first assertion of the following proposition.

Proposition 9. *The quantum determinant $\mathcal{D}_q$ belongs to the center of the algebra $\mathcal{O}(M_q(N))$. If q is not a root of unity, then the center of $\mathcal{O}(M_q(N))$ is generated by $\mathcal{D}_q$.*

The second assertion of Proposition 9 is proved in [NYM], Sect. 1.6. Note that the center of $\mathcal{O}(M_q(N))$ is much larger than the polynomial algebra in $\mathcal{D}_q$ if q is a root of unity.

More properties of quantum determinants can be found in [PW] and [NYM].

9.2.3 The Quantum Groups $GL_q(N)$ and $SL_q(N)$

The coordinate algebra of the quantum group $GL_q(N)$ is obtained by adjoining the inverse $\mathcal{D}_q^{-1}$ of the quantum determinant $\mathcal{D}_q$ to the algebra $\mathcal{O}(M_q(N))$. To be precise, $\mathcal{O}(GL_q(N))$ is the quotient algebra of the algebra $\mathcal{O}(M_q(N))[t]$ of polynomials in t over $\mathcal{O}(M_q(N))$ by the two-sided ideal $\langle t\mathcal{D}_q - 1\rangle$ generated by the element $t\mathcal{D}_q-1$. That is, $\mathcal{O}(GL_q(N))$ is the algebra with N^2+1 generators u_j^i, $i,j=1,2,\cdots,N$, and t and defining relations (17) and

$$\mathcal{D}_q t = t\mathcal{D}_q = 1, \qquad u_j^i t = t u_j^i, \quad i,j=1,2,\cdots,N.$$

In the algebra $\mathcal{O}(GL_q(N))$ we have $t = \mathcal{D}_q^{-1}$. From the very definition

$$\mathcal{O}(GL_q(N)) = \mathcal{O}(M_q(N))[t]/\langle t\mathcal{D}_q - 1\rangle$$

we see that $\mathcal{O}(M_q(N))$ can be regarded as a subalgebra of $\mathcal{O}(GL_q(N))$ and

$$\mathcal{O}(GL_q(N)) = \bigoplus_{m=0}^{\infty} \mathcal{D}_q^{-m} \mathcal{O}(M_q(N)). \tag{26}$$

The coordinate algebra of the quantum group $SL_q(N)$ is defined as the quotient

$$\mathcal{O}(SL_q(N)) = \mathcal{O}(M_q(N))/\langle \mathcal{D}_q - 1\rangle = \mathcal{O}(GL_q(N))/\langle \mathcal{D}_q - 1\rangle$$

of the algebra $\mathcal{O}(M_q(N))$ or, equivalently, of $\mathcal{O}(GL_q(N))$ by the two-sided ideal $\langle \mathcal{D}_q - 1\rangle$ generated by the element $\mathcal{D}_q - 1$.

Proposition 10. *There are unique Hopf algebra structures on the algebras $\mathcal{O}(GL_q(N))$ and $\mathcal{O}(SL_q(N))$ with comultiplications Δ and counits ε such that $\Delta(u^i_j) = u^i_k \otimes u^k_j$ and $\varepsilon(u^i_j) = \delta_{ij}$. The antipodes S of these Hopf algebras are given by $S(u^i_j) = \tilde{u}^i_j \mathcal{D}_q^{-1}$ and $S(\mathcal{D}_q^{-1}) = \mathcal{D}_q$ for $\mathcal{O}(GL_q(N))$ and by $S(u^i_j) = \tilde{u}^i_j$ for $\mathcal{O}(SL_q(N))$. Moreover, we have $S^2(u^i_j) = \mathsf{D}_i^{-1} u^i_j \mathsf{D}_j$ for both Hopf algebras, where $\mathsf{D}_i := q^{-2i}$.*

Proof. By (26), we extend the comultiplication Δ and the counit ε of the bialgebra $\mathcal{O}(M_q(N))$ to algebra homomorphism $\Delta : \mathcal{O}(GL_q(N)) \to \mathcal{O}(GL_q(N)) \otimes \mathcal{O}(GL_q(N))$ and $\varepsilon : \mathcal{O}(GL_q(N)) \to \mathbb{C}$ by setting $\Delta(\mathcal{D}_q^{-1}) = \mathcal{D}_q^{-1} \otimes \mathcal{D}_q^{-1}$ and $\varepsilon(\mathcal{D}_q^{-1}) = 1$. One easily verifies that then $\mathcal{O}(GL_q(N))$ becomes a bialgebra.

Recall that by (25), $\tilde{\mathbf{u}}\mathbf{u} = \mathbf{u}\tilde{\mathbf{u}} = \mathcal{D}_q \cdot I_N$. Multiplying the defining relations $\mathsf{R}\mathbf{u}_1\mathbf{u}_2 = \mathbf{u}_2\mathbf{u}_1\mathsf{R}$ for $\mathcal{O}(M_q(N))$ by $\tilde{\mathbf{u}}_2\tilde{\mathbf{u}}_1\mathcal{D}_q^{-2}$ from the right and by $\tilde{\mathbf{u}}_1\tilde{\mathbf{u}}_2\mathcal{D}_q^{-2}$ from the left we obtain $\mathsf{R}(\tilde{\mathbf{u}}\mathcal{D}_q^{-1})_2(\tilde{\mathbf{u}}\mathcal{D}_q^{-1})_1 = (\tilde{\mathbf{u}}\mathcal{D}_q^{-1})_1(\tilde{\mathbf{u}}\mathcal{D}_q^{-1})_2\mathsf{R}$. That is, the entries of the matrix $\tilde{\mathbf{u}}\mathcal{D}_q^{-1}$ satisfy the defining relations of the algebra $\mathcal{O}(M_q(N))^{\mathrm{op}}$. Hence, there exists an algebra anti-homomorphism $S : \mathcal{O}(M_q(N)) \to \mathcal{O}(M_q(N))$ such that $S(u^i_j) = \tilde{u}^i_j\mathcal{D}_q^{-1}$. Consider the equation $m(\mathrm{id}\otimes S)\Delta(a) = \varepsilon(a)1$ for $a \in \mathcal{O}(M_q(N))$. Clearly, it holds for a product $a{\cdot}b$ if it holds for a and for b. It is satisfied for $a = 1$ and for $a = u^i_j$, since $\mathbf{u}\tilde{\mathbf{u}} = \mathcal{D}_q{\cdot}I_N$ and $\varepsilon(\mathcal{D}_q) = 1$. Thus, the equation is true for all $a \in \mathcal{O}(M_q(N))$. Setting $a = \mathcal{D}_q$, it yields $\mathcal{D}_q S(\mathcal{D}_q) = 1$. Hence, $S(\mathcal{D}_q) = \mathcal{D}_q^{-1}$ and S extends to an anti-homomorphism $S : \mathcal{O}(GL_q(N)) \to \mathcal{O}(GL_q(N))$ by setting $S(\mathcal{D}_q^{-1}) = \mathcal{D}_q$. Since the relations (1.26) hold for the generators $a = 1, u^i_j, \mathcal{D}_q^{-1}$ of $\mathcal{O}(GL_q(N))$ as is easily seen, the mapping S is an antipode and $\mathcal{O}(GL_q(N))$ is a Hopf algebra by Proposition 1.8.

Since $S(u^i_j) = \tilde{u}^i_j\mathcal{D}_q^{-1} = (-q)^{i-j}A^j_i$, the second formula of (24) yields $\sum_n(-q)^{2(j-n)}S(u^n_j)u^k_n = \delta_{jk}$. Applying S, we get $\sum_n(-q)^{2(j-n)}S(u^k_n)S^2(u^n_j) = \delta_{jk}$. Multiplying this equation by u^i_k on the left and summing over k, we obtain $(-q)^{2(j-i)}S^2(u^i_j) = u^i_j$, which gives the formula for S^2.

Now we turn to the algebra $\mathcal{O}(SL_q(N))$. From the equations $\Delta(\mathcal{D}_q-1) = \mathcal{D}_q\otimes\mathcal{D}_q-1\otimes 1 = (\mathcal{D}_q-1)\otimes 1+\mathcal{D}_q\otimes(\mathcal{D}_q-1)$, $\varepsilon(\mathcal{D}_q-1)=0$ and $S(\mathcal{D}_q-1) = \mathcal{D}_q^{-1}-1 = -\mathcal{D}_q^{-1}(\mathcal{D}_q-1)$ it follows that $\langle\mathcal{D}_q-1\rangle$ is a Hopf ideal of the Hopf algebra $\mathcal{O}(GL_q(N))$. Hence, the quotient $\mathcal{O}(SL_q(N)) = \mathcal{O}(GL_q(N))/\langle\mathcal{D}_q-1\rangle$ is also a Hopf algebra with the structure induced from $\mathcal{O}(GL_q(N))$. Since $\Delta(\mathcal{D}_q) = \mathcal{D}_q\otimes\mathcal{D}_q$ and $\varepsilon(\mathcal{D}_q)=1$, the uniqueness assertions are clear. □

Definition 6. *The Hopf algebras $\mathcal{O}(GL_q(N))$ and $\mathcal{O}(SL_q(N))$ defined above are called the* coordinate algebras *of the quantum groups $GL_q(N)$ and $SL_q(N)$, respectively.*

Recall that the quantum vector space $\mathcal{O}(\mathbb{C}_q^N)$ is a left and right quantum space for the bialgebra $\mathcal{O}(M_q(N))$. Since $\mathcal{O}(M_q(N))$ is a subbialgebra of $\mathcal{O}(GL_q(N))$ and $\mathcal{O}(SL_q(N))$ is a quotient of the bialgebra $\mathcal{O}(M_q(N))$, $\mathcal{O}(\mathbb{C}_q^N)$ is also a left and right quantum space for $\mathcal{O}(GL_q(N))$ and $\mathcal{O}(SL_q(N))$ by Proposition 4. The corresponding coactions φ_L and φ_R are determined by the formulas $\varphi_L(x_i) = u^i_j\otimes x_j$ and $\varphi_R(x_i) = x_j\otimes u^j_i$.

The next proposition says that the left (resp. right) quantum space $\mathcal{O}(\mathbb{C}_q^N)$ for $\mathcal{O}(GL_q(N))$ can be considered as a subalgebra of $\mathcal{O}(GL_q(N))$ by identifying x_i and u^i_1 (resp. x_i and u^1_i), where the left (resp. right) coaction on x_i corresponds to the comultiplication of u^i_1 (resp. u^1_i).

Proposition 11. *Define $\theta(x_i) = u^i_1$ and $\vartheta(x_i) = u^1_i$, $i=1,2,\cdots,N$. Then θ and ϑ can be uniquely extended to injective algebra homomorphisms of $\mathcal{O}(\mathbb{C}_q^N)$ into $\mathcal{O}(GL_q(N))$ such that $\theta\in\mathrm{Mor}\,(\varphi_L,\Delta)$ and $\vartheta\in\mathrm{Mor}\,(\varphi_R,\Delta)$.*

Proof. We carry out the proof for θ. Since $u^i_1u^j_1 = qu^j_1u^i_1$, $i<j$, by (17a), the u^i_1 satisfy the defining relations of $\mathcal{O}(\mathbb{C}_q^N)$, so there exists an algebra homomorphism $\theta:\mathcal{O}(\mathbb{C}_q^N)\to\mathcal{O}(GL_q(N))$ such that $\theta(x_i)=u^i_1$. By Proposition 6 and formula (26), $\{x_1^{k_1}\cdots x_N^{k_N}\mid k_1,\cdots,k_N\in\mathbb{N}_0\}$ is a vector space basis of $\mathcal{O}(\mathbb{C}_q^N)$ and $\{(u^1_1)^{k_1}(u^2_1)^{k_2}\cdots(u^N_1)^{k_N}\mid k_1,\cdots,k_N\in\mathbb{N}_0\}$ is a subset of a basis of $\mathcal{O}(GL_q(N))$. This implies that θ is injective. The assertion $\theta\in\mathrm{Mor}\,(\varphi_L,\Delta)$ means that $(\mathrm{id}\otimes\theta)\varphi_L(a) = \Delta(\theta(a))$ for all $a\in\mathcal{O}(\mathbb{C}_q^N)$. Since both $(\mathrm{id}\otimes\theta)\circ\varphi_L$ and $\Delta\circ\theta$ are algebra homomorphisms, it suffices to check this equality on the generators x_i. But then we have by (10) that $(\mathrm{id}\otimes\theta)\varphi_L(x_i) = u^i_j\otimes\theta(x_j) = u^i_j\otimes u^j_1 = \Delta(u^i_1) = \Delta(\theta(x_i))$. □

The algebras $\mathcal{X}(f,g,h,\gamma;R)$ from Subsect. 9.1.2 provide another source of quantum spaces for $SL_q(N)$ and also for $GL_q(N)$. An important particular case is the algebra $\mathcal{X}_{\gamma,\delta}$, where $\delta\in\mathbb{C}$, obtained by setting $f(t)=g(t)=t-q, h(t)=qt+\delta(1-q^2)$. Inserting the values (13) into (11) we see that the defining relations for the generators $x_1,\cdots,x_N,y_1,\cdots,y_N$ of $\mathcal{X}_{\gamma,\delta}$ can be written as follows:

$$x_ix_j = qx_jx_i,\quad y_iy_j = q^{-1}y_jy_i,\quad i<j, \tag{27}$$

$$y_ix_j = qx_jy_i,\quad i\neq j, \tag{28}$$

$$y_ix_i - q^2x_iy_i = \gamma + (q^2-1)\Big(\sum\nolimits_{j>i} x_jy_j - \delta\sum\nolimits_{j=1}^N x_jy_j\Big). \tag{29}$$

The algebra $\mathcal{X}_{\gamma,\delta}$ is a right quantum space for $SL_q(N)$ and for $GL_q(N)$ with coactions given by (12). We will meet this algebra in the case $\delta = 0, \gamma = 1$ as the quantum Weyl algebra in Subsect. 12.3.3.

9.2.4 Real Forms of $GL_q(N)$ and $SL_q(N)$ and $*$-Quantum Spaces

Let $\mathcal{A} = \mathcal{O}(G_q)$ be the coordinate Hopf algebra of a quantum group G_q. If $* : \mathcal{A} \to \mathcal{A}$ is an involution such that $(\mathcal{A}, *)$ becomes a Hopf $*$-algebra, then we say that this Hopf $*$-algebra structure defines a *real form* of the quantum group G_q or briefly it is a *real quantum group.* A quantum space $\mathcal{X}$ for a Hopf $*$-algebra $\mathcal{A}$ is called a $*$-*quantum space* if $\mathcal{X}$ is a $*$-algebra and if the coaction φ of $\mathcal{A}$ on $\mathcal{X}$ satisfies $\varphi(x^*) = \varphi(x)^*$ for all $x \in \mathcal{X}$.

In order to discuss real forms of $GL_q(N)$ and $SL_q(N)$ we consider two cases of values for the parameter q.

Case 1: $|q| = 1$. Then we have $\overline{\mathsf{R}(q)} = \mathsf{R}(q^{-1}) = \mathsf{R}(q)^{-1}$, that is, the R-matrix R is inverse real. There exist unique involutions of the algebras $\mathcal{O}(GL_q(N))$ and $\mathcal{O}(SL_q(N))$ such that $(u^i_j)^* = u^i_j$, $i,j = 1,2,\cdots,N$. Equipped with these involutions, $\mathcal{O}(GL_q(N))$ and $\mathcal{O}(SL_q(N))$ are Hopf $*$-algebras denoted by $\mathcal{O}(GL_q(N;\mathbb{R}))$ and $\mathcal{O}(SL_q(N;\mathbb{R}))$, respectively. They are called the *coordinate algebras of the real quantum groups* $GL_q(N;\mathbb{R})$ and $SL_q(N;\mathbb{R})$, respectively.

To prove the assertions in the preceding paragraph, we restrict ourselves to the case of $\mathcal{O}(GL_q(N))$. By Proposition 3(ii), $\mathcal{A}(\mathsf{R})$ is a $*$-bialgebra with $(u^i_j)^* = u^i_j, i,j = 1,\cdots,N$. Since $|q| = 1$, one verifies that $\mathcal{D}_q^* = \mathcal{D}_q$. Because of (26), there is a unique extension of the involution to the algebra $\mathcal{O}(GL_q(N))$ by setting $(\mathcal{D}_q^{-m})^* = \mathcal{D}_q^{-m}$, $m \in \mathbb{N}$.

The algebra $\mathcal{O}(\mathbb{C}_q^N)$ with the involution defined by $(x_i)^* = x_i$, $i = 1,2,\cdots,N$, becomes a $*$-algebra which is denoted by $\mathcal{O}(\mathbb{R}_q^N)$. (It suffices to note that the relations of $\mathbb{C}_q^N$ are invariant under this involution.) One can easily see that $\mathcal{O}(\mathbb{R}_q^N)$ is a (left and right) $*$-quantum space for the Hopf $*$-algebras $\mathcal{O}(GL_q(N;\mathbb{R}))$ and $\mathcal{O}(SL_q(N;\mathbb{R}))$.

Case 2: $q \in \mathbb{R}$. Then the R-matrix R is real. There is a unique involution on the algebra $\mathcal{O}(GL_q(N))$ such that $(u^i_j)^* = S(u^j_i)$, $i,j = 1,2,\cdots,N$. Equipped with this involution, $\mathcal{O}(GL_q(N))$ becomes a Hopf $*$-algebra, denoted by $\mathcal{O}(U_q(N))$, which is called the *coordinate algebra of the quantum unitary group* $U_q(N)$. In the $*$-algebra $\mathcal{O}(U_q(N))$ we have $\mathcal{D}_q^*\mathcal{D}_q = \mathcal{D}_q\mathcal{D}_q^* = 1$, that is, the quantum determinant $\mathcal{D}_q$ is a unitary element.

Let $\nu = (\nu_1,\cdots,\nu_N)$ be an N-tuple of numbers $\nu_i \in \{1,-1\}$. The Hopf algebra $\mathcal{O}(SL_q(N))$ becomes a Hopf $*$-algebra, denoted by $\mathcal{O}(SU_q(N;\nu))$, with the involution determined by the formulas $(u^i_j)^* = \nu_i\nu_j S(u^j_i)$, $i,j = 1,2,\cdots,N$. We call this Hopf $*$-algebra the *coordinate algebra of the real* *quantum group* $SU_q(N;\nu)$. If $\nu_1 = \cdots = \nu_N = 1$, then the Hopf $*$-algebra $\mathcal{O}(SU_q(N;\nu))$ is called the *coordinate algebra of the real quantum group* $SU_q(N)$ and is denoted simply by $\mathcal{O}(SU_q(N))$.

We prove the preceding assertion for $\mathcal{O}(SL_q(N))$ by applying Proposition 3(i). Let ψ denote the Hopf algebra automorphism of $\mathcal{O}(SL_q(N))$ defined by $\psi(u^i_j) = \nu_i\nu_j u^i_j$. Let $I : \mathbb{C}\langle u^i_j\rangle \to \mathcal{A}(\mathsf{R}) = \mathbb{C}\langle u^i_j\rangle/\mathcal{J}(R)$ be the canonical map. If $\mathcal{D}_q$ denotes an element of $\mathbb{C}\langle u^i_j\rangle$ such that $I(\mathcal{D}_q) = \mathcal{D}_q$, then we have $\theta(\mathcal{D}_q) - \mathcal{D}_q \in \mathcal{J}(R)$. Thus Proposition 3(i) applies with $\mathcal{I} = \{\mathcal{D}_q - 1\}$ and shows that $\mathcal{O}(SL_q(N)) = \mathbb{C}\langle u^i_j\rangle/\langle\mathcal{J}(R), \mathcal{D}_q - 1\rangle$ is indeed a Hopf $*$-algebra.

It is clear from the definitions of the involutions that for both quantum groups $U_q(N)$ and $SU_q(N)$ the fundamental matrix $\mathbf{u} = (u^i_j)$ is unitary, that is, we have $\mathbf{u}\mathbf{u}^* = \mathbf{u}^*\mathbf{u} = I$, where $(\mathbf{u}^*)^i_j := (u^j_i)^* = S(u^i_j)$. Because of this property, we call $U_q(N)$ and $SU_q(N)$ *compact quantum groups*. An extensive treatment of such quantum groups will be given in Sect. 11.3.

Let γ and δ be real. Then the algebra $\mathcal{X}_{\gamma,\delta}$ defined by the relations (27)–(29) becomes a $*$-algebra, denoted by $\mathcal{X}_{\gamma,\delta}(\nu)$, with involution determined by $(x_i)^* = \nu_i y_i$. (In order to prove this, it suffices to check that the relations (27)–(29) are invariant under this involution.) Then we have $\varphi_R(x_i^*) = \nu_i y_j \otimes S(u^i_j) = \nu_j y_j \otimes (u^j_i)^* = \varphi_R(x_i)^*$ by (12). Therefore, the $*$-algebra $\mathcal{X}_{\gamma,\delta}(\nu)$ is a right $*$-quantum space for the Hopf $*$-algebras $\mathcal{O}(U_q(N))$ and $\mathcal{O}(SU_q(N;\nu))$. If $\gamma = 0$ and $\nu_1 = \cdots = \nu_N = 1$, the $*$-algebra $\mathcal{X}_{\gamma,\delta}(\nu)$ for arbitrary $\delta \in \mathbb{R}$ might be considered as the *quantized algebra of polynomials* on $\mathbb{C}^N$. The elements $x_1, \cdots, x_N$ and $y_1 = x_1^*, \cdots, y_N = x_N^*$ are then interpreted as holomorphic and anti-holomorphic generators, respectively.

9.3 The Quantum Groups $O_q(N)$ and $Sp_q(N)$

Throughout this section, R denotes the R-matrix for the vector representation of one of the Drinfeld–Jimbo algebras $U_q(\mathfrak{g})$, $\mathfrak{g} = \mathrm{so}_{2n}, \mathrm{sp}_N$, and $U_{q^{1/2}}(\mathrm{so}_{2n+1})$; see Sect. 8.4 for details. Let us collect some notation and facts (repeated from Subsect. 8.4.3) that will often be used in the rest of the book.

We abbreviate $i' := N + 1 - i$. Let $N = 2n$ if N is even and $N = 2n + 1$ if N is odd. We define numbers ρ_i, ϵ and ϵ_i by setting

$$\rho_i = \tfrac{N}{2} - i \ \text{ if } \ i < i', \quad \rho_{i'} = -\rho_i \ \text{ if } \ i \le i'; \quad \epsilon = \epsilon_1 = \cdots = \epsilon_N = 1$$

for $\mathfrak{g} = \mathrm{so}_N$ and

$$\rho_i = \tfrac{N}{2} + 1 - i, \quad \rho_{i'} = -\rho_i \ \text{ if } \ i < i';$$

$$\epsilon_1 = \cdots = \epsilon_n = 1, \quad \epsilon = \epsilon_{n+1} = \cdots = \epsilon_N = -1$$

for $\mathfrak{g} = \mathrm{sp}_N$.

The matrix $\mathsf{C} = (\mathsf{C}^i_j)$ of the metric is defined by $\mathsf{C}^i_j := \epsilon_i \delta_{ij'} q^{-\rho_i}$. Then we have $\mathsf{C}^{-1} = \epsilon \mathsf{C}$ and $\mathsf{C}^i_{i'} = \epsilon(\mathsf{C}^{i'}_i)^{-1}$. The matrix R has been described by formula (8.61) and investigated in Propositions 8.25 for $\mathfrak{g} = \mathrm{so}_{2n+1}, \mathrm{so}_{2n}$ and 8.26 for $\mathfrak{g} = \mathrm{sp}_N$. A unique formula for the matrix entries of R and $\hat{\mathsf{R}} = \tau \circ \mathsf{R}$ which applies to all three cases is given by

$$\mathsf{R}^{ij}_{mn} = \hat{\mathsf{R}}^{ji}_{mn} = q^{\delta_{ij}-\delta_{ij'}}\delta_{im}\delta_{jn} + (q-q^{-1})\theta(i-m)(\delta_{jm}\delta_{in} - \mathsf{K}^{ji}_{mn}), \quad (30)$$

where $\mathsf{K}^{ji}_{mn} := \epsilon \mathsf{C}^j_i \mathsf{C}^m_n$ and $i, j, m, n = 1, 2, \cdots, N$. In particular we see that $\hat{\mathsf{R}}^{ij}_{mn} = \hat{\mathsf{R}}^{mn}_{ij}$ and $\mathsf{R}^{ji}_{mn} = \mathsf{R}^{nm}_{ij}$. Moreover, R obeys the property

$$\mathsf{R} = \mathsf{C}_1(\mathsf{R}^{t_1})^{-1}\mathsf{C}_1^{-1} = \mathsf{C}_2(\mathsf{R}^{-1})^{t_2}\mathsf{C}_2^{-1}, \quad (31)$$

where as usual $\mathsf{C}_1 = \mathsf{C} \otimes I, \mathsf{C}_2 = I \otimes \mathsf{C}$ and t_1 and t_2 mean the transpositions with respect to the first and second tensor factors, respectively. The matrix $\hat{\mathsf{R}}$ satisfies the cubic equation

$$(\hat{\mathsf{R}} - qI)(\hat{\mathsf{R}} + q^{-1}I)(\hat{\mathsf{R}} - \epsilon q^{\epsilon - N}I) = 0. \quad (32)$$

We suppose that $(q+q^{-1})(q - \epsilon q^{\epsilon-N})(q^{-1} + \epsilon q^{\epsilon - N}) \neq 0$. This assumption implies in particular that $q - q^{-1} \neq 0$. Except for Definition 7 below we also assume that $N > 2$. Then the transformations P_+, P_- and P_0 defined in Subsect. 8.4.3 are mutually orthogonal projections (that is, $\mathsf{P}_\nu^2 = \mathsf{P}_\nu$ and $\mathsf{P}_\nu \mathsf{P}_\mu = 0$ for $\nu \neq \mu$ and $\nu, \mu \in \{+, -, 0\}$) such that

$$\hat{\mathsf{R}} = q\mathsf{P}_+ - q^{-1}\mathsf{P}_- + \epsilon q^{\epsilon - N}\mathsf{P}_0. \quad (33)$$

Moreover, we have $\mathsf{K} = (1 + \epsilon(q-q^{-1})^{-1}(q^{N-\epsilon} - q^{\epsilon - N}))\mathsf{P}_0$.

9.3.1 The Hopf Algebras $\mathcal{O}(O_q(N))$ and $\mathcal{O}(Sp_q(N))$

Recall that the algebra $\mathcal{O}(SL_q(N))$ is obtained as the quotient of the algebra $\mathcal{O}(M_q(N))$ by the additional relation $\mathcal{D}_q = 1$. The algebras $\mathcal{O}(O_q(N))$ and $\mathcal{O}(Sp_q(N))$ will be defined as quotients of the algebra $\mathcal{A}(R)$, with $R = \mathsf{R}$ given by (30), by the metric condition

$$\mathbf{u}\mathsf{C}\mathbf{u}^t\mathsf{C}^{-1} = \mathsf{C}\mathbf{u}^t\mathsf{C}^{-1}\mathbf{u} = I. \quad (34)$$

In the classical case $q = 1$ the matrix C takes the form

$$\begin{pmatrix} 0 & 0 & \cdots & 0 & 1 \\ 0 & 0 & \cdots & 1 & 0 \\ . & . & \cdots & . & . \\ . & . & \cdots & . & . \\ 0 & 1 & \cdots & 0 & 0 \\ 1 & 0 & \cdots & 0 & 0 \end{pmatrix} \quad \text{and} \quad \begin{pmatrix} 0 & . & . & . & . & 1 \\ . & . & . & . & . & . \\ 0 & . & . & 1 & . & 0 \\ 0 & . & -1 & . & . & 0 \\ . & . & . & . & . & . \\ -1 & . & . & . & . & 0 \end{pmatrix}$$

for $O_q(N)$ and $Sp_q(N)$, respectively, so in this case (34) is just the metric condition which defines the complex Lie groups $O(N, \mathbb{C})$ and $Sp(N, \mathbb{C})$.

A heuristic motivation for relation (34) is provided by the following reasoning. Applying the transposition t_1 to the defining relation $\mathsf{R}\mathbf{u}_1\mathbf{u}_2 = \mathbf{u}_2\mathbf{u}_1\mathsf{R}$ for $\mathcal{A}(\mathsf{R})$, we derive $(\mathbf{u}^t)_1\mathsf{R}^{t_1}\mathbf{u}_2 = \mathbf{u}_2\mathsf{R}^{t_1}(\mathbf{u}^t)_1$. Since $\mathsf{R}^{t_1} = \mathsf{C}_1^{-1}\mathsf{R}^{-1}\mathsf{C}_1$ by the first equality of (31), we get $(\mathbf{u}^t)_1\mathsf{C}_1^{-1}\mathsf{R}^{-1}\mathsf{C}_1\mathbf{u}_2 = \mathbf{u}_2\mathsf{C}_1^{-1}\mathsf{R}^{-1}\mathsf{C}_1(\mathbf{u}^t)_1$. Multiplying this relation by $\mathbf{u}_1\mathsf{C}_1$ on the left and by $\mathsf{C}_1^{-1}\mathbf{u}_1$ on the right we obtain

$$(\mathbf{u}\mathsf{C}\mathbf{u}^t\mathsf{C}^{-1})_1\mathsf{R}^{-1}\mathbf{u}_2\mathbf{u}_1 = \mathbf{u}_1\mathbf{u}_2\mathsf{R}^{-1}(\mathsf{C}\mathbf{u}^t\mathsf{C}^{-1}\mathbf{u})_1. \tag{35}$$

On the other hand, the relation $\mathsf{R}\mathbf{u}_1\mathbf{u}_2 = \mathbf{u}_2\mathbf{u}_1\mathsf{R}$ yields

$$\mathsf{R}^{-1}\mathbf{u}_2\mathbf{u}_1 = \mathbf{u}_1\mathbf{u}_2\mathsf{R}^{-1}. \tag{36}$$

A comparison of (35) and (36) suggests that it is natural to impose the conditions (34).

Let $\mathcal{J} \equiv \langle \mathbf{u}\mathsf{C}\mathbf{u}^t\mathsf{C}^{-1} - I, \mathsf{C}\mathbf{u}^t\mathsf{C}^{-1}\mathbf{u} - I\rangle$ denote the two-sided ideal of the algebra $\mathcal{A}(\mathsf{R})$ generated by the entries of the matrices $\mathbf{u}\mathsf{C}\mathbf{u}^t\mathsf{C}^{-1} - I$ and $\mathsf{C}\mathbf{u}^t\mathsf{C}^{-1}\mathbf{u} - I$. The algebras $\mathcal{O}(O_q(N))$ and $\mathcal{O}(Sp_q(N))$ are defined now as quotient algebras $\mathcal{A}(\mathsf{R})/\mathcal{J}$, where R is the corresponding matrix given by (30). By Lemma 12 below, the two-sided ideal $\mathcal{J}$ has a single generator $\mathcal{Q}_q - 1$, so the algebras $\mathcal{O}(O_q(N))$ and $\mathcal{O}(Sp_q(N))$ can also be defined as the quotients $\mathcal{A}(\mathsf{R})/\langle\mathcal{Q}_q - 1\rangle$. This definition closely resembles that of the algebra $\mathcal{O}(SL_q(N))$.

Lemma 12. *In the algebra $\mathcal{A}(\mathsf{R})$ we have the equality*

$$\mathcal{Q}_q := \sum\nolimits_n \epsilon\mathsf{C}_j^{j'}\mathsf{C}_{n'}^{n}u_j^n u_{j'}^{n'} = \sum\nolimits_n \epsilon\mathsf{C}_i^{i'}\mathsf{C}_{n'}^{n}u_n^i u_{n'}^{i'} \quad \textit{for} \quad i,j = 1,2,\cdots,N. \tag{37}$$

The element $\mathcal{Q}_q - 1$ generates the two-sided ideal $\mathcal{J}$. Moreover, $\mathcal{Q}_q$ is group-like and belongs to the center of the algebra $\mathcal{A}(\mathsf{R})$.

Proof. Since P_0 and so K is a polynomial of $\hat{\mathsf{R}}$, the matrix equation $\hat{\mathsf{R}}\mathbf{u}_1\mathbf{u}_2 = \mathbf{u}_1\mathbf{u}_2\hat{\mathsf{R}}$ (by (3)) implies that $\mathsf{K}\mathbf{u}_1\mathbf{u}_2 = \mathbf{u}_1\mathbf{u}_2\mathsf{K}$. In terms of matrix entries, the latter reads as $\mathsf{C}_k^i\mathsf{C}_m^n u_j^n u_l^m = u_n^i u_m^k \mathsf{C}_m^n\mathsf{C}_l^j$. Since $\mathsf{C}_l^j = 0, j' \neq l$ and $(\mathsf{C}_{i'}^i)^{-1} = \epsilon\mathsf{C}_i^{i'}$, this gives the equality (37). Setting $k = i', l = j'$ and using the relations $\epsilon\mathsf{C}_{n'}^n = (\mathsf{C}^{-1})_n^{n'}$ and $\mathsf{C}_l^j = 0, j' \neq l$, we see that the diagonal entries of the matrices $\mathsf{C}\mathbf{u}^t\mathsf{C}^{-1}\mathbf{u} - I$ and $\mathbf{u}\mathsf{C}\mathbf{u}^t\mathsf{C}^{-1} - I$ are all equal to $\mathcal{Q}_q - 1$. Setting $k \neq i'$, $l = j'$ and $k = i'$, $l \neq j'$, it follows that the off-diagonal entries of these matrices vanish. Hence the element $\mathcal{Q}_q - 1$ generates the ideal $\mathcal{J}$.

Using formula (37) it is straightforward to show that $\Delta(\mathcal{Q}_q) = \mathcal{Q}_q \otimes \mathcal{Q}_q$. Set $\tilde{\mathbf{u}} := \mathsf{C}\mathbf{u}^t\mathsf{C}^{-1}$. By the preceding computations, we have $\tilde{\mathbf{u}}\mathbf{u} = \mathbf{u}\tilde{\mathbf{u}} = \mathcal{Q}_q \cdot I$. As in the case of the quantum determinant (see Proposition 9), the latter implies that $\mathcal{Q}_q$ is in the center of $\mathcal{A}(\mathsf{R})$. □

Proposition 13. *There are unique Hopf algebra structures on the algebras $\mathcal{O}(O_q(N))$ and $\mathcal{O}(Sp_q(N))$ such that $\Delta(u_j^i) = u_k^i \otimes u_j^k$ and $\varepsilon(u_j^i) = \delta_{ij}$. For the antipode S of both Hopf algebras we have $S(u_j^i) = \mathsf{C}_{i'}^i u_{i'}^{j'}(\mathsf{C}^{-1})_j^{j'} = \epsilon_i\epsilon_j q^{\rho_j - \rho_i} u_{i'}^{j'}$ and $S^2(u_j^i) = \mathsf{D}_i^{-1}u_j^i\mathsf{D}_j$ with $\mathsf{D}_i := (\mathsf{C}_i^{i'})^2 = q^{2\rho_i}, i,j = 1,\cdots,N$.*

Proof. Since $\mathcal{Q}_q$ is group-like, $\mathcal{J} = \langle \mathcal{Q}_q - 1 \rangle$ is a biideal, so that the quotient $\mathcal{A}(\mathsf{R})/\mathcal{J}$ is again a bialgebra. Arguing similarly as in the proof of Proposition 10 it can be shown that there is an algebra anti-homomorphism S of $\mathcal{A}(\mathsf{R})/\mathcal{J}$ such that $S(u^i_j) = \mathsf{C}^i_{i'} u^{j'}_{i'} (\mathsf{C}^{-1})^{j'}_j$. By (34), S fulfills the conditions (1.26) of the antipode on the generators u^i_j. Hence $\mathcal{A}(\mathsf{R})/\mathcal{J}$ is a Hopf algebra. The formula for $S^2(u^i_j)$ follows at once from that for $S(u^i_j)$. □

The matrix $\mathsf{D} := \mathsf{C}^t\mathsf{C}$ has the entries $\mathsf{D}^i_j := \delta_{ij}\mathsf{D}_i$. In matrix notation the formulas for S and S^2 in Proposition 13 can be written as

$$S(\mathbf{u}) = \mathsf{C}\mathbf{u}^t\mathsf{C}^{-1} \quad \text{and} \quad S^2(\mathbf{u}) = \mathsf{D}^{-1}\mathbf{u}\mathsf{D}.$$

For any Hopf algebra structure satisfying $\Delta(\mathbf{u}) = \mathbf{u} \otimes \mathbf{u}$ and $\varepsilon(\mathbf{u}) = I$ the matrix $S(\mathbf{u})$ must be the inverse of $\mathbf{u}$. Thus the formula $S(\mathbf{u}) = \mathsf{C}\mathbf{u}^t\mathsf{C}^{-1}$ is, of course, expected from the metric condition (34).

Definition 7. *The Hopf algebras $\mathcal{O}(O_q(N))$ and $\mathcal{O}(Sp_q(N))$ are called the* coordinate algebras of the quantum groups *$O_q(N)$ and $Sp_q(N)$, respectively.*

By writing down the defining relations for the algebras $\mathcal{O}(Sp_q(2))$ and $\mathcal{O}(SL_{q^2}(2))$ one observes that they are equivalent, so the mapping $u^i_j \rightarrow u^i_j$ defines an isomorphism of the Hopf algebras $\mathcal{O}(Sp_q(2))$ and $\mathcal{O}(SL_{q^2}(2))$. We express this fact by saying that the quantum groups $Sp_q(2)$ and $SL_{q^2}(2)$ are isomorphic.

9.3.2 The Quantum Vector Space for the Quantum Group $O_q(N)$

Throughout this subsection R denotes the matrix (30) in the orthogonal case.

Definition 8. *Let $f_s(t) = t^2 - (q + q^{1-N})t + q^{2-N}$ and $f_a(t) = t + q^{-1}$. The algebras $\mathcal{O}(O^N_q) := \mathcal{X}(f_s; \mathsf{R})$ and $\Lambda(O^N_q) := \mathcal{X}(f_a; \mathsf{R})$ are called the* coordinate algebra *and the* exterior algebra of the quantum Euclidean space *O^N_q, respectively.*

By (30), the matrix $\hat{\mathsf{R}}$ is symmetric, so that $\mathcal{X}_L(f; R) = \mathcal{X}_R(f; R) = \mathcal{X}(f; R)$ for any polynomial f. Hence, by Proposition 4, both algebras $\mathcal{O}(O^N_q)$ and $\Lambda(O^N_q)$ are left and right $\mathcal{O}(O_q(N))$-quantum spaces.

Since $f_s(\hat{\mathsf{R}}) = \lambda_+(q^{-1} + q^{1-N})\mathsf{P}_-$ by (8.75) and $\lambda_+(q^{-1} + q^{1-N}) \neq 0$ by assumption, the defining relations $f_s(\hat{\mathsf{R}})\mathbf{x}_1\mathbf{x}_2 = 0$ for the algebra $\mathcal{O}(O^N_q)$ can also be expressed as $\mathsf{P}_-\mathbf{x}_1\mathbf{x}_2 = 0$. Clearly, the entries of the $N^2 \times 1$ matrix $\mathsf{P}_-\mathbf{x}_1\mathbf{x}_2$ span $\operatorname{im}(\mathsf{P}_-)^t$. But $\operatorname{im}(\mathsf{P}_-)^t$ coincides with $\operatorname{im}\mathsf{P}_-$, because the matrix P_- is symmetric. Hence $\mathcal{O}(O^N_q)$ is also the quotient of the tensor algebra $T(V) \cong \mathbb{C}\langle x_i \rangle$ by the two-sided ideal generated by $\operatorname{im}\mathsf{P}_- = \ker(\hat{\mathsf{R}} + q^{-1}I)$.

We also have $f_a(\hat{\mathsf{R}}) = \hat{\mathsf{R}} + q^{-1}I = \lambda_+\mathsf{P}_+ + (q^{-1} + q^{1-N})\mathsf{P}_0$, so the algebra $\Lambda(O^N_q)$ has the defining relations $\mathsf{P}_+\mathbf{y}_1\mathbf{y}_2 = 0$ and $\mathsf{P}_0\mathbf{y}_1\mathbf{y}_2 = 0$. Thus, $\Lambda(O^N_q)$ is the quotient of the tensor algebra $T(V) \cong \mathbb{C}\langle y_i \rangle$ by the two-sided ideal generated by $\operatorname{im}\mathsf{P}_+ = \ker(\hat{\mathsf{R}} - qI)$ and $\operatorname{im}\mathsf{P}_0 = \ker(\hat{\mathsf{R}} - q^{1-N}I)$.

Remark 2. Our exterior algebra $\Lambda(O_q^N)$ differs from the one in [RTF], where $\Lambda(O_q^N)$ is defined by the relations $\mathsf{P}_+\mathbf{y}_1\mathbf{y}_2 = 0$ only. △

Some algebraic properties of these algebras are collected in the following two propositions.

Proposition 14. (i) *The algebra $\mathcal{O}(O_q^N)$ for even $N = 2n$ has the following defining relations:*

$$\begin{aligned} x_i x_j &= q x_j x_i, \quad i<j, \ i \neq j', \\ x_{i'} x_i - x_i x_{i'} &= (q-q^{-1}) \sum\nolimits_{i<j\leq n} q^{j-i-1} x_j x_{j'}, \quad i < i'. \end{aligned}$$

(ii) *The algebra $\mathcal{O}(O_q^N)$ for odd $N = 2n+1$ has the defining relations*

$$x_i x_j = q x_j x_i, \quad i<j, \ i \neq j',$$

$$x_{i'} x_i - x_i x_{i'} = q^{n-i}(q^{1/2} - q^{-1/2}) x_{n+1}^2 + (q-q^{-1}) \sum_{i<j\leq n} q^{j-i-1} x_j x_{j'}, \quad i < i'.$$

(iii) *The element $z := \mathbf{x}^t \mathsf{C} \mathbf{x} \equiv \sum_i x_i x_{i'} \mathsf{C}^i_{i'} = \sum_i q^{-\rho_i} x_i x_{i'}$ belongs to the center of the algebra $\mathcal{O}(O_q^N)$ and satisfies $\varphi_L(z) = 1 \otimes z$ and $\varphi_R(z) = z \otimes 1$.*

(iv) *$\{x_1^{k_1} x_2^{k_2} \cdots x_N^{k_N} \mid k_1, \cdots, k_N \in \mathbb{N}_0\}$ is a vector space basis of $\mathcal{O}(O_q^N)$.*

Proof. (i) and (ii): Set $\alpha := (1+q^{N-2})^{-1}$. Since $-\lambda_+ \mathsf{P}_- = \hat{\mathsf{R}} - qI + \alpha(q-q^{-1})\mathsf{K}$, the defining relations $\mathsf{P}_-\mathbf{x}_1\mathbf{x}_2 = 0$ of $\mathcal{O}(O_q^N)$ can be stated as

$$\hat{\mathsf{R}}^{ij}_{kl} x_k x_l - q x_i x_j + \alpha(q-q^{-1}) \mathsf{C}^i_j x_k \mathsf{C}^k_{k'} x_{k'} = 0, \quad i,j = 1, \cdots, N. \tag{38}$$

We insert the values $\hat{\mathsf{R}}^{ij}_{kl}$ from (30) in this equation. If $i < j$, $i \neq j'$, then we get $x_i x_j = q x_j x_i$. First we suppose that N is even. Subtracting the equations (38) for $i = n, j = i'$ and for $i = n+1, j = i'$ we conclude that $x_n x_{n'} = x_{n'} x_n$. Applying (38) with $j = i', i < i'$, we obtain

$$x_{i'} x_i - x_i x_{i'} = q^{i-n+1}(q-q^{-1}) \Big(\sum\nolimits_{j<i'} x_j x_{j'} q^{-\rho_j} - \alpha \sum\nolimits_k x_k x_{k'} q^{-\rho_k} \Big),$$

so that $x_{i'+1} x_{i-1} - x_{i-1} x_{i'+1} = q(x_{i'} x_i - x_i x_{i'}) + (q-q^{-1}) x_i x_{i'}$ and the above relations follow. Now suppose that N is odd. From (38) applied with $i = j = n+1$ and with $i = n, j = i'$ we conclude that $x_{n'} x_n - x_n x_{n'} = (q^{1/2} - q^{-1/2}) x_{n+1}^2$. Similarly as for even N, we derive that $x_{i'} x_i - x_i x_{i'} = q(x_{i'-1} x_{i+1} - x_{i+1} x_{i'-1}) + (q-q^{-1}) x_{i+1} x_{i'-1}$ which implies the above relations.

On the other hand, the above relations correspond to an $N(N-1)/2$-dimensional subspace of $V \otimes V$ which is contained in im P_- by the preceding proof. Since dim (im P_-) $= N(N-1)/2$, this subspace coincides with im P_-. Hence the above relations define the algebra $\mathcal{O}(O_q^N)$.

(iii): That z is in the center of $\mathcal{O}(O_q^N)$ follows from the relations in (i) and (ii). From the definition (10) of φ_L and the metric condition (34) we obtain

$$\varphi_L(z) = \sum\nolimits_{i,k,l} \mathsf{C}^i_{i'} u^i_k u^{i'}_l \otimes x_k x_l = \sum\nolimits_{k,l} \mathsf{C}^k_l \cdot 1 \otimes x_k x_l = 1 \otimes z.$$

The proof of $\varphi_R(z) = z \otimes 1$ is similar.

(iv): Using the relations from (i) and (ii) the assertion can be derived from the diamond Lemma 4.8. □

Proposition 15. (i) *The algebra $\Lambda(O_q^N)$ for odd $N = 2n+1$ has the following set of defining relations:*

$$y_i^2 = 0,\ i \neq i', \quad y_i y_j = -q^{-1} y_j y_i,\ \ i < j,\ i \neq j',$$

$$y_{i'} y_i + y_i y_{i'} = (q - q^{-1}) \sum\nolimits_{1 \leq j < i} q^{j-i+1} y_j y_{j'},\ \ i < i',$$

$$y_{n+1}^2 = (q^{1/2} - q^{-1/2}) \sum\nolimits_{1 \leq j \leq n} q^{j-n} y_j y_{j'}.$$

(ii) *The first two lines of equations of* (i) *are the defining relations of the algebra $\Lambda(O_q^N)$ for even $N = 2n$.*

(iii) *$\{y_1^{i_1} y_2^{i_2} \cdots y_N^{i_N} \mid i_1, \cdots, i_N \in \{0,1\}\}$ is a basis of the vector space $\Lambda(O_q^N)$.*

Proof. The proof is similar to that of Proposition 14. □

Note that for even N there is no index i such that $i = i'$ and hence we have $y_i^2 = 0$ for all $i = 1, 2, \cdots, N$. However, for odd $N = 2n+1$ it follows from Proposition 15 that the element y_{n+1}^2 of $\Lambda(O_q^N)$ is nonzero in contrast to the classical situation.

Definition 9. *The quotient of the algebra $\mathcal{O}(O_q^N)$ by the two-sided ideal $\langle z-1 \rangle$ generated by $z-1$ is called the* coordinate algebra of the quantum orthogonal complex unit sphere $S_q^{O,N-1}$.

By Proposition 14(iii), we have $\varphi_L(\langle z-1 \rangle) \subseteq \mathcal{O}(O_q^N) \otimes \langle z-1 \rangle$, so the left coaction φ_L of $O_q(N)$ on O_q^N passes to the quotient algebra $\mathcal{O}(S_q^{O,N-1}) = \mathcal{O}(O_q^N)/\langle z-1 \rangle$ and defines a left coaction there. This turns $S_q^{O,N-1}$ into a left quantum space for the quantum group $O_q(N)$. In a similar way, $S_q^{O,N-1}$ becomes a right quantum space for $O_q(N)$. The central element $z = \sum_i x_i x_{i'} q^{-\rho_i}$ can be regarded as a quantum analog of the $O(N,\mathbb{C})$-invariant quadratic form on $\mathbb{C}^N$.

Finally, we consider the classical limit $q \to 1$ of the three algebras $\mathcal{O}(O_q^N)$, $\Lambda(O_q^N)$ and $\mathcal{O}(S_q^{O,N-1})$. Recall that we excluded the classical case $q = 1$ since we assumed that $1 - q^{-N} \neq 0$. In the limit $q \to 1$, the algebra $\mathcal{O}(O_q^N)$ is the polynomial algebra in N commuting indeterminates (that is, the coordinate algebra of $\mathbb{C}^N$), the element z is $\sum_i x_i x_{i'}$ and $\Lambda(O_q^N)$ is the exterior algebra of the vector space $\mathbb{C}^N$. On changing variables to v_i defined by $\sqrt{2} x_i = v_i + \sqrt{-1} v_{i'}$, $\sqrt{2} x_{i'} = v_i - \sqrt{-1} v_{i'}$, for $i = 1, 2, \cdots, n$ and $x_{n+1} = v_{n+1}$ if $N = 2n+1$ is odd, the defining equation $z = 1$ of $S_q^{O,N-1}$ becomes the equation $\sum_i v_i^2 = 1$ of the unit sphere in $\mathbb{C}^N$.

9.3.3 The Quantum Group $SO_q(N)$

In order to define the quantum determinant for the quantum group $O_q(N)$ we use the same idea as in the case of $SL_q(N)$ (see Subsect. 9.2.2). By Proposition 15(iii), the vector space $\Lambda(O_q^N)_N$ of elements of degree N of $\Lambda(O_q^N)$ has dimension 1. Let $\varphi_{R,N}$ denote the restriction of the right coaction φ_R of $\mathcal{O}(O_q(N))$ on $\Lambda(O_q^N)$ to the invariant subspace $\Lambda(O_q^N)_N$. There exists a unique element $\mathcal{D}_q$ of $\mathcal{O}(O_q(N))$ such that $\varphi_{R,N}(\zeta) = \zeta\otimes\mathcal{D}_q$ for $\zeta \in \Lambda(O_q^N)_N$. Being the matrix coefficient of the one-dimensional corepresentation $\varphi_{R,N}$, the element $\mathcal{D}_q$ satisfies $\Delta(\mathcal{D}_q) = \mathcal{D}_q\otimes\mathcal{D}_q$ and $\varepsilon(\mathcal{D}_q) = 1$. The latter facts imply that $\mathcal{D}_q$ is invertible in the algebra $\mathcal{O}(O_q(N))$ and $S(\mathcal{D}_q) = \mathcal{D}_q^{-1}$. Further, it follows that the two-sided ideal $\langle\mathcal{D}_q - 1\rangle$ of the algebra $\mathcal{O}(O_q(N))$ is a Hopf ideal, so the quotient $\mathcal{O}(O_q(N))/\langle\mathcal{D}_q - 1\rangle$ is also a Hopf algebra. It will be denoted by $\mathcal{O}(SO_q(N))$.

Definition 10. *The element $\mathcal{D}_q$ of $\mathcal{O}(O_q(N))$ is called the* quantum determinant *of the matrix $\mathbf{u} = (u_j^i)$ for the quantum group $O_q(N)$. The Hopf algebra $\mathcal{O}(SO_q(N))$ is called the* coordinate algebra of the quantum group $SO_q(N)$.

Let us emphasize that the element $\mathcal{D}_q$ defined above is different from the element $\mathcal{D}_q$ occurring in Subsect. 9.2.2. They are quantum determinants of different series of quantum groups and belong to different algebras.

Example 3. An explicit expression of the quantum determinant for $O_q(3)$ is

$$\begin{aligned}\mathcal{D}_q &= u_1^1u_2^2u_3^3 + qu_3^1u_1^2u_2^3 + qu_2^1u_3^2u_1^3 - qu_2^1u_1^2u_3^3 - qu_1^1u_3^2u_2^3 \\ &\quad -q^2u_3^1u_2^2u_1^3 - q(q^{1/2} - q^{-1/2})u_2^1u_2^2u_2^3 \\ &= u_1^1u_2^2u_3^3 + (q+q^2)u_2^1u_3^2u_1^3 - (1+q)u_1^1u_3^2u_2^3 - q^2u_3^1u_2^2u_1^3.\end{aligned}$$

Here the first equality is derived from the definition of $\mathcal{D}_q$, while the second one follows from the commutation rules of the u_j^i. △

A remarkable peculiarity of quantum group theory is that embeddings or homomorphisms of classical Lie groups (for instance, the group inclusions $SO(N) \subseteq SL(N)$ and $Sp(N) \subseteq SL(N)$ for $N \geq 3$) have no quantum analog in general. The technical reason is that quantum groups corresponding to different series of simple Lie groups are derived from different FRT bialgebras $\mathcal{A}(R)$. There is a general principle behind this which says that "quantization removes degeneracy". This means that from the quantum group point of view most of the classical group homomorphisms are particular cases. However, at least two relations between classical simple Lie groups carry over to the quantum case. That the quantum groups $SL_{q^2}(2)$ and $Sp_q(2)$ are isomorphic was already mentioned in Subsect. 9.3.1. The second one is the result $SU(2)/\{I,-I\} \simeq SO(3)$. Our next aim is to give a quantum version of this classical fact. The following notions are also used in Sects. 11.5 and 11.6.

Definition 11. *Let $\mathcal{A}$ be a Hopf algebra. A* quantum subgroup *of $\mathcal{A}$ is a Hopf algebra $\mathcal{B}$ together with a surjective Hopf algebra homomorphism $\pi : \mathcal{A} \to \mathcal{B}$.*

The elements of the space $\mathcal{A}/\mathcal{B} := \{a \in \mathcal{A} \mid (\mathrm{id} \otimes \pi)\Delta(a) = a \otimes 1\}$ *are then called* right $\mathcal{B}$-invariant.

If the Hopf algebras $\mathcal{A}$ and $\mathcal{B}$ in Definition 11 are coordinate Hopf algebras of quantum groups G_q and H_q, respectively, then we say that H_q is a quantum subgroup of G_q.

The classical picture for these concepts is clear: Let $\mathrm{i} : H \to G$ be the inclusion for a subgroup H of a group G. If f is a function on G, then $\pi(f)$ is the function on H given by $\pi(f)(h) := f(\mathrm{i}(h))$, $h \in H$, and the right invariance of f means that $f(g\mathrm{i}(h)) = f(g)$ for $h \in H$ and $g \in G$.

Now let $\mathcal{A} = \mathcal{O}(SL_q(2))$ and let $\mathcal{B} = \mathbb{C}\mathbb{Z}_2$ be the group algebra of $\mathbb{Z}_2 = \mathbb{Z}/2\mathbb{Z}$. The two elements of $\mathbb{Z}_2$ are denoted by $\bar{0}$ and $\bar{1}$. Let $\mathcal{A}_e$ and $\mathcal{A}_o$ be the linear spaces spanned by the basis elements $a^ib^jc^k, d^ib^jc^k$ of $\mathcal{A}$ with even and odd degrees $i+j+k$, respectively. Clearly, there is a Hopf algebra homomorphism $\pi : \mathcal{A} \to \mathcal{B}$ such that $\pi(a_e) = \varepsilon(a_e)\bar{0}$, $a_e \in \mathcal{A}_e$, and $\pi(a_o) = \varepsilon(a_o)\bar{1}$, $a_o \in \mathcal{A}_o$. Hence $\mathcal{B}$ is a quantum subgroup of $\mathcal{A}$. Obviously, we have $\mathcal{A}/\mathcal{B} \simeq \mathcal{A}_e$. Since $\mathcal{A}_e$ is also spanned by the matrix coefficients of odd-dimensional irreducible corepresentations T_l of $SL_q(2)$, $\mathcal{A}_e$ is a Hopf subalgebra of $\mathcal{A}$. Let t_{ij}, $i,j = 1,2,3$, be the elements of $\mathcal{O}(SL_q(2))$ defined by

$$(t_{ij}) = \begin{pmatrix} a^2 & (1+q^{-2})^{1/2}ab & -b^2 \\ (1+q^{-2})^{1/2}ac & 1+(q+q^{-1})bc & -(1+q^{-2})^{1/2}bd \\ -c^2 & -(1+q^{-2})^{1/2}cd & d^2 \end{pmatrix}.$$

Then (t_{ij}) is a matrix for the corepresentation T_1 of the Hopf algebra $\mathcal{O}(SL_q(2))$. It is obtained from the matrix described by formula (4.75) when the third basis vector is multiplied by -1. From the explicit form of these matrix entries we see that $\mathcal{A}_e$ is precisely the unital subalgebra of $\mathcal{A}$ generated by the elements t_{ij}. With some computational work (see [Dij1] for details) it can be shown that the t_{ij} satisfy all defining relations of $SO_{q^{1/2}}(3)$ and no other relations than those. Hence there is an isomorphism ϑ of the algebras $\mathcal{A}_e$ and $\mathcal{O}(SO_{q^{1/2}}(3))$ which sends the element t_{ij} to the generator u^i_j of $SO_{q^{1/2}}(3)$. Since the t_{ij} are matrix coefficients of a corepresentation, it follows that ϑ is also an isomorphism of the Hopf algebras $\mathcal{A}_e$ and $\mathcal{O}(SO_{q^{1/2}}(3))$. Summarizing the preceding, as a quantum anolog of the classical relation $SU(2)/\{I,-I\} \cong SO(3)$ we have the Hopf algebra isomorphism

$$\mathcal{O}(SL_q(2))/\mathbb{C}\mathbb{Z}_2 \simeq \mathcal{O}(SO_{q^{1/2}}(3)).$$

9.3.4 The Quantum Vector Space for the Quantum Group $Sp_q(N)$

In this subsection let R be the R-matrix from (30) for $U_q(\mathrm{sp}_N)$.

Definition 12. *Set* $f_s(t) = t - q$ *and* $f_a(t) = t^2 + (q^{-1} + q^{-1-N})t + q^{-2-N}$. *We call* $\mathcal{O}(Sp^N_q) := \mathcal{X}(f_s; \mathsf{R})$ *and* $\Lambda(Sp^N_q) := \mathcal{X}(f_a; \mathsf{R})$ *the* coordinate algebra *and the* exterior algebra of the quantum symplectic space Sp^N_q, *respectively.*

We state some properties of these algebras which can be derived in a similar manner as for the corresponding quantum spaces of $O_q(N)$.

Because the matrix $\hat{\mathsf{R}}$ is symmetric, both algebras $\mathcal{O}(Sp_q^N)$ and $\Lambda(Sp_q^N)$ are left and right quantum spaces for the Hopf algebra $\mathcal{O}(Sp_q(N))$ by Proposition 4. From the equation $f_s(\hat{\mathsf{R}}) = -\lambda_+\mathsf{P}_- - (q+q^{-1-N})\mathsf{P}_0$ we conclude that the algebra $\mathcal{O}(Sp_q^N)$ has the defining relations $\mathsf{P}_-\mathbf{x}_1\mathbf{x}_2 = 0$ and $\mathsf{P}_0\mathbf{x}_1\mathbf{x}_2 = 0$ and that $\mathcal{O}(Sp_q^N)$ is the quotient of $T(V) \cong \mathbb{C}\langle x_i\rangle$ by the two-sided ideal generated by im $\mathsf{P}_- = \ker(\hat{\mathsf{R}} + q^{-1}I)$ and im $\mathsf{P}_0 = \ker(\hat{\mathsf{R}} + q^{-1-N}I)$. Since the element $\sum_i x_i x_{i'} \mathsf{C}^i_{i'} = \sum_i x_i x_{i'} \epsilon_i q^{-\rho_i}$ belongs to im P_0, it vanishes in the algebra $\mathcal{O}(Sp_q^N)$.

Since $f_a(\mathsf{R}) = \lambda_+(q+q^{-1-N})\mathsf{P}_+$, the defining relations of $\Lambda(Sp_q^N)$ can be rewritten as $\mathsf{P}_+\mathbf{y}_1\mathbf{y}_2 = 0$ or, equivalently, $\Lambda(Sp_q^N)$ is the quotient of $T(V) \cong \mathbb{C}\langle y_i\rangle$ by the two-sided ideal generated by im $\mathsf{P}_+ = \ker(\hat{\mathsf{R}} - qI)$.

Proposition 16. (i) *$\mathcal{O}(Sp_q^N)$ is the algebra with generators $x_1, \cdots, x_N$ and defining relations*

$$x_i x_j = q x_j x_i, \quad i < j, \quad i \neq j',$$

$$x_{i'} x_i - q^{-2} x_i x_{i'} = -(q - q^{-1}) \sum\nolimits_{j<i} q^{j-i-1} x_j x_{j'}, \quad i < i'.$$

(ii) *$\{x_1^{k_1} \cdots x_N^{k_N} \,|\, k_1, \cdots, k_N \in \mathbb{N}_0\}$ is a vector space basis of $\mathcal{O}(Sp_q^N)$.*

Proposition 17. (i) *$\Lambda(Sp_q^N)$ has the defining relations*

$$y_i^2 = 0,$$

$$y_i y_j = -q^{-1} y_j y_i, \quad i < j, \quad i \neq j',$$

$$y_{i'} y_i + q^2 y_i y_{i'} = -(q - q^{-1}) \sum\nolimits_{i<j\leq n} q^{j-i+1} y_j y_{j'}, \quad i < i'.$$

(ii) *$\{y_1^{i_1} \cdots y_N^{i_N} \mid i_1, \cdots, i_N \in \{0, 1\}\}$ is a basis of the vector space $\Lambda(Sp_q^N)$.*

Proof. The proofs of both propositions are similar to that of the corresponding assertions for $\mathcal{O}(O_q^N)$ and $\Lambda(O_q^N)$. □

9.3.5 Real Forms of $O_q(N)$ and $Sp_q(N)$ and *-Quantum Spaces

As in Subsect. 9.2.4 we distinguish two cases of parameter values.

Case 1: $|q| = 1$. Similarly as in Subsect. 9.2.4, we then have $\overline{\mathsf{R}} = \mathsf{R}^{-1}$, that is, R is inverse real, and $\bar{\mathsf{C}}\mathsf{C}^t = I$. From these facts and Proposition 3(ii) it follows that the Hopf algebras $\mathcal{O}(O_q(N))$ and $\mathcal{O}(Sp_q(N))$ equipped with the involutions determined by $(u^i_j)^* = u^i_j$ are Hopf $*$-algebras. These Hopf $*$-algebras are denoted by $\mathcal{O}(O_q(n,n))$ if $N = 2n$, $\mathcal{O}(O_q(n,n+1))$ if $N = 2n+1$ and $\mathcal{O}(Sp_q(N;\mathbb{R}))$, respectively. They are called the *coordinate algebras of the real quantum groups* $O_q(n,n), O_q(n,n+1)$ and $Sp_q(N;\mathbb{R})$, respectively.

The algebras $\mathcal{O}(O_q^N)$ and $\mathcal{O}(Sp_q^N)$ equipped with the involutions given by $x_i^* := x_i$ become $*$-algebras. They are left and right $*$-quantum spaces for the Hopf $*$-algebras $\mathcal{O}(O_q(n,n))$, $\mathcal{O}(O_q(n,n+1))$ and $\mathcal{O}(Sp_q(N;\mathbb{R}))$, respectively.

Case 2: $q \in \mathbb{R}$. Then the R-matrix R is real. Let $\nu = (\nu_1, \cdots, \nu_N)$ be an N-tuple of numbers $\nu_i \in \{1,-1\}$ such that $\nu_{i'} = \epsilon\nu_i$, $i = 1,\cdots,N$. Moreover, we assume that $\nu_{n+1} = 1$ and $q>0$ in the case of $O_q(2n+1)$. Then there are involutions on $\mathcal{O}(O_q(N))$ and $\mathcal{O}(Sp_q(N))$ given by $(u_j^i)^* := \nu_i\nu_j S(u_i^j) = \nu_i\nu_j\epsilon_i\epsilon_j q^{\rho_i-\rho_j} u_{j'}^{i'}$, $i,j = 1,\cdots,N$, turning these algebras into Hopf $*$-algebras. (This assertion can be derived from Proposition 3(i) using formula (37) and arguing similarly as in Subsect. 9.2.4, Case 2.) We denote these Hopf $*$-algebras by $\mathcal{O}(O_q(N;\nu))$ and $\mathcal{O}(Sp_q(N;\nu))$ and call the *coordinate algebras of the real quantum groups* $O_q(N;\nu)$ and $Sp_q(N;\nu)$, respectively.

Let ν be as above and let $\delta \in \{1,-1\}$. The algebras $\mathcal{O}(O_q^N)$, $\mathcal{O}(Sp_q^N)$, $\Lambda(O_q^N)$ and $\Lambda(Sp_q^N)$ equipped with the involutions determined by the formulas $x_i^* := \epsilon_i\nu_i q^{-\delta\rho_i} x_{i'}$ and $y_i^* := \epsilon_i\nu_i q^{-\delta\rho_i} y_{i'}, i = 1,\cdots,N$, become $*$-algebras, denoted by $\mathcal{O}(O_q^N(\delta,\nu))$, $\mathcal{O}(Sp_q^N(\delta,\nu))$, $\Lambda(O_q^N(\delta,\nu))$ and $\Lambda(Sp_q^N(\delta,\nu))$, respectively. (The defining relations listed in Propositions 14–17 are indeed invariant under these involutions.) Then one verifies that $\mathcal{O}(O_q^N(1,\nu))$, $\Lambda(O_q^N(1,\nu))$ and $\mathcal{O}(Sp_q^N(1,\nu))$, $\Lambda(Sp_q^N(1,\nu))$ are right $*$-quantum spaces for the Hopf $*$-algebras $\mathcal{O}(O_q(N;\nu))$ and $\mathcal{O}(Sp_q(N;\nu))$, respectively, the coactions being given by (10). Likewise, $\mathcal{O}(O_q^N(-1,\nu)), \Lambda(O_q^N(-1,\nu))$ and $\mathcal{O}(Sp_q^N(-1,\nu))$, $\Lambda(Sp_q^N(-1,\nu))$ are left $*$-quantum spaces for the Hopf $*$-algebras $\mathcal{O}(O_q(N;\nu))$ and $\mathcal{O}(Sp_q(N;\nu))$, respectively.

Recall that the quantum determinant $\mathcal{D}_q$ for $O_q(N)$ is defined by the property $\varphi_R(\zeta) = \zeta\otimes\mathcal{D}_q$, $\zeta \in \Lambda(O_q^N)_N$. For the $*$-algebras $\Lambda(O_q^N(1,\nu))$ and $\mathcal{O}(O_q(N;\nu))$ we have $\zeta^*\otimes\mathcal{D}_q^* = \varphi_R(\zeta)^* = \varphi_R(\zeta^*) = \zeta^*\otimes\mathcal{D}_q$ and so $\mathcal{D}_q^* = \mathcal{D}_q$. Hence the quotient Hopf algebra $\mathcal{O}(SO_q(N)) = \mathcal{O}(O_q(N))/\langle\mathcal{D}_q - 1\rangle$ with the $*$-structure inherited from $\mathcal{O}(O_q(N;\nu))$ is a Hopf $*$-algebra $\mathcal{O}(SO_q(N;\nu))$.

In the case $\nu_1 = \cdots = \nu_N = 1$ the corresponding Hopf $*$-algebras defined above are denoted by $\mathcal{O}(O_q(N;\mathbb{R})), \mathcal{O}(SO_q(N;\mathbb{R}))$ and $\mathcal{O}(USp_q(N))$. (The latter notation is suggested by the fact that the compact form of the classical symplectic group is just the intersection of $Sp(N,\mathbb{C})$ with the unitary group $U(N)$.) For each of these three Hopf $*$-algebras the fundamental matrix $\mathbf{u} = (u_j^i)$ is unitary, that is, $\mathbf{u}^*\mathbf{u} = \mathbf{u}\mathbf{u}^* = I$, where $(\mathbf{u}^*)_j^i := (u_i^j)^* = S(u_j^i)$. Hence we call $O_q(N;\mathbb{R})$, $SO_q(N;\mathbb{R})$ and $USp_q(N)$ *compact quantum groups*.

Let us return to the $*$-algebra $\mathcal{O}(O_q^N(\delta,\nu))$. Its central element $z = \sum_i x_i x_{i'} q^{-\rho_i}$ is hermitean (that is, $z = z^*$) and can be written as $z = \sum_i \nu_i q^{(\delta-1)\rho_i} x_i x_i^*$. Thus, the quotient algebra $\mathcal{O}(S_q^{N-1}) = \mathcal{O}(O_q^N)/\langle z-1\rangle$ becomes a $*$-algebra, denoted by $\mathcal{O}(S_q^{N-1}(\delta,\nu))$, with $*$-structure induced from $\mathcal{O}(O_q^N(\delta,\nu))$. By construction, $S_q^{N-1}(1,\nu)$ and $S_q^{N-1}(-1,\nu)$ are right and left $*$-quantum spaces, respectively, for the Hopf $*$-algebra $\mathcal{O}(O_q(N;\nu))$. In the case $\nu_1 = \cdots = \nu_N = 1$ the $*$-algebra $\mathcal{O}(S_q^{N-1}(\delta,\nu))$ is called the *coordinate algebra of the quantum Euclidean real unit sphere* $S_{q,\delta}^{N-1}(\mathbb{R})$ and is denoted by $\mathcal{O}(S^{N-1}_{q,\delta}(\mathbb{R}))$. Clearly, in the limit $q \to 1$ the $*$-algebra $\mathcal{O}(S^{N-1}_{q,\delta}(\mathbb{R}))$ gives the coordinate algebra of the unit sphere S^{N-1} of $\mathbb{R}^N$.

9.4 Dual Pairings of Drinfeld–Jimbo Algebras and Coordinate Hopf Algebras

The duality between Drinfeld–Jimbo algebras and coordinate Hopf algebras supports the common viewpoint that both Hopf algebras provide dual approaches to the same "quantum object". It appears to be a powerful tool for the study of quantum groups (see, for instance, Sects. 11.2.3 and 11.5).

First let us repeat some notation from Chap. 8. Let T_1 be the vector representation for one of the algebras $U_q(\mathfrak{g}), U_q^{\mathrm{ext}}(\mathfrak{g})$, or $\tilde{U}_q(\mathrm{so}_N)$ (see Subsects. 8.4.1, 8.5.3, and 8.6.1) and let t_{ij} be the matrix elements of T_1, that is, $T_1(f)e_j = t_{ij}(f)e_i$. Let $R_{1,1} = (R^{ik}_{jl})$ be the R-matrix (see Subsects. 8.3.3 and 8.4.2) for the vector representation, $R^{-1}_{1,1} = ((R^{-1})^{ik}_{jl})$ its inverse and $l^{\pm}_{ij}$ the L-functionals associated with T_1 (see Subsect. 8.5.4).

Theorem 18. *There exist unique dual pairings $\langle\cdot,\cdot\rangle$ of the pairs of Hopf algebras $U_q(\mathrm{gl}_N)$ and $\mathcal{O}(GL_q(N))$, $U_q^{\mathrm{ext}}(\mathrm{sl}_N)$ and $\mathcal{O}(SL_q(N))$, $U_{q^{1/2}}(\mathrm{so}_{2n+1})$ and $\mathcal{O}(O_q(2n+1))$, $U_{q^{1/2}}(\mathrm{so}_{2n+1})$ and $\mathcal{O}(SO_q(2n+1))$, $U_q^{\mathrm{ext}}(\mathrm{so}_{2n})$ and $\mathcal{O}(O_q(2n))$, $U_q^{\mathrm{ext}}(\mathrm{so}_{2n})$ and $\mathcal{O}(SO_q(2n))$, $U_q^{\mathrm{ext}}(\mathrm{sp}_{2n})$ and $\mathcal{O}(Sp_q(2n))$ such that*

$$\langle f, u^k_l\rangle = t_{kl}(f), \quad k,l = 1,2,\cdots,N. \tag{39}$$

The dual pairing $\langle\cdot,\cdot\rangle$ satisfies the relations

$$\langle l^+_{ij}, u^k_l\rangle = R^{ki}_{lj} \text{ and } \langle l^-_{ij}, u^k_l\rangle = (R^{-1})^{ik}_{jl}, \quad i,j,k,l = 1,2,\cdots,N, \tag{40}$$

and it is uniquely determined by these equations.

Proof. Let us write U_q and $\mathcal{O}_q$ for the corresponding Drinfeld–Jimbo algebra and coordinate Hopf algebra, respectively. Let $\mathcal{B}$ be the subalgebra of the dual of U_q generated by the functionals t_{ij}. Since $t_{ij}(ab) = t_{ik}(a)t_{kj}(b)$, $\mathcal{B}$ is a subbialgebra of the dual Hopf algebra U_q°. Clearly, the evaluation of $b \in \mathcal{B}$ at $f \in U_q$ defines a dual pairing $\langle\cdot,\cdot\rangle$ of the bialgebras U_q and $\mathcal{B}$.

Since $\hat{R}_{VV}(T_V \otimes T_V) = (T_V \otimes T_V)\hat{R}_{VV}$ by Proposition 8.19, the matrix $\mathbf{t} = (t_{ij})$ satisfies the equation $\hat{\mathsf{R}}\mathbf{t}_1\mathbf{t}_2 = \mathbf{t}_1\mathbf{t}_2\hat{\mathsf{R}}$. (Recall that $\mathsf{R} = q^{1/N}R_{1,1}$ for $\mathfrak{g} = \mathrm{sl}_N$ and $\mathsf{R} = R_{1,1}$ for $\mathfrak{g} = \mathrm{so}_N, \mathrm{sp}_N$.) Consequently, there is an algebra homomorphism $\psi : \mathcal{A}(\mathsf{R}) \to \mathcal{B}$ such that $\psi(u^i_j) = t_{ij}$. Since $\Delta(u^i_j) = u^i_k \otimes u^k_l$ and $\Delta(t_{ij}) = t_{ik} \otimes t_{kl}$, ψ is a bialgebra homomorphism. Therefore, the definition $\langle f, a\rangle := \langle f, \psi(a)\rangle \equiv \psi(a)(f)$, $f \in U_q$, $a \in \mathcal{A}(\mathsf{R})$, gives a dual pairing of the bialgebras U_q and $\mathcal{A}(\mathsf{R})$. Next we show that the pairing $\langle\cdot,\cdot\rangle$ of U_q and $\mathcal{A}(\mathsf{R})$ passes to a dual pairing of U_q and $\mathcal{O}_q$.

We carry out the proof in the case where $U_q = U_q^{\mathrm{ext}}(\mathrm{sl}_N)$ and $\mathcal{O}_q = \mathcal{O}(SL_q(N)))$. The other cases can be treated similarly. From Subsect. 9.2.2

we recall the relation $\varphi_{R,N}(\zeta) = \zeta \otimes \mathcal{D}_q$ for $\zeta \in \Lambda(\mathbb{C}_q^N)_N$. Setting $\zeta = y_1 \cdots y_N$, we have $\varphi_{R,N}(y_1 \cdots y_N) = y_{i_1} \cdots y_{i_N} \otimes u_1^{i_1} \cdots u_N^{i_N}$. By (1.64), we obtain

$$\hat{\varphi}_{R,N}(K_i) y_1 \cdots y_N = \langle K_i, u_1^{i_1} \cdots u_N^{i_N} \rangle \, y_{i_1} \cdots y_{i_N}$$

$$= \langle K_i, u_1^{i_1} \rangle \cdots \langle K_i, u_N^{i_N} \rangle \, y_{i_1} \cdots y_{i_N} = \langle K_i, \mathcal{D}_q \rangle \, y_1 \cdots y_N.$$

From the particular form of the representation T_1 we see that $\langle K_i, u_l^k \rangle = t_{kl}(K_i) = 0$ if $k \neq l$ and $\langle K_i, u_1^1 \rangle \cdots \langle K_i, u_N^N \rangle = 1$. Therefore, by the above equality, $\langle K_i, \mathcal{D}_q \rangle = 1$. Since $\mathcal{D}_q$ is group-like, the linear functional $\langle \cdot, \mathcal{D}_q \rangle$ on U_q is a character. Hence we get $\langle E_i, \mathcal{D}_q \rangle = \langle F_i, \mathcal{D}_q \rangle = 0$ and so $\langle f, \mathcal{D}_q - 1 \rangle = 0$ for all $f \in U_q$. Thus, the dual pairing $\langle \cdot, \cdot \rangle$ of U_q and $\mathcal{A}(\mathsf{R})$ induces a dual pairing of the Hopf algebras U_q and $\mathcal{O}_q = \mathcal{A}(\mathsf{R})/\langle \mathcal{D}_q - 1 \rangle$. Formula (39) holds by definition and (40) follows immediately from (8.88).

By Theorem 8.33, the elements $l_{ij}^{\pm}$ generate U_q as an algebra. Except for $GL_q(N)$, the algebra $\mathcal{O}_q$ is generated by the elements u_j^i. Hence the pairing $\langle \cdot, \cdot \rangle$ is uniquely determined by the values $\langle f, u_l^k \rangle$ resp. $\langle l_{ij}^{\pm}, u_l^k \rangle$. □

The restrictions in the first variables of the dual pairings for the extended Hopf algebras $U_q^{\mathrm{ext}}(\mathrm{sl}_N)$, $U_q^{\mathrm{ext}}(\mathrm{sp}_{2n})$, $U_q^{\mathrm{ext}}(\mathrm{so}_{2n})$ yield corresponding dual pairings for the Hopf subalgebras $U_q(\mathrm{sl}_N)$, $U_q(\mathrm{sp}_{2n})$, $U_q(\mathrm{so}_{2n})$, respectively. Let us collect the values of these dual pairings on the generators u_j^i of $\mathcal{O}(G_q)$ and E_i, F_i, K_i of $U_q(\mathfrak{g})$ resp. $U_{q^{1/2}}(\mathrm{so}_{2n+1})$. From the formulas for the vector representations in Subsect. 8.4.1 we read off the following list:

$U_q(\mathrm{gl}_N)$:

$$\langle K_i, u_i^i \rangle = q^{-1}, \quad \langle K_i, u_j^j \rangle = 1, \quad j \neq i, \tag{41}$$

$$\langle E_i, u_i^{i+1} \rangle = \langle F_i, u_{i+1}^i \rangle = 1; \tag{42}$$

$U_q(\mathrm{sl}_N)$:

$$\langle K_i^{-1}, u_i^i \rangle = \langle K_i, u_{i+1}^{i+1} \rangle = q, \quad \langle K_i, u_j^j \rangle = 1, \quad j \neq i, i+1,$$

$$\langle E_i, u_i^{i+1} \rangle = \langle F_i, u_{i+1}^i \rangle = 1;$$

$U_{q^{1/2}}(\mathrm{so}_{2n+1})$:

$$\langle K_i^{-1}, u_i^i \rangle = \langle K_i, u_{i+1}^{i+1} \rangle = \langle K_i^{-1}, u_{(i+1)'}^{(i+1)'} \rangle = \langle K_i, u_{i'}^{i'} \rangle = q, \quad i \neq n,$$

$$\langle K_i, u_j^j \rangle = 1 \quad \text{for} \quad i \neq n, \ j \neq i, i+1, i', (i+1)',$$

$$\langle K_n^{-1}, u_n^n \rangle = \langle K_n, u_{n+2}^{n+2} \rangle = q, \quad \langle K_n, u_j^j \rangle = 1 \ \text{ for } \ j \neq n, n+2,$$

$$\langle E_i, u_i^{i+1} \rangle = -\langle E_i, u_{(i+1)'}^{i'} \rangle = \langle F_i, u_{i+1}^i \rangle = -\langle F_i, u_{i'}^{(i+1)'} \rangle = 1, \quad i \neq n,$$

$$\langle E_n, u_n^{n+1} \rangle = (q^{1/2} + q^{-1/2})^{1/2}, \quad \langle E_n, u_{n+1}^{n+2} \rangle = -q^{1/2}(q^{1/2} + q^{-1/2})^{1/2},$$

$$\langle F_n, u_{n+1}^n \rangle = (q^{1/2} + q^{-1/2})^{1/2}, \quad \langle F_n, u_{n+2}^{n+1} \rangle = -q^{-1/2}(q^{1/2} + q^{-1/2})^{1/2};$$

$U_q(\mathrm{sp}_{2n})$:

$$\langle K_i^{-1}, u_i^i\rangle = \langle K_i, u_{i+1}^{i+1}\rangle = \langle K_i^{-1}, u_{(i+1)'}^{(i+1)'}\rangle = \langle K_i, u_{i'}^{i'}\rangle = q, \quad i \neq n,$$

$$\langle K_i^{-1}, u_j^j\rangle = 1 \quad \text{for} \quad i \neq n, \; j \neq i, i+1, i', (i+1)',$$

$$\langle K_n^{-1}, u_n^n\rangle = \langle K_n, u_{n+1}^{n+1}\rangle = q^2, \quad \langle K_n, u_j^j\rangle = 1 \; \text{ for } \; j \neq n, n+1,$$

$$\langle E_i, u_i^{i+1}\rangle = -\langle E_i, u_{(i+1)'}^{i'}\rangle = \langle F_i, u_{i+1}^i\rangle = -\langle F_i, u_{i'}^{(i+1)'}\rangle = 1, \quad i \neq n,$$

$$\langle E_n, u_n^{n+1}\rangle = \langle F_n, u_{n+1}^n\rangle = 1;$$

$U_q(\mathrm{so}_{2n})$:

$$\langle K_i^{-1}, u_i^i\rangle = \langle K_i, u_{i+1}^{i+1}\rangle = \langle K_i^{-1}, u_{(i+1)'}^{(i+1)'}\rangle = \langle K_i, u_{i'}^{i'}\rangle = q, \quad i \neq n,$$

$$\langle K_i^{-1}, u_j^j\rangle = 1 \quad \text{for} \quad i \neq n, \; j \neq i, i+1, i', (i+1)',$$

$$\langle K_n^{-1}, u_{n-1}^{n-1}\rangle = \langle K_n^{-1}, u_n^n\rangle = \langle K_n, u_{n+1}^{n+1}\rangle = \langle K_n, u_{n+2}^{n+2}\rangle = q,$$

$$\langle K_n, u_j^j\rangle = 1 \; \text{ for } \; j \neq n-1, n, n+1, n+2,$$

$$\langle E_i, u_i^{i+1}\rangle = -\langle E_i, u_{(i+1)'}^{i'}\rangle = \langle F_i, u_{i+1}^i\rangle = -\langle F_i, u_{i'}^{(i+1)'}\rangle = 1, \quad i \neq n,$$

$$-\langle E_n, u_n^{n+2}\rangle = \langle E_n, u_{n-1}^{n+1}\rangle = -\langle F_n, u_{n+2}^n\rangle = \langle F_n, u_{n+1}^{n-1}\rangle = 1.$$

The values $\langle Y, u_l^k\rangle$, $Y = E_i, F_i, K_i$, for all other pairs of generators vanish.

In Subsect. 11.2.3 we shall also need a dual pairing of the Hopf algebras $\tilde{U}_q(\mathrm{so}_N)$ (see Subsect. 8.6.1 for its definition) and $\mathcal{O}(O_q(N))$.

Proposition 19. *There exists a unique dual pairing $\langle\cdot,\cdot\rangle$ of the Hopf algebras $\tilde{U}_q(\mathrm{so}_N)$ and $\mathcal{O}(O_q(N))$ satisfying the relations (39) for $f \in \tilde{U}_q(\mathrm{so}_N)$. In particular, we have $\langle \chi, \mathcal{D}_q\rangle = -1$. Moreover, $\langle \chi, u_l^k\rangle = -\delta_{kl}$ for $N = 2n+1$ and $\langle \chi, u_{n+1}^n\rangle = \langle \chi, u_n^{n+1}\rangle = 1$, $\langle \chi, u_i^i\rangle = 1$ for $i = 1, 2, \cdots, n-1, n+2, \cdots, N$ and $\langle \chi, u_j^i\rangle = 0$ otherwise for $N = 2n$.*

Proof. Since $\mathsf{R}\mathbf{t}_1(\chi)\mathbf{t}_2(\chi) = \mathbf{t}_1(\chi)\mathbf{t}_2(\chi)\mathsf{R}$ as easily seen from the form of R and $\tilde{T}_1(\chi)$, there is a dual pairing $\langle\cdot,\cdot\rangle$ of the bialgebras $\tilde{U}_q(\mathrm{so}_N)$ and $\mathcal{A}(\mathsf{R})$ such that $\langle f, u_l^k\rangle = t_{kl}(f)$, $f \in \tilde{U}_q(\mathrm{so}_N)$. An easy computation shows that

$$\langle \chi, \mathcal{Q}_q\rangle = \sum\nolimits_n \mathsf{C}_j^{j'} \mathsf{C}_{n'}^n t_{nj}(\chi) t_{n'j'}(\chi) = 1.$$

Therefore, since $\langle g, \mathcal{Q}_q - 1\rangle = 0$ for $g \in U_q(\mathrm{so}_{2n})$ resp. $g \in U_{q^{1/2}}(\mathrm{so}_{2n+1})$ by Theorem 18, we get $\langle f, \mathcal{Q}_q - 1\rangle = 0$ for all $f \in \tilde{U}_q(\mathrm{so}_N)$. Hence the pairing $\langle\cdot,\cdot\rangle$ of $\tilde{U}_q(\mathrm{so}_N)$ and $\mathcal{A}(\mathsf{R})$ passes to a dual pairing $\langle\cdot,\cdot\rangle$ of $\tilde{U}_q(\mathrm{so}_N)$ and $\mathcal{O}(O_q(N))$. The values $\langle \chi, u_l^k\rangle$ follow from the particular form of the operator $\tilde{T}_1(\chi)$ (see Subsect. 8.6.1). They easily imply that $\langle \chi, \mathcal{D}_q\rangle = -1$. In case $N = 2n$ one has to use also the relation $y_n y_{n+1} = -y_{n+1} y_n$ which holds by Proposition 15(ii). □

Let U_q and $\mathcal{O}_q$ denote one of the dual pairs of Hopf algebras in Theorem 18 or Proposition 19. Then, by Proposition 1.15, any corepresentation φ of $\mathcal{O}_q$ yields a representation $\hat{\varphi}$ of U_q. If φ_1 is the fundamental corepresentation of $\mathcal{O}_q$ given by (6), it follows from (39) and (1.64) that $\hat{\varphi}_1(f)x_i = x_j\langle f, u^j_i\rangle = t_{ji}(f)x_j$, $f \in U_q$. That is, if we take in the carrier space of φ_1 the same basis as for T_1, then *$\hat{\varphi}_1$ is the vector representation T_1 of U_q*. In fact, the dual pairing $\langle\cdot,\cdot\rangle$ of U_q and $\mathcal{O}_q$ was just constructed such that the latter holds. Clearly, the preceeding is also true for the dual pairs of U_q and $\mathcal{A}(\mathsf{R})$.

We conclude this section with another duality result. For an arbitrary matrix $R = (R^{ik}_{jl})_{i,j,k,l=1,\cdots,N}$ with complex entries, let $\mathfrak{U}(R)$ denote the algebra with $2N^2$ generators ℓ^+_{ij}, ℓ^-_{ij}, $i,j = 1,2,\cdots,N$, subject to the relations

$$\mathfrak{L}^+_1\mathfrak{L}^+_2 R = R\mathfrak{L}^+_2\mathfrak{L}^+_1, \quad \mathfrak{L}^-_1\mathfrak{L}^-_2 R = R\mathfrak{L}^-_2\mathfrak{L}^-_1, \quad \mathfrak{L}^-_1\mathfrak{L}^+_2 R = R\mathfrak{L}^+_2\mathfrak{L}^-_1, \tag{43}$$

where $\mathfrak{L}^\pm = (\ell^\pm_{ij})$. Arguing as in Subsect. 8.5.5, it follows that $\mathfrak{U}(R)$ is a bialgebra with comultiplication and counit given by

$$\Delta(\ell^\pm_{ij}) = \ell^\pm_{ik} \otimes \ell^\pm_{kj} \quad \text{and} \quad \varepsilon(\ell^\pm_{ij}) = \delta_{ij}.$$

Proposition 20. *Suppose that R is an invertible solution of the QYBE. Then there is a dual pairing $\langle\cdot,\cdot\rangle$ of the bialgebras $\mathfrak{U}(R)$ and $\mathcal{A}(R)$ such that*

$$\langle\mathfrak{L}^+_1, \mathbf{u}_2\rangle = R_{21} \quad \textit{and} \quad \langle\mathfrak{L}^-_1, \mathbf{u}_2\rangle = R^{-1}_{12}. \tag{44}$$

Outline of proof. First we note that there is a unique dual pairing $\langle\cdot,\cdot\rangle$ of the free bialgebras $\mathbb{C}\langle\ell^+_{ij}, \ell^-_{ij}\rangle$ and $\mathbb{C}\langle u^i_j\rangle$ satisfying (44). In order to obtain a well-defined dual pairing of $\mathfrak{U}(R)$ and $\mathcal{A}(R)$ one has to check that the corresponding consistency conditions in both variables hold. As a sample, we shall do this for the relation $\mathfrak{L}^-_1\mathfrak{L}^+_2 R = R\mathfrak{L}^+_2\mathfrak{L}^-_1$ and compute

$$\begin{aligned}\langle\mathfrak{L}^-_1\mathfrak{L}^+_3 R_{13}, \mathbf{u}_2\rangle &= \langle\mathfrak{L}^-_1, \mathbf{u}_2\rangle\langle\mathfrak{L}^+_3, \mathbf{u}_2\rangle R_{13} = R^{-1}_{12}R_{23}R_{13} = R_{13}R_{23}R^{-1}_{12} \\ &= R_{13}\langle\mathfrak{L}^+_3, \mathbf{u}_2\rangle\langle\mathfrak{L}^-_1, \mathbf{u}_2\rangle = \langle R_{13}\mathfrak{L}^+_3\mathfrak{L}^-_1, \mathbf{u}_2\rangle. \qquad \square\end{aligned}$$

9.5 Notes

The FRT approach and the quantum deformations of the four series of classical simple Lie groups appeared in the fundamental paper [RTF] of L. D. Faddeev, N. Yu. Reshetikhin, and L. A. Takhtajan. These quantum groups were also constructed by M. Takeuchi [Ta2] and investigated in [H3]. The quantum group $SU_q(N)$ was invented and studied by S. L. Woronowicz [Wor2], [Wor3]. Another influential early reference is [Man]. The bialgebra $\mathcal{M}(\mathcal{C},\mathbf{r})$ from Remark 1 is due to [Doi]. Multiparameter quantum groups appear (for instance) in [AST], [Res3] and [Sud1].

10. Coquasitriangularity and Crossed Product Constructions

This chapter is centered around various fundamental notions (coquasitriangular bialgebras, crossed product and coproducts, braided Hopf algebras) from the advanced theory of Hopf algebras. They are useful tools for the deeper study of quantum groups and for the construction of new quantum groups.

10.1 Coquasitriangular Hopf Algebras

Coquasitriangularity is probably the most important general Hopf algebra concept for the second approach to quantum groups. The coordinate Hopf algebras of the quantum groups from the previous chapter and the FRT bialgebra $\mathcal{A}(R)$ for any invertible solution R of the QYBE have this property.

10.1.1 Definition and Basic Properties

Recall that a bialgebra $\mathcal{B}$ is quasitriangular if it admits a universal R-matrix $\mathcal{R} \in \mathcal{B} \otimes \mathcal{B}$. In order to motivate the subsequent definition of the dual notion, let us consider a bialgebra $\mathcal{A}$ which is dually paired with $\mathcal{B}$. Define a linear form $\mathbf{r}$ on $\mathcal{A} \otimes \mathcal{A}$ by $\mathbf{r}(a \otimes a') = \langle a \otimes a', \mathcal{R} \rangle$. Remembering that the role of multiplication and comultiplication is interchanged by the dual pairing, it is not difficult to verify that $\mathbf{r}$ is a universal r-form of $\mathcal{A}$ according to the following

Definition 1. *A* coquasitriangular bialgebra *is a bialgebra $\mathcal{A}$ equipped with a linear form $\mathbf{r} : \mathcal{A} \otimes \mathcal{A} \to \mathbb{C}$ such that the following conditions hold:*

(i) *$\mathbf{r}$ is invertible with respect to the convolution, that is, there exists another linear form $\bar{\mathbf{r}} : \mathcal{A} \otimes \mathcal{A} \to \mathbb{C}$ such that $\mathbf{r}\bar{\mathbf{r}} = \bar{\mathbf{r}}\mathbf{r} = \varepsilon \otimes \varepsilon$ on $\mathcal{A} \otimes \mathcal{A}$,*

(ii) *$m_{\mathcal{A}^{\mathrm{op}}} = \mathbf{r} * m_{\mathcal{A}} * \bar{\mathbf{r}}$ on $\mathcal{A} \otimes \mathcal{A}$,*

(iii) *$\mathbf{r}(m_{\mathcal{A}} \otimes \mathrm{id}) = \mathbf{r}_{13}\mathbf{r}_{23}$ and $\mathbf{r}(\mathrm{id} \otimes m_{\mathcal{A}}) = \mathbf{r}_{13}\mathbf{r}_{12}$ on $\mathcal{A} \otimes \mathcal{A} \otimes \mathcal{A}$,*

where $m_{\mathcal{A}^{\mathrm{op}}}(a \otimes b) = ba$, $m_{\mathcal{A}}(a \otimes b) = ab$, $\mathbf{r}_{12}(a \otimes b \otimes c) = \mathbf{r}(a \otimes b)\varepsilon(c)$, $\mathbf{r}_{23}(a \otimes b \otimes c) = \varepsilon(a)\mathbf{r}(b \otimes c)$ and $\mathbf{r}_{13}(a \otimes b \otimes c) = \varepsilon(b)\mathbf{r}(a \otimes c)$, $a, b, c \in \mathcal{A}$. Recall that the convolution product $*$ is defined by formula (1.21). A linear form $\mathbf{r}$ on $\mathcal{A} \otimes \mathcal{A}$ with the properties (i)–(iii) is called a universal r-form *of $\mathcal{A}$. A bialgebra $\mathcal{A}$ is called* cotriangular *if it has a universal r-form $\mathbf{r}$ such* *that* $\mathbf{r} = \bar{\mathbf{r}}_{21}$, *where* $\bar{\mathbf{r}}_{21}(a \otimes b) := \bar{\mathbf{r}}(b \otimes a)$. *A* coquasitriangular Hopf algebra *is a Hopf algebra which is a coquasitriangular bialgebra.*

Remark 1. Coquasitriangular bialgebras appear under a number of different names in the literature. They are called dual quasitriangular bialgebras in [Maj7], braided bialgebras in [LTo], [Doi] and cobraided bialgebras in [Kas]. The above name is used (for instance) in [H2], [S1], [Mon]. △

The above form of Definition 1 is directly dual to Definition 8.1. However, the following reformulations of conditions (i)–(iii) in Sweedler's notation are more convenient to work with. Since linear forms on $\mathcal{A} \otimes \mathcal{A}$ correspond to bilinear forms on $\mathcal{A} \times \mathcal{A}$, we can consider any linear form $\mathbf{r} : \mathcal{A} \otimes \mathcal{A} \to \mathbb{C}$ as a bilinear form on $\mathcal{A} \times \mathcal{A}$ and write $\mathbf{r}(a,b) := \mathbf{r}(a \otimes b)$, $a,b \in \mathcal{A}$. Now the equations $\mathbf{r}\bar{\mathbf{r}} = \bar{\mathbf{r}}\mathbf{r} = \varepsilon \otimes \varepsilon$ and the conditions (ii) and (iii) read as

$$\sum \mathbf{r}(a_{(1)}, b_{(1)})\bar{\mathbf{r}}(a_{(2)}, b_{(2)}) = \sum \bar{\mathbf{r}}(a_{(1)}, b_{(1)})\mathbf{r}(a_{(2)}, b_{(2)}) = \varepsilon(a)\varepsilon(b), \tag{1}$$

$$ba = \sum \mathbf{r}(a_{(1)}, b_{(1)})a_{(2)}b_{(2)}\bar{\mathbf{r}}(a_{(3)}, b_{(3)}), \tag{2}$$

$$\mathbf{r}(ab, c) = \sum \mathbf{r}(a, c_{(1)})\mathbf{r}(b, c_{(2)}), \tag{3}$$

$$\mathbf{r}(a, bc) = \sum \mathbf{r}(a_{(1)}, c)\mathbf{r}(a_{(2)}, b) \tag{4}$$

with $a,b,c \in \mathcal{A}$. Multiplying condition (ii) in Definition 1 by $\mathbf{r}$ we see that it is equivalent to the requirement

$$\sum \mathbf{r}(a_{(1)}, b_{(1)})a_{(2)}b_{(2)} = \sum \mathbf{r}(a_{(2)}, b_{(2)})b_{(1)}a_{(1)}, \quad a,b \in \mathcal{A}. \tag{5}$$

Condition (ii) means that the universal $\mathbf{r}$-form $\mathbf{r}$ expresses in some sense the deviation of $\mathcal{A}$ from being commutative. Clearly, a bialgebra $\mathcal{A}$ is commutative if and only if $\mathbf{r} = \varepsilon \otimes \varepsilon$ is a universal $\mathbf{r}$-form of $\mathcal{A}$.

From the equations (3) and (4) and from Proposition 2(ii) below we see that the map $a \times b \to \mathbf{r}(a,b)$ defines a dual pairing of the bialgebras $\mathcal{A}$ and $\mathcal{A}^{\mathrm{op}}$ and hence a skew-pairing (see Definition 8.3) of $\mathcal{A}$ and $\mathcal{A}$ itself. Conversely, a skew-pairing $\mathbf{r}$ of $\mathcal{A}$ and $\mathcal{A}$ is a universal r-form of $\mathcal{A}$ if and only if $\mathbf{r}$ satisfies (2) or its equivalent version (5). By Proposition 2(iii), the map $a \times b \to \bar{\mathbf{r}}(a,b)$ is a dual pairing of $\mathcal{A}^{\mathrm{op}}$ and $\mathcal{A}$.

Recall from Proposition 8.4 that the universal R-matrix of a quasitriangular Hopf algebra induces an equivalence of the tensor product representations $T_V \otimes T_W$ and $T_W \otimes T_V$. Our next proposition shows that there is a similar result for comodules of a coquasitriangular Hopf algebra.

Let $\mathcal{A}$ be a coquasitriangular bialgebra with universal r-form $\mathbf{r}$. For right $\mathcal{A}$-comodules V and W we define a linear mapping $\mathbf{r}_{VW} : V \otimes W \to W \otimes V$ by

$$\mathbf{r}_{VW}(v \otimes w) = \sum \mathbf{r}(v_{(1)}, w_{(1)})w_{(0)} \otimes v_{(0)}, \quad v \in V, w \in W. \tag{6}$$

If $V = W$ is the right $\mathcal{A}$-comodule $\mathcal{A}$ with respect to the comultiplication, then we obviously have $(\varepsilon \otimes \varepsilon)\mathbf{r}_{VW} = \mathbf{r}$, that is, the universal r-form $\mathbf{r}$ can be recovered from the map $\mathbf{r}_{VW}$ in this case.

Proposition 1. *If U, V and W are right $\mathcal{A}$-comodules, then we have:*

(i) $\mathbf{r}_{VW}$ *is an isomorphism of the right $\mathcal{A}$-comodules $V \otimes W$ and $W \otimes V$.*
(ii) $\mathbf{r}_{U\otimes V,W} = (\mathbf{r}_{UW} \otimes \mathrm{id})(\mathrm{id} \otimes \mathbf{r}_{VW})$ *and* $\mathbf{r}_{U,V\otimes W} = (\mathrm{id} \otimes \mathbf{r}_{UW})(\mathbf{r}_{UV} \otimes \mathrm{id})$.
(iii) $(\mathbf{r}_{VW} \otimes \mathrm{id})(\mathrm{id} \otimes \mathbf{r}_{UW})(\mathbf{r}_{UV} \otimes \mathrm{id}) = (\mathrm{id} \otimes \mathbf{r}_{UV})(\mathbf{r}_{UW} \otimes \mathrm{id})(\mathrm{id} \otimes \mathbf{r}_{VW})$.

Proof. (i): Define a linear mapping $\tilde{\mathbf{r}}_{WV} : W \otimes V \to V \otimes W$ by $\tilde{\mathbf{r}}_{WV}(w \otimes v) = \sum \bar{\mathbf{r}}(v_{(1)}, w_{(1)}) v_{(0)} \otimes w_{(0)}$, $v \in V, w \in W$. Using (1), we obtain

$$\begin{aligned}
(\tilde{\mathbf{r}}_{WV} \circ \mathbf{r}_{VW})(v \otimes w) &= \tilde{\mathbf{r}}_{WV}\Big(\sum \mathbf{r}(v_{(1)}, w_{(1)}) w_{(0)} \otimes v_{(0)}\Big) \\
&= \sum \mathbf{r}(v_{(2)}, w_{(2)}) \bar{\mathbf{r}}(v_{(1)}, w_{(1)}) v_{(0)} \otimes w_{(0)} \\
&= \sum \varepsilon(v_{(1)}) \varepsilon(w_{(1)}) v_{(0)} \otimes w_{(0)} = v \otimes w
\end{aligned}$$

and similarly $\mathbf{r}_{VW} \circ \tilde{\mathbf{r}}_{WV}(w \otimes v) = w \otimes v$. Hence $\mathbf{r}_{VW}$ is invertible.

Let $\varphi_{V\otimes W}$ and $\varphi_{W\otimes V}$ denote the coactions of $\mathcal{A}$ on the right comodules $V \otimes W$ and $W \otimes V$, respectively. By (6), (1.61) and (5), we get

$$\begin{aligned}
\varphi_{W\otimes V} \circ \mathbf{r}_{V,W}(v \otimes w) &= \varphi_{W\otimes V}\Big(\sum \mathbf{r}(v_{(1)}, w_{(1)}) w_{(0)} \otimes v_{(0)}\Big) \\
&= \sum w_{(0)} \otimes v_{(0)} \otimes w_{(1)} v_{(1)} \mathbf{r}(v_{(2)}, w_{(2)}) \\
&= \sum w_{(0)} \otimes v_{(0)} \otimes v_{(2)} w_{(2)} \mathbf{r}(v_{(1)}, w_{(1)}) \\
&= (\mathbf{r}_{VW} \otimes \mathrm{id}) \circ \varphi_{V\otimes W}(v \otimes w).
\end{aligned}$$

That is, $\mathbf{r}_{VW} \in \mathrm{Mor}\,(\varphi_{V\otimes W}, \varphi_{W\otimes V})$.

Using (3) and (4) the assertions of (ii) are proved in a similar manner. (iii) follows easily from (ii). □

Of course, the preceding results hold for left $\mathcal{A}$-comodules as well.

By Proposition 1(iii), for each right $\mathcal{A}$-comodule V the linear mapping $\mathbf{r}_{VV}$ defined by (6) satisfies braid relation, hence $\tau \circ \mathbf{r}_{VV}$ is a solution of the QYBE. That is, coquasitriangular bialgebras induce solutions of the QYBE on their comodules just as quasitriangular bialgebras do on their modules.

A large part of the theory of coquasitriangular bialgebras can be developed by dualizing the corresponding theory for quasitriangular bialgebras. Some basic facts on coquasitriangular bialgebras are collected in

Proposition 2. *For any coquasitriangular bialgebra $\mathcal{A}$ with universal r-form $\mathbf{r}$ we have:*

(i) $\mathbf{r}_{12}\mathbf{r}_{13}\mathbf{r}_{23} = \mathbf{r}_{23}\mathbf{r}_{13}\mathbf{r}_{12}$ *on* $\mathcal{A} \otimes \mathcal{A} \otimes \mathcal{A}$.
(ii) $\mathbf{r}(1, a) = \mathbf{r}(a, 1) = \varepsilon(a)$ for $a \in \mathcal{A}$.
(iii) $\bar{\mathbf{r}}$ *is a universal r-form for the bialgebras $\mathcal{A}^{\mathrm{op}}$ and $\mathcal{A}^{\mathrm{cop}}$ (see Subsect. 1.2.3 for the definitions of these bialgebras).*

(iv) $\bar{\mathbf{r}}_{21}$ *is also a universal r-form for* $\mathcal{A}$*, where* $\bar{\mathbf{r}}_{21}(a \otimes b) := \bar{\mathbf{r}}(b \otimes a)$.
(v) *If* $\mathcal{A}$ *is a coquasitriangular Hopf algebra, then for* $a, b \in \mathcal{A}$ *we have*

$$\mathbf{r}(S(a), b) = \bar{\mathbf{r}}(a, b), \quad \bar{\mathbf{r}}(a, S(b)) = \mathbf{r}(a, b), \quad \mathbf{r}(S(a), S(b)) = \mathbf{r}(a, b).$$

Proof. All proofs are straightforward verifications. As a sample, we show the first relation of (v). Set $\mathbf{r}'(a \otimes b) = \mathbf{r}(S(a) \otimes b)$. From (3) and (ii) we obtain

$$\begin{aligned}\sum \mathbf{r}(a_{(1)}, b_{(1)})\mathbf{r}'(a_{(2)}, b_{(2)}) &= \sum \mathbf{r}(a_{(1)}, b_{(1)})\mathbf{r}(S(a_{(2)}), b_{(2)}) = \\ &= \sum \mathbf{r}(a_{(1)}S(a_{(2)}), b) = \varepsilon(a)\mathbf{r}(1, b) = \varepsilon(a)\varepsilon(b),\end{aligned}$$

so that $\mathbf{r}\mathbf{r}' = \varepsilon \otimes \varepsilon$. Similarly, $\mathbf{r}'\mathbf{r} = \varepsilon \otimes \varepsilon$. Hence $\mathbf{r}' = \bar{\mathbf{r}}$. □

The next result describes the square and the inverse of the antipode of a coquasitriangular Hopf algebra. It is dual to the corresponding assertion for quasitriangular Hopf algebras (see Proposition 8.5).

Proposition 3. *Let* $\mathcal{A}$ *be a coquasitriangular Hopf algebra with universal r-form* $\mathbf{r}$*. Define linear functionals* f *and* $\bar{f}$ *on* $\mathcal{A}$ *by*

$$f(a) = \sum \mathbf{r}(a_{(1)}, S(a_{(2)})) \quad \text{and} \quad \bar{f}(a) = \sum \bar{\mathbf{r}}(S(a_{(1)}), a_{(2)}), \qquad a \in \mathcal{A}.$$

Then we have $f\bar{f} = \bar{f}f = \varepsilon$ *in* $\mathcal{A}'$*. The antipode* S *of* $\mathcal{A}$ *is bijective and satisfies* $S^2 = \bar{f} * \mathrm{id} * f$ *and* $S^{-1} = f * S * \bar{f}$*. That is, for* $a \in \mathcal{A}$ *we have*

$$S^2(a) = \sum \bar{f}(a_{(1)})a_{(2)}f(a_{(3)}), \quad S^{-1}(a) = \sum f(a_{(1)})S(a_{(2)})\bar{f}(a_{(3)}).$$

Proof. Applying (5) with $a \otimes b$ replaced by $\sum a_{(1)} \otimes S(a_{(2)})$ we get $f(a)1 = \sum S(a_{(3)})a_{(1)}f(a_{(2)})$. This yields

$$\begin{aligned}\sum S^2(a_{(2)})f(a_{(1)}) &= \sum S^2(a_{(4)})S(a_{(3)})a_{(1)}f(a_{(2)}) \\ &= \sum S(a_{(3)}S(a_{(4)}))a_{(1)}f(a_{(2)}) = \sum a_{(1)}f(a_{(2)}). \quad (7)\end{aligned}$$

Using the first and the third equality of Proposition 2(v), formulas (7) and (4) and Proposition 2(ii) we compute

$$\begin{aligned}\bar{f}f(a) &= \sum \bar{\mathbf{r}}(S(a_{(1)}), a_{(2)})f(a_{(3)}) = \sum \mathbf{r}(S^2(a_{(1)}), a_{(2)}f(a_{(3)})) \\ &= \sum \mathbf{r}(S^2(a_{(1)}), S^2(a_{(3)})f(a_{(2)})) = \sum \mathbf{r}(a_{(1)}, a_{(3)})f(a_{(2)}) \\ &= \sum \mathbf{r}(a_{(1)}, a_{(4)})\mathbf{r}(a_{(2)}, S(a_{(3)})) = \sum \mathbf{r}(a_{(1)}, S(a_{(2)})a_{(3)}) \\ &= \mathbf{r}(a, 1) = \varepsilon(a).\end{aligned}$$

Similarly, $f\bar{f} = \varepsilon$. From the relations $\bar{f}f = \varepsilon$ and (7) we obtain

$$\sum \bar{f}(a_{(1)})a_{(2)}f(a_{(3)}) = \sum \bar{f}(a_{(1)})f(a_{(2)})S^2(a_{(3)}) = S^2(a).$$

Since $S^2 = \bar{f} * \mathrm{id} * f$ as just shown, we have $S^2 * \bar{f} = \bar{f} * \mathrm{id}$, that is,

$$\sum S^2(a_{(1)})\bar{f}(a_{(2)}) = \sum \bar{f}(a_{(1)})a_{(2)}. \tag{8}$$

Put $T := f * S * \bar{f}$. Using (7) and the relation $f\bar{f} = \varepsilon$, we get

$$\begin{aligned}\sum T(a_{(2)})a_{(1)} &= \sum f(a_{(2)})S(a_{(3)})\bar{f}(a_{(4)})a_{(1)}\\ &= \sum S(a_{(3)})\bar{f}(a_{(4)})S^2(a_{(2)})f(a_{(1)})\\ &= \sum S(S(a_{(2)})a_{(3)})f(a_{(1)})\bar{f}(a_{(4)}) = \varepsilon(a)1.\end{aligned}$$

Similarly, (8) and the equation $f\bar{f} = \varepsilon$ imply that $\sum a_{(2)}T(a_{(1)}) = \varepsilon(a)1$. This shows that T is an antipode for the bialgebra $\mathcal{A}^{\rm op}$. By Proposition 1.6, $T = S^{-1}$. □

Another interesting feature of a coquasitriangular bialgebra $\mathcal{A}$ is that its comodules carry (left and right) module structures.

Proposition 4. *The equation $v \triangleleft a := \sum v_{(0)}\mathbf{r}(v_{(1)}, a)$, $a \in \mathcal{A}$, $v \in V$, turns any right $\mathcal{A}$-comodule V into a right $\mathcal{A}$-module. If V is a right $\mathcal{A}$-comodule algebra, then V becomes a right $\mathcal{A}$-module algebra in this manner.*

Proof. Proposition 2(ii) yields $v \triangleleft 1 = v$. By (4), we have

$$\begin{aligned}(v \triangleleft a) \triangleleft b &= \left(\sum v_{(0)}\mathbf{r}(v_{(1)}, a)\right) \triangleleft b = \sum v_{(0)}\mathbf{r}(v_{(1)}, b)\mathbf{r}(v_{(2)}, a)\\ &= \sum v_{(0)}\mathbf{r}(v_{(1)}, ab) = v \triangleleft ab.\end{aligned}$$

Thus V is a right $\mathcal{A}$-module. If V is a right $\mathcal{A}$-comodule algebra, (3) implies that $vw \triangleleft a = \sum(v \triangleleft a_{(1)})(w \triangleleft a_{(2)})$, so V is a right $\mathcal{A}$-module algebra. □

Similarly, $a \triangleright v := \sum v_{(0)}\mathbf{r}(a, v_{(1)})$ defines a left $\mathcal{A}$-module structure on V. Also, the inverse form $\bar{\mathbf{r}}$ induces left and right $\mathcal{A}$-module structures on V by $a \triangleright v = \sum \bar{\mathbf{r}}(v_{(1)}, a)v_{(0)}$ and $v \triangleleft a := \sum \bar{\mathbf{r}}(a, v_{(1)})v_{(0)}$.

Next we introduce some important bilinear forms. For a coquasitriangular Hopf algebra $\mathcal{A}$ with universal r-form $\mathbf{r}$, we define two linear forms $\mathbf{q}$ and $\mathbf{k}_R$ on $\mathcal{A} \otimes \mathcal{A}$ by setting

$$\mathbf{q}(a \otimes b) := \mathbf{r}_{21}\mathbf{r}(a \otimes b) \equiv \sum \mathbf{r}(b_{(1)}, a_{(1)})\mathbf{r}(a_{(2)}, b_{(2)}), \tag{9}$$

$$\mathbf{k}_R(a \otimes b) := \mathbf{q}(a \otimes S(b)) \equiv \sum \mathbf{r}(S(b_{(2)}), a_{(1)})\mathbf{r}(a_{(2)}, S(b_{(1)})). \tag{10}$$

As usual we consider $\mathbf{q}$ and $\mathbf{k}_R$ as bilinear forms on $\mathcal{A}\times\mathcal{A}$ as well.

Recall that a linear functional f on a vector space V is called invariant with respect to a right coaction φ of $\mathcal{A}$ on V if $(f\otimes \mathrm{id})\varphi = f1$. A linear form $\mathbf{k}$ on $\mathcal{A}\otimes\mathcal{A}$ is called Ad_R-*invariant* if $\mathbf{k}$ is invariant with respect to the right coaction $\mathrm{Ad}_R \otimes \mathrm{Ad}_R$ of $\mathcal{A}$ on $\mathcal{A}\otimes\mathcal{A}$ or equivalently if

$$\sum \mathbf{k}(a_{(2)} \otimes b_{(2)})\ S(a_{(1)})a_{(3)}S(b_{(1)})b_{(3)} = \mathbf{k}(a \otimes b)1 \quad \text{for} \quad a, b \in \mathcal{A}.$$

Proposition 5. *The linear form* $\mathbf{k}_R$ *on* $\mathcal{A} \otimes \mathcal{A}$ *is* Ad_R*-invariant.*

Proof. Applying condition (5) twice, we obtain

$$\begin{aligned}
&\sum \mathbf{k}_R(a_{(2)} \otimes b_{(2)}) S(a_{(1)}) a_{(3)} S(b_{(1)}) b_{(3)} \\
&= \sum \mathbf{r}(S(b_{(3)}), a_{(2)}) S(a_{(1)}) [\mathbf{r}(a_{(3)}, S(b_{(2)})) a_{(4)} S(b_{(1)})] b_{(4)} \\
&= \sum \mathbf{r}(S(b_{(3)}), a_{(2)}) S(a_{(1)}) [S(b_{(2)}) a_{(3)} \mathbf{r}(a_{(4)}, S(b_{(1)}))] b_{(4)} \\
&= \sum S(a_{(1)}) [\mathbf{r}(S(b_{(3)}), a_{(2)}) S(b_{(2)}) a_{(3)})] b_{(4)} \mathbf{r}(a_{(4)}, S(b_{(1)})) \\
&= \sum S(a_{(1)}) [a_{(2)} S(b_{(3)}) \mathbf{r}(S(b_{(2)}), a_{(3)})] b_{(4)} \mathbf{r}(a_{(4)}, S(b_{(1)})) \\
&= \sum S(a_{(1)}) a_{(2)} S(b_{(3)}) b_{(4)} \mathbf{r}(S(b_{(2)}), a_{(3)}) \mathbf{r}(a_{(4)}, S(b_{(1)})) \\
&= \mathbf{k}_R(a \otimes b) 1.
\end{aligned}$$ □

In a similar way, the linear form $\mathbf{k}_L$ on $\mathcal{A} \otimes \mathcal{A}$ defined by

$$\mathbf{k}_L(a \otimes b) = \mathbf{q}(S(a) \otimes b) \equiv \sum \mathbf{r}(b_{(1)}, S(a_{(2)})) \mathbf{r}(S(a_{(1)}), b_{(2)})$$

is Ad_L-invariant. Further, we note some useful identities:

$$\mathbf{k}_R(a \otimes S^{-1}(b)) = \mathbf{q}(a \otimes b) = \mathbf{q}(S(b) \otimes S(a)) = \mathbf{k}_L(b \otimes S(a)).$$

Next we turn to the relationship between $*$-structures and universal r-forms.

Proposition 6. *Let* $\mathbf{r}$ *be a universal* r*-form of a* $*$*-bialgebra* $\mathcal{A}$*. Define linear forms* $\mathbf{r}_1$ *and* $\mathbf{r}_2$ *on* $\mathcal{A} \otimes \mathcal{A}$ *by* $\mathbf{r}_1(a \otimes b) := \overline{\mathbf{r}(b^* \otimes a^*)}$ *and* $\mathbf{r}_2(a \otimes b) := \overline{\bar{\mathbf{r}}(a^* \otimes b^*)}$*,* $a, b \in \mathcal{A}$*. Then* $\mathbf{r}_1$ *and* $\mathbf{r}_2$ *are both universal* r*-forms of the bialgebra* $\mathcal{A}$*.*

Proof. As a sample, we verify condition (3) for $\mathbf{r}_2$. By Proposition 2(iv), $\bar{\mathbf{r}}_{21}$ is a universal r-form of $\mathcal{A}$. Therefore, by (4) applied to $\bar{\mathbf{r}}_{21}$, we have

$$\begin{aligned}
\mathbf{r}_2(ab, c) &= \overline{\bar{\mathbf{r}}_{21}(c^*, b^* a^*)} = \sum \overline{\bar{\mathbf{r}}_{21}(c_{(1)}{}^*, a^*)}\ \overline{\bar{\mathbf{r}}_{21}(c_{(2)}{}^*, b^*)} \\
&= \sum \mathbf{r}_2(a, c_{(1)}) \mathbf{r}_2(b, c_{(2)}).
\end{aligned}$$ □

Definition 2. *A universal* r*-form* $\mathbf{r}$ *of a* $*$*-bialgebra* $\mathcal{A}$ *is called* real *if* $\mathbf{r} = \mathbf{r}_1$ *(that is,* $\mathbf{r}(a \otimes b) = \overline{\mathbf{r}(b^* \otimes a^*)}$*) and* inverse real *if* $\mathbf{r} = \mathbf{r}_2$ *(that is,* $\mathbf{r}(a \otimes b) = \overline{\bar{\mathbf{r}}(a^* \otimes b^*)}$*).*

Examples of both cases will be seen in the next subsection. We close this subsection with two examples of coquasitriangular Hopf algebras.

Example 1. Let $\mathcal{A} = \mathbb{C}G$ be the group algebra of an *abelian* group G. Then the universal r-forms on $\mathcal{A}$ are in one-to-one correspondence to the bicharacters $\mathbf{r}(\cdot, \cdot)$ of G, that is, to mappings $\mathbf{r}(\cdot, \cdot) : G \times G \to \mathbb{C}$ such that $\mathbf{r}(gh, k) = \mathbf{r}(g, k)\mathbf{r}(h, k)$ and $\mathbf{r}(g, hk) = \mathbf{r}(g, h)\mathbf{r}(g, k)$ for $g, h, k \in G$. For instance, let G be the additive group $\mathbb{Z}$ of integers. Then, for any nonzero $\lambda \in \mathbb{C}$ there is a universal r-form of $\mathcal{A} = \mathbb{C}\mathbb{Z}$ given by $\mathbf{r}_\lambda(n, m) = \lambda^{-nm}$, $n, m \in \mathbb{Z}$. △

Example 2 (*Sweedler's Hopf algebra*). Let $\mathcal{A}$ be as in Example 1.9. One verifies that the Hopf algebra $\mathcal{A}$ is self-dual, that is, there is a dual pairing $\langle\cdot,\cdot\rangle$ of $\mathcal{A}$ and $\mathcal{A}$ such that $\langle x,x\rangle = -\langle g,g\rangle = 1$ and $\langle x,g\rangle = \langle g,x\rangle = 0$. If $\mathcal{R}(\alpha)$, $\alpha\in\mathbb{C}$, is the universal R-matrix from Example 8.2, then $\mathbf{r}_\alpha(a\otimes b) := \langle a\otimes b, \mathcal{R}(\alpha)\rangle$, $a,b\in\mathcal{A}$, defines a universal r-form $\mathbf{r}_\alpha$ on $\mathcal{A}$ such that $\mathcal{A}$ is cotriangular. $\triangle$

10.1.2 Coquasitriangularity of FRT Bialgebras $\mathcal{A}(R)$ and Coordinate Hopf Algebras $\mathcal{O}(G_q)$

The main source of examples of coquasitriangular bialgebras stems from the following

Theorem 7. *Let V be a finite-dimensional vector space. Suppose that R is an invertible linear mapping of $V\otimes V$ to itself which satisfies the QYBE $R_{12}R_{13}R_{23} = R_{23}R_{13}R_{12}$. Then the bialgebra $\mathcal{A}(R)$ is coquasitriangular. There exists a unique universal r-form $\mathbf{r}$ of $\mathcal{A}(R)$ such that $\mathbf{r}_{VV} = \hat{R}$, where V is considered as a right $\mathcal{A}(R)$-comodule by (9.6). Then $\mathbf{r}$ is called the canonical universal r-form of $\mathcal{A}(R)$. For $i,j,k,l = 1,2,\cdots,N$, we have*

$$\mathbf{r}(u^i_j\otimes u^k_l) = R^{ik}_{jl} \quad \textit{and} \quad \bar{\mathbf{r}}(u^i_j\otimes u^k_l) = (R^{-1})^{ik}_{jl}. \tag{11}$$

Proof. We first construct two linear forms $\mathbf{r}$ and $\bar{\mathbf{r}}$ on $\mathbb{C}\langle u^i_j\rangle\otimes\mathbb{C}\langle u^i_j\rangle$. In order to do so, we define $\mathbf{r}(u^i_j\otimes u^k_l)$ and $\bar{\mathbf{r}}(u^i_j\otimes u^k_l)$ by (11) and

$$\mathbf{r}(1\otimes u^i_j) = \mathbf{r}(u^i_j\otimes 1) = \bar{\mathbf{r}}(1\otimes u^i_j) = \bar{\mathbf{r}}(u^i_j\otimes 1) = \delta_{ij}. \tag{12}$$

Since $\mathbb{C}\langle u^i_j\rangle$ is the free algebra with generators u^i_j and a coalgebra (see the proof of Proposition 9.1), there is a unique extension of $\mathbf{r}$ to a linear form $\mathbf{r}:\mathbb{C}\langle u^i_j\rangle\otimes\mathbb{C}\langle u^i_j\rangle\to\mathbb{C}$ such that the equations (3) and (4) are satisfied. Similarly, $\bar{\mathbf{r}}$ extends uniquely to a linear form $\bar{\mathbf{r}}:\mathbb{C}\langle u^i_j\rangle\otimes\mathbb{C}\langle u^i_j\rangle\to\mathbb{C}$ such that (3) and (4) hold for the coalgebra $\mathbb{C}\langle u^i_j\rangle^{\mathrm{op}}$.

The crucial step of this proof is to show that $\mathbf{r}$ and $\bar{\mathbf{r}}$ pass to linear forms on $\mathcal{A}(R)\otimes\mathcal{A}(R)$. This is where the QYBE for R comes in. Recall that $\mathcal{A}(R)$ is the quotient of the algebra $\mathbb{C}\langle u^i_j\rangle$ by the two-sided ideal $\mathcal{J}(R)$ generated by the elements $I^{ij}_{nm} := R^{ji}_{rs}u^r_n u^s_m - u^i_r u^j_s R^{sr}_{nm}$. By (3) and the first equality of (11), we compute

$$\begin{aligned}
\mathbf{r}(I^{ij}_{nm}\otimes u^k_l) &= R^{ji}_{rs}\mathbf{r}(u^r_n u^s_m\otimes u^k_l) - (u^i_r u^j_s\otimes u^k_l)R^{sr}_{nm}\\
&= R^{ji}_{rs}\mathbf{r}(u^r_n\otimes u^k_x)\mathbf{r}(u^s_m\otimes u^x_l) - \mathbf{r}(u^i_r\otimes u^k_x)\mathbf{r}(u^j_s\otimes u^x_l)R^{sr}_{nm}\\
&= R^{ji}_{rs}R^{rk}_{nx}R^{sx}_{ml} - R^{ik}_{rx}R^{jx}_{sl}R^{sr}_{nm}\\
&= (R_{12}R_{13}R_{23})^{jik}_{nml} - (R_{23}R_{13}R_{12})^{jik}_{nml} = 0.
\end{aligned}$$

Similarly, $\mathbf{r}(u^k_l\otimes I^{ij}_{nm}) = 0$ by (4) and (11). By (3) and (4), these equations imply that $\mathbf{r}(\mathcal{J}(R)\otimes u^k_l) = \mathbf{r}(u^k_l\otimes\mathcal{J}(R)) = \{0\}$. Further, by (12), we have $\mathbf{r}(\mathcal{J}(R)\otimes 1) = \mathbf{r}(1\otimes\mathcal{J}(R)) = \{0\}$. Applying once more (3) and (4) we conclude that $\mathbf{r}(\mathcal{J}(R)\otimes\mathbb{C}\langle u^i_j\rangle) = \mathbf{r}(\mathbb{C}\langle u^i_j\rangle\otimes\mathcal{J}(R)) = \{0\}$. Therefore, $\mathbf{r}$ induces a well-defined linear form, again denoted by $\mathbf{r}$, on $\mathcal{A}(R)\otimes\mathcal{A}(R)$ by setting $\mathbf{r}(a\otimes b) := \mathbf{r}((a+\mathcal{J}(R))\otimes(b+\mathcal{J}(R)))$, $a,b\in\mathcal{A}(R)$. Now we repeat the preceding reasoning with $\mathbf{r}$ replaced by $\bar{\mathbf{r}}$ and $\mathbb{C}\langle u^i_j\rangle$ by $\mathbb{C}\langle u^i_j\rangle^{\mathrm{op}}$. Using again the QYBE for R we obtain a well-defined linear form $\bar{\mathbf{r}}$ on $\mathcal{A}(R)\otimes\mathcal{A}(R)$.

Next we show that $\mathbf{r}$ and $\bar{\mathbf{r}}$ satisfy the conditions of Definition 1 or equivalently the equations (1), (3), (4) and (5). By the construction of $\mathbf{r}$, (3) and (4) are fulfilled. Using (3), (4) and (11), one verifies that both equations (1) and (5) hold for $a=a'$, $b=b'c'$ and for $a=a'b'$, $b=c'$ provided that they hold for $a=a'$, $b=b'$, for $a=a'$, $b=c'$ and for $a=b'$, $b=c'$. Thus it suffices to show (1) and (5) for the elements 1 and u^i_j. If a or b is 1, then (1) and (5) clearly hold by (12). Now let $a=u^j_n$ and $b=u^i_m$. Then, by (11), the equations (1) and (5) are equivalent to the relations $RR^{-1}=R^{-1}R=I$ and the equation (9.1), respectively. Hence (1) and (5) hold for arbitrary $a,b\in\mathcal{A}(R)$. This completes the proof that $\mathcal{A}(R)$ is coquasitriangular.

We prove that $\mathbf{r}_{VV}=\hat{R}$. Since $\varphi_V(x_i)=x_j\otimes u^j_i$ by (9.6), we have

$$\mathbf{r}_{VV}(x_k\otimes x_l)=\mathbf{r}(u^i_k,u^j_l)x_j\otimes x_i=R^{ij}_{kl}x_j\otimes x_i=\hat{R}^{ji}_{kl}x_j\otimes x_l=\hat{R}(x_k\otimes x_l) \quad (13)$$

by (6) and (11). Thus, $\mathbf{r}_{VV}=\hat{R}$. Conversely, if $\tilde{\mathbf{r}}$ is another universal r-form on $\mathcal{A}(R)$ such that $\tilde{\mathbf{r}}_{VV}=\hat{R}$, then we conclude from $\tilde{\mathbf{r}}_{VV}(x_k\otimes x_l)=\tilde{\mathbf{r}}(u^i_k,u^j_l)x_j\otimes x_i$ by (6) that $\tilde{\mathbf{r}}(u^i_k,u^j_l)=R^{ij}_{kl}=\mathbf{r}(u^i_k,u^j_l)$. Moreover, $\mathbf{r}(1,u^i_j)=\mathbf{r}(u^i_j,1)=\tilde{\mathbf{r}}(1,u^i_j)=\tilde{\mathbf{r}}(u^i_j,1)=\delta_{ij}$. Since $\tilde{\mathbf{r}}$ and $\mathbf{r}$ both satisfy (3) and (4), it follows that $\tilde{\mathbf{r}}=\mathbf{r}$ on the whole of $\mathcal{A}(R)\otimes\mathcal{A}(R)$. This proves the uniqueness assertion. □

Remark 2. Theorem 7 can be generalized to the following more abstract result. Let $\mathcal{M}(\mathcal{C},\mathbf{r})$ be the bialgebra which was defined in Remark 1 of Subsect. 9.1.1. Suppose that the linear form $\mathbf{r}:\mathcal{C}\otimes\mathcal{C}\to\mathbb{C}$ is invertible and satisfies the QYBE $\mathbf{r}_{12}\mathbf{r}_{13}\mathbf{r}_{23}=\mathbf{r}_{23}\mathbf{r}_{13}\mathbf{r}_{12}$, that is, for any $x,y,z\in\mathcal{C}$ we have

$$\sum\mathbf{r}(x_{(1)},y_{(1)})\mathbf{r}(x_{(2)},z_{(1)})\mathbf{r}(y_{(2)},z_{(2)})=\sum\mathbf{r}(y_{(1)},z_{(1)})\mathbf{r}(x_{(1)},z_{(2)})\mathbf{r}(x_{(2)},y_{(2)})$$

Then $\mathbf{r}$ *admits a unique extension to a universal r-form of the bialgebra* $\mathcal{M}(\mathcal{C},\mathbf{r})$ *which becomes then a coquasitriangular bialgebra.* (The proof of this assertion uses the same reasoning as above. Note that the inverse $\bar{\mathbf{r}}$ of $\mathbf{r}$ also satisfies the QYBE as $\mathbf{r}$ does.)

Recall that the FRT bialgebras $\mathcal{A}(R)$ are a special case of the bialgebras $\mathcal{M}(\mathcal{C},\mathbf{r})$ with $\mathcal{C}=M_N(\mathbb{C})'$ and $\mathbf{r}(u^i_n\otimes u^j_m)=R^{ij}_{nm}$. If R is invertible and satisfies the QYBE, one easily checks that the linear form $\mathbf{r}$ has the same properties. Thus, Theorem 7 is a special case of the preceding result for the bialgebra $\mathcal{M}(\mathcal{C},\mathbf{r})$. △

As a supplement to Theorem 7 we state another related result which will be needed later in Subsect. 10.3.1. Let us say that a solution $R=(R^{ik}_{jl})$ of the QYBE is *regular* if R^{t_2} is invertible (t_2 means the transposition of the second index) or equivalently if there exists another matrix $\tilde{R} = (\tilde{R}^{ik}_{jl})$ such that

$$R^{ik}_{lj}\tilde{R}^{nj}_{im} = \tilde{R}^{ik}_{lj}R^{nj}_{im} = \delta_{ln}\delta_{km}, \quad k,l,n,m = 1,\cdots,N. \tag{14}$$

Proposition 8. *Retain the assumptions and the notation of Theorem 7 and suppose in addition that R is regular. Then there exists a unique linear form* $\mathbf{s} : \mathcal{A}(R) \otimes \mathcal{A}(R) \to \mathbb{C}$ *such that* $\mathbf{s}(\cdot,\cdot) = \mathbf{s}(\cdot \otimes \cdot)$ *is a dual pairing of the bialgebras* $\mathcal{A}(R)^{\mathrm{op}}$ *and* $\mathcal{A}(R)$ *and*

$$\mathbf{s}(u^i_j \otimes u^k_l) = \tilde{R}^{ik}_{jl}, \quad i,j,k,l = 1,2,\cdots,N. \tag{15}$$

Moreover, $\mathbf{s}$ *is the convolution inverse of* $\mathbf{r}$ *when both are considered as linear forms on* $\mathcal{A}(R)^{\mathrm{op}} \otimes \mathcal{A}(R)$.

Proof. The proof is quite similar to the proof of Theorem 7. The crucial steps are again the two consistency conditions $\mathbf{s}(I^{ij}_{nm} \otimes u^k_l) = \mathbf{s}(u^k_l \otimes I^{ij}_{nm}) = 0$ which are needed in order to pass from $\mathbb{C}\langle u^i_j\rangle$ to $\mathcal{A}(R)$. It turns out that both conditions are equivalent to the QYBE. That $\mathbf{s}$ is the convolution inverse of $\mathbf{r} : \mathcal{A}(R)^{\mathrm{op}} \otimes \mathcal{A}(R) \to \mathbb{C}$ follows immediately from (14) and (15). □

Now we apply Theorem 7 in order to inherit coquasitriangular structures on the coordinate Hopf algebras $\mathcal{O}(G_q)$, where G_q is one of the quantum groups $GL_q(N)$, $SL_q(N)$, $O_q(N)$ or $Sp_q(N)$. Let $\mathsf{R} = \mathsf{R}(q)$ be the corresponding R-matrix given by (9.13) resp. (9.30). Since R is an invertible solution of the QYBE (in fact, $\mathsf{R}(q)^{-1} = \mathsf{R}(q^{-1})$) by Proposition 8.19, Theorem 7 applies to the bialgebra $\mathcal{A}(R)$ with $R = z\mathsf{R}$ for any nonzero $z \in \mathbb{C}$. The following theorem says that for certain values of the parameter z the canonical universal r-form $\mathbf{r}_z$ of $\mathcal{A}(R)$ induces a universal r-form on $\mathcal{O}(G_q)$.

Theorem 9. $\mathcal{O}(GL_q(N)), \mathcal{O}(SL_q(N), \mathcal{O}(O_q(N))$ *and* $\mathcal{O}(Sp_q(N))$ *are coquasitriangular Hopf algebras with universal r-forms* $\mathbf{r}_z$ *uniquely determined by*

$$\mathbf{r}_z(u^i_j \otimes u^k_l) = z\mathsf{R}^{ik}_{jl}, \quad i,j,k,l = 1,2,\cdots,N. \tag{16}$$

Here z is a complex number such that $z \neq 0$ *for* $GL_q(N)$, $z^N = q^{-1}$ *for* $SL_q(N)$ *and* $z^2 = 1$ *for* $O_q(N)$ *and* $Sp_q(N)$. *The matrix* $\mathsf{R} = (\mathsf{R}^{ik}_{jl})$ *is given by (9.13) for* $GL_q(N)$ *and* $SL_q(N)$ *and by (9.30) for* $O_q(N)$ *and* $Sp_q(N)$.

Proof. We begin with the quantum groups $GL_q(N)$ and $SL_q(N)$. Let $z \in \mathbb{C}$, $z \neq 0$, and set $R = z\mathsf{R}$. Our first aim is to show that

$$\mathbf{r}_z(\mathcal{D}_q, u^i_j) = \mathbf{r}_z(u^i_j, \mathcal{D}_q) = z^N q\delta_{ij}. \tag{17}$$

For this proof we use the L-functionals $l^{\pm}_{ij} \in U^{\mathrm{ext}}_q(\mathrm{sl}_N)$ defined in Subsect. 8.5.4 and the dual pairing $\langle\cdot,\cdot\rangle$ of the Hopf algebras $U^{\mathrm{ext}}_q(\mathrm{sl}_N)$ and $\mathcal{O}(SL_q(N))$ from Theorem 9.18. Since $\mathcal{D}_q = 1$ in $\mathcal{O}(SL_q(N))$ and $\varepsilon(l^+_{ij}) = \delta_{ij}$, we have $\langle l^+_{ij}, \mathcal{D}_q\rangle = \delta_{ij}$. Recall from Subsects. 8.4.2 and 8.5.3 that $q^{-1/N}\mathsf{R}$ is the R-matrix for the vector representation of $U^{\mathrm{ext}}_q(\mathrm{sl}_N)$. Using these facts, the first equality of (9.21) and the formulas (8.89) and (9.40), we get

$$z^N q\delta_{ij} = z^N q \langle l^+_{ij}, \mathcal{D}_q\rangle = \sum_{p\in\mathcal{P}_N} (-q)^{l(p)-l(r)} z^N q \langle l^+_{ij}, u^{p(1)}_{r(1)} \cdots u^{p(N)}_{r(N)}\rangle$$

$$= \sum_{p\in\mathcal{P}_N} (-q)^{l(p)-l(r)} z^N (\mathsf{R}_{1,N+1}\mathsf{R}_{2,N+1}\cdots \mathsf{R}_{N,N+1})^{p(1)\cdots p(N)i}_{r(1)\cdots r(N)j} \tag{18}$$

for any permutation $r \in \mathcal{P}_N$. Applying the formulas (3) and (11) a similar computation for the term $\mathbf{r}_z(\mathcal{D}_q, u^i_j)$ leads to the expression (18) as well. This proves that $\mathbf{r}_z(\mathcal{D}_q, u^i_j) = z^N q\delta_{ij}$. Using the relation $\mathsf{R}^{kl}_{mn} = \mathsf{R}^{nm}_{lk}$ (which holds by (9.13)) and formula (4) instead of (3), we easily derive that $\mathbf{r}_z(u^i_j, \mathcal{D}_q)$ is equal to (18) with upper indices $p(1)\cdots p(N)i$ and lower indices $r(1)\cdots r(N)j$ interchanged. The latter expression is also obtained if we compute the term $z^N q\delta_{ij} = z^N q \langle l^+_{ji}, \mathcal{D}_q\rangle$ by means of the second equality of (9.21). Therefore, we have $\mathbf{r}_z(u^i_j, \mathcal{D}_q) = z^N q\delta_{ij}$. This completes the proof of (17).

Let $\mathcal{A}_n(R)$ be the linear span of monomials $u^{i_1}_{j_1} \ldots u^{i_n}_{j_n}$. By (3) and (4), equation (17) implies that

$$\mathbf{r}_z(\mathcal{D}_q a, b) = (z^N q)^n \mathbf{r}_z(a,b) \quad \text{and} \quad \mathbf{r}_z(b, \mathcal{D}_q a) = (z^N q)^n \mathbf{r}_z(b,a)$$

for $a \in \mathcal{A}(R)$ and $b \in \mathcal{A}_n(R)$. Therefore, by (9.26), we extend $\mathbf{r}_z$ uniquely to $\mathcal{O}(GL_q(N))$ by setting

$$\mathbf{r}_z(\mathcal{D}_q^{-k} a, \mathcal{D}_q^{-l} b) := (z^N q)^{km+ln+klN} \mathbf{r}_z(a,b), \;\; a \in \mathcal{A}_n(R), b \in \mathcal{A}_m(R), k,l \in \mathbb{N}.$$

For the quantum group $SL_q(N)$ we assume that $z^N = q^{-1}$. Then, we have that $\mathbf{r}_z(\mathcal{D}_q - 1, u^i_j) = \mathbf{r}_z(u^i_j, \mathcal{D}_q - 1) = 0$ by formula (17) and hence $\mathbf{r}_z(\langle \mathcal{D}_q - 1\rangle, \mathcal{A}(R)) = \mathbf{r}_z(\mathcal{A}(R), \langle \mathcal{D}_q - 1\rangle) = \{0\}$ by (3) and (4). Therefore, $\mathbf{r}_z$ passes to the quotient $\mathcal{O}(SL_q(N)) = \mathcal{A}(R)/\langle \mathcal{D}_q - 1\rangle$.

Now we turn to the quantum groups $O_q(N)$ and $Sp_q(N)$ and suppose that $z = \pm 1$. Using the first (resp. second) expression for $\mathcal{Q}_q$ in (9.37) and the second (resp. first) formula for R in (9.31), one immediately checks that $\mathbf{r}_z(u^i_j, \mathcal{Q}_q) = \delta_{ij}$ (resp. $\mathbf{r}_z(\mathcal{Q}_q, u^i_j) = \delta_{ij}$). Hence $\mathbf{r}_z(\langle \mathcal{Q}_q - 1\rangle, \mathcal{A}(R)) = \mathbf{r}_z(\mathcal{A}(R), \langle \mathcal{Q}_q - 1\rangle) = \{0\}$ and $\mathbf{r}_z$ passes to the quotient Hopf algebra $\mathcal{A}(R)/\langle \mathcal{Q}_q - 1\rangle$. By Lemma 9.12, the latter is $\mathcal{O}(O_q(N))$ resp. $\mathcal{O}(Sp_q(N))$.

The above reasoning can be repeated with $R = z\mathsf{R}$ replaced by $R^{-1} = z^{-1}\mathsf{R}^{-1}$ and $\mathbf{r}_z$ by $\overline{\mathbf{r}_z}$. (In the proof of (17) for $\overline{\mathbf{r}_z}$ one has to use l^-_{ij} instead of l^+_{ij}.) We then obtain a linear form $\overline{\mathbf{r}_z}$ on $\mathcal{O}(G_q) \otimes \mathcal{O}(G_q)$ which is the convolution inverse of $\mathbf{r}_z$. Since $\mathbf{r}_z$ is derived from a universal r-form of $\mathcal{A}(R)$, it is clearly a universal r-form of $\mathcal{O}(G_q)$. □

Finally, we consider the universal r-forms $\mathbf{r}_z$ from Theorem 9 for the corresponding real forms of the quantum groups (see Subsects. 9.2.4 and 9.3.5).

Proposition 10. (i) *If the numbers q and z are real, the universal r-forms $\mathbf{r}_z$ of the coordinate Hopf $*$-algebras of $U_q(N), SU_q(N;\nu), O_q(N;\nu)$ and $Sp_q(N;\nu)$ are real.*

(ii) *If* $|q| = |z| = 1$, *the universal r-forms* $\mathbf{r}_z$ *of the coordinate Hopf* $*$-*algebras of* $GL_q(N;\mathbb{R}), SL_q(N;\mathbb{R}), O_q(n,n), O_q(n,n+1)$, *and* $Sp_q(N;\mathbb{R})$ *are inverse real.*

Proof. We prove the assertion for $O_q(N;\nu)$ and $Sp_q(N;\nu)$ and use the corresponding notation from Subsect. 9.3.5, Case 2. The proofs for the other cases are similar. Let $\mathbf{r}_z$ be as in Theorem 9. Since $q \in \mathbb{R}$, the matrix entries R^{lj}_{ki} are real. By (9.30), they satisfy the equations $\nu_i\nu_j\nu_k\nu_l\mathsf{R}^{lj}_{ki} = \mathsf{R}^{lj}_{ki} = \mathsf{R}^{ik}_{jl}$. Thus, we obtain

$$(\mathbf{r}_z)_1(u^i_j, u^k_l) = \overline{\mathbf{r}_z((u^k_l)^*, (u^i_j)^*)} = \nu_i\nu_j\nu_k\nu_l\ \overline{\mathbf{r}_z(S(u^l_k), S(u^j_i))}$$

$$= \nu_i\nu_j\nu_k\nu_l\ \overline{\mathbf{r}_z(u^l_k, u^j_i)} = \nu_i\nu_j\nu_k\nu_l\bar{z}\ \overline{\mathsf{R}^{lj}_{ki}} = z\mathsf{R}^{ik}_{jl} = \mathbf{r}_z(u^i_j, u^k_l).$$

That is, the universal r-forms $\mathbf{r}_z$ and $(\mathbf{r}_z)_1$ (by Proposition 6) are equal on the generators. Hence they coincide everywhere. □

We close this subsection with an example concerning Proposition 3.

Example 3 (*Square of the antipode*). Let $\mathcal{A}$ be an arbitrary coquasitriangular Hopf algebra with universal r-form $\mathbf{r}$ and let $u = (u^i_j)_{i,j=1,\cdots,d}$, $d \in \mathbb{N}$, be a matrix of elements $u^i_j \in \mathcal{A}$ such that $\Delta(u^i_j) = u^i_k \otimes u^k_j$ and $\varepsilon(u^i_j) = \delta_{ij}$, $i,j = 1,2,\cdots,d$. In the notation of Proposition 3, we set

$$D^i_j := f(u^i_j) \equiv \mathbf{r}(u^i_k, S(u^k_j)),\quad (D^{-1})^i_j := \bar{f}(u^i_j) \equiv \mathbf{r}(S^2(u^i_k), u^k_j). \tag{19}$$

Since $f\bar{f} = \bar{f}f = \varepsilon$, the matrix $D^{-1} = ((D^{-1})^i_j)$ is indeed the inverse of the matrix $D = (D^i_j)$. Then Proposition 3 states that

$$S^2(u^i_j) = (D^{-1})^i_n u^n_m D^m_j. \tag{20}$$

Now let $\mathcal{A}$ be one of the Hopf algebras $\mathcal{O}(G_q)$ in Theorem 9 and let $\mathbf{u}=(u^i_j)$ be the fundamental corepresentation of $\mathcal{A}$. By Propositions 9.10 and 9.13, we then have $S^2(u^i_j) = \mathsf{D}^{-1}_i u^i_j \mathsf{D}_j$. Set $\mathbf{r} := \mathbf{r}_z$. By (17) and (18), we get $(D^{-1})^i_j = \mathsf{D}^{-1}_i \mathsf{D}_k \mathbf{r}_z(u^i_k, u^k_j) = \sum_k z\mathsf{D}^{-1}_i\mathsf{D}_k\hat{\mathsf{R}}^{ki}_{kj}$. From the latter and the formulas (9.13) and (9.30) one derives

$$D^i_j = z^{-1}q\mathsf{D}_i\delta_{ij} = z^{-1}q^{-2i+1}\delta_{ij}\ \text{ for }\ G_q = GL_q(N), SL_q(N),$$

$$D^i_j = z^{-1}\epsilon q^{N-\epsilon}\mathsf{D}_i\delta_{ij} = z^{-1}\epsilon q^{2\rho_i+N-\epsilon}\delta_{ij}\ \text{ for }\ G_q = O_q(N), Sp_q(N).$$

For the universal r-form $\mathbf{r} = (\overline{\mathbf{r}_z})_{21}$ of $\mathcal{A}$ we obtain

$$D^i_j = zq^{-2i+2N+1}\delta_{ij}\ \text{ for }\ G_q = GL_q(N), SL_q(N),$$

$$D^i_j = z\epsilon q^{2\rho_i-N+\epsilon}\delta_{ij}\ \text{ for }\ G_q = O_q(N), Sp_q(N). \qquad \triangle$$

10.1.3 L-Functionals of Coquasitriangular Hopf Algebras

In this subsection we investigate some linear functionals on coquasitriangular bialgebras which are important tools for a number of applications. The constructions of bicovariant differential calculi in Sects. 14.5 and 14.6 and of quantum homogeneous spaces in Subsect. 11.6.6 will use these functionals. We suppose that $\mathcal{A}$ is a coquasitriangular bialgebra with universal r-form $\mathbf{r}$.

For $a \in \mathcal{A}$, we introduce linear functionals $l^{\pm}(a)$, $l(a)$ and $\tilde{l}(a)$ on $\mathcal{A}$ by

$$l^+(a) := \mathbf{r}(\cdot, a), \quad l^-(a) := \bar{\mathbf{r}}(a, \cdot), \quad l(a) := \mathbf{q}(\cdot, a), \quad \tilde{l}(a) := \mathbf{q}(a, \cdot), \tag{21}$$

where $\mathbf{q}$ is the bilinear form on $\mathcal{A} \times \mathcal{A}$ defined by (9). If $\mathcal{A}$ is a Hopf algebra, it follows easily from Proposition 2(v) that

$$l(a) = \sum S(l^-(a_{(1)}))l^+(a_{(2)}) \quad \text{and} \quad \tilde{l}(a) = \sum l^+(a_{(1)})S(l^-(a_{(2)})). \tag{22}$$

Moreover, since $\mathbf{q}(S(a), S(b)) = \mathbf{q}(b, a)$, the functionals $l(a)$ and $\tilde{l}(a)$ are related by the formulas

$$\tilde{l}(a) = S(l(S(a))) \quad \text{and} \quad l(a) = S(\tilde{l}(S(a))).$$

Let $\mathcal{U}$ denote the subalgebra of the algebra $\mathcal{A}'$ generated by the functionals $l^{\pm}(a)$, $a \in \mathcal{A}$. If $\mathcal{A}$ is a Hopf algebra, then we have $S(l^{\pm}(a)) = l^{\pm}(S_{\mathcal{A}}^{-1}(a))$ (by Proposition 11(ii) below) and hence $l(a) \in \mathcal{U}$ and $\tilde{l}(a) \in \mathcal{U}$ for any $a \in \mathcal{A}$ by (22).

Proposition 11. (i) *$\mathcal{U}$ is a subbialgebra of the dual bialgebra $\mathcal{A}^\circ$. The maps $\pi_{\pm} : a \to l^{\pm}(a)$ are bialgebra homomorphisms of $\mathcal{A}^{\mathrm{op}}$ to $\mathcal{U}$. For $a, b, c \in \mathcal{A}$, we have*

$$l^{\pm}(ab)(c) = \sum l^{\pm}(b)(c_{(1)})l^{\pm}(a)(c_{(2)}), \tag{23}$$

$$\Delta(l^{\pm}(a)) = \sum l^{\pm}(a_{(1)}) \otimes l^{\pm}(a_{(2)}). \tag{24}$$

(ii) *Suppose that $\mathcal{A}$ is a Hopf algebra. Then $\mathcal{U}$ is a Hopf subalgebra of $\mathcal{A}^\circ$ with antipode determined by the relation $S(l^{\pm}(a)) = l^{\pm}(S_{\mathcal{A}}^{-1}(a))$, $a \in \mathcal{A}$. Moreover, for all $a, b \in \mathcal{A}$ and $g \in \mathcal{A}^\circ$ we have the identities*

$$\Delta(l(a)) = \sum l(a_{(2)}) \otimes S(l^-(a_{(1)}))l^+(a_{(3)}), \tag{25}$$

$$\Delta(\tilde{l}(a)) = \sum l^+(a_{(1)})S(l^-(a_{(3)})) \otimes \tilde{l}(a_{(2)}), \tag{26}$$

$$\mathrm{ad}_R(g)(l(a)) = \sum l(a_{(2)})\; g(a_{(1)}S(a_{(3)})), \tag{27}$$

$$\mathrm{ad}_L(g)(\tilde{l}(a)) = \sum \tilde{l}(a_{(2)})\; g(a_{(1)}S(a_{(3)})), \tag{28}$$

$$l(a)l(b) = l(a_{(1)}b_{(2)})\mathbf{r}(b_{(1)}, S(a_{(3)}))\mathbf{r}(b_{(3)}, a_{(2)}), \tag{29}$$

$$\tilde{l}(a)\tilde{l}(b) = \tilde{l}(a_{(2)}b_{(3)})\mathbf{r}(a_{(1)}, b_{(2)})\mathbf{r}(a_{(3)}, S(b_{(1)})). \tag{30}$$

Proof. (i): Since $l^-(a)(\cdot) = \bar{\mathbf{r}}_{21}(\cdot, a)$ and $\bar{\mathbf{r}}_{21}$ is also a universal r-form on $\mathcal{A}$ by Proposition 2(iv), it suffices to prove the assertions for l^+. By (3), we have

$$l^+(a)(bc) = \mathbf{r}(bc, a) = \sum \mathbf{r}(b, a_{(1)})\mathbf{r}(c, a_{(2)}) = \sum l^+(a_{(1)})(b)l^+(a_{(2)})(c).$$

By the definition of $\mathcal{A}^\circ$ (see Subsect. 1.2.8), this implies that $l^+(a) \in \mathcal{A}^\circ$ and hence $\mathcal{U} \subseteq \mathcal{A}^\circ$. The preceding equation rewrites as (24) using the definition of the comultiplication of $\mathcal{A}^\circ$. By (24), $\mathcal{U}$ is a subbialgebra of $\mathcal{A}^\circ$. Formula (23) follows from (4). The equations (23) and (24) combined with the relations $l^\pm(1) = \varepsilon$ mean that the linear maps $\pi_\pm$ are bialgebra homomorphisms of $\mathcal{A}^{\mathrm{op}}$ to $\mathcal{U}$.

(ii): The first assertion of (ii) follows at once from (i). Recall that the Hopf algebra $\mathcal{A}^{\mathrm{op}}$ has the antipode $S_{\mathcal{A}}^{-1}$. Formula (25) is obtained from (22), (24) and the fact that $S(l^-(a)) = l^-(S_{\mathcal{A}}^{-1}(a))$. Next we prove (27). Using condition (5) twice, we compute

$$\begin{aligned}
(l(a)g)(b) &= \sum \mathbf{q}(b_{(1)}, a)g(b_{(2)}) \\
&= \sum \mathbf{r}(a_{(1)}, b_{(1)})\mathbf{r}(b_{(2)}, a_{(2)})g_{(1)}(b_{(3)}a_{(3)})g_{(2)}(S(a_{(4)})) \\
&= \sum \mathbf{r}(a_{(1)}, b_{(1)})\mathbf{r}(b_{(3)}, a_{(3)})g_{(1)}(a_{(2)}b_{(2)})g_{(2)}(S(a_{(4)})) \\
&= \sum \mathbf{r}(a_{(2)}, b_{(2)})\mathbf{r}(b_{(3)}, a_{(3)})g_{(1)}(b_{(1)}a_{(1)})g_{(2)}(S(a_{(4)})) \\
&= \sum l(a_{(2)})(b_{(2)})g_{(1)}(b_{(1)})g_{(2)}(a_{(1)}S(a_{(3)})).
\end{aligned}$$

That is, we have

$$l(a)g = \sum g_{(1)}l(a_{(2)})\ g_{(2)}(a_{(1)}S(a_{(3)})) \tag{31}$$

and hence

$$\begin{aligned}
\mathrm{ad}_R(g)(l(a)) &= \sum S(g_{(1)})l(a)g_{(2)} = \sum S(g_{(1)})g_{(2)}l(a_{(2)})\ g_{(3)}(a_{(1)}S(a_{(3)})) \\
&= \sum l(a_{(2)})\ g(a_{(1)}S(a_{(3)})).
\end{aligned}$$

Now we verify equation (29). First we apply the QYBE $\mathbf{r}_{12}\mathbf{r}_{13}\mathbf{r}_{23} = \mathbf{r}_{23}\mathbf{r}_{13}\mathbf{r}_{12}$ to $x \otimes \cdot \otimes y$ for $x, y \in \mathcal{A}$. We then obtain the identity

$$\sum S(l^-(x_{(1)}))\mathbf{r}(x_{(2)}, y_{(1)})l^+(y_{(2)}) = \sum l^+(y_{(1)})\mathbf{r}(x_{(1)}, y_{(2)})S(l^-(x_{(2)})). \tag{32}$$

Using formulas (22), (23), (5), (32) and (4) we compute

$$\begin{aligned}
&\sum l(a_{(1)}b_{(2)})\mathbf{r}(b_{(1)}, S(a_{(3)}))\mathbf{r}(b_{(3)}, a_{(2)}) \\
&\quad = \sum \mathbf{r}(b_{(1)}, S(a_{(4)}))S(l^-(a_{(1)}))S(l^-(b_{(2)}))[\mathbf{r}(b_{(4)}, a_{(3)}))l^+(a_{(2)}b_{(3)})] \\
&\quad = \sum \mathbf{r}(b_{(1)}, S(a_{(4)}))S(l^-(a_{(1)}))S(l^-(b_{(2)}))\mathbf{r}(b_{(3)}, a_{(2)})l^+(b_{(4)}a_{(3)})
\end{aligned}$$

$$= \sum \mathbf{r}(b_{(1)}, S(a_{(4)}))S(l^-(a_{(1)}))[S(l^-(b_{(2)}))\mathbf{r}(b_{(3)}, a_{(2)})l^+(a_{(3)})]l^+(b_{(4)})$$
$$= \sum \mathbf{r}(b_{(1)}, S(a_{(4)}))S(l^-(a_{(1)}))l^+(a_{(2)})\mathbf{r}(b_{(2)}, a_{(3)})S(l^-(b_{(3)}))l^+(b_{(4)})$$
$$= \sum \mathbf{r}(b_{(1)}, a_{(2)}S(a_{(3)}))l(a_{(1)})l(b_{(2)}) = l(a)l(b).$$

The formulas for the functionals $\tilde{l}(a)$ are derived in a similar manner. Instead of (31) we use the equation $g\tilde{l}(a) = \sum \tilde{l}(a_{(2)})g_{(2)}\ g_{(1)}(S(a_{(1)})a_{(3)})$. □

Equation (30) gives the motivation for the definition of the covariantized product and the transmutation theory of coquasitriangular Hopf algebras, see Proposition 34(ii) below. An immediate and perhaps surprising consequence of formulas (29) and (30) is

Corollary 12. *If $\mathcal{A}$ is a coquasitriangular Hopf algebra, then the sets $l(\mathcal{A}) = S(\tilde{l}(\mathcal{A}))$ and $\tilde{l}(\mathcal{A}) = S(l(\mathcal{A}))$ are subalgebras of the Hopf dual $\mathcal{A}^\circ$.*

In general the algebras $l(\mathcal{A})$ and $\tilde{l}(\mathcal{A})$ are not Hopf subalgebras of $\mathcal{A}^\circ$.

Let us recall the notation $\bar{a} = a - \varepsilon(a)1$. From the formulas (25) and (27) we easily obtain the following

Corollary 13. *Let V be a linear subspace of a coquasitriangular Hopf algebra $\mathcal{A}$ such that $\Delta^{(2)}(V) \subseteq \mathcal{A} \otimes V \otimes \mathcal{A}$. Then, $\mathcal{T}(V) := \{l(\bar{a}) \mid a \in V\}$ is an ad_R-invariant linear subspace of $\mathcal{A}^\circ$ satisfying $X(1) = 0$ and $\Delta(X) - \varepsilon \otimes X \in \mathcal{T}(V) \otimes \mathcal{A}^\circ$ for $X \in \mathcal{T}(V)$.*

We shall see in Subsect. 14.2.3 that (under some technical assumptions) the subspaces of $\mathcal{A}^\circ$ with the above properties are just the quantum Lie algebras of bicovariant first order differential calculi on $\mathcal{A}$.

Next we consider $*$-structures. Recall from Corollary 1.11 that the algebra $\mathcal{A}^\circ$ for a Hopf $*$-algebra $\mathcal{A}$ becomes a $*$-algebra with respect to the involution $f \to f^*(a) := \overline{f(S(a)^*)}$, $a \in \mathcal{A}$.

Proposition 14. *Suppose that $\mathcal{A}$ is a Hopf $*$-algebra.*

(i) *If the universal r-form $\mathbf{r}$ is real, then*

$$l^\pm(a)^* = l^\mp(S^{-2}(a^*)) \quad \text{and} \quad l(a)^* = l(S(a)^*). \tag{33}$$

(ii) *If the universal r-form $\mathbf{r}$ is inverse real, then*

$$l^\pm(a)^* = l^\pm(S^{-2}(a^*)) \quad \text{and} \quad l(a)^* = \sum \mathbf{r}({a^*}_{(1)}, S({a^*}_{(3)}))\bar{f}({a^*}_{(4)})l({a^*}_{(2)}), \tag{34}$$

where $\bar{f}$ is the linear functional from Proposition 3.

Proof. As samples, we prove the formulas for $l^+(a)^*$ in (33) and for $l(a)^*$ in (34). Using the assumption that $\mathbf{r}$ is real and Proposition 2(v), we get

$$l^+(a)^*(b) = \overline{l^+(a)(S(b)^*)} = \overline{\mathbf{r}(S(b)^*, a)} = \mathbf{r}(a^*, S(b))$$
$$= \mathbf{r}(S^{-2}(a^*), S^{-1}(b)) = \bar{\mathbf{r}}(S^{-2}(a^*), b) = l^-(S^{-2}(a^*))(b).$$

The second proof will be only sketched. Let $b, c \in \mathcal{A}$. Using the first formula of (34) and the relation $* \circ S^{-1} = S \circ *$ (see (1.39)), we compute

$$S(l^-(b))^* = l^-(S^{-1}(b))^* = l^-(S^{-2}(S^{-1}(b)^*)) = l^-(S^{-2}(S(b^*))) = S(l^-(b^*))$$

and hence

$$l(b)^* = \left(\sum S(l^-(b_{(1)}))l^+(b_{(2)})\right)^* = \sum l^+(S^{-2}(b^*{}_{(2)}))S(l^-(b^*{}_{(1)})). \quad (35)$$

Further, by Propositions 3 and 2(v), we have

$$b = \sum \bar{f}(b_{(1)})S^{-2}(b_{(2)})f(b_{(3)}) = \sum \mathbf{r}(b_{(1)}, S^{-2}(b_{(2)}))S^{-2}(b_{(3)})f(b_{(4)}). \quad (36)$$

Using the formulas (22), (36), (32), (35) and (7) in this order, we get

$$\begin{aligned} l(c) &= S(l^-(c_{(1)}))l^+(c_{(2)}) \\ &= \sum S(l^-(c_{(1)}))\mathbf{r}(c_{(2)}, S^{-2}(c_{(3)}))l^+(S^{-2}(c_{(4)}))f(c_{(5)}) \\ &= \sum l^+(S^{-2}(c_{(3)}))\mathbf{r}(c_{(1)}, S^{-2}(c_{(4)}))S(l^-(c_{(2)}))f(c_{(5)}) \\ &= \sum \mathbf{r}(c_{(1)}, S^{-2}(c_{(3)}))f(c_{(4)})l((c_{(2)})^*)^* \\ &= \sum \mathbf{r}(c_{(1)}, c_{(4)})f(c_{(3)})l((c_{(2)})^*)^*. \end{aligned}$$

Inserting the latter formula into the expression $\sum \mathbf{r}(c_{(1)}, S(c_{(3)}))\bar{f}(c_{(4)})l(c_{(2)})$, it simplifies to $l(c^*)^*$. Setting $c = a^*$, this gives the assertion. □

Now let $\mathbf{u}=(u^i_j)_{i,j=1,\cdots,d}$ be a fixed matrix corepresentation of $\mathcal{A}$. Setting

$$R^{in}_{jm} = \mathbf{r}(u^i_j, u^n_m) \quad (37)$$

we obtain a matrix $R = (R^{in}_{jm})$ with inverse $R^{-1} = ((R^{-1})^{in}_{jm})$ given by

$$(R^{-1})^{in}_{jm} = \bar{\mathbf{r}}(u^i_j, u^n_m).$$

Comparing (37) and (6) we recognize $\hat{R} = \tau \circ R$ as the matrix of the map $\mathbf{r}_{VV}$, when V is the right $\mathcal{A}$-comodule $\mathbb{C}^d$ with coaction determined by the matrix corepresentation $\mathbf{u} = (u^i_j)$. Therefore, by Proposition 1(iii), $\hat{R}$ satisfies the braid relation and so R fulfills the QYBE. From Proposition 1(i) we obtain

$$R^{ji}_{kl}u^k_n u^l_m = u^i_k u^j_l R^{lk}_{nm}, \quad i, j = 1, \cdots, d. \quad (38)$$

We define $d \times d$ matrices $\mathsf{L}^\pm = (l^{\pm i}_j)$, $\mathsf{L} = (l^i_j)$ and $\tilde{\mathsf{L}} = (\tilde{l}^i_j)$ of functionals on $\mathcal{A}$ by

$$l^{+i}_j = l^+(u^i_j), \quad l^{-i}_j = l^-(u^i_j), \quad l^i_j = l(u^i_j), \quad \tilde{l}^i_j = \tilde{l}(u^i_j). \quad (39)$$

The linear functionals $l^{\pm i}_j$ on $\mathcal{A}$ are called *L-functionals associated with the matrix corepresentation* $\mathbf{u}$. Since $\Delta(u^i_j) = u^i_k \otimes u^k_j$ and $\varepsilon(u^i_j) = \delta_{ij}$, (24) and Proposition 2(ii) imply that

$$\Delta(l^{\pm i}_j) = l^{\pm i}_k \otimes l^{\pm k}_j \quad \text{and} \quad \varepsilon(l^{\pm i}_j) = \delta_{ij}. \tag{40}$$

The latter relations mean that the mappings $\mathsf{L}^{\pm}(\cdot) : \mathcal{A} \to M_d(\mathbb{C})$ are algebra homomorphisms. Further, if $\mathcal{A}$ is a Hopf algebra, then the matrices L and $\tilde{\mathsf{L}}$ are just the matrix products $\mathsf{L} = S(\mathsf{L}^-)\mathsf{L}^+$ and $\tilde{\mathsf{L}} = \mathsf{L}^+S(\mathsf{L}^-)$.

From the corresponding definitions we see that the evaluation of the functionals l^{+i}_j at the matrix entries u^n_m are given by

$$l^{+i}_j(u^n_m) = R^{ni}_{mj}, \quad l^{-i}_j(u^n_m) = (R^{-1})^{in}_{jm}, \quad l^i_j(u^n_m) = (\hat{R}^2)^{ni}_{mj}, \quad \tilde{l}^i_j(u^n_m) = (\hat{R}^2)^{in}_{jm}.$$

In matrix notation these formulas can be expressed as

$$\mathsf{L}^+_1(\mathbf{u}_2) = R_{21}, \quad \mathsf{L}^-_1(\mathbf{u}_2) = R^{-1}_{12}, \quad \mathsf{L}_1(\mathbf{u}_2) = (\hat{R}^2)_{21}, \quad \tilde{\mathsf{L}}_1(\mathbf{u}_2) = (\hat{R}^2)_{12}. \tag{41}$$

Proposition 15. *The matrices* $\mathsf{L}^{\pm}$ *satisfy the relations*

$$\mathsf{L}^+_1\mathsf{L}^+_2R = R\mathsf{L}^+_2\mathsf{L}^+_1, \quad \mathsf{L}^-_1\mathsf{L}^-_2R = R\mathsf{L}^-_2\mathsf{L}^-_1, \quad \mathsf{L}^-_1\mathsf{L}^+_2R = R\mathsf{L}^+_2\mathsf{L}^-_1. \tag{42}$$

Proof. Using the formulas (21), (4) and (38) we conclude that

$$\begin{aligned}
(l^{+j}_k l^{+i}_l)(a)R^{kl}_{nm} &= \sum l^{+j}_k(a_{(1)})l^{+i}_l(a_{(2)})R^{kl}_{nm} \\
&= \sum \mathbf{r}(a_{(1)}, u^j_k)\mathbf{r}(a_{(2)}, u^i_l)R^{kl}_{nm} \\
&= \mathbf{r}(a, u^i_l u^j_k)R^{kl}_{nm} = R^{ji}_{kl}\mathbf{r}(a, u^k_n u^l_m) \\
&= R^{ji}_{kl}l^{+l}_m l^{+k}_n(a),
\end{aligned}$$

that is, $\mathsf{L}^+_1\mathsf{L}^+_2R = R\mathsf{L}^+_2\mathsf{L}^+_1$. Replacing $\mathbf{r}$ by the universal r-form $\bar{\mathbf{r}}_{21}$, the same reasoning shows that $\mathsf{L}^-_1\mathsf{L}^-_2R = R\mathsf{L}^-_2\mathsf{L}^-_1$. The QYBE for $\mathbf{r}$ (by Proposition 2(i)) implies that $\bar{\mathbf{r}}_{12}\mathbf{r}_{23}\mathbf{r}_{13} = \mathbf{r}_{13}\mathbf{r}_{23}\bar{\mathbf{r}}_{12}$. Applying this equation to $u^i_n \otimes a \otimes u^j_m$ and using formula (37) we obtain

$$\sum \bar{\mathbf{r}}(u^i_k, a_{(1)})\mathbf{r}(a_{(2)}, u^j_l)R^{kl}_{nm} = \sum R^{ij}_{kl}\mathbf{r}(a_{(1)}, u^l_m)\bar{\mathbf{r}}(u^k_n, a_{(2)}).$$

This means that $l^{-i}_k l^{+j}_l R^{kl}_{nm} = R^{ij}_{kl} l^{+l}_m l^{-k}_n$. Thus, $\mathsf{L}^-_1\mathsf{L}^+_2R = R\mathsf{L}^+_2\mathsf{L}^-_1$. □

Proposition 16. *Suppose that* $\mathcal{A}$ *is a coquasitriangular Hopf algebra and* $\mathbf{u}=(u^i_j)_{i,j=1,\cdots,d}$ *is a matrix corepresentation of* $\mathcal{A}$.

(i) *The matrices* $\mathsf{L} = S(\mathsf{L}^-)\mathsf{L}^+$ *and* $\tilde{\mathsf{L}} = \mathsf{L}^+S(\mathsf{L}^-)$ *defined above satisfy the reflection equations*

$$\mathsf{L}_2R_{12}\mathsf{L}_1R_{21} = R_{12}\mathsf{L}_1R_{21}\mathsf{L}_2, \tag{43}$$

$$\tilde{\mathsf{L}}_2R_{21}\tilde{\mathsf{L}}_1R_{12} = R_{21}\tilde{\mathsf{L}}_1R_{12}\tilde{\mathsf{L}}_2. \tag{44}$$

(ii) *The elements* $\mathrm{Tr}\ \mathsf{L}^kD^{-1}$, $k \in \mathbb{N}$, *belong to the center of* $\mathcal{A}^\circ$, *where* D^{-1} *is the matrix given by (19) and* $\mathsf{L} = S(\mathsf{L}^-)\mathsf{L}^+$.

Proof. (i): By the third equality of (42), we have $\mathsf{L}^+_2R_{12}S(\mathsf{L}^-_1) = S(\mathsf{L}^-_1)R_{12}\mathsf{L}^+_2$. Using this relation and the first equality of (42) we get

$$
\begin{aligned}
\mathsf{L}_2 R_{12} \mathsf{L}_1 &= S(\mathsf{L}_2^-)\mathsf{L}_2^+ R_{12} S(\mathsf{L}_1^-)\mathsf{L}_1^+ \\
&= S(\mathsf{L}_2^-)S(\mathsf{L}_1^-) R_{12} \mathsf{L}_2^+ \mathsf{L}_1^+ \\
&= S(\mathsf{L}_1^- \mathsf{L}_2^-)\mathsf{L}_1^+ \mathsf{L}_2^+ R_{12}.
\end{aligned} \tag{45}
$$

Interchanging the lower indices yields

$$\mathsf{L}_1 R_{21} \mathsf{L}_2 = S(\mathsf{L}_2^- \mathsf{L}_1^-)\mathsf{L}_2^+ \mathsf{L}_1^+ R_{21}. \tag{46}$$

Multiplying (46) by R_{12} on the left and using the first two relations of (42) we obtain $R_{12}\mathsf{L}_1 R_{21}\mathsf{L}_2 = S(\mathsf{L}_1^- \mathsf{L}_2^-)\mathsf{L}_1^+ \mathsf{L}_2^+ R_{12}R_{21}$. By (45), the right hand side is equal to $\mathsf{L}_2 R_{12}\mathsf{L}_1 R_{21}$. Thus we obtain formula (43). The proof of the second reflection equation (44) is similar.

(ii): From formula (31) we derive by induction on k the identity

$$l(a^1)\cdots l(a^k)g = \sum g_{(1)} l(a^1_{(2)})\cdots l(a^k_{(2)}) g_{(2)}(a^1_{(1)}S(a^1_{(3)})\cdots a^k_{(1)}S(a^k_{(3)}))$$

for arbitrary elements $a^1,\cdots,a^k \in \mathcal{A}$ and $g \in \mathcal{A}^\circ$. Using this equation and the relation $S(u^m_j)(D^{-1})^j_i = (D^{-1})^m_j S^{-1}(u^j_i)$ by (20), we obtain

$$
\begin{aligned}
&\sum\nolimits_i (\mathsf{L}^k D^{-1})^i_i g = \sum\nolimits_i l^i_{i_2} l^{i_2}_{i_3}\cdots l^{i_k}_j g(D^{-1})^j_i \\
&= \sum g_{(1)} l^n_{n_2} l^{r_2}_{n_3}\cdots l^{r_k}_m \; g_{(2)}(u^i_n S(u^{n_2}_{i_2}) u^{i_2}_{r_2} S(u^{n_3}_{i_3})\cdots u^{i_k}_{r_k} S(u^m_j))(D^{-1})^j_i \\
&= \sum g_{(1)} l^n_{n_2} l^{n_2}_{n_3}\cdots l^{n_k}_m \; g_{(2)}(u^i_n S(u^m_j)(D^{-1})^j_i) \\
&= \sum g_{(1)} (L^k)^n_m \; g_{(2)}(S^{-1}(u^j_i S(u^i_n)))(D^{-1})^m_j \\
&= g\sum\nolimits_j (\mathsf{L}^k D^{-1})^j_j.
\end{aligned}
$$

□

We now assume that $\mathcal{A}$ is a Hopf $*$-algebra. We shall express the conditions in Proposition 14 in terms of the functionals $l^{\pm i}_j$ and l^i_j.

First suppose that $\mathbf{r}$ is real and $(u^i_j)^* = S(u^j_i)$ for $i,j = 1,2,\cdots,d$. Since then $S^{-2}((u^i_j)^*) = S^{-1}(u^j_i)$ and $S(u^i_j)^* = u^j_i$, the equations (33) yield

$$(l^{\pm i}_j)^* = S(l^{\mp j}_i) \quad \text{and} \quad (l^i_j)^* = l^j_i. \tag{47}$$

Next we assume that $\mathbf{r}$ is inverse real and $(u^i_j)^* = u^i_j$ for $i,j = 1,2,\cdots,d$. By (20), we have $S^{-2}(u^i_j) = D^i_n u^n_m (D^{-1})^m_j$, where the numbers D^i_j and $(D^{-1})^i_j$ are given by (19). Inserting this into (34) and setting $\tilde{R}^{in}_{km} := \mathbf{r}(u^i_k, S(u^n_m))$, we get

$$(l^{\pm i}_j)^* = D^i_n l^{\pm n}_m (D^{-1})^m_j \quad \text{and} \quad (l^i_j)^* = \tilde{R}^{in}_{km}(D^{-1})^m_j l^k_n. \tag{48}$$

It might be necessary to emphasize that the matrices $\mathsf{L}^\pm$, L and $\tilde{\mathsf{L}}$ of functionals depend on the choices of the universal r-form $\mathbf{r}$ and of the matrix corepresentation $\mathbf{u}$ of $\mathcal{A}$. If we take the universal r-form $\bar{\mathbf{r}}_{21}$ (by Proposition

2(iv)) instead of $\mathbf{r}$, then l^{+i}_{j} and l^{-i}_{j} interchange. Further, if $\mathcal{A}$ is a coquasitriangular Hopf algebra and $\mathbf{u}$ is replaced by $\mathbf{v} = (v^i_j)$ with $v^i_j := S^{-1}(u^j_i)$, then the representations $\mathsf{L}^{\pm}$ of $\mathcal{A}$ go into the contragredient representations $\mathsf{L}^{\pm,c}$, where $(\mathsf{L}^{\pm,c})^i_j := S(l^{\pm j}_{i})$.

Let us specialize the preceding to coquasitriangular FRT bialgebras. Suppose that R is an invertible solution of the QYBE. Then, by Theorem 7, the FRT bialgebra $\mathcal{A}(R)$ is coquasitriangular and its canonical r-form $\mathbf{r}$ of $\mathcal{A}(R)$ satisfies equation (37), when $\mathbf{u} = (u^i_j)$ denotes the fundamental corepresentation of $\mathcal{A}(R)$. Thus, all the above facts are valid in this case. Let $\mathcal{U}(R)$ denote the bialgebra $\mathcal{U}$ defined above for $\mathcal{A} = \mathcal{A}(R)$. That is, $\mathcal{U}(R)$ is the subalgebra of $\mathcal{A}(R)^\circ$ generated by the functionals l^{+i}_{j} and l^{-i}_{j}, $i,j = 1,2,\cdots,N$.

Now let us suppose that G_q is one of the quantum groups $GL_q(N)$, $SL_q(N)$, $O_q(N)$ or $Sp_q(N)$ and that $\mathcal{A}$ is the coquasitriangular Hopf algebra $\mathcal{O}(G_q)$ with universal r-form $\mathbf{r}_z$ (see Theorem 9) and fundamental corepresentation $\mathbf{u} = (u^i_j)$. Let $\mathcal{U}(G_q)$ denote the bialgebra $\mathcal{U}$ in the case $\mathcal{A} = \mathcal{O}(G_q)$. By Proposition 11(ii), $\mathcal{U}(G_q)$ is a Hopf algebra. All the above formulas remain valid with $R = z\mathsf{R}(q)$, where z and $\mathsf{R}(q)$ are as in Theorem 9. In addition, the L-functionals l^{+i}_{j} associated with the fundamental corepresentation $\mathbf{u}$ of $\mathcal{O}(G_q)$ have the following properties.

Proposition 17. (i) $l^{+i}_{j} = l^{-j}_{i} = 0$, *$i > j$, and* $l^{+k}_{k} l^{-k}_{k} = \varepsilon$, $k = 1,2,\cdots,N$.

(ii) $l^{+1}_{1} l^{+2}_{2} \cdots l^{+N}_{N} = \varepsilon$ *for* $G_q = SL_q(N)$.

(iii) $\mathsf{L}^{\pm}\mathsf{C}^t(\mathsf{L}^{\pm})^t(\mathsf{C}^{-1})^t = \mathsf{C}^t(\mathsf{L}^{\pm})^t(\mathsf{C}^{-1})^t\mathsf{L}^{\pm} = \varepsilon{\cdot}I$ *for* $G_q = O_q(N), Sp_q(N)$.

Proof. (i): Suppose that $i > j$. The matrices R given by (9.13) and (9.30) are lower triangular. Thus, by (41), $l^{+i}_{j}(u^n_m) = z\mathsf{R}^{ni}_{mj} = 0$ for all $n, m = 1,2,\cdots,N$. Since $l^{+i}_{j}(ab) = l^{+i}_{k}(a) l^{+k}_{j}(b)$ and $l^{+i}_{j}(1) = \delta_{ij}$ by (40), we conclude that $l^{+i}_{j} = 0$ on the whole of $\mathcal{A}$ for $G_q = SL_q(N), O_q(N), Sp_q(N)$. In the case of $GL_q(N)$ this is also true, since $\mathsf{L}^+(\mathcal{D}_q^{-1})$ is the inverse of the upper triangular matrix $\mathsf{L}^+(\mathcal{D}_q)$. The proof of $l^{-j}_{i} = 0$ is similar. Since $l^{+i}_{j} = l^{-j}_{i} = 0$ as just shown, we have $\Delta(l^{+i}_{i} l^{-i}_{i}) = l^{+i}_{i} l^{-i}_{i} \otimes l^{+i}_{i} l^{-i}_{i}$ again by (40). Hence $l^{+i}_{i} l^{-i}_{i}$ is a character of the algebra $\mathcal{A}$. By (41), the characters $l^{+i}_{i} l^{-i}_{i}$ and ε are equal on the generators and hence on $\mathcal{A}$.

(ii) follows from the fact that $l^{+1}_{1} \cdots l^{+N}_{N}$ and ε are both characters which coincide on the generators u^n_m by (41) and the assumption $z^N = q^{-1}$. (One may also verify that $\varepsilon = l^{\pm}(\mathcal{D}_q) = l^{\pm 1}_{1} \cdots l^{\pm N}_{N}$).

(iii) is obtained from (3) and the relations (9.34). □

Let $\mathcal{A}$ be $\mathcal{A}(R)$ or $\mathcal{O}(G_q)$. Then the L-functionals $l^{\pm i}_{j}$ for the fundamental corepresentation $\mathbf{u}$ of $\mathcal{A}$ are uniquely determined by the equations (40) and (41) or equivalently by the requirement that the matrices $\mathsf{L}^{\pm} = (l^{\pm i}_{j})$ are algebra homomorphisms of $\mathcal{A}$ to $M_N(\mathbb{C})$ satisfying (41). For most applications this characterization of the functionals $l^{\pm i}_{j}$ is sufficient to work with.

The theory of L-functionals for $\mathcal{O}(G_q)$ developed above is closely related to that for $U_h(\mathfrak{g})$ and $U_q(\mathfrak{g})$ in Sect. 8.5. Both L-functionals have similar properties (see, Propositions 17 and 8.28 and formulas (42), (40), (41) and (8.83), (8.84), (8.89), (9.40), respectively) and even some proofs are analogous (compare, for instance, the proofs of Propositions 15 and 17 with those of Propositions 8.27 and 8.28). However, the L-functionals $l^{\pm i}_{j} \in \mathcal{U}(G_q)$ are defined as linear functionals on $\mathcal{O}(G_q)$, while the L-functionals $l^{\pm}_{ij}$ and $\ell^{\pm}_{ij}$ in 8.5 belong to the abstract algebras $U_h(\mathfrak{g})$, $U_q^{\rm ext}(\mathfrak{g})$ and $U_q^L(\mathfrak{g})$, respectively.

By Propositions 15 and 17 and formulas (40), there is a surjective Hopf algebra homomorphism $\vartheta : U_q^L(\mathfrak{g}) \rightarrow \mathcal{U}(G_q)$ such that $\vartheta(\ell^+_{ij}) = l^{+i}_{j}$ and $\vartheta(\ell^-_{ji}) = l^{-j}_{i}$, $i \leq j$, where $\mathfrak{g} = \mathrm{gl}_N, \mathrm{sl}_N, \mathrm{so}_N, \mathrm{sp}_N$ and $G_q = GL_q(N), SL_q(N), O_q(N), Sp_q(N)$, respectively. Likewise, there is a Hopf algebra homomorphism θ of the algebra $U_q^{\rm ext}(\mathfrak{g})$ generated by l^+_{ij}, l^-_{ji} onto $\mathcal{U}(G_q)$ which maps l^+_{ij} to l^{+i}_{j} and l^-_{ji} to l^{-j}_{i}, $i \leq j$. In general, both homomorphisms ϑ and θ are not injective. (For instance, if $\mathfrak{g} = \mathrm{sl}_2$ and q is a root of unity, one can verify that $(l^{+1}_{1})^n = \varepsilon$ in $\mathcal{U}(SL_q(2))$ for some $n \in \mathbb{N}$.) Obviously, if we identify $U_q^{\rm ext}(\mathfrak{g})$ and $U_q^L(\mathfrak{g})$ by the isomorphism from Theorem 8.33, ϑ and θ coincide.

10.2 Crossed Product Constructions of Hopf Algebras

Crossed products or semidirect products are useful tools for the study of covariance problems and for the construction of new objects from old ones. Most of such constructions occurring in the literature can be thought of as generalizations or combinations of two fundamental concepts developed in this section: crossed product algebras or coalgebras and twisted cocycle algebras or coalgebras. In the course of this the quantum double will be treated as a guiding example.

If not specified otherwise, $\mathcal{A}$ and $\mathcal{B}$ are bialgebras in this section.

10.2.1 Crossed Product Algebras

Recall from Subsect. 1.3.3 that an algebra $\mathcal{X}$ is a *left $\mathcal{A}$-module algebra* if $\mathcal{X}$ is a left $\mathcal{A}$-module (that is, there is a bilinear map $\alpha : \mathcal{A}\times\mathcal{X} \ni (a,x) \rightarrow a\triangleright x \in \mathcal{X}$ satisfying $a \triangleright (b \triangleright x) = (ab) \triangleright x$ and $1 \triangleright x = x$ for $a, b \in \mathcal{A}$, $x \in \mathcal{X}$) such that

$$a \triangleright (xy) = \sum (a_{(1)} \triangleright x)(a_{(2)} \triangleright y), \quad a \triangleright 1 = \varepsilon(a)1. \tag{49}$$

Proposition 18. *Let $\mathcal{X}$ be a left $\mathcal{A}$-module algebra. Then the vector space $\mathcal{X} \otimes \mathcal{A}$ is a (unital associative) algebra, called the* left crossed product algebra *and denoted by $\mathcal{X} \rtimes_\alpha \mathcal{A}$ or simply $\mathcal{X} \rtimes \mathcal{A}$, with multiplication defined by*

$$(x \otimes a)(y \otimes b) = \sum x(a_{(1)} \triangleright y) \otimes a_{(2)}b, \quad x, y \in \mathcal{X}, \quad a, b \in \mathcal{A}. \tag{50}$$

Proof. It suffices to show that this product is associative. Using the formulas (49) and (50) and the left $\mathcal{A}$-module property of $\mathcal{X}$ we obtain

$$\begin{aligned}(x\otimes a)((y\otimes b)(z\otimes c)) &= \sum x(a_{(1)}\triangleright(y(b_{(1)}\triangleright z)))\otimes a_{(2)}b_{(2)}c\\ &= \sum x(a_{(1)}\triangleright y)((a_{(2)}b_{(1)})\triangleright z)\otimes a_{(3)}b_{(2)}c\\ &= \left(\sum x(a_{(1)}\triangleright y)\otimes a_{(2)}b\right)(z\otimes c)\\ &= ((x\otimes a)(y\otimes b))(z\otimes c).\end{aligned}$$ □

Clearly, the maps $\mathcal{X}\ni x\to x\otimes 1\in\mathcal{X}\rtimes\mathcal{A}$ and $\mathcal{A}\ni a\to 1\otimes a\in\mathcal{X}\rtimes\mathcal{A}$ are injective algebra homomorphisms. Therefore, by identifying x with $x\otimes 1$ and a with $1\otimes a$, we can consider $\mathcal{X}$ and $\mathcal{A}$ as subalgebras of the algebra $\mathcal{X}\rtimes\mathcal{A}$. Then definition (50) of the product yields the commutation relations

$$ax=\sum(a_{(1)}\triangleright x)a_{(2)},\quad a\in\mathcal{A},\ x\in\mathcal{X}. \tag{51}$$

The algebra $\mathcal{X}\rtimes\mathcal{A}$ may be thought of as the universal algebra generated by the algebras $\mathcal{X}$ and $\mathcal{A}$ with respect to the commutation relations (51).

If φ is a representation of the algebra $\mathcal{X}\rtimes\mathcal{A}$ on a vector space V, it follows at once from (51) that the restrictions $\varphi_{\mathcal{X}}:=\varphi\lceil\mathcal{X}$ and $\varphi_{\mathcal{A}}:=\varphi\lceil\mathcal{A}$ are representations of $\mathcal{X}$ and $\mathcal{A}$ on V such that the compatibility condition

$$\varphi_{\mathcal{A}}(a)\varphi_{\mathcal{X}}(x)=\sum\varphi_{\mathcal{X}}(a_{(1)}\triangleright x)\varphi_{\mathcal{A}}(a_{(2)}),\quad a\in\mathcal{A},\ x\in\mathcal{X},$$

holds. Conversely, if $\varphi_{\mathcal{X}}$ and $\varphi_{\mathcal{A}}$ are representations of $\mathcal{X}$ and $\mathcal{A}$ on the same vector space V satisfying the latter condition, then the equation $\varphi(ax)=\varphi_{\mathcal{A}}(a)\varphi_{\mathcal{X}}(x)$ defines unambiguously a representation of $\mathcal{X}\rtimes\mathcal{A}$ on V. For instance, by (49), the representations $\varphi_{\mathcal{X}}(x)y=xy$ and $\varphi_{\mathcal{A}}(a)y=a\triangleright y$ of $\mathcal{X}$ and $\mathcal{A}$ on $V=\mathcal{X}$ have this property. In this case the corresponding representation φ of $\mathcal{X}\rtimes\mathcal{A}$ on $\mathcal{X}$ acts as $\varphi(xa)y=x(a\triangleright y)$, $a\in\mathcal{A}$, $x,y\in\mathcal{X}$. It is called the *Heisenberg representation* of the crossed product algebra $\mathcal{X}\rtimes\mathcal{A}$.

Before we discuss a number of examples, we briefly mention the corresponding right handed version $\mathcal{A}\ltimes\mathcal{X}$. Let $\mathcal{X}$ be a *right $\mathcal{A}$-module algebra*, that is, $\mathcal{X}$ is an algebra equipped with a bilinear map $\alpha:\mathcal{X}\times\mathcal{A}\ni(x,a)\to x\triangleleft a\in\mathcal{X}$ such that $(x\triangleleft a)\triangleleft b=x\triangleleft(ab), x\triangleleft 1=x$, $(xy)\triangleleft a=\sum(x\triangleleft a_{(1)})(y\triangleleft a_{(2)})$ and $1\triangleleft a=\varepsilon(a)1$ for $a,b\in\mathcal{A}$ and $x,y\in\mathcal{X}$. Then the vector space $\mathcal{A}\otimes\mathcal{X}$ is an algebra with product defined by the formula

$$(a\otimes x)(b\otimes y)=\sum ab_{(1)}\otimes(x\triangleleft b_{(2)})y,\quad a,b\in\mathcal{A},\ x,y\in\mathcal{X}.$$

This algebra is the *right crossed product algebra* and denoted by $\mathcal{A}_{\alpha}\ltimes\mathcal{X}$ or simply $\mathcal{A}\ltimes\mathcal{X}$.

Remark 3. In the Hopf algebra literature (see, for instance, [Sw] or [Mon], 4.1.1 and 10.6.1) crossed product algebras and coalgebras are usually called *smash products* and *coproducts*, respectively. △

Example 4. Any algebra $\mathcal{X}$ is a left $\mathcal{A}$-module algebra with respect to the trivial action $a\triangleright x=\varepsilon(a)x$. In this case $\mathcal{X}\rtimes\mathcal{A}$ is the "ordinary" tensor product algebra $\mathcal{X}\otimes\mathcal{A}$ with product $(x\otimes a)(y\otimes b)=xy\otimes ab$. △

Example 5. Each Hopf algebra $\mathcal{A}$ is a left $\mathcal{A}$-module algebra with respect to the left adjoint action $a \triangleright b := \mathrm{ad}_L(a)b = \sum a_{(1)}bS(a_{(2)})$. Then the corresponding left crossed product algebra $\mathcal{A} \rtimes_{\mathrm{ad}_L} \mathcal{A}$ has the product

$$(a \otimes b)(c \otimes d) = \sum ab_{(1)}cS(b_{(2)}) \otimes b_{(3)}d, \quad a, b, c, d \in \mathcal{A}.$$

If $\mathcal{A}$ is quasitriangular, then the algebra $\mathcal{A} \rtimes_{\mathrm{ad}_L} \mathcal{A}$ is even a Hopf algebra for some appropriate coproduct, see Example 25 in Subsect. 10.3.6 below. △

Example 6. Let $\mathcal{A}$ be the universal enveloping algebra $U(\mathfrak{g})$ of a Lie algebra $\mathfrak{g}$, see Example 1.6. Suppose that π is a homomorphism of the Lie algebra $\mathfrak{g}$ to the Lie algebra of smooth vector fields on a C^∞-manifold $\mathcal{M}$. (An important special case is when $\mathcal{M}$ is a Lie group G with Lie algebra $\mathfrak{g}$ and the elements of $\mathfrak{g}$ act as left invariant vector fields on G.) Then π extends uniquely to an algebra homomorphism of $U(\mathfrak{g})$. For simplicity let us assume that this extension is injective and identify $\mathcal{A} = U(\mathfrak{g})$ with its image. Then the algebra $\mathcal{X} = C^\infty(\mathcal{M})$ (or any other subalgebra which is invariant under the action of $\mathfrak{g}$) is a left $\mathcal{A}$-module algebra. Indeed, it suffices to check (49) for elements a of $\mathfrak{g}$. Since $\Delta(a) = a \otimes 1 + 1 \otimes a$, the first equation of (49) reduces to the Leibniz rule which holds because $a \in \mathfrak{g}$ acts as a vector field by assumption. In this case the elements of the crossed product algebra $\mathcal{X} \rtimes \mathcal{A}$ act as differential operators on $\mathcal{M}$ with coefficients in $\mathcal{X}$.

Let us specialize to the case where $\mathfrak{g} = \mathbb{R}^n$, $\mathcal{M} = \mathbb{R}^n$, $\mathcal{X} = \mathbb{C}[x_1, \cdots, x_n]$ and $\pi(e_i) = \partial/\partial x_i$, $i = 1, 2, \cdots, n$, for some basis $\{e_i\}$ of $\mathfrak{g}$. Then $\mathcal{X} \rtimes \mathcal{A}$ is just the Weyl algebra and its action on $\mathbb{R}^n$ is the realization as the algebra of differential operators with polynomial coefficients. △

Example 7. Let G and H be groups such that H acts on G by group automorphisms, that is, $h \triangleright (gg') = (h \triangleright g)(h \triangleright g')$ and $1 \triangleright g = g$. The semidirect product $G \rtimes H$ of G and H is the set $G \times H$ with group operation $(g, h)(g', h') := (g(h \triangleright g'), hh')$. Extending the actions of H and G by linearity to the group algebras $\mathcal{X} = \mathbb{C}G$ and $\mathcal{A} = \mathbb{C}H$, $\mathcal{X}$ becomes a left $\mathcal{A}$-module algebra and the crossed product algebra $\mathcal{X} \rtimes \mathcal{A}$ is just the group algebra $\mathbb{C}(G \rtimes H)$ of the semidirect product $G \rtimes H$. △

Important crossed product algebras are the Heisenberg doubles. Let us suppose that $\langle \cdot, \cdot \rangle$ is a dual pairing of two bialgebras $\mathcal{B}$ and $\mathcal{A}$. Define

$$b \triangleright a := \sum \langle b, a_{(2)} \rangle a_{(1)}, \quad a \in \mathcal{A},\ b \in \mathcal{B}. \tag{52}$$

Using the properties of the pairing (1.31) and (1.32) it is easy to check that the map $(b, a) \to b \triangleright a$ defines an action of $\mathcal{B}$ on $\mathcal{A}$ such that $\mathcal{A}$ is a left $\mathcal{B}$-module algebra. The corresponding crossed product algebra $\mathcal{A} \rtimes \mathcal{B}$ is called the *Heisenberg double* of the pair $\mathcal{A}, \mathcal{B}$ and denoted by $H(\mathcal{A}, \mathcal{B})$. When the bialgebra $\mathcal{B}$ is the dual $\mathcal{A}^\circ$ of $\mathcal{A}$, then $H(\mathcal{A}) := H(\mathcal{A}, \mathcal{B})$ is said to be the *Heisenberg double* of $\mathcal{A}$. Inserting the definition (52) of the action into (50), we get the following expression for the product of the algebra $H(\mathcal{A}, \mathcal{B})$:

$$(a \otimes b)(a' \otimes b') = \sum \langle b_{(1)}, a'_{(2)} \rangle a a'_{(1)} \otimes b_{(2)} b', \quad a, a' \in \mathcal{A}, \ b, b' \in \mathcal{B}.$$

We consider this general construction in two examples.

Example 8. Let G be a Lie group with Lie algebra $\mathfrak{g}$, $\mathcal{B} = U(\mathfrak{g})$ and let $\mathcal{A}$ be a bialgebra of C^∞-functions on G. It is well-known that any $b \in \mathcal{B}$ acts as a left invariant differential operator $\tilde{b}$ on G. The evaluation of the function $\tilde{b}(a)$, $a \in \mathcal{A}$, at the unit element of G defines a dual pairing of the bialgebras $\mathcal{B}$ and $\mathcal{A}$ (see Example 1.6). The corresponding Heisenberg double $H(\mathcal{A}, \mathcal{B})$ is an abstract algebra of differential operators $\sum_i a_i b_i$, $a_i \in \mathcal{A}$, $b_i \in \mathcal{B}$, and the Heisenberg representation of $H(\mathcal{A}, \mathcal{B}) = \mathcal{A} \rtimes \mathcal{B}$ describes the action $\sum_i a_i \tilde{b}_i$ of these operators on functions of $\mathcal{A}$. That is, the elements of $\mathcal{B}$ and $\mathcal{A}$ act as differential operators and multiplication operators on G, respectively. This picture is the reason for the name "Heisenberg double". It also motivates the interpretation of (52) for a general Heisenberg double as the action of the "generalized left derivation" $b \in \mathcal{B}$ on the "function" $a \in \mathcal{A}$. △

Example 9. Let R be an invertible solution of the QYBE and let $\mathcal{A}(R)$ and $\mathfrak{U}(R)$ be the corresponding FRT bialgebras defined in Sects. 9.1 and 9.4. By Proposition 9.20, these bialgebras are dually paired with pairing described by (9.44). Suppose that $\mathcal{A}$ is a quotient bialgebra of $\mathcal{A}(R)$ and $\mathcal{B}$ is a quotient bialgebra of $\mathfrak{U}(R)$ such that the pairing of $\mathfrak{U}(R)$ and $\mathcal{A}(R)$ passes to the quotients. Then, by (9.44), the actions of the generators $\ell^{\pm i}_j$ of $\mathcal{B}$ on the generators u^n_m of $\mathcal{A}$ are written in matrix form as $\mathfrak{L}^+_1 \triangleright \mathbf{u}_2 = \mathbf{u}_2 R_{21}$ and $\mathfrak{L}^-_1 \triangleright \mathbf{u}_2 = \mathbf{u}_2 R^{-1}_{12}$. Hence the commutation relations (51) between both sets of generators in the algebra $H(\mathcal{A}, \mathcal{B})$ are

$$\mathfrak{L}^+_1 \mathbf{u}_2 = \mathbf{u}_2 R_{21} \mathfrak{L}^+_1 \quad \text{and} \quad \mathfrak{L}^-_1 \mathbf{u}_2 = \mathbf{u}_2 R^{-1}_{12} \mathfrak{L}^-_1. \tag{53}$$

That is, the Heisenberg double $H(\mathcal{A}, \mathcal{B})$ has $3N^2$ generators $\ell^{\pm i}_j, u^n_m$, $i, j, n, m = 1, 2, \cdots, N$. The defining relations are those of $\mathcal{A}$ and $\mathcal{B}$ and the cross relations (53). △

10.2.2 Crossed Coproduct Coalgebras

This subsection is concerned with the dual notion to crossed product algebras.

A coalgebra $\mathcal{X}$ is called a *right $\mathcal{A}$-comodule coalgebra* if $\mathcal{X}$ is a right $\mathcal{A}$-comodule such that the comultiplication $\Delta_\mathcal{X}$ and the counit $\varepsilon_\mathcal{X}$ are comodule maps. The latter means that

$$\sum (x^{(0)})_{(1)} \otimes (x^{(0)})_{(2)} \otimes x^{(1)} = \sum (x_{(1)})^{(0)} \otimes (x_{(2)})^{(0)} \otimes (x_{(1)})^{(1)} (x_{(2)})^{(1)},$$

$$\sum \varepsilon_\mathcal{X}(x^{(0)}) x^{(1)} = \varepsilon_\mathcal{X}(x) 1.$$

Here and in the following we use the Sweedler notation $\beta(x) = \sum x^{(0)} \otimes x^{(1)}$, $\Delta_\mathcal{X}(x) = \sum x_{(1)} \otimes x_{(2)}$ and $\Delta_\mathcal{A}(a) = \sum a_{(1)} \otimes a_{(2)}$, where $\beta : \mathcal{X} \to \mathcal{X} \otimes \mathcal{A}$ is the coaction of $\mathcal{A}$ on $\mathcal{X}$.

Proposition 19. *Let $\mathcal{X}$ be a right $\mathcal{A}$-comodule coalgebra. Then the vector space $\mathcal{A}\otimes\mathcal{X}$ becomes a coalgebra, called the* right crossed coproduct coalgebra *and denoted by $\mathcal{A}^{\beta}\ltimes\mathcal{X}$ or $\mathcal{A}\ltimes\mathcal{X}$, with comultiplication and counit given by*

$$\Delta(a\otimes x)=\sum a_{(1)}\otimes(x_{(1)})^{(0)}\otimes a_{(2)}(x_{(1)})^{(1)}\otimes x_{(2)}, \tag{54}$$

$$\varepsilon(a\otimes x)=\varepsilon_{\mathcal{A}}(a)\varepsilon_{\mathcal{X}}(x),\quad a\in\mathcal{A},\ x\in\mathcal{X}.$$

Proof. The proof of the coassociativity of Δ is an advanced exercise in the use of the Sweedler notation. We omit the details, see [Mon], 10.6.3. □

Let us turn to the corresponding left handed notion. A *left $\mathcal{A}$-comodule coalgebra* is a coalgebra $\mathcal{X}$ which is a left $\mathcal{A}$-comodule such that $\Delta_{\mathcal{X}}$ and $\varepsilon_{\mathcal{X}}$ are comodule maps, that is,

$$\sum x^{(-1)}\otimes(x^{(0)})_{(1)}\otimes(x^{(0)})_{(2)}=\sum(x_{(1)})^{(-1)}(x_{(2)})^{(-1)}\otimes(x_{(1)})^{(0)}\otimes(x_{(2)})^{(0)},$$

$$\sum x^{(-1)}\varepsilon_{\mathcal{X}}(x^{(0)})=\varepsilon_{\mathcal{X}}(x)1,$$

where we write $\beta(x)=\sum x^{(-1)}\otimes x^{(0)}$ for the coaction $\beta:\mathcal{X}\to\mathcal{A}\otimes\mathcal{X}$. For such a left $\mathcal{A}$-comodule coalgebra $\mathcal{X}$, the vector space $\mathcal{X}\otimes\mathcal{A}$ becomes a coalgebra with comultiplication and counit defined by

$$\Delta(x\otimes a)=\sum x_{(1)}\otimes(x_{(2)})^{(-1)}a_{(1)}\otimes(x_{(2)})^{(0)}\otimes a_{(2)},$$

$$\varepsilon(x\otimes a)=\varepsilon_{\mathcal{X}}(x)\varepsilon_{\mathcal{A}}(a),\quad x\in\mathcal{X},\ a\in\mathcal{A}.$$

This coalgebra is called the *left crossed coproduct coalgebra*. It is denoted by $\mathcal{X}\rtimes^{\beta}\mathcal{A}$ or $\mathcal{X}\rtimes\mathcal{A}$.

Example 10. Each coalgebra $\mathcal{X}$ is a right $\mathcal{A}$-comodule coalgebra with respect to the trivial coaction $\beta(x)=x\otimes 1$. Then $\mathcal{A}^{\beta}\ltimes\mathcal{X}$ is just the tensor product coalgebra $\mathcal{A}\otimes\mathcal{X}$ with coproduct $\Delta(a\otimes x)=\sum a_{(1)}\otimes x_{(1)}\otimes a_{(2)}\otimes x_{(2)}$. △

Example 11. The right adjoint coaction $\mathrm{Ad}_R(a)=\sum a_{(2)}\otimes S(a_{(1)})a_{(3)}$ turns each Hopf algebra $\mathcal{A}$ into a right $\mathcal{A}$-comodule coalgebra. Then the coproduct of $\mathcal{A}^{\mathrm{Ad}_R}\ltimes\mathcal{A}$ is given by $\Delta(a\otimes x)=\sum a_{(1)}\otimes x_{(2)}\otimes a_{(2)}S(x_{(1)})x_{(3)}\otimes x_{(4)}$. For a coquasitriangular Hopf algebra $\mathcal{A}$, we shall show in Subsect. 10.3.6 (see Example 24) that the coalgebra $\mathcal{A}^{\mathrm{Ad}_R}\ltimes\mathcal{A}$ equipped with a suitable product becomes a Hopf algebra. △

10.2.3 Twisting of Algebra Structures by 2-Cocycles and Quantum Doubles

In this subsection we construct new bialgebras from old ones by twisting the product by 2-cocycles and keeping the coproduct unchanged.

Definition 3. *A bilinear mapping* $\gamma : \mathcal{A}\times\mathcal{A}\to\mathbb{C}$ *is said to be a* left 2-cocycle *on* $\mathcal{A}$ *(with values in* $\mathbb{C}$*) if it satisfies the condition*

$$\sum \gamma(a_{(1)}, b_{(1)})\gamma(a_{(2)}b_{(2)}, c) = \sum \gamma(b_{(1)}, c_{(1)})\gamma(a, b_{(2)}c_{(2)}) \tag{55}$$

for $a, b, c \in \mathcal{A}$. *It is called a* right 2-cocycle *on* $\mathcal{A}$ *if it satisfies the equation*

$$\sum \gamma(a_{(1)}b_{(1)}, c)\gamma(a_{(2)}, b_{(2)}) = \sum \gamma(a, b_{(1)}c_{(1)})\gamma(b_{(2)}, c_{(2)}) \tag{56}$$

for $a, b, c \in \mathcal{A}$. *The mapping* γ *is said to be* unital *if* $\gamma(a, 1) = \gamma(1, a) = \varepsilon(a)$, $a \in \mathcal{A}$.

If $\gamma : \mathcal{A}\times\mathcal{A}\to\mathbb{C}$ is a bilinear mapping which is convolution invertible, then (55) for γ is equivalent to (56) for its inverse $\bar{\gamma}$. That is, γ is a left 2-cocycle on $\mathcal{A}$ if and only if $\bar{\gamma}$ is a right 2-cocycle on $\mathcal{A}$.

Example 12. Any universal r-form $\mathbf{r}$ of a bialgebra $\mathcal{A}$ is a unital left 2-cocycle on $\mathcal{A}$ and its inverse $\bar{\mathbf{r}}$ is a unital right 2-cocycle on $\mathcal{A}$. Let us verify (55) for $\mathbf{r}$. Indeed, if we apply (3) to the term $\mathbf{r}(a_{(2)}b_{(2)}, c)$ and (4) to $\mathbf{r}(a, b_{(2)}c_{(2)})$, then (55) reduces to the QYBE which holds by Proposition 2(i). Thus $\mathbf{r}$ is a left 2-cocycle. By the preceding remark, $\bar{\mathbf{r}}$ is a right 2-cocycle. △

Proposition 20. (i) *Let* γ *be a unital left 2-cocycle on a bialgebra* $\mathcal{A}$ *which is convolution invertible. Let* $\bar{\gamma}$ *be its inverse. Define a new product on* $\mathcal{A}$ *by*

$$a\cdot_\gamma b := \sum \gamma(a_{(1)}, b_{(1)})a_{(2)}b_{(2)}\bar{\gamma}(a_{(3)}, b_{(3)}), \qquad a, b \in \mathcal{A}. \tag{57}$$

Then the vector space $\mathcal{A}$ *equipped with the product* $\cdot_\gamma$ *and with the coalgebra structure of* $\mathcal{A}$ *becomes a bialgebra. It is denoted* $\mathcal{A}(\gamma)$ *and called the* twist of $\mathcal{A}$ by the cocycle γ.

(ii) *If* $\mathcal{A}$ *is coquasitriangular with universal* r*-form* $\mathbf{r}$, *then* $\mathbf{r}_\gamma := \gamma_{21}\mathbf{r}\bar{\gamma}$ *and* $\tilde{\mathbf{r}}_\gamma := \gamma_{21}\bar{\mathbf{r}}_{21}\bar{\gamma}$ *are universal* r*-forms of the bialgebra* $\mathcal{A}(\gamma)$. *In particular,* $\mathcal{A}(\gamma)$ *is also coquasitriangular.*

(iii) *If* $\mathcal{A}$ *is a Hopf algebra, then so is* $\mathcal{A}(\gamma)$ *with antipode given by* $S_{\mathcal{A}(\gamma)}(a) = \sum f(a_{(1)})S_{\mathcal{A}}(a_{(2)})\bar{f}(a_{(3)})$, *where* $f(a) := \sum \gamma(a_{(1)}, S(a_{(2)}))$ *and* $\bar{f}(a) := \sum \bar{\gamma}(S(a_{(1)}), a_{(2)})$, $a \in \mathcal{A}$. *Moreover,* $f\bar{f} = \bar{f}f = \varepsilon$ *in* $\mathcal{A}'$.

Proof. All assertions follow by direct verifications of the corresponding axioms. As an example, we prove the associativity of the product $\cdot_\gamma$. Inserting the definition (57) of the product $\cdot_\gamma$, we obtain

$$(a \cdot_\gamma b) \cdot_\gamma c = \sum \gamma(a_{(1)}, b_{(1)}) \gamma(a_{(2)} b_{(2)}, c_{(1)}) a_{(3)} b_{(3)} c_{(2)} \times \overline{\gamma}(a_{(4)} b_{(4)}, c_{(3)}) \overline{\gamma}(a_{(5)}, b_{(5)}),$$

$$a \cdot_\gamma (b \cdot_\gamma c) = \sum \gamma(b_{(1)}, c_{(1)}) \gamma(a_{(1)}, b_{(2)} c_{(2)}) a_{(2)} b_{(3)} c_{(3)} \times \overline{\gamma}(a_{(3)}, b_{(4)} c_{(4)}) \overline{\gamma}(b_{(5)}, c_{(5)}).$$

Since γ is a left 2-cocycle, $\overline{\gamma}$ is a right 2-cocycle. Using (55) for γ and (56) for $\overline{\gamma}$, it follows that the right hand sides of the preceding relations coincide. Hence the product $\cdot_\gamma$ is associative. For the last assertion, see also the proof of Proposition 3. □

Example 12 (Continued). If the left 2-cocycle γ in Proposition 20 is the universal r-form of a coquasitriangular bialgebra, then, comparing (57) and (2), we see that $\mathcal{A}(\gamma)$ is just the opposite bialgebra $\mathcal{A}^{\mathrm{op}}$. △

Let us retain the assumptions and the notation of Proposition 20. Then the twisting procedure can be extended to $\mathcal{A}$-comodule algebras and to $\mathcal{A}$-comodule coalgebras as well. Let $\mathcal{X}$ be a left $\mathcal{A}$-comodule algebra. Define a new product ${}_\gamma\cdot$ by

$$x {}_\gamma\cdot y = \sum \gamma(x_{(-1)}, y_{(-1)}) x_{(0)} y_{(0)}, \quad x, y \in \mathcal{X}.$$

The product ${}_\gamma\cdot$ makes the vector space $\mathcal{X}$ into an algebra ${}_\gamma\mathcal{X}$ which is a left comodule algebra for the twisted bialgebra $\mathcal{A}(\gamma)$. (To prove the associativity of the product ${}_\gamma\cdot$, we note that the scalar factors appearing in the expressions $(x {}_\gamma\cdot y) {}_\gamma\cdot z$ and $x {}_\gamma\cdot (y {}_\gamma\cdot z)$ are just the two sides of left cocycle condition (55). Since γ is unital, the unit of the algebra $\mathcal{X}$ is also a unit for the product ${}_\gamma\cdot$.)

Similarly, a right $\mathcal{A}$-comodule algebra $\mathcal{X}$ is a right $\mathcal{A}(\gamma)$-comodule algebra with respect to the new product $\cdot_\gamma$ defined by

$$x \cdot_\gamma y = \sum x_{(0)} y_{(0)} \overline{\gamma}(x_{(1)}, y_{(1)}), \quad x, y \in \mathcal{X}.$$

If $\mathcal{X}$ is a right $\mathcal{A}$-comodule coalgebra, then

$$\Delta_\gamma(x) = \sum (x_{(1)})^{(0)} \otimes (x_{(2)})^{(0)} \gamma((x_{(1)})^{(1)}, (x_{(2)})^{(1)}), \quad x \in \mathcal{X},$$

gives a new coproduct Δ_γ on the vector space $\mathcal{X}$ such that $\mathcal{X}$ becomes a right $\mathcal{A}(\gamma)$-comodule coalgebra.

Next let us consider cohomologous cocycles. Let u be an invertible element of the dual bialgebra $\mathcal{A}^\circ$ with inverse $\overline{u}$. If γ is a left 2-cocycle on $\mathcal{A}$, then one verifies that $\gamma' := (\overline{u} \otimes \overline{u}) \gamma \Delta(u)$, that is,

$$\gamma'(a, b) = \sum \overline{u}(a_{(1)}) \overline{u}(b_{(1)}) \gamma(a_{(2)}, b_{(2)}) u(a_{(3)} b_{(3)}), \quad a, b \in \mathcal{A},$$

is also a left 2-cocycle on $\mathcal{A}$. Then we say that γ and γ' are *cohomologous*. A left 2-cocycle γ is called a *coboundary* if γ is cohomologous to the trivial cocycle $\gamma_0(a,b) = \varepsilon(a)\varepsilon(b)$, that is, if γ is of the form $(\overline{u}\otimes\overline{u})\gamma_0\Delta(u)$ for some convolution invertible element $u \in \mathcal{A}^\circ$.

Proposition 21. *If γ and γ' are two cohomologous invertible unital left 2-cocycles on $\mathcal{A}$ with $\gamma' = (\overline{u}\otimes\overline{u})\gamma\Delta(u)$, then the map $\theta = u * \mathrm{id} * \overline{u}$ (that is, $\theta(a) = \sum u(a_{(1)})a_{(2)}\overline{u}(a_{(3)})$) is an isomorphism of the twisted bialgebras $\mathcal{A}(\gamma)$ and $\mathcal{A}(\gamma')$. In particular, if γ is a coboundary, then the twisted bialgebra $\mathcal{A}(\gamma)$ is isomorphic to $\mathcal{A}$.*

Proof. Using the relations $(u\otimes u)\gamma' = \gamma\Delta(u)$ and $\overline{\gamma'}(\overline{u}\otimes\overline{u}) = \Delta(\overline{u})\overline{\gamma}$, we get

$$\begin{aligned}
\theta(a)\cdot_{\gamma'}\theta(b) &= \left(\sum u(a_{(1)})a_{(2)}\overline{u}(a_{(3)})\right)\cdot_{\gamma'}\left(\sum u(b_{(1)})b_{(2)}\overline{u}(b_{(3)})\right)\\
&= \sum u(a_{(1)})u(b_{(1)})\gamma'(a_{(2)},b_{(2)})a_{(3)}b_{(3)}\overline{\gamma'}(a_{(4)},b_{(4)})\\
&\qquad\qquad\qquad\qquad \times\overline{u}(a_{(5)})\overline{u}(b_{(5)})\\
&= \sum \gamma(a_{(1)},b_{(1)})u(a_{(2)}b_{(2)})a_{(3)}b_{(3)}\overline{u}(a_{(4)}b_{(4)})\overline{\gamma}(a_{(5)},b_{(5)})\\
&= \sum \gamma(a_{(1)},b_{(1)})\theta(a_{(2)}b_{(2)})\overline{\gamma}(a_{(3)},b_{(3)}) = \theta((a\cdot_\gamma b)).
\end{aligned}$$

It is easy to check that θ preserves the coalgebra structure as well. □

We conclude this subsection by showing how the quantum double (see Subsect. 8.2.1) can be obtained by the twisting procedure.

Proposition 22. *Let $\mathcal{A}$ and $\mathcal{B}$ be bialgebras which admit a skew-pairing $\sigma : \mathcal{A}\times\mathcal{B}\to\mathbb{C}$. Suppose that σ has a convolution inverse $\overline{\sigma}$. We define bilinear mappings $\gamma : \mathcal{B}\otimes\mathcal{A}\times\mathcal{B}\otimes\mathcal{A}\to\mathbb{C}$ and $\hat{\gamma} : \mathcal{A}\otimes\mathcal{B}\times\mathcal{A}\otimes\mathcal{B}\to\mathbb{C}$ by*

$$\gamma(b\otimes a, b'\otimes a') = \varepsilon(b)\overline{\sigma}(a,b')\varepsilon(a') \quad \text{and} \quad \hat{\gamma}(a\otimes b, a'\otimes b') = \varepsilon(a)\sigma(a',b)\varepsilon(b).$$

Then γ and $\hat{\gamma}$ are invertible unital left 2-cocycles on the tensor product bialgebras $\mathcal{B}\otimes\mathcal{A}$ and $\mathcal{A}\otimes\mathcal{B}$, respectively. The corresponding twisted bialgebras are the quantum doubles $\mathcal{D}(\mathcal{A},\mathcal{B};\sigma)$ and $\mathcal{D}(\mathcal{B},\mathcal{A};\overline{\sigma}_{21})$, respectively. That is,

$$\mathcal{D}(\mathcal{A},\mathcal{B};\sigma) = (\mathcal{B}\otimes\mathcal{A})(\gamma) \quad \text{and} \quad \mathcal{D}(\mathcal{B},\mathcal{A};\overline{\sigma}_{21}) = (\mathcal{A}\otimes\mathcal{B})(\hat{\gamma}).$$

Proof. We carry out the proof for γ. The assertion concerning $\hat{\gamma}$ follows then by interchanging the role of $\mathcal{A}$ and $\mathcal{B}$ and of σ and $\overline{\sigma}_{21}$. The left 2-cocycle condition (55) for γ is equivalent to the equality

$$\sum \overline{\sigma}(a_{(1)},b')\overline{\sigma}(a_{(2)}a',b) = \sum \overline{\sigma}(a',b_{(1)})\overline{\sigma}(a,b'b_{(2)}).$$

Therefore, since $\overline{\sigma}_{21}$ is an invertible skew-pairing of $\mathcal{B}$ and $\mathcal{A}$, γ is an invertible unital left 2-cocycle on $\mathcal{B}\otimes\mathcal{A}$. Comparing the formulas (8.21) and (57) we see that the products of the algebras $\mathcal{D}(\mathcal{A},\mathcal{B};\sigma)$ and $(\mathcal{B}\otimes\mathcal{A})(\gamma)$ on the vector space $\mathcal{B}\otimes\mathcal{A}$ coincide. By definition, the coalgebra structures are the same. □

Corollary 23. *If $\mathcal{A}$ and $\mathcal{B}$ are coquasitriangular bialgebras with universal r-forms $\mathbf{r}_\mathcal{A}$ and $\mathbf{r}_\mathcal{B}$, then the quantum double $\mathcal{D}(\mathcal{A},\mathcal{B};\sigma)$ is coquasitriangular with universal r-form $\hat{\mathbf{r}} = \overline{\sigma}_{41}(\mathbf{r}_\mathcal{B})_{13}(\mathbf{r}_\mathcal{A})_{24}\sigma_{23}$, that is,*

$$\hat{\mathbf{r}}(b\otimes a, b'\otimes a') = \sum \bar{\sigma}(a'_{(1)}, b_{(1)})\mathbf{r}_{\mathcal{B}}(b_{(2)}, b'_{(1)})\mathbf{r}_{\mathcal{A}}(a_{(1)}, a'_{(2)})\sigma(a_{(2)}, b'_{(2)}).$$

Proof. Obviously, $\mathbf{r} = (\mathbf{r}_{\mathcal{B}})_{13}(\mathbf{r}_{\mathcal{A}})_{24}$ is a universal r-form for the bialgebra $\mathcal{B}\otimes\mathcal{A}$. Since $\mathcal{D}(\mathcal{A},\mathcal{B};\sigma) = (\mathcal{B}\otimes\mathcal{A})(\gamma)$, $\hat{\mathbf{r}} = \mathbf{r}_\gamma = \gamma_{21}\mathbf{r}\bar{\gamma} = \bar{\sigma}_{41}(\mathbf{r}_{\mathcal{B}})_{13}(\mathbf{r}_{\mathcal{A}})_{24}\sigma_{23}$ is a universal r-form of $\mathcal{D}(\mathcal{A},\mathcal{B};\sigma)$ by Proposition 20(ii). □

Example 13. Let $\mathcal{A}$ be either the FRT bialgebra $\mathcal{A}(R)$ for some invertible solution of the QYBE or the coordinate Hopf algebra $\mathcal{O}(G_q)$ for $G_q = GL_q(N), SL_q(N), O_q(N), Sp_q(N)$. By Theorems 7 and 9, $\mathcal{A}$ is coquasitriangular with universal r-form $\mathbf{r}$ such that $\mathbf{r}(\mathbf{u}_1,\mathbf{u}_2) = R$. Therefore, by Corollary 23, the quantum double $\mathcal{D}(\mathcal{A},\mathcal{A};\mathbf{r})$ with respect to the skew-pairing $\mathbf{r}$ of $\mathcal{A}$ and $\mathcal{A}$ admits a universal r-form $\hat{\mathbf{r}}$ such that

$$\hat{\mathbf{r}}(\mathbf{u}_1\otimes\mathbf{u}_2, \mathbf{u}_3\otimes\mathbf{u}_4) = R_{41}^{-1}R_{13}R_{24}R_{23}. \tag{58}$$

Since $\bar{\mathbf{r}}_{21}$ is a universal r-form of $\mathcal{A}$ as well, we also get three other universal r-forms of $\mathcal{D}(\mathcal{A},\mathcal{A};\mathbf{r})$ in this manner. They are obtained if the term $R_{13}R_{24}$ in (58) is replaced by $R_{31}^{-1}R_{24}$, $R_{13}R_{42}^{-1}$ and $R_{31}^{-1}R_{42}^{-1}$, respectively. △

Remark 4. There is a more general concept of crossed product algebras that contains the constructions from Propositions 17 and 19 as special cases. Suppressing technical details, it is obtained if the action of the Hopf algebra $\mathcal{A}$ on the $\mathcal{A}$-module algebra $\mathcal{X}$ is twisted by an $\mathcal{X}$-valued cocycle $\sigma : \mathcal{A}\otimes\mathcal{A}\to\mathcal{X}$. Such crossed products have been introduced and studied in [DT1] and [BCM], see also [Mon], Chap. 7. △

10.2.4 Twisting of Coalgebra Structures by 2-Cocycles and Quantum Codoubles

In this subsection we dualize the main considerations of the preceding subsection. That is, we construct new bialgebras by twisting the comultiplication and preserving the multiplication.

Definition 4. *An element $\gamma\in\mathcal{A}\otimes\mathcal{A}$ is called a* 2-cocycle *for $\mathcal{A}$ if*

$$\gamma_{12}\cdot(\Delta\otimes\mathrm{id})(\gamma) = \gamma_{23}\cdot(\mathrm{id}\otimes\Delta)(\gamma). \tag{59}$$

The element γ is called counital *if $(\varepsilon\otimes\mathrm{id})\gamma = (\mathrm{id}\otimes\varepsilon)\gamma = 1$ and* invertible *if it has an inverse, denoted by γ^{-1}, in the algebra $\mathcal{A}\otimes\mathcal{A}$.*

Example 14. The universal R-matrix $\mathcal{R}$ of a quasitriangular bialgebra $\mathcal{A}$ is an invertible counital 2-cocycle for $\mathcal{A}$. (Formula (8.7) means that $\mathcal{R}$ is counital and (8.2) and (8.6) imply (59).) △

The next proposition is the counterpart to Proposition 20.

Proposition 24. (i) *Let γ be an invertible counital 2-cocycle for a bialgebra $\mathcal{A}$. The algebra $\mathcal{A}$ equipped with the comultiplication $\Delta_\gamma(a) := \gamma\cdot\Delta_{\mathcal{A}}(a)\cdot\gamma^{-1}$,*

$a \in \mathcal{A}$, and the counit $\varepsilon_{\mathcal{A}}$ becomes a bialgebra, denoted by $\mathcal{A}(\gamma)$ and called the twist of $\mathcal{A}$ by the cocycle γ.

(ii) *If $\mathcal{A}$ is quasitriangular with universal R-matrix $\mathcal{R}$, then $\mathcal{R}_\gamma := \gamma_{21}\mathcal{R}\gamma^{-1}$ is a universal R-matrix of $\mathcal{A}(\gamma)$ and so $\mathcal{A}(\gamma)$ is also quasitriangular.*

(iii) *If $\mathcal{A}$ is a Hopf algebra, then the bialgebra $\mathcal{A}(\gamma)$ is also a Hopf algebra with antipode given by $S_\gamma(a)=vS_{\mathcal{A}}(a)v^{-1}$, where $v := m_{\mathcal{A}}(\mathrm{id} \otimes S_{\mathcal{A}})(\gamma) \in \mathcal{A}$ and $v^{-1} := m_{\mathcal{A}}(S_{\mathcal{A}} \otimes \mathrm{id})(\gamma^{-1}) \in \mathcal{A}$ is the inverse of v in the algebra $\mathcal{A}$.*

Proof. We only sketch the proof of the coassociativity of Δ_γ. From the definition of Δ_γ we obtain

$$(\Delta_\gamma \otimes \mathrm{id})\Delta_\gamma(a) = [\gamma_{12}(\Delta \otimes \mathrm{id})(\gamma)] \cdot [(\Delta \otimes \mathrm{id})\Delta(a)] \cdot [(\Delta \otimes \mathrm{id})(\gamma^{-1})\gamma_{12}^{-1}],$$

$$(\mathrm{id} \otimes \Delta_\gamma)\Delta_\gamma(a) = [\gamma_{23}(\mathrm{id} \otimes \Delta)(\gamma)] \cdot [(\mathrm{id} \otimes \Delta)\Delta(a)] \cdot [(\mathrm{id} \otimes \Delta)(\gamma^{-1})\gamma_{23}^{-1}].$$

The corresponding expressions in the squared brackets of both equations coincide, the first and the third ones by the cocycle condition (59) and the middle ones by the coassociativity of $\Delta = \Delta_{\mathcal{A}}$. □

Remark 5. If $\gamma \in \mathcal{A} \otimes \mathcal{A}$ does not fulfill condition (59), then the comultiplication $\Delta_\gamma = \gamma \cdot \Delta_{\mathcal{A}} \cdot \gamma^{-1}$ is no longer coassociative in general. However,

$$\Phi := \gamma_{12} \cdot (\Delta \otimes \mathrm{id})(\gamma) \cdot (\mathrm{id} \otimes \Delta)(\gamma^{-1}) \cdot \gamma_{23}^{-1}$$

is an invertible element of $\mathcal{A} \otimes \mathcal{A} \otimes \mathcal{A}$ such that

$$(\mathrm{id} \otimes \Delta_\gamma)\Delta_\gamma(a) = \Phi^{-1} \cdot ((\Delta_\gamma \otimes \mathrm{id})\Delta_\gamma(a)) \cdot \Phi.$$

That is, Δ_γ is coassociative up to conjugation by Φ. Such a weakening of the coassociativity axiom leads to the notion of a *quasi-Hopf algebra*, see [Dr5]. Note that (59) implies that $\Phi = 1$ and hence that Δ_γ is coassociative, but the latter property is weaker than (59) in general. △

Our next aim is to define a dual notion to the quantum double. In order to do so, we first dualize the notion of a skew-pairing (see Definition 8.3).

Definition 5. *A* skew-copairing *of two bialgebras $\mathcal{A}$ and $\mathcal{B}$ is an element $\sigma \in \mathcal{A} \otimes \mathcal{B}$ such that*

$$(\varepsilon_{\mathcal{A}} \otimes \mathrm{id})\sigma = 1, \quad (\mathrm{id} \otimes \varepsilon_{\mathcal{B}})\sigma = 1, \tag{60}$$

$$(\Delta_{\mathcal{A}} \otimes \mathrm{id})(\sigma) = \sigma_{13}\sigma_{23}, \quad (\mathrm{id} \otimes \Delta_{\mathcal{B}})(\sigma) = \sigma_{13}\sigma_{12}. \tag{61}$$

Note that if $\sigma \in \mathcal{A} \otimes \mathcal{B}$ is invertible, then the two equations (61) already imply (60) (see the proof of formula (8.7) in Subsect. 8.1.1).

Example 15. The universal R-matrix $\mathcal{R}$ of a quasitriangular bialgebra $\mathcal{A}$ is a skew-copairing of $\mathcal{A}$ and $\mathcal{A}$ itself (see (8.7) and (8.2)). △

Proposition 25. *Let $\sigma \in \mathcal{A} \otimes \mathcal{B}$ be an invertible skew-copairing of two bialgebras $\mathcal{A}$ and $\mathcal{B}$. Then the element $\gamma := \sigma_{23}^{-1} \in \mathcal{B} \otimes \mathcal{A} \otimes \mathcal{B} \otimes \mathcal{A}$ is an invertible* *counital 2-cocycle for the tensor product bialgebra $\mathcal{B} \otimes \mathcal{A}$. The corresponding twisted bialgebra $(\mathcal{B} \otimes \mathcal{A})(\gamma)$ has the algebra structure of the tensor product $\mathcal{B} \otimes \mathcal{A}$, the counit $\varepsilon(b \otimes a) = \varepsilon_{\mathcal{B}}(b)\varepsilon_{\mathcal{A}}(a)$ and the comultiplication*

$$\Delta(b \otimes a) = \sum b_{(1)} \otimes \sigma^{-1}(a_{(1)} \otimes b_{(2)})\sigma \otimes a_{(2)}. \tag{62}$$

If $\mathcal{A}$ and $\mathcal{B}$ are Hopf algebras, then so is $(\mathcal{B} \otimes \mathcal{A})(\gamma)$ with antipode

$$S(b \otimes a) = \sigma_{21}(S_{\mathcal{B}}(b) \otimes S_{\mathcal{A}}(a))\sigma_{21}^{-1}. \tag{63}$$

Proof. From (61) one easily derives that γ is indeed a counital 2-cocycle for $\mathcal{B} \otimes \mathcal{A}$. Thus, Proposition 24 applies. The formula $\Delta_\gamma = \gamma \cdot \Delta_{\mathcal{B}\otimes\mathcal{A}} \cdot \gamma^{-1}$ for the twisted bialgebra $(\mathcal{B} \otimes \mathcal{A})(\gamma)$ now reads as (62). To prove the formula (63), it suffices to show that the element v in Proposition 24 is σ_{21}. Let $\sigma^{-1} = \sum_i a^i \otimes b^i$. Then we have

$$\begin{aligned} v &= m_{\mathcal{B}\otimes\mathcal{A}}(\mathrm{id} \otimes \mathrm{id} \otimes S_{\mathcal{B}\otimes\mathcal{A}})(\gamma) = \sum\nolimits_i (1 \otimes a^i)(S_{\mathcal{B}}(b^i) \otimes 1) \\ &= (S_{\mathcal{B}} \otimes \mathrm{id})(\sigma_{21}^{-1}) = \sigma_{21}, \end{aligned}$$

where the last equality follows similarly as the second formula in (8.8). □

Definition 6. *The bialgebra $(\mathcal{B} \otimes \mathcal{A})(\gamma)$ in Proposition 25 is called the* (generalized) quantum codouble *of the bialgebras $\mathcal{A}$ and $\mathcal{B}$ with respect to the skew-copairing σ and denoted by $\mathcal{D}^{\mathrm{co}}(\mathcal{A}, \mathcal{B}; \sigma)$.*

10.2.5 Double Crossed Product Bialgebras and Quantum Doubles

In this subsection we are concerned with a generalization of the crossed product algebra construction by considering two bialgebras which act on each other. This situation is precisely described by the following definition.

Definition 7. *A pair $\{\mathcal{B}, \mathcal{A}\}$ of two bialgebras is called* matched *if $\mathcal{A}$ is a right $\mathcal{B}$-module and $\mathcal{B}$ is a left $\mathcal{A}$-module with actions $\alpha(a \otimes b) = a \triangleleft b$ of $\mathcal{B}$ on $\mathcal{A}$ and $\beta(a \otimes b) = a \triangleright b$ of $\mathcal{A}$ on $\mathcal{B}$ satisfying the following conditions:*

$$\Delta_{\mathcal{A}}(a \triangleleft b) = \sum a_{(1)} \triangleleft b_{(1)} \otimes a_{(2)} \triangleleft b_{(2)}, \quad \varepsilon_{\mathcal{A}}(a \triangleleft b) = \varepsilon_{\mathcal{A}}(a)\varepsilon_{\mathcal{B}}(b), \tag{64}$$

$$\Delta_{\mathcal{B}}(a \triangleright b) = \sum a_{(1)} \triangleright b_{(1)} \otimes a_{(2)} \triangleright b_{(2)}, \quad \varepsilon_{\mathcal{B}}(a \triangleright b) = \varepsilon_{\mathcal{A}}(a)\varepsilon_{\mathcal{B}}(b), \tag{65}$$

$$(aa') \triangleleft b = \sum (a \triangleleft (a'_{(1)} \triangleright b_{(1)}))(a'_{(2)} \triangleleft b_{(2)}), \quad 1 \triangleleft b = \varepsilon_{\mathcal{B}}(b), \tag{66}$$

$$a \triangleright (bb') = \sum (a_{(1)} \triangleright b_{(1)})((a_{(2)} \triangleleft b_{(2)}) \triangleright b'), \quad a \triangleright 1 = \varepsilon_{\mathcal{A}}(a), \tag{67}$$

$$\sum a_{(1)} \triangleleft b_{(1)} \otimes a_{(2)} \triangleright b_{(2)} = \sum a_{(2)} \triangleleft b_{(2)} \otimes a_{(1)} \triangleright b_{(1)}. \tag{68}$$

Equations (64) and (65) mean that the actions α and β are coalgebra homomorphisms. They can be expressed by saying that $\mathcal{A}$ is a right $\mathcal{B}$-module *coalgebra* and $\mathcal{B}$ is a left $\mathcal{A}$-module *coalgebra*. Equations (66)–(68) are compatibility conditions of the two actions. Note that (68) is always fulfilled if $\mathcal{A}$ and $\mathcal{B}$ are both cocommutative.

Proposition 26. (i) *Let $\{\mathcal{B},\mathcal{A}\}$ be a matched pair of bialgebras. Then the vector space $\mathcal{B}\otimes\mathcal{A}$ becomes a bialgebra with the tensor product coalgebra structure of $\mathcal{B}\otimes\mathcal{A}$ and with the product*

$$(b\otimes a)(b'\otimes a') = \sum b(a_{(1)}\triangleright b'_{(1)})\otimes(a_{(2)}\triangleleft b'_{(2)})a'. \tag{69}$$

This bialgebra is called the double crossed product bialgebra *of the pair $\{\mathcal{B},\mathcal{A}\}$. It is denoted by $\mathcal{B}_\alpha\bowtie_\beta\mathcal{A}$ or simply by $\mathcal{B}\bowtie\mathcal{A}$.*

(ii) *If $\mathcal{A}$ and $\mathcal{B}$ are Hopf algebras, then $\mathcal{B}\bowtie\mathcal{A}$ is also a Hopf algebra with antipode $S(b\otimes a) = (1\otimes S_{\mathcal{A}}(a))(S_{\mathcal{B}}(b)\otimes 1)$, $a\in\mathcal{A}, b\in\mathcal{B}$.*

Proof. The proof follows by direct verification using the conditions in Definition 7. □

As in the case of similar constructions (see Subsects. 8.2.1 or 10.2.1), we can consider $\mathcal{B}$ and $\mathcal{A}$ as subbialgebras of the double crossed product bialgebra $\mathcal{B}\bowtie\mathcal{A}$ by identifying $b\in\mathcal{B}$ with $b\otimes 1\in\mathcal{B}\bowtie\mathcal{A}$ and $a\in\mathcal{A}$ with $1\otimes a\in\mathcal{B}\bowtie\mathcal{A}$. We then have the following cross commutation relations between elements $a\in\mathcal{A}$ and $b\in\mathcal{B}$ in the algebra $\mathcal{B}\bowtie\mathcal{A}$:

$$ab = \sum a_{(1)}\triangleright b_{(1)}\otimes a_{(2)}\triangleleft b_{(2)} \equiv \sum a_{(2)}\triangleright b_{(2)}\otimes a_{(1)}\triangleleft b_{(1)}.$$

We now show that the quantum double $\mathcal{D}(\mathcal{A},\mathcal{B};\sigma)$ can be considered as an example of double crossed product bialgebras. Let us assume that $\mathcal{A}$ and $\mathcal{B}$ are bialgebras equipped with an invertible skew-pairing σ (see Subsect. 8.2.1). For $a\in\mathcal{A}$ and $b\in\mathcal{B}$, we set

$$a\triangleleft b = \sum a_{(2)}\,\bar\sigma(a_{(1)},b_{(1)})\sigma(a_{(3)},b_{(2)}), \tag{70}$$

$$a\triangleright b = \sum b_{(2)}\,\bar\sigma(a_{(1)},b_{(1)})\sigma(a_{(2)},b_{(3)}), \tag{71}$$

where $\bar\sigma$ denotes the inverse of σ. With some work one verifies that (70) and (71) define actions of $\mathcal{B}$ on $\mathcal{A}$ and of $\mathcal{A}$ on $\mathcal{B}$ such that $\{\mathcal{B},\mathcal{A}\}$ is a matched pair of bialgebras. Let $\mathcal{B}\bowtie_\sigma\mathcal{A}$ denote the corresponding double crossed product bialgebra. By (70) and (71), we have

$$\begin{aligned}\sum & a_{(1)}\triangleright b'_{(1)}\otimes a_{(2)}\triangleleft b'_{(2)}\\ &= \sum b'_{(2)}\bar\sigma(a_{(1)},b'_{(1)})\sigma(a_{(2)},b'_{(3)})\otimes a_{(4)}\,\bar\sigma(a_{(3)},b'_{(4)})\sigma(a_{(5)},b'_{(5)})\\ &= \sum b'_{(2)}\otimes a_{(2)}\,\bar\sigma(a_{(1)},b'_{(1)})\sigma(a_{(3)},b'_{(3)}).\end{aligned}$$

This shows that the product (69) of the algebra $\mathcal{B}\bowtie_\sigma\mathcal{A}$ on the vector space $\mathcal{B}\otimes\mathcal{A}$ is just the product (8.21) of the algebra $\mathcal{D}(\mathcal{A},\mathcal{B};\sigma)$. Recall that both

$\mathcal{B} \bowtie_\sigma \mathcal{A}$ and $\mathcal{D}(\mathcal{A},\mathcal{B};\sigma)$ have the tensor product coalgebra structures. Thus we have proved the following equality of bialgebras:

$$\mathcal{D}(\mathcal{A},\mathcal{B};\sigma) = \mathcal{B} \bowtie_\sigma \mathcal{A}. \tag{72}$$

Let us suppose in addition that $\mathcal{A}$ and $\mathcal{B}$ are Hopf algebras such that the antipode $S_\mathcal{B}$ of $\mathcal{B}$ is invertible. Then the two actions $\triangleleft$ and $\triangleright$ defined above can be nicely expressed as duals to the right adjoint coactions of the Hopf algebras $\mathcal{A}$ and $\mathcal{B}^{\mathrm{op}}$. First a few preliminaries are needed. Using the formulas (8.20) and (8.19), we rewrite the equations (70) and (71) as

$$a \triangleleft b = \sum a_{(2)}\sigma(S_\mathcal{A}(a_{(1)})a_{(3)}, b), \tag{73}$$

$$a \triangleright b = \sum b_{(2)}\sigma(a, b_{(3)}S_\mathcal{B}^{-1}(b_{(1)})). \tag{74}$$

Let $\langle\cdot,\cdot\rangle$ be a dual pairing of two Hopf algebras $\mathcal{G}$ and $\mathcal{H}$. By Proposition 1.15, any right $\mathcal{H}$-comodule V becomes a left $\mathcal{G}$-module with action given by $g \triangleright v := \sum v_{(0)}\langle g, v_{(1)}\rangle$, $g \in \mathcal{G}$, $v \in V$. Likewise a right $\mathcal{G}$-module is a left $\mathcal{H}$-module with action $h \triangleright v = \sum v_{(0)}\langle v_{(1)}, h\rangle$. For the right adjoint coactions $\mathrm{Ad}_R(a) = \sum a_{(2)} \otimes S(a_{(1)})a_{(3)}$ of $\mathcal{G}$ and $\mathcal{H}$ on itself, the associated left actions of $\mathcal{G}$ and $\mathcal{H}$ on each other read as

$$h \triangleright g = \sum g_{(2)}\langle S(g_{(1)})g_{(3)}, h\rangle, \tag{75}$$

$$g \triangleright h = \sum h_{(2)}\langle g, S(h_{(1)})h_{(3)}\rangle. \tag{76}$$

We now apply the preceding to the dual pairing $\langle\cdot,\cdot\rangle \equiv \sigma(\cdot,\cdot)$ of $\mathcal{A}$ and $\mathcal{B}^{\mathrm{op}}$. Then the left action (75) of $\mathcal{B}^{\mathrm{op}}$ on $\mathcal{A}$ yields the right action (73) of $\mathcal{B}$ on $\mathcal{A}$. In terms of the product and antipode of $\mathcal{B}$ the right adjoint coaction of the Hopf algebra $\mathcal{B}^{\mathrm{op}}$ can be written as $\mathrm{Ad}_R(b) = \sum b_{(2)} \otimes b_{(3)}S_\mathcal{B}^{-1}(b_{(1)})$, $b \in \mathcal{B}^{\mathrm{op}}$. Hence the left action (76) of $\mathcal{A}$ on $\mathcal{B}^{\mathrm{op}}$ gives just the left action (74).

We close our study of the quantum double by treating an instructive example. It is based on the FRT bialgebra $\mathcal{A}(R)$ which has always been our guiding example.

Example 16 (*The quantum double* $\mathcal{D}(\mathcal{A}(R),\mathcal{A}(R);\mathbf{r})$). Let R be an invertible linear transformation on $\mathbb{C}^N$ satisfying the QYBE. Since the canonical universal r-form $\mathbf{r}$ of the FRT bialgebra $\mathcal{A}(R)$ (see Theorem 7) is a skew-pairing of $\mathcal{A}(R)$ and $\mathcal{A}(R)$ itself, the quantum double $\mathcal{D}(\mathcal{A}(R),\mathcal{A}(R);\mathbf{r})$ is a well-defined bialgebra.

Let $\mathbf{u}$ and $\mathbf{v}$ denote the fundamental matrices of two copies of $\mathcal{A}(R)$. By (11), we have $\mathbf{r}(\mathbf{u}_1,\mathbf{v}_2) = R_{12}$ and $\bar{\mathbf{r}}(\mathbf{u}_1,\mathbf{v}_2) = R_{12}^{-1}$. Inserting these expressions into the formulas (70) and (71) we see that in matrix notation the actions $\triangleleft$ and $\triangleright$ of $\mathcal{A}(R)$ on $\mathcal{A}(R)$ are written as

$$\mathbf{u}_1 \triangleleft \mathbf{v}_2 = R_{12}^{-1}\mathbf{u}_1 R_{12} \quad \text{and} \quad \mathbf{u}_1 \triangleright \mathbf{v}_2 = R_{12}^{-1}\mathbf{v}_2 R_{12}.$$

Similarly, the commutation relations (8.22) of the entries of $\mathbf{u}$ and $\mathbf{v}$ in the algebra $\mathcal{D}(\mathcal{A}(R),\mathcal{A}(R);\mathbf{r})$ are $\mathbf{u}_1\mathbf{v}_2 = R^{-1}\mathbf{v}_2\mathbf{u}_1 R$. Hence, by the discussion after Definition 8.4, the algebra $\mathcal{D}(\mathcal{A}(R),\mathcal{A}(R);\mathbf{r})$ is generated by the entries of the matrices $\mathbf{u}$ and $\mathbf{v}$ subject to the defining relations

$$R\mathbf{u}_1\mathbf{u}_2 = \mathbf{u}_2\mathbf{u}_1 R, \quad R\mathbf{v}_1\mathbf{v}_2 = \mathbf{v}_2\mathbf{v}_1 R, \quad \mathbf{u}_1\mathbf{v}_2 = R^{-1}\mathbf{v}_2\mathbf{v}_1 R.$$

On the other hand, the algebra $\mathfrak{U}(R^{-1})$ has the defining relations (see (9.43))

$$R\mathfrak{L}_1^+\mathfrak{L}_2^+ = \mathfrak{L}_1^+\mathfrak{L}_2^+ R, \quad R\mathfrak{L}_1^-\mathfrak{L}_2^- = \mathfrak{L}_1^-\mathfrak{L}_2^- R, \quad R\mathfrak{L}_1^-\mathfrak{L}_2^+ = \mathfrak{L}_2^+\mathfrak{L}_1^- R.$$

Therefore, the assignments $\mathbf{v} \equiv \mathbf{v}\otimes 1 \to \mathfrak{L}^+$ and $\mathbf{u} \equiv 1 \otimes \mathbf{u} \to \mathfrak{L}^-$ define a bialgebra isomorphism

$$\mathcal{D}(\mathcal{A}(R),\mathcal{A}(R);\mathbf{r}) = \mathcal{A}(R) \bowtie_{\mathbf{r}} \mathcal{A}(R) \simeq \mathfrak{U}(R^{-1}) = \mathfrak{U}(R)^{\mathrm{op}}. \tag{77}$$

We will give two other interpretations of the latter result. Let $\mathfrak{U}_\epsilon(R)$, $\epsilon = +,-$, be the algebra with matrix generators $\mathfrak{L}^\epsilon$ and defining relations $\mathfrak{L}_1^\epsilon\mathfrak{L}_2^\epsilon R = R\mathfrak{L}_2^\epsilon\mathfrak{L}_1^\epsilon$. It is clear that the map $\mathfrak{L}^\epsilon \to \mathbf{u}$ is an algebra isomorphism of $\mathfrak{U}_\epsilon(R)$ to $\mathcal{A}(R^{-1})$, so $\mathfrak{U}_\epsilon(R)$ is a bialgebra with comultiplication $\Delta(\mathfrak{L}^\epsilon) = \mathfrak{L}^\epsilon \otimes \mathfrak{L}^\epsilon$ and counit $\varepsilon(\mathfrak{L}^\epsilon) = I$. Under the isomorphisms $\mathfrak{U}_-(R) \simeq \mathcal{A}(R^{-1})$ and $\mathfrak{U}_+(R) \simeq \mathcal{A}(R^{-1})$ the universal r-form $\mathbf{r}$ of $\mathcal{A}(R^{-1})$ goes into a skew-pairing $\tilde{\mathbf{r}}$ of the bialgebras $\mathfrak{U}_-(R)$ and $\mathfrak{U}_+(R)$ determined by the equation $\tilde{\mathbf{r}}(\mathfrak{L}_1^-,\mathfrak{L}_2^+) = R_{12}^{-1}$. Thus, (77) yields the bialgebra isomorphism

$$\mathcal{D}(\mathfrak{U}_-(R),\mathfrak{U}_+(R);\tilde{\mathbf{r}}) = \mathfrak{U}_+(R) \bowtie_{\tilde{\mathbf{r}}} \mathfrak{U}_-(R) \simeq \mathfrak{U}(R). \tag{78}$$

This representation of the bialgebra $\mathfrak{U}(R)$ as the quantum double of $\mathfrak{U}_-(R)$ and $\mathfrak{U}_+(R)$ explains the source of the third matrix relation $\mathfrak{L}_1^-\mathfrak{L}_2^+R = R\mathfrak{L}_2^+\mathfrak{L}_1^-$. It is just the cross relation of the quantum double. Likewise, we could have expressed $\mathfrak{U}(R)$ also as the quantum double of $\mathfrak{U}_+(R)$ and $\mathfrak{U}_-(R)$ built over the vector space $\mathfrak{U}_-(R)\otimes\mathfrak{U}_+(R)$ rather than $\mathfrak{U}_+(R)\otimes\mathfrak{U}_-(R)$.

To give still another view of the isomorphism (77), we recall from Proposition 9.20 that there is a dual pairing $\langle\cdot,\cdot\rangle$ of the bialgebras $\mathcal{A}(R)$ and $\mathfrak{U}_+(R)$ such that $\langle \mathbf{u}_1, \mathfrak{L}_2^+\rangle = R_{12}$. The corresponding skew-pairing σ of $\mathcal{A}(R)$ and $\mathfrak{U}_+(R)^{\mathrm{op}}$ is just the canonical universal r-form $\mathbf{r}$ of the bialgebra $\mathcal{A}(R)$ under the isomorphism $\mathfrak{U}_+(R)^{\mathrm{op}} = \mathfrak{U}_+(R^{-1}) \simeq \mathcal{A}(R)$, so (77) means also that

$$\mathcal{D}(\mathcal{A}(R),\mathfrak{U}_+(R)^{\mathrm{op}};\sigma) \simeq \mathfrak{U}(R^{-1}). \qquad \triangle$$

10.2.6 Double Crossed Coproduct Bialgebras and Quantum Codoubles

In this very short subsection we only state dual versions to the main results of Subsect. 10.2.5. We consider two bialgebras which coact on each other.

Definition 8. *A pair $\{\mathcal{A},\mathcal{B}\}$ of bialgebras is said to be* comatched *if $\mathcal{B}$ is a right $\mathcal{A}$-comodule algebra with right coaction $\alpha : \mathcal{B} \to \mathcal{B}\otimes\mathcal{A}$ and $\mathcal{A}$ is a left*

$\mathcal{B}$-comodule algebra with left coaction $\beta : \mathcal{A} \to \mathcal{B} \otimes \mathcal{A}$ such that for $a \in \mathcal{A}$ and $b \in \mathcal{B}$ the following compatibility conditions hold:

$$\begin{aligned}(\Delta \otimes \mathrm{id}) \circ \alpha(b) &= \sum((\mathrm{id} \otimes \beta) \circ \alpha(b_{(1)}))(1 \otimes \alpha(b_{(2)})),\\ (\mathrm{id} \otimes \Delta) \circ \beta(a) &= \sum(\beta(a_{(1)}) \otimes 1)((\alpha \otimes \mathrm{id}) \circ \beta(a_{(2)})),\\ \alpha(b)\beta(a) &= \beta(a)\alpha(b).\end{aligned}$$

Proposition 27. *Suppose that $\{\mathcal{A}, \mathcal{B}\}$ is a comatched pair of bialgebras. Then the tensor product algebra $\mathcal{A} \otimes \mathcal{B}$ becomes a bialgebra, called the* double crossed coproduct bialgebra *and denoted by $\mathcal{A}^{\alpha} \bowtie^{\beta} \mathcal{B}$ or $\mathcal{A} \bowtie \mathcal{B}$, with counit $\varepsilon = \varepsilon_{\mathcal{A}} \otimes \varepsilon_{\mathcal{B}}$ and comultiplication*

$$\Delta(a \otimes b) = \sum a_{(1)} \otimes \alpha(b_{(1)})\beta(a_{(2)}) \otimes b_{(2)}, \quad a \in \mathcal{A},\ b \in \mathcal{B}.$$

Proposition 28. *Let $\sigma \in \mathcal{A} \otimes \mathcal{B}$ be an invertible skew-copairing of the bialgebras $\mathcal{A}$ and $\mathcal{B}$. Then the pair $\{\mathcal{B}, \mathcal{A}\}$ is comatched with respect to the coactions $\alpha(a) = \sigma^{-1}(a \otimes 1)\sigma$ and $\beta(b) = \sigma^{-1}(1 \otimes b)\sigma$ and the corresponding double crossed coproduct bialgebra $\mathcal{B} \bowtie \mathcal{A}$ is just the generalized quantum codouble $\mathcal{D}^{\mathrm{co}}(\mathcal{A}, \mathcal{B}; \sigma)$.*

10.2.7 Realifications of Quantum Groups

Let $\mathcal{A}$ be the coordinate Hopf algebra of a quantum group G_q. Roughly speaking, a realification of G_q should be a Hopf $*$-algebra $\mathcal{B}$ which contains $\mathcal{A}$ as a Hopf subalgebra and which is generated as an algebra by $\mathcal{A}$ and $\mathcal{A}^*$. Here $\mathcal{A}^*$ denotes the set of adjoint elements a^* for $a \in \mathcal{A}$ with respect to the involution of the larger $*$-algebra $\mathcal{B}$. The elements of $\mathcal{A}$ are then interpreted as "holomorphic polynomials" and the elements of $\mathcal{A}^*$ as "anti-holomorphic polynomials" on G_q. The purpose of this subsection is to show that the quantum double $\mathcal{D}(\mathcal{A}, \mathcal{A}; \mathbf{r}) = \mathcal{A} \bowtie_{\mathbf{r}} \mathcal{A}$ provides a general procedure for the construction of a realification of a coquasitriangular Hopf $*$-algebra $\mathcal{A}$ equipped with a real universal r-form $\mathbf{r}$. The involution of $\mathcal{A}$ will be used in order to inherit an involution on the algebra $\mathcal{D}(\mathcal{A}, \mathcal{A}; \mathbf{r})$.

Definition 9. *A Hopf $*$-algebra $\mathcal{B}$ is called a* realification *of a Hopf algebra $\mathcal{A}$ if $\mathcal{A}$ is a Hopf subalgebra of $\mathcal{B}$ and if the linear map $\mu : \mathcal{A} \otimes \mathcal{A}^* \to \mathcal{B}$ defined by $\mu(a \otimes b^*) = ab^*$ is bijective, where $\mathcal{A}^*$ is the set $\{a^* \mid a \in \mathcal{A}\}$ in the $*$-algebra $\mathcal{B}$.*

Remark 6. Some authors call the passage from $\mathcal{A}$ to $\mathcal{B}$ a *complexification* and $\mathcal{B}$ a *complex* quantum group. △

Suppose that $\mathcal{A}$ is a coquasitriangular Hopf algebra with universal r-form $\mathbf{r}$. By Proposition 8.8 and Corollary 23, the quantum double $\mathcal{D}(\mathcal{A}, \mathcal{A}; \mathbf{r})$ is also a coquasitriangular Hopf algebra. Because this Hopf algebra is essentially needed in the following, we briefly repeat its structure (see Sects. 8.2,

10.2.3, and 10.2.5 for details). As a coalgebra, $\mathcal{D}(\mathcal{A}, \mathcal{A}; \mathbf{r})$ is the tensor product coalgebra $\mathcal{A} \otimes \mathcal{A}$. As an algebra, $\mathcal{D}(\mathcal{A}, \mathcal{A}; \mathbf{r})$ is the vector space $\mathcal{A} \otimes \mathcal{A}$ with product

$$(a \otimes b)(c \otimes d) = \sum a\, c_{(2)} \otimes b_{(2)} d\, \bar{\mathbf{r}}(b_{(1)}, c_{(1)}) \mathbf{r}(b_{(3)}, c_{(3)}). \qquad (79)$$

Further, $\mathcal{D}(\mathcal{A}, \mathcal{A}; \mathbf{r})$ is the free algebra generated by two copies, denoted by $\mathcal{A}$ and $\tilde{\mathcal{A}}$, of the algebra $\mathcal{A}$ with cross commutations relations

$$\sum \mathbf{r}(a_{(1)}, \tilde{b}_{(1)}) a_{(2)} \tilde{b}_{(2)} = \sum \tilde{b}_{(1)} a_{(1)} \mathbf{r}(a_{(2)}, \tilde{b}_{(2)}), \quad a \in \mathcal{A},\ \tilde{b} \in \tilde{\mathcal{A}}. \qquad (80)$$

Here $a \in \mathcal{A}$ is identified with $a \otimes 1$ and $\tilde{b} \in \tilde{\mathcal{A}}$ with $1 \otimes \tilde{b}$.

Proposition 29. *Suppose that $\mathcal{A}$ is a Hopf $*$-algebra and $\mathbf{r}$ is a real universal r-form of $\mathcal{A}$. Then the quantum double $\mathcal{B} := \mathcal{D}(\mathcal{A}, \mathcal{A}; \mathbf{r})$ is a Hopf $*$-algebra with involution defined by*

$$(a \otimes b)^* := b^* \otimes a^*, \quad a, b \in \mathcal{A}. \qquad (81)$$

$\mathcal{B}$ is a realification of the Hopf algebra $\mathcal{A}$. There is a homomorphism θ of the Hopf $*$-algebra $\mathcal{B}$ to the Hopf $*$-algebra $\mathcal{A}$ such that $\theta(a \otimes b) = ab$, $a, b \in \mathcal{A}$. Moreover, $\mathbf{r}' := \bar{\mathbf{r}}_{41}\mathbf{r}_{13}\mathbf{r}_{24}\mathbf{r}_{23}$ is a real universal r-form and $\mathbf{r}'' := \bar{\mathbf{r}}_{41}\bar{\mathbf{r}}_{31}\mathbf{r}_{24}\mathbf{r}_{23}$ is an inverse real universal r-form of $\mathcal{B}$.

Proof. Since $\mathbf{r}$ is a real universal r-form of $\mathcal{A}$, $\bar{\mathbf{r}}$ is a real universal r-form of $\mathcal{A}^{\mathrm{op}}$. Using these facts and equations (79) and (81), we obtain

$$\begin{aligned} ((a \otimes b)(c \otimes d))^* &= (b_{(2)}d)^* \otimes (ac_{(2)})^* \, \overline{\bar{\mathbf{r}}(b_{(1)}, c_{(1)}) \mathbf{r}(b_{(3)}, c_{(3)})} \\ &= d^* {b^*}_{(2)} \otimes {c^*}_{(2)} a^* \, \bar{\mathbf{r}}({c^*}_{(1)}, {b^*}_{(1)}) \mathbf{r}({c^*}_{(3)}, {b^*}_{(3)}) \\ &= (d^* \otimes c^*)(b^* \otimes a^*) = (c \otimes d)^* (a \otimes b)^*. \end{aligned}$$

The other requirements for $\mathcal{B} = \mathcal{D}(\mathcal{A}, \mathcal{A}; \mathbf{r})$ being a Hopf $*$-algebra are clear. Obviously, $\mathcal{A}$ becomes a Hopf subalgebra of $\mathcal{B}$ if we identify $a \in \mathcal{A}$ with $a \otimes 1 \in \mathcal{B}$. Then, because of (79) and (81), it follows that $\sum_i a_i \otimes b_i^* = \sum_i (a_i \otimes 1)(b_i \otimes 1)^* = \sum_i a_i b_i^*$. Hence the mapping $\mu : \mathcal{A} \otimes \mathcal{A}^* \to \mathcal{B}$ is bijective and $\mathcal{B}$ is a realification of $\mathcal{A}$.

By Corollary 23, $\mathbf{r}'$ and $\mathbf{r}''$ are universal r-forms of $\mathcal{B}$. The reality of $\mathbf{r}'$, the inverse reality of $\mathbf{r}''$ and the fact that θ is a homomorphism of Hopf $*$-algebras are easily verified. □

Let us emphasize that $\mathcal{A}$ is not a $*$-subalgebra of $\mathcal{B}$ because the adjoint a^* in $\mathcal{B}$ of an element $a \equiv a \otimes 1 \in \mathcal{A} \subseteq \mathcal{B}$ is the element $1 \otimes a^*$.

We now apply Proposition 29 in order to construct realifications of the coordinate Hopf algebras $\mathcal{O}(G_q)$, $G_q = GL_q(N)$, $SL_q(N)$, $O_q(N)$, $Sp_q(N)$, for real q. In doing so we use the Hopf $*$-algebras $\mathcal{A} = \mathcal{O}(G_q^{\mathrm{Re}})$ for the corresponding real quantum groups $G_q^{\mathrm{Re}} = U_q(N)$, $SU_q(N;\nu)$, $O_q(N;\nu)$, $Sp_q(N;\nu)$ and the real universal r-forms $\mathbf{r}_z$ from Proposition 10(i). If we consider G_q as a complex quantum group (which means that we "forget" the involution of $\mathcal{O}(G_q^{\mathrm{Re}})$), then Proposition 29 yields a realification of G_q. By the discussion before Proposition 29 the structure of this realification $\mathcal{B} = \mathcal{D}(\mathcal{A}, \mathcal{A}; \mathbf{r}_z)$ can be explicitly described as follows (see also Example 16 in Subsect. 10.2.5).

Let $\mathbf{u} = (u^i_j)$ and $\mathbf{v} = (v^i_j)$ be the fundamental matrices of two copies of the Hopf algebra $\mathcal{A} = \mathcal{O}(G_q)$. Then, the algebra $\mathcal{B} = \mathcal{D}(\mathcal{A}, \mathcal{A}; \mathbf{r}_z)$ admits the generators u^i_j, v^i_j, $i, j = 1, 2, \cdots, N$. The defining relations of $\mathcal{B}$ are the defining relations of $\mathcal{O}(G_q)$ for each of the sets $\{u^i_j\}$ and $\{v^i_j\}$ together with the cross relations $\mathsf{R}^{ij}_{kl} u^k_n v^l_m = v^j_l u^i_k \mathsf{R}^{kl}_{nm}$ or in matrix form $\mathsf{R}\mathbf{u}_1\mathbf{v}_2 = \mathbf{v}_2\mathbf{u}_1\mathsf{R}$. Note that the latter are just the relations (80) for the generators of the two copies of $\mathcal{A}$. The involution of $\mathcal{B}$ is determined by the equations $(u^i_j)^* = v^i_j$. As noted above, we consider the u^i_j as holomorphic generators and the v^i_j as anti-holomorphic generators (or as complex conjugates to u^i_j) of the algebra $\mathcal{B}$. The coalgebra structure of $\mathcal{B}$ is that of the tensor product $\mathcal{A}\otimes\mathcal{A}$. It is given by the formulas $\Delta(u^i_j) = u^i_k \otimes u^k_j$, $\Delta(v^i_j) = v^i_k \otimes v^k_j$ and $\varepsilon(u^i_j) = \varepsilon(v^i_j) = \delta_{ij}$.

Clearly, one can also verify directly that the preceding description gives a realification $\mathcal{B}$ of $\mathcal{A}$. However, an advantage of the above approach is the explanation of the cross commutation relations $\mathsf{R}\mathbf{u}_1\mathbf{v}_2 = \mathbf{v}_2\mathbf{u}_1\mathsf{R}$ as the defining relations for the quantum double. Further, the general theory of the quantum double provides the universal r-forms $\mathbf{r}'$ and $\mathbf{r}''$ of $\mathcal{B}$.

10.3 Braided Hopf Algebras

A braided Hopf algebra $\mathcal{X}$ is in general not a Hopf algebra according to Definition 1.4. The main new feature is the requirement that the comultiplication is an algebra homomorphism of $\mathcal{X}$ to $\mathcal{X} \otimes \mathcal{X}$, where the latter carries a particular product, the so-called braided product.

10.3.1 Covariantized Products for Coquasitriangular Bialgebras

The right adjoint coaction $\mathrm{Ad}_R(a) = \sum a_{(2)} \otimes S(a_{(1)})a_{(3)}$ of a commutative Hopf algebra $\mathcal{A}$ is obviously an algebra homomorphism of $\mathcal{A}$ to $\mathcal{A} \otimes \mathcal{A}$. In the general case this is no longer true, so the algebra $\mathcal{A}$ is not necessarily a comodule algebra for the Hopf algebra $\mathcal{A}$ with respect to the coaction Ad_R. For coquasitriangular Hopf algebras one can remedy this defect by introducing a new "covariantized" product $\underline{\cdot}$ on $\mathcal{A}$ such that Ad_R is a homomorphism of the corresponding algebra $B(\mathcal{A}) := (\mathcal{A}, \underline{\cdot})$ to $B(\mathcal{A}) \otimes B(\mathcal{A})$. Further, there is another "natural" product $\bullet$ on $B(\mathcal{A}) \otimes B(\mathcal{A})$ such that the comultiplication $\Delta_\mathcal{A}$ is an algebra homomorphism of $B(\mathcal{A})$ to $(B(\mathcal{A}) \otimes B(\mathcal{A}), \bullet)$. In this manner $B(\mathcal{A})$ becomes a braided Hopf algebra.

In this subsection we consider a similar passage from $\mathcal{A}$ to $B(\mathcal{A})$ for bialgebras. This requires some additional technical preliminaries. The Hopf algebra case will be treated in the next subsection.

Definition 10. *A universal r-form* $\mathbf{r}$ *of a coquasitriangular bialgebra* $\mathcal{A}$ *is called* regular *if there is a linear form* $\mathbf{s} : \mathcal{A} \otimes \mathcal{A} \to \mathbb{C}$ *such that for* $a, b \in \mathcal{A}$,

$$\sum \mathbf{r}(a_{(1)} \otimes b_{(2)})\mathbf{s}(a_{(2)} \otimes b_{(1)}) = \sum \mathbf{s}(a_{(1)} \otimes b_{(2)})\mathbf{r}(a_{(2)} \otimes b_{(1)}) = \varepsilon(a)\varepsilon(b). \quad (82)$$

Recall that $\mathcal{A}^{\mathrm{op}}$ (resp. $\mathcal{A}^{\mathrm{cop}}$, $\mathcal{A}^{\mathrm{op,cop}}$) is the bialgebra built on the vector space $\mathcal{A}$ with opposite product (resp. coproduct, product and coproduct) of $\mathcal{A}$. Then equation (82) says that $\mathbf{s}$ is the convolution inverse of $\mathbf{r}$ when $\mathbf{r}$ and $\mathbf{s}$ are considered as linear forms on $\mathcal{A}^{\mathrm{cop}} \otimes \mathcal{A}$.

The universal r-form $\mathbf{r}$ of a coquasitriangular Hopf algebra $\mathcal{A}$ is always regular. Indeed, one easily checks that the linear form $\mathbf{s}(a \otimes b) := \mathbf{r}(a \otimes S(b))$ satisfies (82). In the bialgebra considerations below the form $\mathbf{s}$ is used, roughly speaking, as a substitute for the possibly missing antipode.

Example 17 (*The FRT bialgebra* $\mathcal{A}(R)$). Let R be an invertible solution of the QYBE. Then the canonical universal r-form (see Theorem 7) of the FRT bialgebra $\mathcal{A}(R)$ is regular if and only if R is regular, that is, if R^{t_2} is invertible. (The sufficiency was stated in Proposition 8. The necessity follows from (82) applied to $a = u^i_j, b = u^k_l$ and combined with (14).) The R-matrices for the quantum matrix groups $SL_q(N), O_q(N)$ and $Sp_q(N)$ are all regular, because they lead to coquasitriangular Hopf algebras by Theorem 9. The flip $R = \tau$ is a solution of the QYBE which is not regular. △

Throughout this subsection we assume that $\mathcal{A}$ is a coquasitriangular bialgebra and $\mathbf{r}$ is a regular universal r-form of $\mathcal{A}$.

Since $\mathbf{r}$ is a dual pairing of the bialgebras $\mathcal{A}$ and $\mathcal{A}^{\mathrm{op}}$ (see Subsect. 10.1.1), it follows easily that $\mathbf{s}$ is a dual pairing of $\mathcal{A}^{\mathrm{op}}$ and $\mathcal{A}$ and of $\mathcal{A}$ and $\mathcal{A}^{\mathrm{cop}}$. Hence $\mathbf{s}$ is a skew-pairing of the bialgebras $\mathcal{A}$ and $\mathcal{A}^{\mathrm{op,cop}}$, so the quantum double $\mathcal{D}(\mathcal{A}, \mathcal{A}^{\mathrm{op,cop}}; \mathbf{s})$ is well-defined by Proposition 8.8. By (72), it coincides with the double crossed product bialgebra $\mathcal{A}^{\mathrm{op,cop}} \bowtie_{\mathbf{s}} \mathcal{A}$ from 10.2.5. In order to avoid notational confusion, let us denote the identity map from $\mathcal{A}$ to $\mathcal{A}^{\mathrm{op,cop}}$ by i and assume that the Sweedler notation always refers to the comultiplication of $\mathcal{A}$. That is, $\mathrm{i}(ab) = \mathrm{i}(b)\mathrm{i}(a)$ and $\Delta_{\mathcal{A}^{\mathrm{op,cop}}}(\mathrm{i}(b)) = \sum \mathrm{i}(b_{(2)}) \otimes \mathrm{i}(b_{(1)})$ for $a, b \in \mathcal{A}$. Then, by Proposition 8.8, the bialgebra

$$D(\mathcal{A}) := \mathcal{D}(\mathcal{A}, \mathcal{A}^{\mathrm{op,cop}}; \mathbf{s}) = \mathcal{A}^{\mathrm{op,cop}} \bowtie_{\mathbf{s}} \mathcal{A} \quad (83)$$

is built on the vector space $\mathcal{A}^{\mathrm{op,cop}} \otimes \mathcal{A}$ with product

$$(\mathrm{i}(b) \otimes a)(\mathrm{i}(b') \otimes a') = \sum \mathrm{i}(b'_{(2)}b) \otimes a_{(2)}a' \; \mathbf{r}(a_{(1)}, b'_{(3)})\mathbf{s}(a_{(3)}, b'_{(1)}) \quad (84)$$

and coproduct

$$\Delta_{D(\mathcal{A})}(\mathrm{i}(b) \otimes a) = \sum \mathrm{i}(b_{(2)}) \otimes a_{(1)} \otimes \mathrm{i}(b_{(1)}) \otimes a_{(2)}.$$

Proposition 30. *Let* $\mathcal{A}$ *be a bialgebra equipped with a regular universal r-form* $\mathbf{r}$. *The vector space* $\mathcal{A}$ *equipped with the new product*

$$a\underline{\cdot}b := \sum a_{(2)}b_{(3)}\mathbf{r}(a_{(1)},b_{(2)})\mathbf{s}(a_{(3)},b_{(1)}) = \sum b_{(2)}a_{(1)}\mathbf{r}(a_{(2)},b_{(3)})\mathbf{s}(a_{(3)},b_{(1)}) \tag{85}$$

is an algebra, denoted $B(\mathcal{A})$. The algebra $B(\mathcal{A})$ is a right comodule algebra for the bialgebra $D(\mathcal{A})$ with respect to the coaction $\varphi: B(\mathcal{A}) \to B(\mathcal{A})\otimes D(\mathcal{A})$ given by $\varphi(a) := \sum a_{(2)} \otimes \mathrm{i}(a_{(1)}) \otimes a_{(3)}$, $a \in \mathcal{A}$. The product of $\mathcal{A}$ can be recovered from that of $B(\mathcal{A})$ by the formula

$$ab = \sum a_{(2)}\underline{\cdot}b_{(3)}\ \bar{\mathbf{r}}(a_{(1)},b_{(1)})\mathbf{r}(a_{(3)},b_{(2)}), \quad a,b \in \mathcal{A}. \tag{86}$$

Proof. We begin with the associativity of the new product. Using first the definition (85) of $\underline{\cdot}$, then formula (5) and finally the facts that $\mathbf{r}(\cdot,\cdot)$ and $\mathbf{s}(\cdot,\cdot)$ are dual pairings of $\mathcal{A}$ and $\mathcal{A}^{\mathrm{op}}$ resp. $\mathcal{A}^{\mathrm{op}}$ and $\mathcal{A}$, we compute

$$\begin{aligned}
(a\underline{\cdot}b)\underline{\cdot}c &= \sum a_{(3)}b_{(4)}c_{(3)}\mathbf{r}(a_{(2)}b_{(3)},c_{(2)})\mathbf{s}(a_{(4)}b_{(5)},c_{(1)})\mathbf{r}(a_{(1)},b_{(2)})\mathbf{s}(a_{(5)},b_{(1)})\\
&= \sum a_{(3)}b_{(4)}c_{(3)}\mathbf{r}(b_{(2)}a_{(1)},c_{(2)})\mathbf{s}(a_{(4)}b_{(5)},c_{(1)})\mathbf{r}(a_{(2)},b_{(3)})\mathbf{s}(a_{(5)},b_{(1)})\\
&= \sum a_{(3)}b_{(4)}c_{(5)}\mathbf{r}(b_{(2)},c_{(3)})\mathbf{r}(a_{(1)},c_{(4)})\mathbf{s}(a_{(4)},c_{(2)})\mathbf{s}(b_{(5)},c_{(1)})\\
&\qquad\qquad \times\mathbf{r}(a_{(2)},b_{(3)})\mathbf{s}(a_{(5)},b_{(1)}),\\
a\underline{\cdot}(b\underline{\cdot}c) &= \sum a_{(2)}b_{(4)}c_{(5)}\mathbf{r}(a_{(1)},b_{(3)}c_{(4)})\mathbf{s}(a_{(3)},b_{(2)}c_{(3)})\mathbf{r}(b_{(1)},c_{(2)})\mathbf{s}(b_{(5)},c_{(1)})\\
&= \sum a_{(2)}b_{(4)}c_{(5)}\mathbf{r}(a_{(1)},b_{(3)}c_{(4)})\mathbf{s}(a_{(3)},c_{(2)}b_{(1)})\mathbf{r}(b_{(2)},c_{(3)})\mathbf{s}(b_{(5)},c_{(1)})\\
&= \sum a_{(3)}b_{(4)}c_{(5)}\mathbf{r}(a_{(1)},c_{(4)})\mathbf{r}(a_{(2)},b_{(3)})\mathbf{s}(a_{(4)},c_{(2)})\mathbf{s}(a_{(5)},b_{(1)})\\
&\qquad\qquad \times\mathbf{r}(b_{(2)},c_{(3)})\mathbf{s}(b_{(5)},c_{(1)}).
\end{aligned}$$

Proposition 2(ii) implies that $1\underline{\cdot}a = a\underline{\cdot}1 = a$, $a \in \mathcal{A}$. Thus, we have shown that $B(\mathcal{A})=(\mathcal{A},\underline{\cdot})$ is indeed an associative unital algebra.

It is clear that φ is a coaction of $D(\mathcal{A})$ on $B(\mathcal{A})$. Using first the definitions (85) and (84) of the products and then the identity (82) we get

$$\begin{aligned}
\varphi(a)\underline{\cdot}\varphi(b) &= \sum a_{(2)}\underline{\cdot}b_{(2)} \otimes (\mathrm{i}(a_{(1)}) \otimes a_{(3)})(\mathrm{i}(b_{(1)}) \otimes b_{(3)})\\
&= \sum a_{(3)}b_{(6)}\ \mathbf{r}(a_{(2)},b_{(5)})\mathbf{s}(a_{(4)},b_{(4)}) \otimes \mathrm{i}(b_{(2)}a_{(1)}) \otimes a_{(6)}b_{(7)}\\
&\qquad\qquad \times\mathbf{r}(a_{(5)},b_{(3)})\mathbf{s}(a_{(7)},b_{(1)})\\
&= \sum a_{(3)}b_{(4)} \otimes \mathrm{i}(b_{(2)}a_{(1)}) \otimes a_{(4)}b_{(5)}\ \mathbf{r}(a_{(2)},b_{(3)})\mathbf{s}(a_{(5)},b_{(1)}),
\end{aligned}$$

where $\underline{\cdot} = \underline{\cdot}\otimes\cdot$. On the other hand, applying (85) and (5) we obtain

$$\begin{aligned}
\varphi(a\underline{\cdot}b) &= \sum a_{(3)}b_{(4)} \otimes \mathrm{i}(a_{(2)}b_{(3)}) \otimes a_{(4)}b_{(5)}\ \mathbf{r}(a_{(1)},b_{(2)})\mathbf{s}(a_{(5)},b_{(1)})\\
&= \sum a_{(3)}b_{(4)} \otimes \mathrm{i}(b_{(2)}a_{(1)}) \otimes a_{(4)}b_{(5)}\ \mathbf{r}(a_{(2)},b_{(3)})\mathbf{s}(a_{(5)},b_{(1)}).
\end{aligned}$$

Moreover, $\varphi(1) = 1\otimes 1\otimes 1$. This proves that φ is an algebra homomorphism of $B(\mathcal{A})$ to $B(\mathcal{A})\otimes D(\mathcal{A})$.

Inserting (85) into the right hand side of (86) and using (82) one obtains the equality (86). The second equality in (85) follows from (5). □

The product $\underline{\cdot}$ is called the *covariantized product* and the algebra $B(\mathcal{A})$ the *covariantized algebra* of the coquasitriangular bialgebra $\mathcal{A}$, because they turn the right $D(\mathcal{A})$-comodule $\mathcal{A}$ with coaction φ into a right $D(\mathcal{A})$-comodule *algebra.* Note that the algebra $\mathcal{A}$ itself does not have this property in general.

Let us describe the covariantized algebra for the FRT bialgebra $\mathcal{A}(R)$.

Example 18 (*The FRT bialgebra $\mathcal{A}(R)$ – continued*). Let R be a regular invertible solution of the QYBE. By Proposition 8, the canonical universal r-form $\mathbf{r}$ of $\mathcal{A}(R)$ is regular. Recall that $\mathbf{r}(u^i_j, u^k_l) = R^{ik}_{jl}$ and $\mathbf{s}(u^i_j, u^k_l) = \tilde{R}^{ik}_{jl}$ by (11) and (15), where $\tilde{R}^{t_2} = (R^{t_2})^{-1}$. Hence formula (85) for the new product $\underline{\cdot}$ yields $u^i_j \underline{\cdot} u^k_l = u^n_m u^s_l \tilde{R}^{mk}_{jr} R^{ir}_{ns}$ and so $u^i_j \underline{\cdot} u^k_l R^{jr}_{mk} = u^n_m u^s_l R^{ir}_{ns}$. In matrix notation the latter reads as

$$\mathbf{u}_1 \underline{\cdot} R_{12} \mathbf{u}_2 = R_{12} \mathbf{u}_1 \mathbf{u}_2. \tag{87}$$

Each of the three last formulas expresses the products of $B(\mathcal{A}(R))$ and of $\mathcal{A}(R)$ by each other. From (87) we obtain

$$\mathbf{u}_2 \underline{\cdot} R_{21} \mathbf{u}_1 R_{12} = R_{21} \mathbf{u}_2 \mathbf{u}_1 R_{12} \quad \text{and} \quad R_{21} \mathbf{u}_1 \underline{\cdot} R_{12} \mathbf{u}_2 = R_{21} R_{12} \mathbf{u}_1 \mathbf{u}_2. \tag{88}$$

Therefore, since $R_{12}\mathbf{u}_1\mathbf{u}_2 = \mathbf{u}_2\mathbf{u}_1 R_{12}$ by (9.2), the matrix $\mathbf{u}$ satisfies the equation

$$\mathbf{u}_2 \underline{\cdot} R_{21} \mathbf{u}_1 R_{12} = R_{21} \mathbf{u}_1 \underline{\cdot} R_{12} \mathbf{u}_2, \tag{89}$$

which is called the *reflection equation.* Recall that we have met this equation already as formula (44); see also (43). An equivalent form of (89) is

$$\mathbf{u}_2 \underline{\cdot} \hat{R}_{12} \mathbf{u}_2 \hat{R}_{12} = \hat{R}_{12} \mathbf{u}_2 \underline{\cdot} \hat{R}_{12} \mathbf{u}_2.$$

Since R is invertible, it follows from (88) that the reflection equation for the new product is equivalent to the equation $R_{12}\mathbf{u}_1\mathbf{u}_2 = \mathbf{u}_2\mathbf{u}_1 R_{12}$ for the old one. Thus, the covariantized algebra $B(\mathcal{A}(R))$ for the FRT bialgebra $\mathcal{A}(R)$ has generators u^i_j, $i, j = 1, \cdots, N$, with defining relations (89). This algebra is denoted $\mathcal{B}(R)$ and called the *braided matrix algebra associated with R.* △

Example 19. Let R be the matrix (3.40). Using the notation $a = u^1_1$, $b = u^1_2$, $c = u^2_1$, $d = u^2_2$ for the matrix entries of $\mathbf{u}$, the defining equation (89) of the braided matrix algebra $\mathcal{B}(R)$ is equivalent to the following six relations:

$$a \underline{\cdot} b - b \underline{\cdot} a = (q^{-2}-1) b \underline{\cdot} d,\ a \underline{\cdot} c - c \underline{\cdot} a = (1-q^{-2}) d \underline{\cdot} c,\ b \underline{\cdot} c - c \underline{\cdot} b = (q^{-2}-1) d \underline{\cdot} (a-d),$$

$$a \underline{\cdot} d = d \underline{\cdot} a, \quad b \underline{\cdot} d = q^{-2} d \underline{\cdot} b, \quad c \underline{\cdot} d = q^2 d \underline{\cdot} c. \qquad \triangle$$

The comultiplication $\Delta_{\mathcal{A}}$ is not necessarily an algebra homomorphism of $B(\mathcal{A})$ to $B(\mathcal{A}) \otimes B(\mathcal{A})$ if the latter carries the "ordinary" tensor product multiplication. However, there is another product $\bullet$ on the vector space

$B(\mathcal{A}) \otimes B(\mathcal{A})$ such that $\Delta_{\mathcal{A}}$ is a homomorphism of $B(\mathcal{A})$ to $(B(\mathcal{A}) \otimes B(\mathcal{A}), \bullet)$. This product is derived from the following more general lemma.

Lemma 31. *Let $\sigma : \mathcal{A} \times \mathcal{B} \to \mathbb{C}$ be a skew-pairing of two bialgebras $\mathcal{A}$ and $\mathcal{B}$ and let $\mathcal{X}$ and $\mathcal{Y}$ be right comodule algebras for $\mathcal{A}$ and $\mathcal{B}$, respectively. Then the vector space $\mathcal{X} \otimes \mathcal{Y}$ is an algebra with product defined by*

$$(x \otimes y) \bullet (x' \otimes y') := \sum x x'_{(0)} \otimes y_{(0)} y' \; \sigma(y_{(1)}, x'_{(1)}), \quad x, x' \in \mathcal{X}, \; y, y' \in \mathcal{Y}. \tag{90}$$

If $\mathcal{A} = \mathcal{B}$ and σ is a universal r-form of $\mathcal{A}$, then this algebra is a right $\mathcal{A}$-comodule algebra.

Proof. Using the assumption that $\mathcal{X}$ and $\mathcal{Y}$ are comodule algebras and the properties of the skew-pairing σ we get

$$\begin{aligned} ((x \otimes y) \bullet (x' \otimes y')) \bullet (x'' \otimes y'') &= \sum (x x'_{(0)} \otimes y_{(0)} y') \bullet (x'' \otimes y'') \; \sigma(y_{(1)}, x'_{(1)}) \\ &= \sum x x'_{(0)} x''_{(0)} \otimes y_{(0)} y'_{(0)} y'' \; \sigma(y_{(1)} y', x'') \; \sigma(y_{(2)}, x'_{(1)}) \\ &= \sum x x'_{(0)} x''_{(0)} \otimes y_{(0)} y'_{(0)} y'' \; \sigma(y_{(1)}, x''_{(1)}) \; \sigma(y', x''_{(2)}) \; \sigma(y_{(2)}, x'_{(1)}). \end{aligned}$$

The same expression is obtained for $(x \otimes y) \bullet ((x' \otimes y') \bullet (x'' \otimes y''))$. Hence, the product $\bullet$ is associative. Clearly, $(x \otimes y) \bullet (1 \otimes 1) = (1 \otimes 1) \bullet (x \otimes y) = x \otimes y$.

We prove the last assertion of the lemma. By property (5) of the universal r-form σ, the expressions

$$\begin{aligned} &\varphi_{\mathcal{X}} \otimes \varphi_{\mathcal{Y}}((x \otimes y) \bullet (x' \otimes y')) \\ &\qquad = \sum x_{(0)} x'_{(0)} \otimes y_{(0)} y'_{(0)} \otimes x_{(1)} x'_{(1)} y_{(1)} y'_{(1)} \; \sigma(y_{(2)}, x'_{(2)}), \\ &\varphi_{\mathcal{X}} \otimes \varphi_{\mathcal{Y}}(x \otimes y) \varphi_{\mathcal{X}} \otimes \varphi_{\mathcal{Y}}(x' \otimes y') \\ &\qquad = \sum x_{(0)} x'_{(0)} \otimes y_{(0)} y'_{(0)} \otimes x_{(1)} y_{(2)} x'_{(2)} y'_{(1)} \; \sigma(y_{(1)}, x'_{(1)}) \end{aligned}$$

coincide, so $\varphi_{\mathcal{X}} \otimes \varphi_{\mathcal{Y}}$ is indeed an algebra homomorphism. □

Proposition 32. *Let $\mathcal{A}$ and $B(\mathcal{A})$ be as in Proposition 30. The equation*

$$\sigma(\mathfrak{i}(b) \otimes a, \mathfrak{i}(b') \otimes a') := \sum \bar{\mathbf{r}}(b_{(2)}, a'_{(1)}) \mathbf{r}(b_{(1)}, b'_{(2)}) \mathbf{r}(a_{(1)}, a'_{(2)}) \mathbf{s}(a_{(2)}, b'_{(1)}) \tag{91}$$

defines a skew-pairing of the bialgebra $D(\mathcal{A})$ with itself. With the product

$$(a \otimes b) \bullet (c \otimes d) = \sum a \underline{\cdot} c_{(2)} \otimes b_{(2)} \underline{\cdot} d \, \sigma(\mathfrak{i}(b_{(1)}) \otimes b_{(3)}, \mathfrak{i}(c_{(1)}) \otimes c_{(3)}) \tag{92}$$

the vector space $B(\mathcal{A}) \otimes B(\mathcal{A})$ becomes an algebra, denoted $B(\mathcal{A}) \underline{\otimes} B(\mathcal{A})$, such that $\Delta_{\mathcal{A}}$ is an algebra homomorphism of $B(\mathcal{A})$ to $B(\mathcal{A}) \underline{\otimes} B(\mathcal{A})$.

Proof. We omit the lengthy verification that σ is a skew-paring. By Proposition 30, $B(\mathcal{A})$ is a right $D(\mathcal{A})$-comodule algebra with respect to the coaction $\varphi(a) = \sum a_{(2)} \otimes \mathfrak{i}(a_{(1)}) \otimes a_{(3)}$, $a \in \mathcal{A}$. The product (90) specializes to (92) in the case when $\mathcal{A}$ and $\mathcal{B}$ are the bialgebra $D(\mathcal{A})$ and $\mathcal{X}$ and $\mathcal{Y}$ are the comodule algebra $B(\mathcal{A})$. Therefore, by Lemma 31, $(B(\mathcal{A})\otimes B(\mathcal{A}), \bullet)$ is an algebra.

It remains to prove that $\Delta_{\mathcal{A}}$ is an algebra homomorphism of $B(\mathcal{A})$ to $B(\mathcal{A})\underline{\otimes} B(\mathcal{A})$. Applying first the definitions (92) (combined with (91)) and (85) of the products of $B(\mathcal{A})\underline{\otimes} B(\mathcal{A})$ resp. $B(\mathcal{A})$ and finally (82), we get

$$\Delta_{\mathcal{A}}(a) \bullet \Delta_{\mathcal{A}}(b) = \sum (a_{(1)} \otimes a_{(2)}) \bullet (b_{(1)} \otimes b_{(2)})$$

$$= \sum a_{(2)}b_{(5)} \otimes a_{(7)}b_{(10)}\ \mathbf{r}(a_{(1)}, b_{(4)})\mathbf{s}(a_{(3)}, b_{(3)})\mathbf{r}(a_{(6)}, b_{(9)})\mathbf{s}(a_{(8)}, b_{(8)})$$
$$\times \bar{\mathbf{r}}(a_{(5)}, b_{(6)})\mathbf{r}(a_{(9)}, b_{(7)})\mathbf{r}(a_{(4)}, b_{(2)})\mathbf{s}(a_{(10)}, b_{(1)})$$

$$= \sum a_{(2)}b_{(3)} \otimes a_{(3)}b_{(4)}\ \mathbf{r}(a_{(1)}, b_{(2)})\mathbf{s}(a_{(4)}, b_{(1)}) = \Delta_{\mathcal{A}}(a \underline{\cdot} b). \qquad \square$$

Note that the skew-pairing σ defined by (91) is not a universal r-form of the bialgebra $D(\mathcal{A})$.

Example 20 (The FRT bialgebra $\mathcal{A}(R)$ – continued). Recall that the covariantized algebra $B(\mathcal{A}(R))$ of $\mathcal{A}(R)$ is the algebra $\mathcal{B}(R)$ with product $\underline{\cdot}$. Then the product $\bullet$ on the vector space $\mathcal{B}(R) \otimes \mathcal{B}(R)$ is given by

$$(a \otimes u^i_j) \bullet (u^k_l \otimes b) = a \underline{\cdot} u^n_m \otimes u^r_s \underline{\cdot} b\ (R^{-1})^{pm}_{rx} R^{iy}_{pn} R^{sx}_{zl} \tilde{R}^{zk}_{jy}, \quad a, b \in \mathcal{B}(R).$$

In matrix notation this may be rewritten as

$$(a \otimes R^{-1}_{12}\mathbf{u}_1) \bullet (R_{12}\mathbf{u}_2 \otimes b) = a \underline{\cdot} \mathbf{u}_2 R^{-1}_{12} \otimes \mathbf{u}_1 \underline{\cdot} b R_{12}. \qquad \triangle$$

10.3.2 Braided Hopf Algebras Associated with Coquasitriangular Hopf Algebras

We now assume that $\mathcal{A}$ is a coquasitriangular Hopf algebra and show that then the above results take a much simpler form. Some necessary technical facts for that are collected in the next proposition. The first three assertions give the motivation for the use of the forms $\mathbf{s}$ and σ and of the bialgebra $D(\mathcal{A})$ in the preceding subsection.

Proposition 33. *For any coquasitriangular Hopf algebra $\mathcal{A}$ with universal r-form $\mathbf{r}$, we have:*

(i) *$\mathbf{r}$ is regular and $\mathbf{s}(a, b) = \mathbf{r}(a, S(b))$ for $a, b \in \mathcal{A}$.*

(ii) *$\sigma(\imath(b)\otimes a, \imath(b')\otimes a') = \mathbf{r}(S(b)a,\ S(b')a')$ for $a, b, a', b' \in \mathcal{A}$.*

(iii) *There is a Hopf algebra homomorphism $\theta : D(\mathcal{A}) \to \mathcal{A}$ such that $\theta(\imath(b) \otimes a) = S(b)a$.*

(iv) *Define $\underline{S}(a) := \sum S(a_{(2)})\mathbf{r}(S^2(a_{(3)})S(a_{(1)}), a_{(4)})$ for $a \in \mathcal{A}$. Then the linear mapping $\underline{S} : \mathcal{A} \to \mathcal{A}$ intertwines the right adjoint coaction Ad_R of $\mathcal{A}$ (that is, $\underline{S} \in \mathrm{Mor}\,(\mathrm{Ad}_R)$) and satisfies the condition*

$$\sum a_{(1)} \underline{\cdot}\ \underline{S}(a_{(2)}) = \sum \underline{S}(a_{(1)}) \underline{\cdot} a_{(2)} = \varepsilon(a)1, \quad a \in \mathcal{A}. \tag{93}$$

Proof. (i) and (ii) are verified in a straightforward manner using the properties of $\mathbf{r}$. In the case of (ii) we apply (i) and the identity $\mathbf{r}(S(a),b)=\bar{\mathbf{r}}(a,b)$.

(iii): Using the formulas (3) and (5) and the assertion (i), we obtain

$$\begin{aligned}\theta((\mathrm{i}(b)\otimes a)(\mathrm{i}(b')\otimes a')) &= \sum \theta(\mathrm{i}(b'_{(2)}b)\otimes a_{(2)}a')\ \mathbf{r}(a_{(1)},b'_{(3)})\mathbf{s}(a_{(3)},b'_{(1)})\\ &= \sum S(b)(S(b'_{(2)})a_{(2)})a'\ \mathbf{r}(a_{(1)},b'_{(3)})\mathbf{r}(a_{(3)},S(b'_{(1)}))\\ &= \sum S(b)a_{(3)}S(b'_{(1)})a'\ \mathbf{r}(a_{(1)},b'_{(3)})\mathbf{r}(a_{(2)},S(b'_{(2)}))\\ &= \sum S(b)a_{(2)}S(b'_{(1)})a'\ \mathbf{r}(a_{(1)},S(b'_{(2)})b'_{(3)})\\ &= S(b)aS(b')a'=\theta(\mathrm{i}(b)\otimes a)\ \theta(\mathrm{i}(b')\otimes a').\end{aligned}$$

It is easy to check that $(\theta\otimes\theta)\circ\Delta_{D(\mathcal{A})}=\Delta_{\mathcal{A}}\circ\theta$. Thus, θ is a Hopf algebra homomorphism.

(iv): Using the functionals $\bar{f}$ and f from Proposition 3 we compute

$$\begin{aligned}&\sum a_{(1)}\underline{\cdot}\ \underline{S}(a_{(2)})\\ &= \sum a_{(2)}S(a_{(5)})\ \mathbf{r}(a_{(1)},S(a_{(6)}))\mathbf{r}(a_{(3)},S^2(a_{(7)}))\mathbf{r}(S^2(a_{(8)}),a_{(9)})\\ &\qquad\qquad\qquad\qquad\qquad\qquad\times\mathbf{r}(S(a_{(4)}),a_{(10)})\\ &= \sum a_{(2)}S(a_{(5)})\ \mathbf{r}(a_{(1)},S(a_{(6)}))\mathbf{r}(a_{(3)},S^2(a_{(7)})\bar{f}(a_{(8)}))\mathbf{r}(S(a_{(4)}),a_{(9)})\\ &= \sum a_{(2)}S(a_{(5)})\ \mathbf{r}(a_{(1)},S(a_{(6)}))\mathbf{r}(a_{(3)},S^2(a_{(7)}))\\ &\qquad\qquad\times\mathbf{r}(S(a_{(4)}),S^2(a_{(8)}))\bar{f}(a_{(9)})\\ &= \sum a_{(2)}S(a_{(4)})\ \mathbf{r}(a_{(1)},S(a_{(5)}))\mathbf{r}(a_{(3)},S(a_{(7)})S^2(a_{(6)}))\bar{f}(a_{(8)})\\ &= \sum a_{(2)}S(a_{(3)})\ \mathbf{r}(a_{(1)},S(a_{(4)}))\bar{f}(a_{(5)})\\ &= \sum \mathbf{r}(a_{(1)},S(a_{(2)}))\bar{f}(a_{(3)})1=\sum f(a_{(1)})\bar{f}(a_{(2)})1=\varepsilon(a)1.\end{aligned}$$

Here the first equality follows from the definitions of $\underline{\cdot}$ and $\underline{S}$ and assertion (i). The second and third equalities are obtained by inserting $\bar{f}$ and applying formula (8), respectively.

Similar (slightly simpler) reasoning yields the second equality of (93). In order to prove that $\underline{S}\in\mathrm{Mor}\,(\mathrm{Ad}_R)$, we compute

$$\begin{aligned}\mathrm{Ad}_R(\underline{S}(a)) &= \sum S(a_{(3)})\otimes S^2(a_{(4)})S(a_{(2)})\ \mathbf{r}(S^2(a_{(5)})S(a_{(1)}),a_{(6)})\\ &= \sum S(a_{(3)})\otimes S^2(a_{(4)})S(a_{(2)})\ \bar{f}(a_{(5)})\mathbf{r}(S(a_{(1)}),a_{(6)})\\ &= \sum S(a_{(3)})\otimes a_{(5)}S(a_{(2)})\ \bar{f}(a_{(4)})\mathbf{r}(S(a_{(1)}),a_{(6)})\\ &= \sum S(a_{(3)})\otimes S(a_{(1)})a_{(6)}\ \bar{f}(a_{(4)})\mathbf{r}(S(a_{(2)}),a_{(5)})\\ &= \sum S(a_{(3)})\otimes S(a_{(1)})a_{(6)}\mathbf{r}(S^2(a_{(4)})S(a_{(2)}),a_{(5)})\\ &= \sum \underline{S}(a_{(2)})\otimes S(a_{(1)})a_{(3)}=(\underline{S}\otimes\mathrm{id})\mathrm{Ad}_R(a).\end{aligned}$$

Here we used the formulas (8), (4) and (5). □

The next two propositions restate some key results and formulas from the preceding subsection in the case of a coquasitriangular Hopf algebra $\mathcal{A}$ with universal r-form $\mathbf{r}$.

Proposition 34. (i) *The vector space $\mathcal{A}$ equipped with the product*

$$a\underline{\cdot}b = \sum a_{(2)}b_{(2)}\mathbf{r}(S(a_{(1)})a_{(3)}, S(b_{(1)})) = \sum b_{(2)}a_{(1)}\mathbf{r}(a_{(2)}, S(b_{(1)})b_{(3)})$$

is a right $\mathcal{A}$-comodule algebra, denoted by $B(\mathcal{A})$, with respect to the right adjoint coaction Ad_R.

(ii) *Recall that $\tilde{l}(a)$ is the linear functional on $\mathcal{A}$ defined by (21). Then the map $a \to \tilde{l}(a)$ is a homomorphism of the algebra $B(\mathcal{A})$ to the subalgebra $\tilde{l}(\mathcal{A})$ of the dual Hopf algebra $\mathcal{A}^\circ$.*

Proof. (i): From Propositions 32(i) and 2(v) and formula (3) we get

$$\mathbf{r}(S(a)a', S(b)) = \sum \mathbf{r}(S(a), S(b_{(2)}))\mathbf{r}(a', S(b_{(1)})) = \sum \mathbf{r}(a, b_{(2)})\mathbf{s}(a', b_{(1)})$$

and $\mathbf{r}(a, S(b)b') = \sum \mathbf{r}(a_{(1)}, b')\mathbf{s}(a_{(2)}, b)$, so the product (85) takes the above form. Since obviously $\mathrm{Ad}_R = (\mathrm{id}\otimes\theta)\circ\varphi$, $B(\mathcal{A})$ is a right $\mathcal{A}$-comodule algebra with respect to the coaction Ad_R by Propositions 29 and 32(iii).

(ii): Since $\mathbf{s}(a,b) = \mathbf{r}(a, S(b))$ by Proposition 33(i), a comparison of the formulas (30) and (85) shows that $\tilde{l}(a\underline{\cdot}b) = \tilde{l}(a)\tilde{l}(b)$ for $a, b \in \mathcal{A}$. □

Let us say that a coquasitriangular Hopf algebra $\mathcal{A}$ (or more precisely a pair $(\mathcal{A}, \mathbf{r})$ of a coquasitriangular Hopf algebra $\mathcal{A}$ and a universal r-form $\mathbf{r}$ on $\mathcal{A}$) is *factorizable* if the bilinear form $\mathbf{q}$ given by (9) is nondegenerate. Since $\mathbf{q}(S(a), S(b)) = \mathbf{q}(b, a)$ and the antipode of a coquasitriangular Hopf algebra is bijective, it is clear that $\mathcal{A}$ is factorizable if and only if $l(a) = 0$ implies that $a = 0$ or equivalently if $\tilde{l}(a) = 0$ is only possible for $a = 0$. Therefore, by Proposition 34(ii), for a factorizable Hopf algebra $\mathcal{A}$ the map $a \to \tilde{l}(a)$ is an algebra *isomorphism* of the covariantized algebra $B(\mathcal{A})$ to the subalgebra $\tilde{l}(\mathcal{A})$ of $\mathcal{A}^\circ$.

Note that cotriangular Hopf algebras are those coquasitriangular Hopf algebras for which the form $\mathbf{q}$ is trivial, that is, $\mathbf{q} = \varepsilon\otimes\varepsilon$. The factorizability of a coquasitriangular Hopf algebra is just the other extreme where the form $\mathbf{q}$ is far from being trivial.

The following proposition is only a reformulation of Lemma 31.

Proposition 35. *Let $\mathcal{X}$ and $\mathcal{Y}$ be right $\mathcal{A}$-comodule algebras. Then the vector space $\mathcal{X}\otimes\mathcal{Y}$ becomes a right $\mathcal{A}$-comodule algebra, denoted by $\mathcal{X}\underline{\otimes}\mathcal{Y}$, with respect to the tensor product coaction and the product*

$$(x\otimes y)\bullet(x'\otimes y') := (x\otimes 1)\mathbf{r}_{\mathcal{Y}\mathcal{X}}(y\otimes x')(1\otimes y'), \quad x, x' \in \mathcal{X},\ y, y' \in \mathcal{Y}, \tag{94}$$

where the product on the right hand side is the tensor multiplication and

$$\mathbf{r}_{\mathcal{Y}\mathcal{X}}(y \otimes x') := \sum x'_{(0)} \otimes y_{(0)} \; \mathbf{r}(y_{(1)}, x'_{(1)}). \tag{95}$$

The product (94) on $\mathcal{X} \otimes \mathcal{Y}$ is called the *braided product* and the mapping $\mathbf{r}_{\mathcal{Y}\mathcal{X}} : \mathcal{Y} \otimes \mathcal{X} \to \mathcal{X} \otimes \mathcal{Y}$ is called the *braiding* of the $\mathcal{A}$-comodule algebras $\mathcal{Y}$ and $\mathcal{X}$. Recall that $\mathbf{r}_{\mathcal{Y}\mathcal{X}}$ is just the mapping defined by formula (6). Some properties of the braiding maps can be found in Proposition 1. In particular, it was shown therein that each map $\mathbf{r}_{\mathcal{X}\mathcal{X}}$ satisfies the braid relation.

Applied to the case when $\mathcal{X} = \mathcal{Y}$ is the right $\mathcal{A}$-comodule algebra $B(\mathcal{A})$ with coaction Ad_R, the product (94) of the algebra $B(\mathcal{A})\underline{\otimes} B(\mathcal{A})$ writes as

$$(a \otimes b) \bullet (c \otimes d) = \sum a\underline{\cdot} c_{(2)} \otimes b_{(2)}\underline{\cdot} d \; \mathbf{r}(S(b_{(1)})b_{(3)}, S(c_{(1)})c_{(3)}). \tag{96}$$

By Proposition 33(ii), it coincides with the product (92) of the algebra $B(\mathcal{A})\underline{\otimes} B(\mathcal{A})$.

Now let us summarize the main facts established above. By Proposition 34, $B(\mathcal{A})$ is a right $\mathcal{A}$-comodule algebra. Obviously, $B(\mathcal{A})$ is a right $\mathcal{A}$-comodule coalgebra with comultiplication $\Delta_{\mathcal{A}}$ and counit $\varepsilon_{\mathcal{A}}$ inherited from $\mathcal{A}$. By Proposition 32, $\Delta_{\mathcal{A}}$ is an algebra homomorphism of $B(\mathcal{A})$ to $B(\mathcal{A})\underline{\otimes} B(\mathcal{A})$. It is obvious that $\varepsilon_{\mathcal{A}} : B(\mathcal{A}) \to \mathbb{C}$ is an algebra homomorphism. Proposition 33(iv) says that the mapping $\underline{S} : B(\mathcal{A}) \to B(\mathcal{A})$ defined therein intertwines the right adjoint coaction Ad_R and satisfies the antipode axiom. Thus, the algebra $B(\mathcal{A})$ and the mappings $\Delta_{\mathcal{A}}, \varepsilon_{\mathcal{A}}$ and $\underline{S}$ fulfill all Hopf algebra axioms except for one modification: the algebra $B(\mathcal{A})\underline{\otimes} B(\mathcal{A})$ carries the braided product (94) rather than the usual tensor multiplication. We shall express all that by saying that $B(\mathcal{A})$ is a *braided Hopf algebra* associated with the coquasitriangular Hopf algebra $\mathcal{A}$. The precise definition of this concept is

Definition 11. *Let $\mathcal{A}$ be a coquasitriangular bialgebra equipped with a fixed universal r-form $\mathbf{r}$. A right $\mathcal{A}$-comodule algebra and right $\mathcal{A}$-comodule coalgebra $\mathcal{X}$ is called a* braided bialgebra associated with $\mathcal{A}$ *(briefly, a* braided bialgebra*) if $\varepsilon_{\mathcal{X}} : \mathcal{X} \to \mathbb{C}$ and $\Delta_{\mathcal{X}} : \mathcal{X} \to \mathcal{X}\underline{\otimes}\mathcal{X}$ are algebra homomorphisms, where $\mathcal{X}\underline{\otimes}\mathcal{X}$ carries the braided product (94). A braided bialgebra $\mathcal{X}$ is a* braided Hopf algebra associated with $\mathcal{A}$ *(briefly, a* braided Hopf algebra*) if $\mathcal{X}$ admits an antipode that intertwines the coaction of $\mathcal{A}$ on $\mathcal{X}$.*

In other words, a braided bialgebra $\mathcal{X}$ associated with $\mathcal{A}$ is a right $\mathcal{A}$-comodule $\mathcal{X}$ which admits a product and a coproduct such that:

(i) $\mathcal{X}$ satisfies the bialgebra axioms with the modification that the vector space $\mathcal{X}\underline{\otimes}\mathcal{X}$ is equipped with the *braided product* (94);

(ii) the product and the coproduct of $\mathcal{X}$ intertwine the corresponding coactions of $\mathcal{A}$, that is, $\mathcal{X}$ is a $\mathcal{A}$-comodule *algebra* and *coalgebra*.

It should be emphasized that the definitions of the braided product (94) and of a braided bialgebra depend on the choice of the universal r-form of $\mathcal{A}$.

Recall that the antipode of a Hopf algebra is an anti-homomorphism of the algebra and of the coalgebra. For a braided Hopf algebra $\mathcal{X}$ there is a similar result which says that

$$S_{\mathcal{X}} \circ m_{\mathcal{X}} = m_{\mathcal{X}} \circ \mathbf{r}_{\mathcal{X}\mathcal{X}} \circ (S_{\mathcal{X}} \otimes S_{\mathcal{X}}), \quad S_{\mathcal{X}}(1) = 1, \tag{97}$$

$$\Delta_{\mathcal{X}} \circ S_{\mathcal{X}} = (S_{\mathcal{X}} \otimes S_{\mathcal{X}}) \circ \mathbf{r}_{\mathcal{X}\mathcal{X}} \circ \Delta_{\mathcal{X}}, \quad \varepsilon_{\mathcal{X}} \circ S_{\mathcal{X}} = \varepsilon_{\mathcal{X}}. \tag{98}$$

Let us turn to examples of braided Hopf algebras. Each "ordinary" Hopf algebra $\mathcal{X}$ is, of course, a braided Hopf algebra associated with any coquasitriangular bialgebra with respect to the trivial coaction $\beta(x) = x \otimes 1$, $x \in \mathcal{X}$. A nontrivial simple but still instructive example is the following

Example 21 (*The braided line*). Let $\mathcal{A} = \mathbb{C}\mathbb{Z}$ be the group Hopf algebra (see Example 1.7) of the additive group $\mathbb{Z}$ of integers. That is, $\mathcal{A}$ is the Hopf algebra of all complex polynomials in a single group-like generator, say χ, and its inverse χ^{-1}. As noted in Example 1 above, for any nonzero complex number q there exists a universal r-form $\mathbf{r} = \mathbf{r}_q$ on $\mathcal{A}$ such that

$$\mathbf{r}_q(\chi^n, \chi^m) = q^{-mn}, \quad m, n \in \mathbb{Z}. \tag{99}$$

The algebra $\mathcal{X} = \mathbb{C}[z]$ of all polynomials in one variable is a right $\mathcal{A}$-comodule algebra with coaction β given by $\beta(z^n) = z^n \otimes \chi^n$, $n \in \mathbb{Z}$. By (95) and (99), the corresponding braiding map $\mathbf{r}_{\mathcal{X}\mathcal{X}}$ acts on the vector space $\mathcal{X} \otimes \mathcal{X}$ as

$$\mathbf{r}_{\mathcal{X}\mathcal{X}}(z^n \otimes z^m) = q^{-mn} z^m \otimes z^n, \quad m, n \in \mathbb{N}_0.$$

Therefore, the braided product (94) of the algebra $\mathcal{X} \otimes \mathcal{X}$ is given by

$$(z^i \otimes z^j) \bullet (z^n \otimes z^m) = q^{-jn} z^{i+n} \otimes z^{j+m}.$$

In particular, we see that the two generators $x := z \otimes 1$ and $y := 1 \otimes z$ of the algebra $\mathcal{X} \otimes \mathcal{X}$ satisfy the relation $yx = q^{-1}xy$. Hence the algebras $\mathcal{X} \otimes \mathcal{X}$ and $\mathcal{O}(\mathbb{C}_q^2)$ are isomorphic.

Now we define algebra homomorphisms $\Delta_{\mathcal{X}} : \mathcal{X} \to \mathcal{X} \otimes \mathcal{X}$ and $\varepsilon_{\mathcal{X}} : \mathcal{X} \to \mathbb{C}$ such that $\Delta_{\mathcal{X}}(z) = z \otimes 1 + 1 \otimes z$ and $\varepsilon_{\mathcal{X}}(z) = 0$. From Proposition 2.2 we obtain

$$\Delta_{\mathcal{X}}(z^n) = \sum\nolimits_{k=0}^{n} \begin{bmatrix} n \\ k \end{bmatrix}_{q^{-1}} z^{n-k} \otimes z^k, \quad n \in \mathbb{N}_0.$$

Then the vector space $\mathcal{X}$ becomes a right $\mathcal{A}$-comodule coalgebra (see Subsect. 10.2.2) with comultiplication $\Delta_{\mathcal{X}}$, counit $\varepsilon_{\mathcal{X}}$ and coaction β as given above. (Indeed, since all structure maps are algebra homomorphisms, it suffices to check the axioms on the generator z which is easily done.) Further, one verifies that the linear map $S_{\mathcal{X}} : \mathcal{X} \to \mathcal{X}$ defined by

$$S_{\mathcal{X}}(z^n) := (-1)^n q^{-(n-1)n/2} z^n, \quad n \in \mathbb{N}_0,$$

satisfies the antipode axiom. Obviously, we have $\beta \circ S_{\mathcal{X}} = (S_{\mathcal{X}} \otimes \mathrm{id}) \circ \beta$. Thus, we have shown that $\mathcal{X}$ is a braided Hopf algebra associated with the coquasitriangular Hopf algebra $\mathcal{A}$. It is called the *braided line algebra.*

In this example the first two formulas of (97) and (98) read as

$$S_{\mathcal{X}}(z^{m+n}) = q^{-mn} S_{\mathcal{X}}(z^m) S_{\mathcal{X}}(z^n),$$

$$\Delta_{\mathcal{X}}(S_{\mathcal{X}}(z^n)) = \sum\nolimits_{k=0}^{n} \begin{bmatrix} n \\ k \end{bmatrix}_{q^{-1}} q^{-k(n-k)} S_{\mathcal{X}}(z^{n-k}) \otimes S_{\mathcal{X}}(z^k). \qquad \triangle$$

Obviously, Example 21 remains valid if the Hopf algebra $\mathcal{A} = \mathbb{C}\mathbb{Z}$ is replaced by the subbialgebra $\mathbb{C}[\chi]$ of polynomials in χ. If we interpret this bialgebra $\mathbb{C}[\chi]$ as the FRT bialgebra $\mathcal{A}(R)$ (with R being the 1×1 matrix (q^{-1})), then the braided line algebra becomes a special case of an important class of examples which will be constructed in Subsect. 10.3.5. Nevertheless, the braided Hopf algebras $B(\mathcal{A})$ developed above are still a main source of interesting examples. In fact, the same arguments of proof as in the case of $B(\mathcal{A})$ yield the following more general result.

Proposition 36. *Let $p : \mathcal{H} \rightarrow \mathcal{A}$ be a Hopf algebra homomorphism from an arbitrary Hopf algebra $\mathcal{H}$ to a coquasitriangular Hopf algebra $\mathcal{A}$. Then the vector space $\mathcal{H}$ becomes a braided Hopf algebra associated with $\mathcal{A}$, denoted by $B(\mathcal{H}; \mathcal{A}, p)$, with structures given as follows. The product of the algebra $B(\mathcal{H}; \mathcal{A}, p)$ is defined by*

$$h \underline{\cdot} g = \sum h_{(2)} g_{(2)} \, \mathbf{r}(p(S(h_{(1)})h_{(3)}), p(S(g_{(1)}))), \quad h, g \in \mathcal{H}.$$

The coalgebra structure and the unit element of $B(\mathcal{H}; \mathcal{A}, p)$ coincide with those of $\mathcal{H}$ and the antipode $\underline{S}$ of $B(\mathcal{H}; \mathcal{A}, p)$ acts as

$$\underline{S}(h) = \sum S(h_{(2)}) \, \mathbf{r}(p(S^2(h_{(3)}) S(h_{(1)})), p(h_{(4)})).$$

Further, $B(\mathcal{H}; \mathcal{A}, p)$ is a right $\mathcal{A}$-comodule algebra with respect to the coaction

$$\beta(h) := \sum h_{(2)} \otimes p(S(h_{(1)}) h_{(3)}), \quad h \in \mathcal{H}.$$

Clearly, if $\mathcal{H} = \mathcal{A}$ and p is the identity map, then $B(\mathcal{H}; \mathcal{A}, p)$ is just the braided Hopf algebra $B(\mathcal{A})$ defined above. At the other extreme, if p is the trivial Hopf algebra homomorphism (that is, $p(h) = \varepsilon(h)1$ for $h \in \mathcal{H}$), then obviously $B(\mathcal{H}; \mathcal{A}, p) = \mathcal{H}$. In this case the coaction β of $\mathcal{A}$ on $\mathcal{H}$ is trivial, so $\mathcal{H}$ is a braided Hopf algebra with respect to the trivial coaction.

Following S. Majid, the passage from the Hopf algebra $\mathcal{A}$ resp. the Hopf algebra homomorphism $p : \mathcal{H} \rightarrow \mathcal{A}$ to the braided Hopf algebras $B(\mathcal{A})$ resp. $B(\mathcal{H}; \mathcal{A}, p)$ will be called *transmutation*.

10.3.3 Braided Hopf Algebras Associated with Quasitriangular Hopf Algebras

In this subsection let $\mathcal{B}$ be a quasitriangular Hopf algebra with universal R-matrix $\mathcal{R} = \sum_i x_i \otimes y_i$. We briefly state (without giving proofs) the dual versions to the main results of the preceding subsection.

Proposition 37. *Let $a \triangleright b := \mathrm{ad}_L(a)b = \sum a_{(1)} b S(a_{(2)})$, $a, b \in \mathcal{B}$, denote the left adjoint action of $\mathcal{B}$ on itself. With the new comultiplication $\underline{\Delta}$ defined by*

$$\underline{\Delta}(b) = \sum_i \sum b_{(1)} S(y_i) \otimes x_i \triangleright b_{(2)}, \qquad b \in \mathcal{B},$$

and the counit $\varepsilon_{\mathcal{B}}$, the vector space $\mathcal{B}$ becomes a coalgebra, denoted by $B(\mathcal{B})$, such that $B(\mathcal{B})$ is a left $\mathcal{B}$-module coalgebra. Moreover, $\underline{\Delta}(1) = 1 \otimes 1$.

Proposition 38. *If $\mathcal{X}$ and $\mathcal{Y}$ are left $\mathcal{B}$-module algebras, then the vector space $\mathcal{X} \otimes \mathcal{Y}$ is a left $\mathcal{B}$-module algebra, denoted by $\mathcal{X} \underline{\otimes} \mathcal{Y}$, with the tensor product action of $\mathcal{B}$ and the braided product $\bullet$ defined by*

$$(x \otimes y) \bullet (x' \otimes y') = \sum (x \otimes 1) \mathcal{R}_{\mathcal{Y},\mathcal{X}}(y \otimes x')(1 \otimes y'), \;\; x, x' \in \mathcal{X}, \; y, y' \in \mathcal{Y}, \quad (100)$$

where $\mathcal{R}_{\mathcal{Y},\mathcal{X}}(y \otimes x') := \sum_i y_i \triangleright x' \otimes x_i \triangleright y$.

The algebra $\mathcal{B}$ itself is also a left $\mathcal{B}$-module algebra with respect to the left adjoint action. Therefore, the braided product (100) makes the vector space $\mathcal{B} \otimes \mathcal{B}$ into an algebra denoted by $\mathcal{B} \underline{\otimes} \mathcal{B}$. It can be shown that $\underline{\Delta} : \mathcal{B} \to \mathcal{B} \underline{\otimes} \mathcal{B}$ is an algebra homomorphism. Further, the mapping $\underline{S} : \mathcal{B} \to \mathcal{B}$ defined by

$$\underline{S}(b) := \sum_i y_i S_{\mathcal{B}}(x_i \triangleright b), \qquad b \in \mathcal{B},$$

intertwines the left adjoint action (that is, $a \triangleright \underline{S}(b) = \underline{S}(a \triangleright b)$) and satisfies the equation $\sum b_{(\underline{1})} \underline{S}(b_{(\underline{2})}) = \sum \underline{S}(b_{(\underline{1})}) b_{(\underline{2})} = \varepsilon(b)1$, $b \in \mathcal{B}$. Here $\underline{\Delta}(b) = \sum b_{(\underline{1})} \otimes b_{(\underline{2})}$ is the Sweedler notation for the new comultiplication $\underline{\Delta}$. The left $\mathcal{B}$-module coalgebra $B(\mathcal{B})$ is also a left $\mathcal{B}$-module algebra with the algebra structure inherited from $\mathcal{B}$. We call $B(\mathcal{B})$ equipped with these structures a *braided Hopf algebra* associated with the quasitriangular Hopf algebra $\mathcal{B}$.

The general definition is easily guessed from Definition 11 above. A *braided bialgebra associated with* $\mathcal{B}$ is a left $\mathcal{B}$-module algebra and a left $\mathcal{B}$-module coalgebra $\mathcal{X}$ such that $\varepsilon_{\mathcal{X}} : \mathcal{X} \to \mathbb{C}$ and $\Delta_{\mathcal{X}} : \mathcal{X} \to \mathcal{X} \underline{\otimes} \mathcal{X}$ are algebra homomorphisms, where $\mathcal{X} \underline{\otimes} \mathcal{X}$ is equipped with the braided product (100). If $\mathcal{X}$ has an antipode which intertwines the action of $\mathcal{B}$ on $\mathcal{X}$, then $\mathcal{X}$ is called a *braided Hopf algebra*.

Finally, we want to make more precise the sense in which the preceding constructions of $B(\mathcal{A})$ and $B(\mathcal{B})$ are dual to each other. In order to do so, we consider in addition to the quasitriangular Hopf algebra $\mathcal{B}$ a coquasitriangular Hopf algebra $\mathcal{A}$ with universal r-form $\mathbf{r}$. We suppose that there is a dual pairing $\langle \cdot, \cdot \rangle$ of the Hopf algebras $\mathcal{A}$ and $\mathcal{B}$ such that $\mathbf{r}(a \otimes a') = \langle a \otimes a', \mathcal{R} \rangle$

for $a, a' \in \mathcal{A}$. The next proposition shows that there is a close link between the two new structures (the product $\underline{\cdot}$ of $B(\mathcal{A})$ and the coproduct $\underline{\Delta}$ of $B(\mathcal{B})$) and the two old ones (the product $\cdot$ of $\mathcal{B}$ and the coproduct $\Delta_{\mathcal{A}}$ of $\mathcal{A}$).

Proposition 39. *Retain the preceding assumptions on $\mathcal{A}$ and $\mathcal{B}$ and define a linear mapping $Q : \mathcal{A} \to \mathcal{B}$ by $Q(a) = \sum_i \langle a, a_i \rangle b_i$, $a \in \mathcal{A}$, where $\mathcal{R}_{21}\mathcal{R}_{12} = \sum_i a_i \otimes b_i \in \mathcal{B} \otimes \mathcal{B}$. Then, Q is an algebra homomorphism of $B(\mathcal{A})$ to $\mathcal{B}$ (that is, $Q(a \underline{\cdot} b) = Q(a) \cdot Q(b)$) and a coalgebra homomorphism of $\mathcal{A}$ to $B(\mathcal{B})$ (that is, $\underline{\Delta} Q(a) = (Q \otimes Q)\Delta_{\mathcal{A}}(a)$).*

Idea of proof. Some computations show that $Q(a \underline{\cdot} b)$ and $Q(a) \cdot Q(b)$ are both equal to $\langle a \otimes b, \mathcal{R}_{12}\mathcal{R}_{21}\mathcal{R}_{13}\mathcal{R}_{31} \rangle$ and that the expressions $\underline{\Delta} Q(a)$ and $(Q \otimes Q)\Delta_{\mathcal{A}}(a)$ coincide with $\langle a, \mathcal{R}_{13}\mathcal{R}_{31}\mathcal{R}_{23}\mathcal{R}_{32} \rangle$. See the proof of Proposition 7.4.3 in [Maj] for the details. □

10.3.4 Braided Tensor Categories and Braided Hopf Algebras

In this subsection we shall see how both braided products (94) and (100) and both braided Hopf algebras $B(\mathcal{A})$ and $B(\mathcal{B})$ can be considered from a unique point of view. The common notion behind these concepts is that of a braided tensor category. In the following treatment we shall concentrate on the essential points and suppress mathematical subtleties.

A *category* $\mathcal{C}$ is a collection of objects $X, Y, Z, \cdots$ and of sets $\mathrm{Mor}\,(X, Y)$ of morphisms between two objects X, Y such that a composition of morphisms is defined which has similar properties to the composition of maps. That is, given morphisms $f \in \mathrm{Mor}\,(X, Y)$ and $g \in \mathrm{Mor}\,(Y, Z)$ there always exists a morphism $g \circ f \in \mathrm{Mor}\,(X, Z)$ and the composition of three morphisms has to satisfy the associativity law. Moreover, each set $\mathrm{Mor}\,(X, X)$ has to contain a morphism id_X such that $f \circ \mathrm{id}_X = f$ and $\mathrm{id}_X \circ g = g$ for any $f \in \mathrm{Mor}\,(X, Y)$ and $g \in \mathrm{Mor}\,(Y, X)$. A mapping $f \in \mathrm{Mor}\,(X, Y)$ is called an isomorphism if there is another morphism $g \in \mathrm{Mor}\,(Y, X)$ such that $g \circ f = \mathrm{id}_X$ and $f \circ g = \mathrm{id}_Y$. An obvious example of a category is obtained by taking the vector spaces as objects and the set of all linear mappings $f : X \to Y$ as $\mathrm{Mor}\,(X, Y)$. For our purpose the categories $\mathcal{M}^{\mathcal{A}}$ and $\mathcal{M}_{\mathcal{A}}$ of a bialgebra $\mathcal{A}$ are most important. The objects of $\mathcal{M}^{\mathcal{A}}$ and $\mathcal{M}_{\mathcal{A}}$ are the right $\mathcal{A}$-comodules resp. the right $\mathcal{A}$-modules and the morphisms are the linear intertwiners of the corresponding coactions resp. actions.

In what follows we consider *tensor* (or *monoidal*) *categories*. These are categories $\mathcal{C}$ that have a product, denoted $\otimes$ and called the tensor product, which admits several "natural" properties such as associativity and existence of a unit object denoted $\mathbf{1}$. The associativity of the tensor product requires that there is an isomorphism $\phi_{X,Y,Z} : (X \otimes Y) \otimes Z \to X \otimes (Y \otimes Z)$ for any triple X, Y, Z of objects in $\mathcal{C}$. We will not list the axioms of a tensor category here (see, for instance, [Kas], Sect. XI.2). For our subsequent discussion it suffices to know that the categories $\mathcal{M}^{\mathcal{A}}$ and $\mathcal{M}_{\mathcal{A}}$ for any bialgebra $\mathcal{A}$ become such tensor categories if we take the tensor product of comodules resp. modules as the tensor product $\otimes$ and the trivial one-dimensional comodule resp. module as the unit object $\mathbf{1}$ in the category.

The categories we are interested in are tensor categories which are equipped with some kind of transposition of tensor products. A tensor category $(\mathcal{C}, \otimes)$ is called a *braided tensor* (or *monoidal*) *category* if for any pair X, Y of objects in $\mathcal{C}$ there is an isomorphism $\Psi_{X,Y} : X \otimes Y \to Y \otimes X$ such that $(g \otimes f) \circ \Psi_{X,Y} = \Psi_{X',Y'} \circ (f \otimes g)$ for arbitrary morphisms $f \in \operatorname{Mor}(X, X')$ and $g \in \operatorname{Mor}(Y, Y')$ and the hexagon axiom holds. For simplicity we will suppress writing the isomorphisms $\phi_{X,Y,Z}$ which govern the associativity of the tensor product. Then the *hexagon axiom* is the validity of the two conditions

$$\Psi_{X,Z} \circ \Psi_{Y,Z} = \Psi_{X \otimes Y, Z}, \quad \Psi_{X,Y} \circ \Psi_{X,Z} = \Psi_{X, Y \otimes Z}$$

for all objects X, Y, Z in $\mathcal{C}$. A family Ψ of mappings $\Psi_{X,Y}$ having the above properties is called a *braiding* (or a *quasisymmetry*) in the tensor category $(\mathcal{C}, \otimes)$. Obviously, the flip gives a braiding in the category of vector spaces. But in contrast to this example, the mapping $\Psi_{Y,X}$ need not be the inverse of $\Psi_{X,Y}$ in general. The two hexagon identities imply the relation

$$(\Psi_{Y,Z} \otimes \mathrm{id}_X) \circ (\mathrm{id}_Y \otimes \Psi_{X,Z}) \circ (\Psi_{X,Y} \otimes \mathrm{id}_Z)$$
$$= (\mathrm{id}_Z \otimes \Psi_{X,Y}) \circ (\Psi_{X,Z} \otimes \mathrm{id}_Y) \circ (\mathrm{id}_X \otimes \Psi_{Y,Z}).$$

Indeed, there are two ways to go from $(X \otimes Y) \otimes Z$ to $Z \otimes (Y \otimes X)$ by applying Ψ, one over $Z \otimes (X \otimes Y)$ and another one over $(Y \otimes X) \otimes Z$. The hexagon conditions for these ways give the above relation. In the case $X{=}Y{=}Z$ it is just the braid relation for $\Psi_{X,X}$ and this explains why Ψ is called a braiding.

From now on let $\mathcal{A}$ be a coquasitriangular bialgebra with universal r-form $\mathbf{r}$ and $\mathcal{B}$ a quasitriangular bialgebra with universal R-matrix $\mathcal{R}$. Then, by Propositions 1 and 8.4, the families of mappings $\Psi_{X,Y} := \mathbf{r}_{XY}$ and $\Psi_{X,Y} := R_{XY}$ are braidings in the tensor categories $\mathcal{M}^{\mathcal{A}}$ and $\mathcal{M}_{\mathcal{B}}$, respectively. Thus, $\mathcal{M}^{\mathcal{A}}$ and $\mathcal{M}_{\mathcal{B}}$ are braiding tensor categories.

Let $\mathcal{C}$ be a braided tensor category. By an *algebra in the category* $\mathcal{C}$ we mean an object $\mathcal{X}$ in $\mathcal{C}$ together with morphisms $m_{\mathcal{X}} : \mathcal{X} \otimes \mathcal{X} \to \mathcal{X}$ and $\eta : \mathbf{1} \to \mathcal{X}$ from $\mathcal{C}$ (!) satisfying the usual associativity and unit properties. From the corresponding definitions we conclude that the algebras in the categories $\mathcal{M}^{\mathcal{A}}$ and $\mathcal{M}_{\mathcal{B}}$ are precisely the right $\mathcal{A}$-comodule algebras and the right $\mathcal{B}$-module algebras, respectively. Now let $\mathcal{X}$ and $\mathcal{Y}$ be two algebras in the category $\mathcal{C}$. Then it can be shown that the object $\mathcal{X} \otimes \mathcal{Y}$ in $\mathcal{C}$ becomes an algebra in $\mathcal{C}$, denoted by $\mathcal{X} \underline{\otimes} \mathcal{Y}$ and called the *braided product algebra* of $\mathcal{X}$ and $\mathcal{Y}$, with product morphism

$$m_{\mathcal{X} \underline{\otimes} \mathcal{Y}} := (m_{\mathcal{X}} \otimes m_{\mathcal{Y}}) \circ (\mathrm{id}_{\mathcal{X}} \otimes \Psi_{\mathcal{Y},\mathcal{X}} \otimes \mathrm{id}_{\mathcal{Y}})$$

and the tensor product unit morphism. If $\Psi_{\mathcal{Y},\mathcal{X}}$ is the flip, then $\mathcal{X} \underline{\otimes} \mathcal{Y}$ is just the tensor product algebra $\mathcal{X} \otimes \mathcal{Y}$. In the case of the category $\mathcal{M}^{\mathcal{A}}$ with braiding $\Psi_{\mathcal{X},\mathcal{Y}} = \mathbf{r}_{\mathcal{X}\mathcal{Y}}$, the algebra $\mathcal{X} \underline{\otimes} \mathcal{Y}$ in the category $\mathcal{M}^{\mathcal{A}}$ is nothing but the right $\mathcal{A}$-comodule algebra $\mathcal{X}\underline{\otimes}\mathcal{Y}$ from Proposition 35 with braided product (94). Likewise, for the category $\mathcal{M}_{\mathcal{B}}$ the algebra $\mathcal{X}\underline{\otimes}\mathcal{Y}$ in $\mathcal{M}_{\mathcal{B}}$ is the right $\mathcal{B}$-module algebra $\mathcal{X}\underline{\otimes}\mathcal{Y}$ from Proposition 38 with product (100).

In a similar way, a number of other concepts from "ordinary" Hopf algebra theory make sense in the category $\mathcal{C}$. In addition to the usual axioms one has to require that the corresponding object and all structure maps belong to the category $\mathcal{C}$. In particular, a Hopf algebra $\mathcal{X}$ in the category $\mathcal{C}$ is defined in this manner, whereby the comultiplication $\Delta_{\mathcal{X}}$ has to be an algebra homomorphism of $\mathcal{X}$ to the braided product algebra $\mathcal{X}\underline{\otimes}\mathcal{X}$. It is clear that the Hopf algebras in the braided tensor categories $\mathcal{M}^{\mathcal{A}}$ and $\mathcal{M}_{\mathcal{B}}$ are precisely the braided Hopf algebras associated with $\mathcal{A}$ and $\mathcal{B}$, respectively.

Let us illustrate the preceding by a simple example. It can be generalized to any abelian group rather than $\mathbb{Z}_2$ equipped with a fixed bicharacter.

Example 22 ($\mathcal{A} = \mathbb{C}\mathbb{Z}_2$). Let $\mathcal{A}$ be the coquasitriangular group Hopf algebra $\mathbb{C}\mathbb{Z}_2$ of the group $\mathbb{Z}_2 = \mathbb{Z}/2\mathbb{Z} \simeq \{\bar{0}, \bar{1}\}$ with universal r-form $\mathbf{r}(\bar{i}, \bar{j}) = (-1)^{ij}$, $i, j \in \{0, 1\}$. Then any right $\mathcal{A}$-comodule X with coaction φ splits into a direct sum of an even part $X_{\bar{0}} := \{x \in X \mid \varphi(x) = x\otimes\bar{0}\}$ and an odd part $X_{\bar{1}} = \{x \in X \mid \varphi(x) = x\otimes\bar{1}\}$. Therefore, the objects of the category $\mathcal{M}^{\mathcal{A}}$ are the $\mathbb{Z}_2$-graded (or super) vector spaces and the morphisms are the $\mathbb{Z}_2$-graded linear mappings. The braiding obtained from $\mathbf{r}$ is determined by the formula

$$\Psi_{X,Y}(x_{\bar{i}} \otimes y_{\bar{j}}) = (-1)^{ij} y_{\bar{j}} \otimes x_{\bar{i}}, \quad x_{\bar{i}} \in X_{\bar{i}},\ y_{\bar{j}} \in Y_{\bar{j}},\ i, j \in \{0, 1\}. \tag{101}$$

As noted above, the algebras in $\mathcal{M}^{\mathcal{A}}$ are the right $\mathcal{A}$-comodule algebras. If $\mathcal{X}$ is an algebra equipped with a coaction φ of $\mathcal{A}$ on $\mathcal{X}$, then $\varphi : \mathcal{X} \to \mathcal{X} \otimes \mathcal{A}$ is an algebra homomorphism if and only if

$$\mathcal{X}_{\bar{0}}\mathcal{X}_{\bar{0}} \subseteq \mathcal{X}_{\bar{0}}, \quad \mathcal{X}_{\bar{0}}\mathcal{X}_{\bar{1}} \subseteq \mathcal{X}_{\bar{1}}, \quad \mathcal{X}_{\bar{1}}\mathcal{X}_{\bar{0}} \subseteq \mathcal{X}_{\bar{1}} \ \text{ and } \ \mathcal{X}_{\bar{1}}\mathcal{X}_{\bar{1}} \subseteq \mathcal{X}_{\bar{0}}.$$

Thus, the algebras in $\mathcal{M}^{\mathcal{A}}$ are precisely the $\mathbb{Z}_2$-graded algebras. Similarly, the Hopf algebras in $\mathcal{M}^{\mathcal{A}}$ or equivalently the braided Hopf algebras associated with $\mathcal{A}$ are just the super Hopf algebras (see Subsect. 1.2.9). $\triangle$

We close this subsection with some comments on the physical interpretation of the braided product algebras. For this purpose we assume that the right $\mathcal{A}$-comodule (or $\mathcal{B}$-module) algebras $\mathcal{X}$ and $\mathcal{Y}$ are algebras of observables of two physical systems. Let us consider the braided product algebra $\mathcal{X}\underline{\otimes}\mathcal{Y}$ as an algebra of observables of a joint system which is built from the two systems. First let $\mathcal{A} = \mathbb{C}$. Then the braiding of the category $\mathcal{M}^{\mathcal{A}}$ is the flip and the algebra $\mathcal{X}\underline{\otimes}\mathcal{Y}$ carries the product $(x\otimes y) \bullet (x'\otimes y') = xx'\otimes yy'$, so the subalgebras $\mathcal{X}$ and $\mathcal{Y}$ of $\mathcal{X}\underline{\otimes}\mathcal{Y}$ mutually commute. This means that the two systems are independent in the joint system. Next we set $\mathcal{A} = \mathbb{C}\mathbb{Z}_2$. As usual in "super physics", we consider $\mathcal{X}_{\bar{0}}$ and $\mathcal{Y}_{\bar{0}}$ as algebras of observables of bosonic systems and $\mathcal{X}_{\bar{1}}$ and $\mathcal{Y}_{\bar{1}}$ as algebras of observables of fermionic systems. From the formula (101) of the braiding it is clear that bosons and bosons as well as bosons and fermions of the two systems commute, while fermions and fermions of the two systems anticommute. That is, in this case the braiding contains the statistics between the two systems. In the general case the braiding Ψ describes the commutation relations between observables of the two systems. Having the above example $\mathcal{A} = \mathbb{C}\mathbb{Z}_2$ in mind, one says that the systems obey the *braid statistics* Ψ. This picture opens an important conceptual view of the braiding and the braided Hopf algebras and their possible future role in physics.

10.3.5 Braided Vector Algebras

From Proposition 9.4 we know that the algebra $\mathcal{X}_R(f;R)$ is a right quantum space (that is, a right comodule algebra) for any quotient bialgebra $\mathcal{A}$ of the FRT bialgebra $\mathcal{A}(R)$. Recall that $\mathcal{X}_R(f;R)$ is the unital algebra with generators $x_1, x_2, \ldots, x_N$ and relations $f(\hat{R})^{kl}_{ij} x_k x_l = 0$, $i,j = 1,2,\cdots,N$, and that the coaction φ_R of $\mathcal{A}$ on $\mathcal{X}_R(f;R)$ is given by the equations $\varphi_R(x_i) = x_j \otimes u^j_i$, see (9.10). The purpose of this subsection is to show that under suitable assumptions the algebra $\mathcal{X}_R(f;R)$ becomes a braided Hopf algebra.

Proposition 40. *Let R be an invertible solution of the QYBE and let f be a complex polynomial such that*

$$(\hat{R} + I)f(\hat{R}) = 0. \tag{102}$$

Suppose that $\mathcal{A}$ is a quotient bialgebra of $\mathcal{A}(R)$ such that the canonical universal r-form $\mathbf{r}$ of $\mathcal{A}(R)$ passes to a universal r-form of $\mathcal{A}$. Then the right $\mathcal{A}$-comodule algebra $\mathcal{X} := \mathcal{X}_R(f;R)$ is a braided Hopf algebra associated with the coquasitriangular bialgebra $\mathcal{A}$ with comultiplication $\Delta_{\mathcal{X}}$, counit $\varepsilon_{\mathcal{X}}$, antipode $S_{\mathcal{X}}$ and braiding $\mathbf{r}_{\mathcal{X}\mathcal{X}}$ determined by the formulas

$$\Delta_{\mathcal{X}}(x_i) = x_i \otimes 1 + 1 \otimes x_i, \quad \varepsilon_{\mathcal{X}}(x_i) = 0, \quad S_{\mathcal{X}}(x_i) = -x_i, \tag{103}$$

$$\mathbf{r}_{\mathcal{X}\mathcal{X}}(x_i \otimes x_j) = x_k \otimes x_l \, \hat{R}^{kl}_{ij}. \tag{104}$$

Proof. As in similar earlier proofs (for instance, of Proposition 9.1), we first show the corresponding result for the free algebra $\mathbb{C}\langle x_i \rangle$ and then we pass to the quotient algebra $\mathcal{X}$ of $\mathbb{C}\langle x_i \rangle$. Because $\mathbb{C}\langle x_i \rangle$ is a right $\mathcal{A}$-comodule algebra with coaction $\varphi_R(x_i) = x_j \otimes u^j_i$, the braided product algebra $\mathbb{C}\langle x_i \rangle \underline{\otimes} \mathbb{C}\langle x_i \rangle$ is well-defined by Proposition 35. Since $\mathbb{C}\langle x_i \rangle$ is the free algebra with generators x_i, there are unique algebra homomorphisms $\Delta : \mathbb{C}\langle x_i \rangle \to \mathbb{C}\langle x_i \rangle \underline{\otimes} \mathbb{C}\langle x_i \rangle$ and $\varepsilon : \mathbb{C}\langle x_i \rangle \to \mathbb{C}$ such that $\Delta(x_i) = x_i \otimes 1 + 1 \otimes x_i$ and $\varepsilon(x_i) = 0$. The set $\{1, x_{i_1} \cdots x_{i_n}\}$ is a vector space basis of $\mathbb{C}\langle x_i \rangle$. Hence there is a unique linear map S of $\mathbb{C}\langle x_i \rangle$ such that $S(1) = 1$, $S(\mathbf{x}) = -\mathbf{x}$ and

$$S(\mathbf{x}_1 \cdots \mathbf{x}_n) = (-1)^n \mathbf{x}_n \ldots \mathbf{x}_1 R_{12} \cdots R_{1n} \, R_{23} \cdots R_{2n} \cdots R_{n-1,n}, \quad n \geq 2, \tag{105}$$

in the usual matrix notation. That is, we define $S(1) = 1$, $S(x_i) = -x_i$ and extend S to a linear map of $\mathbb{C}\langle x_i \rangle$ according to the formula (97). One easily verifies that $\mathbb{C}\langle x_i\rangle$ equipped with these mappings becomes a braided Hopf algebra associated with $\mathcal{A}$.

In order to complete the proof we have to show that these structures pass to the quotient algebra $\mathcal{X}_R(f;R)=\mathbb{C}\langle x_i\rangle/\mathcal{J}$, where $\mathcal{J}$ denotes the two-sided ideal of $\mathbb{C}\langle x_i\rangle$ generated by the N^2 elements $I_{ij}:=f(\hat{R})^{kl}_{ij}x_kx_l$. For the coaction φ_R this holds by Proposition 9.4. Formula (104) for the braiding $\mathbf{r}_{\mathcal{X}\mathcal{X}}$ is obtained by repeating the reasoning of (13). By the definition of ε we have $\varepsilon(\mathcal{J})=\{0\}$. Using the formulas (103), (94), (104) and finally (102), we obtain

$$\begin{aligned}\Delta(I_{ij}) &= f(\hat{R})^{kl}_{ij}\Delta(x_kx_l)=f(\hat{R})^{kl}_{ij}(x_k\otimes 1+1\otimes x_k)\bullet(x_l\otimes 1+1\otimes x_l)\\ &= f(\hat{R})^{kl}_{ij}x_kx_l\otimes 1+f(\hat{R})^{kl}_{ij}x_k\otimes x_l+f(\hat{R})^{kl}_{ij}x_m\otimes x_n\ \hat{R}^{mn}_{kl}\\ &\qquad +1\otimes f(\hat{R})^{kl}_{ij}x_kx_l\\ &= I_{ij}\otimes 1+((\hat{R}+I)f(\hat{R}))^{mn}_{ij}x_m\otimes x_n+1\otimes I_{ij}=I_{ij}\otimes 1+1\otimes I_{ij}\end{aligned}$$

and hence

$$\Delta(aI_{ij}b)=\Delta(a)\bullet(I_{ij}\otimes 1+1\otimes I_{ij})\bullet\Delta(b)\in\mathcal{J}\otimes\mathbb{C}\langle x_i\rangle+\mathbb{C}\langle x_i\rangle\otimes\mathcal{J}$$

for $a,b\in\mathbb{C}\langle x_i\rangle$. From equation (105) it follows that

$$S(I_{ij})=f(\hat{R})^{kl}_{ij}S(x_kx_l)=f(\hat{R})^{kl}_{ij}x_mx_nR^{nm}_{kl}=\hat{R}^{kl}_{ij}I_{kl}\in\mathcal{J}.$$

By (97), this implies that $S(aI_{ij}b)\in\mathcal{J}$ for $a,b\in\mathbb{C}\langle x_i\rangle$. Thus we have shown that $\mathcal{J}$ is a Hopf ideal of the braided Hopf algebra $\mathbb{C}\langle x_i\rangle$. Hence the quotient $\mathcal{X}=\mathbb{C}\langle x_i\rangle/\mathcal{J}$ is also a braided Hopf algebra. □

Note that (105) also hold for the antipode $S_{\mathcal{X}}$ of the braided Hopf algebra $\mathcal{X}_R(f;R)$ and allows us to compute $S_{\mathcal{X}}$ on products of generators.

In order to discuss the applicability of Proposition 40, let us assume in addition that the transformation $\hat{R}$ admits a spectral decomposition $\hat{R}=\sum_i\lambda_iP_i$, where P_i are nonzero projections such that $\sum_i P_i=I$ and $P_iP_j=0$ and $\lambda_i\neq\lambda_j$ if $i\neq j$. It is well-known from linear algebra that the projections P_j are then polynomials in $\hat{R}$, say, $P_i=f_i(\hat{R})$. If an eigenvalue λ_i is -1, then f_i satisfies (102) and hence, by Proposition 40, $\mathcal{X}_R(f_i;R)$ is a braided Hopf algebra associated with $\mathcal{A}$. In the case $\mathcal{A}=\mathcal{A}(R)$ the transformation R can be replaced by any nonzero complex multiple, so each eigenspace of $\hat{R}$ gives rise to a braided Hopf algebra. However, if $\mathcal{A}$ is one of the coquasitriangular Hopf algebras $\mathcal{O}(G_q)$, $G_q=SL_q(N),O_p(N),Sp_q(N)$, and $R=z\mathsf{R}$ as in Theorem 9, such a scaling is impossible and $\hat{R}$ cannot have the eigenvalue -1 unless q is a root of unity. Our next proposition shows how to overcome this difficulty: $\mathcal{X}_R(f;R)$ becomes a braided Hopf algebra associated with the coquasitriangular bialgebra $\tilde{\mathcal{A}}$ which is obtained from $\mathcal{A}$ by adjoining an additional central invertible group-like generator.

Recall that the group algebra $\mathbb{C}\mathbb{Z}$ of the group $\mathbb{Z}$ of integers is the algebra of Laurent polynomials in a single generator χ. It is a Hopf algebra with comultiplication $\Delta(\chi^n) = \chi^n \otimes \chi^n$, $n \in \mathbb{Z}$. For each nonzero $\lambda \in \mathbb{C}$, there exists a universal r-form $\mathbf{r}_\lambda$ on $\mathbb{C}\mathbb{Z}$ such that $\mathbf{r}_\lambda(\chi^n \otimes \chi^m) = \lambda^{-nm}$ (see Example 1 in Subsect. 10.1.1). One easily verifies that the tensor product bialgebra $\tilde{\mathcal{A}} := \mathcal{A} \otimes \mathbb{C}\mathbb{Z}$ is coquasitriangular with universal r-form $\tilde{\mathbf{r}}$ defined by

$$\tilde{\mathbf{r}}(a \otimes \chi^n \otimes b \otimes \chi^m) = \mathbf{r}(a \otimes b)\lambda^{-nm}, \quad a, b \in \mathcal{A}, \quad n, m \in \mathbb{Z}. \tag{106}$$

Clearly, each algebra $\mathcal{X}_R(f; R)$ is also a right $\tilde{\mathcal{A}}$-comodule algebra with coaction given by $\varphi_R(x_j) = x_j \otimes u^j_i \otimes \chi$.

Proposition 41. *Let R, $\mathcal{A}$ and $\mathbf{r}$ be as in Proposition 40. Suppose that $\lambda \in \mathbb{C}$, $\lambda \neq 0$, and f is a polynomial such that*

$$(\hat{R} + \lambda I) f(\hat{R}) = 0. \tag{107}$$

Then $\mathcal{X} := \mathcal{X}_R(f; R)$ is a braided Hopf algebra associated with the coquasitriangular bialgebra $\tilde{\mathcal{A}}$ defined above. The Hopf algebra structure of $\mathcal{X}$ is determined by (103) and the braiding map $\tilde{\mathbf{r}}_{\mathcal{X}\mathcal{X}}$ is given by

$$\tilde{\mathbf{r}}_{\mathcal{X}\mathcal{X}}(x_i \otimes x_j) = \lambda^{-1} x_k \otimes x_l \hat{R}^{kl}_{ij}. \tag{108}$$

Proof. Clearly, (106) implies (108). The term $(\hat{R} + I) f(\hat{R})$ of the expression $\Delta(I_{ij})$ in the proof of Proposition 40 now becomes $(\lambda^{-1}\hat{R} + I) f(\hat{R})$ which vanishes by (107). The rest of the proof follows by the same reasoning. $\square$

The braided Hopf algebras $\mathcal{X}_R(f; R)$ in Propositions 40 and 41 are called *braided vector algebras.*

Note that Propositions 40 and 41 are also valid for the coquasitriangular Hopf algebra $\mathcal{O}(GL_q(N))$ with the same proofs verbatim.

Finally, let us consider the case when the number λ in (107) is a root of unity, say, $\lambda^k = 1$ for $k \in \mathbb{N}$. Then, $\mathbf{r}_\lambda$ gives a universal r-form of the group algebra $\mathbb{C}(\mathbb{Z}/k\mathbb{Z})$ of the group $\mathbb{Z}/k\mathbb{Z}$. Hence the assertion of Proposition 41 holds also for the coquasitriangular bialgebra $\tilde{\mathcal{A}} := \mathcal{A} \otimes \mathbb{C}(\mathbb{Z}/k\mathbb{Z})$. In particular, in the case $\lambda = 1$ we can take $k = 1$ and recover Proposition 40.

10.3.6 Bosonization of Braided Hopf Algebras

In this subsection we show that the crossed product of a *braided* Hopf algebra $\mathcal{X}$ by the coquasitriangular (or quasitriangular) Hopf algebra $\mathcal{A}$ which $\mathcal{X}$ is associated with becomes an *ordinary* Hopf algebra. In order to avoid possible confusion with the comultiplications of $\mathcal{X}$ and $\mathcal{A}$, let us write the Sweedler notation for the coaction β of $\mathcal{A}$ on $\mathcal{X}$ as $\beta(x) = \sum x^{(0)} \otimes x^{(1)}$. This notation will be kept in the rest of this section.

To begin with, suppose that $\mathcal{X}$ is a braided bialgebra associated with a coquasitriangular bialgebra $\mathcal{A}$. Then, by Definition 11, $\mathcal{X}$ is in particular a right $\mathcal{A}$-comodule coalgebra with respect to the coaction β of $\mathcal{A}$ on $\mathcal{X}$. Therefore, by Proposition 19, we have the *crossed coproduct coalgebra* $\mathcal{A}^\beta \ltimes \mathcal{X}$ built on the vector space $\mathcal{A}\otimes\mathcal{X}$. On the other hand, $\mathcal{X}$ is also a right $\mathcal{A}$-comodule algebra by Definition 11. Hence, by Proposition 4, the right coaction β induces a right action β' of $\mathcal{A}$ on $\mathcal{X}$ by $x\triangleleft a=\sum x^{(0)}\mathbf{r}(x^{(1)},a)$, $a\in\mathcal{A}$, $x\in\mathcal{X}$, such that $\mathcal{X}$ becomes a right $\mathcal{A}$-module algebra, where $\beta(x)=\sum x^{(0)}\otimes x^{(1)}$. Thus the vector space $\mathcal{A}\otimes\mathcal{X}$ carries also the structure of the *crossed product algebra* $\mathcal{A}_{\beta'}\ltimes\mathcal{X}$ (see Subsect. 10.2.1). The next proposition says that these two structures (algebra and coalgebra) on the vector space $\mathcal{A}\otimes\mathcal{X}$ indeed fit nicely together.

Proposition 42. *Let $\mathcal{X}$ be a braided bialgebra associated with a coquasitriangular bialgebra $\mathcal{A}$. Then the vector space $\mathcal{A}\otimes\mathcal{X}$ is a bialgebra, denoted by $\mathcal{A}^{\beta}\ltimes\mathcal{X}$ or $\mathcal{A}\ltimes\mathcal{X}$, with the coalgebra structure of $\mathcal{A}^{\beta}\ltimes\mathcal{X}$ and the algebra structure of $\mathcal{A}_{\beta'}\ltimes\mathcal{X}$. That is, the bialgebra $\mathcal{A}^{\beta}\ltimes\mathcal{X}$ has the coproduct (54) and the product*

$$(a\otimes x)(b\otimes y)=\sum ab_{(1)}\otimes x^{(0)}y\,\mathbf{r}(x^{(1)},b_{(2)}),\quad a,b\in\mathcal{A},\ x,y\in\mathcal{X}.\quad(109)$$

The subalgebra $1\otimes\mathcal{X}$ of $\mathcal{A}\ltimes\mathcal{X}$ is a right $\mathcal{A}\ltimes\mathcal{X}$-comodule algebra with coaction

$$\varphi_R(1\otimes x):=\Delta_{\mathcal{A}\ltimes\mathcal{X}}(1\otimes x)=\sum 1\otimes(x_{(1)})^{(0)}\otimes(x_{(1)})^{(1)}\otimes x_{(2)},\quad x\in\mathcal{X}.\quad(110)$$

If $\mathcal{X}$ is a braided Hopf algebra associated with the coquasitriangular Hopf algebra $\mathcal{A}$, then the bialgebra $\mathcal{A}\ltimes\mathcal{X}$ is a Hopf algebra with antipode

$$S(a\otimes x)=\sum(1\otimes S_{\mathcal{X}}(x^{(1)}))(S_{\mathcal{A}}(ax^{(0)})\otimes 1),\quad a\in\mathcal{A},\ x\in\mathcal{X}.\quad(111)$$

Proof. Being the restriction of the comultiplication to an invariant subspace, φ_R is a coaction. One has to show that φ_R, Δ and ε are algebra homomorphisms and that S satisfies the antipode axiom. These technical verifications are carried out in [Wei] and in the dual situation of quasitriangular Hopf algebras in [Maj10]. Formula (110) follows from (54). □

The construction of the "ordinary" Hopf algebra $\mathcal{A}\ltimes\mathcal{X}$ from the braided Hopf algebra $\mathcal{X}$ is a process called *bosonization* by S. Majid. It is in a sense dual to the process of transmutation which converts the Hopf algebra $\mathcal{A}$ or more generally any Hopf algebra homomorphism $p:\mathcal{H}\to\mathcal{A}$ into the braided Hopf algebras $B(\mathcal{A})$ resp. $B(\mathcal{H};\mathcal{A};p)$, see Subsect. 10.3.2.

Let us see how $\mathcal{A}$ and $\mathcal{X}$ are related to $\mathcal{A}\ltimes\mathcal{X}$. From the formulas (54), (109) and (111) it is clear that the map $a\to a\otimes 1$ is an injective Hopf algebra homomorphism of $\mathcal{A}$ to $\mathcal{A}\ltimes\mathcal{X}$, so we may consider $\mathcal{A}$ as a Hopf subalgebra of $\mathcal{A}\ltimes\mathcal{X}$ by identifying a with $a\otimes 1$. Also, the map $x\to 1\otimes x$ is an algebra isomorphism of $\mathcal{X}$ to the subalgebra $1\otimes\mathcal{X}$ of $\mathcal{A}\ltimes\mathcal{X}$, but as seen from (110) and (111) it does not preserve the comultiplication and the antipode. Being a crossed product algebra (see Subsect. 10.2.1), $\mathcal{A}\ltimes\mathcal{X}$ is the universal algebra generated by the algebras $\mathcal{A}$ and $\mathcal{X}$ with cross commutation relations

$$xa=\sum a_{(1)}x^{(0)}\,\mathbf{r}(x^{(1)},a_{(2)}),\quad a\in\mathcal{A},\ x\in\mathcal{X}.\quad(112)$$

We now illustrate the bosonization procedure by two examples. The Hopf algebras $\mathcal{A} \ltimes \mathcal{X}$ appearing in these examples are also of interest in themselves without referring to braided Hopf algebras.

Example 23 (*The braided line – continued*). Let us retain the notation of Example 21. By Proposition 42, $\mathcal{A}^\beta \ltimes \mathcal{X}$ is a Hopf algebra. It is generated by the elements $\chi \equiv \chi \otimes 1$ and $z \equiv 1 \otimes z$. Since $\beta(z) = z \otimes \chi$ and $\mathbf{r}_q(\chi \otimes \chi) = q^{-1}$, the cross relation (112) with $x := z$ and $a := \chi$ reads as $z\chi = \chi z q^{-1}$. Since $\Delta_{\mathcal{X}}(z) = z \otimes 1 + 1 \otimes z$, formula (54) for the coproduct of $\mathcal{A} \ltimes \mathcal{X}$ yields

$$\Delta(\chi) = \chi \otimes \chi, \quad \Delta(z) = z \otimes \chi + 1 \otimes z. \tag{113}$$

From these formulas (or from (97)) we easily derive

$$\varepsilon(\chi) = 1, \quad \varepsilon(z) = 0, \quad S(\chi) = \chi^{-1}, \quad S(z) = -z\chi^{-1}. \tag{114}$$

That is, the Hopf algebra $\mathcal{A} \ltimes \mathcal{X}$ has two generators χ and z with defining relation $\chi z = qz\chi$ and structure maps determined by (113) and (114). For $q = 1$ the Hopf algebra $(\mathcal{A} \ltimes \mathcal{X})^{\mathrm{cop}}$ is just the coordinate algebra of the group of affine transformations of the real line (the so-called $ax + b$-group). This suggests the use of the bosonization for the construction of inhomogeneous quantum groups. This idea will be realized in the next subsection. △

Example 24 (*The Hopf algebra* $\mathcal{A}^{\mathrm{Ad}_R} \ltimes B(\mathcal{A})$). Let $\mathcal{X}$ be the braided Hopf algebra $B(\mathcal{A})$ associated with a coquasitriangular Hopf algebra $\mathcal{A}$. Recall from Subsect. 10.3.2 that $B(\mathcal{A})$ has the same coalgebra structure as $\mathcal{A}$ and that the product $\underline{\cdot}$ of $B(\mathcal{A})$ is given by (85). The coaction β of $\mathcal{A}$ on $B(\mathcal{A})$ is the right adjoint coaction $\mathrm{Ad}_R(a) = \sum a_{(2)} \otimes S(a_{(1)})a_{(3)}$. Inserting these data into the formulas (109) and (54), the product and coproduct of $\mathcal{A} \bowtie B(\mathcal{A})$ read as

$$(a \otimes x)(b \otimes y) = \sum ab_{(1)} \otimes x_{(2)} \underline{\cdot} y\, \mathbf{r}(S(x_{(1)})x_{(3)}, b_{(2)})$$

$$= \sum ab_{(1)} \otimes x_{(3)}y_{(2)}\, \mathbf{r}(S(x_{(2)})x_{(4)}, S(y_{(1)}))\mathbf{r}(S(x_{(1)})x_{(5)}, b_{(2)}), \tag{115}$$

$$\Delta(a \otimes x) = \sum a_{(1)} \otimes x_{(2)} \otimes a_{(2)}S(x_{(1)})x_{(3)} \otimes x_{(4)}. \tag{116}$$

The following interesting result emphasizes once more the fundamental importance of Drinfeld's double construction.

Statement: *The linear map θ defined by $\theta(a \otimes x) = \sum ax_{(1)} \otimes x_{(2)}$ is a Hopf algebra isomorphism of the quantum double $\mathcal{D}(\mathcal{A}, \mathcal{A}; \mathbf{r}) = \mathcal{A} \bowtie_{\mathbf{r}} \mathcal{A}$ (see Subsects. 8.2.1 and 10.2.5) to the Hopf algebra $\mathcal{A}^{\mathrm{Ad}_R} \ltimes B(\mathcal{A})$.*

Proof. Using the definitions (115) and (8.21) of the corresponding products and some properties of the universal r-form $\mathbf{r}$ we compute

$$
\begin{aligned}
\theta(a\otimes x)\theta(b\otimes y) &= \sum (ax_{(1)} \otimes x_{(2)})(by_{(1)} \otimes y_{(2)}) \\
&= \sum ax_{(1)}b_{(1)}y_{(1)} \otimes x_{(4)}y_{(4)}\, \mathbf{r}(S(x_{(3)})x_{(5)}, S(y_{(3)})) \\
&\qquad\qquad \times \mathbf{r}(S(x_{(2)})x_{(6)}, b_{(2)}y_{(2)}) \\
&= \sum ax_{(1)}b_{(1)}y_{(1)} \otimes x_{(5)}y_{(4)}\, \mathbf{s}(S(x_{(4)})x_{(6)}, y_{(3)}) \\
&\qquad\qquad \times \mathbf{r}(S(x_{(3)})x_{(7)}, y_{(2)})\mathbf{r}(S(x_{(2)})x_{(8)}, b_{(2)}) \\
&= \sum ax_{(1)}b_{(1)}y_{(1)} \otimes x_{(3)}y_{(2)}\, \mathbf{r}(S(x_{(2)})x_{(4)}, b_{(2)}) \\
&= \sum ax_{(1)}b_{(1)}y_{(1)} \otimes x_{(3)}y_{(2)}\, \bar{\mathbf{r}}(x_{(2)}, b_{(2)})\mathbf{r}(x_{(4)}, b_{(3)}) \\
&= \sum ab_{(2)}x_{(2)}y_{(1)} \otimes x_{(3)}y_{(2)}\, \bar{\mathbf{r}}(x_{(1)}, b_{(1)})\mathbf{r}(x_{(4)}, b_{(3)}) \\
&= \sum \theta(ab_{(2)} \otimes x_{(2)}y)\, \bar{\mathbf{r}}(x_{(1)}, b_{(1)})\mathbf{r}(x_{(3)}, b_{(3)}) \\
&= \theta((a \otimes x)(b \otimes y)).
\end{aligned}
$$

Here the third equality follows from Proposition 33(i), the fourth from (82) and the sixth from (5). Recall that the quantum double has the tensor product coalgebra structure. Hence, by (116), we obtain

$$
\begin{aligned}
\Delta \circ \theta(a \otimes x) &= \sum \Delta(ax_{(1)} \otimes x_{(2)}) \\
&= \sum a_{(1)}x_{(1)} \otimes x_{(4)} \otimes a_{(2)}x_{(2)}S(x_{(3)})x_{(5)} \otimes x_{(6)} \\
&= \sum a_{(1)}x_{(1)} \otimes x_{(2)} \otimes a_{(2)}x_{(3)} \otimes x_{(4)} = (\theta \otimes \theta) \circ \Delta(a \otimes x).
\end{aligned}
$$

Since obviously $\theta(1\otimes 1) = 1\otimes 1$ and $\varepsilon\circ\theta = \varepsilon$, θ is an algebra and a coalgebra homomorphism. It is easy to check that $\theta^{-1}(a \otimes x) := \sum aS(x_{(1)}) \otimes x_{(2)}$ is the inverse of θ. Therefore, θ is a Hopf algebra isomorphism. □ △

We briefly turn to the dual situation of a quasitriangular Hopf algebra $\mathcal{B}$ and state the counterpart to Proposition 42 as

Proposition 43. *Suppose that $\mathcal{X}$ is a braided Hopf algebra associated with a quasitriangular Hopf algebra $\mathcal{B}$. Then, by definition, $\mathcal{X}$ is a left $\mathcal{B}$-module algebra. Let $\alpha : (b \otimes x) \to b \triangleright x$ denote the corresponding left action of $\mathcal{B}$ on $\mathcal{X}$ and let $\mathcal{R} = \sum_i x_i \otimes y_i$ be the universal R-matrix of $\mathcal{B}$. Then $\mathcal{X}$ is a left $\mathcal{B}$-comodule coalgebra with respect to the coaction $\alpha'(x) := \sum_i y_i \otimes x_j \triangleright x$.*

The vector space $\mathcal{X} \otimes \mathcal{B}$ equipped with the left crossed product algebra structure of $\mathcal{X} \rtimes_\alpha \mathcal{B}$ (see Subsect. 10.2.1) and with the left crossed coproduct coalgebra structure of $\mathcal{X} \rtimes^{\alpha'} \mathcal{B}$ is a Hopf algebra, denoted by $\mathcal{X} \rtimes_\alpha \mathcal{B}$ or simply $\mathcal{X} \rtimes \mathcal{B}$. The product, the coproduct and the antipode of this Hopf algebra are given by

$$
(x \otimes a)(y \otimes b) = \sum x(a_{(1)} \triangleright y) \otimes a_{(2)}b,
$$

$$
\Delta(x \otimes a) = \sum\nolimits_i \sum x_{(1)} \otimes y_i a_{(1)} \otimes x_i \triangleright x_{(2)} \otimes a_{(2)},
$$

$$S(x \otimes a) = \sum\nolimits_{i,j} \sum S_{\mathcal{B}}(y_j a_{(2)}) x_j (x_i \triangleright S_{\mathcal{X}}(x)) \otimes S_{\mathcal{B}}(y_i a_{(1)}),$$

where $\Delta_{\mathcal{A}}(a) = \sum a_{(1)} \otimes a_{(2)}$ *and* $\Delta_{\mathcal{X}}(x) = \sum x_{(1)} \otimes x_{(2)}$ *are the Sweedler notations for the comultiplications of* $\mathcal{A}$ *and* $\mathcal{X}$.

Example 25 (*The Hopf algebra* $B(\mathcal{B}) \rtimes_{\mathrm{ad}_L} \mathcal{B}$). Let $\mathcal{B}$ and $\mathcal{R}$ be as in Proposition 43 and let $\mathcal{X}$ be the braided Hopf algebra $B(\mathcal{B})$ from Subsect. 10.3.3. In this case α is the left adjoint action $\mathrm{ad}_L(b)x = \sum b_{(1)} x S_{\mathcal{B}}(b_{(2)})$. The product and the coproduct of the Hopf algebra $B(\mathcal{B}) \rtimes_{\mathrm{ad}_L} \mathcal{B}$ are determined by

$$(x \otimes a)(y \otimes b) = \sum x a_{(1)} y S_{\mathcal{B}}(a_{(2)}) \otimes a_{(3)} b,$$

$$\Delta(x \otimes a) = \sum\nolimits_i \sum x_{(1)} \otimes y_i a_{(1)} \otimes x_{i(1)} x_{(2)} S_{\mathcal{B}}(x_{i(2)}) \otimes a_{(2)}. \qquad \triangle$$

10.3.7 $*$-Structures on Bosonized Hopf Algebras

The definition of $*$-structures on quantum spaces of a coquasitriangular $*$-bialgebra depends essentially on the reality type of the universal r-form. In this subsection we treat only the inverse real case which is much simpler than the real case.

Proposition 44. *Suppose that* $\mathbf{r}$ *is an inverse real universal* r*-form of a* $*$*-bialgebra* $\mathcal{A}$. *Let* $\mathcal{X}$ *and* $\mathcal{Y}$ *be right* $\mathcal{A}$*-comodule* $*$*-algebras. Then the vector space* $\mathcal{X} \otimes \mathcal{Y}$ *equipped with the braided product (94) and the involution*

$$(x \otimes y)^* := \sum x^{(0)*} \otimes y^{(0)*} \mathbf{r}(y^{(1)*}, x^{(1)*}), \quad x \in \mathcal{X},\ y \in \mathcal{Y}, \tag{117}$$

becomes a right $\mathcal{A}$*-comodule* $*$*-algebra, denoted by* $\mathcal{X} \underline{\otimes} \mathcal{Y}$.

Proof. By Proposition 35, $\mathcal{X} \underline{\otimes} \mathcal{Y}$ is a right $\mathcal{A}$-comodule algebra, so only the assertions concerning the involution have to be proved. Using the assumption that $\mathbf{r}$ is inverse real, we compute

$$\begin{aligned}
((x \otimes y)(z \otimes w))^* &= \sum (x z^{(0)} \otimes y^{(0)} w\, \mathbf{r}(y^{(1)}, z^{(1)}))^* \\
&= \sum z^{(0)*} x^{(0)*} \otimes w^{(0)*} y^{(0)*}\, \mathbf{r}(w^{(1)*} y^{(1)*}, z^{(1)*} x^{(1)*})\, \overline{\mathbf{r}(y^{(2)}, z^{(2)})} \\
&= \sum z^{(0)*} x^{(0)*} \otimes w^{(0)*} y^{(0)*}\, \mathbf{r}(w^{(1)*}, x^{(1)*})\, \mathbf{r}(w^{(2)*}, z^{(1)*}) \\
&\qquad \times \mathbf{r}(y^{(1)*}, x^{(2)*}) \mathbf{r}(y^{(2)*}, z^{(2)*}) \bar{\mathbf{r}}(y^{(3)*}, z^{(2)*}) \\
&= \sum (z^{(0)*} \otimes w^{(0)*})(x^{(0)*} \otimes y^{(0)*})\, \mathbf{r}(w^{(1)*}, z^{(1)*})\, \mathbf{r}(y^{(1)*}, x^{(1)*}) \\
&= (z \otimes w)^* (x \otimes y)^*
\end{aligned}$$

and

$$((x \otimes y)^*)^* = \sum x^{(0)} \otimes y^{(0)} \mathbf{r}(y^{(1)}, x^{(1)}) \bar{\mathbf{r}}(y^{(2)}, y^{(2)}) = x \otimes y.$$

Hence we have an involution of the algebra $\mathcal{X} \underline{\otimes} \mathcal{Y}$. Next we show that the coaction φ_R of $\mathcal{A}$ on $\mathcal{X} \underline{\otimes} \mathcal{Y}$ is $*$-preserving. Indeed, by (117) and (5), we get

$$\begin{aligned}\varphi_R((x\otimes y)^*) &= \sum x^{(0)*}\otimes y^{(0)*}\otimes x^{(1)*}y^{(1)*}\,\mathbf{r}(y^{(2)*},x^{(2)*})\\ &= \sum x^{(0)*}\otimes y^{(0)*}\otimes y^{(2)*}x^{(2)*}\,\mathbf{r}(y^{(1)*},x^{(1)*})\\ &= \sum (x^{(0)}\otimes y^{(0)})^*\otimes (x^{(1)}y^{(1)})^* = \varphi_R(x\otimes y)^*. \qquad \square\end{aligned}$$

The next proposition shows that the involution (117) turns the bosonized Hopf algebra $\mathcal{A}\ltimes\mathcal{X}$ from Proposition 42 into a Hopf $*$-algebra.

Proposition 45. *Let* $\mathbf{r}$ *be an inverse real universal r-form of a $*$-bialgebra* $\mathcal{A}$. *Let* $\mathcal{X}$ *be a braided bialgebra associated with* $\mathcal{A}$ *such that* $\mathcal{X}$ *is a* $\mathcal{A}$*-comodule $*$-algebra and* $\Delta_{\mathcal{X}}(x^*)=\Delta_{\mathcal{X}}(x)^*$ *for* $x\in\mathcal{X}$, *where* $\mathcal{X}\underline{\otimes}\mathcal{X}$ *carries the involution (117). Then the bialgebra* $\mathcal{A}\ltimes\mathcal{X}$ *from Proposition 42 equipped with the involution*

$$(a\otimes x)^* = \sum (a_{(1)})^*\otimes x^{(0)*}\,\mathbf{r}(x^{(1)*},(a_{(2)})^*),\quad a\in\mathcal{A},\ x\in\mathcal{X}, \qquad (118)$$

becomes a $*$-bialgebra, where $\Delta_{\mathcal{A}}(a)=\sum a_{(1)}\otimes a_{(2)}$.

Proof. Since the involution (118) is a special case of (117) (by considering $\mathcal{A}$ as a right $\mathcal{A}$-comodule $*$-algebra with respect to the comultiplication), $\mathcal{A}\ltimes\mathcal{X}$ is a $*$-algebra by Proposition 44. It only remains to show that the counit and the comultiplication of the bialgebra $\mathcal{A}\ltimes\mathcal{X}$ are $*$-preserving. For the counit this is obvious. For the comultiplication this requires some technical computations which will be omitted here. They are carried out in [Wei]. □

As usual, let us identify $x\in\mathcal{X}$ with $x\otimes 1\in\mathcal{X}\underline{\otimes}\mathcal{Y}$ and $y\in\mathcal{Y}$ with $1\otimes y\in\mathcal{X}\underline{\otimes}\mathcal{Y}$. Then it is clear from the definition (117) of the involution that $\mathcal{X}$ and $\mathcal{Y}$ are $*$-subalgebras of the $*$-algebra $\mathcal{X}\underline{\otimes}\mathcal{Y}$ from Proposition 44. Conversely, formula (117) already follows from that property, since

$$\begin{aligned}(x\otimes y)^* &= ((x\otimes 1)(1\otimes y))^* = (1\otimes y)^*(x\otimes 1)^* = (1\otimes y^*)(x^*\otimes 1)\\ &= \sum x^{(0)*}\otimes y^{(0)*}\,\mathbf{r}(y^{(1)*},x^{(1)*}).\end{aligned}$$

In other words, *there exists a unique involution of the algebra* $\mathcal{X}\underline{\otimes}\mathcal{Y}$ *which extends the involutions of its subalgebras* $\mathcal{X}$ *and* $\mathcal{Y}$.

The last assertion is no longer true if the universal r-form $\mathbf{r}$ is real rather than inverse real. More precisely, if $\mathcal{A}$ is a Hopf $*$-algebra with real universal r-form $\mathbf{r}$ and $\mathcal{X}$ is as in Proposition 45, then there is no well-defined(!) involution of the algebra $\mathcal{A}\ltimes\mathcal{X}$ which coincides with the involutions on its subalgebras $\mathcal{A}$ and $\mathcal{X}$. An example will be seen in Subsect. 10.3.9 below. One possibility of handling this difficulty is to work with the realification $\mathcal{B}=\mathcal{D}(\mathcal{A},\mathcal{A};\mathbf{r})$ of $\mathcal{A}$ (see Subsect. 10.2.7) rather than with $\mathcal{A}$ itself. Recall that, by Proposition 29, $\mathcal{B}$ always admits an *inverse real* universal r-form $\mathbf{r}''$, so that Propositions 43 and 44 apply to $\mathcal{B}$ and $\mathbf{r}''$. Details can be found in [Wei].

10.3.8 Inhomogeneous Quantum Groups

In this subsection we combine Propositions 41 and 42 in order to construct coordinate Hopf algebras of inhomogeneous quantum groups.

In what follows let G_q be one of the quantum groups $GL_q(N)$, $SL_q(N)$, $O_q(N)$ or $Sp_q(N)$. By Theorem 9, the Hopf algebra $\mathcal{A} = \mathcal{O}(G_q)$ possesses a universal r-form $\mathbf{r}_z$ such that $\mathbf{r}_z(\mathbf{u}_1, \mathbf{u}_2) = R_{12}$ with $R := z\mathsf{R}$, where the parameter z and the matrix R are as stated there. Let us suppose that $q^2+1 \neq 0$ for $G_q = GL_q(N), SL_q(N)$ and $(q^2+1)(\epsilon + q^{N-1-\epsilon}) \neq 0$ for $G_q = O_q(N), Sp_q(N)$. Then the spectral projector P_- of the matrix $\hat{\mathsf{R}}$ to the eigenvalue $-q^{-1}$ is of the form $f(\hat{R})$ for some polynomial f and the right quantum space $\mathcal{X} := \mathcal{X}(f; R)$ of $\mathcal{A}$ has the defining relations $\mathsf{P}_-\mathbf{x}_1\mathbf{x}_2 = 0$. Note that $\mathcal{X} = \mathcal{O}(\mathbb{C}_q^N)$ if $G_q = GL_q(N), SL_q(N)$ and $\mathcal{X} = \mathcal{O}(O_q^N)$ if $G_q = O_q(N)$.

Since $(\hat{R} + zq^{-1})f(\hat{R}) = (z\hat{\mathsf{R}} + zq^{-1})\mathsf{P}_- = -zq^{-1}\mathsf{P}_- + zq^{-1}\mathsf{P}_- = 0$, Proposition 41 applies to $\mathcal{X}$ with $\lambda = zq^{-1}$. Hence $\mathcal{X}$ is a braided Hopf algebra associated with the coquasitriangular Hopf algebra $\tilde{\mathcal{A}} = \mathcal{A} \otimes \mathbb{C}\mathbb{Z}$. If $(zq^{-1})^k = 1$ for some $k \in \mathbb{N}$, it is possible to take the quotient group $\mathbb{Z}/k\mathbb{Z}$ instead of $\mathbb{Z}$ (see the remarks at the end of Subsect. 10.3.5). By Proposition 42, there exists the (ordinary) Hopf algebra $\tilde{\mathcal{A}} \ltimes \mathcal{X}$.

Definition 12. *The Hopf algebra $\tilde{\mathcal{A}} \ltimes \mathcal{X}$ is called the* coordinate algebra of the inhomogeneous quantum group *IG_q and denoted by $\mathcal{O}(IG_q)$.*

We are now going to desribe the structure of the Hopf algebra $\mathcal{O}(IG_q)$ explicitly in terms of generators. We shall do this first in the case where $\tilde{\mathcal{A}} = \mathcal{A} \otimes \mathbb{C}\mathbb{Z}$ and recall briefly the structure of the Hopf algebra $\tilde{\mathcal{A}} \ltimes \mathcal{X}$. As a vector space, $\tilde{\mathcal{A}} \ltimes \mathcal{X}$ is $\mathcal{A} \otimes \mathbb{C}\mathbb{Z} \otimes \mathcal{X}$. We can consider $\mathcal{A}$ as a Hopf subalgebra of $\tilde{\mathcal{A}} \ltimes \mathcal{X}$ by identifying $a \in \mathcal{A}$ with $a \otimes 1 \otimes 1 \in \tilde{\mathcal{A}} \ltimes \mathcal{X}$. Likewise, $\mathbb{C}\mathbb{Z}$ becomes a Hopf subalgebra of $\tilde{\mathcal{A}} \ltimes \mathcal{X}$ and $\mathcal{X}$ a subalgebra of $\tilde{\mathcal{A}} \ltimes \mathcal{X}$ by identifying $g \in \mathbb{C}\mathbb{Z}$ with $1 \otimes g \otimes 1$ and $x \in \mathcal{X}$ with $1 \otimes 1 \otimes x$. As an algebra, $\tilde{\mathcal{A}} \ltimes \mathcal{X}$ is a crossed product algebra of $\tilde{\mathcal{A}}$ and $\mathcal{X}$. Therefore, as noted in Subsect. 10.3.6, it is the algebra generated by the algebras $\tilde{\mathcal{A}}$ (and so by $\mathcal{A}$ and $\mathbb{C}\mathbb{Z}$) and $\mathcal{X}$ satisfying the cross relations (112). The coalgebra structure of $\tilde{\mathcal{A}} \ltimes \mathcal{X}$ is that of a crossed coproduct coalgebra of $\tilde{\mathcal{A}}$ and $\mathcal{X}$ and described by Proposition 19. Finally, the coaction φ_R and the antipode of $\tilde{\mathcal{A}} \ltimes \mathcal{X}$ are given by (110) and (111), respectively. Inserting the data of the coaction ($\beta(x_i) = x_j \otimes u^j_i\chi$), of the universal r-form ($\tilde{\mathbf{r}}(\mathbf{u}_1, \mathbf{u}_2) = z\mathsf{R}_{12}$, $\tilde{\mathbf{r}}(u^j_i\chi, \chi) = \mathbf{r}(u^j_i, 1)\lambda^{-1} = \delta_{ij}qz^{-1}$) and of the braided Hopf algebra ($\Delta_{\mathcal{X}}(x_i) = x_i \otimes 1 + 1 \otimes x_i$, $\varepsilon_{\mathcal{X}}(x_i) = 0$, $S_{\mathcal{X}}(x_i) = -x_i$) in all these formulas, we obtain the following description of the Hopf algebra $\mathcal{O}(IG_q)$ for the quantum groups $G_q = SL_q(N), O_q(N), Sp_q(N)$.

Algebra structure of $\mathcal{O}(IG_q)$:

$\mathcal{O}(IG_q)$ has N^2+N+2 generators u^i_j, x_i, χ, χ^{-1}, $i,j = 1,2,\cdots,N$. The defining relations are the defining relations of $\mathcal{O}(G_q)$ for the generators u^i_j (see Sects. 9.2 and 9.3), the defining relations of $\mathcal{X}$ for the generators x_i (that is, $(\mathsf{P}_-)^{kl}_{ij}x_kx_l = 0$ for $i,j = 1,2,\cdots,N$) and the following relations:

$$x_n u^i_j = z\hat{\mathsf{R}}^{kl}_{nj} u^i_k x_l, \quad \chi u^i_j = u^i_j \chi, \quad x_i \chi = q z^{-1} \chi x_i, \quad \chi\chi^{-1} = \chi^{-1}\chi = 1.$$

Coalgebra structure of $\mathcal{O}(IG_q)$:

$$\Delta(u^i_j) = u^i_k \otimes u^k_j, \quad \Delta(x_i) = 1 \otimes x_i + x_j \otimes u^j_i \chi, \quad \Delta(\chi) = \chi \otimes \chi,$$

$$\varepsilon(u^i_j) = \delta_{ij}, \quad \varepsilon(x_i) = 0, \quad \varepsilon(\chi) = 1.$$

Antipode of $\mathcal{O}(IG_q)$:

$$S(u^i_j) = S_{\mathcal{O}(G_q)}(u^i_j), \quad S(x_i) = -x_j S_{\mathcal{O}(G_q)}(u^j_i)\chi^{-1}, \quad S(\chi) = \chi^{-1}.$$

Coaction φ_R of $\mathcal{O}(IG_q)$:

$\mathcal{X}$ is a right $\mathcal{O}(IG_q)$-comodule algebra with coaction $\varphi_R(x_i) = \Delta(x_i)$.

Let us emphasize that the preceding formulas already give a complete description of the structures of the Hopf algebra $\mathcal{O}(IG_q)$. One may also work alone with these formulas combined with the fact that they define a Hopf algebra and ignore the braided Hopf algebra approach presented above.

For the quantum group $G_q = GL_q(N)$ it suffices to add a generator $\mathcal{D}_q^{-1}$ and to require that it is the inverse of the quantum determinant $\mathcal{D}_q$. The corresponding formulas for $\mathcal{D}_q^{-1}$ then follow from the above structures. For example, the commutation relations for x_n and u^i_j imply that $x_n\mathcal{D}_q = z^N q\mathcal{D}_q x_n$, so we obtain $\mathcal{D}_q^{-1} x_n = z^N q x_n \mathcal{D}_q^{-1}$.

The new phenomenon for the quantum inhomogenous groups IG_q is the appearance of the additional element χ which is not present in the classical case. It can be interpreted as a scaling (or dilation) generator which is adjoined as a central element to the homogeneous part. The generator χ cannot be avoided if the parameter $\lambda = zq^{-1}$ is different from 1. Let us give two arguments for its necessity when $\lambda \neq 1$. First, tracing back the above approach without χ or more precisely with $\chi = 1$, it would require that $\mathcal{X}$ is a braided Hopf algebra associated with $\mathcal{A}$. However, the proof of Proposition 41 is not valid for $\lambda \neq 1$, because the term $(R+I)f(R) = (1-\lambda)\mathsf{P}_-$ of $\Delta(I_{ij})$ therein does not vanish. Secondly, having the above set of formulas, one can set $\chi = 1$ and try to verify directly that they define a Hopf algebra. Computing $(\mathsf{P}_-)^{kl}_{ij}\Delta(x_k x_l) = 0$ by using the formulas $\Delta(x_i) = 1 \otimes x_i + x_j \otimes u^j_i$ (recall that $\chi = 1$) and $x_n u^i_j = z\hat{\mathsf{R}}^{kl}_{nj} u^i_k x_l$ leads to the relations $(\lambda - 1)(\mathsf{P}_-)^{kl}_{ij} u^n_k x_l = 0$ for all $i, j, n = 1, 2, \cdots, N$. This is not reasonable when $\lambda \neq 1$.

Next let us discuss the case when $\lambda^k \equiv z^k q^{-k} = 1$ for some $k \in \mathbb{Z}$. Then, in our above approach, the group $\mathbb{Z}$ can be replaced by $\mathbb{Z}/k\mathbb{Z}$. In terms of generators this means that we have to add the equation $\chi^k = 1$ to the defining relations. For the quantum groups $SL_q(N)$, $O_q(N)$ and $Sp_q(N)$ an equation $\lambda^k = 1$ with $k \in \mathbb{Z}$ is only possible if q is a root of unity. However, in the case of $GL_q(N)$ the parameter $z \neq 0$ is arbitrary, so it may be choosen as q. Then we have $\lambda = 1$ and $\mathcal{X}$ is a braided Hopf algebra associated with $\mathcal{A} = \mathcal{O}(GL_q(N))$ by Proposition 40. Thus, in this case the generator χ can indeed be avoided.

That is, for $\mathcal{A} = \mathcal{O}(GL_q(N))$ there exists a Hopf algebra with $N^2 + N + 1$ generators $u^i_j, x_i, \mathcal{D}_q^{-1}$ and the defining relations and structure equations as listed above with $\chi := 1$. This Hopf algebra can be taken as the coordinate algebra of the inhomogeneous quantum group $IGL_q(N)$.

The Hopf algebra structure of $\mathcal{O}(IG_q)$ can be nicely kept in mind if we use the matrix

$$\tilde{\mathbf{u}} = \begin{pmatrix} \mathbf{u}\chi & 0 \\ \mathbf{x} & 1 \end{pmatrix}.$$

The above formulas for the comultiplication are obtained from the entries of the product of $\tilde{\mathbf{u}}$ by itself and for the antipode from the entries of the inverse matrix

$$\tilde{\mathbf{u}}^{-1} = \begin{pmatrix} S(\mathbf{u}) & 0 \\ -\mathbf{x}S(\mathbf{u})\chi^{-1} & 1 \end{pmatrix}.$$

10.3.9 $*$-Structures for Inhomogeneous Quantum Groups

As in earlier subsections, we shall distinguish between two cases of parameter values and reality types.

First we assume that $|q| = |z| = 1$. Let G_q be one of the real forms $GL_q(N;\mathbb{R})$, $SL_q(N;\mathbb{R})$, $O_q(n,n)$, $O_q(n,n+1)$ or $Sp_q(N;\mathbb{R})$ from Case 1 in Subsects. 9.2.4 or 9.3.5. Then, by Proposition 10(i), the universal r-form $\mathbf{r}_z$ of the Hopf $*$-algebra $\mathcal{A} = \mathcal{O}(G_q)$ is inverse real. It is easily checked that the Hopf algebra $\tilde{\mathcal{A}} = \mathcal{A} \otimes \mathbb{C}\mathbb{Z}$ becomes a Hopf $*$-algebra with the involution defined by $(a \otimes \chi^n)^* := a^* \otimes \chi^n$, $a \in \mathcal{A}$, $n \in \mathbb{Z}$, and that $\mathcal{X}$ is a right $\tilde{\mathcal{A}}$-comodule $*$-algebra. Since $|\lambda| = |zq^{-1}| = 1$, the universal r-form $\mathbf{r}_\lambda$ of $\tilde{\mathcal{A}}$ is also inverse real. Therefore, by Proposition 45 and the subsequent remarks, *the Hopf algebra* $\mathcal{O}(IG_q) = \tilde{\mathcal{A}} \ltimes \mathcal{X}$ *is a Hopf* $*$*-algebra with respect to the involution determined by the equations* $(u^i_j)^* = u^i_j$, $(x_i)^* = x_i$ *and* $\chi^* = \chi$.

From now on we suppose that q and z are real and we turn to the real forms $U_q(N)$, $SU_q(N;\nu)$, $O_q(N;\nu)$ and $Sp_q(N;\nu)$ treated in Case 2 of Subsects. 9.2.4 and 9.3.5. Then the universal r-form $\mathbf{r}_z$ of $\mathcal{A}$ is real (by Proposition 10(ii)) and Proposition 45 does not apply in general. In order to point out the difficulties that occur in this case, let us consider the Hopf $*$-algebra $\mathcal{A} = \mathcal{O}(O_q(N;\nu))$ with $\nu_1 = \cdots = \nu_N = 1$ and its right comodule $*$-algebra $\mathcal{X} = \mathcal{O}(O_q^N(1;\nu))$; see Subsect. 9.3.5, Case 2. As in the inverse real case above, we define the involution on $\tilde{\mathcal{A}}$ by $(a \otimes \chi^n)^* := a^* \otimes \chi^n$. Then $\tilde{\mathcal{A}}$ is a Hopf $*$-algebra and $\mathcal{X}$ is a right $\tilde{\mathcal{A}}$-comodule $*$-algebra. The universal r-form $\mathbf{r}_\lambda$ of $\tilde{\mathcal{A}}$ is real, because $\lambda = zq^{-1}$ is real. However, *there is no involution of the algebra* $\tilde{\mathcal{A}} \ltimes \mathcal{X}$ *which extends the involutions of* $\tilde{\mathcal{A}}$ *and* $\mathcal{X}$. Assume on the contrary that such an involution exists. Recall that by definition $(u^i_j)^* = S(u^j_i)$ and $(x_n)^* = q^{-\rho_n} x_{n'}$. Applying the involution to the equation $x_n u^i_j = z\hat{\mathsf{R}}^{kl}_{nj} u^i_k x_l$, then multiplying by u^m_j on the left and by u^i_s on the right and finally summing over j and i, we get $x_{n'} q^{-\rho_n} u^m_s = z\hat{\mathsf{R}}^{sl}_{nj} u^m_j x_{l'} q^{-\rho_l}$. Comparing the latter with the relation

$x_{n'}u^m_s = z\hat{\mathsf{R}}^{jl'}_{n's}u^m_j x_{l'}$ (note that the elements $u^m_j x_{l'} \simeq u^m_j \otimes 1 \otimes x_{l'} \in \tilde{\mathcal{A}} \otimes \mathcal{X}$ are linearly independent), we obtain $\hat{\mathsf{R}}^{sl}_{nj} q^{\rho_n - \rho_l} = \hat{\mathsf{R}}^{jl'}_{n's}$ for all j, l, n, s. This contradicts formula (9.30). Thus, the involutions of the $*$-algebras $\tilde{\mathcal{A}}$ and $\mathcal{X}$ cannot be extended to an involution of the algebra $\tilde{\mathcal{A}} \ltimes \mathcal{X}$. One way to circumvent this difficulty and to still get a Hopf $*$-algebra is to "double" the generators of the translation part. This procedure will be carried out in the rest of this subsection.

Let $\mathcal{X}_{\mathrm{Re}}$ denote the algebra $\mathcal{X}(f, g, h, \gamma; \mathsf{R})$ from Subsect. 9.1.2 with $f(t) = g(t) = t - q$, $h(t) = qt$, $\gamma = 0$. That is, $\mathcal{X}_{\mathrm{Re}}$ has $2N$ generators $x_1, \ldots, x_N$, $y_1, \cdots, y_N$ and the defining relations (9.11) of $\mathcal{X}_{\mathrm{Re}}$ read as

$$\hat{\mathsf{R}}^{kl}_{ij} x_k x_l = q x_i x_j, \quad \hat{\mathsf{R}}^{lk}_{ji} y_k y_l = q y_i y_j, \quad \hat{\mathsf{R}}^{ik}_{jl} x_k y_l = q^{-1} y_i x_j. \tag{119}$$

Let $\mathcal{X}$ and $\mathcal{Y}$ denote the algebras with generators $x_1, \cdots, x_N$ resp. $y_1, \cdots, y_N$ and defining relations $\hat{\mathsf{R}}^{kl}_{ij} x_k x_l = q x_i x_j$ resp. $\hat{\mathsf{R}}^{lk}_{ji} y_k y_l = q y_i y_j$. Clearly, $\mathcal{X}$ and $\mathcal{Y}$ are right $\tilde{\mathcal{A}}$-comodule algebras with coactions given by the formulas $\varphi_R(x_i) = x_j \otimes u^j_i \otimes \chi$ and $\varphi_R(y_i) = y_j \otimes S(u^i_j) \otimes \chi^{-1}$. Then the algebra $\mathcal{X}_{\mathrm{Re}}$ is nothing but the braided product algebra $\mathcal{X} \underline{\otimes} \mathcal{Y}$ with product defined by (94) with respect to the universal r-form $(\bar{\tilde{\mathbf{r}}})_{21}$ of $\tilde{\mathcal{A}}$, where $\tilde{\mathbf{r}}$ is given by (106) and $\bar{\tilde{\mathbf{r}}}$ is the convolution inverse of $\tilde{\mathbf{r}}$. Indeed, the cross commutation relations of the braided product algebra are

$$y_i x_j = x_k y_l \, (\bar{\tilde{\mathbf{r}}})_{21}(S(u^i_l) \otimes \chi^{-1}, u^k_j \otimes \chi) = q z^{-1} x_k y_l \, \mathbf{r}(u^k_j, u^i_l) = q \hat{\mathsf{R}}^{ik}_{jl} x_k y_l.$$

This is just the third group of relations in (119).

The algebra $\mathcal{X}_{\mathrm{Re}}$ becomes a $*$-algebra with involution determined by $x^*_i := \nu_i y_i$, where ν_i are the numbers in the definitions of $SU_q(N; \nu)$, $O_q(N; \nu)$ and $Sp_q(N; \nu)$ and $\nu_i := 1$ in the case of $U_q(N)$. To prove this assertion, it suffices to check that the relations (119) are invariant under this involution. Using the facts that $\hat{\mathsf{R}}^{kl}_{ij} = \hat{\mathsf{R}}^{kl}_{ij} \nu_i \nu_j \nu_k \nu_l = \hat{\mathsf{R}}^{ij}_{kl}$ this is easily done. In the special case $\nu_1 = \cdots = \nu_N = 1$ the elements x_i and y_i are interpreted as holomorphic and anti-holomorphic generators, respectively, and the $*$-algebra $\mathcal{X}_{\mathrm{Re}}$ may be considered as a realification of $\mathcal{X}$ (see Subsect. 10.2.7).

Obviously, the algebra $\tilde{\mathcal{A}}$ is also a $*$-algebra with involution $(a \otimes \chi^n)^* := a^* \otimes \chi^{-n}$, $a \in \mathcal{A}$, $n \in \mathbb{Z}$. Recall that, by Proposition 9.5, $\mathcal{X}_{\mathrm{Re}}$ is a right quantum space of $\mathcal{A}$. Hence $\mathcal{X}_{\mathrm{Re}}$ is also a right quantum space of $\tilde{\mathcal{A}}$ with coaction φ_R given by

$$\varphi_R(x_i) = x_j \otimes u^j_i \otimes \chi, \quad \varphi_R(y_i) = y_j \otimes S(u^i_j) \otimes \chi^{-1}.$$

From the relations $x^*_i = \nu_i y_i$, $(u^j_i)^* = \nu_i \nu_j S(u^i_j)$ and $\chi^* = \chi^{-1}$ we conclude that $\varphi_R(x^*_i) = \varphi_R(x_i)^*$. Therefore, $\mathcal{X}_{\mathrm{Re}}$ is a right $\tilde{\mathcal{A}}$-comodule $*$-algebra.

Next we turn $\mathcal{X}_{\mathrm{Re}}$ into a right $\tilde{\mathcal{A}}$-module algebra. Since $\tilde{\mathbf{r}}$ and $(\bar{\tilde{\mathbf{r}}})_{21}$ are universal r-forms of $\tilde{\mathcal{A}}$, it follows from Proposition 4 that $\mathcal{X}$ and $\mathcal{Y}$ are right $\tilde{\mathcal{A}}$-module algebras with actions defined by

$$x \triangleleft \tilde{a} := \sum x^{(0)} \tilde{\mathbf{r}}(x^{(1)}, \tilde{a}) \text{ and } y \triangleleft \tilde{a} := \sum y^{(0)} \bar{\tilde{\mathbf{r}}}(\tilde{a}, y^{(1)}), \quad \tilde{a} \in \tilde{\mathcal{A}}, \tag{120}$$

respectively. Hence the vector space $\mathcal{X} \otimes \mathcal{Y}$ is a right $\tilde{\mathcal{A}}$-module with respect to the action

$$(x \otimes y) \triangleleft \tilde{a} := \sum x \triangleleft \tilde{a}_{(1)} \otimes y \triangleleft \tilde{a}_{(2)}. \tag{121}$$

We shall show below that the algebra $\mathcal{X}_{\mathrm{Re}} \simeq \mathcal{X} \underline{\otimes} \mathcal{Y}$ even becomes a right $\tilde{\mathcal{A}}$-module *algebra* in this manner.

Now we consider the right crossed product algebra $\tilde{\mathcal{A}} \ltimes \mathcal{X}_{\mathrm{Re}}$ of the Hopf algebra $\tilde{\mathcal{A}}$ with its right module algebra $\mathcal{X}_{\mathrm{Re}}$. Recall from Subsect. 10.2.1 that $\tilde{\mathcal{A}} \ltimes \mathcal{X}_{\mathrm{Re}}$ is generated by the algebras $\tilde{\mathcal{A}}$ and $\mathcal{X}_{\mathrm{Re}}$ subject to the cross commutation relations $x\tilde{a} = \sum \tilde{a}_{(1)}(x \triangleleft \tilde{a}_{(2)})$. From (120) we get

$$x_n \triangleleft u^i_j = x_l \tilde{\mathbf{r}}(u^l_n \otimes \chi, u^i_j \otimes 1) = x_l\, \mathbf{r}(u^l_n, u^i_j) = x_l z \hat{\mathsf{R}}^{il}_{nj},$$

$$y_n \triangleleft u^i_j = y_l\, \bar{\tilde{\mathbf{r}}}(u^i_j \otimes 1, S(u^n_l) \otimes \chi^{-1}) = y_l\, \mathbf{r}(u^i_j, u^n_l) = y_l z \hat{\mathsf{R}}^{ni}_{jl},$$

$$x_n \triangleleft \chi = x_n \lambda^{-1} = x_n q z^{-1}, \quad y_n \triangleleft \chi = y_n q z^{-1}.$$

Hence $\tilde{\mathcal{A}} \ltimes \mathcal{X}_{\mathrm{Re}}$ is generated by the algebras $\tilde{\mathcal{A}}$ and $\mathcal{X}_{\mathrm{Re}}$ with cross relations

$$x_n u^i_j = z \hat{\mathsf{R}}^{kl}_{nj} u^i_k x_l, \quad y_n u^i_j = z \hat{\mathsf{R}}^{nk}_{jl} u^i_k y_l, \quad x_n \chi = q z^{-1} \chi x_n, \quad y_n \chi = q z^{-1} \chi y_n \tag{122}$$

for the generators. The involutions of $\tilde{\mathcal{A}}$ and $\mathcal{X}_{\mathrm{Re}}$ extend uniquely to an involution of $\tilde{\mathcal{A}} \ltimes \mathcal{X}_{\mathrm{Re}}$ such that $\tilde{\mathcal{A}} \ltimes \mathcal{X}_{\mathrm{Re}}$ becomes a $*$-algebra. Indeed, it is not difficult to show that the cross relations (122) are invariant under the involutions.

Proposition 46. *The $*$-algebra $\tilde{\mathcal{A}} \ltimes \mathcal{X}_{\mathrm{Re}}$ defined above is a Hopf $*$-algebra with comultiplication, counit and antipode determined by*

$$\Delta(u^i_j) = u^i_k \otimes u^k_j, \quad \Delta(\chi) = \chi \otimes \chi,$$

$$\Delta(y_i) = 1 \otimes y_i + y_j \otimes S(u^i_j)\chi^{-1}, \quad \Delta(x_i) = 1 \otimes x_i + x_j \otimes u^j_i \chi,$$

$$\varepsilon(u^i_j) = \delta_{ij}, \quad \varepsilon(x_i) = \varepsilon(y_i) = 0, \quad \varepsilon(\chi) = 1,$$

$$S(u^i_j) = S_{\mathcal{O}(G_q)}(u^i_j), \quad S(\chi) = \chi^{-1},$$

$$S(y_i) = -y_j S^2_{\mathcal{O}(G_q)}(u^i_j)\chi, \quad S(x_i) = -x_j S_{\mathcal{O}(G_q)}(u^j_i)\chi^{-1}.$$

Proof. The proof follows by similar reasoning as Proposition 9.1. We do not carry out the details. □

The Hopf $*$-algebra $\tilde{\mathcal{A}} \ltimes \mathcal{X}_{\mathrm{Re}}$ from Proposition 46 is called the coordinate Hopf $*$-algebra $\mathcal{O}(IG_q)$ of the *real inhomogeneous quantum group IG_q*, where G_q is one of the real quantum groups $U_q(N)$, $SU_q(N;\nu)$, $O_q(N;\nu)$ and $Sp_q(N;\nu)$. The structures of the Hopf $*$-algebra $\mathcal{O}(IG_q)$ can be summarized as follows:

As an algebra, $\mathcal{O}(IG_q)$ admits the generators $u^i_j, x_i, y_i, \chi, \chi^{-1}$. The defining relations are the defining relations of $\mathcal{O}(G_q)$ for u^i_j, the relations (119) and (122) for x_i and y_j and $\chi\chi^{-1} = \chi^{-1}\chi = 1$. The involution is given by the equations $(u^i_j)^* = \nu_i\nu_j S_{\mathcal{O}(G_q)}(u^j_i)$, $x^*_i = \nu_i y_i$ and $\chi^* = \chi^{-1}$. The coalgebra structure and the antipode are determined by the formulas in Proposition 46.

Knowing that $\tilde{\mathcal{A}} \ltimes \mathcal{X}_{\mathrm{Re}} = \mathcal{O}(IG_q)$ is a Hopf $*$-algebra (by Proposition 46), some of these formulas can be derived from others by means of axioms or properties of a Hopf $*$-algebra. For instance, by applying the involution the second relation of (119) leads to the first one, the first and the third relations of (122) yield the second and the fourth one, respectively, and the formulas of $\Delta(x_i), \varepsilon(x_i)$ and $S(x_i)$ imply the corresponding formulas for y_i.

The preceding constructions become clearer if they are reviewed in terms of the realification $\tilde{\mathcal{B}} = \mathcal{D}(\tilde{\mathcal{A}}, \tilde{\mathcal{A}}; \tilde{\mathbf{r}})$ of the Hopf $*$-algebra $\tilde{\mathcal{A}}$ with respect to the real universal r-form $\tilde{\mathbf{r}}$ (see Subsect. 10.2.7). This will be sketched below.

First recall that, by Proposition 29, $\tilde{\mathcal{B}}$ admits an inverse real universal r-form $\mathbf{r}'' = \bar{\tilde{\mathbf{r}}}_{41}\bar{\tilde{\mathbf{r}}}_{31}\tilde{\mathbf{r}}_{24}\tilde{\mathbf{r}}_{23}$. Obviously, $\mathcal{X}$ and $\mathcal{Y}$ are right $\tilde{\mathcal{B}}$-comodule algebras with coactions given by $\psi(x) = \sum x^{(0)} \otimes 1 \otimes x^{(1)}$ and $\psi(y) = \sum y^{(0)} \otimes y^{(1)} \otimes 1$, where $\varphi(x) = \sum x^{(0)} \otimes x^{(1)}$ and $\varphi(y) = \sum y^{(0)} \otimes y^{(1)}$ denote the coactions of $\tilde{\mathcal{A}}$ on $\mathcal{X}$ and $\mathcal{Y}$, respectively. Since $\mathbf{r}''(\tilde{a} \otimes 1, 1 \otimes \tilde{b}) = \bar{\tilde{\mathbf{r}}}_{21}(\tilde{a}, \tilde{b})$ for $\tilde{a}, \tilde{b} \in \tilde{\mathcal{A}}$, the braided product algebra $\mathcal{X}_{\mathrm{Re}} = \mathcal{X} \underline{\otimes} \mathcal{Y}$ with respect to $\tilde{\mathcal{A}}$ and $\bar{\tilde{\mathbf{r}}}_{21}$ coincides with the braided product algebra $\mathcal{X} \underline{\otimes} \mathcal{Y}$ with respect to $\tilde{\mathcal{B}}$ and $\mathbf{r}''$. Therefore, by Propositions 34 and 4, $\mathcal{X}_{\mathrm{Re}}$ is a right $\tilde{\mathcal{B}}$-comodule algebra with coaction

$$\psi(x \otimes y) = \sum x^{(0)} \otimes y^{(0)} \otimes (1 \otimes x^{(1)})(y^{(1)} \otimes 1) \tag{123}$$

and so a right $\tilde{\mathcal{B}}$-module algebra with action

$$(x \otimes y) \triangleleft (\tilde{a} \otimes \tilde{b}) := \sum x^{(0)} \otimes y^{(0)} \mathbf{r}''((1 \otimes x^{(1)})(y^{(1)} \otimes 1), \tilde{a} \otimes \tilde{b}).$$

Hence $\mathcal{X}_{\mathrm{Re}}$ is a right $\tilde{\mathcal{A}}$-module algebra with action $(x \otimes y) \triangleleft \tilde{a} := (x \otimes y) \triangleleft (\tilde{a} \otimes 1)$. For elements $x \equiv x \otimes 1 \in \mathcal{X}$ and $y \equiv 1 \otimes y \in \mathcal{Y}$ this is just the action of $\tilde{\mathcal{A}}$ on $\mathcal{X}$ and $\mathcal{Y}$ given by (121). Therefore, it follows that this action of $\tilde{\mathcal{A}}$ on $\mathcal{X}_{\mathrm{Re}}$ coincides with the action of $\tilde{\mathcal{A}}$ on $\mathcal{X} \otimes \mathcal{Y}$ defined by (122). In particular, as promised above, we have proved that *$\mathcal{X}_{\mathrm{Re}}$ is a right $\tilde{\mathcal{A}}$-module algebra.*

As we have seen above the involutions of $\tilde{\mathcal{A}}$ and $\mathcal{X}$ do not extend to a well-defined involution of the algebra $\tilde{\mathcal{A}} \ltimes \mathcal{X}$, while the involutions of $\tilde{\mathcal{A}}$ and $\mathcal{X}_{\mathrm{Re}}$ do for $\tilde{\mathcal{A}} \ltimes \mathcal{X}_{\mathrm{Re}}$. The deeper reason why $\tilde{\mathcal{A}} \ltimes \mathcal{X}_{\mathrm{Re}}$ becomes a $*$-algebra lies in the facts that $\mathcal{X}_{\mathrm{Re}}$ is a $\tilde{\mathcal{B}}$-comodule $*$-algebra with coaction ψ and

$$(\tilde{a} \otimes w)^* = \sum (\tilde{a}_{(1)})^* \otimes w^{(0)*} \mathbf{r}''(w^{(1)*}, (\tilde{a}_{(2)})^* \otimes 1), \quad \tilde{a} \in \tilde{\mathcal{A}}, \quad w \in \mathcal{X}_{\mathrm{Re}}, \tag{124}$$

where $\psi(w) = \sum w^{(0)} \otimes w^{(1)}$ is the Sweedler notation for the coaction ψ of $\tilde{\mathcal{B}}$ on $\mathcal{X}_{\mathrm{Re}}$ given by (123). Formula (124) shows that the involution of $\tilde{\mathcal{A}} \ltimes \mathcal{X}_{\mathrm{Re}}$ can be obtained in a similar manner as in Proposition 44 by means the *inverse real* universal r-form $\mathbf{r}''$ of the larger Hopf $*$-algebra $\tilde{\mathcal{B}}$. In order to prove formula (124), one first checks that it is true for $\tilde{a} = \tilde{a}'\tilde{a}''$ if it holds for $\tilde{a}'$ and for $\tilde{a}''$ and that it is valid for $z = z'z''$ if it holds for z' and for z''. Then it suffices to show (124) for the generators of $\tilde{\mathcal{A}}$ and $\mathcal{X}_{\mathrm{Re}}$. We omit these verifications. In the cases where $\tilde{a} = u^i_j \otimes 1$ and $z = x_n \otimes 1$ resp. $z = 1 \otimes y_n$ equation (124) turns out to be equivalent to the first two defining relations (122).

10.4 Notes

Coquasitriangular Hopf algebras appeared implicitly in the announcement [Ly1] and more explicitly in [Maj6]. They were first studied in detail by R. G. Larson and J. Towber in [LTo] and by T. Hayashi in [H2]. Theorem 7 can be found therein. Other early papers on coquasitriangular Hopf algebras are [Maj7], [S1], [Doi]. The theory of L-functionals for the quantized simple Lie groups was developed in [RTF]. Our Subsect. 10.1.3 deals with the general case of coquasitriangular Hopf algebras and contains a large amount of new material.

Crossed products are standard constructions in the Hopf algebra literature (see [Mon] and the references therein). The twisting of Hopf algebras was studied by V.G. Drinfeld [Dr5]. The approach to the quantum double via twisting is taken from [DT2]. Double crossed products of Hopf algebras were introduced in [Maj2]. Realifications of quantum groups have been investigated in [Pod4], [Zak], [DSWZ], [Maj8], [WoZ] and others.

Most of the constructions and results of Sect. 10.3 are due to S. Majid [Maj5], [Maj8–10]. The results on braided Hopf algebras and on transmutation of Hopf algebras can be found in [Maj9], [Maj8], while the bosonization theory was developed in [Maj10]. Hopf algebras in categories occurred much earlier (see [MM], [Pa]).

Inhomogeneous quantum groups were introduced and studied in [SWW1], [SWW2], [Maj8], [OSWZ], [Dt2]. Our main reference for Subsect. 10.3.9 is [Wei].

11. Corepresentation Theory and Compact Quantum Groups

The material in this chapter is related, directly or indirectly, to the corepresentation theory of Hopf algebras. In Sects. 11.2 and 11.3 we investigate two classes of Hopf algebras resp. Hopf $*$-algebras, called cosemisimple Hopf algebras resp. compact quantum group algebras, which are determined by their irreducible resp. unitary corepresentations. The second class leads at once to Woronowicz' compact quantum group C^*-algebras studied in Sect. 11.4. In Sect. 11.5 the irreducible representations of the quantum group $GL_q(N)$ when q is not a root of unity are classified. The final Sect. 11.6 gives an introduction to quantum homogeneous spaces.

In this chapter we retain the summation convention to sum over repeated indices.

11.1 Corepresentations of Hopf Algebras

In this section the basics of general corepresentation theory are developed. For the convenience of the reader, this section is self-contained and some notions which have already appeared in Chaps. 1 and 4 are repeated.

If not stated explicitly otherwise, $\mathcal{A}$ denotes a coalgebra.

11.1.1 Corepresentations

First we recall the following definition.

Definition 1. *A* corepresentation *of $\mathcal{A}$ on a complex vector space V is a linear map $\varphi : V \to V \otimes \mathcal{A}$ such that* $(\mathrm{id} \otimes \Delta) \circ \varphi = (\varphi \otimes \mathrm{id}) \circ \varphi$ *and* $(\mathrm{id} \otimes \varepsilon) \circ \varphi = \mathrm{id}$.

In other words, that φ is a corepresentation of $\mathcal{A}$ on V means that V is a right $\mathcal{A}$-comodule with right coaction φ.

We want to give some reformulations of the conditions in Definition 1 and assume that the vector space V is *finite-dimensional*. First we introduce some more notation. Recall that $\mathcal{L}(V)$ and $\mathcal{L}(V, V \otimes \mathcal{A})$ denote the vector spaces of linear mappings of V to V and of V to $V \otimes \mathcal{A}$, respectively. For an element $\check{\varphi} = \sum_i T_i \otimes a^i \in \mathcal{L}(V) \otimes \mathcal{A}$ with $T_i \in \mathcal{L}(V)$ and $a^i \in \mathcal{A}$, we define

$\varphi(.) = \sum_i T_i(.) \otimes a^i \in \mathcal{L}(V, V \otimes \mathcal{A})$. Since V is finite-dimensional, the mapping $\breve{\varphi} \to \varphi$ is an isomorphism of the vector spaces $\mathcal{L}(V) \otimes \mathcal{A}$ and $\mathcal{L}(V, V \otimes \mathcal{A})$.

Let $\varphi \in \mathcal{L}(V, V \otimes \mathcal{A})$ and let e_i, $i = 1, 2, \cdots, n$, be a basis of V. Then there exist uniquely determined elements $v_{ij} \in \mathcal{A}$, $i, j = 1, 2, \cdots, n$, such that $\varphi(e_j) = e_i \otimes v_{ij}$, $j = 1, 2, \cdots, n$. The elements v_{ij} are called the *matrix coefficients* or the *matrix elements* of φ with respect to the basis $\{e_i\}$. For $f \in \mathcal{A}'$, we define a linear mapping $\hat{\varphi}(f) \in \mathcal{L}(V)$ by setting $\hat{\varphi}(f) = (\mathrm{id} \otimes f)\varphi$. Recall that the dual $\mathcal{A}'$ of $\mathcal{A}$ is an algebra with product $fg = (f \otimes g)\Delta$ and unit element ε.

Proposition 1. *Let V be a finite-dimensional vector space and let $\varphi \in \mathcal{L}(V, V \otimes \mathcal{A})$. Then the following conditions are equivalent:*

(i) $(\mathrm{id} \otimes \Delta) \circ \varphi = (\varphi \otimes \mathrm{id}) \circ \varphi$.
(ii) $(\mathrm{id} \otimes \Delta) \circ \breve{\varphi} = \breve{\varphi}_{12}\breve{\varphi}_{13}$.
(iii) $\hat{\varphi}(fg) = \hat{\varphi}(f)\hat{\varphi}(g)$ *for* $f, g \in \mathcal{A}'$.
(iv) $\Delta(v_{ij}) = v_{ik} \otimes v_{kj}$ *for* $i, j = 1, 2, \cdots, n$.

Proof. The proof follows easily from the corresponding definitions combined with the coassociativity axiom. □

Proposition 2. *If V is a finite-dimensional vector space and $\varphi \in \mathcal{L}(V, V \otimes \mathcal{A})$ satisfies the conditions in Proposition 1, then the following are equivalent:*

(i) $(\mathrm{id} \otimes \varepsilon) \circ \varphi = \mathrm{id}$.
(ii) $\ker \varphi = \{0\}$.
(iii) $\hat{\varphi}(\varepsilon) = \mathrm{id}$.
(iv) $\varepsilon(v_{ij}) = \delta_{ij}$ *for* $i, j = 1, 2, \cdots, n$.

If $\mathcal{A}$ is a Hopf algebra, then these conditions are also equivalent to:

(v) $\breve{\varphi}$ *admits a left inverse in the algebra* $\mathcal{L}(V) \otimes \mathcal{A}$.
(vi) $\breve{\varphi}$ *is invertible in the algebra* $\mathcal{L}(V) \otimes \mathcal{A}$ *and* $\breve{\varphi}^{-1} = (\mathrm{id} \otimes S)\breve{\varphi}$.

Proof. (ii)→(iii): Since $\hat{\varphi}(\varepsilon)\hat{\varphi}(\varepsilon) = \hat{\varphi}(\varepsilon)$ by Proposition 1(iii), $\hat{\varphi}(\varepsilon)$ is a projection. If $x \in \ker \hat{\varphi}(\varepsilon)$, then $\hat{\varphi}(f)(x) = \hat{\varphi}(f\varepsilon)(x) = \hat{\varphi}(f)\hat{\varphi}(\varepsilon)(x) = 0$ for any $f \in \mathcal{A}'$, hence $\varphi(x) = 0$. Thus, $\ker \hat{\varphi}(\varepsilon) = \{0\}$ and $\hat{\varphi}(\varepsilon) = \mathrm{id}$.

(iii)→(vi) for a Hopf algebra $\mathcal{A}$: Let $\breve{\varphi} = \sum_i T_i \otimes a^i$ with $T_i \in \mathcal{L}(V)$ and $a^i \in \mathcal{A}$. The equation $(\mathrm{id} \otimes \Delta) \circ \breve{\varphi} = \breve{\varphi}_{12}\breve{\varphi}_{13}$ by Proposition 1(ii) yields the identity $\sum_i \sum T_i \otimes a^i_{(1)} \otimes a^i_{(2)} = \sum_{i,j} T_i T_j \otimes a^i \otimes a^j$. Applying $\mathrm{id} \otimes m(\mathrm{id} \otimes S)$ to both sides, we obtain $\hat{\varphi}(\varepsilon) \otimes 1 = \sum_{i,j} T_i T_j \otimes a^i S(a^j)$. Since $\hat{\varphi}(\varepsilon) = \mathrm{id}$ by (iii), this means that $(\mathrm{id} \otimes S)\breve{\varphi}$ is a right inverse of $\breve{\varphi}$ in the algebra $\mathcal{L}(V) \otimes \mathcal{A}$. Replacing the mapping $\mathrm{id} \otimes m(\mathrm{id} \otimes S)$ by $\mathrm{id} \otimes m(S \otimes \mathrm{id})$ in the preceding, it follows that $(\mathrm{id} \otimes S)\breve{\varphi}$ is also a left inverse of $\breve{\varphi}$.

The implications (i) ↔ (iv) → (ii) and (v) → (ii) are easily verified. The equivalence (i) ↔ (iii) is just the definition of $\hat{\varphi}(\varepsilon)$. (vi) → (v) is trivial. □

A matrix $v = (v_{ij})_{i,j=1,\cdots,n}$ of n^2 elements $v_{ij} \in \mathcal{A}$ satisfying

$$\Delta(v_{ij}) = v_{ik} \otimes v_{kj} \quad \text{and} \quad \varepsilon(v_{ij}) = 1, \quad i, j = 1, 2, \cdots, n,$$

is called a *matrix corepresentation* of $\mathcal{A}$. By Propositions 1 and 2, such a matrix v defines a corepresentation φ on an n-dimensional vector space V with basis e_i, $i = 1, 2, \cdots, n$, by $\varphi(e_j) = e_i \otimes v_{ij}$. Sometimes we abuse language and shall not distinguish between corepresentations and corresponding matrix corepresentations.

11.1.2 Intertwiners

Let φ and ψ be corepresentations of $\mathcal{A}$ on finite-dimensional vector spaces V and W, respectively, and let $v = (v_{ij})$ and $w = (w_{ij})$ be the corresponding matrix corepresentations of $\mathcal{A}$ with respect to bases e_i, $i = 1, 2, \cdots, n$, and f_k, $k = 1, 2, \cdots, m$, of V and W, respectively. Let $T : V \to W$ be a linear mapping. We write $T(e_i) = \sum_k T_{ki} f_k$.

Proposition 3. *In the above notation, the following are equivalent:*

(i) $(T \otimes \mathrm{id}) \circ \varphi = \psi \circ T$.

(ii) $(T \otimes \mathrm{id}) \circ \check{\varphi} = \check{\psi} \circ (T \otimes \mathrm{id})$.

(iii) $T\hat{\varphi}(f) = \hat{\psi}(f)T$ *for* $f \in \mathcal{A}'$.

(iv) $\sum_i T_{ki} v_{ij} = \sum_l w_{kl} T_{lj}$ *for* $j = 1, 2, \cdots, n$ *and* $k = 1, 2, \cdots, m$.

Proof. The proof is straightforward and is omitted. □

Definition 2. *A linear mapping $T : V \to W$ satisfying the conditions in Proposition 3 is called an* intertwiner *of φ and ψ. The vector space of all such operators T is denoted by* Mor (φ, ψ). *We set* Mor $(\varphi) :=$ Mor (φ, φ).

The space of matrices satisfying condition (iv) is denoted by Mor (v, w). We write Mor $(v) :=$ Mor (v, w) if $V = W$, $e_i = f_i$ and $v_{ij} = w_{ij}$. Note that Mor (φ) and Mor (v) are subalgebras of $\mathcal{L}(V)$ and $M_n(\mathbb{C})$, respectively.

Two corepresentations φ and ψ of $\mathcal{A}$ are said to be *equivalent* if there exists an invertible intertwiner of φ and ψ. In this case we write $\varphi \simeq \psi$.

Remark 1. Those parts of Propositions 1–3 which do not contain the mapping $\check{\varphi}$ remain valid for infinite-dimensional vector spaces. The corresponding matrices (v_{ij}), (w_{kl}) and (T_{ki}) have the property that for a fixed second index all but finitely many terms vanish, so all occurring sums are still finite. △

11.1.3 Constructions of New Corepresentations

The direct sum $\bigoplus_\nu \varphi_\nu$ of a family $\{\varphi_\nu\}$ of corepresentations of $\mathcal{A}$ is defined in the obvious manner as the direct sum of the linear mappings φ_ν.

a) *Tensor product of corepresentations*: Let φ and ψ be corepresentations of a bialgebra $\mathcal{A}$ on vector spaces V and W, respectively. Then the mapping

$$\varphi \otimes \psi := (\mathrm{id} \otimes \mathrm{id} \otimes m)(\mathrm{id} \otimes \tau \otimes \mathrm{id})(\varphi \otimes \psi) : \; V \otimes W \to V \otimes W \otimes \mathcal{A}$$

defines a corepresentation of $\mathcal{A}$ on $V \otimes W$. It is called the *tensor product* of φ and ψ. In Sweedler's notation, we have

$$(\varphi \otimes \psi)(x \otimes y) = \sum x_{(0)} \otimes y_{(0)} \otimes x_{(1)} y_{(1)}, \qquad x \in V,\ y \in W.$$

Retaining the notation of Subsect. 11.1.2, the matrix corepresentation of the tensor product $\varphi \otimes \psi$ with respect to the basis $\{e_i \otimes f_k\}$ of $V \otimes W$ is given by $v \otimes w \equiv \big((v \otimes w)_{ik,jl}\big) = (v_{ij} w_{kl})$.

Clearly, the tensor product of corepresentations is associative. If $\mathcal{A}$ is noncommutative, it is no longer commutative in the usual sense that the flip provides an equivalence of $\varphi \otimes \psi$ and $\psi \otimes \varphi$. (This may be seen by comparing the corresponding matrix elements.) However, if $\mathcal{A}$ is a coquasitriangular bialgebra, Proposition 10.1(i) states that $\varphi \otimes \psi$ and $\psi \otimes \varphi$ are equivalent.

b) *Conjugate corepresentation*: Let φ be a corepresentation of a $*$-coalgebra $\mathcal{A}$ (that is, $\mathcal{A}$ is a coalgebra and a $*$-algebra such that Δ and ε are $*$-preserving) on a vector space V. Let $\overline{V}$ be the complex conjugate vector space of V, that is, $\overline{V}$ consists of elements $\overline{x}$, where $x \in V$, and the vector space structure of $\overline{V}$ is determined by $\overline{x} + \overline{y} := \overline{x+y}$ and $\lambda\overline{x} := \overline{\overline{\lambda} x}$, $x, y \in V$, $\lambda \in \mathbb{C}$. Put $\overline{\varphi} := (- \otimes *)\varphi$, where $-$ and $*$ denote the mappings $x \to \overline{x}$ and $a \to a^*$, respectively. Then $\overline{\varphi} : \overline{V} \to \overline{V} \otimes \mathcal{A}$ is a corepresentation of $\mathcal{A}$ on $\overline{V}$ and the corresponding matrix corepresentation is $\overline{v} \equiv (\overline{v}_{ij}) = ((v_{ij})^*)$.

c) *Contragredient corepresentation*: Let φ be a finite-dimensional corepresentation of a Hopf algebra $\mathcal{A}$ on V. For $T \in \mathcal{L}(V)$, let $T^t \in \mathcal{L}(V')$ denote the transposed mapping of T, that is, $(T^t x')(x) = x'(Tx)$ for $x' \in V'$, $x \in V$. If $\check{\varphi} = \sum_i T_i \otimes a^i$ with $T_i \in \mathcal{L}(V)$ and $a^i \in \mathcal{A}$, define $\check{\varphi}^c := \sum_i T_i^t \otimes S(a^i) \in \mathcal{L}(V') \otimes \mathcal{A}$. The corresponding map $\varphi^c \in \mathcal{L}(V', V' \otimes \mathcal{A})$ is uniquely determined by the requirement $\varphi^c(x')(x \otimes \mathrm{id}) = (x' \otimes \mathrm{id})\varphi(x)$, $x' \in V'$, $x \in V$. If φ is represented by the matrix $v = (v_{ij})$ with respect to a basis of V, then φ^c is obviously given by the matrix $v^c \equiv \big((v^c)_{ij}\big) = \big(S(v_{ji})\big)$ with respect to the dual basis of V'. Then φ^c is a corepresentation of $\mathcal{A}$ on V'. (The easiest way to see this might be to check the conditions (iv) in Propositions 1 and 2.) We call φ^c the *contragredient corepresentation* of φ.

11.1.4 Irreducible Corepresentations

Let φ be a corepresentation of $\mathcal{A}$ on a vector space V. A linear subspace W of V is called *φ-invariant* if $\varphi(W) \subseteq W \otimes \mathcal{A}$ or, equivalently, if $\hat{\varphi}(f)(W) \subseteq W$ for all functionals $f \in \mathcal{A}'$. Then the restriction of φ to W is obviously a corepresentation of $\mathcal{A}$ on W, called a *subcorepresentation* of φ. An element $x \in V$ is said to be *φ-invariant* if $\varphi(x) = x \otimes 1$ or, equivalently, if $\hat{\varphi}(f)(x) = f(1)x$ for all $f \in \mathcal{A}'$.

Definition 3. *A corepresentation φ of $\mathcal{A}$ on a (not necessarily finite-dimensional) vector space V is said to be* irreducible *if φ has only the trivial invariant subspaces $W = \{0\}$ and $W = V$. A coalgebra $\mathcal{A}$ is called* cosimple *if it has no nonzero proper subcoalgebra.*

Remark 2. In the Hopf algebra literature, the irreducible corepresentations are the simple right comodules and a cosimple coalgebra is usually called a simple coalgebra. However, the name "cosimple coalgebra" seems to be more appropriate for the latter, because it can be shown that a coalgebra $\mathcal{A}$ is cosimple if and only if $\mathcal{A}'$ is a finite-dimensional simple algebra. $\triangle$

Next we prove two finiteness results.

Proposition 4. (i) *If φ is a corepresentation of $\mathcal{A}$ on V, then any element $x \in V$ is contained in a finite-dimensional φ-invariant subspace.*

(ii) *Any element $a \in \mathcal{A}$ belongs to a finite-dimensional subcoalgebra of $\mathcal{A}$.*

Proof. (i): Let $\{c_i\}$ be a basis of the vector space $\mathcal{A}$. Write $\varphi(x) = \sum_i x_i \otimes c_i$ and $\Delta(c_i) = \sum_{j,k} \gamma_{ijk} c_j \otimes c_k$ with $x_i \in V$ and $\gamma_{ijk} \in \mathbb{C}$. (All sums in this proof are finite.) Comparing the coefficients of c_k in the identity

$$\sum_k \varphi(x_k) \otimes c_k = (\varphi \otimes \mathrm{id})\varphi(x) = (\mathrm{id} \otimes \Delta)\varphi(x) = \sum_{i,j,k} x_i \otimes \gamma_{ijk} c_j \otimes c_k$$

we conclude that $\varphi(x_k) = \sum_{i,j} x_i \otimes \gamma_{ijk} c_j$. Thus, $W := \mathrm{Lin}\,\{x, x_i\}$ is a finite-dimensional φ-invariant subspace.

(ii): By (i) applied to $V = \mathcal{A}$, $\varphi = \Delta$, there is a finite-dimensional subspace W of $\mathcal{A}$ such that $a \in W$ and $\Delta(W) \subseteq W \otimes \mathcal{A}$. Let $\{e_i\}$ be a basis of W and let $\Delta(e_j) = e_i \otimes w_{ij}$. Since $\Delta(w_{ij}) = w_{ik} \otimes w_{kj}$ by Proposition 1, $\mathcal{B} = \mathrm{Lin}\,\{e_i, w_{ij}\}$ satisfies $\Delta(\mathcal{B}) \subseteq \mathcal{B} \otimes \mathcal{B}$ and $a \in \mathcal{B}$. $\square$

Corollary 5. (i) *Each irreducible corepresentation is finite-dimensional.*

(ii) *Any cosimple coalgebra is finite-dimensional.*

The following very simple but very useful result, called *Schur's lemma*, will often be used in the following.

Proposition 6. *Let φ and ψ be irreducible corepresentations of $\mathcal{A}$. Then:*

(i) *If φ and ψ are not equivalent, then* $\mathrm{Mor}\,(\varphi, \psi) = \{0\}$.

(ii) *If φ and ψ are equivalent, then* $\dim \mathrm{Mor}\,(\varphi, \psi) = 1$ *and any* $T \in \mathrm{Mor}\,(\varphi, \psi)$ *is either 0 or invertible. In particular,* $\mathrm{Mor}\,(\varphi) = \mathbb{C} \cdot I$.

Proof. The proof follows immediately from the facts that $\ker T$ is φ-invariant and $\mathrm{im}\, T$ is ψ-invariant for any $T \in \mathrm{Mor}\,(\varphi, \psi)$. $\square$

Proposition 7. *Any irreducible corepresentation φ of $\mathcal{A}$ is equivalent to a subcorepresentation of the corepresentation Δ on $\mathcal{A}$.*

Proof. Take a functional $f \in V'$, $f \neq 0$. One easily verifies that $T := (f \otimes \mathrm{id})\varphi$ belongs to $\mathrm{Mor}\,(\varphi, \Delta)$. Since $\varepsilon \circ T = f$, we have $T \neq 0$. Since $\ker T$ is a φ-invariant subspace, T is injective and provides an equivalence of φ and the restriction of Δ to its invariant subspace $\mathrm{im}\, T$. $\square$

Let φ be a corepresentation of $\mathcal{A}$ on an n-dimensional vector space V and let $v = (v_{ij})$ be the corresponding matrix corepresentation with respect to some basis of V. Then we have

$$\begin{aligned} \mathrm{Lin}\,\{(x' \otimes \mathrm{id})\varphi(x) \mid x' \in V', x \in V\} &= \mathrm{Lin}\,\{(f \otimes \mathrm{id})\check{\varphi} \mid f \in \mathcal{L}(V)'\} \\ &= \mathrm{Lin}\,\{v_{ij} \mid i, j = 1, 2, \cdots, n\}. \end{aligned}$$

From the latter and Proposition 1(iv) we see that this vector space is a subcoalgebra of $\mathcal{A}$. It will be denoted by $\mathcal{C}(\varphi)$ or likewise by $\mathcal{C}(v)$.

Definition 4. *$\mathcal{C}(\varphi)$ (resp. $\mathcal{C}(v)$) is called the* coefficient coalgebra *of the corepresentation φ (resp. of the matrix corepresentation v).*

If φ and ψ are equivalent corepresentations, then $\mathcal{C}(\varphi) = \mathcal{C}(\psi)$. Further, we have

$$\mathcal{C}(\varphi \oplus \psi) = \mathcal{C}(\varphi) + \mathcal{C}(\psi), \quad \mathcal{C}(\varphi \otimes \psi) = \mathcal{C}(\varphi)\mathcal{C}(\psi),$$
$$\mathcal{C}(\overline{\varphi}) = \mathcal{C}(\varphi)^*, \quad \mathcal{C}(\varphi^c) = S(\mathcal{C}(\varphi)).$$

For the second formula we have to assume that $\mathcal{A}$ is a bialgebra, for the third that $\mathcal{A}$ is $*$-coalgebra and for the fourth that $\mathcal{A}$ is a Hopf algebra.

The next proposition shows that the cosimple subcoalgebras of $\mathcal{A}$ are precisely the coefficient coalgebras of irreducible corepresentations of $\mathcal{A}$.

Proposition 8. (i) *A corepresentation φ of $\mathcal{A}$ is irreducible if and only if the coalgebra $\mathcal{C}(\varphi)$ is cosimple. Any cosimple subcoalgebra of $\mathcal{A}$ is of the form $\mathcal{C}(\varphi)$ for some irreducible corepresentation φ of $\mathcal{A}$.*

(ii) *Let φ be an irreducible corepresentation of $\mathcal{A}$ on an n-dimensional vector space V. Then the corepresentation $\Delta_{\mathcal{C}(\varphi)} : \mathcal{C}(\varphi) \to \mathcal{C}(\varphi) \otimes \mathcal{A}$ of $\mathcal{A}$ on $\mathcal{C}(\varphi)$ is equivalent to a direct sum of n corepresentations φ. In particular,* dim $\mathcal{C}(\varphi) = n^2$ *and the matrix elements of φ with respect to any basis of V form a vector space basis of $\mathcal{C}(\varphi)$.*

Proof. (i): If ψ is the restriction of φ to a nontrivial invariant subspace of φ, then $\mathcal{C}(\psi)$ is a nontrivial subcoalgebra of $\mathcal{C}(\varphi)$. So $\mathcal{C}(\varphi)$ is not cosimple if φ is not irreducible. The converse will follow from the proof of (ii) given below.

Let $\mathcal{B}$ be a cosimple subcoalgebra of $\mathcal{A}$. Since $\mathcal{B}$ is finite-dimensional by Corollary 5, the corepresentation $\Delta_{\mathcal{B}} : \mathcal{B} \to \mathcal{B} \otimes \mathcal{A}$ of $\mathcal{A}$ on $\mathcal{B}$ admits a nonzero invariant subspace $\mathcal{E}$ which does not contain any proper invariant subspace. Then the restriction φ of $\Delta_{\mathcal{B}}$ to $\mathcal{E}$ is an irreducible corepresentation of $\mathcal{A}$ and $\mathcal{C}(\varphi)$ is a nonzero subcoalgebra of $\mathcal{B}$. Since $\mathcal{B}$ is cosimple, we have $\mathcal{C}(\varphi) = \mathcal{B}$.

(ii): The proof consists of showing that $V' \otimes V$ considered as a two-sided (that is, right and left) comodule of $\mathcal{A}$ is irreducible. Let $e_1, \cdots, e_n$ and $e'_1, \cdots, e'_n$ be bases of V and V', respectively, and let $\varphi(e_j) = e_i \otimes v_{ij}$. Define right and left $\mathcal{A}$-comodule structures on $V' \otimes V$ by $\varphi_R(e'_i \otimes e_j) = e'_i \otimes e_k \otimes v_{kj}$ and $\varphi_L(e'_i \otimes e_j) = v_{ik} \otimes e'_k \otimes e_j$, respectively. Since $\Delta_{\mathcal{C}(\varphi)}(v_{ij}) = v_{ik} \otimes v_{kj}$, the linear mapping $T : V' \otimes V \to \mathcal{C}(\varphi)$ given by $T(e'_i \otimes e_j) = v_{ij}$ intertwines φ_R (resp. φ_L) and the comultiplication $\Delta_{\mathcal{C}(\varphi)}$. Hence $\mathcal{K} :=$ ker T is an invariant subspace for φ_R and φ_L.

We prove that $\mathcal{K} = \{0\}$. Assume on the contrary that $\mathcal{K} \neq \{0\}$. Take a nonzero φ_R-invariant subspace $\mathcal{K}_0$ of $\mathcal{K}$ such that the restriction ψ of φ_R to $\mathcal{K}_0$ is irreducible. Define linear mappings $A_i : V' \otimes V \to V$ by $A_i(e'_j \otimes x) = \delta_{ij} x$, $x \in V$. Then the restriction B_i of A_i to $\mathcal{K}_0$ belongs to Mor (ψ, φ), so B_i is either zero or bijective by Schur's lemma. Since $\mathcal{K}_0 \neq \{0\}$, $B_k \neq 0$ for some k. Then $B_i B_k^{-1} \in$ Mor (φ), so that $B_i B_k^{-1} = \lambda_i I$ with $\lambda_i \in \mathbb{C}$ again by

Schur's lemma. Hence any nonzero $y \in \mathcal{K}_0$ is of the form $y = \sum_i e'_i \otimes B_i(y) = (\sum_i \lambda_i e'_i) \otimes B_k(y) =: x' \otimes x$ with $x' \neq 0$ and $x \neq 0$. Since $\mathcal{K}$ is invariant under φ_R and φ_L, it follows that $\mathcal{K} \supseteq x' \otimes V$ and $\mathcal{K} \supseteq V' \otimes x$, so $\mathcal{K} = V' \otimes V$ and $T = 0$. The latter implies that $\varphi = 0$, which is a contradiction. Thus, we have proved that $\mathcal{K} = \ker T = \{0\}$.

Hence T is bijective and provides an equivalence of a direct sum of n corepresentations φ and the corepresentation $\Delta_{\mathcal{C}(\varphi)}$. Moreover, the elements v_{ij} constitute a basis of $\mathcal{C}(\varphi)$. Therefore, the dual algebra $\mathcal{C}(\varphi)'$ is isomorphic to $M_n(\mathbb{C})$ (in fact, the dual basis to $\{v_{ij}\}$ corresponds to the matrix units). Thus, $\mathcal{C}(\varphi)'$ is simple and hence $\mathcal{C}(\varphi)$ is cosimple. □

Corollary 9. *If φ and ψ are irreducible corepresentations of $\mathcal{A}$, then either $\mathcal{C}(\varphi) = \mathcal{C}(\psi)$ and φ and ψ are equivalent or $\mathcal{C}(\varphi) \cap \mathcal{C}(\psi) = \{0\}$ and φ and ψ are not equivalent.*

Proof. Assume that $\mathcal{C}(\varphi) \cap \mathcal{C}(\psi) \neq \{0\}$. Since $\mathcal{C}(\varphi) \cap \mathcal{C}(\psi)$ is a subcoalgebra of the cosimple coalgebras $\mathcal{C}(\varphi)$ and $\mathcal{C}(\psi)$, it follows that $\mathcal{C}(\varphi) = \mathcal{C}(\psi)$. Thus, $\Delta_{\mathcal{C}(\varphi)}$ and $\Delta_{\mathcal{C}(\psi)}$ are equivalent corepresentations, hence φ and ψ are equivalent by Proposition 8(ii). □

Corollary 10. *Let $\{\varphi^\alpha \mid \alpha \in I\}$ be a family of pairwise nonequivalent irreducible corepresentations of $\mathcal{A}$ with matrix corepresentations $\{v^\alpha = (v^\alpha_{ij})_{i,j=1,\cdots,d_\alpha} \mid \alpha \in I\}$. Then the set $\{v^\alpha_{ij} \mid i,j = 1,2,\cdots,d_\alpha,\ \alpha \in I\}$ of matrix elements is linearly independent.*

Proof. The proof follows from Proposition 8(ii) and Corollary 9. □

11.1.5 Unitary Corepresentations

In this subsection we suppose that $\mathcal{A}$ is a Hopf $*$-algebra.

Let V be a finite-dimensional Hilbert space with scalar product $\langle \cdot, \cdot \rangle$ conjugate linear in the first variable and linear in the second one. Then $\mathcal{L}(V)$ and $\mathcal{L}(V) \otimes \mathcal{A}$ are $*$-algebras. Let $\varphi : V \to V \otimes \mathcal{A}$ be a corepresentation of $\mathcal{A}$ on V and let $v = (v_{ij})$ be the corresponding matrix corepresentation with respect to an orthonormal basis $\{e_i\}$ of V. Set $v^* \equiv ((v^*)_{ij}) = ((v_{ji})^*)$.

Proposition 11. *The following conditions are equivalent:*

(i) $\sum \langle x, y_{(0)} \rangle y_{(1)} = \sum \langle x_{(0)}, y \rangle S(x_{(1)})^*, \quad x, y \in V.$
(ii) $\sum \langle x_{(0)}, y_{(0)} \rangle x_{(1)}{}^* y_{(1)} = \langle x, y \rangle 1, \quad x, y \in V.$
(iii) $\breve{\varphi}^* \breve{\varphi} = \breve{\varphi} \breve{\varphi}^* = I$ *in the $*$-algebra* $\mathcal{L}(V) \otimes \mathcal{A}$.
(iv) $\breve{\varphi}^* \breve{\varphi} = I$ *(resp.* $\breve{\varphi} \breve{\varphi}^* = I$*) in the $*$-algebra* $\mathcal{L}(V) \otimes \mathcal{A}$.
(v) $v^* v = v v^* = I$ *(that is,* $S(v_{ij}) = (v_{ji})^*$ *for all* i, j*).*
(vi) $v^* v = I$ *(resp.* $v v^* = I$*).*

Proof. If $e_{kl} \in \mathcal{L}(V)$ are the matrix units defined by $e_{kl}(e_i) = \delta_{ik} e_l$, then $\breve{\varphi}$ can be written as $\breve{\varphi} = \sum_{i,j} e_{ji} \otimes v_{ij}$. From this we immediately obtain that (iii) $\leftrightarrow$ (v) and (iv) $\leftrightarrow$ (vi). Setting $x = e_j, y = e_i$ into (i) yields $v_{ij} = S(v_{ji})^*$, so $v^* v = v v^* = I$. Conversely, the relations $v_{ij} = S(v_{ji})^*$ imply (i). Thus,

(i) $\leftrightarrow$ (v). Similarly, (ii) is equivalent to the equality $v^*v = I$. Multiplying $v^*v = I$ (resp. $vv^* = I$) by $S(v)$ on the right (resp. on the left), we obtain $v^* = S(v)$ and so $v^*v = vv^* = I$. This proves (v) $\leftrightarrow$ (vi). □

Definition 5. *A corepresentation φ of $\mathcal{A}$ on a finite-dimensional Hilbert space V is called* unitary *if it satisfies the conditions in Proposition 11. A corepresentation φ on a finite-dimensional vector space V is said to be* unitarizable *if there exists a scalar product on V such that φ is unitary.*

Let $v = (v_{ij})_{i,j=1,\cdots,n}$ be the matrix corepresentation of a corepresentation φ of $\mathcal{A}$ on a vector space V. Then φ is unitarizable if and only if there exists an invertible matrix $A \in M_n(\mathbb{C})$ such that $w := AvA^{-1}$ satisfies the equality $S(w) = w^*$ or, equivalently, if $BS(v)B^{-1} = v^*$ with $B = \overline{A}^t A$.

It is clear from Proposition 11 that direct sums, tensor products and subcorepresentations of unitary (resp. unitarizable) corepresentations are again unitary (resp. unitarizable). However, contragredient and conjugate corepresentations of unitary corepresentations are not necessarily unitary.

Proposition 12. *Any unitarizable (in particular, any unitary) corepresentation is completely reducible, that is, it is a direct sum of irreducible corepresentations.*

Proof. It suffices to show that for a unitary corepresentation φ the orthogonal complement $W^\perp$ of a φ-invariant subspace W in V is also φ-invariant. To prove this, let $x \in W$ and $y \in W^\perp$. Write $\varphi(y) = \sum_i y_i \otimes a^i$ such that $\{a^i\}$ is a linearly independent set. Since W is φ-invariant and $y \in W^\perp$, the right hand side of the equation in Proposition 11(i) vanishes, so $\sum\langle x, y_i\rangle a^i = 0$. Hence, $\langle x, y_i\rangle = 0$ for arbitrary $x \in W$ which implies $y_i \in W^\perp$. □

11.2 Cosemisimple Hopf Algebras

This section is concerned with Hopf algebras which are spanned by the matrix elements of their irreducible corepresentations.

11.2.1 Definition and Characterizations

First we put the subsequent considerations in the proper algebraic context and give

Definition 6. *A coalgebra is called* cosemisimple *if it is the sum of its cosimple subcoalgebras (that is, of coalgebras which have no nonzero proper subcoalgebras).*

Example 1. Any group algebra $\mathbb{C}G$ is obviously cosemisimple. △

Example 2. The universal enveloping algebra $U(\mathfrak{g})$ of a Lie algebra $\mathfrak{g} \neq \{0\}$ is not cosemisimple, because its only cosimple subcoalgebra is $\mathbb{C}\cdot 1$. △

Example 3. For the Hopf algebra of polynomial functions on an algebraic group G cosemisimplicity means that G is *linearly reductive*, that is, the finite-dimensional polynomial representations of G are completely reducible (see [Ho], Sect. V.3). A large number of classical groups such as $SL(N, \mathbb{C})$, $SO(N, \mathbb{C})$, $Sp(N, \mathbb{C})$ has this property. △

Example 4. If q is transcendental, the Hopf algebras $\mathcal{O}(GL_q(N))$, $\mathcal{O}(SL_q(N))$, $\mathcal{O}(O_q(N))$, $\mathcal{O}(Sp_q(N))$ are cosemisimple (see Subsects. 11.2.3 and 11.5.3). △

Let $\mathcal{A}$ be a coalgebra. Denote by $\hat{\mathcal{A}}$ the set of equivalence classes of irreducible corepresentations of $\mathcal{A}$. For any $\alpha \in \hat{\mathcal{A}}$, we fix a corepresentation φ^α from the class α and a basis of the underlying space. Then φ^α is given by a matrix corepresentation $u^\alpha = (u^\alpha_{ij})_{i,j=1,\cdots,d_\alpha}$. Recall from Subsect. 11.1.4 that irreducible corepresentations are always finite-dimensional and that the coefficient coalgebra $\mathcal{C}(\varphi^\alpha) = \mathcal{C}(u^\alpha) = \mathrm{Lin}\,\{u^\alpha_{ij} \mid i,j = 1,2,\cdots,d_\alpha\}$ depends only on the class α, but not on the choice of the representative $\varphi^\alpha \in \alpha$ and of the basis.

Definition 7. *A linear functional h on $\mathcal{A}$ is called* left-invariant *(resp.* right-invariant*) if for all $a \in \mathcal{A}$,*

$$(\mathrm{id} \otimes h)\Delta(a) = h(a)1 \quad (resp. \quad (h \otimes \mathrm{id})\Delta(a) = h(a)1). \tag{1}$$

Equation (1) means that h intertwines the left (resp. right) comodule $\mathcal{A}$ with respect to the comultiplication and the trivial comodule $\mathbb{C} \cdot 1$.

Clearly, a functional h on $\mathcal{A}$ is left-invariant (resp. right-invariant) if and only if $fh = f(1)h$ (resp. $hf = f(1)h$) for all functionals $f \in \mathcal{A}'$. In the Hopf algebra literature such functionals h are called *left* (resp. *right*) *integrals on* $\mathcal{A}$ (see [Mon], Definition 2.4.4).

A number of characterizations and properties of cosemisimple Hopf algebras are collected in the following theorem.

Theorem 13. *For any Hopf algebra $\mathcal{A}$ the following statements are equivalent:*

(i) *$\mathcal{A}$ is cosemisimple.*

(ii) *Every corepresentation of $\mathcal{A}$ is a direct sum of irreducible corepresentations.*

(iii) *There exists a left-invariant linear functional h on $\mathcal{A}$ such that $h(1)=1$.*

(iv) *There exists a unique left- and right-invariant linear functional h on $\mathcal{A}$ satisfying $h(1) = 1$.*

(v) *$\mathcal{A}$ is equal to the sum of all $\mathcal{C}(u^\alpha)$, $\alpha \in \hat{\mathcal{A}}$.*

(vi) *The set $\{u^\alpha_{ij} \mid \alpha \in \hat{\mathcal{A}},\ i,j=1,2,\cdots,d_\alpha\}$ is a vector space basis of $\mathcal{A}$.*

If $\mathcal{A}$ is a cosemisimple Hopf algebra, then the antipode S of $\mathcal{A}$ is bijective and we have $S^2(\mathcal{C}(u^\alpha)) = \mathcal{C}(u^\alpha)$ for any $\alpha \in \hat{\mathcal{A}}$.

Proof. The equivalence of (i)–(iv) is Theorem 14.0.3 in [Sw]. By Proposition 8(i), the cosimple subcoalgebras of $\mathcal{A}$ are precisely the coefficient spaces

$\mathcal{C}(u^\alpha)$, $\alpha \in \hat{\mathcal{A}}$. Hence we have (i)$\leftrightarrow$(v). (v)$\rightarrow$(vi) follows from Corollary 10. (vi)$\rightarrow$(v) is obvious. The last assertion is proved in [La], Theorem 3.3. The crucial implications of the theorem will be proved in the next subsection. □

Let $\mathcal{A}$ be a cosemisimple Hopf algebra. The unique left- and right-invariant linear functional h such that $h(1) = 1$ from Theorem 13(iv) is called the *Haar functional* of $\mathcal{A}$. By (v) and (vi), we have

$$\mathcal{A} = \bigoplus\nolimits_{\alpha \in \hat{\mathcal{A}}} \mathcal{C}(u^\alpha). \tag{2}$$

The decomposition (2) is referred to as the *Peter–Weyl decomposition* of $\mathcal{A}$.

11.2.2 The Haar Functional of a Cosemisimple Hopf Algebra

The existence and uniqueness of the Haar functional is a crucial property of a cosemisimple Hopf algebra. Because of its importance we carry out the proof of the relevant implication (vi) $\rightarrow$ (iv) of Theorem 13 and obtain additional information in this way. (In the above proof we referred only to [Sw].)

Suppose that (vi) is fulfilled. Then $\mathcal{A}$ is the *direct sum* of the coefficient coalgebras $\mathcal{C}(u^\alpha)$. Let $\alpha = 1$ denote the trivial one-dimensional corepresentation of $\mathcal{A}$ with matrix element 1. We define a linear functional h on $\mathcal{A}$ by

$$h(1) = 1 \quad \text{and} \quad h(a) = 0 \quad \text{for} \quad a \in \mathcal{C}(u^\alpha),\ \alpha \in \hat{\mathcal{A}},\ \alpha \neq 1. \tag{3}$$

Since $\Delta(\mathcal{C}(u^\alpha)) \subseteq \mathcal{C}(u^\alpha) \otimes \mathcal{C}(u^\alpha)$ for each $\alpha \in \hat{\mathcal{A}}$, h is obviously left- and right-invariant. This proves the existence assertion of (iv). For the uniqueness, let $\tilde{h}$ denote a *left*-invariant linear functional on $\mathcal{A}$ such that $\tilde{h}(1) = 1$. Then we have $\sum a_{(1)}\tilde{h}(a_{(2)}) = \tilde{h}(a)1$ for $a \in \mathcal{C}(u^\alpha)$. Since $\Delta(a) \in \mathcal{C}(u^\alpha) \otimes \mathcal{C}(u^\alpha)$ and $1 \notin \mathcal{C}(u^\alpha)$ if $\alpha \neq 1$, it follows that $\tilde{h}(a) = 0$ for $a \in \mathcal{C}(u^\alpha)$, $\alpha \neq 1$. That is, $\tilde{h}$ is of the form stated in (3). The same reasoning proves the uniqueness of a *right*-invariant linear functional h on $\mathcal{A}$ satisfying $h(1) = 1$. This finishes the proof of the implication (vi) $\rightarrow$ (iv) in Theorem 13.

Since the antipode S of $\mathcal{A}$ is injective, one easily verifies that $\tilde{h}(\cdot) := h(S(\cdot))$ is also a left-invariant functional on $\mathcal{A}$. Since $\tilde{h}(1) = 1$, the uniqueness of h yields $\tilde{h} = h$. That is, we have

$$h(S(a)) = h(a), \quad a \in \mathcal{A}. \tag{4}$$

Note that the condition $h(1) = 1$ in Theorem 13(iv) is essential. Indeed, for any finite-dimensional Hopf algebra there exists a unique (up to complex multiples) left-invariant linear functional h_l (see [Mon], 2.1.3), but h_l is not right-invariant in general (see Example 5). But if $h_l(1) \neq 0$, then the Hopf algebra is cosemisimple by Theorem 13(iii) and h_l is right-invariant as well.

Example 5 (*Sweedler's Hopf algebra*). Let us retain the notation of Examples 1.9 and 10.2. One easily verifies that the functionals $h_l(\cdot) := \langle x + gx, \cdot\rangle$ and $h_r(\cdot) := \langle x - gx, \cdot\rangle$ are left- resp. right-invariant on $\mathcal{A}$. Note that $h_l(1) =$

$h_r(1)=0$. If $\mathcal{A}$ would possess a Peter–Weyl decomposition (2), then we have $h_l(a)=h_r(a)$ for $a\in\mathcal{C}(u^\alpha)$ by the above reasoning and hence $h_l=h_r$. Therefore, $\mathcal{A}$ is not cosemisimple, because $h_l\neq h_r$. △

In the rest of this subsection we assume that $\mathcal{A}$ *is a cosemisimple Hopf algebra*. Let h denote the Haar functional of $\mathcal{A}$. Our next aim are the Schur type orthogonality relations (7) and (8) below. A key step for this is

Lemma 14. *If $v=(v_{ij})$ and $w=(w_{ij})$ are matrix corepresentations of $\mathcal{A}$, then*

$$\sum\nolimits_n h(v_{ij}S(w_{kn}))w_{nl}=\sum\nolimits_n v_{in}h(v_{nj}S(w_{kl})), \tag{5}$$

$$\sum\nolimits_n h(S(w_{kl})v_{in})v_{nj}=\sum\nolimits_n w_{kn}h(S(w_{nl})v_{ij})). \tag{6}$$

Setting $A^{(j,k)}_{in}:=h(v_{ij}S(w_{kn}))$ and $B^{(l,i)}_{kn}:=h(S(w_{kl})v_{in})$, the equations (5) and (6) can be written as $A^{(j,k)}w=vA^{(j,k)}$ and $B^{(l,i)}v=wB^{(l,i)}$, respectively. That is, $A^{(j,k)}\in\mathrm{Mor}\,(w,v)$ and $B^{(l,i)}\in\mathrm{Mor}\,(v,w)$.

Proof. By the left invariance (1) of h, we have

$$h(v_{ij}S(w_{kn}))1=(\mathrm{id}\otimes h)\Delta(v_{ij}S(w_{kn}))=\sum\nolimits_{r,s}h(v_{rj}S(w_{ks}))v_{ir}S(w_{sn}).$$

Inserting this into the left hand side of (5) and using the identity $S(w_{sn})w_{nl}=\delta_{sl}1$, we obtain (5). Applying the right invariance (1) of h with $a=S(w_{kl})v_{in}$, (6) is derived in a similar manner. □

Proposition 15. *Let $v=(v_{ij})$ and $w=(w_{kl})$ be irreducible matrix corepresentations of $\mathcal{A}$. Then we have:*

(i) *If v and w are not equivalent, then for any i,j,k,l,*

$$h(v_{ij}S(w_{kl}))=h(S(w_{kl})v_{ij})=0. \tag{7}$$

(ii) *w is equivalent to its bicontragredient corepresentation w^{cc}. If $F=(F_{in})$ is an invertible intertwiner of w and w^{cc}, then $\mathrm{Tr}\,F\neq 0$, $\mathrm{Tr}\,F^{-1}\neq 0$ and*

$$h(w_{kl}S(w_{ij}))=\delta_{kj}\frac{F_{il}}{\mathrm{Tr}\,F}\quad\textit{and}\quad h(S(w_{kl})w_{ij})=\delta_{kj}\frac{(F^{-1})_{il}}{\mathrm{Tr}\,F^{-1}}. \tag{8}$$

Proof. (i): Since v and w are not equivalent, their intertwiners $A^{(j,k)}$ and $B^{(l,i)}$ are zero by Schur's lemma (Proposition 6(i)). This gives (7).

(ii): First we apply Lemma 14 in the case $v=w$. By Schur's lemma (Proposition 6(ii)), the intertwiners $A^{(j,k)}$ and $B^{(l,i)}$ are scalar multiples of the identity, so there are complex numbers α_{jk} and β_{li} such that

$$A^{(j,k)}_{il}=h(w_{ij}S(w_{kl}))=\alpha_{jk}\delta_{il}\quad\text{and}\quad B^{(l,i)}_{kj}=h(S(w_{kl})w_{ij})=\beta_{li}\delta_{kj}. \tag{9}$$

Setting $i=l,\ j=k$ and summing over j and i, respectively, we obtain

$$\sum_j \alpha_{jj} = \sum_i \beta_{ii} = 1. \tag{10}$$

Next we apply Lemma 14 with $v = w^{cc}$. Then $v_{ij} = S^2(w_{ij})$. Combining (5) and (6) with (4) yields

$$\sum_n h(w_{kn}S(w_{ij}))w_{nl} = \sum_n S^2(w_{in})h(w_{kl}S(w_{nj})), \tag{11}$$

$$\sum_n h(S(w_{in}w_{kl})S^2(w_{nj}) = \sum_n w_{kn}h(S(w_{ij})w_{nl}). \tag{12}$$

With $C^{(j,k)}_{in} := h(w_{kn}S(w_{ij}))$ and $D^{(l,i)}_{kn} := h(S(w_{in})w_{kl})$ the equations (11) and (12) mean that $C^{(j,k)} \in \text{Mor}\,(w, w^{cc})$ and $D^{(l,i)} \in \text{Mor}\,(w^{cc}, w)$. Moreover, by (9), we have $C^{(j,k)}_{in} = \alpha_{ni}\delta_{kj}$ and $D^{(l,i)}_{kn} = \beta_{nk}\delta_{li}$ which gives (8) for $k \neq j$. From (10) together with the preceding expressions we obtain $\text{Tr}\, C^{(j,j)} = \text{Tr}\, D^{(i,i)} = 1$. Since the antipode S is bijective by Theorem 13 and w is irreducible, w^{cc} is irreducible as well. Because $C^{(j,j)} \in \text{Mor}\,(w, w^{cc})$ is nonzero, w and w^{cc} are equivalent by Schur's lemma. Let F be an invertible element of Mor (w, w^{cc}). Since such an F is unique up to a complex factor (again by Schur's lemma), there are complex numbers α_j and β_i such that $\alpha_j F = C^{(j,j)}$ and $\beta_i F^{-1} = D^{(i,i)}$. Thus, $\alpha_j \text{Tr}\, F = \text{Tr}\, C^{(j,j)} = 1$ and $\beta_i \text{Tr}\, F^{-1} = \text{Tr}\, B^{(i,i)} = 1$, and (8) for $k = j$ follows. □

Let us restate formula (8) in the case where $S^2 = \text{id}$. Recall that, generally speaking, all Hopf algebras coming from groups have this property. Then we have $w_{ij} = (w^{cc})_{ij}$, so (8) applies with $F = I$. If d_w is the dimension of the representation space of w, then (8) reads as

$$h(w_{kl}S(w_{ij})) = h(S(w_{kl})w_{ij}) = \delta_{kj}\delta_{il}d_w^{-1}.$$

Next we apply the Schur type orthogonality relations (7) and (8) to the study of characters.

Definition 8. *Let φ be a corepresentation of $\mathcal{A}$ on a finite-dimensional vector space. The element $\chi_\varphi := (\text{Tr} \otimes \text{id})\varphi$ of $\mathcal{A}$ is called the* character *of φ.*

If (v_{ij}) is a matrix corepresentation of φ , then $\chi_\varphi = \sum_i v_{ii}$. For any $f \in \mathcal{A}'$, we have $f.\chi_\varphi = \chi_\varphi.f$, where $f.\chi_\varphi$ and $\chi_\varphi.f$ are given by (1.65). Note that $\tau \circ \Delta(\chi_\varphi) = \Delta(\chi_\varphi)$, $\chi_{\varphi\oplus\psi} = \chi_\varphi + \chi_\psi$, $\chi_{\varphi\otimes\psi} = \chi_\varphi\chi_\psi$ and $\chi_{\varphi^c} = S(\chi_\varphi)$. Moreover, $\chi_\varphi = \chi_\psi$ if $\varphi \simeq \psi$. Of course, Definition 8 and the preceding simple facts are valid for any Hopf algebra $\mathcal{A}$.

Let χ_α, $\alpha \in \hat{\mathcal{A}}$, denote the character of u^α. An immediate consequence of the formulas (7) and (8) is

Corollary 16. *$h(\chi_\alpha S(\chi_\beta)) = h(S(\chi_\alpha)\chi_\beta) = \delta_{\alpha\beta}$ for $\alpha, \beta \in \hat{\mathcal{A}}$.*

Proposition 17. *Let φ be a corepresentation of $\mathcal{A}$ on a finite-dimensional vector space V. For $\alpha \in \hat{\mathcal{A}}$, we set $n_\alpha := h(S(\chi_\alpha)\chi_\varphi)$ and $P_\alpha := \hat{\varphi}(h_\alpha) \equiv (\text{id}\otimes h_\alpha)\varphi$, where h_α is the linear functional on $\mathcal{A}$ defined by $h_\alpha(u^\beta_{ij}) = \delta_{\alpha\beta}\delta_{ij}$, $\beta \in \hat{\mathcal{A}}$, $i, j = 1, 2, \cdots, d_\beta$. Then we have*

$$\varphi \simeq \bigoplus_{\alpha \in \hat{\mathcal{A}}} n_\alpha u^\alpha.$$

P_α is a projection of V onto the φ-invariant subspace $V_\alpha := P_\alpha(V)$ such that the restriction of φ to V_α is equivalent to the direct sum of n_α copies of u^α. Moreover, $\chi_\varphi = \sum_\alpha n_\alpha \chi_\alpha$ and $h(S(\chi_\varphi)\chi_\varphi) = \sum_\alpha n_\alpha^2$.

Proof. From the definition of h_α we immediately obtain that $fh_\alpha = h_\alpha f$ for $f \in \mathcal{A}'$ and $h_\alpha h_\beta = \delta_{\alpha\beta} h_\alpha$. Hence $\hat{\varphi}(f)\hat{\varphi}(h_\alpha) = \hat{\varphi}(h_\alpha)\hat{\varphi}(f)$ and $\hat{\varphi}(h_\alpha)\hat{\varphi}(h_\beta) = \delta_{\alpha\beta}\hat{\varphi}(h_\alpha)$ by Proposition 1(iii). Therefore, by Proposition 3(iii), $P_\alpha = \hat{\varphi}(h_\alpha) \in \mathrm{Mor}\ (\varphi)$ and $P_\alpha P_\beta = \delta_{\alpha\beta} P_\alpha$, $\alpha, \beta \in \hat{\mathcal{A}}$. Thus, $\{P_\alpha \mid \alpha \in \hat{\mathcal{A}}\}$ is a family of projections such that $\sum_\alpha P_\alpha = I$ on V and $P_\alpha P_\beta = 0$ if $\alpha \neq \beta$. Let φ_α denote the restriction of φ to $V_\alpha = P_\alpha(V)$ and let $v^\alpha = (v^\alpha_{nm})$ be the matrix corepresentation of φ_α with respect to a basis $\{x^\alpha_n\}$ of V_α. For $f \in \mathcal{A}'$, we have $\sum_n x^\alpha_n f(v^\alpha_{nm}) = \hat{\varphi}(f)(x^\alpha_m) = \hat{\varphi}(f)(\hat{\varphi}(h_\alpha)(x^\alpha_m)) = \sum_n x^\alpha_n (fh_\alpha)(v^\alpha_{nm})$, so $f(v^\alpha_{nm}) = fh_\alpha(v^\alpha_{nm})$. Since fh_α vanishes on $\mathcal{C}(u^\beta)$, $\alpha \neq \beta$, this implies that $v^\alpha_{nm} \in \mathcal{C}(u^\alpha)$, so there are complex numbers Λ^{ni}_{mj} (depending also on α) such that $v^\alpha_{nm} = \sum_{i,j} \Lambda^{ni}_{mj} u^\alpha_{ij}$. Applying the comultiplication to both sides and comparing the coefficients of u^α_{rs}, we obtain $\sum_k \Lambda^{ni}_{kr}\Lambda^{ks}_{mj} = \Lambda^{ni}_{mj}\delta_{rs}$. Using the latter we compute that for fixed m and i the elements $y_j := \sum_n \Lambda^{ni}_{mj} x^\alpha_n$ satisfy $\varphi(y_i) = y_j \otimes u^\alpha_{ji}$. Thus, either $W_{\alpha mi} := \mathrm{Lin}\,\{y_j = \sum_n \Lambda^{ni}_{mj} x^\alpha_n\}$ is $\{0\}$ or the restriction of φ to $W_{\alpha mi}$ is equivalent to u^α. Since obviously $V_\alpha = \sum_{m,i} W_{\alpha mi}$, it follows that φ_α is equivalent to a direct sum of copies of u^α. Since $\varphi = \bigoplus_\alpha \varphi_\alpha$, the remaining assertions follows at once from Corollary 16. □

Proposition 17 implies in particular that each finite-dimensional corepresentation of $\mathcal{A}$ is completely reducible. Now let φ be an arbitrary corepresentation of $\mathcal{A}$ on V. By Proposition 4(i), any element of V belongs to some finite-dimensional φ-invariant subspace of V. Thus, by Proposition 17, V is the sum and hence the direct sum of φ-invariant subspaces such that the corresponding restrictions of φ are irreducible. This proves that *any corepresentation of $\mathcal{A}$ is completely reducible.*

Corollary 18. *Two finite-dimensional corepresentations of $\mathcal{A}$ are equivalent if and only if their characters coincide.*

Corollary 19. *For any finite-dimensional corepresentation φ of $\mathcal{A}$ on V, $\varphi_h \equiv (\mathrm{id} \otimes h)\varphi$ is a projection of V to the subspace of φ-invariant elements.*

Proof. The proof follows from Proposition 17 applied to $\alpha = 1$, since then $h_\alpha = h$. □

Of course, the assertion of Corollary 19 can also be verified directly without using the orthogonality relations (7) and (8). It is then obtained for an arbitrary (not necessarily finite-dimensional) corepresentation of $\mathcal{A}$.

11.2.3 Peter–Weyl Decomposition of Coordinate Hopf Algebras

Throughout this subsection we suppose that G_q is one of the quantum groups $SL_q(N)$, $O_q(N)$, $SO_q(N)$, $Sp_q(N)$ and that the complex number q is *transcendental*. Our aim is to prove that then the Hopf algebra $\mathcal{O}(G_q)$ is cosemisimple and to determine its Peter–Weyl decomposition. The main tools for that are the Brauer–Schur–Weyl duality (see Subsect. 8.6.3) and the representation theory of the corresponding Drinfeld–Jimbo algebras $U_q(\mathfrak{g})$.

First let us collect some notation and facts needed later. Recall that $\tilde{U}_q(\mathrm{so}_N)$ is the Hopf algebra defined in Subsect. 8.6.1. For notational simplicity, we also set $\tilde{U}_q(\mathrm{sl}_N) := U_q(\mathrm{sl}_N)$ and $\tilde{U}_q(\mathrm{sp}_N) := U_q(\mathrm{sp}_N)$. Let $\tilde{T}_1$ denote the vector representation of $\tilde{U}_q(\mathfrak{g})$ with matrix coefficients t_{ij} (see (8.87)). By Theorem 9.18 and Proposition 9.19, there exists a dual pairing $\langle\cdot,\cdot\rangle$ of the Hopf algebras $\tilde{U}_q(\mathfrak{g})$, $\mathfrak{g} = \mathrm{sl}_N, \mathrm{so}_N, \mathrm{sp}_N$, and $\mathcal{O}(G_q)$, $G_q = SL_q(N), O_q(N), Sp_q(N)$, of $U_q(\mathrm{so}_{2n})$ and $\mathcal{O}(SO_q(2n))$ and of $U_{q^{1/2}}(\mathrm{so}_{2n+1})$ and $\mathcal{O}(SO_q(2n+1))$ such that $\langle\cdot, u^i_j\rangle = t_{ij}(\cdot)$. Let $\mathcal{A}(\mathsf{R})$ be the FRT bialgebra corresponding to $\mathcal{O}(G_q)$, where R is given by (9.13) and (9.30), respectively. Since $\mathcal{O}(G_q)$ is a quotient of $\mathcal{A}(\mathsf{R})$, the pairing of $\tilde{U}_q(\mathfrak{g})$ and $\mathcal{O}(G_q)$ induces a dual pairing of $\tilde{U}_q(\mathfrak{g})$ and $\mathcal{A}(\mathsf{R})$. Therefore, by Proposition 1.15, any corepresentation φ of $\mathcal{O}(G_q)$ or $\mathcal{A}(\mathsf{R})$ yields a representation $\hat{\varphi}$ of $\tilde{U}_q(\mathfrak{g})$. Let φ_1 denote the fundamental corepresentation of $\mathcal{O}(G_q)$ or $\mathcal{A}(\mathsf{R})$ defined by (9.6) and let V_1 be its carrier space. As noted in Sect. 9.4, we have $\hat{\varphi}_1 = \tilde{T}_1$.

Proposition 20. *Suppose that $r \in \mathbb{N}$. The assignment $\varphi \to \hat{\varphi}$ gives a one-to-one correspondence between subcorepresentations of the corepresentation $\varphi_1^{\otimes r} : V_1^{\otimes r} \to V_1^{\otimes r} \otimes \mathcal{A}(\mathsf{R})$ of the bialgebra $\mathcal{A}(\mathsf{R})$ and subrepresentations of the representation $\tilde{T}_1^{\otimes r}$ of the algebra $\tilde{U}_q(\mathfrak{g})$. Two subcorepresentations φ and ψ are equivalent if and only if the subrepresentations $\hat{\varphi}$ and $\hat{\psi}$ are. A subcorepresentation φ is irreducible if and only if $\hat{\varphi}$ is. The corepresentation $\varphi_1^{\otimes r}$ of $\mathcal{A}(\mathsf{R})$ is completely reducible.*

Proof. Since $\hat{\varphi}_1 = \tilde{T}_1$, we have $(\varphi_1^{\otimes r})\hat{} = \tilde{T}_1^{\otimes r}$ and any subcorepresentation φ of $\varphi_1^{\otimes r}$ leads to a subrepresentation $\hat{\varphi}$ of $\tilde{T}_1^{\otimes r}$. Conversely, let T be a subrepresentation of $\tilde{T}_1^{\otimes r}$ realized on a subspace V of the carrier space $V_1^{\otimes r}$ of $\tilde{T}_1^{\otimes r}$. Since the representation $\tilde{T}_1^{\otimes r}$ is completely reducible by Theorem 7.8, there exists an invariant subspace W for $\tilde{T}_1^{\otimes r}$ such that $V_1^{\otimes r} = V \oplus W$. Obviously, the projection P of $V_1^{\otimes r}$ onto V is in the commutant $\tilde{T}_1^{\otimes r}(\tilde{U}_q(\mathfrak{g}))'$. Therefore, by Theorem 8.38, P belongs to the algebra $B(r)$ generated by the transformations $\hat{\mathsf{R}}_{i,i+1}$ for $\mathfrak{g} = \mathrm{sl}_N$ and by $\hat{\mathsf{R}}_{i,i+1}$ and $\mathsf{K}_{i,i+1}$ for $\mathfrak{g} = \mathrm{so}_N, \mathrm{sp}_N$. Since $\hat{\mathsf{R}}_{i,i+1}\mathbf{u}_i\mathbf{u}_{i+1} = \mathbf{u}_i\mathbf{u}_{i+1}\hat{\mathsf{R}}_{i,i+1}$ and $\mathsf{K}_{i,i+1}\mathbf{u}_i\mathbf{u}_{i+1} = \mathbf{u}_i\mathbf{u}_{i+1}\mathsf{K}_{i,i+1}$, we have $P \in \mathrm{Mor}\,(\mathbf{u}^{\otimes r})$ and so $P \in \mathrm{Mor}\,(\varphi_1^{\otimes r})$ by Proposition 3. Hence $PV_1^{\otimes r} = V$ is also an invariant subspace for $\varphi_1^{\otimes r}$. If φ denotes the restriction of $\varphi_1^{\otimes r}$ to V, we clearly have $\hat{\varphi} = T$.

It is obvious that any equivalence of subcorepresentations φ and ψ yields an equivalence of $\hat{\varphi}$ and $\hat{\psi}$ and that any invariant subspace for φ is invariant for $\hat{\varphi}$. The opposite direction, as well as the last assertion of the proposition, can be verified as in the preceding paragraph using Theorem 8.38. □

By Corollary 8.37, any representation $\tilde{T}(\mathbf{n})$, $\mathbf{n} \in \mathcal{P}_r(\mathfrak{g})$, of $\tilde{U}_q(\mathfrak{g})$ is an irreducible subrepresentation of $\tilde{T}_1^{\otimes r}$. Hence, by Proposition 20, there exists a unique irreducible subcorepresentation, denoted $\varphi_{\mathbf{n},r}$, of the corepresentation $\varphi_1^{\otimes r}$ of $\mathcal{A}(\mathsf{R})$ such that $\hat{\varphi}_{\mathbf{n},r} = \tilde{T}(\mathbf{n})$.

Theorem 21. *The bialgebra $\mathcal{A}(\mathsf{R})$ decomposes as a direct sum of coefficient coalgebras $\mathcal{C}(\varphi_{\mathbf{n},r})$, $\mathbf{n} \in \mathcal{P}_r(\mathfrak{g})$, $r \in \mathbb{N}_0$, that is,*

$$\mathcal{A}(\mathsf{R}) = \bigoplus\nolimits_{r\geq 0} \bigoplus\nolimits_{\mathbf{n}\in\mathcal{P}_r(\mathfrak{g})} \mathcal{C}(\varphi_{\mathbf{n},r}). \tag{13}$$

Any irreducible corepresentation of $\mathcal{A}(\mathsf{R})$ is equivalent to one of the corepresentations $\varphi_{\mathbf{n},r}$.

Proof. Let $r \in \mathbb{N}_0$. We denote by $\mathcal{A}_r(\mathsf{R})$ the homogeneous part of $\mathcal{A}(\mathsf{R})$ of homogeneity degree r. Clearly, $\mathcal{A}_r(\mathsf{R})$ is just the coefficient coalgebra of the corepresentation $\varphi_1^{\otimes r}$ of $\mathcal{A}(\mathsf{R})$. By Proposition 20, the decomposition (8.117) of the representation $\tilde{T}_1^{\otimes r}$ leads to the direct sum decomposition

$$\varphi_1^{\otimes r} = \bigoplus\nolimits_{\mathbf{n}\in\mathcal{P}_r(\mathfrak{g})} m_{\mathbf{n}}\varphi_{\mathbf{n},r}, \qquad m_{\mathbf{n}} > 0,$$

of the corepresentation $\varphi_1^{\otimes r}$. The coefficient coalgebras of both sides must coincide, so we get

$$\mathcal{A}_r(\mathsf{R}) = \sum\nolimits_{\mathbf{n}\in\mathcal{P}_r(\mathfrak{g})} \mathcal{C}(\varphi_{\mathbf{n},r}). \tag{14}$$

If $\mathbf{n} \neq \mathbf{n}'$, $\mathbf{n}, \mathbf{n}' \in \mathcal{P}_r(\mathfrak{g})$, then the representations $\tilde{T}(\mathbf{n})$ and $\tilde{T}(\mathbf{n}')$ and so the corepresentations $\varphi_{\mathbf{n},r}$ and $\varphi_{\mathbf{n}',r}$ (by Proposition 20) are not equivalent. Hence the sum (14) is direct by Corollary 9. Since $\mathcal{A}(\mathsf{R}) = \bigoplus_r \mathcal{A}_r(\mathsf{R})$, this yields the decomposition (13).

Let φ be an irreducible corepresentation of $\mathcal{A}(\mathsf{R})$. By Proposition 8, $\mathcal{C}(\varphi)$ is a cosimple subcoalgebra of $\mathcal{A}(\mathsf{R})$. Therefore, by (13), $\mathcal{C}(\varphi) = \mathcal{C}(\varphi_{\mathbf{n},r})$ for some $\mathbf{n} \in \mathcal{P}_r(\mathfrak{g})$ and $r \in \mathbb{N}_0$. Hence $\varphi \simeq \varphi_{\mathbf{n},r}$ by Corollary 9. □

Our next step is to formulate the Peter–Weyl decomposition theorem for the Hopf algebras $\mathcal{O}(G_q)$. Recall that these Hopf algebras are defined as

$$\mathcal{O}(SL_q(N)) = \mathcal{A}(\mathsf{R})/\langle \mathcal{D}_q - 1\rangle, \quad \mathcal{O}(Sp_q(N)) = \mathcal{A}(\mathsf{R})/\langle \mathcal{Q}_q - 1\rangle,$$

$$\mathcal{O}(O_q(N)) = \mathcal{A}(\mathsf{R})/\langle \mathcal{Q}_q - 1\rangle, \quad \mathcal{O}(SO_q(N)) = \mathcal{A}(\mathsf{R})/\langle \mathcal{Q}_q - 1, \mathcal{D}_q - 1\rangle,$$

where $\mathcal{A}(\mathsf{R})$ is the corresponding FRT bialgebra, $\mathcal{Q}_q$ is the element of $\mathcal{A}(\mathsf{R})$ defined by (9.37) and $\mathcal{D}_q$ is the unique element of $\mathcal{A}(\mathsf{R})$ such that $\varphi_{R,N}(\zeta) = \zeta \otimes \mathcal{D}_q$ for $\zeta \in \Lambda(G_q)_N$. That is, $\mathcal{D}_q$ is the quantum determinant from Definition 9.5 for $G_q = SL_q(N)$. In the case $G_q = O_q(N)$ the quantum determinant from Definition 9.10 belongs to the algebra $\mathcal{O}(O_q(N))$, but here $\mathcal{D}_q$ denotes the corresponding element from the FRT bialgebra $\mathcal{A}(\mathsf{R})$.

In the proof of Theorem 22 it will be shown that the irreducible corepresentations of $\mathcal{O}(G_q)$, $G_q = SL_q(N), SO_q(N), Sp_q(N)$, are characterized by the highest weights of those irreducible representations of the corresponding Drinfeld–Jimbo algebra $U_q(\mathfrak{g})$, $\mathfrak{g} = \mathrm{sl}_N, \mathrm{so}_N, \mathrm{sp}_N$, which are contained in tensor products of some number of vector representations. These are precisely the weights of the set $\mathfrak{P}(\mathfrak{g})$ from Proposition 7.20. We write $P(G_q) := \mathfrak{P}(\mathfrak{g})$ in this case. Further, it will be proved that the irreducible corepresentations of $\mathcal{O}(O_q(N))$ are parametrized by the set $\mathcal{P}(\mathrm{so}_N)$ described in Subsect. 8.6.2. We set $P(O_q(N)) := \mathcal{P}(\mathrm{so}_N)$. In both cases the corresponding irreducible corepresentation $\varphi_{\mathbf{n}}, \mathbf{n} \in P(G_q)$, of $\mathcal{O}(G_q)$ has the property that $\hat{\varphi}_{\mathbf{n}}$ is just the irreducible representation $T_{\mathbf{n}}$ resp. $\tilde{T}(\mathbf{n}) \equiv \tilde{T}_{\mathbf{n}}$ of $U_q(\mathfrak{g})$ resp. $\tilde{U}_q(\mathfrak{g})$.

Theorem 22. *Let G_q be one of the quantum groups $SL_q(N)$, $O_q(N)$, $SO_q(N)$ or $Sp_q(N)$. Suppose that q is transcendental. Then the Hopf algebra $\mathcal{O}(G_q)$ is cosemisimple and it decomposes as a direct sum*

$$\mathcal{O}(G_q) \simeq \bigoplus\nolimits_{\mathbf{n} \in P(G_q)} \mathcal{C}(\varphi_{\mathbf{n}}) \tag{15}$$

of coefficient coalgebras $\mathcal{C}(\varphi_{\mathbf{n}})$ of irreducible corepresentations $\varphi_{\mathbf{n}}, \mathbf{n} \in P(G_q)$, of $\mathcal{O}(G_q)$. Every irreducible corepresentation of $\mathcal{O}(G_q)$ is equivalent to one of the corepresentations $\varphi_{\mathbf{n}}, \mathbf{n} \in P(G_q)$.

Formula (15) is the Peter–Weyl decomposition of the cosemisimple Hopf algebra $\mathcal{O}(G_q)$. Before we begin with the proof of Theorem 22, let us derive an important corollary concerning the dual pairing $\langle\cdot,\cdot\rangle$ of the Hopf algebras $\tilde{U}_q(\mathfrak{g})$, resp. $U_q(\mathfrak{g})$, and $\mathcal{O}(G_q)$ defined in Sect. 9.4.

Corollary 23. *The dual pairings $\langle\cdot,\cdot\rangle$ of the pairs of Hopf algebras $\tilde{U}_q(\mathfrak{g})$, $\mathfrak{g} = \mathrm{sl}_N, \mathrm{so}_N, \mathrm{sp}_N$, and $\mathcal{O}(G_q)$, $G_q = SL_q(N), O_q(N), Sp_q(N)$, of $U_q(\mathrm{so}_{2n})$ and $\mathcal{O}(SO_q(2n))$ and of $U_{q^{1/2}}(\mathrm{so}_{2n+1})$ and $\mathcal{O}(SO_q(2n+1))$ are nondegenerate.*

Proof. We give the proof for the Hopf algebras $\tilde{U}_q(\mathfrak{g})$ and $\mathcal{O}(G_q)$, $G_q = SL_q(N), O_q(N), Sp_q(N)$. The proof in the other cases is completely similar.

Let $\mathcal{N}$ be the set of $a \in \mathcal{O}(G_q)$ such that $\langle f, a\rangle = 0$ for all $f \in \tilde{U}_q(\mathfrak{g})$. Let $\varphi = \Delta$ be the corepresentation of $\mathcal{O}(G_q)$ on itself. By Theorem 22 and Proposition 8, we have $\varphi \simeq \bigoplus_{\mathbf{n}} d_{\mathbf{n}}\varphi_{\mathbf{n}}$, where $d_{\mathbf{n}} = \dim \varphi_{\mathbf{n}}$. Since $\hat{\varphi}_{\mathbf{n}} = \tilde{T}_{\mathbf{n}}$ by the definition of $\varphi_{\mathbf{n}}$, this yields $\hat{\varphi} \simeq \bigoplus_{\mathbf{n}} d_{\mathbf{n}}\tilde{T}_{\mathbf{n}}$. If $a \in \mathcal{N}$ and $f \in \tilde{U}_q(\mathfrak{g})$, then, by (1.65), we get $\langle g, \hat{\varphi}(f)a\rangle = \sum\langle g, a_{(1)}\rangle\langle f, a_{(2)}\rangle = \langle gf, a\rangle = 0$ for all $g \in \tilde{U}_q(\mathfrak{g})$, so that $\hat{\varphi}(f)a \in \mathcal{N}$. That is, $\mathcal{N}$ is $\hat{\varphi}$-invariant. Assume that $\mathcal{N} \neq \{0\}$. Since $\hat{\varphi} \simeq \bigoplus_{\mathbf{n}} d_{\mathbf{n}}\tilde{T}_{\mathbf{n}}$, there exists a $\hat{\varphi}$-invariant subspace $\mathcal{N}_0 \neq \{0\}$ of $\mathcal{N}$ such that $\mathcal{N}_0 \subseteq \mathcal{C}(\varphi_{\mathbf{n}})$ for some $\mathbf{n}$ and $\hat{\varphi}\lceil\mathcal{N}_0 \simeq \tilde{T}_{\mathbf{n}}$. Hence we can find a basis $\{e_i\}$ of $\mathcal{N}_0$ such that $\hat{\varphi}(f)e_i = \sum_j \langle f, v^{\mathbf{n}}_{ji}\rangle e_j$, where $(v^{\mathbf{n}}_{ij})$ is the matrix of the corepresentation $\varphi_{\mathbf{n}}$ for this basis. We write the vector e_1 as $e_1 = \sum_{i,j} \alpha_{ij} v^{\mathbf{n}}_{ij}$ with $\alpha_{ij} \in \mathbb{C}$. By (1.65), we have $\langle g, f_1.a.f_2\rangle = \langle f_2 g f_1, a\rangle = 0$ for $a \in \mathcal{N}$ and $f_1, f_2, g \in \tilde{U}_q(\mathfrak{g})$. Therefore, $f_1.a.f_2 \in \mathcal{N}$ for arbitrary $f_1, f_2 \in \tilde{U}_q(\mathfrak{g})$. The matrix coefficients $\langle\cdot, v^{\mathbf{n}}_{ij}\rangle$ of the irreducible representation $\hat{\varphi}\lceil\mathcal{N}_0 \simeq \tilde{T}_{\mathbf{n}}$ of

$\tilde{U}_q(\mathfrak{g})$ are linearly independent. Hence, for any $k, l, r, s = 1, 2, \cdots, \dim \mathcal{N}_0$, we can find elements $f_1, f_2 \in \tilde{U}_q(\mathfrak{g})$ such that $\langle f_1, v_{ij}^{\mathbf{n}} \rangle = \delta_{ki}\delta_{lj}$ and $\langle f_2, v_{ij}^{\mathbf{n}} \rangle = \delta_{ri}\delta_{sj}$. Then, we obtain $f_1.e_1.f_2 = \alpha_{rl} v_{sk}^{\mathbf{n}} \in \mathcal{N}$. Since k, l, r, s are arbitrary, any matrix element $v_{ij}^{\mathbf{n}}$ belongs to $\mathcal{N}$, that is, $\langle f, v_{ij}^{\mathbf{n}} \rangle = 0$ for all $f \in \tilde{U}_q(\mathfrak{g})$. This in turn yields that $\hat{\varphi}(f)a = 0$ for any $f \in \tilde{U}_q(\mathfrak{g})$ and $a \in \mathcal{N}_0$. Since $\mathcal{N}_0 \neq \{0\}$ by construction, this is a contradiction. Thus, $\mathcal{N} = \{0\}$. This means that $\langle f, a \rangle = 0$ for all $f \in \tilde{U}_q(\mathfrak{g})$ implies that $a = 0$.

Now suppose that $\langle f, a \rangle = 0$ for all $a \in \mathcal{O}(G_q)$. Since $\langle f, a \rangle = \psi(a)(f)$ by the definition of the pairing $\langle \cdot, \cdot \rangle$ (see the proof of Theorem 9.18), it follows that $t_{ij}(f) = 0$ for all matrix coefficients t_{ij} of subrepresentations T of an r-fold tensor product $\tilde{T}_1^{\otimes r}$ of the vector representation $\tilde{T}_1$ of $\tilde{U}_q(\mathfrak{g})$. If $\mathfrak{g} = \mathrm{sl}_N$ or $\mathfrak{g} = \mathrm{sp}_N$, then $\tilde{U}_q(\mathfrak{g}) = U_q(\mathfrak{g})$ and so $f = 0$ by Proposition 7.21. Combining the results of Subsect. 8.6.1 and Proposition 7.21 one can easily verify that if an element $b \in \tilde{U}_q(\mathrm{so}_N)$ is annihilated by all irreducible representations $\tilde{T}$ contained in some r-fold tensor product $\tilde{T}_1^{\otimes r}$, $r \in \mathbb{N}$, then $b = 0$. □

Corollary 23 implies that the maps $a \to \langle \cdot, a \rangle$ and $f \to \langle f, \cdot \rangle$ are injective Hopf algebra homomorphisms of $\mathcal{O}(G_q)$ to $\tilde{U}_q(\mathfrak{g})^\circ$ and of $\tilde{U}_q(\mathfrak{g})$ to $\mathcal{O}(G_q)^\circ$. The injections $\mathcal{O}(G_q) \to \tilde{U}_q(\mathfrak{g})^\circ$ are not surjective, because not every irreducible representation of $\tilde{U}_q(\mathfrak{g})$ is of type 1.

Recall from Sect. 7.1 that each finite-dimensional representation of $U_q(\mathfrak{g})$ is a direct sum of (irreducible) representations of the form $T_{\omega\mathbf{n}}$, where $\mathbf{n} \in P_+$ and $\omega \in \{1, -1\}^n$, $n = \operatorname{rank} \mathfrak{g}$. Since P_+ is precisely the set $P(G_q)$ for $G_q = SL_q(n+1), Sp_q(2n)$, it follows from the preceding and Theorem 22 that there are Hopf algebra isomorphisms

$$U_q(\mathrm{sl}_{n+1})^\circ = \mathcal{O}(SL_q(n+1)) \rtimes \mathbb{C}\mathbb{Z}_2^n, \quad U_q(\mathrm{sp}_{2n})^\circ = \mathcal{O}(Sp_q(2n)) \rtimes \mathbb{C}\mathbb{Z}_2^n.$$

Let us briefly recall the structures of the Hopf algebras on the right hand sides. As a coalgebra, $\mathcal{O}(G_q) \rtimes \mathbb{C}\mathbb{Z}_2^n$ is the tensor product coalgebra $\mathcal{O}(G_q) \otimes \mathbb{C}\mathbb{Z}_2^n$. As an algebra, it is a left crossed product algebra (see Subsect. 10.2.1). The generators $\chi_k, k = 1, 2, \ldots, n$, of $\mathbb{C}\mathbb{Z}_2^n$ act as Hopf algebra automorphisms χ_k of $\mathcal{O}(G_q)$ which are determined by the following formulas: $\chi_k(u_j^i) = -u_j^i$ if $i \le k < j$ or $j \le k < i$ for $G_q = SL_q(n+1)$, and if $i \le k < j < k'$ or $k < j < k' \le i$ or $j \le k < i < k'$ or $k < i < k' \le j$ for $G_q = Sp_q(2n)$ and $\chi_k(u_j^i) = u_j^i$ otherwise. Thus the algebra $\mathcal{O}(G_q) \rtimes \mathbb{C}\mathbb{Z}_2^n$ is generated by the algebra $\mathcal{O}(G_q)$ and n additional pairwise commuting generators χ_k such that $\chi_k^2 = 1$ and $\chi_k(a) = \chi_k a \chi_k^{-1}$ for $a \in \mathcal{O}(G_q)$ and $k = 1, 2, \cdots, n$.

In the case $\mathfrak{g} = \mathrm{so}_N$ there is still an injective Hopf algebra homomorphism of $\mathcal{O}(SO_q(N)) \rtimes \mathbb{C}\mathbb{Z}_2^n$ into $U_q(\mathrm{so}_{2n})^\circ$ resp. $U_{q^{1/2}}(\mathrm{so}_{2n+1})^\circ$, where as usual $N = 2n$ or $N = 2n+1$. However, this homomorphism is not surjective, because its image does not contain the matrix coefficients of the spinor representations of $U_q(\mathrm{so}_{2n})$ resp. $U_{q^{1/2}}(\mathrm{so}_{2n+1})$.

The rest of this subsection is devoted to an *outline of the proof of Theorem 22*. The decomposition (15) will be derived from (13). In order to do so, we have to find the parts of the decomposition (13) that cancel when we pass to the quotient $\mathcal{O}(G_q)$.

We begin with some notation. The right $\mathcal{A}(\mathsf{R})$-comodule corresponding to the corepresentation $\varphi_{\mathbf{n},r}$ is denoted by $M(\mathbf{n},r)$. Let M be a right comodule of $\mathcal{A}(\mathsf{R})$ or $\mathcal{O}(G_q)$. Then, by Proposition 1.15, M is also a left module of the corresponding algebra $\tilde{U}_q(\mathfrak{g})$ resp. $U_q(\mathfrak{g})$, denoted by $M_{\tilde{U}}$ resp. M_U.

Since $\Delta(\mathcal{D}_q) = \mathcal{D}_q \otimes \mathcal{D}_q$ and $\Delta(\mathcal{Q}_q) = \mathcal{Q}_q \otimes \mathcal{Q}_q$, the one-dimensional spaces $\mathbb{C}\mathcal{Q}_q$ and $\mathbb{C}\mathcal{D}_q$ are right $\mathcal{A}(\mathsf{R})$-comodules. They are described in

Lemma 24.

$$\mathbb{C}\mathcal{D}_q \simeq M((N)', N) \quad \textit{for} \quad \mathfrak{g} = \mathrm{sl}_N, \tag{16}$$

$$\mathbb{C}\mathcal{D}_q \simeq M(\mathbf{0}^\circ, N), \quad \mathbb{C}\mathcal{Q}_q \simeq M(\mathbf{0}, 2) \quad \textit{for} \quad \mathfrak{g} = \mathrm{so}_N, \tag{17}$$

$$\mathbb{C}\mathcal{D}_q \simeq M(\mathbf{0}, N), \quad \mathbb{C}\mathcal{Q}_q \simeq M(\mathbf{0}, 2) \quad \textit{for} \quad \mathfrak{g} = \mathrm{sp}_N, \tag{18}$$

$$\mathcal{D}_q^2 = \mathcal{Q}_q^N \quad \textit{for} \quad \mathfrak{g} = \mathrm{so}_N \quad \textit{and} \quad \mathcal{D}_q = \mathcal{Q}_q^n \quad \textit{for} \quad \mathfrak{g} = \mathrm{sp}_{2n}.$$

Proof. We carry out the proof of (17). Clearly, $(\mathbb{C}\mathcal{Q}_q)_{\tilde{U}}$ and $(\mathbb{C}\mathcal{D}_q)_{\tilde{U}}$ are one-dimensional $\tilde{U}_q(\mathrm{so}_N)$-modules. But there are precisely two one-dimensional type 1 representations of $\tilde{U}_q(\mathrm{so}_N)$. They correspond to the partitions $\mathbf{0}$ and $\mathbf{0}^\circ$. Since $\langle f, \mathcal{Q}_q\rangle = \langle f, 1\rangle$ for $f \in \tilde{U}_q(\mathrm{so}_N)$ and $\langle \chi, \mathcal{D}_q\rangle = -1$ by Proposition 9.19, it follows that $(\mathbb{C}\mathcal{D}_q)_{\tilde{U}} = (M(\mathbf{0}^\circ, N))_{\tilde{U}}$ and $(\mathbb{C}\mathcal{Q}_q)_{\tilde{U}} = (M(\mathbf{0}, 2))_{\tilde{U}}$. By Proposition 20, the latter implies that $\mathbb{C}\mathcal{D}_q \simeq M(\mathbf{0}^\circ, N)$ and $\mathbb{C}\mathcal{Q}_q \simeq M(\mathbf{0}, 2)$. The formulas (16) and (18) are verified similarly.

Let us prove the equation $\mathcal{D}_q^2 = \mathcal{Q}_q^N$ for $\mathfrak{g} = \mathrm{so}_N$. The equality $\mathcal{D}_q = \mathcal{Q}_q^n$ for $\mathfrak{g} = \mathrm{sp}_{2n}$ is proved similarly. We have $\mathcal{D}_q^2 \in \mathcal{A}_{2N}(\mathsf{R})$ and $\mathcal{Q}_q^N \in \mathcal{A}_{2N}(\mathsf{R})$. Both $\mathcal{D}_q^2$ and $\mathcal{Q}_q^N$ are group-like elements belonging to the one-dimensional coalgebra $\mathcal{C}(\varphi_{\mathbf{0},2N})$ in the decomposition (13). Hence we get $\mathcal{D}_q^2 = \mathcal{Q}_q^N$. □

Lemma 25. *There are the following $\mathcal{A}(\mathsf{R})$-comodule isomorphisms:*

$$M(\mathbf{n}, r) \otimes \mathbb{C}\mathcal{D}_q \simeq M(\mathbf{n} + (N)', r + N) \quad \textit{for} \quad \mathfrak{g} = \mathrm{sl}_N, \tag{19}$$

$$M(\mathbf{n}, r) \otimes \mathbb{C}\mathcal{D}_q \simeq M(\mathbf{n}, r + N) \quad \textit{for} \quad \mathfrak{g} = \mathrm{sp}_N, \tag{20}$$

$$M(\mathbf{n}, r) \otimes \mathbb{C}\mathcal{D}_q \simeq M(\mathbf{n}^\circ, r + N) \quad \textit{for} \quad \mathfrak{g} = \mathrm{so}_N, \tag{21}$$

$$M(\mathbf{n}, r) \otimes \mathbb{C}\mathcal{Q}_q \simeq M(\mathbf{n}, r + 2) \quad \textit{for} \quad \mathfrak{g} = \mathrm{sp}_N, \mathrm{so}_N. \tag{22}$$

Proof. The relations (19)–(22) are derived from the formulas (16)–(18). In order to prove the relation (19), we first note that both sides of (19) correspond to $\mathcal{A}_{r+N}(\mathsf{R})$-subcorepresentations of $\varphi_1^{\otimes(r+N)}$. Thus, by Proposition 20, it is sufficient to show that $(M(\mathbf{n}, r) \otimes \mathbb{C}\mathcal{D}_q)_{\tilde{U}} \simeq M(\mathbf{n} + (N)', r + N)_{\tilde{U}}$. But the latter follows at once from (16).

Let us prove also the relations (21) and (22) for $\mathfrak{g} = \mathrm{so}_{2n}$. Using again the dual pairing of $\tilde{U}_q(\mathrm{so}_{2n})$ and $\mathcal{A}(\mathsf{R})$ we find that $\chi \in \tilde{U}(\mathrm{so}_{2n})$ acts on $\mathcal{D}_q$ and $\mathcal{Q}_q$ as $\chi(\mathcal{D}_q) = -\mathcal{D}_q$ and $\chi(\mathcal{Q}_q) = \mathcal{Q}_q$. Therefore, by the definition of the irreducible representations $\tilde{T}(\mathbf{n})$ and $\tilde{T}(\mathbf{n}^\circ)$ (see Subsect. 8.6.2), we have

$$(M(\mathbf{n},r)\otimes\mathbb{C}\mathcal{D}_q)_{\tilde{U}}\simeq M(\mathbf{n}^\circ,r+N)_{\tilde{U}},\quad (M(\mathbf{n},r)\otimes\mathbb{C}\mathcal{Q}_q)_{\tilde{U}}\simeq M(\mathbf{n},r+2)_{\tilde{U}},$$

which proves the assertion for $\mathfrak{g}=\mathrm{so}_{2n}$. □

Now we turn to the corresponding $\mathcal{O}(G_q)$-comodules. Clearly, any $\mathcal{A}(\mathsf{R})$-comodule M gives rise to an $\mathcal{O}(G_q)$-comodule by composing the coaction with the canonical mapping of $\mathcal{A}(\mathsf{R})$ to its quotient $\mathcal{O}(G_q)$. It will be denoted by $M_{\mathcal{O}}$ if $G_q=SL_q(N),SO_q(N),Sp_q(N)$ and by $M_{\tilde{\mathcal{O}}}$ if $G_q=O_q(N)$.

Recall that the irreducible representations of $\tilde{U}(\mathfrak{g})$ are described by the partitions $\mathbf{n}$ from Subsect. 8.6.2 and also by the highest weights $\lambda\equiv\lambda(\mathbf{n})=(\lambda_1,\lambda_2,\cdots,\lambda_n)$, where $\lambda_i=2(\lambda,\alpha_i)/(\alpha_i,\alpha_i)$, as in Subsect. 7.1.2.

Lemma 26. (i) *There are the following comodule isomorphisms:*

$$M(\mathbf{n},r)_{\tilde{\mathcal{O}}}\simeq M(\mathbf{n},s)_{\tilde{\mathcal{O}}},\quad M(\mathbf{n},r)_{\mathcal{O}}\simeq M(\mathbf{n},s)_{\mathcal{O}}\ \text{ if }\ \mathbf{n}\in\mathcal{P}_r(\mathfrak{g})\cap\mathcal{P}_s(\mathfrak{g}),$$

$$M(\mathbf{n},r)_{\mathcal{O}}\simeq M(\mathbf{n}^\circ,s)_{\mathcal{O}}\ \text{ if }\ \mathbf{n}\in\mathcal{P}_r(\mathrm{so}_N),\ \ \mathbf{n}^\circ\in\mathcal{P}_s(\mathrm{so}_N),$$

$$M(\mathbf{n},r)_{\mathcal{O}}\simeq M(\mathbf{n}+(N)',r+N)_{\mathcal{O}}\ \text{ if }\ \mathbf{n}\in\mathcal{P}_r(\mathrm{sl}_N).$$

(ii) *If* $\mathbf{n}\in\mathcal{P}_r(\mathfrak{g})$, $\mathbf{m}\in\mathcal{P}_s(\mathfrak{g})$ *and* $\lambda(\mathbf{n})=\lambda(\mathbf{m})$, *then* $M(\mathbf{n},r)_{\mathcal{O}}\simeq M(\mathbf{m},s)_{\mathcal{O}}$.

(iii) *Let* $\mathbf{n}\in\mathcal{P}(\mathrm{so}_{2n})$ *be such that* $n_1'=n$, *where* $\mathbf{n}'=(n_1',n_2',\cdots,n_s')$ *is the transpose of the partition* $\mathbf{n}$. *Then the comodule* $M(\mathbf{n},r)_{\mathcal{O}}$ *is the direct sum of two irreducible comodules* M' *and* M'' *such that* $M_U'\simeq M(\lambda(\mathbf{n}))$ *and* $M_U''\simeq M(\sigma(\lambda(\mathbf{n})))$ *are irreducible* $U_q(\mathrm{so}_{2n})$*-modules with highest weights* $\lambda(\mathbf{n})$ *and* $\sigma(\lambda(\mathbf{n}))$, *respectively (see Proposition 8.34).*

(iv) *Except for the case* $\mathfrak{g}=\mathrm{so}_{2n},n_1'=n$ *(considered in* (iii)*) the comodule* $M(\mathbf{n},r)_{\mathcal{O}}$ *is irreducible for any partition* $\mathbf{n}\in\mathcal{P}_r(\mathfrak{g})$.

Proof. (i): We prove the first two lines of relations for $\mathfrak{g}=\mathrm{so}_{2n}$. The cases $\mathfrak{g}=\mathrm{sl}_N,\mathrm{so}_{2n+1},\mathrm{sp}_N$ are treated similarly. By Proposition 8.36, if $\mathbf{n}\in\mathcal{P}_r(\mathrm{so}_{2n})\cap\mathcal{P}_s(\mathrm{so}_{2n})$, then $r-s\in2\mathbb{Z}$. Since the comodules $(\mathbb{C}\mathcal{Q}_q)_{\tilde{\mathcal{O}}}$ and $(\mathbb{C}\mathcal{Q}_q)_{\mathcal{O}}$ are trivial, the first line of relations follows from (22). By the definition of $\mathbf{n}^\circ$, the assumptions $\mathbf{n}\in\mathcal{P}_r(\mathrm{so}_{2n})$ and $\mathbf{n}^\circ\in\mathcal{P}_s(\mathrm{so}_{2n})$ imply that $r-s\in2n\mathbb{Z}$. Since $(\mathbb{C}\mathcal{D}_q)_{\mathcal{O}}$ is the trivial comodule, the relation $M(\mathbf{n},r)_{\mathcal{O}}\simeq M(\mathbf{n}^\circ,s)_{\mathcal{O}}$ follows from (21). Similarly, the last relation in (i) follows from (19).

(ii): By the definition of $\lambda(\mathbf{n})$, the equality $\lambda(\mathbf{n})=\lambda(\mathbf{m})$ yields that $\mathbf{n}=\mathbf{m}$ for $\mathfrak{g}=\mathrm{sp}_N$, $\mathbf{n}=\mathbf{m}$ or $\mathbf{n}=\mathbf{m}\pm k(N)'$, $k\in\mathbb{N}_0$, for $\mathfrak{g}=\mathrm{sl}_N$ and $\mathbf{n}=\mathbf{m}$ or $\mathbf{n}=\mathbf{m}^\circ$ for $\mathfrak{g}=\mathrm{so}_N$. Hence the assertion follows from (i).

(iii): Since $\mathbf{n}=\mathbf{n}^\circ$ in the present case, it follows from (21) and (22) that there is an $\mathcal{A}_{r+2n}(\mathsf{R})$-comodule isomorphism

$$\Phi: M(\mathbf{n},r)\otimes\mathbb{C}\mathcal{Q}_q^n\to M(\mathbf{n},r)\otimes\mathbb{C}\mathcal{D}_q.$$

Let v be a nonzero highest weight vector of $M(\mathbf{n},r)_U$. Since Φ is also a $\tilde{U}_q(\mathrm{so}_{2n})$-module isomorphism, we may suppose that $\Phi(v\otimes\mathcal{Q}_q^n)=v\otimes\mathcal{D}_q$. Since $\chi(\mathcal{Q}_q^n)=\mathcal{Q}_q^n$ and $\chi(\mathcal{D}_q)=-\mathcal{D}_q$, we have $\Phi(\chi(v)\otimes\mathcal{Q}_q^n)=-\chi(v)\otimes\mathcal{D}_q$. On the other hand, since $\mathcal{O}(SO_q(2n))=\mathcal{A}(\mathsf{R})/\langle\mathcal{Q}_q-1,\mathcal{D}_q-1\rangle$, the linear map Φ' defined by $\Phi' : u \otimes \mathcal{Q}_q^n \to u \otimes \mathcal{D}_q$, $u \in M(\mathbf{n}, r)$, gives a comodule isomorphism $(M(\mathbf{n},r)\otimes\mathbb{C}\mathcal{Q}_q^n)_{\mathcal{O}} \simeq (M(\mathbf{n},r)\otimes\mathbb{C}\mathcal{D}_q)_{\mathcal{O}}$. We define $\mathcal{O}(SO_q(2n))$-comodule isomorphisms $\Phi_\pm := \Phi' \pm \Phi$. Then we have

$$\begin{aligned}
\mathrm{Im}\ \Phi_+ &= \Phi_+(U_q(\mathrm{so}_{2n})(v \otimes \mathcal{Q}_q^n) + U_q(\mathrm{so}_{2n})(\chi(v) \otimes \mathcal{Q}_q^n)) \\
&= U_q(\mathrm{so}_{2n})\Phi_+(v \otimes \mathcal{Q}_q^n) + U_q(\mathrm{so}_{2n})\Phi_+(\chi(v) \otimes \mathcal{Q}_q^n) \\
&= U_q(\mathrm{so}_{2n})(v \otimes \mathcal{D}_q), \\
\mathrm{Im}\ \Phi_- &= U_q(\mathrm{so}_{2n})(\chi(v) \otimes \mathcal{D}_q).
\end{aligned}$$

Thus, Im Φ_+ and Im Φ_- are distinct irreducible components of the $U_q(\mathrm{so}_{2n})$-module $(M(\mathbf{n},r) \otimes \mathbb{C}\mathcal{D}_q)_U$. This implies the assertion.

(iv): The $\mathcal{O}(G_q)$-comodule $M(\mathbf{n},r)_{\mathcal{O}}$ is irreducible, because the corresponding $U_q(\mathfrak{g})$-module $M(\mathbf{n},r)_U$ is so. □

By Lemma 26(i), the irreducible corepresentations of $\mathcal{O}(O_q(N))$ which are derived from irreducible corepresentations of the bialgebra $\mathcal{A}(\mathsf{R})$ are parametrized by the set of partitions $\mathcal{P}(\mathrm{so}_N)$ from Subsect. 8.6.2.

Now let $G_q = SL_q(N), SO_q(N), Sp_q(N)$. We define an irreducible $\mathcal{O}(G_q)$-comodule $M_{\mathcal{O}}(\lambda)$ for any dominant integral weight $\lambda = (\lambda_1, \lambda_2, \cdots, \lambda_n) \in \mathfrak{P}(\mathfrak{g})$, $\lambda_i = 2(\lambda,\alpha_i)/(\alpha_i,\alpha_i)$, of the corresponding Drinfeld–Jimbo algebra $U_q(\mathfrak{g})$. First we consider the case where $\mathfrak{g} = \mathrm{so}_{2n}$ and $\lambda_{n-1} \neq \lambda_n$. Then there exists a partition $\mathbf{n} \in \mathcal{P}(\mathrm{so}_{2n})$ such that $\lambda = \lambda(\mathbf{n})$ and $n_1' = n$. (Recall that $(n_1', n_2', \cdots, n_s')$ is the transpose of the partition $\mathbf{n}$ and $\lambda(\mathbf{n})$ is given by the formulas in Subsect. 8.6.2). In this case we set $M_{\mathcal{O}}(\lambda) := M'$ and $M_{\mathcal{O}}(\sigma(\lambda)) := M''$, where M' and M'' are the irreducible $\mathcal{O}(SO_q(2n))$-comodules from Lemma 26(iii). In all other cases we set $M_{\mathcal{O}}(\lambda) := M(\mathbf{n},r)_{\mathcal{O}}$, where $\mathbf{n} \in \mathcal{P}_r(\mathfrak{g})$ is such that $\lambda(\mathbf{n}) = \lambda$. Thus we obtain irreducible $\mathcal{O}(G_q)$-comodules $M_{\mathcal{O}}(\lambda)$ or equivalently irreducible corepresentations of $\mathcal{O}(G_q)$ parametrized by the elements λ of the set $\mathfrak{P}(\mathfrak{g})$ from Proposition 7.20. If $\lambda \neq \mu$, then these corepresentations of $\mathcal{O}(G_q)$ are not equivalent, because the corresponding representations of the Drinfeld–Jimbo algebras are not.

Proof of Theorem 22. Taking the coefficient coalgebras of both sides of the equivalent $\mathcal{A}(\mathsf{R})$-comodules in (19)–(22) we obtain the relations

$$\begin{aligned}
\mathcal{C}(\varphi_{\mathbf{n}+(N)',r+N}) &= \mathcal{C}(\varphi_{\mathbf{n},r}) \cdot \mathcal{D}_q \quad \text{for} \quad \mathfrak{g} = \mathrm{sl}_N, \\
\mathcal{C}(\varphi_{\mathbf{n},r+N}) &= \mathcal{C}(\varphi_{\mathbf{n},r}) \cdot \mathcal{D}_q \quad \text{for} \quad \mathfrak{g} = \mathrm{sp}_N, \\
\mathcal{C}(\varphi_{\mathbf{n}^{\mathrm{o}},r+N}) &= \mathcal{C}(\varphi_{\mathbf{n},r}) \cdot \mathcal{D}_q \quad \text{for} \quad \mathfrak{g} = \mathrm{so}_N, \\
\mathcal{C}(\varphi_{\mathbf{n},r+2}) &= \mathcal{C}(\varphi_{\mathbf{n},r}) \cdot \mathcal{Q}_q \quad \text{for} \quad \mathfrak{g} = \mathrm{sp}_N, \mathrm{so}_N.
\end{aligned}$$

The direct sum decomposition (15) will be derived from the decomposition (13) and the preceding formulas. Let us carry out the proof in the case $G_q = Sp_q(N)$. Since $\mathcal{C}(\varphi_{\mathbf{n},r+2}) = \mathcal{C}(\varphi_{\mathbf{n},r}) \cdot \mathcal{Q}_q$, the direct sum (13) is now

$$\mathcal{A}(\mathsf{R}) = \bigoplus_{\mathbf{n}\in\mathcal{P}(\mathrm{sp}_N),|\mathbf{n}|\in 2\mathbb{Z}} \mathcal{C}(\varphi_{\mathbf{n},0}) \oplus \bigoplus_{\mathbf{n}\in\mathcal{P}(\mathrm{sp}_N),|\mathbf{n}|\in 2\mathbb{Z}+1} \mathcal{C}(\varphi_{\mathbf{n},1}) \oplus (\mathcal{Q}_q - 1)\mathcal{A}(\mathsf{R}).$$

Since the union of $\{\mathbf{n} \in \mathcal{P}(\mathrm{sp}_N) \mid |\mathbf{n}| \in 2\mathbb{Z}\}$ and $\{\mathbf{n} \in \mathcal{P}(\mathrm{sp}_N) \mid |\mathbf{n}| \in 2\mathbb{Z}+1\}$ is just the set $P(Sp_q(N))$ and $\mathcal{O}(Sp_q(N)) = \mathcal{A}(\mathsf{R})/\langle \mathcal{Q}_q - 1\rangle$ by the definition, we get

$$\mathcal{O}(Sp_q(N)) \simeq \bigoplus\nolimits_{\lambda \in P(Sp_q(N))} \mathcal{C}(\varphi_\lambda),$$

which proves (15). Using the above results the other cases are treated similarly. The last assertion is proved as in the case of Theorem 21. □

Remarks 3. That q is transcendental was only used in order to apply Theorem 8.38. If the latter theorem could be shown for q not a root of unity, then all results of this subsection remain valid under this assumption.

4. The representations $T_{\mathbf{n}}$ and $\tilde{T}(\mathbf{n})$ of $U_q(\mathfrak{g})$ resp. $\tilde{U}_q(\mathfrak{g})$ and the corepresentations $\varphi_{\mathbf{n}}$ of $\mathcal{O}(G_q)$ are parametrized by highest weights with respect to the simple roots $\alpha_1, \cdots, \alpha_l$. Likewise we could have used highest weights with respect to the simple roots $-\alpha_1, \cdots, -\alpha_l$. Then we obtain the same correspondence of irreducible corepresentations with Young frames as for the classical groups. For $G_q = GL_q(N)$ this is carried out in Subsect. 11.5.3. △

11.3 Compact Quantum Group Algebras

This section is devoted to a detailed study of the class of Hopf $*$-algebras which are linear spans of the matrix coefficients of their finite-dimensional unitary corepresentations.

11.3.1 Definitions and Characterizations of CQG Algebras

Let G be a compact topological group and let $\mathcal{A} = \mathrm{Rep}\ (G)$ be the linear span of matrix elements v_{ij} of unitary finite-dimensional continuous representations of G (see Example 1.2). Then $\mathcal{A}$ is a (commutative) Hopf $*$-algebra with $*$-structure and comultiplication given by $(v_{ij})^* = \overline{v_{ij}}$ (the bar means complex conjugation) and $\Delta(v_{ij}) = v_{ik} \otimes v_{kj}$. Further, the algebra $\mathcal{A}$ is finitely generated if and only if G is a compact Lie group or, equivalently, if G is isomorphic to a closed subgroup of the group $U(N)$ of $N \times N$ unitary matrices. Such groups are called compact matrix groups. These facts provide the motivation for the following two definitions.

Definition 9. *A Hopf $*$-algebra $\mathcal{A}$ is called a* compact quantum group algebra *(abbreviated, a* CQG algebra*) if $\mathcal{A}$ is the linear span of all matrix elements of finite-dimensional unitary corepresentations of $\mathcal{A}$.*

Definition 10. *A* compact matrix quantum group algebra *(abbreviated, a* CMQG algebra*) is a CQG algebra which is generated as an algebra by finitely many elements.*

We give first two slight reformulations of Definition 9. Let $\mathcal{A}$ be a Hopf $*$-algebra. Since finite-dimensional unitary corepresentations are completely reducible by Proposition 12, $\mathcal{A}$ is a CQG algebra if and only if it is spanned by the matrix elements of *irreducible* (hence finite-dimensional) unitary corepresentations. Combining the latter characterization with Theorem 13(v) it follows that $\mathcal{A}$ is a CQG algebra if and only if $\mathcal{A}$ is cosemisimple and each corepresentation φ_α, $\alpha \in \hat{\mathcal{A}}$, is unitarizable.

Since CQG algebras are cosemisimple, the theory developed in Subsects. 11.2.1 and 11.2.2 applies in particular to any CQG algebra.

Further characterizations and basic properties of these algebras are contained in the following two results.

Theorem 27. *For any Hopf $*$-algebra $\mathcal{A}$ the following statements are equivalent:*

(i) *$\mathcal{A}$ is a CQG algebra.*

(ii) *$\mathcal{A}$ is generated as an algebra by the matrix elements of all finite-dimensional unitarizable corepresentations of $\mathcal{A}$.*

(iii) *Any finite-dimensional corepresentation of $\mathcal{A}$ is unitarizable (and hence a direct sum of irreducible corepresentations).*

(iv) *There is a linear functional h on $\mathcal{A}$ such that*

$$(h \otimes \mathrm{id})\Delta(a) = h(a)1 \quad \textit{and} \quad h(a^*a) > 0 \quad \textit{for all} \quad a \in \mathcal{A},\ a \neq 0.$$

Proof. (i) $\rightarrow$ (ii) is trivial.

(ii) $\rightarrow$ (i): Let $\{u^i\}$ be a set of finite-dimensional unitarizable corepresentations such that their matrix elements generate $\mathcal{A}$ as an algebra. Let $\mathcal{R}$ be the smallest family of corepresentations of $\mathcal{A}$ which contains all u^i and is closed under direct sums and tensor products. These operations preserve unitarizability, so all corepresentations in $\mathcal{R}$ are unitarizable. Since $\mathcal{C}(\varphi \oplus \psi) = \mathcal{C}(\varphi) + \mathcal{C}(\psi)$ and $\mathcal{C}(\varphi \otimes \psi) = \mathcal{C}(\varphi)\mathcal{C}(\psi)$, the union of $\mathcal{C}(\varphi)$, $\varphi \in \mathcal{R}$, is an algebra which contains all $\mathcal{C}(u^i)$. By assumption (ii), this algebra is $\mathcal{A}$. For any $\varphi \in \mathcal{R}$ there is a unitary corepresentation $\tilde{\varphi}$ such that $\mathcal{C}(\varphi) = \mathcal{C}(\tilde{\varphi})$. Hence $\mathcal{A}$ is a CQG algebra.

(iv) $\rightarrow$ (iii): The proof is an adaption of the proof for the corresponding result on representations of compact groups. Let φ be a corepresentation of $\mathcal{A}$ on a finite-dimensional vector space V. Let $(\cdot,\cdot)$ be an arbitrary scalar product on V. Define another Hermitian sesquilinear form $\langle\cdot,\cdot\rangle$ on $V \times V$ by

$$\langle x, y\rangle := \sum (x_{(0)}, y_{(0)}) h(x_{(1)}{}^* y_{(1)}), \quad x, y \in V.$$

Let v_{ij} be the matrix elements of φ with respect to an orthonormal basis $\{x_i\}$ of $\big(V, (\cdot,\cdot)\big)$. Then, we have

$$\langle x_i, x_j\rangle = \sum\nolimits_{k,l} (x_k, x_l) h((v_{ki})^* v_{lj}) = \sum\nolimits_k h((v_{ki})^* v_{kj}),$$

so

$$\Big\langle \sum\nolimits_i \alpha_i x_i, \sum\nolimits_j \alpha_j x_j \Big\rangle = \sum\nolimits_k h\Big(\Big(\sum\nolimits_i \alpha_i v_{ki}\Big)^* \Big(\sum\nolimits_j \alpha_j v_{kj}\Big)\Big) \geq 0$$

for any $\alpha_i \in \mathbb{C}$. Since $h(a^*a) > 0$ for $a \neq 0$, the left hand side vanishes only when $\sum_i \alpha_i v_{ki} = 0$ for all k. But then $\alpha_k = \varepsilon(\sum_i \alpha_i v_{ki}) = 0$ for all k. Hence $\langle \cdot, \cdot \rangle$ is a scalar product on V.

Using the right invariance of the functional h, we conclude that

$$\sum \langle x_{(0)}, y_{(0)} \rangle x_{(1)}{}^* y_{(1)} = \sum (x_{(0)}, y_{(0)}) h(x_{(1)}{}^* y_{(1)}) x_{(2)}{}^* y_{(2)}$$

$$= \sum (x_{(0)}, y_{(0)})(h \otimes \mathrm{id})\Delta(x_{(1)}{}^* y_{(1)}) = \sum (x_{(0)}, y_{(0)}) h(x_{(1)}{}^* y_{(1)}) 1 = \langle x, y \rangle 1.$$

Therefore, by Proposition 11, the corepresentation φ is unitary with respect to the scalar product $\langle \cdot, \cdot \rangle$.

(iii) $\rightarrow$ (i): Let $a \in \mathcal{A}$. By Proposition 4, there is a finite-dimensional subcoalgebra $\mathcal{C}$ containing a. The restriction φ of Δ to $\mathcal{C}$ is a finite-dimensional corepresentation of $\mathcal{A}$ and hence is unitary with respect to a certain scalar product on $\mathcal{C}$. Since $(\varepsilon \otimes \mathrm{id})\varphi(a) = a$, we have $a \in \mathcal{C}(\varphi)$.

The proof of the implication (i) $\rightarrow$ (iv) is given in Subsect. 11.3.2 along with the proof of Proposition 29 below. □

Proposition 28. *For a Hopf $*$-algebra $\mathcal{A}$, the following are equivalent:*

(i) *$\mathcal{A}$ is a CMQG algebra.*

(ii) *There is a finite-dimensional unitary corepresentation u of $\mathcal{A}$ whose matrix elements generate $\mathcal{A}$ as an algebra.*

(iii) *There is a finite-dimensional corepresentation u of $\mathcal{A}$ such that u and u^c are both unitarizable and the matrix elements of u and u^c generate $\mathcal{A}$ as an algebra.*

Proof. (i)$\rightarrow$(ii): Take a finite set of generators of the algebra $\mathcal{A}$. Since $\mathcal{A}$ is a CQG algebra, each generator is a linear combination of matrix elements of finite-dimensional unitary corepresentations of $\mathcal{A}$. The direct sum of these corepresentations has the desired property. (ii)$\rightarrow$(i) follows by the proof of implication (ii)$\rightarrow$(i) of Theorem 27 applied to a singleton $\{u\}$.

(ii)$\rightarrow$(iii): By the implication (ii)$\rightarrow$(i), $\mathcal{A}$ is a CQG algebra, so u^c is unitarizable by Theorem 27(iii). (Here the more involved part of Theorem 27, which contains the implication (i) $\rightarrow$ (iii), is needed. This can be avoided by an alternative proof based on Proposition 12.) (iii)$\rightarrow$(i) is obvious. □

Let $\mathcal{A}$ be a CMQG algebra and let u be a corepresentation of $\mathcal{A}$ as in Proposition 28(iii). By choosing an appropriate basis, we can assume that u is given by a unitary matrix corepresentation $u = (u_{ij})_{i,j=1,\cdots,N}$. Then we have $u^*u = uu^* = I$ and hence $u^c = \bar{u}$. Since u^c is unitarizable by Proposition 28(iii), there is an invertible matrix $A \in M_N(\mathbb{C})$ such that $v := Au^cA^{-1}$ is a unitary matrix over $\mathcal{A}$. Setting $E := A^*A$, one immediately checks that the relations $v^*v = vv^* = I$ are equivalent to the matrix equations

$$u^t E \bar{u} E^{-1} = E \bar{u} E^{-1} u^t = I. \tag{23}$$

That is, *any CMQG algebra $\mathcal{A}$ is generated as a $*$-algebra (!) by the entries of a unitary matrix $u = (u_{ij})_{i,j=1,\cdots N}$ which satisfies the equations (23).* Further, if $V \in M_N(\mathbb{C})$ is a unitary matrix, then $\tilde{u} := V^t u \overline{V}$ is another unitary matrix corepresentation of $\mathcal{A}$ such that the entries of $\tilde{u}$ generate $\mathcal{A}$ as a $*$-algebra and $\tilde{u}$ satisfies (23) with E replaced by $\tilde{E} := VEV^{-1}$ and u by $\tilde{u}$. By the spectral theorem, the matrix $\tilde{E}$ is diagonal with positive numbers on the diagonal for some suitable choice of V.

We close this subsection by considering some examples.

Example 6 (*Universal CMQG algebras*). Let $E \in M_N(\mathbb{C})$ be an invertible positive Hermitian matrix. Let $\mathcal{A}_{\rm un}(E)$ denote the free $*$-algebra with generators u_{ij}, $i, j = 1, 2, \cdots, N$, subject to the $4N^2$ relations $u^*u = uu^* = I$ and (23), where $u := (u_{ij})$, $\bar{u} \equiv (\bar{u}_{ij}) := ((u_{ij})^*)$ and $u^* := \bar{u}^t$. *Then $\mathcal{A}_{\rm un}(E)$ is a CMQG algebra with comultiplication given by $\Delta(u_{ij}) = u_{ik} \otimes u_{kj}$.*

Proof. The free algebra $\mathbb{C}\langle u_{ij}, v_{ij}\rangle$ with $2N^2$ generators u_{ij} and v_{ij} is a bialgebra with comultiplication determined by $\Delta(u_{ij}) = u_{ik} \otimes u_{kj}$ and $\Delta(v_{ij}) = v_{ik} \otimes v_{kj}$. Let $\mathcal{J}$ be the two-sided ideal of $\mathbb{C}\langle u_{ij}, v_{ij}\rangle$ generated by the elements corresponding to the relations $v^t u = uv^t = u^t E v E^{-1} = E v E^{-1} u^t = I$, where $u = (u_{ij})$ and $v = (v_{ij})$. As in the proof of Proposition 9.1 we verify that $\mathcal{J}$ is a biideal, so that $\mathcal{A} := \mathbb{C}\langle u_{ij}, v_{ij}\rangle / \mathcal{J}$ is a bialgebra. One easily checks that the elements $u'_{ij} := v_{ji}$, $v'_{ij} := (E^{-1}u^t E)_{ij}$ of $\mathcal{A}$ satisfy the defining relations of the opposite algebra $\mathcal{A}^{\rm op}$. Hence there exists an algebra anti-homomorphism $S : \mathcal{A} \to \mathcal{A}$ such that $S(u_{ij}) = u'_{ij}$ and $S(v_{ij}) = v'_{ij}$. Since $S(u)u = uS(u) = I$ and $S(v)v = vS(v) = I$ by construction, it follows from Proposition 1.8 that S is an antipode for the bialgebra $\mathcal{A}$. Thus, $\mathcal{A}$ is a Hopf algebra.

Clearly, $\mathbb{C}\langle u_{ij}, v_{ij}\rangle$ is a $*$-bialgebra with respect to the involution given by $u^*_{ij} = v_{ij}$. Since the matrix E is Hermitian, $\mathcal{J}$ is a $*$-ideal and hence $\mathcal{A}$ is a Hopf $*$-algebra. Obviously, as a $*$-algebra, $\mathcal{A}$ is just the $*$-algebra $\mathcal{A}_{\rm un}(E)$ defined above. From the defining relations $u^*u = uu^* = I$ and (23) it follows that u and $w := E^{1/2} u^c E^{-1/2} \equiv E^{1/2} \bar{u} E^{-1/2}$ are unitary matrix corepresentations. Since $w \simeq u^c$, the corepresentation u^c is unitarizable. Since the matrix entries of u and $u^c = \bar{u}$ generate $\mathcal{A}$ as an algebra, the Hopf algebra $\mathcal{A}$ is a CMQG algebra by Proposition 28(iii). □

From the remarks preceding this example it follows that any CMQG algebra is a quotient of some CMQG algebra $\mathcal{A}_{\rm un}(E)$ for a certain diagonal matrix E with positive diagonal entries. That is, the CMQG algebras $\mathcal{A}_{\rm un}(E)$ are universal CMQG algebras. Note that the CMQG algebras $\mathcal{A}_{\rm un}(E)$ are very large in general, because no relations among the entries u_{ij} are required. △

Example 7. Each of the Hopf $*$-algebras $\mathcal{O}(U_q(N))$, $\mathcal{O}(SU_q(N))$, $\mathcal{O}(O_q(N;\mathbb{R}))$, $\mathcal{O}(SO_q(N;\mathbb{R}))$ and $\mathcal{O}(USp_q(N))$ (see Subsects. 9.2.4 and 9.3.5 for their definitions) for $q \in \mathbb{R}$ is a CMQG algebra. In these cases condition (ii) in Proposition 28 applies, since the fundamental corepresentations $\mathbf{u}$ are unitary (recall that $\mathbf{u}\mathbf{u}^* = I$ by the definitions of the involutions). Except for $U_q(N)$, the matrix elements u^i_j generate the corresponding algebras. The algebra $\mathcal{O}(U_q(N))$ is generated by the matrix elements of the unitary corepresentation $\mathbf{u} \oplus (\mathcal{D}_q^{-1})$. Recall that $\mathcal{D}_q^{-1}$ is a unitary element of the $*$-algebra $\mathcal{O}(U_q(N))$. △

11.3.2 The Haar State of a CQG Algebra

Because any CQG algebra is a cosemisimple Hopf algebra, it possesses a unique Haar functional. In this and the next subsections we are mainly concerned with the Haar functional h of a CQG algebra $\mathcal{A}$.

Recall that a linear functional f on a $*$-algebra $\mathcal{A}$ with unit is called a *state* if $f(1) = 1$ and $f(a^*a) \geq 0$ for $a \in \mathcal{A}$. A state f on $\mathcal{A}$ is called *faithful* if $f(a^*a) > 0$ for all nonzero $a \in \mathcal{A}$.

Proposition 29. *The Haar functional h of a CQG algebra $\mathcal{A}$ is a faithful state on the $*$-algebra $\mathcal{A}$.*

The key step for the proof of this proposition is the following

Lemma 30. *Let w be an irreducible unitary corepresentation of $\mathcal{A}$ on a (finite-dimensional) Hilbert space W. Then there exists a unique invertible intertwiner $F \in \mathrm{Mor}\ (w, w^{cc})$ such that $\mathrm{Tr}\ F = \mathrm{Tr}\ F^{-1} > 0$. This operator F is positive definite, that is, $\langle Fx, x\rangle > 0$ for all $x \in W$, $x \neq 0$.*

Proof. We fix an orthonormal basis of W and consider w and w^{cc} as matrix corepresentations. Since w^c is also irreducible, there exist an irreducible unitary corepresentation v of $\mathcal{A}$ and an invertible complex matrix A such that $w^c = A^{-1}vA$. Let $\bar{A}$ denote the matrix with entries $\overline{A_{ij}}$. Since v and w are unitary, we have $(v^c)_{ij} = S(v_{ji}) = v_{ij}{}^*$, $(w^c)_{ij} = w_{ij}{}^*$ and $(w^{cc})_{ij} = S((w^c)_{ji})$. Therefore, we conclude from $vA = Aw^c$ that $v^c\bar{A} = \bar{A}w$ and $A^tv^c = w^{cc}A^t$ and so $A^t\bar{A}w = w^{cc}A^t\bar{A}$. That is, $A^t\bar{A} \in \mathrm{Mor}\ (w, w^{cc})$. Since A is invertible, $A^t\bar{A}$ is positive definite. Since $\dim \mathrm{Mor}\,(w, w^{cc}) = 1$, $F \in \mathrm{Mor}\,(w, w^{cc})$ is uniquely determined by requiring that $\mathrm{Tr}\,F = \mathrm{Tr}\,F^{-1} > 0$. □

Let $\mathcal{A}$ be a CQG algebra. By the second reformulation of Definition 9 given in Subsect. 11.3.1, we can choose a unitary corepresentation u^α in each equivalence class $\alpha \in \hat{\mathcal{A}}$. Moreover, let us fix $F_\alpha \in \mathrm{Mor}\ (u^\alpha, (u^\alpha)^{cc})$ such that $M_\alpha := \mathrm{Tr}\ F_\alpha = \mathrm{Tr}\ F_\alpha^{-1} > 0$. Since $(u^\beta_{ji})^* = S(u^\beta_{ij})$ and $(u^\alpha_{lk})^* = S(u^\alpha_{kl})$ by the unitarity of u^β and u^α, (7) and (8) can be rewritten as

$$h(u^\alpha_{kl}(u^\beta_{ji})^*) = \delta_{\alpha\beta}\delta_{kj}M_\alpha^{-1}(F_\alpha)_{il}, \tag{24}$$

$$h((u^\alpha_{lk})^*u^\beta_{ij}) = \delta_{\alpha\beta}\delta_{kj}M_\alpha^{-1}(F_\alpha^{-1})_{il}. \tag{25}$$

These two equations are referred to as the *Schur orthogonality relations* for the CQG algebra $\mathcal{A}$.

Proof of Proposition 29. Since $\mathcal{A}$ is a CQG algebra, any element $a \in \mathcal{A}$ is a finite sum $\sum_{\alpha,i,j} \lambda_{\alpha ij}u^\alpha_{ij}$ with $\lambda_{\alpha ij} \in \mathbb{C}$. By (25), we have

$$h(a^*a) = \sum\nolimits_{\alpha,l,k,i} \lambda_{\alpha ik}\overline{\lambda_{\alpha lk}}M_\alpha^{-1}(F_\alpha^{-1})_{il}.$$

Since F_α^{-1} is positive definite as well, $h(a^*a) \geq 0$ and $h(a^*a) = 0$ implies that all coefficients $\lambda_{\alpha ik}$ are zero, so $a = 0$. □

Because of Proposition 29, the Haar functional of a CQG algebra will be called the *Haar state.* The Haar state is a fundamental tool for the study of CQG algebras. It was essentially used in the previous subsection in order to prove the important result that any finite-dimensional corepresentation of a CQG algebra is unitarizable (and hence completely reducible). Also, the character theory in Subsect. 11.2.2 was based on the Haar functional. Other applications will follow below.

Since the Haar state h is a faithful state on $\mathcal{A}$, the equation

$$\langle a, b\rangle := h(a^*b), \qquad a, b \in \mathcal{A},$$

defines a scalar product $\langle\cdot,\cdot\rangle$ on $\mathcal{A}$. Let $\mathcal{H}(\mathcal{A})$ denote the Hilbert space completion of the pre-Hilbert space $(\mathcal{A}, \langle\cdot,\cdot\rangle)$. The left and right invariance of h are expressed in terms of the scalar product $\langle\cdot,\cdot\rangle$ as

$$\langle a, b\rangle = \sum a_{(1)}{}^*b_{(1)}\langle a_{(2)}, b_{(2)}\rangle = \sum \langle a_{(1)}, b_{(1)}\rangle a_{(2)}{}^*b_{(2)}, \quad a, b \in \mathcal{A}.$$

Equation (25) shows that for $\alpha \neq \beta$, $\alpha, \beta \in \hat{\mathcal{A}}$, the coefficient coalgebras $\mathcal{C}(u^\alpha)$ and $\mathcal{C}(u^\beta)$ are orthogonal subspaces and the characters χ_α and χ_β are orthogonal elements of $\mathcal{H}(\mathcal{A})$. Therefore, for any CQG algebra $\mathcal{A}$ the Peter–Weyl decomposition (2) is an *orthogonal sum* in the Hilbert space $\mathcal{H}(\mathcal{A})$.

Example 8. Let G be a compact topological group, μ the Haar measure of G and $\mathcal{A}$ the CQG algebra Rep (G), see Example 1.2. Then the Haar state h on $\mathcal{A}$ is given by $h(a) = \int_G a(g)d\mu(g)$, $a \in \mathcal{A}$, and $\mathcal{H}(\mathcal{A})$ is just the Hilbert space $L^2(G, \mu)$ of square integrable functions on G with respect to μ. The latter follows from the Peter–Weyl theorem which asserts that Rep (G) is dense in $L^2(G, \mu)$. △

11.3.3 C^*-Algebra Completions of CQG Algebras

In this subsection we show that each CQG algebra $\mathcal{A}$ admits a completion to a C^*-algebra. In order to do so, we extend the considerations of the first two paragraphs of Subsect. 4.3.4 to general CQG algebras.

First let us recall some standard terminology (see, for instance, [BR] or [Mu]). A *C^*-seminorm* on a $*$-algebra A is a seminorm p which has the so-called C^*-property $p(a^*a) = p(a)^2$, $a \in A$. It can be shown that any C^*-seminorm p satisfies $p(a^*) = p(a)^*$ and $p(ab) \leq p(a)p(b)$ for $a, b \in A$ and $p(1) = 1$ if $p \not\equiv 0$. A *C^*-norm* is a C^*-seminorm which is a norm. A *C^*-algebra* is a $*$-algebra A equipped with a C^*-norm $\|\cdot\|$ such that the normed linear space $(A, \|\cdot\|)$ is complete. A *$*$-representation* of a $*$-algebra A on a Hilbert space $\mathcal{H}$ is a $*$-homomorphism of A into the $*$-algebra of bounded linear operators on $\mathcal{H}$.

We need a simple technical lemma.

Lemma 31. *Let V be a dense linear subspace of a Hilbert space $\mathcal{H}$ with scalar product $\langle\cdot,\cdot\rangle$. Suppose that π is an algebra homomorphism of a CQG algebra*

$\mathcal{A}$ into the algebra of (not necessarily bounded) linear operators on V such that $\langle \pi(a)x, y\rangle = \langle x, \pi(a^*)y\rangle$ for all $a \in \mathcal{A}$ and $x, y \in V$.

(i) *Let $v = (v_{ij})$ be a unitary matrix corepresentation of $\mathcal{A}$. Then each linear operator $\pi(v_{ij})$ on V is bounded and $\|\pi(v_{ij})\| \le 1$.*

(ii) *π extends uniquely to a $*$-representation of $\mathcal{A}$ on $\mathcal{H}$.*

Proof. (ii) follows immediately from (i). To prove (i), recall that $v^*v = I$ and so $\sum_k v_{kj}{}^* v_{kj} = 1$ by Proposition 11(v). Hence, for $x \in V$ we have

$$\|\pi(v_{ij})x\|^2 \le \sum_k \langle \pi(v_{kj})x, \pi(v_{kj})x\rangle = \sum_k \langle \pi(v_{kj}{}^* v_{kj})x, x\rangle = \|x\|^2. \quad \square$$

We now apply Lemma 31 to the case $V = \mathcal{A}$, $\mathcal{H} = \mathcal{H}(\mathcal{A})$, $\pi = \pi_h$, where π_h is defined by $\pi_h(a)b = ab$, $a, b \in \mathcal{A}$. Since $\langle \pi_h(a)b, c\rangle = h((ab)^*c) = h(b^*a^*c) = \langle b, \pi_h(a^*)c\rangle$ for $a, b \in \mathcal{A}$, π_h satisfies the assumptions of Lemma 31. Hence there exists a unique extention of π_h to a $*$-representation, denoted again by π_h, of the CQG algebra $\mathcal{A}$ on the Hilbert space $\mathcal{H}(\mathcal{A})$. The $*$-representation π_h of $\mathcal{A}$ on $\mathcal{H}(\mathcal{A})$ is called the *GNS representation* of the state h (see [BR] or [Mu]). From the definition it is clear that $\pi_h(\mathcal{A})1$ is dense in the Hilbert space $\mathcal{H}(\mathcal{A})$ and that $h(a) = \langle \pi_h(a)1, 1\rangle$ for $a \in \mathcal{A}$. These two properties characterize the GNS representation up to unitary equivalence. Moreover, $\pi_h(a) = 0$ implies that $a = 0$. (Indeed, $\pi_h(a) = 0$ yields $h(a^*a) = \|\pi_h(a)1\|^2 = 0$, so $a = 0$ by the faithfulness of h.) Hence $\|\pi_h(\cdot)\|$ is a C^*-norm on $\mathcal{A}$.

Proposition 32. *Let $\mathcal{A}$ be a CQG algebra. For $a \in \mathcal{A}$, let $\|a\|_\infty$ denote the supremum of $p(a)$ for all C^*-seminorms p on the $*$-algebra $\mathcal{A}$. Then, $\|\cdot\|_\infty$ is a C^*-norm on $\mathcal{A}$ and the completion of $\mathcal{A}$ with respect to this norm is a C^*-algebra A with unit element. The counit ε and the Haar state h of $\mathcal{A}$ have unique continuous extensions, denoted by the same symbols, to the C^*-algebra A. The comultiplication of $\mathcal{A}$ can be uniquely extended to a $*$-homomorphism $\Delta : A \to A\bar{\otimes}A$. Here $A\bar{\otimes}A$ denotes the C^*-algebra obtained by completing the algebraic tensor product $A \otimes A$ in the least C^*-norm on $A \otimes A$. These extensions satisfy the relations $(\Delta \otimes \mathrm{id}) \circ \Delta = (\mathrm{id} \otimes \Delta) \circ \Delta$ on $A\bar{\otimes}A\bar{\otimes}A$, $(\varepsilon \otimes \mathrm{id}) \circ \Delta = (\mathrm{id} \otimes \varepsilon) \circ \Delta = \mathrm{id}$ and $(h \otimes \mathrm{id})\Delta = (\mathrm{id} \otimes h)\Delta = h1$ on A.*

Proof. First we have to show that $\|a\|_\infty$ is finite for all $a \in \mathcal{A}$. Since $\mathcal{A}$ is a CQG algebra, it suffices to do so for $a = v_{ij}$, where v_{ij} is a matrix element of some unitary matrix corepresentation of $\mathcal{A}$. Let p be a C^*-seminorm on $\mathcal{A}$ and let i_p denote the canonical mapping of $\mathcal{A}$ to the quotient $*$-algebra $\mathcal{A}/\ker p$. Then $\bar{p}(i_p(\cdot)) := p(\cdot)$ defines a C^*-norm on $\mathcal{A}/\ker p$. Let $\bar{\mathcal{A}}_p$ be the C^*-algebra which is the completion of $\mathcal{A}/\ker p$ with respect to $\bar{p}$. The C^*-algebra $\bar{\mathcal{A}}_p$ has a faithful norm preserving $*$-representation (see [BR], 2.3.4), say $\bar{\pi}_p$, on some Hilbert space $\mathcal{H}_p$. Then $\pi_p = \bar{\pi}_p \circ i_p$ is a $*$-representation of $\mathcal{A}$ on $\mathcal{H}_p$. By Lemma 31(i), $\|\pi_p(v_{ij})\| \le 1$ and hence $p(v_{ij}) = \bar{p}(i_p(v_{ij})) = \|\bar{\pi}_p(i_p(v_{ij}))\| = \|\pi_p(v_{ij})\| \le 1$. Thus, $\|v_{ij}\|_\infty \le 1$ and $\|\cdot\|_\infty$ is finite on $\mathcal{A}$.

Since $\|\pi_h(\cdot)\|$ is a C^*-norm on $\mathcal{A}$ as noted above, $\|\cdot\|_\infty$ is a norm, hence a C^*-norm, on $\mathcal{A}$. Clearly, the completion of $(\mathcal{A}, \|\cdot\|_\infty)$ becomes a unital

C^*-algebra A. Let $\|\cdot\|_{\min}$ denote the least C^*-norm of $A \otimes A$ (see [Mu]). Then $\|\Delta(\cdot)\|_{\min}$ is a C^*-seminorm on $\mathcal{A}$. Also, $|\varepsilon(\cdot)|$ is a C^*-seminorm on $\mathcal{A}$ and $|h(\cdot)| \le \|\pi_h(\cdot)\|$. Therefore, $\|\Delta(\cdot)\|_{\min} \le \|\cdot\|_\infty$, $|\varepsilon(\cdot)| \le \|\cdot\|_\infty$ and $|h(\cdot)| \le \|\cdot\|_\infty$ by the definition of $\|\cdot\|_\infty$. Since Δ, ε and h are continuous with respect to the norm $\|\cdot\|_\infty$, they have unique continuous extensions to A. The algebraic relations are preserved by continuity. □

Remarks: 5. The antipode of $\mathcal{A}$ is not necessarily bounded with respect to the norm $\|\cdot\|_\infty$, so it cannot be extended to the C^*-algebra A in general.

6. Since $\|\cdot\|_\infty$ is the largest C^*-seminorm on $\mathcal{A}$, each $*$-representation of $\mathcal{A}$ on a Hilbert space has a continuous extension to a $*$-representation of the C^*-algebra A. This property characterizes the C^*-algebra A in Proposition 32. However, the largest C^*-seminorm is not the only possible choice of a C^*-norm on $\mathcal{A}$ for getting a C^*-algebra completion $\tilde{\mathcal{A}}$ of $\mathcal{A}$ such that the comultiplication of $\mathcal{A}$ extends continuously to $\tilde{\mathcal{A}}$ (see also Subsect. 11.4.1). Further, it should be mentioned that in general there are several inequivalent C^*-norms on the algebraic tensor product $A \otimes A$ (see [Mu]). △

11.3.4 Modular Properties of the Haar State

Throughout this subsection let $\mathcal{A}$ be a CQG algebra with Haar state h.

A linear functional f on $\mathcal{A}$ is called *central* if $f(ab) = f(ba)$ for $a, b \in \mathcal{A}$. It is called a *character* if $f(1) = 1$ and $f(ab) = f(a)f(b)$ for $a, b \in \mathcal{A}$. We shall see below (Corollary 35) that the Haar state h on $\mathcal{A}$ is not central if the square of the antipode of $\mathcal{A}$ is not the identity. In this subsection we investigate a family of characters and of related automorphisms of $\mathcal{A}$ which provide qualitative measures of the extent to which both properties fail.

First we state a preliminary result on holomorphic functions. Let us say that an entire function g is of exponential growth on the right half-plane if there exist constants $M > 0$ and $\mu \in \mathbb{R}$ such that $|g(z)| \le M e^{\mu \operatorname{Re} z}$ for any $z \in \mathbb{C}$, $\operatorname{Re} z \ge 0$.

Lemma 33. *If g_1 and g_2 are entire functions of exponential growth on the right half-plane such that $g_1(z) = g_2(z)$ for $z \in \mathbb{N}$, then $g_1(z) = g_2(z)$ for all $z \in \mathbb{C}$.*

Proof. This is in fact Carlson's theorem (see, for instance, [Tit]). □

Proposition 34. *There exists a family $\{f_z\}_{z \in \mathbb{C}}$ of characters on the CQG algebra $\mathcal{A}$ such that for all $z, z' \in \mathbb{C}$ and $a, b \in \mathcal{A}$ the following conditions are fulfilled:*

(i) *The function $z \to f_z(a)$ is an entire function of exponential growth on the right half-plane.*

(ii) *$f_z f_{z'} = f_{z+z'}$ and $f_0 = \varepsilon$.*

(iii) *$f_z(1) = 1$.*

(iv) *$f_z(S(a)) = f_{-z}(a)$ and $f_z(a^*) = \overline{f_{-\bar{z}}(a)}$.*

(v) $S^2(a) = f_{-1}.a.f_1$, *where* $f_{-1}.a.f_1$ *is defined by (1.65).*
(vi) $h((f_{z-1}.a.f_{z'-1})b) = h(b(f_z.a.f_{z'}))$. *In particular,*

$$h(ab) = h(b(f_1.a.f_1)). \tag{26}$$

The linear functionals f_z, $z \in \mathbb{C}$, *are uniquely determined by the conditions* (i), (ii) *and equation (26).*

Proof. Recall that by Lemma 30 the intertwiners $F_\alpha \in \text{Mor}\ (u^\alpha, (u^\alpha)^{cc})$ defined in Subsect. 11.3.2 are positive definite. Hence F_α has positive eigenvalues, say $\lambda_1, \cdots, \lambda_{d_\alpha}$. For $z \in \mathbb{C}$, let D_α^z denote the diagonal matrix with diagonal entries $\lambda_1^z, \cdots, \lambda_{d_\alpha}^z$. By the finite-dimensional spectral theorem, there is a unitary matrix U_α such that $F_\alpha = U_\alpha D_\alpha^1 U_\alpha^*$. Define the complex power F_α^z, $z \in \mathbb{C}$, of F_α by $F_\alpha^z := U_\alpha D_\alpha^z U_\alpha^*$. We write F_α^z as the matrix $(F_\alpha^z)_{ij}$ with respect to the fixed basis of the representation space of u^α. Now we define the linear functional f_z, $z \in \mathbb{C}$, on the elements of the vector space basis $\{u_{ij}^\alpha \mid \alpha \in \hat{\mathcal{A}}, i,j = 1,2,\cdots,d_\alpha\}$ of $\mathcal{A}$ (see Theorem 13(vi)) by setting $f_z(u_{ij}^\alpha) := (F_\alpha^z)_{ij}$. Then the properties (i)–(iii) are clear. For (ii) we use the relation $F_\alpha^z F_\alpha^{z'} = F_\alpha^{z+z'}$.

By (ii), f_{-z} is the inverse of f_z in the algebra $\mathcal{A}'$. This implies the first equality of (iv). To verify the second one, it suffices to take $a = u_{ij}^\alpha$. Using the facts that $F_\alpha^{-z} = (F_\alpha^{-\bar{z}})^*$ and u^α is unitary, we compute

$$f_z((u_{ij}^\alpha)^*) = f_z(S(u_{ji}^\alpha)) = f_{-z}(u_{ji}^\alpha) = (F_\alpha^{-z})_{ji} = \overline{(F_\alpha^{-\bar{z}})_{ij}} = \overline{f_{-\bar{z}}(u_{ij}^\alpha)}.$$

Thus, (iv) is proved.

Since $F_\alpha \in \text{Mor}\ (u^\alpha, (u^\alpha)^{cc})$, we have $S^2(u_{ij}^\alpha) = (F_\alpha)_{ik} u_{kl}^\alpha (F_\alpha^{-1})_{lj} = (f_1 \otimes \text{id} \otimes f_{-1})(\Delta \otimes \text{id}) \circ \Delta(u_{ij}^\alpha) = f_{-1}.u_{ij}^\alpha.f_1$. This proves (v) for $a = u_{ij}^\alpha$ and hence for all $a \in \mathcal{A}$.

It suffices to prove (vi) for elements $a = u_{ij}^\alpha$ and $b = (u_{kl}^\beta)^*$. If $\alpha \neq \beta$, then both sides of (vi) vanish by (24) and (25). If $\alpha = \beta$, then using once more the Schur orthogonality relations (24) and (25) we compute that both sides of the equation in (vi) are equal to $M_\alpha^{-1}(F_\alpha^{z-1})_{ik}(F_\alpha^{z'})_{lj}$.

Next we show that the functionals f_z are characters. Put $\rho(a) := f_1.a.f_1$. By (26), we have $h(c\rho(ab)) = h(abc) = h(bc\rho(a)) = h(c\rho(a)\rho(b))$ and hence $h(c(\rho(ab) - \rho(a)\rho(b))) = 0$ for any $c \in \mathcal{A}$. Setting $c = (\rho(ab) - \rho(a)\rho(b))^*$ and using the faithfulness of h (by Proposition 29), we conclude that $\rho(ab) = \rho(a)\rho(b)$, so that $f_1.ab.f_1 = (f_1.a.f_1)(f_1.b.f_1)$. Applying ε to this equality and using the relation $f_1 f_1 = f_2$, we get $f_2(ab) = f_2(a)f_2(b)$. Since the convolution product of characters is again a character as is easily seen, the latter implies that $f_n(ab) = f_n(a)f_n(b)$ for $n = 2,4,\cdots$. From Lemma 33 applied to the functions $g_1(z) = f_{2z}(ab)$ and $g_2(z) = f_{2z}(a)f_{2z}(b)$ we derive that $f_z(ab) = f_z(a)f_z(b)$ for all $z \in \mathbb{C}$.

To prove the uniqueness assertion, let $\{\tilde{f}_z\}_{z\in\mathbb{C}}$ be an arbitrary family of linear functionals satisfying (i), (ii) and (26). By a similar argument to that used in the preceding paragraph, (26) implies that $f_1.a.f_1 = \tilde{f}_1.a.\tilde{f}_1$.

Therefore $f_2(a) = \tilde{f}_2(a)$ and hence $f_n(a) = \tilde{f}_n(a)$ by (ii) for all $n = 2, 4, \cdots$. Finally, Lemma 33 yields that $f_z(a) = \tilde{f}_z(a)$ for any $z \in \mathbb{C}$. □

Corollary 35. *The Haar state h of $\mathcal{A}$ is central if and only if $S^2 = \mathrm{id}$. This holds if and only if $f_z = \varepsilon$ for all $z \in \mathbb{C}$.*

Proof. If $S^2 = \mathrm{id}$, then $u^\alpha = (u^\alpha)^{cc}$ and hence $F_\alpha = I$ for $\alpha \in \hat{\mathcal{A}}$. Therefore, $f_z = \varepsilon$ for all $z \in \mathbb{C}$. By (26), the latter implies that h is central. Conversely, if h is central, then $f_1.a.f_1 = a$ by (26) and so $f_n = \varepsilon$ for $n = 2, 4, \cdots$ by the reasoning used in the above proof. Hence $f_z = \varepsilon$ for all $z \in \mathbb{C}$ by Proposition 34(i) and Lemma 33. □

In terms of the functionals f_z the Schur orthogonality relations (24) and (25) can be expressed as

$$h(u^\alpha_{kl}(u^\beta_{ji})^*) = \delta_{\alpha\beta}\delta_{kj}M_\alpha^{-1}f_1(u^\alpha_{il}), \quad h((u^\alpha_{lk})^*u^\beta_{ij}) = \delta_{\alpha\beta}\delta_{kj}M_\alpha^{-1}f_{-1}(u^\alpha_{il}).$$

Since the functionals f_z, $z \in \mathbb{C}$, are characters, they belong to the Hopf dual $\mathcal{A}^\circ$ of $\mathcal{A}$. Thus, the first equality in (iv) can be rewritten as $S(f_z) = f_{-z}$. Further, for any $z, z' \in \mathbb{C}$ the equation

$$\rho_{z,z'}(a) = f_z.a.f_{z'}, \qquad a \in \mathcal{A}, \tag{27}$$

defines an automorphism $\rho_{z,z'}$ of the algebra $\mathcal{A}$. From Proposition 34 we immediately obtain the following formulas for these automorphisms:

$$\rho_{z,z'} \circ \rho_{w,w'} = \rho_{z+w,z'+w'} \quad \text{and} \quad \rho_{0,0} = \mathrm{id}, \tag{28}$$

$$(\rho_{z,w} \otimes \rho_{-w,z'}) \circ \Delta = \Delta \circ \rho_{z,z'}, \tag{29}$$

$$\rho_{z,z'}(a^*) = \rho_{-\bar{z},-\bar{z}'}(a)^*, \tag{30}$$

$$\rho_{z,z'}(S(a)) = S(\rho_{-z',-z}(a)), \tag{31}$$

$$h(\rho_{z,z'}(a)) = h(a). \tag{32}$$

Set $\theta_z := \rho_{\mathrm{i}z,\mathrm{i}z}$ and $\vartheta_z := \rho_{\mathrm{i}z,-\mathrm{i}z}$, $z \in \mathbb{C}$, where $\mathrm{i} = \sqrt{-1}$. Then $\{\theta_t\}_{t\in\mathbb{R}}$ and $\{\vartheta_t\}_{t\in\mathbb{R}}$ are commuting one-parameter groups of $*$-automorphisms of $\mathcal{A}$ by (30). They are important tools for the study of the CQG algebra $\mathcal{A}$.

By condition (vi) in Proposition 34, for any $a, b \in \mathcal{A}$ we have

$$h(ab) = h(b\theta_{-\mathrm{i}}(a)) \quad \text{and} \quad h(\theta_z(a)b) = h(b\theta_{z-\mathrm{i}}(a)), \qquad z \in \mathbb{C}. \tag{33}$$

That is, $\{\theta_z\}$ measures the extent to which h fails to be central. Recall that, by the proof of Corollary 35, h is central if and only if $\theta_z = \mathrm{id}$ for all $z \in \mathbb{C}$. Further, by Proposition 34(i), the function $z \to h(\theta_z(a)b)$ is entire and bounded on any strip $\{z \mid \alpha \leq \mathrm{Im}\ z \leq \beta\}$, $\alpha, \beta \in \mathbb{R}$. Thus, by (33), for any $a, b \in \mathcal{A}$ there exists a bounded function f on the strip $\{z \mid 0 \leq \mathrm{Im}\ z \leq 1\}$ which is holomorphic in its interior and satisfies the boundary conditions $f(t) = h(\theta_t(a)b)$ and $f(t + \mathrm{i}) = h(b\theta_t(a))$, $t \in \mathbb{R}$. The latter is expressed by saying that $\{\theta_t\}_{t\in\mathbb{R}}$ satisfies the *KMS condition* relative to the Haar state

h. The KMS condition plays an important role in the theory of operator algebras and in statistical mechanics (see [BR]). It characterizes the modular automorphism group of a faithful normal state on a von Neumann algebra. We call $\{\theta_t\}_{t\in\mathbb{R}}$ the *modular automorphism group* of the CQG algebra $\mathcal{A}$.

The second automorphism group $\{\vartheta_z\}$ is related to the behavior of the antipode S of $\mathcal{A}$. First we note that each ϑ_z commutes with S by (31). Iterating condition (v) in Proposition 34 we obtain

$$S^{2n} = \rho_{-n,n} = \vartheta_{\mathrm{i}n}, \quad n \in \mathbb{Z}.$$

Set $U(a) := S(\vartheta_{-\mathrm{i}/2}(a))$, $a \in \mathcal{A}$. From this definition it is clear that

$$S = U \circ \vartheta_{\mathrm{i}/2} = \vartheta_{\mathrm{i}/2} \circ U. \tag{34}$$

The map U is a $*$-preserving anti-automorphism of $\mathcal{A}$ such that $U \circ U = \mathrm{id}$. It is called the *unitary antipode* of the CQG algebra $\mathcal{A}$. Note that $U \circ \vartheta_z = \vartheta_z \circ U$, $U \circ \theta_z = \theta_{-z} \circ U$, $US = SU$ and $h(U(a)) = h(a)$ for $z \in \mathbb{C}$ and $a \in \mathcal{A}$.

We close this subsection by considering once more the coordinate Hopf $*$-algebras of compact quantum matrix groups.

Example 9. Let G_q denote one of the compact quantum groups $U_q(N)$, $SU_q(N)$, $O_q(N;\mathbb{R})$ or $USp_q(N)$, $q \in \mathbb{R}$. As noted in Example 7, the Hopf $*$-algebra $\mathcal{A} := \mathcal{O}(G_q)$ is a CQMG algebra. In order to treat all four quantum groups G_q at once, we set

$$\hat{\imath} := 2i - N - 1 \text{ for } U_q(N), SU_q(N), \quad \hat{\imath} := -2\rho_i \text{ for } O_q(N;\mathbb{R}), USp_q(N),$$

where ρ_i are the numbers defined in Subsect. 8.4.2. Then the value of the character f_z at the entry u^i_j of the fundamental matrix $\mathbf{u} = (u^i_j)$ of G_q is given by

$$f_z(u^i_j) = q^{z\hat{\imath}}\delta_{ij} \;\text{ for }\; i,j = 1,2,\cdots,N \;\text{ and }\; z \in \mathbb{C}. \tag{35}$$

Let us prove this formula. By the definition of the involution in $\mathcal{O}(G_q)$, the fundamental corepresentation u is unitary. As noted in Sect. 9.4, the vector representation T_1 of $U_q(\mathfrak{g})$ is the associated representation with $\mathbf{u}$ by Proposition 1.15. Therefore, since T_1 is irreducible, so is $\mathbf{u}$ and we can assume that $\mathbf{u}$ is one of the corepresentations u^α, $\alpha \in \hat{A}$. Set $F = (F_{ij}) := (q^{\hat{\imath}}\delta_{ij})$. By Propositions 9.10 and 9.13, $S^2(u^i_j) = q^{\hat{\imath}}u^i_j q^{-\hat{\jmath}}$. Hence $F \in \mathrm{Mor}\,(\mathbf{u}, \mathbf{u}^{cc})$. Since $\hat{\imath} = -\hat{\imath'}$ with $i' := N + 1 - i$ by the definition of $\hat{\imath}$, we have $\mathrm{Tr}\, F = \mathrm{Tr}\, F^{-1}$. Thus, F is the intertwiner from Lemma 30 and formula (35) follows from the definition of the functionals f_z given in the proof of Proposition 34.

Further, let $\langle \cdot,\cdot \rangle$ denote the dual pairing of the Hopf algebras $U_q(\mathfrak{g})$, $\mathfrak{g} = \mathrm{gl}_N, \mathrm{sl}_N, \mathrm{so}_{2n}, \mathrm{sp}_N$, resp. $U_{q1/2}(\mathrm{so}_{2n+1})$ and $\mathcal{O}(G_q)$ from Theorem 9.18. If ρ is the half-sum of positive roots of $\mathfrak{g}$ and $K_{2\rho}$ the element of $U_q(\mathfrak{g})$ resp. $U_{q1/2}(\mathrm{so}_{2n+1})$ defined by (6.26), then

$$f_1(a) = \langle K_{2\rho}, a \rangle, \quad a \in \mathcal{O}(G_q). \tag{36}$$

Because f_1 is a character and $K_{2\rho}$ is group-like, it is sufficient to prove (36) for $a = u^i_j$. By (35) and (9.39), formula (36) is then equivalent to the equation $t_{ij}(K_{2\rho}) = q^{\hat{\imath}}\delta_{ij}$. The latter can be verified by using the explicit forms of the vector representations T_1 (see Subsect. 8.4.1). Under the additional assumption that $U_q(\mathfrak{g})$ resp. $U_{q^{1/2}}(\mathrm{so}_{2n+1})$ separates the elements of $\mathcal{O}(G_q)$ (see Corollaries 23 and 54), there is a more instructive proof of (36) given as follows. Since $S^2(x) = K_{2\rho}xK_{2\rho}^{-1}$ for x in $U_q(\mathfrak{g})$ resp. $U_{q^{1/2}}(\mathrm{so}_{2n+1})$ by Proposition 6.6, we have

$$\langle x, S^2(u^i_n)\rangle\langle K_{2\rho}, u^n_j\rangle = \langle S^2(x)K_{2\rho}, u^i_j\rangle = \langle K_{2\rho}, u^i_n\rangle\langle x, u^n_j\rangle$$

and hence $S^2(u^i_n)\langle K_{2\rho}, u^n_j\rangle = \langle K_{2\rho}, u^i_n\rangle u^n_j$. Therefore, the linear transformation $F = (F_{ij}) := (\langle K_{2\rho}, u^i_j\rangle)$ is the intertwiner of $\mathbf{u}$ and $\mathbf{u}^{cc}$ from Lemma 30. By the definition of f_1, (36) follows for $a = u^i_j$.

Finally, combining the formulas (27) and (35) we obtain the action of the algebra automorphism $\rho_{z,z'}$ of $\mathcal{O}(G_q)$ on the generators u^i_j as

$$\rho_{z,z'}(u^i_j) = q^{z'\hat{\imath}+z\hat{\jmath}}u^i_j. \qquad \triangle$$

11.3.5 Polar Decomposition of the Antipode

With formula (34) a decomposition of the antipode as an anti-automorphism of $\mathcal{A}$ was given. The following proposition describes the polar decomposition of the closure of the antipode in the Hilbert space $\mathcal{H}(\mathcal{A})$.

Proposition 36. *Let $\mathcal{A}$ be a CQG algebra. The antipode S of $\mathcal{A}$, considered as a linear operator in the Hilbert space $\mathcal{H}(\mathcal{A})$, is closable. Let $\bar{S} = U_0|\bar{S}|$ be the polar decomposition of the closure $\bar{S}$ of the operator S. Then U_0 is a unitary self-adjoint operator on $\mathcal{H}(\mathcal{A})$ and we have*

$$U_0(a) = f_1.S(a) \quad and \quad |\bar{S}|(a) = a.f_1, \quad a \in \mathcal{A}.$$

Proof. Using the formulas from Subsect. 11.3.4 and formula (4), we derive that $\langle b, S(a)\rangle = \langle \rho_{1,1}(S(b)), a\rangle$, $a, b \in \mathcal{A}$. Therefore, the domain of the adjoint operator of S contains $\mathcal{A}$, hence it is dense and S is closable. Define $U_0(a) = f_1.S(a)$ and $S_0(a) = a.f_1$, $a \in \mathcal{A}$. Then, by (31), we have

$$U_0S_0(a) = \rho_{1,0}(S(\rho_{0,1}(a))) = S(\rho_{0,-1}\circ\rho_{0,1}(a)) = S(a), \quad a \in \mathcal{A}.$$

Since $S^2 = \rho_{-1,1}$ by Proposition 34(v) and (27) and $S\circ * = *\circ S^{-1}$ by (1.39), we obtain $\rho_{1,0}(S(a))^* = \rho_{0,-1}(S(a^*))$ by (31). Using Proposition 34(vi) and the formulas (32) and (4) we get

$$\begin{aligned}\langle U_0(a), U_0(b)\rangle &= h\big(\rho_{1,0}(S(a))^*\rho_{1,0}(S(b))\big) = h\big(\rho_{0,-1}(S(a^*))\rho_{1,0}(S(b))\big)\\ &= h\big(\rho_{1,0}(S(b))\rho_{1,0}(S(a^*))\big) = h\big(\rho_{1,0}(S(a^*b))\big)\\ &= h(S(a^*b)) = h(a^*b) = \langle a, b\rangle.\end{aligned}$$

Therefore, since $U_0(\mathcal{A}) = \mathcal{A}$, U_0 extends uniquely to a unitary operator, denoted again by U_0, on $\mathcal{H}(\mathcal{A})$. Further, by (32) and (30), we have

$$\begin{aligned}\langle a, S_0(a)\rangle &= h(a^*\rho_{0,1}(a)) = h\big(\rho_{0,1/2}(\rho_{0,-1/2}(a^*)\rho_{0,1/2}(a))\big)\\ &= h(\rho_{0,1/2}(a)^*\rho_{0,1/2}(a)) \geq 0.\end{aligned}$$

That is, S_0 is a positive operator on $\mathcal{A}$. Thus, by the uniqueness of the polar decomposition (see [RS], Sect. VIII.9), U_0 and the closure of S_0 are the two factors of the polar decomposition of $\bar{S}$. From Proposition 34(v) and (31) it follows that $U_0^2 = I$. Hence U_0 is unitary and self-adjoint. □

11.3.6 Multiplicative Unitaries of CQG Algebras

First let us motivate the definitions given below. Let G be a unimodular locally compact group and let $\mathcal{H} = L^2(G)$ be the Hilbert space of square integrable functions on G with respect to the Haar measure of G. The unitary operator V on the tensor product Hilbert space $\mathcal{H}\otimes\mathcal{H} = L^2(G\times G)$ defined by $(Vf)(s,t) = f(s,st)$, $s,t \in G$, $f \in L^2(G\times G)$, plays a fundamental role in the duality theory of locally compact groups. This operator, originally invented by W. F. Stinespring, is usually called *multiplicative unitary* or the *Kac–Takesaki operator*. In fact, V allows one to introduce a Hopf algebra structure on the von Neumann algebra generated by the left regular representation of G from which the group G itself can be recovered (see [ES] and [Tk] for treatments of this topic). The unitary operator W on $\mathcal{H}\otimes\mathcal{H}$ given by $(Wf)(s,t) = f(st,t)$ is used in a similar manner. It is easy to check that V and W satisfy the following equations on $\mathcal{H}\otimes\mathcal{H}\otimes\mathcal{H}$:

$$V_{12}V_{23} = V_{23}V_{13}V_{12}, \tag{37}$$

$$W_{23}W_{12} = W_{12}W_{13}W_{23}. \tag{38}$$

Equations (37) and (38) are called *pentagon equations*. Obviously, V satisfies (37) if and only if V^{-1} satisfies (38).

Now let $\mathcal{A}$ be a Hopf algebra with invertible antipode. We define linear mappings V, W: $\mathcal{A}\otimes\mathcal{A} \to \mathcal{A}\otimes\mathcal{A}$ by

$$V(a\otimes b) = \Delta(b)\cdot(a\otimes 1) \quad \text{and} \quad W(a\otimes b) = \Delta(a)\cdot(1\otimes b). \tag{39}$$

Clearly, these are generalizations of the mappings V and W mentioned above, but because of the commutativity of the function algebras we could have also taken $\tilde{V}(a\otimes b) = (a\otimes 1)\cdot\Delta(b)$ and $\tilde{W}(a\otimes b) = (1\otimes b)\cdot\Delta(a)$.

Proposition 37. (i) *V and W are bijections of $\mathcal{A}\otimes\mathcal{A}$. Their inverses act as $V^{-1}(a\otimes b) = ((S^{-1}\otimes \mathrm{id})\Delta(b))\cdot(a\otimes 1)$ and $W^{-1}(a\otimes b) = ((\mathrm{id}\otimes S)\Delta(a))\cdot(1\otimes b)$.*

(ii) *V and W satisfy the pentagon equations (37) and (38), respectively.*

(iii) *The comultiplication, the counit and the antipode of $\mathcal{A}$ can be recovered from the mappings V and W by*

$$\Delta(a)=V(1\otimes a)V^{-1},\ \varepsilon(a)1=m\circ\tau\circ V^{-1}(1\otimes a),\ S^{-1}(a)=m(\mathrm{id}\otimes\varepsilon)V^{-1}(1\otimes a),$$

$$\Delta(a)=W(a\otimes 1)W^{-1},\ \varepsilon(a)1=mW^{-1}(a\otimes 1),\ S(a)=m(\varepsilon\otimes\mathrm{id})W^{-1}(a\otimes 1),$$

where $\Delta(a)$ is considered as a multiplication operator on $\mathcal{A}\otimes\mathcal{A}$.

Proof. All assertions follow by straightforward algebraic manipulations. We verify (for instance) the pentagon equation (38) for W. In the Sweedler notation, we have $W(a\otimes b)=\sum a_{(1)}\otimes a_{(2)}b$ and hence

$$\begin{aligned}
&W_{12}W_{13}W_{23}(a\otimes b\otimes c)=W_{12}W_{13}\Big(\sum a\otimes b_{(1)}\otimes b_{(2)}c\Big)\\
&\quad=W_{12}\Big(\sum a_{(1)}\otimes b_{(1)}\otimes a_{(2)}b_{(2)}c\Big)=\sum a_{(1)}\otimes a_{(2)}b_{(1)}\otimes a_{(3)}b_{(2)}c,\\
&W_{23}W_{12}(a\otimes b\otimes c)=W_{23}\Big(\sum a_{(1)}\otimes a_{(2)}b\otimes c\Big)\\
&\quad=\sum a_{(1)}\otimes a_{(2)}b_{(1)}\otimes a_{(3)}b_{(2)}c.
\end{aligned}$$

□

Proposition 38. *Let $\mathcal{A}$ be a CQG algebra. Then the mappings V and W defined by (39) have unique extensions to unitary operators of the Hilbert space $\mathcal{H}(\mathcal{A})\otimes\mathcal{H}(\mathcal{A})$.*

Proof. By Proposition 37(i), V and W are bijective mappings of $\mathcal{A}\otimes\mathcal{A}$. Therefore, since $\mathcal{A}\otimes\mathcal{A}$ is dense in $\mathcal{H}(\mathcal{A})\otimes\mathcal{H}(\mathcal{A})$, it suffices to show that V and W preserve the scalar product of $\mathcal{A}\otimes\mathcal{A}$. We prove this in the case of W. Let $a,b,c,d\in\mathcal{A}$. Using the definitions of the scalar product on $\mathcal{A}\otimes\mathcal{A}$ and $\mathcal{A}$ and the right invariance of the Haar state h, we compute

$$\begin{aligned}
\langle W(a\otimes b),W(c\otimes d)\rangle &=\sum\langle a_{(1)}\otimes a_{(2)}b,c_{(1)}\otimes c_{(2)}d\rangle\\
&=\sum\langle a_{(1)},c_{(1)}\rangle\langle a_{(2)}b,c_{(2)}d\rangle=\sum h(a_{(1)}{}^*c_{(1)})h((a_{(2)}b)^*c_{(2)}d)\\
&=h\Big(b^*\Big(\sum h((a^*c)_{(1)})(a^*c)_{(2)}\Big)d\Big)=h(b^*h(a^*c)1)d)=\langle a,c\rangle\langle b,d\rangle\\
&=\langle a\otimes b,c\otimes d\rangle.
\end{aligned}$$

Let us denote the continuous extensions of V and W to $\mathcal{H}(\mathcal{A})\otimes\mathcal{H}(\mathcal{A})$ again by V and W, respectively. These extensions (which also satisfy the pentagon equations (37) and (38)) are called *multiplicative unitaries* of the CQG algebra $\mathcal{A}$. The unitary operators V and W are linked by the relation

$$V^*=\tau(U_0\otimes 1)W(U_0\otimes 1)\tau, \tag{40}$$

where τ is the flip and U_0 is the unitary self-adjoint operator from the polar decomposition of the closure of the antipode (see Proposition 36). Indeed, (40) is equivalent to the equation $V^{-1}\tau(U_0\otimes 1)=\tau(U_0\otimes 1)W$, which can be verified by inserting the corresponding formulas for V^{-1}, U_0 and W.

Multiplicative unitaries play a crucial role in the C^*-algebra approach to noncompact quantum groups (see [BS], [B], [Wor8]).

11.4 Compact Quantum Group C^*-Algebras

In Proposition 32 we constructed a C^*-algebra completion of a CQG algebra. In this section we take the C^*-algebra of a compact quantum group as the starting point and show that it always contains a unique dense CQG subalgebra.

11.4.1 CQG C^*-Algebras and Their CQG Algebras

Definition 11. *A* compact quantum group C^*-algebra *(briefly, a* CQG C^*-algebra*) is a C^*-algebra A with unit element such that the following two conditions hold:*

(i) *There is a unital $*$-homomorphism $\Delta : A \to A\bar{\otimes}A$ which satisfies the relation $(\Delta \otimes \mathrm{id})\Delta = (\mathrm{id} \otimes \Delta)\Delta$ on $A\bar{\otimes}A\bar{\otimes}A$.*

(ii) *There is a family $\{u^\nu = (u^\nu_{ij})_{i,j=1,\cdots,d_\nu} \mid \nu \in \mathsf{I}\}$ (I is an index set and $d_\nu \in \mathbb{N}$) of matrices with entries in A such that:*

(ii.1) $\Delta(u^\nu_{ij}) = u^\nu_{ik} \otimes u^\nu_{kj}$ *for* $\nu \in \mathsf{I}$, $i,j = 1,2,\cdots,d_\nu$.

(ii.2) u^ν *and its transpose* $(u^\nu)^t$, $\nu \in \mathsf{I}$, *are invertible matrices over* A.

(ii.3) *The $*$-subalgebra $\mathcal{A}$ of A generated by the entries u^ν_{ij} is dense in A.*

If the family $\{u^\nu \mid \nu \in \mathsf{I}\}$ consists only of a singleton, then A is called a compact quantum matrix group C^*-algebra *(a* CQMG C^*-algebra*).*

As in Subsect. 11.3.3, $\mathcal{A}\bar{\otimes}A$ denotes the completion of the algebraic tensor product $A \otimes A$ in the least C^*-norm (see [Mu], p.190). The invertibility of a matrix $v = (v_{ij})$ over A means that there is another matrix $w = (w_{ij})$ with entries in A such that $\sum_k v_{ik}w_{kj} = \sum_k w_{ik}v_{kj} = \delta_{ij}1$ for all i,j.

Let A be a CQG C^*-algebra. Note that the $*$-homomorphism Δ maps A to the completion $A\bar{\otimes}A$ rather than to the algebraic tensor product $A \otimes A$. Let us still call Δ the *comultiplication* of A. However, Δ maps the $*$-algebra $\mathcal{A}$ to the algebraic tensor product $\mathcal{A}\otimes\mathcal{A}$ by (ii.1) and satisfies the "ordinary" coassociativity axiom $(\Delta\otimes\mathrm{id})\circ\Delta = (\mathrm{id}\otimes\Delta)\circ\Delta$ on $\mathcal{A}\otimes\mathcal{A}\otimes\mathcal{A}$. By Theorem 39 stated below, $\mathcal{A}$ is even a Hopf $*$-algebra, but at the moment we do not yet have a counit and an antipode on $\mathcal{A}$.

Next we define finite-dimensional corepresentations of A.

Definition 12. *Let V be a finite-dimensional vector space. A linear mapping $\varphi : V \to V \otimes A$ is called a* corepresentation *of the CQG C^*-algebra A on V if* $(\mathrm{id} \otimes \Delta) \circ \varphi = (\varphi \otimes \mathrm{id}) \circ \varphi$ *and* $\ker \varphi = \{0\}$.

By Propositions 1 and 2, the two conditions in Definition 12 characterize a corepresentation of a coalgebra. Let us emphasize here once more that a CQG C^*-algebra is in general not a coalgebra according to Definition 1.2.

The concepts and results on finite-dimensional corepresentations of coalgebras developed in Sect. 11.1 go over almost verbatim to corepresentations of CQG C^*-algebras (except for the few statements where the counit occurs). In particular, the first condition $(\mathrm{id} \otimes \Delta) \circ \varphi = (\varphi \otimes \mathrm{id}) \circ \varphi$ in Definition 12 is equivalent to the relation $(\mathrm{id} \otimes \Delta) \circ \breve{\varphi} = \breve{\varphi}_{12}\breve{\varphi}_{13}$, where $\breve{\varphi}$ is the associated element to φ of $\mathcal{L}(V) \otimes A$ (see Subsect. 11.1.1). Let us recall the notion of a unitary corepresentation which plays a crucial role in the following. A corepresentation φ of a *CQG* C^*-algebra A on a finite-dimensional Hilbert space V is called *unitary* if $\breve{\varphi}$ is a unitary element of the $*$-algebra $\mathcal{L}(V) \otimes A$, that is, $\breve{\varphi}^*\breve{\varphi} = \breve{\varphi}\breve{\varphi}^* = I_V \otimes 1_A$. For a CQG C^*-algebra A it is more important to treat finite-dimensional corepresentations of A as elements $\breve{\varphi} \in \mathcal{L}(V) \otimes A$ rather than as mappings $\varphi : V \to V \otimes A$. The technical reason is that $\mathcal{L}(V) \otimes A$ is itself a C^*-algebra, so the C^*-algebra theory applies to $\breve{\varphi}$. On the conceptual side, this view paves the way for the definition of *infinite*-dimensional corepresentations of CQG C^*-algebras (see Subsect. 11.4.3 below).

There is a rather close link between CQG C^*-algebras and CQG algebras which we will now discuss. On the one hand, for each CQG algebra $\mathcal{A}_0$ the C^*-algebra completion A of $\mathcal{A}_0$ from Proposition 32 obviously satisfies the conditions of Definition 11 (with $\mathcal{A} = \mathcal{A}_0$ and $\mathsf{I} = \hat{\mathcal{A}}_0$), so A is a CQG C^*-algebra and $\Delta_{\mathcal{A}_0} = \Delta_A \lceil \mathcal{A}_0$. Thus, each example of a CQG algebra $\mathcal{A}_0$ gives rise to an example of a CQG C^*-algebra A in this way. On the other hand, for each CQG C^*-algebra A there exists a unique CQG algebra $\mathcal{A}_0$ such that $\mathcal{A}_0$ is a dense $*$-subalgebra of A and $\Delta_{\mathcal{A}_0} = \Delta_A \lceil \mathcal{A}_0$. The latter assertion is the main content of the following theorem.

Theorem 39. *Let A be a CQG C^*-algebra and let $\mathcal{A}$ be the $*$-algebra from Definition* 11(ii.3). *Then $\mathcal{A}$ is the linear span of matrix elements of all finite-dimensional unitary corepresentations of A and $\mathcal{A}$ is a CQG algebra with comultiplication $\Delta_{\mathcal{A}} := \Delta_A \lceil \mathcal{A}$. It is the unique CQG algebra which is a dense $*$-subalgebra of A and whose comultiplication is the restriction of the comultiplication of A.*

The proof of Theorem 39 will be given in Subsect. 11.4.3. It is essentially based on the existence of a Haar state on a CQG C^*-algebra which will be proved in Subsect. 11.4.2.

Let us continue the above discussion concerning the connections between the CQG C^*-algebra A and the associated CQG algebra $\mathcal{A}$. By Lemma 45 below and Theorem 27(iii), finite-dimensional corepresentations of A and $\mathcal{A}$ are unitarizable. Therefore, the first assertion of Theorem 39 implies that there is a one-to-one correspondence between finite-dimensional corepresentations of A and finite-dimensional corepresentations of $\mathcal{A}$. Clearly, this correspondence preserves all standard operations and concepts from corepresentation theory such as direct sums, tensor products, equivalence, irreducibility and unitarity. That is, the CQG algebra $\mathcal{A}$ carries the full information on the finite-dimensional corepresentation theory of the CQG C^*-algebra A.

It might be necessary to emphasize that despite the close relations explained above the correspondence between CQG C^*-algebras A and their associated CQG algebras $\mathcal{A}$ is not one-to-one. As already mentioned in Subsect. 11.3.3, for a given CQG algebra $\mathcal{A}$ there is in general no *unique* C^*-norm on $\mathcal{A}$ such that the corresponding completion is a CQG C^*-algebra.

11.4.2 Existence of the Haar State of a CQG C^*-Algebra

The main aim of this subsection is to prove the following

Theorem 40. *Let A be a CQG C^*-algebra. Then there exists a unique state h of A such that* $(\mathrm{id} \otimes h)\Delta(a) = (h \otimes \mathrm{id})\Delta(a) = h(a)1$ *for all $a \in A$.*

The state h in Theorem 40 is called the *Haar state* of the CQG C^*-algebra A.

The proof of Theorem 40 is divided into several steps stated as lemmas.

Lemma 41. *If A is a CQG C^*-algebra, then the sets* $\Delta(A)(A \otimes 1) \equiv \mathrm{Lin}\,\{\Delta(a)(b \otimes 1) \,|\, a, b \in A\}$ *and* $\Delta(A)(1 \otimes A) \equiv \mathrm{Lin}\,\{\Delta(a)(1 \otimes b) \,|\, a, b \in A\}$ *are dense in $A \bar{\otimes} A$.*

Proof. Let $v = (v_{ij})_{i,j=1,\cdots,d}$ be a matrix with entries in A such that $\Delta(v_{ij}) = v_{ik} \otimes v_{kj}$. Let V be a d-dimensional vector space with basis $\{e_i \,|\, i = 1, 2, \cdots, d\}$ and define $\varphi \in \mathcal{L}(V, V \otimes A)$ by $\varphi(e_j) = e_i \otimes v_{ij}$. If v^t is invertible, then $\ker \varphi = \{0\}$ and hence φ is a corepresentation and v is a matrix corepresentation of A. If v^t is invertible, then $\bar{v}$ is also invertible and hence $\bar{v}$ is also a matrix corepresentation. From these remarks and conditions (ii.1) and (ii.2) it follows that u^ν and $\overline{u^\nu}$ are matrix corepresentations of A.

Now let $\mathcal{R}$ be the smallest set of matrix corepresentations of A that contains all u^ν and $\overline{u^\nu}$, $\nu \in \mathsf{I}$, and is closed under direct sums and tensor products. Then $\mathcal{A}$ is the linear span of matrix coefficients of corepresentations in $\mathcal{R}$. Let $u = (u_{ij})$ be an arbitrary matrix corepresentation in $\mathcal{R}$. By induction one can easily verify that the matrices u and $\bar{u}$ are invertible. Hence u^t is also invertible. For $v := u^{-1}$ and $w := (u^t)^{-1}$, we have $\sum_i \Delta(u_{ni})(1 \otimes v_{im}) = \sum_{i,j} u_{nj} \otimes u_{ji} v_{im} = u_{nm} \otimes 1$ and $\sum_i \Delta(u_{in})(w_{im} \otimes 1) = 1 \otimes u_{mn}$. Therefore, $\Delta(A)(1 \otimes A) \supseteq \mathcal{A} \otimes \mathcal{A}$ and $\Delta(A)(A \otimes 1) \supseteq \mathcal{A} \otimes \mathcal{A}$. Since $\mathcal{A}$ is dense in A by condition (ii.3), the assertion follows. □

For the following three lemmas we assume that A is a unital C^*-algebra such that condition (i) of Definition 11 is satisfied and $\Delta(A)(A \otimes 1)$ in dense in $A \bar{\otimes} A$.

Lemma 42. *For any state f on A there is a state g such that $gf = fg = g$.*

Proof. Define a state f_n on A by the Cesaro sum $f_n := \frac{1}{n}(f + f^2 + \cdots + f^n)$, $n \in \mathbb{N}$. Since the state space of a unital C^*-algebra is weakly compact (see [Mu] or [BR]), there exists a weak accumulation point g of the sequence $\{f_n\}$. From the inequality $\|f_n f - f_n\| = \frac{1}{n}\|f^{n+1} - f\| \le \frac{2}{n}$ we obtain $gf = g$ in the limit. Similarly, $fg = g$. □

Lemma 43. *Let f and g be states on A such that $fg = g$. If φ is another positive linear functional on A such that $\varphi \le f$, then $\varphi g = \varphi(1)g$.*

Proof. Let $a \in A$ and put $b := (\mathrm{id} \otimes g)\Delta(a)$. Then we have

$$(\mathrm{id} \otimes f)\Delta(b) = (\mathrm{id} \otimes fg)\Delta(a) = (\mathrm{id} \otimes g)\Delta(a) = b$$

and so

$$\begin{aligned}&(\mathrm{id}\otimes f)(\Delta(b)-b\otimes 1)^*(\Delta(b)-b\otimes 1)\\&\quad=(\mathrm{id}\otimes f)\Delta(b^*b)+b^*b-(\mathrm{id}\otimes f)\Delta(b^*)(b\otimes 1)-(\mathrm{id}\otimes f)(b^*\otimes 1)\Delta(b)\\&\quad=(\mathrm{id}\otimes f)\Delta(b^*b)-b^*b.\end{aligned}$$

Now let ψ be a state on A such that $\psi f{=}\psi$. Such a state always exists by Lemma 42. Applying ψ to the preceding equation, we obtain

$$(\psi\otimes f)(\Delta(b)-b\otimes 1)^*(\Delta(b)-b\otimes 1)=0.$$

By the Cauchy–Schwarz inequality, this yields $(\psi\otimes f)((c\otimes d)(\Delta(b)-b\otimes 1))=0$ for all $c,d\in A$. Inserting the definition of b, we get

$$(\psi\otimes f\otimes g)((c\otimes d\otimes 1)(\Delta\otimes\mathrm{id})\Delta(a))=f(d)(\psi\otimes g)((c\otimes 1)\Delta(a)).$$

We have

$$\begin{aligned}(c\otimes d\otimes 1)(\Delta\otimes\mathrm{id})\Delta(a)&=(c\otimes d\otimes 1)(\mathrm{id}\otimes\Delta)\Delta(a)\\&=(1\otimes d\otimes 1)(\mathrm{id}\otimes\Delta)((c\otimes 1)\Delta(a)).\end{aligned}$$

Applying the involution to $\Delta(A)(A\otimes 1)$ we see that $(A\otimes 1)\Delta(A)$ is dense in $A\bar{\otimes}A$. Therefore, we can replace $(c\otimes 1)\Delta(a)$ by $1\otimes x$, $x\in A$, on the right hand side of the preceding equalities and obtain

$$(\psi\otimes f\otimes g)((1\otimes d\otimes 1)(\mathrm{id}\otimes\Delta)(1\otimes x))=f(d)(\varphi\otimes g)(1\otimes x).$$

Setting $y_x:=(\mathrm{id}\otimes g)\Delta(x)$ and using the equality $\psi(1)=1$, this reads as $f(dy_x)=f(d)g(x)$, $d,x\in A$.

Now let π_f denote the GNS representation of f on the Hilbert space $\mathcal{H}_f$ (see [Mu] or [BR] for this concept). Since $f(a)=\langle\zeta_f,\pi_f(a)\zeta_f\rangle$, $a\in A$, by the definition of π_f, we have

$$f(dy_x)=\langle\pi_f(d^*)\zeta_f,\pi_f(y_x)\zeta_f\rangle=g(x)\langle\pi_f(d^*)\zeta_f,\zeta_f\rangle=g(x)f(d),\quad d,x\in A.$$

Because $\pi_f(A)\zeta_f$ is dense in $\mathcal{H}_f$, the latter implies $\langle\eta,\pi_f(y_x)\zeta_f\rangle=g(x)\langle\eta,\zeta_f\rangle$ for all $\eta\in\mathcal{H}_f$. From the inequality $\varphi\leq f$ it follows that there exists a vector $\eta_\varphi\in\mathcal{H}_f$ such that $\varphi(a)=\langle\eta_\varphi,\pi_f(a)\zeta_f\rangle$ for all $a\in A$. Hence $\varphi(y_x)=g(x)\varphi(1)$. But $\varphi(y_x)=(\varphi g)(x)$ by the definition of y_x. Thus, $\varphi g(x)=g(x)\varphi(1)$ for $x\in A$. □

Lemma 44. *There exists a state h_l on A such that $(id\otimes h_l)\Delta(a)=h_l(a)1$ for all $a\in A$.*

Proof. For a positive linear functional f on A, let K_f be the set of all states g on A satisfying $fg=f(1)g$. Clearly, K_f is a weakly compact subset of the dual space of the C^*-algebra A. By Lemma 42, K_f is not empty. By Lemma 43 we have $K_f\subseteq K_\varphi$ if $\varphi\leq f$. Hence $K_{f_1+f_2}\subseteq K_{f_1}\cap K_{f_2}$ for arbitrary positive linear functionals f_1 and f_2 on A. Since the state space of A is weakly compact, the intersection of all K_f is nonempty. If h_l is an element of this intersection, we have $fh_l = f(1)h_l$ for all positive and hence for all continous linear functionals f on A. This implies the assertion. $\square$

Proof of Theorem 40. By Lemma 41, the sets $\Delta(A)(A \otimes 1)$ and $\Delta(A)(1 \otimes A)$ are dense in $A\bar{\otimes}A$, so the preceding lemmas apply to the CQG C^*-algebra A. Let h_l be as in Lemma 44. By similar arguments (or by applying Lemma 44 to A equipped with the opposite comultiplication $\Delta^{\rm op} = \tau \circ \Delta$), there is a state h_r on A such that $(h_r \otimes {\rm id})\Delta(a) = h_r(a)1$. Applying h_r to the formula for h_l and h_l to the formula for h_r, we get $h_r h_l(a) = h_l(a)$ and $h_r h_l(a) = h_r(a)$ for $a \in A$. Thus, $h_l = h_r =: h$ and the existence of h is proved. The latter reasoning proves also the uniqueness assertion. $\square$

Remark 7. Let A be a C^*-algebra with unit element such that $A \neq \mathbb{C}{\cdot}1$. Define a linear mapping $\Delta : A \to A\bar{\otimes}A$ by $\Delta(a) = 1 \otimes a$. Obviously, Δ satisfies the coassociativity axiom and $\Delta(A)(A \otimes 1)$ is dense in $A\bar{\otimes}A$. In this case, each state h on A is left-invariant, but there is no right-invariant state on A. Hence A is not a CQG C^*-algebra. (If A is the algebra of the continuous functions on a compact topological Hausdorff space, the above comultiplication Δ corresponds to the semigroup structure on X defined by $x{\cdot}y = y$ for all $x, y \in X$.) $\triangle$

11.4.3 Proof of Theorem 39

The following lemma is used in the proof of Theorem 39. It is the counterpart to assertion (iii) of Theorem 27 for CQG C^*-algebras.

Lemma 45. *Let φ be a corepresentation of the CQG C^*-algebra A on a finite-dimensional vector space V. If $\breve{\varphi}$ is an invertible element of $\mathcal{L}(V) \otimes A$, then φ is unitarizable.*

Proof. Let $\langle \cdot, \cdot \rangle$ be a scalar product on V. Then $\mathcal{L}(V) \otimes A$ becomes a C^*-algebra. Since $\breve{\varphi}$ is invertible, $\breve{\varphi}^*\breve{\varphi} \geq \delta(I \otimes 1)$ for some $\delta > 0$. Put $E := ({\rm id} \otimes h)(\breve{\varphi}^*\breve{\varphi})$, where h is the Haar state of A. Then E is a self-adjoint operator on the Hilbert space V such that $E \geq \delta I$.

Since φ is a corepresentation, we have $({\rm id} \otimes \Delta)(\breve{\varphi}^*\breve{\varphi}) = \breve{\varphi}^*_{13}\breve{\varphi}^*_{12}\breve{\varphi}_{12}\breve{\varphi}_{13}$ (see Proposition 1(ii)). Applying $({\rm id} \otimes h \otimes {\rm id})$ to both sides of this identity and using the right invariance of h we obtain $\breve{\varphi}^*(E \otimes 1)\breve{\varphi} = E \otimes 1$ and hence $\breve{\varphi}(E^{-1} \otimes 1)\breve{\varphi}^* = \breve{\varphi}(E^{-1} \otimes 1)[\breve{\varphi}^*(E \otimes 1)\breve{\varphi}]\breve{\varphi}^{-1}(E^{-1} \otimes 1) = E^{-1} \otimes 1$.

Let $\breve{\psi} := (E^{1/2} \otimes 1)\breve{\varphi}(E^{-1/2} \otimes 1)$. Then ψ is a corepresentation of A which is equivalent to φ (via $E^{1/2} \in {\rm Mor}\,(\varphi, \psi)$). By the two equalities at the end of the preceding paragraph we have $\breve{\psi}^*\breve{\psi} = \breve{\psi}\breve{\psi}^* = I \otimes 1$, that is, ψ is unitary. Therefore, φ is unitarizable. $\square$

Now we begin the proof of Theorem 39. Suppose that A is a CQG C^*-algebra. Let $\mathcal{R}_0$ be the class of all finite-dimensional corepresentations φ of A which have a matrix v such that v and $\bar{v}$ are invertible. Since this property is preserved under direct sums, tensor products and conjugations, the linear span $\mathcal{A}_0$ of the entries of all such matrices v is a $*$-algebra. Any corepresentation from $\mathcal{R}_0$ is unitary with respect to some scalar product by Lemma 45 and hence a direct sum of irreducible unitary corepresentations by Proposition 12. Since unitary corepresentations of A belong to $\mathcal{R}_0$, $\mathcal{A}_0$ is spanned by the matrix coefficients of all irreducible unitary corepresentations of A. Let $\hat{A}$ denote the equivalence classes of such corepresentations. For each $\alpha \in \hat{A}$, we fix a unitary corepresentation φ^α of the class α and an orthonormal basis of the underlying Hilbert space, so that φ^α is given by a unitary matrix $v^\alpha = (v^\alpha_{ij})_{i,j=1,\dots,d_\alpha}$. By Corollary 10, the set $\{v^\alpha_{ij} \mid i,j = 1,2,\cdots,d_\alpha,\ \alpha \in \hat{A}\}$ is a vector space basis of $\mathcal{A}_0$. (Note that in the proof of Corollary 10 the counit was not used.)

We define two linear mappings $\varepsilon : \mathcal{A}_0 \to \mathbb{C}$ and $S : \mathcal{A}_0 \to \mathcal{A}_0$ by $\varepsilon(v^\alpha_{ij}) = \delta_{ij}$ and $S(v^\alpha_{ij}) = (v^\alpha_{ji})^*$ for $\alpha \in \hat{A}$, $i,j = 1,2,\cdots,d_\alpha$. Using the relations $\Delta(v^\alpha_{ij}) = v^\alpha_{ik} \otimes v^\alpha_{kj}$ and $\sum_i (v^\alpha_{ij})^* v^\alpha_{ik} = \sum_i v^\alpha_{ji}(v^\alpha_{ki})^* = \delta_{jk}$ one easily checks that ε and S satisfy the counit and antipode axioms, respectively. Therefore, $\mathcal{A}_0$ is a Hopf algebra. Since the conjugate $\overline{\varphi^\alpha}$ of φ^α is irreducible and unitarizable by Lemma 45, $\overline{\varphi^\alpha}$ belongs also to some class of $\hat{A}$, so that $\mathcal{A}_0$ is a $*$-algebra. By Definition 11(i), Δ is a $*$-homomorphism. Hence $\mathcal{A}_0$ is a Hopf $*$-algebra. Since $\mathcal{A}_0$ fulfills condition (ii) in Theorem 27, $\mathcal{A}_0$ is a CQG algebra.

Now let $\mathcal{B}$ be an arbitrary CQG algebra such that $\mathcal{B}$ is a dense $*$-subalgebra of A and $\Delta_{\mathcal{B}} = \Delta_{\mathcal{A}} \lceil \mathcal{B}$. Then the irreducible unitary corepresentations of $\mathcal{B}$ are also irreducible unitary corepresentations of A. Hence $\mathcal{B} \subseteq \mathcal{A}_0$ by the Peter–Weyl decomposition (2) of $\mathcal{B}$ and the definition of $\mathcal{A}_0$. We prove that $\mathcal{B} = \mathcal{A}_0$. Assume on the contrary that $\mathcal{B} \neq \mathcal{A}_0$. Then there exists a $\beta \in \hat{A}$ such that $\mathcal{C}(u^\beta) \cap \mathcal{B} = \{0\}$. By the uniqueness of the Haar functional (see Subsect. 11.2.2), the restriction of the Haar state h of A to $\mathcal{A}_0$ is the Haar state of $\mathcal{A}_0$. By (25), $h(b^*a) = 0$ for all $b \in \mathcal{B}$ and $a \in \mathcal{C}(u^\beta)$. Approaching a^* by elements of $\mathcal{B}$ (because $\mathcal{B}$ is dense in A) and using the continuity of the state h on A, we get $h(a^*a) = 0$ for $a \in \mathcal{C}(u^\beta)$. This contradicts Proposition 29. Thus we have proved that $\mathcal{B} = \mathcal{A}_0$.

Because the matrices u^ν and $\overline{u^\nu}$, $\nu \in I$, are invertible by Definition 11(ii.2), the class $\mathcal{R}$ defined in the proof of Lemma 41 is contained in $\mathcal{R}_0$, so $\mathcal{A}$ is a $*$-subalgebra of $\mathcal{A}_0$. Repeating the above reasoning, it follows that $\mathcal{A}$ is also a CQG algebra such that $\Delta_{\mathcal{A}} = \Delta_{\mathcal{A}} \lceil \mathcal{A}$ and $\mathcal{A}$ is dense in A by Definition 11(ii.3). As shown in the preceding paragraph, this implies that $\mathcal{A} = \mathcal{A}_0$. This completes the proof of Theorem 39. □

11.4.4 Another Definition of CQG C*-Algebras

An equivalent definition of CQG C^*-algebras is the following

Definition 11′. *A* compact quantum group C^*-algebra *is a unital C^*-algebra A which satisfies the conditions* (i) *of Definition 11 and*

(ii)′: *The linear spaces $\Delta(A)(A \otimes 1)$ and $\Delta(A)(1 \otimes A)$ are dense in $A \bar{\otimes} A$.*

Lemma 41 above says that Definition 11 implies Definition 11′. From Theorem 1.2 in [Wor7] it follows that Definition 11′ implies Definition 11, so that both definitions of a CQG C^*-algebra are equivalent.

In the proof of Theorem 40 only the conditions (i) and (ii)′ are essentially used. Therefore, this proof shows also that each CQG C^*-algebra in the sense of Definition 11′ admits a unique Haar state.

Definition 11 (with condition (ii) rather than (ii)′) is better adapted to the construction of examples of CQG C^*-algebras, because the standard compact quantum matrix groups (see Subsects. 9.2.4 and 9.3.5) are built from a unitary fundamental corepresentation $\mathbf{u} = (u^i_j)$. But in contrast to that, Definition 11′ gives an elegant axiomatic approach to CQG C^*-algebras.

For the next definition, let $M(B)$ denote the multiplier C^*-algebra of a C^*-algebra B (see [Mu], 2.1.5). Let $\mathfrak{L}(\mathcal{H})$ and $\mathcal{C}(\mathcal{H})$ be the C^*-algebras of bounded resp. compact operators on a Hilbert space $\mathcal{H}$.

Definition 13. *A* unitary corepresentation *of a CQG C^*-algebra A on a Hilbert space $\mathcal{H}$ is a unitary element v of the multiplier algebra $M(\mathcal{C}(\mathcal{H})\bar{\otimes}A)$ such that* $(\mathrm{id}\otimes\Delta)v = v_{12}v_{13}$.

In particular, each unitary element v of the C^*-algebra $\mathfrak{L}(\mathcal{H})\bar{\otimes}A$ satisfying $(\mathrm{id}\otimes\Delta)v = v_{12}v_{13}$ is a unitary corepresentation of A, because $\mathfrak{L}(\mathcal{H})\bar{\otimes}A$ is a C^*-subalgebra of the multiplier algebra $M(\mathcal{C}(\mathcal{H})\bar{\otimes}A)$. Clearly, if the Hilbert space $\mathcal{H}$ is finite-dimensional, then $M(\mathcal{C}(\mathcal{H})\bar{\otimes}A) = \mathfrak{L}(\mathcal{H})\otimes A$ and Definition 13 reduces to the definition of a unitary corepresentation as used above.

11.5 Finite-Dimensional Representations of $GL_q(N)$

In classical representation theory (see, for instance, [Zhe], Chap. XVI) all finite-dimensional analytic irreducible representations of the group $GL(N,\mathbb{C})$ can be constructed by means of characters of the subgroup of diagonal matrices and the invariance with respect to the Borel subgroups of triangular matrices. In this section we extend this procedure to the quantum group $GL_q(N)$ and show that $\mathcal{O}(GL_q(N))$ is a cosemisimple Hopf algebra when q is not a root of unity.

Throughout this section we assume that q is not a root of unity.

11.5.1 Some Quantum Subgroups of $GL_q(N)$

Let D_N be the diagonal subgroup of $GL(N,\mathbb{C})$. Its coordinate Hopf algebra $\mathcal{O}(D_N)$ is the commutative algebra $\mathbb{C}[t_1,t_1^{-1},\cdots,t_N,t_N^{-1}]$ of all Laurent polynomials in N indeterminates $t_1,t_2,\cdots,t_N$ with comultiplication $\Delta(t_i)=t_i\otimes t_i$ and counit $\varepsilon(t_i)=1$. Obviously, there exists a surjective Hopf algebra homomorphism $\pi_D:\mathcal{O}(GL_q(N))\rightarrow\mathcal{O}(D_N)$ determined by $\pi_D(u^i_j)=\delta_{ij}t_i$, $i,j=1,2,\cdots,N$. Therefore, by Definition 9.11, D_N is a quantum subgroup of $GL_q(N)$, called the *diagonal subgroup*.

Next we define the quantum Borel subgroups B_q^+ and B_q^- of $GL_q(N)$. Let $\mathcal{I}^+$ (resp. $\mathcal{I}^-$) be the two-sided ideal of the algebra $\mathcal{O}(GL_q(N))$ generated by the elements u_j^i, $i > j$ (resp. $i < j$). It is easily checked that $\mathcal{I}^\pm$ is a Hopf ideal of $\mathcal{O}(GL_q(N))$. Hence the quotient $\mathcal{O}(B_q^\pm) := \mathcal{O}(GL_q(N))/\mathcal{I}^\pm$ is again a Hopf algebra. Its structure is described a little more explicitly as follows. As algebras, $\mathcal{O}(B_q^+)$ and $\mathcal{O}(B_q^-)$ have $\frac{1}{2}N(N+3)$ generators z_{ij}, $i \le j$, z_{ii}^{-1} and z_{ij}, $i \ge j$, z_{ii}^{-1}, respectively. The defining relations are $z_{ii}z_{ii}^{-1} = z_{ii}^{-1}z_{ii} = 1$ and the relations (9.17) for $\mathcal{O}(M_q(N))$ when we set $u_j^i = 0$, $i > j$ (resp. $u_j^i = 0$, $i < j$) therein. Under the latter replacement the Hopf algebra structure of $\mathcal{O}(B_q^+)$ (resp. of $\mathcal{O}(B_q^-)$) is derived from that of $\mathcal{A} = \mathcal{O}(GL_q(N))$. There are surjective Hopf algebra homomorphisms $\pi_{B_q^\pm} : \mathcal{O}(GL_q(N)) \to \mathcal{O}(B_q^\pm)$ such that $\pi_{B_q^\pm}(\mathcal{D}_q^{-1}) = z_{11}^{-1} \cdots z_{NN}^{-1}$ and

$$\pi_{B_q^+}(u_j^i) = z_{ij}, \quad i \le j, \quad \pi_{B_q^+}(u_j^i) = 0, \quad i > j, \quad \text{for} \quad \mathcal{O}(B_q^+),$$

$$\pi_{B_q^-}(u_j^i) = z_{ij}, \quad i \ge j, \quad \pi_{B_q^-}(u_j^i) = 0, \quad i < j, \quad \text{for} \quad \mathcal{O}(B_q^-).$$

Hence B_q^+ and B_q^- are quantum subgroups of the quantum group $GL_q(N)$, called the *quantum Borel subgroups*. From (9.17) it follows in particular that the elements $z_{11}, \cdots, z_{NN}, z_{11}^{-1}, \cdots, z_{NN}^{-1}$ pairwise commute.

11.5.2 Submodules of Relative Invariant Elements

Here we introduce some general notions which will be used for the construction of irreducible representations of $GL_q(N)$.

Let $\mathcal{O}(G_q)$ be the coordinate Hopf algebra of a quantum group G_q and let $\mathcal{O}(X_q)$ be the coordinate algebra of a *left* (resp. *right*) *G_q-quantum space X_q*. Recall that the latter means that there exists a left coaction $\phi_L : \mathcal{O}(X_q) \to \mathcal{O}(G_q) \otimes \mathcal{O}(X_q)$ (resp. a right coaction $\phi_R : \mathcal{O}(X_q) \to \mathcal{O}(X_q) \otimes \mathcal{O}(G_q)$) which is an algebra homomorphism.

Definition 14. *Let χ be an element of $\mathcal{O}(G_q)$. An element $a \in \mathcal{O}(X_q)$ is called* left (*resp.* right) relative G_q-invariant *with respect to χ if $\phi_L(a) = \chi \otimes a$ (resp. $\phi_R(a) = a \otimes \chi$). The set of all left (resp. right) relative G_q-invariant elements with respect to χ is denoted by $\mathcal{O}(G_q \backslash X_q; \chi)$ (resp. $\mathcal{O}(X_q / G_q; \chi)$).*

In the case $\chi = 1$ the spaces $\mathcal{O}(G_q \backslash X_q; \chi)$ and $\mathcal{O}(X_q / G_q; \chi)$ are just the algebras of left- and right-invariant elements of $\mathcal{O}(X_q)$, respectively.

Let X_q be a left G_q-quantum space with left coaction $\phi_L : \mathcal{O}(X_q) \to \mathcal{O}(G_q) \otimes \mathcal{O}(X_q)$ and a right H_q-quantum space with right coaction $\phi_R : \mathcal{O}(X_q) \to \mathcal{O}(X_q) \otimes \mathcal{O}(H_q)$. We say that X_q is a *two-sided (G_q, H_q)-quantum space* if the coactions ϕ_L and ϕ_R commute, that is, if

$$(\phi_L \otimes \mathrm{id}) \circ \phi_R = (\mathrm{id} \otimes \phi_R) \circ \phi_L. \tag{41}$$

The following rather simple facts are crucial for what follows.

Proposition 46. *Let X_q be a two-sided (G_q, H_q)-quantum space and let $\chi \in \mathcal{O}(G_q)$ and $\chi' \in \mathcal{O}(H_q)$. Then $\mathcal{O}(G_q\backslash X_q; \chi)$ is a right $\mathcal{O}(H_q)$-subcomodule of $\mathcal{O}(X_q)$ and $\mathcal{O}(X_q/H_q; \chi')$ is a left $\mathcal{O}(G_q)$-subcomodule of $\mathcal{O}(X_q)$.*

Proof. For $a \in \mathcal{O}(G_q\backslash X_q; \chi)$, we have $(\phi_L \otimes \mathrm{id}) \circ \phi_R(a) = (\mathrm{id} \otimes \phi_R) \circ \phi_L(a) = \chi \otimes \phi_R(a)$ by (41), hence $\phi_R(a) \in \mathcal{O}(G_q\backslash X_q; \chi) \otimes \mathcal{O}(H_q)$. The proof for $\mathcal{O}(X_q/H_q; \chi')$ is similar. □

Now let K_q be a quantum subgroup of G_q with surjective Hopf algebra homomorphism $\pi_{K_q} : \mathcal{O}(G_q) \to \mathcal{O}(K_q)$. Then the formulas

$$L_{K_q} := (\pi_{K_q} \otimes \mathrm{id}) \circ \Delta \quad \text{and} \quad R_{K_q} := (\mathrm{id} \otimes \pi_{K_q}) \circ \Delta \tag{42}$$

define a left coaction L_{K_q} and a right coaction R_{K_q} of $\mathcal{O}(K_q)$ on the algebra $\mathcal{O}(G_q)$ such that G_q becomes a left and right K_q-quantum space. In this way, G_q is a two-sided (K_q, G_q)-quantum space with coactions $\phi_L = L_{K_q}$, $\phi_R = \Delta$ and a two-sided (G_q, K_q)-quantum space with coactions $\phi_L = \Delta$ and $\phi_R = R_{K_q}$. Applying Proposition 46 to these particular cases we obtain

Corollary 47. *For any element $\chi \in \mathcal{O}(K_q)$, the subspace $\mathcal{O}(K_q\backslash G_q; \chi)$ is a right $\mathcal{O}(G_q)$-subcomodule of the right comodule $\mathcal{O}(G_q)$ with respect to the comultiplication. Likewise, $\mathcal{O}(G_q/K_q; \chi)$ is a left $\mathcal{O}(G_q)$-subcomodule of $\mathcal{O}(G_q)$.*

11.5.3 Irreducible Representations of $GL_q(N)$

In this subsection we apply the above notions and facts to the quantum group $G_q = GL_q(N)$ and its quantum subgroups $K_q = D_N, B_q^-$.

Let $P = \{\lambda = \lambda_1\epsilon_1 + \cdots + \lambda_N\epsilon_N \mid \lambda_1, \cdots, \lambda_N \in \mathbb{Z}\}$ be the set of integral weights of the Lie algebra $\mathrm{gl}(N, \mathbb{C})$ and let $P_+ = \{\lambda = \lambda_1\epsilon_1 + \cdots + \lambda_N\epsilon_N \in P \mid \lambda_1 \geq \cdots \geq \lambda_N\}$ denote the subset of dominant integral weights. For any $\lambda = \lambda_1\epsilon_1 + \cdots + \lambda_N\epsilon_N \in P$, we define group-like elements

$$t^\lambda = t_1^{\lambda_1} \cdots t_N^{\lambda_N} \in \mathcal{O}_q(D_N) \qquad \text{and} \qquad z^\lambda = z_{11}^{\lambda_1} \cdots z_{NN}^{\lambda_N} \in \mathcal{O}(B_q^-).$$

Clearly, we have $t^\lambda t^\mu = t^{\lambda+\mu}$ and $z^\lambda z^\mu = z^{\lambda+\mu}$ for $\lambda, \mu \in P$. From Corollary 47 we know that the spaces

$$\mathcal{O}(D_N\backslash GL_q(N); t^\lambda) \quad \text{and} \quad \mathcal{O}(B_q^-\backslash GL_q(N); z^\lambda)$$

are right $\mathcal{O}(GL_q(N))$-subcomodules of the comodule $\mathcal{O}(GL_q(N))$.

Next we describe some elements of these spaces. For any matrix $A = (a_{ij}) \in M_N(\mathbb{N}_0)$ the monomial

$$u^A := (u_1^1)^{a_{11}}(u_2^1)^{a_{12}} \cdots (u_N^1)^{a_{1n}}(u_1^2)^{a_{21}}(u_2^2)^{a_{22}} \cdots (u_N^N)^{a_{NN}}$$

satisfies the condition $L_{D_n}(u^A) = t^{\alpha(A)} \otimes u^A$ with $\alpha(A) := \sum_i \left(\sum_j a_{ij}\right) \epsilon_i \in P$, so that $u^A \in \mathcal{O}(D_N\backslash GL_q(N); t^{\alpha(A)})$.

Relative invariants for the Borel subgroup B_q^- are obtained by means of quantum minor determinants. Let Ω_n be the set of n-tuples $I = \{i_1, i_2, \cdots, i_n\}$ of integers from $\{1, 2, \cdots, N\}$ such that $i_1 < i_2 < \cdots < i_n$. Recall that the quantum n-minor $\mathcal{D}_J^I$ for $I, J \in \Omega_n$ was defined by formula (9.18). In the case $I = \{1, 2, \cdots, n\}$ we write $\mathcal{D}_J := \mathcal{D}_J^I$. Since $\Delta(\mathcal{D}_J) = \sum_I \mathcal{D}_I \otimes \mathcal{D}_J^I$ by Proposition 9.7(ii) and $\pi_{B_q^-}(u_j^i) = 0$ for $i < j$, we get $L_{B_q^-}(\mathcal{D}_J) = \sum_I \pi_{B_q^-}(\mathcal{D}_I) \otimes \mathcal{D}_J^I = z^{\Lambda_n} \otimes \mathcal{D}_J$ and so $\mathcal{D}_J \in \mathcal{O}(B_q^- \backslash GL_q(N); z^{\Lambda_n})$, where $\Lambda_n = \epsilon_1 + \epsilon_2 + \cdots + \epsilon_n$ is the fundamental weight. Since $L_{B_q^-}$ is an algebra homomorphism, it follows that

$$\mathcal{D}_{I_1} \mathcal{D}_{I_2} \cdots \mathcal{D}_{I_r} \in \mathcal{O}(B_q^- \backslash GL_q(N); z^\lambda), \tag{43}$$

for arbitrary m_i-tuples $I_i \in \Omega_{m_i}$, $i = 1, 2, \cdots, r$, where $\lambda = \Lambda_{m_1} + \cdots + \Lambda_{m_r}$.

Clearly, an integral weight $\lambda \in P$ is of the form $\lambda = \Lambda_{m_1} + \Lambda_{m_2} + \cdots + \Lambda_{m_s}$ with $N \geq m_1 \geq m_2 \geq \cdots \geq m_s \geq 0$ if and only if it can be written as $\lambda = \lambda_1 \epsilon_1 + \cdots + \lambda_N \epsilon_N$ with $\lambda_1 \geq \lambda_2 \geq \cdots \geq \lambda_N \geq 0$. The set $(\lambda_1, \cdots, \lambda_N)$ is obtained from $(m_1, \cdots, m_s)$ by transposing the corresponding Young diagram. For such $\lambda \in P$, let $\mathbf{T} = \{T_{rp} \mid 1 \leq r \leq N,\ 1 \leq p \leq \lambda_r\}$ be a tableau of elements from the set $\{1, 2, \cdots, N\}$. The tableau $\mathbf{T}$ is called *semistandard of shape* λ if

$$T_{r-1,p} < T_{rp}, \qquad 1 \leq p \leq s, \quad 2 \leq r \leq m_p,$$

$$T_{r,p-1} \leq T_{rp}, \qquad 1 \leq r \leq N, \quad 2 \leq p \leq \lambda_r.$$

The set of all semistandard tableaux of shape λ is denoted by $\mathrm{Tab}_N(\lambda)$. For each $\mathbf{T} \in \mathrm{Tab}_N(\lambda)$ we define the *standard monomial* $\mathcal{D}_\mathbf{T}$ by setting

$$\mathcal{D}_\mathbf{T} := \mathcal{D}_{J_1} \cdots \mathcal{D}_{J_s} \in \mathcal{O}(GL_q(N, \mathbb{C})),$$

where $J_p = \{T_{1p}, \cdots, T_{m_p p}\}$, $1 \leq p \leq s$. By (43), $\mathcal{D}_\mathbf{T} \in \mathcal{O}(B_q^- \backslash GL_q(N); z^\lambda)$.

Proposition 48. (i) *Suppose that $\lambda \in P_+$ and s is an integer such that $\lambda_N \geq -s$. Then the set $\{\mathcal{D}_q^{-s} \mathcal{D}_\mathbf{T} \mid \mathbf{T} \in \mathrm{Tab}_N(\lambda + s(\epsilon_1 + \cdots + \epsilon_N))\}$ is a basis of the vector space $\mathcal{O}(B_q^- \backslash GL_q(N, \mathbb{C}); z^\lambda)$.*

(ii) *If $\lambda \in P$ is not in P_+, then $\mathcal{O}(B_q^- \backslash GL_q(N, \mathbb{C}); z^\lambda) = \{0\}$.*

Proof. The proof is given in [NYM], Theorem 2.5. □

By Theorem 9.18, there is a dual pairing $\langle \cdot, \cdot \rangle$ of the Hopf algebras $U_q(\mathrm{gl}_N)$ and $\mathcal{O}(GL_q(N))$. By Proposition 1.16, such a pairing turns the algebra $\mathcal{O}(GL_q(N))$ into a $U_q(\mathrm{gl}_N)$-bimodule with actions defined by (1.65).

Proposition 49. *Let $\lambda \in P$. An element $a \in \mathcal{O}(D_n \backslash GL_q(N); t^\lambda)$ belongs to $\mathcal{O}(B_q^- \backslash GL_q(N); z^\lambda)$ if and only if a is annihilated by the right action of all generators $E_k \in U_q(\mathrm{gl}_N)$, $k = 1, 2, \cdots, N-1$, that is, $a.E_k \equiv \sum \langle E_k, a_{(1)} \rangle a_{(2)} = 0$ for $k = 1, 2, \cdots, N-1$.*

Proof. The proof follows from Theorem 2.2 in [NYM]. (Note that our dual pairing of $U_q(\mathrm{gl}_N)$ and $\mathcal{O}(GL_q(N))$ differs from that in [NYM].) □

For $\lambda \in P_+$, let T_λ denote the corepresentation of $\mathcal{O}(GL_q(N))$ on the right $\mathcal{O}(GL_q(N))$-comodule $\mathcal{O}(B_q^- \backslash GL_q(N); z^\lambda)$. Then, by Proposition 1.15, the equation $\hat{T}_\lambda(f) := (\mathrm{id} \otimes f) \circ T_\lambda$, $f \in U_q(\mathrm{gl}_N)$, defines a representation $\hat{T}_\lambda$ of the algebra $U_q(\mathrm{gl}_N)$ on the finite-dimensional vector space $\mathcal{O}(B_q^- \backslash GL_q(N); z^\lambda)$. In the next proposition and in the proof of Theorem 51 below we shall work with highest weight representations of $U_q(\mathrm{gl}_N)$ with respect to the sequence of simple roots $-\alpha_1, \cdots, -\alpha_N$ (see also Remark 7 in Subsect. 6.2.4).

Proposition 50. *For any $\lambda \in P_+$, T_λ is an irreducible corepresentation of $\mathcal{O}(GL_q(N))$ and $\hat{T}_\lambda$ is an irreducible representation of $U_q(\mathrm{gl}_N)$ with highest weight λ with respect to the simple roots $-\alpha_1, \cdots, -\alpha_N$ of gl_N.*

Proof. It is sufficient to prove the assertion for λ such that $\lambda_1 \geq \lambda_2 \geq \cdots \geq \lambda_N \geq 0$ (otherwise we multiply all elements of $\mathcal{O}(B_q^- \backslash GL_q(N); z^\lambda)$ by $\mathcal{D}_q^{-s}$, $s \in \mathbb{N}$, and apply Proposition 48(i)). Let $\lambda = r_1 \Lambda_1 + r_2 \Lambda_2 + \cdots + r_N \Lambda_N$, $r_1, r_2, \cdots, r_N \in \mathbb{N}_0$. Let us consider the standard monomial $a := \mathcal{D}^{r_N}_{\{1,\cdots,N\}} \mathcal{D}^{r_{N-1}}_{\{1,\cdots,N-1\}} \cdots \mathcal{D}^{r_2}_{\{1,2\}} \mathcal{D}^{r_1}_{\{1\}}$ and set $V := \hat{T}_\lambda(U_q(\mathrm{gl}_N))a$. The restriction of $\hat{T}_\lambda$ to V is a finite-dimensional representation and hence a weight representation of $U_q(\mathrm{gl}_N)$. Since $\langle K_i^{-1}, u_l^k \rangle = \delta_{kl} q^{\delta_{ki}}$ by the formulas from Sect. 9.4, we obtain $\hat{T}_\lambda(K_i^{-1})a \equiv \sum a_{(1)} \langle K_i^{-1}, a_{(2)} \rangle = q^{\lambda_i} a$. From the relation $\Delta(F_i) = F_i \otimes 1 + K_i^{-1} K_{i+1} \otimes F_i$ (see Subsect. 6.1.2) we get

$$\Delta^{(n)}(F_i) = \sum\nolimits_{j=1}^{n+1} (K_i^{-1} K_{i+1})^{\otimes(n-j+1)} \otimes F_i \otimes 1^{\otimes(j-1)}.$$

Since $\langle F_i, u_k^k \rangle = 0$ and $\langle K_j^{-1}, u_l^k \rangle = 0, k \neq l$, by the formulas from Sect. 9.4, we derive that $\hat{T}_\lambda(F_i)a = 0$. Hence a is a highest weight vector for the representation $\hat{T}_\lambda \lceil V$ with weight λ with respect to the simple roots $-\alpha_1, \cdots, -\alpha_N$.

By Proposition 48(i), the representation space of $\hat{T}_\lambda$ has a basis labeled by the standard tableaux of shape λ. From classical representation theory it is known that the same is true for the irreducible representation of the Lie algebra $\mathrm{gl}(N, \mathbb{C})$ with highest weight λ. As noted in Subsect. 7.3.3, the dimensions of irreducible representations of $\mathrm{gl}(N, \mathbb{C})$ and $U_q(\mathrm{gl}_N)$ with the same highest weight coincide. Since $\hat{T}_\lambda$ contains a subrepresentation with highest weight λ as shown above, it follows that $\hat{T}_\lambda$ is irreducible. The irreducibility of $\hat{T}_\lambda$ obviously implies the irreducibility of T_λ. □

11.5.4 Peter–Weyl Decomposition of $\mathcal{O}(GL_q(N))$

The main result of this subsection is the following.

Theorem 51. *Suppose that q is not a root of unity. Then the coordinate Hopf algebra $\mathcal{O}(GL_q(N))$ decomposes into a direct sum of coefficient coalgebras $\mathcal{C}(T_\lambda)$ of irreducible corepresentations T_λ, $\lambda \in P_+$, of $\mathcal{O}(GL_q(N))$:*

$$\mathcal{O}(GL_q(N)) = \bigoplus\nolimits_{\lambda \in P_+} \mathcal{C}(T_\lambda). \tag{44}$$

Proof. Throughout this proof highest weights always refer to the simple roots $-\alpha_1, \cdots, -\alpha_N$. From the defining relations for $U_q(\mathrm{gl}_N)$ we see that there is an algebra anti-homomorphism θ of $U_q(\mathrm{gl}_N)$ such that $\theta(E_i) = F_i$, $\theta(F_i) = E_i$, $i = 1, 2, \cdots, N-1$, and $\theta(K_i) = K_i$, $i = 1, 2, \cdots, N$. We define a representation ρ of $U_q(\mathrm{gl}_N)$ on $\mathcal{O}(GL_q(N))$ by $\rho(f)a = a.\theta(f)$, $f \in U_q(\mathrm{gl}_N)$, $a \in \mathcal{O}(GL_q(N))$. Let $a \in \mathcal{O}(GL_q(N))$ be a highest weight vector for ρ. By definition we have $\rho(F_k)a = a.\theta(F_k) = a.E_k = 0$ for $k = 1, \cdots, N-1$. From the definition of $\mathcal{O}(D_N \backslash GL_q(N); t^\lambda)$ it follows that an element $b \in \mathcal{O}(GL_q(N))$ belongs to $\mathcal{O}(D_N \backslash GL_q(N); t^\lambda)$ if and only if $b.K_i^{-1} = q^{\lambda_i} b$, $i = 1, 2, \cdots, N$. Therefore, $a \in \mathcal{O}(D_N \backslash GL_q(N); t^\lambda)$ for some $\lambda \in P_+$ and hence $a \in \mathcal{O}(B_q^- \backslash GL_q(N); z^\lambda)$ by Proposition 49. Clearly, we have $\mathcal{O}(B_q^- \backslash GL_q(N); z^\lambda) \subseteq \mathcal{C}(T_\lambda)$, so that $a \in \mathcal{C}(T_\lambda)$.

On the other hand, let $\mathcal{O}(M_q(N))_s$ be the span of monomials in u^i_j of degree less than or equal to s. The algebra $\mathcal{O}(GL_q(N))$ is the sum of finite-dimensional left $\mathcal{O}(GL_q(N))$-subcomodules $\mathcal{D}_q^{-r} \mathcal{O}(M_q(N))_s$, $r, s \in \mathbb{N}_0$, and so of finite-dimensional ρ-invariant subspaces. Therefore, by Theorem 7.8, ρ is a direct sum of finite-dimensional irreducible representations. Each irreducible component contains a highest weight vector. Since such vectors belong to some $\mathcal{C}(T_\lambda)$ as shown in the preceding paragraph and $\mathcal{C}(T_\lambda)$ is ρ-invariant, it follows that $\mathcal{O}(GL_q(N)) = \sum_{\lambda \in P_+} \mathcal{C}(T_\lambda)$. If $\lambda, \mu \in P_+$ and $\lambda \neq \mu$, the representations $\hat{T}_\lambda$ and $\hat{T}_\mu$ are inequivalent (because they have different highest weights). So the corepresentations T_λ and T_μ are also inequivalent. Hence, by Corollary 9, the sum $\sum_\lambda \mathcal{C}(T_\lambda)$ is actually a direct sum. □

Theorem 51 has a number of important corollaries. Recall that, as always in this section, q is not a root of unity.

Corollary 52. *Any irreducible corepresentation of $\mathcal{O}(GL_q(N))$ is equivalent to one of the corepresentations T_λ, $\lambda \in P_+$.*

Proof. If T is an irreducible corepresentation not equivalent to some T_λ, then we have $\mathcal{C}(T_\lambda) \bigcap \mathcal{C}(T) = \{0\}$ for all $\lambda \in P_+$ by Corollary 9. Because of (44), this is impossible. □

Corollary 53. *The Hopf algebra $\mathcal{O}(GL_q(N))$ is a cosemisimple Hopf algebra and the direct sum (44) is its Peter–Weyl decomposition.*

Proof. Combine Theorem 51 and the implication (i)→(vi) of Theorem 13. □

Corollary 54. *The dual pairing $\langle \cdot, \cdot \rangle$ of $U_q(\mathrm{gl}_N)$ and $\mathcal{O}(GL_q(N))$ from Theorem 9.18 is non-degenerate.*

Proof. The proof is analogous that of Corollary 23. □

Another immediate consequence of Theorem 51 is the existence of a unique Haar functional h on $\mathcal{O}(GL_q(N))$. From formula (11.3) we know that h is given by $h(1) = 1$ and $h(a) = 0$ for $a \in \mathcal{C}(T_\lambda)$, $\lambda \in P_+$, $\lambda \neq 0$.

11.5.5 Representations of the Quantum Group $U_q(N)$

In this subsection we assume that q is a real number such that $q \neq 0, \pm 1$.

Recall from Subsect. 9.2.4 that the Hopf $*$-algebra $\mathcal{O}(U_q(N))$ is just the Hopf algebra $\mathcal{O}(GL_q(N))$ equipped with the involution defined by $(u^i_j)^* = S(u^j_i)$, $i,j = 1,2,\cdots,N$, and $(\mathcal{D}_q^{-1})^* = \mathcal{D}_q$. Further, as noted in Example 11.7, $\mathcal{O}(U_q(N))$ is a CMQG algebra. Therefore, all results on the quantum group $GL_q(N)$ obtained in the previous subsections and the theory of CQG algebras developed in Sect. 11.3 are valid for the quantum group $U_q(N)$. Most of the corepresentation theory of $U_q(N)$ can be derived by combining these two sources of results. Some of them will be sketched in the following.

From Proposition 29, the Haar functional h of the cosemisimple Hopf algebra $\mathcal{O}(GL_q(N))$ is a faithful state of the $*$-algebra $\mathcal{A} := \mathcal{O}(U_q(N))$. Hence the equation $\langle a, b\rangle := h(a^* b)$, $a, b \in \mathcal{A}$, defines a scalar product on $\mathcal{A}$. Formula (36) says that the character f_1 of $\mathcal{A}$ is given by $f_1(\cdot) = \langle K_{2\rho}, \cdot\rangle$, where

$$2\rho = \sum\nolimits_{k=1}^{N} (N - 2k + 1)\epsilon_k.$$

Now let $\lambda \in P_+$ be a dominant integral weight. We choose an orthonormal basis $\{e_i\}$ of the space $V_\lambda := \mathcal{O}(B_q^- \backslash GL_q(N), z^\lambda)$ with respect to the scalar product $\langle\cdot,\cdot\rangle$ on $\mathcal{A}$ such that e_i is a weight vector with weight $\lambda(i) \in P$ for the representation $\hat{T}_\lambda$ of $U_q(\mathrm{gl}_N)$ on V_λ. Let $\{u^\lambda_{ij}\}$ be the matrix elements of the corepresentation T_λ of $\mathcal{A}$ with respect to the basis $\{e_i\}$. Then, $u^\lambda = (u^\lambda_{ij})$ is an irreducible unitary matrix corepresentation of $\mathcal{A}$. The unique intertwiner $F_\lambda = ((F_\lambda)_{ij}) \in \mathrm{Mor}\,(u^\lambda, (u^\lambda)^{cc})$ such that $\mathrm{Tr}\, F_\lambda = \mathrm{Tr}\, F_\lambda^{-1}$ (by Lemma 30) is given by

$$(F_\lambda)_{ij} = q^{2(\rho,\lambda(i))}\delta_{ij}. \tag{45}$$

We shall prove formula (45). Using Proposition 6.6 and (45) we get

$$\langle f, S^2(u^\lambda_{in})(F_\lambda)_{nj}\rangle = \langle S^2(f), u^\lambda_{ij}\rangle q^{2(\rho,\lambda(j))}$$

$$= \langle K_{2\rho} f K_{2\rho}^{-1}, u^\lambda_{ij}\rangle q^{2(\rho,\lambda(j))} = \langle K_{2\rho} f, u^\lambda_{ij}\rangle = \langle f, (F_\lambda)_{in} u^\lambda_{nj}\rangle$$

for $f \in U_q(\mathrm{gl}_N)$ and so $(F_\lambda)_{in} u^\lambda_{nj} = S^2(u^\lambda_{in})(F_\lambda)_{nj}$ by Corollary 54. Thus, $F_\lambda \in \mathrm{Mor}\,(u^\lambda, (u^\lambda)^{cc})$. Since $\mathrm{Tr}\, F_\lambda = \mathrm{Tr}\, F_\lambda^{-1}$, the matrix F_λ defined by (45) is indeed the intertwiner from Lemma 30.

Next we combine formula (45) with the Schur orthogonality relation (25) from Subsect. 11.3.2. Then we conclude that the set $\{u^\lambda_{ij} \,|\, \lambda \in P_+, i,j = 1,2,\cdots,\dim V_\lambda\}$ is an orthogonal vector space basis of $\mathcal{A}$ and

$$\|u^\lambda_{ij}\|^2 = d_\lambda^{-1} q^{-2(\rho,\lambda(i))}.$$

The number

$$d_\lambda = \mathrm{Tr}\, F_\lambda = \sum\nolimits_{\lambda(i)\in P(\lambda)} q^{2(\rho,\lambda(i))} = \sum\nolimits_{\lambda(i)\in P(\lambda)} q^{-2(\rho,\lambda(i))} \tag{46}$$

may be considered as a quantum analog of the dimension of V_λ. It coincides with the quantum dimension (see Subsect. 7.1.6) of the corresponding representation of $U_q(\mathrm{gl}_N)$. Here $P(\lambda)$ is the set of all weights of the representation $\hat{T}_\lambda$ of $U_q(\mathrm{gl}_N)$ taken with multiplicities.

Finally, let us turn to the character $\chi(\lambda) := \sum_i u^\lambda_{ii}$ of the corepresentation T_λ, $\lambda_1 \geq \cdots \geq \lambda_N \geq 0$. It is an element of the algebra $\mathcal{A} = \mathcal{O}(U_q(N))$. Since $\pi_D(u^\lambda_{ii}) = t^{\lambda(i)} \equiv t_1^{\lambda_1} \cdots t_N^{\lambda_N}$ for $\lambda(i) = \lambda_1\epsilon_1 + \cdots + \lambda_N\epsilon_N$, $\pi_D(\chi_\lambda)$ coincides with the corresponding expression in the classical case. The latter is known to be the *Schur function* $S_\lambda(t) \equiv S_\lambda(t_1, \cdots, t_N)$ (see [Mac]). Thus, we have

$$\pi_D(\chi(\lambda)) = S_\lambda(t) \;\; \text{and} \;\; d_\lambda = S_\lambda(q^{n-1}, q^{n-3}, \cdots, q^{-n+1}).$$

11.6 Quantum Homogeneous Spaces

11.6.1 Definition of a Quantum Homogeneous Space

Recall that a quantum space for a Hopf algebra may be viewed as a generalization to quantum groups of a space on which a group acts. If the action is transitive, such a space is usually called a homogeneous space. The following definition gives a generalization of the latter notion to quantum groups.

Let G_q be a quantum group given by its coordinate Hopf algebra $\mathcal{O}(G_q)$ and let $\mathcal{O}(X_q)$ be the coordinate algebra of a right G_q-space with coaction $\varphi : \mathcal{O}(X_q) \rightarrow \mathcal{O}(X_q) \otimes \mathcal{O}(G_q)$.

Definition 15. *We say that X_q is a* right quantum homogeneous G_q-space *if there exist a subalgebra $\mathcal{X}$ of $\mathcal{O}(G_q)$ such that $\Delta(\mathcal{X}) \subseteq \mathcal{X} \otimes \mathcal{O}(G_q)$ and an algebra isomorphism $\theta : \mathcal{O}(X_q) \rightarrow \mathcal{X}$ such that $\Delta \circ \theta = (\theta \otimes \mathrm{id}) \circ \varphi$.*

Any subalgebra $\mathcal{X}$ of $\mathcal{O}(G_q)$ such that $\Delta(\mathcal{X}) \subseteq \mathcal{X} \otimes \mathcal{O}(G_q)$ is obviously a right quantum G_q-space with right coaction $\Delta\lceil\mathcal{X}$. Definition 15 says that up to isomorphisms the right quantum G_q-spaces of this form are precisely the right quantum homogeneous G_q-spaces. If X_q is such a quantum homogeneous G_q-space, then $f := \varepsilon \circ \theta$ is a character of the algebra $\mathcal{O}(X_q)$, that is, $f(xy) = f(x)f(y)$ for $x, y \in \mathcal{O}(X_q)$ and $f(1) = 1$. This means that any quantum homogeneous G_q-space contains at least one "classical point".

Left quantum homogeneous G_q-spaces are defined similarly (by requiring that $\Delta(\mathcal{X}) \subseteq \mathcal{O}(G_q) \otimes \mathcal{X}$ and $\Delta \circ \theta = (\mathrm{id} \otimes \theta) \circ \varphi$).

Example 10. The quantum vector space $\mathbb{C}_q^N$ is a right and left quantum homogeneous $GL_q(N)$-space by Proposition 9.11. △

Example 11. Let q be not a root of unity. By Proposition 4.31, the quantum 2-spheres $S^2_{q\rho}$ are right quantum homogeneous $SL_q(2)$-spaces. △

11.6.2 Quantum Homogeneous Spaces Associated with Quantum Subgroups

In differential geometry it is well-known that a homogeneous manifold for a Lie group G is isomorphic to the coset space $H\backslash G$ (resp. G/H) of G with respect to the stabilizer subgroup H (see, for example, [Wa], 3.58). Then, omitting technical details, functions on $H\backslash G$ (resp. G/H) can be identified with those functions on G which are invariant under the left (resp. right) action of the subgroup H on G. Thus, it is natural to construct quantum homogeneous G_q-spaces in a similar manner.

Let K_q be a quantum subgroup of G_q (see Definition 9.11) with coordinate Hopf algebra $\mathcal{O}(K_q)$ and surjective Hopf algebra homomorphism $\pi_K : \mathcal{O}(G_q) \to \mathcal{O}(K_q)$. For notational simplicity we abbreviate $\mathcal{O}_G := \mathcal{O}(G_q)$ and $\mathcal{O}_K := \mathcal{O}(K_q)$. Let L_{K_q} and R_{K_q} be the left and right coactions of $\mathcal{O}_K$ on $\mathcal{O}_G$, respectively, defined by $L_{K_q} := (\pi_K \otimes \mathrm{id}) \circ \Delta$ and $R_{K_q} := (\mathrm{id} \otimes \pi_K) \circ \Delta$.

Definition 16. *The elements of the sets*

$$\mathcal{O}_K\backslash\mathcal{O}_G := \{a \in \mathcal{O}_G \mid L_{K_q}(a) = 1 \otimes a\},$$

$$\mathcal{O}_G/\mathcal{O}_K := \{a \in \mathcal{O}_G \mid R_{K_q}(a) = a \otimes 1\},$$

$$\mathcal{O}_K\backslash\mathcal{O}_G/\mathcal{O}_K := \mathcal{O}_K\backslash\mathcal{O}_G \cap \mathcal{O}_G/\mathcal{O}_K$$

are said to be left-invariant, right-invariant *and* biinvariant, *respectively, with respect to the quantum subgroup K_q. The sets $\mathcal{O}_K\backslash\mathcal{O}_G$ and $\mathcal{O}_G/\mathcal{O}_K$ are called coordinate algebras of the* right *resp.* left quantum homogeneous G_q-spaces *$K_q\backslash G_q$ resp. G_q/K_q.*

Since L_{K_q} and R_{K_q} are algebra homomorphisms, $\mathcal{O}_K\backslash\mathcal{O}_G$, $\mathcal{O}_G/\mathcal{O}_K$ and $\mathcal{O}_K\backslash\mathcal{O}_G/\mathcal{O}_K$ are subalgebras of $\mathcal{O}_G$. Proposition 55(ii) says that $K_q\backslash G_q$ and G_q/K_q are indeed right and left quantum homogeneous G_q-spaces.

Proposition 55. (i) $S(\mathcal{O}_K\backslash\mathcal{O}_G) \subseteq \mathcal{O}_G/\mathcal{O}_K$, $S(\mathcal{O}_G/\mathcal{O}_K) \subseteq \mathcal{O}_K\backslash\mathcal{O}_G$.

(ii) $\Delta(\mathcal{O}_K\backslash\mathcal{O}_G) \subseteq \mathcal{O}_K\backslash\mathcal{O}_G \otimes \mathcal{O}_G$, $\Delta(\mathcal{O}_G/\mathcal{O}_K) \subseteq \mathcal{O}_G \otimes \mathcal{O}_G/\mathcal{O}_K$.

(iii) *Setting* $\mathcal{C} := \{a \in \mathcal{O}_G \mid \pi_K(a) = \varepsilon(a)1\}$, *we have*

$$\mathcal{O}_K\backslash\mathcal{O}_G = \{a \in \mathcal{C} \mid \Delta(a) \in \mathcal{C} \otimes \mathcal{O}_G\},\ \mathcal{O}_G/\mathcal{O}_K = \{a \in \mathcal{C} \mid \Delta(a) \in \mathcal{O}_G \otimes \mathcal{C}\},$$

$$\mathcal{O}_K\backslash\mathcal{O}_G/\mathcal{O}_K = \{a \in \mathcal{C} \mid \Delta(a) \in \mathcal{C} \otimes \mathcal{C}\}.$$

Proof. (i): Let $a \in \mathcal{O}_K\backslash\mathcal{O}_G$, that is, $L_{K_q}(a) = 1 \otimes a$. Since π_K is a Hopf algebra homomorphism, $\pi_K \circ S = S \circ \pi_K$. Thus, we have

$$\begin{aligned} R_{K_q}(S(a)) &= (\mathrm{id} \otimes \pi_K) \circ \Delta(S(a)) = \sum S(a_{(2)}) \otimes \pi_K(S(a_{(1)})) \\ &= \sum (S \otimes S)(a_{(2)} \otimes \pi_K(a_{(1)})) = (S \otimes S) \circ \tau \circ L_{K_q}(a) \\ &= (S \otimes S) \circ \tau(1 \otimes a) = S(a) \otimes 1, \end{aligned}$$

that is, $S(a) \in \mathcal{O}_G/\mathcal{O}_K$. The second inclusion is proved similarly.

(ii) follows from Corollary 47 applied with $\chi = 1$.

(iii): We verify the first equality. Let $a \in \mathcal{O}_K \backslash \mathcal{O}_G$. Applying $\mathrm{id} \otimes \varepsilon$ to $(\pi_K \otimes \mathrm{id}) \circ \Delta(a) = 1 \otimes a$ yields $\pi_K(a) = \varepsilon(a)1$. Hence, $\mathcal{O}_K \backslash \mathcal{O}_G \subseteq \mathcal{C}$. By (ii), $\Delta(a) \in \mathcal{O}_K \backslash \mathcal{O}_G \otimes \mathcal{O}_G$. Conversely, suppose that $a \in \mathcal{C}$ and $\Delta(a) \in \mathcal{C} \otimes \mathcal{O}_G$. Then $(\pi_K \otimes \mathrm{id}) \circ \Delta(a) = \sum \varepsilon(a_{(1)})1 \otimes a_{(2)} = 1 \otimes a$. Thus, $a \in \mathcal{O}_K \backslash \mathcal{O}_G$. □

We illustrate the above notions by restating some results from Chap. 4.

Example 12 ($G_q = SL_q(2)$). Throughout this example we assume that q is not a root of unity. Recall that the group K_q of diagonal matrices in $SL(2, \mathbb{C})$ is a quantum subgroup of $G_q = SL_q(2)$. From Subsect. 4.2.2 we know that

$$\mathcal{O}_K \backslash \mathcal{O}_G = \bigoplus\nolimits_n \mathcal{A}[0, n], \quad \mathcal{O}_G / \mathcal{O}_K = \bigoplus\nolimits_n \mathcal{A}[n, 0], \quad \mathcal{O}_K \backslash \mathcal{O}_G / \mathcal{O}_K = \mathbb{C}[bc].$$

That is, the algebra $\mathcal{O}_K \backslash \mathcal{O}_G$ is generated by the elements bc, ac, db, the algebra $\mathcal{O}_G / \mathcal{O}_K$ is generated by bc, ab, cd and the algebra $\mathcal{O}_K \backslash \mathcal{O}_G / \mathcal{O}_K$ possesses the single generator bc. Hence $\mathcal{O}_K \backslash \mathcal{O}_G$ coincides with the algebra generated by the elements $\tilde{x}_i$, $i = -1, 0, 1$, from (4.80). Therefore, by Proposition 4.31, the right quantum homogeneous G_q-space $\mathcal{O}_K \backslash \mathcal{O}_G$ is just the quantum 2-sphere $S^2_{q\infty}$. It is not difficult to show that the other quantum 2-spheres $S^2_{q\rho}$, $\rho \neq \infty$, are not of the form $\mathcal{O}_H \backslash \mathcal{O}_G$ for some quantum subgroup H_q of G_q.

Let us note that $\Delta(bc)$ is neither contained in $\mathcal{O}_K \backslash \mathcal{O}_G / \mathcal{O}_K \otimes \mathcal{O}_G$ nor in $\mathcal{O}_G \otimes \mathcal{O}_K \backslash \mathcal{O}_G / \mathcal{O}_K$, so the subalgebra $\mathcal{O}_K \backslash \mathcal{O}_G / \mathcal{O}_K$ of $\mathcal{O}_G$ is neither a left nor a right quantum homogeneous G_q-space. △

Proposition 56. *Let $\mathcal{O}(K_q)$ be a cosemisimple Hopf algebra with Haar functional h_K. Then the linear mappings $P_l := (h_K \circ \pi_K \otimes \mathrm{id}) \circ \Delta$ and $P_r := (\mathrm{id} \otimes h_K \circ \pi_K) \circ \Delta$ are projections of $\mathcal{O}_G$ onto $\mathcal{O}_K \backslash \mathcal{O}_G$ and $\mathcal{O}_G / \mathcal{O}_K$, respectively. The projections P_l and P_r commute and $P_l P_r$ is a projection of $\mathcal{O}_G$ onto $\mathcal{O}_K \backslash \mathcal{O}_G / \mathcal{O}_K$.*

Proof. Since $\pi_K : \mathcal{O}_G \to \mathcal{O}_K$ is a Hopf algebra homomorphism, we obtain

$$\begin{aligned} P_l P_l &= (h_K \circ \pi_K \otimes \mathrm{id}) \circ \Delta \circ (h_K \circ \pi_K \otimes \mathrm{id}) \circ \Delta \\ &= (h_K \circ \pi_K \otimes h_K \circ \pi_K \otimes \mathrm{id}) \circ (\Delta \otimes \mathrm{id}) \circ \Delta \\ &= (h_K \otimes h_K \otimes \mathrm{id})(\Delta \otimes \mathrm{id}) \circ (\pi_K \otimes \mathrm{id}) \circ \Delta = P_l. \end{aligned}$$

It is obvious that $P_l(a) = a$ if $a \in \mathcal{O}_K \backslash \mathcal{O}_G$. For $a \in \mathcal{O}_G$, we have

$$\begin{aligned} (\pi_K \otimes \mathrm{id}) \circ \Delta(P_l(a)) &= (\pi_K \otimes \mathrm{id}) \circ \Delta \circ (h_K \circ \pi_K \otimes \mathrm{id}) \circ \Delta(a) \\ &= (h_K \otimes \pi_K \otimes \mathrm{id})(\pi_K \otimes \mathrm{id} \otimes \mathrm{id}) \circ (\Delta \otimes \mathrm{id}) \circ \Delta(a) \\ &= (h_K \otimes \mathrm{id} \otimes \mathrm{id})(\Delta \otimes \mathrm{id}) \circ (\pi_K \otimes \mathrm{id}) \circ \Delta(a) \\ &= 1 \otimes (h_K \circ \pi_K \otimes \mathrm{id}) \circ \Delta(a) = 1 \otimes P_l(a), \end{aligned}$$

so $P_l(a) \in \mathcal{O}_K \backslash \mathcal{O}_G$. Thus, we have shown that P_l is a projection of $\mathcal{O}_G$ onto $\mathcal{O}_K \backslash \mathcal{O}_G$. Similar reasoning proves the assertions concerning P_r. One verifies that $P_l P_r$ and $P_r P_l$ are both equal to $(h_K \circ \pi_K \otimes \mathrm{id} \otimes h_K \circ \pi_K)(\mathrm{id} \otimes \Delta) \circ \Delta$. Hence $P_l P_r = P_r P_l$. The remaining assertions follow easily. □

11.6.3 Quantum Gel'fand Pairs

Let us first recall the corresponding classical notion. A pair (G, K) of a locally compact group G and a compact subgroup K is called a Gel'fand pair if the algebra $L^1(K\backslash G/K)$ of integrable K-biinvariant functions with respect to the convolution product is commutative. If G is compact, this is equivalent to the requirement that the dimension of the subspace of K-biinvariant elements in the coefficient space of each irreducible representation of G is at most one. In the following we shall generalize this concept to quantum groups. Throughout this subsection we assume that K_q is a quantum subgroup of a quantum group G_q such that the Hopf algebras $\mathcal{O}(K_q)$ and $\mathcal{O}(G_q)$ are cosemisimple.

As we have seen in Example 12, the comultiplication Δ of $\mathcal{O}_G$ does not map $\mathcal{O}_K\backslash\mathcal{O}_G/\mathcal{O}_K$ to $\mathcal{O}_K\backslash\mathcal{O}_G/\mathcal{O}_K \otimes \mathcal{O}_K\backslash\mathcal{O}_G/\mathcal{O}_K$. For this reason we introduce a new comultiplication Δ_{bi} on $\mathcal{O}_K\backslash\mathcal{O}_G/\mathcal{O}_K$ by

$$\Delta_{\mathrm{bi}} := (\mathrm{id} \otimes h_K \circ \pi_K \otimes \mathrm{id}) \circ (\Delta \otimes \mathrm{id}) \circ \Delta = (\mathrm{id} \otimes P_l) \circ \Delta = (P_r \otimes \mathrm{id}) \circ \Delta.$$

Proposition 57. *The vector space $\mathcal{C}_{\mathrm{bi}} := \mathcal{O}_K\backslash\mathcal{O}_G/\mathcal{O}_K$ is a coalgebra with comultiplication Δ_{bi} and counit $\varepsilon\lceil\mathcal{C}_{\mathrm{bi}}$. The restriction to $\mathcal{C}_{\mathrm{bi}}$ of a left- (resp. right-) invariant linear functional on $\mathcal{O}_G$ is also left- (resp. right-) invariant with respect to the comultiplication Δ_{bi}. On $\mathcal{C}_{\mathrm{bi}}$ we have the identity $\Delta_{\mathrm{bi}} \circ S = \tau \circ (S \otimes S) \circ \Delta_{\mathrm{bi}}$.*

Proof. We sketch the proof of the first assertion. Let $a \in \mathcal{C}_{\mathrm{bi}}$. By Proposition 56, $P_r P_l(a) = a$. From the relations $\Delta \circ P_r = (\mathrm{id} \otimes P_r) \circ \Delta$ and $\Delta \circ P_l = (P_l \otimes \mathrm{id}) \circ \Delta$ and the fact that P_r and P_l are commuting projections we get

$$\Delta_{\mathrm{bi}}(a) = (P_r \otimes \mathrm{id}) \circ \Delta \circ P_r P_l(a) = (P_r P_l \otimes P_r) \circ \Delta(a),$$

$$\Delta_{\mathrm{bi}}(a) = (\mathrm{id} \otimes P_l) \circ \Delta \circ P_r P_l(a) = (P_l \otimes P_l P_r) \circ \Delta(a).$$

These identities imply that $\Delta_{\mathrm{bi}}(a) \in \mathcal{C}_{\mathrm{bi}} \otimes \mathcal{C}_{\mathrm{bi}}$. Some computations show that both expressions $(\Delta_{\mathrm{bi}} \otimes \mathrm{id}) \circ \Delta_{\mathrm{bi}}$ and $(\mathrm{id} \otimes \Delta_{\mathrm{bi}}) \circ \Delta_{\mathrm{bi}}$ are equal to $(\mathrm{id} \otimes h_K \circ \pi_K \otimes \mathrm{id} \otimes h_K \circ \pi_K \otimes \mathrm{id}) \circ \Delta^{(4)}$, so Δ_{bi} is coassociative. We have

$$\begin{aligned}(\varepsilon \otimes \mathrm{id}) \circ \Delta_{\mathrm{bi}}(a) &= (\varepsilon \otimes \mathrm{id})(\mathrm{id} \otimes h_K \circ \pi_K \otimes \mathrm{id}) \circ (\Delta \otimes \mathrm{id}) \circ \Delta(a) \\ &= (h_K \circ \pi_K \otimes \mathrm{id})(\varepsilon \otimes \mathrm{id} \otimes \mathrm{id}) \circ (\Delta \otimes \mathrm{id}) \circ \Delta(a) \\ &= (h_K \circ \pi_K \otimes \mathrm{id}) \circ \Delta(a) = (h_K \otimes \mathrm{id})(1 \otimes a) = a\end{aligned}$$

for $a \in \mathcal{O}_K\backslash\mathcal{O}_G/\mathcal{O}_K$. Similarly, $(\mathrm{id} \otimes \varepsilon) \circ \Delta_{\mathrm{bi}}(a) = a$. Hence $\varepsilon\lceil\mathcal{C}_{\mathrm{bi}}$ is a counit for $\mathcal{C}_{\mathrm{bi}}$. This proves that $\mathcal{C}_{\mathrm{bi}}$ is a coalgebra. □

The comultiplication Δ_{bi} is not an algebra homomorphism in general.

Proposition 58. *The following conditions are equivalent:*

(i) *The coalgebra $\mathcal{O}_K\backslash\mathcal{O}_G/\mathcal{O}_K$ with comultiplication Δ_{bi} is cocommutative.*

(ii) dim $(\mathcal{O}_K\backslash\mathcal{O}_G/\mathcal{O}_K \bigcap \mathcal{C}(\varphi)) \leq 1$ *for all irreducible corepresentations φ of $\mathcal{O}_G$.*

In the proof of Proposition 58 we will use the following

Lemma 59. *Let φ be a corepresentation of $\mathcal{O}(G_q)$ on a finite-dimensional vector space V such that $\mathcal{O}_K\backslash\mathcal{O}_G/\mathcal{O}_K \bigcap \mathcal{C}(\varphi) \neq \{0\}$. Then there exists a basis of V such that the corresponding matrix $v = (v_{ij})$ of φ satisfies the condition $\mathcal{O}_K\backslash\mathcal{O}_G/\mathcal{O}_K \bigcap \mathcal{C}(\varphi) = \mathrm{Lin}\,\{v_{ij} \mid i,j = 1,2,\cdots,r\}$ for some $r \leq \dim V$.*

Proof. Clearly, $\varphi_K := (\mathrm{id} \otimes \pi_K) \circ \varphi$ is a corepresentation of $\mathcal{O}(K_q)$ on V. Put $V_1 := \{v \in V \mid \varphi_K(v) = v \otimes 1\}$. Since $\mathcal{O}(K_q)$ is cosemisimple by assumption, φ_K is a direct sum of irreducible corepresentations. Hence there exists a φ_K-invariant subspace V_2 of V such that $V = V_1 \oplus V_2$ and $\varphi_K \lceil V_2$ does not contain the trivial corepresentation. Take a basis $\{e_1, \cdots, e_n\}$ of V such that $e_i \in V_1$, $i \leq r$, and $e_i \in V_2$, $i > r$, and let (v_{ij}) be the matrix corepresentation of φ with respect to this basis. Since $\varphi_K(e_i) = \sum_j e_j \otimes \pi_K(v_{ji})$ and V_1 and V_2 are φ_K-invariant, we get $\pi_K(v_{ji}) = 0$ if $j \leq r, i > r$ and if $j > r, i \leq r$. Moreover, $\pi_K(v_{ji}) = \delta_{ij} \cdot 1$ if $i,j \leq r$, because the elements of V_1 are φ_K-invariant. From these facts it follows that $v_{ij} \in \mathcal{O}_K\backslash\mathcal{O}_G/\mathcal{O}_K$ for all $i,j \leq r$. Since $\varphi_K \lceil V_2$ does not contain the trivial corepresentation, $P_l P_r(v_{ij}) = \sum_{m,n} h_K(\pi_K(v_{im}))v_{mn}h_K(\pi_K(v_{nj})) = 0$ for all $i,j > r$ by (3). Since $P_l P_r$ is a projection of $\mathcal{O}_G$ onto $\mathcal{O}_K\backslash\mathcal{O}_G/\mathcal{O}_K$ by Proposition 56, we conclude that $\mathcal{O}_K\backslash\mathcal{O}_G/\mathcal{O}_K \bigcap \mathrm{Lin}\,\{v_{ij} \mid i,j > r\} = \{0\}$. □

Proof of Proposition 58. (i)→(ii): Assume that $\mathcal{O}_K\backslash\mathcal{O}_G/\mathcal{O}_K \bigcap \mathcal{C}(\varphi) \neq \{0\}$. We then choose a matrix (v_{ij}) as in Lemma 59. For $i,j \leq r$ we then have $v_{ij} \in \mathcal{O}_K\backslash\mathcal{O}_G/\mathcal{O}_K$ and $\pi_K(v_{ij}) = \delta_{ij} \cdot 1$, so we obtain

$$\Delta_{\mathrm{bi}}(v_{ij}) = \sum_{m,n} h_K(\pi_K(v_{mn}))v_{im} \otimes v_{nj} = \sum_{n=1}^{r} v_{in} \otimes v_{nj}.$$

Since the matrix elements v_{mn} are linearly independent and $\tau \circ \Delta_{\mathrm{bi}} = \Delta_{\mathrm{bi}}$ by (i), the latter is only possible if $r = 1$.

(ii)→(i): Since the Hopf algebra $\mathcal{O}_G$ is cosemisimple, $\mathcal{O}_K\backslash\mathcal{O}_G/\mathcal{O}_K$ is the sum of subspaces $\mathcal{O}_K\backslash\mathcal{O}_G/\mathcal{O}_K \bigcap \mathcal{C}(\varphi)$ for all irreducible corepresentations φ of $\mathcal{O}_G$. By (ii) and Lemma 59, $\mathcal{O}_K\backslash\mathcal{O}_G/\mathcal{O}_K \bigcap \mathcal{C}(\varphi) = \mathbb{C} \cdot v_{11}$ and hence $\Delta_{\mathrm{bi}}(v_{11}) = \sum_{m,n} v_{1m}h_K(\pi_K(v_{mn})) \otimes v_{n1} = v_{11} \otimes v_{11}$. Hence $\tau \circ \Delta_{\mathrm{bi}} = \Delta_{\mathrm{bi}}$. □

Definition 17. *The pair (G_q, K_q) is called a* quantum Gel'fand pair *if the equivalent conditions in Proposition 58 are satisfied. If in addition the algebra $\mathcal{O}(K_q)\backslash\mathcal{O}(G_q)/\mathcal{O}(K_q)$ is commutative, then the quantum Gel'fand pair is called* strict.

A condition for (G_q, K_q) being a quantum Gel'fand pair is contained in

Proposition 60. *Suppose that there exist a bijective coalgebra anti-homomorphism $\delta_G : \mathcal{O}(G_q) \to \mathcal{O}(G_q)$ and an injective coalgebra anti-homomorphism $\delta_K : \mathcal{O}(K_q) \to \mathcal{O}(K_q)$ such that $\pi_K \circ \delta_G = \delta_K \circ \pi_K$ and $\delta_G(a) = a$ for all $a \in \mathcal{O}_K\backslash\mathcal{O}_G/\mathcal{O}_K$. Then (G_q, K_q) is a quantum Gel'fand pair.*

Proof. The proof is given in [Fl]. □

Example 13 ($G_q = SL_q(2)$). Let ϑ be the algebra automorphism of $\mathcal{O}(SL_q(2))$ determined by $\theta(a) = a$, $\theta(b) = c$, $\theta(c) = b$, $\theta(d) = d$ (see Proposition 4.5(i)). Then the mappings $\delta_G := \theta$ and $\delta_K := \mathrm{id}$ have the properties required in Proposition 60, where K_q denotes the diagonal subgroup of G_q. If q is not a root of unity, then the Hopf algebra $\mathcal{O}(SL_q(2))$ is cosemisimple by Theorem 4.17. Therefore, since $\mathcal{O}_K \backslash \mathcal{O}_G / \mathcal{O}_K = \mathbb{C}[bc]$ (see Example 12) is commutative, (G_q, K_q) is a strict quantum Gel'fand pair if q is not a root of unity. $\triangle$

11.6.4 The Quantum Homogeneous Space $U_q(N-1)\backslash U_q(N)$

Let us briefly recall the corresponding classical situation. It is well-known that the unitary group $U(N)$ acts transitively on the unit sphere $S_{\mathbb{C}}^{N-1}$ of the complex vector space $\mathbb{C}^N$ and that the stabilizer of the point $\mathbf{e}_N = (0, \cdots, 0, 1) \in S_{\mathbb{C}}^{N-1}$ in $U(N)$ is just the subgroup $U(N-1)$. Hence the right coset space $U(N-1)\backslash U(N)$ can be identified with the unit sphere $S_{\mathbb{C}}^{N-1}$.

In the following we study a quantum analog of the homogeneous space $U(N-1)\backslash U(N)$. In the course of this we use the representation theory of the quantum group $GL_q(N)$ developed in Sect. 11.5. Throughout this subsection we suppose that q is real and $q \neq 0, \pm 1$. Whenever we speak about highest weights for representations of $U_q(\mathrm{gl}_n)$, we refer to the simple roots $-\alpha_1, \cdots, -\alpha_n$ (see also Remark 6.7 and Proposition 50).

Let us abbreviate $G_q := U_q(N)$ and $K_q := U_q(N-1)$. We shall denote the generators of $\mathcal{O}(K_q)$ by y^i_j, $1 \leq i, j \leq N-1$, and its quantum determinant by $\mathcal{D}'_q$. There exists a unique surjective Hopf algebra homomorphism $\pi_K : \mathcal{O}(G_q) \to \mathcal{O}(K_q)$ such that

$$\pi_K(u^i_j) = y^i_j,\ i, j \leq N-1, \quad \pi_K(u^k_N) = \pi_K(u^N_k) = \delta_{kN}1,\ k = 1, 2, \cdots, N.$$

(Indeed, it suffices to check that the corresponding elements satisfy the defining relations of $U_q(N)$.) Thus, K_q is a quantum subgroup of G_q.

From Sect. 11.5 we recall that for any $\lambda \in P_+$ there exists a unique irreducible corepresentation T_λ of $\mathcal{O}(G_q)$. Let t^λ_{ij} be the matrix elements of T_λ with respect to some basis of the carrier space. We fix an index i and set $V_R(\lambda) = \mathrm{Lin}\,\{t^\lambda_{ij} \mid j = 1, 2, \cdots, d_\lambda\}$ and $V_L(\lambda) = \mathrm{Lin}\,\{t^\lambda_{ji} \mid j = 1, 2, \cdots, d_\lambda\}$, $d_\lambda = \dim T_\lambda$. Then $V_R(\lambda)$ and $V_L(\lambda)$ are right and left $\mathcal{O}(K_q)$-comodules with coactions $R^\lambda_{K_q} = (\mathrm{id} \otimes \pi_K) \circ \Delta$ and $L^\lambda_{K_q} = (\pi_K \otimes \mathrm{id}) \circ \Delta$, respectively.

Proposition 61. *The right $\mathcal{O}(K_q)$-comodule $V_R(\lambda)$ and the left $\mathcal{O}(K_q)$-comodule $V_L(\lambda)$ decompose into a direct sum of irreducible subcomodules as*

$$V_R(\lambda) = \bigoplus\nolimits_\mu V_R(\lambda, \mu), \quad V_L(\lambda) = \bigoplus\nolimits_\mu V_L(\lambda, \mu), \tag{47}$$

where in both cases the summations are over all dominant integral weights $\mu = \mu_1 \epsilon_1 + \cdots + \mu_{N-1} \epsilon_{N-1}$ such that

$$\lambda_1 \geq \mu_1 \geq \lambda_2 \geq \mu_2 \geq \cdots \geq \lambda_{N-1} \geq \mu_{N-1} \geq \lambda_N. \tag{48}$$

$V_R(\lambda,\mu)$ and $V_L(\lambda,\mu)$ are right and left irreducible $\mathcal{O}(K_q)$-subcomodules of $V_R(\lambda)$ and $V_L(\lambda)$, respectively, with highest weights μ.

Outline of Proof. Clearly, $\pi_K(\mathcal{D}_q) = \mathcal{D}'_q$ and the one-dimensional $\mathcal{O}(K_q)$-comodule $\mathbb{C}\cdot\mathcal{D}_q^l$ is equivalent to $\mathbb{C}\cdot(\mathcal{D}'_q)^l$, $l\in\mathbb{N}_0$. Therefore, by Proposition 48(i), we may assume that $\lambda_N \ge 0$. For each dominant integral weight μ satisfying condition (48) we introduce the standard monomials

$$v_\mu = \mathcal{D}^{\lambda_N}_{\{1,\cdots,N\}}\mathcal{D}^{b_{N-1}}_{\{1,\cdots,N-1\}}\mathcal{D}^{a_{N-1}}_{\{1,\cdots,N-2,N\}}\cdots\mathcal{D}^{b_2}_{\{1,2\}}\mathcal{D}^{a_2}_{\{1,N\}}\mathcal{D}^{b_1}_{\{1\}}\mathcal{D}^{a_1}_{\{N\}}, \tag{49}$$

where $a_k = \lambda_k - \mu_k$ and $b_k = \mu_k - \lambda_{k+1}$, $k = 1,2,\cdots,N-1$. One verifies that v_μ is a highest weight vector of weight $\mu + (\sum_{k=1}^{N-1} a_k)\epsilon_N$ for the corresponding representation of $U_q(\mathrm{gl}_{N-1})$ on $V_R(\lambda)$. Thus, there exists a unique irreducible $\mathcal{O}(K_q)$-subcomodule $V_R(\lambda,\mu)$ in $V_R(\lambda)$ containing v_μ as a highest weight vector of weight μ. This proves that $\bigoplus_\mu V_R(\lambda,\mu) \subseteq V_R(\lambda)$. Since both sides have the same dimensions, the first equality in (47) follows. The second equality is proved similarly. □

We shall say that the right (resp. left) $\mathcal{O}(G_q)$-comodule $V_R(\lambda)$ (resp. $V_L(\lambda)$) is of *class 1 with respect to the quantum subgroup K_q* if it contains a nonzero K_q-fixed vector, that is, $R^\lambda_{K_q}(v) = v\otimes 1$ (resp. $L^\lambda_{K_q}(v) = 1\otimes v$).

Corollary 62. *The left $\mathcal{O}(G_q)$-comodule $V_L(\lambda)$ is of class 1 with respect to K_q if and only if $\lambda = l\epsilon_1 - l'\epsilon_N$, $l,l'\in\mathbb{N}_0$. In this case, the standard monomial*

$$v_0 = (\mathcal{D}_q)^{-l'}(\mathcal{D}_{\{1,\cdots,N-1\}})^{l'}(u^N_1)^l \equiv S(u^N_N)^{l'}(u^N_1)^l$$

is up to a complex multiple the unique K_q-fixed vector in $V_L(\lambda)$.

Proof. The proof follows from Proposition 61 and formula (49). □

Proposition 63. *The algebra $\mathcal{O}(K_q\backslash G_q)$ of left K_q-invariant elements of $\mathcal{O}(G_q)$ decomposes into a direct sum of irreducible right $\mathcal{O}(G_q)$-comodules as*

$$\mathcal{O}(K_q\backslash G_q) = \bigoplus\nolimits_{l,l'=0}^{\infty} C_1(T_{l\epsilon_1-l'\epsilon_N}).$$

Here $C_1(T_{l\epsilon_1-l'\epsilon_N})$ is the irreducible right $\mathcal{O}(G_q)$-subcomodule of the coefficient coalgebra $C(T_{l\epsilon_1-l'\epsilon_N})$ which contains $S(u^N_N)^{l'}(u^N_1)^l$ as its highest weight vector of weight $l\epsilon_1 - l'\epsilon_N$.

Proof. Since the Hopf algebra $\mathcal{O}(G_q)$ is cosemisimple by Corollary 53, the right $\mathcal{O}(G_q)$-comodule $\mathcal{O}(K_q\backslash G_q)$ decomposes into a direct sum of irreducible components. In order to determine this decomposition explicitly, we first note that $\mathcal{O}(K_q\backslash G_q) \subseteq \bigoplus_{l,l'=0}^{\infty} C(T_{l\epsilon_1-l'\epsilon_N})$ by Corollary 62. For $\lambda = l\epsilon_1 - l'\epsilon_N$ we choose a basis $e_1,\cdots,e_k$ of the carrier space of T_λ such that the vector e_1 is $U_q(N-1)$-invariant and the space $\mathrm{Lin}\,\{e_2,\cdots,e_k\}$ is $U_q(N-1)$-invariant. Then one easily verifies that a matrix coefficient t^λ_{ij} of T_λ is left $U_q(N-1)$-invariant if and only if $i = 1$. Moreover, each left $U_q(N-1)$-invariant element in $C(T_\lambda)$ is a linear combination of t^λ_{1j}, $j = 1,2,\cdots,k$. Let $C_1(T_\lambda)$,

$\lambda = l\epsilon_1 - l'\epsilon_N$, be the set of left $U_q(N-1)$-invariant elements in $\mathcal{C}(T_\lambda)$. Then $\mathcal{O}(K_q\backslash G_q) = \bigoplus_{l,l'=0}^{\infty} \mathcal{C}_1(T_{l\epsilon_1 - l'\epsilon_N})$. The comultiplication Δ realizes the irreducible corepresentation $T_{l\epsilon_1-l'\epsilon_N}$ of $\mathcal{O}(G_q)$ on $\mathcal{C}_1(T_{l\epsilon_1-l'\epsilon_N})$. By Corollary 62, $S(u_N^N)^{l'}(u_1^N)^l$ is the unique highest weight vector in $\mathcal{C}_1(T_{l\epsilon_1-l'\epsilon_N})$. □

All results developed so far are still valid for the quantum groups $G_q = GL_q(N)$ and $K_q = GL_q(N-1)$ if q is not a root of unity. From now on we shall use the $*$-structures of $\mathcal{O}(U_q(N))$ and $\mathcal{O}(U_q(N-1))$ too.

Since $\pi_K : \mathcal{O}(G_q) \to \mathcal{O}(K_q)$ is a $*$-algebra homomorphism, $\mathcal{O}(K_q\backslash G_q)$ is a $*$-subalgebra of $\mathcal{O}(G_q)$ and a right $*$-quantum space for the Hopf $*$-algebra $\mathcal{O}(U_q(N))$. By construction, $\mathcal{O}(K_q\backslash G_q)$ is a right quantum homogeneous $U_q(N)$-space.

Definition 18. *The $*$-algebra $\mathcal{O}(K_q\backslash G_q)$ is called the* coordinate algebra of the quantum sphere *related to $U_q(N)$.*

Our next aim is to describe the $*$-algebra $\mathcal{O}(K_q\backslash G_q)$ in terms of generators and relations. Let us abbreviate $z_i = u_i^N$, $i = 1, 2, \cdots, N$. Since $L_{K_q}(z_i) = \pi_K(u_j^N) \otimes u_i^j = 1 \otimes z_i$, the elements z_i and $z_i^* = (u_i^N)^* = S(u_N^i)$ belong to $\mathcal{O}(K_q\backslash G_q)$. The right coaction φ_R of $\mathcal{O}(G_q)$ on $z_i \in \mathcal{O}(K_q\backslash G_q)$ is given by the comultiplication, that is, we have

$$\varphi_R(z_j) = z_i \otimes u_j^i,\ j = 1, 2, \cdots, N.$$

Proposition 64. (i) *$\mathcal{O}(K_q\backslash G_q)$ is the $*$-subalgebra of $\mathcal{O}(G_q)$ generated by the elements z_i, $i = 1, 2, \cdots, N$.*

(ii) *The generators $z_1, z_2, \cdots, z_N$ satisfy the relations*

$$z_i z_j = q z_j z_i,\quad i < j; \qquad z_j^* z_i = q z_i z_j^*,\ i \neq j; \tag{50}$$

$$z_k^* z_k = z_k z_k^* + (1-q^2) \sum\nolimits_{j<k} z_j z_j^*,\quad k = 1, \cdots, N; \quad \sum\nolimits_{k=1}^{N} z_k z_k^* = 1. \tag{51}$$

As a $*$-algebra, $\mathcal{O}(K_q\backslash G_q)$ is generated by N elements $z_1, z_2, \cdots, z_N$ with defining relations (50) and (51).

(iii) *The set $\{z_1^{s_1} \cdots z_N^{s_N} (z_1^*)^{m_1} \cdots (z_N^*)^{m_N} \mid s_i, m_i \in \mathbb{N}_0\}$ and the set $\{(z_1^*)^{m_1} \cdots (z_N^*)^{m_N} z_1^{s_1} \cdots z_N^{s_N} \mid s_i, m_i \in \mathbb{N}_0\}$ are bases of the vector space $\mathcal{O}(K_q\backslash G_q)$.*

Proof. (i): The elements $(z_N^*)^{l'} z_1^l = S(u_N^N)^{l'}(u_1^N)^l$ are highest weight vectors for the $\mathcal{O}(G_q)$-comodules $\mathcal{C}_1(T_{l\epsilon_1-l'\epsilon_N})$. Hence these comodules and so $\mathcal{O}(K_q\backslash G_q)$ by Proposition 63 are contained in the $*$-subalgebra generated by z_i, $i = 1, 2, \cdots, N$.

(ii): The relations (50) and (51) follow easily from the commutation relations for the elements u_j^i in the $*$-algebra $\mathcal{O}(U_q(N))$. That the relations (50) and (51) are already the defining relations of the $*$-algebra $\mathcal{O}(K_q\backslash G_q)$ can be shown by using appropriate $*$-representations of $\mathcal{O}(U_q(N))$ (see [VS2]).

(iii) can be proved by means of the diamond Lemma 4.8. □

Note that because of (50) the relations (51) can be rewritten as

$$z_k z_k^* = z_k^* z_k - (1-q^2)\sum\nolimits_{j<k} q^{2(k-j-1)} z_j^* z_j, \quad \sum\nolimits_{k=1}^{N} q^{2(N-k)} z_k^* z_k = 1.$$

The restriction of the Haar functional of $\mathcal{O}(G_q)$ to the subalgebra $\mathcal{O}(K_q\backslash G_q)$ determines a right-invariant linear functional h on $\mathcal{O}(K_q\backslash G_q)$ such that $h(1) = 1$. From Proposition 63 it follows immediately that h is the only functional with this property.

We want to describe the functional h more explicitly. First we note that the diagonal subgroup D_N of $GL_q(N)$ becomes a quantum subgroup of $U_q(N)$ if we introduce a $*$-structure on $\mathcal{O}(D_N)$ by $(t_i)^* := t_i^{-1}$. We have

$$\mathcal{O}(K_q\backslash G_q) = \bigoplus\nolimits_{\lambda\in P} \mathcal{O}(K_q\backslash G_q)_\lambda, \tag{52}$$

where $\mathcal{O}(K_q\backslash G_q)_\lambda = \{a \in \mathcal{O}(K_q\backslash G_q) \mid R_{D_N}(a) = a \otimes t^\lambda\}$ are the weight subspaces. Clearly, $\mathcal{O}(K_q\backslash G_q)_{\lambda=0}$ is just the algebra $\mathcal{O}(K_q\backslash G_q/D_N)$ of D_N-invariant elements in $\mathcal{O}(K_q\backslash G_q)$. The functional h on $\mathcal{O}(K_q\backslash G_q)$ vanishes on all subspaces $\mathcal{O}(K_q\backslash G_q)_\lambda$ from (52) with $\lambda \neq 0$. Thus, it suffices to determine the functional h on the subalgebra $\mathcal{O}(K_q\backslash G_q/D_N)$. It is easily seen that the subalgebra $\mathcal{O}(K_q\backslash G_q/D_N)$ is spanned by the monomials $(z_1^*)^{m_1}\cdots(z_N^*)^{m_N} z_N^{m_N}\cdots z_1^{m_1}$, $m_k \in \mathbb{N}_0$.

Proposition 65. (i) *The algebra $\mathcal{O}(K_q\backslash G_q/D_N)$ is commutative and generated by the elements $\zeta_k = \sum_{i=1}^k z_i z_i^*$, $k = 1, 2, \cdots, N-1$.*

(ii) *On $\mathcal{O}(K_q\backslash G_q/D_N)$, the functional h is given by*

$$h((z_1^*)^{m_1}\cdots(z_N^*)^{m_N} z_N^{m_N}\cdots z_1^{m_1}) = \frac{(q^2;q^2)_{m_1}\cdots(q^2;q^2)_{m_N}(q^2;q^2)_{N-1}}{(q^2;q^2)_m},$$

where $m := m_1 + \cdots + m_N + N - 1$.

(iii) *For any polynomial $p(\zeta_1, \cdots, \zeta_{N-1}) \in \mathcal{O}(K_q\backslash G_q/D_N)$ the value $h(p)$ can be expressed in terms of q-integrals as*

$$h(p) = \frac{(q^2;q^2)_{N-1}}{(1-q^2)^{N-1}} \int_0^1 \int_0^{\zeta_{N-1}} \cdots \int_0^{\zeta_2} p(\zeta_1, \cdots, \zeta_{N-1}) d_{q^2}\zeta_1 \cdots d_{q^2}\zeta_{N-1}.$$

Proof. The proof is given in [NYM]. □

From Propositions 58(ii) and 63 one can derive

Corollary 66. *(G_q, K_q) is a quantum Gel'fand pair and we have*

$$\mathcal{O}(K_q\backslash G_q/K_q) = \bigoplus\nolimits_{l,l'=0}^{\infty} \mathbb{C}\cdot t_{11}^{l\epsilon_1 - l'\epsilon_N}.$$

The matrix coefficients $\phi^{ll'} := t_{11}^{l\epsilon_1 - l'\epsilon_N}$ are called *quantum zonal spherical functions* on $K_q\backslash G_q$. In the next proposition they are explictly described in terms of the little q-Jacobi polynomials (see Subsect. 2.3.3).

Proposition 67. *The quantum zonal spherical functions $\phi^{ll'}$ are given by*

$$\phi^{ll'} = z_N^{l-l'} p_{l'}(\zeta_{N-1}; q^{N-2}, q^{l-l'} \,|\, q^2) \qquad if \qquad l \geq l',$$

$$\phi^{ll'} = p_l(\zeta_{N-1}; q^{N-2}, q^{l'-l} \,|\, q^2) {z^*}_N^{l'-l} \qquad if \qquad l' \geq l,$$

where $\zeta_{N-1} = \sum_{i=1}^{N-1} z_i z_i^* = 1 - z_N z_N^*$ and p_l are the little q-Jacobi polynomials.

Proof. The proof is given in [NYM]. □

Proposition 67 is a quantum analog of classical results on zonal spherical functions for the unitary group $U(N)$ (see [VK1], Chap. 11).

11.6.5 Quantum Homogeneous Spaces of Infinitesimally Invariant Elements

The quantum homogeneous spaces treated in the preceding subsections were constructed by means of quantum subgroups. However, this method does not yield quantum deformations of all compact Riemannian symmetric spaces, since quantum groups do not have as many quantum subgroups as the corresponding Lie groups have Lie subgroups (see [H4]). In particular, quantum analogs of the classical symmetric spaces $SU(N)/SO(N)$ and $SU(N)/USp(N)$ cannot be obtained by means of quantum subgroups. Moreover, as we know from the study of quantum 2-spheres (see Example 12 above and Sect. 4.5), even if there exists a corresponding quantum subgroup, we do not get all interesting deformations of the classical space in this manner.

In this brief subsection we sketch the "infinitesimal" method for the construction of quantum homogeneous spaces. Let us first explain the underlying idea in the classical situation. Functions on a homogeneous space G/K are usually considered as functions on the Lie group G which are right-invariant with respect to shifts by elements from K. But they can also be defined as those functions on G which are annihilated by the right action of elements from $U(\mathfrak{k})$, where $\mathfrak{k}$ is the Lie algebra of K.

We assume that $\mathcal{U}$ and $\mathcal{A}$ are Hopf algebras equipped with a dual pairing $\langle\cdot,\cdot\rangle : \mathcal{U} \times \mathcal{A} \to \mathbb{C}$. By Proposition 1.16, the algebra $\mathcal{A}$ becomes then a $\mathcal{U}$-bimodule with left and right actions of $\mathcal{U}$ on $\mathcal{A}$ given by $X.a = (\mathrm{id} \otimes X)\Delta(a)$ and $a.X = (X \otimes \mathrm{id})\Delta(a)$, $X \in \mathcal{U}$, $a \in \mathcal{A}$. Recall also that a coideal of $\mathcal{U}$ is a a linear subspace $\mathcal{I}$ of $\mathcal{U}$ such that $\Delta(\mathcal{I}) \subseteq \mathcal{I} \otimes \mathcal{U} + \mathcal{U} \otimes \mathcal{I}$ and $\varepsilon(\mathcal{I}) = \{0\}$.

Definition 19. *Let $\mathcal{I}$ be a coideal of $\mathcal{U}$. The elements of the sets*

$${}_{\mathcal{I}}\mathcal{A} := \{a \in \mathcal{A} \mid X.a = 0,\ X \in \mathcal{I}\}, \quad \mathcal{A}_{\mathcal{I}} := \{a \in \mathcal{A} \mid a.X = 0,\ X \in \mathcal{I}\},$$

$${}_{\mathcal{I}}\mathcal{A}_{\mathcal{I}} := {}_{\mathcal{I}}\mathcal{A} \cap \mathcal{A}_{\mathcal{I}}$$

are called infinitesimally left-invariant, right-invariant *and* biinvariant *with respect to $\mathcal{I}$, respectively.*

Proposition 68. *Let $\mathcal{I}$ be a coideal of $\mathcal{U}$. Then the sets ${}_{\mathcal{I}}\mathcal{A}$, $\mathcal{A}_{\mathcal{I}}$ and ${}_{\mathcal{I}}\mathcal{A}_{\mathcal{I}}$ are subalgebras of $\mathcal{A}$ and we have*

$$\Delta({}_{\mathcal{I}}\mathcal{A}) \subseteq \mathcal{A} \otimes {}_{\mathcal{I}}\mathcal{A}, \quad \Delta(\mathcal{A}_{\mathcal{I}}) \subseteq \mathcal{A}_{\mathcal{I}} \otimes \mathcal{A}.$$

Moreover, $_{\mathcal{U}}.{}_{\mathcal{I}}\mathcal{A} = {}_{\mathcal{I}}\mathcal{A}$ *and* $\mathcal{A}_{\mathcal{I}}.{}_{\mathcal{U}} = \mathcal{A}_{\mathcal{I}}$.

Proof. The fact that $\mathcal{A}_{\mathcal{I}}$ is a subalgebra follows immediately from the relation $ab.X = \sum(a.X_{(1)})(b.X_{(2)})$, $a, b \in \mathcal{A}$, $X \in \mathcal{I}$, combined with the assumption $\Delta(\mathcal{I}) = \mathcal{I} \otimes \mathcal{U} + \mathcal{U} \otimes \mathcal{I}$. For $a \in \mathcal{A}_{\mathcal{I}}$ and $X \in \mathcal{I}$ we have $0 = \Delta(a.X) = \sum a_{(1)}.X \otimes a_{(2)}$. Hence $\Delta(a) \in \mathcal{A}_{\mathcal{I}} \otimes \mathcal{A}$. The equation $a.XY = (a.X).Y$ implies the equality $\mathcal{A}_{\mathcal{I}}.{}_{\mathcal{U}} = \mathcal{A}_{\mathcal{I}}$. The assertions for ${}_{\mathcal{I}}\mathcal{A}$ and ${}_{\mathcal{I}}\mathcal{A}_{\mathcal{I}}$ are derived similarly. □

If $\mathcal{A}$ is a coordinate Hopf algebra $\mathcal{O}(G_q)$, then by Proposition 68 the algebras ${}_{\mathcal{I}}\mathcal{A}$ and $\mathcal{A}_{\mathcal{I}}$ are left resp. right quantum homogeneous G_q-spaces.

Example 14 ($G_q = SL_q(2)$). Let $\mathcal{A} = \mathcal{O}(SL_q(2))$ and $\mathcal{U} = \breve{U}_q(\mathrm{sl}_2)$. Let X_ρ, $\rho \in \mathbb{C}\bigcup\{\infty\}$, be the element of $\breve{U}_q(\mathrm{sl}_2)$ defined by (4.84) resp. (4.85). Since $\Delta(X_\rho) = K^{-1} \otimes X_\rho + X_\rho \otimes K$, the one-dimensional space $\mathcal{I}_\rho := \mathbb{C}{\cdot}X_\rho$ is a coideal of $\mathcal{U}$. If q is not a root of unity, then Proposition 4.31 says that the right $\mathcal{A}$-quantum space $\mathcal{A}_{\mathcal{I}_\rho}$ is isomorphic to the coordinate algebra $\mathcal{O}(S^2_{q\rho})$. Let us emphasize once more (see Example 12) that only the quantum 2-sphere $S^2_{q\infty}$ can be obtained by means of a quantum subgroup of $SL_q(2)$. △

11.6.6 Quantum Projective Spaces

In this subsection we outline how a family of quantum homogeneous $U_q(N)$-spaces can be constructed by using the coideal approach. These spaces are quantum analogs of the complex projective space $(U(N-1) \times U(1))\backslash U(N)$. Throughout this subsection q is a real number such that $q \neq 0, \pm 1$.

Let $\mathcal{A}$ be the Hopf $*$-algebra $\mathcal{O}(U_q(N))$ and let $\mathcal{U}$ be the Hopf algebra $\mathcal{U}(GL_q(N))$ of L-functionals on the coquasitriangular Hopf algebra $\mathcal{A}$ (see Subsect. 10.1.3 for details). Since q is real, $\mathcal{U}(GL_q(N))$ becomes a Hopf $*$-algebra, denoted $\mathcal{U}(U_q(N))$, with involution determined by $(l^{\pm i}_{j})^* = S(l^{\mp j}_{i})$ (see (10.48)). By construction, $\mathcal{U}$ is a Hopf subalgebra of the dual Hopf algebra $\mathcal{A}^\circ$. Hence the evaluation of functionals $f \in \mathcal{U}$ at elements $a \in \mathcal{A}$ gives a dual pairing of Hopf $*$-algebras $\mathcal{U} = \mathcal{U}(U_q(N))$ and $\mathcal{A} = \mathcal{O}(U_q(N))$.

Suppose that r and s are nonnegative real numbers such that $(r, s) \neq (0, 0)$. Let $\mathfrak{k}^{(r,s)}$ denote the subalgebra of $\mathcal{U}$ generated by the elements

$$l^{+1}_{1} - l^{-N}_{N},\ l^{-1}_{1} - l^{+N}_{N},\ \sqrt{rs}\, l^{+1}_{N} + \sqrt{rs}\, l^{-N}_{1} - (r-s)(l^{+1}_{1} - l^{-1}_{1}), \tag{53}$$

$$l^{+k}_{k} - l^{-k}_{k},\ \sqrt{r}\, l^{+1}_{k} + \sqrt{s}\, l^{-N}_{k},\ \sqrt{r}\, l^{+k}_{N} + \sqrt{s}\, l^{-k}_{1},\quad 2 \leq k \leq N-1, \tag{54}$$

$$l^{+i}_{j},\ l^{-j}_{i},\quad 2 \leq i < j \leq N-1. \tag{55}$$

Lemma 69. $\mathfrak{k}^{(r,s)}$ *is a coideal of* $\mathcal{U}$ *such that* $S(f)^* \in \mathfrak{k}^{(r,s)}$ *for* $f \in \mathfrak{k}^{(r,s)}$.

Proof. The assertion is easily obtained from (10.40) and (10.47). □

Note that the coideal $\mathfrak{k}^{(r,s)}$ depends only on the ratio s/r. Let us write $\sqrt{s/r} = q^\sigma$, $\sigma \in \mathbb{R}$, if $r > 0$ and $s > 0$. We also set $\sigma = \infty$ if $s = 0$ and $\sigma = -\infty$ if $r = 0$. The coideal $\mathfrak{k}^{(r,s)}$ is also denoted by $\mathfrak{k}^\sigma$. For arbitrary σ, $-\infty \le \sigma \le +\infty$, we define

$$\mathcal{B}_q^\sigma \equiv \mathcal{B}_q^{r,s} := \{a \in \mathcal{O}(U_q(N)) \mid a.X = 0 \ \text{ for all } \ X \in \mathfrak{k}^\sigma\}.$$

By property (1.41) of the dual pairing of the Hopf $*$-algebras $\mathcal{U}$ and $\mathcal{A}$ we have $a^*.f = (a.S(f)^*)^*$, $a \in \mathcal{A}, f \in \mathcal{U}$. Therefore, since $\mathfrak{k}^\sigma$ is invariant under the mapping $* \circ S$ by Lemma 59, the algebra $\mathcal{B}_q^\sigma$ is a $*$-subalgebra of $\mathcal{O}(U_q(N))$. From Proposition 68 it follows that $\mathcal{B}_q^\sigma \equiv \mathcal{A}_{\mathfrak{k}^\sigma}$ is a right quantum homogeneous $U_q(N)$-space. The harmonic analysis on the quantum homogeneous space $\mathcal{B}_q^\sigma$ has been studied extensively in the paper [DN], where the proofs of the following two propositions may be found.

Proposition 70. (i) *The elements*

$$x_{ij} := s(u_i^1)^* u_j^1 + r(u_i^N)^* u_j^N + \sqrt{rs}\,(u_i^N)^* u_j^1 + \sqrt{rs}\,(u_i^1)^* u_j^N$$

of $\mathcal{O}(U_q(N))$ belong to $\mathcal{B}_q^{r,s}$. They generate the algebra $\mathcal{B}_q^{r,s}$ and satisfy the relation $(x_{ij})^* = x_{ij}$.

(ii) *The right $\mathcal{O}(U_q(N))$-comodule $\mathcal{B}_q^\sigma$ decomposes as a direct sum*

$$\mathcal{B}_q^\sigma = \bigoplus\nolimits_{l=0}^{\infty} V(l\epsilon_1 - l\epsilon_N),$$

where $V(l\epsilon_1 - l\epsilon_N)$ is the irreducible right $\mathcal{O}(U_q(N))$-subcomodule generated by the highest weight vector x_{N1}^l.

The irreducible decomposition of the comodule $\mathcal{B}_q^\sigma$ is analogous to the irreducible decomposition of the space of functions on the complex projective space $U(N)/(U(N-1) \times U(1))$ (see [VK1], Chap. 11). For this reason we call $\mathcal{B}_q^\sigma$ the coordinate algebra of the *quantum projective space* $\mathbb{CP}_{q,\sigma}^{N-1}$.

We now consider biinvariant elements of $\mathcal{O}(U_q(N))$ with respect to the coideal $\mathfrak{k}^\sigma$ and define

$$\mathcal{H}_q^\sigma := \{a \in \mathcal{O}(U_q(N)) \mid X.a = a.X = 0 \ \text{ for all } \ X \in \mathfrak{k}^\sigma\}.$$

Then, $\mathcal{H}_q^\sigma$ is a $*$-subalgebra of $\mathcal{O}(U_q(N))$. An element of $\mathcal{H}_q^\sigma$ is called a *quantum zonal spherical function* of $\mathcal{B}_q^\sigma$ if it belongs to the coefficient coalgebra $\mathcal{C}(T_\lambda)$ for some $\lambda \in P_+$.

Proposition 71. (i) *If we set $\mathcal{H}_q^\sigma(\lambda) := \mathcal{H}_q^\sigma \bigcap \mathcal{C}(T_\lambda)$, then we have*

$$\mathcal{H}_q^\sigma = \bigoplus\nolimits_{l=0}^{\infty} \mathcal{H}_q^\sigma(l\epsilon_1 - l\epsilon_N).$$

The algebra $\mathcal{H}_q^\sigma$ is commutative and the spaces $\mathcal{H}_q^\sigma(l\epsilon_1 - l\epsilon_N)$ are one-dimensional.

(ii) *If $\sigma \in \mathbb{R}$, then the algebra $\mathcal{H}_q^\sigma$ is generated by the element*

$$x^{(\sigma)} := \tilde{x}_{1N} + \tilde{x}_{N1} + q^{\sigma+1}\tilde{x}_{11} + q^{-\sigma-1}\tilde{x}_{NN} - (q^{2\sigma+1} + q^{-2\sigma-1})1,$$

where $\tilde{x}_{ij} = q^\sigma (u_i^1)^* u_j^1 + q^{-\sigma}(u_i^N)^* u_j^N + (u_i^N)^* u_j^1 + (u_i^1)^* u_j^N \in \mathcal{O}(U_q(N))$.

The assertion of Proposition 71(i) can be expressed by saying that the pair $(\mathcal{U}, \mathfrak{k}^\sigma)$ is an *infinitesimal quantum Gel'fand pair.*

The preceding exposition and the treatment of the quantum 2-spheres in Subsect. 4.5.4 have shown how one-parameter families of quantum symmetric spaces with the same classical counterpart can be obtained by the infinitesimal approach. This method makes it possible to construct quantum analogs for most of the classical compact Riemannian symmetric spaces (see [Dij2] and the references therein). The L-functionals and the reflection equation are essentially used in these constructions.

11.7 Notes

Cosemisimple Hopf algebras occurred in [Sw], Chaps. 11 and 14, see also [La]. The important Theorem 22 on the cosemisimplicity of the coordinate Hopf algebras $\mathcal{O}(G_q)$ is due to T. Hayashi [H3].

The theory of compact quantum (matrix) groups and their representations was developed by S. L. Woronowicz in his pioneering papers [Wor1] and [Wor3]. In Sect. 11.3 we followed the algebraic approach of M. S. Dijkhuizen and T. H. Koornwinder [DK1] thus avoiding the use of C^*-algebras. The discrete quantum groups in [ER] and [VD4] are in fact equivalent to the CQG algebras (see [Ks] for a comparison). The universal CQMG algebras $A_{\rm un}(E)$ are in [VDW]. The polar decomposition of the antipode is in [BS].

Both approaches to compact quantum groups C^*-algebras are due to S. L. Woronowicz [Wor5], [Wor7]. Our proof of the existence of the Haar state follows [VD3]. There exists an extensive literature on compact quantum group algebras and C^*-algebras, see [An], [Le], [LS2], [MN], [Nag], [Rie1], [Rie2], [Soi1], [Wan1], [Wan2], [Y]. Multiplicative unitaries have been studied in [BS], [B], [Wor8].

The representation theory of $GL_q(N)$ is taken from the paper [NYM].

The first well-studied quantum homogeneous spaces were the quantum spheres of P. Podleś [Pod1] and of L. L. Vaksman and Y. S. Soibelman [VS2]. In our treatment of the quantum spheres in Subsect. 11.6.4 and of the quantum projective spaces in Subsect. 11.6.6 we have followed the papers [NYM] and [DN], respectively. Quantum Gel'fand pairs were introduced in [Ko3] and studied in [Vai1], [Vai2], [Fl]. The comultiplication $\Delta_{\rm bi}$ appeared in [Vai2]. As mentioned in the Notes to Chap. 4, the infinitesimal method for the construction of quantum homogeneous spaces was first used by T. H. Koornwinder [Ko4], [DK2] in the case of the quantum 2-spheres. It was considerably generalized in [N] and [DN] (see also [Sag] and [Dij2]).

Part IV

Noncommutative Differential Calculus

12. Covariant Differential Calculus on Quantum Spaces

Nowadays differential forms on manifolds have entered the formulation of a number of physical theories such as Maxwell's theory, mechanics, the theory of relativity and others. There are various physical ideas and considerations (quantum gravity, discrete space-time structures, models of elementary particle physics) that strongly motivate the replacement of the commutative algebra of C^∞-functions on a manifold by an appropriate noncommutative algebra and the study of "noncommutative geometry" there. Differential forms also appear to be a proper framework for doing this. The basic concept in this context is that of a "differential calculus" of an algebra. It allows us to introduce differential geometric notions and carries in this sense the geometry of the "noncommutative space" which may be thought to be behind the algebra.

The three final chapters of this book deal with covariant differential calculi on coordinate algebras of quantum spaces and quantum groups. In the first two sections of this chapter we develop general notions and facts on covariant differential calculi on quantum spaces, while the final section is concerned with the construction of such calculi. In particular, a covariant differential calculus on the quantum vector space $\mathbb{C}_q^N$ is treated in detail.

In Subsects. 12.1.1, 12.2.1 and 12.2.2, $\mathcal{X}$ denotes an arbitrary algebra. In the rest of this chapter we assume that $\mathcal{X}$ is a left quantum space (or equivalently, a left $\mathcal{A}$-comodule algebra) for a Hopf algebra $\mathcal{A}$ with left coaction $\varphi : \mathcal{X} \to \mathcal{A} \otimes \mathcal{X}$. We retain the convention to sum over repeated indices.

12.1 Covariant First Order Differential Calculus

12.1.1 First Order Differential Calculi on Algebras

It is well-known that the exterior derivative d of a C^∞- manifold $\mathcal{M}$ satisfies the Leibniz rule and maps the algebra $C^\infty(\mathcal{M})$ to the 1-forms on $\mathcal{M}$. If $\mathcal{M}$ is compact, any 1-form is a finite sum of 1-forms $f \cdot \mathrm{d}g \cdot h$ with $f, g, h \in C^\infty(\mathcal{M})$. These facts give the motivation for the following general definition.

Definition 1. *A* first order differential calculus *(abbreviated, a* FODC*) over an algebra $\mathcal{X}$ is an $\mathcal{X}$-bimodule Γ with a linear mapping* $\mathrm{d} : \mathcal{X} \to \Gamma$ *such that*

(i) d *satisfies the Leibniz rule* $\mathrm{d}(xy) = x{\cdot}\mathrm{d}y + \mathrm{d}x{\cdot}y$ *for any* $x, y \in \mathcal{X}$,
(ii) Γ *is the linear span of elements* $x{\cdot}\mathrm{d}y{\cdot}z$ *with* $x, y, z \in \mathcal{X}$.

We shall say that two FODC Γ_1 *and* Γ_2 *over* $\mathcal{X}$ *are* isomorphic *if there exists a bijective linear mapping* $\psi : \Gamma_1 \to \Gamma_2$ *such that* $\psi(x{\cdot}\mathrm{d}_1y{\cdot}z) = x{\cdot}\mathrm{d}_2y{\cdot}z$ *for* $x, y, z \in \mathcal{X}$.

We shall use self-explanatory notation such as $\mathcal{X}{\cdot}\mathrm{d}\mathcal{X} := \mathrm{Lin}\{x{\cdot}\mathrm{d}y \mid x, y \in \mathcal{X}\}$, $\mathcal{X}{\cdot}\mathrm{d}\mathcal{X}{\cdot}\mathcal{X} := \mathrm{Lin}\{x{\cdot}\mathrm{d}y{\cdot}z \mid x, y, z \in \mathcal{X}\}$, etc. If Γ is a FODC, then by (i) we have $x{\cdot}\mathrm{d}y{\cdot}z = x{\cdot}\mathrm{d}(yz) - xy{\cdot}\mathrm{d}z$ for $x, y, z \in \mathcal{X}$ and hence $\Gamma = \mathcal{X}{\cdot}\mathrm{d}\mathcal{X} = \mathrm{d}\mathcal{X}{\cdot}\mathcal{X}$. Sometimes we omit the dot and write $x\mathrm{d}y$ for $x{\cdot}\mathrm{d}y$.

Condition (i) means that d is a derivation of the algebra $\mathcal{X}$ with values in the bimodule Γ. Conversely, if d is a derivation of $\mathcal{X}$ with values in an $\mathcal{X}$-bimodule Γ_0, then $\Gamma := \mathcal{X}{\cdot}\mathrm{d}\mathcal{X}{\cdot}\mathcal{X}$ is obviously a FODC over $\mathcal{X}$.

Let us emphasize that in contrast to the case of classical differential forms for a general FODC the 1-forms $\mathrm{d}x{\cdot}y$ and $y{\cdot}\mathrm{d}x$ are not necessarily equal, even when the underlying algebra $\mathcal{X}$ is commutative.

As noted above, the 1-forms on a compact C^∞-manifold $\mathcal{M}$ constitute a FODC over the algebra $C^\infty(\mathcal{M})$. Having this classical picture in mind, we consider the bimodule Γ in Definition 1 as a variant of the space of 1-forms over the algebra $\mathcal{X}$. We refer to elements of Γ simply as *1-forms* and to the mapping d as the *differentiation.*

We illustrate the generality of the notion of a FODC by a simple but instructive example.

Example 1 ($\mathcal{X} = \mathbb{C}[x]$). Let $\mathcal{X}$ be the algebra $\mathbb{C}[x]$ of all polynomials in one variable x. Fix a polynomial $p \in \mathcal{X}$. Let Γ be the free right $\mathcal{X}$-module with a single basis element denoted $\mathrm{d}x$. That is, the elements of Γ are expressions $\mathrm{d}x{\cdot}f$ with $f \in \mathcal{X}$ and the right action of $\mathcal{X}$ is given by the multiplication of $\mathcal{X}$. There is a unique $\mathcal{X}$-bimodule structure on Γ such that $x\mathrm{d}x = \mathrm{d}x{\cdot}p$. We denote this $\mathcal{X}$-bimodule by Γ_p. It is easily seen that Γ_p becomes a FODC over $\mathcal{X}$ with differentiation d defined by

$$\mathrm{d}\Big(\sum_n \alpha_n x^n\Big) = \sum_n \sum_{i+j=n-1} \alpha_n x^i{\cdot}\mathrm{d}x{\cdot}x^j.$$

Two such FODC Γ_p and $\Gamma_{\tilde{p}}$ are isomorphic only if $p = \tilde{p}$. Since $\mathrm{d}x$ is a right $\mathcal{X}$-module basis of Γ, for any $f \in \mathcal{X}$ there is a unique element $\partial(f) \in \mathcal{X}$ such that $\mathrm{d}f = \mathrm{d}x{\cdot}\partial(f)$. Because $f(x)\mathrm{d}x = \mathrm{d}x{\cdot}f(p(x))$ by definition, we obtain

$$\partial(x^n) = \sum_{i+j=n-1} p(x)^i x^j.$$

Let us consider the special case $p(x) = qx$, where $q \in \mathbb{C}$, $q \neq 1$. Then the $\mathcal{X}$-bimodule Γ_p is characterized by the equation $f(x)\mathrm{d}x = \mathrm{d}xf(qx)$ and $\partial(f)$ is just the q-derivative $D_q(f) = (f(qx) - f(x))/(qx - x)$ (see Subsect. 2.2.1). Another interesting situation is obtained when $p(x) = x + c$, where $c \in \mathbb{C}[x]$,

$c \neq 0$. Then we have $f(x)\mathrm{d}x = \mathrm{d}x \cdot f(x+c)$ and $\partial(f) = \frac{1}{c}(f(x+c) - f(x))$. Note that in both cases $\partial(f)$ becomes the ordinary derivative of f when $q \to 1$ and $c \to 0$, respectively.

Take a fixed 1-form $\eta \neq 0$ of Γ and define $\mathrm{d}_\eta f = f\eta - \eta f$, $f \in \mathcal{X}$. Then $\mathcal{X}_{p,\eta} := \mathcal{X} \cdot \mathrm{d}_\eta \mathcal{X} \cdot \mathcal{X}$ is a FODC over $\mathcal{X}$ with differentiation d_η. If $\eta = r\mathrm{d}s$ with $r, s \in \mathcal{X}$, then we have $\mathrm{d}_\eta f = \mathrm{d}x \cdot r(p(x))\partial(s)(f(p(x)) - f(x))$. △

Definition 2. *A FODC Γ over a $*$-algebra $\mathcal{X}$ is called a* $*$-calculus *if there exists an involution $\rho \to \rho^*$ of the vector space Γ such that $(x \cdot \mathrm{d}y \cdot z)^* = z^* \cdot \mathrm{d}(y^*) \cdot x^*$ for $x, y, z \in \mathcal{X}$.*

Proposition 1. *A FODC Γ over a $*$-algebra $\mathcal{X}$ is a $*$-calculus if and only if $\sum_i x_i \cdot \mathrm{d}y_i = 0$ with $x_i, y_i \in \mathcal{X}$ always implies that $\sum_i \mathrm{d}(y_i^*) \cdot x_i^* = 0$.*

Proof. The only if part is trivial. Conversely, if the latter condition is fulfilled, then $(\sum_i x_i \mathrm{d}y_i)^* := \sum_i \mathrm{d}(y_i^*) \cdot x_i^*, x_i, \ y_i \in \mathcal{X}$, gives a well-defined(!) antilinear mapping of Γ which has the desired properties. □

12.1.2 Covariant First Order Calculi on Quantum Spaces

If the algebra $\mathcal{X}$ is a quantum space for a Hopf algebra $\mathcal{A}$, it is natural to look for covariant FODC. The precise definition of this notion is given as follows.

Definition 3. *A first order differential calculus Γ over a left quantum space $\mathcal{X}$ with left coaction $\varphi : \mathcal{X} \to \mathcal{A} \otimes \mathcal{X}$ is called* left-covariant *with respect to $\mathcal{A}$ if there exists a left coaction $\phi : \Gamma \to \mathcal{A} \otimes \Gamma$ of $\mathcal{A}$ on Γ such that*

(i) $\phi(x\rho y) = \varphi(x)\phi(\rho)\varphi(y)$ *for all $x, y \in \mathcal{X}$ and $\rho \in \Gamma$,*
(ii) $\phi(\mathrm{d}x) = (\mathrm{id} \otimes \mathrm{d})\varphi(x)$ *for all $x \in \mathcal{X}$.*

Conditions (i) and (ii) mean that the coaction ϕ of $\mathcal{A}$ on Γ is compatible with the coaction φ of $\mathcal{A}$ on $\mathcal{X}$ and with the differentiation d, respectively. Condition (ii) says that the mapping d intertwines the coactions φ and ϕ, that is, $\mathrm{d} \in \mathrm{Mor}(\varphi, \phi)$, or equivalently that the following diagram is commutative:

$$\begin{array}{ccc} \mathcal{X} & \xrightarrow{\ \mathrm{d}\ } & \Gamma \\ \varphi \big\downarrow & & \big\downarrow \phi \\ \mathcal{A} \otimes \mathcal{X} & \xrightarrow{\ \mathrm{id} \otimes \mathrm{d}\ } & \mathcal{A} \otimes \Gamma \end{array}$$

For instance, if $x_1, \cdots, x_n$ are elements of $\mathcal{X}$ such that $\varphi(x_i) = u^i_j \otimes x_j$, then $\phi(\mathrm{d}x_i) = u^i_j \otimes \mathrm{d}x_j$ by (ii). That is, (ii) implies that the differentials $\mathrm{d}x_i$ transform under ϕ just as the elements x_i do under φ. However, the differentials $\mathrm{d}x_i$ need not be linearly independent if the x_i are.

Example 2 ($\mathcal{X} = \mathbb{C}[x]$ – continued). The algebra $\mathbb{C}[x]$ can be interpreted as the enveloping algebra $U(\mathfrak{g})$ with $\mathfrak{g} = \mathbb{R}$. Hence, by Example 1.6, $\mathcal{X}$ is a Hopf algebra with comultiplication given by $\Delta(x) = x\otimes 1 + 1\otimes x$. If we consider $\mathcal{X}$ as a left quantum space for the Hopf algebra $\mathcal{X}$ itself with respect to the comultiplication, then the FODC Γ_p from Example 1 is left-covariant if and only if $p(x) = x$. (In order to prove this, apply ϕ to the equation $x\mathrm{d}x = \mathrm{d}x \cdot p(x)$ and use conditions (i) and (ii).)

Let $\mathcal{A} = \mathbb{C}\mathbb{Z}$ be the group Hopf algebra of $\mathbb{Z}$ (see Example 1.7). We write $\mathcal{A}$ as the algebra $\mathbb{C}[z, z^{-1}]$ of Laurent polynomials in z with comultiplication $\Delta(z) = z\otimes z$. Clearly, $\mathcal{X}$ is a left quantum space of $\mathcal{A}$ with coaction determined by $\varphi(x) = z\otimes x$. Then the FODC Γ_p is left-covariant if and only if $p(x) = qx$ for some $q \in \mathbb{C}$. Thus we see that in both cases the covariance requirement is an essential restriction of the wealth of possible FODC. △

Note that given a FODC Γ over $\mathcal{X}$ there is at most one linear mapping $\phi : \Gamma \to \mathcal{A}\otimes\Gamma$ as in Definition 3. Indeed, if such a ϕ exists, then by (i) and (ii), we have

$$\phi\left(\sum\nolimits_i x_i \mathrm{d}y_i\right) = \sum\nolimits_i \varphi(x_i)(\mathrm{id}\otimes\mathrm{d})\varphi(y_i)\ . \tag{1}$$

The right-hand side of (1) and so ϕ is uniquely determined by φ and d.

An intrinsic characterization of the left-covariance of a FODC which avoids the use of the coaction ϕ is given by

Proposition 2. *For any FODC Γ over $\mathcal{X}$ the following assertions are equivalent:*

(i) *Γ is left-covariant with respect to $\mathcal{A}$.*

(ii) *There is a linear mapping $\phi : \Gamma \to \mathcal{A}\otimes\Gamma$ such that for all $x, y \in \mathcal{X}$ we have $\phi(x\mathrm{d}y) = \varphi(x)(\mathrm{id}\otimes\mathrm{d})\varphi(y)$.*

(iii) *$\sum_i x_i\mathrm{d}y_i = 0$ in Γ implies that $\sum_i \varphi(x_i)(\mathrm{id}\otimes\mathrm{d})\varphi(y_i) = 0$ in $\mathcal{A}\otimes\Gamma$.*

Proof. (i)→(ii) is true by definition and (ii)→(iii) is trivial.

(iii)→(i): By (iii), equation (1) defines unambiguously(!) a linear mapping $\phi : \Gamma \to \mathcal{A}\otimes\Gamma$ which obviously satisfies condition (ii) of Definition 3. By the properties of the coaction φ, we have

$$\begin{aligned}
(\Delta\otimes\mathrm{id})\phi(x\mathrm{d}y) &= (\Delta\otimes\mathrm{id})[\varphi(x)(\mathrm{id}\otimes\mathrm{d})\varphi(y)]\\
&= [(\Delta\otimes\mathrm{id})\varphi(x)][(\mathrm{id}\otimes\mathrm{id}\otimes\mathrm{d})(\Delta\otimes\mathrm{id})\varphi(y)]\\
&= [(\mathrm{id}\otimes\varphi)\varphi(x)][(\mathrm{id}\otimes\mathrm{id}\otimes\mathrm{d})(\mathrm{id}\otimes\varphi)\varphi(y)]\\
&= (\mathrm{id}\otimes\phi)[\varphi(x)(\mathrm{id}\otimes\mathrm{d})\varphi(y)] = (\mathrm{id}\otimes\phi)\phi(x\mathrm{d}y),\\
(\varepsilon\otimes\mathrm{id})\phi(x\mathrm{d}y) &= (\varepsilon\otimes\mathrm{id})[\varphi(x)(\mathrm{id}\otimes\mathrm{d})\varphi(y)]\\
&= (\varepsilon\otimes\mathrm{id})\varphi(x)\cdot(\varepsilon\otimes\mathrm{d})\varphi(y) = x\mathrm{d}y,
\end{aligned}$$

that is, ϕ is a left coaction of $\mathcal{A}$ on Γ. To prove condition (i) of Definition 3, let $\rho = \sum_i x_i\mathrm{d}y_i$. Using the Leibniz rule for d and the fact that φ is an algebra homomorphism (because $\mathcal{X}$ is a quantum space), we obtain

$$\begin{aligned}\phi(x\rho y) &= \sum\nolimits_i \phi(xx_i \mathrm{d}(y_i y)) - \sum\nolimits_i \phi(xx_i y_i \mathrm{d}y)\\ &= \sum\nolimits_i \varphi(xx_i)(\mathrm{id}\otimes \mathrm{d})\varphi(y_i y) - \sum\nolimits_i \varphi(xx_i y_i)(\mathrm{id}\otimes \mathrm{d})\varphi(y)\\ &= \varphi(x)\left(\sum\nolimits_i \varphi(x_i)(\mathrm{id}\otimes \mathrm{d})\varphi(y_i)\right)\varphi(y) = \varphi(x)\phi(\rho)\varphi(y).\end{aligned}$$ □

12.2 Covariant Higher Order Differential Calculus

12.2.1 Differential Calculi on Algebras

The de Rham complex of a C^∞-manifold is formed by extending the exterior derivative to the algebra of differential forms. A differential calculus may be viewed as a substitute of the de Rham complex for arbitrary algebras.

Definition 4. *A* differential calculus *(abbreviated, a* DC*) over* $\mathcal{X}$ *is a graded algebra* $\Gamma^\wedge = \bigoplus_{n=0}^\infty \Gamma^{\wedge n}$ *(that is,* $\Gamma^\wedge$ *is a direct sum of vector spaces* $\Gamma^{\wedge n}$ *and the product* $\wedge$ *of* $\Gamma^\wedge$ *maps* $\Gamma^{\wedge n}\times\Gamma^{\wedge m}$ *into* $\Gamma^{\wedge(n+m)}$*) with a linear mapping* $\mathrm{d}:\Gamma^\wedge\to\Gamma^\wedge$ *of degree one (that is,* $\mathrm{d}:\Gamma^{\wedge n}\to\Gamma^{\wedge(n+1)}$*) such that:*

(i) $\mathrm{d}^2=0$,

(ii) $\mathrm{d}(\rho\wedge\zeta)=\mathrm{d}\rho\wedge\zeta+(-1)^n\rho\wedge\mathrm{d}\zeta$ *for* $\rho\in\Gamma^{\wedge n}$ *and* $\zeta\in\Gamma^\wedge$,

(iii) $\Gamma^{\wedge 0}=\mathcal{X}$ *and* $\Gamma^{\wedge n}=\mathrm{Lin}\,\{x_0\wedge \mathrm{d}x_1\wedge\cdots\wedge \mathrm{d}x_n \mid x_0,\cdots,x_n\in\mathcal{X}\}$ *for* $n\in\mathbb{N}$.

If condition (iii) of Definition 4 is dropped, then we obtain the definition of a *differential graded algebra.*

The notation $\rho\wedge\xi$ reminds us of the cap product of forms in differential geometry. For the product by elements $x\in\Gamma^{\wedge 0}=\mathcal{X}$ we shall write simply $x\rho$ and ρx, $\rho\in\Gamma^\wedge$. Condition (ii) is called the *graded Leibniz rule.* By induction on n one proves that condition (iii) in Definition 4 can be replaced by

(iii)′ $\Gamma^{\wedge 0}=\mathcal{X}$ and $\Gamma^{\wedge n}=\mathcal{X}\cdot \mathrm{d}\Gamma^{\wedge(n-1)}$ for $n\in\mathbb{N}$.

By a *differential ideal* of a DC $\Gamma^\wedge$ over $\mathcal{X}$ we mean a two-sided ideal $\mathcal{J}$ of the algebra $\Gamma^\wedge$ such that $\mathcal{J}\cap\Gamma^{\wedge 0}=\{0\}$ and $\mathcal{J}$ is invariant under the differentiation d. Suppose that $\mathcal{J}$ is a differential ideal of a DC $\Gamma^\wedge$ over $\mathcal{X}$. Let $\pi:\Gamma^\wedge\to\Gamma^\wedge/\mathcal{J}$ denote the canonical map of $\Gamma^\wedge$ to the quotient algebra $\tilde\Gamma^\wedge:=\Gamma^\wedge/\mathcal{J}$. We define $\tilde{\mathrm{d}}(\pi(\rho)):=\pi(\mathrm{d}\rho)$, $\rho\in\Gamma^\wedge$. The condition $\mathcal{J}\cap\Gamma^{\wedge 0}=\{0\}$ ensures that $\tilde\Gamma^{\wedge 0}=\Gamma^{\wedge 0}=\mathcal{X}$. Then $\tilde\Gamma^\wedge$ is again a DC over $\mathcal{X}$ with differentiation $\tilde{\mathrm{d}}$, called the *quotient of* $\Gamma^\wedge$ *by the differential ideal* $\mathcal{J}$.

For any DC over the algebra $\mathcal{X}$ the following identities hold:

$$\mathrm{d}(x_0\mathrm{d}x_1\wedge\cdots\wedge\mathrm{d}x_n)=\mathrm{d}x_0\wedge\mathrm{d}x_1\wedge\cdots\wedge\mathrm{d}x_n, \tag{2}$$

$$(x_0\mathrm{d}x_1\wedge\cdots\wedge\mathrm{d}x_n)\wedge(x_{n+1}\mathrm{d}x_{n+2}\wedge\cdots\wedge\mathrm{d}x_{n+k})$$

$$=(-1)^n x_0x_1\mathrm{d}x_2\wedge\cdots\wedge\mathrm{d}x_{n+k}+\sum_{r=1}^{n}(-1)^{n-r}x_0\mathrm{d}x_1\wedge\cdots\wedge\mathrm{d}(x_rx_{r+1})\wedge\cdots\wedge\mathrm{d}x_{n+k}. \tag{3}$$

Indeed, (2) and (3) are easily verified by induction using the Leibniz rule and the equation $\mathrm{d}^2 = 0$.

Definition 2′. *A DC $\Gamma^\wedge$ over a $*$-algebra $\mathcal{X}$ is a* $*$-calculus *if there exists an involution $\rho \to \rho^*$ of the vector space $\Gamma^\wedge$ which coincides with the involution of the $*$-algebra $\mathcal{X}$ on $\Gamma^{\wedge 0} = \mathcal{X}$ and satisfies the relations $(\rho_n \wedge \rho_k)^* = (-1)^{nk}\rho_k^* \wedge \rho_n^*$ and $\mathrm{d}(\rho^*) = (\mathrm{d}\rho)^*$ for any $\rho_n \in \Gamma^{\wedge n}, \rho_k \in \Gamma^{\wedge k}$ and $\rho \in \Gamma^\wedge$.*

Note that the involution of a $*$-calculus $\Gamma^\wedge$ (if it exists) is always uniquely determined by the involution of the $*$-algebra $\mathcal{X}$, because $\Gamma^\wedge$ is generated as an algebra by $\Gamma^{\wedge 0} = \mathcal{X}$ and $\mathrm{d}\mathcal{X}$.

12.2.2 The Differential Envelope of an Algebra

The aim of this subsection is to define a DC $\Omega(\mathcal{X})$ over $\mathcal{X}$ which is universal in the sense that any other DC over $\mathcal{X}$ is isomorphic to some quotient of $\Omega(\mathcal{X})$. The calculus $\Omega(\mathcal{X})$ itself is very large and so is not of interest. We shall need it only as a technical tool in order to construct differential calculi as its quotients (just as free algebras are used to construct algebras in terms of generators and relations). Roughly speaking, the main idea behind the definition of the DC $\Omega(\mathcal{X})$ can be described as follows: $\Omega(\mathcal{X})$ is the free algebra generated by elements $x \in \mathcal{X}$ and symbols $\mathrm{d}x, x \in \mathcal{X}$, subject to the Leibniz rule $\mathrm{d}x{\cdot}y = \mathrm{d}(xy) - x{\cdot}\mathrm{d}y$. By the latter relation, any element of $\Omega(\mathcal{X})$ can be written as a sum of monomials $x_0\mathrm{d}x_1 \wedge \cdots \wedge \mathrm{d}x_n$ with $x_0, x_1, \cdots, x_n \in \mathcal{X}$, where $\wedge$ denotes the product of the algebra $\Omega(\mathcal{X})$. Thus it suffices to define the product and the differentiation of $\Omega(\mathcal{X})$ for such monomials. Since we want $\Omega(\mathcal{X})$ to be a DC, the formulas (2) and (3) must hold. They will give us the corresponding definitions for monomials.

The above idea is realized by the following precise mathematical construction. Let $\bar{\mathcal{X}} := \mathcal{X}/\mathbb{C}{\cdot}1$ be the quotient vector space of $\mathcal{X}$ by $\mathbb{C}{\cdot}1$ and let $\bar{x}$ denote the equivalence class $x + \mathbb{C}{\cdot}1$ of $x \in \mathcal{X}$. We set

$$\Omega(\mathcal{X}) := \bigoplus_{n=0}^{\infty} \Omega^n(\mathcal{X}), \quad \text{where } \Omega^0(\mathcal{X}) := \mathcal{X}, \quad \Omega^n(\mathcal{X}) := \mathcal{X} \otimes \bar{\mathcal{X}}^{\otimes n}, \ n \in \mathbb{N}.$$

We write $x_0\mathrm{d}x_1 \wedge \cdots \wedge \mathrm{d}x_n$ for $x_0 \otimes \bar{x}_1 \otimes \cdots \otimes \bar{x}_n$. It can be shown (see, for instance, [Con], [CQ] or [Ka]) that the formulas (2) and (3) (with the obvious interpretations for $n = 0$ and $k = 1$) define a differentiation d and a product $\wedge$, respectively, such that $\Omega(\mathcal{X})$ becomes a differential calculus over $\mathcal{X}$. Clearly, $\Omega(\mathcal{X})$ is a $*$-calculus if $\mathcal{X}$ is a $*$-algebra.

Let $\Gamma^\wedge$ be another DC over $\mathcal{X}$ with differentiation $\tilde{\mathrm{d}}$. Define a linear map $\psi : \Omega(\mathcal{X}) \to \Gamma^\wedge$ by $\psi(x) := x$ and $\psi(x_0\mathrm{d}x_1 \wedge \cdots \wedge \mathrm{d}x_n) := x_0\tilde{\mathrm{d}}x_1 \wedge \cdots \wedge \tilde{\mathrm{d}}x_n$, $n \in \mathbb{N}$. (The definition of $\Omega^n(\mathcal{X})$ as a tensor product ensures that ψ is well-defined!) From the identities (2) and (3) it follows that $\mathcal{N} := \ker \psi$ is a differential ideal of $\Omega(\mathcal{X})$ and that $\Gamma^\wedge$ is isomorphic to the quotient DC $\Omega(\mathcal{X})/\mathcal{N}$.

The first order part $\Omega^1(\mathcal{X})$ of $\Omega(\mathcal{X})$ is obviously a FODC over $\mathcal{X}$ and an arbitrary FODC over $\mathcal{X}$ is isomorphic to a quotient of $\Omega^1(\mathcal{X})$.

The DC $\Omega(\mathcal{X})$ is called the *differential envelope* or the *universal differential calculus* of the algebra $\mathcal{X}$. Likewise, the FODC $\Omega^1(\mathcal{X})$ is called the *universal first order differential calculus* of $\mathcal{X}$.

12.2.3 Covariant Differential Calculi on Quantum Spaces

First we extend Definition 3 and Proposition 2 to higher order calculi.

Definition 3′. *A differential calculus $\Gamma^\wedge$ over the left quantum space $\mathcal{X}$ is called* left-covariant *with respect to $\mathcal{A}$ if there exists an algebra homomorphism $\phi^\wedge : \Gamma^\wedge \to \mathcal{A}\otimes\Gamma^\wedge$ which is a left coaction of $\mathcal{A}$ on $\Gamma^\wedge$ such that $\phi^\wedge(x) = \varphi(x)$ for $x \in \Gamma^{\wedge 0} = \mathcal{X}$ and $\phi^\wedge(\mathrm{d}\rho) = (\mathrm{id}\otimes\mathrm{d})\phi^\wedge(\rho)$ for $\rho \in \Gamma^\wedge$.*

Proposition 2′. *For any DC $\Gamma^\wedge$ over $\mathcal{X}$ the following statements are equivalent:*
(i) *$\Gamma^\wedge$ is left-covariant with respect to $\mathcal{A}$.*
(ii) *There exists a linear mapping $\phi^\wedge : \Gamma^\wedge \to \mathcal{A}\otimes\Gamma^\wedge$ such that the restriction of $\phi^\wedge$ to $\mathcal{X}$ is equal to φ and for any $x_0, x_1, \cdots, x_n \in \mathcal{X}$ we have*

$$\phi^\wedge(x_0\mathrm{d}x_1 \wedge \cdots \wedge \mathrm{d}x_n) = \varphi(x_0)(\mathrm{id}\otimes\mathrm{d})\varphi(x_1)\cdots(\mathrm{id}\otimes\mathrm{d})\varphi(x_n).$$

(iii) *$\sum_i x_0^i\mathrm{d}x_1^i \wedge \cdots \wedge \mathrm{d}x_n^i = 0$ in $\Gamma^\wedge$ always implies that*

$$\sum\nolimits_i \varphi(x_0^i)(\mathrm{id}\otimes\mathrm{d})\varphi(x_1^i)\cdots(\mathrm{id}\otimes\mathrm{d})\varphi(x_n^i) = 0 \text{ in } \mathcal{A}\otimes\Gamma^\wedge.$$

Using formulas (2) and (3) with $k = 1$, Proposition 2′ is proved in a similar manner as Proposition 2. For the proof of the main implication (iii) $\to$ (i) the coaction $\phi^\wedge$ is defined as in (ii). From the latter we see in particular that the mapping $\phi^\wedge$ in Definition 3′ is uniquely determined by the coaction φ of $\mathcal{A}$ on $\mathcal{X}$ and the differentiation d.

Since $(\mathrm{id}\otimes\mathrm{d})\varphi(x) = (\mathrm{id}\otimes\mathrm{d})\varphi(\bar{x})$ for $x \in \mathcal{X}$, conditions (iii) of Propositions 2 and 2′ are satisfied for the DC $\Omega(\mathcal{X})$ and the FODC $\Omega^1(\mathcal{X})$. Thus, we obtain

Corollary 3′. *The universal DC $\Omega(\mathcal{X})$ and the universal FODC $\Omega^1(\mathcal{X})$ over $\mathcal{X}$ are left-covariant.*

Now suppose that Γ is a FODC over $\mathcal{X}$ with differentiation d_1. In general there are many higher order calculi with first order part Γ. (A trivial one is obtained by setting $\Gamma^{\wedge n} = \{0\}$ for $n \geq 2$.) We want to define the "most free" differential calculus over $\mathcal{X}$ whose first order calculus is the given Γ. For this purpose we set $\mathcal{N} := \{\sum_i x_i\mathrm{d}y_i \in \Omega^1(\mathcal{X}) \mid \sum_i x_i\mathrm{d}_1 y_i = 0 \text{ in } \Gamma\}$. From the Leibniz rule for d_1 it follows easily that $\mathcal{N}$ is a $\mathcal{X}$-subbimodule of $\Omega^1(\mathcal{X})$ and the FODC Γ is isomorphic to the FODC $\Omega^1(\mathcal{X})/\mathcal{N}$. Let $\mathcal{J}_\Gamma := \Omega(\mathcal{X})\wedge\mathcal{N}\wedge\Omega(\mathcal{X})+\Omega(\mathcal{X})\wedge\mathrm{d}\mathcal{N}\wedge\Omega(\mathcal{X})$ be the differential ideal of the DC $\Omega(\mathcal{X})$ generated by $\mathcal{N}$. Then $\Gamma_u^\wedge := \Omega(\mathcal{X})/\mathcal{J}_\Gamma$ is a DC over $\mathcal{X}$ whose first order part is $\Omega^1(\mathcal{X})/\mathcal{N}$ and so is isomorphic to Γ. It is clear from the construction that the DC $\Gamma_u^\wedge$ has the following universal property: *If $\Gamma^\wedge = \bigoplus_n \Gamma^{\wedge n}$ is another DC over $\mathcal{X}$ such that $\Gamma^{\wedge 1}$ is isomorphic to Γ, then $\Gamma^\wedge$ is isomorphic to a quotient of $\Gamma_u^\wedge$.* Because of the latter property, $\Gamma_u^\wedge = \Omega(\mathcal{X})/\mathcal{J}_\Gamma$ is called the *universal differential calculus of the first order differential calculus Γ.*

Proposition 4. (i) *If Γ is left-covariant, then so is $\Gamma_u^\wedge$.*

(ii) *If $\mathcal{X}$ is a $*$-algebra and Γ is a $*$-calculus, then $\Gamma_u^\wedge$ is also a $*$-calculus.*

Proof. (i): Let $\phi^\wedge$ denote the coaction of $\mathcal{A}$ for the left-covariant DC $\Omega(\mathcal{X})$. The characterization of left-covariance of Γ given in Proposition 2(iii) means precisely that $\phi^\wedge(\mathcal{N}) \subseteq \mathcal{A} \otimes \mathcal{N}$. By Definition 3′ applied to the left-covariant DC $\Omega(\mathcal{X})$ we have $\phi^\wedge \mathrm{d} = (\mathrm{id} \otimes \mathrm{d})\phi^\wedge$ and so $\phi^\wedge(\mathrm{d}\mathcal{N}) \subseteq \mathcal{A} \otimes \mathrm{d}\mathcal{N}$. Since $\phi^\wedge$ is an algebra homomorphism, it follows from the preceding that $\phi^\wedge(\mathcal{J}_\Gamma) \subseteq \mathcal{A} \otimes \mathcal{J}_\Gamma$. Therefore, $\phi^\wedge$ passes to the quotient $\Gamma_u^\wedge = \Omega(\mathcal{X})/\mathcal{J}_\Gamma$ and defines a left coaction of $\mathcal{A}$ on $\Gamma_u^\wedge$. Hence $\Gamma_u^\wedge$ is left-covariant.

(ii): Using the Leibniz rule and Definition 2 for the $*$-calculi $\Omega^1(\mathcal{X})$ and Γ, we conclude that $\mathcal{N}$ is $*$-invariant. This implies that the differential ideal $\mathcal{J}_\Gamma$ is $*$-invariant. Therefore, the involution of the $*$-calculus $\Omega(\mathcal{X})$ passes to $\Gamma_u^\wedge$ and hence $\Gamma_u^\wedge$ is also a $*$-calculus. □

Remark. All notions and results concerning left-covariant FODC and DC on a left quantum space established above have their counterparts for right-covariant FODC and DC on a right quantum space. They are obtained if we replace "left" by "right", "id ⊗ d" by "d ⊗ id", "$\mathcal{A} \otimes \Gamma$" by "$\Gamma \otimes \mathcal{A}$" and "$\mathcal{A} \otimes \Gamma^\wedge$" by "$\Gamma^\wedge \otimes \mathcal{A}$" in Definitions 3 and 3′, Propositions 2, 2′, and 4, and Corollary 3′. △

12.3 Construction of Covariant Differential Calculi on Quantum Spaces

12.3.1 General Method

The purpose of this subsection is to develop and discuss a general result (Proposition 5) concerning the existence of left-covariant FODC on quantum spaces.

Let $\{x_i\}_{i \in I}$ be a family of linearly independent generators of the algebra $\mathcal{X}$ and let $\{x_{ij}^k\}_{i,j,k \in I}$ be a set of elements of $\mathcal{X}$.

Proposition 5. *Under the assumptions (5) and (7) stated below, there exists a unique left-covariant FODC Γ over $\mathcal{X}$ such that $\{\mathrm{d}x_i\}_{i \in I}$ is a free right $\mathcal{X}$-module basis of Γ and the $\mathcal{X}$-bimodule structure of Γ is given by the relations*

$$x_i \cdot \mathrm{d}x_j = \mathrm{d}x_k \cdot x_{ij}^k, \quad i, j \in I. \tag{4}$$

Assumption (5) expresses the covariance of the defining relations (4) of the FODC Γ and (7) requires two consistency conditions, one for the $\mathcal{X}$-bimodule structure of Γ and one for the differentiation d. Note that the structure of the FODC Γ is completely described by the equations (4) and the property that $\{dx_i\}_{i\in I}$ is a free right $\mathcal{X}$-module basis of Γ.

Proof. Let $\mathcal{X}_0 = \mathrm{Lin}\{x_i \mid i \in I\}$ and $\varphi_0 := \varphi\lceil\mathcal{X}_0$. Let $T : \mathcal{X}_0 \otimes \mathcal{X}_0 \to \mathcal{X}_0 \otimes \mathcal{X}$ denote the linear mapping defined by $T(x_i \otimes x_j) = x_k \otimes x^k_{ij}, i, j \in I$. We assume that

$$\varphi(\mathcal{X}_0) \subseteq \mathcal{A} \otimes \mathcal{X}_0 \quad \text{and} \quad T \in \mathrm{Mor}\,(\varphi_0 \otimes \varphi_0, \varphi_0 \otimes \varphi). \tag{5}$$

To begin the construction of Γ, let $\mathcal{Y}$ be the free algebra with generators $x_i, i \in I$. Let $d\mathcal{X}_0$ denote a vector space with a basis indexed by the set I. The basis elements are denoted by $\tilde{d}x_i$, $i \in I$. The vector space $\tilde{\Gamma} := \mathcal{Y} \otimes d\mathcal{X}_0 \otimes \mathcal{Y}$ is a $\mathcal{Y}$-bimodule with the obvious right and left actions of $\mathcal{Y}$ on $\tilde{\Gamma}$. For simplicity we write $y \cdot \tilde{d}x_i \cdot z$ for $y \otimes \tilde{d}x_i \otimes z$. Since $\mathcal{Y}$ is the free algebra generated by $\{x_i\}_{i\in I}$, there are a unique algebra homomorphism $\psi_0 : \mathcal{Y} \to \mathcal{A} \otimes \mathcal{Y}$ which extends the linear map $\varphi_0 : \mathcal{X}_0 \to \mathcal{A} \otimes \mathcal{X}_0$ and a well-defined linear mapping $\tilde{d} : \mathcal{Y} \to \tilde{\Gamma}$ such that $\tilde{d}(1) = 0$, $\tilde{d}(x_i) = \tilde{d}x_i$ and

$$\tilde{d}(x_{i_1} \cdots x_{i_n}) = \sum_{m=1}^{n} x_{i_1} \cdots x_{i_{m-1}} \cdot \tilde{d}x_{i_m} \cdot x_{i_{m+1}} \cdots x_{i_n}$$

for $i, i_1, \cdots, i_n \in I$. Then $\mathcal{Y}$ is a left quantum space of $\mathcal{A}$ with coaction ψ_0. Define a coaction $\phi_0 : d\mathcal{X}_0 \to \mathcal{A} \otimes d\mathcal{X}_0$ by $\phi_0(\tilde{d}x_i) := (\mathrm{id} \otimes \tilde{d})\varphi_0(x_i)$. It is easily seen that $\tilde{d}$ satisfies the Leibniz rule and $\psi := \psi_0 \otimes \phi_0 \otimes \psi_0 : \tilde{\Gamma} \to \mathcal{A} \otimes \tilde{\Gamma}$ is a left coaction of $\mathcal{A}$ on $\tilde{\Gamma}$ such that $\tilde{d} \in \mathrm{Mor}\,(\psi_0, \psi)$. Hence $\tilde{\Gamma}$ is a left-covariant FODC over $\mathcal{Y}$ with differentiation $\tilde{d}$.

Now we suppose that $\mathcal{J}$ is a two-sided ideal of $\mathcal{Y}$ such that the algebra $\mathcal{X}$ is (isomorphic to) the quotient algebra $\mathcal{Y}/\mathcal{J}$ and $\psi_0(\mathcal{J}) \subseteq \mathcal{A} \otimes \mathcal{J}$. Note that such an $\mathcal{J}$ always exists, since we assumed that the algebra $\mathcal{X}$ is generated by the set $\{x_i\}_{i\in I}$. Let $\pi : \mathcal{Y} \to \mathcal{X}$ be the corresponding quotient map and take elements $y^k_{ij} \in \mathcal{Y}$ such that $\pi(y^k_{ij}) = x^k_{ij}$. Using once again that $\mathcal{Y}$ is the free algebra with generators $\{x_i\}_{i\in I}$, it follows that there exist linear mappings $\rho^i_j : \mathcal{Y} \to \mathcal{Y}$, $i, j \in I$, such that $\rho^i_j(yz) = \rho^i_k(y)\rho^k_j(z)$ for $y, z \in \mathcal{Y}$, $\rho^i_j(1) = \delta_{ij}1$ and $\rho^k_j(x_i) = y^k_{ij}$. Then $\mathcal{N}_1 := \mathrm{Lin}\{z \cdot \tilde{d}x_i \cdot y - \tilde{d}x_k \cdot \rho^k_i(z)y \mid z, y \in \mathcal{Y}, i \in I\}$ is a $\mathcal{Y}$-subbimodule of $\tilde{\Gamma}$. Hence $\Gamma_1 := \tilde{\Gamma}/\mathcal{N}_1$ is a $\mathcal{Y}$-bimodule. That is, as a vector space Γ_1 is isomorphic to $d\mathcal{X}_0 \otimes \mathcal{Y}$ and the $\mathcal{Y}$-bimodule structure of Γ_1 is determined by the relations

$$x_i \cdot \tilde{d}x_j = \tilde{d}x_k \cdot y^k_{ij}, \quad i,\ j \in I. \tag{6}$$

Let $\mathcal{N}_0 := d\mathcal{X}_0 \cdot \mathcal{J} \equiv \mathrm{Lin}\{\tilde{d}x_i \cdot z \mid z \in \mathcal{J}, i \in I\}$ and let $\mathcal{J}_0$ be a set of generators of the two-sided ideal $\mathcal{J}$. We assume that

$$z_0 \cdot \tilde{d}x_i \in d\mathcal{X}_0 \cdot \mathcal{J} \quad \text{and} \quad \tilde{d}z_0 \in d\mathcal{X}_0 \cdot \mathcal{J} \quad \text{for} \quad z_0 \in \mathcal{J}_0 \quad \text{and} \quad i \in I. \tag{7}$$

Clearly, $\mathcal{N}_0$ is a $\mathcal{Y}$-subbimodule of Γ_1. The first condition of (7) implies that $\mathcal{J}\Gamma_1 \subseteq \mathcal{N}_0$. It is obvious that $\Gamma_1\mathcal{J} \subseteq \mathcal{N}_0$. Hence $\Gamma := \Gamma_1/\mathcal{N}_0$ becomes a bimodule for the quotient algebra $\mathcal{X} = \mathcal{Y}/\mathcal{J}$. Let $\pi_1 : \Gamma_1 \to \Gamma$ be the quotient map. The second condition of (7) and the Leibniz rule for $\tilde{\mathrm{d}}$ imply that $\mathrm{d}(\pi(y)) := \pi_1(\tilde{\mathrm{d}}(y))$, $y \in \mathcal{Y}$, defines unambiguously a linear mapping $\mathrm{d} : \mathcal{X} \to \Gamma$. Then Γ is a FODC over $\mathcal{X}$ with differentiation d.

We can consider $\mathcal{N}_0$ also as a subset of $\tilde{\Gamma}$ rather than of Γ_1, since $\mathcal{N}_1 \cap \mathcal{N}_0 = \{0\}$. From assumption (5) and the facts that $\pi(y^k_{ij}) = x^k_{ij}$ and $\psi_0(\mathcal{J}) \subseteq \mathcal{A} \otimes \mathcal{J}$ we conclude that $\psi(x_i{\cdot}\tilde{\mathrm{d}}x_j - \tilde{\mathrm{d}}x_k{\cdot}y^k_{ij}) \subseteq \mathcal{A} \otimes (\mathcal{N}_0 + \mathcal{N}_1)$. Since the $\mathcal{Y}$-subbimodule $\mathcal{N}_1$ is generated by the elements $x_i{\cdot}\tilde{\mathrm{d}}x_j - \tilde{\mathrm{d}}x_k{\cdot}y^k_{ij}$, $i, j \in I$, and $\mathcal{N}_0$ is a $\mathcal{Y}$-subbimodule of Γ_1,we obtain $\psi(\mathcal{N}_1) \subseteq \mathcal{A} \otimes (\mathcal{N}_0 + \mathcal{N}_1)$. Since also $\psi(\mathcal{N}_0) \subseteq \mathcal{A} \otimes \mathcal{N}_0$, we have $\psi(\mathcal{N}_1 + \mathcal{N}_0) \subseteq \mathcal{A} \otimes (\mathcal{N}_0 + \mathcal{N}_1)$. Hence the coaction ψ of $\mathcal{A}$ on $\tilde{\Gamma}$ passes to the quotient $\tilde{\Gamma}/(\mathcal{N}_1 + \mathcal{N}_0) \cong \Gamma$ to define a coaction ϕ there. The map ϕ satisfies the conditions of Definition 3 for Γ, since ψ does for $\tilde{\Gamma}$. Thus we have shown that Γ is a left-covariant FODC over $\mathcal{X}$. From the preceding construction it is clear that $\{\mathrm{d}x_i\}_{i \in I}$ is a free right $\mathcal{X}$-module basis of Γ and the relations (4) hold. □

Let Γ be the left-covariant FODC from Proposition 5. Then, for any $x \in \mathcal{X}$ there are uniquely determined elements $\partial_i(x) \in \mathcal{X}$ such that

$$\mathrm{d}x = \sum\nolimits_i \mathrm{d}x_i{\cdot}\partial_i(x). \tag{8}$$

The mappings $\partial_i : \mathcal{X} \to \mathcal{X}$ are called the *partial derivatives* of the FODC Γ.

Let us turn to the universal higher order DC $\Gamma_u^\wedge$ of Γ. Recall from Subsect. 12.2.3 that $\Gamma_u^\wedge$ is the quotient of the universal DC $\Omega(\mathcal{X})$ over $\mathcal{X}$ by the two-sided ideal $\mathcal{J}_\Gamma$ generated by the defining elements $x_i{\cdot}\mathrm{d}x_j - \mathrm{d}x_k{\cdot}x^k_{ij}$ for the bimodule Γ and their differentials $\mathrm{d}(x_i{\cdot}\mathrm{d}x_j - \mathrm{d}x_k{\cdot}x^k_{ij}) = \mathrm{d}x_i \wedge \mathrm{d}x_j$ $+\mathrm{d}x_k \wedge \mathrm{d}x^k_{ij}$, $i, j \in I$. That is, the algebra $\Gamma_u^\wedge$ has the generators x_i and $\mathrm{d}x_i, i \in I$, and the defining relations of $\Gamma_u^\wedge$ are those of the algebra $\mathcal{X}$, the relations (4) and

$$\mathrm{d}x_i \wedge \mathrm{d}x_j = -\mathrm{d}x_k \wedge \mathrm{d}x^k_{ij}, \quad i, j \in I. \tag{9}$$

Because of the applications given in the next subsection, we specialize the above procedure. Suppose that the quantum space $\mathcal{X}$ is an algebra with generators $x_i, i = 1, \cdots, N$, and defining relations written in matrix form as

$$B_{12}\mathbf{x}_1\mathbf{x}_2 = 0, \tag{10}$$

where B is a complex $N^2 \times N^2$ matrix and $\mathbf{x}$ denotes the column vector $(x_1, \cdots, x_N)^t$ (see Subsect. 9.1.2 for the corresponding notation). Further, we assume that the elements x^k_{ij} are *linear* in the generators x_i, that is, we have $y^k_{ij} = x^k_{ij} = A^{ij}_{kl}x_l$ for some complex matrix $A = (A^{ij}_{kl})$. Then (4) reads as $x_i{\cdot}\mathrm{d}x_j = A^{ij}_{kl}\,\mathrm{d}x_k{\cdot}x_l$ or in matrix notation as

$$\mathbf{x}_1 \cdot \mathrm{d}\mathbf{x}_2 = A_{12}\, \mathrm{d}\mathbf{x}_1 \cdot \mathbf{x}_2 \ . \tag{11}$$

Now we set $\mathcal{J}_0 := \{B^{nk}_{ij} x_i x_j \mid n, k = 1, 2, \cdots, N\}$ and reformulate the key assumption (7) in this case. From the equations (6) and (11) we obtain

$$(B_{12}\mathbf{x}_1\mathbf{x}_2)\, \tilde{\mathrm{d}}\mathbf{x}_3 = B_{12}A_{23}\, \mathbf{x}_1 \cdot \tilde{\mathrm{d}}\mathbf{x}_2 \cdot \mathbf{x}_3 = B_{12}A_{23}A_{12} \tilde{\mathrm{d}}\mathbf{x}_1 \cdot \mathbf{x}_2\mathbf{x}_3,$$

$$\tilde{\mathrm{d}}(B_{12}\mathbf{x}_1\mathbf{x}_2) = B_{12}(\tilde{\mathrm{d}}\mathbf{x}_1 \cdot \mathbf{x}_2 + \mathbf{x}_1 \cdot \tilde{\mathrm{d}}\mathbf{x}_2) = (B_{12} + B_{12}A_{12})\tilde{\mathrm{d}}\mathbf{x}_1 \cdot \mathbf{x}_2,$$

respectively. Therefore, by (10), the assumption (7) is certainly fulfilled if there exists a complex $N^3 \times N^3$ matrix T such that

$$B_{12}A_{23}A_{12} = T_{123}B_{23}, \tag{12}$$

$$B_{12} + B_{12}A_{12} = 0. \tag{13}$$

12.3.2 Covariant Differential Calculi on Quantum Vector Spaces

In this subsection we shall apply the method of the preceding subsection to the following situation: $\mathcal{A}$ is the Hopf algebra $\mathcal{O}(G_q)$ with $G_q = GL_q(N), SL_q(N), O_q(N), Sp_q(N)$ and $\mathcal{X}$ is the left quantum space with defining relations $\mathsf{P}_-\mathbf{x}_1\mathbf{x}_2 = 0$. Here P_- is the spectral projector of $\hat{\mathsf{R}}$ to the eigenvalue $-q^{-1}$ (see Sects. 9.2 and 9.3). We suppose that $q^2 + 1 \neq 0$ for $GL_q(N)$ and $SL_q(N)$, $(q^2+1)(1+q^{2-N}) \neq 0$ for $O_q(N)$ and $(q^2+1)(1-q^{-2N}) \neq 0$ for $Sp_q(N)$. These assumptions ensure that $\mathsf{P}_- = f(\hat{\mathsf{R}})$ for some polynomial f, so $\mathcal{X} = \mathcal{X}_L(f; \mathsf{R})$ is indeed a left quantum space of $\mathcal{A}$ by Proposition 9.4. In fact, we have $\mathcal{X} = \mathcal{O}(\mathbb{C}^N_q)$ if $G_q = GL_q(N), SL_q(N)$ and $\mathcal{X} = \mathcal{O}(O^N_q)$ if $G_q = O_q(N)$.

By (9.10), the left coaction $\varphi \equiv \varphi_L$ of $\mathcal{A}$ on $\mathcal{X}$ is given by $\varphi(x_i) = u^i_j \otimes x_j$, where $\mathbf{u} = (u^i_j)$ is the fundamental matrix of $\mathcal{A}$ and $\mathbf{x} = (x_1, \cdots, x_N)^t$ is the vector of coordinate functions of $\mathcal{X}$. Then we have $\varphi(\mathcal{X}_0) \subseteq \mathcal{A} \otimes \mathcal{X}_0$ and

$$(\mathrm{id} \otimes T)(\varphi_0 \otimes \varphi_0)(x_i \otimes x_j) = (\mathrm{id} \otimes T)(u^i_k u^j_l \otimes x_k \otimes x_l) = u^i_k u^j_l A^{kl}_{nm} \otimes x_n \otimes x_m,$$

$$(\varphi_0 \otimes \varphi)T(x_i \otimes x_j) = (\varphi_0 \otimes \varphi)(x_k \otimes A^{ij}_{kl} x_l) = A^{ij}_{kl} u^k_n u^l_m \otimes x_n \otimes x_m.$$

Hence the assumption (5) is fulfilled if $A = A_+ := q\hat{\mathsf{R}}$ and if $A = A_- := q^{-1}\hat{\mathsf{R}}^{-1}$. For $A = A_+$ and $A = A_-$ both equations (12) and (13) are satisfied with $T_{123} := A_{23}A_{12}$. The relation (13) follows from the equation $\mathsf{P}_-\hat{\mathsf{R}}^{\pm 1} = -q^{\mp 1}\mathsf{P}_-$, while (12) is an immediate consequence of the braid relation for $\hat{\mathsf{R}}$. Therefore, by the previous considerations, we proved the first assertion of

Theorem 6. *There exist two left-covariant FODC Γ_+ and Γ_- over $\mathcal{X}$ with respect to $\mathcal{A}$ such that the set $\{\mathrm{d}x_i\}_{i=1,\cdots,N}$ is a free right $\mathcal{X}$-module basis of $\Gamma_\pm$ and*

$$x_i \cdot \mathrm{d}x_j = q^{\pm 1}(\hat{\mathsf{R}}^{\pm 1})^{ij}_{kl} \mathrm{d}x_k \cdot x_l. \tag{14}$$

Moreover, we have the following formulas of commutation relations for $\Gamma_\pm$:

$$\partial_i x_j = \delta_{ij} + q^{\pm 1}(\hat{\mathsf{R}}^{\pm 1})^{jl}_{ik} x_k \partial_l, \tag{15}$$

$$(\mathsf{P}_-)^{lk}_{ij} \partial_k \partial_l = 0, \tag{16}$$

$$\mathrm{d}x_i \wedge \mathrm{d}x_j = -q^{\pm 1}(\hat{\mathsf{R}}^{\pm 1})^{ij}_{kl} \mathrm{d}x_k \wedge \mathrm{d}x_l. \tag{17}$$

Proof. It remains to verify the formulas (15)–(17). Using (8) and (4) we get

$$\begin{aligned} \mathrm{d}x_i \cdot \partial_i(x_j y) &= \mathrm{d}(x_j y) = \mathrm{d}x_j \cdot y + x_j \cdot \mathrm{d}y = \mathrm{d}x_j \cdot y + x_j \cdot \mathrm{d}x_l \cdot \partial_l(y) \\ &= \mathrm{d}x_j \cdot y + q^{\pm 1}(\hat{\mathsf{R}}^{\pm 1})^{jl}_{ik} \mathrm{d}x_i \cdot x_k \partial_l(y). \end{aligned}$$

Since the set $\{\mathrm{d}x_i\}_{i=1,\cdots,N}$ is a free right $\mathcal{X}$-module basis of $\Gamma_\pm$, the coefficients of $\mathrm{d}x_i$ on both sides must coincide. This gives (15).

Next we turn to (16). Let $y \in \mathcal{X}$ be such that $(\mathsf{P}_-)^{lk}_{ij}\partial_k\partial_l(y) = 0$ for all $i, j=1, \ldots, N$. By induction it suffices to prove that $(\mathsf{P}_-)^{lk}_{ij}\partial_k\partial_l(x_p y) = 0$ for $p = 1, 2, \cdots, N$. The braid relation for $\hat{\mathsf{R}}$ implies that $\hat{\mathsf{R}}^{\pm 1}_{12}\hat{\mathsf{R}}^{\pm 1}_{23}(\mathsf{P}_-)_{12} = (\mathsf{P}_-)_{23}\hat{\mathsf{R}}^{\pm 1}_{12}\hat{\mathsf{R}}^{\pm 1}_{23}$. Applying first equation (15) twice, then the identity $\hat{\mathsf{R}}\mathsf{P}_- = -q^{-1}\mathsf{P}_-$ and finally the preceding equality we obtain

$$\begin{aligned} &(\mathsf{P}_-)^{lk}_{ij}\partial_k\partial_l(x_p y) = (\mathsf{P}_-)^{pk}_{ij}\partial_k(y) \\ &\qquad + q^{\pm 1}(\mathsf{P}_-)^{lm}_{ij}(\hat{\mathsf{R}}^{\pm 1})^{pn}_{lm}\partial_n(y) + q^{\pm 2}(\mathsf{P}_-)^{lk}_{ij}(\hat{\mathsf{R}}^{\pm 1})^{pn}_{lm}(\hat{\mathsf{R}}^{\pm 1})^{mr}_{ks} x_s \partial_r \partial_n(y) \\ &= (\mathsf{P}_-)^{pk}_{ij}\partial_k(y) + q^{\pm 1}(\hat{\mathsf{R}}^{\pm 1}\mathsf{P}_-)^{pn}_{ij}\partial_n(y) + q^{\pm 2}(\hat{\mathsf{R}}^{\pm 1}_{12}\hat{\mathsf{R}}^{\pm 1}_{23}(\mathsf{P}_-)_{12})^{pnr}_{ijs} x_s \partial_r \partial_n(y) \\ &= (\mathsf{P}_-)^{pk}_{ij}\partial_k(y) + q^{\pm 1}(-q^{\mp 1}\mathsf{P}_-)^{pn}_{ij}\partial_n(y) + q^{\pm 2}((\mathsf{P}_-)_{23}\hat{\mathsf{R}}^{\pm 1}_{12}\hat{\mathsf{R}}^{\pm 1}_{23})^{pnr}_{ijs} x_s \partial_r \partial_n(y) \\ &= q^{\pm 2}(\mathsf{P}_-)^{nr}_{kl}(\hat{\mathsf{R}}^{\pm 1})^{pk}_{in}(\hat{\mathsf{R}}^{\pm 1})^{nl}_{js} x_s \partial_r \partial_n(y) = 0. \end{aligned}$$

Finally, note that (17) follows by differentiation of (14). □

12.3.3 Covariant Differential Calculus on $\mathbb{C}_q^N$ and the Quantum Weyl Algebra

In this subsection we treat the covariant FODC $\Gamma_\pm$ on $\mathcal{X} = \mathcal{O}(\mathbb{C}_q^N)$ in detail.

First we insert the values (9.13) for the numbers $\hat{\mathsf{R}}^{ij}_{kl}$ into the formulas (14)–(17) in Theorem 6. Then the commutation relations for the FODC Γ_+ and the defining relations of the algebra $\mathcal{O}(\mathbb{C}_q^N)$ take the following explicit form:

$$x_i \cdot \mathrm{d}x_j = q\mathrm{d}x_j \cdot x_i + (q^2 - 1)\mathrm{d}x_i \cdot x_j, \quad i < j, \tag{18}$$

$$x_i \cdot \mathrm{d}x_i = q^2 \mathrm{d}x_i \cdot x_i, \tag{19}$$

$$x_j \cdot \mathrm{d}x_i = q\, \mathrm{d}x_i \cdot x_j, \quad i < j, \tag{20}$$

$$\mathrm{d}x_i \wedge \mathrm{d}x_j = -q^{-1}\mathrm{d}x_j \wedge \mathrm{d}x_i, \quad i < j, \tag{21}$$

$$\mathrm{d}x_i \wedge \mathrm{d}x_i = 0, \tag{22}$$

$$x_i x_j = q\, x_j x_i, \quad i < j, \tag{23}$$

$$\partial_i \partial_j = q^{-1} \partial_j \partial_i, \quad i < j, \tag{24}$$

$$\partial_i x_j = q x_j \partial_i, \quad i \neq j, \tag{25}$$

$$\partial_i x_i - q^2 x_i \partial_i = 1 + (q^2 - 1) \sum_{j>i} x_j \partial_j. \tag{26}$$

The corresponding relations for the second FODC Γ_- over $\mathcal{X} = \mathcal{O}(\mathbb{C}_q^N)$ are obtained if we replace q by q^{-1} and the inequality $i < j$ by $j < i$ in the formulas (18)–(26). Of course, the equations (21)–(25) remain unchanged in this manner. In particular, we see that for $q \neq 1$ the FODC Γ_+ and Γ_- are not isomorphic and that for $q = 1$ both FODC give the "ordinary" differential calculus on the commutative polynomial algebra $\mathbb{C}[x_1, \cdots, x_N]$.

The above equations (18)–(20) completely describe the bimodule structure of Γ_+. The algebra $(\Gamma_+)^\wedge_u$ of the universal DC associated with Γ_+ admits the $2N$ generators x_i and $\mathrm{d}x_i, i = 1, \cdots, N$, with defining relations (18)–(23).

In the case $q = 1$ the formulas (23)–(26) reduce to the defining relations of the Weyl algebra. This motivates the following

Definition 5. *The* quantum Weyl algebra $\mathcal{A}_q(N)$ *is the unital algebra with $2N$ generators $x_1, \cdots, x_N, \partial_1, \cdots, \partial_N$ and defining relations (23)–(26).*

We mention a few algebraic properties of this algebra $\mathcal{A}_q(N)$.

1. The Weyl algebra $\mathcal{A}_q(N)$ with $q = 1$ is simple, that is, it has no nontrivial two-sided ideals. However, the algebra $\mathcal{A}_q(N)$ is no longer simple if $q^2 \neq 1$. Indeed, the kernel of the one dimensional representation given by $x_1 \to 1,\ \partial_1 \to (1 - q^2)^{-1},\ x_i \to 0,\ \partial_i \to 0,\ i \geq 2$, is a nontrivial ideal.

2. Let $\mathfrak{D}$ be the unital algebra with N generators $\partial_1, \partial_2, \cdots, \partial_N$ and defining relations (24) and let $\mathcal{X} := \mathcal{O}(\mathbb{C}_q^N)$. Then the linear map of $\mathcal{X} \otimes \mathfrak{D}$ to $\mathcal{A}_q(N)$ defined by $x \otimes y \to x{\cdot}y$ is a vector space isomorphism. (A proof can be given by using the diamond Lemma 4.8.) Therefore, we can consider $\mathcal{X}$ and $\mathfrak{D}$ as subalgebras of $\mathcal{A}_q(N)$ and the set of monomials

$$\{x_1^{k_1} \cdots x_N^{k_N} \partial_1^{n_1} \cdots \partial_N^{n_N} \mid k_1, \cdots, k_N, n_1, \cdots, n_N \in \mathbb{N}_0\}$$

is a vector space basis of $\mathcal{A}_q(N)$.

3. The element $\mathsf{D} := \sum_i x_i \partial_i$ of $\mathcal{A}_q(N)$ is called the *Euler derivation.* If $f_n \in \mathcal{X}$ and $g_n \in \mathfrak{D}$ are homogeneous elements of degree n, then we have the identities

$$\mathsf{D} f_n = [[n]]_{q^2} f_n + q^{2n} f_n \mathsf{D} \quad \text{and} \quad g_n \mathsf{D} = [[n]]_{q^2} g_n + q^{2n} \mathsf{D} g_n,$$

where we have set $[[n]]_{q^2} := (q^2 - 1)^{-1}(q^{2n} - 1)$.

4. The partial derivatives ∂_i of the FODC Γ_+ act on the algebra $\mathcal{X}$ by

$$\partial_i(f_N(x_N) \cdots f_1(x_1)) = f_N(q x_N) \cdots f_{i+1}(q x_{i+1}) D_{q^2} f_i(x_i) f_{i-1}(x_{i-1}) \cdots f_1(x_1),$$

where $f_1, \cdots, f_N$ are polynomials in one variable and D_{q^2} is the q^2-derivative (see (2.43)). It suffices to prove this formula for monomials $f_j(x_j) = x_j^{k_j}$.

First we shift $x_i^{k_i}$ to the right by using (23) and ∂_i in front of $x_i^{k_i}$ by applying (25). Then we compute $\partial_i(x_i^{k_i}) = [[k_i]]_{q^2} x_i^{k_i-1}$ and shift $x_i^{k_i-1}$ back.

It is well-known that the "ordinary" canonical commutation relations are invariant under symplectic linear transformations. The next proposition contains a quantum analog of this fact.

Proposition 7. *Let $\hat{\mathsf{R}} = (\hat{\mathsf{R}}^{ij}_{kl})$ be the R-matrix and $\mathsf{C} = (\mathsf{C}^i_j)$ the matrix of the metric for the quantum group $Sp_q(2N)$ (see Sect. 9.3). Let $\alpha \in \mathbb{C}$, $\alpha \neq 0$. Set $y_i := \alpha q^i \partial_{N+1-i}$ and $y_{N+i} = x_i$ for $i = 1, 2, \cdots, N$. Then the generators $x_1, \cdots, x_N, \partial_1, \cdots, \partial_N$ satisfy the defining relations (23)–(26) of the algebra $\mathcal{A}_q(N)$ if and only if*

$$\sum_{k,l=1}^{2N} \hat{\mathsf{R}}^{ij}_{kl} y_k y_l - q y_i y_j - \alpha q^{-N} \mathsf{C}^i_j = 0 \quad for \quad i, j = 1, 2, \cdots, 2N. \tag{27}$$

The quantum Weyl algebra $\mathcal{A}_q(N)$ is a left quantum space for the Hopf algebra $\mathcal{O}(Sp_q(2N))$ with coaction φ given by $\varphi(y_i) = u^i_j \otimes y_j$, $i = 1, 2, \cdots, 2N$.

Proof. For $i \neq j'$ we have $\mathsf{C}^i_j = 0$ and (27) reduces to the equations (23)–(25). Now we show that the equations (27) for $i = j'$ are equivalent to the equations (26). We first rewrite the relations (26) recursively as

$$\partial_{i+1} x_{i+1} - x_{i+1} \partial_{i+1} = \partial_i x_i - q^2 x_i \partial_i, \quad i = 1, 2, \cdots, N-1,$$

$$\partial_N x_N - q^2 x_N \partial_N = 1.$$

These equations are easily verified to be equivalent to

$$y_{j'} y_j - q^{-2} y_j y_{j'} = -q^{-j}\alpha + (q^{-2} - 1) \sum_{k<j} q^{k-j} y_k y_{k'}, \; j = 1, 2, \cdots, N. \tag{28}$$

On the other hand, inserting the values of $\mathsf{C}^i_{i'}$ and $\hat{\mathsf{R}}^{ii'}_{kl}$ from (9.30) into (27), we see that the equations (27) for $i = j'$ are also equivalent to the relations (28).

For the second assertion we proceed as in the proof of Proposition 9.4. There exists a unique algebra homomorphism $\varphi : \mathbb{C}\langle y_i \rangle \rightarrow \mathcal{O}(Sp_q(2N)) \otimes \mathbb{C}\langle y_i \rangle$ such that $\varphi(y_i) = u^i_j \otimes y_j$. Let $\mathcal{J}$ be the two-sided ideal of the free algebra $\mathbb{C}\langle y_i \rangle$ which is generated by the elements I^{ij} from the left hand side of (27). Using the relations $\mathbf{u}\mathsf{C}\mathbf{u}^t = \mathsf{C}$ and $\hat{\mathsf{R}}\mathbf{u}_1\mathbf{u}_2 = \mathbf{u}_1\mathbf{u}_2\hat{\mathsf{R}}$ (see (9.34) and (9.3)) we obtain that $\varphi(I^{ij}) = u^i_k u^j_l \otimes I^{kl}$. Therefore, φ passes to the quotient algebra $\mathcal{A}_q(N) = \mathbb{C}\langle y_i \rangle / \mathcal{J}$ to define a left coaction of $\mathcal{O}(Sp_q(2N))$ on $\mathcal{A}_q(N)$. □

There are two important special cases, where the quantum Weyl algebra $\mathcal{A}_q(N)$ admits a natural involution.

Case 1: $q \in \mathbb{R}$. Then there is a unique algebra involution on $\mathcal{A}_q(N)$ such that $\partial_i^* := x_i$, $i = 1, 2, \cdots, N$. Indeed, since q is real, the defining relations (23)–(26) are obviously invariant under this involution. Hence $\mathcal{A}_q(N)$ becomes a

$*$-algebra. The generators of $\mathcal{A}_q(N)$ can be interpreted as q-boson creation operators $a_i^* = x_i$ and q-boson annihilation operators $a_i = \partial_i$ in this way.

Case 2: $|q| = 1$. In this case the algebra $\mathcal{O}(\mathbb{C}_q^N)$ becomes a $*$-algebra $\mathcal{O}(\mathbb{R}_q^N)$ with involution defined by $x_i^* = x_i$, $i = 1, 2, \cdots, N$, and the FODC Γ_+ is a $*$-calculus for the $*$-algebra $\mathcal{O}(\mathbb{R}_q^N)$. The involution of Γ_+ induces the involution for the generators ∂_i given by $\partial_i^* = -q^{2(N+1-i)}\partial_i, i = 1, 2, \cdots, N$. Using the assumption $|q| = 1$ one verifies that the relations (23)–(26) are preserved under this involution. Therefore, the algebra $\mathcal{A}_q(N)$ becomes a $*$-algebra, denoted $\mathcal{A}_q(N;\mathbb{R})$, such that $x_i^* = x_i$ and $\partial_i^* = -q^{2(N+1-i)}\partial_i$. The hermitian elements x_i and $p_i := -\sqrt{-1}\; q^{N+1-i}\; \partial_i$ of this $*$-algebra can be viewed as q-analogs of the position and momentum operators, respectively. If one adopts this point of view, it is natural to consider the $*$-algebra $\mathcal{A}_q(N;\mathbb{R})$ as a q-deformation of the real quantum mechanical phase space. Let us set $\alpha = -\sqrt{-1}$ in Proposition 7. Then we have $y_i = p_{N+1-i}$ and $y_{N+i} = x_i$ therein and the last assertion of Propostion 7 means that the quantum group $Sp_q(2N)$ acts on the q-deformation $\mathcal{A}_q(N;\mathbb{R})$ of the phase space.

We close this subsection by stating a uniqueness result for the two covariant FODC $\Gamma_\pm$ on $\mathcal{O}(\mathbb{C}_q^N)$. Recall from Theorem 6 that the set $\{\mathrm{d}x_i\}_{i=1,\cdots,N}$ is a free right module basis of $\Gamma_\pm$ (and also a free left module basis as follows from the formulas (18)–(20)). The next proposition shows that for $N \geq 3$ and for q not a root of unity the latter property characterizes Γ_+ and Γ_- among all possible left-covariant FODC on $\mathcal{O}(\mathbb{C}_q^N)$.

Proposition 8. *Suppose that $N \geq 3$ and $q^n \neq 1$ for all $n \in \mathbb{N}$, $n \geq 2$. If Γ is a left-covariant FODC over $\mathcal{X} = \mathcal{O}(\mathbb{C}_q^N)$ with respect to the Hopf algebra $\mathcal{A} = \mathcal{O}(SL_q(N))$ such that the set $\{\mathrm{d}x_i\}_{i=1,\cdots,N}$ is a free right $\mathcal{X}$-module basis of Γ, then Γ is isomorphic to Γ_+ or to Γ_-.*

Proof. The proof can be found in [PuW]. (In the paper [PuW], covariant FODC with respect to $SU_q(N), q \in (-1, 1)$, are considered, but the proof therein works for $SL_q(N)$ if q is not a nontrivial root of unity.) □

In the case $N = 2$ the assertion of Proposition 8 is not true. In this case there are two one-parameter families of such FODC (see [SS1] for details).

12.3.4 Covariant Differential Calculi on the Quantum Hyperboloid

For a nonzero complex number γ, let $\mathcal{X}_{q,\gamma}$ denote the algebra with generators x_1, x_2 and defining relation

$$x_1x_2 - qx_2x_1 = \gamma{\cdot}1.$$

It is not difficult to check that there is an algebra homomorphism $\varphi : \mathcal{X}_{q,\gamma} \rightarrow \mathcal{O}(SL_q(2))\otimes\mathcal{X}_{q,\gamma}$ defined by $\varphi(x_i) = u_j^i \otimes x_j$, $i = 1, 2$, such that $\mathcal{X}_{q,\gamma}$ becomes a left quantum space for the quantum group $SL_q(2)$ with coaction φ. The algebra $\mathcal{X}_{q,\gamma}$ is called the coordinate algebra of the *quantum hyperboloid.* Obviously, $\mathcal{X}_{q^2,1}$ is just the quantum Weyl algebra $\mathcal{A}_q(1)$. The following proposition says that there exist two distinguished covariant FODC on the quantum space $\mathcal{X}_{q,\gamma}$.

Proposition 9. *Suppose that $\gamma \neq 0$ and $q^n \neq 1$ for $n \in \mathbb{N}$, $n \geq 2$. Then there are precisely two nonisomorphic left-covariant FODC Γ over $\mathcal{X}_{q,\gamma}$ with respect to the Hopf algebra $\mathcal{O}(SL_q(2))$ such that $\{\mathrm{d}x_1, \mathrm{d}x_2\}$ is a free right $\mathcal{X}_{q,\gamma}$-module basis of Γ. The bimodule structures for these two calculi are described by the formulas*

$$\begin{aligned} x_1{\cdot}\mathrm{d}x_2 &= (p-p^{-1}q^{-1})\mathrm{d}x_1{\cdot}x_2 + p^{-1}\mathrm{d}x_2{\cdot}x_1 + (1-p^{-1})\gamma^{-1}\omega_{\mathrm{inv}}\, x_1x_2, \\ x_2{\cdot}\mathrm{d}x_1 &= p^{-1}\mathrm{d}x_1{\cdot}x_2 + (p-qp^{-1})\mathrm{d}x_2{\cdot}x_1 + (1-p^{-1})\gamma^{-1}\omega_{\mathrm{inv}}\, x_2x_1, \\ x_j{\cdot}\mathrm{d}x_j &= p\mathrm{d}x_j{\cdot}x_j + (1-p^{-1})\gamma^{-1}\omega_{\mathrm{inv}}(x_j)^2 \quad \text{for} \quad j=1,2, \end{aligned}$$

where $p := \pm(q^{1/2}+q^{-1/2})-1$ and $\omega_{\mathrm{inv}} := \mathrm{d}x_1{\cdot}x_2 - q\,\mathrm{d}x_2{\cdot}x_1$.

This proposition is stated in [SS1]. We shall not carry out its proof.

The uniqueness assertion of Proposition 9 is no longer true if $q^3=1$ and $q \neq 1$. Indeed, let q be either $\frac{1}{2}(-1+\sqrt{-3})$ or $\frac{1}{2}(-1-\sqrt{-3})$. Then it can be shown that for any nonzero complex number α there exists a left-covariant FODC Γ_α over $\mathcal{X}_{q,\gamma}$ with respect to $\mathcal{O}(SL_q(2))$ such that the set $\{\mathrm{d}x_1, \mathrm{d}x_2\}$ is a free right module basis of Γ_α. The corresponding formulas for the bimodule Γ_α are

$$\begin{aligned} x_1{\cdot}\mathrm{d}y_2 &= (\alpha-\alpha^{-1}q^{-1})\mathrm{d}x_1{\cdot}x_2 + \alpha^{-1}\mathrm{d}x_2{\cdot}x_1 + \tilde{\alpha}\;\omega_{\mathrm{inv}}\; x_1x_2, \\ x_2{\cdot}\mathrm{d}x_1 &= \alpha^{-1}\mathrm{d}x_1{\cdot}x_2 + (\alpha-q\,\alpha^{-1})\mathrm{d}x_2{\cdot}x_1 + \tilde{\alpha}\;\omega_{\mathrm{inv}}\; x_2x_1, \\ x_j{\cdot}\mathrm{d}x_j &= \alpha\;\mathrm{d}x_j{\cdot}x_j + \tilde{\alpha}\;\omega_{\mathrm{inv}}\;(x_j)^2 \quad \text{for} \quad j=1,2, \end{aligned}$$

where $\tilde{\alpha} := -(\alpha+\alpha^{-1}+1)\gamma^{-1}$ and $\omega_{\mathrm{inv}} := \mathrm{d}x_1{\cdot}x_2 - q\,\mathrm{d}x_2{\cdot}x_1$.

12.4 Notes

Differential forms on noncommutative algebras and the differential envelope of an algebra appeared in the work of A. Connes, M. Karoubi, D. Kastler and others (see the references in [Con], [CQ] and [Ka]). The covariant differential calculus on the quantum vector space $\mathbb{C}_q^N$ was obtained independently in [WZ] and [PuW]. These papers were very influential for the study of noncommutative differential calculi, see (for instance) [CSW], [Sch], [Og], [Z], [BKO], [Sc1], [GZ]. First steps towards a q-deformed quantum mechanics have been made in [SW] and [HSSWW]. Apart from quantum vector spaces and quantum groups themselves, covariant differential calculi have been investigated on quantum 2-spheres (see [Pod2], [Pod3], [AS]). A theory of such calculi on general quantum spaces (if possible) is still at the very beginning.

13. Hopf Bimodules and Exterior Algebras

This chapter is devoted to some concepts (left-covariant bimodules, bicovariant bimodules) and constructions (tensor algebras, exterior algebras) which are basic tools for the covariant differential calculus on quantum groups in the next chapter. They are also of interest in themselves.

The structure theory of covariant bimodules is developed coordinate free in abstract Hopf algebra language. This allows us elegant statements and short proofs of the results. At the end of the subsections the results are reformulated in coordinate form which is more convenient for computations.

Throughout this chapter $\mathcal{A}$ is a Hopf algebra with invertible antipode S.

13.1 Covariant Bimodules

Covariant bimodules can be considered as quantum group analogs of vector fiber bundles over Lie groups which are endowed with left or/and right actions of the groups.

13.1.1 Left-Covariant Bimodules

Definition 1. *A* left-covariant bimodule *over $\mathcal{A}$ is an $\mathcal{A}$-bimodule Γ which is a left comodule of $\mathcal{A}$, with coaction $\Delta_L : \Gamma \to \mathcal{A} \otimes \Gamma$, such that*

$$\Delta_L(a\rho b) = \Delta(a)\Delta_L(\rho)\Delta(b) \quad \text{for} \quad a, b \in \mathcal{A} \quad \text{and} \quad \rho \in \Gamma. \tag{1}$$

In Sweedler's notation, the last condition can be written as

$$\sum (a\rho b)_{(-1)} \otimes (a\rho b)_{(0)} = \sum a_{(1)}\rho_{(-1)}b_{(1)} \otimes a_{(2)}\rho_{(0)}b_{(2)}. \tag{2}$$

An element ρ of a left-covariant bimodule Γ is called *left-invariant* if $\Delta_L(\rho) = 1 \otimes \rho$. The vector space of left-invariant elements of Γ is denoted by ${}_{\text{inv}}\Gamma$.

Lemma 1. *Let Γ be a left-covariant bimodule over $\mathcal{A}$. There exists a unique linear projection $P_L : \Gamma \to {}_{\text{inv}}\Gamma$ such that*

$$P_L(a\rho) = \varepsilon(a)P_L(\rho), \quad a \in \mathcal{A}, \ \rho \in \Gamma. \tag{3}$$

The map P_L is given by $P_L(\rho) = \sum S(\rho_{(-1)})\rho_{(0)}$. For $\rho \in \Gamma$ and $a \in \mathcal{A}$, we have

$$\rho = \sum \rho_{(-1)} P_L(\rho_{(0)}), \tag{4}$$

$$P_L(\rho a) = \sum S(a_{(1)}) P_L(\rho) a_{(2)} \equiv \mathrm{ad}_R(a)(P_L(\rho)), \tag{5}$$

where ad_R is the right adjoint action of $\mathcal{A}$ on the bimodule Γ (see Remark 2 in Subsect. 1.3.4).

Proof. Using the formulas (2), (1.30), (1.27), and (1.20) we have

$$\begin{aligned} \Delta_L(P_L(\rho)) &= \sum \Delta_L(S(\rho_{(-1)})\rho_{(0)}) \\ &= \sum S(\rho_{(-2)})\rho_{(-1)} \otimes S(\rho_{(-3)})\rho_{(0)} \\ &= \sum \varepsilon(\rho_{(-1)})1 \otimes S(\rho_{(-2)})\rho_{(0)} = 1 \otimes P_L(\rho), \end{aligned}$$

so that $P_L(\rho) \in {}_{\mathrm{inv}}\Gamma$. Using the formulas (2), (1.28), and (1.27) we get

$$\begin{aligned} P_L(a\rho) &= \sum S((a\rho)_{(-1)})(a\rho)_{(0)} = \sum S(a_{(1)}\rho_{(-1)})a_{(2)}\rho_{(0)} \\ &= \sum S(\rho_{(-1)})S(a_{(1)})a_{(2)}\rho_{(0)} \\ &= \sum S(\rho_{(-1)})\varepsilon(a)\rho_{(0)} = \varepsilon(a)P_L(\rho). \end{aligned}$$

The relations (4) and (5) follow in a similar way. The second equality in (5) is just the definition of $\mathrm{ad}_R(a)$. If $\rho \in {}_{\mathrm{inv}}\Gamma$, then $\Delta_L(\rho) = 1 \otimes \rho$ and hence $P_L(\rho) = S(1)\rho = \rho$, that is, P_L is a projection of Γ onto ${}_{\mathrm{inv}}\Gamma$. If P is another projection $P : \Gamma \to {}_{\mathrm{inv}}\Gamma$ satisfying (3), then by (4) and (3),

$$P(\rho) = \sum \varepsilon(\rho_{(-1)})P(P_L(\rho_{(0)})) = \sum P_L(\varepsilon(\rho_{(-1)})\rho_{(0)}) = P_L(\rho),$$

which proves the uniqueness assertion. □

We now begin to describe the structure of left-covariant bimodules. Let Γ_0 be an arbitrary right $\mathcal{A}$-module, that is, there exists a bilinear map $\Gamma_0 \times \mathcal{A} \ni (\omega, a) \to \omega \triangleleft a \in \Gamma_0$ such that $\omega \triangleleft (ab) = (\omega \triangleleft a) \triangleleft b$ and $\omega \triangleleft 1 = \omega$ for $a, b \in \mathcal{A}$ and $\omega \in \Gamma_0$. It is easy to check that the equations

$$a(b \otimes \omega)c = \sum abc_{(1)} \otimes \omega \triangleleft c_{(2)}, \tag{6}$$

$$\Delta_L(b \otimes \omega) = \sum b_{(1)} \otimes b_{(2)} \otimes \omega \tag{7}$$

define $\mathcal{A}$-bimodule and left $\mathcal{A}$-comodule structures on the vector space $\Gamma_L := \mathcal{A} \otimes \Gamma_0$ such that Γ_L becomes a left-covariant bimodule over $\mathcal{A}$. As an example, we verify the right $\mathcal{A}$-module property of Γ_L by computing

$$\begin{aligned} (a \otimes \omega)bc &= \sum a(bc)_{(1)} \otimes \omega \triangleleft ((bc)_{(2)}) = \sum ab_{(1)}c_{(1)} \otimes (\omega \triangleleft b_{(2)}) \triangleleft c_{(2)} \\ &= \sum (ab_{(1)} \otimes \omega \triangleleft b_{(2)})c = ((a \otimes \omega)b)c. \end{aligned}$$

Obviously, we have ${}_{\mathrm{inv}}(\Gamma_L) = 1 \otimes \Gamma_0$. Moreover, from the formulas (6) and (3) we obtain $P_L((1 \otimes \omega)b) = \sum \varepsilon(b_{(1)})(1 \otimes \omega \triangleleft b_{(2)}) = 1 \otimes \omega \triangleleft b$, that is, the right $\mathcal{A}$-module structure on Γ_0 can be recovered from the left-covariant bimodule Γ_L.

Conversely, suppose that Γ is an arbitrary left-covariant bimodule over $\mathcal{A}$. We proceed in the opposite direction and show that up to isomorphism Γ is of the form just described. For $a \in \mathcal{A}$ and $\omega \in {}_{\mathrm{inv}}\Gamma$ we put

$$\omega \triangleleft a := P_L(\omega a) = \mathrm{ad}_R(a)\omega, \tag{8}$$

where the second equality holds by (5). By Proposition 1.14(i) and Remark 2 after it, this defines a right $\mathcal{A}$-module structure on ${}_{\mathrm{inv}}\Gamma$. Let $\Gamma_L := \mathcal{A} \otimes {}_{\mathrm{inv}}\Gamma$ denote the left-covariant bimodule over $\mathcal{A}$ which is given by (6) and (7) with respect to this right $\mathcal{A}$-module action. We assert that the linear mapping $\vartheta := (\mathrm{id} \otimes P_L) \circ \Delta_L : \Gamma \to \Gamma_L$ is an isomorphism of the left-covariant bimodules Γ and Γ_L. To prove this, we let $a, b \in \mathcal{A}$ and $\omega \in {}_{\mathrm{inv}}\Gamma$ and compute

$$\vartheta((a\omega)b) = (\mathrm{id} \otimes P_L)\Delta(a)\Delta_L(\omega)\Delta(b) = \sum a_{(1)}b_{(1)} \otimes P_L(a_{(2)}\omega b_{(2)})$$

$$= \sum a_{(1)}b_{(1)} \otimes \varepsilon(a_{(2)})P_L(\omega b_{(2)}) = \sum ab_{(1)} \otimes \omega \triangleleft b_{(2)} = (a \otimes \omega)b, \tag{9}$$

where we used (1), (1.20) and (6). Recall that $\Gamma = \mathcal{A} \cdot {}_{\mathrm{inv}}\Gamma$ by (4). Therefore, applying (9) with $b = 1$, we see that ϑ is a bijective linear mapping of Γ and Γ_L. It is obvious that ϑ preserves the left $\mathcal{A}$-module structures and that $\Delta_L \circ \vartheta = (\mathrm{id} \otimes \vartheta) \circ \Delta_L$. By (9), ϑ preserves also the right $\mathcal{A}$-module structures of Γ and Γ_L. That is, ϑ provides an isomorphism of the left-covariant bimodules Γ and Γ_L. For notational simplicity we shall identify Γ and Γ_L via ϑ, that is, we set $a\omega = a \otimes \omega$.

Summarizing, the preceding considerations have shown that *there is a one-to-one correspondence, given by the formulas (6)–(8), between left-covariant bimodules Γ and right $\mathcal{A}$-module structures on the vector space ${}_{\mathrm{inv}}\Gamma$*. More precisely, we have proved that Γ is isomorphic to the left-covariant bimodule $\mathcal{A} \otimes {}_{\mathrm{inv}}\Gamma$ with structures defined by the formulas (6)–(7).

Now we use the assumption that the antipode S is invertible. For $a \in \mathcal{A}$ and $\omega \in {}_{\mathrm{inv}}\Gamma$, we then have

$$a\omega = \sum (\omega \triangleleft S^{-1}(a_{(2)}))a_{(1)}. \tag{10}$$

Indeed, setting $b = S^{-1}(a)$ and using (6), we obtain

$$\begin{aligned}
\sum (\omega \triangleleft S^{-1}(a_{(2)}))a_{(1)} &= \sum a_{(1)}((\omega \triangleleft S^{-1}(a_{(3)})) \triangleleft a_{(2)}) \\
&= \sum a_{(1)}\omega \triangleleft (S^{-1}(a_{(3)})a_{(2)}) \\
&= \sum S(b_{(3)})\omega \triangleleft (b_{(1)}S(b_{(2)})) \\
&= \sum S(b_{(2)})\omega \triangleleft (\varepsilon(b_{(1)})1) \\
&= S(b)\omega = a\omega.
\end{aligned}$$

In particular, (10) implies that $\Gamma = {}_{\rm inv}\Gamma \cdot \mathcal{A}$.

Similar considerations as above can be done with $\mathcal{A} \otimes {}_{\rm inv}\Gamma$ replaced by ${}_{\rm inv}\Gamma \otimes \mathcal{A}$. For any right $\mathcal{A}$-module Γ_0 with right action $(\omega, a) \to \omega \triangleleft a$, one can verify that the vector space $\Gamma_R := \Gamma_0 \otimes \mathcal{A}$ becomes a left-covariant bimodule by defining

$$a(\omega \otimes b)c := \sum \omega \triangleleft S^{-1}(a_{(2)}) \otimes a_{(1)}bc, \tag{11}$$

$$\Delta_L(\omega \otimes b) := \sum b_{(1)} \otimes \omega \otimes b_{(2)}. \tag{12}$$

Conversely, let Γ be a left-covariant bimodule over $\mathcal{A}$. Let $\Gamma_R := {}_{\rm inv}\Gamma \otimes \mathcal{A}$ be the left-covariant bimodule with structures given by (11) and (12), where $\omega \triangleleft a := P_L(\omega a), \omega \in {}_{\rm inv}\Gamma$. Then the mapping $\theta : \Gamma \to \Gamma_R$ defined by

$$\theta(\rho) = \sum P_L(\rho_{(0)}) \triangleleft S^{-1}(\rho_{(-1)}) \otimes \rho_{(-2)} \tag{13}$$

provides an isomorphism of the left-covariant bimodules Γ and Γ_R. We only prove that θ is bijective. Define a linear mapping $\theta^{-1} : \Gamma_R \to \Gamma$ by $\theta^{-1}(\omega \otimes a) = \omega a$. Then θ^{-1} is indeed the inverse of θ, since by (13), (6), (4), and (8) we have

$$\begin{aligned}
\theta^{-1}(\theta(\rho)) &= \sum (P_L(\rho_{(0)}) \triangleleft S^{-1}(\rho_{(-1)}))\rho_{(-2)} \\
&= \sum \rho_{(-3)}(P_L(\rho_{(0)}) \triangleleft S^{-1}(\rho_{(-1)}) \triangleleft \rho_{(-2)}) \\
&= \sum \rho_{(-1)} P_L(\rho_{(0)}) = \rho, \\
\theta(\theta^{-1}(\omega \otimes b)) &= \theta(\omega b) = \sum P_L(\omega b_{(3)}) \triangleleft S^{-1}(b_{(2)}) \otimes b_{(1)} \\
&= \sum (\omega \triangleleft b_{(3)}) \triangleleft S^{-1}(b_{(2)}) \otimes b_{(1)} = \omega \otimes b.
\end{aligned}$$

We close this subsection by restating some of the above results in coordinate form. Let I be an index set. We call matrices $(v^i_j)_{i,j \in I}$ and $(f^i_j)_{i,j \in I}$ of elements $v^i_j \in \mathcal{A}$ and functionals $f^i_j \in \mathcal{A}'$ *pointwise finite* if for any $i \in I$, resp. $i \in I$ and $a \in \mathcal{A}$, all but finitely many terms v^i_j resp. $f^i_j(a)$ vanish. This ensures that all sums over the index set I occurring below are in fact finite sums. Recall the notation $f.a = \sum a_{(1)} f(a_{(2)})$ for $a \in \mathcal{A}$ and $f \in \mathcal{A}'$.

Proposition 2. *Let Γ be a left-covariant bimodule over $\mathcal{A}$ and let $\{\omega_i\}_{i \in I}$ be a basis of the vector space ${}_{\rm inv}\Gamma$. Then the set $\{\omega_i\}_{i \in I}$ is a free left $\mathcal{A}$-module basis of Γ and there exists a pointwise finite matrix $(f^i_j)_{i,j \in I}$ consisting of functionals $f^i_j \in \mathcal{A}'$ such that for any $a, b \in \mathcal{A}$ and $i, j \in I$,*

$$f^i_j(ab) = f^i_k(a) f^k_j(b), \quad f^i_j(1) = \delta_{ij}, \tag{14}$$

$$\omega_i a = (f^i_k.a)\omega_k. \tag{15}$$

The set $\{\omega_i\}_{i \in I}$ is also a free right $\mathcal{A}$-module basis of Γ and we have

$$a\omega_i = \omega_k((f^i_k \circ S^{-1}).a) . \tag{16}$$

Proof. Since $\omega_i \triangleleft a \in {}_{\mathrm{inv}}\Gamma$, there exist complex numbers $f^i_j(a), i, j \in I$, such that all but finitely many $f^i_j(a)$ vanish and

$$\omega_i \triangleleft a = f^i_j(a)\omega_j \ . \tag{17}$$

Clearly, $(f^i_j)_{i,j\in I}$ is a pointwise finite matrix with entries $f^i_j \in \mathcal{A}'$. The right $\mathcal{A}$-module property of ${}_{\mathrm{inv}}\Gamma$ implies (14). Equations (6) and (10) yield (15) and (16). Since $\vartheta : \Gamma \to \mathcal{A} \otimes {}_{\mathrm{inv}}\Gamma$ and $\theta : \Gamma \to {}_{\mathrm{inv}}\Gamma \otimes \mathcal{A}$ are bijective mappings, $\{\omega_i\}_{i\in I}$ is a free left $\mathcal{A}$-module basis and a free right $\mathcal{A}$-module basis of Γ. □

13.1.2 Right-Covariant Bimodules

Definition 2. *A* right-covariant bimodule *over $\mathcal{A}$ is an $\mathcal{A}$-bimodule Γ equipped with a right coaction $\Delta_R : \Gamma \to \Gamma \otimes \mathcal{A}$ of $\mathcal{A}$ on Γ such that*

$$\Delta_R(a\rho b) = \Delta(a)\Delta_R(\rho)\Delta(b) \quad \textit{for any} \quad a, b \in \mathcal{A} \quad \textit{and} \quad \rho \in \Gamma.$$

For a right-covariant bimodule Γ, the elements of the vector space

$$\Gamma_{\mathrm{inv}} := \{\rho \in \Gamma \mid \Delta_R(\rho) = \rho \otimes 1\}$$

are called *right-invariant.* For right-covariant bimodules a similar theory as in Subsect. 13.1.1 can be developed. We will not carry out the details and mention only some necessary modifications. The canonical projection $P_R : \Gamma \to \Gamma_{\mathrm{inv}}$ is defined by $P_R(\rho) = \sum \rho_{(0)} S(\rho_{(1)})$ and satisfies

$$P_R(\rho a) = \varepsilon(a) P_R(\rho), \quad P_R(a\rho) = \sum a_{(1)} P_R(\rho) S(a_{(2)}) \equiv \mathrm{ad}_L(a)(P_R(\rho))$$

for $a \in \mathcal{A}$ and $\rho \in \Gamma$. The left $\mathcal{A}$-module structure on Γ_{inv} defined by the left adjoint action of $\mathcal{A}$ on $\Gamma_{\mathrm{inv}}, a \triangleright \eta := P_R(a\eta) = \mathrm{ad}_L(a)(\eta)$, determines completely the structure of the right-covariant bimodule Γ, since $\Gamma = \Gamma_{\mathrm{inv}} \cdot \mathcal{A}$ and $a(\eta b) = \sum (a_{(1)} \triangleright \eta) a_{(2)} b$ for $a, b \in \mathcal{A}$ and $\eta \in \Gamma_{\mathrm{inv}}$.

13.1.3 Bicovariant Bimodules (Hopf Bimodules)

The main notion of this chapter is introduced by the following

Definition 3. *A* bicovariant bimodule *(or a* Hopf bimodule*) over $\mathcal{A}$ is an $\mathcal{A}$-bimodule Γ together with linear mappings $\Delta_L : \Gamma \to \mathcal{A} \otimes \Gamma$ and $\Delta_R : \Gamma \to \Gamma \otimes \mathcal{A}$ such that:*

(i) *Γ is a left-covariant bimodule over $\mathcal{A}$ with left coaction Δ_L,*
(ii) *Γ is a right-covariant bimodule over $\mathcal{A}$ with right coaction Δ_R,*
(iii) $(\mathrm{id} \otimes \Delta_R) \circ \Delta_L = (\Delta_L \otimes \mathrm{id}) \circ \Delta_R$.

Condition (iii) means that the left coaction Δ_L and the right coaction Δ_R of $\mathcal{A}$ on Γ commute. This condition justifies the use of the Sweedler notation $\sum \rho_{(-1)} \otimes \rho_{(0)} \otimes \rho_{(1)}$ for $(\mathrm{id} \otimes \Delta_R) \circ \Delta_L(\rho) = (\Delta_L \otimes \mathrm{id}) \circ \Delta_R(\rho), \ \rho \in \Gamma$.

The subsequent considerations will determine the structure of bicovariant bimodules. Let Γ_0 be a vector space carrying the structures of a right $\mathcal{A}$-module, with right action denoted by $\triangleleft$, and of a right comodule, with right coaction $\delta_R : \Gamma_0 \to \Gamma_0 \otimes \mathcal{A}$, such that the compatibility condition

$$\sum \omega_{(0)} \triangleleft a_{(1)} \otimes \omega_{(1)} a_{(2)} = \sum (\omega \triangleleft a_{(2)})_{(0)} \otimes a_{(1)} (\omega \triangleleft a_{(2)})_{(1)} \qquad (18)$$

is fulfilled for $\omega \in \Gamma_0$ and $a \in \mathcal{A}$. Equation (18) is called the *Yetter–Drinfeld condition.* A vector space with these properties is usually called a *Yetter–Drinfeld module.*

As discussed in Subsect. 13.1.1, the right $\mathcal{A}$-module action on Γ_0 induces the structure of a left-covariant bimodule over $\mathcal{A}$ on $\Gamma := \mathcal{A} \otimes \Gamma_0$ given by (6) and (7). We now define a mapping $\Delta_R : \Gamma \to \Gamma \otimes \mathcal{A}$ by setting

$$\Delta_R(a \otimes \omega) = \sum a_{(1)} \otimes \omega_{(0)} \otimes a_{(2)} \omega_{(1)} \quad \text{for} \quad a \in \mathcal{A} \quad \text{and} \quad \omega \in \Gamma_0 \, . \quad (19)$$

Then Γ becomes a right $\mathcal{A}$-comodule. It is obvious that $\Delta_R(a\rho) = \Delta(a)\Delta_R(\rho)$ and $(\mathrm{id} \otimes \Delta_R) \circ \Delta_L(\rho) = (\Delta_L \otimes \mathrm{id}) \circ \Delta_R(\rho)$ for $\rho \in \Gamma$ and $a \in \mathcal{A}$. Thus Γ is a bicovariant bimodule over $\mathcal{A}$ provided we have verified the condition $\Delta_R(\rho a) = \Delta_R(\rho)\Delta(a)$. In fact, the latter is equivalent to the Yetter–Drinfeld condition (18). By (6) and (19), we have

$$\begin{aligned} \Delta_R((b \otimes \omega)a) &= \sum \Delta_R(ba_{(1)} \otimes \omega \triangleleft a_{(2)}) \\ &= \sum b_{(1)} a_{(1)} \otimes (\omega \triangleleft a_{(3)})_{(0)} \otimes b_{(2)} a_{(2)} (\omega \triangleleft a_{(3)})_{(1)} \quad (20) \\ \Delta_R(b \otimes \omega)\Delta(a) &= \sum b_{(1)} \otimes \omega_{(0)}) a_{(1)} \otimes b_{(2)} \omega_{(1)} a_{(2)} \\ &= \sum b_{(1)} a_{(1)} \otimes \omega_{(0)} \triangleleft a_{(2)} \otimes b_{(2)} \omega_{(1)} a_{(2)} \quad (21) \end{aligned}$$

for $a, b \in \mathcal{A}$ and $\omega \in {}_{\mathrm{inv}}\Gamma$. By (18), the right hand sides of (20) and (21) coincide. Therefore, Γ is a bicovariant bimodule over $\mathcal{A}$.

Conversely, let Γ be a bicovariant bimodule over $\mathcal{A}$. Recall from Subsect. 13.1.1 that $\omega \triangleleft a := P_L(\omega a)$ defines a right $\mathcal{A}$-module structure on ${}_{\mathrm{inv}}\Gamma$. By condition (iii) in Definition 3, we have

$$(\Delta_L \otimes \mathrm{id}) \circ \Delta_R(\omega) = (\mathrm{id} \otimes \Delta_R) \circ \Delta_L(\omega) = 1 \otimes \Delta_R(\omega) \quad \text{for} \quad \omega \in {}_{\mathrm{inv}}\Gamma,$$

so that $\Delta_R({}_{\mathrm{inv}}\Gamma) \subseteq {}_{\mathrm{inv}}\Gamma \otimes \mathcal{A}$. Let δ_R denote the restriction of Δ_R to ${}_{\mathrm{inv}}\Gamma$. As in Subsect. 13.1.1 we identify Γ and $\mathcal{A} \otimes {}_{\mathrm{inv}}\Gamma$ by setting $a\omega = a \otimes \omega$ for $a \in \mathcal{A}$ and $\omega \in {}_{\mathrm{inv}}\Gamma$. Since Γ is a bicovariant bimodule, the left-hand sides of (20) and (21) coincide. Applying $\varepsilon \otimes \mathrm{id}$ to the expressions of the right-hand sides of (20) and (21), we obtain the Yetter–Drinfeld condition (18).

Thus, we have proved that *there is a one-to-one correspondence, given by the formulas (6)–(8) and (19), between bicovariant bimodules Γ over $\mathcal{A}$ and pairs of a right $\mathcal{A}$-module and a right $\mathcal{A}$-comodule structures on the vector space ${}_{\mathrm{inv}}\Gamma$ satisfying the Yetter–Drinfeld condition (18).*

Next we show that the linear mapping $\Phi : \Gamma \to \Gamma$ defined by

$$\Phi(\rho) = \sum S(\rho_{(-1)})\rho_{(0)}S(\rho_{(1)}) \tag{22}$$

is bijective. Indeed, putting $\Psi(\rho) = \sum S^{-1}(\rho_{(1)})\rho_{(0)}S^{-1}(\rho_{(-1)})$, we compute

$$\sum \Phi(\rho)_{(-1)} \otimes \Phi(\rho)_{(0)} \otimes \Phi(\rho)_{(1)} = \sum S(\rho_{(2)}) \otimes S(\rho_{(-1)})\rho_{(0)}S(\rho_{(1)}) \otimes S(\rho_{(-2)}),$$

so

$$\begin{aligned} \Psi(\Phi(\rho)) &= \sum S^{-1}(\Phi(\rho)_{(1)})\Phi(\rho)_{(0)}S^{-1}(\Phi(\rho)_{(-1)}) \\ &= \sum S^{-1}(S(\rho_{(-2)}))(S(\rho_{(-1)})\rho_{(0)}S(\rho_{(1)}))S^{-1}(S(\rho_{(2)})) \\ &= \sum \rho_{(-2)}S(\rho_{(-1)})\rho_{(0)}S(\rho_{(1)})\rho_{(2)} = \rho. \end{aligned}$$

Similarly, $\Phi(\Psi(\rho)) = \rho$ for $\rho \in \Gamma$. Hence Φ is bijective with inverse Ψ.

For $\omega \in {}_{\mathrm{inv}}\Gamma$ and $\eta \in \Gamma_{\mathrm{inv}}$, we have $\Phi(\omega) = \sum \omega_{(0)}S(\omega_{(1)}) = P_R(\omega)$ and $\Psi(\eta) = \sum \eta_{(0)}S^{-1}(\eta_{(-1)}) \in {}_{\mathrm{inv}}\Gamma$, so $\Phi : {}_{\mathrm{inv}}\Gamma \to \Gamma_{\mathrm{inv}}$ and $\Psi : \Gamma_{\mathrm{inv}} \to {}_{\mathrm{inv}}\Gamma$. That is, Φ is an isomorphism of the vector spaces ${}_{\mathrm{inv}}\Gamma$ and Γ_{inv}.

We now summarize and reformulate the structure theory of bicovariant bimodules developed above in the coordinate version by the following

Theorem 3 (*Fundamental theorem for bicovariant bimodules*). *Suppose that Γ is a bicovariant bimodule over $\mathcal{A}$ and let $\{\omega_i\}_{i\in I}$ be a basis of the vector space ${}_{\mathrm{inv}}\Gamma$ of left-invariant forms from Γ. Then there exist pointwise finite matrices $v = (v^i_j)_{i,j\in I}$ and $f = (f^i_j)_{i,j\in I}$ of elements $v^i_j \in \mathcal{A}$ and functionals $f^i_j \in \mathcal{A}'$ such that for $a, b \in \mathcal{A}$ and $i, j \in I$ we have*

(i) $\omega_i a = (f^i_k.a)\omega_k$,

(ii) $\Delta_R(\omega_i) = \omega_k \otimes v^k_i$,

(iii) *$f = (f^i_j)$ is an algebra representation of $\mathcal{A}$ (that is, $f^i_j(ab) = f^i_k(a)f^k_j(b)$ and $f^i_j(1) = \delta_{ij}$),*

(iv) *$v = (v^i_j)$ is a matrix corepresentation of $\mathcal{A}$ (that is, $\Delta(v^i_j) = v^i_k \otimes v^k_j$ and $\varepsilon(v^i_j) = \delta_{ij}$),*

(v) *the following equality holds:*

$$\sum\nolimits_k v^k_i(a.f^k_j) = \sum\nolimits_k (f^i_k.a)v^j_k. \tag{23}$$

Moreover, $\{\eta_i := \omega_j S(v^j_i)\}_{i\in I}$ is a basis of the vector space Γ_{inv} of right-invariant forms from Γ. Both sets $\{\omega_i\}_{i\in I}$ and $\{\eta_i\}_{i\in I}$ are free left $\mathcal{A}$-module bases and free right $\mathcal{A}$-module bases of Γ.

Conversely, if $\{\omega_i\}_{i\in I}$ is a basis of a vector space Γ_0 and if $v = (v^i_j)_{i,j\in I}$ and $f = (f^i_j)_{i,j\in I}$ are pointwise finite matrices with $v^i_j \in \mathcal{A}$ and $f^i_j \in \mathcal{A}'$ satisfying (iii)–(v), *then there exists a unique bicovariant bimodule Γ, denoted by the symbol (v, f), such that $\Gamma_0 = \Gamma_{\mathrm{inv}}$ and* (i) *and* (ii) *hold.*

Proof. Clearly, there exists a one-to-one correspondence, given by (17), between right $\mathcal{A}$-module structures on ${}_{\rm inv}\Gamma$ and pointwise finite matrices $f = (f^i_j)_{i,j\in I}$ satisfying (iii). As noted in the proof of Proposition 2, condition (i) follows from (6). Moreover, there is a one-to-one correspondence, given by (ii), between right comodule structures on ${}_{\rm inv}\Gamma$ and pointwise finite matrices $v = (v^i_j)$ satisfying (iv). From (22) we obtain $\eta_i = \Phi(\omega_i)$. To complete the proof it suffices to check that the Yetter–Drinfeld condition (18) is equivalent to the compatibility condition (23). For $i \in I$ and $a \in \mathcal{A}$, we compute

$$\begin{aligned}\sum (\omega_i)_{(0)} \triangleleft a_{(1)} \otimes (\omega_i)_{(1)} a_{(2)} &= \sum\nolimits_k \omega_j \otimes v^k_i (a.f^k_j),\\ \sum (\omega_i \triangleleft a_{(2)})_{(0)} \otimes a_{(1)} (\omega_i \triangleleft a_{(2)})_{(1)} &= \sum\nolimits_k \omega_j \otimes (f^i_k.a) v^j_k\,.\end{aligned}$$

Hence (18) and (23) are equivalent. □

The next corollary shows that in the case of the quantum matrix groups the compatibility condition (23) reduces to an intertwining property.

Corollary 4. *Let $\mathcal{A}$ be the coordinate Hopf algebra $\mathcal{O}(G_q)$ for one of the quantum groups $G_q = GL_q(N), SL_q(N), O_q(N), Sp_q(N)$ and let $\mathbf{u} = (u^m_n)$ denote the fundamental corepresentation of $\mathcal{A}$. Suppose that $v^i_j \in \mathcal{A}$ and $f^i_j \in \mathcal{A}', i,j \in I$, are given such that the matrices $v = (v^i_j)$ and $f = (f^i_j)$ satisfy the conditions* (iii) *and* (iv) *of Theorem 3. Set $A^{mj}_{in} := f^i_j(u^m_n)$ for $i,j \in I$ and $m,n = 1,2,\cdots,N$. Then the matrices v and f fulfill equation (23) (and hence they define a bicovariant bimodule (v,f) over $\mathcal{A}$ by Theorem 3) if and only if $A \equiv (A^{mj}_{in})$ belongs to* Mor $(v \otimes \mathbf{u}, \mathbf{u} \otimes v)$.

Proof. Using (iii) it is easily seen that (23) is true for ab if it is valid for a and b. Since $f^i_j(1) = \delta_{ij}$, condition (23) holds for $a = 1$. Thus it suffices to check it for the generators u^r_s of $\mathcal{A}$ and in addition for the inverse $\mathcal{D}_q^{-1}$ of the quantum determinant $\mathcal{D}_q$ in the case $G_q = GL_q(N)$. From the equations

$$v^k_i (u^r_s.f^k_j) = A^{rj}_{km} v^k_i u^m_s \quad \text{and} \quad (f^i_k.u^r_s) v^j_k = u^r_m v^j_k A^{mk}_{is}$$

we see that (23) holds for all u^r_s if and only if $A \in$ Mor $(v \otimes \mathbf{u}, \mathbf{u} \otimes v)$. If $G_q = GL_q(N)$, then, as we have just shown, the relation $A \in$ Mor $(v\otimes\mathbf{u}, \mathbf{u}\otimes v)$ implies in particular that (23) holds for $a = \mathcal{D}_q$. Multiplying equation (23) for $a = \mathcal{D}_q$ by $f^j_r(\mathcal{D}_q^{-1}) f^s_i(\mathcal{D}_q^{-1}) \mathcal{D}_q^{-2}$, summing over i and j and using (iii) we obtain condition (23) for $a = \mathcal{D}_q^{-1}$. □

13.1.4 Woronowicz' Braiding of Bicovariant Bimodules

One of the most interesting features of bicovariant bimodules is that they possess a natural braiding of tensor products. The definition and properties of this braiding are developed in this subsection.

Let $\Gamma_1, \cdots, \Gamma_n$ be bicovariant bimodules over $\mathcal{A}$. It is easy to check that the tensor product $\Gamma_1 \otimes_{\mathcal{A}} \cdots \otimes_{\mathcal{A}} \Gamma_n$ of $\mathcal{A}$-bimodules is also a bicovariant bimodule Γ over $\mathcal{A}$ with bimodule structure, left coaction Δ_L and right coaction Δ_R defined by

$$\begin{aligned} a(\rho^1 \otimes_{\mathcal{A}} \cdots \otimes_{\mathcal{A}} \rho^n)b &= (a\rho^1) \otimes_{\mathcal{A}} \cdots \otimes_{\mathcal{A}} (\rho^n b), \\ \Delta_L(\rho^1 \otimes_{\mathcal{A}} \cdots \otimes_{\mathcal{A}} \rho^n) &= \sum \rho^1_{(-1)} \cdots \rho^n_{(-1)} \otimes (\rho^1_{(0)} \otimes_{\mathcal{A}} \cdots \otimes_{\mathcal{A}} \rho^n_{(0)}), \\ \Delta_R(\rho^1 \otimes_{\mathcal{A}} \cdots \otimes_{\mathcal{A}} \rho^n) &= \sum (\rho^1_{(0)} \otimes_{\mathcal{A}} \cdots \otimes_{\mathcal{A}} \rho^n_{(0)}) \otimes \rho^1_{(1)} \cdots \rho^n_{(1)} \end{aligned}$$

for $\rho^1 \in \Gamma_1, \cdots, \rho^n \in \Gamma_n$ and $a, b \in \mathcal{A}$. Clearly, ${}_{\rm inv}\Gamma = {}_{\rm inv}\Gamma_1 \otimes_{\mathcal{A}} \cdots \otimes_{\mathcal{A}} {}_{\rm inv}\Gamma_n$. This bicovariant bimodule Γ is called the *tensor product of the bicovariant bimodules* $\Gamma_1, \cdots, \Gamma_n$ and is denoted by $\Gamma_1 \otimes_{\mathcal{A}} \cdots \otimes_{\mathcal{A}} \Gamma_n$. In the case $\Gamma_1 = \cdots = \Gamma_n$ we simply write (with a slight abuse of notation) $(\Gamma_1)^{\otimes n}$ for Γ.

By a *homomorphism of bicovariant bimodules* Γ_1 and Γ_2 over $\mathcal{A}$ we mean an $\mathcal{A}$-bimodule homomorphism $T : \Gamma_1 \to \Gamma_2$ which intertwines the left and the right coactions (that is, $T(a\rho b) = aT(\rho)b$ for $a, b \in \mathcal{A}$, $\rho \in \Gamma$; $({\rm id} \otimes T) \circ \Delta_L = \Delta_L \circ T$ and $(T \otimes {\rm id}) \circ \Delta_R = \Delta_R \circ T$). If T is also bijective, it is called an *isomorphism* of Γ_1 and Γ_2.

Definition 4. *The linear mapping* $\sigma : \Gamma_1 \otimes_{\mathcal{A}} \Gamma_2 \to \Gamma_2 \otimes_{\mathcal{A}} \Gamma_1$ *given by*

$$\begin{aligned} \sigma(\rho^1 \otimes_{\mathcal{A}} \rho^2) &:= \sum \rho^1_{(-1)} P_R(\rho^2_{(0)}) \otimes_{\mathcal{A}} P_L(\rho^1_{(0)}) \rho^2_{(1)} \qquad (24) \\ &= \sum \rho^1_{(-2)} \rho^2_{(0)} S(\rho^2_{(1)}) \otimes_{\mathcal{A}} S(\rho^1_{(-1)}) \rho^1_{(0)} \rho^2_{(2)} \end{aligned}$$

for $\rho^1 \in \Gamma_1$ *and* $\rho^2 \in \Gamma_2$ *is called the* braiding *of the bicovariant bimodules* Γ_1 *and* Γ_2.

This terminology stems from the facts that the bicovariant bimodules over $\mathcal{A}$ form a braided tensor category (see Subsect. 10.3.4 for this notion) and σ is the braiding in this category. We shall not carry out the details of the proof, because we will not need this in the following.

One verifies that the linear mapping $\sigma^{-1} : \Gamma_2 \otimes_{\mathcal{A}} \Gamma_1 \to \Gamma_1 \otimes_{\mathcal{A}} \Gamma_2$ given by

$$\sigma^{-1}(\rho^2 \otimes_{\mathcal{A}} \rho^1) := \sum \rho^2_{(2)} \rho^1_{(0)} S^{-1}(\rho^1_{(-1)}) \otimes_{\mathcal{A}} S^{-1}(\rho^2_{(1)}) \rho^2_{(0)} \rho^1_{(-2)}$$

is the inverse of σ. Thus, in particular, σ is bijective.

We give another form of σ and σ^{-1} which is more convenient for applications. For $a \in \mathcal{A}$, $\omega^1 \in {}_{\rm inv}\Gamma_1$ and $\omega^2 \in {}_{\rm inv}\Gamma_2$, we have

$$\begin{aligned} \sigma(a\omega^1 \otimes_{\mathcal{A}} \omega^2) &= \sum a\omega^2_{(0)} \otimes_{\mathcal{A}} (\omega^1 \triangleleft \omega^2_{(1)}), \qquad (25) \\ \sigma^{-1}(a\omega^2 \otimes_{\mathcal{A}} \omega^1) &= \sum a(\omega^1 \triangleleft S^{-1}(\omega^2_{(1)})) \otimes_{\mathcal{A}} \omega^2_{(0)}, \end{aligned}$$

where $\triangleleft$ is the right adjoint action of $\mathcal{A}$ on ${}_{\rm inv}\Gamma_k$, $k = 1, 2$, (see (8)). We verify the formula (25) and set $\rho^1 = a\omega^1$ and $\rho^2 = \omega^2$ in (24). Then we have

$$\sum \rho^1_{(-2)} \otimes \rho^1_{(-1)} \otimes \rho^1_{(0)} = \sum a_{(1)} \otimes a_{(2)} \otimes a_{(3)}\omega^1,$$

so that

$$\begin{aligned}\sigma(\rho^1 \otimes_{\mathcal{A}} \rho^2) &= \sum a_{(1)}\omega^2_{(0)} S(\omega^2_{(1)}) \otimes_{\mathcal{A}} S(a_{(2)})a_{(3)}\omega^1\omega^2_{(2)} \\ &= \sum a_{(1)}\omega^2_{(0)} \otimes_{\mathcal{A}} S(\omega^2_{(1)})\varepsilon(a_{(2)})\omega^1\omega^2_{(2)} \\ &= \sum a\omega^2_{(0)} \otimes_{\mathcal{A}} (\omega^1 \triangleleft \omega^2_{(1)}).\end{aligned}$$

Similar reasoning shows that for $a \in \mathcal{A}$, $\eta^1 \in (\Gamma_1)_{\rm inv}$ and $\eta^2 \in (\Gamma_2)_{\rm inv}$,

$$\sigma(\eta^1 \otimes_{\mathcal{A}} \eta^2 a) = \sum (\eta^1_{(-1)} \triangleright \eta^2) \otimes_{\mathcal{A}} \eta^1_{(0)} a, \tag{26}$$

where $b \triangleright \eta^2 := P_R(b\eta^2) = {\rm ad}_L(b)(\eta^2)$ for $b \in \mathcal{A}$ (see Subsect. 13.1.2).

Proposition 5. *The mapping σ defined by (24) is an isomorphism of the bicovariant bimodules $\Gamma_1 \otimes_{\mathcal{A}} \Gamma_2$ and $\Gamma_2 \otimes_{\mathcal{A}} \Gamma_1$ over $\mathcal{A}$. It is the unique linear map $\sigma : \Gamma_1 \otimes_{\mathcal{A}} \Gamma_2 \to \Gamma_2 \otimes_{\mathcal{A}} \Gamma_1$ such that*

$$\sigma(a\omega \otimes_{\mathcal{A}} \eta) = a\eta \otimes_{\mathcal{A}} \omega \quad \text{for} \quad a \in \mathcal{A},\ \omega \in {}_{\rm inv}\Gamma_1 \ \text{ and } \ \eta \in (\Gamma_2)_{\rm inv}\,. \tag{27}$$

The map σ satisfies the braid relation on $\Gamma_1 \otimes_{\mathcal{A}} \Gamma_2 \otimes_{\mathcal{A}} \Gamma_3$, that is,

$$(\sigma \otimes {\rm id})({\rm id} \otimes \sigma)(\sigma \otimes {\rm id}) = ({\rm id} \otimes \sigma)(\sigma \otimes {\rm id})({\rm id} \otimes \sigma)\,. \tag{28}$$

Proof. From Subsect. 13.1.3 it is clear that $\Gamma_1 \otimes_{\mathcal{A}} \Gamma_2$ is linearly spanned by each of the sets $\{a\omega^1 \otimes_{\mathcal{A}} \eta^2\}$, $\{a\omega^1 \otimes_{\mathcal{A}} \omega^2\}$ and $\{\eta^1 \otimes_{\mathcal{A}} \eta^2 b\}$ with $a, b \in \mathcal{A}$, $\omega^k \in {}_{\rm inv}\Gamma_k$ and $\eta^k \in (\Gamma_k)_{\rm inv}$ for $k = 1, 2$. Hence σ is uniquely determined by (27) and it suffices to verify all required relations on $\Gamma_1 \otimes_{\mathcal{A}} \Gamma_2$ for such elements.

We conclude from (25) that $({\rm id}\otimes\sigma)\circ\Delta_L(\rho) = \Delta_L\circ\sigma(\rho)$ and $\sigma(a\rho) = a\sigma(\rho)$ for $a \in \mathcal{A}$ and $\rho \in \Gamma_1 \otimes_{\mathcal{A}} \Gamma_2$. Formula (26) in turn implies that $(\sigma \otimes {\rm id}) \circ \Delta_R(\rho) = \Delta_R \circ \sigma(\rho)$ and $\sigma(\rho a) = \sigma(\rho)a$. We have already noted that σ is bijective. Thus, σ is an isomorphism of bicovariant bimodules.

For $\omega \in {}_{\rm inv}\Gamma_1$ and $\eta \in (\Gamma_2)_{\rm inv}$, we have $\sum \omega_{(-2)} \otimes \omega_{(-1)} \otimes \omega_{(0)} = 1 \otimes 1 \otimes \omega$ and $\sum \eta_{(0)} \otimes \eta_{(1)} \otimes \eta_{(2)} = \eta \otimes 1 \otimes 1$. Inserting this into (24) we obtain (27).

Since the elements $\rho = \omega^1 \otimes_{\mathcal{A}} \omega^2 \otimes_{\mathcal{A}} \eta^3$ with $\omega^1 \in {}_{\rm inv}\Gamma_1$, $\omega^2 \in {}_{\rm inv}\Gamma_2$ and $\eta^3 \in (\Gamma_3)_{\rm inv}$ generate $\Gamma_1 \otimes_{\mathcal{A}} \Gamma_2 \otimes_{\mathcal{A}} \Gamma_3$ as a left $\mathcal{A}$-module, it suffices to prove the equality (28) for such elements ρ. Using (25) and (27), we get

$$\begin{aligned}(\sigma \otimes {\rm id})({\rm id} \otimes \sigma)(\sigma \otimes {\rm id})(\rho) &= \sum (\sigma \otimes {\rm id})({\rm id} \otimes \sigma)(\omega^2_{(0)} \otimes_{\mathcal{A}} \omega^1 \triangleleft \omega^2_{(1)} \otimes_{\mathcal{A}} \eta^3) \\ &= \sum (\sigma\otimes{\rm id})(\omega^2_{(0)} \otimes_{\mathcal{A}} \eta^3 \otimes_{\mathcal{A}} \omega^1 \triangleleft \omega^2_{(1)}) \\ &= \sum \eta^3 \otimes_{\mathcal{A}} \omega^2_{(0)} \otimes_{\mathcal{A}} \omega^1 \triangleleft \omega^2_{(1)}, \\ ({\rm id} \otimes \sigma)(\sigma \otimes {\rm id})({\rm id} \otimes \sigma)(\rho) &= \sum ({\rm id} \otimes \sigma)(\sigma \otimes {\rm id})(\omega^1 \otimes_{\mathcal{A}} \eta^3 \otimes_{\mathcal{A}} \omega^2) \\ &= \sum ({\rm id} \otimes \sigma)(\eta^3 \otimes_{\mathcal{A}} \omega^1 \otimes_{\mathcal{A}} \omega^2) \\ &= \sum \eta^3 \otimes_{\mathcal{A}} \omega^2_{(0)} \otimes_{\mathcal{A}} \omega^1 \triangleleft \omega^2_{(1)},\end{aligned}$$

which proves the braid relation (28). □

We close this subsection by giving the coordinate versions for some of the above notions and results.

Let $\{\omega_i^1\}_{i\in I}$ and $\{\omega_j^2\}_{j\in J}$ be bases of the vector spaces ${}_{\rm inv}\Gamma_1$ and ${}_{\rm inv}\Gamma_2$, respectively. In the notation of Theorem 3, we write $\Gamma_1 = (v, f)$ and $\Gamma_2 = (w, g)$ and set $\eta_i^1 = \omega_k^1 S(v_i^k)$ and $\eta_j^2 = \omega_k^2 S(w_j^k)$. Then each of the four sets $\{\omega_i^1 \otimes_{\mathcal{A}} \omega_j^2\}, \{\omega_i^1 \otimes_{\mathcal{A}} \eta_j^2\}, \{\eta_i^1 \otimes_{\mathcal{A}} \omega_j^2\}, \{\eta_i^1 \otimes_{\mathcal{A}} \eta_j^2\}$ is a free left $\mathcal{A}$-module basis and also a free right $\mathcal{A}$-module basis of $\Gamma_1 \otimes_{\mathcal{A}} \Gamma_2$. The bicovariant bimodule $\Gamma_1 \otimes_{\mathcal{A}} \Gamma_2$ can be written as $(v\otimes w, f\otimes g)$, where $v\otimes w$ is the corepresentation of $\mathcal{A}$ with matrix elements $(v \otimes w)_{kn}^{ij} := v_k^i w_n^j$ and $f \otimes g$ is the algebra representation of $\mathcal{A}$ with matrix elements $(f \otimes g)_{kn}^{ij} := f_k^i g_n^j$. Having these data, the formulas (i) and (ii) in Theorem 3 provide an explicit description of the bicovariant bimodule $\Gamma_1 \otimes_{\mathcal{A}} \Gamma_2$ in terms of the basis $\{\omega_i^1 \otimes_{\mathcal{A}} \omega_j^2\}$ of the vector space ${}_{\rm inv}(\Gamma_1 \otimes_{\mathcal{A}} \Gamma_2)$. Since $\Delta_R(\omega_j^2) = \omega_n^2 \otimes w_j^n$, it follows at once from (25) and (17) that the braiding map σ acts on elements of this basis by

$$\sigma(\omega_i^1 \otimes_{\mathcal{A}} \omega_j^2) = \sigma_{ij}^{nk} \omega_n^2 \otimes_{\mathcal{A}} \omega_k^1 \quad \text{with} \quad \sigma_{ij}^{nk} := f_k^i(w_j^n)\ . \tag{29}$$

The equality $(\sigma \otimes {\rm id}) \circ \Delta_R(\omega_i^1 \otimes_{\mathcal{A}} \omega_j^2) = \Delta_R \circ \sigma(\omega_i^1 \otimes_{\mathcal{A}} \omega_i^2)$ implies that σ belongs to the space Mor $(v \otimes w, w \otimes v)$, that is,

$$\sigma_{rs}^{nk} v_i^r w_j^s = w_s^n v_r^k \sigma_{ij}^{sr}, \quad i,k \in I, \quad j,n \in J. \tag{30}$$

By (17) we have the identity $\omega_i^1 \otimes_{\mathcal{A}} \omega_j^2 a = (f_r^i g_s^j.a)\, \omega_r^1 \otimes_{\mathcal{A}} \omega_s^2$. Applying σ to both sides, we get

$$\sigma_{ij}^{rs} g_n^r f_k^s = f_r^i g_s^j \sigma_{rs}^{nk}, \quad i,k \in I, \quad j,n \in J. \tag{31}$$

Let $\tilde{\sigma} : \Gamma_2 \otimes_{\mathcal{A}} \Gamma_1 \to \Gamma_1 \otimes_{\mathcal{A}} \Gamma_2$ denote the braiding of Γ_2 and Γ_1. Inserting the expression $\Delta_L(\eta_j^2) = \Delta_L(\omega_r^2)\Delta(S(w_j^r)) = S(w_j^k) \otimes \eta_k^2$ into formula (26) applied to $\tilde{\sigma}(\eta_j^2 \otimes_{\mathcal{A}} \eta_i^1)$ and using first (16) and then (29) and (30) we derive

$$\tilde{\sigma}(\eta_j^2 \otimes_{\mathcal{A}} \eta_i^1) = \sigma_{ij}^{nk} \eta_k^1 \otimes_{\mathcal{A}} \eta_n^2, \quad i \in I, \quad j \in J. \tag{32}$$

13.1.5 Bicovariant Bimodules and Representations of the Quantum Double

This subsection relates the Yetter–Drinfeld modules to representations of the quantum double. Combined with the one-to-one correspondence between Yetter–Drinfeld modules and bicovariant bimodules shown in Subsect. 13.1.3, this characterizes bicovariant bimodules by representations of the quantum double.

Let $\mathcal{U}$ be another Hopf algebra and let $\langle \cdot, \cdot \rangle$ be a dual pairing of the Hopf algebras $\mathcal{U}$ and $\mathcal{A}$. As noted in Subsect. 8.2.1, $\langle \cdot, \cdot \rangle$ is an invertible skew-pairing of $\mathcal{U}$ and $\mathcal{A}^{\rm op}$. Hence the corresponding quantum double $\mathcal{D}(\mathcal{U}, \mathcal{A}^{\rm op})$ is a well-defined Hopf algebra (see Proposition 8.8). Recall that as an algebra,

$\mathcal{D}(\mathcal{U}, \mathcal{A}^{\mathrm{op}})$ is generated by the algebras $\mathcal{U}$ and $\mathcal{A}^{\mathrm{op}}$ with cross relations (see (8.23))

$$\sum \langle f_{(1)}, a_{(1)} \rangle f_{(2)} a_{(2)} = \sum a_{(1)} f_{(1)} \langle f_{(2)}, a_{(2)} \rangle, \quad f \in \mathcal{U}, \ a \in \mathcal{A}^{\mathrm{op}}. \tag{33}$$

By Proposition 1.15, for any corepresentation φ of $\mathcal{A}$, the equation $\hat{\varphi}(f) = (\mathrm{id} \otimes f) \circ \varphi$, $f \in \mathcal{U}$, defines a representation $\hat{\varphi}$ of $\mathcal{U}$. Representations of $\mathcal{U}$ of this form are called *rational.*

Now let Γ_0 be a right $\mathcal{A}$-module with action $\triangleleft$ and a right $\mathcal{A}$-comodule with coaction δ. Then the equations

$$\pi_{\mathcal{A}^{\mathrm{op}}}(a)\omega := \omega \triangleleft a \quad \text{and} \quad \pi_{\mathcal{U}}(f)\omega := \hat{\delta}(f)\omega = \sum \omega_{(0)} \langle f, \omega_{(1)} \rangle$$

define representations $\pi_{\mathcal{A}^{\mathrm{op}}}$ and $\pi_{\mathcal{U}}$ of the algebras $\mathcal{A}^{\mathrm{op}}$ and $\mathcal{U}$, respectively. By construction, $\pi_{\mathcal{U}}$ is rational.

Proposition 6. (i) *If Γ_0 is a Yetter–Drinfeld module of $\mathcal{A}$, then $\pi_{\mathcal{A}^{\mathrm{op}}}$ and $\pi_{\mathcal{U}}$ are restrictions of a representation of the algebra $\mathcal{D}(\mathcal{U}, \mathcal{A}^{\mathrm{op}})$ on Γ_0.*

(ii) *Suppose that $\mathcal{U}$ separates the elements of $\mathcal{A}$. If π is a representation of $\mathcal{D}(\mathcal{U}, \mathcal{A}^{\mathrm{op}})$ on a vector space Γ_0 such that the restriction $\pi \lceil \mathcal{U}$ of π to $\mathcal{U}$ is rational, then there is a unique Yetter–Drinfeld module structure on Γ_0 such that $\pi \lceil \mathcal{A}^{\mathrm{op}} = \pi_{\mathcal{A}^{\mathrm{op}}}$ and $\pi \lceil \mathcal{U} = \pi_{\mathcal{U}}$.*

Proof. Let π be a representation of $\mathcal{D}(\mathcal{U}, \mathcal{A}^{\mathrm{op}})$ on Γ_0 such that $\pi \lceil \mathcal{U}$ is rational, say $\pi \lceil \mathcal{U} = \hat{\delta}$ for a right coaction δ of $\mathcal{A}$ on Γ_0. Let $\omega \triangleleft a := \pi(a)\omega$, $a \in \mathcal{A}$. Applying π for the element in (33) to $\omega \in \Gamma_0$, we get

$$\sum \langle f_{(1)}, a_{(1)} \rangle (\omega \triangleleft a_{(2)})_{(0)} \langle f_{(2)}, (\omega \triangleleft a_{(2)})_{(1)} \rangle$$

$$= \sum \langle f_{(1)}, \omega_{(1)} \rangle \omega_{(0)} \triangleleft a_{(1)} \langle f_{(2)}, a_{(2)} \rangle \tag{34}$$

or equivalently

$$\sum (\mathrm{id} \otimes f)((\omega \triangleleft a_{(2)})_{(0)} \otimes a_{(1)} (\omega \triangleleft a_{(2)})_{(1)}) = \sum (\mathrm{id} \otimes f)(\omega_{(0)} \triangleleft a_{(1)} \otimes \omega_{(1)} a_{(2)}). \tag{35}$$

If $\mathcal{U}$ separates the points of $\mathcal{A}$, then equation (35) implies the Yetter–Drinfeld condition (18) and δ is uniquely determined by $\pi \lceil \mathcal{U}$. This proves (ii).

Conversely, if Γ_0 is a Yetter–Drinfeld module, then (35) and hence (34) hold. Therefore, it follows that $\pi(fa) := \pi_{\mathcal{U}}(f) \pi_{\mathcal{A}^{\mathrm{op}}}(a)$ gives a well-defined(!) representation π of the algebra $\mathcal{D}(\mathcal{U}, \mathcal{A}^{\mathrm{op}})$ on Γ_0. □

For a representation π of the quantum double $\mathcal{D}(\mathcal{U}, \mathcal{A}^{\mathrm{op}})$ as in (ii), the associated bicovariant bimodule Γ can be explicitly given as follows. Let $\{\omega_i\}$ be a basis of the vector space Γ_0. There exist pointwise finite matrices $f = (f_i^j)$ and (F_i^j) of functionals $f_i^j \in \mathcal{A}'$ and $F_i^j \in \mathcal{U}'$ such that $\pi(a)\omega_i = f_j^i(a)\omega_j, a \in \mathcal{A}$, and $\pi(f)\omega_i = F_i^j(f)\omega_j, f \in \mathcal{U}$. Since $\pi \lceil \mathcal{U}$ is rational, there are elements $v_i^j \in \mathcal{A}$ such that $F_i^j(f) = \langle f, v_i^j \rangle, f \in \mathcal{U}$. Then, $\Gamma = (v, f)$ is the corresponding bicovariant bimodule by Theorem 3, where $v = (v_j^i)$.

13.2 Tensor Algebras and Exterior Algebras of Bicovariant Bimodules

From Examples 1.8 and 1.13 we know that the tensor algebra and the exterior algebra of a vector space are a Hopf algebra and an $\mathbb{N}_0$-graded super Hopf algebra, respectively. In this section we will show that these results and the corresponding formulas obtained in both examples have generalizations to the tensor algebra and the exterior algebra of a bicovariant bimodule.

If not stated otherwise, Γ denotes a bicovariant bimodule over $\mathcal{A}$ and σ is the braiding of $\Gamma^{\otimes 2} = \Gamma \otimes_{\mathcal{A}} \Gamma$.

13.2.1 The Tensor Algebra of a Bicovariant Bimodule

First let $\mathcal{A}$ be an arbitrary algebra and Γ a bimodule for $\mathcal{A}$. We set $\Gamma^{\otimes 0} := \mathcal{A}, \Gamma^{\otimes 1} := \Gamma$ and $\Gamma^{\otimes n} := \Gamma \otimes_{\mathcal{A}} \cdots \otimes_{\mathcal{A}} \Gamma$ (n times). Then the direct sum of vector spaces

$$T_{\mathcal{A}}(\Gamma) := \bigoplus_{n=0}^{\infty} \Gamma^{\otimes n}$$

becomes an algebra with multiplication defined by $x_n y_k := x_n \otimes_{\mathcal{A}} y_k$ for $x_n \in \Gamma^{\otimes n}$ and $y_k \in \Gamma^{\otimes k}$. Here we identify $x_0 \otimes_{\mathcal{A}} y_0$ and $x_0 y_0$ for $x_0, y_0 \in \Gamma^{\otimes 0} = \mathcal{A}$. The latter means that the multiplication of zeroth components in $T_{\mathcal{A}}(\Gamma)$ is just the multiplication in $\mathcal{A}$. The algebra $T_{\mathcal{A}}(\Gamma)$ is called the *tensor algebra* of Γ. Just as for the ordinary tensor algebra of a vector space, the tensor algebra $T_{\mathcal{A}}(\Gamma)$ has the following important *universal property: if Φ^1 is a linear mapping of $\mathcal{A} \oplus \Gamma$ into an algebra $\mathcal{B}$ for which $\Phi^1 \lceil \mathcal{A}$ is an algebra homomorphism and $\Phi^1(a\rho b) = \Phi^1(a)\Phi^1(\rho)\Phi^1(b)$ for any $a, b \in \mathcal{A}$ and $\rho \in \Gamma$, then there exists a unique algebra homomorphism $\Phi : T_{\mathcal{A}}(\Gamma) \to \mathcal{B}$ such that Φ^1 is the restriction of Φ to $\mathcal{A} \oplus \Gamma$.* In particular, if $\Phi_1 : \Gamma \to \tilde{\Gamma}$ is a bimodule homomorphism of the $\mathcal{A}$-bimodule Γ into another $\mathcal{A}$-bimodule $\tilde{\Gamma}$, then there is a unique algebra homomorphism $\Phi : T_{\mathcal{A}}(\Gamma) \to T_{\mathcal{A}}(\tilde{\Gamma})$ such that $\Phi(a) = a$ and $\Phi(\rho) = \Phi_1(\rho)$ for $a \in \mathcal{A}$ and $\rho \in \Gamma$.

From now on we suppose that Γ is a bicovariant bimodule over the Hopf algebra $\mathcal{A}$. Then all $\Gamma^{\otimes n}$ are bicovariant bimodules over $\mathcal{A}$. Obviously, $\Gamma^{\otimes 0} = \mathcal{A}$ is a bicovariant bimodule over $\mathcal{A}$ with $\Delta_R = \Delta_L := \Delta$. Therefore, $T_{\mathcal{A}}(\Gamma)$ is also a *bicovariant bimodule* over $\mathcal{A}$ with bimodule operations and right and left coactions defined componentwise.

Proposition 7. *The $\mathbb{N}_0$-graded algebra $T_{\mathcal{A}}(\Gamma)$ becomes a Hopf algebra, denoted again by $T_{\mathcal{A}}(\Gamma)$, and an $\mathbb{N}_0$-graded super Hopf algebra, denoted by $T^s_{\mathcal{A}}(\Gamma)$, with structure maps Δ_T, ε_T, S_T and Δ^s_T, ε^s_T, S^s_T, respectively, such that*

$$\Delta_T(a) = \Delta^s_T(a) = \Delta_{\mathcal{A}}(a), \quad \Delta_T(\rho) = \Delta^s_T(\rho) = \Delta_R(\rho) + \Delta_L(\rho), \tag{36}$$

$$\varepsilon_T(a) = \varepsilon^s_T(a) = \varepsilon_{\mathcal{A}}(a), \quad \varepsilon_T(\rho) = \varepsilon^s_T(\rho) = 0, \tag{37}$$

$$S_T(\rho) = S_T^s(\rho) = -\sum S_{\mathcal{A}}(\rho_{(-1)})\rho_{(0)}S_{\mathcal{A}}(\rho_{(1)}) \tag{38}$$

for $a \in \Gamma^{\otimes 0} = \mathcal{A}$ *and* $\rho \in \Gamma^{\otimes 1} = \Gamma$.

Proof. We prove the assertion for $T_{\mathcal{A}}^s(\Gamma)$. By the universal property of $T_{\mathcal{A}}(\Gamma)$, the linear mappings Δ_T^s, ε_T^s, S_T^s of $\mathcal{A}\oplus\Gamma$ into the algebras $T_{\mathcal{A}}(\Gamma)\otimes T_{\mathcal{A}}(\Gamma)$, $\mathbb{C}$ and $T_{\mathcal{A}}(\Gamma)^{\mathrm{op}}$, respectively, defined by (36)–(38), extend uniquely to algebra homomorphisms of $T_{\mathcal{A}}(\Gamma)$. Here it is understood that $T_{\mathcal{A}}(\Gamma)\otimes T_{\mathcal{A}}(\Gamma)$ carries the $\mathbb{N}_0$-graded product $\bullet$ given by formula (1.42). By Proposition 1.8 (more precisely, by its $\mathbb{N}_0$-graded version), it suffices to check that the Hopf algebra axioms are satisfied for the generating set $\mathcal{A}\cup\Gamma$. We omit these straightforward verifications. □

In order to generalize the formulas (1.37), (1.38), (1.43) and (1.44) to the present situation, some further notation and preliminaries are necessary.

First let us recall Artin's braid groups $\mathfrak{B}_n$ corresponding to the Lie algebra $\mathrm{sl}(n,\mathbb{C})$ (see Subsect. 6.2.1). Fix $n \in \mathbb{N}$, $n \geq 2$. Then $\mathfrak{B}_n$ is the group with $n-1$ generators $s_1, s_2, \cdots, s_{n-1}$ and defining relations

$$\begin{aligned} s_k s_{k+1} s_k &= s_{k+1} s_k s_{k+1}, \quad k = 1,\cdots,n-2,\\ s_i s_j &= s_j s_i, \qquad |i-j| \geq 2. \end{aligned}$$

The group $\mathfrak{B}_n$ plays an important role in knot theory (see, for instance, [Ka], Chap. X). It is the group of equivalence classes of braids on n strands with composition of braids as group multiplication.

For $k = 1, 2, \cdots, n-1$, we define an automorphism σ_k of the bicovariant bimodule $\Gamma^{\otimes n}$ by $\sigma_k = \mathrm{id}\otimes_{\mathcal{A}}\cdots\otimes_{\mathcal{A}}\sigma\otimes_{\mathcal{A}}\cdots\otimes_{\mathcal{A}}\mathrm{id}$, where the braiding σ from Proposition 5 stands in the place $(k, k+1)$. Since $\sigma_k\sigma_{k+1}\sigma_k = \sigma_{k+1}\sigma_k\sigma_{k+1}$ by (28), $\sigma_1, \sigma_2, \cdots, \sigma_{n-1}$ satisfy the defining relations of the braid group $\mathfrak{B}_n$. So there exists a unique group homomorphism $x \to \pi(x)$ of $\mathfrak{B}_n$ into the group of automorphisms of the bicovariant bimodule $\Gamma^{\otimes n}$ such that $\pi(e) = \mathrm{id}$ and $\pi(s_k) = \sigma_k$, $k = 1, 2, \cdots, n-1$. We give a more explicit description of the homomorphism π. Let t_k denote the permutation from the symmetric group $\mathcal{P}_n$ which interchanges k and $k+1$ and leaves all other numbers fixed. Each permutation $p \in \mathcal{P}_n$ can be written as a product $p = t_{i_1}\cdots t_{i_{l(p)}}$, where $l(p)$ denotes the length of the permutation p. Then we set $\pi(p) := \sigma_{i_1}\cdots\sigma_{i_{l(p)}}$. It can be shown that the latter is independent of the particular representation $p = t_{i_1}\cdots t_{i_{l(p)}}$ of p. Moreover, we have

$$\pi(p_1p_2) = \pi(p_1)\pi(p_2) \ \text{ if } \ l(p_1p_2) = l(p_1) + l(p_2) \ \text{ and } \ p_1, p_2 \in \mathcal{P}_n \,. \tag{39}$$

For $n \in \mathbb{N}$ and $k = 0, 1, \cdots, n$ we define

$$B_{nk}^{\pm} := \sum_{p^{-1}\in\mathcal{P}_{nk}} (\pm 1)^{l(p)}\pi(p)\,, \tag{40}$$

where $\mathcal{P}_{nk}$ is the set of $(k, n-k)$-shuffles, that is, the set of permutations $p \in \mathcal{P}_n$ such that $p(1) < \cdots < p(k)$ and $p(k+1) < \cdots < p(n)$.

Let $\{\omega_i\}_{i\in I}$ be a fixed basis of the vector space ${}_{\rm inv}\Gamma$. Let η_i and v^j_i, $i,j\in I$, be as in Theorem 3. Recall that $\Delta_R(\omega_i)=\omega_j\otimes v^j_i$ and $\{\eta_i:=\omega_jS(v^j_i)\}_{i\in I}$ is a vector space basis of $\Gamma_{\rm inv}$. For $\mathfrak{i}=(i_1,\cdots,i_n)\in I^n$ we abbreviate the product $\omega_{i_1}\cdots\omega_{i_n}$ in the algebra $T_{\mathcal{A}}(\Gamma)$ by $\omega_{\mathfrak{i}}$. In a similar way we use the multi-index notation $\eta_{\mathfrak{i}}$ and $v^{\mathfrak{j}}_{\mathfrak{i}}$, $\mathfrak{i},\mathfrak{j}\in I^n$. Recall from Subsect. 13.1.4 that $\{\omega_{\mathfrak{i}}\}_{\mathfrak{i}\in I^n}$ is a basis of the vector space ${}_{\rm inv}(\Gamma^{\otimes n})$ and that the braiding σ leaves ${}_{\rm inv}(\Gamma^{\otimes 2})$ invariant. Thus any operator $\pi(p)$, $p\in\mathcal{P}_n$, leaves ${}_{\rm inv}(\Gamma^{\otimes n})$ invariant. Hence for each $B\in \operatorname{Lin}\{\pi(p)\,|\,p\in\mathcal{P}_n\}$ there exists a pointwise finite matrix $(B^{\mathfrak{j}}_{\mathfrak{i}})_{\mathfrak{i},\mathfrak{j}\in I^n}$ of complex entries such that $B(\omega_{\mathfrak{i}})=B^{\mathfrak{j}}_{\mathfrak{i}}\omega_{\mathfrak{j}}$. For $\mathfrak{i}=(i_1,\cdots,i_k)\in I^k$ and $\mathfrak{j}=(j_1,\cdots,j_m)\in I^m$, we put $\bar{\mathfrak{i}}:=(i_k,\cdots,i_1)$ and $(\mathfrak{i},\mathfrak{j}):=(i_1,\cdots,i_k,j_1,\cdots,j_m)$. Now we are able to give generalizations of the formulas in Examples 1.8 and 1.13.

Proposition 8. *For $n\in\mathbb{N}$ and $\mathfrak{i}\in I^n$, we have*

$$\Delta_T(\omega_{\mathfrak{i}})=\sum_{k=0}^{n}\left(B^+_{nk}\right)^{(\mathfrak{j}',\mathfrak{j}'')}_{\mathfrak{i}}\omega_{\mathfrak{r}}\otimes v^{\mathfrak{r}}_{\mathfrak{j}'}\omega_{\mathfrak{j}''},\quad S_T(\omega_{\mathfrak{i}})=(-1)^n\eta_{\bar{\mathfrak{i}}},\tag{41}$$

$$\Delta^s_T(\omega_{\mathfrak{i}})=\sum_{k=0}^{n}(B^-_{nk})^{(\mathfrak{j}',\mathfrak{j}'')}_{\mathfrak{i}}\omega_{\mathfrak{r}}\otimes v^{\mathfrak{r}}_{\mathfrak{j}'}\omega_{\mathfrak{j}''},\qquad S^s_T(\omega_{\mathfrak{i}})=(-1)^{\binom{n+1}{2}}\eta_{\bar{\mathfrak{i}}}.\tag{42}$$

Note that there are multi-index summations over $\mathfrak{j}',\mathfrak{r}\in I^k$ and $\mathfrak{j}''\in I^{n-k}$ in (41) and (42). For $k=0$ and $k=n$ the summands should be interpreted as $1\otimes\omega_{\mathfrak{i}}$ and $\omega_{\mathfrak{r}}\otimes v^{\mathfrak{r}}_{\mathfrak{i}}$, respectively.

Proof. We carry out the proof of (42). In order to prove the formula for Δ^s_T, we proceed by induction on n. For $n=1$ this equation is obviously true. Assume that it holds for all $\mathfrak{i}\in I^n$. Let $\tilde{\mathfrak{i}}=(i_1,\cdots,i_{n+1})\in I^{n+1}$. Put $\mathfrak{i}=(i_1,\cdots,i_n)$. By (42) and Theorem 3(i), $\Delta^s_T(\omega_{i_{n+1}})=1\otimes\omega_{i_{n+1}}+\omega_m\otimes v^m_{i_{n+1}}$. Therefore, by the induction hypothesis, equation (42) and the definition (1.42) of the $\mathbb{N}_0$-graded product $\bullet$, we obtain

$$\begin{aligned}\Delta^s_T(\omega_{\tilde{\mathfrak{i}}}) &= \Delta^s_T(\omega_{\mathfrak{i}})\bullet\Delta^s_T(\omega_{i_{n+1}})\\ &= \left(\sum_{k=0}^{n}(B^-_{nk})^{(\mathfrak{j}',\mathfrak{j}'')}_{\mathfrak{i}}\omega_{\mathfrak{r}}\otimes v^{\mathfrak{r}}_{\mathfrak{j}'}\omega_{\mathfrak{j}''}\right)\bullet(1\otimes\omega_{i_{n+1}}+\omega_m\otimes v^m_{i_{n+1}})=\end{aligned}$$

$$\sum_{k=0}^{n}\sum(-1)^{l(p)}\pi(p)^{(\mathfrak{j}',\mathfrak{j}'')}_{\mathfrak{i}}\left(\omega_{\mathfrak{r}}\otimes v^{\mathfrak{r}}_{\mathfrak{j}'}\omega_{\mathfrak{j}''}\omega_{i_{n+1}}+(-1)^{n-k}\omega_{\mathfrak{r}}\omega_{r_{k+1}}\otimes v^{\mathfrak{r}}_{\mathfrak{j}'}\omega_{\mathfrak{j}''}v^{r_{k+1}}_{i_{n+1}}\right),\tag{43}$$

where the second sum is over all permutations $p\in\mathcal{P}_n$ such that $p^{-1}\in\mathcal{P}_{nk}$. The first summands of the interior sum in (43) correspond to summands $(-1)^{l(\tilde{p})}\pi(\tilde{p})$ of $B^-_{n+1,k}$ with $\tilde{p}^{-1}\in\mathcal{P}_{n+1,k}$, where $\tilde{p}(i)=p(i)$ for $i=1,\cdots,n$ and $\tilde{p}(n+1)=n+1$. In order to treat the second summands of the interior sum in (43), we observe that $\omega_jv^r_i=v^r_sf^j_m(v^s_i)\omega_m=v^r_s\sigma^{sm}_{ji}\omega_m$ by (29) and hence $\omega_{\mathfrak{j}''}v^{r_{k+1}}_{i_{n+1}}=v^{r_{k+1}}_{m_{k+1}}\sigma^{m_{k+1}s_{k+1}}_{j_{k+1}m_{k+2}}\sigma^{m_{k+2}s_{k+2}}_{j_{k+2}m_{k+3}}\cdots\sigma^{m_ns_n}_{j_ni_{n+1}}\omega_{\mathfrak{s}''}$ with

$\mathfrak{j}'' = (j_{k+1}, \cdots, j_n)$ and $\mathfrak{s}'' = (s_{k+1}, \cdots, s_n)$. Putting $\tilde{\mathfrak{j}}' := (j_1, \cdots, j_k, m_{k+1})$ and $\tilde{\mathfrak{r}} = (r_1, \cdots, r_k, r_{k+1})$, it follows that the second summands are equal to $(-1)^{l(p)+k}(\sigma_{k+1} \cdots \sigma_n \pi(p))_{\tilde{\mathfrak{i}}}^{(\tilde{\mathfrak{j}}', \mathfrak{s}'')} \omega_{\tilde{\mathfrak{r}}} \otimes v^{\tilde{\mathfrak{r}}}_{\tilde{\mathfrak{j}}'} \omega_{\mathfrak{s}''}$. Let $\tilde{p} := p_1 p' \in \mathcal{P}_{n+1}$, where $p_1 := t_{k+1} \cdots t_n$ and the restriction to $\{1, 2, \cdots, n\}$ of $p' \in \mathcal{P}_{n+1}$ is p. It is easy to check that $\tilde{p}^{-1}(i) = p^{-1}(i)$ for $i = 1, 2, \cdots, k$, $\tilde{p}^{-1}(k+1) = n+1$ and $\tilde{p}^{-1}(i+1) = \tilde{p}^{-1}(i)$ for $i = k+1, \cdots, n$. Hence $\tilde{p}^{-1}$ is a shuffle permutation of $\mathcal{P}_{n+1,k+1}$ and we have $l(\tilde{p}) = n - k + l(p) = l(p_1) + l(p')$. Therefore, by (42), $(-1)^{l(p)+n-k} \sigma_{k+1} \cdots \sigma_n \pi(p) = (-1)^{l(\tilde{p})} \pi(\tilde{p})$ and the latter is a summand of $B^-_{n+1,k+1}$. The preceding shows that there is a one-to-one correspondence between summands of (43) and summands of (42) for $\mathfrak{i} \in I^{n+1}$. By induction, this completes the proof of the first formula of (42).

Formula (38) yields $S^s_T(\omega_i) = -\eta_i$ for $i \in I$. Since the antipode of the $\mathbb{N}_0$-graded super Hopf algebra $T^s_{\mathcal{A}}(\Gamma)$ satisfies $S^s_T(x_k y_m) = (-1)^{km} S^s_T(y_m) S^s_T(x_k)$ for $x_k \in \Gamma^{\otimes k}$ and $y_m \in \Gamma^{\otimes m}$, the second formula of (42) follows. □

13.2.2 The Exterior Algebra of a Bicovariant Bimodule

In this subsection we study quantum analogs of the antisymmetrizer and the exterior algebra for bicovariant bimodules. The basic idea is to replace the flip operator in the corresponding classical constructions by the braiding map σ.

For $n \in \mathbb{N}$ and $k = 0, 1, \cdots, n$, we define

$$A_n := \sum_{p \in \mathcal{P}_n} (-1)^{l(p)} \pi(p) \quad \text{and} \quad A_{nk} := \sum_{p \in \mathcal{P}_{nk}} (-1)^{l(p)} \pi(p).$$

The map A_n is called the *quantum antisymmetrizer* of $\Gamma^{\otimes n}$.

Let $k \in \{1, 2, \cdots, n-1\}$. Each permutation $p \in \mathcal{P}_n$ admits unique representations $p = r p_1 p_2$ and $p = p_3 p_4 s$, where $r, s^{-1} \in \mathcal{P}_{nk}$ and $p_1, p_3 \in \mathcal{P}_n$ (resp. $p_2, p_4 \in \mathcal{P}_n$) leave the numbers $k+1, \cdots, n$ (resp. $1, \cdots, k$) fixed. Then we have $l(p) = l(r) + l(p_1) + l(p_2) = l(p_3) + l(p_4) + l(s)$ and hence $\pi(p) = \pi(r)\pi(p_1)\pi(p_2) = \pi(p_3)\pi(p_4)\pi(s)$ by (39). From these facts we conclude that

$$A_n = A_{nk}(A_k \otimes A_{n-k}) = (A_k \otimes A_{n-k}) B^-_{nk} \,. \tag{44}$$

We shall use (44) also for $k = 0$ and $k = n$, where it is interpreted as $A_n = A_{n0} A_n = A_n B^-_{n0}$ and $A_n = A_{nn} A_n = A_n B^-_{nn}$, respectively.

Let $\mathfrak{S} := \bigoplus_{n=2}^{\infty} \mathfrak{S}^n$, where $\mathfrak{S}^n := \ker A_n$. The elements of $\mathfrak{S}^n$ can be considered as quantum analogs of symmetric tensors of degree n. From the first equality in (44) we see that $\mathfrak{S}$ is a two-sided ideal of the tensor algebra $T_{\mathcal{A}}(\Gamma)$.

The quotient algebra $\Gamma^\wedge_e := T_{\mathcal{A}}(\Gamma)/\mathfrak{S} \equiv T^s_{\mathcal{A}}(T)/\mathfrak{S}$ is called the *exterior algebra* of the bicovariant bimodule Γ over $\mathcal{A}$. Clearly, $\Gamma^\wedge_e$ is again a graded algebra, that is,

$$\Gamma_e^{\wedge}=\bigoplus_{n=0}^{\infty}\Gamma^{\wedge n}\quad\text{with}\quad \Gamma^{\wedge 0}:=\mathcal{A},\ \ \Gamma^{\wedge 1}:=\Gamma,\ \ \Gamma^{\wedge n}:=\Gamma^{\otimes n}/\mathfrak{S}^n,\ n\geq 2.$$

The quantum antisymmetrizer A_n is a homomorphism of the bicovariant bimodule $\Gamma^{\otimes n}$ to itself as σ is of $\Gamma^{\otimes 2}$ by Proposition 5. Therefore, $\mathfrak{S}$ is a bicovariant subbimodule of $T_{\mathcal{A}}(\Gamma)$. Hence, the quotient $\Gamma_e^{\wedge}=T_{\mathcal{A}}(\Gamma)/\mathfrak{S}$ is also a *bicovariant bimodule* over $\mathcal{A}$.

Taking Proposition 10 stated below for granted, $\mathfrak{S}$ is a Hopf ideal of the $\mathbb{N}_0$-graded super Hopf algebra $T_{\mathcal{A}}^{\mathfrak{s}}(\Gamma)$ defined in Proposition 7. Therefore, the quotient $\Gamma_e^{\wedge}=T_{\mathcal{A}}^{\mathfrak{s}}(\Gamma)/\mathfrak{S}$ is also an *$\mathbb{N}_0$-graded super Hopf algebra* with structures induced from $T_{\mathcal{A}}^{\mathfrak{s}}(\Gamma)$.

Let $\langle\mathfrak{S}^2\rangle$ denote the two-sided ideal of the algebra $T_{\mathcal{A}}(\Gamma)$ generated by $\mathfrak{S}^2=\ker(\sigma-I)$. Since $\langle\mathfrak{S}^2\rangle$ is a bicovariant subbimodule of $T_{\mathcal{A}}(\Gamma)$ and a Hopf ideal of $T_{\mathcal{A}}^{\mathfrak{s}}(\Gamma)$ by the proof of Proposition 10 below, the quotient $T_{\mathcal{A}}(\Gamma)/\langle\mathfrak{S}^2\rangle$ is also a bicovariant bimodule over $\mathcal{A}$ and an $\mathbb{N}_0$-graded super Hopf algebra.

The following result is needed later in Subsect. 14.4.1.

Proposition 9. *Let $\tilde{\Gamma}$ be a bicovariant bimodule over $\mathcal{A}$ and let Γ be a bicovariant subbimodule of $\tilde{\Gamma}$ (that is, Γ is an $\mathcal{A}$-subbimodule of $\tilde{\Gamma}$ and the left and right coactions of $\mathcal{A}$ on Γ are restrictions of the corresponding coactions on $\tilde{\Gamma}$). Then there exists an embedding $\Gamma_e^{\wedge}\subseteq\tilde{\Gamma}_e^{\wedge}$, which is the identity map on $\Gamma^{\wedge 0}=\tilde{\Gamma}^{\wedge 0}=\mathcal{A}$ and the inclusion $\Gamma\subseteq\tilde{\Gamma}$ on $\Gamma^{\wedge 1}=\Gamma$, such that $\Gamma_e^{\wedge}$ is an $\mathbb{N}_0$-graded super Hopf subalgebra and a bicovariant subbimodule of $\tilde{\Gamma}_e^{\wedge}$.*

Proof. By the universal property of the tensor algebra (see Subsect. 13.2.1), there exists an injective algebra homomorphism $I:T_{\mathcal{A}}(\Gamma)\to T_{\mathcal{A}}(\tilde{\Gamma})$ such that $I(a)=a$ and $I(\rho)=\rho$ for $a\in\mathcal{A}$ and $\rho\in\Gamma$. We identify $I(\zeta)$ and ζ for $\zeta\in T_{\mathcal{A}}(\Gamma)$, so $T_{\mathcal{A}}(\Gamma)$ becomes a subalgebra of $T_{\mathcal{A}}(\tilde{\Gamma})$. Since Γ is a bicovariant subbimodule of $\tilde{\Gamma}$, we have $\sigma=\tilde{\sigma}\lceil\Gamma^{\otimes 2}$ and hence $A_n=\tilde{A}_n\lceil\Gamma^{\otimes n}$. (The tilde always refers to the corresponding objects for $\tilde{\Gamma}$.) Thus, $\mathfrak{S}=\tilde{\mathfrak{S}}\cap T_{\mathcal{A}}(\Gamma)$ and the assertion follows immediately from the constructions of $\Gamma_e^{\wedge}$ and $\tilde{\Gamma}_e^{\wedge}$. □

Proposition 10. *$\mathfrak{S}$ is a Hopf ideal of $T_{\mathcal{A}}^{\mathfrak{s}}(\Gamma)$, that is, $\mathfrak{S}$ is a two-sided ideal of the algebra $T_{\mathcal{A}}^{\mathfrak{s}}(\Gamma)$ such that $\varepsilon(\mathfrak{S})=\{0\}$, $\Delta_T^{\mathfrak{s}}(\mathfrak{S})\subseteq\mathfrak{S}\otimes T_{\mathcal{A}}(\Gamma)+T_{\mathcal{A}}(\Gamma)\otimes\mathfrak{S}$ and $S_T^{\mathfrak{s}}(\mathfrak{S})\subseteq\mathfrak{S}$.*

Proof. Suppose that $\zeta\in\mathfrak{S}^n=\ker A_n$. Since $\{\omega_{\mathfrak{i}}\}_{\mathfrak{i}\in I^n}$ is a free left $\mathcal{A}$-module basis of $\Gamma^{\otimes n}$, ζ can be written as a finite sum $\zeta=a^{\mathfrak{i}}\omega_{\mathfrak{i}}$ with $a^{\mathfrak{i}}\in\mathcal{A}$ and $A_n\zeta=0$ implies that $(A_n)^{\mathfrak{j}}_{\mathfrak{i}}a^{\mathfrak{i}}=0$ for all $\mathfrak{j}\in I^n$. Let $\xi_{k\mathfrak{i}}$ denote the k-th summand of the sum for $\Delta_T^{\mathfrak{s}}(\omega_{\mathfrak{i}})$ in (42). Since $\sigma\in\mathrm{Mor}\,(v\otimes v)$ by (30), it follows that $(A_k)^{\mathfrak{s}}_{\mathfrak{r}'}v^{\mathfrak{r}'}_{\mathfrak{j}'}=v^{\mathfrak{s}}_{\mathfrak{r}'}(A_k)^{\mathfrak{r}'}_{\mathfrak{j}'}$. Using these facts and the second equality of (44) we obtain

$$\begin{aligned}(A_k\otimes A_{n-k})\Delta(a^{\mathfrak{i}})\xi_{k\mathfrak{i}} &= \Delta(a^{\mathfrak{i}})(B_{nk}^{-})_{\mathfrak{i}}^{(\mathfrak{j}',\mathfrak{j}'')}(A_k)^{\mathfrak{s}}_{\mathfrak{r}'}\omega_{\mathfrak{s}}\otimes v^{\mathfrak{r}'}_{\mathfrak{j}'}(A_{n-k})^{\mathfrak{r}''}_{\mathfrak{j}''}\omega_{\mathfrak{r}''}\\ &= \Delta(a^{\mathfrak{i}})((A_k\otimes A_{n-k})B_{nk}^{-})_{\mathfrak{i}}^{(\mathfrak{r}',\mathfrak{r}'')}\omega_{\mathfrak{s}}\otimes v^{\mathfrak{s}}_{\mathfrak{r}'}\omega_{\mathfrak{r}''}\\ &= \Delta((A_n)_{\mathfrak{i}}^{(\mathfrak{r}',\mathfrak{r}'')}a^{\mathfrak{i}})\omega_{\mathfrak{s}}\otimes v^{\mathfrak{s}}_{\mathfrak{r}'}\omega_{\mathfrak{r}''}=0,\end{aligned}$$

so $\Delta(a^{\mathfrak{i}})\xi_{k\mathfrak{i}} \in \ker(A_k \otimes A_{n-k}) = \ker A_k \otimes \Gamma^{\otimes(n-k)} + \Gamma^{\otimes k} \otimes \ker A_{n-k} \subseteq \mathfrak{S}^k \otimes T_{\mathcal{A}}(\Gamma) + T_{\mathcal{A}}(\Gamma) \otimes \mathfrak{S}^{n-k}$. Since $\Delta^s_T(\zeta) = \Delta(a^{\mathfrak{i}})\Delta^s_T(\omega_{\mathfrak{i}}) = \sum_{k=0}^n \Delta(a^{\mathfrak{i}})\xi_{k\mathfrak{i}}$ by (42), this shows that $\Delta^s_T(\mathfrak{S}) \subseteq \mathfrak{S} \otimes T_{\mathcal{A}}(\Gamma) + T_{\mathcal{A}}(\Gamma) \otimes \mathfrak{S}$.

Let $p_0 \in \mathcal{P}_n$ denote the permutation with $p_0(i) = n - i + 1$. Since $\sigma(\eta_j\eta_i) = \sigma^{rs}_{ij}\eta_s\eta_r$ by (32), we obtain that $\sigma_k(\eta_{\bar{\mathfrak{i}}}) = (\sigma_{n-k})^{\mathfrak{j}}_{\mathfrak{i}}\eta_{\bar{\mathfrak{j}}}$ for any $\mathfrak{i} \in I^n$ and $k = 1, 2, \cdots, n-1$. Since $\sigma_k = \pi(t_k)$ and $\sigma_{n-k} = \pi(t_{n-k}) = \pi(p_0 t_k p_0^{-1})$ by definition, it follows that $\pi(p)(\eta_{\bar{\mathfrak{i}}}) = \pi(p_0 p p_0^{-1})^{\mathfrak{j}}_{\mathfrak{i}}\eta_{\bar{\mathfrak{j}}}$ for all $p \in \mathcal{P}_n$ and hence $A_n(\eta_{\bar{\mathfrak{i}}}) = (A_n)^{\mathfrak{j}}_{\mathfrak{i}}\eta_{\bar{\mathfrak{j}}}$ for $\mathfrak{i} \in I^n$. From the latter and (42) we get

$$\begin{aligned} A_n S^s_T(\zeta) &= A_n S^s_T(a^{\mathfrak{i}}\omega_{\mathfrak{i}}) = A_n(S^s_T(\omega_{\mathfrak{i}})S(a^{\mathfrak{i}})) = (-1)^{\binom{n+1}{2}} A_n(\eta_{\bar{\mathfrak{i}}})S(a^{\mathfrak{i}}) \\ &= (-1)^{\binom{n+1}{2}} (A_n)^{\mathfrak{j}}_{\mathfrak{i}}\eta_{\bar{\mathfrak{j}}}S(a^{\mathfrak{i}}) = (-1)^{\binom{n+1}{2}} \eta_{\bar{\mathfrak{j}}}S((A_n)^{\mathfrak{j}}_{\mathfrak{i}}a^{\mathfrak{i}}) = 0, \end{aligned}$$

so $S^s_T(\zeta) \in \ker A_n = \mathfrak{S}^n$. Thus, $S^s_T(\mathfrak{S}) \subseteq \mathfrak{S}$. Since $\mathfrak{S}$ is a two-sided ideal of $T^s_{\mathcal{A}}(\Gamma)$, as noted above, we have proved that $\mathfrak{S}$ is a Hopf ideal of $T^s_{\mathcal{A}}(\Gamma)$. □

The construction of the exterior algebra and the above results have their counterparts for the symmetric algebra over Γ. Let $S_n := \sum_p \pi(p)$ and $\mathfrak{A}^n := \ker S_n$. By similar reasoning as above, $\mathfrak{A} := \bigoplus_{n=2}^{\infty} \mathfrak{A}^n$ is a Hopf ideal of $T_{\mathcal{A}}(\Gamma)$. Hence the quotient $T_{\mathcal{A}}(\Gamma)/\mathfrak{A}$ is a Hopf algebra. This Hopf algebra $T_{\mathcal{A}}(\Gamma)/\mathfrak{A}$ is called the *symmetric algebra* of the bicovariant bimodule Γ. In the case $\mathcal{A} = \mathbb{C}$ it is just the symmetric algebra of the vector space Γ. However, in the important case when Γ comes from a bicovariant differential calculus as in Sect. 14.6 we have $\mathfrak{A}^2 = \ker(\sigma + I) = \{0\}$ if q is not a root of unity (see Subsect. 14.6.2), so the algebra $T_{\mathcal{A}}(\Gamma)/\mathfrak{A}$ is not of interest in this particular situation.

13.3 Notes

The main reference for this chapter is [Wor4]. Yetter–Drinfeld modules were introduced in [Yt] where they are called crossed bimodules. Their connections to bicovariant bimodules were first observed in [S2]. The relation to quantum double representations (in different forms) appeared in [PoW], [Maj4] and [Tk4]. Propositions 8 and 10 were found in [Dt1] and independently by one of the authors (K.S.).

14. Covariant Differential Calculus on Quantum Groups

This chapter contains the main concepts and results of the general theory of covariant differential calculi on quantum groups. The underlying Hopf algebra structure allows us to develop a rich theory of such calculi which is suggested by ideas from classical Lie theory.

In the first two sections left-covariant and bicovariant first order calculi are studied. To each left-covariant FODC Γ we associate a right ideal $\mathcal{R}_\Gamma$ of $\ker \varepsilon$ and a quantum tangent space $\mathcal{T}_\Gamma$. If Γ is bicovariant, the quantum tangent space $\mathcal{T}_\Gamma$ carries an analog of the classical Lie bracket and is called the quantum Lie algebra of Γ. Sections 14.3 and 14.4 deal with higher order left-covariant and bicovariant calculi, respectively. One of the main results is that any bicovariant FODC admits a unique extension to a bicovariant DC on the exterior algebra which becomes then a differential Hopf algebra. Section 14.5 develops a general method for the construction of bicovariant DC on coquasitriangular Hopf algebras. In Sect. 14.6 this method is elaborated for the coordinate algebras of the quantum groups $GL_q(N)$, $SL_q(N)$, $O_q(N)$ and $Sp_q(N)$.

Throughout this chapter $\mathcal{A}$ is a Hopf algebra with *invertible* antipode.

14.1 Left-Covariant First Order Differential Calculi

14.1.1 Left-Covariant First Order Calculi and Their Right Ideals

Definition 12.3 can also be applied to the Hopf algebra $\mathcal{A}$ considered as a left quantum space with respect to the comultiplication.

Definition 1. *A FODC Γ over the Hopf algebra $\mathcal{A}$ is called* left-covariant *if Γ is a left-covariant FODC Γ over the left quantum space $\mathcal{X} = \mathcal{A}$ with left coaction $\varphi = \Delta$ according to Definition 12.3.*

By Proposition 12.2, a FODC Γ over $\mathcal{A}$ is left-covariant if and only if there is a linear mapping $\Delta_L : \Gamma \rightarrow \mathcal{A} \otimes \Gamma$ such that

$$\Delta_L(a\mathrm{d}b) = \Delta(a)(\mathrm{id} \otimes \mathrm{d})\Delta(b) \quad \text{for all} \quad a, b \in \mathcal{A}. \tag{1}$$

If such a mapping Δ_L exists, then, by the proof of Proposition 12.2, Δ_L is a left coaction of $\mathcal{A}$ on Γ which satisfies $\Delta_L(a\rho b) = \Delta(a)\Delta_L(\rho)\Delta(b)$ for

$a, b \in \mathcal{A}$ and $\rho \in \Gamma$. Thus, according to Definition 13.1, Γ is a left-covariant bimodule over $\mathcal{A}$.

Let Γ be a left-covariant FODC over $\mathcal{A}$. First we restate some results obtained in Subsect. 13.1.1 in the present context. We define a linear mapping $\omega_\Gamma : \mathcal{A} \to {}_{\rm inv}\Gamma$ by setting $\omega_\Gamma(a) = P_L(\mathrm{d}a)$. If no confusion can arise, we omit the subscript Γ and write simply $\omega(a)$. From Lemma 13.1 we get $\omega(\mathcal{A}) = {}_{\rm inv}\Gamma$. As shown in Subsect. 13.1.1, $\Gamma = \mathcal{A}\omega(\mathcal{A}) = \omega(\mathcal{A})\,\mathcal{A}$ and any basis of the vector space $\omega(\mathcal{A})$ is a left $\mathcal{A}$-module basis and a right $\mathcal{A}$-module basis of Γ. Since $\Delta_L(\mathrm{d}a) = \sum a_{(1)} \otimes \mathrm{d}a_{(2)}$ by (1), it follows immediately from the definition of P_L and (13.4) that

$$\omega(a) = \sum S(a_{(1)})\mathrm{d}a_{(2)} \quad \text{and} \quad \mathrm{d}a = \sum a_{(1)}\omega(a_{(2)}) \quad \text{for} \quad a \in \mathcal{A}. \tag{2}$$

From the formulas (13.8), (2), and (13.3) and the Leibniz rule we obtain

$$\begin{aligned} \omega(a) \triangleleft b &= P_L\left(\sum S(a_{(1)})\mathrm{d}a_{(2)}{\cdot}b\right) = \sum \varepsilon(S(a_{(1)}))P_L(\mathrm{d}a_{(2)}{\cdot}b) = P_L(\mathrm{d}a{\cdot}b) \\ &= P_L(\mathrm{d}(ab)) - P_L(a\mathrm{d}b) = \omega(ab) - \varepsilon(a)\omega(b) = \omega(\bar{a}b). \end{aligned}$$

Recall the notation $\bar{a} := a - \varepsilon(a)1$. Hence, by (13.5) and (13.6), we have

$$\mathrm{ad}_R(b)(\omega(a)) = \omega(a) \triangleleft b = \omega(\bar{a}b), \tag{3}$$

$$b\omega(a)c = \sum bc_{(1)}\omega(\bar{a}c_{(2)}). \tag{4}$$

In order to characterize left-covariant FODC's as quotients of the universal FODC, we use another realization of the universal FODC over $\mathcal{A}$ which is better adapted to the Hopf algebra structure of $\mathcal{A}$.

Example 1 (The universal first order differential calculus over $\mathcal{A}$ – revised). Let $\Omega^1(\mathcal{A}) := \mathcal{A} \otimes \ker \varepsilon$. Let us write $a\omega(b)$ instead of $a \otimes \bar{b}$, where $a, b \in \mathcal{A}$. In particular, $\omega(b)$ denotes the element $1 \otimes \bar{b}$ for $b \in \mathcal{A}$. Taking into account the formulas (2) and (4), we introduce an $\mathcal{A}$-bimodule structure of $\Omega^1(\mathcal{A})$ and a linear mapping $\mathrm{d} : \mathcal{A} \to \Omega^1(\mathcal{A})$ by $c(b\omega(a)) := cb\omega(a)$, $(b\omega(a))c := \sum bc_{(1)}\omega(\bar{a}c_{(2)})$ and $\mathrm{d}a := \sum a_{(1)}\omega(a_{(2)})$ for $a, b, c \in \mathcal{A}$. From the relations

$$a{\cdot}\mathrm{d}b + \mathrm{d}a{\cdot}b = \sum ab_{(1)}\omega(b_{(2)}) + \sum a_{(1)}\omega(a_{(2)})b$$

$$= \sum ab_{(1)}\omega(b_{(2)}) + \sum a_{(1)}b_{(1)}\omega(a_{(2)}b_{(2)}) - \sum a_{(1)}\varepsilon(a_{(2)})b_{(1)}\omega(b_{(2)})$$

$$= \sum a_{(1)}b_{(1)}\omega(a_{(2)}b_{(2)}) = \mathrm{d}(ab),$$

$$\omega(a) = \sum \varepsilon(a_{(1)})\omega(a_{(2)}) = \sum S(a_{(1)})a_{(2)}\omega(a_{(3)}) = \sum S(a_{(1)})\mathrm{d}a_{(2)} \tag{5}$$

we see that d satisfies the Leibniz rule and that $\Omega^1(\mathcal{A}) = \mathcal{A}\mathrm{d}\mathcal{A}$, that is, $\Omega^1(\mathcal{A})$ is a FODC over the Hopf algebra $\mathcal{A}$.

We prove that $\Omega^1(\mathcal{A})$ is the universal FODC over $\mathcal{A}$. Let Γ be another FODC over $\mathcal{A}$ with differentiation $\tilde{\mathrm{d}}$. We first show that the linear mapping $\psi : \Omega^1(\mathcal{A}) \to \Gamma$ given by $\psi(a{\cdot}\mathrm{d}b) = a{\cdot}\tilde{\mathrm{d}}b$ is well-defined. For this we suppose that $\sum_i a_i{\cdot}\mathrm{d}b^i = 0$ in $\Omega^1(\mathcal{A})$ with $a_i, b^i \in \mathcal{A}$. By the definitions of d and $\Omega^1(\mathcal{A})$, we have $\sum\sum_i a_i b^i_{(1)}\omega(b^i_{(2)}) = \sum\sum_i a_i b^i_{(1)}\otimes\overline{b^i_{(2)}} = 0$ in $\mathcal{A}\otimes\ker\varepsilon$. Applying the mapping $(m\otimes\mathrm{id})(\mathrm{id}\otimes S\otimes\mathrm{id})(\mathrm{id}\otimes\Delta)$ and using the Hopf algebra axioms, we compute $\sum_i(a_i\otimes b^i - a_i b^i\otimes 1) = 0$. Hence $\sum_i a_i{\cdot}\tilde{\mathrm{d}}b^i = 0$ in Γ and ψ is well-defined. It is straightforward to verify that the FODC Γ is isomorphic to the quotient FODC $\Omega^1(\mathcal{A})/\mathcal{N}$, where $\mathcal{N}$ is the $\mathcal{A}$-subbimodule $\ker\psi$ of $\Omega^1(\mathcal{A})$. Therefore, $\Omega^1(\mathcal{A})$ is indeed the universal FODC over $\mathcal{A}$.

Of course, the universal FODC over $\mathcal{A}$ is unique up to isomorphism. Hence the universal FODC $\Omega^1(\mathcal{A})$ defined above and the universal FODC constructed in Subsect. 12.2.2 are isomorphic. By Corollary 12.3′, the FODC $\Omega^1(\mathcal{A})$ is left-covariant. Moreover, formula (5) shows that $\omega(a)$ is nothing but the 1-form $P_L(\mathrm{d}a) = \omega_{\Omega^1(\mathcal{A})}(a)$, $a\in\mathcal{A}$, in accordance with the notation introduced above. △

For a left-covariant FODC Γ over $\mathcal{A}$ we define

$$\mathcal{R}_\Gamma := \{a\in\ker\varepsilon \,|\, \omega_\Gamma(a) = 0\}. \tag{6}$$

The next proposition shows that the mapping $\Gamma \to \mathcal{R}_\Gamma$ gives a bijection between left-covariant FODC over $\mathcal{A}$ and right ideals of $\ker\varepsilon$.

Proposition 1. (i) *Let $\mathcal{R}$ be a right ideal of* $\ker\varepsilon$. *Then $\mathcal{N} := \mathcal{A}\,\omega_{\Omega^1(\mathcal{A})}(\mathcal{R})$ is an $\mathcal{A}$-subbimodule of $\Omega^1(\mathcal{A})$ and the quotient $\Gamma := \Omega^1(\mathcal{A})/\mathcal{N}$ is a left-covariant FODC over $\mathcal{A}$ such that $\mathcal{R}_\Gamma = \mathcal{R}$.*

(ii) *We suppose that Γ is a left-covariant FODC over $\mathcal{A}$. Then $\mathcal{R}_\Gamma$ is a right ideal of* $\ker\varepsilon$ *and the FODC Γ is isomorphic to the quotient FODC $\Omega^1(\mathcal{A})/\mathcal{A}\,\omega_{\Omega^1(\mathcal{A})}(\mathcal{R}_\Gamma)$.*

Proof. (i): Throughout the proof of (i) we write $\omega(\cdot)$ for $\omega_{\Omega^1(\mathcal{A})}(\cdot)$. Since $\mathcal{R}$ is also a right ideal of $\mathcal{A}$, we have $\omega(a)c = \sum c_{(1)}\omega(ac_{(2)}) \in \mathcal{A}\omega(\mathcal{R})$ by (4) for $a\in\mathcal{R}$ ($\subseteq\ker\varepsilon$) and $c\in\mathcal{A}$. That is, $\mathcal{N} = \mathcal{A}\omega(\mathcal{R})$ is an $\mathcal{A}$-subbimodule of $\Omega^1(\mathcal{A})$. Hence $\Gamma = \Omega^1(\mathcal{A})/\mathcal{N}$ is a FODC over $\mathcal{A}$.

Let $\Delta_L : \Omega^1(\mathcal{A}) \to \mathcal{A}\otimes\Omega^1(\mathcal{A})$ denote the coaction for the left-covariant FODC $\Omega^1(\mathcal{A})$. Since $\Delta_L(a\omega(b)) = \Delta(a)(1\otimes\omega(b))$, we have $\Delta_L(\mathcal{N}) \subseteq \mathcal{A}\otimes\mathcal{N}$. Therefore, Δ_L passes to a left coaction of $\mathcal{A}$ on the quotient $\Gamma = \Omega^1(\mathcal{A})/\mathcal{N}$ which obviously has all the properties required in Definition 12.3. This shows that the FODC Γ is left-covariant.

By the definition of d on the quotient FODC it is clear that $\mathcal{R}\subseteq\mathcal{R}_\Gamma$. Conversely, suppose that $\omega_\Gamma(a) = 0$ for some $a\in\ker\varepsilon$. Then $\omega(a)\in\mathcal{N} = \mathcal{A}\omega(\mathcal{R})$. By the definition of $\Omega^1(\mathcal{A})$, this means that $1\otimes a\in\mathcal{A}\otimes\mathcal{R}$ which in turn implies $a\in\mathcal{R}$. Thus, $\mathcal{R} = \mathcal{R}_\Gamma$.

(ii): Since Γ is left-covariant, (3) holds. Therefore, if $a\in\mathcal{R}_\Gamma$ and $b\in\ker\varepsilon$, then $\omega_\Gamma(a) = 0$ and hence $\omega_\Gamma(ab) = \mathrm{ad}_R(b)(\omega_\Gamma(a)) = 0$ by (3). Thus, $\mathcal{R}_\Gamma$ is a right ideal of $\ker\varepsilon$. The second assertion of (ii) follows from (i). □

Proposition 2. *Suppose that $\mathcal{A}$ is a Hopf $*$-algebra and Γ is a left-covariant FODC over $\mathcal{A}$. Then Γ is a $*$-calculus if and only if $S(\mathcal{R}_\Gamma)^* \subseteq \mathcal{R}_\Gamma$.*

Proof. Let Γ be a $*$-calculus and let $a \in \mathcal{R}_\Gamma$. Using the identity $\Delta(S(a)^*) = \sum S(a_{(2)})^* \otimes S(a_{(1)})^*$, the formulas (2) and (1.39), properties of the involution, and finally the Leibniz rule, we compute

$$\begin{aligned}\omega_\Gamma(S(a)^*) &= \sum S(S(a_{(2)})^*)\mathrm{d}(S(a_{(1)})^*) = \sum a_{(2)}{}^*(\mathrm{d}S(a_{(1)}))^* \\ &= \sum((\mathrm{d}S(a_{(1)}))a_{(2)})^* = \sum(\mathrm{d}(S(a_{(1)})a_{(2)}) - S(a_{(1)})\mathrm{d}a_{(2)})^* \\ &= (\varepsilon(a)\mathrm{d}1 - \omega_\Gamma(a))^* = -\omega_\Gamma(a)^* = 0, \end{aligned} \tag{7}$$

so $S(a)^* \in \mathcal{R}_\Gamma$. Thus, $S(\mathcal{R}_\Gamma)^* \subseteq \mathcal{R}_\Gamma$.

Conversely, suppose that $S(\mathcal{R}_\Gamma)^* \subseteq \mathcal{R}_\Gamma$ and write ω for $\omega_{\Omega^1(\mathcal{A})}$. Recall from Subsect. 12.2.2 that the universal FODC over $\mathcal{A}$ is a $*$-calculus. Therefore, by the calculation (7) applied to the $*$-calculus $\Omega^1(\mathcal{A})$ and by the assumption, we have $\omega(a)^* = -\omega(S(a)^*) \in \omega(\mathcal{R}_\Gamma)$ for $a \in \mathcal{R}_\Gamma$, so that $\omega(\mathcal{R}_\Gamma)^* \subseteq \omega(\mathcal{R}_\Gamma)$. Since $\mathcal{R}_\Gamma$ is a right ideal, $\omega(\mathcal{R}_\Gamma)\mathcal{A} \subseteq \mathcal{A}\omega(\mathcal{R}_\Gamma)$ by (4). Hence we get $(\mathcal{A}\omega(\mathcal{R}_\Gamma))^* = \omega(\mathcal{R}_\Gamma)^*\mathcal{A}^* \subseteq \omega(\mathcal{R}_\Gamma)\mathcal{A} \subseteq \mathcal{A}\omega(\mathcal{R}_\Gamma)$, so the involution on $\Omega^1(\mathcal{A})$ passes to the quotient $\Omega^1(\mathcal{A})/\mathcal{A}\omega(\mathcal{R}_\Gamma)$ which therefore becomes a $*$-calculus. Since this quotient FODC is isomorphic to Γ by Proposition 1, Γ is also a $*$-calculus. □

14.1.2 The Quantum Tangent Space

If Γ is the "ordinary" differential calculus on a Lie group G, then we have $\mathcal{R}_\Gamma = (\ker\varepsilon)^2 = \{f \in \mathcal{A} \,|\, f(e) = (\mathrm{d}f)(e) = 0\}$ and the tangent space at the unit element can be considered as the vector space of all linear functionals on $C^\infty(G)$ annihilating $\mathcal{R}_\Gamma$ and the constant functions. This motivates

Definition 2. *For a left-covariant FODC Γ over $\mathcal{A}$, the vector space*

$$\mathcal{T}_\Gamma := \{X \in \mathcal{A}' \mid X(1) = 0 \;\; \textit{and} \;\; X(a) = 0 \;\; \textit{for} \;\; a \in \mathcal{R}_\Gamma\} \tag{8}$$

is called the quantum tangent space *to Γ.*

In what follows we will omit the subscripts Γ in $\mathcal{T}_\Gamma$ and $\mathcal{R}_\Gamma$.

Propositions 3 and 4 below show that $\mathcal{T}$ admits properties analogous to those of the classical tangent space at the identity of a Lie group. However, the commutation relations between elements of $\mathcal{T}$ are not necessarily quadratic as in the classical case (see Subsect. 14.1.4 for an example).

Proposition 3. *There is a unique bilinear form $(.,.) : \mathcal{T} \times \Gamma \to \mathbb{C}$ such that*

$$(X, a{\cdot}\mathrm{d}b) = \varepsilon(a)X(b) \quad \textit{for} \quad a, b \in \mathcal{A}, \;\; X \in \mathcal{T}. \tag{9}$$

The vector spaces ${}_{\mathrm{inv}}\Gamma = \omega(\mathcal{A})$ and $\mathcal{T}$ form a nondegenerate dual pairing with respect to this bilinear form (that is, $(\omega, X) = 0$ for all $X \in \mathcal{T}$ implies $\omega = 0$ and $(\omega, X) = 0$ for all $\omega \in {}_{\mathrm{inv}}\Gamma$ implies $X = 0$), and we have

$$(\omega(a), X) = X(a) \quad for \quad a \in \mathcal{A}, \quad X \in \mathcal{T}. \tag{10}$$

Proof. For $\zeta = \sum_i a_i \mathrm{d}b^i \in \Gamma$ and $X \in \mathcal{T}$, we define $(X, \zeta) := X(\sum_i \varepsilon(a_i)b^i)$. We have to prove that this value does not depend on the particular representation $\sum_i a_i \mathrm{d}b^i$ of the form ζ. For this it suffices to check that the right-hand side of the defining equation vanishes if $\zeta = 0$. Indeed, by (13.3), $\zeta = 0$ yields $0 = P_L(\zeta) = \sum_i \varepsilon(a_i)P_L(\mathrm{d}b^i) = \omega(\sum_i \varepsilon(a_i)b^i)$, so $\sum_i \varepsilon(a_i)\overline{b^i} \in \mathcal{R}$ and hence $X(\sum_i \varepsilon(a_i)b^i) = 0$ by the definition of $\mathcal{T}$. Thus, the bilinear form $(\cdot,\cdot)$ is well-defined.

Since $\Gamma = \mathcal{A}\,\mathrm{d}\mathcal{A}$, the bilinear form $(\cdot,\cdot)$ is uniquely determined by (9). Formula (10) follows at once from (9) and (2). The assertions concerning the dual pairing are easily verified. □

It is clear from (8) and the fact that ${}_{\mathrm{inv}}\Gamma$ and $\mathcal{T}$ form a nondegenerate dual pairing that

$$\dim {}_{\mathrm{inv}}\Gamma = \dim \mathcal{T} = \dim \ker \varepsilon/\mathcal{R}. \tag{11}$$

We call this (cardinal) number the *dimension* of the left-covariant FODC Γ.

Let us fix the following notation which will always be kept in the following: $\{X_i\}_{i\in I}$ denotes a basis of the vector space $\mathcal{T}$, $\{\omega_i\}_{i\in I}$ is the dual basis of ${}_{\mathrm{inv}}\Gamma$ (that is, $(X_i, \omega_j) = \delta_{ij}$ for $i, j \in I$) and $\{f_i^j\}_{i,j\in I}$ is the family of functionals from Proposition 13.2.

Proposition 4. *With the above notation, we have for a and b in* $\mathcal{A}$ *that*

$$\mathrm{d}a = \sum\nolimits_i (X_i.a)\omega_i, \tag{12}$$

$$X_i(ab) = \varepsilon(a)X_i(b) + \sum\nolimits_j X_j(a)f_i^j(b). \tag{13}$$

Proof. Since $(X_i, \omega(a)) = X_i(a)$ by (10), we have $\omega(a) = \sum_i X_i(a)\omega_i$ and hence $\mathrm{d}a = \sum a_{(1)}\omega(a_{(2)}) = \sum\sum_i a_{(1)}X_i(a_{(2)})\omega_i = \sum_i (X_i.a)\omega_i$ which proves (12). Using (12) and the Leibniz rule, we get

$$\sum\nolimits_i (X_i.ab)\omega_i = \mathrm{d}(ab) = a{\cdot}\mathrm{d}b + \mathrm{d}a{\cdot}b = \sum\nolimits_i a(X_i.b)\omega_i + \sum\nolimits_j (X_j.a)\omega_j b$$

$$= \sum\nolimits_i a(X_i.b)\omega_i + \sum\nolimits_{i,j} (X_j.a)(f_i^j.b)\omega_i.$$

Equating the coefficients of ω_i one obtains

$$X_i.ab = a(X_i.b) + \sum\nolimits_j (X_j.a)(f_i^j.b), \ i \in I.$$

Applying ε to this equation and using the identity $\varepsilon(f.c) = f(c)$ for $f \in \mathcal{A}'$ and $c \in \mathcal{A}$, we obtain (13). □

Formula (12) indicates that the linear mapping $L_{X_i} : \mathcal{A} \to \mathcal{A}$ defined by $L_{X_i}(a) := X_i.a$ might be regarded as a quantum analog of a left-invariant vector field in classical Lie theory (see Subsect. 14.4.2 for more details).

Suppose now that the dimension of the vector space $\mathcal{T}$ is *finite*. Then we conclude from (13) and (13.14) that the functionals X_i and f^r_s belong to the dual Hopf algebra $\mathcal{A}^\circ$ (see Subsect. 1.2.8) and we have

$$\Delta(X_i) = \varepsilon \otimes X_i + X_k \otimes f^k_i, \quad \Delta(f^r_s) = f^r_k \otimes f^k_s \tag{14}$$

in $\mathcal{A}^\circ$. Hence the linear span of functionals ε, X_i and f^r_s, $i, r, s \in I$, is a subcoalgebra of $\mathcal{A}^\circ$. The antipode of X_i in the dual Hopf algebra $\mathcal{A}^\circ$ is given by

$$S(X_i) = -X_k S(f^k_i). \tag{15}$$

Indeed, by (1.26) and (14), we have the relation

$$\mathrm{m}_{\mathcal{A}^\circ}(\mathrm{id} \otimes S)\Delta(X_i) = \mathrm{m}_{\mathcal{A}^\circ}(\varepsilon \otimes S(X_i) + X_k \otimes S(f^k_i)) = 0 = \varepsilon(X_i),$$

which in turn implies (15).

Proposition 5. *A finite-dimensional vector space $\mathcal{T}$ of linear functionals on a Hopf algebra $\mathcal{A}$ is a quantum tangent space of a left-covariant FODC if and only if $X(1) = 0$ and $\Delta(X) - \varepsilon \otimes X \in \mathcal{T} \otimes \mathcal{A}^\circ$ for all $X \in \mathcal{T}$.*

Proof. The necessity of the second condition follows at once from (14). To prove the sufficiency part, let $\mathcal{R} = \{a \in \ker\varepsilon \mid X(a) = 0 \text{ for all } X \in \mathcal{T}\}$. From the assumption $\Delta(X) - \varepsilon \otimes X \in \mathcal{T} \otimes \mathcal{A}^\circ$ for $X \in \mathcal{T}$ it follows that $\mathcal{R}$ is a right ideal of $\ker\varepsilon$. By Proposition 1, there exists a left-covariant FODC Γ over $\mathcal{A}$ such that $\mathcal{R} = \mathcal{R}_\Gamma$. Since $\mathcal{T}$ is finite-dimensional, the definition of $\mathcal{R}$ implies that $\mathcal{T} = \{X \in \mathcal{A}' \mid X(1) = 0 \text{ and } X(a) = 0 \text{ for } a \in \mathcal{R}\}$. That is, $\mathcal{T}$ is the quantum tangent space of Γ. □

Proposition 6. *Let Γ be a finite-dimensional left-covariant FODC over a Hopf $*$-algebra $\mathcal{A}$. Then Γ is a $*$-calculus if and only if its quantum tangent space $\mathcal{T} \subseteq \mathcal{A}'$ is $*$-invariant, where $\mathcal{A}'$ is endowed with the involution (1.40).*

Proof. By (1.40), the functional f^* is defined by $f^*(\cdot) = \overline{f(S(\cdot)^*)}$ for $f \in \mathcal{A}'$. If Γ is a $*$-calculus and $X \in \mathcal{T}$, we have $X^*(a) = \overline{X(S(a)^*)} = 0$ for all $a \in \mathcal{R}$ by Proposition 2, so that $X^* \in \mathcal{T}$. Conversely, if $\mathcal{T}$ is $*$-invariant and $a \in \mathcal{R}$, then $X(S(a)^*) = \overline{X^*(a)} = 0$ for all $X \in \mathcal{T}$. Since $\mathcal{T}$ is finite-dimensional, this implies that $S(a)^* \in \mathcal{R}$, so Γ is a $*$-calculus again by Proposition 2. □

14.1.3 An Example: The 3D-Calculus on $SL_q(2)$

Throughout this and the next subsection, $\mathcal{A}$ is the Hopf algebra $\mathcal{O}(SL_q(2))$ and we assume that $q^2 \neq 1$. Let $a = u^1_1$, $b = u^1_2$, $c = u^2_1$, $d = u^2_2$ denote the entries of the fundamental matrix of $\mathcal{A}$ and let E, F, K be the generators of $\breve{U}_q(\mathrm{sl}_2)$, see Subsect. 3.1.2. Recall from Subsect. 4.4.1 that there is a dual pairing of the Hopf algebras $\breve{U}_q(\mathrm{sl}_2)$ and $\mathcal{A} = \mathcal{O}(SL_q(2))$ determined on the generators by the equations

$$\langle K, a\rangle = q^{-1/2}, \ \langle K, d\rangle = q^{1/2}, \langle E, \ c\rangle = \langle F, b\rangle = 1 \text{ and zero otherwise.} \tag{16}$$

We define three linear functionals X_0, X_1, X_2 on $\mathcal{A}$ by

$$X_0 := q^{-1/2}FK, \quad X_2 := q^{1/2}EK, \quad X_1 := (1-q^{-2})^{-1}(\varepsilon - K^4).$$

By the definition (3.12) of the comultiplication of $\breve{U}_q(\mathrm{sl}_2)$ we have

$$\Delta X_j = \varepsilon \otimes X_j + X_j \otimes K^2 \text{ for } j{=}0,2 \text{ and } \Delta X_1 = \varepsilon \otimes X_1 + X_1 \otimes K^4. \quad (17)$$

Therefore, by Proposition 5, $\mathcal{T} := \mathrm{Lin}\{X_0, X_1, X_2\}$ is the quantum tangent space of a (unique up to isomorphism) left-covariant FODC Γ over $\mathcal{A}$. This is the so-called 3D-calculus of S. L. Woronowicz. Let $\{\omega_0, \omega_1, \omega_2\}$ be a basis of ${}_{\mathrm{inv}}\Gamma$ which is dual to the basis $\{X_0, X_1, X_2\}$ of $\mathcal{T}$. Then, by (12), we can write

$$\mathrm{d}x = \sum_{j=0}^{2}(X_j.x)\omega_j, \quad x \in \mathcal{A}. \quad (18)$$

Since $X_0(b) = X_2(c) = X_1(a) = 1$ and $X_0(a) = X_0(c) = X_2(a) = X_2(b) = X_1(b) = X_2(c) = 0$ because of (16), we obtain

$$\omega_0 = \omega(b), \quad \omega_2 = \omega(c), \quad \omega_1 = \omega(a) = -q^{-2}\omega(d), \quad (19)$$

$$\mathrm{d}a = b\omega_2 + a\omega_1, \mathrm{d}b = a\omega_0 - q^2 b\omega_1, \mathrm{d}c = c\omega_1 + d\omega_2, \mathrm{d}d = -q^2 d\omega_1 + c\omega_0. \quad (20)$$

Comparing (17) with (14) we see that $f^j_j = K^2$ for $j = 0,2$, $f^1_1 = K^4$ and $f^r_s = 0$ otherwise. This leads to the following commutation rules between matrix entries and 1-forms:

$$\omega_j a = q^{-1}a\omega_j, \; \omega_j b = qb\omega_j, \; \omega_j c = q^{-1}c\omega_j, \; \omega_j d = qd\omega_j \text{ for } j = 0,2, \quad (21)$$

$$\omega_1 a = q^{-2}a\omega_1, \; \omega_1 b = q^2 b\omega_1, \; \omega_1 c = q^{-2}c\omega_1, \; \omega_1 d = q^2 d\omega_1. \quad (22)$$

Let $\mathcal{R} := \mathcal{B}\cdot\mathcal{A}$ be the right ideal of $\ker\varepsilon$, where $\mathcal{B}$ is the vector space generated by the six elements

$$a + q^{-2}d - (1+q^{-2})1, \quad b^2, \quad c^2, \quad bc, \quad (a-1)b, \quad (a-1)c. \quad (23)$$

Using (16) one easily verifies that the functionals X_0, X_2, X_1 annihilate these elements, hence $\mathcal{R} \subseteq \mathcal{R}_\Gamma$. Since $\dim(\ker\varepsilon)/\mathcal{R} \le 3$ by the definition of $\mathcal{R}$ and $\dim(\ker\varepsilon)/\mathcal{R}_\Gamma = \dim\mathcal{T} = 3$ by (11), we conclude that $\mathcal{R} = \mathcal{R}_\Gamma$, that is, $\mathcal{R}$ is the right ideal of $\ker\varepsilon$ associated with the FODC Γ.

The six generators x from (23) satisfy the condition $S(x)^* \in \mathcal{R}$ for the Hopf $*$-algebras $\mathcal{O}(SL_q(2,\mathbb{R}))$, $|q| = 1$, $\mathcal{O}(SU_q(2))$ and $\mathcal{O}(SU_q(1,1))$, $q \in \mathbb{R}$. Hence, by Proposition 2, Γ is a $*$-calculus for all three Hopf $*$-algebras.

From the commutation rules in the algebra $\breve{U}_q(\mathrm{sl}_2)$ it follows that the basis elements of the quantum tangent space $\mathcal{T}$ satisfy the relations

$$q^2 X_1 X_0 - q^{-2} X_0 X_1 = (1+q^2)X_0, \quad (24)$$

$$q^2 X_2 X_1 - q^{-2} X_1 X_2 = (1+q^2)X_2\,, \quad (25)$$

$$qX_2X_0 - q^{-1}X_0X_2 = -q^{-1}X_1. \quad (26)$$

In the limit $q \to 1$ all the above formulas give the corresponding formulas for the ordinary first order differential calculus on $SL(2)$.

14.1.4 Another Left-Covariant Differential Calculus on $SL_q(2)$

We put

$$X_0 := q^{-5/2}FK^5, \quad X_2 := q^{5/2}EK^5, \quad X_1 := (1-q^{-2})^{-1}(\varepsilon - K^4).$$

Note that X_1 is the same as in Subsect. 14.1.3. Now we have

$$\Delta X_j = \varepsilon \otimes X_j + X_j \otimes K^6 + (q^{-2}-1)X_1 \otimes X_j, \quad j = 0, 2.$$

By Proposition 5, there exists a left-covariant FODC Γ over $SL_q(2)$ with quantum tangent space $\mathcal{T} := \mathrm{Lin}\,\{X_0, X_1, X_2\}$. This FODC Γ can be developed in a similar manner as in Subsect. 14.1.3, so we only mention the necessary modifications. Formulas (18)–(20), (24) and (25) remain valid without change. In (21) we replace q by q^3 and q^{-1} by q^{-3}, and in (23) the two last generators of $\mathcal{R}$ must be replaced by $(a-q^{-2})b$ and $(a-q^{-2})c$. Instead of (22), we now have

$$\omega_1 a = q^{-2}a\omega_1 + (q^{-2}-1)b\omega_2, \quad \omega_1 c = q^{-2}c\omega_1 + (q^{-2}-1)d\omega_2,$$
$$\omega_1 b = q^2 b\omega_1 + (q^{-2}-1)a\omega_0, \quad \omega_1 d = q^2 d\omega_1 + (q^{-2}-1)c\omega_0.$$

The commutation relation (26) becomes the equation

$$q^5 X_2 X_0 - q^{-5}X_0X_2 = -q^{-1}X_1 + 2q^{-2}(q-q^{-1})X_1^2 - q^{-3}(q-q^{-1})^2 X_1^3.$$

As in Subsect. 14.1.3, the classical limit $q \to 1$ of the FODC Γ is just the ordinary first order calculus on $SL(2)$. Note that in the case $q = 1$ the commutation relations of X_0, X_1 and X_2 for both FODC in Subsects. 14.1.3 and 14.1.4 give the relations of the classical Lie algebra $\mathrm{sl}(2,\mathbb{C})$.

14.2 Bicovariant First Order Differential Calculi

14.2.1 Right-Covariant First Order Differential Calculi

At the end of Subsect. 12.2.3 we explained how Definition 12.3 and Proposition 12.2 have to be changed in the case of a right quantum space $\mathcal{X}$ and a right-covariant FODC Γ on $\mathcal{X}$. We apply these modified versions in

Definition 3. *A FODC Γ over $\mathcal{A}$ is said to be* right-covariant *if Γ is a right-covariant FODC over the right quantum space $\mathcal{X} = \mathcal{A}$ with right coaction $\varphi = \Delta$ or equivalently if there exists a linear mapping $\Delta_R : \Gamma \to \Gamma \otimes \mathcal{A}$ such that*

$$\Delta_R(a\mathrm{d}b) = \Delta(a)(\mathrm{d}\otimes \mathrm{id})\Delta(b) \quad \textit{for} \quad a, b \in \mathcal{A}. \tag{27}$$

Suppose that Γ is a right-covariant FODC over $\mathcal{A}$. Then Γ is in particular a right-covariant bimodule over $\mathcal{A}$. We introduce a linear mapping $\eta = \eta_\Gamma : \mathcal{A} \to \Gamma_{\mathrm{inv}}$ by setting $\eta(a) := P_R(\mathrm{d}a)$. Clearly, $\eta(\mathcal{A}) = \Gamma_{\mathrm{inv}}$. Since $\Delta_R(\mathrm{d}a) = \sum \mathrm{d}a_{(1)} \otimes a_{(2)}$ by (27), it follows that

$$\eta(a) = \sum \mathrm{d}a_{(1)} \cdot S(a_{(2)}) \quad \text{and} \quad \mathrm{d}a = \sum \eta(a_{(1)})a_{(2)} \quad \text{for} \quad a \in \mathcal{A}.$$

14.2.2 Bicovariant First Order Differential Calculi

For a Hopf algebra it is quite natural to look for FODC which are compatible with the comultiplication from the left and from the right. These are the bicovariant FODC's.

Definition 4. *A first order differential calculus over a Hopf algebra $\mathcal{A}$ is called* bicovariant *if it is both left-covariant and right-covariant.*

By Subsects. 14.1.1 and 14.2.1, a FODC Γ over $\mathcal{A}$ is bicovariant if and only if there exist linear mappings $\Delta_L : \Gamma \to \mathcal{A} \otimes \Gamma$ and $\Delta_R : \Gamma \to \Gamma \otimes \mathcal{A}$ such that the equations (1) and (27) are satisfied.

Suppose that Γ is a bicovariant FODC over $\mathcal{A}$ and let Δ_L and Δ_R be the corresponding mappings. From (1) and (27) we obtain

$$\begin{aligned}(\mathrm{id} \otimes \Delta_R) \circ \Delta_L(a\mathrm{d}b) &= \sum (\mathrm{id} \otimes \Delta_R)(a_{(1)}b_{(1)} \otimes a_{(2)}\mathrm{d}b_{(2)}) \\ &= \sum a_{(1)}b_{(1)} \otimes a_{(2)}\mathrm{d}b_{(2)} \otimes a_{(3)}b_{(3)}, \\ (\Delta_L \otimes \mathrm{id}) \circ \Delta_R(a\mathrm{d}b) &= \sum (\Delta_L \otimes \mathrm{id})(a_{(1)}\mathrm{d}b_{(1)} \otimes a_{(2)}b_{(2)}) \\ &= \sum a_{(1)}b_{(1)} \otimes a_{(2)}\mathrm{d}b_{(2)} \otimes a_{(3)}b_{(3)}\end{aligned}$$

for $a, b \in \mathcal{A}$. Since $\Gamma = \mathcal{A}{\cdot}\mathrm{d}\mathcal{A}$, the preceding shows that $(\mathrm{id} \otimes \Delta_R) \circ \Delta_L = (\Delta_L \otimes \mathrm{id}) \circ \Delta_R$, so condition (iii) in Definition 13.3 is fulfilled. The validity of the two other conditions in Definition 13.3 have been already mentioned in Subsects. 14.1.1 and 14.2.1. Therefore, Γ is a *bicovariant bimodule* over $\mathcal{A}$ and the whole theory developed in Chap. 13 applies to Γ.

The following formulas describe the right coaction Δ_R on a left-invariant form $\omega(a)$ and the left coaction Δ_L on a right invariant form $\eta(a)$ in terms of the right and left adjoint coactions of $\mathcal{A}$ on itself, respectively:

$$\Delta_R(\omega(a)) = (\omega \otimes \mathrm{id})(\mathrm{Ad}_R(a)), \quad \Delta_L(\eta(a)) = (\mathrm{id} \otimes \eta)(\mathrm{Ad}_L(a)), \quad a \in \mathcal{A}. \tag{28}$$

We verify (for instance) the first equality and compute

$$\begin{aligned}\Delta_R(\omega(a)) &= \sum \Delta_R(S(a_{(1)})\mathrm{d}a_{(2)}) = \sum \Delta(S(a_{(1)}))(\mathrm{d} \otimes \mathrm{id})\Delta(a_{(2)}) \\ &= \sum S(a_{(2)})\mathrm{d}a_{(3)} \otimes S(a_{(1)})a_{(4)} = \sum \omega(a_{(2)}) \otimes S(a_{(1)})a_{(3)} \\ &= (\omega \otimes \mathrm{id})(\mathrm{Ad}_R(a)).\end{aligned}$$

Next we characterize bicovariant FODC among left-covariant FODC in terms of their right ideals.

Proposition 7. *Let Γ be a left-covariant FODC over $\mathcal{A}$ with associated right ideal $\mathcal{R}$. The FODC Γ is bicovariant if and only if $\mathcal{R}$ is invariant under the right adjoint coaction (that is, $\mathrm{Ad}_R(\mathcal{R}) \subseteq \mathcal{R} \otimes \mathcal{A}$).*

Proof. If Γ is bicovariant, then (28) holds. Since $\mathcal{R} = \{a \in \ker \varepsilon \mid \omega(a) = 0\}$, (28) implies that $\mathrm{Ad}_R(\mathcal{R}) \subseteq \mathcal{R} \otimes \mathcal{A}$. Conversely, assume that $\mathrm{Ad}_R(\mathcal{R}) \subseteq$

$\mathcal{R} \otimes \mathcal{A}$. By Corollary 12.3′, the universal FODC $\Omega^1(\mathcal{A})$ is left-covariant. Similarly, $\Omega^1(\mathcal{A})$ is right-covariant (see the Remark at the end of Subsect. 12.2.3) and hence bicovariant. From formula (28) applied to the bicovariant FODC $\Omega^1(\mathcal{A})$ and from the Ad_R-invariance of $\mathcal{R}$ it follows that $\Delta_R(\mathcal{A}\omega_{\Omega^1(\mathcal{A})}(\mathcal{R})) \subseteq \mathcal{A}\omega_{\Omega^1(\mathcal{A})}(\mathcal{R})\otimes\mathcal{A}$. Therefore, the right coaction Δ_R of the FODC $\Omega^1(\mathcal{A})$ passes to the quotient $\Omega^1(\mathcal{A})/\mathcal{N}$, where $\mathcal{N} := \mathcal{A}\omega_{\Omega^1(\mathcal{A})}(\mathcal{R})$, and the quotient FODC $\Omega^1(\mathcal{A})/\mathcal{N}$ is right-covariant. Since Γ is isomorphic to $\Omega^1(\mathcal{A})/\mathcal{N}$ by Proposition 1(ii), Γ is also right-covariant. □

Let Γ be a bicovariant FODC over $\mathcal{A}$. We state two useful formulas for the braiding σ of $\Gamma \otimes_{\mathcal{A}} \Gamma$ which will be needed later. For any $a, b \in \mathcal{A}$, we have

$$\sigma(\omega(a) \otimes_{\mathcal{A}} \omega(b)) = \sum \omega(b_{(2)}) \otimes_{\mathcal{A}} \omega(\bar{a}S(b_{(1)})b_{(3)}), \tag{29}$$

$$(\omega \otimes_{\mathcal{A}} \omega)\mathrm{Ad}_R(a) = (\mathrm{id} - \sigma)(\omega \otimes_{\mathcal{A}} \omega)\Delta(a). \tag{30}$$

Formula (29) follows at once from (13.25) combined with (28) and (3). Using the Hopf algebra axioms, (30) is easily derived from (29).

14.2.3 Quantum Lie Algebras of Bicovariant First Order Calculi

Unless stated otherwise we suppose in this subsection that Γ is a bicovariant FODC over $\mathcal{A}$ with right ideal $\mathcal{R}$ and quantum tangent space $\mathcal{T}$ such that $\dim \mathcal{T} < \infty$.

For X and Y in $\mathcal{T}$, we define a linear functional $[X, Y]$ on $\mathcal{A}$ by setting

$$[X, Y](a) := (X \otimes Y)(\mathrm{Ad}_R(a)), \quad a \in \mathcal{A}. \tag{31}$$

Since the right ideal $\mathcal{R}$ is Ad_R-invariant by Proposition 7 and the elements of $\mathcal{T}$ annihilate $\mathcal{R}$ because of (8), it follows from (31) that $[X, Y]$ also annihilates $\mathcal{R}$. Since $X(1) = Y(1) = 0$, we get $[X, Y](1) = 0$. Hence we have $[X, Y] \in \mathcal{T}$ by (8). Thus, $C : X \otimes Y \to [X, Y]$ defines a linear mapping of $\mathcal{T} \otimes \mathcal{T}$ to $\mathcal{T}$. The assertions of Proposition 8 below show that the mapping C can be viewed as a quantum analog of the classical Lie commutator bracket.

First we need some notation. Let $(\cdot,\cdot)$ be the bilinear form on $\mathcal{T} \times {}_{\mathrm{inv}}\Gamma$ from Proposition 3. There exists a unique bilinear form $(\cdot,\cdot)_2$ on $(\mathcal{T} \otimes \mathcal{T}) \times {}_{\mathrm{inv}}(\Gamma \otimes_{\mathcal{A}} \Gamma)$ such that

$$(X \otimes Y, \omega^1 \otimes_{\mathcal{A}} \omega^2)_2 = (X, \omega^1)(Y, \omega^2) \quad \text{for } \omega^1, \omega^2 \in {}_{\mathrm{inv}}\Gamma,\ X, Y \in \mathcal{T}. \tag{32}$$

Clearly, the bilinear form $(\cdot,\cdot)_2$ is nondegenerate as $(\cdot,\cdot)$ is. Since the braiding σ maps ${}_{\mathrm{inv}}(\Gamma \otimes_{\mathcal{A}} \Gamma)$ into itself, its transpose σ^t with respect to the nondegenerate bilinear form $(\cdot,\cdot)_2$ is a well-defined map of $\mathcal{T} \otimes \mathcal{T}$ into itself. Let $m_{\mathcal{A}'} : \mathcal{A}' \otimes \mathcal{A}' \to \mathcal{A}'$ denote the multiplication map of the algebra $\mathcal{A}'$, that is, $m_{\mathcal{A}'}(f \otimes g)(a) = fg(a) = \sum f(a_{(1)})g(a_{(2)})$ for $f, g \in \mathcal{A}'$ and $a \in \mathcal{A}$. Recall from Subsect. 14.1.2 that $\mathcal{T}$ is contained in the dual Hopf algebra $\mathcal{A}^\circ$, because we have assumed that $\dim \mathcal{T} < \infty$.

Proposition 8. *For arbitrary elements $X, Y, Z \in \mathcal{T}$ we have:*

(i) $[X,Y] = \mathrm{ad}_R(Y)X = XY - m_{\mathcal{A}'}\sigma^t(X \otimes Y)$.

(ii) $[X,[Y,Z]] = [[X,Y],Z] - \sum_i [[X,Z_i],Y^i]$ *with* $\sum_i Z_i \otimes Y^i := \sigma^t(Y \otimes Z)$.

(iii) *If* $\sigma^t(\xi) = \xi$ *for some* $\xi \in \mathcal{T} \otimes \mathcal{T}$, *then* $C(\xi) = 0$.

Proof. (i): First let $f, g \in \mathcal{A}^\circ$ and $a \in \mathcal{A}$. By the definition of the Hopf algebra structure of $\mathcal{A}^\circ$ (see Subsect. 1.2.8), we have

$$(\mathrm{ad}_R(g)f)(a) = \sum S(g_{(1)}) f g_{(2)}(a) = \sum g_{(1)}(S(a_{(1)})) f(a_{(2)}) g_{(2)}(a_{(3)})$$
$$= (f \otimes g)\mathrm{Ad}_R(a). \tag{33}$$

Comparing with (31), we get $[X,Y] = \mathrm{ad}_R(Y)X$. Let us write $\sigma^t(X \otimes Y)$ as a finite sum $\sum_i Y_i \otimes X^i$ with $X^i, Y_i \in \mathcal{T}$. Using (10), (29) and (32) we compute

$$\begin{aligned}
\sigma^t(X \otimes Y)(a \otimes b) &= \sum\nolimits_i Y_i(a) X^i(b) = \sum\nolimits_i (Y_i, \omega(a))(X^i, \omega(b)) \\
&= \Big(\sum\nolimits_i Y_i \otimes X^i, \omega(a) \otimes_{\mathcal{A}} \omega(b)\Big)_2 = (X \otimes Y, \sigma(\omega(a) \otimes_{\mathcal{A}} \omega(b)))_2 \\
&= \sum (X, \omega(b_{(2)}))(Y, \omega(\bar{a} S(b_{(1)}) b_{(3)})) = \sum Y_{(1)}(\bar{a}) X(b_{(2)}) Y_{(2)}(S(b_{(1)}) b_{(3)}) \\
&= \sum \overline{Y_{(1)}}(a) S(Y_{(2)}) X Y_{(3)}(b)
\end{aligned}$$

for $a, b \in \mathcal{A}$. That is, we have $\sigma^t(X \otimes Y) = \sum \overline{Y_{(1)}} \otimes \mathrm{ad}_R(Y_{(2)})X$ and so

$$m_{\mathcal{A}'}\sigma^t(X \otimes Y) = \sum \overline{Y_{(1)}} \mathrm{ad}_R(Y_{(2)})X$$

$$= \sum (Y_{(1)} S(Y_{(2)}) X Y_{(3)} - \varepsilon(Y_{(1)}) S(Y_{(2)}) X Y_{(3)}) = XY - \mathrm{ad}_R(Y)X, \tag{34}$$

which yields the second equality of (i). (iii) follows at once from (i).

(ii): Since ad_R is a right action of $\mathcal{A}^\circ$ by Proposition 1.14, we have

$$[[X,Y],Z] = (\mathrm{ad}_R(Z) \circ \mathrm{ad}_R(Y))X = \mathrm{ad}_R(YZ)X = [X, YZ]. \tag{35}$$

Inserting the expression $YZ = [Y,Z] + \sum_i Z_i Y^i$ from (34) and using the relation $[X, Z_i Y^i] = [[X, Z_i], Y^i]$ from (35) we obtain the desired identity. □

If the braiding σ is the flip operator, then Proposition 8 says that $\mathcal{T}$ is an "ordinary" complex Lie algebra with bracket $[X,Y] = XY - YX$. In the general case, assertions (ii) and (iii) are quantum versions of the Jacobi identity and the antisymmetry, respectively. For this reason, we call $\mathcal{T}$ the *quantum Lie algebra* of the bicovariant FODC Γ.

Corollary 9. (i) *For any $g \in \mathcal{A}^\circ$, we have* $\mathrm{ad}_R(g)\mathcal{T} \subseteq \mathcal{T}$.

(ii) *Suppose that $\mathcal{B}_0$ is a subset of $\mathcal{A}^\circ$ such that the subalgebra $\mathcal{B}$ generated by $\mathcal{B}_0$ separates the elements of $\mathcal{A}$, that is, $g(a) = 0$ for all $g \in \mathcal{B}$ implies $a = 0$. If Γ is a left-covariant FODC over $\mathcal{A}$ such that $\dim \mathcal{T}_\Gamma < \infty$ and $\mathrm{ad}_R(g)\mathcal{T}_\Gamma \subseteq \mathcal{T}_\Gamma$ for any $g \in \mathcal{B}_0$, then Γ is a bicovariant FODC.*

Proof. (i): Since $\mathrm{Ad}_R(\mathcal{R}) \subseteq \mathcal{R}\otimes\mathcal{A}$ by Proposition 7, it follows from (33) that $\mathrm{ad}_R(g)X(a) = 0$ for $X \in \mathcal{T}$ and $a \in \mathcal{A}$, so that $\mathrm{ad}_R(g)X \in \mathcal{T}$.

(ii): The assumptions and formula (33) imply that $\mathrm{Ad}_R(\mathcal{R}_\Gamma) \subseteq \mathcal{R}_\Gamma \otimes \mathcal{A}$, so Γ is bicovariant by Proposition 7. □

Example 2. The 3D-calculus from Subsect. 14.1.3 is not bicovariant by Corollary 9(i), since $\mathrm{ad}_R(F)X_1 = (q^{-1}+q^{-3})FK^5$ is not in $\mathcal{T}$. △

For the conclusion of Corollary 9(ii) it does *not* suffice that the assumption $\mathrm{ad}_R(g)\mathcal{T} \subseteq \mathcal{T}$ holds for all functionals $g = X_i$ and $g = f^r_s$, $i,r,s \in I$. For instance, the left-covariant FODC over $SL_q(2)$ with quantum tangent space $\mathcal{T} = \mathrm{Lin}\,\{\varepsilon - K^2\}$ has this property, but it is not bicovariant.

Corollary 10. *Suppose that the dual Hopf algebra $\mathcal{A}^\circ$ separates the elements of $\mathcal{A}$. Then a finite-dimensional linear subspace $\mathcal{T}$ of $\mathcal{A}^\circ$ is the quantum Lie algebra of a bicovariant FODC over $\mathcal{A}$ if and only if $\mathrm{ad}_R(g)\mathcal{T} \subseteq \mathcal{T}$ for $g \in \mathcal{A}^\circ$ and $X(1) = 0$ and $\Delta(X) - \varepsilon\otimes X \in \mathcal{T}\otimes\mathcal{A}^\circ$ for $X \in \mathcal{T}$.*

Proof. Combine Proposition 5 and Corollary 9. □

As an interesting application of Corollary 10 we prove

Proposition 11. *If the Hopf dual $\mathcal{A}^\circ$ separates the elements of $\mathcal{A}$, then for any central element c of $\mathcal{A}^\circ$ there exists a bicovariant FODC over $\mathcal{A}$ with quantum Lie algebra*

$$\mathcal{T}(c) := \mathrm{Lin}\,\{\chi_a := \textstyle\sum c_{(2)}(a)c_{(1)} - c(a)\varepsilon \mid a \in \mathcal{A}\}. \tag{36}$$

Proof. Let us write $\Delta(c) = \sum_i c_i \otimes b_i$ such that the set $\{b_i\}$ is linearly independent. From its very definition we see that the vector space $\mathcal{T}(c)$ is the linear span of elements $c_i - c_i(1)\varepsilon$. Hence $\mathcal{T}(c)$ is a finite-dimensional subspace of $\mathcal{A}^\circ$ and we have $\mathcal{T}(c) = \{f \in \mathcal{A}^\circ \mid f(b) = 0 \text{ for } b \in \mathcal{R}(c) \oplus \mathbb{C}{\cdot}1\}$, where $\mathcal{R}(c) := \{b \in \ker\varepsilon \mid \chi_a(b) = 0 \text{ for } a \in \mathcal{A}\}$. For $a, x \in \mathcal{A}$ and $b \in \mathcal{R}(c) \oplus \mathbb{C}{\cdot}1$, we have

$$\langle \Delta(\chi_a) - \varepsilon\otimes\chi_a, b\otimes x\rangle$$

$$= \sum c_{(3)}(a)c_{(1)}(b)c_{(2)}(x) - c(a)\varepsilon(b)\varepsilon(x) - \varepsilon(b)\left(\sum c_{(2)}(a)c_{(1)}(x) - c(a)\varepsilon(x)\right)$$

$$= \sum c_{(1)}(b)c_{(2)}(xa) - c(xa)\varepsilon(b) = \chi_{xa}(b) = 0,$$

so that $\Delta(\chi_a) - \varepsilon\otimes\chi_a \in \mathcal{T}(c)\otimes\mathcal{A}^\circ$ for any $a \in \mathcal{A}$.

Next we show that $\mathcal{T}(c)$ is ad_R-invariant. Let $g \in \mathcal{A}^\circ$. Since c is central in $\mathcal{A}^\circ$, we have $\Delta^{(2)}(g)(1\otimes c\otimes 1) = (1\otimes c\otimes 1)\Delta^{(2)}(g)$. Applying the mapping $(m_{12}\otimes m_{34})(S\otimes \mathrm{id}\otimes\mathrm{id}\otimes S)(\mathrm{id}\otimes\Delta\otimes\mathrm{id})$ to this equality we get

$$\sum S(g_{(1)})c_{(1)}g_{(2)} \otimes c_{(2)} = \sum c_{(1)} \otimes g_{(1)}c_{(2)}S(g_{(2)}).$$

Using the latter formula, some properties of the dual pairing of $\mathcal{A}^\circ$ and $\mathcal{A}$ and again the centrality of c we compute

$$\begin{aligned}
\mathrm{ad}_R(g)\chi_a &= \sum c_{(2)}(a)S(g_{(1)})c_{(1)}g_{(2)} - c(a)g(1)\varepsilon \\
&= \sum g_{(1)}c_{(2)}S(g_{(2)})(a)c_{(1)} - c(a)g(1)\varepsilon \\
&= \sum g(a_{(1)}S(a_{(3)}))c_{(2)}(a_{(2)})c_{(1)} - c(a)g(1)\varepsilon \\
&= \sum g(a_{(1)}S(a_{(3)}))\chi_{a_{(2)}} + \sum g(a_{(1)}S(a_{(3)}))c(a_{(2)}) - c(a)g(1)\varepsilon \\
&= \sum g(a_{(1)}S(a_{(3)}))\chi_{a_{(2)}} \in \mathcal{T}(c).
\end{aligned}$$

Thus $\mathcal{T}(c)$ is ad_R-invariant. Since obviously $X(1) = 0$ for all $X \in \mathcal{T}(c)$, we have shown that $\mathcal{T}(c)$ has the properties required in Corollary 10. □

The quantum Lie algebras of the bicovariant FODC $\Gamma_{\pm,z}$ on the Hopf algebras $\mathcal{O}(G_q)$ constructed in Subsect. 14.6.1 will be of the form (36).

Let us make the structure of the quantum Lie algebra more transparent.

Proposition 12. *The basis elements X_i, $i \in I$, of $\mathcal{T}$ satisfy the commutation relations*

$$[X_i, X_j] = X_iX_j - \sigma^{ij}_{nm}X_nX_m = C^k_{ij}X_k, \tag{37}$$

where

$$\sigma^{nm}_{ij} = f^i_m(v^n_j) \quad \textit{and} \quad C^k_{ij} = X_j(v^i_k) \quad \textit{for} \quad i,j,n,m,k \in I. \tag{38}$$

Proof. The first equalities in (37) and (38) follow immediately from Proposition 8(i) and (13.29). We prove the formula for C^k_{ij} and compute

$$\sum a_{(1)}X_i(a_{(2)})\omega_i \otimes a_{(3)} = \sum \mathrm{d}a_{(1)} \otimes a_{(2)} = (\mathrm{d} \otimes \mathrm{id})\Delta(a) = \Delta_R(\mathrm{d}a)$$

$$= \sum \Delta_R(a_{(1)}X_m(a_{(2)})\omega_m) = \sum a_{(1)}X_m(a_{(3)})\omega_i \otimes a_{(2)}v^i_m$$

for $a \in \mathcal{A}$. Applying $P_L \otimes \mathrm{id}$ to both sides of this equality, using formula (13.3) and equating the coefficients of ω_i, we obtain

$$\sum X_i(a_{(1)})a_{(2)} = \sum a_{(1)}v^i_m X_m(a_{(2)}), \quad a \in \mathcal{A}. \tag{39}$$

Applying X_j to (39) and then using (13) and the formula for σ^{ij}_{nm}, we get

$$\begin{aligned}
X_iX_j(a) &= \sum X_i(a_{(1)})X_j(a_{(2)}) = \sum X_j(a_{(1)}v^i_m)X_m(a_{(2)}) \\
&= \sum \varepsilon(a_{(1)})X_j(v^i_m)X_m(a_{(2)}) + \sum X_n(a_{(1)})f^n_j(v^i_m)X_m(a_{(2)}) \\
&= X_j(v^i_k)X_k(a) + \sigma^{ij}_{nm}X_nX_m(a),
\end{aligned}$$

which gives the above expression for C^k_{ij} and completes the proof. □

Moreover, applying the functional f^k_j to (39) and taking (13.14) and (13.29) into account, we obtain

$$X_if^k_j = \sigma^{ij}_{nm}f^k_nX_m. \tag{40}$$

We specialize (33) to the functionals X_i and f^r_s. From the formulas (40), (37), (14) and (15) we easily derive

$$\mathrm{ad}_R(f^i_m)X_n = \sigma^{nm}_{ij} X_j \quad \text{and} \quad \mathrm{ad}_R(X_j)X_i = C^k_{ij} X_k. \tag{41}$$

By (41), the matrix coefficients σ^{nm}_{ij} of the braiding σ and the structure constants C^k_{ij} of the quantum Lie algebra $\mathcal{T}$ are computed from the right adjoint action of the dual Hopf algebra $\mathcal{A}^\circ$ on $\mathcal{T}$.

14.2.4 The 4D$_+$- and the 4D$_-$-Calculus on $SL_q(2)$

As in Subsect. 14.1.3 we use the Hopf algebra $\breve{U}_q(\mathrm{sl}_2)$ and the dual pairing of $\breve{U}_q(\mathrm{sl}_2)$ and $\mathcal{A} := \mathcal{O}(SL_q(2))$ given by (16). In order to treat both calculi Γ_+ and Γ_- at once, we introduce the character ε_- of the algebra $\mathcal{A}$ defined by $\varepsilon_-(u^i_j) = -\delta_{ij}$, $i,j = 1,2$. It is easy to check that $\Delta(\varepsilon_-) = \varepsilon_- \otimes \varepsilon_-$, $\varepsilon_-^2 = \varepsilon$ and that ε_- belongs to the center of the dual algebra $\mathcal{A}'$. Moreover, we set $\varepsilon_+ := \varepsilon$.

Let $\mathcal{T}_\pm$ be the vector subspace of $\mathcal{A}'$ spanned by the four linear functionals

$$X_1 := \varepsilon_\pm K^{-2} - \varepsilon, \quad X_2 := q^{1/2}\varepsilon_\pm F K^{-1}, \quad X_3 := q^{-1/2}\varepsilon_\pm E K^{-1},$$
$$X_4 := \varepsilon_\pm K^2 + (q-q^{-1})^2 q^{-1}\varepsilon_\pm FE - \varepsilon.$$

Note that $q^{-1}X_1 + qX_4 + (q+q^{-1})1 = \varepsilon_\pm (q-q^{-1})^2 C_q$, where

$$C_q := FE + (q-q^{-1})^{-2}(qK^2 + q^{-1}K^{-2})$$

is the quantum Casimir element of $\breve{U}_q(\mathrm{sl}_2)$. In particular, $\mathcal{T}_+$ contains the element $C_q - \varepsilon(C_q)1$. From Proposition 5 and the formulas

$$\begin{aligned}
\Delta X_1 &= \varepsilon \otimes X_1 + X_1 \otimes \varepsilon_\pm K^{-2},\\
\Delta X_j &= \varepsilon \otimes X_j + X_j \otimes \varepsilon_\pm + X_1 \otimes X_j, \quad j = 2,3,\\
\Delta X_4 &= \varepsilon \otimes X_4 + (q-q^{-1})^2 q^{-1} X_1 \otimes \varepsilon_\pm FE + X_4 \otimes \varepsilon_\pm K^2\\
&\quad + (q-q^{-1})^2 q^{-1/2}(X_2 \otimes \varepsilon_\pm EK + X_3 \otimes \varepsilon_\pm KF)
\end{aligned}$$

we conclude that $\mathcal{T}_\pm$ is the quantum tangent space of a left-covariant FODC $\Gamma_\pm$ over $\mathcal{A}$. The FODC $\Gamma_\pm$ is called the 4D$_\pm$-calculus on $SL_q(2)$.

Let $\mathcal{R}_\pm := \mathcal{B}_\pm \cdot \mathcal{A}$ be the right ideal of $\ker\varepsilon$ generated by the vector space $\mathcal{B}_\pm$ spanned by the following nine elements:

$$b^2, \quad c^2, b(a-d), \quad c(a-d), \quad a^2 + q^2 d^2 - (1+q^2)(ad + q^{-1}bc),$$
$$x_\pm b, \quad x_\pm c, \quad x_\pm(a-d), \quad x_\pm(a + q^{-2}d - (1+q^{-2})1).$$

Here we abbreviated $x_\pm := q^2 a + d \mp (q^{-1} + q^3)1$. Using (16) one can check that the functionals X_1, X_2, X_3, X_4 annihilate these nine elements, so $\mathcal{R}_\pm \subseteq \mathcal{R}_{\Gamma_\pm}$. Since $\dim(\ker\varepsilon)/\mathcal{R}_\pm \le 4$ and $\dim \mathcal{T}_\pm = 4$, it follows that $\mathcal{R}_\pm$ is the right ideal $\mathcal{R}_{\Gamma_\pm}$ of the FODC $\Gamma_\pm$.

The vector space $\mathcal{B}_\pm$ is Ad_R-invariant. (Indeed, it is not difficult to verify that the generators listed above span an Ad_R-invariant subspace.) From the formula $\mathrm{Ad}_R(xy) = \sum x_{(2)}y_{(2)} \otimes S((xy)_{(1)})(xy)_{(3)}$ we see that the right ideal $\mathcal{R}_\pm = \mathcal{B}_\pm \cdot \mathcal{A}$ is Ad_R-invariant as $\mathcal{B}_\pm$ is. Therefore, by Proposition 7, $\Gamma_\pm$ is a bicovariant FODC.

From Proposition 2 we derive that $\Gamma_\pm$ is a $*$-calculus for the three real forms $SL_q(2,\mathbb{R})$, $|q|=1$, $SU_q(2)$ and $SU_q(1,1)$, $q \in \mathbb{R}$.

Remark. Let $U(\mathcal{T}_+)$ be the subalgebra of $\breve{U}_q(\mathrm{sl}_2)$ generated the elements of the quantum Lie algebra $\mathcal{T}_+$ and the unit element. Contrary to what is occasionally asserted in the literature, $U(\mathcal{T}_+)$ is different from $\breve{U}_q(\mathrm{sl}_2)$. In fact, the set of monomials $F^iK^jE^l$ such that $i+j+l$ is odd is linearly independent modulo $U(\mathcal{T}_+)$. But $U(\mathcal{T}_+)$ still separates the points of $\mathcal{O}(SL_q(2))$ when q is not a root of unity. $\triangle$

14.2.5 Examples of Bicovariant First Order Calculi on Simple Lie Groups

In this subsection we shall construct two examples of bicovariant FODC on simple Lie groups which are different from the ordinary differential calculus. For $SL(N,\mathbb{C})$ it can be shown that the N bicovariant FODC obtained in this manner are just the classical limits of the $2N$ bicovariant FODC for the quantum group $SL_q(N)$ developed in Sect. 14.6.

Suppose that G is one of the Lie groups $SL(N,\mathbb{C})$, $O(N,\mathbb{C})$ or $Sp(N,\mathbb{C})$. Let $\mathcal{A} = \mathcal{O}(G)$ be the coordinate Hopf algebra of G (see Example 1.3), $\mathfrak{g}$ the complex Lie algebra of G, and $\{X_1,\cdots,X_n\}$ a basis of $\mathfrak{g}$. We shall use the dual pairing of the Hopf algebras $U(\mathfrak{g})$ and $\mathcal{O}(G)$ described in Example 1.6.

Example 3. Let $C = \sum_{i,j=1}^n \alpha_{ij}X_iX_j$ be a nonzero fixed quadratic element of the enveloping algebra $U(\mathfrak{g})$. Then we have $\Delta(X_i) = \varepsilon \otimes X_i + X_i \otimes \varepsilon$ and

$$\Delta(C) = \varepsilon \otimes C + C \otimes \varepsilon + \sum\nolimits_{i,j} X_i \otimes X_j(\alpha_{ij}+\alpha_{ji}). \tag{42}$$

Hence, by Proposition 5, the $(n+1)$-dimensional subspace $\mathcal{T} := \mathfrak{g} \oplus \mathbb{C}\cdot C$ of $\mathcal{A}^\circ$ is the quantum tangent space of a left-covariant FODC Γ over $\mathcal{A}$. Because of Proposition 4, the FODC Γ can be easily determined explicitly. For notational simplicity, let us set $X_0 := C$. Then we can write

$$\mathrm{d}a = \sum\nolimits_{i=0}^n (X_i.a)\omega_i, \quad a \in \mathcal{A}. \tag{43}$$

For the commutation rules of forms and functions we need the functionals f_i^j from Proposition 13.2. Comparing the formulas (14) and (42), we obtain $f_0^j = \sum_{k=1}^n(\alpha_{jk}+\alpha_{kj})X_k$, $f_i^0 = \delta_{0i}\varepsilon$ and $f_i^j = \delta_{ij}\varepsilon$ for $i=1,2,\cdots,n$. Therefore, by (13.15),

$$\omega_0 a = a\omega_0, \quad \omega_i a = a\omega_i + (f_0^i.a)\omega_0 \ \text{ for } \ i=1,2,\ldots,n. \tag{44}$$

Suppose now that C is a Casimir element of $U(\mathfrak{g})$. Since C lies in the center of $U(\mathfrak{g})$, it is clear that $\mathrm{ad}_R(y)C = \varepsilon(y)C$ and $\mathrm{ad}_R(y)x \in \mathfrak{g}$ for $x \in \mathfrak{g}$ and $y \in U(\mathfrak{g})$. This implies that $\mathcal{T}$ is ad_R-invariant. Therefore, by Corollary 9(ii), Γ is bicovariant. Let us emphasize that in contrast to the ordinary differential calculus on G the differentiation (43) involves the second order differential operator $X_0 = C$, and forms and functions do not commute as seen from (44). △

Example 4. For $G = SL(N, \mathbb{C})$, let $\zeta \in \mathbb{C}$ be such that $\zeta \neq 1$ and $\zeta^N = 1$. For $G = O(N, \mathbb{C})$, $Sp(N, \mathbb{C})$, we set $\zeta := -1$. Let ε_ζ denote the multiplicative linear functional on $\mathcal{A} := \mathcal{O}(G)$ defined by $\varepsilon_\zeta(u^i_j) = \delta_{ij}\zeta$. Set $X'_0 := \varepsilon_\zeta - \varepsilon$, $X'_i := \varepsilon_\zeta X_i$, $i = 1, 2, \ldots, n$, and $\mathcal{T} = \mathrm{Lin}\,\{X'_i \mid i = 0, 1, \cdots, n\}$. We have

$$\Delta(X'_0) = \varepsilon \otimes X'_0 + X'_0 \otimes \varepsilon_\zeta, \quad \Delta(X'_i) = \varepsilon \otimes X'_i + X'_i \otimes \varepsilon_\zeta + X'_0 \otimes X'_i \tag{45}$$

for $i = 1, 2, \cdots, n$. Since the functional ε_ζ belongs to the center of $\mathcal{A}^\circ$, the linear subspace $\mathcal{T}$ of $\mathcal{A}^\circ$ is ad_R-invariant. Thus, $\mathcal{T}$ is the quantum Lie algebra of a bicovariant FODC Γ over $\mathcal{A}$ by Corollary 9(ii).

Let us describe this FODC Γ explicitly. From (14) and (45) we see that $f^j_0 = \delta_{0j}\varepsilon_\zeta$ and $f^0_i = X'_i$, $f^i_i = \varepsilon_\zeta$ for $i = 1, 2, \cdots, n$. Let $\mathcal{A}_k$ be the linear span of elements $u^{i_1}_{j_1} \cdots u^{i_k}_{j_k}$. For $a \in \mathcal{A}_k$ we have $\varepsilon_\zeta . a = \zeta^k a$ and so we obtain

$$\omega_0 a = \zeta^k a\omega_0 + \sum\nolimits_{i=1}^{n} (X'_i . a)\omega_i \quad \text{and} \quad \omega_i a = \zeta^k a\omega_i \quad \text{for} \quad i = 1, 2, \cdots, n. \tag{46}$$

Combining the first equality of (46) with (12), we conclude that

$$\mathrm{d}a = \omega_0 a - a\omega_0, \quad a \in \mathcal{A}.$$

The quantum Lie derivatives $L_i(a) := X'_i . a$ satisfy the twisted Leibniz rule

$$L_i(ab) = \zeta^k a L_i(b) + \zeta^m L_i(a) b, \quad a \in \mathcal{A}_k, \ b \in \mathcal{A}_m. \qquad \triangle$$

14.3 Higher Order Left-Covariant Differential Calculi

In this section Γ is a left-covariant FODC over the Hopf algebra $\mathcal{A}$ with associated right ideal $\mathcal{R}$.

14.3.1 The Maurer–Cartan Formula

The Maurer–Cartan equations for left-invariant differential forms on Lie groups have the following quantum analog.

Proposition 13. *If* $\Gamma^\wedge = \bigoplus_{n=0}^\infty \Gamma^{\wedge n}$ *is an arbitrary* DC *over* $\mathcal{A}$ *such that* $\Gamma^{\wedge 1} = \Gamma$, *then we have*

$$\mathrm{d}\omega(a) = -\sum \omega(a_{(1)}) \wedge \omega(a_{(2)}), \quad a \in \mathcal{A}. \tag{47}$$

Proof. Differentiation of the identity $\sum S(a_{(1)})a_{(2)} = \varepsilon(a)1$ yields the equation $\sum \mathrm{d}S(a_{(1)})\cdot a_{(2)} + \sum S(a_{(1)})\mathrm{d}a_{(2)} = 0$ and so

$$\mathrm{d}S(a) = -\sum S(a_{(1)})\mathrm{d}a_{(2)}\cdot S(a_{(3)}).$$

Hence, by the graded Leibniz rule and the relation $\mathrm{d}^2 = 0$ in $\Gamma^\wedge$, we get

$$\begin{aligned}\mathrm{d}\omega(a) &= \sum \mathrm{d}(S(a_{(1)})\mathrm{d}a_{(2)}) = \sum \mathrm{d}S(a_{(1)}) \wedge \mathrm{d}a_{(2)} \\ &= -\sum S(a_{(1)})\mathrm{d}a_{(2)} \wedge S(a_{(3)})\mathrm{d}a_{(4)} = -\sum \omega(a_{(1)}) \wedge \omega(a_{(2)}). \qquad \square\end{aligned}$$

Equation (47) is called the *Maurer–Cartan formula* of the DC $\Gamma^\wedge$. Let us give two coordinate versions of this equation. From (10) it follows that $\omega(a) = \sum_i X_i(a)\omega_i$ and hence $\sum \omega(a_{(1)}) \wedge \omega(a_{(2)}) = \sum_{i,j} X_iX_j(a)\omega_i \wedge \omega_j$. Therefore, the Maurer–Cartan formula (47) can be expressed as

$$\sum_i X_i(a)\mathrm{d}\omega_i = -\sum_{i,j} X_iX_j(a)\omega_i \wedge \omega_j, \quad a \in \mathcal{A}. \tag{48}$$

For the second variant of formula (47), let us take elements $a_j \in \mathcal{A}$ such that $X_i(a_j) = \delta_{ij}$ and set $c^k_{ij} = X_iX_j(a_k)$, $i,j,k \in I$. Inserting $a = a_k$ into (48) and applying (37) to a_k, we then obtain

$$\mathrm{d}\omega_k = -c^k_{ij}\omega_i \wedge \omega_j, \ k \in I, \ \text{ and } \ c^k_{ij} - c^k_{nm}\sigma^{ij}_{nm} = C^k_{ij}, \ \ k,i,j \in I. \tag{49}$$

Moreover, since $\omega(a) = 0$ for $a \in \mathcal{R}$, formula (47) implies also that

$$\sum \omega(a_{(1)}) \wedge \omega(a_{(2)}) = 0, \quad a \in \mathcal{R}\,. \tag{50}$$

Note that the right ideal $\mathcal{R}$ depends only on the FODC Γ, while the cap product $\wedge$ in (50) is taken in $\Gamma^\wedge$. We shall see in Subsect. 14.3.3 that (50) is the only obstruction for the construction of a higher order DC which contains Γ as its FODC.

14.3.2 The Differential Envelope of a Hopf Algebra

In Subsect. 14.1.1 we gave a second construction of the universal first order calculus for a Hopf algebra. In this subsection we extend this idea to the universal higher order calculus.

Let $\Omega^0(\mathcal{A}) := \mathcal{A}$ and $\Omega^n(\mathcal{A}) := \mathcal{A} \otimes (\ker \varepsilon)^{\otimes n}$ for $n \in \mathbb{N}$. We shall write $a\omega(a^1) \wedge \cdots \wedge \omega(a^n)$ for the tensor $a \otimes a^1 \otimes \cdots \otimes a^n \in \Omega^n(\mathcal{A})$. As in Example 1, we set $\omega(a) := 1 \otimes \bar{a}$ for $a \in \mathcal{A}$.

On the vector space $\Omega(\mathcal{A}) := \bigoplus_{n=0}^\infty \Omega^n(\mathcal{A})$ we define a product $\wedge$ and a linear mapping $\mathrm{d} : \Omega(\mathcal{A}) \to \Omega(\mathcal{A})$ by

$$\begin{aligned}&(a\omega(a^1)\wedge \ldots \wedge\omega(a^n)) \wedge (b\omega(b^1)\wedge\ldots\wedge\omega(b^m)) \\ &\qquad := \sum ab_{(1)}\, \omega(a^1b_{(2)})\wedge\cdots\wedge\omega(a^nb_{(n+1)}) \wedge \omega(b^1)\wedge\cdots\wedge\omega(b^m), \end{aligned}\tag{51}$$

$$\mathrm{d}(a\omega(a^1)\wedge\ldots\wedge\omega(a^n)) := \sum a_{(1)}\omega(a_{(2)})\wedge\omega(a^1)\wedge\ldots\wedge\omega(a^n)$$

$$+\sum_{k=1}^{n}(-1)^k a\omega(a^1)\wedge\ldots\wedge\omega(a^{k-1})\wedge\omega(a^k_{(1)})\wedge\omega(a^k_{(2)})\wedge\omega(a^{k+1})\wedge\ldots\wedge\omega(a^n) \quad (52)$$

for $a^1,\cdots,a^n,b^1,\cdots,b^m \in \ker\varepsilon$ and $a,b \in \mathcal{A}$. (In order to motivate these formulas, let us recall that for any left-covariant DC over $\mathcal{A}$ we have $\omega(b)c = \sum c_{(1)}\omega(bc_{(2)})$, $\mathrm{d}a = \sum a_{(1)}\omega(a_{(2)})$, $\mathrm{d}\omega(a) = -\sum \omega(a_{(1)})\wedge\omega(a_{(2)})$ by the formulas (4), (2), and (47), respectively, for $a,c\in\mathcal{A}$ and $b\in\ker\varepsilon$.)

Proposition 14. *With the above definitions, $\Omega(\mathcal{A})$ is a bicovariant differential calculus over $\mathcal{A}$ and the differential envelope of $\mathcal{A}$.*

Proof. It is easily seen that $\Omega(\mathcal{A})$ is an algebra. The associativity of the product follows from the fact that Δ is an algebra homomorphism. The unit element 1 of $\mathcal{A} = \Omega^0(\mathcal{A})$ is the unit of $\Omega(\mathcal{A})$.

It suffices to verify the Leibniz rule for elementary tensors $x_n = a\omega(a^1)\wedge\cdots\wedge\omega(a^n)$ and $y_m = b\omega(b^1)\wedge\cdots\wedge\omega(b^m)$. In order to do this we fix x_n. From (51) and (52) we see that $\mathrm{d}(x_n b) = (\mathrm{d}x_n)b + (-1)^n x_n\wedge \mathrm{d}b_n$. By induction on m we prove that $\mathrm{d}(x_n\wedge y_m) = \mathrm{d}x_n\wedge y_m + (-1)^n x_n \wedge \mathrm{d}y_m$. Using the Leibniz rule, another simple induction argument shows that $\mathrm{d}^2(x_n) = 0$ for all x_n, so $\mathrm{d}^2 = 0$. Thus, $\Omega(\mathcal{A})$ is a differential calculus over $\mathcal{A}$.

Next we prove that $\Omega(\mathcal{A})$ is the universal DC over $\mathcal{A}$. Let $\Gamma^\wedge$ be an arbitrary DC over $\mathcal{A}$ with differentiation $\tilde{\mathrm{d}}$. Each $\rho\in\Omega^n(\mathcal{A})$ is a finite sum $\rho = \sum_i a^i \mathrm{d}a^i_1\cdots \mathrm{d}a^i_n$. First let us show that $\rho = 0$ in $\Omega^n(\mathcal{A})$ implies that $\sum_i a^i\tilde{\mathrm{d}}a^i_1\wedge\cdots\wedge\tilde{\mathrm{d}}a^i_n = 0$. Indeed, arguing as in Example 1, it follows that

$$\sum_i a^i(1\otimes a^i_1 - a^i_1\otimes 1)_{12}(1\otimes a^i_2 - a^i_2\otimes 1)_{23}\cdots(1\otimes a^i_n - a^i_n\otimes 1)_{n,n+1} = 0.$$

The element on the left-hand side is a sum of $\sum_i a^i\otimes a^i_1\otimes\cdots\otimes a^i_n$ and an element from the set $\mathcal{A}\otimes 1\otimes\mathcal{A}\otimes\cdots\otimes\mathcal{A}+\cdots+\mathcal{A}\otimes\cdots\otimes\mathcal{A}\otimes 1$. This in turn implies that $\sum_i a^i\tilde{\mathrm{d}}a^i_1\wedge\cdots\wedge\tilde{\mathrm{d}}a^i_n = 0$.

By the fact proved in the preceding paragraph, there is a *well-defined* linear map $\psi : \Omega(\mathcal{A}) \to \Gamma^\wedge$ such that $\psi(a) = a$ and $\psi(a\mathrm{d}a_1\wedge\cdots\wedge\mathrm{d}a_n) = a\tilde{\mathrm{d}}a_1\wedge\cdots\wedge\tilde{\mathrm{d}}a_n$. Using the formulas (12.2) and (12.3) applied to both differential calculi $\Omega(\mathcal{A})$ and $\Gamma^\wedge$ we derive that $\Omega(\mathcal{A})/\ker\psi$ is a DC which is isomorphic to $\Gamma^\wedge$. Hence $\Omega(\mathcal{A})$ is the universal DC over $\mathcal{A}$.

As noted in Subsect. 12.2.3, the universal DC of a left (resp. right) quantum space of $\mathcal{A}$ is left- (resp. right-) covariant. Therefore, the FODC $\Omega(\mathcal{A})$ is bicovariant. □

In what follows, the symbol Ω will always refer to the *universal* DC of $\mathcal{A}$.

14.3.3 The Universal DC of a Left-Covariant FODC

In Subsect. 12.2.3 we defined the universal DC $\Omega/\mathcal{J}_\Gamma$ of a FODC Γ over an arbitrary algebra. In this subsection we realize this DC for a Hopf algebra $\mathcal{A}$ as a quotient of the tensor algebra $T_{\mathcal{A}}(\Gamma)$.

Recall from Subsect. 12.2.3 that $\mathcal{J}_\Gamma$ is the differential ideal of the universal DC Ω generated by the $\mathcal{A}$-subbimodule $\mathcal{N}$ of Ω^1 such that Γ is isomorphic to $\Omega^1/\mathcal{N}$. Since $\mathcal{N} = \mathcal{A}\omega_{\Omega^1}(\mathcal{R})$ by Proposition 1, we have $\mathcal{J}_\Gamma := \Omega\,\omega_{\Omega^1}(\mathcal{R})\Omega + \Omega\,\mathrm{d}\omega_{\Omega^1}(\mathcal{R})\Omega$. Let $\mathcal{I}_\Gamma := \Omega\,\omega_{\Omega^1}(\mathcal{R})\Omega$ be the two-sided ideal of Ω generated by $\omega_{\Omega^1}(\mathcal{R})$. It is not difficult to verify that the mapping

$$\sum\nolimits_i a^i\omega_{\Omega^1}(a_1^i)\wedge\cdots\wedge\omega_{\Omega^1}(a_n^i)+\mathcal{I}_\Gamma \to \sum\nolimits_i a^i\omega_\Gamma(a_1^i)\cdots\omega_\Gamma(a_n^i)$$

defines an isomorphism of the quotient algebra $\Omega/\mathcal{I}_\Gamma$ and the tensor algebra $T_\mathcal{A}(\Gamma)$ of the $\mathcal{A}$-bimodule Γ. By (47), under this isomorphism the quotient $\mathcal{J}_\Gamma/\mathcal{I}_\Gamma$ goes into the two-sided ideal $\mathcal{S}_\mathcal{R}$ of the algebra $T_\mathcal{A}(\Gamma)$ generated by the elements

$$\mathcal{S}(x) := \sum \omega_\Gamma(x_{(1)})\omega_\Gamma(x_{(2)}),\quad x\in\mathcal{R}. \tag{53}$$

Since $\Omega/\mathcal{J}_\Gamma$ is isomorphic to $(\Omega/\mathcal{I}_\Gamma)/(\mathcal{J}_\Gamma/\mathcal{I}_\Gamma)$, we obtain a canonical isomorphism π of $\Omega/\mathcal{J}_\Gamma$ onto the quotient algebra $\Gamma_u^\wedge := T_\mathcal{A}(\Gamma)/\mathcal{S}_\mathcal{R}$. Clearly, $\Gamma_u^\wedge = \bigoplus_{n=0}^\infty \Gamma^{\wedge n}$, where $\Gamma^{\wedge 1} := \Gamma$, $\Gamma^{\wedge n} := \Gamma^{\otimes n}/\mathcal{S}_\mathcal{R}^n$, and $\mathcal{S}_\mathcal{R}^n := \mathcal{S}_\mathcal{R}\cap\Gamma^{\otimes n}$. We define $\mathrm{d}\rho := \mathrm{d}\pi^{-1}(\rho)$ for $\rho\in\Gamma_u^\wedge$. Then $\Gamma_u^\wedge$ is a DC over $\mathcal{A}$ which is isomorphic to $\Omega/\mathcal{J}_\Gamma$ by construction. Thus, $\Gamma_u^\wedge$ is the *universal differential calculus of the first order calculus* Γ. By Proposition 12.4, $\Gamma_u^\wedge$ is left-covariant as Γ is.

In terms of bases, the mapping $\mathcal{S}:\mathcal{R}\to {}_{\mathrm{inv}}(\Gamma^{\otimes 2})$ defined by (53) can be written as

$$\mathcal{S}(x) = \sum\nolimits_{i,j} X_iX_j(x)\omega_i\omega_j,\quad x\in\mathcal{R}. \tag{54}$$

Lemma 15. *For $x\in\mathcal{R}$ and $a\in\mathcal{A}$ we have*

$$\mathcal{S}(x)a = \sum a_{(1)}\mathcal{S}(xa_{(2)}),\quad \mathcal{S}(xa) = \sum S(a_{(1)})\mathcal{S}(x)a_{(2)}. \tag{55}$$

Proof. Using formula (4) combined with the fact that $\omega\equiv\omega_\Gamma$ vanishes on the right ideal $\mathcal{R}\subseteq\ker\varepsilon$, we obtain

$$\begin{aligned}
\mathcal{S}(x)a &= \sum \omega(x_{(1)})\omega(x_{(2)})a = \sum a_{(1)}\omega(\overline{x_{(1)}}a_{(2)})\omega(\overline{x_{(2)}}a_{(3)})\\
&= \sum a_{(1)}[\omega(x_{(1)}a_{(2)})\omega(x_{(2)}a_{(3)}) - \omega(a_{(2)})\omega(\varepsilon(x_{(1)})x_{(2)}a_{(3)})\\
&\quad -\omega(\varepsilon(x_{(2)})x_{(1)}a_{(2)})\omega(a_{(3)}) + \varepsilon(x_{(1)})\varepsilon(x_{(2)})\omega(a_{(2)})\omega(a_{(3)})]\\
&= \sum a_{(1)}[\mathcal{S}(xa_{(2)}) - \omega(a_{(2)})\omega(xa_{(3)})\\
&\qquad -\omega(xa_{(2)})\omega(a_{(3)}) + \varepsilon(x)\omega(a_{(1)})\omega(a_{(2)})]\\
&= \sum a_{(1)}\mathcal{S}(xa_{(2)})
\end{aligned}$$

and so

$$\mathcal{S}(xa) = \sum S(a_{(1)})a_{(2)}\mathcal{S}(xa_{(3)}) = \sum S(a_{(1)})\mathcal{S}(x)a_{(2)}. \qquad\square$$

Recall that $\mathcal{S}_\mathcal{R}^2 = \mathcal{A}\mathcal{S}(\mathcal{R})\mathcal{A}$ by definition. By (55), $\mathcal{S}_\mathcal{R}^2 = \mathcal{A}\mathcal{S}(\mathcal{R})$ is an $\mathcal{A}$-subbimodule of $\Gamma^{\otimes 2}$ and hence a left-covariant bimodule of $\mathcal{A}$. Clearly,

$\mathcal{S}(\mathcal{R})$ is just the left-invariant part ${}_{\rm inv}(\mathcal{S}^2_{\mathcal{R}})$ of $\mathcal{S}^2_{\mathcal{R}}$. Therefore, the results of Subsect. 13.1.1 and formula (55) imply the following

Proposition 16. (i) *Each basis of the vector space $\mathcal{S}(\mathcal{R})$ is a free left $\mathcal{A}$-module basis of $\mathcal{S}^2_{\mathcal{R}}$.*

(ii) *Let $\mathcal{B}$ be a linear subspace of $\mathcal{A}$ such that $\mathcal{R} = \mathcal{B}{\cdot}\mathcal{A}$. Then we have $\mathcal{S}^2_{\mathcal{R}} = \mathcal{A}\mathcal{S}(\mathcal{B})\mathcal{A}$. If for any $a \in \mathcal{A}$ and $b \in \mathcal{B}$ there exist finitely many $a_i \in \mathcal{A}$ and $c_i \in \mathcal{B}$ such that $\mathcal{S}(b)a = \sum_i a_i\mathcal{S}(c_i)$, then $\mathcal{S}(\mathcal{B}) = \mathcal{S}(\mathcal{R})$.*

The previous construction becomes explicit in the following example.

Example 5 (The 3D-calculus on $SL_q(2)$). Let $\mathcal{A} = \mathcal{O}(SL_q(2))$ and let Γ be the FODC over $\mathcal{A}$ from Subsect. 14.1.3. Assume that $q^2 \neq -1$. We compute the elements $\mathcal{S}(x)$ (see (54)) for the six generators of the right ideal $\mathcal{R}$ listed in (23). Let $x = bc$. Then we have $\mathcal{S}(x) = \sum_{i,j} X_i(u^1_r u^2_s)X_j(u^r_2 u^s_1)\omega_i\omega_j$. By the dual pairing (16) of $\breve{U}_q(\mathrm{sl}_2)$ and $\mathcal{A}$, the only nonvanishing terms $X_i(u^1_r u^2_s)$ are $X_0(u^1_2 u^2_2) = q$ and $X_2(u^1_1 u^2_1) = 1$. There are only two corresponding nonzero terms $X_j(u^r_2 u^s_1)$, namely $X_2(u^2_2 u^2_1) = 1$ and $X_0(u^1_2 u^1_1) = q^{-1}$. Thus we get $\mathcal{S}(x) = q\omega_0\omega_2 + q^{-1}\omega_2\omega_0$. Similar computations for the other five elements in (23) show that $\mathcal{S}(\mathcal{B})$ is spanned by the elements

$$\omega_0^2,\ \omega_1^2,\ \omega_2^2,\ \omega_2\omega_0 + q^2\omega_0\omega_2,\ \omega_1\omega_0 + q^4\omega_0\omega_1,\ \omega_2\omega_1 + q^4\omega_1\omega_2 \tag{56}$$

of the tensor algebra $\mathcal{T}_{\mathcal{A}}(\Gamma)$ over Γ. From the relations (21) and (22) we see that for each element $\mathcal{S}(x)$ from (56) and any $a \in \mathcal{A}$ we have $\mathcal{S}(x)a = \alpha a\mathcal{S}(x)$ with some $\alpha \in \mathbb{C}$. Therefore, it follows from Proposition 16 that the six elements in (56) form a basis of the vector space $\mathcal{S}(\mathcal{R})$ and a free left $\mathcal{A}$-module basis of $\mathcal{S}^2_{\mathcal{R}}$. Hence $\Gamma^{\wedge 2} = \Gamma^{\otimes 2}/\mathcal{S}^2_{\mathcal{R}}$ has the free left $\mathcal{A}$-module basis $\{\omega_0 \wedge \omega_2, \omega_0 \wedge \omega_1, \omega_1 \wedge \omega_2\}$. From (56) we get the equations

$$\omega_j \wedge \omega_j = 0 \ \text{ for } \ j = 0,1,2, \qquad \omega_2 \wedge \omega_0 = -q^2\omega_0 \wedge \omega_2, \tag{57}$$

$$\omega_1 \wedge \omega_0 = -q^4\omega_0 \wedge \omega_1, \qquad \omega_2 \wedge \omega_1 = -q^4\omega_1 \wedge \omega_2 \tag{58}$$

in $\Gamma^{\wedge 2}$. The 3-form $\omega_0 \wedge \omega_1 \wedge \omega_2$ is a free left-$\mathcal{A}$-module basis of $\Gamma^{\wedge 3}$. Clearly, $\Gamma^{\wedge n} = \{0\}$ for $n \geq 4$. From the equations (19), (47), (57), and (58) we derive the Maurer–Cartan formulas

$$\mathrm{d}\omega_0 = (q^2 + q^4)\omega_0 \wedge \omega_1,\quad \mathrm{d}\omega_1 = -\omega_0 \wedge \omega_2,\quad \mathrm{d}\omega_2 = (q^2 + q^4)\omega_1 \wedge \omega_2.$$

Summarizing the preceding, we have obtained a complete description of the universal higher order differential calculus $\Gamma^{\wedge}_u$ of the FODC Γ. In particular, we see that its classical limit $q \to 1$ is just the ordinary DC on $SL(2)$. △

Example 6 (The FODC on $SL_q(2)$ from Subsect. 14.1.4). Let Γ and $\mathcal{R}$ be as in Subsect. 14.1.4 and suppose that $q^{12} \neq 1$. The elements $x_1 := a + q^{-2}d - (1 + q^{-2})1$, $x_2 := a^2 - (1 + q^{-2})a + q^{-2}1$ and $x_3 := (a - q^{-2})bc$ belong to the right ideal $\mathcal{R}$. One computes $\mathcal{S}(q^2x_1 - q^6x_2) = (q^6 + q^4 - 1)\omega_0\omega_2 + \omega_2\omega_0$ and $\mathcal{S}(x_3) = (q^{-2} - 1)\ (q^3\omega_0\omega_2 + q^{-7}\omega_2\omega_0)$. Therefore, we have $\omega_0\omega_2 \in \mathcal{S}(\mathcal{R})$ and hence $\omega_0 \wedge \omega_2 = 0$ in $\Gamma^{\wedge 2}$. One also verifies that $\omega_j \wedge \omega_j = 0$, $j = 0,1,2$. Hence all 3-forms of $\Gamma^{\wedge}_u$ vanish, so the higher order calculus $\Gamma^{\wedge}_u$ does *not* give the ordinary higher order calculus on $SL(2)$ when $q \to 1$. △

14.4 Higher Order Bicovariant Differential Calculi

14.4.1 Bicovariant Differential Calculi and Differential Hopf Algebras

Let Γ be a FODC over $\mathcal{A}$. In general, there are many differential calculi over $\mathcal{A}$ which have Γ as their first order parts. For instance, if $\Gamma^\wedge = \bigoplus_{n=0}^\infty \Gamma^{\wedge n}$ is a DC with $\Gamma^{\wedge 1} = \Gamma$, then so is any quotient $\Gamma^\wedge/\mathcal{J}$ of $\Gamma^\wedge$ by a differential ideal $\mathcal{J}$ such that $\mathcal{J} \cap (\mathcal{A} \oplus \Gamma) = \{0\}$ (for example, $\mathcal{J} = \bigoplus_{n=k}^\infty \Gamma^{\wedge n}$ for some $k \geq 2$). A natural higher order calculus with first order part Γ is the universal DC $\Gamma_u^\wedge$ considered in Subsect. 14.3.3. If Γ is bicovariant, then there is another distinguished DC $\Gamma_e^\wedge$ with FODC Γ built on the exterior algebra $\Gamma_e^\wedge$ (see Subsect. 13.2.2) over the bicovariant bimodule Γ. The aim of this subsection is to construct this differential calculus $\Gamma_e^\wedge$ and to prove that both bicovariant calculi $\Gamma_e^\wedge$ and $\Gamma_u^\wedge$ give rise to differential Hopf algebras.

Definition 5. *A* differential Hopf algebra *is a differential graded algebra* $\Gamma^\wedge = \bigoplus_{n=0}^\infty \Gamma^{\wedge n}$ *with differentiation* d *which is an* $\mathbb{N}_0$*-graded super Hopf algebra with comultiplication* Δ *satisfying the condition* $\mathrm{d}^\otimes \circ \Delta = \Delta \circ \mathrm{d}$*, where* $\mathrm{d}^\otimes : \Gamma^\wedge \otimes \Gamma^\wedge \to \Gamma^\wedge \otimes \Gamma^\wedge$ *is defined by* $\mathrm{d}^\otimes(\rho_1 \otimes \rho_2) = \mathrm{d}\rho_1 \otimes \rho_2 + (-1)^n \rho_1 \otimes \mathrm{d}\rho_2$ *for* $\rho_1 \in \Gamma^{\wedge n}$ *and* $\rho_2 \in \Gamma^\wedge$.

Note that $\mathrm{d}^\otimes \circ \Delta = \Delta \circ \mathrm{d}$ is the "dual" relation to the graded Leibniz rule $m \circ \mathrm{d}^\otimes = \mathrm{d} \circ m$, where m denotes the multiplication map for the algebra $\Gamma^\wedge$. By differentiating the identity $\sum \rho_{(1)} S(\rho_{(2)}) = \sum S(\rho_{(1)})\rho_{(2)} = \varepsilon(\rho)1$ for $\rho \in \Gamma^{\wedge n}$ it follows easily that $\mathrm{d}S = S\mathrm{d}$ in any differential Hopf algebra.

The bicovariant DC's $\Gamma_e^\wedge$ and $\Gamma_{e,2}^\wedge$ defined below and related constructions (see the last paragraph in Subsect. 14.6.2) appear to be the most important DC's with given bicovariant FODC Γ. A formal reason is that the definition of the exterior algebra $\Gamma_e^\wedge$ over Γ is completely similar to the classical case if the permutation group is replaced by the braid group. More deeper, these DC's are based on the braiding σ which is a fundamental concept in quantum group theory.

Theorem 17. *Let* Γ *be a bicovariant FODC over* $\mathcal{A}$ *and let* $\Gamma_e^\wedge = \bigoplus_{n=0}^\infty \Gamma^{\wedge n}$ *be the exterior algebra over the bicovariant bimodule* Γ *with the* $\mathbb{N}_0$*-graded super Hopf algebra structure defined in Subsect. 13.2.2. Then there exists a unique linear mapping* $\mathrm{d} : \Gamma_e^\wedge \to \Gamma_e^\wedge$ *whose restriction to* $\mathcal{A}$ *is the differentiation* d_1 *of* Γ *such that* $\Gamma_e^\wedge$ *is a bicovariant differential calculus over* $\mathcal{A}$ *with differentiation* d *and a differential Hopf algebra.*

Proof. A convenient way to define the differentiation d on the exterior algebra is Woronowicz' method of extending the bicovariant bimodule Γ. Let Γ_0 be the free left $\mathcal{A}$-module with one generator, denoted Θ, and let $\tilde{\Gamma}$ be the left $\mathcal{A}$-module $\Gamma_0 \oplus \Gamma$. Any element $\tilde{\rho} \in \tilde{\Gamma}$ can be written as $\tilde{\rho} = b\Theta + \rho$ with $\rho \in \Gamma$ and $b \in \mathcal{A}$ uniquely determined by $\tilde{\rho}$. For such $\tilde{\rho}$ and $a \in \mathcal{A}$ we define

$$\tilde{\rho}a = ba\Theta + (b\mathrm{d}_1 a + \rho a), \tag{59}$$

$$\tilde{\Delta}_L(\tilde{\rho}) = \Delta(b)(1 \otimes \Theta) + \Delta_L(\rho), \quad \tilde{\Delta}_R(\tilde{\rho}) = \Delta(b)(\Theta \otimes 1) + \Delta_R(\rho). \qquad (60)$$

Then one easily verifies that $\tilde{\Gamma}$ becomes a bicovariant bimodule over $\mathcal{A}$ with right $\mathcal{A}$-module action of $\mathcal{A}$ on $\tilde{\Gamma}$ given by (59) and left and right coactions $\tilde{\Delta}_L$ and $\tilde{\Delta}_R$ defined by (60), respectively. From (59) and (60) applied with $\tilde{\rho} = \rho$ and $b = 0$ we see that Γ is a bicovariant subbimodule of $\tilde{\Gamma}$. Hence $\Gamma_e^\wedge \subseteq \tilde{\Gamma}_e^\wedge$ by Proposition 13.9.

We define a linear mapping $\mathrm{d} : \tilde{\Gamma}_e^\wedge \to \tilde{\Gamma}_e^\wedge$ by $\mathrm{d}\tilde{\rho} := \Theta \wedge \tilde{\rho} - (-1)^n \tilde{\rho} \wedge \Theta$ for $\tilde{\rho} \in \tilde{\Gamma}^{\wedge n}$. Obviously, d satisfies the graded Leibniz rule. Since Θ is left- and right-invariant by (60), we have $\tilde{\sigma}(\Theta \otimes_{\mathcal{A}} \Theta) = \Theta \otimes_{\mathcal{A}} \Theta$ in $\tilde{\Gamma}^{\otimes 2}$ by (13.27) and hence $\Theta \otimes_{\mathcal{A}} \Theta \in \ker A_2 = \tilde{\mathfrak{S}}^2$, so that $\Theta \wedge \Theta = 0$ in $\tilde{\Gamma}_e^\wedge$. Since $\mathrm{d}(\mathrm{d}\tilde{\rho}) = \Theta \wedge \Theta \wedge \tilde{\rho} - \tilde{\rho} \wedge \Theta \wedge \Theta$ for any $\tilde{\rho} \in \tilde{\Gamma}$, it follows that $\mathrm{d}^2 = 0$. Thus, $\tilde{\Gamma}_e^\wedge$ is a differential calculus over $\mathcal{A}$ with differentiation d. Using the fact that Θ is left- and right-invariant, it is easy to check that this DC is bicovariant. From formula (59) applied with $\rho = 0$ and $b = 1$, we have $\mathrm{d}_1 a = \Theta a - a\Theta$ for $a \in \mathcal{A}$. Therefore, $\mathrm{d}\lceil \mathcal{A} = \mathrm{d}_1$. The proof of the fact that $\Gamma_e^\wedge$ is a bicovariant DC is complete as soon as we have shown that $\mathrm{d}\rho \in \Gamma_e^\wedge$ for any $\rho \in \Gamma_e^\wedge$. It suffices to prove this for $\rho \in \Gamma_e^\wedge$ of the form $\rho = a\mathrm{d}_1 a_1 \wedge \cdots \wedge \mathrm{d}_1 a_n$ with $a, a_1, \cdots, a_n \in \mathcal{A}$, because $\Gamma_e^\wedge$ is spanned by such elements. Using the facts that $\mathrm{d}\lceil \mathcal{A} = \mathrm{d}_1$, $\mathrm{d}^2 = 0$ and that d satisfies the graded Leibniz rule, we derive that $\mathrm{d}\rho = \mathrm{d}a \wedge \mathrm{d}a_1 \wedge \cdots \wedge \mathrm{d}a_n = \mathrm{d}_1 a \wedge \mathrm{d}_1 a_1 \wedge \cdots \wedge \mathrm{d}_1 a_n \in \Gamma_e^\wedge$. The latter formula proves also the uniqueness of d.

Finally, it remains to show that $\mathrm{d}^\otimes \circ \Delta = \Delta \circ \mathrm{d}$. Since Γ is a bicovariant FODC with differentiation $\mathrm{d}_1 = \mathrm{d}\lceil \mathcal{A}$, we have $\Delta(\mathrm{d}(a)) = \Delta_L(\mathrm{d}a) + \Delta_R(\mathrm{d}a) = \sum a_{(1)} \otimes \mathrm{d}a_{(2)} + \sum \mathrm{d}a_{(1)} \otimes a_{(2)} = \mathrm{d}^\otimes(\Delta(a))$ for $a \in \mathcal{A}$. Since Δ is an $\mathbb{N}_0$-graded homomorphism and d satisfies the graded Leibniz rule, the equality $\mathrm{d}^\otimes \circ \Delta = \Delta \circ \mathrm{d}$ on $\Gamma^{\wedge n}$ follows by induction on n. The relation $\mathrm{d}^\otimes \circ \Delta = \Delta \circ \mathrm{d}$ can also be derived from the definition $\mathrm{d}\rho = \Theta \wedge \rho - (-1)^n \rho \wedge \Theta$, $\rho \in \Gamma^{\wedge n}$, combined with the fact that $\Delta(\Theta) = 1 \otimes \Theta + \Theta \otimes 1$. □

As noted in Subsect. 13.2.2, the two-sided ideal $\langle \mathfrak{S}^2 \rangle$ generated by $\mathfrak{S}^2 = \ker(\sigma - I)$ is a Hopf ideal of $T_{\mathcal{A}}^s(\Gamma)$. Therefore, the quotient $\Gamma_{e,2}^\wedge := T_{\mathcal{A}}^s(\Gamma)/\langle \mathfrak{S}^2 \rangle$ is also an $\mathbb{N}_0$-graded super Hopf algebra. The preceding proof applies to $\Gamma_{e,2}^\wedge$ as well and shows that $\Gamma_{e,2}^\wedge$ is also a bicovariant DC over $\mathcal{A}$ and a differential Hopf algebra. It is obvious that the DC $\Gamma_e^\wedge$ is a quotient of $\Gamma_{e,2}^\wedge$ and that $\Gamma_{e,2}^\wedge$ has Γ as its first order part. The bicovariant DC $\Gamma_{e,2}^\wedge$ is often more convenient to work with rather than $\Gamma_e^\wedge$. The reason is that it might be difficult to describe $\mathfrak{S}^n = \ker A_n$ for general n, while it is usually easier to determine $\mathfrak{S}^2 = \ker(\sigma - I)$.

Now we turn to the universal DC $\Gamma_u^\wedge$ of the bicovariant FODC Γ defined in Subsect. 14.3.3. Let $\mathcal{R}$ denote the right ideal $\mathcal{R}_\Gamma$.

Theorem 18. *The universal differential calculus $\Gamma_u^\wedge$ of a bicovariant first order calculus Γ over $\mathcal{A}$ is bicovariant. The ideal $\mathcal{S}_\mathcal{R}$ is a Hopf ideal of the $\mathbb{N}_0$-graded super Hopf algebra $T_{\mathcal{A}}^s(\Gamma)$, so the quotient $\Gamma_u^\wedge = T_{\mathcal{A}}^s(\Gamma)/\mathcal{S}_\mathcal{R}$ is also an $\mathbb{N}_0$-graded super Hopf algebra. With these structures, $\Gamma_u^\wedge$ is a differential Hopf algebra.*

Proof. By Proposition 12.4(i) and the Remark at the end of Subsect. 12.2.3, $\Gamma_u^\wedge$ is left- and right-covariant and hence bicovariant.

Now we prove that $\mathcal{S}_\mathcal{R}$ is a Hopf ideal of $T_\mathcal{A}^s(\Gamma)$. Let $x \in \mathcal{R}$. Recall that

$$\Delta_T^s(\omega(a)) = 1 \otimes \omega(a) + \sum \omega(a_{(2)}) \otimes S(a_{(1)})a_{(3)}, \quad a \in \mathcal{A}, \tag{61}$$

by (13.36) and (28). Since $\mathrm{Ad}_R(\mathcal{R}) \subseteq \mathcal{R} \otimes \mathcal{A}$, we have

$$(\mathrm{id} \otimes \Delta)\mathrm{Ad}_R(x) = \sum x_{(3)} \otimes S(x_{(2)})x_{(4)} \otimes S(x_{(1)})x_{(5)} \in \mathcal{R} \otimes \mathcal{A} \otimes \mathcal{A}.$$

Hence, since $\omega(.)$ vanishes on $\mathcal{R}$, we obtain

$$\sum \omega(x_{(3)}) \otimes S(x_{(2)})x_{(4)}\omega(S(x_{(1)})x_{(5)}) = 0, \tag{62}$$

$$\sum \omega(x_{(3)})\omega(S(x_{(2)})x_{(4)})S(S(x_{(1)})x_{(5)}) = 0. \tag{63}$$

Using definition (1.42) of the $\mathbb{N}_0$-graded product $\bullet$ and equations (61), (4) and finally (62) we compute

$$\begin{aligned}
\Delta_T^s(\mathcal{S}(x)) &= \sum \Delta_T^s(\omega(x_{(1)})) \bullet \Delta_T^s(\omega(x_{(2)})) \\
&= \sum 1 \otimes \omega(x_{(1)})\omega(x_{(2)}) - \sum \omega(x_{(3)}) \otimes \omega(x_{(1)})S(x_{(2)})x_{(4)} \\
&\quad + \sum \omega(x_{(2)}) \otimes S(x_{(1)})x_{(3)}\omega(x_{(4)}) \\
&\quad + \sum \omega(x_{(2)})\omega(x_{(5)}) \otimes S(x_{(1)})x_{(3)}S(x_{(4)})x_{(6)} \\
&= 1 \otimes \mathcal{S}(x) - \sum \omega(x_{(4)}) \otimes S(x_{(3)})x_{(5)}\omega(\overline{x_{(1)}}S(x_{(2)})x_{(6)}) \\
&\quad + \sum \omega(x_{(2)}) \otimes S(x_{(1)})x_{(3)}\omega(x_{(4)}) \\
&\quad + \sum \omega(x_{(2)})\omega(x_{(3)}) \otimes S(x_{(1)})x_{(4)} \\
&= 1 \otimes \mathcal{S}(x) + \sum \omega(x_{(3)}) \otimes S(x_{(2)})x_{(4)}\omega(S(x_{(2)})x_{(5)}) \\
&\quad + (\mathcal{S} \otimes \mathrm{id})\mathrm{Ad}_R(x) \\
&= 1 \otimes \mathcal{S}(x) + (\mathcal{S} \otimes \mathrm{id})\mathrm{Ad}_R(x) \in \mathcal{A} \otimes \mathcal{S}(\mathcal{R}) + \mathcal{S}(\mathcal{R}) \otimes \mathcal{A}.
\end{aligned}$$

Thus, it follows that $\Delta_T^s(\mathcal{S}_\mathcal{R}) \subseteq T_\mathcal{A}^s(\Gamma) \otimes \mathcal{S}_\mathcal{R} + \mathcal{S}_\mathcal{R} \otimes T_\mathcal{A}^s(\Gamma)$.

From (61) one easily derives that

$$S_T^s(\omega(a)) = -\sum \omega(a_{(2)})S(a_{(3)})S^2(a_{(1)}), \quad a \in \mathcal{A}. \tag{64}$$

Using the Hopf algebra axioms we compute

$$\begin{aligned}
S^s_T(\mathcal{S}(x)) &= \sum S^s_T(\omega(x_{(1)})\omega(x_{(2)})) = -\sum S^s_T(\omega(x_{(2)}))S^s_T(\omega(x_{(1)}))\\
&= -\sum \omega(x_{(5)})S(x_{(6)})S^2(x_{(4)})\omega(x_{(2)})S(x_{(3)})S^2(x_{(1)})\\
&= -\sum \omega(x_{(6)})\omega(\overline{x_{(2)}}S(x_{(5)})x_{(7)})S(x_{(8)})S^2(x_{(4)})S(x_{(3)})S^2(x_{(1)})\\
&= -\sum \omega(x_{(4)})\omega(\overline{x_{(2)}}S(x_{(3)})x_{(5)})S(x_{(6)})S^2(x_{(1)})\\
&= +\sum \omega(x_{(3)})\omega(S(x_{(2)})x_{(4)})S(S(x_{(1)})x_{(5)})\\
&\quad -\sum \omega(x_{(2)})\omega(x_{(3)})S(x_{(4)})S^2(x_{(1)})\\
&= 0-\sum \mathcal{S}(x_{(2)})S(S(x_{(1)})x_{(3)}) = -m(\mathcal{S}\otimes S)\mathrm{Ad}_R(x) \in \mathcal{S}(\mathcal{R})\mathcal{A}.
\end{aligned}$$

For the third equality we used (64) and for the fourth we applied (13.10) combined with (3) twice. The seventh equality follows from (63) and (53). The preceding formula implies that $S^s_T(\mathcal{S}_\mathcal{R}) \subseteq \mathcal{S}_\mathcal{R}$. Therefore, $\mathcal{S}_\mathcal{R}$ is a Hopf ideal of $T^s_\mathcal{A}(\Gamma)$ and hence $\Gamma^\wedge_u = T^s_\mathcal{A}(\Gamma)/\mathcal{S}_\mathcal{R}$ is an $\mathbb{N}_0$-graded super Hopf algebra. Arguing as in the proof of Theorem 17, it follows that $\mathrm{d}^\otimes \circ \Delta = \Delta \circ \mathrm{d}$, so that $\Gamma^\wedge_u$ is indeed a differential Hopf algebra. □

Finally, let us note that the DC $\Gamma^\wedge_e$ and $\Gamma^\wedge_u$ do not coincide in general as shown by the following artificial but simple example.

Example 7. Let Γ be the universal FODC over $\mathcal{A}$. Then $\mathcal{R} = \{0\}$ and hence $\Gamma^\wedge_u = T^s_\mathcal{A}(\Gamma)$. Since $\ker(\sigma - I) \neq \{0\}$ in general (for example, if $x \in \mathcal{A}$ is group-like, then we have $(\sigma - I)(\omega(x)\otimes_\mathcal{A}\omega(x)) = 0$ by (29)), $\Gamma^\wedge_e$ is different from $\Gamma^\wedge_u$. △

14.4.2 Quantum Lie Derivatives and Contraction Operators

In this subsection we investigate in detail two basic operations on differential forms (Lie derivatives and interior products) for the bicovariant differential calculus $\Gamma^\wedge_e = T^s_\mathcal{A}(\Gamma)/\mathfrak{S}$ defined in the preceding subsection.

First we introduce a quantum analog of the Lie derivative of differential forms along a vector field. Since $\Gamma^\wedge_e$ is a right comodule for $\mathcal{A}$, we know from Proposition 1.15 that

$$L_f(\rho) \equiv f.\rho := (\mathrm{id}\otimes f)\Delta_R(\rho), \quad \rho \in \Gamma^\wedge_e, \quad f \in \mathcal{A}', \tag{65}$$

defines a left action of the algebra $\mathcal{A}'$ on $\Gamma^\wedge_e$. Note that for $a \in \mathcal{A} = \Gamma^{\wedge 0} \subseteq \Gamma^\wedge_e$, we have $f.a = (\mathrm{id}\otimes f)\Delta_\mathcal{A}(a)$ which is in accordance with the notation from (1.65). Thus, $\Gamma^\wedge_e$ becomes a left $\mathcal{A}'$-module and we have $L_\varepsilon = I$ and $L_f L_g = L_{fg}$ for $f, g \in \mathcal{A}'$. Combined with (37) the latter yields

$$L_{X_i}L_{X_j} - \sigma^{ij}_{rs}L_{X_r}L_{X_s} = L_{[X_i,X_j]} = C^k_{ij}L_{X_k} \quad \text{for} \quad i,j \in I,$$

that is, the map $X \to L_X$ preserves the commutator of the quantum Lie algebra $\mathcal{T}$. Moreover, we have $L_f(\rho_1 \wedge \rho_2) = \sum L_{f_{(1)}}(\rho_1) \wedge L_{f_{(2)}}(\rho_2)$ for f in the dual Hopf algebra $\mathcal{A}^\circ$. By (14), this gives

$$L_{X_i}(\rho_1 \wedge \rho_2) = \rho_1 \wedge L_{X_i}(\rho_2) + L_{X_k}(\rho_1) \wedge L_{f_i^k}(\rho_2), \quad \rho_1, \rho_2 \in \Gamma_e^\wedge. \tag{66}$$

(Strictly speaking, (14) applies only when $\dim \mathcal{T}$ is finite, since then $X_i \in \mathcal{A}^\circ$. But (66) is also valid in the general case and follows then from (13).)

Definition 6. *The map* $L_X : \Gamma_e^\wedge \to \Gamma_e^\wedge$ *defined by (65) is called the* quantum Lie derivative *along* $X \in \mathcal{T}$.

Equation (66) can be viewed as a twisted Leibniz rule for L_{X_i}. It is easy to check that for any $f \in \mathcal{A}'$ the map L_f commutes with the right action Δ_R, the left action Δ_L and the differentiation d of $\Gamma_e^\wedge$, that is, we have

$$(\mathrm{id} \otimes L_f) \circ \Delta_R = \Delta_R \circ L_f, \quad (\mathrm{id} \otimes L_f) \circ \Delta_L = \Delta_L \circ L_f, \quad L_f \mathrm{d} = \mathrm{d} L_f.$$

For instance, we verify the latter by computing $L_f \mathrm{d}\rho = (\mathrm{id} \otimes f)\Delta_R(\mathrm{d}\rho) = (\mathrm{id} \otimes f)(\mathrm{d} \otimes \mathrm{id})\Delta_R(\rho) = \sum \mathrm{d}\rho_{(0)} f(\rho_{(1)}) = \sum \mathrm{d}(\rho_{(0)} f(\rho_{(1)})) = \mathrm{d}L_f(\rho)$.

Moreover, since $\Delta_R(\omega_k) = \omega_i \otimes v_k^i$, (38) yields

$$L_{X_j}(\omega_k) = C_{ij}^k \omega_i \quad \text{and} \quad L_{f_s^r}(\omega_k) = \sigma_{rk}^{is} \omega_i. \tag{67}$$

Next we generalize the notion of the interior product of a differential form by a vector field from differential geometry to the calculus $\Gamma_e^\wedge$.

Proposition 19. *For any* $r \in I$ *there exists a unique linear mapping* $i(X_r) : \Gamma_e^\wedge \to \Gamma_e^\wedge$ *of degree* -1 *(that is,* $i(X_r) : \Gamma^{\wedge n} \to \Gamma^{\wedge(n-1)}$ *for* $n \in \mathbb{N}_0$*, where* $\Gamma^{\wedge(-1)} := \{0\}$*) such that* $i(X_r)(\omega) = (X_r, \omega)$ *for all* $\omega \in {}_{\mathrm{inv}}\Gamma$ *and*

$$i(X_r)(\rho_1 \wedge \rho_2) = (-1)^n \rho_1 \wedge i(X_r)(\rho_2) + i(X_s)(\rho_1) \wedge L_{f_r^s}(\rho_2) \tag{68}$$

for $\rho_1 \in \Gamma^{\wedge n}$ *and* $\rho_2 \in \Gamma_e^\wedge$*. On* $\Gamma_e^\wedge$*, the mappings* d, L_{X_r} *and* $i(X_r)$ *satisfy the identity*

$$L_{X_r} = i(X_r)\mathrm{d} + \mathrm{d}i(X_r). \tag{69}$$

Proof. In order to get well-defined maps $i(X_r)$ on $\Gamma_e^\wedge$, we first define these mappings for the tensor algebra $T_{\mathcal{A}}(\Gamma)$ over Γ and then pass to the quotient $\Gamma_e^\wedge = T_{\mathcal{A}}(\Gamma)/\mathfrak{S}$. We freely use the multi-index notation established in Sect. 13.2. Let us set $f_{\mathfrak{k}}^{\mathfrak{j}} := f_{k_1}^{j_1} \cdots f_{k_n}^{j_n}$ for $\mathfrak{j} = (j_1, \cdots, j_n)$ and $\mathfrak{k} = (k_1, \cdots, k_n)$ in I^n. Further, recall that $T_{\mathcal{A}}(\Gamma)$ is a left $\mathcal{A}'$-module by Proposition 1.15 with action given by $f.\rho := (\mathrm{id} \otimes f)\Delta_R(\rho)$, $f \in \mathcal{A}'$, $\rho \in T_{\mathcal{A}}(\Gamma)$.

Let $B_1 := I$ and $B_n := (-1)^{n+1} B_{n,n-1}^-$ for $n \in \mathbb{N}$, $n \geq 2$, where $B_{n,n-1}^-$ is given by (13.40). We define a linear mapping $i_r : T_{\mathcal{A}}(\Gamma) \to T_{\mathcal{A}}(\Gamma)$ of degree -1 by $i_r(\zeta) := a^{\mathfrak{j}} (B_n)_{\mathfrak{j}}^{(\mathfrak{k},r)} \omega_{\mathfrak{k}}$ for $\zeta = a^{\mathfrak{j}} \omega_{\mathfrak{j}} \in \Gamma^{\otimes n}$, $a^{\mathfrak{j}} \in \mathcal{A}$, $r \in I$. (Note that there are summations over $\mathfrak{j} \in I^n$ and $\mathfrak{k} \in I^{n-1}$.) Since $A_{n-1} B_n = (-1)^{n+1} A_n$ by the second equality of (13.44), we have

$$A_{n-1} i_r(\zeta) = a^{\mathfrak{j}} (B_n)_{\mathfrak{j}}^{(\mathfrak{k},r)} (A_{n-1})_{\mathfrak{k}}^{\mathfrak{l}} \omega_{\mathfrak{l}} = (-1)^n (A_n)_{\mathfrak{j}}^{(\mathfrak{l},r)} a^{\mathfrak{j}} \omega_{\mathfrak{l}}. \tag{70}$$

Therefore, if $\zeta \in \mathfrak{S}^n$, then $(A_n)_{\mathfrak{j}}^{(\mathfrak{l},r)} a^{\mathfrak{j}} = 0$ and hence $i_r(\zeta) \in \mathfrak{S}^{n-1}$ by (70). Hence $i_r(\mathfrak{S}) \subseteq \mathfrak{S}$ and i_r passes to the quotient $\Gamma_e^\wedge = T_{\mathcal{A}}(\Gamma)/\mathfrak{S}$

to define a linear mapping $i(X_r) : \Gamma_e^\wedge \to \Gamma_e^\wedge$. Since obviously $\mathcal{P}_{n,n-1} = \{e, t_{n-1}, t_{n-2}t_{n-1}, \dots, t_1 \cdots t_{n-1}\}$, the definition (13.40) of $B^-_{n,n+1}$ yields

$$B_n = (-1)^{n+1} I + (-1)^n \sigma_{n-1} + \ldots - \sigma_{n-1}\sigma_{n-2} \cdots \sigma_2 + \sigma_{n-1}\sigma_{n-2} \cdots \sigma_1$$

and hence $B_{1,\cdots,n+k} = (-1)^n B_{n+1,\cdots,n+k} + \sigma_{n+k-1} \cdots \sigma_n B_{1,\dots,n}$ for $k, n \in \mathbb{N}$, where the lower indices of B denote the variables on which B acts. Formula (13.31) implies that $(B_n)^{\mathfrak{k}}_{\mathfrak{j}} f^{\mathfrak{l}}_{\mathfrak{j}} = (B_n)^{\mathfrak{j}}_{\mathfrak{l}} f^{\mathfrak{j}}_{\mathfrak{k}}$ for $\mathfrak{k}, \mathfrak{l} \in I^n$. Let $r, s \in I$. Since $f^r_s.\omega_j = \omega_l f^r_s(v^l_j) = \sigma^{ls}_{rj}\omega_l$, $j \in I$, by (13.29) we obtain the relation $f^r_s.\omega_{\mathfrak{k}} = (\sigma_k \cdots \sigma_1)^{(\mathfrak{k}',s)}_{(r,\mathfrak{k})}\omega_{\mathfrak{k}'}$ for $\mathfrak{k} \in I^k$. Let $a^{\mathfrak{n}}\omega_{\mathfrak{n}}$ and $b^{\mathfrak{k}}\omega_{\mathfrak{k}}$ be finite sums with $a^{\mathfrak{n}}, b^{\mathfrak{k}} \in \mathcal{A}$ for $\mathfrak{n} \in I^n$, $\mathfrak{k} \in I^k$ and $n, k \in \mathbb{N}$. Using the preceding facts and the definition of i_r we compute

$$\begin{aligned}
&(-1)^n a^{\mathfrak{n}}\omega_{\mathfrak{n}} i_r(b^{\mathfrak{k}}\omega_{\mathfrak{k}}) + i_s(a^{\mathfrak{n}}\omega_{\mathfrak{n}}) f^s_r.(b^{\mathfrak{k}}\omega_{\mathfrak{k}}) \\
&\quad = (-1)^n a^{\mathfrak{n}}\omega_{\mathfrak{n}} i_r(b^{\mathfrak{k}}\omega_{\mathfrak{k}}) + a^{\mathfrak{n}}(B_n)^{(\mathfrak{l},s)}_{\mathfrak{n}}\omega_{\mathfrak{l}}(f^s_x.b^{\mathfrak{k}})(f^x_r.\omega_{\mathfrak{k}}) \\
&\quad = (-1)^n a^{\mathfrak{n}}\omega_{\mathfrak{n}} b^{\mathfrak{k}} i_r(\omega_{\mathfrak{k}}) + a^{\mathfrak{n}}(B_n)^{(\mathfrak{l},s)}_{\mathfrak{n}}(f^{(\mathfrak{l},s)}_{(\mathfrak{p},x)}.b^{\mathfrak{k}})\omega_{\mathfrak{p}}(\sigma_k \cdots \sigma_1)^{(\mathfrak{k}',r)}_{(x,\mathfrak{k})}\omega_{\mathfrak{k}'} \\
&\quad = (-1)^n a^{\mathfrak{n}}(f^{\mathfrak{n}}_{\mathfrak{n}'}.b^{\mathfrak{k}})\omega_{\mathfrak{n}'}(B_k)^{(\mathfrak{l},r)}_{\mathfrak{k}}\omega_{\mathfrak{l}} + a^{\mathfrak{n}}(\sigma_k \cdots \sigma_1)^{(\mathfrak{k}',r)}_{(x,\mathfrak{k})}(B_n)^{(\mathfrak{p},x)}_{\mathfrak{n}'}(f^{\mathfrak{n}}_{\mathfrak{n}'}.b^{\mathfrak{l}})\omega_{\mathfrak{p}}\omega_{\mathfrak{k}'} \\
&\quad = a^{\mathfrak{n}}(f^{\mathfrak{n}}_{\mathfrak{n}'}.b^{\mathfrak{k}})[(-1)^n(B_{n+1,\dots,n+k})^{(\mathfrak{n}',\mathfrak{l},r)}_{(\mathfrak{n}',\mathfrak{k})}\omega_{(\mathfrak{n}',\mathfrak{l})} \\
&\qquad\qquad + (\sigma_{n+k-1} \cdots \sigma_n B_{1,\dots,n})^{(\mathfrak{p},\mathfrak{k}',r)}_{(\mathfrak{n}',\mathfrak{k})}\omega_{(\mathfrak{p},\mathfrak{k}')}] \\
&\quad = a^{\mathfrak{n}}(f^{\mathfrak{n}}_{\mathfrak{n}'}.b^{\mathfrak{k}})(B_{1,\dots,n+k})^{(\mathfrak{j},r)}_{(\mathfrak{n}',\mathfrak{k})}\omega_{\mathfrak{j}} = a^{\mathfrak{n}}(f^{\mathfrak{n}}_{\mathfrak{n}'}.b^{\mathfrak{k}}) i_r(\omega_{\mathfrak{n}'}\omega_{\mathfrak{k}}) = i_r(a^{\mathfrak{n}}\omega_{\mathfrak{n}} b^{\mathfrak{k}}\omega_{\mathfrak{k}}).
\end{aligned}$$

Passing to the quotient in the preceding equation, we obtain (68) for arbitrary elements $\rho_1 \in \Gamma^{\wedge n}$ and $\rho_2 \in \Gamma^{\wedge k}$. With obvious modifications this proof works for $k = 0$ and $n = 0$ as well. By linearity, (68) holds for all $\rho_2 \in \Gamma_e^\wedge$. The uniqueness assertion is clear.

Using (68), (66) and the graded Leibniz rule for d, it is straightforward to check that (69) holds for the product $\rho_1 \wedge \rho_2$ as soon as it holds for ρ_1 and for ρ_2. So it suffices to prove the identity (69) applied to $a \in \mathcal{A}$ and ω_k, $k \in I$, separately. Since $i(X_r)(\omega_i) = \delta_{ir}$ by definition, (69) is obviously true for $a \in \mathcal{A}$. From the definition of $i(X_r)$ and the formulas (49), (67), and (68) we obtain

$$\begin{aligned}
\mathrm{d}i(X_r)(\omega_k) + i(X_r)(\mathrm{d}\omega_k) &= \mathrm{d}(\delta_{kr}1) + i(X_r)(-c^k_{ij}\omega_i \wedge \omega_j) \\
&= -c^k_{ij}(\sigma - I)^{lr}_{ij}\omega_l = C^k_{lr}\omega_l = L_{X_r}(\omega_k). \qquad \square
\end{aligned}$$

We define $i(X)$ for any X in the quantum Lie algebra $\mathcal{T}$ of Γ by requiring that the map $X \to i(X)$ is linear. Then (69) is still valid for $X \in \mathcal{T}$.

Definition 7. *The form* $i(X)(\rho) \in \Gamma_e^\wedge$ *is called the* contraction *of the form* $\rho \in \Gamma_e^\wedge$ *by* $X \in \mathcal{T}$.

Equation (68) can be regarded as a twisted Leibniz rule for $i(X_r)$. This formula is useful for computations. For instance, it implies that $i(X_r)(a\rho) = ai(X_r)(\rho)$ for $a \in \mathcal{A}$ and $\rho \in \Gamma_e^\wedge$, $i(X_r)(\omega_i \wedge \omega_j) = \sigma^{lr}_{ij}\omega_l - \delta_{rj}\omega_i$, etc.

14.5 Bicovariant Differential Calculi on Coquasitriangular Hopf Algebras

In this section $\mathcal{A}$ denotes a coquasitriangular Hopf algebra equipped with two universal r-forms $\mathbf{r}$ and $\mathbf{s}$. Further, let $\mathbf{u}=(u^i_j)_{i,j=1,\cdots,d}$ be a matrix corepresentation of $\mathcal{A}$, that is, $\mathbf{u}$ is a matrix of elements $u^i_j \in \mathcal{A}$ such that $\Delta(u^i_j)=u^i_k \otimes u^k_j$ and $\varepsilon(u^i_j)=\delta_{ij}$. Our aim is to construct bicovariant differential calculi over $\mathcal{A}$ by using the L-functionals associated with $\mathbf{r}$ resp. $\mathbf{s}$ and $\mathbf{u}$.

First let us repeat some notation and facts from Sect. 10.1. For this reason, let $\mathbf{r}$ be an arbitrary universal r-form on $\mathcal{A}$. The L-functionals associated with $\mathbf{r}$ and $\mathbf{u}$ are the linear functionals $l^{\pm i}_j$ on $\mathcal{A}$ defined by

$$l^{+i}_j(a) = \mathbf{r}(a, u^i_j) \quad \text{and} \quad l^{-i}_j(a) = \bar{\mathbf{r}}(u^i_j, a), \quad a \in \mathcal{A}.$$

We introduce matrices $R = (R^{in}_{jm})$, $\tilde{R} = (\tilde{R}^{in}_{jm})$ and $D = (D^i_j)$ by

$$R^{ij}_{nm} := \mathbf{r}(u^i_n, u^j_m), \quad \tilde{R}^{ij}_{nm} := \mathbf{r}(u^i_n, S(u^j_m)), \quad D^i_j := \sum\nolimits_k \tilde{R}^{ik}_{kj}. \tag{71}$$

That is, R and D are the matrices appearing in (10.37) and (10.19), respectively. Hence, by (10.20), we have $S^2(u^i_j) = (D^{-1})^i_n u^n_m D^m_j$. From the latter equality and the properties of $\mathbf{r}$ and $\bar{\mathbf{r}}$ we obtain the relations

$$l^{+i}_j(u^k_l) = \hat{R}^{ik}_{lj}, \quad l^{-i}_j(u^k_l) = (\hat{R}^{-1})^{ik}_{lj}, \tag{72}$$

$$S(l^{+i}_j)(u^k_l) = (\hat{R}^{-1})^{ki}_{jl}, \quad S(l^{-i}_j)(u^k_l) = \hat{R}^{ki}_{jl}, \tag{73}$$

$$R^{ik}_{lj}\tilde{R}^{nj}_{im} = \tilde{R}^{ik}_{lj}R^{nj}_{im} = \delta_{ln}\delta_{km}, \tag{74}$$

$$\tilde{R}^{ij}_{nm} = (D^{-1})^j_k(\hat{R}^{-1})^{ik}_{ln}D^l_m, \tag{75}$$

where as usual $\hat{R} = \tau \circ R$. By (74), $\tilde{R}$ is just the matrix $\tilde{R}$ defined by (10.14). Recall from Subsect. 10.1.3 that the matrix $\mathsf{L}^{\pm} = (l^{\pm i}_j)$ gives a representation of the algebra $\mathcal{A}$. Let $\mathsf{L}^{\pm,c}$ be the contragredient representation of $\mathsf{L}^{\pm}$ and let $\mathbf{u}^c$ be the contragredient corepresentation of $\mathbf{u}$. That is, $\mathsf{L}^{\pm,c}$ and $\mathbf{u}^c$ have the matrix elements $(\mathsf{L}^{\pm,c})^i_j = S(l^{\pm j}_i)$ and $(\mathbf{u}^c)^i_j = S(u^j_i)$, respectively.

The starting point for the construction of bicovariant DC is the crucial observation that the pairs

$$\Gamma_{\mathbf{r}} := (\mathbf{u}, \mathsf{L}^{-,c}) \quad \text{and} \quad \Gamma^c_{\mathbf{r}} := (\mathbf{u}^c, \mathsf{L}^+)$$

(see Theorem 13.3 for this notation) are bicovariant bimodules over $\mathcal{A}$. In order to prove that they are indeed bicovariant bimodules, by Theorem 13.3 it remains to verify the compatibility condition (13.23). Setting $f^k_j = S(l^{-j}_k)$ and using equation (10.5) and Proposition 10.2(v), we compute

$$\begin{aligned}\sum\nolimits_k u_i^k(a.f_j^k) &= \sum S(l^{-j}_{\ k})(a_{(1)})u_i^k a_{(2)} = \sum \bar{\mathbf{r}}(u_k^j, S(a_{(1)}))u_i^k a_{(2)} \\ &= \sum \mathbf{r}(u_k^j, a_{(1)})u_i^k a_{(2)} = \sum a_{(1)} u_k^j \mathbf{r}(u_i^k, a_{(2)}) \\ &= \sum a_{(1)} u_k^j S(l^{-k}_{\ i})(a_{(2)}) = \sum\nolimits_k (f_k^i.a)u_k^j.\end{aligned}$$

This proves condition (13.23) for $\Gamma_{\mathbf{r}}$. The proof for $\Gamma_{\mathbf{r}}^c$ is similar.

In the following we will need the matrices $\mathsf{L}^{\pm} = (l^{\pm i}_{\ j})$, $R = (R^{in}_{jm})$ and $\tilde{R} = (\tilde{R}^{in}_{jm})$ for the matrix corepresentation $\mathbf{u}$ and for both universal r-forms $\mathbf{r}$ and $\mathbf{s}$. We shall express the dependence on the universal r-forms by adding $\mathbf{r}$ resp. $\mathbf{s}$ as a lower index.

Since $\Gamma_{\mathbf{r}}^c$ and $\Gamma_{\mathbf{s}}$ are bicovariant bimodules as just shown,

$$\Gamma_{12} := \Gamma_{\mathbf{r}}^c \otimes_{\mathcal{A}} \Gamma_{\mathbf{s}} = (\mathbf{u}^c \otimes \mathbf{u}, \mathsf{L}_{\mathbf{r}}^{+} \otimes \mathsf{L}_{\mathbf{s}}^{-,c})$$

is also a bicovariant bimodule over $\mathcal{A}$. By Theorem 13.3, the structure of the bicovariant bimodule Γ_{12} is described as follows. The vector space ${}_{\mathrm{inv}}\Gamma_{12}$ of left-invariant elements of Γ_{12} has a basis $\{\theta_{ij} \mid i,j = 1,2,\dots,d\}$ such that the right and left actions of $\mathcal{A}$ on the $\mathcal{A}$-bimodule Γ_{12} satisfy the relations

$$\theta_{ij}a = (f^{ij}_{nm}.a)\theta_{nm}, \quad \text{where} \quad f^{ij}_{nm} = l^{+i}_{\mathbf{r}n} S(l^{-m}_{\mathbf{s}\ j}), \tag{76}$$

and the right coaction of $\mathcal{A}$ on this basis is given by

$$\Delta_R(\theta_{ij}) = \theta_{nm} \otimes v^{nm}_{ij}, \quad \text{where} \quad v^{nm}_{ij} = (\mathbf{u}^c \otimes \mathbf{u})^{nm}_{ij} = S(u_n^i)u_j^m. \tag{77}$$

We set $\theta := D_i^j \theta_{ij}$, where the matrix $D = (D_j^i)$ is defined by (71) for one of the forms $\mathbf{r}$ or $\mathbf{s}$. Using again condition (10.5) one easily shows that $\Delta_R(\theta) = \theta \otimes 1$. Since $\theta \in {}_{\mathrm{inv}}\Gamma_{12}$ is left-invariant, θ is a biinvariant (that is, left- and right-invariant) element of Γ_{12}. We now define

$$\mathrm{d}a = \theta a - a\theta, \quad a \in \mathcal{A}. \tag{78}$$

Further, let $\mathcal{T}$ be the span of linear functionals

$$X_{ij} := D_n^m f^{nm}_{ij} - D_i^j \varepsilon, \quad i,j = 1,2,\cdots,d.$$

Proposition 20. *$\Gamma := \mathcal{A}{\cdot}\mathrm{d}\mathcal{A}{\cdot}\mathcal{A}$ is a bicovariant FODC over the Hopf algebra $\mathcal{A}$ with differentiation* d *and quantum Lie algebra $\mathcal{T}$.*

Proof. It is clear that Γ is an $\mathcal{A}$-subbimodule of Γ_{12} which is invariant under the left and right coactions of $\mathcal{A}$ and that d satisfies the Leibniz rule. Hence Γ is a FODC over $\mathcal{A}$. Since θ is biinvariant, we have

$$\Delta_R(\mathrm{d}a) = \sum \theta a_{(1)} \otimes a_{(2)} - \sum a_{(1)}\theta \otimes a_{(2)} = (\mathrm{d} \otimes \mathrm{id})\Delta(a)$$

and similarly $\Delta_L(\mathrm{d}a) = (\mathrm{id} \otimes \mathrm{d})\Delta(a)$ by (78). Thus, Γ is bicovariant. Let $\{Y_n\}$ be a basis of the vector space $\mathcal{T}$. Let $X_{ij} = \alpha^n_{ij} Y_n$. Setting $\theta_n := \alpha^n_{ij}\theta_{ij}$, we derive from (78) that $\mathrm{d}a = \sum_{i,j}(X_{ij}.a)\theta_{ij} = \sum_n (Y_n.a)\theta_n$ and so $\omega(a) =$

$\sum_n Y_n(a)\theta_n$ for $a \in \mathcal{A}$. By (8) and (10), this implies that $\{\theta_n\}$ is a basis of ${}_{\rm inv}\Gamma$ and that $\{Y_n\}$ is the corresponding dual basis of $\mathcal{T}_\Gamma$. In particular, $\mathcal{T} = {\rm Lin}\,\{Y_n\}$ is the quantum Lie algebra of Γ. □

Of course, the dimension of the quantum Lie algebra $\mathcal{T}$ may be less than d^2. For instance, if $\mathcal{A}$ is commutative and $\mathbf{r}$ and $\mathbf{s}$ are the trivial universal r-form $\varepsilon \otimes \varepsilon$ on $\mathcal{A}$, then all X_{ij} are zero. *Let us assume in the remainder of this section that the linear functionals X_{ij}, $i,j = 1,2,\cdots,d$, on $\mathcal{A}$ are linearly independent.* Then the quantum Lie algebra $\mathcal{T}$ and the FODC Γ have dimension d^2. Since $\omega(a) = \sum S(a_{(1)}){\rm d}a_{(2)} = \sum_{i,j} X_{ij}(a)\theta_{ij}$ for $a \in \mathcal{A}$, it follows then that $\theta_{ij} \in \mathcal{A}{\cdot}{\rm d}\mathcal{A} = \Gamma$. Hence we have $\Gamma = \Gamma_{12}$ and the set $\{\theta_{ij}\}$ is a free left $\mathcal{A}$-module basis of Γ. Moreover, it is the dual basis to $\{X_{ij}\}$.

Let σ denote the braiding of $\Gamma \otimes_\mathcal{A} \Gamma$ (see Subsect. 13.1.4). We denote by $\sigma^{nm,rs}_{ij,kl}$ the matrix coefficients of σ with respect to the basis $\{\theta_{ij}\otimes_\mathcal{A}\theta_{kl}\}$ of ${}_{\rm inv}(\Gamma\otimes_\mathcal{A}\Gamma)$ (that is, $\sigma(\theta_{ij}\otimes_\mathcal{A}\theta_{kl}) = \sigma^{nm,rs}_{ij,kl}\theta_{nm}\otimes_\mathcal{A}\theta_{rs}$) and by $C^{kl}_{nm,rs}$ the structure constants of the quantum Lie algebra $\mathcal{T}$ with respect to the basis $\{X_{ij}\}$ (see Subsect. 14.2.3). These quantities are determined by

Proposition 21.

$$\sigma^{nm,rs}_{ij,kl} = (\hat{R}^{-1}_{\mathbf{r}})^{pi}_{xn}(\hat{R}_{\mathbf{r}})^{xm}_{yr}(\hat{R}_{\mathbf{s}})^{ys}_{wl}(\tilde{R}_{\mathbf{s}})^{wk}_{jp},$$

$$C^{kl}_{nm,rs} = D^j_i(\hat{R}^{-1}_{\mathbf{r}})^{pi}_{xn}(\hat{R}_{\mathbf{r}})^{xm}_{yr}(\hat{R}_{\mathbf{s}})^{ys}_{wl}(\tilde{R}_{\mathbf{s}})^{wk}_{jp} - D^s_r\delta_{kn}\delta_{lm}.$$

Proof. From the second formula of (71) and Proposition 10.2(v) we obtain

$$S(l^{-i}_{\mathbf{s}\ j})(S(u^n_m)) = l^{-i}_{\mathbf{s}\ j}(S^2(u^n_m)) = \bar{\mathbf{s}}(u^i_j, S^2(u^n_m)) = \mathbf{s}(u^i_j, S(u^n_m)) = (\tilde{R}_{\mathbf{s}})^{in}_{jm}.$$

Using this fact and the formulas (13.29), (77), (13.14), (76), (72), and (73) we compute

$$\begin{aligned}
\sigma^{nm,rs}_{ij,kl} &= f^{ij}_{rs}(v^{nm}_{kl}) = f^{ij}_{xw}(S(u^k_n))f^{xw}_{rs}(u^m_l)\\
&= l^{+i}_{\mathbf{r}x}(S(u^p_n))S(l^{-w}_{\mathbf{s}j})(S(u^k_p))l^{+x}_{\mathbf{r}r}(u^m_y)S(l^{-s}_{\mathbf{s}w})(u^y_l)\\
&= (\hat{R}^{-1}_{\mathbf{r}})^{pi}_{xn}(\tilde{R}_{\mathbf{s}})^{wk}_{jp}(\hat{R}_{\mathbf{r}})^{xm}_{yr}(\hat{R}_{\mathbf{s}})^{ys}_{wl}.
\end{aligned}$$

Since $C^{kl}_{nm,rs} = X_{rs}(v^{nm}_{kl}) = D^j_i f^{ij}_{rs}(v^{nm}_{kl}) - D^s_r\varepsilon(v^{nm}_{kl})$ by (38), the second formula follows at once from the previous computation. □

Under the assumption the matrix $R_{\mathbf{r}}$ is a scalar multiple of $R_{\mathbf{s}}$, we now determine the spectral decomposition of the braiding map σ. Let us write R instead of $R_{\mathbf{r}}$. For $A = (A^{nm}_{ij}), B = (B^{nm}_{ij}) \in \mathcal{L}(\mathbb{C}^d \times \mathbb{C}^d)$ we define

$$\Phi(A,B)^{nm,rs}_{ij,kl} := A^{pi}_{xn}\hat{R}^{xm}_{yr}B^{ys}_{wl}\tilde{R}^{wk}_{jp}.$$

Clearly, if $R_{\mathbf{s}} = \alpha R_{\mathbf{r}}$, then $\tilde{R}_{\mathbf{s}} = \alpha^{-1}\tilde{R}_{\mathbf{r}}$ by (74). Hence the formula for $\sigma^{nm,rs}_{ij,kl}$ remains unchanged if we set $R_{\mathbf{s}} = R_{\mathbf{r}} = R$ therein. Therefore, we have $\sigma = \Phi(\hat{R}^{-1},\hat{R})$. Using formula (74) it follows that

$$\Phi(A_1,B_1)\Phi(A_2,B_2)=\Phi(A_2A_1,B_1B_2) \quad \text{and} \quad \Phi(I,I)=I. \tag{79}$$

Let $\hat{R}=\sum_i \lambda_i P_i$ be the spectral decomposition of $\hat{R}$ (that is, $\sum_i P_i = I$, $P_i^2=P_i$ and $P_iP_j=0$ for $i\neq j$). By (79), $P_{ij}:=\Phi(P_i,P_j)$ are also projections with $\sum_{i,j}P_{ij}=I$ and $P_{ij}P_{nm}=0$ if $(i,j)\neq(n,m)$. Hence the spectral decomposition of σ is given by

$$\sigma=\Phi(\hat{R}^{-1},\hat{R})=\sum\nolimits_{i,j}\lambda_i^{-1}\lambda_jP_{ij}. \tag{80}$$

We briefly turn to another variant of the preceding construction. In order to define a bicovariant FODC Γ', we consider the bicovariant bimodule

$$\Gamma_{21}=\Gamma_{\mathbf{s}}\otimes_{\mathcal{A}}\Gamma_{\mathbf{r}}^{c}=(\mathbf{u}\otimes\mathbf{u}^{c},\mathsf{L}_{\mathbf{s}}^{-,c}\otimes\mathsf{L}_{\mathbf{r}}^{+}).$$

Let us denote all structure quantities for Γ_{21} and Γ' by the same symbols as for Γ_{12} and Γ but equipped with a prime. It is obvious that $\theta':=\sum\theta_{ii}$ is a biinvariant element of Γ_{21}. Setting $\mathrm{d}'a:=\theta'a-a\theta'$ for $a\in\mathcal{A}$, we obtain a bicovariant FODC $\Gamma':=\mathcal{A}{\cdot}\mathrm{d}'\mathcal{A}{\cdot}\mathcal{A}$ over $\mathcal{A}$. For the FODC Γ' we have

$$f'^{ij}_{\ nm}=S(l^{-n}_{\mathbf{s}\ i})l^{+j}_{\mathbf{r}\ m} \quad \text{and} \quad X'_{ij}=S(l^{-i}_{\mathbf{s}\ n})l^{+n}_{\mathbf{r}j}-\delta_{ij}\varepsilon. \tag{81}$$

Assume that the functionals X'_{ij}, $i,j=1,2,\cdots,d$, are linearly independent. Then the matrix coefficients of the braiding map σ' with respect to the basis $\{\theta'_{ij}\otimes_{\mathcal{A}}\theta'_{kl}\}$ and the structure constants of the quantum Lie algebra with respect to the basis $\{X'_{ij}\}$ are given by the formulas

$$\sigma'^{nm,rs}_{\ ij,kl}=(\hat{R}_{\mathbf{s}})^{np}_{ix}(\hat{R}_{\mathbf{r}})^{jx}_{ky}(\hat{R}_{\mathbf{r}}^{-1})^{ly}_{sw}(\tilde{R}_{\mathbf{s}})^{rw}_{pm},$$

$$C'^{kl}_{\ nm,rs}=(\hat{R}_{\mathbf{s}})^{np}_{ix}(\hat{R}_{\mathbf{r}})^{ix}_{ky}(\hat{R}_{\mathbf{r}}^{-1})^{ly}_{sw}(\tilde{R}_{\mathbf{s}})^{rw}_{pm}-\delta_{rs}\delta_{nk}\delta_{ml}.$$

An advantage of the FODC Γ over Γ' is the fact that the 1-forms θ_{ij} transform under the right coaction Δ_R just as the left-invariant 1-forms $\omega_{ij}:=\omega(u^i_j)\equiv P_L(\mathrm{d}u^i_j)\equiv S(u^i_k)\mathrm{d}u^k_j$ do. Indeed, by (28), we have for any bicovariant FODC over $\mathcal{A}$ that $\Delta_R(\omega_{ij})=\omega_{nm}\otimes S(u^i_n)u^m_j$.

The bicovariant bimodules Γ_{12} and Γ_{21} and hence the bicovariant FODC Γ and Γ' over $\mathcal{A}$ are always isomorphic. It can be shown that there is an isomorphism of Γ_{21} to Γ_{12} and so of Γ' to Γ which maps $(\hat{R}_{\mathbf{r}})^{in}_{jm}\theta'_{nm}$ to θ_{ij}.

Now let $\mathcal{A}$ be a Hopf $*$-algebra. If both universal r-forms $\mathbf{r}$ and $\mathbf{s}$ are real and $(u^i_j)^*=S(u^j_i)$, $i,j=1,2,\ldots,d$, then the FODC Γ and Γ' are $*$-calculi. In order to prove this, first note that the reality of $\mathbf{r}$ resp. $\mathbf{s}$ implies that $\overline{D^i_j}=D^j_i$. From this fact and the formulas (10.47) we obtain $(X_{ij})^*=X_{ji}$ and $(X'_{ij})^*=X'_{ji}$, so Γ and Γ' are indeed $*$-calculi by Proposition 6.

Finally, let us turn to the higher order calculi. If we know the braiding map σ, we can form the exterior algebra $\Gamma^{\wedge}_e=T_{\mathcal{A}}(\Gamma)/\mathfrak{S}$ over the bicovariant bimodule Γ (see Subsect. 13.2.2). In the present case the differentiation d of the bicovariant DC $\Gamma^{\wedge}_e$ from Theorem 17 can be expressed by the $\mathbb{N}_0$-graded commutator

$$\mathrm{d}\rho = \theta \wedge \rho - (-1)^n \rho \wedge \theta \quad \text{for} \quad \rho \in \Gamma^{\wedge n},\ n \in \mathbb{N}_0. \tag{82}$$

Indeed, since the latter defines a bicovariant DC on $\Gamma_e^\wedge$ with first order part Γ, it must be the DC from Theorem 17. The differentiation of the bicovariant DC $\Gamma_{e,2}^\wedge = T_{\mathcal{A}}(\Gamma)/\langle \mathfrak{S}_2 \rangle$ is also given by formula (82). All assertions of this paragraph remain valid when Γ is replaced by Γ'.

Summarizing the preceding, we have developed a general method for the construction of bicovariant differential calculi over a coquasitriangular Hopf algebra $\mathcal{A}$. These DC depend on a matrix corepresentation $\mathbf{u} = (u^i_j)$ and on two universal r-forms $\mathbf{r}$ and $\mathbf{s}$ of $\mathcal{A}$. For many quantum groups there is a natural fundamental corepresentation $\mathbf{u}$ of $\mathcal{A}$. But even if $\mathbf{u}$ is fixed, different choices of the universal r-forms $\mathbf{r}$ and $\mathbf{s}$ may still give nonisomorphic calculi in general.

14.6 Bicovariant Differential Calculi on Quantized Simple Lie Groups

In what follows, $\mathcal{A}$ denotes the Hopf algebra $\mathcal{O}(G_q)$ for one of the quantum groups $G_q = GL_q(N), SL_q(N), O_q(N), Sp_q(N)$ and $\mathbf{u} = (u^i_j)_{i,j=1,\dots,N}$ is the fundamental corepresentation. In this section we elaborate the method from the previous section in this situation. Complete proofs of most of the unproven assertions in the following treatment can be found in the papers [SS2], [SS3].

14.6.1 A Family of Bicovariant First Order Differential Calculi

Let us say that a complex number z is *admissible* if $z \neq 0$ for $G_q = GL_q(N)$, $z^N = q^{-1}$ for $G_q = SL_q(N)$ and $z^2 = 1$ for $G_q = O_q(N), Sp_q(N)$. By Theorem 10.9, the Hopf algebra $\mathcal{A} = \mathcal{O}(G_q)$ is coquasitriangular with universal r-form $\mathbf{r}_z$ and

$$\mathbf{r}_z(u^i_j, u^n_m) = z\mathsf{R}^{in}_{jm}, \quad i,j,n,m = 1,2,\cdots,N,$$

where $z \in \mathbb{C}$ is admissible and the matrix $\mathsf{R} = (\mathsf{R}^{in}_{jm})$ is given by the formulas (9.13) resp. (9.30).

Let x and y be admissible parameters and set $z := xy$. We denote the FODC Γ defined in Subsect. 14.5 in the case $\mathbf{r} = \mathbf{r}_x$, $\mathbf{s} = \mathbf{r}_y$ by $\Gamma_{+,z}$ and in the case $\mathbf{r} = (\overline{\mathbf{r}_x})_{21}$, $\mathbf{s} = (\overline{\mathbf{r}_y})_{21}$ by $\Gamma_{-,z}$. From the description given below it will be clear that the FODC $\Gamma = \Gamma_{\pm,z}$ depends only on the product xy, so this notation is justified. It can be shown that the corresponding generators X_{ij}, $i,j = 1,2,\cdots,N$, of the quantum Lie algebra $\mathcal{T}$ are linearly independent. Hence $\Gamma_{\pm,z}$ is an N^2-dimensional bicovariant FODC over $\mathcal{A}$.

First we describe the bicovariant bimodule $\Gamma_{\pm,z}$ more explicitly. Recall from Subsect. 10.1.3 that the functional $l^{\pm i}_j$ goes into $l^{\mp i}_j$ if the universal r-form $\mathbf{r}$ is replaced by $\bar{\mathbf{r}}_{21}$. Thus, we have the following equality of bicovariant bimodules

$$\Gamma_{\pm,z} = (\mathbf{u}^c \otimes \mathbf{u}, \mathsf{L}^{\pm}_{\mathbf{r}_x} \otimes \mathsf{L}^{\mp,c}_{\mathbf{r}_y}).$$

Let us repeat the main structure data of the bicovariant bimodule $\Gamma_{\pm,z}$ according to Theorem 13.3. The matrix entries of $\mathsf{L}^{\pm}_{\mathbf{r}_x}$ and $\mathsf{L}^{\mp,c}_{\mathbf{r}_y}$ are the functionals determined by the formulas (72) resp. (73) if $\hat{R}$ is replaced by $x\hat{\mathsf{R}}$ resp. $y\hat{\mathsf{R}}$ therein. Since $xy = z$, these formulas yield the commutation relations

$$\theta_{ij} u^r_s = z^{\pm 1} (\hat{\mathsf{R}}^{\pm 1})^{ik}_{pn} (\hat{\mathsf{R}}^{\pm 1})^{pm}_{js} u^r_k \theta_{nm} \tag{83}$$

between the basis elements θ_{ij}, $i, j = 1, 2, \cdots, N$, of the vector space of left-invariant elements of $\Gamma_{\pm,z}$ and the generators u^r_s. The right coaction of $\mathcal{A}$ on this basis is given by

$$\Delta_R(\theta_{ij}) = \theta_{nm} \otimes S(u^i_n) u^m_j.$$

A convenient form of the corresponding functionals f^{ij}_{nm} is obtained as follows. Let ε_ζ be the character of the algebra $\mathcal{A}$ such that $\varepsilon_\zeta(u^i_j) = \zeta \delta_{ij}$ for $i, j = 1, 2, \cdots, N$, where $\zeta \neq 0$ for $GL_q(N)$, $\zeta^N = 1$ for $SL_q(N)$ and $\zeta^2 = 1$ for $O_q(N)$ and $Sp_q(N)$. Obviously, ε_ζ belongs to the center of the algebra $\mathcal{A}'$. Further, let $l^{\pm i}_{\ j}$ denote the L-functionals with respect to $\mathbf{r}_{z_0}$ and $\mathbf{u}$ for some fixed admissible parameter z_0. Then the bimodule structure of $\Gamma_{\pm,z}$ is determined by the equations

$$\theta_{ij} a = (f^{ij}_{nm} . a) \theta_{nm}, \quad \text{where} \quad f^{ij}_{nm} = \varepsilon_\zeta l^{\pm i}_{\ n} S(l^{\mp m}_{\ j}), \quad \zeta := z z_0^{-2}. \tag{84}$$

Next we turn to the differentiation d of the FODC $\Gamma_{\pm,z}$. From Example 10.3 we know that the matrix $D = (D^i_j)$ from (10.19) is a scalar multiple of the diagonal matrix $(\mathsf{D}_i \delta_{ij})$, where $\mathsf{D}_i = q^{-2i}$ for $G_q = GL_q(N), SL_q(N)$ and $\mathsf{D}_i = q^{2\rho_i}$ for $G_q = O_q(N), Sp_q(N)$. Therefore, after scaling the basis elements θ_{ij}, the biinvariant element θ of $\Gamma_{\pm,z}$ is $\theta := \sum_i \mathsf{D}_i \theta_{ii}$. Thus the differentiation d of the FODC $\Gamma_{\pm,z}$ is defined by

$$\mathrm{d}a = \theta a - a\theta = \sum\nolimits_{i,j=1}^{N} (X_{ij} . a) \theta_{ij},$$

where $\{X_{ij} \mid i, j = 1, 2, \cdots, N\}$ is a basis of the quantum Lie algebra of $\Gamma_{\pm,z}$. From (84) we immediately derive that

$$X_{ij} = \sum\nolimits_n \mathsf{D}_n \varepsilon_\zeta l^{\pm n}_{\ i} S(l^{\mp j}_{\ n}) - \mathsf{D}_i \delta_{ij} \varepsilon.$$

Before we continue the study of the bicovariant FODC $\Gamma_{\pm,z}$, we briefly comment on isomorphic calculi among them. First let us note that the FODC Γ' of Sect. 14.5 is isomorphic to $\Gamma = \Gamma_{\pm,z}$. (The map $(\hat{\mathsf{R}}^{\pm 1})^{in}_{jm} \theta_{nm} \to \theta'_{ij}$ extends to an isomorphism of Γ and Γ'.) For the quantum groups $O_q(N)$ and $Sp_q(N)$ the bicovariant bimodules $\Gamma_{+,z}$ and $\Gamma_{-,z}$ and so the corresponding FODC are isomorphic. (There is an isomorphism of $\Gamma_{+,z}$ and $\Gamma_{-,z}$ such that $(\mathsf{C}^{-1})^r_n \hat{\mathsf{R}}^{rm}_{sj} \mathsf{C}^s_i \theta_{nm} \to \theta_{ij}$.) Similarly, the FODC $\Gamma_{+,z}$ and $\Gamma_{-,z}$ for $GL_q(2)$ and $SL_q(2)$ are isomorphic. Let us write $\Gamma_+ := \Gamma_{+,1}$ and $\Gamma_- := \Gamma_{+,-1}$ for the quantum groups $O_q(N)$, $Sp_q(N)$ and $SL_q(2)$. (It can be shown that the FODC $\Gamma_{+,1}$ and $\Gamma_{+,-1}$ on $SL_q(2)$ defined above are indeed isomorphic to the FODC Γ_+ and Γ_-, respectively, defined in Subsect. 14.2.4, so the notation $\Gamma_\pm$ is justified.)

Summarizing, we have constructed so far the following bicovariant FODC:

- *one-parameter families* $\Gamma_{+,z}$ *and* $\Gamma_{-,z}$, $z \in \mathbb{C}\backslash\{0\}$, *for* $GL_q(N), N{\geq}3$;
- *a one-parameter family* $\Gamma_{+,z}$, $z \in \mathbb{C}\backslash\{0\}$, *for* $GL_q(2)$;
- $2N$ *calculi* $\Gamma_{\pm,z}$, $z^N = q^{-2}$, *for* $SL_q(N)$, $N \geq 3$;
- *two calculi* Γ_+ *and* Γ_- *for* $O_q(N)$, $Sp_q(N)$ *and* $SL_q(2)$.

It can be shown that different FODC of this list are not isomorphic at least if q is not a root of unity. Note that all these FODC have dimension N^2. We close this subsection with some remarks about these FODC $\Gamma = \Gamma_{\pm,z}$.

As noted in the paragraph before last, the FODC Γ and Γ' (see Sect. 14.5) are isomorphic. Hence their quantum Lie algebras coincide and the functionals X'_{ij} from (81) give another basis of the quantum Lie algebra of $\Gamma = \Gamma_{\pm,z}$. In the above notation, these functionals are written as

$$X'_{ij} = \varepsilon_\zeta S(l^{\mp i}_{\ n})l^{\pm n}_{\ j} - \delta_{ij}\varepsilon, \quad i,j = 1,2,\cdots,N. \tag{85}$$

Note that $\varepsilon_\zeta = \varepsilon$ if $z = z_0^2$. Since $S(l^{-i}_{\ n})l^{+n}_{\ j}$ is equal to the functional l^i_j defined by (10.39) for the universal r-form $\mathbf{r} = \mathbf{r}_{z_0}$, the basis elements X'_{ij} of the quantum Lie algebra of the bicovariant FODC $\Gamma_{+,z}$ are just the entries of the matrix $\varepsilon_\zeta \mathsf{L} - \varepsilon I$. This shows the importance of the matrix $\mathsf{L} = S(\mathsf{L}^-)\mathsf{L}^+$ introduced in Subsect. 10.1.3 for the study of bicovariant differential calculi. Further, by Proposition 10.16(ii),

$$c := \varepsilon_\zeta \operatorname{Tr}\ \mathsf{L}D^{-1} = \varepsilon_\zeta l^i_j (D^{-1})^j_i \tag{86}$$

is a central element of $\mathcal{A}^\circ$. From formula (85) we see that $c - c(1)\varepsilon$ belongs to the quantum Lie algebra of $\Gamma_{+,z}$ and

$$\Delta(c) = \varepsilon_\zeta l^k_n \otimes Y_{kn}, \quad \text{where} \quad Y_{kn} := \varepsilon_\zeta S(l^{-i}_{\ k})l^{+n}_{\ j}(D^{-1})^j_i. \tag{87}$$

If q is not a root of unity, it can be shown that the functionals $Y_{kn}, k, n = 1,\cdots,N$, are linearly independent. Therefore, by (87) and the first paragraph of the proof of Proposition 11, the vector space $\mathcal{T}(c)$ defined by (36) coincides with the linear span of functionals $X'_{ij} = \varepsilon_\zeta l^i_j - \delta_{ij}\varepsilon$. That is, the quantum Lie algebra of the bicovariant FODC $\Gamma_{+,z}$ is of the form $\mathcal{T}(c)$, where c is the central element of $\mathcal{A}^\circ$ given by (86). An analogous result holds also for the FODC $\Gamma_{-,z}$ and the universal r-form $(\overline{\mathbf{r}_{z_0}})_{21}$, because the passage from $\mathbf{r}$ to $\bar{\mathbf{r}}_{21}$ interchanges the functionals $l^{+i}_{\ j}$ and $l^{-i}_{\ j}$.

Finally, we turn to the question of $*$-calculi. Suppose that the parameters q, x and y are real and $\mathcal{A}$ is one of the Hopf $*$-algebras in Proposition 10.10(i). Then the universal r-forms $\mathbf{r}$ and $\mathbf{s}$ are real and hence, as discussed in Sect. 14.5, the corresponding FODC $\Gamma_{\pm,z}$, $z = xy$, are $*$-calculi.

Next we assume that $|q| = |z| = |z_0| = 1$, $z^2 = z_0^4$ and $\mathcal{A}$ denotes one of the Hopf $*$-algebras from Proposition 10.10(ii). Then the universal r-forms $\mathbf{r}_{z_0}$ and $(\overline{\mathbf{r}_{z_0}})_{21}$ on $\mathcal{A}$ are inverse real. Since $z^2 = z_0^4$, we have $\zeta = \bar{\zeta}$ and hence $\varepsilon_\zeta{}^* = \varepsilon_\zeta$. Therefore, it follows from the formulas (10.48) and (85) that all functionals $(X'_{ij})^* = (\varepsilon_\zeta l^i_j)^* - \delta_{ij}\varepsilon = \tilde{R}^{in}_{km}(D^{-1})^m_j X'_{kn}$ again belong to the quantum Lie algebra of $\Gamma_{\pm,z}$. Therefore, by Proposition 6, $\Gamma_{\pm,z}$ is a $*$-calculus. In particular, $\Gamma_{\pm,z_0^2}$ is a $*$-calculus by applying the preceding with $z = z_0^2$. Setting $z = \pm 1$ and $z_0 = \pm 1$ we conclude that both FODC Γ_+ and Γ_- are $*$-calculi for $O_q(n,n)$, $O_q(n,n+1)$ and $Sp_q(N;\mathbb{R})$.

14.6.2 Braiding and Structure Constants of the FODC $\Gamma_{\pm,z}$

The matrix coefficients $(\sigma_\pm)^{nm,rs}_{ij,kl}$ of the braiding $\sigma_\pm$ and the structure constants $(C^\pm)^{kl}_{nm,rs}$ of the quantum Lie algebra of the FODC $\Gamma_{\pm,z}$ with respect to the bases $\{\theta_{ij} \otimes_{\mathcal{A}} \theta_{kl}\}$ and $\{X_{ij}\}$ are given by

Proposition 22.

$$(\sigma_\pm)^{nm,rs}_{ij,kl} = (\hat{\mathsf{R}}^{\mp 1})^{pi}_{wn}(\hat{\mathsf{R}}^{\pm 1})^{wm}_{xr}(\hat{\mathsf{R}}^{\pm 1})^{xs}_{yl}(\hat{\mathsf{R}}^{\mp 1})^{yk}_{pj}\mathsf{D}_k^{-1}\mathsf{D}_p,$$
$$(C_\pm)^{kl}_{nm,rs} = (\hat{\mathsf{R}}^{\mp 2})^{yk}_{wn}(\hat{\mathsf{R}}^{\pm 1})^{wm}_{xr}(\hat{\mathsf{R}}^{\pm 1})^{xs}_{yl}\mathsf{D}_y - \mathsf{D}_r\delta_{kn}\delta_{lm}\delta_{rs}.$$

Proof. As noted in Example 10.3, $D^i_j = c\mathsf{D}_i\delta_{ij}$ for some constant $c \neq 0$. Since $S^2(u^n_m) = \mathsf{D}_n^{-1}u^n_m\mathsf{D}_m$ by Propositions 9.10 and 9.13, we get

$$z_0(\hat{\mathsf{R}}^{-1})^{wk}_{pj}\mathsf{D}_w^{-1}\mathsf{D}_j\mathsf{D}_k^{-1}\mathsf{D}_p = \bar{\mathbf{r}}_{z_0}(u^w_j, u^k_p)\mathsf{D}_w^{-1}\mathsf{D}_j\mathsf{D}_k^{-1}\mathsf{D}_p$$
$$= \bar{\mathbf{r}}_{z_0}(S^2(u^w_j), S^2(u^k_p)) = \bar{\mathbf{r}}_{z_0}(u^w_j, u^k_p) = z_0(\hat{\mathsf{R}}^{-1})^{wk}_{pj}.$$

Moreover, for $\Gamma_{+,z}$ we have $\hat{R}_{\mathbf{r}} = x\hat{\mathsf{R}}$ and $\hat{R}_{\mathbf{s}} = y\hat{\mathsf{R}}$. Inserting these facts and equation (75) into the formulas in Proposition 21 we obtain the above formulas in the case of $\Gamma_{+,z}$. The proof for $\Gamma_{-,z}$ is similar. □

Let us turn to the spectral decomposition of the braiding map $\sigma_\pm$. Since $R_{\mathbf{r}} = (x/y)^{\pm 1}R_{\mathbf{s}}$, the assumption made in Sect. 14.5 is fulfilled and formula (80) gives the spectral decomposition of $\sigma_\pm$ when the spectral decomposition of $\hat{\mathsf{R}}^{\pm 1}$ is known. Recall that $\sigma_\pm$ then has precisely the eigenvalues $\lambda_i^{-1}\lambda_j$, $i,j \in I$, if $\hat{\mathsf{R}}^{\pm 1}$ possesses the eigenvalues λ_i, $i \in I$.

In the case $G_q = GL_q(N), SL_q(N)$ the eigenvalues of $\hat{\mathsf{R}}$ are q and $-q^{-1}$, so $\sigma := \sigma_\pm$ has the eigenvalues $1, -q^2, -q^{-2}$ and satisfies the equation

$$(\sigma - I)(\sigma + q^2 I)(\sigma + q^{-2}I) = 0.$$

For the quantum groups $G_q = O_q(N), Sp_q(N)$ the matrix $\hat{\mathsf{R}}$ has the eigenvalues $q, -q^{-1}, \epsilon q^{\epsilon - N}$, hence $\sigma := \sigma_\pm$ has the eigenvalues $1, -q^2, -q^{-2}, q^N$, $q^{-N}, -q^{\epsilon N-2}, -q^{-\epsilon N+2}$ and we have the equation

$$(\sigma - I)(\sigma + q^2 I)(\sigma + q^{-2}I)(\sigma - q^N I)(\sigma - q^{-N}I)(\sigma + q^{\epsilon N-2}I)(\sigma + q^{-\epsilon N+2}I) = 0.$$

In both cases the spectra of σ_+ and σ_- coincide and for q not a root of unity the operator $\sigma_\pm + I$ is always invertible in contrast to the classical situation.

The last formula shows another peculiarity that appears for the quantum groups $O_q(N)$ and $Sp_q(N)$. Recall that the space of 2-forms for the DC $\Gamma_e^\wedge$ from Theorem 17 is $(\Gamma \otimes_{\mathcal{A}} \Gamma)/\ker(\sigma - I)$. Let Γ denote one of the FODC $\Gamma_\pm$ for $O_p(N)$ or $Sp_q(N)$. In the classical limit $q \to 1$ the eigenspaces of $\sigma_\pm$ for the eigenvalues q^N and q^{-N} belong also to the symmetric part of $\Gamma \otimes_{\mathcal{A}} \Gamma$. Thus, it seems to be natural in this case to take the quotient $\Gamma^{\wedge 2} = (\Gamma \otimes_{\mathcal{A}} \Gamma)/\mathcal{J}_2$ as the space of 2-forms, where $\mathcal{J}_2 := \ker(\sigma - I) \oplus \ker(\sigma - q^N I) \oplus \ker(\sigma - q^{-N} I)$. This definition has the advantage that the dimension of the space of left-invariant 2-forms ${}_{\mathrm{inv}}\Gamma^{\wedge 2}$ is $\binom{N^2}{2}$ just as in the case of an N^2-dimensional *classical* first order calculus. But $\Gamma^\wedge = \bigoplus_n \Gamma^{\wedge n}$ is then no longer an $\mathbb{N}_0$-graded super Hopf algebra, because the comultiplication of $T^s_{\mathcal{A}}(\Gamma)$ does not map $\mathcal{J}_2$ to $\mathcal{J}_2 \otimes \mathcal{A} + \mathcal{A} \otimes \mathcal{J}_2$ when $q^N \neq 1$. Another disadvantage of this definition is that the dimensions of the spaces of higher order left-invariant forms ${}_{\mathrm{inv}}\Gamma^{\wedge n}$, $n \geq 3$, are smaller than in the classical case.

14.6.3 A Canonical Basis for the Left-Invariant 1-Forms

In the preceding subsections we worked with the artificially chosen basis $\{\theta_{ij}\}$ of the vector space ${}_{\mathrm{inv}}(\Gamma_{\pm,z})$. A more natural and intrinsic subset of this space is formed by the projections

$$\omega_{ij} := \omega(u^i_j) \equiv P_L(\mathrm{d}u^i_j) \equiv S(u^i_k)\mathrm{d}u^k_j$$

of the differentials $\mathrm{d}u^i_j$ of the generators u^i_j. The purpose of this subsection is to decide when the set $\{\omega_{ij}\}$ is a basis of the vector space ${}_{\mathrm{inv}}(\Gamma_{\pm,z})$ and to determine the basis transformation $\{\theta_{ij}\} \to \{\omega_{ij}\}$ in this case.

Since $\{\theta_{ij}\}$ is a basis of ${}_{\mathrm{inv}}(\Gamma_{\pm,z})$, there is a complex $N^2 \times N^2$ matrix $A_{\pm,z} = ((A_{\pm,z})^{nm}_{ij})$ such that

$$\omega_{ij} = (A_{\pm,z})^{nm}_{ij}\theta_{nm}.$$

It is clear that the set $\{\omega_{ij}\}$ is a basis of ${}_{\mathrm{inv}}(\Gamma_{\pm,z})$ if and only if the matrix $A_{\pm,z}$ is invertible. We shall use the notation

$$\lambda := q - q^{-1}, \quad \lambda_+ := q + q^{-1} \quad \text{and} \quad B = (B^{nm}_{ij}) := (\mathsf{D}_n \delta_{nm} \delta_{ij}),$$

where as earlier $\mathsf{D}_i = q^{-2i}$ for $G_q = GL_q(N), SL_q(N)$ and $\mathsf{D}_i = q^{2\rho_i}$ for $G_q = O_q(N)$, $Sp_q(N)$.

Case 1 ($G_q = GL_q(N), SL_q(N)$). Let us abbreviate $\mathfrak{s} := \sum_{i=1}^N q^{-2i}$, $\mathfrak{s}_+ := \mathfrak{s} + 1 - q^{-2}$ and $\mathfrak{s}_- := \mathfrak{s} - q^{-2N} + q^{-2N-2}$. The transformations $A_{\pm,z}$ are

$$A_{+,z} = \lambda q^{-1} z I + (z-1)B, \quad A_{-,z} = -\lambda q^{-2N-1} z^{-1} I + (z^{-1} - 1)B.$$

The set $\{\omega_{ij} \mid i,j = 1,2,\cdots,N\}$ is a basis of the vector space ${}_{\mathrm{inv}}(\Gamma_{\pm,z})$ if and only if $z^{\pm 1}\mathfrak{s}_\pm \neq \mathfrak{s}$. If this is fulfilled, the inverses of $A_{\pm,z}$ are given by

$$A_{+,z}^{-1} = qz^{-1}\lambda^{-1}(I+(1-z)\alpha_{+,z}B), A_{-,z}^{-1} = -q^{2N+1}z\lambda^{-1}(I+(1-z^{-1})\alpha_{-,z}B),$$

where $\alpha_{\pm,z} := (z^{\pm 1}\mathfrak{s}_\pm - \mathfrak{s})^{-1}$.

Case 2 ($G_q = O_q(N), Sp_q(N)$). Setting $A_\pm := A_{\pm,1}$ and $\mu := \epsilon q^{N-\epsilon} - \epsilon q^{\epsilon-N}$, we have

$$A_+ = \lambda\epsilon q^{N-\epsilon}I - \lambda(\mathsf{C}^{-1})_1^t\hat{\mathsf{R}}\mathsf{C}_1^t, \quad A_- = -\lambda\epsilon q^{N-\epsilon}I - 2B + \lambda(\mathsf{C}^{-1})_1^t\hat{\mathsf{R}}\mathsf{C}_1^t.$$

The set $\{\omega_{ij} \mid i,j = 1,2,\cdots,N\}$ is a vector space basis of ${}_{\rm inv}\Gamma_+$, resp. ${}_{\rm inv}\Gamma_-$, if and only if $q^{2N-2\epsilon} \neq 1$, resp. $2+\mu\lambda+2\mu\lambda^{-1} \neq 0$. In these cases the inverses $A_\pm^{-1}$ are

$$\begin{aligned} A_+^{-1} &= \gamma((\epsilon q^{N-\epsilon} - \lambda)I + (\mathsf{C}^{-1})_1^t\hat{\mathsf{R}}\mathsf{C}_1^t - \lambda\alpha_+ B), \\ A_-^{-1} &= -\gamma((\epsilon q^{N-\epsilon} - \lambda)I + (\mathsf{C}^{-1})_1^t\hat{\mathsf{R}}\mathsf{C}_1^t + (\lambda - \lambda_+^2\epsilon q^{N-\epsilon}\alpha_-)B), \end{aligned}$$

where

$$\alpha_+ := \mu^{-1}\epsilon q^{\epsilon-N}, \quad \alpha_- = (2+\mu\lambda+2\mu\lambda^{-1})^{-1}, \quad \gamma := \epsilon q^{\epsilon-N}\lambda^{-1}(\mu-\lambda)^{-1}.$$

We close this subsection by considering the two FODC over $\mathcal{O}(GL_q(N))$ which are obtained from the family $\Gamma_{\pm,z}$ in the particular case $z = 1$.

Example 8 (*The calculi $\Gamma_{+,1}$ and $\Gamma_{-,1}$ for $GL_q(N)$*). Let $G_q = GL_q(N)$ and $z = 1$. In this case (83) holds with θ_{ij} replaced by ω_{ij}, because the above transformation $A_{\pm,1}$ is a complex multiple of I. Therefore, we have

$$\begin{aligned} {\rm d}u_j^i \cdot u_s^r &= u_n^i\omega_{nj}u_s^r = u_n^i u_m^r(\hat{\mathsf{R}}^{\pm1})_{kx}^{nm}(\hat{\mathsf{R}}^{\pm})_{js}^{kl}\omega_{xl} \\ &= (\hat{\mathsf{R}}^{\pm1})_{nm}^{ir}u_k^n u_x^m(\hat{\mathsf{R}}^{\pm1})_{js}^{kl}S(u_y^x){\rm d}u_l^y = (\hat{\mathsf{R}}^{\pm1})_{nm}^{ir}u_k^n{\rm d}u_l^m(\hat{\mathsf{R}}^{\pm1})_{js}^{kl} \end{aligned} \tag{88}$$

or in matrix notation

$$\mathrm{d}\mathbf{u}_1 \cdot \mathbf{u}_2 = \hat{\mathsf{R}}_{12}^{\pm1}\mathbf{u}_1 \cdot \mathrm{d}\mathbf{u}_2\hat{\mathsf{R}}_{12}^{\pm1}.$$

Thus for both calculi $\Gamma_{\pm,1}$ we have *linear* commutation relations between the matrix entries u_s^r and their differentials ${\rm d}u_j^i$. For any higher order DC of $\Gamma_{\pm,1}$ it follows from (88) by applying d that

$$\mathrm{d}\mathbf{u}_1 \wedge \mathrm{d}\mathbf{u}_2 = -\hat{\mathsf{R}}_{12}^{\pm1}\mathrm{d}\mathbf{u}_1 \wedge \mathrm{d}\mathbf{u}_2\hat{\mathsf{R}}_{12}^{\pm1}.$$

Recall that by Proposition 9.11 the quantum space $\mathcal{O}(\mathbb{C}_q^N)$ can be considered as a subalgebra of $\mathcal{O}(GL_q(N))$ by identifying the generators x_i and u_1^i. Since $(\hat{\mathsf{R}}^{\pm1})_{11}^{kl} = q^{\pm1}\delta_{k1}\delta_{l1}$, under this identification (88) yields

$$\mathrm{d}x_i \cdot x_j = q^{\pm1}(\hat{\mathsf{R}}^{\pm1})_{nm}^{ij}x_n\mathrm{d}x_m.$$

That is, on the subalgebra $\mathcal{O}(\mathbb{C}_q^N)$ the FODC $\Gamma_{\pm,1}$ of $GL_q(N)$ induces the left-covariant FODC $\Gamma_\mp$ from Theorem 12.6. △

14.6.4 Classification of Bicovariant First Order Differential Calculi

It might be natural to find all bicovariant FODC over $\mathcal{A}$ for which the differentials du^i_j of the matrix entries u^i_j generate the left module of 1-forms. The "ordinary" differential calculus on the coresponding Lie groups obviously possesses this property. We first give some equivalent formulations of this condition and then we state the classification theorem.

Lemma 23. *For any left-covariant FODC Γ over $\mathcal{A}$ the following three conditions are equivalent:*

(i) $\Gamma = \mathrm{Lin}\,\{a\,du^i_j \mid a \in \mathcal{A},\ i,j = 1,2,\cdots,N\}$,

(ii) ${}_{\mathrm{inv}}\Gamma = \mathrm{Lin}\,\{\omega_{ij} \mid i,j = 1,2,\cdots,N\}$,

(iii) $\ker\varepsilon = \mathcal{R}_\Gamma + \mathrm{Lin}\,\{u^i_j - \delta_{ij}1 \mid i,j = 1,2,\cdots,N\}$.

Proof. (i)$\rightarrow$(ii) is clear, since $P_L(a du^i_j) = \varepsilon(a)P_L(du^i_j) = \varepsilon(a)\omega_{ij}$ by (13.3).

(ii)$\rightarrow$(iii): Let $a \in \ker\varepsilon$. By (ii), $\omega_\Gamma(a) = \alpha^{ij}\omega_{ij}$ with $\alpha^{ij} \in \mathbb{C}$. Thus we have $\omega_\Gamma(a - \alpha^{ij}(u^i_j - \delta_{ij}1)) = 0$ and so $a - \alpha^{ij}(u^i_j - \delta_{ij}1) \in \mathcal{R}_\Gamma$, which proves (iii).

(iii)$\rightarrow$(i): Let $a \in \mathcal{A}$. By (iii), $\bar{a}$ is of the form $\bar{a} = \alpha^{kj}(u^k_j - \delta_{kj}1) + x$ with $\alpha^{kj} \in \mathbb{C}$ and $x \in \mathcal{R}_\Gamma$. Thus, $\omega_\Gamma(a) = \omega_\Gamma(\bar{a}) = \alpha^{kj}\omega_{kj} = \alpha^{kj}S(u^k_i)du^i_j$ and hence $\Gamma = \mathcal{A}\,\omega_\Gamma(\mathcal{A}) = \mathrm{Lin}\,\{a\,du^i_j\}$. □

Theorem 24. *Suppose that q is not a root of unity. Then there exist up to isomorphism precisely the following bicovariant FODC Γ over $\mathcal{A}$ such that* $\dim\,\Gamma \geq 2$ *and* $\Gamma = \mathrm{Lin}\,\{a\,du^i_j \mid a \in \mathcal{A},\ i,j = 1,2,\cdots,N\}$*:*

- $\mathcal{A} = \mathcal{O}(GL_q(N)), N{\geq}3:\ \Gamma_{\pm,z}$ *with* $z^{\pm1}\mathfrak{s}_\pm \neq \mathfrak{s}, z \in \mathbb{C}\backslash\{0\}$;
- $\mathcal{A} = \mathcal{O}(GL_q(2)):\ \Gamma_{+,z}$ *with* $z^{-1}(1+q^2) \neq 1+q^4,\ z \in \mathbb{C}\backslash\{0\}$;
- $\mathcal{A} = \mathcal{O}(SL_q(N)),\ N \geq 3:\ \Gamma_{\pm,z}$ *with* $z^{\pm1}\mathfrak{s}_\pm \neq \mathfrak{s},\ z^N = q^{-2}$;
- $\mathcal{A} = \mathcal{O}(SL_q(2))$ *and* $\mathcal{A} = \mathcal{O}(Sp_q(2)):\ \Gamma_+$ *and* Γ_-;
- $\mathcal{A} = \mathcal{O}(O_q(N))$ *and* $\mathcal{A} = \mathcal{O}(Sp_q(N)),\ N \geq 4:\ \Gamma_+$ *and* Γ_- *with* $2 + \nu\lambda + 2\nu\lambda^{-1} \neq 0$.

Two FODC over $\mathcal{A}$ of this list with different parameters are not isomorphic.

For the quantum group $O_q(3)$ under the assumptions of Theorem 24 there are precisely four such bicovariant FODC. These are the two 9-dimensional FODC $\Gamma_\pm$ from Sect. 14.6.1 and two additional 4-dimensional FODC developed in [SS3]. Theorem 24 and the preceding assertion show in particular that under the above assumptions for the quantum groups $SL_q(N)$, $O_q(N)$ and $Sp_q(N)$ there is no *bicovariant* FODC whose dimension coincides with the dimension of the corresponding classical Lie group. Further, all calculi occurring in Theorem 24 have dimension N^2 and belong to the family of FODC $\Gamma_{\pm,z}$ which was constructed by the general method of Sect. 14.5.

The conditions $z^{\pm1}\mathfrak{s}_\pm \neq \mathfrak{s}$ for $GL_q(N)$ and $SL_q(N), N{\geq}3$, and $2 + \mu\lambda + 2\mu\lambda^{-1} \neq 0$ for $O_q(N)$ and $Sp_q(N)$, $N \geq 4$, are only needed to ensure that the corresponding FODC $\Gamma_{\pm,z}$ satisfy the assumption $\Gamma = \mathrm{Lin}\{a\, du^i_j\}$ or equivalently ${}_{\mathrm{inv}}\Gamma = \mathrm{Lin}\{\omega_{ij}\}$ (see Subsect. 14.6.3). If N is fixed, then except for the quantum group $GL_q(N)$ there exist only finitely many q for which these conditions apply. For all other values of q the list in Theorem 24 contains exactly $2N$ FODC for $SL_q(N)$, $N \geq 3$, and 2 FODC for $O_q(N)$ and $Sp_q(N)$, $N \geq 4$.

14.7 Notes

The poineering work on covariant differential calculi on quantum groups is due to S. L. Woronowicz [Wor2], [Wor4]. Many results (but not all) in Sects. 14.1–3 appeared in the fundamental paper [Wor4]. The 3D-calculus in [Wor2] was the first example of a noncommutative differential calculus on quantum groups. The short approach given in Subsect. 14.1.3 follows the paper [SS4]. The results in Subsects. 14.3.2 and 14.3.3 on higher order left-covariant calculi are also from [SS4]. Proposition 11 is taken from [BM1].

The higher order bicovariant calculus $\Gamma^\wedge_e$ was obtained in [Wor4]. The fact that it yields a differential Hopf algebra was first observed in [Brz], but the proof therein works only for $\Gamma^\wedge_{e,2}$. Complete proofs have been provided since then by [Dt1], one of the authors (K.S.) and probably others. Note that our proof in the text was based on Proposition 13.10. Theorem 18 seems to be new.

Quantum Lie derivatives and contraction operators appeared first in [SWZ1] and [AC], but a complete proof of Proposition 19 seems to be still missing in the literature. The method for the construction of bicovariant differential calculi developed in Sect. 14.5 is new at least in the general setting of coquasitriangular Hopf algebras. In the special situation of Subsect. 14.6.1 it gives the procedure used in [Jur] and [CSWW], where bicovariant calculi for the standard quantized simple Lie groups have been constructed in this manner. The two bicovariant calculi in Example 8 of Subsect. 14.6.2 have been found in [Man1] and also in [Mal]. Theorem 24 was proved in [SS2] and [SS3].

There is now an extensive literature on differential calculi on quantum groups. In addition to the papers mentioned above, an (incomplete) list includes [Bd], [BMDS], [Ct], [CW], [DSWZ], [HS], [IV], [MHR], [Sud2], [Sud3], [SWZ2], [Ts].

Bibliography

1. Books

[A] Abe, E.: *Hopf Algebras*, Cambridge Univ. Press, Cambridge, 1980

[Bax] Baxter, R. J.: *Exactly Solved Models in Statistical Mechanics*, Academic Press, New York, 1982

[BL] Biedenharn, L. C., Lohe, M. A.: *Quantum Group Symmetry and q-Tensor Algebras*, World Scientific, Singapore, 1995

[BM] Berger, J., Montgomery, S. (eds.): *Advances in Hopf Algebras*, Marcel Dekker, New Jork, 1996

[Bou1] Bourbaki, N.: *Algébre Commutative*, Hermann, Paris, 1961

[Bou2] Bourbaki, N.: *Groupes et Algébres de Lie, Chaps. 4, 5 et 6*, Herman, Paris, 1968

[BR] Bratteli, O., Robinson, D. W.: *Operator Algebras and Quantum Statistical Mechanics*, Springer, New York, 1979

[BPZ] Budzyński, R., Pusz, W., Zakrewski, S. (eds.): *Quantum Groups and Quantum Spaces*, Banach Center Publications, vol. 40, Warsaw, 1997

[Con] Connes, A.: *Non-Commutative Geometry*, Academic Press, New York, 1994

[CP] Chari, V., Pressley, A. N.: *A Guide to Quantum Groups*, Cambridge Univ. Press, Cambridge, 1994

[CR] Curtis, Ch. W., Reiner, I.: *Representation Theory of Finite Groups and Associative Algebras*, Wiley, New York, 1962

[DH] Doebner, H.-D., Hennig, J.-D. (eds.): *Quantum Groups*, Lecture Notes Phys. **370**, Springer, Berlin, 1989

[Dix] Dixmier, J.: *Enveloping Algebras*, North-Holland, Amsterdam, 1977

[ES] Enock, M., Schwartz, J. M.: *Kac Algebras and Duality of Locally Compact Groups*, Springer, Berlin, 1992

[FK] Fröhlich, J., Kerler, T.: *Quantum Groups, Quantum Categories and Quantum Field Theory*, Lecture Notes Math. **1542**, Springer, Berlin, 1993

[GR] Gasper, G., Rahman, M.: *Basic Hypergeometric Series*, Cambridge Univ. Press, Cambridge, 1990

[Ge] Ge, Mo-Lin (ed.): *Quantum Groups and Quantum Integrable Systems*, World Scientific, Singapore, 1992

[GV] Ge, Mo-Lin, de Vega, H. J. (eds.): *Quantum Groups, Integrable Statistical Models and Knot Theory*, World Scientific, Singapore, 1993

[GS] Gerstenhaber, M., Stasheff, J. (eds.): *Deformation Theory and Quantum Groups with Applications to Mathematical Physics*, Amer. Math. Soc., Providence, RI, 1992

[GLP] Gielerak, R., Lukierski, J., Popowicz, Z. (eds.): *Quantum Groups and Related Topics*, Kluwer, Dordrecht, 1992

[Gr] Gruber, B. (ed.): *Symmetries in Science VI. From the Rotation Group to Quantum Algebras*, Plenum Press, New York, 1993

[Ho] Hochschild, G.: *Basic Theory of Algebraic Groups and Lie Algebras*, Springer, New York, 1981

[Hum] Humphreys, J. E.: *Introduction to Lie Algebras and Representation Theory*, Springer, New York, 1972

[J] Jimbo, M. (ed.): *Yang–Baxter Equation in Integrable Systems*, World Scientific, Singapore, 1989

[Jan] Jantzen, J. C.: *Lectures on Quantum Groups*, Amer. Math. Soc., Providence, RI, 1996

[Jos] Joseph, A.: *Quantum Groups and Their Primitive Ideals*, Springer, Berlin, 1995

[Ka] Kastler, D.: *Cyclic Cohomology within the Differential Envelope*, Hermann, Paris, 1988

[Kas] Kassel, Ch.: *Quantum Groups*, Springer, New York, 1994

[Ku] Kulish, P. P. (ed.): *Quantum Groups*, Lecture Notes Math. **1510**, Springer, Berlin, 1992

[Lus] Lusztig, G.: *Introduction to Quantum Groups*, Birkhäuser, Boston, 1993

[LV] LePourneux, J., Vinet, L. (eds.): *Quantum Groups, Integrable Models and Statistical Systems*, World Scientific, Singapore, 1993

[Mac] Macdonald, I. G.: *Symmetric Functions and Hall Polynomials*, Clarendon Press, Oxford, 1979

[Mad] Madore, J.: *An Introduction to Noncommutative Differential Geometry and its Physical Applications*, Cambridge Univ. Press, Cambridge, 1995

[Maj] Majid, S.: *Foundations of Quantum Group Theory*, Cambridge Univ. Press, Cambridge, 1995

[Man] Manin, Yu. I.: *Quantum Groups and Non-Commutative Geometry*, CRM, Montreal, 1988

[Mon] Montgomery, S.: *Hopf Algebras and Their Actions on Rings*, Amer. Math. Soc., Providence, RI, 1993

[Mu] Murphy, G. J.: *C^*-Algebra and Operator Theory*, Academic Press, Boston, 1990

[PW] Parshall, B., Wang, J.: *Quantum Linear Groups*, Memoirs Amer. Math. Soc. **439**, Providence, RI, 1991

[Pit] Pittner, L.: *Algebraic Foundations of Non-Commutative Differential Geometry and Quantum Groups*, Lecture Notes Phys. **M39**, Springer, Berlin, 1996

[RS] Reed, M., Simon, B.: *Methods of Modern Mathematical Physics*, vol. 1, Academic Press, New York, 1983

[Rie] Rieffel, M.: *Quantum Deformations for Actions of $\mathbb{R}^d$*, Memoirs Amer. Math. Soc. **506**, Providence, RI, 1993

[Sc] Schmüdgen, K.: *Unbounded Operator Algebras and Representation Theory*, Birkhäuser, Basel, 1990

[Ser] Serre, J.-P.: *Lie Algebras and Lie Groups*, Benjamin, Reading, MA, 1965

[ShS] Shnider, S., Sternberg, S.: *Quantum Groups*, International Press, Boston, 1993

[Sw] Sweedler, M. E.: *Hopf Algebras*, Benjamin, New York, 1969

[Tit] Titchmarsh, E.: *The Theory of Functions*, Oxford Univ. Press, 1939

[Tur] Turaev, V. G.: *Quantum Invariants of Knots and 3-Manifolds*, W. de Gruyter, Berlin, 1994

[VK1] Vilenkin, N. Ya., Klimyk, A. U.: *Representation of Lie Groups and Special Functions*, vols. 1–3, Kluwer, Dordrecht, 1991–1993

[VK2] Vilenkin, N. Ya., Klimyk, A. U.: *Representation of Lie Groups and Special Functions: Recent Advances*, Kluwer, Dordrecht, 1995

[Wa] Warner, F.: *Foundations of Differentiable Manifolds and Lie Groups*, Springer, New York, 1983

[Wey] Weyl, H.: *The Classical Groups*, Princeton Univ. Press, Princeton, 1939

[Zhe] Zhelobenko, D. P.: *Compact Lie Groups and Their Representations*, Amer. Math. Soc., Providence, RI, 1973

2. Articles

[AlC] Alev, J., Chamarie, M.: Dérivations et Automorphismes de quelques algèbres quantique, *Commun. Algebra* **20** (1992), 1787–1802

[AlS] Ališauskas, S., Smirnov, Yu. F.: Multiplicity-free $u_q(n)$ coupling coefficients, *J. Phys. A* **27** (1994), 5925–5939

[AGS] Alvarez-Gaumé, L., Gomez, C., Sierra, G.: Duality and quantum groups, *Nucl. Phys. B* **330** (1990), 347–398

[APW] Andersen, H., Polo, P., Wen Kexin: Representations of quantum algebras, *Invent. Math.* **104** (1991), 1–59

[An] Andruskiewitsch, N.: Some exceptional compact matrix pseudogroups, *Bull. Soc. Math. France* **120** (1992), 297-325

[AS] Apel, J., Schmüdgen, K.: Classification of three-dimensional covariant differential calculi on Podleś quantum spheres and on related spaces, *Lett. Math. Phys.* **32** (1994), 25–36

[ACh1] Arnaudon, D., Chakrabarti, A.: Flat periodic representations of $U_q(\mathfrak{G})$, *Commun. Math. Phys.* **139** (1991), 605–617

[ACh2] Arnaudon, D., Chakrabarti, A.: Periodic and partially periodic representations of $SU(N)_q$, *Commun. Math. Phys.* **139** (1991) 461–478

[Ar] Artin, E.: Theory of braids, *Ann. Math.* **48** (1947), 101–126

[AST] Artin, M., Schelter, W., Tate, J.: Quantum deformations of GL_n, *Commun. Pure Appl. Math.* **44** (1991), 879-895

[AC] Aschieri, P., Castellani, L.: An introduction to noncommutative differential geometry on quantum groups, *Intern. J. Modern Phys. A* **8** (1993), 1667–1706

[B] Baaj, S.: Représentation reguliere du groupe quantique des deplacements de Woronowicz, *Astérisque* **232** (1994), 11–48

[BS] Baaj, S., Skandalis, G.: Unitaries multiplicatifs et dualité pour les produits croisés de C^*-algébres, *Ann. Sci. École Norm. Sup.* **26** (1993), 425–488

[BK] Bazhanov, V. V., Kashaev, R. M.: Cyclic L-operators related with 3-state R-matrix, *Commun. Math. Phys.* **136** (1991), 607–623

[BSt] Bazhanov, V. V., Stroganov, Yu. G.: Chiral Potts models as a descendant of the six-vertex model, *J. Stat. Phys.* **51** (1990), 799–817

[Bec] Beck, J.: Representations of quantum groups at even roots of unity, *J. Algebra* **107** (1994), 29–57

[Ber] Bergman, G. M.: The diamond lemma for ring theory, *Adv. Math.* **29** (1978), 178–218

[Bd] Bernard, D.: Quantum Lie algebras and differential calculus on quantum groups, *Progr. Theor. Phys. Suppl.* **102** (1990), 49–66

[Bes] Bespalov, Yu. N.: Crossed modules and quantum groups in braiding categories, *Applied Categorical Structures* **5** (1997), 155–204

[Bid] Biedenharn, L. C.: The quantum group $SU_q(2)$ and a q-analogue of the boson operators, *J. Phys. A* **22** (1989), L873–L878

[BT] Biedenharn, L. C., Tarlini, M.: On q-tensor operators for quantum groups, *Lett. Math. Phys.* **20** (1990), 271–278

[BW] Birman, J. S., Wenzl, H.: Braids, links and a new algebra, *Trans. Amer. Math. Soc.* **313** (1989), 239–245

[BCM] Blattner, R. J., Cohen, M., Montgomery, S.: Crossed products and inner actions of Hopf algebras, *Trans. Amer. Math. Soc.* **298** (1986), 671–711

[BDKL] Bonatsos, D., Daskaloyannis, C., Kolokotronis, P., Lenis, D.: Quantum algebras in nuclear structure, *Preprint*, Trento, Italy, 1995

[BFGP] Bonneau, P., Flato, M., Gerstenhaber, M., Pinczon, G.: The hidden group structure of quantum groups: strong duality, rigidity and preferred deformations, *Commun. Math. Phys.* **161** (1994), 125–156

[BKO] Borowiec, A., Kharchenko, V.K., Oziewicz, Z.: On free differentials on associative algebras, in *Non-Associative Algebras and its Applications* (S.Gonzalez, ed.), Kluwer, Dordrecht, 1994, pp. 46–53

[Br1] Bragiel, K.: On the Wigner–Eckart theorem for tensor operators connected with compact matrix quantum groups, *Lett. Math. Phys.* **21** (1991), 181–191

[Br2] Bragiel, K.: On the spherical and zonal functions on the compact quantum groups, *Lett. Math. Phys.* **22** (1991), 195–202

[Bra] Brauer, R.: On algebras which are connected with the semisimple continuous groups, *Ann. Math.* **38** (1937), 857–872

[BMDS] Bresser, K., Müller-Hoissen, F., Dimakis, A., Sitarz, A.: Noncommutative geometry of finite groups, *J. Phys. A* **29** (1996), 2705-2735

[Brz] Brzesinski, T.: Remarks on bicovariant differential calculi and exterior Hopf algebras, *Lett. Math. Phys.* **27** (1993), 287–300

[BM1] Brzezinski, T., Majid, S.: A class of bicovariant differential calculi on Hopf algebras, *Lett. Math. Phys.* **26** (1992), 67–78

[BM2] Brzesinski, T., Majid, S.: Quantum group gauge theory on quantum spaces, *Commun. Math. Phys.* **157** (1993), 591–638

[BHV] Burdik, C., Havliček, M., Vancura, A.: Irreducible highest weight representations of quantum group $U_q(gl(n,\mathbb{C}))$, *Commun. Math. Phys.* **148** (1992), 417–423

[Bur] Burroughs, N.: Relating the approaches to quantized algebras and quantum groups, *Commun. Math. Phys.* **133** (1990), 91–117

[CSW] Carow-Watamura, U., Schlieker, M., Watamura, S.: $SO_q(N)$ covariant differential calculus on quantum space and quantum deformation of Schrödinger equation, *Z. Phys. C* **49** (1991), 439–446

[CSWW] Carow-Watamura, U., Schlieker, M., Watamura, S, Weich, W.: Bicovariant differential calculus on quantum groups $SU_q(N)$ and $SO_q(N)$, *Commun. Math. Phys.* **142** (1991), 605–641

[CW] Carow-Watamura, U., Watamura, S.: Complex quantum group, dual algebra, and bicovariant differential calculus, *Commun. Math. Phys.* **151** (1993), 487–514

[Ct] Castellani, L.: Gauge theories of quantum groups, *Phys. Lett.* B **292** (1992), 93-98

[CGST] Celeghini, E., Giachetti, R., Sorace, E., Tarlini, M.: The quantum Heisenberg group $H(1)_q$, *J. Math. Phys.* **32** (1991), 1155–1158

[CP1] Chari, V., Pressley, A. N.: Quantum affine algebras, *Commun. Math. Phys.* **142** (1991), 261–283

[CP2] Chari, V., Pressley, A. N.: Minimal cyclic representations of quantum groups at root of 1, *C. R. Acad. Sci. Paris, Ser. I* **313** (1991), 429–434

[CP3] Chari, V., Pressley, A. N.: Representation of quantum $SO(8)$ and related quantum algebras, *Commun. Math. Phys.* **159** (1994), 29–49

[ChK] Chung, Won-Sang, Klimyk, A. U.: On position and momentum operators in the q-oscillator algebra, *J. Math. Phys.* **37** (1996), 917–932

[CQ] Cuntz, J., Quillen, D.: Algebraic extensions and nonsingularity, *J. Amer. Math. Soc.* **8** (1995), 251–289

[DKu] Damaskinskii, E. V., Kulish, P. P.: Deformed oscillators and their applications, *Zap. Nauchn. Sem. LOMI* **189** (1991), 37–74

[DJM] Date, M., Jimbo, M., Miwa, T.: Representations of $U_q(\hat{gl}(n,\mathbb{C}))$ at $q = 0$ and the Robinson–Schensted correspondence, in *Physics and Mathematics of Strings* (L. Brink, D. Frieden and A. M. Polyakov, eds.), World Scientific, Singapore, 1990, pp. 185–211

[DJMM1] Date, M., Jimbo, M., Miki, K., Miwa, T.: Generalized chiral Potts models and minimal cyclic representations of $U_q(\hat{gl}(n,\mathbb{C}))$, *Commun. Math. Phys.* **137** (1991), 133–147

[DJMM2] Date, M., Jimbo, M., Miki, K., Miwa, T.: Cyclic representations of $U_q(sl(n+1))$ at $q^N = 1$, *Publ. RIMS Kyoto Univ.* **27** (1991), 347–366

[DCK] De Concini, C., Kac, V. G.: Representations of quantum groups at root of 1, in *Operator Algebras, Unitary Representations, Enveloping Algebras and Invariant Theory* (A. Connes, M. Duflo and R. Rentschler, eds.), Birkhäuser, Boston, 1990, pp. 472–506

[DCKP] De Concini, C., Kac, V. G., Procesi, C.: Quantum coadjoint action, *J. Amer. Math. Soc.* **5** (1992), 151–189

[DCL] De Concini, C., Lyubashenko, V.: Quantum function algebras at roots of 1, *Adv. Math.* **108** (1995), 205–262

[DCP] De Concini, C., Procesi, C.: Quantum Groups, *Lecture Notes Math.* **1565**, Springer, Berlin, 1993, pp. 31–140

[DV] De Vega, H. J.: Yang–Baxter algebras, integrable systems and quantum groups, *Intern. J. Modern Phys. A* **4** (1989), 2371–2463

[Dij1] Dijkhuizen, M. S.: The double covering of the quantum group $SO_q(3)$, *Rend. Circ. Math. Palermo* Serie II **37** (1994), 47–57

[Dij2] Dijkhuizen, M. S.: Some remarks on the construction of quantum symmetric spaces, *Acta Appl. Math.* **44** (1996), 59–80

[DK1] Dijkhuizen, M. S., Koornwinder, T. H.: CQG algebras: A direct algebraic approach to compact quantum groups, *Lett. Math. Phys.* **32** (1994), 315–330

[DK2] Dijkhuizen, M. S., Koornwinder, T. H.: Quantum homogeneous spaces, duality and quantum 2-spheres, *Geom. Dedicata* **52** (1994), 291–315

[DN] Dijkhuizen, M. S., Noumi, M.: A family of quantum projective spaces and related q-hypergeometric orthogonal polynomials, *Trans. Amer. Math. Soc.*

[DF] Ding, J., Frenkel, I. B.: Isomorphism of two realizations of quantum affine algebra $U_q(\hat{gl}(n))$, *Commun. Math. Phys.* **156** (1993), 277–300

[DD] Dipper, R., Donkin, S.: Quantum GL_n, *Proc. London Math. Soc.* **63** (1991), 165–211

[Dob1] Dobrev, V. K.: Singular vectors of representations of quantum groups, *J. Phys. A* **25** (1992), 149–160

[Dob2] Dobrev, V. K.: Duality for the matrix quantum group $GL_{p,q}(2,\mathbb{C})$, *J. Math. Phys.* **33** (1992), 3419–3426

[Doi] Doi, Y.: Braided bialgebras and quadratic bialgebras, *Commun. Algebra* **21** (1993), 1731–1749

[DT1] Doi, Y., Takeuchi, M.: Cleft comodule algebras for a bialgebra, *Commun. Algebra* **14** (1986), 801–818

[DT2] Doi, Y., Takeuchi, M.: Multiplication alteration by two-cocycles, *Commun. Algebra* **22** (1994), 5715–5732

[Dt1] Drabant, B.: Hopf bimodules and differential calculus, Preprint, University of Amsterdam, 1994

[Dt2] Drabant, B.: Braided bosonization and inhomogeneous quantum groups, *Acta Appl. Math.* **44** (1996), 117–132

[DSWZ] Drabant, B., Schlieker, M., Weich, W., Zumino, B.: Complex quantum groups and their quantum universal enveloping algebras, *Commun. Math. Phys.* **147** (1992), 625–633

[Dr1] Drinfeld, V. G.: Hopf algebras and the quantum Yang–Baxter equation, *Soviet Math. Dokl.* **32** (1985), 254-258

[Dr2] Drinfeld, V. G.: Quantum groups, in *Proceedings of the International Congress of Mathematicians* (A. M. Gleason, ed.), Amer. Math. Soc., Providence, RI, 1986, pp. 798–820

[Dr3] Drinfeld, V. G.: A new realization of Yangians and quantized affine algebras, *Soviet Math. Dokl.* **41** (1988), 212–216

[Dr4] Drinfeld, V. G.: On almost cocommutative Hopf algebras, *Leningrad Math. J.* **1** (1989), 321–342

[Dr5] Drinfeld, V.G.: Quasi-Hopf algebras, *Leningrad Math. J.* **1** (1990), 1419–1457

[ER] Effros, E., Ruan, Z.-J.: Dicrete quantum groups, I. The Haar measure, *Intern. J. Math.* **5** (1994), 681–723

[F1] Faddeev, L. D.: Integrable models in (1+1)-dimensional quantum field theory, in *Recent Advances in Field Theory and Statistical Mechanics* (Zuber, J.-B. and R. Stora, eds.), North-Holland, Amsterdam, 1984, pp. 561–608

[F2] Faddeev, L. D.: From integrable models to conformal field theory via quantum groups, in *Integrable Systems, Quantum Groups, and Quantum Field Theory* (L. A. Ibort and M. A. Rodriguez, eds.), Kluwer, Dordrecht, 1993, pp. 1–24

[FL] Fateev, V. A., Lukyanov, S. L.: Vertex operators and representations of quantum universal enveloping algebras, *Intern. J. Modern Phys. A* **7** (1992), 1325–1359

[Fe] Feinsilver, P.: Elements of q-harmonic analysis, *J. Math. Anal. Appl.* **141** (1989), 509–526

[FT] Feng, P., Tsygan, B.: Hochschild and cyclic homology of quantum groups, *Commun. Math. Phys.* **140** (1991), 481–521

[FS] Flato, M., Sternheimer, D.: On a possible origin of quantum groups, *Lett. Math. Phys.* **22** (1991), 155–160

[Fl] Floris, P. G. A.: Gel'fand pair criteria for compact quantum groups, *Indag. Math.* **6** (1995), 83–98

[FGP] Furlan, P., Ganchev, A. Ch., Petkova, V. B.: Quantum groups and fusion rules multiplicities, *Nucl. Phys. B* **343** (1990), 205–227

[Gav] Gavrilik A. M.: q-Serre relations in $U_q(u_n)$ and q-deformed meson mass sum, *J. Phys. A* **27** (1994), L91–L94

[GZ] Giaquinto, A., Zhang, I. J.: Quantum Weyl algebras, *J. Algebra* **176** (1995), 861–881

[GoS] Gomez, C., Sierra, G.: The quantum group symmetry of rational conformal field theories, *Nucl. Phys. B* **352** (1991), 791–828

[Gou] Gould, M. D.: Reduced Wigner coefficients for $U_q(gl(n))$, *J. Math. Phys.* **33** (1992), 1023–1031

[GLB] Gould, M. D., Links, J., Bracken, A. J.: Matrix elements and Wigner coefficients for $U_q(sl(n))$, *J. Math. Phys.* **33** (1992), 1008–1022

[GL] Grojnowski, I., Lusztig, G.: A comparison of bases of quantized enveloping algebras, *Contemp. Math.* **153** (1993), 11–19

[GKK] Groza, V. A., Kachurik, I. I., Klimyk, A. U.: On Clebsch–Gordan coefficients and matrix elements of the quantum algebra $U_q(su_2)$, *J. Math. Phys.* **31** (1990), 2769–2780

[Gur] Gurevich, D. I.: Algebraic aspects of the quantum Yang–Baxter equation, *Leningrad Math. J.* **2** (1991), 801–828

[HKW] Harpe, P., Kervaire, M., Weber, C.: On the Jones polynomial, *Enseign. Math.* **32** (1986), 271–335

[H1] Hayashi, T.: q-Analogue of Clifford and Weyl algebras – spinor and oscillator representations of quantum enveloping algebras, *Commun. Math. Phys.* **127** (1990), 129–144

[H2] Hayashi, T.: Quantum groups and quantum determinants, *J. Algebra* **152** (1992), 146–165

[H3] Hayashi, T.: Quantum deformations of classical groups, *Publ. RIMS Kyoto Univ.* **28** (1992), 57–81

[H4] Hayashi, T.: Non-existence of homomorphisms between quantum groups, *Tokyo J. Math.* **15** (1992), 431–435

[HSSWW] Hebecker, A., Schreckenberg, S., Schwenk, J. Weich, W., Wess, J.: Representations of a q-deformed Heisenberg algebra, *Z. Phys. C* **64** (1994), 355–359

[HS] Heckenberger, I., Schmüdgen, K.: Levi–Civita connections on the quantum groups $SL_q(N), O_q(N)$ and $Sp_q(N)$, *Commun. Math. Phys.* **182** (1997), 1–20

[Hi] Hietarinta, J.: All solutions to the constant quantum Yang–Baxter equation in two dimensions, *Phys. Lett A* **165** (1992), 245–251

[HR] Hinrichsen, H., Rittenberg, V.: Quantum groups, correlation functions and infrared divergences, *Phys. Lett. B* **304** (1993), 115–120

[Hop] Hopf, H.: Über die Topologie der Gruppenmannigfaltigkeiten und ihre Verallgemeinerungen, *Ann. Math.* **42** (1941), 22–52

[Is1] Isaev, A. P.: Interrelation between quantum groups and reflection equation (braided) algebras, *Lett. Math. Phys.* **34** (1995), 333–341

[Is2] Isaev, A. P.: Quantum groups and Yang–Baxter equations, *Phys. Part. Nucl.* **26** (1995), 501–526

[IV] Isaev, A. P., Vladimirov, A. A.: $GL_q(n)$-Covariant braided differential bialgebras, *Lett. Math. Phys.* **33** (1995), 297–303

[Jk] Jackson, F. H.: q-Difference equations, *Amer. J. Math.* **32** (1910), 305–314

[Jim1] Jimbo, M.: A q-analogue of $U(\mathfrak{g})$ and the Yang–Baxter equation, *Lett. Math. Phys.* **10** (1985), 63–69

[Jim2] Jimbo, M.: A q-analogue of $U(gl(N+1))$, Hecke algebra and the Yang–Baxter equation, *Lett. Math. Phys.* **11** (1986), 247–252

[Jim3] Jimbo, M.: Quantum R-matrix related to the generalized Toda system: an algebraic approach, *Lecture Notes Phys.* **246**, Springer, Berlin, 1986, pp. 335–361

[Jim4] Jimbo, M.: Quantum R-matrix for the generalized Toda system, *Commun. Math. Phys.* **102** (1986), 537–547

[Jim5] Jimbo, M.: Introduction to the Yang–Baxter equation, *Intern. J. Modern Phys.* **4** (1989), 3759–3777

[JMM] Jimbo, M., Misra, K. C., Miwa, T.: Crystal basis for the basic representation of $U_q(\hat{sl}(n))$, *Commun. Math. Phys.* **136** (1991), 79–88

[JMMO] Jimbo, M., Misra, K. C., Miwa, T., Okado, M.: Combinatorics of representations of $U_q(\hat{sl}(n))$ at $q=0$, *Commun. Math. Phys.* **136** (1991), 543–566

[JL] Joseph, A., Letzter, G.: Local finiteness of the adjoint action for quantized enveloping algebras, *J. Algebra* **153** (1992), 289–318

[JS] Joyal, A., Street, R.: Braided monoidal categories, Macquaria Univ. Report 860081, 1986

[Jur] Jurčo, B.: Differential calculus on quantized simple Lie groups, *Lett. Math. Phys.* **22** (1991), 177–186

[JS] Jurčo, B., Stoviček, A.: Quantum dressing orbits on compact groups, *Commun. Math. Phys.* **152** (1993), 97-126

[K] Kac, G. I.: Ring groups and the duality principle, *Proc. Moscow Math. Soc.* **12** (1963), 259–301

[KP] Kac, G. I., Palytkin, V. G.: An example of a ring group generated by Lie groups, *Ukrain. Math. J.* **16** (1964), 99–105

[KKl] Kachurik, I. I., Klimyk, A. U.: On Racah coefficients of the quantum algebra $U_q(su_2)$, *J. Phys. A* **23** (1990), 2717–2728

[KKMMNN] Kang, S. J., Kashiwara, M., Misra, K. C., Miwa, T., Nakashima, T., Nakayashiki, A.: Vertex models and crystals, *C. R. Acad. Sci. Paris, Ser. I* **315** (1992), 375–380

[Kas1] Kashiwara, M.: Crystalizing the q-analogue of universal enveloping algebras, *Commun. Math. Phys.* **133** (1990), 249–260

[Kas2] Kashiwara, M.: On crystal bases of the q-analogue of universal enveloping algebras, *Duke. Math. J.* **63** (1991), 465–516

[Kas3] Kashiwara, M.: Crystallization of quantized enveloping algebras, *Sugaku* **7** (1994), 99–115

[KN] Kashiwara, M., Nakashima, T.: Crystal graphs for representations of the q-analogue of classical Lie algebras, *J. Algebra* **165** (1994), 295–345

[KS] Kazhdan, D., Soibelman, Y.: Representations of quantum affine algebras, *Selecta Math.* **1** (1995), 537–595

[KM] Kempf, A., Majid, S.: Algebraic q-integration and Fourier theory on quantum and braided spaces, *J. Math. Phys.* **35** (1994), 6802–6837

[KT] Khoroshkin, S. M., Tolstoy, V. N.: Universal R-matrix for quantized (super) algebras, *Commun. Math. Phys.* **141** (1991), 599–617

[KR1] Kirillov, A.N., Reshetikhin, N.Yu.: Representations of the algebra $U_q(sl(2))$, q-orthogonal polynomials and invariants of links, in *Infinite-Dimensional Lie Algebras and Groups* (V. G. Kac, ed.), World Scientific, Singapore, 1989, pp. 285–339

[KR2] Kirillov, A. N., Reshetikhin, N. Yu.: q-Weyl group and a multiplicative formula for universal R-matrices, *Commun. Math. Phys.* **134** (1990), 421–431

[KL1] Klimek, S., Lesniewski, A.: Quantum Riemann surfaces I. The unit disk, *Commun. Math. Phys.* **146** (1992), 103–122

[KL2] Klimek, S., Lesniewski, A.: Quantum Riemann surfaces II. The discrete series, *Lett. Math. Phys.* **24** (1992), 125–127

[Kl1] Klimyk, A. U.: Decomposition of a tensor product of irreducible representations of a semisimple Lie algebra into a direct sum of irreducible representations, *Amer. Math. Soc. Transl. Ser. 2* **76** (1968), 63–73

[Kl2] Klimyk, A. U.: Wigner–Eckart theorem for tensor operators of the quantum group $U_q(n)$, *J. Phys. A* **25** (1992), 2919–2927

[Kl3] Klimyk, A. U.: Symmetries of Clebsch–Gordan coefficients of the quantum group $U_q(n)$, *J. Math. Phys.* **36** (1995), 508–523

[KlK] Klimyk, A. U., Kachurik, I. I.: Spectra, eigenvectors and overlap functions for representations of q-deformed algebras, *Commun. Math. Phys.* **175** (1996), 89–111

[Koe] Koelink, E.: 8 Lectures on quantum groups and q-special functions, *Report 96-10*, University of Amsterdam, 1996

[KK] Koelink, H. T., Koornwinder, T. H.: The Clebsch–Gordan coefficients for the quantum group $S_\mu U(2)$ and q-Hahn polynomials, *Nederl. Akad. Wetensch. Proc. Ser. A* **92** (1989), 443–456

[KV] Koelink, H. T., Verding, J.: Spectral analysis and the Haar functional on the quantum $SU(2)$ group, *Commun. Math. Phys.* **1** (1995), 1-14

[Ko1] Koornwinder, T. H.: Representations of the twisted $SU(2)$ quantum group and some q-hypergeometric orthogonal polynomials, *Nederl. Akad. Wetensch. Proc. Ser. A* **92** (1989), 97–117

[Ko2] Koornwinder, T. H.: Orthogonal polynomials in connection with quantum groups, in *Orthogonal Polynomials: Theory and Practice* (P. Nevai, ed.), Kluwer, Dordrecht, 1990, pp. 257–292

[Ko3] Koornwinder, T. H.: Positive convolution structures associated with quantum groups, in *Probability measures on groups X* (H. Heyer, ed.), Plenum Press, New York, 1991, pp. 249–286

[Ko4] Koornwinder, T. H.: Askey–Wilson polynomials as zonal spherical functions on the $SU(2)$ quantum group, *SIAM J. Math. Anal.* **24** (1993), 795–813

[Ko5] Koornwinder, T. H.: Special functions and q-commuting variables, *Preprint*, Institute Mittag–Leffler 1-95/96, 1995

[KoS] Koornwinder, T. H., Swarttow, R. F.: On q-analogues of the Fourier and Hankel transforms, *Trans. Amer. Math. Soc.* **333** (1992), 445–461

[Kul] Kulish, P. P.: Contraction of quantum algebras and q-oscillators, *Theoret. Math. Phys.* **86** (1991), 108–110

[KuR] Kulish, P. P., Reshetikhin, N. Yu.: Quantum linear problem for the sine-Gordon equation and higher representations, *Zap. Nauchn. Sem. LOMI* **101** (1981), 101–110

[KSS] Kulish, P. P., Sasaki, R., Schwiebert, C.: Constant solutions of reflection equations and quantum groups, *J. Math. Phys.* **34** (1993), 286–304

[Ks] Kustermans, J.: Two notions of discrete quantum groups, *Preprint*, Katholic University of Leuven, 1994

[LR] Lambe, L. A., Radford, D. E.: Algebraic aspects of the quantum Yang–Baxter equation, *J. Algebra* **154** (1993), 228–288

[La] Larson, R.G., Characters of Hopf algebras, *J. Algebra* **17** (1971), 352–368

[LaR] Larson, R. G., Radford, D. E.: Finite-dimensional cosemisimple Hopf algebras in characteristic 0 are semisimple, *J. Algebra* **117** (1988), 267–289

[LT] Larson, R. G., Taft, E. J.: The algebraic structure of linearly recursive sequences under Hadamard product, *Israel J. Math.* **72** (1990), 118–132

[LTo] Larson, R. G., Towber, J.: Two dual classes of bialgebras related to the concepts of "quantum group" and "quantum Lie algebra", *Commun. Algebra* **19** (1991), 3295–3345

[Le] Levendorskii, S. Z.: Twisted function algebras on a compact quantum group and their representations, *St. Petersburg Math. J.* **3** (1992), 405–423

[LS1] Levendorskii, S. Z., Soibelman, Ya. S.: Some applications of quantum Weyl groups, *J. Geom. Phys.* **7** (1990), 241–254

[LS2] Levendorskii, S. Z., Soibelman, Ya. S.: Algebras of functions on compact quantum groups, Schubert cells and quantum tori, *Commun. Math. Phys.* **139** (1991), 141–170

[L1] Lusztig, G.: Quantum deformations of certain simple modules over enveloping algebras, *Adv. Math.* **70** (1988), 237–249

[L2] Lusztig, G.: Finite-dimensional Hopf algebras arising from quantized universal enveloping algebras, *J. Amer. Math. Soc.* **3** (1990), 257–296

[L3] Lusztig, G.: Quantum groups at root of 1, *Geom. Dedicata* **35** (1990), 89–113

[L4] Lusztig, G.: Canonical bases arising from quantized enveloping algebras, *J. Amer. Math. Soc.* **3** (1990), 447–498

[L5] Lusztig, G.: Canonical bases arising from quantized enveloping algebras II, *Progr. Theor. Phys. Suppl.* **102** (1990), 175–201

[Ly1] Lyubashenko, V. V.: Hopf algebras and vector symmetries, *Sov. Math. Surveys* **41** (1986), 153–154

[Ly2] Lyubashenko, V. V.: Real and imaginary forms of quantum groups, *Lecture Notes Math.* **1510**, Springer, Berlin, 1992, pp. 67–78

[Macf] Macfarlane, A.: On q-analogues of the quantum harmonic oscillator and the quantum group $SU(2)_q$, *J. Phys. A* **22** (1989), 4581–4588

[MS] Mack, C. , Schomerus, V.: Quasi Hopf Symmetry in Quantum Theory, *Nucl. Phys. B* **370** (1992), 185–230

[Maj1] Majid, S.: Quasitriangular Hopf algebras and Yang–Baxter equations, *Intern. J. Modern Phys. A* **5** (1990), 1-91

[Maj2] Majid, S.: Physics for algebraists: noncommutative and noncocommutative Hopf algebras by a bicrossproduct construction, *J. Algebra* **130** (1990), 17–64

[Maj3] Majid, S.: More examples of bicrossproducts and double cross product Hopf algebras, *Isreal J. Math.* **72** (1990), 133–148

[Maj4] Majid, S.: Doubles of quasitriangular Hopf algebras, *Commun. Algebra* **19** (1991), 3061–3073

[Maj5] Majid, S.: Examples of braided groups and braided matrices, *J. Math. Phys.* **32** (1991) 3246–3253

[Maj6] Majid, S.: Rank of quantum groups and braided groups in dual form, *Lecture Notes Math.* **1510**, Springer, Berlin, 1992, pp. 79–89

[Maj7] Majid, S.: Braided groups, *J. Pure Appl. Algebra* **86** (1993), 187–221

[Maj8] Majid, S.: Braided momentum in the q-Poincare group, *J. Math. Phys.* **34** (1993), 2045–2058

[Maj9] Majid, S.: Transmutation theory and rank for quantum braided groups, *Math. Proc. Camb. Phil. Soc.* **113** (1993), 45–70

[Maj10] Majid, S.: Cross products by braided groups and bosonization, *J. Algebra* **163** (1994), 165–190

[Maj11] Majid, S.: Quantum and braided Lie algebras, *J. Geom. Phys.* **13** (1994), 307–356

[Mal] Maltsiniotis, G.: Calcul différential sur le groupe linéare quantique, *Preprint*, ENS, Paris, 1990

[Man1] Manin, Yu. I.: Notes on quantum groups and quantum de Rham complexes, *Theoret. Math. Phys.* **92** (1992), 997–1019

[MMN] Masuda, T., Mimachi, K., Nakagami, Y., Noumi, M., Ueno, K.: Representations of the quantum group $SU_q(2)$ and the little q-Jacobi polynomials, *J. Funct. Anal.* **99** (1991), 357–386

[MN] Masuda, T., Nakagami, Y.: A von Neumann algebra framework for the duality of the quantum groups, *Publ. RIMS Kyoto Univ.* **30** (1994), 799–850

[MM] Milnor, J. W., Moore, J. C.: On the structure of Hopf algebras, *Ann. Math.* **81** (1965), 211–264

[MR] Moore, G., Reshetikhin, N. Yu.: A comment on quantum group symmetry in conformal field theory, *Nucl. Phys. B* **328** (1989), 557–574

[MHR] Müller-Hoissen, F., Reuten, C.: Bicovariant differential calculi on $GL_{p,q}(2)$ and quantum subgroups, *J. Phys. A* **26** (1993), 2955–2975

[Mur] Murakami, J.: The representations of the q-analogue of Brauer's centralizer algebras and the Kauffman polynomials of links, *Publ. RIMS Kyoto Univ.* **26** (1990), 935–945

[Nag] Nagy, G.: On the Haar measure of the quantum $SU(n)$ groups, *Commun. Math. Phys.* **153** (1993), 217–228

[Na] Nakashima, T.: Crystal base and a generalization of the Littlewood–Richardson rule for the classical Lie algebras, *Commun. Math. Phys.* **154** (1993), 215–243

[N] Noumi, M.: Macdonald's symmetric polynomials as zonal spherical functions on quantum homogeneous spaces, *Adv. Math.* **123** (1996), 16–77

[NM] Noumi, M., Mimachi, K.: Quantum 2-spheres and big q-Jacobi polynomials, *Commun. Math. Phys.* **128** (1990), 521–531

[NYM] Noumi, M., Yamada, H., Mimachi, K.: Finite dimensional representations of the quantum group $GL_q(n, \mathbb{C})$ and the zonal spherical functions on $U_q(n)/U_q(n-1)$, *Jap. J. Math.* **19** (1993), 31–80

[Og] Ogievetsky, O.: Reality in the differential calculus on q-Euclidean spaces, *Lett. Math. Phys.* **25** (1992), 121–130

[OSWZ] Ogievetsky, O., Schmidke, W. B., Wess, J., Zumino, B.: q-deformed Poincaré algebra, *Commun. Math. Phys.* **150** (1992), 495–518

[OS] Ostrovskyi, V. L., Samoilenko, Yu. S.: Representations of quadratic $*$-algebras by bounded and unbounded operators, *Rep. Math. Phys.* **35** (1995), 283–301

[Pa] Pareigis, B.: A non-commutative, non-cocommutative Hopf algebra in "nature", *J. Algebra* **70** (1981), 356–374

[PT] Peterson, B., Taft, E.: The Hopf algebra of linearly recursive sequences, *Aequationes Math.* **20** (1980), 1–17

[Pod1] Podleś, P.: Quantum spheres, *Lett. Math. Phys.* **14** (1987), 193–202

[Pod2] Podleś, P.: Differential calculus on quantum spheres, *Lett. Math. Phys.* **18** (1989), 107–119

[Pod3] Podleś, P.: The classification of differential structures on quantum 2-spheres, *Commun. Math. Phys.* **150** (1992), 177–180

[Pod4] Podleś, P.: Complex quantum groups and their real representations, *Publ. RIMS Kyoto Univ.* **28** (1992), 709–745

[PoW] Podleś, P., Woronowicz, S. L.: Quantum deformation of Lorentz group, *Commun. Math. Phys.* **130** (1990), 381–431

[Pus] Pusz, W.: Irreducible unitary representations of quantum Lorentz group, *Commun. Math. Phys.* **152** (1993), 591–626

[PuW] Pusz, W., Woronowicz, S. L.: Twisted second quantization, *Rep. Math. Phys.* **27** (1989), 231–257

[Rad1] Radford, D. E.: The order of the antipode of a finite dimensional Hopf algebra is finite, *Amer. J. Math.* **98** (1976), 333–355

[Rad2] Radford, D. E.: On the antipode of a quasitriangular Hopf algebra, *J. Algebra* **151** (1992), 1–11

[Rad3] Radford, D. E.: Solutions of the quantum Yang–Baxter equation and the Drinfel'd double, *J. Algebra* **161** (1993), 20–32

[Res1] Reshetikhin, N. Yu.: Quantized universal enveloping algebras, the Yang–Baxter equation and invariants of links, I and II, *Preprints*, LOMI, Leningrad, 1988

[Res2] Reshetikhin, N. Yu.: Multiparameter quantum groups and twisted quasitriangular Hopf algebras, *Lett. Math. Phys.* **20** (1990), 331–335

[Res3] Reshetikhin, N. Yu.: Quasitriangularity of quantum groups at roots of 1, *Commun. Math. Phys.* **170** (1995), 79–99

[ReS] Reshetikhin, N. Yu., Semenov-Tianshansky, M. A.: Quantum R-matrices and factorization problems, *J. Geom. Phys.* **5** (1988), 533–550

[RTF] Reshetikhin, N. Yu., Takhtajan, L. A., Faddeev, L. D.: Quantization of Lie groups and Lie algebras, *Leningrad Math. J.* **1** (1990), 193–225

[Rid] Rideau, G.: On the representations of the quantum oscillator algebra, *Lett. Math. Phys.* **24** (1992), 147–154

[Rie1] Rieffel, M. A.: Deformation quantization and operator algebras, *Proc. Symp. Pure Math.* **51**, Amer. Math. Soc., Providence, RI, 1990, pp. 411–423

[Rie2] Rieffel,M.: Compact quantum groups associated with toral subgroups, *Contemp. Math.* **145** (1993), 465–491

[Rin] Ringel, C. M.: PBW–bases of quantum groups, *J. Reine Angew. Math.* **470** (1996), 51–88

[RiS] Rittenberg, V., Scheunert, M.: Tensor operators for quantum groups and applications, *J. Math. Phys.* **33** (1992), 436–445

[RA] Roche, P., Arnaudon, D.: Irreducible representations of the quantum analogue of $SU(2)$, *Lett. Math. Phys.* **17** (1989), 295–300

[Ros1] Rosso, M.: Finite dimensional representations of the quantum analog of the enveloping algebra of a complex simple Lie algebra, *Commun. Math. Phys.* **117** (1988), 581–593

[Ros2] Rosso, M.: An analogue of P.B.W. theorem and the universal R-matrix for $U_h(sl(N+1))$, *Commun. Math. Phys.* **124** (1989), 307–318

[Ros3] Rosso, M.: Analogues de la forme de Killing et du théoreme d'Harish-Chandra pour les groupes quantiques, *Ann. Sci. École Norm. Sup.* **23** (1990), 445–467

[Sag] Sagutani, T.: Harmonic analysis on quantum spheres associated with the representations of $U_q(\mathfrak{so}_N)$ and q-Jacobi polynomials, *Comp. Math.* **99** (1995), 249–281

[S1] Schauenburg, P.: On coquasitriangular Hopf algebras and the quantum Yang–Baxter equation, *Algebra-Berichte* **67**, Verlag R. Fischer, Munich, 1992

[S2] Schauenburg, P.: Hopf modules and Yetter–Drinfeld Modules, *J. Algebra* **169** (1994), 874–890

[Sch] Schirrmacher, A.: The multiparametric deformation of $GL(n)$ and the covariant differential calculus on the quantum vector space, *Z. Phys. C* **50** (1991), 321–327

[SWW1] Schlieker, M., Weich, W., Weixler, R.O.: Inhomogeneous quantum groups, *Z. Phys. C.* **53** (1992), 79–82

[SWW2] Schlieker, M., Weich, W. and Weixler, R.O.: R-matrix approach to inhomogeneous quantum groups, *Preprint*, LMU-TPW 1992-18, Munich

[SWZ] Schmidke, W. B., Wess, J., Zumino, B.: A q-deformed Lorentz algebra, *Z. Phys. C* **52** (1991), 471–476

[Sc1] Schmüdgen, K.: Operator representations of the real twisted canonical commutation relations, *J. Math. Phys.* **35** (1994), 3211–3229

[Sc2] Schmüdgen, K.: Integrable operator representations of $\mathbb{R}_q^2$, $X_{q,\gamma}$ and $SL_q(2,\mathbb{R})$, *Commun. Math. Phys.* **159** (1994), 217–237

[Sc3] Schmüdgen, K.: Operator representations of $U_q(sl_2(\mathbf{R}))$, *Lett. Math. Phys.* **37** (1996), 211–222

[SS1] Schmüdgen, K., Schüler, A.: Covariant differential calculi on quantum spaces and on quantum groups, *C. R. Acad. Sci. Paris Ser I* **316** (1993), 1155–1160

[SS2] Schmüdgen, K., Schüler, A.: Classification of bicovariant differential calculi on quantum groups of type A, B, C and D, *Commun. Math. Phys.* **167** (1995), 635–670

[SS3] Schmüdgen, K., Schüler, A.: Classification of bicovariant differential calculi on quantum groups, *Commun. Math. Phys.* **170** (1995), 315–335

[SS4] Schmüdgen, K., Schüler, A.: Left-covariant differential calculi on $SL_q(2)$ and $SL_q(3)$, *J. Geom. Phys.* **20** (1996), 87–105

[Schn] Schneider, H.-J.: Principal homogeneous spaces for arbitrary Hopf algebras, *Israel Math. J.* **72** (1990), 167–195

[Schn1] Schnizer, A. W.: Roots of unity: representations for symplectic and orthogonal quantum groups, *J. Math. Phys.* **34** (1993), 4340–4363

[Schn2] Schnizer, A. W.: Roots of unity: representations of quantum groups, *Commun. Math. Phys.* **163** (1994), 293–306

[SW] Schwenk, J., Wess, J.: A q-deformed quantum mechanical toy model, *Phys. Lett. B* **291** (1992), 273–277

[SWZ1] Schupp, P., Watts, P., Zumino, B.: Differential geometry on linear quantum groups, *Lett. Math. Phys.* **25** (1992), 139–147

[SWZ2] Schupp, P., Watts, P., Zumino, B.: Bicovariant quantum algebras and quantum Lie algebras, *Commun. Math. Phys.* **157** (1993), 305–329

[Skl1] Sklyanin, E. K.: Some algebraic structures connected with the Yang–Baxter equation. Representations of quantum algebras, *Funct. Anal. Appl.* **17** (1983), 273–284

[Skl2] Sklyanin, E. K.: On an algebra generated by quadratic relations, *Uspekhi Mat. Nauk* **40** (1985), 214

[Soi1] Soibelman, Ya. S.: The algebra of functions on a compact quantum group and its representations, *Leningrad Math. J.* **2** (1991), 161–178

[Soi2] Soibelman, Ya. S.: Selected topics in quantum groups, in *Infinite Analysis* (A. Tsuchiya, T. Eguchi and M. Jimbo, eds.), World Scientific, Singapore, 1992, pp. 859–888

[SV] Soibelman, Ya. S., Vaksman, L. L.: On some problems in the theory of quantum groups, *Adv. Soviet Math.* **9** (1992), 3–55

[Sud1] Sudbery, A.: Consistent multiparametric quantisation of $GL(n)$, *J. Phys. A* **23** (1990), L697–L704

[Sud2] Sudbery, A.: Canonical differential calculus on quantum general linear groups and supergroups, *Phys. Lett. B* **284** (1992), 61–65

[Sud3] Sudbery, A.: The algebra of differential forms on a full matric bialgebra, *Math. Proc. Camb. Phil. Soc.* **114** (1993), 111–130

[Sud4] Sudbery, A.: Matrix-element bialgebras determined by quadratic coordinate algebras, *J. Algebra* **158** (1993), 375–399

[Tk] Takesaki, M.: Duality and von Neumann algebras, *Lecture Notes Math.* **247**, Springer-Verlag, Heidelberg, 1972, pp. 665–779

[Ta1] Takeuchi, M.: The #-product of group sheaf extensions applied to Long's theory of dimodule algebras, *Algebra-Berichte* **34**, Verlag R. Fischer, Munich, 1977

[Ta2] Takeuchi, M.: Matric bialgebras and quantum groups, *Israel J. Math.* **72** (1990), 232–251

[Ta3] Takeuchi, M.: Some topics on $GL_q(n)$, *J. Algebra* **147** (1992), 379–410

[Ta4] Takeuchi, M.: Finite dimensional representations of the quantum Lorentz group, *Commun. Math. Phys.* **144** (1992), 557–580

[Ta5] Takeuchi, M.: Hopf algebra techniques applied to the quantum group $U_q(sl(2))$, *Contemp. Math.* **134** (1992), 309–323

[Tan1] Tanisaki, T.: Harish-Chandra isomorphisms for quantum algebras, *Commun. Math. Phys.* **127** (1990), 555–571

[Tan2] Tanisaki, T.: Finite dimensional representations of quantum groups, *Osaka J. Math.* **28** (1991), 37–53

[Tan3] Tanisaki, T.: Killing forms, Harish-Chandra homomorphisms and universal R-matrices for quantum algebras, in *Infinite Analysis* (A. Tsuchiya, T. Eguchi and M. Jimbo, eds.), World Scientific, Singapore, 1992, pp. 941–962

[TV] Truini, P., Varadarajan, V. S.: Quantization of reductive Lie algebras: construction and universality, *Rev. Math. Phys.* **5** (1993), 363–415

[Ts] Tsygan, B.: Notes on differential forms on quantum groups, *Selecta Math.* **12** (1993), 75–103

[Tw] Twietmeyer, E.: Real forms of $U_q(\mathfrak{g})$, *Lett. Math. Phys.* **24** (1992), 49–58

[UTS] Ueno, K., Takebayashi, T., Shibukawa, Y.: Construction of Gel'fand-Zetlin basis for $U_q(gl(N+1))$ modules, *Publ. RIMS Kyoto Univ.* **26** (1990), 667–679

[Vai1] Vainerman, L.: Gel'fand pairs of quantum groups, hypergroups and q-special functions, *Contemp. Math.* **183** (1995), 373–394

[Vai2] Vainerman, L.: Hypergroup structures associated with Gel'fand pairs of compact quantum groups, *Astérisque* **232** (1995), 231–242

[Vak] Vaksman, L. L.: q-Analogues of Clebsch–Gordan coefficients and the algebra of functions on the quantum group $SU(2)$, *Soviet Math. Dokl.* **39** (1989), 467–470

[VK] Vaksman, L. L., Korogodskii, L. I.: Spherical functions on the quantum group $SU(1,1)$ and a q-analogue of the Mehler–Fock formula, *Funct. Anal. Appl.* **25** (1991), 48–49

[VS1] Vaksman, L. L., Soibelman, Ya. S.: Algebra of functions on the quantum group $SU(2)$, *Funct. Anal. Appl.* **22** (1988), 170–181

[VS2] Vaksman, L. L., Soibelman, Ya. S.: Algebra of functions on the quantum group $SU(N+1)$ and odd-dimensional quantum spheres, *Leningrad Math. J.* **2** (1991), 1023–1042

[VD1] Van Daele, A.: Dual pairs of $*$-Hopf algebras, *Bull. London Math. Soc.* **25** (1993), 209–230

[VD2] Van Daele, A.: Multiplier Hopf algebras, *Trans. Amer. Math. Soc.* **342** (1994), 917–932

[VD3] Van Daele, A.: The Haar measure on a compact quantum group, *Proc. Amer. Math. Soc.* **123** (1995), 3125–3128

[VD4] Van Daele, A.: Discrete quantum groups, *J. Algebra,* **180** (1996), 431–444

[VDW] Van Daele, A., Wang, S. Z.: Universal quantum groups, *Intern. J. Math.* **7** (1996), 255–263

[Var] Varchenko, A. N.: Asymptotic solution to the Knizhnik–Zamolodchikov equation and crystal base, *Commun. Math. Phys.* **171** (1995), 99–137

[Wan1] Wang, S. Z.: Free products of compact quantum groups, *Commun. Math. Phys.* **167** (1995), 671–692

[Wan2] Wang, S. Z.: Deformations of compact quantum groups via Rieffel's quantization, *Commun. Math. Phys.* **178** (1996), 747–764

[W] Watanabe, J.: On the function algebras of compact quantum groups $SU_q(N)$, *Preprint*, Tokyo, 1995

[Wei] Weixler, R. O.: *Inhomogene Quantengruppen*, Ph.D. Thesis, Ludwig-Maximilians-Universität, Munich, 1993

[Wen1] Wenzl, H.: Representations of braid groups and the quantum Yang–Baxter equation, *Pacific J. Math.* **145** (1990), 153–180

[Wen2] Wenzl, H.: Quantum groups and subfactors of types B, C, and D, *Commun. Math. Phys.* **133** (1990), 383–432

[WZ] Wess, J., Zumino, B.: Covariant differential calculus on the quantum hyperplane, *Nucl. Phys. B Proc. Suppl.* **18** (1991), 302–312

[Wit] Witten, E.: Gauge theories, vector models and quantum groups, *Nucl. Phys. B* **117** (1990), 285–346

[Wor1] Woronowicz, S. L.: Compact matrix pseudogroups, *Commun. Math. Phys.* **111** (1987), 613–665

[Wor2] Woronowicz, S. L.: Twisted $SU(2)$ group. An example of a noncommutative differential calculus, *Publ. RIMS Kyoto Univ.* **23** (1987), 117–181

[Wor3] Woronowicz, S. L.: Tannaka–Krein duality for compact matrix pseudogroups. Twisted $SU(N)$ group, *Invent. Math.* **93** (1988), 35–76

[Wor4] Woronowicz, S. L.: Differential calculus on compact matrix pseudogroups (quantum groups), *Commun. Math. Phys.* **122** (1989), 125–170

[Wor5] Woronowicz, S. L.: A remark on compact matrix quantum groups, *Lett. Math. Phys.* **21** (1991), 35–39

[Wor6] Woronowicz, S. L.: Unbounded elements affiliated with C^*-algebras and noncompact quantum groups, *Commun. Math. Phys.* **136** (1991), 399–432

[Wor7] Woronowicz, S. L.: Compact Quantum Groups, *Preprint*, Warsaw, 1992

[Wor8] Woronowicz, S. L.: From multiplicative unitaries to quantum groups, *Intern. J. Math.* **7** (1996), 127–149

[WoZ] Woronowicz, S. L., Zakrzewski, S.: Quantum deformations of the Lorentz group. The Hopf $*$-algebra level, *Comp. Math.* **90** (1994), 211–243

[Y] Yamagami, Sh.: On unitary representation theories of compact quantum groups, *Commun. Math. Phys.* **167** (1995), 509–529

[Yam] Yamane, H.: A Poincare–Birkhoff–Witt theorem for quantized enveloping algebras of type A_n, *Publ. RIMS Kyoto Univ.* **25** (1989), 503–520

[Yan] Yang, C. N.: Some exact results for the many-body problem in one dimension with repulsive delta-function integration, *Phys. Rev. Lett.* **19** (1967), 1212–1214

[Yt] Yetter, D. N.: Quantum groups and representations of monoidal categories, *Math. Proc. Cambridge Phil. Soc.* **108** (1990), 261–290

[Zak] Zakrzewski, S.: Realifications of complex quantum groups, in *Groups and Related Topics* (R. Gielerak *et al.*, eds.), Kluwer, Dordrecht, 1992, pp. 83–100

[Z] Zumino, B.: Deformation of the quantum mechanical phase space with bosonic or fermionic coordinates, *Modern Phys. Lett A* **6** (1991), 1225–1235

Index

Subject Index

Index of Symbols

Springer and the environment

At Springer we firmly believe that an international science publisher has a special obligation to the environment, and our corporate policies consistently reflect this conviction.

We also expect our business partners – paper mills, printers, packaging manufacturers, etc. – to commit themselves to using materials and production processes that do not harm the environment. The paper in this book is made from low- or no-chlorine pulp and is acid free, in conformance with international standards for paper permanency.